System Dynamics and Controls

S. Graham Kelly
The University of Akron
Professor Emeritus

CENGAGE

Australia • Brazil • Canada • Mexico • Singapore • United Kingdom • United States

System Dynamics and Controls
S. Graham Kelly

SVP, Product Management: Cheryl Costantini

VP, Product Management: Heather Bradley Cole

Portfolio Product Director: Colin Grover

Senior Portfolio Product Manager: Timothy L. Anderson

Learning Designer: MariCarmen Constable

Senior Content Manager: Alexander Sham

Digital Project Manager: Nikkita Kendrick

Senior Content Acquisition Analyst: Deanna Ettinger

Production Service: MPS Limited

Designer: Chris Doughman

Cover Image Source: Patty Lagera/Moment/Getty Images

Library of Congress Control Number: 2024935800

ISBN: 978-0-357-87920-7

Cengage
5191 Natorp Boulevard
Mason, OH 45040
USA

Cengage is a leading provider of customized learning solutions.
Our employees reside in nearly 40 different countries and serve digital learners in 165 countries around the world. Find your local representative at **www.cengage.com.**

To learn more about Cengage platforms and services, register or access your online learning solution, or purchase materials for your course, visit **www.cengage.com.**

Printed at CLDPC, USA, 06-24

To Seala

Contents

Preface

Engineering systems are interdisciplinary in nature. Engineers often work in multi-disciplinary teams. In order to effectively communicate with their other team members, a mechanical engineer must have some knowledge of electrical and chemical engineering, and an electrical engineer must have some knowledge of mechanical and chemical engineering. All engineers should have some familiarity with the modeling of physical systems of any kind, as well as with the time-dependent responses of these systems. They should be able to incorporate this knowledge into the design of systems which control the behavior of these systems. For example, the design of a robot involves mechanical systems and electrical systems. A control system for the robot, usually an electrical controller, is designed to make the robot respond appropriately to certain inputs. The objective of this book is to provide the reader with a general understanding of how to mathematically model a linear engineering system and how to control that system.

The prerequisites for this study include knowledge of single-variable calculus, including ordinary differential equations, and a course in engineering physics that includes mechanics, electromagnetism, and thermal systems. Some advanced topics in the book require knowledge of multi-variable calculus. Readers should be familiar with concepts studied in the core engineering subjects of statics and dynamics, with knowledge of dynamics of rigid bodies undergoing planar motion. Previous or concurrent study of chemistry, circuit analysis, fluid mechanics, and thermodynamics, while not essential, will enhance understanding of some topics and examples.

Chapter 1 presents introductory material and discusses topics that are common to multiple systems: the linearization of nonlinear models, the unit impulse function and the unit step function, nondimensionalization, and the use of MATLAB. After this introductory chapter, this text can be divided into four parts.

Chapters 2 and 3 set a mathematical framework for the mathematical modeling of any linear physical system. Chapter 2 introduces the Laplace transform method. Chapter 3 explores using a mathematical model for a linear system to develop a framework for analyzing the model. The Laplace transform method is first used to define transfer functions. Then the state-space method is used to develop a matrix formulation for a system. The chapter shows how to go back and forth between the state-space model and the transfer function.

Chapters 4, 5, and 6 focus on mathematical modeling of different types of systems. They also discuss formulating the mathematical models into the transfer function framework or the state-space formulation framework. Chapter 4 deals with mechanical systems; Chapter 5 deals with electrical systems including electromechanical systems; and Chapter 6 focuses on systems that are best modeled using transport phenomena, including thermal systems, chemical systems, and biomedical systems.

Chapters 7 and 8 deal with the system response once the mathematical modeling is performed. Chapter 7 focuses on the transient response; that is, the free response, the impulsive response, and the step response. Chapter 8 explores the steady-state response, which is the response in the frequency domain. The sinusoidal transfer function is also introduced. This chapter includes the graphical representation of the responses such as the frequency response curves, the Bode diagram, and the Nyquist diagram.

Chapters 9 through 12 deal with the control of linear systems. Chapter 9 serves as an introductory chapter detailing categories of controllers (P, PI, PD, PID, and compensators). It also deals with mechanical, electrical, and transport systems which constitute controllers. This chapter introduces concepts of error and offset, and provides examples of the control of first- and second-order plants. Chapter 10 presents the root-locus method and its role in controller design. Chapter 11 deals with the design of controllers in the frequency domain using the Nyquist diagram and the Bode diagram. Finally, Chapter 12 discusses state-space methods and how to design controllers that provide state-variable feedback.

This text frequently uses MATLAB as a tool for the determination of the response of a dynamic system. MATLAB is used as a computational tool, a programming tool, and a graphical tool. Specific applications of functions from the Symbolic Toolbox and the Control Systems Toolbox are used throughout. For example, the Symbolic Toolbox is used to determine inverse Laplace transforms, while the Control Systems Toolbox is used to develop Bode diagrams and Nyquist diagrams for a system. Appendix C provides a short introduction to MATLAB and the commands used in this text.

Simulink, a MATLAB-developed simulation and modeling tool, is also used throughout this text. Simulink allows development of models, using either the transfer function or the state-space formulation, without the programming required for MATLAB.

Instructor Resources

Additional instructor resources for this book are available online. Instructor assets include a Solution and Answer Guide and PowerPoint® slides. Sign up or sign in at www.cengage.com to search for and access this product and its associated resources.

Acknowledgments

The author wishes to thank the editorial staff at Cengage for aiding in the development of this text. Included in this acknowledgment are Timothy Anderson, Senior Product Portfolio Manager; Alexander Sham, Senior Content Manager; and MariCarmen Constable, Learning Designer. The author also wishes to thank Rose Kernan of RPK Editorial Services, Inc., as well as Carol Reitz for her excellent copyediting. The author also acknowledges the contributions made by Stephen Chapman and Aly El-Iraki, who thoroughly reviewed this title. The author also recognizes the many excellent comments and suggestions made by students at the University of Akron throughout the authoring process. Finally, this project could not have been possible without the support and encouragement from my wife, Seala Fletcher-Kelly.

S. Graham Kelly

Digital Resources

MindTap Reader

Available via Cengage Unlimited, **MindTap Reader** is Cengage's next-generation eTextbook for students.

The MindTap Reader provides more than just text learning for the student. It offers a variety of tools to help our future engineers learn chapter concepts in a way that resonates with their workflow and learning styles.

- **Personalize their experience**

Within the MindTap Reader, students can highlight key concepts, add notes, and bookmark pages. These are collected in My Notes, ensuring they will have their own study guide when it comes time to study for exams.

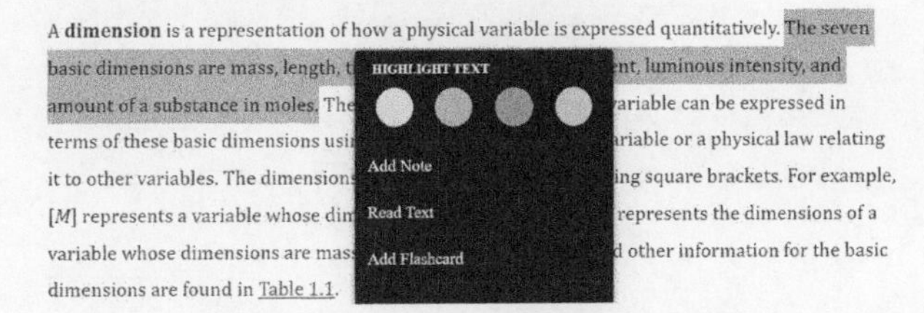

- **Flexibility at their fingertips**

Students can personalize their study experience by creating and collating their own custom flashcards. The ReadSpeaker feature reads text aloud to students so they can learn on the go—wherever they are.

Cengage Read

The Cengage Read app enables students to learn anywhere, anytime with interactive eTextbooks, 24/7 course access, and more.

To learn more and download the mobile app, visit www.cengage.com/mobile-app/.

Cengage Unlimited

All-You-Can-Learn Access with Cengage Unlimited

Cengage Unlimited is the cost-saving student plan that includes access to our entire library of eTextbooks, online platforms, and more—in one place, for one price. For just $129.99 for four months, a student gets online and offline access to Cengage course materials across disciplines, plus hundreds of student success and career readiness skill-building activities. To learn more, visit www.cengage.com/unlimited.

1

Introduction

Learning Objectives

At the conclusion of this chapter, you should be able to:

- Name the terms involved in the formulation of a mathematical model of a dynamic system
- Define a control system
- Distinguish between a dimension and a unit, and define systems of units
- Name the steps in the mathematical modeling of a dynamic system
- Understand the two methods used to linearize a nonlinear differential equation: the small angle assumptions and the perturbation method
- Understand that the unit impulse function is a mathematical model of the force required to impart a unit impulse to a system
- Explain the unit step function as being the integral of the unit impulse function, and describe how it is used in the mathematical modeling of systems that change at discrete values of time
- Define the concepts of stability, neutral stability, and instability
- Know how MATLAB and Simulink are used in the determination of system response and control system design

1.1 Dynamic Systems

1.1.1 Inputs and Outputs

A model for many physical processes is illustrated by Figure 1.1(a). An **input** is provided to a **system** that delivers an **output**. The system is the set of **components** that act together to generate the output from the input. The components are individual units within the system. The system is **static** if the output is dependent only on the instantaneous input. The system is **dynamic** when the output is a function of the history of the input. The output from the system for a given system input is called the **system response**.

The mathematical modeling of dynamic systems is the first step in the process described in this book. The system response, obtained using the mathematical model, is then analyzed and modified so that a desired objective is achieved. The problem of defining the relationship between the input, the system components, and the system response can be posed in two ways:

1. Given the input and the system components, determine the system response.
2. Given the input and the desired output, determine a set of system components that can be used to achieve the desired output.

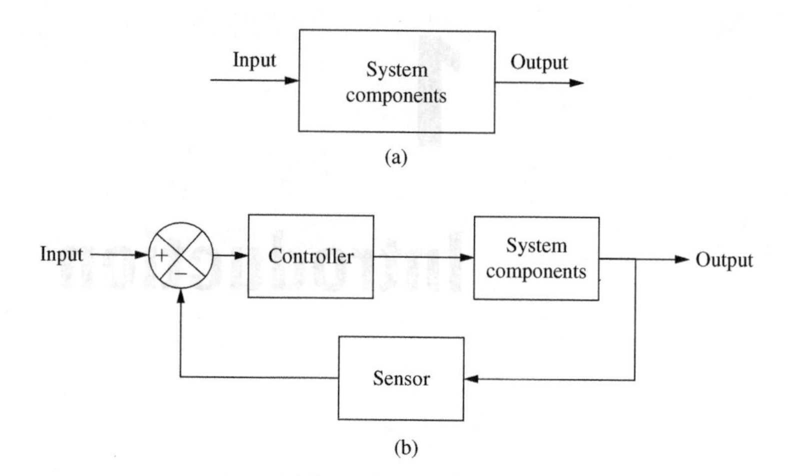

Figure 1.1

(a) System model of a physical process; (b) schematic of a feedback control system.

The first problem is a major focus of this book. A thorough knowledge of dynamic system response is necessary to study the second problem, which is related to the design and synthesis of dynamic systems. The second problem is related to control system design; that is, choosing components such that the behavior of a system is modified.

The output of a dynamic system is dependent on the history of the input, which may be time dependent. Thus, the output varies with time. Time is an independent variable in the study of dynamic system response and is designated by t. The dynamic response of a system is obtained when the system output, represented by the dependent variable x, is determined as a function of time $x(t)$. The measurement of time usually starts at 0. At this instance, the conditions of the system are designated as the system's initial conditions.

Mathematical modeling is the process through which the dynamic response of a system is obtained. Mathematical modeling, described in Section 1.3, leads to the development of mathematical equations that describe the behavior of the system. The behavior of a dynamic system is usually governed by a differential equation, an integral equation, an integrodifferential equation, or a set of differential equations in which time is an independent variable. The dependent variables represent the system outputs. An assumption made in this book is that all system components are discrete and have lumped properties. This assumption implies that time is the only independent variable in the system modeling. Since the dependent variables are functions of a single independent variable, the derived equations are ordinary differential equations.

The dot notation, used in Equations (1.1) and (1.2), is used throughout the book to represent derivatives with respect to time. A single dot over a dependent variable represents a derivative with respect to time, and two dots over a dependent variable represent two derivatives with respect to time.

A differential equation is **linear** if it contains only linear functions of the dependent variables and their time derivatives. A **nonlinear** differential equation contains nonlinear functions of the dependent variables or their time derivatives. Equations (1.1) are examples of linear differential equations. Equation (1.1a) is a linear ordinary differential equation (ODE) with constant coefficients. Equation (1.1b) is linear in the dependent variable x, but it has a variable coefficient, $\sin(t)$.

$$\ddot{x} + 4\dot{x} + 3x = 2\sin(5t) \tag{1.1a}$$

$$\ddot{x} + \sin(t)x = 0 \tag{1.1b}$$

Equations (1.2) are examples of nonlinear differential equations.

$$\ddot{x} + 2\dot{x} + 3x + 4x^3 = 0 \tag{1.2a}$$

$$\ddot{x} + \sin(x) = 2\sin(5t) \tag{1.2b}$$

A system whose mathematical model involves only linear differential equations is called a **linear system**. A system whose mathematical model contains nonlinear differential equations is called a **nonlinear system**. Linear systems are the focus of this book. All systems are inherently nonlinear. Aerodynamic drag and sliding friction lead to nonlinear terms in differential equations governing the response of mechanical systems. Equations used to specify the behavior of system components such as springs and resistors are linearized by neglecting nonlinear terms. Assumptions are often made to eliminate nonlinear terms from the differential equations, thus approximating nonlinear systems using linear equations. Such assumptions include neglecting nonlinear effects such as aerodynamic drag, assuming linear behavior of system components such as springs and resistors, or postulating system operation in a range where nonlinear effects are small. In other cases when a nonlinear differential equation is obtained, it is linearized using the techniques described in Section 1.6.

A system whose input is known at all values of time is a **continuous time system**. The response of a continuous time system is governed by differential equations. A system whose input is known only at discrete times, but at regularly spaced intervals, is called a **discrete time system**. Discrete time systems are governed by difference equations.

1.1.2 Control Systems

Dynamic systems are designed such that their response is predictable when subject to defined input. However, input to a system may change unpredictably. For example, an output variable for a heating and air conditioning system is the temperature of a room it is designed to service. An input to the system is the temperature to which the room is to be heated or cooled, the thermostat setting. When the input is changed, the system responds dynamically to change the room temperature. This is an example of a predictable change in input. However, for the heating and air conditioning system to function, it must also be able to sense the room temperature and respond to changes in room temperature caused by external sources when the thermostat setting is constant. The heating and air conditioning system must be designed so that the current room temperature is a system input, and the system must be able to respond to the unpredictable changes caused by external sources.

The heating and air conditioning system just described must be designed with a sensor to determine the room temperature (the input) and compare it to the thermostat temperature (the desired output). When the two temperatures are different, the system must respond so that the temperature from the heating and air conditioning system dynamically approaches the input temperature. When the desired output temperature is achieved, the heating and air conditioning system stops responding. The differential in temperature between the output and the input is zero.

A system that senses its output and responds to a difference between input and output is called a **feedback control system**. Feedback control systems are designed to provide stable responses to unpredictable changes in system input. A simple model for a feedback control system is illustrated in Figure 1.1(b). This system is closed loop in that the output from the system components is fed to a sensor whose output is compared with the input, thus closing the loop. The closed-loop response differs from that of the open loop. The controller is a system that dynamically changes the output variable in order to achieve this result.

Table 1.1 Basic Dimensions

Quantity	*M-L-T* Dimensions	*F-L-T* Dimensions	SI Unit	English Unit	Conversion
Mass	$[M]$	$[FT^2L^{-1}]^2$	kilogram (kg)	slug[2]	1 kg = 0.00685 slug
Length	$[L]$	$[L]$	meter (m)	foot (ft)	1 m = 3.28 ft
Time	$[T]$	$[T]$	second (s)	second (s)	
Electric current	$[i]$	$[i]$	ampere (A)	ampere (A)	
Number of particles of a substance	[mol]	[mol]	gram mole (gmol)	pound mole (lbmol)	
Temperature	Θ	Θ	kelvin (K)	rankine (R)	1 K = 1.8 R[1]
Luminous intensity	$[I]$	$[I]$	candela (cd)	candela (cd)	
Force	$[MLT^{-2}]^2$	$[F]$	newton (N)[2]	pound (lb)	1 N = 0.225 lb

[1]Valid over a temperature interval; not for absolute temperatures
[2]A derived unit or dimension

1.2 Dimensions and Units

A **dimension** is a representation of how a physical variable is expressed quantitatively. The seven basic dimensions are mass, length, time, temperature, electric current, luminous intensity, and amount of a substance in moles. The dimensions of every physical variable can be expressed in terms of these basic dimensions using either the definition of the variable or a physical law relating it to other variables. The dimensions of a quantity are expressed using square brackets. For example, $[M]$ represents a variable whose dimensions are mass, and $[MT^{-2}]$ represents the dimensions of a variable whose dimensions are mass per time squared. Symbols and other information for the basic dimensions are found in Table 1.1.

The dimensions of any physical variable can be expressed in terms of the basic dimensions. A quantity such as a displacement has the basic dimension of length. Velocity is defined as the time rate of change of displacement; thus its dimensions are $[LT^{-1}]$. Similarly, the dimensions of acceleration are obtained as $[LT^{-2}]$. The dimensions of force, $[MLT^{-2}]$, are obtained using a physical law, Newton's second law, which states that force equals mass times acceleration.

The set of basic dimensions is called the *M-L-T* system. An alternate formulation of the set of basic dimensions, called the *F-L-T* system, uses force, $[F]$, as a basic dimension rather than mass. In the *F-L-T* system, the dimensions of mass are derived as $[FT^2L^{-1}]$. The *F-L-T* system can be used for any physical system, but it is often used for physical systems in which mass is not a system parameter.

Example 1.1

The force, F, acting on the piston in a piston-cylinder viscous damper is proportional to the velocity, v, of the piston. The equation commonly used to state this proportionality is

$$F = cv \tag{a}$$

Determine the dimensions of c, called the viscous damping coefficient. Use both the *M-L-T* system and the *F-L-T* system.

Solution

Solving Equation (a) for c leads to

$$c = \frac{F}{v} \tag{b}$$

Thus, the dimensions of c are equal to the dimensions of force divided by the dimensions of velocity. Using the M-L-T system,

$$\begin{aligned}\text{dimensions of } c &= \frac{[MLT^{-2}]}{[LT^{-1}]} \\ &= [MT^{-1}]\end{aligned} \tag{c}$$

Using the F-L-T system, the dimensions of the damping coefficient are

$$\begin{aligned}\text{dimensions of } c &= \frac{[F]}{[LT^{-1}]} \\ &= [FTL^{-1}]\end{aligned} \tag{d}$$

Even though Equations (c) and (d) are equivalent, Equation (d) provides a better sense of the physical meaning of the damping coefficient.

The basic dimension used for electrical systems is that of electric current. Current results form the motion of charged particles. If i represents electric current and q represents electric charge, then

$$i = \frac{dq}{dt} \tag{1.3}$$

Thus, electric charge has dimensions of $[iT]$. The unit of charge is called a coulomb, which is defined as an ampere·second. The electric potential represents the amount of work required to move an electric charge through an electric field. Usually given the algebraic symbol v, electric potential has the dimensions of work divided by electric charge. In terms of the basic dimensions using the F-L-T system, electric potential has dimensions of $[FLi^{-1}T^{-1}]$.

Example 1.2

The relation between voltage and current in a circuit component called an inductor is

$$v = L\frac{di}{dt} \tag{a}$$

Determine the dimensions of L, the inductance, using the F-L-T system of dimensions.

Solution

Solving Equation (a) for L leads to

$$L = \frac{v}{\left(\frac{di}{dt}\right)} \tag{b}$$

Thus,

$$\begin{aligned}\text{dimensions of } L &= \frac{[FLi^{-1}T^{-1}]}{[iT^{-1}]} \\ &= [FLi^{-2}]\end{aligned} \tag{c}$$

The dimensions for physical quantities used in this book are listed in Table 1.2.

Any equation derived from physical principles satisfies the **principle of dimensional homogeneity**, which states that every additive term in an equation has equal dimensions. Its inverse can be stated in terms of a popular cliché, "You can't add apples and oranges." A corollary to the principle of dimensional homogeneity is that in mathematical expressions, arguments of transcendental functions must be dimensionless.

Example 1.3

The integrodifferential equation derived in Example 5.10 to model the response of a series LRC circuit is

$$L\frac{di}{dt} + Ri + \frac{1}{C}\int_0^t i dt = v(t) \tag{a}$$

where $i(t)$ is the current in the circuit and $v(t)$ is an electric potential.

a. Show that Equation (a) satisfies the principle of dimensional homogeneity.

b. If $v(t) = V_0 \sin(\omega t)$, where t is time, what are the dimensions of ω and V_0?

Solution

a. The dimensions of each of the parameters are given in Table 1.2. Since the right-hand side of the equation is an electric potential, for the principle of dimensional homogeneity to be satisfied, every term in the equation must also have dimensions of electric potential, $[FLi^{-1}T^{-1}]$. Example 1.2 shows that the first term has the same dimensions as electric potential. Checking the remaining terms,

$$\begin{aligned} \text{dimensions of } Ri &= [FLi^{-2}T^{-1}][i] \\ &= [FLi^{-1}T^{-1}] \end{aligned} \tag{b}$$

$$\begin{aligned} \text{dimensions of } \frac{1}{C}\int_0^t i dt &= \frac{[i][T]}{[i^2T^2F^{-1}L^{-1}]} \\ &= [FLi^{-1}T^{-1}] \end{aligned} \tag{c}$$

The results in Equations (b) and (c) verify that all terms in Equation (a) have the same dimensions.

b. Since the argument of the transcendental function $\sin(\omega t)$ must be dimensionless, the dimensions of ω are $[T^{-1}]$. Transcendental functions are dimensionless. Thus, applying the principle of dimensional homogeneity, the dimensions of V_0 must be those of electric potential, which are $[FLi^{-1}T^{-1}]$.

A system of units is a system in which numerical values for basic dimensions are defined. A benchmark quantity for each basic dimension is designated, and all variables with that dimension are referenced to the benchmark quantity.

The **English system** is an F-L-T system that uses the pound as the fundamental unit of force, the foot as the fundamental unit of length, the second as the fundamental unit of time, and degrees Rankine as the fundamental unit of temperature. The SI **(Système International)** system of units is an M-L-T system that uses the kilogram as the fundamental unit of mass, the meter as the fundamental unit of length, the second as the fundamental unit of time, and kelvins as the fundamental unit of temperature. Both systems use amperes as the fundamental unit of current and candelas as the fundamental unit of luminous intensity. Both systems also use the mole as the basic unit of chemical matter. One mole of a substance is the mass

Table 1.2 Derived Dimensions and Units

Quantity (Symbol)	*M-L-T* Dimensions	*F-L-T* Dimensions	SI Units	English Units
Acceleration (a)	$[LT^{-2}]$	$[LT^{-2}]$	m/s^2	ft/s^2
Angular acceleration (α)	$[T^{-2}]$	$[T^{-2}]$	r/s^2	r/s^2
Angular velocity (ω)	$[T^{-1}]$	$[T^{-1}]$	r/s	r/s
Area (A)	$[L^2]$	$[L^2]$	m^2	ft^2
Capacitance (C)	$[i^2T^4M^{-1}L^{-2}]$	$[i^2T^2F^{-1}L^{-1}]$	farad (F) = C/V	F
Damping coefficient (c)	$[MT^{-1}]$	$[FTL^{-1}]$	N·s/m	lb·s/ft
Density (ρ)	$[ML^{-3}]$	$[FT^2L^{-4}]$	kg/m^3	slug/ft^3
Dynamic viscosity (μ)	$[ML^{-1}T^{-1}]$	$[FTL^{-2}]$	N·s/m^2	lb·s/ft^2
Electric charge (q)	$[iT]$	$[iT]$	coulomb (C) = A·s	C
Electric potential (v)	$[ML^2i^{-1}T^{-3}]$	$[FLi^{-1}T^{-1}]$	volt (V) = J/C	V
Energy (E)	$[ML^2T^{-2}]$	$[FL]$	joule (J) = N·m	lb·ft
Film coefficient (h)	$[MT^{-3}\Theta^{-1}]$	$[FL^{-1}T^{-1}\Theta^{-1}]$	J/(m^2·K·s)	lb/(ft·R·s)
Frequency (f)	$[T^{-1}]$	$[T^{-1}]$	hertz (Hz) = cycles/s	Hz
Frequency (ω)	$[T^{-1}]$	$[T^{-1}]$	r/s	r/s
Gas constant (R)	$[L^2T^{-2}\Theta^{-1}]$	$[L^2T^{-2}\Theta^{-1}]$	J/(kg·K)	ft·lb/(lbm·R)
Head (h)	$[L]$	$[L]$	m	ft
Impulse (I)	$[MLT^{-1}]$	$[FT]$	N·s	lb·s
Inductance (L)	$[ML^2i^{-2}T^{-2}]$	$[FLi^{-2}]$	henry (H) = Wb/A	H
Magnetic flux (ψ)	$[ML^2i^{-1}T^{-2}]$	$[FLi^{-1}]$	weber (Wb) = V·s	Wb
Mass flow rate ($\dot{m}$)	$[MT^{-1}]$	$[FTL^{-1}]$	kg/s	slug/s
Moment of inertia (I)	$[ML^2]$	$[FLT^2]$	kg·m^2	slug·ft^2
Mutual inductance (M)	$[ML^2i^{-2}T^{-2}]$	$[FLi^{-2}]$	H	H
Pipe roughness (ε)	$[L]$	$[L]$	m	ft
Power (P)	$[ML^2T^{-3}]$	$[FLT^{-1}]$	watt (W) = J/s	lb·ft/s
Pressure (p)	$[ML^{-1}T^{-2}]$	$[FL^{-2}]$	pascal (Pa) = N/m^2	lb/ft^2
Rate of reaction (k)	$[(\text{mol})L^{-3}T^{-1}]$	$[(\text{mol})L^{-3}T^{-1}]$	gmol/(m^3·s)	lbmol/(ft^3·s)
Resistance (R)	$[ML^2T^{-3}i^{-2}]$	$[FLT^{-1}i^{-2}]$	ohm (Ω) = V/A	Ω
Species concentration (C)	$[(\text{mol})L^{-3}]$	$[(\text{mol})L^{-3}]$	gmol/m^3	lbmol/ft^3
Specific energy (e)	$[ML^{-1}T^{-2}]$	$[FL^{-2}]$	J/m^3	lb/ft^3
Specific heat (c_p)	$[L^2T^{-2}\Theta^{-1}]$	$[L^2T^{-2}\Theta^{-1}]$	J/(kg·K)	ft·lb/(lbm·R)
Stiffness (k)	$[MT^{-2}]$	$[FL^{-1}]$	N/m	lb/ft
Thermal conductivity (k)	$[MLT^{-3}\Theta^{-1}]$	$[FT^{-1}\Theta^{-1}]$	J/(m·K·s)	lb/(s·R)
Torsional damping coefficient (c_t)	$[ML^2T^{-1}]$	$[FLT]$	N·m·s/r	lb·ft·s/r
Torsional stiffness (k_t)	$[ML^2T^{-2}]$	$[FL]$	N·m/r	lb·ft/r
Velocity (v)	$[LT^{-1}]$	$[LT^{-1}]$	m/s	ft/s
Volume (V)	$[L^3]$	$[L^3]$	m^3	ft^3
Volume flow rate (Q)	$[L^3T^{-1}]$	$[L^3T^{-1}]$	m^3/s	ft^3/s

of Avogadro's number (6.022×10^{23}) of fundamental molecular units (atoms, molecules, ions) of the substance. The SI system uses the gram-mole (gmol), which is the mass in grams of Avogadro's number of molecules, whereas the English system uses the pound-mole (lbmol), which is the mass of the same number of molecules measured in pound-mass. These units are summarized in Table 1.1 with conversions between the two systems. In the English system, the unit of mass, called a slug, is a derived unit defined by

$$1 \text{ slug} = 1 \text{ lb}\cdot\text{s}^2/\text{ft} \tag{1.4}$$

Table 1.3 SI Prefixes

Prefix	Symbol	Factor
Terra	T	10^{12}
Giga	G	10^{9}
Mega	M	10^{6}
Kilo	k	10^{3}
Centi	c	10^{-2}
Milli	m	10^{-3}
Micro	μ	10^{-6}
Nano	n	10^{-9}

As noted, an alternate unit of mass is the pound mass (lbm), which is the mass of a substance that exerts a gravitational force of one pound. In the SI system, the unit of force, called a Newton, is a derived unit defined by

$$1 \text{ newton (N)} = 1 \text{ kg}\cdot\text{m/s}^2 \tag{1.5}$$

Derived units for common variables are given in Table 1.2 for both the English and the SI systems. Derived units are often given names, especially in the SI system. For example, the unit for electric potential is called the volt and is defined as

$$1 \text{ volt (V)} = 1 \frac{\text{joule (J)}}{\text{coulomb (C)}} = 1 \text{ N}\cdot\text{m/(A}\cdot\text{s)} \tag{1.6}$$

It is common practice to use scientific notation to express an answer when using the English system. For example, a displacement may be expressed as $x = 1.35 \times 10^{-5}$ ft. In the SI system, it is common practice to use prefixes representing powers of 10 rather than scientific notation. For example, a numerical value for energy may be expressed as $E = 35.6$ MJ. Table 1.3 lists the prefixes used in the SI system. In either system, unless more digits are needed for a calculation, three significant digits are used to express numerical values.

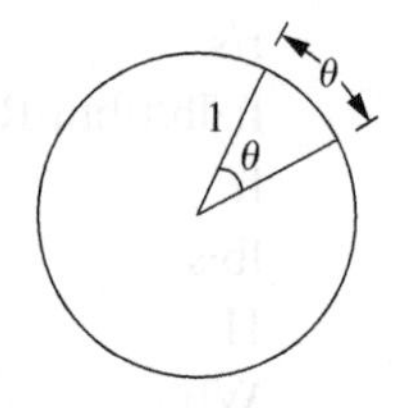

Figure 1.2

Angular measure in radians is equal to the arc length of a sector of a circle of radius 1 formed by two radii making an angle θ.

Angular measure is inherently dimensionless. However, there are several units in which angular measure is expressed. The measure of an angle in **radians**, the fundamental unit of angular measure, is illustrated in Figure 1.2. The angle between the two line segments of unit length, measured in radians, is equal to the length of the arc subtended from the line segments. There are 2π radians in the arc of a full circle. This is equal to the circumference of a circle of radius 1. A protractor is used to measure angles in **degrees**, with 360 degrees in a circle. Thus,

$$2\pi \text{ radians (r)} = 360 \text{ degrees (°)} \tag{1.7}$$

The units of **cycles** and **revolutions** measure the number of times a full circle has been executed. Thus,

$$1 \text{ cycle} = 2\pi r \tag{1.8}$$

If an angular measure is determined using degrees, cycles, or revolutions, it must be converted to radians using Equation (1.7) or (1.8) before it is used in calculations.

Frequency measures the rate at which a cyclic motion is executed. Frequency is often measured in hertz (Hz) defined by

$$1 \text{ Hz} = 1 \text{ cycle/s} \tag{1.9}$$

Frequency in hertz is usually given the symbol f. However, in calculations the frequency in r/s, ω, must be used. The conversion is

$$\omega = 2\pi f \tag{1.10}$$

Since angular measure is inherently dimensionless, frequency has dimensions of $[T^{-1}]$.

1.3 Mathematical Modeling of Dynamic Systems

Mathematical modeling is a process through which the output of a dynamic system, as illustrated in Figure 1.1, is determined given the system input. Mathematical modeling of a system can be used to achieve one of three objectives. **(1) System analysis** is used to determine the output for a specific system. The system components and their parameters are defined. The system may be analyzed for a variety of system inputs. **(2) System design** is used to determine the system components and their parameters such that a specific system output is achieved. **(3)** There is a subtle difference between system design and **system synthesis**, which is the determination of system components and their parameters to achieve a specific performance for a variety of system inputs. System synthesis is used in the analysis and design of control systems.

The procedure for the mathematical modeling of dynamic systems is outlined in this section. The steps are not necessarily followed in the order presented. For example, steps 2–4 may be done simultaneously, or step 4 may be performed before step 2 or 3. Some systems do not require all of the steps. For example, step 7, determining the system's initial conditions, is not necessary when only a steady-state response is required.

Step 1: Define the system to be modeled. Identify the input to the system and what will constitute the output. For example, an electromechanical system may consist of a circuit designed to operate a motor, which in turn provides power to turn a shaft that operates a turbine. This system consists of several subsystems. The output from one subsystem is the input for another. Figure 1.3(a) illustrates subsystems that are uncoupled, whereas Figure 1.3(b) illustrates subsystems that are coupled. In Figure 1.3(b), there is **feedback** from subsystem C into subsystem A; the output from subsystem C affects the input to subsystem A. Each subsystem in Figure 1.3(a) could be modeled as a system. The feedback in the system of Figure 1.3(b) requires that the modeling include the entire system.

Step 2: The assumptions under which the modeling occurs must be identified and stated. **Implicit assumptions** are those made for almost any system. They are taken for granted and rarely stated. Indeed, these assumptions should be explicitly stated when they are not applicable. Examples of implicit assumptions are that the earth is an inertial reference frame, that a continuum model can be used for all matter in the system, that no nuclear reactions are occurring in the system, and that the acceleration due to gravity is a constant. **Explicit assumptions** must be stated and later verified. Examples of explicit assumptions are that displacements are small, specific laws for system components apply, and friction effects are negligible.

Assumptions are made to simplify the modeling. A mathematical model is an approximation of a true physical system. If it were possible to model a physical system exactly, a mathematical analysis would be most likely impossible. Thus approximations, in the form of assumptions, must be made. These approximations reduce the accuracy of the predicted response. Some assumptions are made so that the modeling is not needlessly complicated by including negligible effects. Other assumptions are made to linearize a

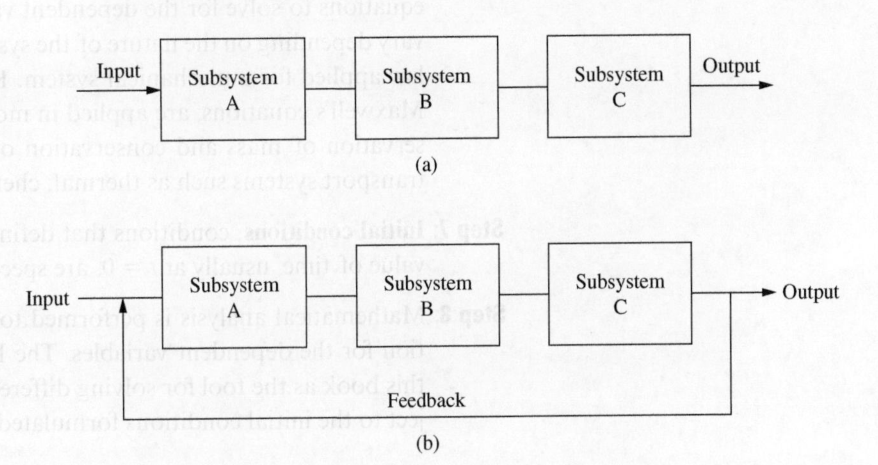

Figure 1.3

(a) The system is composed of uncoupled subsystems; (b) feedback from subsystem C into subsystem A couples the subsystems.

system. Because all systems are inherently nonlinear, some linearizing assumptions are essential. The validity of assumptions can be determined upon completion of the modeling.

This book considers mostly linear systems whose response is governed by linear ordinary differential equations. Thus, all nonlinear effects are assumed to be negligible. In addition, all system components are assumed to be discrete and their properties are lumped. This assumption implies that output variables have no variation with spatial coordinates. Specific assumptions leading to this result include that all bodies in mechanical systems are rigid, all springs and viscous dampers are massless, and the resistance of a wire can be modeled by a single resistor in a circuit. The discrete, lumped parameter system is implicit throughout this study.

Step 3: System components are identified and their behavior quantified. Relations defining the behavior of system components are the result of constitutive equations (equations defining the behavior of materials and specifying their properties) and equations of state (equations defining the physical state in which the component exists). Hooke's law relating stress and strain for an elastic solid is an example of a constitutive equation, while the ideal gas law is an example of an equation of state.

For example, the force-displacement relationships for springs are quantified in mechanical systems. The equations for the voltage drop across a resistor and stored energy in the electric field of a capacitor are defined for an electrical system. Appropriate equations for friction losses in pipes are established for hydraulic systems. Laws for heat transfer mechanisms are established for thermal systems. Rate of reaction equations are determined for components in chemical systems.

Step 4: **Variables** and **parameters** are defined. The independent variable in a dynamic system is time. The dependent variables in a dynamic system are related to the system output. Parameters are properties of system components that are assumed to remain constant as the dependent variables change. Examples of parameters include geometric quantities such as length and area; stiffnesses of springs; properties of electric circuit components such as resistance and capacitance; rates of reactions in chemical systems; and properties such as density, viscosity, and specific heat for fluid and thermal systems.

Step 5: A schematic diagram is drawn for the system. A mechanical system requires a free-body diagram. A circuit diagram may be drawn for an electrical system. A control volume diagram is drawn for the modeling of a transport system.

Step 6: Applicable physical laws are applied, resulting in an equation or a set of equations to solve for the dependent variables. The applicable physical laws vary depending on the nature of the system. Newton's second law is the basic law applied for a mechanical system. Kirchhoff's circuit laws, derived from Maxwell's equations, are applied in modeling discrete electric circuits. Conservation of mass and conservation of energy are often applied to model transport systems such as thermal, chemical, and biological systems.

Step 7: **Initial conditions**, conditions that define the state of the system at an initial value of time, usually at $t = 0$, are specified when required.

Step 8: Mathematical analysis is performed to determine the time-dependent solution for the dependent variables. The Laplace transform method is used in this book as the tool for solving differential equations derived in step 6 subject to the initial conditions formulated in step 7.

Step 9: The system output is determined from the mathematical solution obtained in step 8. The output is analyzed to attain the objectives of the modeling process.

Step 10: The model is validated. Model validation may include checking the validity of the assumptions, comparing numerical results or system performance to benchmark situations, or simulating the system in a laboratory.

Often, especially in design and synthesis, the modeling process is iterative. For example, in a design situation, the system components are not defined in advance. Also, the process may proceed with proposed components. If the desired objective is not achieved, the components may be changed.

1.4 System Components

The **components** of a system consist of anything that stores, dissipates, or adds energy to the system. Examples of system components in a mechanical system include springs, masses, friction elements, and actuators. Examples of components in an electrical system include resistors, capacitors, inductors, and voltage sources. In a transport system, components include storage tanks, piping systems, thermal insulation, and bellows. The system components traditionally have some type of parameter defining them.

A system component is classified as **passive** if it stores or dissipates energy. It is designated as **active** if it adds energy to the system. Energy may take different forms. It may be kinetic, potential, stored in an electric field, or stored in a magnetic field, among other forms.

Consider a mechanical system that has a displacement from equilibrium x and a velocity $\dot{x}$. The system consists of components of a spring, a rigid body, a viscous damper, and an actuator. If the spring is linear, it has a parameter called the stiffness or spring constant k. The spring stores potential energy of $\frac{1}{2}kx^2$ and is a passive component. The rigid body has a mass m and stores kinetic energy of $\frac{1}{2}m\dot{x}^2$. The rigid body is a passive element. Suppose the friction element is a linear viscous damper with the parameter of a viscous damping coefficient c; it then causes an energy dissipation of $c\dot{x}^2$ and it is also a passive component. If the actuator provides a force $F(t)$ to the system, the energy provided by that force is $\int_0^t F(t)\dot{x}dt$ and the actuator is an active component. If the actuator provides motion to the system $y(t)$ at the end of the spring not connected to the mass, it provides potential energy to the system as $\frac{1}{2}ky^2$ and it is an active component.

Consider a series LRC circuit consisting of a linear resistor with resistance R, capacitance C, and inductance L. The circuit is connected to a voltage source $v(t)$. The circuit has a current i flowing through it. The resistor dissipates energy at a rate Ri^2 and is a passive component. The capacitor stores energy of $\frac{1}{2}Cv^2$, where v is the potential difference across the capacitor, in an electric field. It is also a passive component. The inductor stores energy of $\frac{1}{2}Li^2$ in a magnetic field and is a passive component. The voltage source, which adds energy to the system, is an active component.

Consider a liquid-level system consisting of a tank of cross-sectional area A that contains a fluid of mass density ρ. The tank has an outlet into a piping system with a total resistance R. The flow rate into the tank is q. The tank which stores potential energy, $\frac{1}{2}\rho Agh^2$, is a passive component. The piping system, which dissipates energy proportional to the resistance, is also a passive component. The assumption is usually that the velocity of the fluid in the tank is very small and kinetic energy is neglected. The fluid flowing into the tank is a source of energy input and is an active component.

1.5 System Response

The response of a system depends on many factors, including system components and how the system is modeled. It has been established that this book deals with the dynamics of lumped parameter linear systems modeled by ordinary differential equations. The **order** of the system is a key factor in understanding the dynamics of such a system. The term "order" can be defined in a variety of ways, including a definition relating to energy. A definition of order is presented in Chapter 3, which relates order to a mathematical property of a system, called its transfer function. For now, it suffices to define the term as the order of the derivatives appearing in the system's mathematical model. In this context, the order of a system with one output is the highest order derivative appearing in the mathematical model of a system when it is expressed as a single differential equation. A **first-order system** is modeled by a first-order differential equation. A **second-order system** is modeled by a second-order differential equation. When the modeling of a system results in an integrodifferential equation, the order of the system is obtained by differentiating the equation to eliminate the integral. A **higher-order system** is modeled by a set of differential equations; a system modeled by three second-order differential equations is a sixth-order system.

The **free response** of a system is its response due to nonzero initial conditions and occurs in the absence of any other system input. A **forced response** occurs when the system is subject to a nonzero input for $t > 0$. When subject to a nonzero input, a linear system also has a free response caused by the sudden change in conditions. For linear systems, a **general response** is the sum of the forced response and the free response.

The **transient response** of a system refers to either its free response or to the system response shortly after input is changed. The **steady-state** response is the system's response after a long period of time. In many systems, the transient response decays and is insignificant after some time. In such a case, the steady-state response is the forced response. If the input is periodic, the steady-state response is periodic. When the energy added to the system through the input is continually increased, the system may not reach a steady state. The steady-state response of a linear system, when it exists, is independent of initial conditions. The mass-spring-viscous damper system of Figure 1.4(a) is at rest in equilibrium when an external periodic force is applied. Mathematical modeling of the system leads to the dynamic response of Figure 1.4(b), which clearly shows an initial transient response that decays, leaving the steady-state response.

Figure 1.5(a) illustrates the free response of a typical second-order system. Figure 1.5(b) illustrates the general response of a typical first-order system due to a step input. Transient system response is studied in Chapter 7. Figure 1.5(c) illustrates the steady-state response of a typical second-order system due to a periodic input. Steady-state responses are studied in Chapter 8.

Equilibrium is the balance achieved between competing forces. The term is often used to describe the state of a system when system variables do not change with time. A mechanical system is in static equilibrium when the resultant of external forces is zero. When current does not flow through a circuit, the circuit is in equilibrium. When the total heat transfer to or from a thermal system is zero, the system is in equilibrium. A chemical system is in equilibrium when the concentrations of reaction components are constant. In each of these systems, the equilibrium is disturbed when external conditions change. A switch may be closed in a circuit that allows current to flow, additional reactants may be added to a chemical system, or the ambient temperature may change in a thermal system. A transient response occurs when an equilibrium state is disturbed.

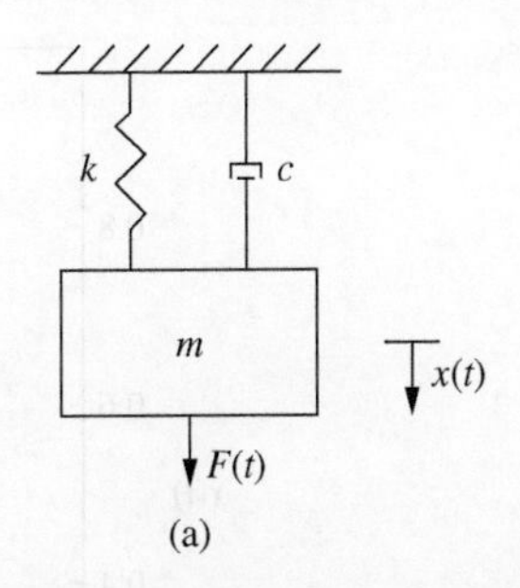

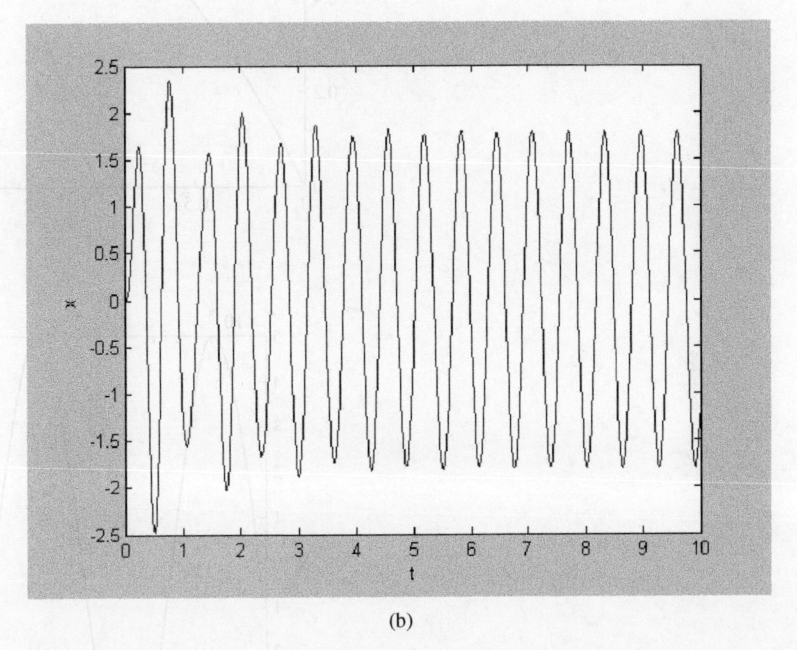

Figure 1.4

(a) A mechanical system is initially at rest in equilibrium, when an external periodic force is applied. (b) The response of a mechanical system at rest in equilibrium when suddenly subject to a periodic input begins with a transient response, which eventually decays, leaving only the steady-state response.

In system dynamics terms, input to the system disturbs an equilibrium position. Mathematical modeling of the system results in the prediction of the system response due to the input. Thus, mathematical models of systems are built around predicting changes from equilibrium. For this reason, dependent variables used in system analysis are often measured from the system's equilibrium position.

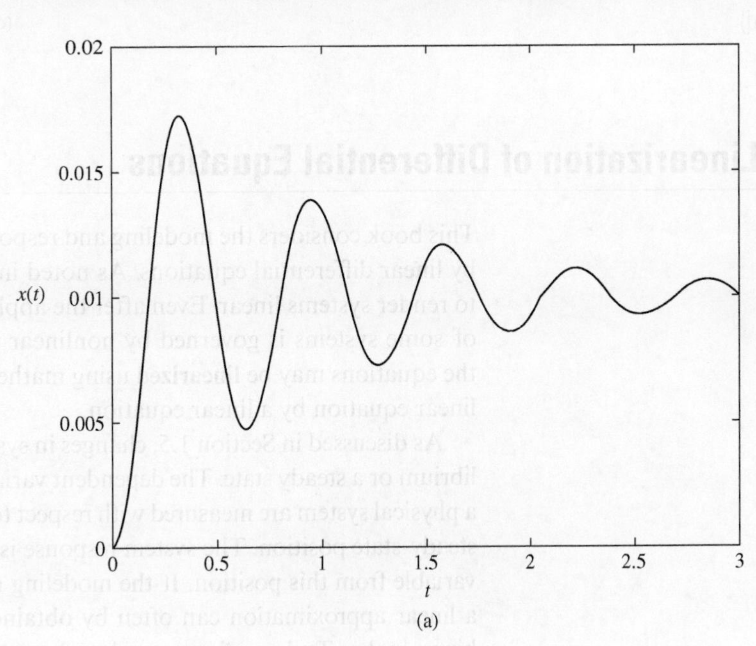

Figure 1.5

(a) Typical free response of a second-order system; (b) typical response of a first-order system due to a step input; (c) typical steady-state response of a second-order system due to a periodic input.

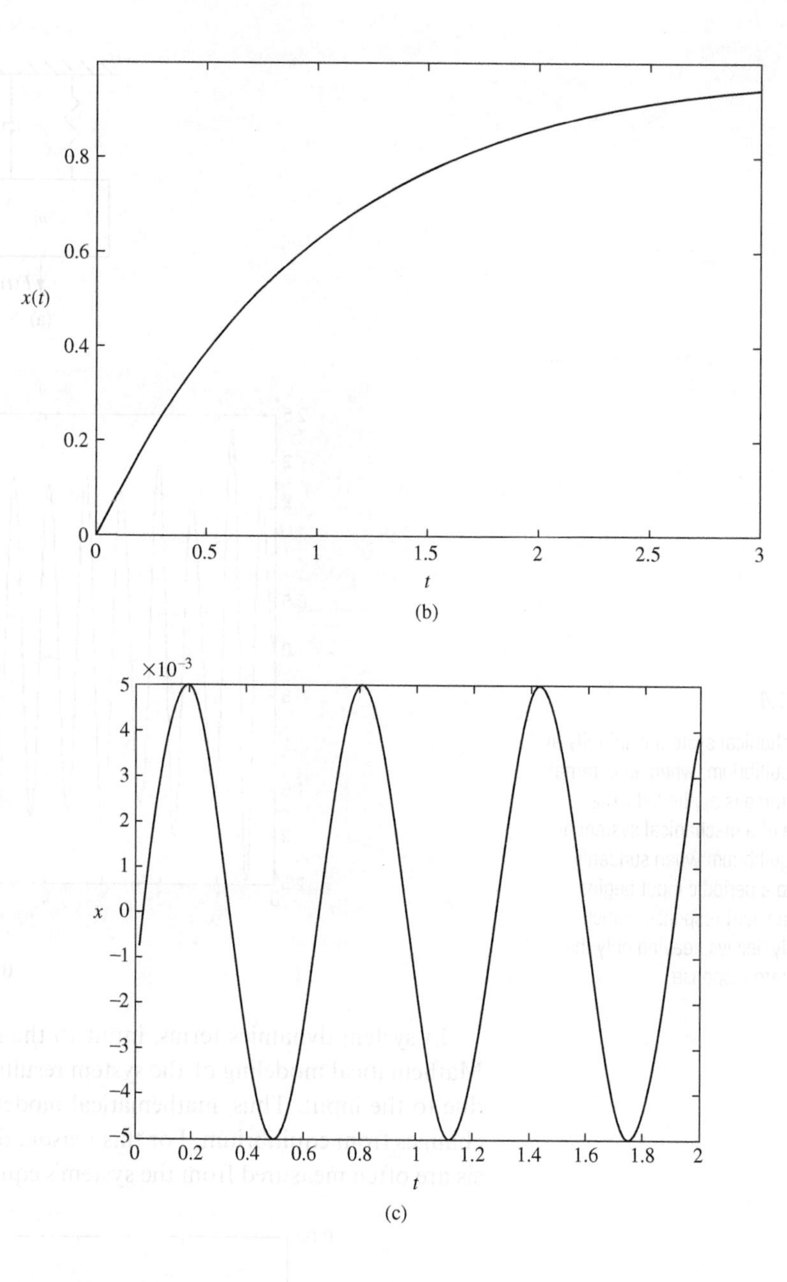

Figure 1.5
(Continued)

1.6 Linearization of Differential Equations

This book considers the modeling and response of linear systems, or systems governed by linear differential equations. As noted in Section 1.3, assumptions are often made to render systems linear. Even after the application of such assumptions, the behavior of some systems is governed by nonlinear differential equations. When appropriate, the equations may be **linearized** using mathematical methods to approximate the nonlinear equation by a linear equation.

As discussed in Section 1.5, changes in system input may disturb an established equilibrium or a steady state. The dependent variables used in the mathematical modeling of a physical system are measured with respect to an equilibrium position or from a defined steady-state position. The system response is a measure of the change in the dependent variable from this position. If the modeling results in a nonlinear differential equation, a linear approximation can often by obtained by expanding the nonlinear terms in a binominal or Taylor series expansion about the equilibrium position or the steady state.

Consider a differential equation of the form

$$\frac{d^2x}{dt^2} + f(x) = g(t) \tag{1.11}$$

Suppose $x = 0$ corresponds to the equilibrium position of the system. A McLaurin series is obtained by making a Taylor series expansion about $x = 0$. If $f(x)$ is continuous and continuously differentiable at $x = 0$, then it has a McLaurin series expansion of the form

$$f(x) = f(0) + x\frac{df}{dx}(0) + \frac{x^2}{2}\frac{d^2f}{dx^2}(0) + \cdots + \frac{x^n}{n!}\frac{d^nf}{dx^n}(0) + \cdots \tag{1.12}$$

In most cases, because x is measured from the system's equilibrium position, $f(0) = 0$. In this case, substituting Equation (1.12) into Equation (1.11) leads to

$$\frac{d^2x}{dt^2} + x\frac{df}{dx}(0) + \frac{x^2}{2}\frac{d^2f}{dx^2}(0) + \cdots + \frac{x^n}{n!}\frac{d^nf}{dx^n}(0) + \cdots = g(t) \tag{1.13}$$

If x is small, then $x^2 << x$, and higher-order terms in the McLaurin series expansion for $f(x)$ are negligible in comparison to the linear term. An approximate linear differential equation is obtained by truncating the McLaurin series expansion for $f(x)$ after the linear term, leading to

$$\frac{d^2x}{dt^2} + \left[\frac{df}{dx}(0)\right]x = g(t) \tag{1.14}$$

Equation (1.14) is an approximation to Equation (1.11) under the assumption of small x. After the solution is obtained, the assumption should be checked to validate the linear approximation.

Example 1.4

The differential equation governing the motion of a compound pendulum is given here as

$$\frac{d^2\theta}{dt^2} + \frac{3g}{2L}\sin\theta = 0 \tag{a}$$

where θ is the counterclockwise angular displacement of the system measured from the system's vertical equilibrium position. Linearize Equation (a).

Solution

The McLaurin series expansion for $f(\theta) = \sin\theta$ about $\theta = 0$ is

$$\sin\theta = \theta - \frac{1}{6}\theta^3 + \frac{1}{120}\theta^5 - \cdots \tag{b}$$

Truncating Equation (b) after the linear term leads to the approximation

$$\sin\theta \approx \theta \tag{c}$$

Using Equation (c) in Equation (a) leads to the linear differential equation

$$\frac{d^2\theta}{dt^2} + \frac{3g}{2L}\theta = 0 \tag{d}$$

Equation (d) is a linear approximation to Equation (a).

The approximation used in Equation (c) of Example 1.4 is called the **small angle approximation** and is often used to linearize differential equations whose dependent variable is an angular displacement. As shown in Table 1.4, the error in using Equation (c) to approximate $\sin\theta$ is only 2.1 percent for $\theta = 20°$ and 11.1 percent for $\theta = 45°$.

Table 1.4 Accuracy of Small Angle Approximation

θ (degrees)	θ (rad)	$\sin\theta$	% error
0	0	0	0
1	0.017453	0.017452	0.005077
3	0.052359	0.052335	0.045706
4	0.069812	0.069756	0.081276
5	0.087266	0.087155	0.127034
7	0.122172	0.121868	0.2492
10	0.174531	0.173646	0.509495
13	0.22689	0.224949	0.863169
15	0.261797	0.258816	1.151492
18	0.314156	0.309014	1.664039
20	0.349062	0.342017	2.059983
25	0.436328	0.422614	3.244952
30	0.523593	0.499995	4.719654
35	0.610859	0.573571	6.500963
40	0.698124	0.642782	8.609822
45	0.78539	0.707101	11.07183
50	0.872656	0.766039	13.91796

As an alternative to linearizing a derived nonlinear differential equation, the small angle approximation can be applied before the application of appropriate conservation laws. Displacements are identified in terms of the dependent variable at an arbitrary instant using the small angle approximation. For example, the change in length of a spring may be $a \sin\theta$, but when the small angle approximation is applied it is written simply as $a\,\theta$. The use of the small angle assumption at the beginning of the modeling often leads to the derivation of a linear differential equation.

The appropriate approximations used for trigonometric functions when the small angle approximation is used are

$$\sin\theta \approx \theta \tag{1.15a}$$

$$\cos\theta \approx 1 \tag{1.15b}$$

$$\tan\theta \approx \theta \tag{1.15c}$$

$$1 - \cos\theta \approx \frac{1}{2}\theta^2 \tag{1.15d}$$

Consider a system that has a steady-state response $x_s(t)$ due to a system input $f_s(t)$. The system input is changed to $F(t)$, and the resulting system response is $X(t)$. However, mathematical modeling of the system with this change in input leads to a nonlinear differential equation for $X(t)$. This system is linearized by considering the **perturbation variables**, $x(t)$ and $f(t)$, defined by

$$X(t) = x_s(t) + x(t) \tag{1.16a}$$

$$F(t) = f_s(t) + f(t) \tag{1.16b}$$

The perturbation variables measure changes from steady-state conditions. Equations (1.16) are substituted into the governing differential equations. Nonlinear terms involving $X(t)$ are linearized assuming small $x(t)$ and using an appropriate binomial or Taylor series expansion truncated after the linear term. The steady-state relation between $x_s(t)$ and $f_s(t)$ is used to simplify the equation. Perturbation variables become the dependent variables used in the system.

Suppose a nonlinearity is of the form of X^n. Use of Equation (1.16a) leads to

$$X^n = (x_s + x)^n$$
$$= x_s^n\left(1 + \frac{x}{x_s}\right)^n \tag{1.17}$$

Recall the binomial expansion

$$(1 + a)^n = 1 + na + \frac{1}{2!}n(n-1)a^2 + \frac{1}{3!}n(n-1)(n-2)a^3 + \cdots \quad |a| < 1 \tag{1.18}$$

The perturbation variable should be much smaller than the steady-state variable, thus $|\frac{x}{x_s}| << 1$ and Equation (1.18) is used in Equation (1.17) to give

$$X^n = x_s^n\left[1 + n\frac{x}{x_s} + \frac{1}{2!}n(n-1)\left(\frac{x}{x_s}\right)^2 + \cdots\right] \tag{1.19}$$

Equation (1.19) is linearized by truncating after the linear term, leading to

$$X^n \approx x_s^n + nx_s^{n-1}x \tag{1.20}$$

If the nonlinearity is not in the form of a power of the dependent variable, a Taylor series expansion is used to linearize the term. Suppose the nonlinear term is of the form $g(X)$. Then a Taylor series expansion for $g(X)$ about $X = x_s$ is

$$g(X) = g(x_s) + \left[\frac{dg}{dX}\right]_{X=x_s} x + \frac{1}{2!}\left[\frac{d^2g}{dX^2}\right]_{X=x_s} x^2 + \cdots \tag{1.21}$$

Truncating Equation (1.21) after the linear term leads to

$$g(X) = g(x_s) + \left[\frac{dg}{dX}\right]_{X=x_s} x \tag{1.22}$$

Example 1.5

The differential equation governing the level of a fluid in a reservoir is

$$A\frac{dH}{dt} + K\sqrt{H} = Q \tag{a}$$

where H is the liquid level in the reservoir, A is the cross-sectional area of the reservoir, Q is the input flow rate, and K is a constant. For a constant input flow rate q_s, the steady-state liquid level is h_s, where

$$K\sqrt{h_s} = q_s \tag{b}$$

Suppose the flow rate is perturbed from the steady state such that

$$Q(t) = q_s + q(t) \tag{c}$$

If $h(t)$ is the perturbation of the liquid level, then

$$H(t) = h_s + h(t) \tag{d}$$

Substitute Equations (b)–(d) into Equation (a), and linearize the system to obtain a linear differential equation for $h(t)$.

Solution

Substituting Equations (c) and (d) into Equation (a) leads to

$$\frac{d}{dt}[h_s + h(t)] + K\sqrt{h_s + h(t)} = q_s + q(t) \tag{e}$$

Since h_s is a constant, $\frac{dh_s}{dt} = 0$. Use of Equation (1.20) in Equation (e) with $n = 1/2$ leads to

$$\frac{dh}{dt} + K\left(\sqrt{h_s} + \frac{h}{2\sqrt{h_s}}\right) = q_s + q(t) \tag{f}$$

Equation (b) is used to simplify Equation (f) to

$$\frac{dh}{dt} + \frac{K}{2\sqrt{h_s}}h = q(t) \tag{g}$$

Example 1.6

In the mathematical modeling of nonisothermal systems with chemical reactions, the Arrhenius law is used to specify the dependence of the rate of reaction k on absolute temperature T:

$$k = \alpha e^{-\frac{E}{RT}} \tag{a}$$

where α is a constant of proportionality, E is the activation energy, and R is the gas constant. Suppose the temperature is perturbed from a steady-state temperature of T_s by a perturbation T_p. Determine a linear relationship between the perturbation in the rate of reaction k_p and the perturbation temperature.

Solution

The rate of reaction at the steady state is

$$k_s = \alpha e^{-\frac{E}{RT_s}} \tag{b}$$

Let k_p be the perturbation in the reaction rate. Using perturbation variables, Equation (a) is written as

$$k_s + k_p = \alpha e^{-\frac{E}{R(T_s + T_p)}} \tag{c}$$

Equation (1.22) is used to linearize the exponential function

$$e^{-\frac{E}{R(T_s + T_p)}} = e^{-\frac{E}{RT_s}} + \left[\frac{d}{dT}\left(e^{-\frac{E}{RT}}\right)\right]_{T=T_s} T_p \tag{d}$$

Noting that

$$\frac{d}{dT}\left(e^{-\frac{E}{RT}}\right) = \frac{E}{RT^2}e^{-\frac{E}{RT}} \tag{e}$$

Equation (d) becomes

$$e^{-\frac{E}{R(T_s+T_p)}} = e^{-\frac{E}{RT_s}} + \frac{E}{RT_s^2}e^{-\frac{E}{RT_s}}T_p \tag{f}$$

Using Equations (b) and (f) in Equation (c) leads to

$$k_p = \alpha\frac{E}{RT_s^2}e^{-\frac{E}{RT_s}}T_p$$

$$= \frac{k_s E}{RT_s^2}T_p \tag{g}$$

1.7 Unit Impulse Function and Unit Step Function

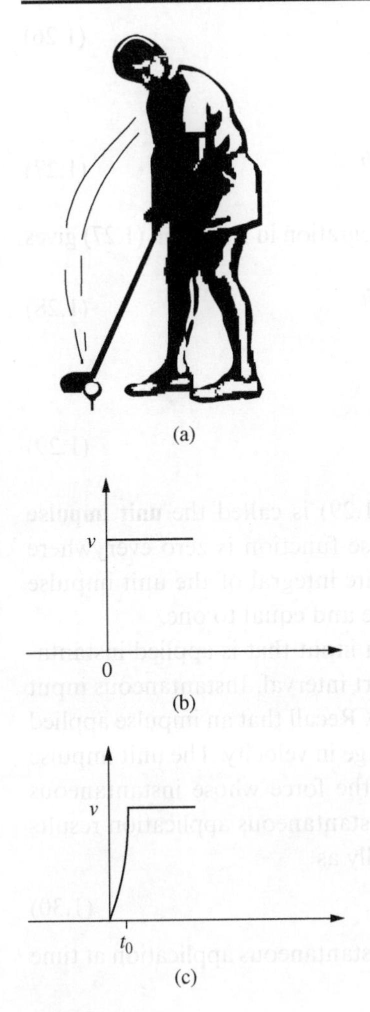

Figure 1.6
(a) Before the golfer strikes the ball, its velocity is zero, but immediately after being struck, the ball appears to have a velocity, v; (b) the discontinuous model of velocity of the ball; (c) the actual time-dependent velocity of the ball.

Nature is continuous; that is, physical systems occurring in nature have continuous properties and their responses are continuous functions of time. True discontinuities rarely occur in nature. However, for convenience, some phenomena are often modeled using discontinuous functions. Consider a golfer about to strike a golf ball, as illustrated in Figure 1.6(a). The ball is at rest on the tee while the club head is approaching and about to strike the ball with some velocity. It appears that the ball instantaneously is given a velocity and begins its trajectory; that is, it appears that the velocity of the ball is discontinuous in time. Immediately before the ball is struck, it has a velocity of zero, and immediately after it is stuck, it has a finite velocity, as illustrated in Figure 1.6(b). In reality, the velocity of the ball is continuous; the club is in contact with the ball for a finite time, after which the ball has attained its initial velocity. During this very short time, the force applied to the ball is very large, causing a large acceleration and leading to a finite velocity after contact. The actual time-dependent velocity may be as in Figure 1.6(c), where t_0 is a very short time.

A very large force applied over a very short interval of time is called an impulsive force, and its application results in the application of an impulse. When the club is in contact with the ball, it is applying an impulse to the ball. The modeling of the ball's subsequent motion is dependent on knowledge of the initial velocity imparted to it. Thus it is necessary to mathematically model impulsive forces.

Figures 1.7 and 1.8 (page 20) illustrate systems in which it is convenient to use discontinuous models for system input. The valve controlling the flow of liquid into the tank of Figure 1.7(a) is opened, changing the flow rate from Q_1 to Q_2. The actual time-dependent inlet flow rate may be given by Figure 1.7(b), whereas a useful discontinuous approximation is given in Figure 1.7(c). When the switch in the circuit of Figure 1.8 is closed, a finite time is required for the battery to reach its full potential. The true potential supplied by the battery may be given by Figure 1.8(b), whereas a useful discontinuous approximation is given in Figure 1.8(c).

These examples show that it is necessary and useful to approximate short continuous inputs with discontinuous approximations. To this end, mathematical functions that model these discontinuities are necessary.

1.7.1 Unit Impulse Function

The mathematical description of the function illustrated in Figure 1.9 (page 21) is

$$F(t; a) = \begin{cases} 0 & t < -\frac{a}{2} \\ \frac{1}{a} & -\frac{a}{2} < t < \frac{a}{2} \\ 0 & t > \frac{a}{2} \end{cases} \tag{1.23}$$

Independent of the value of a, the function defined by Equation (1.23) has the property

$$\int_{-\infty}^{\infty} F(t; a)\, dt = 1 \tag{1.24}$$

The function $\delta(t)$ is defined by

$$\delta(t) = \lim_{a \to 0} F(t; a) \tag{1.25}$$

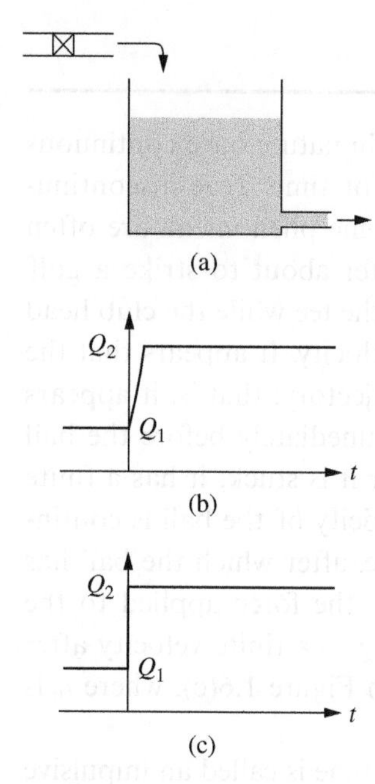

Figure 1.7

(a) Single tank system is at a steady state when the upstream valve is opened, increasing flow rate into the tank; (b) the actual time history of the inlet flow rate; (c) the discontinuous model of the inlet flow rate.

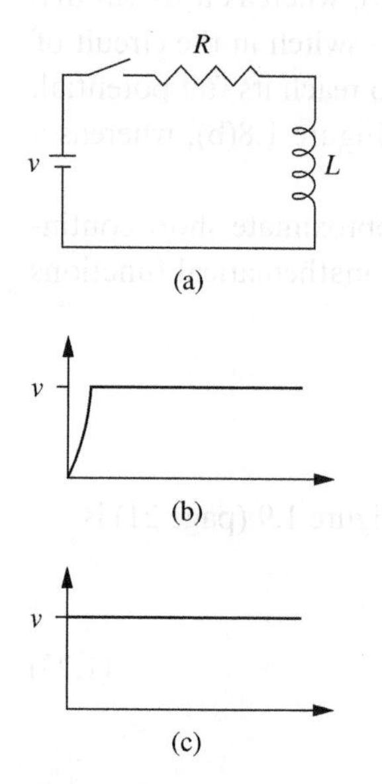

Figure 1.8

(a) When the switch in the LR circuit is closed at $t = 0$, a battery provides potential to the circuit; (b) the actual potential supplied to the circuit; (c) the discontinuous model of potential.

Substitution of Equation (1.23) into Equation (1.25) leads to

$$\delta(t) = \begin{cases} 0 & t \neq 0 \\ \infty & t = 0 \end{cases} \tag{1.26}$$

Integrating Equation (1.25) leads to

$$\int_{-\infty}^{\infty} \delta(t)dt = \int_{-\infty}^{\infty} \lim_{a \to 0} F(t; a)dt \tag{1.27}$$

Interchanging the order of the limit process and the integration in Equation (1.27) gives

$$\int_{-\infty}^{\infty} \delta(t)dt = \lim_{a \to 0} \int_{-\infty}^{\infty} F(t; a)dt \tag{1.28}$$

Using Equation (1.24) in Equation (1.28) yields

$$\int_{-\infty}^{\infty} \delta(t)dt = 1 \tag{1.29}$$

The function defined by Equations (1.26) and (1.29) is called the **unit impulse function** or the **Dirac delta function.** The unit impulse function is zero everywhere except at $t = 0$, where it is infinite. However, a definite integral of the unit impulse function over a region containing the impulse is finite and equal to one.

The unit impulse function is used to model system input that is applied instantaneously. In reality, the input is applied over a very short interval. Instantaneous input leads to an instantaneous change in a physical variable. Recall that an impulse applied to a mechanical system leads to an instantaneous change in velocity. The unit impulse function provides a mathematical representation of the force whose instantaneous application results in a unit impulse. A force whose instantaneous application results in an impulse of magnitude I is modeled mathematically as

$$F(t) = I\delta(t) \tag{1.30}$$

The mathematical representation of the force whose instantaneous application at time t_0 results in an impulse of magnitude I is modeled as

$$F(t) = I\delta(t - t_0) \tag{1.31}$$

The unit impulse function can be used to model the impulsive force applied to the golf ball as it is struck by the club head. Indeed, it can be used to model impulsive forces applied to many mechanical systems.

1.7.2 Unit Step Function

Consider the function defined as

$$u(t - t_0) = \int_{-\infty}^{t} \delta(\tau - t_0)d\tau \tag{1.32}$$

If $t < t_0$, then the integrand is identically zero everywhere over the range of integration and the value of the integral is zero. If $t > t_0$, then the range of integration includes the time where the impulse is applied; thus the value of the integral is one. In summary,

$$u(t - t_0) = \begin{cases} 0 & t < t_0 \\ 1 & t > t_0 \end{cases} \tag{1.33}$$

The function $u(t - t_0)$ defined in Equation (1.33) and illustrated in Figure 1.10 is called the **unit step function**. The unit step function is zero when its argument is

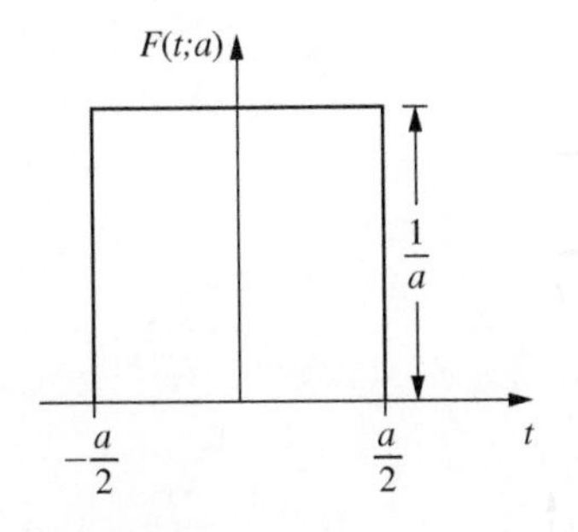

Figure 1.9

The unit impulse function: $\lim_{a \to 0} F(t;a) = \delta(t)$.

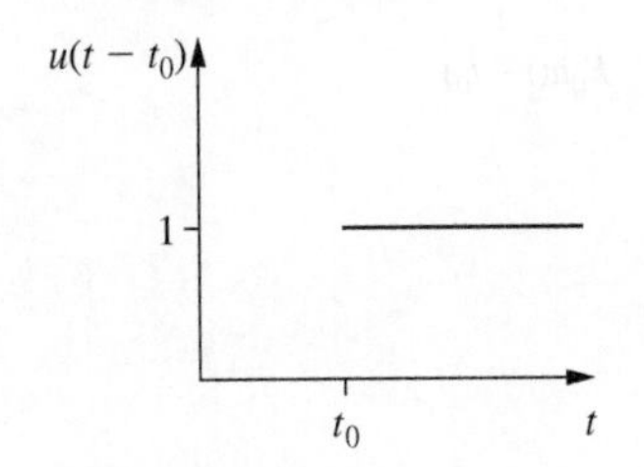

Figure 1.10

Definition of unit step function, $u(t - t_0)$.

negative and one when its argument is positive. The unit step function is used to model input that is discontinuous at time t_0. For example, if a switch in the electric circuit of Figure 1.8 is closed at $t = t_0$ leading to a battery of potential V being connected to the circuit, then the voltage supplied to the circuit through this source is mathematically modeled as $Vu(t - t_0)$. The mathematical model for the time-dependent flow rate into the tank of Figure 1.7 when the upstream valve is opened at $t = 0$ is $Q(t) = Q_1 + (Q_2 - Q_1)u(t)$.

Differentiation of Equation (1.33) with respect to time leads to

$$\frac{d}{dt}[u(t - t_0)] = \delta(t - t_0) \tag{1.34}$$

Thus the unit impulse function is the derivative of the unit step function.

Two useful integral formulas are

$$\int_0^t \delta(\tau - t_0)\, F(\tau) d\tau = F(t_0)\, u(t - t_0) \tag{1.35}$$

$$\int_0^t F(\tau)(\tau - a)\, d\tau = u(t - a) \int_0^t F(\tau) d\tau \tag{1.36}$$

Example 1.7

Develop a mathematical model for each of the forces of Figure 1.11.

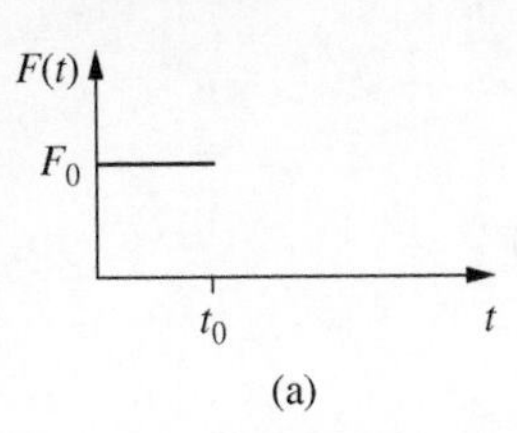

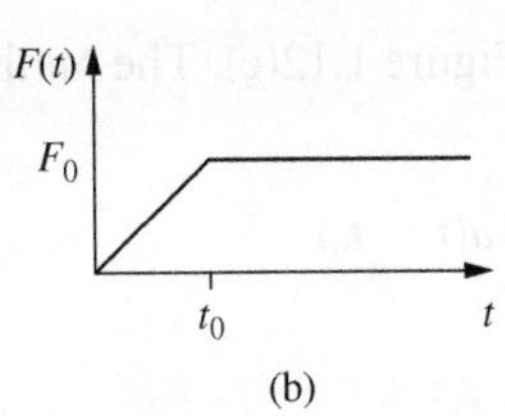

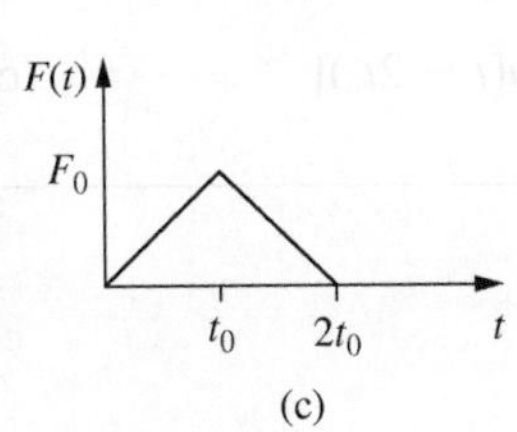

Figure 1.11

Time-dependent forces for Example 1.7: (a) step excitation removed at time t_0; (b) step excitation with rise time t_0; (c) triangular pulse excitation.

Solution

Each of the forces can be written as a linear superposition of forces, as illustrated in Figure 1.12 (page 22). Each force in the superposition has a mathematical representation that is written using the unit step function.

a. The force in Figure 1.11(a) is that of a step excitation applied at $t = 0$ and removed at $t = t_0$. The mathematical representation of the force is obtained using Figure 1.12(a) as

$$F(t) = F_0 u(t) - F_0 u(t - t_0) \tag{a}$$

b. The force of Figure 1.11(b) represents a step excitation, but with a finite rise time of t_0. The appropriate superposition for this force is shown in Figure 1.12(b), leading to

$$F(t) = F_0 \frac{t}{t_0} u(t) - F_0 \frac{t}{t_0} u(t - t_0) + F_0 u(t - t_0)$$

$$= \frac{F_0}{t_0}[tu(t) - (t - t_0)u(t - t_0)] \tag{b}$$

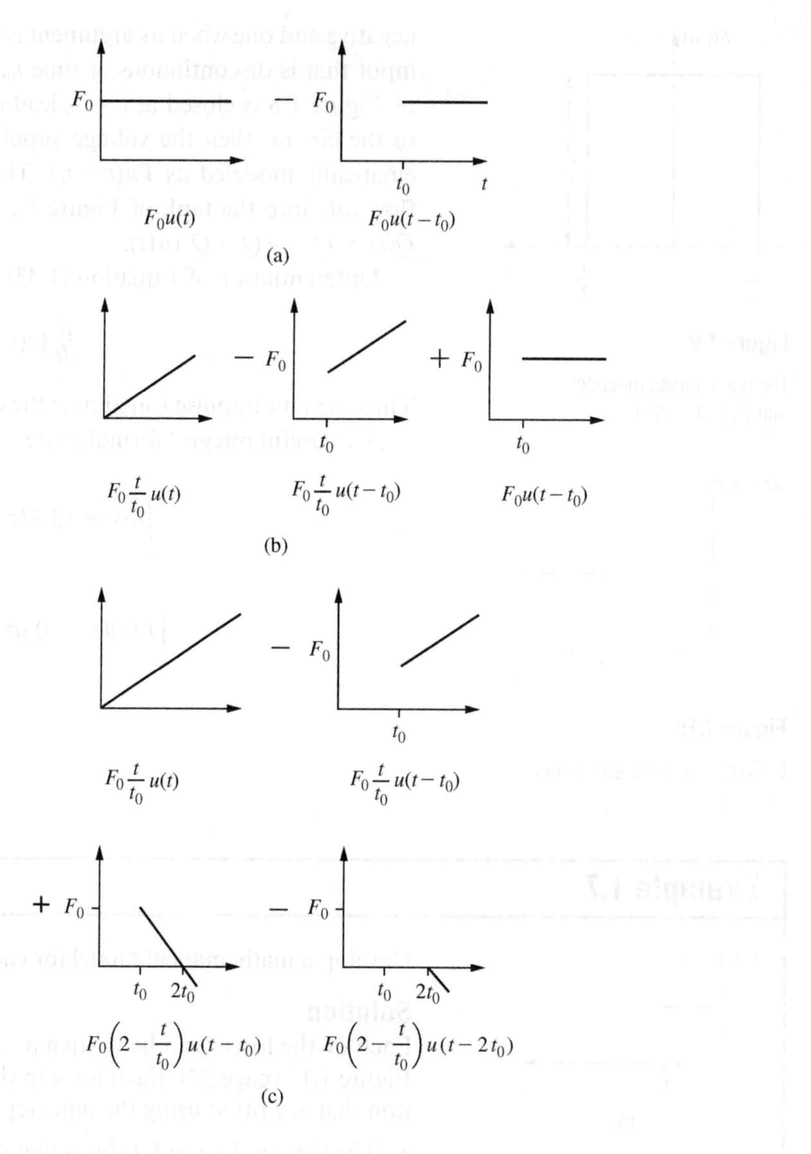

Figure 1.12

Superpositions used to develop mathematical model for forces of Example 1.7.

c. The superposition for the triangular pulse is illustrated in Figure 1.12(c). The mathematical model for this pulse is

$$\begin{aligned} F(t) &= F_0\frac{t}{t_0}u(t) - F_0\frac{t}{t_0}u(t-t_0) + \left(2F_0 - F_0\frac{t}{t_0}\right)u(t-t_0) \\ &\quad - \left(2F_0 - F_0\frac{t}{t_0}\right)u(t-2t_0) \\ &= \frac{F_0}{t_0}[tu(t) - 2(t-t_0)\,u(t-t_0) + (t-2t_0)\,u(t-2t_0)] \end{aligned} \quad \text{(c)}$$

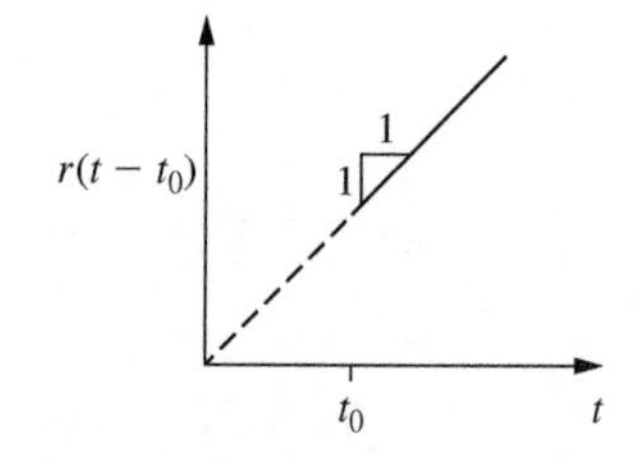

Figure 1.13

Delayed unit ramp function.

The **unit ramp function** is defined by

$$\begin{aligned} r(t) &= \int_0^t u(t)dt \\ &= tu(t) \end{aligned} \quad (1.37)$$

The delayed unit ramp function is

$$r(t-t_0) = (t-t_0)\,u(t-t_0) \quad (1.38)$$

The delayed unit ramp function is illustrated in Figure 1.13.

The mathematical form of the triangular pulse of Example 1.7(c) can be written in terms of ramp functions as

$$F(t) = \frac{F_0}{t_0}[r(t) - 2r(t - t_0) + r(t - 2t_0)] \tag{1.39}$$

1.8 Stability

An equilibrium position may be **stable, unstable**, or **neutrally stable**. When a stable equilibrium position is disturbed, a transient response, if developed, will die out when the disturbance is removed and the system will return to the equilibrium position. The transient response of a system about an unstable equilibrium position will continue when the disturbance is removed and equilibrium will not be reestablished. The response of a linear system about a neutrally stable equilibrium position is such that the system will establish a new equilibrium position when the disturbance is removed. A famous example illustrating the stability of equilibrium positions is illustrated in Figure 1.14 (page 24). In each case the sphere is at rest in equilibrium. If the sphere of Figure 1.14(a) is displaced slightly from this equilibrium position, the action of gravity causes the sphere to oscillate about its original equilibrium position. Friction causes those oscillations to decrease, and the system eventually returns to its equilibrium position. Thus the system is stable. When the sphere in Figure 1.14(b) is displaced from equilibrium, it indefinitely continues to move farther away from equilibrium; the system is unstable. If the sphere of Figure 1.14(c) is displaced from equilibrium, it remains in its displaced position; the equilibrium position is neutrally stable.

In this study, the stability of a dynamic system about an equilibrium position is determined from the response of the system when a unit impulse is applied. This response is called the system's **impulsive response**. If, after application of a unit impulse, the system approaches its equilibrium position in the steady state, then the system is stable. If no steady state exists for the impulsive response, the system is unstable. If, in the steady state, after the application of a unit impulse the response is bounded but does not return to the original equilibrium position, the system is neutrally stable.

Example 1.8

An actuator is often used in feedback control systems to modify transient system response. The system response is dependent on parameters of the actuator, called gains. Consider a feedback system in which an actuator is used such that the time-dependent impulsive responses of the third-order system for different values of the actuator gain are

a. $x(t) = 0.03e^{-0.25t} + 0.05e^{0.15t}\sin(120t + 0.5)$

b. $x(t) = 0.03e^{-0.25t} + 0.05e^{-0.15t}\sin(120t + 0.5)$

c. $x(t) = 0.03e^{-0.25t} + 0.05\sin(120t + 0.5)$

Discuss the stability of the system when each actuator is used.

Solution

a. Due to the presence of the exponential function with a positive exponent, the impulsive response of this system grows without bound, and a steady state does not exist. Thus this system is unstable.

b. Since both terms in the impulsive response contain exponential functions with negative exponents, the impulsive response approaches zero for large t and the system is stable.

c. The steady-state impulsive response is $0.05\sin(120t + 0.5)$, which is bounded but does not approach zero. Thus the system is neutrally stable.

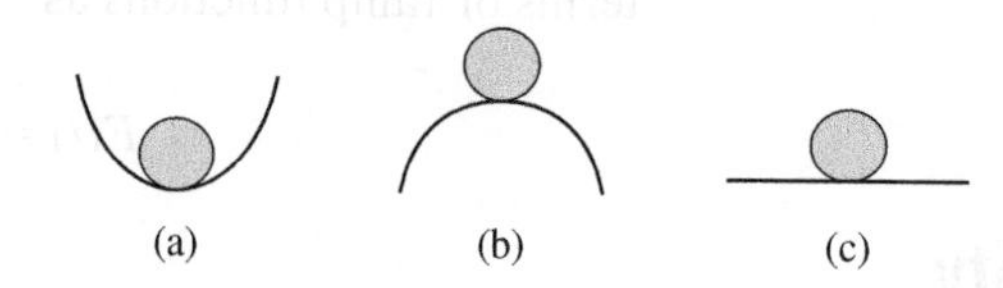

Figure 1.14

(a) Stable equilibrium position; (b) unstable equilibrium position; (c) neutrally stable equilibrium position.

The forced response of a system is unstable if it grows without bound. Instabilities in the forced response occur due to the nature of the system input as well as system parameters. Resonance, explained in Section 8.2, is an example of an instability that occurs in the forced response of an undamped dynamic system when the frequency of a periodic input is equal to a certain value, which depends on system parameters. Instabilities frequently occur in forced responses of nonlinear systems. An example is the "pull-in" instability in many microelectric mechanical (MEMS) devices, as discussed in Section 5.8. However, instabilities in the forced responses of linear systems are the exception rather than the rule.

1.9 MATLAB

MATLAB is a high-performance language with many applications in engineering. The capabilities of MATLAB relevant to this study are computation, programming, and graphics. Advanced capabilities of MATLAB include data acquisition, data analysis, modeling, prototyping, and building interfaces. MATLAB is a command-based language in that it acts on commands provided interactively or from a script.

MATLAB has all the features of any scientific programming language. It accepts formatted input and provides formatted output; variables are declared through arithmetic assignment statements that follow a standard hierarchy of operation; decision-making statements allow for development of structured programs; and subroutines may be written that allow for modular programming.

There are many significant differences between MATLAB and programming languages such as FORTRAN and C. The name MATLAB is a contraction of Matrix Laboratory. MATLAB treats each piece of data as a matrix. A scalar is treated and stored as a one-by-one matrix. However, unlike FORTRAN, MATLAB does not require data stored in matrices to be declared as such and dimensioned. MATLAB is free formatted and less syntax driven than most programming languages.

MATLAB can be used interactively, as in using a calculator or worksheet, or it can be used by writing a script, storing the script in a text file with a .m extension (called an M-file) and executing the M-file. Solutions to problems using MATLAB are presented in this book as M-files. Some M-files are listed as figures; in other cases, only the results obtained through using M-files are presented in the solution to some problems.

Special collections of M-files, called toolboxes, are available for use with MATLAB. The toolboxes extend the capability of MATLAB for a variety of applications. The Symbolic Toolbox allows for symbolic computation, that is, representation of data in terms of symbolic variables. Simulink is an add-on that allows system simulation by building systems from components and using block diagrams. The Control System Toolbox contains many M-files appropriate for computing and plotting dynamic system response. Use of the Symbolic Toolbox is illustrated in Chapter 2 in the Laplace transform solution of differential equations. Simulations of mechanical and fluid systems using Simulink are presented in Chapters 9–12.

MATLAB is best learned through continual usage. The language is introduced in this section. A compendium of commands used in this text is presented in Appendix C.

Each M-file presented is annotated with comment statements so that the meaning of each line is clear. The comment statements serve as a tutorial for the M-file. For further elaboration, the reader is encouraged to read one of the many books published on using MATLAB in the solution of engineering problems. In addition, MATLAB's help function provides valuable assistance with syntax. A major goal of this book is to provide the reader with an understanding of the modeling and response of dynamic systems. MATLAB is a valuable tool used to achieve this goal.

The following examples illustrate MATLAB's capabilities in computation, programming, and graphics that will be used throughout this text. Each example includes comment statements that begin with the percent (%) symbol.

Example 1.9

The steady-state response $x(t)$ of a second-order system of mass m, natural frequency ω_n, and damping ratio ζ due to a sinusoidal input of the form $F_0 \sin(\omega t)$ is obtained in Chapter 8 as

$$x(t) = X \sin(\omega t - \phi) \tag{a}$$

where the steady-state amplitude is

$$X = \frac{F_0}{m\omega_n^2} \frac{1}{\sqrt{(1 - r^2)^2 + (2\zeta r)^2}} \tag{b}$$

the steady-state phase is

$$\phi = \tan^{-1}\left(\frac{2\zeta r}{1 - r^2}\right) \tag{c}$$

and the frequency ratio is defined as

$$r = \frac{\omega}{\omega_n} \tag{d}$$

A special case occurs when $\zeta = 0$ and $r = 1$. In this case, a condition called resonance occurs and the response that grows without bound is given by

$$x(t) = \frac{F_0}{2m\omega_n} t \cos(\omega_n t) \tag{e}$$

Develop a MATLAB M-file that, when executed, will

a. input the values of all parameters;

b. calculate the frequency ratio and check to see if resonance occurs;

c. if resonance occurs, determine and plot the response given by Equation (e);

d. if resonance does not occur, determine and plot the response given by Equations (a)–(c).

Solution

The script of the MATLAB M-file is provided in Figure 1.15 (pages 26–28). The MATLAB screen obtained from execution of the M-file and the resulting plot file is shown in Figure 1.16 (page 29) for the case in which the response is given by Equation (a). Figure 1.17 (page 30) shows the same when resonance occurs, and the response is given by Equation (e).

The M-file is annotated with comment statements to illustrate the syntax of certain statements and the flow of the program. The following additional notes are made:

- The logic in the program is controlled using if loops. A logical statement follows the word if. If the statement is true, the succeeding statements are executed. When the else statement is reached, the control is shifted to the end statement for the loop. If the statement is false, control is shifted to the else statement and the statements following the word else are executed.

```
% Example1_9.m
% m file for Example 1.9
%
clear
disp('Steady-state response of second-order mechanical system')
disp('due to harmonic input of F(t)=F_0sin(omega*t)')
disp('Enter system parameters')
%
% Input for this m file is provided interactively when running the script.
% Text contained in single quotes, ' ' ,is printed on the screen.
% The following statements prompt the user to input parameters
% and assign a numerical value to the stated variable.
%
mass = input('Enter mass in kg m= ');
omegan = input('Enter natural frequency in rad/s omega_n= ');
zeta = input('Enter dimensionless damping ratio zeta= ');
omega = input('Enter frequency in rad/s omega= ');
F0 = input('Enter magnitude of input force in N F_0=  ');
%
% Calculate and display frequency ratio
%
disp('Dimensionless frequency ratio')
r=omega/omegan;
disp(r)
%
% Text required for annotating plot
% Set up strings for identifying system parameters on graph
%
str2(1)={['F(t)=',num2str(F0),'sin(',num2str(omega),'t) N']};
str2(2)={['m=',num2str(mass), ' kg,']};
str2(3)={['\omega_n=',num2str(omegan), ' rad/s']};
str2(4)={['\zeta=',num2str(zeta)]};
%
% Resonance is a special case which occurs when zeta=0 and r=1
%
% If resonance occurs
%
if zeta==0 && r==1
   disp('Undamped system with input frequency equal to natural frequency')
   disp('Resonance occurs and a steady state is not attained')
   %
   % When resonance occurs the response is calculated using Eq.(e) of Example
   % 1.9. The response is calculated for 0<t<tf
   %
   tf=15*pi/omega;
   %
   % The values of time at which the response is calculated are stored in the
   % vector t assigned using the linspace command.  The corresponding
   % displacements at these times are stored in the vector x.
   %
   t=linspace(0,tf,1001);
   c1=-F0/(2*mass*omegan);
```

Figure 1.15

Script of file Example1_9.m used for solution of Example 1.9.

```
    x=c1.*t.*cos(omegan.*t);
    %
    % Plot graph of system response
    %
    plot(t,x)
    %
    % Annotation of graph
    %
    xlabel('time (s)') % Prints label on x axis
    ylabel('x (m)') % Prints label on y axis
    title('Resonance response for undamped system') % Prints title on graph
    %
    % Set up string to identify response
    %
    str2(5)={['x(t)=',num2str(c1),'tcos(',num2str(omegan),'t) m']};
    xmax=max(x);
    %
    % The text will be printed on the graph beginning at the point
    % corresponding to t=0.05*tf, and x=0.7*xmax
    %
    text(0.05*tf,0.7*xmax,str2)
    %
    % If resonance occurs the above is the last statement that will be executed
    % Program control is shifted to the end statement closing this loop. The
    % statements following the else statement are executed only when
    % resonance does not occur
    %

else

    tf=6*pi/omega;
    %
    % For no resonance the steady-state amplitude is calculated using
    % Equation (b) of Example 1.9
    %
    amp=F0/(mass*omegan^2)/((1-r^2)^2+(2*zeta*r)^2)^0.5;
    %
    disp('Steady-state amplitude in m')
    disp(amp)
    %
    % Calculation of phase angle
    %
    % When zeta=0 the phase angle is either zero or pi, depending on the sign
    % of r. If r>0 then phi=0. If r<0 then phi=pi
    %
    if zeta==0
       if r<1
          phi=0;
       else
          phi=pi;
       end
    % If zeta>0 then the phase angle is calculated using
```

Figure 1.15
(Continued)

```
    % Equation (c) of Example 1.9
    else
       if r==1
          phi=pi/2;
       else
          phi=atan(2*zeta*r/(1-r^2));
       end
    end
    disp('Phase angle in rad')
    disp(phi)
    %
    % Steady-state response for non-resonant conditions
    %
    % The vector of times t is assigned using the linspace command.
    % The corresponding displacements are calculated using Equation (a) of
    % Example 1.9
    t=linspace(0,tf,1001);
    x=amp.*sin(omega.*t-phi);
    %
    % Plot of response with t on horizontal axis and x on vertical axis
    %
    plot(t,x)
    %
    % Provide title and labels for the graph
    %
    xlabel('time (s)')
    ylabel('x (m)')
    title('Steady-state response of second-order system')
    %
    % Defining text strings to annotate graph with values of system parameters
    %
    % The case when the phase angle is negative is treated as a special case
    % such that only one sign is printed
    %
    if phi<0
       str2(5)={['x(t)=',num2str(amp),'sin(',num2str(omega),'t+',...
       num2str(-phi), ') m ']};
    else
       str2(5)={['x(t)=',num2str(amp),'sin(',num2str(omega),'t-',...
       num2str(phi), ') m' ]};
    end
    %
    % Placement of text on graph
    % The text is started at t=0.05*tf and x=xp
    %
    C=min(x);
    D=max(x);
    xp=C+0.2*(D-C);
    text(0.05*tf,xp,str2)
end
%
% End of file
```

Figure 1.15

(Continued)

```
Steady-state response of second-order mechanical system
due to harmonic input of F(t)=F_0sin(omega*t)
Enter system parameters
Enter mass in kg m= 2
Enter natural frequency in rad/s omega_n= 100
Enter dimensionless damping ratio zeta= 0.4
Enter frequency in rad/s omega= 150
Enter magnitude of input force in N F_0= 100
Dimensionless frequency ratio
    1.5000

Steady-state amplitude in m
    0.0029

Phase angle in rad
   -0.7650
```

(a)

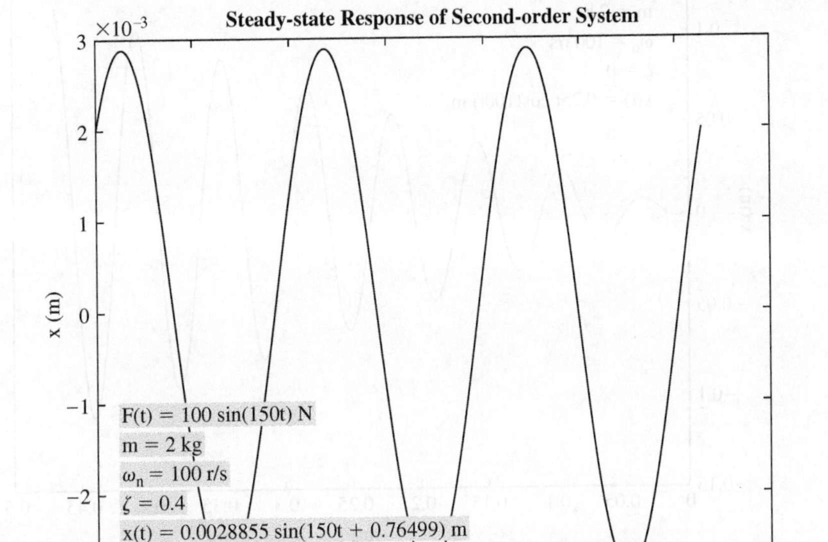

(b)

Figure 1.16

(a) MATLAB screen obtained when running file Example1_9.m for the nonresonant case;
(b) MATLAB-generated plot for system response.

- In the logical sentence within the first if statement, the two equal signs (==) are a relational operator used to test equality, and the double ampersand (&&) is a logical operator meaning "and". Other relational operators that may be used in logical sentences are testing less than (<), testing less than or equal to (<=), testing greater than (>), and testing greater than or equal to (>=). The logical operator meaning "or" is two vertical bars (||).
- The hierarchy of operations, from highest to lowest in arithmetic assignment statements for scalars (1×1 matrix) is

  (i) parentheses ();
  (ii) power (^);
  (iii) multiplication (*) and division (/);
  (iv) addition (+) and subtraction (−).

```
Steady-state response of second-order mechanical system
due to harmonic input of F(t)=F_0sin(omega*t)
Enter system parameters
Enter mass in kg m= 2
Enter natural frequency in rad/s omega_n= 100
Enter dimensionless damping ratio zeta= 0
Enter frequency in rad/s omega= 100
Enter magnitude of input force in N F_0=  100
Dimensionless frequency ratio
     1
Undamped system with input frequency equal to natural frequency
Resonance occurs and a steady state is not attained
```

(a)

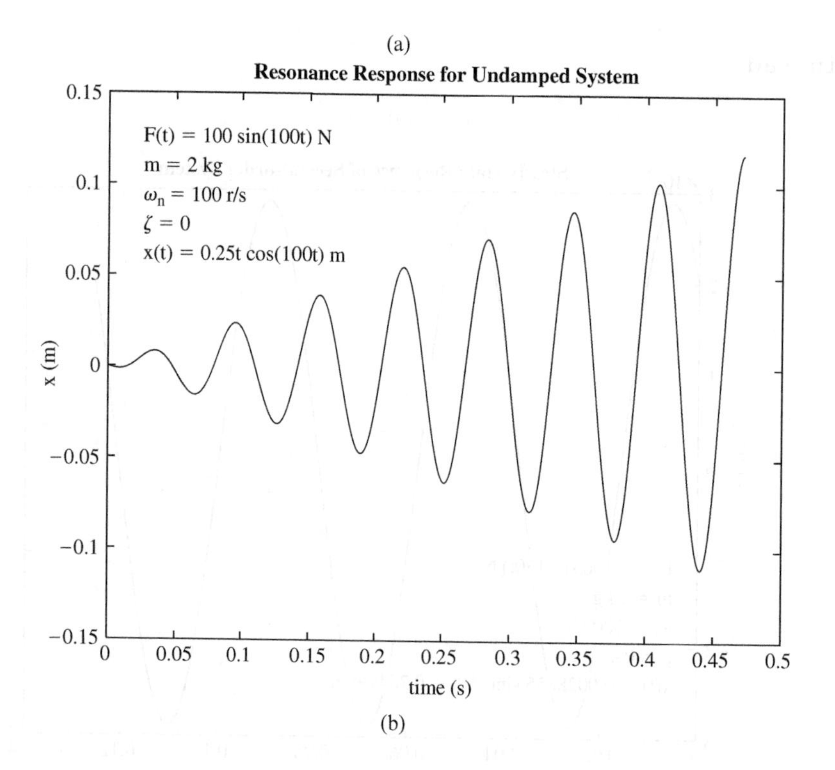

(b)

Figure 1.17

(a) MATLAB screen obtained when running file Example1_9.m for the resonant case;
(b) MATLAB-generated plot for the response of the system with resonance.

In the case of operations at equal level in the hierarchy, evaluation proceeds from left to right. Since MATLAB stores all data in matrices, additional levels are used for matrix operations such as matrix transpose.

- The "linspace" command is used to create a vector of values of the variable "t". Its syntax is "t=linspace(a,b,n)", where "a" is the first element of the vector t_1, "b" is the final element of the vector "t_n", and "n" is the number of elements in the vector. For example, if the goal is to create a vector beginning at 0 and ending at 5 with increments of 0.1 between the elements, the appropriate syntax is "linspace(0,5,51)".
- The ".*", "./", and ".^" operators are used for multiplication, division, and exponentiation on individual terms in a matrix. For example, the command "A.*B" means to take the corresponding elements of matrix A and matrix B and multiply them. The command does not mean matrix multiplication. In the same way, "A.^3" means to take each element of the matrix A and raise it to the third power. It does not mean A*A*A.

- The title, xlabel, ylabel, and text statements are used to annotate the graphs from the execution of an M-file. It is often easier to annotate a graph using the mouse and visible menus when viewing the graph.
- Unless otherwise specified, the value of a variable is printed after the execution of an assignment statement. The use of the semicolon (;) at the end of the statement suppresses the printing.
- An M-file is executed from the MATLAB command prompt by typing its name.

Example 1.10

In Chapter 5, Kirchhoff's Current Law and Kirchhoff's Voltage Law are applied to derive equations whose solutions lead to currents in multiloop circuits. Such a circuit analysis of the three-loop circuit of Figure 1.18 leads to the following equations, formulated in a matrix form:

$$\begin{bmatrix} R_1 + R_2 + R_3 & -R_2 & 0 \\ -R_2 & R_2 + R_4 + R_5 & -R_5 \\ 0 & -R_5 & R_5 + R_6 + R_7 \end{bmatrix} \begin{bmatrix} i_1 \\ i_2 \\ i_3 \end{bmatrix} = \begin{bmatrix} v_1 \\ 0 \\ v_2 \end{bmatrix} \tag{a}$$

Write a MATLAB M-file that inputs the resistances and dc voltages and that uses matrix methods to solve for the currents. Use the program to solve for the currents when $R_1 = 200\,\Omega$, $R_2 = 100\,\Omega$, $R_3 = 400\,\Omega$, $R_4 = 200\,\Omega$, $R_5 = 600\,\Omega$, $R_6 = 300\,\Omega$, $R_7 = 100\,\Omega$, $v_1 = 12\,\text{V}$, and $v_2 = 20\,\text{V}$.

Solution

The file Example1_10.m developed to solve this problem is shown in Figure 1.19 (pages 32–33). The output generated when the file is executed using the stated input values is shown in Figure 1.20 (page 33). Note the following about the programming in this M-file:

- The syntax for defining a matrix in MATLAB is to enclose the elements of the matrix in square brackets with entry by rows. Elements in the same row but in adjacent columns are separated by spaces. A semicolon (;) is used to separate rows.
- Contrary to programming languages such as FORTRAN and BASIC, the size of a matrix is not declared before defining the matrix. MATLAB checks for compatibility in performing arithmetic operations using matrices.
- A while loop is used to check the validity of the entered resistances. If an entered resistance is less than zero, the user is requested to re-enter the resistance. The program will continue to request re-entry of the data until a valid value is entered.
- The use of string variables is not the only method to develop the text to be displayed when requesting input data. The command num2str(i) converts the numerical value of the variable *i* into a text string for the purpose of display.

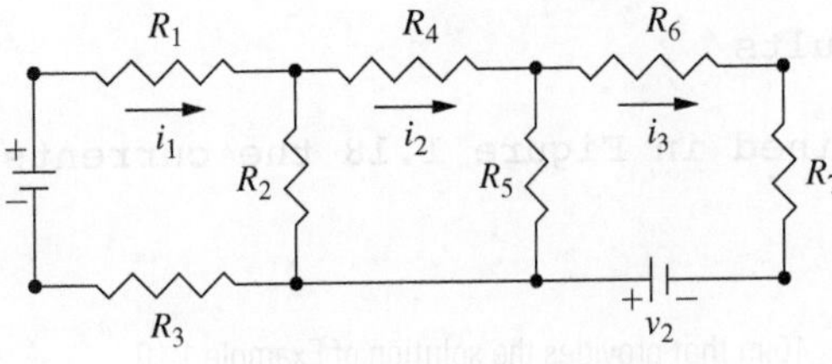

Figure 1.18

Circuit for Example 1.10.

```
% Example1_10.m
%
% Example 1.10
% m file inputs parameter values, sets up matrices, computes currents,
% and outputs values
%
disp('Example 1.10')
disp('Referring to Figure 1.18, input values of resistances, R(i), i=1,2,...,7,
each in ohms')
%
% Input resistances
%
for i=1:7
   str=['Please input R(',num2str(i), ')'];
   disp(str)
   R(i)=input('>> ');
   %
   % Each resistance must have a positive numerical value
   %
   while R(i)<=0
      disp('Invalid entry, resistance must be positive')
      str=['Please reenter R(',num2str(i), ')'  ];
      disp(str)
      R(i)=input(' ');
   end
end
disp('The coefficient matrix of resistances is ')
%
% Develop coefficient matrix
%
A=[R(1)+R(2)+R(3) -R(2) 0;-R(2) R(2)+R(4)+R(5) -R(5);
    0 -R(5) R(5)+R(6)+R(7)]
%
% Input dc voltages
%
disp('Referring to Figure 1.18, input values of dc voltages in V ')
for i=1:2
   str=['Please input v(',num2str(i), ')' ];
   disp(str)
   v(i)=input(' ');
end
disp('The right-hand side vector is')
%
% Develop right-hand side vector
%
y=[v(1);0;v(2)]
%
% The solution vector is obtained by multiplying the inverse of the
% coefficient matrix by the right-hand side vector
%
I=A^-1*y;
%
% Output results
%
disp('As defined in Figure 1.18 the currents are ')
```

Figure 1.19

Script of file Example1_10.m that provides the solution of Example 1.10.

```
for i=1:3
   str=['i(',num2str(i),')= ',num2str(I(i)), ' A' ];
   disp(str)
end
%
% End of file
```

Figure 1.19
(Continued)

```
Example 1.10
Referring to Figure 1.18, input values of resistances, R(i), i=1,2,...,7, each in
ohms
Please input R(1)
>> 200
Please input R(2)
>> 100
Please input R(3)
>> 400
Please input R(4)
>> 200
Please input R(5)
>> 600
Please input R(6)
>> 300
Please input R(7)
>> 100
The coefficient matrix of resistances is

A =

         700        -100           0
        -100         900        -600
           0        -600        1000

Referring to Figure 1.18, input values of dc voltages in V
Please input v(1)
 12
Please input v(2)
 20
The right-hand side vector is

y =

    12
     0
    20

As defined in Figure 1.18, the currents are
i(1)= 0.02087 A
i(2)= 0.026087 A
i(3)= 0.035652 A
>>
```

Figure 1.20
MATLAB output for Example 1.10.

Example 1.11

A transfer function, $G(s)$, a function introduced in Chapter 3, is used to determine dynamic system response. The Nyquist diagram, introduced in Chapter 8, is a diagram used in the analysis of system response and control system design. The Nyquist diagram is a plot of the real part of $G(j\omega)$ on the horizontal axis and the imaginary part of $G(j\omega)$ on the vertical axis, where ω is a parameter that varies such that $-\infty < \omega < \infty$. MATLAB has a command for developing the Nyquist plot from the definition of the transfer function. However, it is instructive to develop an M-file that, given the transfer function, develops and plots the Bode diagram. The script is to be developed so that it can be executed without modification for any transfer function. The transfer function is provided in a separate file called trans.m. Use the file to develop Nyquist diagrams for

$$G(s) = \frac{2s + 10}{s^2 + 2s + 10}$$

and

$$\frac{2}{(s + 2)(s + 5)}$$

Solution

The MATLAB script Example 1_11.m is shown in Figure 1.21. Execution of the script requires the presence of a function subprogram in the trans.m file.

```
% Example 1.11
%
% Program to develop Nyquist diagram for arbitrary transfer function
%
clear
%
% The variable omega is used as a parameter in developing the Nyquist
% diagram. The variable omega is a continuous variable that ranges from
% negative infinity to infinity. For computational purposes it becomes a
% discrete variable ranging between -200 and 200 with an increment of 0.1
% between successive values. The vector is set up using the linspace
% command. The first number is the first element of the vector,
% omega(1)=-200, the second number is the last value of 200. The
% last number is the number of elements in the vector assuming a constant
% difference between elements. With a difference of 0.1 between elements,
% there are 4001 elements between -200 and 200.
%
omega=linspace(-200,200,4001);
%
% The value of G is the transfer function that is taken from the function
% subprogram trans. The input to trans is j*omega where j is the square
% root of -1.
%
   G=trans(j.*omega);
%
% The real and imaginary parts of G are calculated.
%
   x=real(G);
   y=imag(G);
%
```

Figure 1.21

MATLAB script Example1_11.m developed for Example 1.11.

```
% The Nyquist diagram is a plot with real(G) plotted on the horizontal
% scale and imag(G) plotted on the vertical scale.
%
plot(x,y)
%
% The title and axis labels are added to the graph
%
title('Nyquist diagram')
xlabel('Re[G(j\omega]')
ylabel('Im[G(j\omega]')
%
% End of file
```

Figure 1.21

(Continued)

The appropriate scripts for trans.m used to provide the transfer function to Example1_11.m are shown in Figure 1.22. The Nyquist diagrams generated from the execution of Example1_11.m using these files for trans.m are in Figure 1.23 (page 36). Note the following regarding these files:

- The file uses $-200 \leq \omega \leq 200$ for development of the Nyquist diagram.
- MATLAB recognizes either i or j to represent $\sqrt{-1}$.
- MATLAB performs algebra using complex variables without requiring declaring them as complex. The MATLAB functions `real` and `imag` evaluate the real and imaginary parts of a complex number, respectively.
- The syntax used for the function subprogram trans.m is explained in the comments statements in Figure 1.22. When MATLAB encounters a statement such as G=trans(omega*j), it searches for the file trans.m. The first statement in trans.m must be the function statement defining what is transferred between the calling file and the function subprogram. In mathematical terms, in this example, this statement is comparable to G=trans(s), where trans is the name of the function, s is the independent variable, and G is the dependent variable. The argument list for a function statement may contain more than one independent variable whose values are specified from the calling program and more than one dependent variable whose values are calculated in the function subprogram. For example, the statement function [x,y] = h(a,b,c) indicates that $x = f(a, b, c)$ and $y = g(a, b, c)$; that a, b, and c are specified from the calling program; and that x and y are calculated in the function subprogram.

```
% Transfer function required for Example 1.11
% Name of function subprogram is trans
%
function t=trans(s)
%
% The input to the subprogram from the calling program is s, which will be
% used in calculations in the subprogram.
%
% The output from the function subprogram, which is returned to the calling
% program, is the variable t.
%
```

Figure 1.22

Function subprogram trans.m required for execution of MATLAB script Example1_11.m. The function subprogram provides the transfer functions for Example 1.11: (a) $G(s) = (2s + 10)/(s^2 + 2s + 10)$ and (b) $G(s) = 2/[(s + 2)(s + 5)]$.

```
% Calculating the transfer function for an input s
%
t=(2.*s+10)./(s.^2+2.*s+10);
%
% End of function subprogram
```

(a)

```
% Transfer function required for Example 1.11
% Name of function subprogram is trans
%
function t=trans(s)
%
% The input to the subprogram from the calling program is s, which will be
% used in calculations in the subprogram.
%
% The output from the function subprogram, which is returned to the calling
% program, is the variable t.
%
% Calculating the transfer function for an input s
%
t=2./(s+2)./(s+5);
%
% End of function subprogram
```

(b)

Figure 1.22
(Continued)

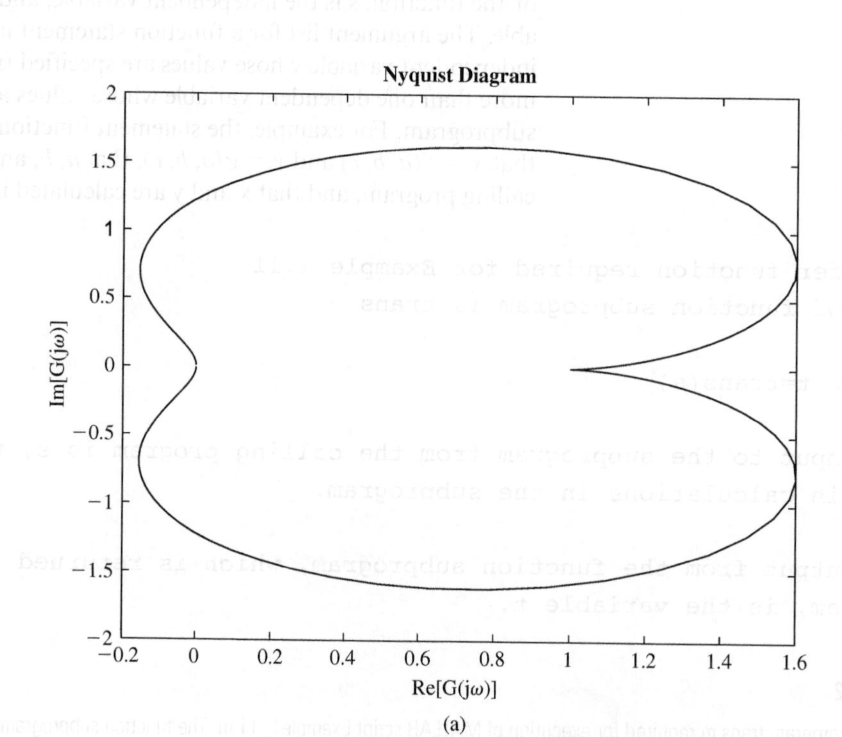

Figure 1.23
Nyquist diagrams obtained from execution of MATLAB script Example1_11.m with (a) $G(s) = (2s + 10)/(s^2 + 2s + 10)$ and (b) $G(s) = 2/[(s + 2)(s + 5)]$.

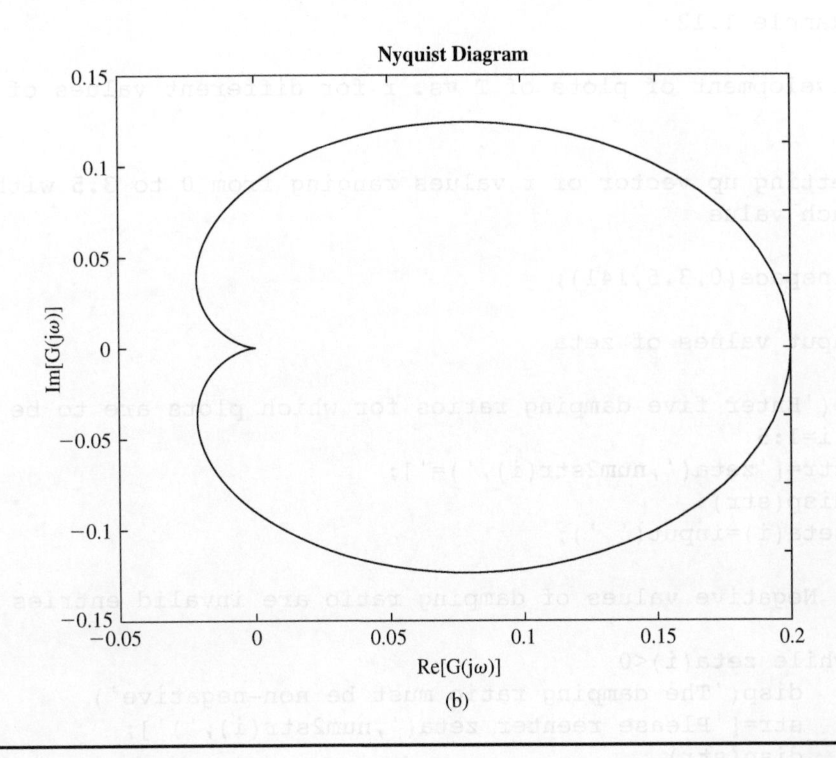

Figure 1.23
(Continued)

Example 1.12

It is often instructive to compare the response of a system for several values of a system parameter and thus be able to plot several responses on the same set of axes. Equation (8.82) presents a function used in the frequency response of mechanical systems

$$T = \sqrt{\frac{1 + (2\zeta r)^2}{(1 - r)^2 + (2\zeta r)^2}} \tag{a}$$

It is instructive to plot T vs. r for several values of the parameter ζ. Write a MATLAB M-file that plots on the same set of axes T vs. r for $0 < r < 3.5$ for $\zeta = 0.1, 0.25, 0.4, 0.6$, and 0.8.

Solution

The script for Example1_12.m is given in Figure 1.24 (page 38), while the plot developed from the execution of Example1_12.m is given in Figure 1.25 (page 39). Note the following regarding this program:

- When executed, the file produces a figure with five plots on the same set of axes. The extra arguments in the plot statement refer to the line style. Other arguments for line width and color can be used. These are listed by using MATLAB's Help function.
- The program uses the variable "T(i,:)" to refer to a double subscripted variable, $T_{i,j}$, but i is held constant while j varies over its entire range.

The preceding programs illustrate most of MATLAB's programming concepts, operations, and plotting functions that are used in the remainder of the book. Appendix C provides a list of important commands and their explanations.

```
% Example 1.12
%
% Development of plots of T vs. r for different values of zeta
%
%
% Setting up vector of r values ranging from 0 to 3.5 with 0.025 between
% each value
%
r=linspace(0,3.5,141);
%
% Input values of zeta
%
disp('Enter five damping ratios for which plots are to be developed')
for i=1:5
   str=['zeta(',num2str(i),')='];
   disp(str)
   zeta(i)=input(' ');
   %
   % Negative values of damping ratio are invalid entries
   %
   while zeta(i)<0
      disp('The damping ratio must be non-negative')
      str=['Please reenter zeta(',num2str(i),')'];
      disp(str)
      zeta(i)=input(' ');
   end
   % Calculating the values of T for each value of zeta
   %
   T(i,:)=((1.+(2*zeta(i).*r).^2)./((1-r.^2).^2+(2*zeta(i).*r).^2)).^0.5;
   %
end
%
% Plot of T vs. r for various values of zeta
%
plot(r,T(1,:),'-',r,T(2,:),'.',r,T(3,:),'-.',r,T(4,:),'--',r,T(5,:),'.')
%
% Annotating graph
%
title('T vs. r')
xlabel('r')
ylabel('T')
%
% Defining string variables to be used in legend
%
str1=['\zeta=',num2str(zeta(1))];
str2=['\zeta=',num2str(zeta(2))];
str3=['\zeta=',num2str(zeta(3))];
str4=['\zeta=',num2str(zeta(4))];
str5=['\zeta=',num2str(zeta(5))];
legend(str1,str2,str3,str4,str5)
%
% End of Example1_12.m
```

Figure 1.24

Script of MATLAB file Example1_12.m.

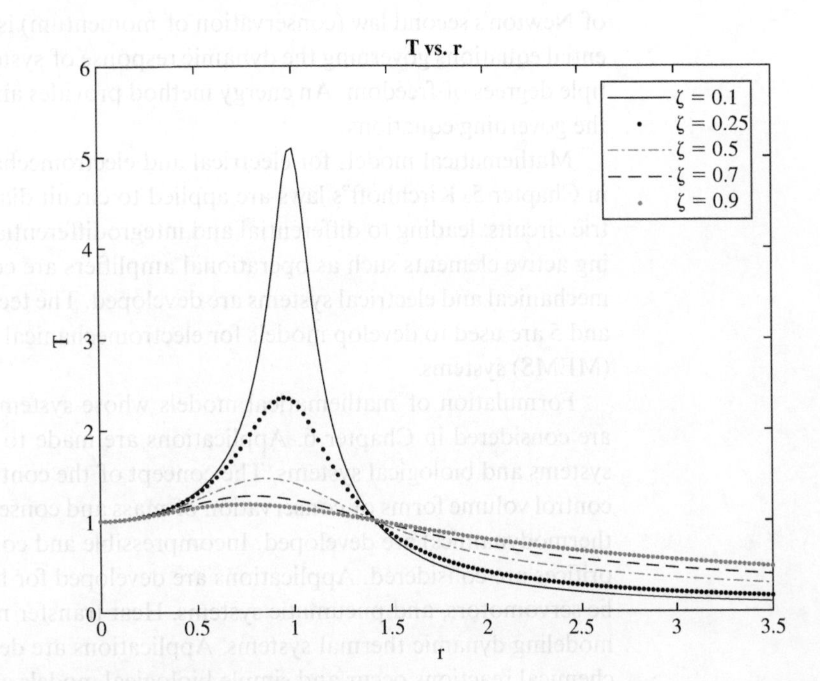

Figure 1.25
Plot generated from execution of Example1_12.m.

1.10 Scope of Study

This text is concerned with the modeling and response of dynamic systems. The text is presented in four distinct parts.

1.10.1 Mathematical Formulation

The mathematical formulation of models is covered in Chapters 2 and 3. Chapter 2 discusses the Laplace transform method and formulation of a problem in the "s" domain. The Laplace transform method provides a convenient way to solve linear ordinary differential and integrodifferential equations with constant coefficients. Chapter 3 introduces the concept of a transfer function for a system that is the Laplace transform of the output of a system to the Laplace transform of the system's input. State variables are introduced in Chapter 3 and are used in the development of the state-space formulation. The transfer function formulation and the state-space formulation of a system are the basic tools used for control system design.

1.10.2 Model Formulation

Model formulation, the development of an appropriate mathematical model for a dynamic system, is discussed in Chapters 4–6. Given a dynamic system, basic assumptions are made to simplify the model without compromising its integrity. Appropriate diagrams defining the system at an arbitrary instant are drawn. Basic conservation laws, constitutive equations, and equations of state are applied to formulate a mathematical model. The resulting model is a differential equation, an integrodifferential equation, or a set of coupled differential and/or integrodifferential equations. The methods of Chapter 3 are applied to develop transfer function models and state-space models.

Mathematical models for mechanical systems are formulated in Chapter 4. Fr body diagrams of the system at an arbitrary instant are drawn. An appropriate of Newton's second law (conservation of momentum) is applied, leading to the differential equations governing the dynamic response of systems with single as well as multiple degrees of freedom. An energy method provides an alternate method of deriving the governing equations.

Mathematical models for electrical and electromechanical systems are formulated in Chapter 5. Kirchhoff's laws are applied to circuit diagrams drawn of discrete electric circuits, leading to differential and integrodifferential equations. Circuits containing active elements such as operational amplifiers are considered. Analogies between mechanical and electrical systems are developed. The techniques studied in Chapters 4 and 5 are used to develop models for electromechanical and microelectric mechanical (MEMS) systems.

Formulation of mathematical models whose systems involve transport processes are considered in Chapter 6. Applications are made to fluid, thermal, and chemical systems and biological systems. The concept of the control volume is introduced, and control volume forms of conservation of mass and conservation of energy (first law of thermodynamics) are developed. Incompressible and compressible flows in pipes and orifices are considered. Applications are developed for liquid level problems, hydraulic servomotors, and pneumatic systems. Heat transfer mechanisms are considered in modeling dynamic thermal systems. Applications are developed for systems in which chemical reactions occur and simple biological models are considered.

1.10.3 System Response

System response is the focus of Chapters 7 and 8. The transfer function, developed from the mathematical model of a system, is used to study system response. Transient responses of first-order, second-order, and higher-order systems are studied in Chapter 7. The free response of each system is considered, as well as the response due to an impulsive input and a unit step input. The transient response is characterized in terms of system parameters.

System response due to periodic input is developed in Chapter 8. The sinusoidal transfer function is used to develop the steady-state response. Frequency response, the dependency of steady-state response on the frequency of periodic input, is studied. Graphical methods are introduced to study frequency response. Applications are made to electrical and mechanical systems, with emphasis on vibration isolation and electronic filter design.

1.10.4 Introduction to Control Systems

An introduction to feedback control systems and their effect on the response of dynamic systems is considered in Chapters 9–11. Chapter 9 focuses on the types of control systems and on the analysis and response of systems with feedback control. Chapter 10 focuses on the root-locus method for controller and compensator design, while Chapter 11 focuses on compensator design using principles of frequency response.

Modern control systems are analyzed using the state-space methods introduced in Chapter 12. The governing equations are transformed into the state space, and matrix methods are used to determine the system response.

MATLAB is used as an analysis tool in Chapters 7–12. Examples are presented using MATLAB to perform numerical computations and graphically illustrate system response. Simulink is used in Chapters 9–12 to build and numerically evaluate models.

Each subsequent chapter has a section titled "Further Examples," which illustrate in some detail the concepts of the current and preceding chapters. Each chapter concludes with a summary of the important points in the chapter and a recap of the most important equations from the chapter. Exercises are presented at the end of each chapter.

1.11 Summary

1.11.1 Chapter Highlights

The important points of Chapter 1 are as follows:

- A system is a set of components working together to achieve a desired output from a specified input. A system is dynamic if its output is a function of the history of the input.
- A dimension is a representation of how a physical quantity is expressed quantitatively. Every numerical representation of a physical quantity must be accompanied by appropriate units.
- The principle of dimensional homogeneity requires that all additive terms in an equation have the same dimensions and that all arguments of transcendental functions are dimensionless.
- Mathematical modeling is a process by which the output of a system is determined by knowing the system input. A formal process is established and applied throughout this book.
- All physical systems are inherently nonlinear. Assumptions are made to approximate systems using linear differential equations. A formal linearization process may be used if certain assumptions are valid.
- The unit impulse function is the mathematical model of a force whose instantaneous application results in the application of a unit impulse.
- The unit step function is used to mathematically model system input that has finite changes at discrete time.
- The stability of a system about an equilibrium position is defined in relation to the impulsive response of the system.
- MATLAB is a valuable tool that can be used to aid in the numerical determination and graphical representation of system response.

1.11.2 Important Equations

- Definition of unit impulse function

$$\delta(t) = \begin{cases} 0 & t \neq 0 \\ \infty & t = 0 \end{cases} \tag{1.26}$$

$$\int_{-\infty}^{\infty} \delta(t)dt = 1 \tag{1.29}$$

- Definition of unit step function

$$u(t - t_0) = \begin{cases} 0 & t < t_0 \\ 1 & t > t_0 \end{cases} \tag{1.33}$$

- Definition of unit ramp function

$$r(t - t_0) = (t - t_0)u(t - t_0) \tag{1.38}$$

Short Answer Problems

SA1.1 What are the English units of mass?

SA1.2 What are the SI units of mass?

SA1.3 In the *F-L-T* system, what are the dimensions of energy?

SA1.4 In the *M-L-T* system, what are the dimensions of energy?

SA1.5 In the *F-L-T* system, what are the dimensions of electric resistance?

SA1.6 What are the English units of electric resistance?

SA1.7 In the *M-L-T* system, what are the dimensions of mass flow rate?

SA1.8 In the *M-L-T* system, what are the dimensions of concentration of a substance?

SA1.9 In the *M-L-T* system, what are the dimensions of dynamic viscosity?

SA1.10 In the *M-L-T* system, what are the dimensions of impulse?

SA1.11 What are the SI units of dynamic viscosity?

SA1.12 What is the definition of the unit impulse function as applied to a mechanical system?

SA1.13 An electrical system has a voltage spike totaling 26 V·s. Express this spike in mathematical form.

SA1.14 A mechanical system is subject to an impulse of 0.4 N·s. Express this in mathematical form.

SA1.15 An electrical system is subject to the voltage input shown in Figure SA1.15. Write a mathematical expression for the voltage.

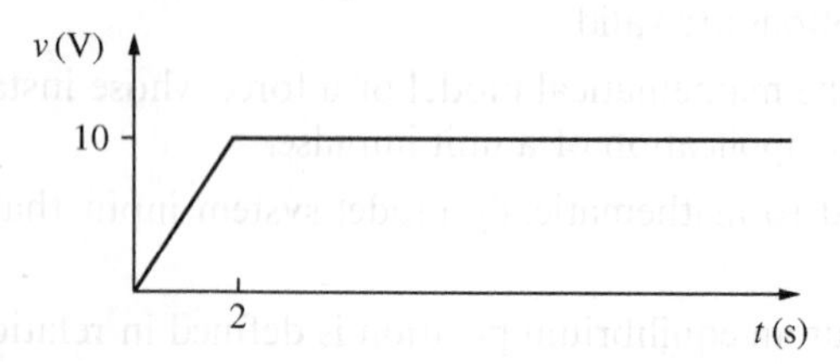

Figure SA1.15

SA1.16 A mechanical system is subject to a force that is illustrated in Figure SA1.16. Write a unified mathematical expression for the force.

SA1.17 What is the total impulse imparted to a mechanical system from the force shown in Figure SA1.16?

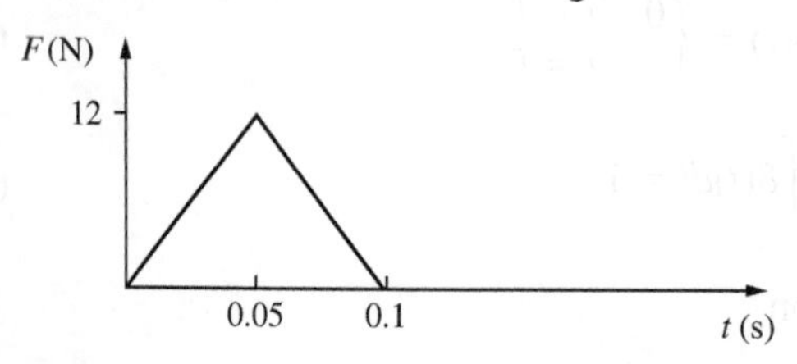

Figure SA1.16

SA1.18 Linearize the function $(1 + 0.2x)^{3/2}$ for small x.

SA1.19 Linearize the function $(1 - 0.2x)^{1/4}$ for small x.

SA1.20 Linearize the function $(1 + 0.3x)^{-0.76}$ for small x.

SA1.21 Linearize the function e^{-2x} for small x.

SA1.22 Linearize the function $\sin\theta \cos 2\theta$ for small θ.

SA1.23 Linearize the function $\tan 0.3\theta$ for small θ.

SA1.24 Linearize the function $\sin\theta\cos^2\theta$ for small θ.

SA1.25 Linearize the differential equation assuming small θ:

$$\frac{d^2\theta}{dt^2} + \cos\theta\frac{d\theta}{dt} + \tan\theta = 6\sin 2t$$

SA1.26 Linearize the differential equation assuming small θ:

$$\frac{d^2\theta}{dt^2} + 4t\cos\theta\frac{d\theta}{dt} + 16\sin\left(\frac{1}{2}\theta\right)\cos(3\theta) = 6\sin 2t$$

SA1.27 Fill in the blanks from the choices in parentheses regarding the differential equation

$$\frac{d^2\theta}{dt^2} + 4t\frac{d\theta}{dt} + 12\theta = 6\sin 2t$$

The equation is a(n) ________________ (ordinary, partial) differential equation. It has an independent variable of ________________ (θ, t) and a dependent variable of ________________ (θ, t). It is a ________________ (linear, nonlinear), ________________ (homogeneous, nonhomogeneous) differential equation with ________________ (constant, variable) coefficients.

SA1.28 Fill in the blanks from the choices in parentheses regarding the differential equation

$$\frac{d^2\theta}{dt^2} + 4\cos\theta\frac{d\theta}{dt} + 12\sin\theta = 0$$

The equation is a(n) ________________ (ordinary, partial) differential equation. It has an independent variable of ________________ (θ, t) and a dependent variable of ________________ (θ, t). It is a ________________ (linear, nonlinear), ________________ (homogeneous, nonhomogeneous) differential equation with ________________ (constant, variable) coefficients.

Problems

1.1 Newton's law of cooling is used to calculate the rate at which heat is transferred by convection to a solid body at a temperature T from a surrounding fluid at a temperature T_∞. The formula is

$$\dot{Q} = hA(T - T_\infty) \quad \text{(a)}$$

where $\dot{Q}$ is the rate of heat transfer (rate of energy transfer), A is the area over which heat is transferred by convection, and h is the heat transfer coefficient (also called the film coefficient). Use Equation (a) to determine the basic dimensions of h. Suggest appropriate units for h using the English system and the SI system.

1.2 The Reynolds number is used as a measure of the ratio of inertia forces to the friction forces in the flow of a fluid in a circular pipe. The Reynolds number (Re) is defined as

$$\text{Re} = \frac{\rho V D}{\mu} \tag{a}$$

where ρ is the mass density of the fluid, V is the average velocity of the flow, D is the diameter of the pipe, and μ is the dynamic viscosity of the fluid. Show that the Reynolds number is dimensionless.

1.3 The relationship between voltage and current in a capacitor is $i = C\frac{dv}{dt}$. Use this relation to determine the basic dimensions of the capacitance C.

1.4 The natural frequency ω_n of a mechanical system of mass m and stiffness k is calculated by $\omega_n = \sqrt{k/m}$. (a) What are the dimensions of ω_n? (b) A mass-spring system of mass 100 g has a natural frequency of 20 Hz. What is the stiffness of the system?

1.5 Carbon nanotubes are new materials that consist of carbon atoms and are of significant interest because of their good conductivity properties, light weight, and high strength. (a) The density of a carbon nanotube is 1300 kg/m^3. A carbon nanotube is a tube with the diameter of 1 carbon atom, $d = 0.68$ nm. Determine the mass of a carbon nanotube whose length is 80 nm. (b) The elastic modulus of a carbon nanotube is $E = 1.1$ TPa. What is its elastic modulus in pounds per square inch? (c) If modeled as a fixed-fixed beam, the fundamental frequency of the nanotube is

$$\omega = 22.37\sqrt{\frac{EI}{\rho A L^4}}$$

where $I = \frac{\pi}{4}r^4$ is the cross-sectional moment of inertia of the beam and A is its cross-sectional area. Calculate the fundamental frequency in hertz of a nanotube of length 80 nm.

1.6 During 24 hours of operation, a motor expends 900 kW·hr of energy. Noting that a horsepower (hp) is a unit of power such that 1 hp = 550 ft·lb/s, determine the power delivered by the motor in horsepower.

1.7 The density of water is 1.94 slugs/ft^3. Express the density of water in kilograms per cubic meter (kg/m^3).

1.8 A train is traveling at a speed of 180 km/hr and is decelerating at 6 m/s^2. Assuming uniform deceleration, how far will the train travel, in miles, before it stops?

1.9 The differential equation governing the angular velocity ω of a shaft of mass moment of inertia J, acted on by a torque T and attached to bearings of torsional damping coefficient c_t, is

$$J\frac{d\omega}{dt} + c_t\omega = T \tag{a}$$

Use Equation (a) to determine the appropriate dimensions of c_t.

1.10 In an armature controlled dc servomotor, the torque applied to the armature due to a magnetic coupling field is given by

$$T = K_a i_a i_f \tag{a}$$

and the back electromotive force (voltage) generated as the shaft rotates through the magnetic field is

$$v = K_f i_f \omega \tag{b}$$

where i_a is the current in the armature circuit, i_f is the current in the field circuit, and ω is the angular velocity of the shaft. Show that the constants K_a and K_f have the same dimensions. Referring to Table 1.2, what physical quantity has the same dimensions as these constants?

1.11 The rate of heat transfer by radiation from one body to another is given by

$$\dot{Q} = \sigma A \epsilon (T^4 - T_b^4) \tag{a}$$

where A is the area of heat transfer, ϵ is the dimensionless emissivity of the body, T is the temperature of the body, T_b is the temperature of the radiating body, and the Stefan-Boltzmann constant is

$$\sigma = 1.73 \times 10^{-7}\ \text{Btu/ft}^2\cdot\text{hr}\cdot\text{R}^4 = 5.67 \times 10^{-8}\ \text{W/m}^2\cdot\text{K}^4 \tag{b}$$

(a) What are the basic dimensions of $\dot{Q}$? (b) The differential equation governing the transient temperature in a body due to radiation heat transfer is

$$\rho c\frac{dT}{dt} + \sigma\epsilon(T^4 - T_b^4) = 0 \tag{c}$$

where ρ is the mass density of the body and c is its specific heat. The body is in a steady state with a uniform temperature defined by a temperature T_s when the temperature of the radiating body is suddenly changed from T_{bs} to T_{b1}. Let T_1 be perturbation in temperature from the steady state such that the temperature T is

$$T = T_s + T_1(t) \tag{d}$$

Substitute Equation (d) into Equation (c), use the binomial theorem to expand the nonlinear term, and derive a linearized differential equation to solve for $T_1(t)$.

1.12 A nonlinear differential equation governing the motion of a mechanical system is

$$\frac{1}{3}mL^2\ddot{\theta} + \frac{1}{4}cL^2\dot{\theta} + kL^2\sin\theta\cos\theta = 0 \tag{a}$$

Assuming small θ, derive a linearized differential equation for the system.

1.13 The differential equation governing the motion of the system of Figure P1.13 is

$$\frac{1}{3}mL^2\ddot{\theta} + L\ddot{y}\sin\theta + L\ddot{x}\cos\theta + mg\frac{L}{2}\sin\theta = 0 \quad \text{(a)}$$

where $x(t)$ and $y(t)$ are known functions of time. Derive a linearized equation governing the motion of the system for small θ.

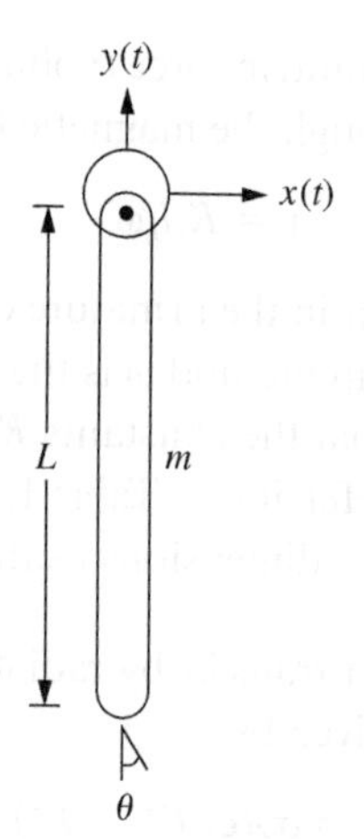

Figure P1.13

1.14 A chemical reaction that converts moles of a reactant A into moles of a substance B occurs in a continuously stirred tank reactor. The reaction is exothermic, and heat is given off during the reaction. The reactor is cooled by adding heat at a rate $\dot{Q}$ to the reactor. The differential equations governing the concentration of the reactant C_A and the temperature in the reactor T are

$$V\frac{dC_A}{dt} + (q + \alpha V e^{-E/(RT)})C_A = qC_{Ai} \quad \text{(a)}$$

$$\rho q c_p T_i - \rho q c_p T - \dot{Q} + \lambda V \alpha e^{-E/(RT)} C_A = \rho V c_p \frac{dT}{dt} \quad \text{(b)}$$

where V is the volume of the reactor, q is the flow rate of the reactant into the reactor, λ is the heat of reaction, E is the activation energy of the reaction, R is the universal gas constant, α is a constant related to the rate of reaction, ρ is the mass density of the inlet stream, c_p is the specific heat of the mixture, C_{Ai} is the concentration of the reactant in the inlet stream, and T_i is the temperature of the inlet stream. The steady-state conditions C_{As} and T_s are determined from Equations (a) and (b) by setting $\frac{dC_A}{dt} = \frac{dT}{dt} = 0$. The reactor is operating at steady state when it is subject to a sudden change in the flow rate of the inlet stream defined so that the flow rate is

$$q = q_s + q_p(t) \quad \text{(c)}$$

where q_s is the flow rate at steady state and q_p is the perturbation in the flow rate. The sudden change in flow rate leads to transient behavior in the reactor with time-dependent changes in reactant concentration and temperature defined by

$$C_A = C_{As} + C_{Ap}(t) \quad \text{(d)}$$

$$T = T_s + T_p(t) \quad \text{(e)}$$

Derive a set of linearized equations that could be solved to obtain the perturbations in reactant concentration and temperature.

1.15 The differential equation governing the transient temperature in a structure is

$$c_p\frac{dT}{dt} + \frac{1}{R}T = \frac{1}{R}T_\infty \quad \text{(a)}$$

where c_p is the specific heat of the material from which the structure is made, R is its thermal resistance, and T_∞ is the ambient temperature of the surrounding medium. The specific heat of a material is a function of temperature. However, in many thermal problems the changes in temperature are small and the specific heat is not greatly affected. It is common to assume a constant value for c_p, perhaps evaluated at T_∞. Consider a situation in which the ambient temperature has a sudden large perturbation $T_{\infty p}(t)$ from its steady-state value of $T_{\infty s}$, resulting in a transient temperature in the structure defined by $T = T_s + T_p(t)$. The temperature-dependent form of the specific heat for the structure is

$$c_p(T) = A_1 + A_2T^{1.5} + A_3T^{2.6} \quad \text{(b)}$$

(a) Use binominal expansions to linearize Equation (b), assuming the perturbation temperature is small compared to the steady-state temperature. (b) Derive a linearized mathematical model for the perturbation temperature.

1.16 An air spring consists of a piston moving in a cylinder of compressed air. When the piston is in equilibrium, the height of the column of air is h and it exerts a pressure p_0 on the face of the piston. The area of the face is A. When the piston moves a distance x downward, as illustrated in Figure P1.16, the air is further compressed. The process is assumed to be isentropic so that the air pressure is related to its density by

$$p = C\rho^\gamma \quad \text{(a)}$$

where C is a constant and γ is the ratio of specific heats. (a) Derive a relation between the force on the piston and the distance the piston moves into the column of air x. Hint: The mass of the air in the cylinder is constant. (b) Linearize the relationship of part (a) so that the force acting on the face of the piston is approximated as

$$F = p_0A + kx \quad \text{(b)}$$

Specify k.

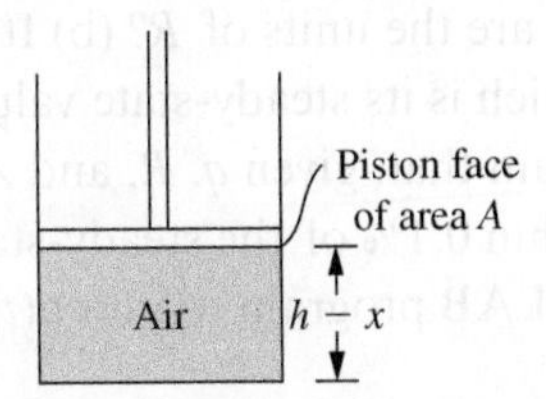

Figure P1.16

1.17 The time-dependent voltage of Figure P1.17 is supplied by a source in a series LRC circuit. Write a unified mathematical representation of the voltage provided by the source using unit step functions.

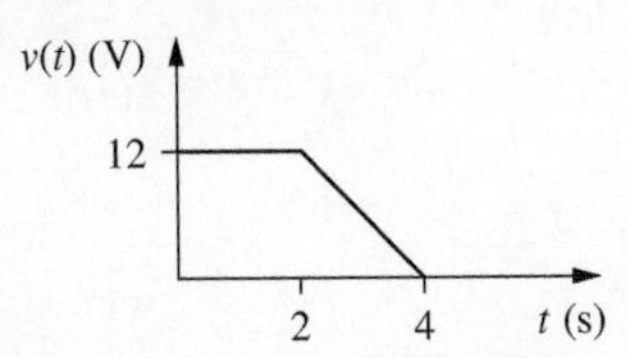

Figure P1.17

1.18 A mechanical system is subject to the time-dependent force shown in Figure P1.18. Develop a unified mathematical representation of the force using unit step functions.

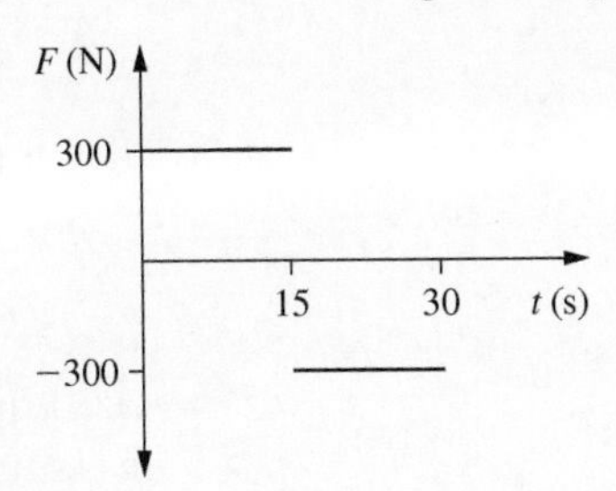

Figure P1.18

1.19 A cam and follower system has been designed to provide a periodic displacement to a valve in a mechanical system. Figure P1.19 illustrates the displacement provided over one period. (a) Use unit step functions to write a unified mathematical representation for the displacement over one period. (b) The displacement provided by the cam and follower system is periodic of period 0.5 s. Generalize the results of part (a) to provide a unified mathematical representation of the displacement for all time. Hint: Use an infinite series representation of the displacement with the index of summation representing one period.

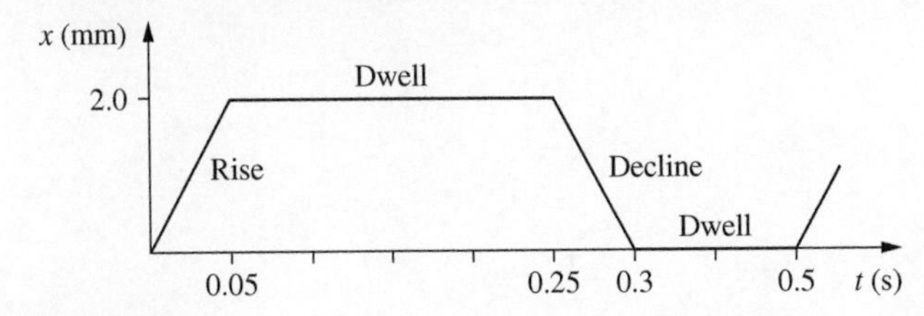

Figure P1.19

1.20 A particle of mass 4 kg is at rest when it is subject to an impulsive force that imparts a total impulse of 12 N·s to the mass. Knowing that Newton's second law can be written as $F = m\frac{dv}{dt}$, determine the velocity of the particle after application of the impulse.

1.21 No current is flowing through a circuit when it is subject to an impulsive voltage such that the total impulse is 20 V·s. Knowing that the relation between voltage and current in an inductor is $v = L\frac{di}{dt}$, where $L = 0.4$ H is the inductance, determine the current through the inductor immediately after application of the impulse.

1.22 A machine is subject to three impulses: an impulse of magnitude 100 N·s applied at $t = 0$, an impulse of magnitude 150 N·s applied at $t = 2.5$ s, and an impulse of magnitude 50 N·s applied at $t = 3.8$ s. What is the mathematical formulation for the force applied to the machine?

1.23 The steady-state current in a series LRC circuit due to a sinusoidal voltage source is

$$i(t) = \frac{V_0\omega}{L}\frac{1}{\sqrt{(\omega^2+\omega_n^2)^2+(2\zeta\omega\omega_n)^2}} \times \sin\left[\omega t + \tan^{-1}\left(\frac{\omega^2-\omega_n^2}{2\zeta\omega\omega_n}\right)\right] \tag{a}$$

where the circuit's natural frequency is $\omega_n = \frac{1}{\sqrt{LC}}$ and the damping ratio is $\zeta = \frac{R}{2}\sqrt{\frac{C}{L}}$. Write a MATLAB M-file that does the following: (a) inputs the values of resistance (R) in ohms, capacitance (C) in farads, inductance (L) in henrys, input frequency (ω) in r/s, and input amplitude (V_0) in volts; (b) evaluates $i(t)$ for 200 discrete values of t in the interval $0 \le t \le \frac{10\pi}{\omega}$; and (c) plots the response with an annotated graph.

1.24 Write a MATLAB M-file that (a) inputs the elements of two 5×5 matrices **A** and **B**; (b) calculates **A**+**B**; (c) calculates **AB**; (d) calculates $\mathbf{A}^{-1}$; (e) calculates det(**A**); and (f) determines the eigenvalues and eigenvectors of **A** (see Appendix B) using the MATLAB function eigs.

1.25 The function

$$\Lambda = \frac{r^2}{\sqrt{(1-r^2)^2+(2\zeta r)^2}} \tag{a}$$

is used in the analysis of the steady-state response of certain mechanical systems. Write a MATLAB M-file that plots, on the same axes, Λ vs. r for five different values of ζ. The plot should be annotated.

1.26 The step response of an underdamped mechanical system of natural frequency ω_n and damping ratio ζ is

$$x(t) = \frac{1}{\omega_n^2}\left[1 - e^{-\zeta\omega_n t}\cos\omega_d t + \frac{\zeta\omega_n}{\omega_d}e^{-\zeta\omega_n t}\sin\omega_d t\right] \quad \text{(a)}$$

where $\omega_d = \omega_n\sqrt{1-\zeta^2}$. Write a MATLAB M-file that (a) inputs the values of natural frequency and damping ratio and (b) evaluates and plots the step response on an annotated graph. The response should be evaluated for at least five cycles, $t = 5\left(\frac{2\pi}{\omega_d}\right)$.

1.27 The perturbation in the liquid level in a tank of area A and hydraulic resistance R due to a flow rate perturbation q is

$$h(t) = qR\left(1 - e^{-t/(RA)}\right) \quad \text{(a)}$$

(a) If A is in square meters (m^2) and t is measured in seconds (s), what are the units of R? (b) It is noted that $\lim_{t\to\infty} h(t) = qR$, which is its steady-state value. Write a MATLAB program that, given q, R, and A, calculates $h(t)$ until it is within 0.1% of the steady-state value. (c) Use the MATLAB program to plot $h(t)$.

2

Laplace Transforms

Learning Objectives

At the conclusion of this chapter, you should be able to:

- Determine if the Laplace transform of a function exists
- Determine how to use the definition of the Laplace transform to determine the Laplace transform of a function
- Use a table of transform pairs
- Use the properties of the Laplace transform
- Define the inverse transform and use properties and known transform pairs to determine inverses
- Use partial fraction decomposition on inverse transforms
- Use the Laplace transform method to solve ordinary differential equations, integrodifferential equations, and systems of differential and integrodifferential equations

Methods of mathematical modeling of time-dependent physical systems are presented in Chapters 4–6. Mathematical modeling of a physical system whose dependent variables are time dependent leads to a differential equation, an integrodifferential equation, or a set of differential and integrodifferential equations. Methods of solution of the derived equations include exact analytical methods, approximate methods, and numerical methods. Exact analytical solutions, when available, are preferable. Approximate and numerical solutions are not exact and deviate from the exact solution at each time. Errors in numerical solutions propagate with time. It is easier to study the effects of parameters on the response of a system when using an analytical solution.

The Laplace transform is the mathematical tool used in this book to solve linear differential equations and integrodifferential equations with constant coefficients. The Laplace transform method may be applied to determine the response of a linear dynamic system due to any physically possible system input. If an exact solution to the differential equation cannot be determined, the application of the Laplace transform method provides a solution in terms of an integral whose integrand is dependent on the system input. The integral could be evaluated numerically.

The Laplace transform method is not chosen arbitrarily. The method is schematically illustrated in Figure 2.1. The mathematical models of Chapters 4–6 are derived

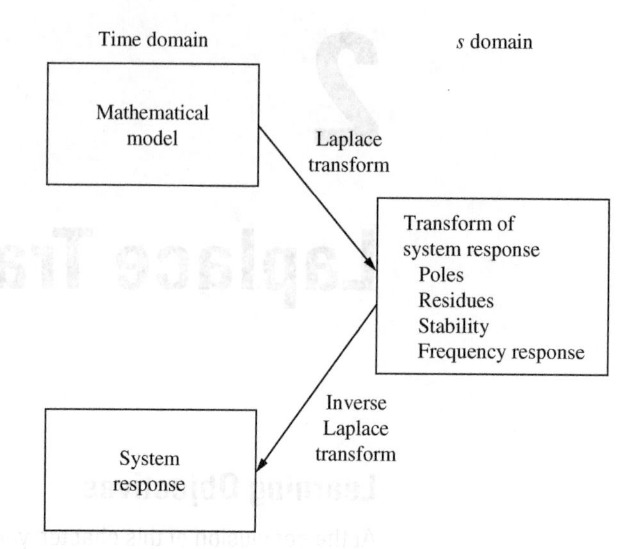

Figure 2.1

The schematic of the Laplace transform method, applied to determine system response from a mathematical model.

in the "time domain," that is, time is the dependent variable. The Laplace transform is a mathematical procedure that transforms a function in the time domain into a function of a complex variable, usually designated as s. The Laplace transform exists in the s domain, which is mathematically related to the physically defined frequency domain. The elegance of the method arises from its properties, which allow a differential equation for the system response in the time domain to be transformed into an algebraic equation for the transform of the system response in the s domain. Algebraic methods are used to determine the transform of the system response. Much can be learned about the system response from the s domain, including end behavior of the response, system stability, and frequency response. The response in the time domain is obtained through the application of an inverse Laplace transform to the function in the s domain.

The schematic application of this method to the solution of a linear differential equation is shown in Figure 2.2. The differential equation is derived in the time domain and then transformed into the s domain. The transform is then inverted to determine the response in the time domain.

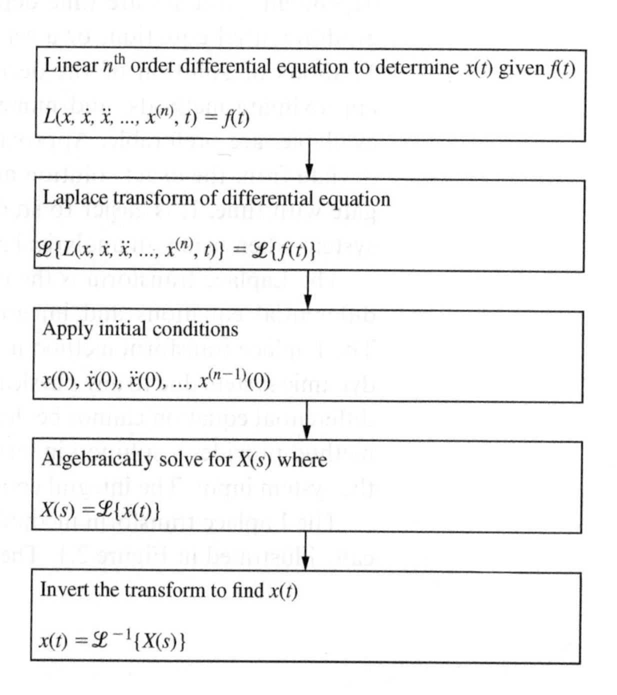

Figure 2.2

Schematic diagram for the solution of a linear n^{th} order differential equation by the Laplace transform method where $L(x, \dot{x}, \ddot{x}, \ldots, x^{(n)}, t)$ is an nth order differential operator.

2.1 Definition and Existence

The **Laplace transform** of a function $f(t)$ is defined by

$$\mathcal{L}\{f(t)\} = \int_0^\infty f(t)e^{-st}\,dt \tag{2.1}$$

where $\mathcal{L}$ is a symbolic operator representing the Laplace transform and s is a complex variable. Since the operation defined by Equation (2.1) results in a function of the complex variable s, the following notation is used to denote the Laplace transform of $f(t)$.

$$F(s) = \mathcal{L}\{f(t)\} \tag{2.2}$$

The Laplace transform of $f(t)$ exists if $f(t)$ is piecewise continuous and of **exponential order**. $f(t)$ is piecewise continuous if on the interval $a \le t \le b$, for any values of a and b, it has a finite number of jump discontinuities (finite jumps). $f(t)$ is of exponential order if there exists an $\alpha > 0$ such that

$$\lim_{t\to\infty} e^{-\alpha t}|f(t)| = 0 \tag{2.3}$$

Functions such as $\sin \omega t$, $\cos \omega t$, $e^{\lambda t}$, and t^n $(n > 0)$ are of exponential order. Any function that can be physically generated is of exponential order.

The **inverse Laplace transform** is an operator denoted by $\mathcal{L}^{-1}$ that operates on a function of the complex variable s, whose result is a function of a real variable t. The operator is defined by

$$\mathcal{L}^{-1}\{G(s)\} = \frac{1}{2\pi j}\int_{\gamma-j\infty}^{\gamma+j\infty} G(s)e^{st}ds \tag{2.4}$$

The integral in Equation (2.4) is carried out in the complex plane. If

$$F(s) = \mathcal{L}\{f(t)\} \tag{2.5}$$

then

$$f(t) = \mathcal{L}^{-1}\{F(s)\} \tag{2.6}$$

where $f(t) = 0$ for all $t < 0$.

For every $f(t)$ there is one and only one $F(s)$ that satisfies Equations (2.5) and (2.6). That is, the Laplace transform of $f(t)$ is unique, as is the inverse Laplace transform of $F(s)$. If $f(t)$ and $F(s)$ satisfy Equations (2.5) and (2.6), they are called a **transform pair**.

2.2 Determination of Transform Pairs

2.2.1 Direct Integration

If $f(t)$ is a piecewise continuous function of exponential order, then its Laplace transform can be determined by direct integration using Equation (2.1). This method works for the simple functions illustrated in this section. These results lead to the development of a **table of transform pairs** (Table 2.1). The properties of Laplace transforms derived in Section 2.3 and the application of tabulated transform pairs are used to add

Table 2.1 Laplace Transform Pairs

	$f(t)$	$F(s) = \mathcal{L}\{f(t)\}$
1	e^{at}	$\dfrac{1}{s-a}$
2	$\sin(\omega t)$	$\dfrac{\omega}{s^2+\omega^2}$
3	$\cos(\omega t)$	$\dfrac{s}{s^2+\omega^2}$
4	t	$\dfrac{1}{s^2}$
5	t^2	$\dfrac{2}{s^3}$
6	t^n	$\dfrac{n!}{s^{n+1}}$
7	$\delta(t-a)$	e^{-as}
8	$u(t-a)$	$\dfrac{e^{-as}}{s}$
9	$\sinh(\omega t)$	$\dfrac{\omega}{s^2-\omega^2}$
10	$\cosh(\omega t)$	$\dfrac{s}{s^2-\omega^2}$
11	te^{at}	$\dfrac{1}{(s-a)^2}$
12	$e^{-\zeta\omega_n t}\sin\left(\omega_n\sqrt{1-\zeta^2}\,t\right)$	$\dfrac{\omega_n\sqrt{1-\zeta^2}}{s^2+2\zeta\omega_n s+\omega_n^2}$
13	$e^{-\zeta\omega_n t}\cos\left(\omega_n\sqrt{1-\zeta^2}\,t\right)$	$\dfrac{s+\zeta\omega_n}{s^2+2\zeta\omega_n s+\omega_n^2}$
14	$t\sin(\omega t)$	$\dfrac{2s\omega}{(s^2+\omega^2)^2}$
15	$t\cos(\omega t)$	$\dfrac{s^2-\omega^2}{(s^2+\omega^2)^2}$
16	$e^{-at}-e^{-bt}$	$\dfrac{(b-a)}{(s+a)(s+b)}$
17	$1-e^{-at}$	$\dfrac{a}{s(s+a)}$
18	$ae^{-at}-be^{-bt}$	$\dfrac{(a-b)s}{(s+a)(s+b)}$

more entries to the table and to determine transform pairs not listed in the table. From here on it is taken for granted that the Laplace transform exists and is unique for any function for which it is attempted.

The Laplace transform of $f(t)$ is a function of the complex variable s. The domain of s is a subset of the complex plane. When a transform pair is specified, the domain of s must also be specified. The integral in Equation (2.4) that defines the inverse Laplace transform is performed only over the domain of s.

The determination of the Laplace transforms for several of the examples in this section and the derivation of several properties in Section 2.3 uses the technique of integration by parts. The standard integration by parts formula is

$$\int u\,dv = uv - \int v\,du \tag{2.7}$$

Equation (2.7) can be applied to aid in the evaluation of both definite and indefinite integrals.

Example 2.1

Determine the Laplace transform of the exponential function $f(t) = e^{at}$.

Solution

Application of Equation (2.1) for the given $f(t)$ leads to

$$\mathscr{L}\{e^{at}\} = \int_0^\infty e^{at}\, e^{-st}\, dt$$

$$= \int_0^\infty e^{-(s-a)t} dt$$

$$= -\frac{1}{s-a} e^{-(s-a)t}\Big|_0^\infty$$

$$= \frac{1}{s-a}\left(1 - \lim_{t\to\infty} e^{-(s-a)t}\right) \quad \text{(a)}$$

The limit in the last line of Equation (a) exists only if $\text{Re}(s) > a$. In this case, the limit is evaluated to zero, leading to

$$\mathscr{L}\{e^{at}\} = \frac{1}{s-a} \qquad \text{Re}(s) > a \quad (2.8)$$

Example 2.2

Determine $\mathscr{L}\{t^n\}$ where n is a non-negative integer.

Solution

First consider the case when $n = 1$. Substituting $f(t) = t$ into Equation (2.1) leads to

$$\mathscr{L}\{t\} = \int_0^\infty te^{-st}\, dt \quad \text{(a)}$$

The application of integration by parts, Equation (2.7), with $u = t$ and $dv = e^{-st}dt$ leads to Equation (a) being rewritten as

$$\mathscr{L}\{t\} = \left[-\frac{1}{s}te^{-st}\Big|_0^\infty + \frac{1}{s}\int_0^\infty e^{-st}\, dt\right]$$

$$= -\frac{1}{s}\lim_{t\to\infty} te^{-st} + \frac{1}{s^2} - \frac{1}{s^2}\lim_{t\to\infty} e^{-st} \quad \text{(b)}$$

The limits in Equation (b) exist only for $\text{Re}(s) > 0$, leading to

$$\mathscr{L}\{t\} = \frac{1}{s^2} \qquad \text{Re}(s) > 0 \quad \text{(c)}$$

Now consider the general case

$$\mathcal{L}\{t^n\} = \int_0^\infty t^n e^{-st} dt \tag{d}$$

Using the integration by parts formula [Equation (2.7)] with $u = t^n$ and $dv = e^{-st}dt$ in evaluation of the integral in Equation (d) and assuming $\mathrm{Re}(s) > 0$ leads to

$$\begin{aligned}\mathcal{L}\{t^n\} &= -\frac{1}{s}t^n e^{-st}\Big|_0^\infty + \frac{n}{s}\int_0^\infty t^{n-1}e^{-st}dt \\ &= \frac{n}{s}\int_0^\infty t^{n-1}e^{-st}dt \\ &= \frac{n}{s}\mathcal{L}\{t^{n-1}\}\end{aligned} \tag{e}$$

Using Equation (e) with $n = 2$

$$\mathcal{L}\{t^2\} = \frac{2}{s}\mathcal{L}\{t\} = \left(\frac{2}{s}\right)\left(\frac{1}{s^2}\right) = \frac{2}{s^3} \tag{f}$$

then with $n = 3$, Equation (e) becomes

$$\mathcal{L}\{t^3\} = \frac{3}{s}\mathcal{L}\{t^2\} = \left(\frac{3}{s}\right)\left(\frac{2}{s^3}\right) = \frac{6}{s^4} \tag{g}$$

This process can be continued, leading to the following formula, which is verified by induction

$$\mathcal{L}\{t^n\} = \frac{n!}{s^{n+1}} \tag{2.9}$$

Example 2.3

Determine $\mathcal{L}\{\sin(\omega t)\}$.

Solution

Substituting $f(t) = \sin(\omega t)$ in Equation (2.1) leads to

$$\mathcal{L}\{\sin(\omega t)\} = \int_0^\infty \sin(\omega t)e^{-st}dt \tag{a}$$

The application of integration by parts [Equation (2.7)] to the integral in Equation (a) with $u = \sin(\omega t)$ and $dv = e^{-st}dt$ leads to

$$\mathcal{L}\{\sin(\omega t)\} = -\frac{1}{s}\sin(\omega t)e^{-st}\Big|_0^\infty + \frac{\omega}{s}\int_0^\infty \cos(\omega t)e^{-st}dt \tag{b}$$

The first term on the right-hand side of Equation (b) is zero if $\mathrm{Re}(s) > 0$. Assuming this is the case and using integration by parts on the integral in Equation (b) with $u = \cos(\omega t)$ and $dv = e^{-st}dt$ leads to

$$\begin{aligned}\mathcal{L}\{\sin(\omega t)\} &= \frac{\omega}{s}\left[-\frac{1}{s}\cos(\omega t)e^{-st}\Big|_0^\infty - \frac{\omega}{s}\int_0^\infty \sin(\omega t)e^{-st}dt\right] \\ &= \frac{\omega}{s^2} - \frac{\omega^2}{s^2}\mathcal{L}\{\sin(\omega t)\}\end{aligned} \tag{c}$$

Rearranging Equation (c) and solving for $\{\sin(\omega t)\}$ leads to

$$\mathcal{L}\{\sin(\omega t)\} = \frac{\omega}{s^2 + \omega^2} \qquad \mathrm{Re}(s) > 0 \tag{2.10}$$

The Laplace transform of cos(ωt) is obtained by a method similar to that used in Example 2.3. The result is

$$\mathcal{L}\{\cos(\omega t)\} = \frac{s}{s^2 + \omega^2} \tag{2.11}$$

Example 2.4

Determine $\mathcal{L}\{u(t - a)\}$, where $u(t)$ is the unit step function.

Solution

Substituting $f(t) = u(t - a)$ into Equation (2.1) and recalling that $u(t - a) = 0$ for $t < a$ and $u(t - a) = 1$ for $t > a$ leads to

$$\mathcal{L}\{u(t - a)\} = \int_0^\infty u(t - a)e^{-st}\,dt$$

$$= \int_a^\infty e^{-st}\,dt$$

$$= -\frac{1}{s}e^{-st}\Big|_a^\infty \tag{a}$$

Assuming $\mathrm{Re}(s) > 0$, Equation (a) leads to

$$\mathcal{L}\{u(t - a)\} = \frac{1}{s}e^{-as} \tag{2.12}$$

Example 2.5

Determine $\mathcal{L}\{\delta(t - a)\}$, where $\delta(t)$ is the unit impulse function.

Solution

Substitution of $f(t) = \delta(t - a)$ in Equation (2.1) leads to

$$\mathcal{L}\{\delta(t - a)\} = \int_0^\infty \delta(t - a)e^{-st}dt \tag{a}$$

Using Equation (1.35) in Equation (a) leads to

$$\mathcal{L}\{\delta(t - a)\} = e^{-as} \tag{2.13}$$

Equations (2.8)–(2.13) provide Laplace transforms for basic functions and are summarized as transform pairs in Table 2.1. These equations and the properties derived in Section 2.3 are used to derive the remaining transform pairs in Table 2.1 as well as many others.

2.2.2 Use of MATLAB

MATLAB's symbolic toolbox has the capability to determine Laplace transforms of time-dependent functions. Since the Laplace transform is in the symbolic toolbox, time and the transform variable must be declared as symbolic variables. The statement

syms s t

declares both s and t as symbolic variables. This allows MATLAB to perform algebraic manipulations on functions involving these variables.

The function to be transformed is defined as a function of t using the standard hierarchy of operations. For example,

$$\mathrm{F} = \exp(-3*t)$$

defines the function $F = e^{-3t}$. The statement

$$\mathrm{H} = \mathrm{laplace}(\mathrm{F})$$

evaluates the Laplace transform of F and assigns it to the function H, which will be displayed as an algebraic function of s.

Example 2.6

Write a MATLAB M-file that determines the Laplace transform of each of the functions in Examples 2.1–2.5. Use t^4 as the function for Example 2.2 and $a = 3$ for the functions in Examples 2.4 and 2.5.

Solution

The MATLAB script file Example2_6.m is illustrated in Figure 2.3, and output from its execution is illustrated in Figure 2.4. The results obtained are identical to those of Examples 2.1–2.5. The following is noted about the use of MATLAB for this example:

- The Heaviside function is another name for the unit step function, the name used by MATLAB: heaviside(t − a) = u(t − a).
- The Dirac function is another name for the unit impulse function, the name used by MATLAB: dirac(t − a) = δ(t − a)

```
% MATLAB program for Example 2.6
% Symbolic computation of Laplace transforms
%
% Declaration of symbolic variables
% t=time
% s=transform variable
% w, a are parameters for symbolic evaluation of transforms
%
syms t s w a
%
% Example 2.1
%
F1=exp(a*t);
```

Figure 2.3
MATLAB script file for Example 2.6.

```
H1=laplace(F1)
%
% Example 2.2
% f(t)=t^4
%
F2=t^4;
H2=laplace(F2)
%
% Example 2.3
% f(t)=sin(omega*t)
%
F3=sin(w*t);
H3=laplace(F3)
%
% Example 2.4
% The heaviside function is equivalent to the unit step function u(t-a)
%
F4=heaviside(t-3);
H4=laplace(F4)
%
% Example 2.5
% The dirac function is the Dirac delta function also known as the unit
% impulse function
%
F5=dirac(t-3);
H5=laplace(F5)
%
% End of script file for Example 2.6
```

Figure 2.3
(Continued)

```
>> Example2_6
H1 =

-1/(a - s)

H2 =

24/s^5

H3 =

w/(s^2 + w^2)

H4 =

exp(-3*s)/s

H5 =

exp(-3*s)
>>
```

Figure 2.4
Output from execution of MATLAB script file Example2_6.m. The results are identical to those obtained in Examples 2.1–2.5.

2.3 Laplace Transform Properties

The properties developed in this section are useful in the derivation of transform pairs as well as in the application of Laplace transforms to the solution of differential and integrodifferential equations. Each property is stated, followed by a short proof. Relevant examples using the properties in determining transform pairs are presented. The application of the properties to the solution of differential equations is pursued in Section 2.5. The properties are summarized in Table 2.2.

The following properties are stated and derived using certain assumptions. When it is stated that $F(s) = \mathcal{L}\{f(t)\}$, it is implicitly assumed that $f(t)$ is of exponential order and its Laplace transform exists.

Table 2.2 Properties of Laplace Transforms

	Name	Equation
1	Definition of transform	$F(s) = \mathcal{L}\{f(t)\} = \int_0^\infty f(t)e^{-st}dt$
2	Inverse transform	$f(t) = \mathcal{L}^{-1}\{F(s)\} = \frac{1}{2\pi j}\int_{\gamma-j\infty}^{\gamma+j\infty} F(s)e^{st}ds$
3	Linearity of transform	$\mathcal{L}\{af(t) + bg(t)\} = aF(s) + bG(s)$
4	Linearity of inverse transform	$\mathcal{L}^{-1}\{aF(s) + bG(s)\} = af(t) + bg(t)$
5	First shifting theorem	$\mathcal{L}\{e^{at}f(t)\} = F(s-a)$
6	Second shifting theorem	$\mathcal{L}\{f(t-a)u(t-a)\} = e^{-as}F(s)$
7	Transform of first derivative	$\mathcal{L}\left\{\frac{df}{dt}\right\} = sF(s) - f(0)$
8	Transform of second derivative	$\mathcal{L}\left\{\frac{d^2f}{dt^2}\right\} = s^2F(s) - sf(0) - \dot{f}(0)$
9	Transform of nth derivative	$\mathcal{L}\left\{\frac{d^nf}{dt^n}\right\} = s^nF(s) - \sum_{k=1}^{n-1} s^{n-k}\frac{d^{k-1}f}{dt^{k-1}}(0)$
10	Transform of integral	$\mathcal{L}\left\{\int_0^t f(t)\right\} = \frac{1}{s}F(s)$
11	Derivatives of transform	$\mathcal{L}\{t^nf(t)\} = (-1)^n\frac{d^nF(s)}{ds^n}$
12	Transform of periodic function	If $f(t)$ is periodic of Period T, then $\mathcal{L}\{f(t)\} = \frac{1}{1-e^{-Ts}}\int_0^T f(t)e^{-st}dt$
13	Convolution	$\mathcal{L}\{f(t)*g(t)\} = F(s)G(s)$
14	Final value theorem	$\lim_{t\to\infty} f(t) = \lim_{s\to 0} sF(s)$
15	Initial value theorem	$\lim_{s\to\infty} sF(s) = f(0)$

Linearity of the transform: If $F(s) = \mathcal{L}\{f(t)\}$, $G(s) = \mathcal{L}\{g(t)\}$, and a and b are any scalars, then

$$\mathcal{L}\{af(t) + bg(t)\} = aF(s) + bG(s) \tag{2.14}$$

Proof: From the definition of the Laplace transform and using the basic properties of integrals,

$$\begin{aligned}\mathcal{L}\{af(t) + bg(t)\} &= \int_0^\infty [af(t) + bg(t)]e^{-st}\,dt \\ &= \int_0^\infty af(t)e^{-st}\,dt + \int_0^\infty bg(t)e^{-st}\,dt \\ &= a\int_0^\infty f(t)e^{-st}\,dt + b\int_0^\infty g(t)e^{-st}\,dt \\ &= aF(s) + bG(s)\end{aligned} \tag{a}$$

Example 2.7

Determine $\mathcal{L}\{\sinh(\omega t)\}$.

Solution

The hyperbolic sine function is written in terms of exponentials as

$$\sinh(\omega t) = \frac{1}{2}(e^{\omega t} - e^{-\omega t}) \tag{a}$$

Taking the Laplace transform of Equation (a) using the linearity property [Equation (2.14)] leads to

$$\mathcal{L}\{\sinh(\omega t)\} = \frac{1}{2}\mathcal{L}\{e^{\omega t}\} - \frac{1}{2}\mathcal{L}\{e^{-\omega t}\} \tag{b}$$

The transforms on the right-hand side of Equation (b) are evaluated using Equation (2.8). This leads to

$$\begin{aligned}\mathcal{L}\{\sinh(\omega t)\} &= \frac{1}{2}\left(\frac{1}{s-\omega}\right) - \frac{1}{2}\left(\frac{1}{s+\omega}\right) \\ &= \frac{1}{2}\left[\frac{s+\omega-(s-\omega)}{(s-\omega)(s+\omega)}\right] \\ &= \frac{\omega}{s^2-\omega^2}\end{aligned} \tag{c}$$

First shifting theorem: If $F(s) = \mathcal{L}\{f(t)\}$, then

$$\mathcal{L}\{e^{at}f(t)\} = F(s-a) \tag{2.15}$$

Proof: By definition

$$\begin{aligned}F(s-a) &= \int_0^\infty f(t)e^{-(s-a)t}\,dt \\ &= \int_0^\infty [e^{at}f(t)]e^{-st}\,dt\end{aligned} \tag{a}$$

The right-hand side of Equation (a) is by definition the Laplace transform of $e^{at}f(t)$ [Equation (2.1)]. Hence Equation (2.15) is proved.

Example 2.8

Determine $\mathcal{L}\{e^{-2t}\cos(4t)\}$.

Solution

Equation (2.11) is used to determine

$$\mathcal{L}\{\cos(4t)\} = \frac{s}{s^2 + 16} \tag{a}$$

The first shifting theorem indicates that the desired transform can be obtained by replacing s by $s + 2$ in Equation (a), leading to

$$\mathcal{L}\{e^{-2t}\cos(4t)\} = \frac{s + 2}{(s + 2)^2 + 16}$$

$$= \frac{s + 2}{s^2 + 4s + 20} \tag{b}$$

Second shifting theorem: If $F(s) = \mathcal{L}\{f(t)\}$, then

$$\mathcal{L}\{f(t - a)u(t - a)\} = e^{-as}F(s) \tag{2.16}$$

Proof: The definition of the Laplace transform, Equation (2.1), is used to obtain

$$\mathcal{L}\{f(t - a)u(t - a)\} = \int_0^\infty f(t - a)u(t - a)e^{-st}\,dt \tag{a}$$

Since $u(t - a) = 0$ for $t < a$ and $u(t - a) = 1$ for $t > a$, Equation (a) can be written as

$$\mathcal{L}\{f(t - a)u(t - a)\} = \int_a^\infty f(t - a)e^{-st}\,dt \tag{b}$$

The following change in variable

$$w = t - a \tag{c}$$

is used to rewrite Equation (b) as

$$\mathcal{L}\{f(t - a)u(t - a)\} = \int_a^\infty f(w)e^{-s(w+a)}\,dw$$

$$= e^{-as}\int_0^\infty f(w)e^{-sw}\,dw \tag{d}$$

The name of the variable of integration is irrelevant in the definition of the Laplace transform. Equation (2.1) could have been written with a w replacing t in the integral. Thus, Equation (d) is the same as Equation (2.16).

Example 2.9

Determine the Laplace transform of the triangular excitation of Example 1.7(c), which has the mathematical representation

$$f(t) = \frac{F_0}{t_0}[tu(t) - 2(t - t_0)u(t - t_0) + (t - 2t_0)u(t - 2t_0)] \tag{a}$$

Solution

The application of linearity of transforms [Equation (2.14)] to Equation (a) leads to

$$\mathcal{L}\{f(t)\} = \frac{F}{t_0}[\mathcal{L}\{tu(t)\} - 2\mathcal{L}\{(t - t_0)u(t - t_0)\} + \mathcal{L}\{(t - 2t_0)u(t - 2t_0)\}] \tag{b}$$

Each of the Laplace transforms in Equation (b) can be evaluated using Equation (2.9), with $n = 1$ and the second shifting theorem, Equation (2.16). For example, using Equation (2.16) with $f(t) = t$, $F(s) = 1/s^2$, and $a = 2t_0$ leads to

$$\mathscr{L}\{(t - 2t_0)u(t - 2t_0)\} = \frac{e^{-2t_0 s}}{s^2} \tag{c}$$

Using Equation (c) and similar evaluations for $\mathscr{L}\{tu(t)\}$ and $\mathscr{L}\{(t - t_0)u(t - t_0)\}$ in Equation (b) leads to

$$\begin{aligned}\mathscr{L}\{f(t)\} &= \frac{F_0}{t_0}\left[\frac{1}{s^2} - 2\frac{e^{-t_0 s}}{s^2} + \frac{e^{-2t_0 s}}{s^2}\right] \\ &= \frac{F_0}{t_0}\frac{1 - 2e^{-t_0 s} + e^{-2t_0 s}}{s^2} \\ &= \frac{F_0}{t_0}\left(\frac{1 - e^{-t_0 s}}{s}\right)^2\end{aligned} \tag{d}$$

Transform of derivatives: If $F(s) = \mathscr{L}\{f(t)\}$, then

$$\mathscr{L}\left\{\frac{d^n f}{dt^n}\right\} = s^n F(s) - s^{n-1}f(0) - s^{n-2}\frac{df}{dt}(0) \cdots - s\frac{d^{n-2}f}{dt^{n-2}}(0) - \frac{d^{n-1}f}{dt^{n-1}}(0) \tag{2.17}$$

Proof: The proof of Equation (2.17) is by induction. First consider $n = 1$. Equation (2.1) is used to give

$$\mathscr{L}\left\{\frac{df}{dt}\right\} = \int_0^\infty \frac{df}{dt}e^{-st}\,dt \tag{a}$$

Integration by parts [Equation (2.7)] is used on Equation (a) with $u = e^{-st}$ and $dv = \frac{df}{dt}dt$, leading to

$$\begin{aligned}\mathscr{L}\left\{\frac{df}{dt}\right\} &= f(t)e^{-st}\Big|_0^\infty + s\int_0^\infty f(t)e^{-st}\,dt \\ &= sF(s) - f(0)\end{aligned} \tag{b}$$

Thus, Equation (2.17) is proved for $n = 1$. Now assume Equation (2.17) is valid for $n = 1, 2, 3, \ldots k - 1$ for an arbitrary k and consider

$$\mathscr{L}\left\{\frac{d^k f}{dt^k}\right\} = \int_0^\infty \frac{d^k f}{dt^k}e^{-st}\,dt \tag{c}$$

Integration by parts [Equation (2.7)] is used on Equation (c) with $u = e^{-st}$ and $dv = \frac{d^k f}{dt^k}dt$, leading to

$$\begin{aligned}\mathscr{L}\left\{\frac{d^k f}{dt^k}\right\} &= e^{-st}\frac{d^{k-1}f}{dt^{k-1}} + s\int_0^\infty \frac{d^{k-1}f}{dt^{k-1}}e^{-st}\,dt \\ &= -\frac{d^{k-1}f}{dt^{k-1}}(0) + s\mathscr{L}\left\{\frac{d^{k-1}f}{dt^{k-1}}\right\}\end{aligned} \tag{d}$$

Use of the assumption that Equation (2.17) is true for $n = k - 1$ in Equation (d) leads to

$$\begin{aligned}\mathscr{L}\left\{\frac{d^k f}{dt^k}\right\} &= -\frac{d^{k-1}f}{dt^{k-1}}(0) + s\left[s^{k-1}F(s) - s^{k-2}f(0) - \cdots - s\frac{d^{k-3}f}{dt^{k-3}}(0) - \frac{d^{k-2}f}{dt^{k-2}}(0)\right] \\ &= s^k F(s) - s^{k-1}f(0) - s^{k-2}\frac{df}{dt}(0) - \cdots - s\frac{d^{k-2}f}{dt^{k-2}}(0) - \frac{d^{k-1}f}{dt^{k-1}}(0)\end{aligned} \tag{e}$$

Equation (b) shows that Equation (2.17) is true for $n = 1$. Equation (e) shows that Equation (2.17) is true for $n = k$ for any value of k, assuming it is true for $n = k - 1$. Thus, by induction, Equation (2.17) is true for all n.

The property of the transform of the derivatives is useful in the solution of linear differential equations. It is convenient to write explicit expressions of Equation (2.17) for $n = 1$ and $n = 2$. If $F(s) = \mathcal{L}\{f(t)\}$ then

$$\mathcal{L}\left\{\frac{df}{dt}\right\} = sF(s) - f(0) \tag{2.18}$$

$$\mathcal{L}\left\{\frac{d^2f}{dt^2}\right\} = s^2F(s) - sf(0) - \frac{df}{dt}(0) \tag{2.19}$$

Example 2.10

Use Equation (2.11) for $\mathcal{L}\{\cos(\omega t)\}$ and Equation (2.18) to derive Equation (2.10).

Solution

Applying Equation (2.18) with $f(t) = \cos(\omega t)$ and noting from Equation (2.11) that $F(s) = \frac{s}{s^2 + \omega^2}$ leads to

$$\mathcal{L}\{-\omega \sin(\omega t)\} = s\left(\frac{s}{s^2 + \omega^2}\right) - 1 \tag{a}$$

Rearranging Equation (a) leads to

$$\begin{aligned} \mathcal{L}\{\sin(\omega t)\} &= \frac{1}{\omega}\left(1 - \frac{s^2}{s^2 + \omega^2}\right) \\ &= \frac{1}{\omega}\left(\frac{s^2 + \omega^2 - s^2}{s^2 + \omega^2}\right) \\ &= \frac{\omega}{s^2 + \omega^2} \end{aligned} \tag{b}$$

Transform of an integral: If $F(s) = \mathcal{L}\{f(t)\}$, then

$$\mathcal{L}\left\{\int_0^t f(\tau)d\tau\right\} = \frac{1}{s}F(s) \tag{2.20}$$

Proof: The application of the definition of the Laplace transform to the left-hand side of Equation (2.20) leads to

$$\mathcal{L}\left\{\int_0^t f(\tau)d\tau\right\} = \int_0^\infty\left(\int_0^t f(\tau)d\tau\right)e^{-st}\,dt \tag{a}$$

The application of integration by parts [Equation (2.7)] to Equation (a) with $u = \int_0^t f(\tau)\,d\tau$ and $dv = e^{-st}\,dt$ leads to

$$\begin{aligned} \left\{\mathcal{L}\int_0^t f(\tau)d\tau\right\} &= -\frac{1}{s}\left[\left(\int_0^t f(\tau)d\tau\right)e^{-st}\right]_{t=0}^{t=\infty} + \frac{1}{s}\int_0^\infty f(t)e^{-st}\,dt \\ &= \frac{1}{s}F(s) \end{aligned} \tag{b}$$

Example 2.11

It can be shown that

$$\mathcal{L}\{t^{-1/2}\} = \sqrt{\frac{\pi}{s}} \tag{a}$$

Use Equation (a) and Equation (2.20) to determine $\mathcal{L}\{t^{1/2}\}$.

Solution

It is noted that

$$\int_0^{t} \tau^{-1/2}\, d\tau = 2\tau^{1/2}\big|_0^t = 2t^{1/2} \tag{b}$$

Thus from Equation (b)

$$\mathcal{L}\{2t^{1/2}\} = \mathcal{L}\left\{\int_0^t \tau^{-1/2}\, d\tau\right\} \tag{c}$$

Application of Equation (2.20) to Equation (c) leads to

$$\mathcal{L}\{t^{1/2}\} = \frac{1}{2}\left(\frac{1}{s}\sqrt{\frac{\pi}{s}}\right)$$

$$= \frac{\sqrt{\pi}}{2s^{3/2}} \tag{d}$$

Final value theorem: If $F(s) = \mathcal{L}\{f(t)\}$, then if $\lim\limits_{t\to\infty} f(t)$ exists,

$$\lim_{t\to\infty} f(t) = \lim_{s\to 0} sF(s) \tag{2.21}$$

Proof: Taking the limit of both sides of Equation (2.18) (transform of the first derivative) as s approaches zero gives

$$\lim_{s\to 0}\int_0^\infty \frac{df}{dt} e^{-st}\, dt = \lim_{s\to 0}[sF(s) - f(0)] \tag{a}$$

Since the integral on the left-hand side of Equation (a) has a finite value, the order of the limit and the integration may be interchanged. This leads to

$$\lim_{s\to 0} sF(s) - f(0) = \int_0^\infty \frac{df}{dt}\left(\lim_{s\to 0} e^{-st}\right) dt$$

$$= \int_0^\infty \frac{df}{dt}\, dt$$

$$= f(t)\big|_0^\infty$$

$$= \lim_{t\to\infty} f(t) - f(0) \tag{b}$$

Equation (2.21) is obtained directly from Equation (b).

The final value theorem is useful in predicting the behavior of a system as time grows large using the Laplace transform of the system response. This is illustrated in the following example.

Example 2.12

The Laplace transform of the time-dependent response of a dynamic system is

$$X(s) = \frac{2(s^2 + 4s + 3)}{s(s^2 + 9)(s^2 + 4)(s + 5)} \tag{a}$$

Determine $\lim_{t \to \infty} x(t)$.

Solution

The application of the final value theorem to Equation (a) leads to

$$\lim_{t \to \infty} x(t) = \lim_{s \to 0} s\left[\frac{2(s^2 + 4s + 3)}{s(s^2 + 9)(s^2 + 4)(s + 5)}\right]$$
$$= \frac{2(3)}{(9)(4)(5)} = \frac{1}{30} \tag{b}$$

The properties proved above are the Laplace transform properties that are used most often in this book. Other useful properties, which follow, are stated without proof.

Initial value theorem: If $F(s) = \mathcal{L}\{f(t)\}$, then

$$\lim_{s \to \infty} sF(s) = f(0) \tag{2.22}$$

Change of scale: If $F(s) = \mathcal{L}\{f(t)\}$ and $a \neq 0$, then

$$\mathcal{L}\left\{f\left(\frac{t}{a}\right)\right\} = aF(as) \tag{2.23}$$

Differentiation of transform: If $F(s) = \mathcal{L}\{f(t)\}$, then

$$\mathcal{L}\{tf(t)\} = -\frac{dF(s)}{ds} \tag{2.24}$$

$$\mathcal{L}\{t^2 f(t)\} = \frac{d^2F(s)}{ds^2} \tag{2.25}$$

$$\mathcal{L}\{t^n f(t)\} = (-1)^n \frac{d^n F(s)}{ds^n} \tag{2.26}$$

Transform of periodic function: If $f(t)$ is a piecewise continuous function of period T, then

$$\mathcal{L}\{f(t)\} = \frac{1}{1 - e^{-Ts}} \int_0^T f(t)e^{-st}\,dt \tag{2.27}$$

Convolution of two functions: The convolution of two functions is defined as

$$f(t)*g(t) = \int_0^t f(t - \tau)g(\tau)d\tau \tag{2.28}$$

If $F(s) = \mathcal{L}\{f(t)\}$ and $G(s) = \mathcal{L}\{g(t)\}$, then

$$\mathcal{L}\{f(t)*g(t)\} = F(s)G(s) \tag{2.29}$$

The applications of these properties are illustrated in the following examples.

Example 2.13

Determine $F(s)$ if $f(t) = te^{-2t}\sin\left(2t + \frac{\pi}{4}\right)$.

Solution

The trigonometric identity for the sine of the sum of angles is used to write

$$\sin\left(2t + \frac{\pi}{4}\right) = \sin(2t)\cos\left(\frac{\pi}{4}\right) + \cos(2t)\sin\left(\frac{\pi}{4}\right)$$

$$= \frac{\sqrt{2}}{2}[\sin(2t) + \cos(2t)] \quad \text{(a)}$$

The application of the first shifting theorem [Equation (2.15)] gives

$$\mathcal{L}\{e^{-2t}\sin 2t\} = \frac{2}{(s+2)^2 + 4} = \frac{2}{s^2 + 4s + 8} \quad \text{(b)}$$

$$\mathcal{L}\{e^{-2t}\cos(2t)\} = \frac{s+2}{(s+2)^2 + 4} = \frac{s+2}{s^2 + 4s + 8} \quad \text{(c)}$$

Using Equation (a) and linearity of the transform,

$$\mathcal{L}\left\{e^{-2t}\sin\left(2t + \frac{\pi}{4}\right)\right\} = \frac{\sqrt{2}}{2}\left[\frac{2}{s^2 + 4s + 8} + \frac{s+2}{s^2 + 4s + 8}\right]$$

$$= \frac{\sqrt{2}}{2}\left[\frac{s+4}{s^2 + 4s + 8}\right] \quad \text{(d)}$$

The property of differentiation of the transform [Equation (2.24)] is used to yield

$$\mathcal{L}\{f(t)\} = -\frac{d}{ds}\left[\frac{\sqrt{2}}{2}\frac{s+4}{(s^2 + 4s + 8)}\right]$$

$$= -\frac{\sqrt{2}}{2}\left[\frac{(1)(s^2 + 4s + 8) - (s+4)(2s+4)}{(s^2 + 4s + 8)^2}\right]$$

$$= \frac{\sqrt{2}}{2}\left[\frac{s^2 + 8s + 8}{(s^2 + 4s + 8)^2}\right] \quad \text{(e)}$$

Example 2.14

Determine the Laplace transform of the periodic function of Figure 2.5.

Solution

The function in Figure 2.5 is periodic of period t_0. Over one period, it is defined as

$$f(t) = \begin{cases} F_0 & 0 \le t \le \frac{2t_0}{3} \\ 0 & \frac{2t_0}{3} < t < t_0 \end{cases} \quad \text{(a)}$$

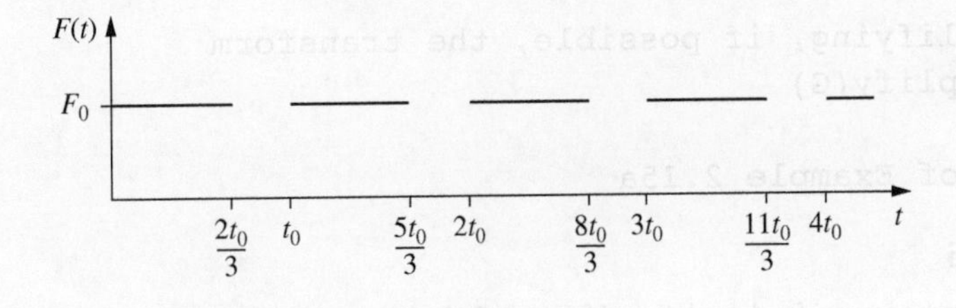

Figure 2.5

The periodic function for Example 2.14.

Equation (2.27) is used to determine the Laplace transform of $f(t)$ as

$$\begin{aligned}\mathcal{L}\{f(t)\} &= \frac{1}{1-e^{-t_0 s}}\int_0^{t_0} f(t)e^{-st}\,dt \\ &= \frac{1}{1-e^{-t_0 s}}\int_0^{2t_0/3} F_0 e^{-st}\,dt \\ &= \frac{F_0}{1-e^{-t_0 s}}\left(-\frac{1}{s}e^{-st}\right)\Big|_0^{2t_0/3} \\ &= \frac{F_0\left(1-e^{-(2t_0/3)s}\right)}{s\left(1-e^{-t_0 s}\right)}\end{aligned} \tag{b}$$

Example 2.15

Use MATLAB to symbolically determine the Laplace transform of **a.** the function of Example 2.9 with $F_0 = 1$ and $t_0 = 1$ and **b.** the function of Example 2.13.

Solution

MATLAB M-files are convenient when a computation is to be performed many times. However, direct interactive work sessions using MATLAB are convenient when the sequence of commands necessary to achieve an objective is not well defined. The MATLAB script files are shown in Figures 2.6(a) and (b). The following is noted about these files:

```
% Solution to Example 2.15a
% Laplace transform of a triangular pulse
% Defining symbolic variables
% t=time
% s=transform variable
%
syms s t
%
% A=u(t-1)    B=u(t-2)
%
A=heaviside(t-1);
B=heaviside(t-2);
%
% F=t-2(t-1)u(t-1)+(t-2)u(t-2)
%
% Laplace transform of F(t)
%
F=t-2*(t-1)*A+(t-2)*B;
G=laplace(F)
%
% Simplifying, if possible, the transform
G1=simplify(G)
%
% End of Example 2.15a
```

Figure 2.6

MATLAB work sessions for the solution of Example 2.15, parts a. and b.

```
G =
exp(-2*s)/s^2 - (2*exp(-s))/s^2 + 1/s^2

G1 =

(exp(-2*s)*(exp(s) - 1)^2)/s^2
```

(a)

```
% Solution to Example 2.15b
% Declaring symbolic variables
% t=time
% s=transform variable
%
syms s t
%
% F=t*e^(-2*t)*sin(2t+pi/4)
%
F=t*exp(-2*t)*sin(2*t+pi/4);
%
% Laplace transform of F
%
G=laplace(F)
%
% Simplifying Laplace transform
%
G1=simplify(G)
%
% Writing transform in a form that is easier to read and use.
% The "3" refers to the number of significant digits that are to be used
%
G2=vpa(G1,3)
%
% End of file

G =

(2^(1/2)*(2*s + 4))/((s + 2)^2 + 4)^2 - (2^(1/2)*(1/((s + 2)^2 + 4) - ((2*s +
4)*(s + 2))/((s + 2)^2 + 4)^2))/2

G1 =

(2^(1/2)*(s^2 + 8*s + 8))/(2*(s^2 + 4*s + 8)^2)

G2 =

(0.707*(s^2 + 8.0*s + 8.0))/(s^2 + 4.0*s + 8.0)^2
```

(b)

Figure 2.6
(Continued)

- MATLAB makes a number of symbolic commands available. When developing a MATLAB M-file, it is difficult to know in advance which symbolic commands may be necessary to put the result in the desired form. During the work session for part b., the command "simplify" resulted in a convenient algebraic form for the result. The result, when written in the form of G1, compares exactly with the result of Example 2.13.
- When MATLAB displays a symbolic result, it displays numerical terms as simple fractions or irrational terms as powers of rational numbers. It is often more convenient to write the result in a decimal form. The command "vpa(G1,3)" rewrites the symbolic expression in a decimal form with three significant digits. The term "vpa" stands for variable precision arithmetic.

2.4 Inversion of Transforms

The process of determining $f(t)$ from its Laplace transform $F(s)$ is called **inversion of the transform**. The relation between $f(t)$ and $F(s)$ is written symbolically in Equation (2.6), repeated here:

$$f(t) = \mathcal{L}^{-1}\{F(s)\} \tag{2.6}$$

There are several methods of transform inversion. The direct method is to use the inversion integral, Equation (2.4). However, its use requires knowledge of integration in a complex plane, including residue analysis. Thus, it is not presented in this book.

The examples in this section are provided to illustrate the use of the table of transform pairs (Table 2.1) and the table of transform properties (Table 2.2). For the purposes of illustration, the examples assume the availability of only transform pairs 1 through 10. The solution of some of the examples would be easier if the remainder of the transform pairs in Table 2.1 were used.

The use of MATLAB to determine the inverse transform is also presented.

2.4.1 Use of Tables and Properties

The most convenient method of transform inversion is to apply the properties of the transform in Table 2.2 and use the transform pairs in Table 2.1. This method requires logic that is the reverse of using the tables and properties to determine transform pairs. For example, if $X(s)$ is of the form $e^{-as}F(s-a)$ where the transform pair for $f(t)$ and $F(s)$ can be located in Table 2.1, then the first shifting property from Table 2.2 is applied, leading to $x(t) = f(t)u(t-a)$.

The property of **linearity of the inverse transform** is that if $f(t) = \mathcal{L}^{-1}\{F(s)\}$ and $g(t) = \mathcal{L}^{-1}\{G(s)\}$ and a and b are scalars, then

$$\mathcal{L}^{-1}\{aF(s) + bG(s)\} = af(t) + bg(t) \tag{2.30}$$

Example 2.16

Determine $x(t)$ if its Laplace transform is

$$X(s) = \frac{4s + 6}{s^2 + 9} \tag{a}$$

Solution

Equation (a) can be written as

$$X(s) = 4\left(\frac{s}{s^2 + 9}\right) + 2\left(\frac{3}{s^2 + 9}\right) \tag{b}$$

The application of linearity for inverse transforms [Equation (2.30)], in addition to transform pairs 3 and 2 from Table 2.1, leads to

$$x(t) = 4\cos(3t) + 2\sin(3t) \tag{c}$$

Example 2.17

Determine $x(t)$ if its Laplace transform is

$$X(s) = \frac{2s}{s^2 + 6s + 13} \tag{a}$$

Solution

The quadratic polynomial in the denominator of Equation (a) does not have real roots. It is convenient to complete the square of the quadratic polynomial, leading to

$$X(s) = \frac{2s}{(s + 3)^2 + 4}$$

$$= 2\left[\frac{s + 3}{(s + 3)^2 + 4} - \frac{3}{(s + 3)^2 + 4}\right] \tag{b}$$

Wherever s appears on the right-hand side of the last step of Equation (b), it appears as $s + 3$. Thus if

$$F(s) = 2\left[\frac{s}{s^2 + 4} - \frac{3}{s^2 + 4}\right] \tag{c}$$

Then $X(s) = F(s + 3)$ and the first shifting theorem is applied, leading to

$$x(t) = e^{-3t}f(t) \tag{d}$$

where $f(t) = \mathcal{L}^{-1}\{F(s)\} = 2\cos(2t) - 3\sin(2t)$ is obtained using linearity, Equation (c), and transform pairs 3 and 2 from Table 2.1. Thus

$$x(t) = e^{-3t}[2\cos(2t) - 3\sin(2t)] \tag{e}$$

Example 2.18

Determine $x(t)$ if its Laplace transform is

$$X(s) = \frac{e^{-3s}}{(s + 2)^2} \tag{a}$$

Solution

The presence of the exponential terms suggests the use of the second shifting theorem. Its application leads to

$$x(t) = f(t - 3)\,u(t - 3) \tag{b}$$

where $f(t)$ is the inverse transform of

$$F(s) = \frac{1}{(s + 2)^2} \tag{c}$$

The inverse transform of Equation (c) is obtained using transform pair 4 from Table 2.1 and the first shifting theorem as

$$f(t) = te^{-2t} \tag{d}$$

Combining Equations (b) and (d) gives

$$x(t) = (t - 3)e^{-2(t-3)}u(t - 3) \tag{e}$$

The preceding examples illustrate the basic procedure for using tables of transform pairs and properties to perform inversion of transforms. These examples also demonstrate some basic strategies for inverting transforms.

- If an exponential function of the transform variable appears in the numerator, use the second shifting theorem.
- If the denominator contains a linear factor raised to an integer power, use transform pair 6 and the first shifting theorem.
- If the denominator contains a quadratic factor that does not have real roots, then complete the square and use the first shifting theorem in conjunction with transform pairs 2 and 3.
- If whenever s appears in the transform it appears as $s - a$, apply the first shifting theorem.

2.4.2 Partial Fraction Decompositions

The inversion of the transform is more complicated when the denominator of the transform contains polynomials in s higher than second order. Polynomials with real coefficients can always be factored into linear and quadratic factors. Consider as a general case a Laplace transform of the form

$$X(s) = \frac{Q(s)}{D(s)} \tag{2.31}$$

where $D(s)$ is a polynomial of order n with a leading coefficient of one.

The function $Q(s)$ can be written as

$$Q(s) = R(s)N(s) \tag{2.32}$$

where $R(s)$ contains everything that is not part of a polynomial, such as exponential terms involving the transform variable. If $Q(s)$ is a polynomial, then $R(s) = 1$. $N(s)$ is a polynomial of order p. If $p \geq n$, then a long division or a synthetic division must be performed such that

$$\frac{N(s)}{D(s)} = W(s) + \frac{B(s)}{D(s)} \tag{2.33}$$

where $W(s)$ is a polynomial in s of order $p - n$ (if $p = n$, $W(s)$ is a constant) and $B(s)$ is a polynomial of order $n - 1$ or less.

Let $s_1, s_2, \ldots, s_n$ be the roots of $D(s)$ such that it has a factorization of

$$D(s) = (s - s_1)(s - s_2)\cdots(s - s_n) \tag{2.34}$$

The roots of $D(s)$ are referred to as the **poles** of the transform.

A **partial fraction decomposition** of $B(s)/D(s)$ can be developed in the form

$$\frac{B(s)}{D(s)} = \sum_{k=1}^{n} \frac{A_k}{s - s_k} \tag{2.35}$$

Table 2.3 Partial Fraction Decompositions

	$X(s)$	Partial Fraction Decomposition for $X(s)$
1	$\frac{1}{s(s+a)}$	$\frac{1}{a}\left(\frac{1}{s}-\frac{1}{s+a}\right)$
2	$\frac{1}{(s+a)(s+b)}$	$\frac{1}{b-a}\left(\frac{1}{s+a}-\frac{1}{s+b}\right)$
3	$\frac{s}{(s+a)(s+b)}$	$\frac{1}{b-a}\left(\frac{b}{s+b}-\frac{a}{s+a}\right)$
4	$\frac{1}{s(s^2+\omega^2)}$	$\frac{1}{\omega^2}\left(\frac{1}{s}-\frac{s}{s^2+\omega^2}\right)$
5	$\frac{1}{(s+a)(s^2+\omega^2)}$	$\frac{1}{\omega^2+a^2}\left(\frac{1}{s+a}-\frac{s-a}{s^2+\omega^2}\right)$
6	$\frac{s}{(s+a)(s^2+\omega^2)}$	$\frac{1}{\omega^2+a^2}\left(\frac{as+\omega^2}{s^2+\omega^2}-\frac{a}{s+a}\right)$
7	$\frac{1}{(s^2+\omega_1^2)(s^2+\omega_2^2)}$	$\frac{1}{\omega_2^2-\omega_1^2}\left(\frac{1}{s^2+\omega_1^2}-\frac{1}{s^2+\omega_2^2}\right)$
8	$\frac{1}{(s+a)(s^2+2\zeta\omega_n s+\omega_n^2)}$	$\frac{1}{a^2+\omega_n^2-2\zeta a\omega_n}\left(\frac{1}{s+a}-\frac{s+2\zeta\omega_n-a}{s^2+2\zeta\omega_n s+\omega_n^2}\right)$
9	$\frac{s}{(s+a)(s^2+2\zeta\omega_n s+\omega_n^2)}$	$\frac{1}{a^2+\omega_n^2-2\zeta a\omega_n+\omega_n^2}\left(\frac{as+\omega_n^2}{s^2+2\zeta\omega_n s+\omega_n^2}-\frac{a}{s+a}\right)$
10	$\frac{1}{(s^2+\omega^2)(s^2+2\zeta\omega_n s+\omega_n^2)}$	$\frac{1}{(\omega_n^2-\omega^2)^2+(2\zeta\omega\omega_n)^2}\left[\frac{-2\zeta\omega_n s+\omega_n^2-\omega^2}{s^2+\omega^2}+\frac{2\zeta\omega_n s-(2\zeta\omega_n)^2+(\omega_n^2-\omega^2)}{s^2+2\zeta\omega_n s+\omega_n^2}\right]$
11	$\frac{s}{(s^2+\omega^2)(s^2+2\zeta\omega_n s+\omega_n^2)}$	$\frac{1}{(\omega_n^2-\omega^2)^2+(2\zeta\omega\omega_n)^2}\left[\frac{(\omega_n^2-\omega^2)s-2\zeta\omega_n+\omega^2}{s^2+\omega^2}-\frac{(\omega_n^2-\omega^2)s-2\zeta\omega_n^3}{s^2+2\zeta\omega_n s+\omega_n^2}\right]$

where A_k, $k = 1, 2, \ldots, n$ are coefficients to be determined. These coefficients are called the **residues** of the transform.

Three cases are considered:

- All poles are real and distinct.
- Some of the poles are imaginary or complex.
- $D(s)$ has repeated roots. [In this case the form of $B(s)/D(s)$ is modified from that of Equation (2.35).]

Real Distinct Poles Multiplying Equation (2.35) by $D(s)$ and using the form of Equation (2.34) for $D(s)$ leads to

$$B(s) = \sum_{i=1}^{n}\left[A_i \prod_{\substack{j=1 \\ j\neq i}}^{n}(s-s_j)\right] \tag{2.36}$$

Equation (2.36) is valid for all s. Consider its evaluation for $s = s_k$ for any k, $1 \le k \le n$. The only nonzero term in the resulting summation corresponds to $i = k$, leading to

$$A_k = \frac{B(s_k)}{\prod_{\substack{j=1 \\ j\neq k}}^{n}(s_k-s_j)} \tag{2.37}$$

An alternate representation of Equation (2.37) is

$$A_k = \lim_{s \to s_k} \frac{(s - s_k)B(s)}{D(s)} \tag{2.38}$$

Equation (2.38) is a mathematical representation of the residue theorem.

Example 2.19

Determine $x(t)$ if

$$X(s) = \frac{3s^2 + 2s + 1}{s^3 + 7s^2 + 14s + 8} \tag{a}$$

Solution

Equation (a) is of the form of Equation (2.31) with

$$Q(s) = B(s) = 3s^2 + 2s + 1 \tag{b}$$

$$D(s) = s^3 + 7s^2 + 14s + 8 \tag{c}$$

The roots of $D(s)$ are determined as $s_1 = -4$, $s_2 = -2$, and $s_3 = -1$, such that

$$D(s) = (s + 4)(s + 2)(s + 1) \tag{d}$$

The coefficients in the partial fraction decomposition of $X(s)$ are determined using Equation (2.37) as

$$\begin{aligned} A_1 &= \frac{3s_1^2 + 2s_1 + 1}{(s_1 + 2)(s_1 + 1)} \\ &= \frac{3(-4)^2 + 2(-4) + 1}{(-4 + 2)(-4 + 1)} \\ &= \frac{41}{6} \end{aligned} \tag{e}$$

$$\begin{aligned} A_2 &= \frac{3s_2^2 + 2s_2 + 1}{(s_2 + 4)(s_2 + 1)} \\ &= \frac{3(-2)^2 + 2(-2) + 1}{(-2 + 4)(-2 + 1)} \\ &= -\frac{9}{2} \end{aligned} \tag{f}$$

$$\begin{aligned} A_3 &= \frac{3s_3^2 + 2s_3 + 1}{(s_3 + 4)(s_3 + 2)} \\ &= \frac{3(-1)^2 + 2(-1) + 1}{(-1 + 4)(-1 + 2)} \\ &= \frac{2}{3} \end{aligned} \tag{g}$$

The partial fraction decomposition of $X(s)$ is

$$X(s) = \frac{41}{6}\left(\frac{1}{s + 4}\right) - \frac{9}{2}\left(\frac{1}{s + 2}\right) + \frac{2}{3}\left(\frac{1}{s + 1}\right) \tag{h}$$

Transform pair 1 of Table 2.1 is used to invert Equation (h), resulting in

$$x(t) = \frac{41}{6}e^{-4t} - \frac{9}{2}e^{-2t} + \frac{2}{3}e^{-t} \tag{i}$$

Complex Poles If $D(s)$ has complex roots, they occur in complex conjugate pairs; that is, if $D(s_k) = 0$, then $D(\bar{s}_k) = 0$ where $\bar{s}_k$ is the complex conjugate of s_k. Equations (2.35) and (2.37) may still be used to determine the partial fraction decomposition of $B(s)/D(s)$, but evaluation of Equation (2.37) leads to complex coefficients when s_k is complex. The use of complex algebra can be avoided by noting that

$$\frac{A_k}{s - s_k} + \frac{\overline{A}_k}{s - \bar{s}_k} = \frac{Cs + D}{(s - s_k)(s - \bar{s}_k)} \tag{2.39}$$

where C and D are real coefficients and $(s - s_k)(s - \bar{s}_k)$ is a quadratic factor with real coefficients

$$\begin{aligned}(s - s_k)(s - \bar{s}_k) &= s^2 - 2\text{Re}(s_k)s + [\text{Re}(s_k)]^2 + [\text{Im}(s_k)]^2 \\ &= s^2 + 2as + b\end{aligned} \tag{2.40}$$

Any polynomial with real coefficients can be written as a product of linear factors with real roots and quadratic factors. A quadratic factor that has real roots can be factored into two linear factors and is called reducible. A quadratic factor that has roots that are complex conjugates of one another cannot be further factored and is called irreducible. The partial fraction decomposition of Equation (2.35) can be modified using Equation (2.39) to handle the quadratic factors. The appropriate term in the partial fraction decomposition corresponding to a quadratic factor is $(Cs + D)/(s^2 + 2as + b)$. This term is easily inverted by completing the square in the denominator, as in Example 2.17, and using the first shifting theorem and transform pairs for $\sin(\omega t)$ and $\cos(\omega t)$.

The coefficients corresponding to the real factors can be determined using Equation (2.37). The coefficients, such as those of Equation (2.39), corresponding to the quadratic factors can be determined by multiplying Equation (2.35) by $N(s)$ and then evaluating the resulting expression at independent values of s that do not coincide with the real roots.

Example 2.20

Determine $x(t)$ if

$$X(s) = \frac{4s + 2}{s^3 + 11s^2 + 44s + 60} \tag{a}$$

Solution

The denominator of Equation (a) can be factored as

$$s^3 + 11s^2 + 44s + 60 = (s + 3)(s^2 + 8s + 20) \tag{b}$$

The quadratic factor of Equation (b) is irreducible and does not have real roots. Thus the appropriate partial fraction decomposition for Equation (a) is

$$\frac{4s + 2}{s^3 + 11s^2 + 44s + 60} = \frac{A}{s + 3} + \frac{Cs + D}{s^2 + 8s + 20} \tag{c}$$

The coefficient of the linear factor is determined using the residue theorem as

$$\begin{aligned}A &= \left.\frac{4s + 2}{s^2 + 8s + 20}\right|_{s=-3} \\ &= \frac{4(-3) + 2}{(-3)^2 + 8(-3) + 20} \\ &= -2\end{aligned} \tag{d}$$

Substituting Equation (d) into Equation (c) and multiplying by the denominator of the left-hand side leads to

$$4s + 2 = -2(s^2 + 8s + 20) + (Cs + D)(s + 3) \tag{e}$$

Evaluation of Equation (e) for $s = 0$ leads to

$$2 = -2(20) + D(3) \Rightarrow D = 14 \tag{f}$$

Evaluation of Equation (e) for $s = 1$ using Equation (f) leads to

$$4(1) + 2 = -2[(1)^2 + 8(1) + 20] + [C(1) + 14](1 + 3) \Rightarrow C = 2 \tag{g}$$

Substituting Equations (d), (f), and (g) into Equation (c) and completing the square of the quadratic factor in the denominator leads to

$$\begin{aligned} X(s) &= -\frac{2}{s+3} + \frac{2s + 14}{s^2 + 8s + 20} \\ &= -\frac{2}{s+3} + \frac{2(s+4)}{(s+4)^2 + 4} + \frac{6}{(s+4)^2 + 4} \end{aligned} \tag{h}$$

The first shifting theorem and transform pairs 1, 3, and 2 respectively from Table 2.1 are used to invert Equation (h), leading to

$$x(t) = -2e^{-3t} + 2e^{-4t}\cos(2t) + 3e^{-4t}\sin(2t) \tag{i}$$

Repeated Poles Consider the case where s_k is a root of $D(s)$ of multiplicity $p > 1$. Equation (2.35) is modified to account for repeated roots by replacing the terms in the equation corresponding to the repeated roots by

$$\frac{C_1}{s - s_k} + \frac{C_2}{(s - s_k)^2} + \dots + \frac{C_p}{(s - s_k)^p} \tag{2.41}$$

Suppose the roots of $D(s)$ are numbered such that $s_1, s_2, \dots, s_{n-p}$ are distinct roots and $s_k = s_{n-p+1}$ is a repeated root of multiplicity p. Using Equation (2.41) in Equation (2.35) leads to

$$\frac{B(s)}{D(s)} = \sum_{i=1}^{n-p} \frac{A_i}{s - s_i} + \sum_{j=1}^{p} \frac{C_j}{(s - s_k)^j} \tag{2.42}$$

Multiplying Equation (2.42) by $(s - s_k)^p$ leads to

$$(s - s_k)^p \frac{B(s)}{D(s)} = \sum_{i=1}^{n-p} \frac{A_i (s - s_k)^p}{s - s_i} + \sum_{j=1}^{p} C_j (s - s_k)^{p-j} \tag{2.43}$$

Evaluation of Equation (2.43) for $s = s_k$ leads to

$$C_p = \left[(s - s_k)^p \frac{B(s)}{D(s)}\right]_{s=s_k} \tag{2.44}$$

Differentiation of Equation (2.43) with respect to s and then evaluating for $s = s_k$ leads to

$$C_{p-1} = \left\{\frac{d}{ds}\left[(s - s_k)^p \frac{B(s)}{D(s)}\right]\right\}_{s=s_k} \tag{2.45}$$

Continued differentiation and evaluation for $s = s_k$ lead to

$$C_{p-\ell} = \frac{1}{\ell!}\left\{\frac{d^\ell}{ds^\ell}\left[(s - s_k)^p \frac{B(s)}{D(s)}\right]\right\}_{s=s_k} \qquad \ell = 0,1,2,\dots,p \tag{2.46}$$

The coefficients in the partial fraction decomposition corresponding to repeated roots are obtained using Equation (2.46).

Example 2.21

Determine $x(t)$ if

$$X(s) = \frac{5s^3 + 2s^2 + 3s - 1}{(s+2)(s+1)^3} \tag{a}$$

Solution

The appropriate partial fraction decomposition for $X(s)$ is

$$\frac{5s^3 + 2s^2 + 3s - 1}{(s+2)(s+1)^3} = \frac{A}{s+2} + \frac{C_1}{s+1} + \frac{C_2}{(s+1)^2} + \frac{C_3}{(s+1)^3} \tag{b}$$

The coefficient A is determined using the residue theorem, Equation (2.37), as

$$A = \left[\frac{5s^3 + 2s^2 + 3s - 1}{(s+1)^3}\right]_{s=-2} = \frac{5(-2)^3 + 2(-2)^2 + 3(-2) - 1}{(-2+1)^3} = 39 \tag{c}$$

The coefficients for the repeated root $s = -1$ are determined using Equation (2.46):

$$C_3 = \left[\frac{5s^3 + 2s^2 + 3s - 1}{s+2}\right]_{s=-1} = \frac{5(-1)^3 + 2(-1)^2 + 3(-1) - 1}{-1+2} = -7 \tag{d}$$

$$\begin{aligned} C_2 &= \left\{\frac{d}{ds}\left[\frac{5s^3 + 2s^2 + 3s - 1}{s+2}\right]\right\}_{s=-1} \\ &= \left[\frac{(s+2)(15s^2 + 4s + 3) - (5s^3 + 2s^2 + 3s - 1)(1)}{(s+2)^2}\right]_{s=-1} \\ &= \left[\frac{10s^3 + 32s^2 + 8s + 7}{(s+2)^2}\right]_{s=-1} \\ &= \frac{10(-1)^3 + 32(-1)^2 + 8(-1) + 7}{(-1+2)^2} = 21 \end{aligned} \tag{e}$$

$$\begin{aligned} C_1 &= \frac{1}{2}\left\{\frac{d^2}{ds^2}\left[\frac{5s^3 + 2s^2 + 3s - 1}{s+2}\right]\right\}_{s=-1} \\ &= \frac{1}{2}\left\{\frac{d}{ds}\left[\frac{10s^3 + 32s^2 + 8s + 7}{(s+2)^2}\right]\right\}_{s=-1} \\ &= \frac{1}{2}\left[\frac{(s+2)^2(30s^2 + 64s + 8) - 2(s+2)(10s^3 + 32s^2 + 8s + 7)}{(s+2)^4}\right]_{s=-1} \\ &= \frac{1}{2}\left[\frac{(-1+2)^2(30(-1)^2 + 64(-1) + 8) - 2(-1+2)(10(-1)^3 + 32(-1)^2 + 8(-1) + 7)}{(-1+2)^4}\right] \\ &= -34 \end{aligned} \tag{f}$$

Using Equations (b)–(f) in Equation (a) leads to

$$X(s) = \frac{39}{s+2} - \frac{34}{s+1} + \frac{21}{(s+1)^2} - \frac{7}{(s+1)^3} \tag{g}$$

The transform of Equation (g) is inverted using the first shifting theorem and transform pairs 1 and 4 from Table 2.1. The result is

$$x(t) = 39e^{-2t} - 34e^{-t} + 21te^{-t} - \frac{7}{2}t^2e^{-t} \tag{h}$$

Brute Force Methods The methods just described are elegant methods for determining residues for partial fraction decomposition. The formulas are easy to forget and can be tedious to apply. The brute force methods described in this section are also tedious, but they do not require special knowledge.

Once the poles are determined, the appropriate partial fraction decomposition is assumed as a sum of fractions of linear factors of s, nonreducible quadratic factors of s, and powers of repeated factors. The numerator for each of the linear factors contains one unknown coefficient, and the numerator for each quadratic factor contains two unknown coefficients. There are n unknown coefficients in the partial fraction decomposition when $D(s)$ is a polynomial of order n.

The assumed partial fraction decomposition is multiplied by $D(s)$. Both sides of the resulting equation are polynomials in s of order $n - 1$. Since s is a variable that can change independently, the equations must be satisfied for all possible values of s. Two methods can be used to determine a set of n simultaneous linear equations to solve for the unknown coefficients.

- Since powers of s are linearly independent, coefficients of like powers of s on each side must be equal. Equating coefficients of s^k for $k = 0, 1, \ldots, n - 1$ leads to n equations.
- The equations may be evaluated for n values of s, which are not poles. Numerical evaluation for each value of s leads to one equation.

Example 2.22

Determine $x(t)$ if

$$X(s) = \frac{s^4 + 8s^3 + 29s^2 + 40s + 36}{(s + 2)^2(s + 3)(s + 4)(s^2 + 4s + 9)} \tag{a}$$

Solution

The appropriate partial fraction decomposition for Equation (a) is

$$\frac{s^4 + 8s^3 + 29s^2 + 40s + 36}{(s + 2)^2(s + 3)(s + 4)(s^2 + 4s + 9)} = \frac{A}{s + 3} + \frac{B}{s + 4} + \frac{C}{s + 2} + \frac{D}{(s + 2)^2} + \frac{Es + F}{s^2 + 4s + 9} \tag{b}$$

Multiplication of Equation (b) by $D(s)$ leads to

$$\begin{aligned} s^4 + 8s^3 + 29s^2 + 40s + 36 = {} & A(s + 4)(s + 2)^2(s^2 + 4s + 9) \\ & + B(s + 3)(s + 2)^2(s^2 + 4s + 9) \\ & + C(s + 3)(s + 4)(s + 2)(s^2 + 4s + 9) \\ & + D(s + 3)(s + 4)(s^2 + 4s + 9) \\ & + (Es + F)(s + 3)(s + 4)(s + 2)^2 \end{aligned} \tag{c}$$

Evaluation of Equation (c) for $s = -1, 0, 1, 2, 3,$ and 4 respectively leads to the following set of equations

$$\begin{bmatrix} 18 & 12 & 36 & 36 & -6 & 6 \\ 144 & 108 & 216 & 108 & 0 & 48 \\ 630 & 504 & 840 & 280 & 180 & 180 \\ 2016 & 1680 & 2520 & 630 & 960 & 480 \\ 5250 & 4500 & 6300 & 1260 & 3150 & 1050 \\ 11{,}808 & 10{,}332 & 13{,}776 & 2296 & 8064 & 2016 \end{bmatrix} \begin{bmatrix} A \\ B \\ C \\ D \\ E \\ F \end{bmatrix} = \begin{bmatrix} 18 \\ 36 \\ 114 \\ 312 \\ 714 \\ 1428 \end{bmatrix} \tag{d}$$

The solution of Equation (d) is $A = -9.80$, $B = 5.13$, $C = 3.70$, $D = -0.480$, $E = 1.01$, $F = 3.05$. Substituting these values into Equation (b) leads to

$$X(s) = -\frac{9.80}{s+3} + \frac{5.13}{s+4} + \frac{3.70}{s+2} - \frac{0.480}{(s+2)^2} + \frac{1.01s + 3.05}{s^2 + 4s + 9} \tag{e}$$

Inversion of Equation (e) yields

$$\begin{aligned} x(t) &= -9.80e^{-3t} + 5.13e^{-4t} + 3.70e^{-2t} - 0.480te^{-2t} \\ &\quad + 1.01e^{-2t}\cos(\sqrt{5}t) + 0.461e^{-2t}\sin(\sqrt{5}t) \end{aligned} \tag{f}$$

2.4.3 Inversion of Transforms of Periodic Functions

The Laplace transform of a periodic function is obtained using property 12 of Table 2.2. From property 12, it is clear that the transform of a periodic function has a term of the form of $1 - e^{-Ts}$ in its denominator. Thus if a transform has this type of term in its denominator, its inverse is a periodic function.

Consider a transform of the form

$$X(s) = \frac{F(s)}{1 - e^{-Ts}} \tag{2.47}$$

The binomial expansion of the denominator is

$$\begin{aligned} (1 - e^{-Ts})^{-1} &= 1 + e^{-Ts} + e^{-2Ts} + e^{-3Ts} + \cdots \\ &= \sum_{k=0}^{\infty} e^{-kTs} \end{aligned} \tag{2.48}$$

Use of Equation (2.48) in Equation (2.47) leads to

$$X(s) = \sum_{k=0}^{\infty} [e^{-kTs}F(s)] \tag{2.49}$$

The second shifting theorem and linearity of the inverse transform are applied to Equation (2.49), leading to

$$x(t) = \sum_{k=0}^{\infty} f(t - kT)u(t - kT) \tag{2.50}$$

where $f(t) = \mathcal{L}^{-1}\{F(s)\}$.

Example 2.23

Determine $x(t)$ if

$$X(s) = \frac{1}{1 - e^{-3s}}\left[\frac{4}{s(s^2 + 4)}\right] \tag{a}$$

Solution

A partial fraction decomposition of Equation (a) leads to

$$X(s) = \frac{1}{1 - e^{-3s}}\left(\frac{1}{s} - \frac{s}{s^2 + 4}\right) \tag{b}$$

Equation (b) is in the form of Equation (2.48) with $T = 3$ and

$$F(s) = \frac{1}{s} - \frac{s}{s^2 + 4} \tag{c}$$

Equation (c) is inverted using the transform pairs 8 and 3 of Table 2.1, leading to

$$f(t) = 1 - \cos 2t \tag{d}$$

Using Equation (d) in Equation (2.50) leads to

$$x(t) = \sum_{k=0}^{\infty} \{[1 - \cos(2t - 6k)]\, u(t - 3k)\} \tag{e}$$

2.4.4 Use of MATLAB

Just as MATLAB is capable of symbolically determining the Laplace transform of a function of a real variable t, it is also capable of determining the inverse transform of a function of a complex variable s. Assuming s is declared as a symbolic variable and G is a defined function of s, the appropriate MATLAB command to determine the inverse transform of $G(s)$ and assign it to the variable x is

x = ilaplace(G)

Example 2.24

Use MATLAB to determine and plot the inverse transforms of the functions in **a.** Example 2.18 and **b.** Example 2.21.

Solution

A MATLAB script file, Example2_24.m, which provides the inverse transform and plots the results as a function of time, is illustrated in Figure 2.7. The output from the execution of the M-file and the resulting plots are shown in Figure 2.8. Note the following regarding Example2_24.m:

```
% Example 2.24
% Determination and plotting of inverse transforms
%
% Declaration of symbolic variables
% s=transform variable
% t=time
%
syms s t
%
% Example 2.24a
% F1 is function in transform domain
% G1 is inverse transform of F1
% G1A is decimal form of G1 with 4 significant figures
%
F1=exp(-2*s)/(s+2)^2;
G1=ilaplace(F1)
G1A=vpa(G1,4)
%
% Example 2.24b
```

Figure 2.7

Script for Example2_24.m.

```
%
% F2 is function in transform domain
% G2 is inverse transform of F1
% G2A is decimal form of G1 with 4 significant figures
%
F2=(5*s^3+2*s^2+3*s-1)/(s+2)/(s+1)^3;
G2=ilaplace(F2)
G2A=vpa(G2,4)
%
% Defining t to range between 0 and 10 in increments of 0.1
%
t1=linspace(0,10,101)
%
% Defining vectors of numerical values for purposes of plotting
% subs(G1A,t1) numerically evaluates the symbolic function G1A at the
% current value of t1
%
x1=subs(G1A,t1);
x2=subs(G2A,t1);
% Since more than one figure is to be plotted the figure statement is
% necessary such that both figures are retained
%
% Plotting and annotating graph of ilaplace(F1)
%
figure
plot(t1,x1)
xlabel('t')
ylabel('x')
title('Inverse transform of e^-^2^s/(s+2)^2')
%
% Plotting and annotating graph of ilaplace(F2)
%
figure
plot(t1,x2)
xlabel('t')
ylabel('x')
title('Inverse transform of (5s^3+2s^2+3s-1)/[(s+2)(s+1)^3]')
%
% End of Example2_24.m
```

Figure 2.7

(Continued)

```
>> Example2_24
G1 =

heaviside(t - 2)*exp(4 - 2*t)*(t - 2)

G1A =
```

Figure 2.8

Output from the execution of Example2_24.m: (a) the screen output, (b) the plot of the inverse transform of part a., and (c) the plot of the inverse transform of part b.

```
heaviside(1.0*t - 2.0)*exp(4.0 - 2.0*t)*(t - 2.0)


G2 =

39*exp(-2*t) - 34*exp(-t) + 21*t*exp(-t) - (7*t^2*exp(-t))/2


G2A =
```

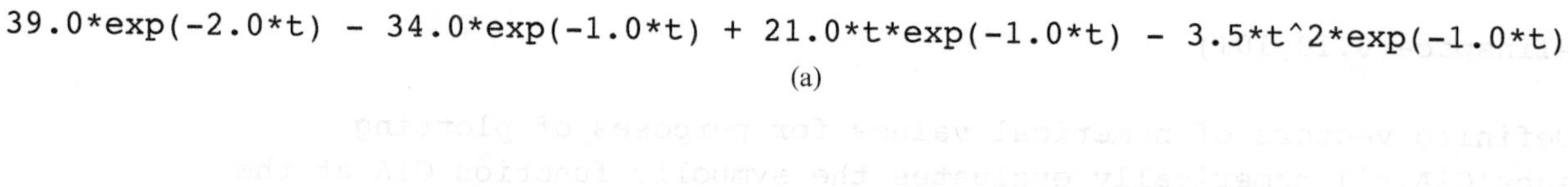

```
39.0*exp(-2.0*t) - 34.0*exp(-1.0*t) + 21.0*t*exp(-1.0*t) - 3.5*t^2*exp(-1.0*t)
```

(a)

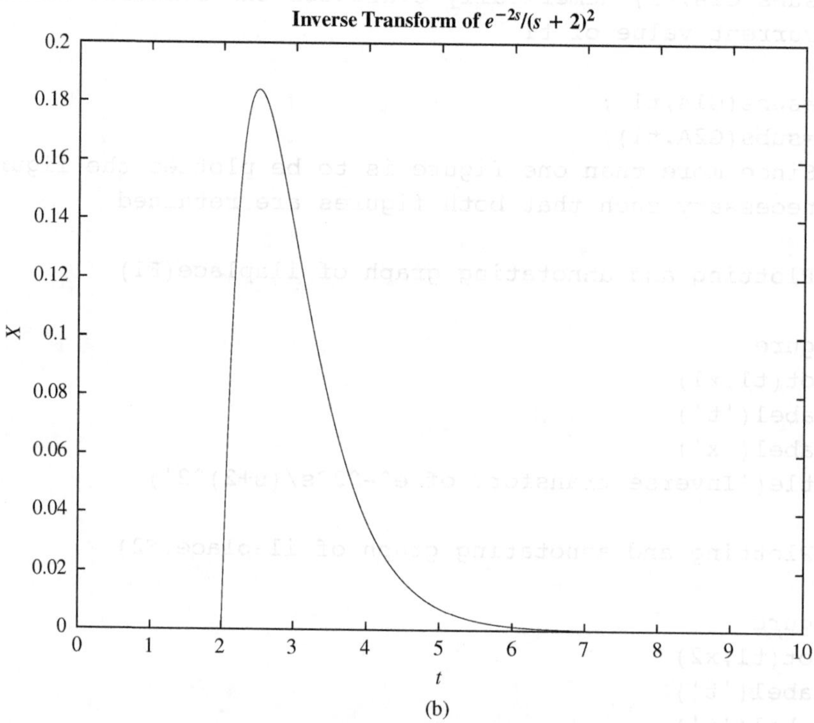

(b)

Figure 2.8
(Continued)

- Since G1A and G2A are symbolic functions, it is necessary to numerically evaluate them at discrete values of t for purposes of plotting. The command "x1=subs (G1A,t)" numerically evaluates the symbolic function G1A at the current value of t and assigns the value to $x1$. Note that this also works for arrays; if t is an array, then $x1$ will be the result of evaluation at every element of the array.
- The "figure" statements are necessary so that both figures are retained as output. Without the second "figure" statement, the first plot would be erased.
- In text annotating graphs, the caret (^) means to use the next character as a superscript and the underscore (_) means to use the next character as a subscript.

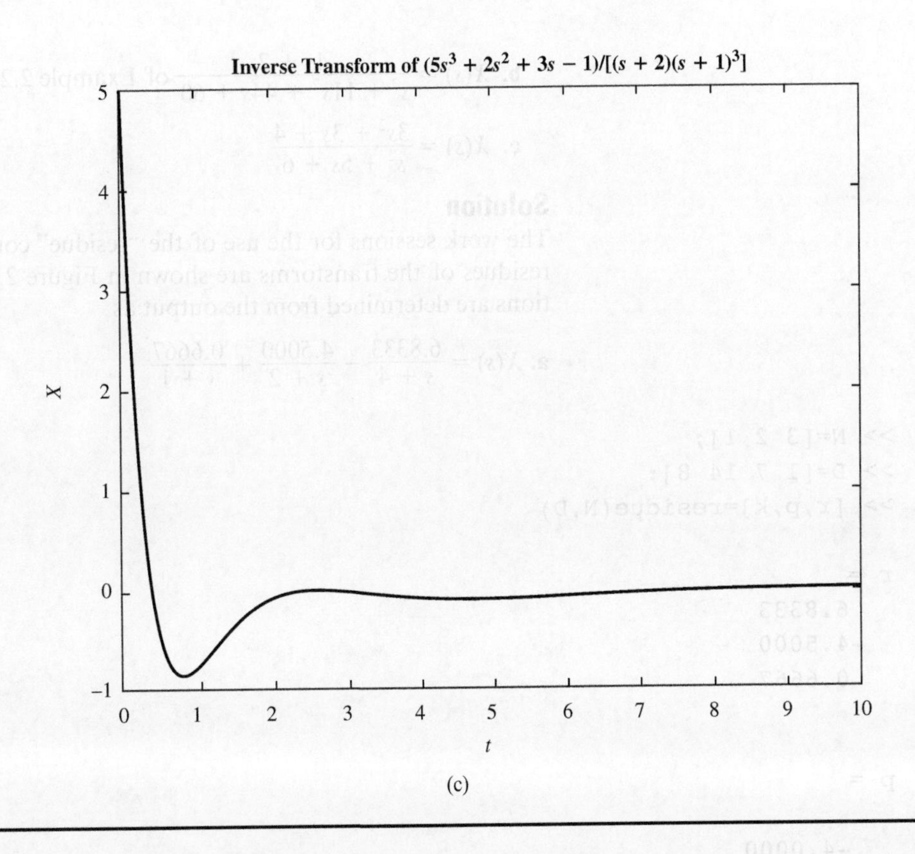

Figure 2.8
(Continued)

MATLAB is also useful in performing partial fraction decompositions on ratios of polynomials of the form of Equation (2.33). The polynomials in the numerator and denominator are defined by their coefficients. The coefficients of a polynomial are stored in a row vector. If $N(s) = a_3s^3 + a_2s^2 + a_1s + a_0$, then $N(s)$ is defined in MATLAB by defining a row vector with four columns according to

$$N = [a_3 \quad a_2 \quad a_1 \quad a_0]$$

Once $N(s)$ and $D(s)$ are defined, MATLAB calculates the poles and residues using the statement

$$[\mathrm{r, p, k}] = \mathrm{residue(N, D)}$$

where r is a column vector of the residues of $N(s)/D(s)$ corresponding to the elements in the vector of poles p. The vector k is returned as the null vector if the order of $N(s)$ is less than the order of $D(s)$. If the order of $N(s)$ is equal to or greater than the order of $D(s)$, then k is the vector of coefficients in the function $W(s)$ defined in Equation (2.33). If the poles are complex, MATLAB provides complex residues. The partial fraction decomposition is built using the elements of r, p, and k. The order of the elements in r corresponds to the order of the elements in p.

Example 2.25

Use the MATLAB "residue" command to help determine the partial fraction decompositions for:

a. $X(s) = \dfrac{3s^2 + 2s + 1}{s^3 + 7s^2 + 14s + 8}$ of Example 2.19

b. $X(s) = \dfrac{4s + 2}{s^3 + 11s^2 + 44s + 60}$ of Example 2.20

c. $X(s) = \dfrac{3s^2 + 3s + 4}{s^2 + 5s + 6}$

Solution

The work sessions for the use of the "residue" command to determine the poles and residues of the transforms are shown in Figure 2.9. The partial fraction decompositions are determined from the output as

$$\text{a. } X(s) = \frac{6.8333}{s+4} - \frac{4.5000}{s+2} + \frac{0.6667}{s+1} \qquad \text{(a)}$$

```
>> N=[3 2 1];
>> D=[1 7 14 8];
>> [r,p,k]=residue(N,D)

r =

   6.8333
  -4.5000
   0.6667


p =

  -4.0000
  -2.0000
  -1.0000


k =

   []
```

(a)

```
>> N=[4 2];
>> D=[1 11 44 60];
>> [r,p,k]=residue(N,D)
r =

  1.0000 - 1.5000i
  1.0000 + 1.5000i
  -2.0000
p =
  -4.0000 + 2.0000i
  -4.0000 - 2.0000i
  -3.0000
```

Figure 2.9

The solution of Example 2.25, parts a., b., and c.: The MATLAB "residue" command aids in partial fraction decompositions.

```
k =

     []
>>
```

(b)

```
>> N=[3 3 4];
>> D=[1 5 6];
>> [r,p,k]=residue(N,D)

r =

  -22.0000
   10.0000


p =
  -3.0000
  -2.0000

k =

     3
>>
```

(c)

Figure 2.9
(Continued)

b. $$X(s) = \frac{1 - 1.5j}{s - (-4 + 2j)} + \frac{1 + 1.5j}{s - (-4 - 2j)} - \frac{2}{s+3}$$

$$= \frac{2s + 14}{s^2 + 8s + 20} - \frac{2}{s+3} \quad \text{(b)}$$

c. $$X(s) = 3 - \frac{22}{s+3} + \frac{10}{s+2} \quad \text{(c)}$$

2.5 Laplace Transform Solution of Differential Equations

Linear differential equations with constant coefficients and integrodifferential equations obtained through the mathematical modeling of dynamic systems can be solved using the **Laplace transform method**. When the Laplace transform of the differential equation is taken, linearity of the transform allows the transform of each term to be taken individually. The property of transform of derivatives property 9 of Table 2.2 allows a differential equation in the time domain to be transformed into an algebraic equation in the s domain. The property of transform of integrals property 10 of Table 2.2 allows integrodifferential equations to be transformed into algebraic equations in the s domain. Once the transform is obtained in the s domain, the methods of Section 2.4 are employed to determine the response in the time domain.

2.5.1 Systems with One Dependent Variable

Given the differential equation and appropriate initial conditions, the following steps summarize the application of the Laplace transform method to a linear differential equation with constant coefficients where $x(t)$ is the dependent variable and defining $X(s) = \mathcal{L}\{x(t)\}$.

1. The property of uniqueness of the transform is used to apply the Laplace transform to both sides of the differential equation.
2. The property of linearity of the transform is applied.
3. If nonzero, the Laplace transform of the nonhomogeneous term is obtained.
4. The property of transform of derivatives is applied using appropriate initial conditions.
5. Algebraic methods are used to determine $X(s)$.
6. $x(t)$ is determined using the inversion methods of Section 2.4.

This procedure is illustrated in the following examples.

Example 2.26

Use the Laplace transform method to solve the differential equation

$$\ddot{x} + 4\dot{x} + 400x = 0 \tag{a}$$

subject to the initial conditions $x(0) = 0.01$ and $\dot{x}(0) = 2$.

Solution

Taking the Laplace transform of both sides of Equation (a) leads to

$$\mathcal{L}\{\ddot{x} + 4\dot{x} + 400x\} = \mathcal{L}\{0\} \tag{b}$$

Linearity of the transform is used on Equation (b) to give

$$\mathcal{L}\{\ddot{x}\} + 4\mathcal{L}\{\dot{x}\} + 400\mathcal{L}\{x\} = 0 \tag{c}$$

Defining $X(s) = \mathcal{L}\{x(t)\}$ and using the property of transforms of derivatives in equation (c) yields

$$s^2X(s) - sx(0) - \dot{x}(0) + 4[sX(s) - x(0)] + 400X(s) = 0 \tag{d}$$

Substituting the given initial conditions into Equation (d) and rearranging leads to

$$(s^2 + 4s + 400)X(s) = 0.01s + 2.04 \tag{e}$$

Solving for $X(s)$ from Equation (e) yields

$$X(s) = \frac{0.01s + 2.04}{s^2 + 4s + 400} \tag{f}$$

The quadratic polynomial in the denominator does not have real roots; it is irreducible. Completing the square in the denominator of Equation (f) gives

$$X(s) = \frac{0.01s + 2.04}{(s + 2)^2 + 396} \tag{g}$$

Equation (g) can be rewritten as

$$X(s) = \frac{0.01(s + 2)}{(s + 2)^2 + 396} + \frac{2.02}{(s + 2)^2 + 396} \tag{h}$$

The inverse transform of Equation (h) is obtained using linearity and transform pairs 12 and 13 of Table 2.1, yielding

$$x(t) = 0.01e^{-2t}\cos(\sqrt{396}t) + \frac{2.02}{\sqrt{396}}e^{-2t}\sin(\sqrt{396}t)$$

$$= 0.01\,e^{-2t}\cos(19.90t) + 0.102\,e^{-2t}\sin(19.90t) \quad \text{(i)}$$

Example 2.27

The differential equation governing the current in a series *RL* circuit with a direct current source connected at $t = 0$ is

$$L\frac{di}{dt} + Ri = V_0 u(t) \quad \text{(a)}$$

where $i(t)$ is the current in the circuit, L is the inductance, R is the resistance, and V_0 is the potential in the source. If there is no current in the circuit when the source is connected at $t = 0$, $i(0) = 0$, determine the time-dependent current in the circuit.

Solution

Taking the Laplace transform of both sides of Equation (a) leads to

$$\mathcal{L}\left\{L\frac{di}{dt} + Ri\right\} = \mathcal{L}\{V_0 u(t)\} \quad \text{(b)}$$

Using linearity of the transform, Equation (b) becomes

$$L\mathcal{L}\left\{\frac{di}{dt}\right\} + R\mathcal{L}\{i(t)\} = \mathcal{L}\{V_0 u(t)\} \quad \text{(c)}$$

Defining $I(s) = \mathcal{L}\{i(t)\}$ and using the transform of derivative property and transform pair 8 of Table 2.1 in Equation (c),

$$L[sI(s) - i(0)] + RI(s) = \frac{V_0}{s} \quad \text{(d)}$$

Applying the initial condition and solving for $I(s)$ yields

$$I(s) = \frac{V_0}{(Ls + R)s}$$

$$= \frac{V_0}{L}\frac{1}{\left(s + \frac{R}{L}\right)s} \quad \text{(e)}$$

A partial fraction decomposition of the right-hand side of Equation (e) leads to

$$I(s) = \frac{V_0}{L}\left[\frac{\frac{L}{R}}{s} - \frac{\frac{L}{R}}{s + \frac{R}{L}}\right]$$

$$= \frac{V_0}{R}\left[\frac{1}{s} - \frac{1}{s + \frac{R}{L}}\right] \quad \text{(f)}$$

The transform in Equation (f) is inverted using transform pairs 8 and 1 of Table 2.1, resulting in

$$i(t) = \frac{V_0}{R}\left[u(t) - e^{-(R/L)t}\right] \quad \text{(g)}$$

Example 2.28

Use the Laplace transform method to determine the solution of the differential equation

$$\ddot{x} + 25x = 0.9 \sin 4t \tag{a}$$

subject to the initial conditions $x(0) = 0$ and $\dot{x}(0) = 0$.

Solution

Taking the Laplace transform of Equation (a) and using the linearity of the transform gives

$$\mathcal{L}\{\ddot{x}\} + 25\mathcal{L}\{x\} = 0.9\mathcal{L}\{\sin(4t)\} \tag{b}$$

Defining $X(s) = \mathcal{L}\{x(t)\}$, using transform pair 2 from Table 2.1 to determine the Laplace transform of the nonhomogeneous term, and applying the property of transform of derivatives to Equation (b) leads to

$$s^2X(s) - sx(0) - \dot{x}(0) + 25X(s) = \frac{3.6}{s^2 + 16} \tag{c}$$

Applying the given initial conditions and solving Equation (c) for $X(s)$ yields

$$X(s) = \frac{3.6}{(s^2 + 16)(s^2 + 25)} \tag{d}$$

A partial fraction decomposition of the right-hand side of Equation (d) results in

$$X(s) = \frac{0.4}{s^2 + 16} - \frac{0.4}{s^2 + 25} \tag{e}$$

The inversion of Equation (e) leads to

$$x(t) = 0.1 \sin(4t) - 0.08 \sin(5t) \tag{f}$$

2.5.2 Systems of Differential Equations

The Laplace transform method can be applied to determine the solution of a system of linear differential equations with constant coefficients. Let $x_1(t)$, $x_2(t)$, ..., $x_n(t)$ be the dependent variables whose Laplace transforms are $X_1(s)$, $X_2(s)$, ..., $X_n(s)$ respectively. A system of n differential equations with appropriate initial conditions has been formulated. Use the following steps to apply the Laplace transform method to solve for the dependent variables.

1. Apply the Laplace transform to both sides of all differential equations in the system.
2. The property of linearity of the transforms is applied to each differential equation.
3. The Laplace transforms of all nonhomogeneous terms are determined.
4. The property of derivatives of transforms is applied to all equations using the initial conditions.
5. A system of simultaneous linear algebraic equations is developed. The unknowns are the transforms of the dependent variables.
6. The simultaneous solution of the linear equations using algebraic methods leads to $X_1(s)$, $X_2(s)$, ..., $X_n(s)$.
7. The inversion methods of Section 2.4 are used to determine $x_1(t)$, $x_2(t)$, ..., $x_n(t)$.

The procedure is illustrated in the following example.

Example 2.29

Determine the solution to the differential equations

$$\ddot{x}_1 + 5x_1 - 2x_2 = 2e^{-t} \tag{a}$$

$$\ddot{x}_2 - 2x_1 + 2x_2 = 0 \tag{b}$$

subject to the initial conditions $x_1(0) = 0$, $x_2(0) = 0$, $\dot{x}_1(0) = 0$, $\dot{x}_2(0) = 0$.

Solution

Taking the Laplace transform of Equations (a) and (b) and using linearity of the transform leads to

$$\mathcal{L}\{\ddot{x}_1\} + 5\mathcal{L}\{x_1\} - 2\mathcal{L}\{x_2\} = 2\mathcal{L}\{e^{-t}\} \tag{c}$$

$$\mathcal{L}\{\ddot{x}_2\} - 2\mathcal{L}\{x_1\} + 2\mathcal{L}\{x_2\} = 0 \tag{d}$$

Using transform pair 1 from Table 2.1 in Equation (c) and the property of transform of derivatives in Equations (c) and (d) leads to

$$s^2 X_1(s) - sx_1(0) - \dot{x}_1(0) + 5X_1(s) - 2X_2(s) = \frac{2}{s+1} \tag{e}$$

$$s^2 X_2(s) - sx_2(0) - \dot{x}_2(0) - 2X_1(s) + 2X_2(s) = 0 \tag{f}$$

The application of the initial conditions to Equations (e) and (f) and rearranging leads to

$$(s^2 + 5)X_1(s) - 2X_2(s) = \frac{2}{s+1} \tag{g}$$

$$-2X_1(s) + (s^2 + 2)X_2(s) = 0 \tag{h}$$

The matrix form of Equations (g) and (h) is

$$\begin{bmatrix} s^2 + 5 & -2 \\ -2 & s^2 + 2 \end{bmatrix} \begin{bmatrix} X_1(s) \\ X_2(s) \end{bmatrix} = \begin{bmatrix} \frac{2}{s+1} \\ 0 \end{bmatrix} \tag{i}$$

Since the solution of the preceding equations leads to functions of the variable s, the algebra is minimized by solving the equations using Cramer's rule. Its application leads to

$$X_1(s) = \frac{\begin{vmatrix} \frac{2}{s+1} & -2 \\ 0 & s^2 + 2 \end{vmatrix}}{\begin{vmatrix} s^2 + 5 & -2 \\ -2 & s^2 + 2 \end{vmatrix}} \tag{j}$$

$$X_2(s) = \frac{\begin{vmatrix} s^2 + 5 & \frac{2}{s+1} \\ -2 & 0 \end{vmatrix}}{\begin{vmatrix} s^2 + 5 & -2 \\ -2 & s^2 + 2 \end{vmatrix}} \tag{k}$$

The determinants in Equations (j) and (k) are expanded, leading to

$$\begin{aligned} X_1(s) &= \frac{\left(\frac{2}{s+1}\right)(s^2 + 2) - (-2)(0)}{(s^2 + 5)(s^2 + 2) - (-2)(-2)} \\ &= \frac{2(s^2 + 2)}{(s+1)(s^4 + 7s^2 + 6)} \\ &= \frac{2(s^2 + 2)}{(s+1)(s^2 + 1)(s^2 + 6)} \end{aligned} \tag{l}$$

$$X_2(s) = \frac{(s^2+5)(0) - \left(\frac{2}{s+1}\right)(-2)}{(s^2+5)(s^2+2) - (-2)(-2)}$$

$$= \frac{4}{(s+1)(s^2+1)(s^2+6)} \tag{m}$$

The appropriate partial fraction decomposition for Equation (l) is

$$\frac{2(s^2+2)}{(s+1)(s^2+1)(s^2+6)} = \frac{A}{s+1} + \frac{Bs+C}{s^2+1} + \frac{Ds+E}{s^2+6} \tag{n}$$

The residue for the linear factor is determined using Equation (2.37):

$$A = \left.\frac{2(s^2+2)}{(s^2+1)(s^2+6)}\right|_{s=-1}$$

$$= \frac{2[(-1)^2+2]}{[(-1)^2+1][(-1)^2+6]}$$

$$= \frac{3}{7} = 0.429 \tag{o}$$

Substituting for A in Equation (n) and multiplying both sides by the denominator of the left-hand side leads to

$$2(s^2+2) = \frac{3}{7}(s^2+1)(s^2+6) + (Bs+C)(s+1)(s^2+6)$$
$$+(Ds+E)(s+1)(s^2+1) \tag{p}$$

Since Equation (p) is valid for all s, four independent equations are developed by giving s four different numerical values. This leads to a set of four simultaneous equations to solve for B, C, D, and E. To this end, setting $s = 0$, 1, 2, and -2 leads to the following equations, respectively:

$$6C + E = \frac{10}{7} \tag{q}$$

$$14B + 14C + 4D + 4E = 0 \tag{r}$$

$$60B + 30C + 30D + 15E = -\frac{66}{7} \tag{s}$$

$$20B - 10C + 10D - 5E = -\frac{66}{7} \tag{t}$$

Simultaneous solution leads to $B = -0.200$, $C = 0.200$, $D = -0.2286$, and $E = 0.2286$. Thus Equation (l) can be written as

$$X_1(s) = \frac{0.429}{s+1} + \frac{-0.200s+0.200}{s^2+1} + \frac{-0.229s+0.229}{s^2+6} \tag{u}$$

A similar partial fraction decomposition for Equation (m) leads to

$$X_2(s) = \frac{0.286}{s+1} + \frac{-0.400s+0.400}{s^2+1} + \frac{0.114s+0.114}{s^2+6} \tag{v}$$

Inversion of the transforms in Equations (u) and (v) using transform pairs 1, 2, and 3 of Table 2.1 yields

$$x_1(t) = 0.429e^{-t} - 0.200\cos t + 0.200\sin t - 0.229\cos(2.45t)$$
$$+0.0935\sin(2.45t) \tag{w}$$

$$x_2(t) = 0.286e^{-t} - 0.400\cos t + 0.400\sin t - 0.114\cos(2.45t)$$
$$-0.0466\sin(2.45t) \tag{x}$$

2.5.3 Integrodifferential Equations

An integrodifferential equation contains derivatives of the dependent variable as well as an indefinite integral whose integrand contains the dependent variable. Integrodifferential equations arise in the study of *LRC* circuits. The Laplace transform method may be applied to determine the solution of a linear integrodifferential equation with constant coefficients. Given the integrodifferential equation and appropriate initial conditions, the following steps are used in the application of the Laplace transform method to a linear integrodifferential equation with constant coefficients, where $x(t)$ is the dependent variable and defining $X(s) = \mathcal{L}\{x(t)\}$:

1. The Laplace transform is applied to both sides of the integrodifferential equation.
2. Linearity of the transform is applied.
3. The Laplace transform of the nonhomogeneous terms are obtained.
4. The properties of transform of derivatives and transform of indefinite integrals are applied, including application of initial conditions.
5. Algebraic methods are applied to determine $X(s)$.
6. $x(t)$ is determined using the inversion methods of Section 2.4.

This procedure is illustrated in the following example.

Example 2.30

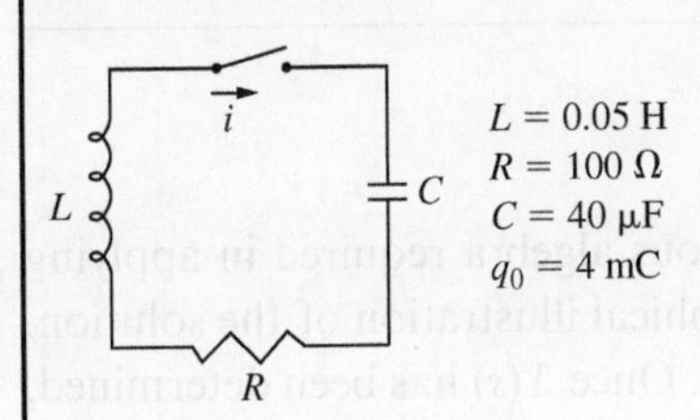

Figure 2.10

The *RLC* circuit of Example 2.30: The capacitor has a charge q_0 before the switch is closed.

The integrodifferential equation governing the current in the series *RLC* circuit of Figure 2.10 is

$$L\frac{di}{dt} + Ri + \frac{1}{C}\int_0^t i(t)dt + v(0) = 0 \tag{a}$$

where $v(0)$ is the voltage on the capacitor at $t = 0$. At $t = 0$, when the switch is closed, the capacitor has a charge of $q_0 = 4$ mC, leading to $v(0) = 100$ V. Determine $i(t)$. Assume $i(0) = 0$.

Solution

The voltage on the capacitor at $t = 0$ is from

$$\begin{aligned} v(0) &= \frac{q_0}{C} \\ &= \frac{4 \text{ mC}}{40\ \mu\text{F}} \\ &= 100 \text{ V} \end{aligned}$$

Taking the Laplace transform of Equation (a) and using linearity of the transform leads to

$$L\mathcal{L}\left\{\frac{di}{dt}\right\} + R\mathcal{L}\{i\} + \frac{1}{C}\mathcal{L}\left\{\int_0^t i(t)dt\right\} + \mathcal{L}\{v(0)\} = 0 \tag{b}$$

Using the properties of transform of the derivative and transform of the integral in Equation (c) leads to

$$L[sI(s) - i(0)] + RI(s) + \frac{1}{Cs}I(s) + \frac{v(0)}{s} = 0 \tag{c}$$

Solving Equation (c) for $I(s)$ leads to

$$I(s) = -\frac{v(0)}{\left(Ls^2 + Rs + \frac{1}{C}\right)} \tag{d}$$

Substituting known values into Equation (d) leads to

$$I(s) = -\frac{100}{\left(0.05s^2 + 100s + \frac{1}{40 \times 10^{-6}}\right)}$$

$$= -\frac{100}{0.05(s^2 + 2000s + 500{,}000)}$$

$$= -\frac{2000}{s^2 + 2000s + 500{,}000} \tag{e}$$

The roots of the polynomial in the denominator are $s_1 = -1.71 \times 10^3$ and $s_2 = -292.9$. Thus, Equation (e) is rewritten as

$$I(s) = -\frac{2000}{(s + 1.71 \times 10^3)(s + 292.9)} \tag{f}$$

A partial fraction decomposition of the right-hand side of Equation (f) leads to

$$I(s) = -\frac{-1.414}{s + 292.9} + \frac{1.414}{s + 1.71 \times 10^3} \tag{g}$$

The inversion of Equation (f) is performed using transform pair 1 from Table 2.1, leading to

$$i(t) = 1.414e^{-1.71\times10^3 t} - 1.414e^{-242.9t}\text{ A} \tag{h}$$

2.5.4 Use of MATLAB

MATLAB is useful in reducing the amount of tedious algebra required in applying the Laplace transform method as well as for the graphical illustration of the solution. MATLAB can be applied as described in Section 2.4. Once $X(s)$ has been determined, the symbolic capabilities of MATLAB may be used to invert the transform. MATLAB is particularly useful for systems with more than one dependent variable that require symbolic solution of a set of algebraic equations.

Example 2.31

Use MATLAB to aid in the solution of Example 2.29. Write an M-file that, beginning with Equation (i) of Example 2.29, solves for the transforms, inverts the transforms, and plots the response.

Solution

The MATLAB file Example2_31.m is shown in Figure 2.11. The output from the execution and the plot are shown in Figure 2.12. Note the following regarding this program:

- The elements of the matrix **A** and the column vector **b** are functions of the symbolic variable *s*.
- The vector xbar is the symbolic solution of the set of algebraic equations.

```
% Example2.31
% Use of MATLAB to aid in the Laplace transform solution of a coupled set
% of differential equations
%
% Defining symbolic variables
% s=transform variable
% t=time
syms s t
%
% Defining coefficient matrix
%
A=[s^2+5 -2;-2 s^2+2];
%
% Defining right-hand side vector
%
b=[2/(s+1);0];
%
% Solving for transforms
%
xbar=A^-1*b
%
% Inverse transform of vector of solutions
%
x=ilaplace(xbar)
x1=vpa(x,3)
%
% Plotting system response
%
t=linspace(0,10,201);
%
% Determining numerical values of response at discrete times
%
y1=subs(x1(1),t);
y2=subs(x1(2),t);
plot(t,y1,'-',t,y2,'.')
xlabel('t')
ylabel('x')
title('Solution of Example 2.31')
legend('x_1(t)','x_2(t)')
%
% End of file
```

Figure 2.11

The file Example2_31.m, which determines inverse transforms for the system of Example 2.29 and plots the responses.

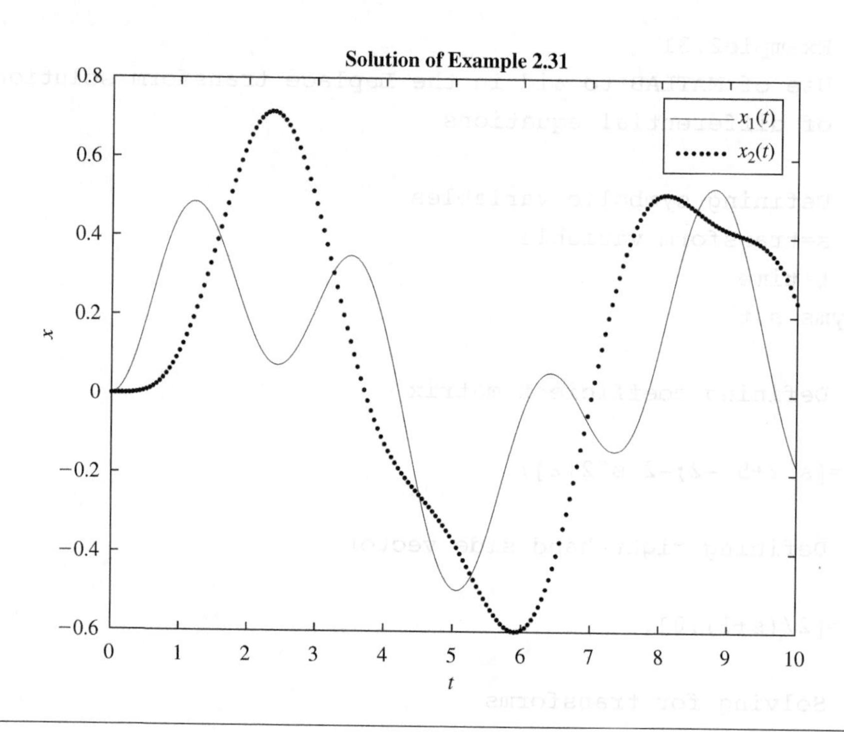

Figure 2.12

The plots of system response. (Note: The screen output from the execution of Example2_31.m shows that the results are identical to those obtained in Example 2.29.)

2.6 Further Examples

The examples in this section illustrate the use of the Laplace transform method in the solution of differential equations or integrodifferential equations obtained through the mathematical modeling of physical systems. The previous examples in this chapter have paid attention to detail. Partial fraction decompositions have been developed in detail. The use of Table 2.1 has been limited to transform pairs 1–10.

The steps followed in these examples have been outlined in Section 2.5. The procedure is repetitive. The examples in the preceding sections provided familiarity and practice with the procedure so that several steps may now be performed simultaneously. Partial fraction decompositions are presented without the detail of previous sections, sometimes using Table 2.3. All entries in Table 2.1 are available for use in determining transforms and inverting transforms.

MATLAB is used to perform computations, numerically and symbolically, and to plot system response.

Example 2.32

The differential equation governing the response of a centrifuge due to an unbalanced rotating mass in the centrifuge is derived as

$$(m + m_0)\ddot{y} + c\dot{y} + ky = m_0 e\omega^2 \sin(\omega t) \tag{a}$$

Use the Laplace transform method to determine the response of the centrifuge using the following values:

Mass of centrifuge	$m = 48$ kg
Rotating mass	$m_0 = 2$ kg
Eccentricity	$e = 0.1$ m

Damping coefficient	$c = 4000$ N·s/m
Stiffness	$k = 2 \times 10^6$ N/m
Angular velocity	$\omega = 100$ r/s
Initial displacement	$y(0) = 0$ m
Initial velocity	$\dot{y}(0) = 0$ m/s

Solution

Substitution of the given values into Equation (a) leads to

$$50\ddot{y} + 4000\dot{y} + 2 \times 10^6 y = 2000\sin(100t) \tag{b}$$

Dividing Equation (b) by 50 leads to

$$\ddot{y} + 80\dot{y} + 40{,}000y = 40\sin(100t) \tag{c}$$

Define $Y(s) = \mathscr{L}\{y(t)\}$. Taking the Laplace transform of both sides of Equation (c) and using linearity of the transform leads to

$$\mathscr{L}\{\ddot{y}\} + 80\mathscr{L}\{\dot{y}\} + 40{,}000\mathscr{L}\{y\} = 40\mathscr{L}\{\sin(100t)\} \tag{d}$$

Applying the property of transform of derivatives and using transform pair 2 from Table 2.1 in Equation (d) leads to

$$[s^2 Y(s) - sy(0) - \dot{y}(0)] + 80[sY(s) - y(0)] + 40{,}000\,Y(s) = \frac{4000}{s^2 + 10{,}000} \tag{e}$$

Using the initial conditions in Equation (e) and solving for $Y(s)$ yields

$$Y(s) = \frac{4000}{(s^2 + 80s + 40{,}000)(s^2 + 10{,}000)} \tag{f}$$

The quadratic polynomials in the denominator do not have real roots. The partial fraction decomposition of the right-hand side of Equation (f) leads to

$$Y(s) = \frac{3.32 \times 10^{-4}s - 9.79 \times 10^{-2}}{s^2 + 80s + 40{,}000} + \frac{-3.32 \times 10^{-4}s + 1.2 \times 10^{-1}}{s^2 + 10{,}000} \tag{g}$$

Equation (g) can be rewritten as

$$Y(s) = \frac{3.32 \times 10^{-4}(s + 40)}{s^2 + 80s + 40{,}000} - \frac{1.11 \times 10^{-1}}{s^2 + 80s + 40{,}000} - \frac{3.32 \times 10^{-4}s}{s^2 + 10{,}000} + \frac{1.2 \times 10^{-1}}{s^2 + 10{,}000} \tag{h}$$

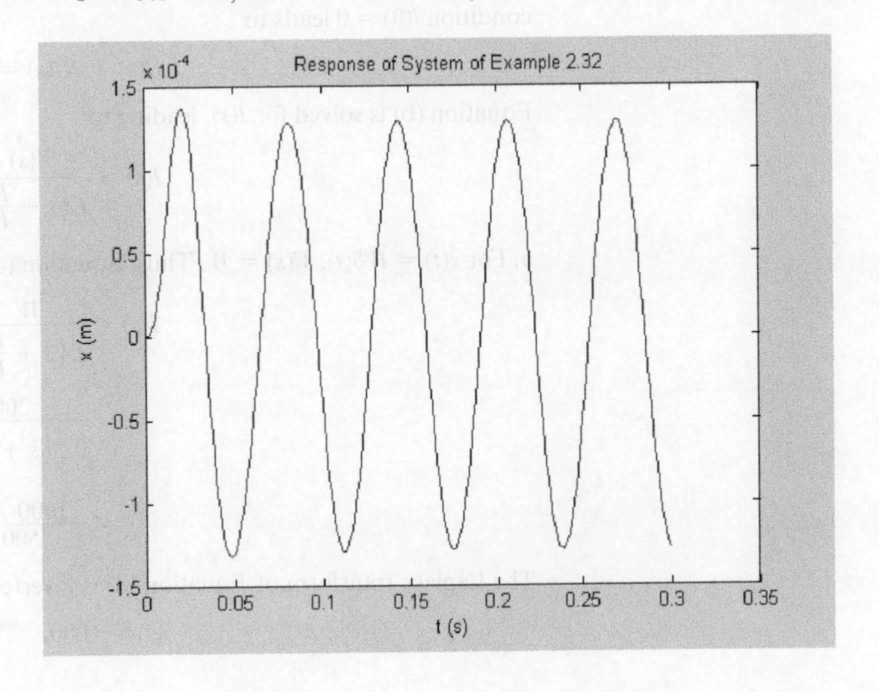

Figure 2.13
The MATLAB-generated plot of response of the system of Example 2.32. The initial transient response decays quickly, leaving only a steady-state response.

Transform pairs 12, 11, 3, and 2 from Table 2.1 are used to invert Equation (h), resulting in

$$y(t) = 3.32 \times 10^{-4} e^{-40t} \cos(1.96 \times 10^3 t) - 5.66 \times 10^{-4} e^{-40t} \sin(1.96 \times 10^3 t)$$
$$-3.32 \times 10^{-4} \cos(100t) + 1.2 \times 10^{-3} \sin(100t) \quad \text{(i)}$$

The first two terms in Equation (i) represent the initial transient response of the centrifuge that occurs because of the movement away from equilibrium. These terms die out quickly, leaving only the last two terms as a steady-state response. The response is illustrated in Figure 2.13.

Example 2.33

The differential equation governing the current through a series *RL* circuit is

$$L\frac{di}{dt} + Ri = v(t) \quad \text{(a)}$$

The parameters for a specific circuit have the following values:

Inductance	$L = 0.2$ H
Resistance	$R = 100\ \Omega$

No current is in the circuit at $t = 0$. Determine the current through the circuit and the voltage across the conductor if

a. $v(t) = W\delta(t)$, where $W = 200$ V·s. The solution to this problem provides the response of the circuit due to an impulsive voltage source.

b. $v(t) = V_0 u(t)$, where $V_0 = 10$ V. The solution of this problem provides the response of the circuit to a 10-V dc source connected at $t = 0$.

c. $v(t) = V_0 \sin \omega t$, where $V_0 = 10$ V and $\omega = 100$ r/s. The solution to this problem provides the response of the *RL* circuit to an alternating voltage source.

Solution

Taking the Laplace transform of Equation (a), using linearity, and applying the initial condition $i(0) = 0$ leads to

$$LsI(s) + RI(s) = V(s) \quad \text{(b)}$$

Equation (b) is solved for $I(s)$, leading to

$$I(s) = \frac{V(s)}{L\left(s + \frac{R}{L}\right)} \quad \text{(c)}$$

a. For $v(t) = W\delta(t)$, $V(s) = W$. Thus, Equation (a) becomes

$$I(s) = \frac{W}{L\left(s + \frac{R}{L}\right)}$$
$$= \frac{200}{0.2\left(s + \frac{100}{0.2}\right)}$$
$$= \frac{1000}{s + 500} \quad \text{(d)}$$

The Laplace transform of Equation (d) is inverted, leading to

$$i(t) = 1000 e^{-500t} \text{ A} \quad \text{(e)}$$

b. For $v(t) = V_0u(t)$, $V(s) = V_0/s$. Thus, Equation (a) becomes

$$I(s) = \frac{V_0}{Ls\left(s + \frac{R}{L}\right)}$$

$$= \frac{10}{0.2s\left(s + \frac{R}{L}\right)}$$

$$= \frac{50}{s(s + 500)} \tag{f}$$

A partial fraction decomposition of Equation (f) leads to

$$I(s) = \frac{0.1}{s} - \frac{0.1}{s + 500} \tag{g}$$

Inversion of the Laplace transform of Equation (g) leads to

$$i(t) = 0.1u(t) - 0.1e^{-500t} \tag{h}$$

c. For $v(t) = V_0 \sin(\omega t)$, $V(s) = \frac{V_0\omega}{s^2 + \omega^2}$. Thus, Equation (a) becomes

$$I(s) = \frac{V_0\omega}{L(s^2 + \omega^2)\left(s + \frac{R}{L}\right)}$$

$$= \frac{10(100)}{0.2(s^2 + 10{,}000)(s + 500)}$$

$$= \frac{5000}{(s^2 + 10{,}000)(s + 500)} \tag{i}$$

A partial fraction decomposition of Equation (i) using entry (5) of Table 2.3 leads to

$$I(s) = \frac{1}{7}\left(\frac{1}{s + 500} - \frac{s - 500}{s^2 + 10{,}000}\right) \tag{j}$$

Inversion of the transform in Equation (j) leads to

$$i(t) = \frac{1}{7}\left[e^{-500t} - \cos(100t) + 5\sin(100t)\right] \text{ A} \tag{k}$$

Example 2.34

The integrodifferential equations governing the current flow through a three-loop circuit are

$$L\frac{di_1}{dt} + Ri_1 + \frac{1}{C}\int_0^t (i_1(t) - i_2(t))dt = v(t) \tag{a}$$

$$L\frac{di_2}{dt} + R(i_2 - i_3) - \frac{1}{C}\int_0^t (i_1 - i_2)dt = 0 \tag{b}$$

$$L\frac{di_3}{dt} - R(i_2 - i_3) + \frac{1}{C}\int_0^t i_3\,dt = 0 \tag{c}$$

A specific circuit has components with the following values:

Inductance	$L = 0.5$ H
Resistance	$R = 3$ kΩ
Capacitance	$C = 2.5$ μF

Determine $I_1(s)$, $I_2(s)$, and $I_3(s)$ if $v(t) = 100(1 - e^{-0.05t})$ V and there is no current in the circuit at $t = 0$. Determine $i_3(t)$.

Solution

Taking the Laplace transform of Equations (a)–(c), using linearity of the transform and the properties of transform of derivatives and transform of integrals, and setting the initial conditions to zero leads to

$$LsI_1(s) + RI_1(s) + \frac{1}{Cs}[I_1(s) - I_2(s)] = V(s) \tag{d}$$

$$LsI_2(s) + R[I_2(s) - I_3(s)] - \frac{1}{Cs}[I_1(s) - I_2(s)] = 0 \tag{e}$$

$$LsI_3(s) + R[I_2(s) - I_3(s)] + \frac{1}{Cs}I_3(s) = 0 \tag{f}$$

Substitution of parameter values and $V(s)$ obtained using transform pairs 1 and 8 from Table 2.1 into Equations (d)–(f), multiplying by s, and collecting terms leads to

$$(0.5s^2 + 3000s + 400{,}000)I_1(s) - 400{,}000I_2(s) = 100s\left(\frac{1}{s} - \frac{1}{s + 0.05}\right) \tag{g}$$

$$-400{,}000I_1(s) + (0.5s^2 + 3000s + 400{,}000)I_2(s) - 3000sI_3(s) = 0 \tag{h}$$

$$-3000sI_2(s) + (0.5s^2 + 3000s + 400{,}000)I_3(s) = 0 \tag{i}$$

Equations (g)–(i) are summarized in matrix form as

$$\begin{bmatrix} 0.5s^2 + 3000s + 400{,}000 & -400{,}000 & 0 \\ -400{,}000 & 0.5s^2 + 3000s + 400{,}000 & -3000s \\ 0 & -3000s & 0.5s^2 + 3000s + 400{,}000 \end{bmatrix} \times \begin{bmatrix} I_1(s) \\ I_2(s) \\ I_3(s) \end{bmatrix} = \begin{bmatrix} \frac{5}{s(s + 0.05)} \\ 0 \\ 0 \end{bmatrix} \tag{j}$$

Cramer's rule is used to solve for $I_1(s)$ as

$$I_1(s) = \frac{\begin{vmatrix} \frac{5}{s(s + 0.05)} & -400{,}000 & 0 \\ 0 & 0.5s^2 + 3000s + 400{,}000 & -3000s \\ 0 & -3000s & 0.5s^2 + 3000s + 400{,}000 \end{vmatrix}}{\begin{vmatrix} 0.5s^2 + 3000s + 400{,}000 & -400{,}000 & 0 \\ -400{,}000 & 0.5s^2 + 3000s + 400{,}000 & -3000s \\ 0 & -3000s & 0.5s^2 + 3000s + 400{,}000 \end{vmatrix}}$$

$$= \frac{5(s^4 + 1.2 \times 10^4 s^3 + 1.6 \times 10^6 s^2 + 9.6 \times 10^9 s + 6.4 \times 10^{11})}{s^2(s + 0.05)(s^3 + 1.2 \times 10^4 s^2 + 1.6 \times 10^6 s + 9.6 \times 10^9)} \tag{k}$$

$I_2(s)$ and $I_3(s)$ are obtained in a similar fashion as

$$I_2(s) = \frac{4 \times 10^6(s^2 + 6 \times 10^3 s + 8 \times 10^5)}{s^2(s + 0.05)(s^3 + 1.2 \times 10^4 s^2 + 1.6 \times 10^6 s + 9.6 \times 10^9)} \tag{l}$$

$$I_3(s) = \frac{2.4 \times 10^{10}}{s(s + 0.05)(s^3 + 1.2 \times 10^4 s^2 + 1.6 \times 10^6 s + 9.6 \times 10^9)} \tag{m}$$

The cubic term in the denominator of Equations (k)–(m) can be factored as $(s + 1.19 \times 10^4)(s^2 + 7.08s + 868.6)$. Thus the partial fraction decomposition for $I_3(s)$ is

$$I_3(s) = \frac{2.5}{s} - \frac{50.0}{s + 0.05} + \frac{169.5}{s + 1.19 \times 10^4} + \frac{149.6s + 5.36 \times 10^3}{s^2 + 7.08s + 868.6} \tag{n}$$

The inversion of Equation (n) leads to

$$i_3(t) = 2.5u(t) - 50.0e^{-0.05t} + 169.5e^{-1.19\times10^4 t} + 149.6e^{-3.54t}\cos(29.3t) + 164.8e^{-3.54t}\sin(29.3t) \tag{o}$$

Example 2.35

The voltage source in an LC circuit is the square waveform of Figure 2.14 (page 96). The specific circuit has components with the following parameters:

Inductance	$L = 0.64$ H
Capacitance	$C = 100\ \mu$F
Maximum voltage	$V_0 = 6.4$ V
Period	$T = 0.06$ s
Initial current	$i(0) = 0$ A

The equation governing the current in the circuit is

$$L\frac{di}{dt} + \frac{1}{C}\int_0^t i\,dt = v(t) \tag{a}$$

Determine and plot $i(t)$.

Solution

Define $I(s) = \mathcal{L}\{i(t)\}$ and $V(s) = \mathcal{L}\{v(t)\}$. Taking the Laplace transform of Equation (a) and using linearity of the transform leads to

$$\mathcal{L}\left\{\frac{di}{dt}\right\} + \frac{1}{C}\mathcal{L}\left\{\int_0^t i\,dt\right\} = V(s) \tag{b}$$

$V(s)$ is determined using the property of transform of periodic functions:

$$\begin{aligned}
V(s) &= \frac{1}{1 - e^{-Ts}}\int_0^T V(t)e^{-st}\,dt \\
&= \frac{1}{1 - e^{-Ts}}\left[\int_0^{T/2} V_0 e^{-st}\,dt + \int_{T/2}^{T}(-V_0)e^{-st}\,dt\right] \\
&= \frac{1}{1 - e^{-Ts}}\left[-\frac{V_0}{s}e^{-st}\Big|_0^{T/2} + \frac{V_0}{s}e^{-st}\Big|_{T/2}^{T}\right] \\
&= \frac{V_0(1 - 2e^{-(T/2)s} + e^{-Ts})}{s(1 - e^{-Ts})} \\
&= \frac{V_0[1 - e^{-(T/2)s}]^2}{s[1 - (e^{-(T/2)s})^2]} \\
&= \frac{V_0[1 - e^{-(T/2)s}]}{s[1 + e^{-(T/2)s}]}
\end{aligned} \tag{c}$$

Using Equation (c) and the properties of transform of derivatives and transform of integrals in Equation (b) and applying the initial condition leads to

$$LsI(s) + \frac{1}{Cs}I(s) = \frac{V_0[1 - e^{-(T/2)s}]}{s[1 + e^{-(T/2)s}]} \tag{d}$$

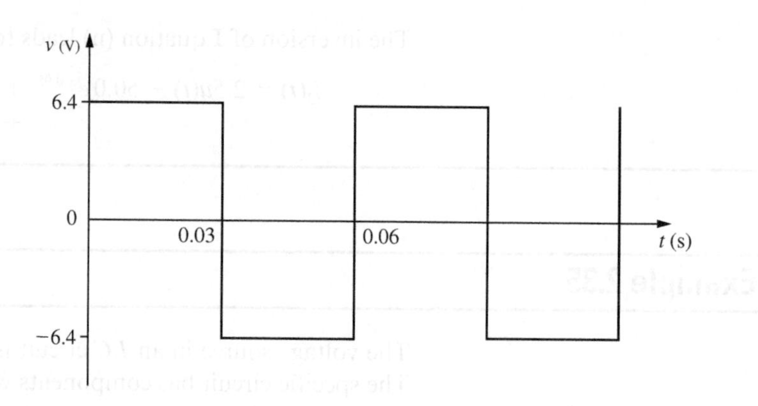

Figure 2.14

The square waveform of Example 2.35.

Solving Equation (d) for $I(s)$ leads to

$$I(s) = \frac{V_0[1 - e^{-(T/2)s}]}{L\left(s^2 + \frac{1}{LC}\right)[1 + e^{-(T/2)s}]} \tag{e}$$

Application of the binomial expansion

$$[1 + e^{-(T/2)s}]^{-1} = \sum_{k=0}^{\infty}(-1)^k e^{-k(T/2)s} \tag{f}$$

in Equation (e) leads to

$$I(s) = \frac{V_0}{L\left(s^2 + \frac{1}{LC}\right)}\left[1 - 2\sum_{k=1}^{\infty}(-1)^k e^{-k(T/2)s}\right] \tag{g}$$

The inversion of Equation (g) using the second shifting theorem and transform pair 2 leads to

$$i(t) = V_0\sqrt{\frac{C}{L}}\sin\left(\sqrt{\frac{1}{LC}}t\right) - 2\sum_{k=1}^{\infty}(-1)^k \sin\left(\sqrt{\frac{1}{LC}}\frac{t - kT}{2}\right)u\frac{t - kT}{2} \tag{h}$$

Substitution of given values leads to

$$i(t) = 0.08\left\{\sin(80t) - 2\sum_{k=1}^{\infty}(-1)^k \sin[80(t - 0.03k)]u(t - 0.03k)\right\} \tag{i}$$

MATLAB is used to plot Equation (i) in Figure 2.15.

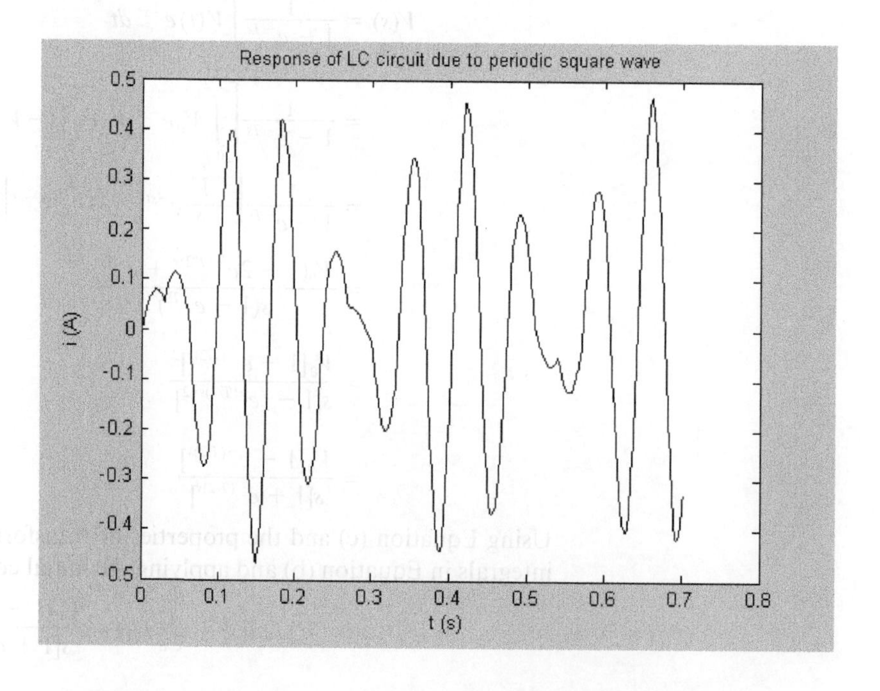

Figure 2.15

The MATLAB-generated plot for the response of system of Example 2.35.

Example 2.36

The differential equations governing the perturbations in liquid levels, h_1 and h_2, in a two-tank system are

$$A_1\frac{dh_1}{dt} + \frac{1}{R_1}h_1 - \frac{1}{R_1}h_2 = q_1(t) \quad \text{(a)}$$

$$A_2\frac{dh_2}{dt} - \frac{1}{R_1}h_1 + \left(\frac{1}{R_1} + \frac{1}{R_2}\right)h_2 = q_2(t) \quad \text{(b)}$$

where A_1 and A_2 are the cross-sectional areas of the tanks, R_1 is the resistance of the pipe connecting the tanks, R_2 is the resistance in the pipe out of the second tank, and q_1 and q_2 are the perturbations in flow rates into the tanks. Determine the response for a system with $A_1 = 900\ \text{m}^2$, $A_2 = 1500\ \text{m}^2$, $R_1 = 4.0\ \text{s/m}^2$, and $R_2 = 5.0\ \text{s/m}^2$ when the flow rate perturbations are

$$q_1(t) = -1.0u(t)\,\text{m}^3/\text{s} \quad \text{(c)}$$

$$q_2(t) = 1.0u(t-5)\,\text{m}^3/\text{s} \quad \text{(d)}$$

Solution

Substitution of the given values and of Equations (c) and (d) into Equations (a) and (b) leads to

$$18{,}000\frac{dh_1}{dt} + 5h_1 - 5h_2 = -20\,u(t) \quad \text{(e)}$$

$$30{,}000\frac{dh_2}{dt} - 5h_1 + 9h_2 = 20\,u(t-5) \quad \text{(f)}$$

Taking the Laplace transform of Equations (e) and (f), noting that $dh_1/dt(0) = dh_2/dt(0) = 0$ m/s, and defining $H_1(s) = \mathcal{L}\{h_1(t)\}$ and $H_2(s) = \mathcal{L}\{h_2(t)\}$ leads to

$$18{,}000s\,H_1(s) + 5H_1(s) - 5H_2(s) = -\frac{20}{s} \quad \text{(g)}$$

$$30{,}000s\,H_2(s) - 5H_1(s) + 9H_2(s) = \frac{20e^{-5s}}{s} \quad \text{(h)}$$

Equations (g) and (h) are summarized in matrix form as

$$\begin{bmatrix} 18{,}000s + 5 & -5 \\ -5 & 30{,}000s + 9 \end{bmatrix}\begin{bmatrix} H_1(s) \\ H_2(s) \end{bmatrix} = \frac{20}{s}\begin{bmatrix} -1 \\ e^{-5s} \end{bmatrix} \quad \text{(i)}$$

Cramer's rule is applied to Equation (i), leading to

$$H_1(s) = \frac{\frac{20}{s}\begin{vmatrix} -1 & -5 \\ e^{-5s} & 30{,}000s + 9 \end{vmatrix}}{\begin{vmatrix} 18{,}000s + 5 & -5 \\ -5 & 30{,}000s + 9 \end{vmatrix}} = \frac{\frac{20}{s}(5e^{-5s} - 30{,}000s - 9)}{5.4 \times 10^8 s^2 + 3.12 \times 10^5 s + 20} \quad \text{(j)}$$

and

$$H_2(s) = \frac{\frac{20}{s}[(18{,}000s + 5)e^{-5s} - 5]}{5.4 \times 10^8 s^2 + 3.12 \times 10^5 s + 20} \quad \text{(k)}$$

Equations (j) and (k) are rewritten as

$$H_1(s) = 3.70 \times 10^{-8}\frac{5e^{-5s} - 30{,}000s - 9}{s(s + 5.04 \times 10^{-4})(s + 7.34 \times 10^{-5})} \quad \text{(l)}$$

$$H_2(s) = 3.70 \times 10^{-8}\frac{(18{,}000s + 5)e^{-5s} - 5}{s(s + 5.04 \times 10^{-4})(s + 7.34 \times 10^{-5})} \quad \text{(m)}$$

Partial fraction decompositions lead to

$$H_1(s) = \frac{-9.0}{s} + \frac{1.04}{s + 5.04 \times 10^{-4}} + \frac{7.96}{s + 7.34 \times 10^{-5}} + e^{-5s}\left(\frac{5.0}{s} + \frac{0.852}{s + 5.04 \times 10^{-4}} - \frac{5.85}{s + 7.34 \times 10^{-5}}\right) \quad \text{(n)}$$

$$H_2(s) = \frac{-5.0}{s} - \frac{0.852}{s + 5.04 \times 10^{-4}} + \frac{5.85}{s + 7.34 \times 10^{-5}} + e^{-5s}\left(\frac{-5.0}{s} - \frac{0.694}{s + 5.04 \times 10^{-4}} - \frac{4.31}{s + 7.34 \times 10^{-5}}\right) \quad \text{(o)}$$

Inversion of the transforms leads to

$$h_1(t) = (-9.0 + 1.04e^{-5.04 \times 10^{-4}t} + 7.96e^{-7.34 \times 10^{-5}t})u(t) + [5 + 0.852e^{-5.04 \times 10^{-4}(t-5)} - 5.85e^{-7.34 \times 10^{-5}(t-5)}]u(t-5) \quad \text{(p)}$$

$$h_2(t) = (-5 - 0.852e^{-5.04 \times 10^{-4}t} + 5.85e^{-7.34 \times 10^{-5}t})u(t) + [5 - 0.694e^{-5.04 \times 10^{-4}(t-5)} - 4.31e^{-7.34 \times 10^{-5}(t-5)}]u(t-5) \quad \text{(q)}$$

2.7 Summary

2.7.1 Mathematical Solutions for Response of Dynamic Systems

The Laplace transform method is applied to solve differential and integrodifferential equations obtained through the mathematical modeling of dynamic systems. Important points regarding Laplace transforms and their application to the solution of differential equations are as follows:

- The Laplace transform of a function of time t is a function of a complex variable s. It is defined in terms of an integral over time, but with the integrand dependent on s. The Laplace transform is unique and exists for any function that can be physically generated.
- The inverse transform of a function of the complex variable s is a function of time. The inverse transform is defined by an integral in the complex plane with t as a parameter. The inverse transform is unique.
- If $F(s) = \mathcal{L}\{f(t)\}$, then $f(t) = \mathcal{L}^{-1}\{F(s)\}$ and $f(t)$ and $F(s)$ form a transform pair. Table 2.1 presents a useful list of transform pairs.
- The Laplace transform of a function may be determined directly from its definition.
- Properties are derived that aid in the determination of transform pairs. These properties are summarized in Table 2.2.
- Inverse transforms are determined using Tables 2.1 and 2.2.
- Partial fraction decompositions are often required to write the inverse transform in a form in which the tables can be used.
- The poles of a transform are the zeroes of its denominator.
- The residues of a transform are the numerators in a partial fraction decomposition of the transform.
- The residue theorem can be used to calculate residues.
- The properties of linearity of the transform and transform of derivatives allow the Laplace transform method to be successful in obtaining solutions of linear differential equations with constant coefficients.
- When taking the Laplace transform of both sides of a differential equation, the application of the transform of derivatives property leads to an algebraic equation that is solved for the transform of the dependent variable. The transform is inverted to obtain the solution of the differential equation.

- The initial conditions are applied during the application of the transform of derivatives property.
- When applying the Laplace transform method to integrodifferential equations, the property of transform of integrals is used.
- Taking the Laplace transform of a coupled set of differential equations results in a set of simultaneous algebraic equations to solve for the transforms of the dependent variables.
- The Symbolic Toolbox of MATLAB can be used to determine symbolically Laplace transforms and inverse Laplace transforms. It is also useful for the symbolic solutions of sets of simultaneous algebraic equations that arise when the method is applied to a coupled set of differential equations. The symbolic equations may be evaluated at discrete values of time to allow plotting the response.

2.7.2 Important Equations

Important equations for Chapter 2 include those contained in Table 2.1, a table of transform pairs, and in Table 2.2, a table of transform properties. Additionally,

- Definition of Laplace transform

$$\mathscr{L}\{f(t)\} = \int_0^\infty f(t)e^{-st}dt \tag{2.1}$$

$$F(s) = \mathscr{L}\{f(t)\} \tag{2.2}$$

- Inversion integral

$$\mathscr{L}^{-1}\{G(s)\} = \frac{1}{2\pi j}\int_{\gamma-j\infty}^{\gamma+j\infty} G(s)\,e^{st}ds \tag{2.4}$$

- Transform after long division:

$$\frac{N(s)}{D(s)} = W(s) + \frac{B(s)}{D(s)} \tag{2.33}$$

- Poles of the transform $s_1, s_2, \ldots, s_n$:

$$D(s) = (s - s_1)(s - s_2)\cdots(s - s_n) \tag{2.34}$$

- Partial fraction decomposition in terms of residues $A_1, A_2, \ldots, A_n$:

$$\frac{B(s)}{D(s)} = \sum_{k=1}^{n} \frac{A_k}{s - s_k} \tag{2.35}$$

- Residue theorem:

$$A_k = \lim_{s\to s_k} \frac{(s - s_k)B(s)}{D(s)} \tag{2.38}$$

Short Answer Problems

For all questions, $F(s) = \mathscr{L}\{f(t)\}$, $G(s) = \mathscr{L}\{g(t)\}$, and $H(s) = \mathscr{L}\{h(t)\}$.

Each of the statements in Problems SA2.1–SA2.25 is either true or false. If the statement if false, rewrite it to make it true.

SA2.1 The Laplace transform of a function $f(t)$ is $F(s) = \int_0^\infty f(t)e^{-st}\,dt$.

SA2.2 The Laplace transform of a function $f(t)$ exists as long as $f(t)$ is of exponential order.

SA2.3 The Laplace transform of a function $f(t)$ is not unique.

SA2.4 A function $f(t)$ and its Laplace transform $F(s)$ form a transform pair.

SA2.5 If $F(s)$ is a Laplace transform, then there corresponds a unique function $f(t)$ such that $F(s) = \mathscr{L}^{-1}\{f(t)\}$.

SA2.6 If $g(t) = 3$, then $G(s) = 3$.

SA2.7 If $g(t) = e^{-2t}f(t)$, then $G(s) = (s + 2)F(s)$.

SA2.8 If $g(t) = f(t - 2)u(t - 2)$, then $G(s) = e^{-2s}F(s)$.

SA2.9 If $g(t) = 3f(t) + 4h(t)$, then $G(s) = 12F(s)H(s)$.

SA2.10 If $g(t) = 3\frac{df}{dt}$ and $f(0) = 2$, then $G(s) = sF(s) - 2$.

SA2.11 If $g(t) = \int_0^t e^{-6t} f(t)dt$, then $G(s) = \frac{1}{s}F(s+6)$.

SA2.12 If $g(t) = e^{-6t}\int_0^t f(t)dt$, then $G(s) = \frac{1}{s+6}F(s)$.

SA2.13 If $g(t) = f(t)h(t)$, then $G(s) = F(s)H(s)$.

SA2.14 If $G(s) = F(s+3)$, then $g(t) = e^{-3t}f(t)$.

SA2.15 If $G(s) = e^{-3s}F(s)$, then $g(t) = f(t)u(t-3)$.

SA2.16 If $G(s) = \frac{1}{s}F(s)$, then $g(t) = \int_0^t f(t)dt$.

SA2.17 If $G(s) = \frac{1}{s+5}F(s)$, then $g(t) = \int_0^t e^{-5t}f(t)dt$.

SA2.18 If $G(s) = s^2F(s) - 2s + 3$, then $g(t) = \frac{d^2f(t)}{dt^2}$, where $f(0) = 2$ and $\frac{df}{dt}(s) = -3$.

SA2.19 If $G(s) = \frac{s+2}{s^2+3s+2}$, then $g(0) = 2$.

SA2.20 If $G(s) = \frac{2s^2+4s+7}{s^3+2s^2+10s+15}$, then $g(0) = 2$.

SA2.21 If $G(s) = \frac{1}{s}F(s)$, then $\lim_{t\to\infty} f(t) = F(0)$.

SA2.22 The poles of a transform are the roots of the numerator of the transform.

SA2.23 A reducible quadratic function is one that has real factors.

SA2.24 The polynomial in the denominator of a transform is the product of linear factors and irreducible quadratic factors.

SA2.25 The Laplace transform method can be used to solve any initial value problem.

Problems SA2.26–SA2.38 require a short answer.

SA2.26 What property of transforms is used in the statement if $f(t) = g(t)$, then $F(s) = G(s)$?

SA2.27 What property of transforms is used in the statement $\mathscr{L}\{af(t)\} = a\mathscr{L}\{f(t)\}$ for any constant a?

SA2.28 What property of transforms is used in the statement if $G(s) = \frac{1}{s}F(s)$, then $\lim_{t\to\infty} f(t) = F(0)$?

SA2.29 What is meant by a transform pair?

SA2.30 Define the term "poles."

SA2.31 Define the term "residues."

SA2.32 If $g(t) = e^{-4t}f(t)$, determine $G(s)$ in terms of $F(s)$.

SA2.33 If $g(t) = e^{-5t}\int_0^t f(t)dt$, determine $G(s)$ in terms of $F(s)$.

SA2.34 If $g(t) = e^{-5t}\int_0^t e^{-4t}f(t)dt$, determine $G(s)$ in terms of $F(s)$.

SA2.35 If $g(t) = t\int_0^t f(t)dt$, determine $G(s)$ in terms of $F(s)$.

SA2.36 If $g(t) = e^{-5t}\int_0^t e^{-2t}\frac{df}{dt}dt$, determine $G(s)$ in terms of $F(s)$ if $f(0) = 1$.

SA2.37 Determine $F(s)$ if $f(t)$ is given in Figure SA2.37.

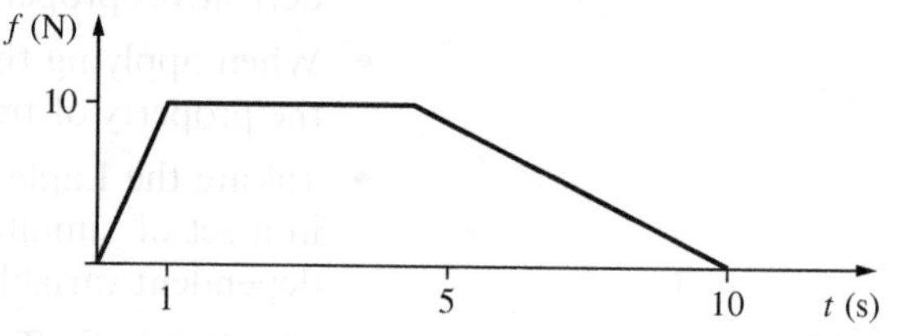

Figure SA2.37

SA2.38 Determine $F(s)$ if $f(t)$ is given in Figure SA2.38.

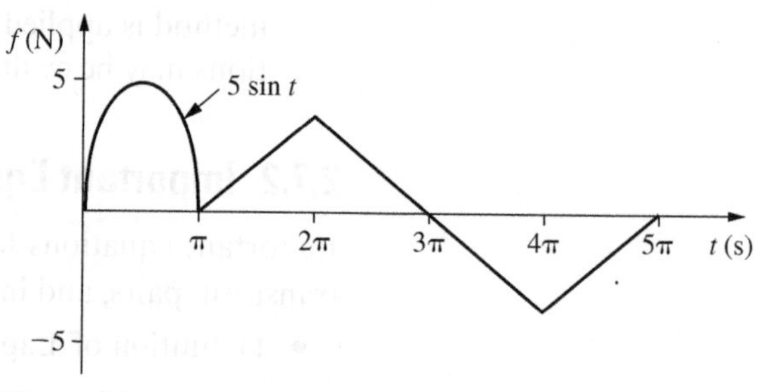

Figure SA2.38

In Problems SA2.39–SA2.52, given $f(t)$, use Table 2.1, a table of transform pairs, and Table 2.2, a table of properties of the Laplace transform, to determine $F(s)$.

SA2.39 $f(t) = 10e^{-t} + 4e^{-2t}$

SA2.40 $f(t) = 10\cos 3t + 14\sin 3t$

SA2.41 $f(t) = e^{-2t}(10\cos 3t + 14\sin 3t)$

SA2.42 $f(t) = te^{-2t}$

SA2.43 $f(t) = t\cos 4t$

SA2.44 $f(t) = te^{-2t}[u(t) - u(t-2)]$

SA2.45 $f(t) = 4\sin 4t\, u(t-3)$

SA2.46 $f(t) = \int_0^t e^{-2t}dt$

SA2.47 $f(t) = \int_0^t e^{-2t}\cos 2t\, dt$

SA2.48 $f(t) = e^{-3t}\int_0^t e^{-2t}\cos 2t\, dt$

SA2.49 $f(t) = t\int_0^t e^{-2t}\cos 2t\, dt$

SA2.50 $f(t) = t\int_0^t e^{-2t}u(t-3)dt$

SA2.51 $f(t) = \frac{d}{dt}(e^{-2t}\cos 2t)$

SA2.52 $f(t) = \int_0^t e^{-3(t-\tau)}\sin 2\tau\, d\tau$

In Problems SA2.53–SA2.72, given $F(s)$, use Table 2.1, a table of transform pairs, and Table 2.2, a table of properties of the Laplace transform, to determine $f(t)$.

SA2.53 $F(s) = \frac{3}{s+6}$

SA2.54 $F(s) = \frac{3s+1}{s^2+9}$

SA2.55 $F(s) = \frac{10}{s^2+25}$

SA2.56 $F(s) = \dfrac{s}{(s+2)(s+5)}$

SA2.57 $F(s) = \dfrac{10}{s^2+7s+12}$

SA2.58 $F(s) = \dfrac{4s+9}{s^2+7s+12}$

SA2.59 $F(s) = \dfrac{4s}{s^2+9}e^{-2s}$

SA2.60 $F(s) = \dfrac{4s}{(s+1)(s+3)(s+6)}e^{-4s}$

SA2.61 $F(s) = \dfrac{3s+1}{s(s^2+9)}$

SA2.62 $F(s) = \dfrac{3s+1}{s^2-9}$

SA2.63 $F(s) = \dfrac{3s+1}{s(s^2-9)}$

SA2.64 $F(s) = \dfrac{3s+1}{s^2+9}(1-e^{-2s})$

SA2.65 $F(s) = \dfrac{3s+1}{s(s^2+9)}(1-e^{-2s})$

SA2.66 $F(s) = \dfrac{s^2+2}{(s^2+9)(s+1)(s+3)}$

SA2.67 $F(s) = \dfrac{5}{(s+9)^2}$

SA2.68 $F(s) = \dfrac{5s}{(s+9)^2}$

SA2.69 If $F(s) = \dfrac{3s^3+3s+10}{(s+2)(s+5)(s+7)(s+10)}$, determine $f(0)$.

SA2.70 If $F(s) = \dfrac{3s^3+3s+10}{(s+2)(s+5)(s+7)(s+10)}$, determine $\lim_{t\to\infty} f(t)$.

SA2.71 If $F(s) = \dfrac{3s^2+s+4}{(s+2)(s+5)(s+7)(s+10)}$, determine $f(0)$.

SA2.72 If $F(s) = \dfrac{3s^2+s+4}{(s+2)(s+5)(s+7)(s+10)}$, determine $\lim_{t\to\infty} f(t)$.

SA2.73 Determine the poles of the transform

$$F(s) = \frac{s+4}{(s+7)(s^2+6s+5)}$$

SA2.74 Determine the poles of the transform

$$F(s) = \frac{s^2+4s+5}{(s+3)(s^2+9s+18)}$$

SA2.75 Determine the residues of the transform by using the residue theorem

$$F(s) = \frac{s+7}{(s+3)(s+4)}$$

SA2.76 Determine the residues of the transform by using the residue theorem

$$F(s) = \frac{s^2+3s+6}{(s+1)(s+4)(s+7)}$$

SA2.77 Perform a partial fraction decomposition on

$$F(s) = \frac{s+4}{s(s+5)^2}$$

SA2.78 Perform a partial fraction decomposition on

$$F(s) = \frac{s+4}{s^2(s+2)}$$

SA2.79 Use the MATLAB command "residue" for the transform

$$F(s) = \frac{s+7}{(s+3)(s+4)}$$

and explain the output.

SA2.80 Use the MATLAB command "residue" for the transform

$$F(s) = \frac{s+4}{s^2(s+2)}$$

and explain the output.

SA2.81 Use the MATLAB command "residue" for the transform

$$F(s) = \frac{4s^2+3}{s^2+7s+6}$$

and explain the output.

Problems

2.1 Use the definition of the transform to determine $\mathcal{L}\{\sinh(\omega t)\}$.

2.2 Use the definition of the transform to determine $\mathcal{L}\{t\sin(\omega t)\}$.

2.3 Noting that $\mathcal{L}\{t^{-1/2}\} = \sqrt{\pi/s}$, evaluate (a) $\mathcal{L}\{t^{1/2}\}$ using the property of differentiation of the transform, (b) $\{t^{3/2}\}$, and (c) $\mathcal{L}\{t^{1/2}e^{-3t}\}$.

2.4 The Laplace transform of the response of a dynamic system is

$$X(s) = \frac{s^2+5}{s(s+3)(s+4)(s^2+15)}$$

(a) Determine $\lim_{t\to\infty} x(t)$ using the final value theorem.
(b) Determine $x(t)$ by using partial fraction decomposition and inverting the transform.
(c) Determine $x(t)$ using MATLAB.

2.5 The Laplace transform of the response of a dynamic system is

$$X(s) = \frac{s+3}{4s^2+6s+17}$$

(a) Determine $x(0)$.
(b) Determine $x(t)$ by inverting the transform by hand.
(c) Use MATLAB to determine $x(t)$.

2.6 The Bessel function of order 0 of t, denoted by $J_0(t)$, is a special function that has many engineering applications. It can be shown that $\mathcal{L}\{J_0(t)\} = 1/\sqrt{s^2+1}$. (a) Evaluate $\mathcal{L}\{tJ_0(t)\}$. (b) Evaluate $\mathcal{L}\{J_0(2t)\}$. (c) It can be shown that the Bessel function of order 1 is related to the Bessel function of order 0 by $J_1(t) = -\frac{d}{dt}(J_0(t))$. Given that $J_0(0) = 1$, evaluate $\mathcal{L}\{J_1(t)\}$.

2.7 Show that

$$\mathcal{L}\{\sin(\omega t)\cosh(\omega t) - \cos(\omega t)\sinh(\omega t)\} = \frac{4\omega^3}{s^2 + 4\omega^2}$$

2.8 Show that

$$\mathcal{L}\left\{\tfrac{2}{t}[1 - \cos(\omega t)]\right\} = \ln\left(\frac{s^2 + \omega^2}{s^2}\right)$$

2.9 Show that

$$\mathcal{L}\left\{\tfrac{1}{t}\sin(\omega t)\right\} = \tan^{-1}\left(\tfrac{\omega}{s}\right)$$

2.10 Prove the change of scale property of Laplace transforms.

2.11 Prove the initial value theorem.

2.12 Prove the property of differentiation of the transform.

2.13 Determine the transform of the periodic function of Figure P2.13.

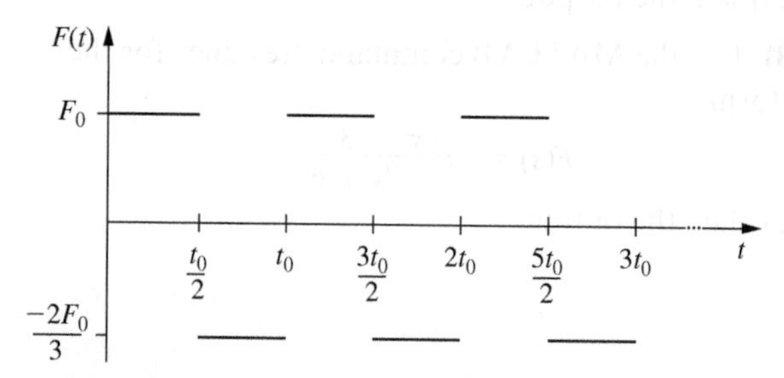

Figure P2.13

2.14 Use the Laplace transform method to solve the differential equation

$$\dot{x} + 3x = 2e^{-3t}$$

subject to $x(0) = 0$.

2.15 Use the Laplace transform method to solve the differential equation

$$\ddot{x} + 2\dot{x} + 17x = 0$$

subject to the initial conditions $x(0) = 0$ and $\dot{x}(0) = 1$.

2.16 Use the Laplace transform method to solve the differential equation

$$\ddot{x} + 25x = 3\sin(5t)$$

subject to the initial conditions $x(0) = 0$ and $\dot{x}(0) = 0$.

2.17 Use the Laplace transform method to solve the differential equation

$$\ddot{x} + 49x = 7e^{-2t}$$

subject to the initial conditions $x(0) = 0$ and $\dot{x}(0) = 0$.

2.18 Use the Laplace transform method to solve the differential equation

$$\dddot{x} + 2\ddot{x} + 25\dot{x} + 50x = 0$$

subject to the initial conditions $x(0) = 1$, $\dot{x}(0) = 0$, and $\ddot{x}(0) = 0.5$.

2.19 Use the Laplace transform method to solve the differential equations

$$\dot{x}_1 + 2x_1 - x_2 = 2e^{-t}$$

$$\dot{x}_2 - x_1 + 3x_2 = 0$$

subject to the initial conditions $x_1(0) = 0$ and $x_2(0) = 0$.

For Problems 2.20–2.24, use MATLAB as needed.

2.20 The integrodifferential equations obtained for the currents in a two-loop circuit are

$$Ri_1 + \frac{1}{C}\int_0^t i_1\,dt - \frac{1}{C}\int_0^t i_2\,dt = v(t) \quad \text{(a)}$$

$$-\frac{1}{C}\int_0^t i_1\,dt + L\frac{di_2}{dt} + Ri_2 + \frac{1}{C}\int_0^t i_2\,dt = 0 \quad \text{(b)}$$

Solve Equations (a) and (b) when $v(t) = 500\ u(t)\ V$ with $i_1(0) = 0$ and $i_2(0) = 0$. The circuit parameters are $R = 500\ \Omega$, $C = 20\ \mu F$, and $L = 0.2$ H. Plot $i_1(t)$ and $i_2(t)$ on the same axes.

2.21 The differential equations obtained for the displacements of some blocks are

$$m_1\ddot{x}_1 + (k_1 + k_2)x_1 - k_2x_2 = 0 \quad \text{(a)}$$

$$m_2\ddot{x}_2 - k_2x_1 + (k_2 + k_3)x_2 = 0 \quad \text{(b)}$$

Use the Laplace transform method to solve Equations (a) and (b) for $x_1(t)$ and $x_2(t)$ with the initial conditions $x_1(0) = 1$ mm, $\dot{x}_1(0) = 0$, $x_2(0) = 0$, and $\dot{x}_2(0) = 0$ when $m_1 = 10$ kg, $m_2 = 20$ kg, $k_1 = 10{,}000$ N/m, $k_2 = 15{,}000$ N/m, and $k_3 = 10{,}000$ N/m. Plot $x_1(t)$ and $x_2(t)$ on the same axes.

2.22 The differential equations for the perturbations in liquid level in a two-tank system due to perturbations in inlet flow rates are

$$A_1\frac{dh_1}{dt} + \frac{1}{R_1}h_1 - \frac{1}{R_1}h_2 = q_{i1}(t) \quad \text{(a)}$$

$$A_2\frac{dh_2}{dt} - \frac{1}{R_1}h_1 + \left(\frac{1}{R_1} + \frac{1}{R_2}\right)h_2 = q_{i2}(t) \quad \text{(b)}$$

Determine the liquid-level perturbation $h_1(t)$ when $q_{i1}(t) = 0$, $q_{i2}(t) = 0.5\ u(t)\ \text{m}^3/\text{s}$ when $A_1 = 100\ \text{m}^2$, $A_2 = 150\ \text{m}^2$, $R_1 = 10\ \text{s/m}^2$, and $R_2 = 8\ \text{s/m}^2$.

2.23 The equations obtained for the perturbation in temperatures of oil and water in a double pipe heat exchanger are

$$\rho_o c_o \pi r_i^2 L\frac{d\theta_o}{dt} + (2\pi r_o LU + c_o\dot{m}_o)\theta_o - 2\pi r_o UL\theta_w = c_o\dot{m}_o\theta_{oi}$$

$$\rho_w c_w \pi(r_2^2 - r_o^2)L\frac{d\theta_w}{dt} - 2\pi r_o LU\theta_o + (2\pi r_o LU + c_w\dot{m}_w)\theta_w = c_w\dot{m}_w\theta_{wi}$$

Determine the perturbations in temperatures of the oil and water when $\theta_{oi}(t) = 0$, $\theta_{wi}(t) = 3u(t)$, and

$\rho_w = 9.60 \times 10^2\,\text{kg/m}^3$, $\rho_o = 7.50 \times 10^2\,\text{kg/m}^3$,
$c_w = 4.19 \times 10^3\,\text{J/kg·C}$, $c_o = 2.4 \times 10^3\,\text{J/kg·C}$,
$U = 8.65 \times 10^2\,\text{W/m}^2\text{·C}$, $\dot{m}_w = 2.51\ \text{kg/s}$,
$\dot{m}_o = 4.85\ \text{kg/s}$, $r_i = 2\ \text{cm}$, $r_o = 2.2\ \text{cm}$, $r_2 = 4.2\ \text{cm}$,
$L = 2.5\ \text{cm}$

Assume $\theta_o(0) = \theta_w(0) = 0$.

2.24 The differential equations obtained for the concentrations of the reactant and product in a constant volume CSTR are

$$V\frac{dC_A}{dt} + (q + kV)C_A = qC_{Ai}$$

$$V\frac{dC_B}{dt} - kVC_A + qC_B = qC_{Bi}$$

Determine the concentrations of the reactant A and the product B when $V = 1.00 \times 10^{-3}\ \text{m}^3$, $q = 2.0 \times 10^{-6}\ \text{m}^3/\text{s}$, $k = 2.0 \times 10^{-3}\ \text{s}^{-1}$, $C_{Ai}(t) = 0.2u(t)$ mol/L, and $C_{Bi}(t) = 0$. Plot $C_{Ai}(t)$ and $C_{Bi}(t)$.

3

Transfer Function and State-Space Modeling

Learning Objectives

At the conclusion of this chapter, you should be able to:

- Determine the order of a system from the number of independent energy storage devices present in the system
- Determine the order of a system from the order of the polynomial in the denominator of the transfer function
- Understand that the number of state variables needed to model a system is equal to the order of the system
- Determine the transfer function of a system as the ratio of the Laplace transform of the system's input to the Laplace transform of the system's output
- Determine the transfer function matrix for a MIMO system
- Determine how to quantify the state variables
- Understand how to develop a state-space formulation and write it in a matrix form
- Determine the transfer function from the state-space formulation for a SISO system
- Determine the state-space formulation from the transfer function or transfer matrix
- Determine the transfer function from the state-space formulation for a MIMO system

A control system is designed using either the transfer function or the state-space formulation. It is prudent to have these formulations ready to apply to the mathematical models obtained through the application of system dynamics.

3.1 Introduction

The mathematical modeling of a system with one independent variable and one dependent variable results in an ordinary differential equation or an integrodifferential equation. Modeling a system with one independent variable and multiple dependent variables results in a system of ordinary differential equations or a system of integrodifferential equations. The number of equations in the system is usually equal to the number of dependent variables. Time, t, is the independent variable for dynamic systems analysis. The dependent variable or variables have physical meanings, such as the displacement of a mass in a mechanical system, the current flowing through a circuit in an electrical system, or the concentration of a reactant in a chemical system.

The derivation of differential equations governing a physical system is the subject of Chapter 4 (mechanical systems), Chapter 5 (electrical systems and electromechanical systems), and Chapter 6 (fluid, thermal, chemical, and biological systems). System dynamics play a major role in the mathematical models derived. The transfer function for a mathematical model contains all the information about the system dynamics. Two systems with the same transfer function have the same system dynamics.

The transfer function is expressed as

$$G(s) = \frac{N(s)}{D(s)} \tag{3.1}$$

where $D(s)$ is a polynomial in s. The order of the transfer function and the system is the order of $D(s)$. A system with an nth-order polynomial in its denominator is called an nth-order system. The numerator is usually a polynomial of order less than or equal to that of the denominator. The numerator is not a polynomial when the system has a time delay.

The order of a system and hence the order of the system's transfer function are equal to the number of independent energy storage elements in the system. For example, consider the discrete mechanical system of Figure 3.1. The system has two masses that store kinetic energy independently of each other. It also has two springs that store potential energy independently of each other, for a total of four independent storage elements. The order of the system is four. Now consider the system of Figure 3.2. Consider the system dynamics. Let x_1 denote the displacement of the leftmost mass, and let x_2 denote the displacement of the rightmost mass. The potential energies developed in the springs from left to right are $\frac{1}{2}k_1x_1^2$, $\frac{1}{2}k_2(x_2 - x_1)^2$, and $\frac{1}{2}k_3x_2^2$. The potential energy of the rightmost spring can be written in terms of the potential energies of the other two springs at any instant. Thus the additional spring is not independent of the other two. This system has four independent storage elements: two masses and two springs.

Now consider the electric circuit of Figure 3.3 (page 106). The system has one inductor that stores energy in a magnetic field and two independent capacitors that store energy in an electric field. The system has three energy storage devices, so the order of the system and its transfer function is three. The resistors dissipate energy rather than store energy.

The assumptions made in modeling a system can affect its order. Consider the single-tank liquid-level system of Figure 3.4 (page 106). The liquid in the tank moves and stores kinetic energy. The liquid is drained through a pipe, and the tank stores

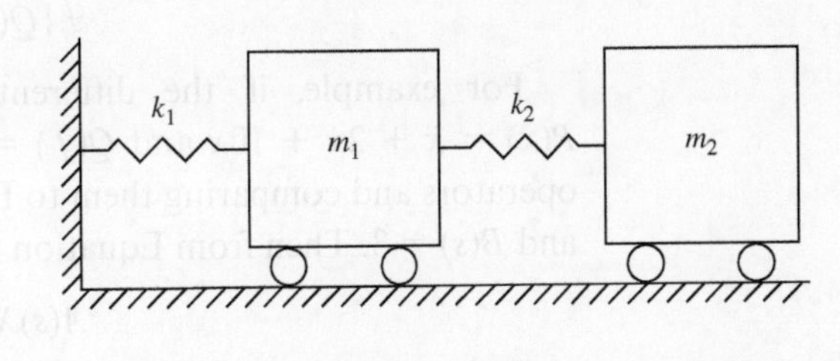

Figure 3.1

The mechanical system has four energy storage elements: two independent springs which store potential energy and two masses which store kinetic energy. The order of the system is four.

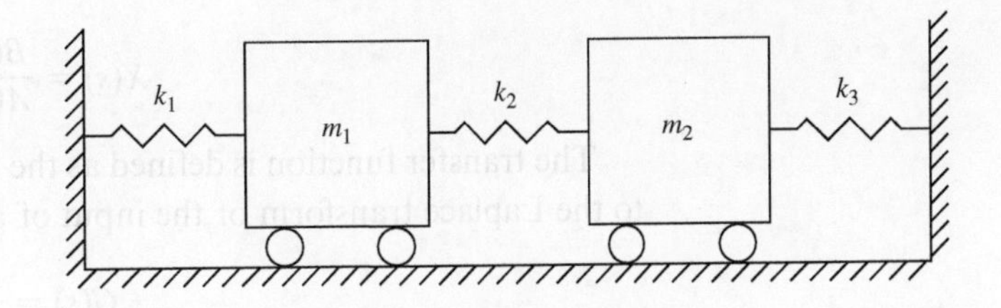

Figure 3.2

The mechanical system has three springs, but only two are independent. Along with the two masses, the system has four independent energy storage elements, thus the order of the system is four.

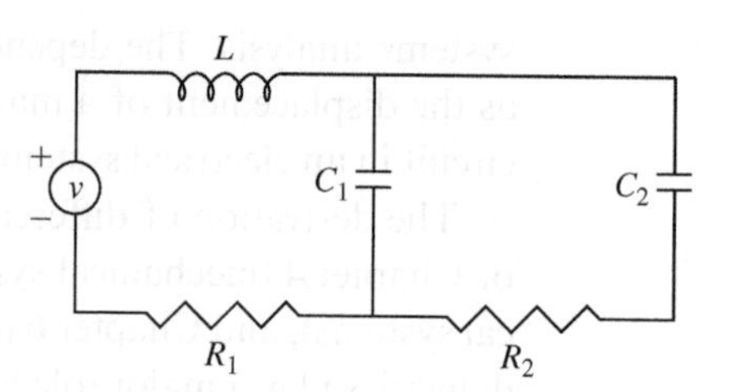

Figure 3.3

As current flows through the circuit, the inductor stores energy in a magnetic field. The two capacitors each store energy independently in the electric field surrounding them. There are three energy storage elements, making the order of the system three.

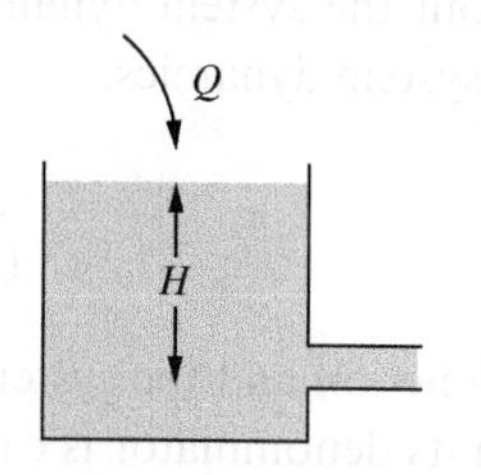

Figure 3.4

The tank stores potential energy due to gravity. The motion of the fluid in the tank is assumed to be negligible, so the kinetic energy associated with motion of the fluid is ignored. The system has one energy storage element, making it a first order system.

potential energy due to the change in the level in the tank. The pipe dissipates energy rather than storing energy. Thus, there are two energy storage elements and the order of the system is two. Modeling the motion of the liquid is difficult; it requires application of the unsteady form of Bernoulli's equation and leads to highly nonlinear terms in the mathematical model. However, the motion of the liquid is slow and its kinetic energy is neglected in comparison to the potential energy. In this case, there is only one energy storage element and the predicted order of the system is one.

The transfer function for a linear system is formulated in what is called the Laplace domain or the s domain. It is formulated from the system's mathematical model, so it is in accordance with assumptions made in the modeling. The transfer function is the ratio of the Laplace transform of the output of a system to the Laplace transform of the input to the system. Therefore, it is independent of the specific system input. The transfer function is formulated assuming all initial conditions are zero. All information about the system dynamics is included in the transfer function.

3.2 Transfer Functions for Systems with One Independent Variable (SISO Systems)

A system with one input and one output is called a single input and single output (SISO) system. Consider such a linear system with input $f(t)$ and output $x(t)$. Given the input, the output is determined by solving a differential equation of the form

$$P(x) = Q(f) \tag{3.2}$$

where P and Q are linear differential operators with constant coefficients. Let $X(s) = \mathcal{L}\{x(t)\}$ and $F(s) = \mathcal{L}\{f(t)\}$. Using the properties of linearity of the transform and transform of derivatives and assuming all initial conditions are zero leads to

$$\mathcal{L}\{P(x)\} = A(s)X(s) \tag{3.3}$$

and

$$\mathcal{L}\{Q(y)\} = B(s)F(s) \tag{3.4}$$

For example, if the differential equation is $\ddot{x} + 2\dot{x} + 10x = 3f$, then $P(x) = \ddot{x} + 2\dot{x} + 10x$ and $Q(f) = 3f$. Then taking the Laplace transform of these operators and comparing them to Equations (3.3) and (3.4) yields $A(s) = s^2 + 2s + 3$ and $B(s) = 3$. Then from Equation (3.2),

$$A(s)X(s) = B(s)F(s) \tag{3.5}$$

Solving for $X(s)$ gives

$$X(s) = \frac{B(s)}{A(s)}F(s) \tag{3.6}$$

The transfer function is defined as the ratio of the Laplace transform of the output to the Laplace transform of the input of a system, or

$$G(s) = \frac{X(s)}{F(s)} \tag{3.7}$$

Equation (3.6) yields

$$G(s) = \frac{B(s)}{A(s)} \tag{3.8}$$

The Laplace transform of the output is written as

$$X(s) = G(s)F(s) \tag{3.9}$$

The convolution property of Laplace transforms is applied to Equation (3.9), leading to

$$x(t) = \int_0^t g(\tau)f(t-\tau)d\tau \tag{3.10}$$

where

$$g(t) = \mathcal{L}^{-1}\{G(s)\} \tag{3.11}$$

A transfer function, given by Equation (3.8), is usually written as the ratio of two polynomials. Without loss of generality, the transfer function can be written such that the leading coefficient in the denominator, $A(s)$ in Equation (3.8), is one.

If $Q(y) = y$ or the differential operator is unity, then $B(s) = 1$ and $G(s) = \frac{1}{A(s)}$, where $A(s)$ is a polynomial in s. If the leading coefficient of $A(s)$ is not one, then $A(s) = \frac{1}{\alpha}D(s)$, where α is a constant and the leading coefficient of the polynomial $D(s)$ is one. For this case,

$$G(s) = \frac{\alpha}{D(s)} \tag{3.12}$$

If the system is governed by an integrodifferential equation, then $A(s)$ is not a polynomial in s. The property of transform of integrals is used, and the result is $A(s) = \frac{1}{\alpha s}D(s)$, where α is chosen to make the leading coefficient of $D(s)$ equal to one. For a system governed by an integrodifferential equation, the transfer function is written as

$$G(s) = \frac{\alpha s B(s)}{D(s)} \tag{3.13}$$

Example 3.1

Determine the transfer function for a system whose mathematical model is

$$30\dot{h} + 15h = q(t) \tag{a}$$

Solution

The input to the system is $q(t)$ and the output is $h(t)$. The transfer function is $G(s) = \frac{H(s)}{Q(s)}$. Taking the Laplace transform of Equation (a) using the property of linearity of the transform leads to

$$30\mathcal{L}\{\dot{h}\} + 15\mathcal{L}\{h\} = \mathcal{L}\{q(t)\} \tag{b}$$

Applying the property of transform of derivatives to Equation (b) leads to

$$30[sH(s) - h(0)] + 15H(s) = Q(s) \tag{c}$$

The transfer function is defined assuming all initial conditions are zero. Thus, Equation (b) becomes

$$(30s + 15)H(s) = Q(s) \tag{d}$$

Rearranging Equation (d),

$$G(s) = \frac{H(s)}{Q(s)} = \frac{1}{30s + 15} \tag{e}$$

Rewriting Equation (e) such that the leading coefficient of its denominator is one leads to

$$G(s) = \frac{\frac{1}{30}}{s + \frac{1}{2}} \tag{f}$$

Example 3.2

Determine the transfer function for a system whose mathematical model is

$$10\ddot{x} + 200\dot{x} + 4000x = f(t) \tag{a}$$

Solution

The input to the system is $f(t)$ and the output is $x(t)$. The transfer function is $G(s) = \frac{X(s)}{F(s)}$. Taking the Laplace transform of Equation (a) using the property of linearity of the transform leads to

$$10\mathcal{L}\{\ddot{x}\} + 200\mathcal{L}\{\dot{x}\} + 4000\mathcal{L}\{x\} = \mathcal{L}\{f(t)\} \tag{b}$$

Applying the property of transform of derivatives to Equation (b) leads to

$$10[s^2X(s) - sx(0) - \dot{x}(0)] + 200[sX(s) - x(0)] + 4000X(s) = F(s) \tag{c}$$

The transfer function is defined assuming all initial conditions are zero. Thus, Equation (b) becomes

$$(10s^2 + 200s + 4000)X(s) = F(s) \tag{d}$$

Rearranging Equation (d),

$$G(s) = \frac{X(s)}{F(s)} = \frac{1}{10s^2 + 200s + 4000} \tag{e}$$

Rewriting Equation (e) such that the leading coefficient of its denominator is one leads to

$$G(s) = \frac{0.1}{s^2 + 20s + 400} \tag{f}$$

Example 3.3

Determine the transfer function for a system whose mathematical model is

$$6\ddot{x} + 24\dot{x} + 120x = 3\dot{y} + 12y \tag{a}$$

The input to the system is $y(t)$ and the output is $x(t)$. Determine the transfer function $G(s) = \frac{X(s)}{Y(s)}$.

Solution

Taking the Laplace transform of the differential equation using the property of linearity of the transform leads to

$$6\mathcal{L}\{\ddot{x}\} + 24\mathcal{L}\{\dot{x}\} + 120\mathcal{L}\{x\} = 3\mathcal{L}\{\dot{y}\} + 12\mathcal{L}\{y\} \tag{b}$$

Applying the property of derivatives of the transform to Equation (b) leads to

$$6[s^2X(s) - sx(0) - \dot{x}(0)] + 24[sX(s) - x(0)] + 120X(s) = 3[sY(s) - y(0)] + 12Y(s) \tag{c}$$

Assuming all initial conditions are zero, $\dot{x}(0) = x(0) = y(0) = 0$, in Equation (c) leads to

$$[6s^2 + 24s + 120]X(s) = [3s + 12]Y(s) \tag{d}$$

Note that in Equation (3.6), $A(s) = 6s^2 + 24s + 120$ and $B(s) = 3s + 12$. Rearranging Equation (d) gives

$$\frac{X(s)}{Y(s)} = \frac{3s + 12}{6s^2 + 24s + 120} \tag{e}$$

Requiring that the leading coefficient in the denominator is one gives

$$G(s) = \frac{0.5s + 2}{s^2 + 4s + 20} \tag{f}$$

Example 3.4

Determine the transfer function for a system whose mathematical model is the integrodifferential equation

$$2\frac{dx}{dt} + 10x + 500\int_0^t x\,dt = f(t) \tag{a}$$

Solution

The input to the system is $f(t)$ and the output is $x(t)$. Determine the transfer function $G(s) = \frac{X(s)}{F(s)}$. Taking the Laplace transform of Equation (a) and using the property of linearity of the transform yields

$$2\mathcal{L}\left\{\frac{dx}{dt}\right\} + 10\mathcal{L}\{x\} + 500\mathcal{L}\left\{\int_0^t x\,dt\right\} = \mathcal{L}\{f\} \tag{b}$$

Applying the property of the transform of the first derivative, assuming the initial condition is zero, and applying the property of the transform of the integral gives

$$2sX(s) + 10X(s) + \frac{500}{s}X(s) = F(s) \tag{c}$$

Multiplying Equation (c) by s and collecting terms on the left-hand side leads to

$$(2s^2 + 10s + 500)X(s) = sF(s) \tag{d}$$

The transfer function is obtained from Equation (d) as

$$G(s) = \frac{s}{2s^2 + 10s + 500} \tag{e}$$

Requiring that the leading coefficient in the denominator is one leads to

$$G(s) = \frac{0.5s}{s^2 + 5s + 250} \tag{f}$$

3.3 Transfer Functions for Systems with Multiple Independent Variables (MIMO Systems)

A system with multiple independent variables is called a multiple input and multiple output (MIMO) system. Consider a system with n independent variables (outputs) that has m inputs. The outputs are described by $x_i(t)$, $i = 1, 2, \ldots, n$. The inputs are denoted by $f_j(t)$, $j = 1, 2, \ldots, m$. The output is summarized by an $n \times 1$ column vector $\mathbf{x}(t)$, while the input is summarized by an $m \times 1$ column vector $\mathbf{f}(t)$. The outputs are related to the inputs through n linear differential equations with variable coefficients

$$\sum_{i=1}^{n} P_{i,j}(x_j) = \sum_{i=1}^{m} Q_{i,j}(f_j) \tag{3.14}$$

where $P_{i,j}$ and $Q_{i,j}$ are linear differential operators.

Taking the Laplace transform of Equation (3.14) using the properties of linearity of the transform and transform of derivatives and assuming all initial conditions are zero leads to

$$\mathbf{Z}(s)\mathbf{X}(s) = \mathbf{R}(s)\mathbf{F}(s) \tag{3.15}$$

In Equation (3.15), $\mathbf{X}(s)$ is a vector whose i^{th} element is $X_i(s)$; $\mathbf{F}(s)$ is a vector whose i^{th} element is $F_i(s)$; $\mathbf{Z}(s)$, the impedance matrix, is an $n \times n$ matrix obtained by taking the Laplace transform of the left-hand side of Equation (3.14); and $\mathbf{R}(s)$ is an $n \times m$ matrix obtained by taking the Laplace transform of the right-hand side of Equation (3.14). Multiplying Equation (3.15) by the inverse of $\mathbf{Z}(s)$ gives

$$\mathbf{X}(s) = \mathbf{G}(s)\mathbf{F}(s) \tag{3.16}$$

The transfer function matrix $\mathbf{G}(s)$ is the $n \times m$ matrix defined as

$$\mathbf{G}(s) = \mathbf{Z}^{-1}(s)\mathbf{R}(s) \tag{3.17}$$

The transfer function $G_{i,j}(s)$ is the transfer function that shows the effect of $f_j(t)$ on $x_i(t)$. That is, $G_{i,j}(s)$ is calculated as

$$G_{i,j}(s) = \frac{X_i(s)}{F_j(s)} \tag{3.18}$$

assuming that $f_k(t) = 0$ for $k = 1, 2, \ldots, m$ with $k \neq j$.

Example 3.5

a. A system is governed by the differential equations

$$\ddot{x}_1 + 3x_1 - 2x_2 = f(t) \tag{a}$$

$$2\ddot{x}_2 - 2x_1 + 4x_2 = 0 \tag{b}$$

Determine the vector of transfer functions.

b. A similar system is governed by the differential equations

$$\ddot{x}_1 + 3x_1 - 2x_2 = 4f_1(t) + f_2(t) \tag{c}$$

$$2\ddot{x}_2 - 2x_1 + 4x_2 = f_1(t) + 3f_2(t) \tag{d}$$

Determine the transfer function matrix for this system.

Solution

a. The system has one input and two outputs. The matrix of transfer functions is therefore 2×1 or a column vector. Define $G_1(s) = \frac{X_1(s)}{F(s)}$ and $G_2(s) = \frac{X_2(s)}{F(s)}$. Taking the Laplace transforms of Equations (a) and (b) and using the properties of linearity and transforms of derivatives, assuming all initial conditions are zero, leads to

$$s^2X_1(s) + 3X_1(s) - 2X_2(s) = F(s) \tag{e}$$

$$2s^2X_2(s) - 2X_1(s) + 4X_2(s) = 0 \tag{f}$$

Summarizing Equations (e) and (f) in matrix form gives

$$\begin{bmatrix} s^2+3 & -2 \\ -2 & 2s^2+4 \end{bmatrix} \begin{bmatrix} X_1(s) \\ X_2(s) \end{bmatrix} = \begin{bmatrix} F(s) \\ 0 \end{bmatrix} \tag{g}$$

Comparing Equation (g) to Equation (3.15), it is seen that the impedance matrix is $\mathbf{Z}(s) = \begin{bmatrix} s^2+3 & -2 \\ -2 & 2s^2+4 \end{bmatrix}$ and $\mathbf{R}(s) = \begin{bmatrix} 1 \\ 0 \end{bmatrix}$. The transfer function vector can be obtained by inverting $\mathbf{Z}(s)$ and multiplying by $\mathbf{R}(s)$, which in this case leads to the first row of the inverse matrix. Another method is to solve the equations simultaneously for $X_1(s)$ and $X_2(s)$ and then divide by $F(s)$ to obtain $G_1(s)$ and $G_2(s)$, respectively. From the latter, Cramer's rule states that an element of the solution vector is obtained as the ratio of two determinants. The determinant in the denominator is the determinant of the coefficient matrix, and the determinant in the numerator is the determinant of the coefficient matrix with the appropriate column replaced by the vector on the right-hand side, $\begin{bmatrix} F(s) \\ 0 \end{bmatrix}$. To determine $X_1(s)$, the first column is replaced, and to determine $X_2(s)$, the second column is replaced. It is noted that

$$\begin{aligned} D(s) &= \begin{vmatrix} s^2+3 & -2 \\ -2 & 2s^2+4 \end{vmatrix} \\ &= (s^2+3)(2s^2+4) - (-2)(-2) \\ &= 2s^4 + 10s^2 + 8 \end{aligned} \tag{h}$$

Then

$$X_1(s) = \frac{\begin{vmatrix} F(s) & -2 \\ 0 & 2s^2+4 \end{vmatrix}}{D(s)} = \frac{2s^2+4}{2s^4+10s^2+8} F(s) \tag{i}$$

$$X_2(s) = \frac{\begin{vmatrix} s^2+3 & F(s) \\ -2 & 0 \end{vmatrix}}{D(s)} = \frac{2}{2s^4+10s^2+8} F(s) \tag{j}$$

and

$$G_1(s) = \frac{X_1(s)}{F(s)} = \frac{2s^2+4}{2s^4+10s^2+8} \tag{k}$$

$$G_2(s) = \frac{X_2(s)}{F(s)} = \frac{2}{2s^4+10s^2+8} \tag{l}$$

b. The system defined by Equations (c) and (d) has two inputs and two outputs. Taking the Laplace transform and using linearity of the transform and the property of transform of derivatives, assuming all initial conditions are zero, leads to

$$s^2X_1(s) + 3X_1(s) - 2X_2(s) = 4F_1(s) + F_2(s) \tag{m}$$

$$2s^2X_2(s) - 2X_1(s) + 4X_2(s) = F_1(s) + 3F_2(s) \tag{n}$$

Summarizing Equations (m) and (n) in matrix form results in

$$\begin{bmatrix} s^2+3 & -2 \\ -2 & 2s^2+4 \end{bmatrix} \begin{bmatrix} X_1(s) \\ X_2(s) \end{bmatrix} = \begin{bmatrix} 4 & 1 \\ 1 & 3 \end{bmatrix} \begin{bmatrix} F_1(s) \\ F_2(s) \end{bmatrix} \tag{o}$$

The impedance matrix is $\mathbf{Z}(s) = \begin{bmatrix} s^2+3 & -2 \\ -2 & 2s^2+4 \end{bmatrix}$ and $\mathbf{R}(s) = \begin{bmatrix} 4 & 1 \\ 1 & 3 \end{bmatrix}$. The inverse of the impedance matrix is

$$\mathbf{Z}^{-1}(s) = \frac{1}{2s^4+10s^2+8} \begin{bmatrix} 2s^2+4 & 2 \\ 2 & s^2+3 \end{bmatrix} \tag{p}$$

Then

$$\mathbf{Z}^{-1}(s)\mathbf{R}(s)\mathbf{F}(s) = \frac{1}{2s^4 + 10s^2 + 8}\begin{bmatrix} 2s^2 + 4 & 2 \\ 2 & s^2 + 3 \end{bmatrix}\begin{bmatrix} 4 & 1 \\ 1 & 3 \end{bmatrix}\begin{bmatrix} F_1(s) \\ F_2(s) \end{bmatrix}$$

$$= \frac{1}{2s^4 + 10s^2 + 8}\begin{bmatrix} 4(2s^2 + 4)F_1(s) + 2F_2(s) \\ 2F_1(s) + 3(s^2 + 3)F_2(s) \end{bmatrix} \quad \text{(q)}$$

The transfer function $G_{1,1}(s)$ is defined as the ratio of $X_1(s)$ to $F_1(s)$ if $F_2(s) = 0$. From Equation (q), this yields

$$G_{1,1}(s) = \frac{4(2s^2 + 4)}{2s^4 + 10s^2 + 8} \quad \text{(r)}$$

$G_{1,2}(s)$ is defined as the ratio of $X_1(s)$ to $F_2(s)$ if $F_1(s) = 0$. From Equation (q), this yields

$$G_{1,2}(s) = \frac{2}{2s^4 + 10s^2 + 8} \quad \text{(s)}$$

$G_{2,1}(s)$ is defined as the ratio of $X_2(s)$ to $F_1(s)$ if $F_2(s) = 0$. From Equation (q), this yields

$$G_{2,1}(s) = \frac{2}{2s^4 + 10s^2 + 8} \quad \text{(t)}$$

$G_{2,2}(s)$ is defined as the ratio of $X_2(s)$ to $F_2(s)$ if $F_1(s) = 0$. From Equation (q), this yields

$$G_{2,2}(s) = \frac{3(s^2 + 3)}{2s^4 + 10s^2 + 8} \quad \text{(u)}$$

The transfer matrix is written as

$$\mathbf{G}(s) = \frac{1}{2s^4 + 10s^2 + 8}\begin{bmatrix} 4(2s^2 + 4) & 2 \\ 2 & 3(s^2 + 3) \end{bmatrix} \quad \text{(v)}$$

3.4 State Variables and the State-Space Method

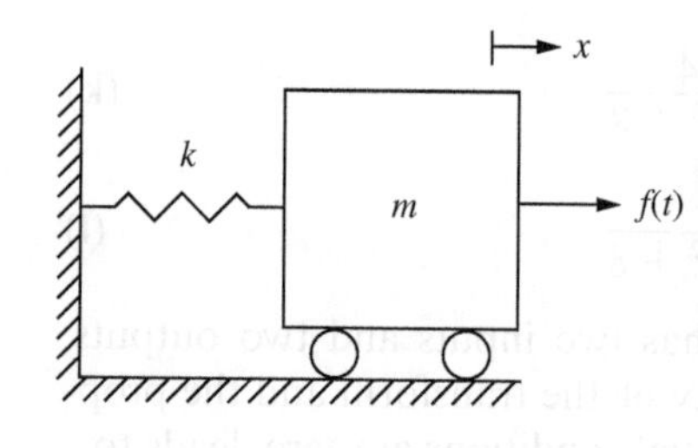

Figure 3.5

Two state variables are necessary to model the mechanical system. They may be taken as $y_1 = x$ and $y_2 = \dot{x}$.

The state of a dynamic system is defined by the smallest set of variables such that, given initial conditions, the time-dependent behavior of the entire system can be determined. State variables are defined as such a set of variables. For example, consider the simple mass-spring system of Figure 3.5. Let x be the time-dependent displacement of the mass measured from the system's equilibrium position. The behavior of the system is defined by the displacement x and the velocity of the mass $v = \dot{x}$. The acceleration $\ddot{x}$ is not necessary to specify separately from the mathematical model to determine the behavior of the system. The state variables for this system are x and $\dot{x}$. There are two state variables.

Consider the electric circuit of Figure 3.6. The input to the system is the potential provided by the source $v(t)$. The state variables are the charge q and the current $i = \frac{dq}{dt}$. The state variables, along with the mathematical model, are used to define the time-dependent behavior of a system given its initial conditions.

The number of state variables used to define the behavior of a system is unique, but the definition of the state variables is not. Consider the mass-spring system connected to a movable support. The input to this system is the displacement of the base y. The obvious state variables are the displacement of the mass x and the velocity $\dot{x}$. However, another definition of state variables would be the displacement of the mass relative to the base $z = y - x$ and the relative velocity $\dot{z} = \dot{y} - \dot{x}$. State variables may be taken to be z and $\dot{z}$.

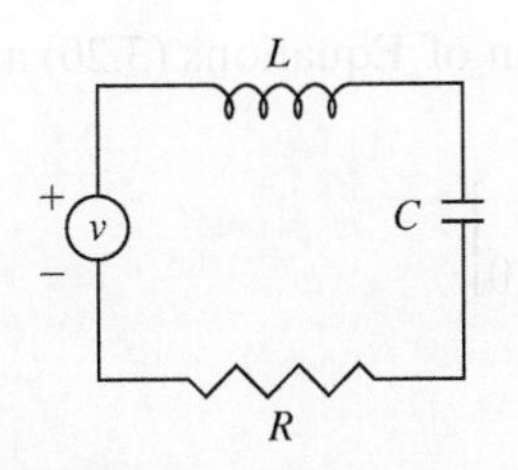

Figure 3.6

Two state variables are necessary to model the series LRC circuit. The state variables are related to the energy stored by the inductor in a magnetic field $\left(\frac{1}{2}Li^2\right)$, and the energy stored by the capacitor in an electric field $\left(\frac{1}{2}\frac{1}{C}q^2\right)$. A possible choice of state variables is $y_1 = q = \int_0^t i\,dt$ and $y_2 = i$.

One set of state variables can be defined to calculate the energy stored by all the independent storage devices in the system. For the system in Figure 3.5, the kinetic energy stored by the mass is $\frac{1}{2}m\dot{x}^2$, and the potential energy stored by the spring is $\frac{1}{2}kx^2$. For the electric circuit of Figure 3.6, the energy stored by the inductor in a magnetic field is $\frac{1}{2}Li^2$, and the energy stored by the capacitor in an electric field is $\frac{1}{2}\frac{1}{C}q^2$.

Consider a system with m outputs and p inputs. Let $\mathbf{u}$ be a $p \times 1$ vector of inputs. Let $\mathbf{x}$ be an $m \times 1$ vector of outputs. Let $y_1, y_2, \ldots, y_n$ be the defined state variables and define the state vector as

$$\mathbf{y} = \begin{bmatrix} y_1 \\ \vdots \\ y_n \end{bmatrix} \tag{3.19}$$

The state-space method uses the state variables to formulate, using the mathematical model, a time domain set of equations that can be used to determine the behavior of the system at any time. The state-space model is summarized in matrix form as

$$\dot{\mathbf{y}} = \mathbf{A}\mathbf{y} + \mathbf{B}\mathbf{u} \tag{3.20}$$

where $\mathbf{A}$ is the $n \times n$ **state matrix** and $\mathbf{B}$ is the $n \times p$ **input matrix**. The state-space formulation defines the output as

$$\mathbf{x} = \mathbf{C}\mathbf{y} + \mathbf{D}\mathbf{u} \tag{3.21}$$

where $\mathbf{C}$ is an $m \times n$ **output matrix** and $\mathbf{D}$ is an $n \times p$ **transmission matrix**. Equations (3.20) and (3.21) represent the state-space formulation of a mechanical system.

Example 3.6

Determine a state-space formulation for a system governed by the differential equation

$$3\ddot{x} + 9\dot{x} + 30x = f(t) \tag{a}$$

Solution

Define the state variables as

$$y_1 = x \tag{b}$$

$$y_2 = \dot{x} \tag{c}$$

Equations (b) and (c) imply that

$$\dot{y}_1 = y_2 \tag{d}$$

Substitution of Equations (b) and (c) into Equation (a) results in

$$3\dot{y}_2 + 9y_2 + 30y_1 = f(t) \tag{e}$$

Equation (e) is rearranged to give

$$\dot{y}_2 = -3y_2 - 10y_1 + \frac{1}{3}f(t) \tag{f}$$

Summarizing Equations (d) and (f) in matrix form leads to

$$\begin{bmatrix} \dot{y}_1 \\ \dot{y}_2 \end{bmatrix} = \begin{bmatrix} 0 & 1 \\ -3 & -10 \end{bmatrix}\begin{bmatrix} y_1 \\ y_2 \end{bmatrix} + \begin{bmatrix} 0 \\ \frac{1}{3} \end{bmatrix} f(t) \tag{g}$$

The relation between the output and the state variables is defined by Equations (b) and (c), which are summarized in matrix form as

$$[x] = [1 \quad 0]\begin{bmatrix} y_1 \\ y_2 \end{bmatrix} + [0][f(t)] \tag{h}$$

From Equations (g) and (h), consistent with the formulation of Equations (3.20) and (3.21), the following matrices are defined:

$$\text{State matrix } \mathbf{A} = \begin{bmatrix} 0 & 1 \\ -3 & -10 \end{bmatrix}$$

$$\text{Input matrix } \mathbf{B} = \begin{bmatrix} 0 \\ \frac{1}{3} \end{bmatrix}$$

$$\text{Output matrix } \mathbf{C} = [1 \quad 0]$$

$$\text{Transmission matrix } \mathbf{D} = [0]$$

Example 3.7

A system is governed by the integrodifferential equation

$$2\frac{dx}{dt} + 10x + 500\int_0^t x\,dt = f(t) \tag{a}$$

Determine a state-space formulation for the system.

Solution

The input is $f(t)$ and the output is $x(t)$. Define as state variables

$$y_1 = \int_0^t x\,dt \tag{b}$$

$$y_2 = x \tag{c}$$

It is noted that

$$\frac{dy_1}{dt} = y_2 \tag{d}$$

Substitution of Equations (b) and (c) into Equation (a) leads to

$$2\frac{dy_2}{dt} + 10y_2 + 500y_1 = f(t) \tag{e}$$

Equation (e) is rearranged to

$$\frac{dy_2}{dt} = -250y_1 - 5y_2 + \frac{1}{2}f(t) \tag{f}$$

Equations (d) and (f) are summarized in matrix form as

$$\begin{bmatrix} \frac{dy_1}{dt} \\ \frac{dy_2}{dt} \end{bmatrix} = \begin{bmatrix} 0 & 1 \\ -250 & -5 \end{bmatrix}\begin{bmatrix} y_1 \\ y_2 \end{bmatrix} + \begin{bmatrix} 0 \\ \frac{1}{2} \end{bmatrix} f(t) \tag{g}$$

The state variables are related to the input and output by Equation (c), which is summarized by

$$[x] = [0 \quad 1]\begin{bmatrix} y_1 \\ y_2 \end{bmatrix} + [0][f(t)] \tag{h}$$

From Equations (g) and (h), consistent with the formulation of Equations (3.20) and (3.21), the following matrices are defined:

$$\text{State matrix } \mathbf{A} = \begin{bmatrix} 0 & 1 \\ -250 & -5 \end{bmatrix}$$

$$\text{Input matrix } \mathbf{B} = \begin{bmatrix} 0 \\ \frac{1}{2} \end{bmatrix}$$

$$\text{Output matrix } \mathbf{C} = [0 \quad 1]$$

$$\text{Transmission matrix } \mathbf{D} = [0]$$

Example 3.8

A system is governed by the differential equations

$$\ddot{x}_1 + 3x_1 - 2x_2 = 4f_1(t) + f_2(t) \tag{a}$$

$$2\ddot{x}_2 - 2x_1 + 4x_2 = f_1(t) + 3f_2(t) \tag{b}$$

Determine a state-space formulation for this system.

Solution

This is a MIMO system with inputs $f_1(t)$ and $f_2(t)$ and outputs $x_1(t)$ and $x_2(t)$. Define as the state variables

$$y_1 = x_1 \tag{c}$$

$$y_2 = x_2 \tag{d}$$

$$y_3 = \frac{dx_1}{dt} \tag{e}$$

$$y_4 = \frac{dx_2}{dt} \tag{f}$$

Using the definition of the state variables,

$$\dot{y}_1 = y_3 \tag{g}$$

$$\dot{y}_2 = y_4 \tag{h}$$

Substitution of Equations (c)–(f) into Equations (a) and (b) leads to

$$\dot{y}_3 + 3y_1 - 2y_2 = 4f_1(t) + f_2(t) \tag{i}$$

$$\dot{y}_4 - 2y_1 + 4y_2 = f_1(t) + 3f_2(t) \tag{j}$$

Equations (i) and (j) are rearranged to

$$\dot{y}_3 = -3y_1 + 2y_2 + 4f_1(t) + f_2(t) \tag{k}$$

$$\dot{y}_4 = 2y_1 - 4y_2 + f_1(t) + 3f_2(t) \tag{l}$$

Equations (k) and (l) are summarized by

$$\begin{bmatrix} \dot{y}_1 \\ \dot{y}_2 \\ \dot{y}_3 \\ \dot{y}_4 \end{bmatrix} = \begin{bmatrix} 0 & 0 & 1 & 0 \\ 0 & 0 & 0 & 1 \\ -3 & 2 & 0 & 0 \\ 2 & -4 & 0 & 0 \end{bmatrix} \begin{bmatrix} y_1 \\ y_2 \\ y_3 \\ y_4 \end{bmatrix} + \begin{bmatrix} 0 & 0 \\ 0 & 0 \\ 4 & 1 \\ 1 & 3 \end{bmatrix} \begin{bmatrix} f_1 \\ f_2 \end{bmatrix} \tag{m}$$

The relation among the output variables, the state variables, and the input variables is

$$\begin{bmatrix} x_1 \\ x_2 \end{bmatrix} = \begin{bmatrix} 1 & 0 & 0 & 0 \\ 0 & 1 & 0 & 0 \end{bmatrix} \begin{bmatrix} y_1 \\ y_2 \\ y_3 \\ y_4 \end{bmatrix} + \begin{bmatrix} 0 & 0 \\ 0 & 0 \end{bmatrix} \begin{bmatrix} f_1 \\ f_2 \end{bmatrix} \tag{n}$$

From Equations (m) and (n), consistent with the formulation of Equations (3.20) and (3.21), the following matrices are defined:

$$\text{State matrix } \mathbf{A} = \begin{bmatrix} 0 & 0 & 1 & 0 \\ 0 & 0 & 0 & 1 \\ -3 & 2 & 0 & 0 \\ 2 & -4 & 0 & 0 \end{bmatrix}$$

$$\text{Input matrix } \mathbf{B} = \begin{bmatrix} 0 & 0 \\ 0 & 0 \\ 4 & 1 \\ 1 & 3 \end{bmatrix}$$

$$\text{Output matrix } \mathbf{C} = \begin{bmatrix} 1 & 0 & 0 & 0 \\ 0 & 1 & 0 & 0 \end{bmatrix}$$

$$\text{Transmission matrix } \mathbf{D} = \begin{bmatrix} 0 & 0 \\ 0 & 0 \end{bmatrix}$$

3.5 State-Space Models for Systems with Derivatives of the Input in the Mathematical Model

A state-space formulation does not allow derivatives of an input. The state variables must be modified from their usual definition. Modifying the definition of the second state variable to include z allows a term proportional to $\dot{z}$ to appear when the state variables are substituted into the mathematical model. The coefficients are adjusted to eliminate terms proportional to the derivative of the input. Systems such as the one shown in Figure 3.7 have derivatives of the input in the mathematical model.

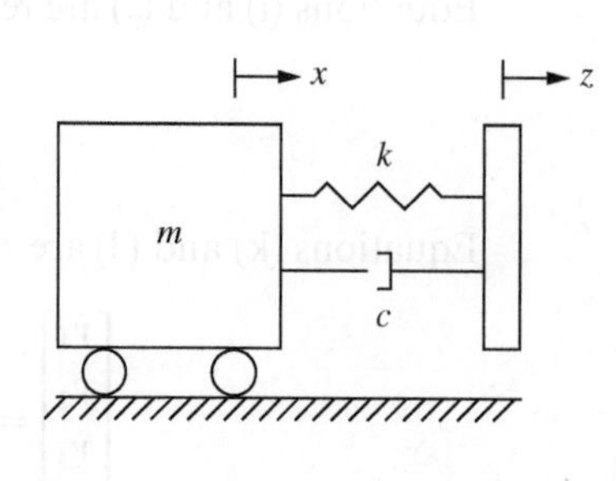

Figure 3.7

The mathematical model for the system with an input $z(t)$ and an output $x(t)$ is $m\ddot{x} + c\dot{x} + kx = c\dot{z} + kz$. The model includes the first derivative of the input.

Example 3.9

A system has the mathematical model

$$6\ddot{x} + 24\dot{x} + 120x = 3\dot{z} + 20z \tag{a}$$

where the input is $z(t)$ and the output is $x(t)$. Determine a state-space model for the system.

Solution

The input is $z(t)$ and the output is $x(t)$. Since the mathematical model contains the derivative of the input, a modification to the state variables must be made. Define the state variables as

$$y_1 = x \tag{b}$$

$$y_2 = \dot{x} + \mu z \tag{c}$$

where μ is to be determined to achieve the desired state-space model. From Equations (b) and (c),

$$y_2 = \dot{y}_1 + \mu z \tag{d}$$

Substitution of Equations (b) and (c) into Equation (a) leads to

$$6(\dot{y}_2 - \mu\dot{z}) + 24(y_2 - \mu z) + 120y_1 = 3\dot{z} + 20z \tag{e}$$

The derivative of the input is eliminated from Equation (e) by choosing

$$-6\mu = 3 \Rightarrow \mu = -\frac{1}{2} \tag{f}$$

Then Equation (e) becomes

$$\dot{y}_2 = -20y_1 - 4y_2 + \frac{4}{3}z \tag{g}$$

From Equations (d) and (f),

$$\dot{y}_1 = y_2 - \mu z = y_2 - \frac{1}{2}z \tag{h}$$

Equations (g) and (h) are summarized in matrix form as

$$\begin{bmatrix} \dot{y}_1 \\ \dot{y}_2 \end{bmatrix} = \begin{bmatrix} 0 & 1 \\ -20 & -4 \end{bmatrix} \begin{bmatrix} y_1 \\ y_2 \end{bmatrix} + \begin{bmatrix} -\frac{1}{2} \\ \frac{4}{3} \end{bmatrix} [z] \tag{i}$$

The output is expressed in terms of state variables, and the input as

$$[x] = [1 \quad 0] \begin{bmatrix} y_1 \\ y_2 \end{bmatrix} + [0][z(t)] \tag{j}$$

From Equations (i) and (j), consistent with the formulation of Equations (3.20) and (3.21), the following matrices are defined:

$$\text{State matrix } \mathbf{A} = \begin{bmatrix} 0 & 1 \\ -20 & -4 \end{bmatrix}$$

$$\text{Input matrix } \mathbf{B} = \begin{bmatrix} -\frac{1}{2} \\ \frac{4}{3} \end{bmatrix}$$

$$\text{Output matrix } \mathbf{C} = [1 \quad 0]$$

$$\text{Transmission matrix } \mathbf{D} = [0]$$

Example 3.10

Determine a state-space formulation for a system whose mathematical model is

$$3\ddot{x} + 12\dot{x} + 120x = 3\ddot{z} \tag{a}$$

Solution

The input to the system is $z(t)$. Its output is $x(t)$. The mathematical model explicitly has a term proportional to the second derivative of the input. Since the state-space formulation does not allow for terms proportional to the derivatives of the input to appear explicitly, modifications to the state variables are necessary. To this end, define

$$y_1 = x + vz \tag{b}$$

$$y_2 = \dot{x} + \mu z + v\dot{z} \tag{c}$$

where v and μ are to be chosen to develop an appropriate state-space formulation. Rearranging Equations (b) and (c) leads to

$$x = y_1 - vz \tag{d}$$

$$\dot{x} = y_2 - \mu z - v\dot{z} \tag{e}$$

$$\ddot{x} = \dot{y}_2 - \mu\dot{z} - v\ddot{z} \tag{f}$$

Substitution of Equations (d), (e), and (f) into Equation (a) leads to

$$3(\dot{y}_2 - \mu\dot{z} - v\ddot{z}) + 12(y_2 - \mu z - v\dot{z}) + 120(y_1 - vz) = 3\ddot{z} \tag{g}$$

In order to eliminate the derivatives of the input from the state-space formulation, choose

$$-3v = 3 \Rightarrow v = -1 \tag{h}$$

and

$$-3\mu - 12v = 0 \Rightarrow \mu = -4 \tag{i}$$

Using Equations (h) and (i) in Equation (g) leads to

$$3\dot{y}_2 + 12y_2 + 120y_1 + 168z = 0 \Rightarrow \dot{y}_2 = -40y_1 - 4y_2 - 56z \tag{j}$$

Equations (b) and (c) are rearranged to give

$$\dot{y}_1 = y_2 + z \tag{k}$$

Equations (j) and (k) are summarized in matrix form by

$$\begin{bmatrix} \dot{y}_1 \\ \dot{y}_2 \end{bmatrix} = \begin{bmatrix} 0 & 1 \\ -40 & -4 \end{bmatrix} \begin{bmatrix} y_1 \\ y_2 \end{bmatrix} + \begin{bmatrix} 1 \\ -56 \end{bmatrix} [z] \tag{l}$$

Equation (d) is written as

$$[x] = [1 \quad 0] \begin{bmatrix} y_1 \\ y_2 \end{bmatrix} + [1][z] \tag{m}$$

From Equations (l) and (m), consistent with the formulation of Equations (3.20) and (3.21), the following matrices are defined:

$$\text{State matrix } \mathbf{A} = \begin{bmatrix} 0 & 1 \\ -40 & -4 \end{bmatrix}$$

$$\text{Input matrix } \mathbf{B} = \begin{bmatrix} 1 \\ -56 \end{bmatrix}$$

$$\text{Output matrix } \mathbf{C} = [1 \quad 0]$$

$$\text{Transmission matrix } \mathbf{D} = [1]$$

3.6 Transfer Functions from the State-Space Model for SISO Systems

The transfer function $G(s) = \dfrac{X(s)}{F(s)}$ for a SISO system is unique. However, the state-space formulation for the system may take many forms and is not unique. It should be possible to determine a unique transfer function from a state-space formulation and to determine a state-space formulation from a transfer function.

Consider first the state-space formulation of an nth-order system described by Equations (3.20) and (3.21). The state variables are $y_1, y_2, \dots, y_n$, and since we are considering a SISO system, let $f(t)$ be the system input and let $x(t)$ be the system output. Taking the Laplace transform of Equation (3.20) leads to

$$s\mathbf{Y}(s) = \mathbf{AY}(s) + \mathbf{B}F(s) \tag{3.22}$$

Equation (3.22) is rearranged to give

$$(s\mathbf{I} - \mathbf{A})\mathbf{Y}(s) = \mathbf{B}F(s) \tag{3.23}$$

Equation (3.23) can be viewed as a system of simultaneous equations to solve for the vector of transforms. Using Cramer's rule to solve for each member of the solution vector individually gives

$$y_k(s) = \frac{|\mathbf{V}_k(s)|}{|s\mathbf{I} - \mathbf{A}|}F(s) \tag{3.24}$$

where $\mathbf{V}_k(s)$ is the matrix formed by replacing the kth column of $s\mathbf{I} - \mathbf{A}$ with the column vector $\mathbf{B}$. The vector of transforms can be written as

$$\mathbf{Y}(s) = \mathbf{\Lambda}(s)F(s) \tag{3.25}$$

where

$$\Lambda_k(s) = \frac{|\mathbf{V}_k(s)|}{|s\mathbf{I} - \mathbf{A}|} \tag{3.26}$$

Then taking the Laplace transform of Equation (3.21), recalling that the system is SISO, yields

$$\begin{aligned} X(s) &= \mathbf{CY}(s) + \mathbf{D}F(s) \\ &= (\mathbf{C\Lambda}(s) + \mathbf{D})F(s) \end{aligned} \tag{3.27}$$

The transfer function is obtained from Equation (3.27) as

$$G(s) = \mathbf{C\Lambda}(s) + \mathbf{D} \tag{3.28}$$

Example 3.11

The state-space formulation of a mathematical model is

$$\begin{bmatrix} \dot{y}_1 \\ \dot{y}_2 \end{bmatrix} = \begin{bmatrix} 0 & 1 \\ -10 & -5 \end{bmatrix}\begin{bmatrix} y_1 \\ y_2 \end{bmatrix} + \begin{bmatrix} 0 \\ 1 \end{bmatrix}[f] \tag{a}$$

$$[x] = [1 \;\; 0]\begin{bmatrix} y_1 \\ y_2 \end{bmatrix} + [0][f] \tag{b}$$

Determine the system's transfer function.

Solution

The matrix $s\mathbf{I} - \mathbf{A}$ is

$$s\mathbf{I} - \mathbf{A} = \begin{bmatrix} s & -1 \\ 10 & s+5 \end{bmatrix} \tag{c}$$

The denominator of the transfer function is

$$D(s) = |s\mathbf{I} - \mathbf{A}| = \begin{vmatrix} s & -1 \\ 10 & s+5 \end{vmatrix} = s(s+5) - (-1)(10) = s^2 + 5s + 10 \tag{d}$$

The numerators of the transforms of the state variables are

$$|\mathbf{V}_1(s)| = \begin{vmatrix} 0 & -1 \\ 1 & s+5 \end{vmatrix} = 1 \tag{e}$$

and

$$|\mathbf{V}_2(s)| = \begin{vmatrix} s & 0 \\ 10 & 1 \end{vmatrix} = s \tag{f}$$

The vector of the transforms is

$$\mathbf{Y}(s) = \frac{1}{s^2 + 5s + 10}\begin{bmatrix} 1 \\ s \end{bmatrix} \tag{g}$$

Then from Equation (3.28), the transfer function becomes

$$G(s) = [1 \quad 0]\frac{1}{s^2 + 5s + 10}\begin{bmatrix} 1 \\ s \end{bmatrix} + [0] = \frac{1}{s^2 + 5s + 10} \tag{h}$$

Example 3.12

A state-space formulation for a mathematical model of a system is

$$\begin{bmatrix} \dot{y}_1 \\ \dot{y}_2 \\ \dot{y}_3 \end{bmatrix} = \begin{bmatrix} -3 & -1 & 1 \\ 1 & 0 & 0 \\ -10 & 3 & -4 \end{bmatrix}\begin{bmatrix} y_1 \\ y_2 \\ y_3 \end{bmatrix} + \begin{bmatrix} 1 \\ 0 \\ 0 \end{bmatrix}[f(t)] \tag{a}$$

$$[x] = [1 \quad 1 \quad 0]\begin{bmatrix} y_1 \\ y_2 \\ y_3 \end{bmatrix} + [0][f(t)] \tag{b}$$

Determine the transfer function for the system.

Solution

The matrix $s\mathbf{I} - \mathbf{A}$ is

$$s\mathbf{I} - \mathbf{A} = \begin{bmatrix} s+3 & 1 & -1 \\ -1 & s & 0 \\ 10 & -3 & s+4 \end{bmatrix} \tag{c}$$

The denominator of the transfer function is

$$D(s) = |s\mathbf{I} - \mathbf{A}| = \begin{vmatrix} s+3 & 1 & -1 \\ -1 & s & 0 \\ 10 & -3 & s+4 \end{vmatrix} = s^3 + 7s^2 + 13s + 31 \tag{d}$$

The numerators of the transforms of the state variables are determined using

$$|\mathbf{V}_1(s)| = \begin{vmatrix} 1 & 1 & -1 \\ 0 & s & 0 \\ 0 & -3 & s+4 \end{vmatrix} = s^2 + 4s \tag{e}$$

$$|\mathbf{V}_2(s)| = \begin{vmatrix} s+3 & 1 & -1 \\ -1 & 0 & 0 \\ 10 & 0 & s+4 \end{vmatrix} = s + 4 \tag{f}$$

and

$$|\mathbf{V}_3(s)| = \begin{vmatrix} s+3 & 1 & 1 \\ -1 & s & 0 \\ 10 & -3 & 0 \end{vmatrix} = -10s + 3 \tag{g}$$

Thus, the vector of the transforms of the state variables is

$$\mathbf{Y}(s) = \frac{1}{s^3 + 7s^2 + 13s + 31} \begin{bmatrix} s^2 + 4s \\ s + 4 \\ -10s + 3 \end{bmatrix} [F(s)]$$

The transfer function for the output is obtained as

$$G(s) = \frac{1}{s^3 + 7s^2 + 13s + 31} [1 \quad 1 \quad 0] \begin{bmatrix} s^2 + 4s \\ s + 4 \\ -10s + 3 \end{bmatrix} = \frac{s^2 + 5s + 4}{s^3 + 7s^2 + 13s + 31} \tag{h}$$

3.7 A State-Space Model from the Transfer Function

Equation (3.28) suggests a method to determine a state-space formulation for a SISO system. The denominator of the transfer function is an nth-order polynomial in s, which can be used to construct a matrix whose determinant is the denominator. Then the vector $\mathbf{B}$ is chosen to adjust the numerator. To this end, assume that the denominator of the transfer function is

$$D(s) = s^n + a_{n-1}s^{n-1} + \cdots + a_1 s + a_0 \tag{3.29}$$

Consider the matrix

$$\mathbf{A} = \begin{bmatrix} 0 & 1 & 0 & 0 & 0 & 0 \\ 0 & 0 & 1 & 0 & 0 & 0 \\ 0 & 0 & 0 & 1 & 0 & 0 \\ \vdots & \vdots & \vdots & \ddots & \vdots & \vdots \\ 0 & 0 & 0 & \cdots & 1 & 0 \\ -a_0 & -a_1 & -a_2 & -a_3 & \cdots & -a_{n-1} \end{bmatrix} \tag{3.30}$$

The determinant of $s\mathbf{I} - \mathbf{A}$ is evaluated by expanding by minors using its last row. This shows that the determinant is of the form of $D(s)$ in Equation (3.29). For example, consider a third-order system in which

$$\mathbf{A} = \begin{bmatrix} 0 & 1 & 0 \\ 0 & 0 & 1 \\ -a_0 & -a_1 & -a_2 \end{bmatrix} \Rightarrow s\mathbf{I} - \mathbf{A} = \begin{bmatrix} s & -1 & 0 \\ 0 & s & -1 \\ a_0 & a_1 & s + a_2 \end{bmatrix} \tag{3.31}$$

Calculation of the determinant of the matrix in Equation (3.31) by expanding using the third row gives

$$|s\mathbf{I} - \mathbf{A}| = a_0\begin{vmatrix} -1 & 0 \\ s & -1 \end{vmatrix} - a_1\begin{vmatrix} s & 0 \\ 0 & -1 \end{vmatrix} + (s + a_2)\begin{vmatrix} s & -1 \\ 0 & s \end{vmatrix}$$

$$= a_0 + a_1 s + (s + a_2)s^2$$

$$= s^3 + a_2 s^2 + a_1 s + a_0 \tag{3.32}$$

Equation (3.32) is the denominator of a third-order transfer function.

When the system output is defined as a state variable, say $x = y_k$, then Equation (3.24) is used to replace the kth column $s\mathbf{I} - \mathbf{A}$ by $\mathbf{B}$ in the evaluation of $|\mathbf{V}_k(s)|$. Then the elements of $\mathbf{B}$ are chosen such that the numerator matches $|\mathbf{V}_k(s)|$.

Example 3.13

The transfer function formulation for a mathematical model is

$$G(s) = \frac{s^2 + 10s + 15}{s^3 + 2s^2 + 10s + 100} \tag{a}$$

The state variables are to be defined such that $y_1 = x$, where x is the system output. Determine a state-space formulation for the model.

Solution

According to Equation (3.30), the state matrix is determined as

$$\mathbf{A} = \begin{bmatrix} 0 & 1 & 0 \\ 0 & 0 & 1 \\ -100 & -10 & -2 \end{bmatrix} \tag{b}$$

It is determined from Equation (b) that

$$s\mathbf{I} - \mathbf{A} = \begin{bmatrix} s & -1 & 0 \\ 0 & s & -1 \\ 100 & 10 & s+2 \end{bmatrix} \tag{c}$$

Since $y_1 = x$, then $|\mathbf{V}_1(s)|$ is set equal to the numerator of the transfer function,

$$|\mathbf{V}_1(s)| = \begin{vmatrix} b_1 & -1 & 0 \\ b_2 & s & -1 \\ b_3 & 10 & s+2 \end{vmatrix} = b_1[s(s+2) + 10] - b_2[-1(s+2)] + b_3$$

$$= b_1 s^2 + (2b_1 + b_2)s + (10b_1 + b_3) \tag{d}$$

Then

$$b_1 s^2 + (2b_1 + b_2)s + (10b_1 + b_3) = s^2 + 10s + 15 \tag{e}$$

The coefficients of like powers of s are set equal in Equation (e), leading to

$$b_1 = 1, \quad b_2 = 8, \quad b_3 = 5 \tag{f}$$

A state-space formulation of the mathematical model of the system is

$$\begin{bmatrix} \dot{y}_1 \\ \dot{y}_2 \\ \dot{y}_3 \end{bmatrix} = \begin{bmatrix} 0 & 1 & 0 \\ 0 & 0 & 1 \\ -100 & -10 & -2 \end{bmatrix}\begin{bmatrix} y_1 \\ y_2 \\ y_3 \end{bmatrix} + \begin{bmatrix} 1 \\ 8 \\ 5 \end{bmatrix}[f(t)] \tag{g}$$

$$[x] = [1 \ 0 \ 0]\begin{bmatrix} y_1 \\ y_2 \\ y_3 \end{bmatrix} + [0][f(t)] \tag{h}$$

3.8 Transfer Functions for a MIMO System from a State-Space Model

For a MIMO system, converting a state-space formulation to the transfer formulation is very similar to the process for SISO systems. The Laplace transform of Equation (3.20) is taken, and the resulting system of algebraic equations becomes

$$(s\mathbf{I} - \mathbf{A})\mathbf{Y}(s) = \mathbf{BF}(s) \tag{3.33}$$

Solving for the vector of transforms of the state variables gives

$$\mathbf{Y}(s) = (s\mathbf{I} - \mathbf{A})^{-1}\mathbf{BF}(s) \tag{3.34}$$

The output vector is determined using Equation (3.27) as

$$\mathbf{X}(s) = \{\mathbf{C}(s\mathbf{I} - \mathbf{A})^{-1}\mathbf{B} + \mathbf{D}\}\mathbf{F}(s) \tag{3.35}$$

Comparing Equation (3.35) with Equation (3.16) shows that the transfer matrix becomes

$$\mathbf{G}(s) = \mathbf{C}(s\mathbf{I} - \mathbf{A})^{-1}\mathbf{B} + \mathbf{D} \tag{3.36}$$

Example 3.14

A state-space formulation of a mathematical model is

$$\begin{bmatrix} \dot{y}_1 \\ \dot{y}_2 \end{bmatrix} = \begin{bmatrix} -1 & 1 \\ 1 & -2 \end{bmatrix}\begin{bmatrix} y_1 \\ y_2 \end{bmatrix} + \begin{bmatrix} 1 & 0 \\ 0 & 1 \end{bmatrix}\begin{bmatrix} f_1 \\ f_2 \end{bmatrix} \tag{a}$$

$$\begin{bmatrix} x_1 \\ x_2 \end{bmatrix} = \begin{bmatrix} 1 & 0 \\ 0 & 1 \end{bmatrix}\begin{bmatrix} y_1 \\ y_2 \end{bmatrix} + \begin{bmatrix} 0 & 0 \\ 0 & 0 \end{bmatrix}\begin{bmatrix} f_1 \\ f_2 \end{bmatrix} \tag{b}$$

Determine the transfer matrix for the mathematical model.

Solution

From Equation (a),

$$s\mathbf{I} - \mathbf{A} = \begin{bmatrix} s+1 & -1 \\ -1 & s+2 \end{bmatrix} \tag{c}$$

From Equation (c),

$$(s\mathbf{I} - \mathbf{A})^{-1} = \frac{1}{s^2 + 3s + 1}\begin{bmatrix} s+2 & 1 \\ 1 & s+1 \end{bmatrix} \tag{d}$$

Using Equation (3.36), the transfer function matrix is

$$\begin{aligned} \mathbf{G}(s) &= \begin{bmatrix} 1 & 0 \\ 0 & 1 \end{bmatrix}\frac{1}{s^2 + 3s + 1}\begin{bmatrix} s+2 & 1 \\ 1 & s+1 \end{bmatrix}\begin{bmatrix} 1 & 0 \\ 0 & 1 \end{bmatrix} + \begin{bmatrix} 0 & 0 \\ 0 & 0 \end{bmatrix} \\ &= \frac{1}{s^2 + 3s + 1}\begin{bmatrix} s+2 & 1 \\ 1 & s+1 \end{bmatrix} \end{aligned} \tag{e}$$

Example 3.15

In Example 3.8, a state-space formulation of a mathematical model is determined as

$$\begin{bmatrix} \dot{y}_1 \\ \dot{y}_2 \\ \dot{y}_3 \\ \dot{y}_4 \end{bmatrix} = \begin{bmatrix} 0 & 0 & 1 & 0 \\ 0 & 0 & 0 & 1 \\ -3 & 2 & 0 & 0 \\ 2 & -4 & 0 & 0 \end{bmatrix}\begin{bmatrix} y_1 \\ y_2 \\ y_3 \\ y_4 \end{bmatrix} + \begin{bmatrix} 0 & 0 \\ 0 & 0 \\ 4 & 1 \\ 1 & 3 \end{bmatrix}\begin{bmatrix} f_1 \\ f_2 \end{bmatrix} \tag{a}$$

$$\begin{bmatrix} x_1 \\ x_2 \end{bmatrix} = \begin{bmatrix} 1 & 0 & 0 & 0 \\ 0 & 1 & 0 & 0 \end{bmatrix} \begin{bmatrix} y_1 \\ y_2 \\ y_3 \\ y_4 \end{bmatrix} + \begin{bmatrix} 0 & 0 \\ 0 & 0 \end{bmatrix} \begin{bmatrix} f_1 \\ f_2 \end{bmatrix} \tag{b}$$

Derive the transfer matrix.

Solution

From Equation (a),

$$s\mathbf{I} - \mathbf{A} = \begin{bmatrix} s & 0 & -1 & 0 \\ 0 & s & 0 & -1 \\ 3 & -2 & s & 0 \\ -2 & 4 & 0 & s \end{bmatrix} \tag{c}$$

From Equation (c),

$$(s\mathbf{I} - \mathbf{A})^{-1} = \frac{1}{s^4 + 7s^2 + 8} \begin{bmatrix} s(s^2+4) & 2s & s^2+4 & 2 \\ 2s & s^2+3 & 2 & s^2+3 \\ -(3s^2+8) & 2s^2 & s(s^2+4) & 2s \\ 2s^2 & -4(s^2+2) & 2s & s(s^2+3) \end{bmatrix} \tag{d}$$

Equation (d) is used in Equation (3.36) to give

$$\mathbf{G}(s) = \begin{bmatrix} 1 & 0 & 0 & 0 \\ 0 & 1 & 0 & 0 \end{bmatrix} \left\{ \frac{1}{s^4 + 7s^2 + 8} \begin{bmatrix} s(s^2+4) & 2s & s^2+4 & 2 \\ 2s & s^2+3 & 2 & s^2+3 \\ -(3s^2+8) & 2s^2 & s(s^2+4) & 2s \\ 2s^2 & -4(s^2+2) & 2s & s(s^2+3) \end{bmatrix} \begin{bmatrix} 0 & 0 \\ 0 & 0 \\ 4 & 1 \\ 1 & 3 \end{bmatrix} \right\} + \begin{bmatrix} 0 & 0 \\ 0 & 0 \end{bmatrix}$$

$$= \frac{1}{s^4 + 7s^2 + 8} \begin{bmatrix} 2(s^2+9) & s^2+10 \\ s^2+11 & 3s^2+11 \end{bmatrix} \tag{e}$$

3.9 Further Examples

Example 3.16

The mathematical model for a mechanical system is

$$5\ddot{x} + 20\dot{x} + 150x = 10\dot{z} + 4z \tag{a}$$

The input to the system is $z(t)$ and the output is $x(t)$.

a. Determine the transfer function $G(s) = \frac{X(s)}{Z(s)}$.

b. Determine a state-space formulation.

c. Determine a state-space formulation from the transfer function.

Solution

a. Taking the Laplace transform of Equation (a) using the properties of linearity and transform of derivatives and assuming all initial conditions are zero leads to

$$5s^2 X(s) + 20sX(s) + 150X(s) = 10sZ(s) + 4Z(s) \quad \text{(b)}$$

Rearranging Equation (b) leads to the transfer function

$$G(s) = \frac{X(s)}{Z(s)} = \frac{10s + 4}{5s^2 + 20s + 150} \quad \text{(c)}$$

b. Since a derivative of the input is expressed in the mathematical model, the state variables must be modified from the input and its derivative. The appropriate definitions of the state variables are

$$y_1 = x \quad \text{(d)}$$

$$y_2 = \dot{x} + \mu z \quad \text{(e)}$$

The definitions of the state variables, Equations (d) and (e), are used to give

$$\dot{y}_1 = y_2 - \mu z \quad \text{(f)}$$

Substituting the state variables into the mathematical model, Equation (a), leads to

$$5(\dot{y}_2 - \mu\dot{z}) + 20(y_2 - \mu z) + 150y_1 = 10\dot{z} + 4z \quad \text{(g)}$$

In order to eliminate the derivative of the input from the state-space model,

$$-5\mu = 10 \Rightarrow \mu = -2 \quad \text{(h)}$$

Equation (g) then becomes

$$5\dot{y}_2 + 20(y_2 + 2z) + 150y_1 = 4z \quad \text{(i)}$$

Equation (i) is rearranged to give

$$\dot{y}_2 = -30y_1 - 4y_2 - 7.2z \quad \text{(j)}$$

A state-space formulation of the mathematical model is

$$\begin{bmatrix} \dot{y}_1 \\ \dot{y}_2 \end{bmatrix} = \begin{bmatrix} 0 & 1 \\ -30 & -4 \end{bmatrix} \begin{bmatrix} y_1 \\ y_2 \end{bmatrix} + \begin{bmatrix} 2 \\ -7.2 \end{bmatrix} [z] \quad \text{(k)}$$

Equation (d) is written as

$$[x] = [1 \;\; 0] \begin{bmatrix} y_1 \\ y_2 \end{bmatrix} + [0][z] \quad \text{(l)}$$

From Equations (k) and (l), consistent with the formulation of Equations (3.20) and (3.21), the following matrices are defined:

$$\text{State matrix } \mathbf{A} = \begin{bmatrix} 0 & 1 \\ -30 & -4 \end{bmatrix}$$

$$\text{Input matrix } \mathbf{B} = \begin{bmatrix} 2 \\ -7.2 \end{bmatrix}$$

$$\text{Output matrix } \mathbf{C} = [1 \;\; 0]$$

$$\text{Transmission matrix } \mathbf{D} = [0]$$

c. Rewriting Equation (a) such that the leading coefficient of the denominator is one yields

$$G(s) = \frac{2s + 0.8}{s^2 + 4s + 30} \quad \text{(m)}$$

Using Equation (3.30), the matrix **A** is determined as

$$\mathbf{A} = \begin{bmatrix} 0 & 1 \\ -30 & -4 \end{bmatrix} \quad \text{(n)}$$

and

$$s\mathbf{I} - \mathbf{A} = \begin{bmatrix} s & -1 \\ 30 & s + 4 \end{bmatrix} \quad \text{(o)}$$

Since $y_1 = x$, $|\mathbf{V}_1|$ is set equal to the numerator of the transfer function. Calculating $|\mathbf{V}_1|$,

$$|\mathbf{V}_1| = \begin{bmatrix} b_1 & -1 \\ b_2 & s + 4 \end{bmatrix} = b_1(s + 4) + b_2 = b_1 s + 4b_1 + b_2 \quad \text{(p)}$$

Setting $|\mathbf{V}_1| = 2s + 0.8$ implies that $b_1 = 2$ and $b_2 = -7.2$. The state matrix and the input matrix are the same as those obtained in part b. The output matrix for $x = y_1$ is the output matrix of part b.

Example 3.17

A system has the mathematical model

$$2\frac{di_1}{dt} + 10i_1 - 5i_2 + 20\int_0^t i_1\,dt - 20\int_0^t i_2\,dt = v(t) \quad \text{(a)}$$

$$\frac{di_2}{dt} - 5i_1 + 10i_2 - 20\int_0^t i_1\,dt + 30\int_0^t i_2\,dt = 0 \quad \text{(b)}$$

a. Determine the elements of the transfer function vector $G_1(s) = \frac{I_1(s)}{V(s)}$ and $G_2(s) = \frac{I_2(s)}{V(s)}$

b. Determine a state-space formulation.

c. Determine the transfer function vector from the state-space formulation determined in part b.

Solution

a. Taking the Laplace transform of Equations (a) and (b) using the properties of linearity and transform of derivatives and assuming that all initial conditions are zero leads to

$$2sI_1(s) + 10I_1(s) - 5I_2(s) + \frac{20}{s}I_1(s) - \frac{20}{s}I_2(s) = V(s) \quad \text{(c)}$$

$$sI_2(s) - 5I_1(s) + 10I_2(s) - \frac{20}{s}I_1(s) + \frac{30}{s}I_2(s) = 0 \quad \text{(d)}$$

Multiplying Equations (c) and (d) by s, rearranging, and writing the resulting equations in matrix form leads to

$$\begin{bmatrix} 2s^2 + 10s + 20 & -5s - 20 \\ -5s - 20 & s^2 + 10s + 30 \end{bmatrix} \begin{bmatrix} I_1(s) \\ I_2(s) \end{bmatrix} = \begin{bmatrix} sV(s) \\ 0 \end{bmatrix} \quad \text{(e)}$$

The solution of Equation (e) is

$$\begin{bmatrix} I_1(s) \\ I_2(s) \end{bmatrix} = \frac{1}{2s^4 + 30s^3 + 155s^2 + 300s + 200} \begin{bmatrix} s^2 + 10s + 30 & 5s + 20 \\ 5s + 20 & 2s^2 + 10s + 20 \end{bmatrix} \begin{bmatrix} sV(s) \\ 0 \end{bmatrix}$$

$$= \frac{1}{2s^4 + 30s^3 + 155s^2 + 300s + 200} \begin{bmatrix} s^2 + 10s + 30 \\ 5s + 20 \end{bmatrix} sV(s) \tag{f}$$

Using Equation (f), it is determined that

$$\begin{bmatrix} G_1(s) \\ G_2(s) \end{bmatrix} = \frac{1}{2s^4 + 30s^3 + 155s^2 + 300s + 200} \begin{bmatrix} s^3 + 10s^2 + 30s \\ 5s^2 + 20s \end{bmatrix} \tag{g}$$

where

$$G_1(s) = \frac{I_1(s)}{V(s)} = \frac{s^3 + 10s^2 + 30s}{2s^4 + 30s^3 + 155s^2 + 300s + 200} \tag{h}$$

$$G_2(s) = \frac{I_2(s)}{V(s)} = \frac{5s^2 + 20s}{2s^4 + 30s^3 + 155s^2 + 300s + 200} \tag{i}$$

b. Define as the state variables

$$y_1 = \int_0^t i_1 dt \tag{j}$$

$$y_2 = \int_0^t i_2 dt \tag{k}$$

$$y_3 = i_1 \tag{l}$$

$$y_4 = i_2 \tag{m}$$

From Equations (j)–(m), it is determined that

$$y_3 = \dot{y}_1 \tag{n}$$

$$y_4 = \dot{y}_2 \tag{o}$$

Substituting Equations (j)–(m) into Equations (a) and (b) leads to

$$2\dot{y}_3 + 10y_3 - 5y_4 + 20y_1 - 20y_2 = v(t) \tag{p}$$

$$\dot{y}_4 - 5y_3 + 10y_4 - 20y_1 + 30y_2 = 0 \tag{q}$$

Equations (o) and (p) can be rewritten as

$$\dot{y}_3 = -10y_1 + 10y_2 - 5y_3 + 2.5y_4 + v(t) \tag{r}$$

$$\dot{y}_4 = 20y_1 - 30y_2 + 5y_3 - 10y_4 \tag{s}$$

Equations (n), (o), (s), and (t) are summarized in matrix form as

$$\begin{bmatrix} \dot{y}_1 \\ \dot{y}_2 \\ \dot{y}_3 \\ \dot{y}_4 \end{bmatrix} = \begin{bmatrix} 0 & 0 & 1 & 0 \\ 0 & 0 & 0 & 1 \\ -10 & 10 & -5 & 2.5 \\ 20 & -30 & 5 & -10 \end{bmatrix} \begin{bmatrix} y_1 \\ y_2 \\ y_3 \\ y_4 \end{bmatrix} + \begin{bmatrix} 0 \\ 0 \\ 1 \\ 0 \end{bmatrix} [v(t)] \tag{t}$$

The output vector is related to the state variables and the input by

$$\begin{bmatrix} i_1 \\ i_2 \end{bmatrix} = \begin{bmatrix} 0 & 0 & 1 & 0 \\ 0 & 0 & 0 & 1 \end{bmatrix} \begin{bmatrix} y_1 \\ y_2 \\ y_3 \\ y_4 \end{bmatrix} + \begin{bmatrix} 0 \\ 0 \end{bmatrix} [v(t)] \tag{u}$$

c. A state-space formulation of the mathematical model is given by Equations (s) and (t). Formulating the matrix $s\mathbf{I} - \mathbf{A}$:

$$s\mathbf{I} - \mathbf{A} = \begin{bmatrix} s & 0 & -1 & 0 \\ 0 & s & 0 & -1 \\ 10 & -10 & s+5 & -2.5 \\ -20 & 30 & -5 & s+10 \end{bmatrix} \tag{v}$$

From Equation (u),

$$(s\mathbf{I} - \mathbf{A})^{-1} = \frac{1}{2s^4 + 30s^3 + 155s^2 + 300s + 200}$$

$$\times \begin{bmatrix} 2s^3 + 30s^2 + 135s + 200 & 20s^2 + 50 & 2s^2 + 20s + 60 & 5s + 20 \\ 40s + 100 & 2s^3 + 30s^2 + 95s + 100 & 10s + 40 & 2s^2 + 10s + 20 \\ -(20s^2 + 100s + 200) & 20s^2 + 50s & 2s^3 + 20s^2 + 60s & 5s^2 + 20s \\ 40s^2 + 100s & -(60s^2 + 200s + 200) & 10s^2 + 4s & 2s^3 + 10s^2 + 20s \end{bmatrix} \tag{w}$$

The transfer function is obtained using Equation (3.36):

$$\mathbf{G}(s) = \mathbf{C}(s\mathbf{I} - \mathbf{A})^{-1}\mathbf{B} + \mathbf{D}$$

$$= \frac{1}{2s^4 + 30s^3 + 155s^2 + 300s + 200}$$

$$\times \begin{bmatrix} 0 & 0 & 1 & 0 \\ 0 & 0 & 0 & 1 \end{bmatrix} \begin{bmatrix} 2s^3 + 30s^2 + 135s + 200 & 20s^2 + 50 & 2s^2 + 20s + 60 & 5s + 20 \\ 40s + 100 & 2s^3 + 30s^2 + 95s + 100 & 10s + 40 & 2s^2 + 10s + 20 \\ -(20s^2 + 100s + 200) & 20s^2 + 50s & 2s^3 + 20s^2 + 60s & 5s^2 + 20s \\ 40s^2 + 100s & -(60s^2 + 200s + 200) & 10s^2 + 4s & 2s^3 + 10s^2 + 20s \end{bmatrix} \begin{bmatrix} 0 \\ 0 \\ 1 \\ 0 \end{bmatrix}$$

$$+ \begin{bmatrix} 0 \\ 0 \end{bmatrix} = \frac{1}{2s^4 + 30s^3 + 155s^2 + 300s + 200} \begin{bmatrix} s^3 + 20s^2 + 60s \\ 5s^2 + 20s \end{bmatrix} \tag{x}$$

Note that the result expressed in Equation (x) is the same as the results in Equation (h) and Equation (i).

3.10 Systems with a Time Delay

3.10.1 Transfer Functions

A dynamic system experiences a time delay when its response at time t is explicitly affected by the system's response at a previous time $t - \tau$ for a fixed value of τ. If $x(t)$ is the output of a SISO system, then the mathematical model for $x(t)$ may include a term of the form $x(t - \tau)$. It is necessary to employ a modified version of the second shifting theorem to determine the transfer function for the system. If $x(t) = 0$ for $t < 0$, then

$$\mathscr{L}\{x(t - \tau)\} = e^{-\tau s}X(s) \tag{3.37}$$

Then the transfer function includes a term proportional to $e^{-\tau s}$ in the denominator. The order of the transfer function is theoretically infinite since the exponential term has an infinite McLaurin series representation in s. In Chapter 7, the McLaurin series is replaced by a Pade approximation. Then the order is the highest power appearing in the denominator after the Pade approximation is applied.

Example 3.18

A dynamic system has the mathematical model

$$\frac{d^2x}{dt^2} + 5\frac{dx}{dt} + 200x(t) - 100x(t-2) = f(t) \tag{a}$$

The input to the system is $f(t)$ and its output is $x(t)$. Determine the system's transfer function.

Solution

Taking the Laplace transform of Equation (a), assuming all the initial conditions are zero, and using linearity of the transform and the transform of derivatives leads to

$$s^2X(s) + 5sX(s) + 200X(s) - 100e^{-2s}X(s) = F(s) \tag{b}$$

Rearranging Equation (b) gives

$$G(s) = \frac{X(s)}{F(s)} = \frac{1}{s^2 + 5s + 200 - 100e^{-2s}} \tag{c}$$

Equation (c) gives the transfer function for the system whose mathematical model is Equation (a). It is not of the form $\frac{N(s)}{D(s)}$, where $D(s)$ is a polynomial.

Example 3.19

A dynamic system has the mathematical model

$$\frac{d^2x}{dt^2} + 3\frac{d}{dt}[x(t-0.1)] + 50x(t-0.4) = f(t) \tag{a}$$

The input to the system is $f(t)$ and its output is $x(t)$. Determine the system's transfer function.

Solution

The Laplace transform of $\frac{d}{dt}[x(t-0.1)]$ is obtained by applying the transform of first derivatives to the transform of $x(t-0.1)$, assuming $x(t) = 0$ for $t < 0$. The result is

$$\mathcal{L}\left\{\frac{d}{dt}[x(t-0.1)]\right\} = se^{-0.1s}X(s) \tag{b}$$

Taking the Laplace transform of Equation (a), assuming all initial conditions are zero, and using linearity of the transform and the transform of derivatives leads to

$$s^2X(s) + 3se^{-0.1s}X(s) + 50e^{-0.4s}X(s) = F(s) \tag{c}$$

Rearranging Equation (b) gives

$$G(s) = \frac{X(s)}{F(s)} = \frac{1}{s^2 + 3se^{-0.1s} + 100e^{-0.4s}} \tag{d}$$

3.10.2 State-Space Models

The state-space model for transfer functions is modified by replacing Equation (3.20) with

$$\dot{\mathbf{y}} = \mathbf{A}\mathbf{y} + \mathbf{L}\mathbf{y}(t-\tau) + \mathbf{B}\mathbf{u} \tag{3.38}$$

where $\mathbf{L}$ is called the delay matrix. It is a matrix of constants that multiplies a vector of delayed state variables.

Example 3.20

Determine a state-space formulation for a system whose mathematical model is

$$\frac{d^2x}{dt^2} + 5\frac{dx}{dt} + 200x(t) - 100x(t-2) = f(t) \tag{a}$$

Solution

Define the state variables as

$$y_1 = x \quad y_2 = \dot{x} \tag{b}$$

From the definition of state variables, Equation (b),

$$\dot{y}_1 = y_2 \tag{c}$$

Using Equation (a) in the mathematical model, Equation (a) leads to

$$\dot{y}_2 = -200y_1 - 5y_2 + 100y_1(t-\tau) + F(t) \tag{d}$$

Summarizing Equations (c) and (d) in matrix form leads to

$$\begin{bmatrix}\dot{y}_1\\ \dot{y}_2\end{bmatrix} = \begin{bmatrix}0 & 1\\ -200 & -5\end{bmatrix}\begin{bmatrix}y_1(t)\\ y_2(t)\end{bmatrix} + \begin{bmatrix}0 & 0\\ 100 & 0\end{bmatrix}\begin{bmatrix}y_1(t-\tau)\\ y_2(t-\tau)\end{bmatrix} + \begin{bmatrix}0\\ 1\end{bmatrix}[F(t)] \tag{e}$$

The output is related to the state variables and the input by

$$[x(t)] = [1 \;\; 0]\begin{bmatrix}y_1(t)\\ y_2(t)\end{bmatrix} + [0][F(t)] \tag{f}$$

Equations (e) and (f) provide a state-space formulation of the mathematical model.

3.11 Use of MATLAB

A transfer function in MATLAB can be defined by its numerator and denominator. Consider the transfer function

$$G(s) = \frac{4s+3}{s^3 + 3s^2 + 10s + 20} \tag{3.39}$$

The numerator and denominator are defined in MATLAB by vectors of their coefficients. For the transfer function in Equation (3.39), the command N=[4 3] defines the numerator and the command D=[1 3 10 20] defines the denominator. These are combined into a transfer function by the command G=tf(N,D). A MATLAB command window is shown in Figure 3.8, which shows these commands and the output in MATLAB.

```
K>> N=[4 3];
K>> D=[1 3 10 20];
K>> G=tf(N,D)

G =

        4 s + 3
  ----------------------
  s^3 + 3 s^2 + 10 s + 20

Continuous-time transfer function.
```

Figure 3.8

MATLAB command window for the specification of the transfer function given in Equation (3.39).

Figure 3.9

MATLAB command window for the specification of the transfer function given in Equation (3.40).

```
K>> N=[1 0 0];
K>> D=[1 0 10 0 100];
K>> G=tf(N,D)

G =

         s^2
  -------------------
  s^4 + 10 s^2 + 100

Continuous-time transfer function.
```

If a vector has n elements, MATLAB assumes for the purposes of defining a transfer function that it represents a polynomial of order $n - 1$. If any power of s is not present in the polynomial, it is represented by a 0 in the vector. For example, consider

$$G(s) = \frac{s^2}{s^4 + 10s^2 + 100} \tag{3.40}$$

The MATLAB commands for defining a transfer function from Equation (3.40) are shown in Figure 3.9.

MATLAB is also used to specify a state-space model for a system. For a system with a state-space model defined by Equations (3.20) and (3.21), enter the matrices **A**, **B**, **C**, and **D** into MATLAB and use the command sys=ss(A,B,C,D). For example, consider the state-space formulation developed in Equations (t) and (u) of Example 3.17, repeated here: tf2ss(G) takes a transfer function and converts it to a state-space model. Remember that the state-space model of a system is not unique. Figure 3.10(a) illustrates the MATLAB formulation of a state-space model given the matrices of Equations (s) and (t) of Example 3.17. Figure 3.10(b) illustrates how MATLAB presents the output of these matrices. Figure 3.10(c) shows the MATLAB output when this state-space model is used to convert to a transfer function. Note that there are two outputs for this model, so the vector **N** has two components. Figure 3.10(d) shows the MATLAB output when the transfer function is used to convert back to a state-space model.

```
A=[0 0 1 0; 0 0 0 1;-10 10 -5 2.5;20 -30 5 -10];
B=[0;0;1;0];
C=[0 0 1 0;0 0 0 1];
D=[0;0];
sys=ss(A,B,C,D)
```

(a)

```
sys =

  A =
         x1    x2    x3    x4
   x1     0     0     1     0
   x2     0     0     0     1
   x3   -10    10    -5   2.5
   x4    20   -30     5   -10
```

Figure 3.10

(a) MATLAB specification of the state-space model of Example 3.17 from the state matrix, the input matrix, the output matrix, and the transmission matrix. (b) MATLAB output once the state-space model has been specified. (c) The MATLAB command ss2tf is used to convert the state-space model to the transfer function. (d) The MATLAB command ss2tf is used to convert the transfer function back into a state-space model.

```
B =
       u1
  x1   0
  x2   0
  x3   1
  x4   0

C =
       x1  x2  x3  x4
  y1   0   0   1   0
  y2   0   0   0   1

D =
       u1
  y1   0
  y2   0

Continuous-time state-space model.
```

(b)

```
K>> [N,D]=ss2tf(A,B,C,D)

N =

         0    1.0000   10.0000   30.0000         0
         0         0    5.0000   20.0000    0.0000

D =

    1.0000   15.0000   77.5000  150.0000  100.0000
```

(c)

```
[A,B,C,D]=tf2ss(N,D)

A =

  -15.0000  -77.5000 -150.0000 -100.0000
    1.0000         0         0         0
         0    1.0000         0         0
         0         0    1.0000         0

B =

     1
     0
     0
     0

C =

    1.0000   10.0000   30.0000         0
         0    5.0000   20.0000    0.0000

D =

     0
     0
```

(d)

Figure 3.10

(Continued)

3.12 Summary

3.12.1 Chapter Highlights

- A transfer function for a linear system is the ratio of the Laplace transform of the system output to the Laplace transform of the system input.
- The transfer function for a system is unique.
- The transfer function is independent of the output; it is the same no matter what the output is.
- The transfer function is defined assuming all initial conditions are zero.
- The transfer function contains all information about the system dynamics.
- Properties of the transform such as linearity and transforms of derivatives are used to determine the transfer function.
- The order of a system is the order of the polynomial in the denominator of the transfer function.
- The order of a system is also the number of independent energy storage devices in the system.
- MIMO systems have a transfer function matrix. The transfer functions in the matrix reflect the effect of an output on an input.
- The state of a dynamic system is defined by the smallest set of variables such that, given initial conditions, the time-dependent behavior of the entire system can be determined. State variables are defined as such a set of variables.
- A state-space formulation uses the defined state variables to express the system dynamics.
- The state-space formulation of a system is not unique.
- If a system has the derivative of the input appear in its mathematical formulation, the state variables must be defined to take this into account.
- A state-space formulation for a system is defined in terms of a state matrix, an input matrix, an output matrix, and a transmission matrix.
- It is possible to obtain a state-space formulation for a system from its transfer function.
- It is possible to obtain the transfer function for a system for a state-space formulation.
- The transfer function for time delay systems have an exponential term in the denominator.
- The state-space formulation for a time delay system involves a delay matrix.
- MATLAB may be used to specify a transfer function in terms of the coefficients in the numerator and denominator.
- MATLAB may be used to specify a state-space formulation in terms of the coefficients in the state matrix, the input matrix, the output matrix, and the transmission matrix.
- MATLAB can be used to determine the unique transfer function from its state-space formulation.
- MATLAB can be used to determine a state-space formulation from the transfer fucntion.

3.12.2 Important Equations

- Definition of the transfer function for a system with input $f(t)$ and output $x(t)$

$$G(s) = \frac{X(s)}{F(s)} \tag{3.7}$$

- Convolution property applied to the definition of a transfer function

$$x(t) = \int_0^t g(\tau)f(t-\tau)d\tau \tag{3.10}$$

where

$$g(t) = \mathcal{L}^{-1}\{G(s)\} \tag{3.11}$$

- Transfer function for a MIMO system

$$\mathbf{X}(s) = \mathbf{G}(s)\mathbf{F}(s) \tag{3.16}$$

The transfer function matrix $\mathbf{G}(s)$ is the $n \times m$ matrix defined as

$$\mathbf{G}(s) = \mathbf{Z}^{-1}(s)\mathbf{R}(s) \tag{3.17}$$

- State-space formulation where **u** is the input vector, **x** is the output vector, and **y** is the vector of state variables

$$\dot{\mathbf{y}} = \mathbf{A}\mathbf{y} + \mathbf{B}\mathbf{u} \tag{3.20}$$

$$\mathbf{x} = \mathbf{C}\mathbf{y} + \mathbf{D}\mathbf{u} \tag{3.21}$$

- Solution for the kth element of the state vector

$$y_k(s) = \frac{|\mathbf{V}_k(s)|}{|s\mathbf{I} - \mathbf{A}|}F(s) \tag{3.24}$$

- Denominator of the transfer function when the leading coefficient is one

$$D(s) = s^n + a_{n-1}s^{n-1} + \ldots + a_1 s + a_0 \tag{3.29}$$

- State matrix from the transfer function

$$\mathbf{A} = \begin{bmatrix} 0 & 1 & 0 & 0 & 0 & 0 \\ 0 & 0 & 1 & 0 & 0 & 0 \\ 0 & 0 & 0 & 1 & 0 & 0 \\ \vdots & \vdots & \vdots & \ddots & \vdots & \vdots \\ 0 & 0 & 0 & \cdots & 1 & 0 \\ -a_0 & -a_1 & -a_2 & -a_3 & \cdots & -a_{n-1} \end{bmatrix} \tag{3.30}$$

- Transfer function matrix from the state-space formulation for a MIMO system

$$\mathbf{G}(s) = \mathbf{C}(s\mathbf{I} - \mathbf{A})^{-1}\mathbf{B} + \mathbf{D} \tag{3.36}$$

Short Answer Problems

For Problems SA3.1–SA3.6, determine whether the statement is true or false. If false, revise the statement so that it is true.

SA3.1 The transfer function of a system is unique.

SA3.2 The state-space formulation of a system is unique.

SA3.3 The order of a system whose transfer function is $G(s) = \dfrac{2s^2 + 7s + 12}{s^4 + 7s^3 + 18s^2 + 30s + 76}$ is two.

SA3.4 A system whose transfer function is $G(s) = \dfrac{2s^2 + 7s + 12}{s^4 + 7s^3 + 18s^2 + 30s + 76}$ has four independent energy storage devices.

SA3.5 If a transfer function for a SISO system is known, that transfer function can be used to determine a state-space formulation of the mathematical model.

SA3.6 If the system of transfer functions for a system with two inputs and three outputs is given, these transfer functions can be used to determine a unique state-space formulation for the mathematical model.

SA3.7 State in words the definition of the transfer function.

SA3.8 If a system has an input $f(t)$ and an output $x(t)$, what is the mathematical definition of the transfer function for the system $G(s)$?

SA3.9 The property of transform of derivatives is used in the determination of most transfer functions. What initial conditions are used when applying this property?

SA3.10 A system is defined with three inputs and five outputs. What is the size of the matrix of transfer functions?

SA3.11 The transfer function for a system has a numerator that is a polynomial of order three and a denominator that is a polynomial of order five. What is the order of the system?

SA3.12 How many independent energy storage devices are in the system of Figure SA3.12?

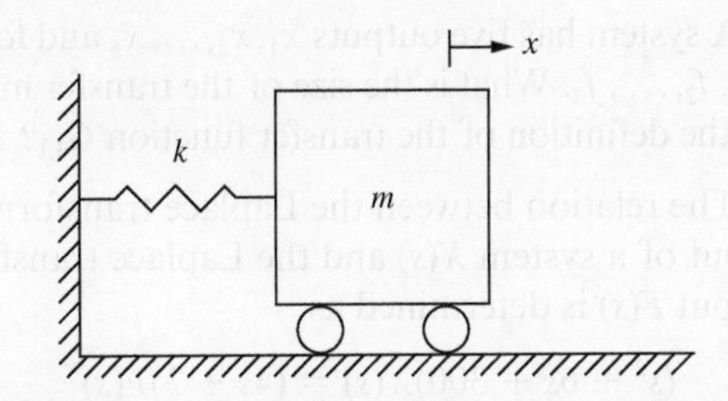

Figure SA3.12

SA3.13 How many independent energy storage devices are in the system of Figure SA3.13?

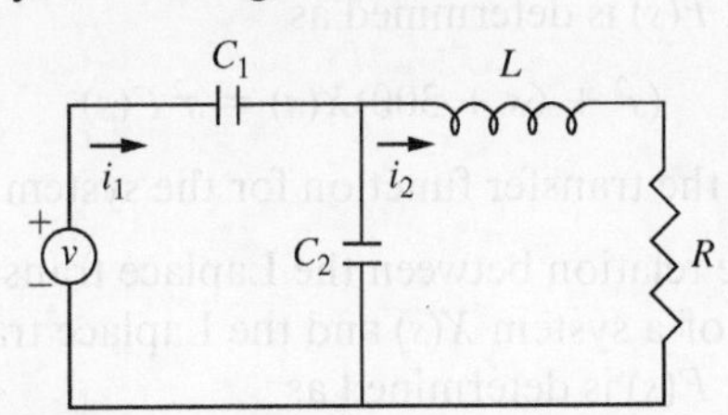

Figure SA3.13

SA3.14 How many independent energy storage devices are in the system of Figure SA3.14?

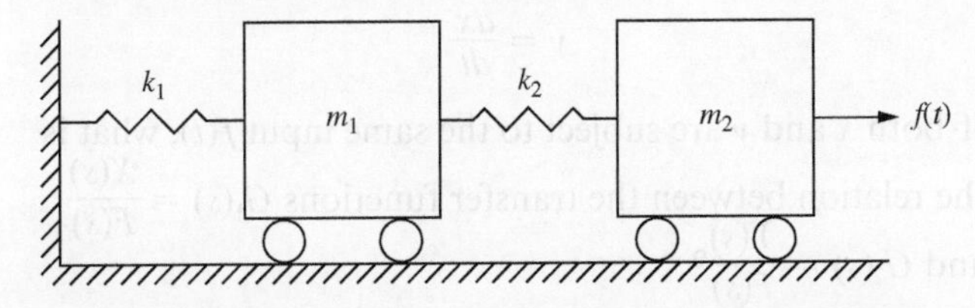

Figure SA3.14

SA3.15 How many independent energy storage devices are in the system of Figure SA3.15?

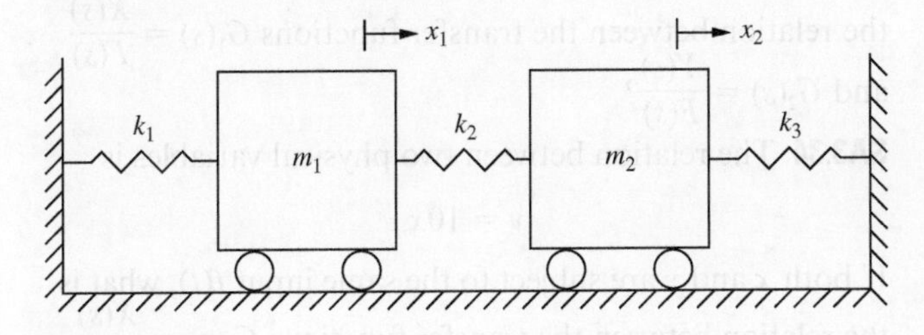

Figure SA3.15

SA3.16 How many independent energy storage devices are in the system of Figure SA3.16?

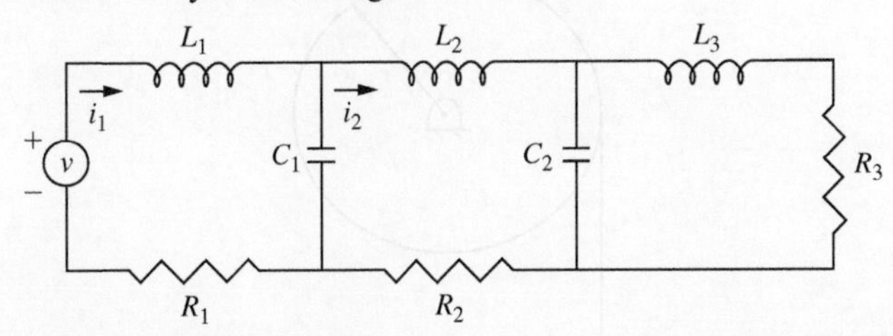

Figure SA3.16

SA3.17 What is the order of the system of Figure SA3.17?

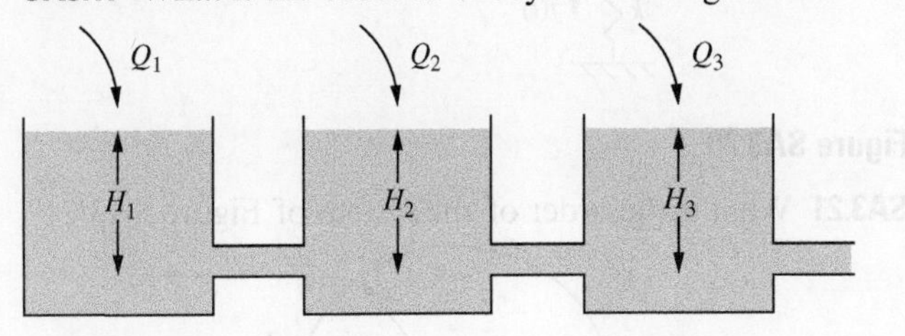

Figure SA3.17

SA3.18 What is the order of the system of Figure SA3.18?

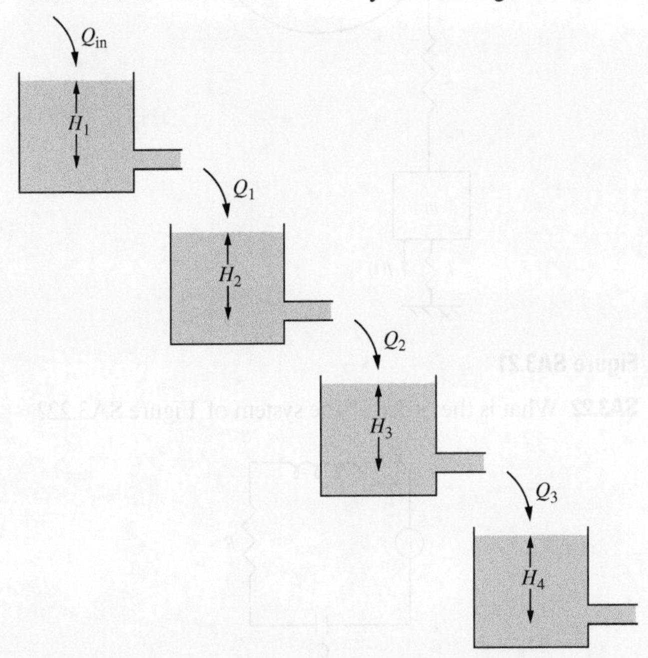

Figure SA3.18

SA3.19 What is the order of the system of Figure SA3.19?

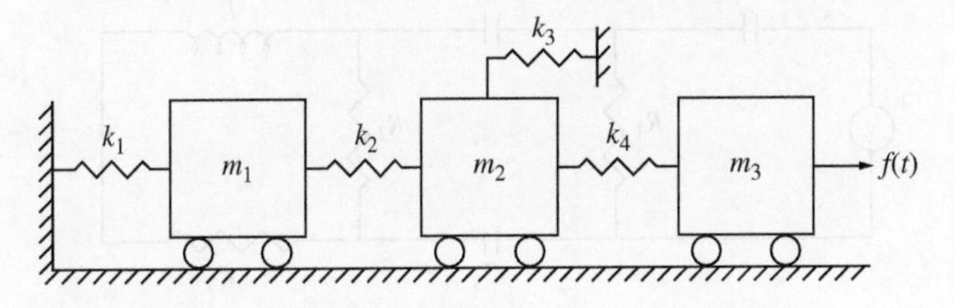

Figure SA3.19

SA3.20 What is the order of the system of Figure SA3.20?

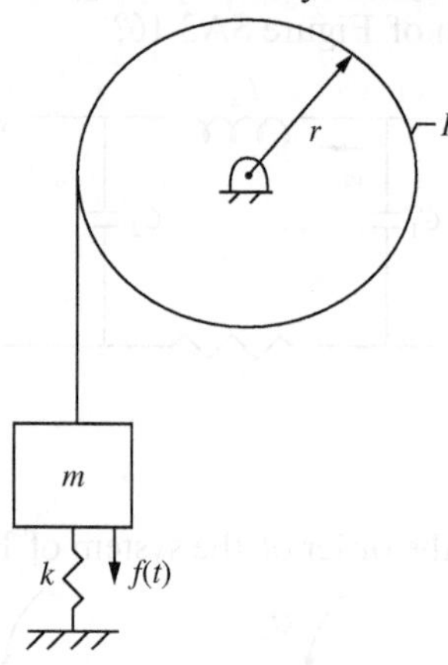

Figure SA3.20

SA3.21 What is the order of the system of Figure SA3.21?

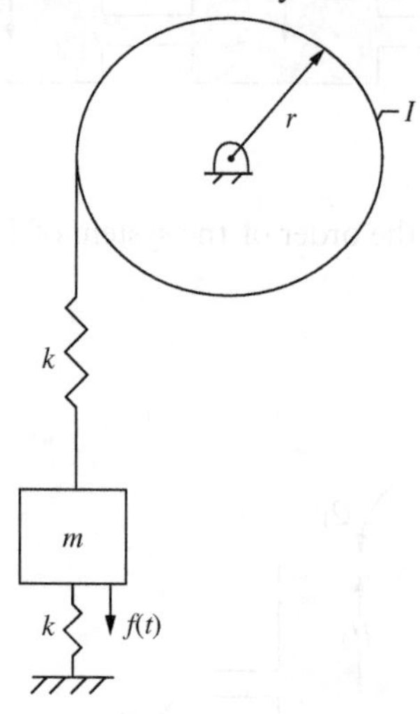

Figure SA3.21

SA3.22 What is the order of the system of Figure SA3.22?

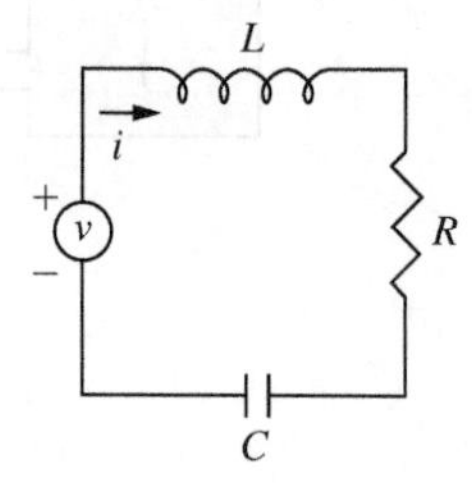

Figure SA3.22

SA3.23 What is the order of the system of Figure SA3.23?

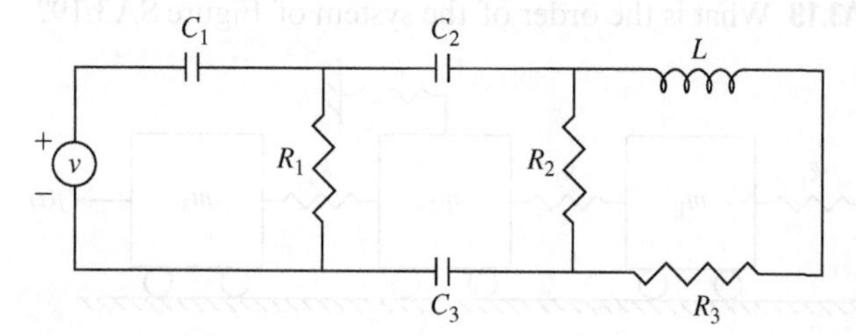

Figure SA3.23

SA3.24 What is the definition of the transfer function matrix?

SA3.25 A system has five outputs $x_1, x_2, \ldots, x_5$ and four inputs $f_1, f_2, \ldots, f_4$. What is the size of the transfer matrix?

SA3.26 A system has five outputs $x_1, x_2, \ldots, x_5$ and four inputs $f_1, f_2, \ldots, f_4$. In the relation $\mathbf{Z}(s)\mathbf{X}(s) = \mathbf{R}(s)\mathbf{F}(s)$, what is the size of the matrix $\mathbf{R}(s)$?

SA3.27 A system has five outputs $x_1, x_2, \ldots, x_5$ and four inputs $f_1, f_2, \ldots, f_4$. What is the size of the transfer matrix? What is the definition of the transfer function $G_{2,2}$?

SA3.28 A system has five outputs $x_1, x_2, \ldots, x_5$ and four inputs $f_1, f_2, \ldots, f_4$. What is the size of the transfer matrix? What is the definition of the transfer function $G_{3,1}$?

SA3.29 The relation between the Laplace transform of the output of a system $X(s)$ and the Laplace transform of its input $F(s)$ is determined as

$$(s^2 + 6s + 300)X(s) = (4s + 3)F(s)$$

Determine the transfer function for the system.

SA3.30 The relation between the Laplace transform of the output of a system $X(s)$ and the Laplace transform of its input $F(s)$ is determined as

$$(s^2 + 6s + 300)X(s) = s^2 F(s)$$

Determine the transfer function for the system.

SA3.31 The relation between the Laplace transform of the output of a system $X(s)$ and the Laplace transform of its input $F(s)$ is determined as

$$(s^2 + 6s + 300)X(s) = \frac{1}{s}F(s)$$

Determine the transfer function for the system.

SA3.32 The relation between two physical variables is

$$y = \frac{dx}{dt}$$

If both x and y are subject to the same input $f(t)$, what is the relation between the transfer functions $G_1(s) = \dfrac{X(s)}{F(s)}$ and $G_2(s) = \dfrac{Y(s)}{F(s)}$?

SA3.33 The relation between two physical variables is

$$y = \int_0^t x(t)\,dt$$

If both x and y are subject to the same input $f(t)$, what is the relation between the transfer functions $G_1(s) = \dfrac{X(s)}{F(s)}$ and $G_2(s) = \dfrac{Y(s)}{F(s)}$?

SA3.34 The relation between two physical variables is

$$y = 10x$$

If both x and y are subject to the same input $f(t)$, what is the relation between the transfer functions $G_1(s) = \dfrac{X(s)}{F(s)}$ and $G_2(s) = \dfrac{Y(s)}{F(s)}$?

SA3.35 The relation between two physical variables is

$$y = 4\frac{dx}{dt} + 2x$$

If both x and y are subject to the same input $f(t)$, what is the relation between the transfer functions $G_1(s) = \frac{X(s)}{F(s)}$ and $G_2(s) = \frac{Y(s)}{F(s)}$?

SA3.36 The transfer function of a system is

$$G(s) = \frac{4s + 3}{s^3 + 6s^2 + 10s + 30}$$

What are the coefficients of the third row of the state matrix?

SA3.37 In converting from the transfer function formulation to a state-space formulation of a third-order system, the output is defined as y_2, the second state variable. How is the output matrix determined?

SA3.38 The MATLAB commands defining the numerator and denominator of a transfer function are

```
num=[1 0 2]
den=[1 5 7 13]
```

What is the transfer function $G(s)$?

SA3.39 The MATLAB commands defining the numerator and denominator of a transfer function are

```
num=[1 0 2]
den=[1 5 7 13]
```

What are the MATLAB commands to determine a state-space formulation for the system?

SA3.40 If the following commands are made in MATLAB to define a transfer function

```
num=[2 10 20]
den=[1 25 140 1000]
```

what is the MATLAB command that defines a state-space formulation for this system?

SA3.41 The transfer function for a system is

$$G(s) = \frac{3s}{(s + 1)(s + 5)}$$

What are the commands that define this transfer function in MATLAB?

SA3.42 The transfer function for a system is defined as

$$G(s) = \frac{3}{(s + 2)(s + 5)}$$

Write the MATLAB commands that define a state-space formulation for this system.

SA3.43 A state-space formulation for a system is

$$\begin{bmatrix} \dot{y}_1 \\ \dot{y}_2 \end{bmatrix} = \begin{bmatrix} 0 & 1 \\ -50 & -3 \end{bmatrix} \begin{bmatrix} y_1(t) \\ y_2(t) \end{bmatrix} + \begin{bmatrix} 0 \\ 1 \end{bmatrix} [F(t)]$$

$$[x(t)] = [1 \;\; 0] \begin{bmatrix} y_1(t) \\ y_2(t) \end{bmatrix} + [0][F(t)]$$

Write the MATLAB commands that are used to define the transfer function for this system.

Problems

3.1 A mechanical system has the mathematical model

$$5\ddot{x} + 12\dot{x} + 100x = f(t)$$

The input is $f(t)$ and the output is $x(t)$. Determine the transfer function defined as $G(s) = \frac{X(s)}{F(s)}$.

3.2 A mechanical system has the mathematical model

$$2\ddot{x} + 7\dot{x} + 50x = 7\dot{y} + 50y$$

The input is $y(t)$ and the output is $x(t)$. Determine the transfer function defined as $G(s) = \frac{X(s)}{Y(s)}$.

3.3 An electrical system has the model

$$10\frac{di}{dt} + 100i + 5000\int_0^t i\,dt = v(t)$$

The input is $v(t)$ and the output is $i(t)$. Determine the transfer function defined as $G(s) = \frac{I(s)}{V(s)}$.

3.4 A single-tank liquid-level problem has the mathematical model

$$6\frac{dh}{dt} + 15h = q$$

The input is $q(t)$ and the output is $h(t)$. Determine the transfer function defined as $G(s) = \frac{H(s)}{Q(s)}$.

3.5 A system has the mathematical model

$$6\dddot{x} + 15\ddot{x} + 25\dot{x} + 70x = f(t)$$

The input is $f(t)$ and the output is $x(t)$. Determine the transfer function defined as $G(s) = \frac{X(s)}{F(s)}$.

3.6 A mechanical system has the mathematical model

$$3\ddot{x}_1 + 10x_1 - 5x_2 = 0$$

$$4\ddot{x}_2 - 5x_1 + 12x_2 = f(t)$$

The input is $f(t)$ and the outputs are $x_1(t)$ and $x_2(t)$. Determine the transfer function defined as $G(s) = \frac{X_2(s)}{F(s)}$.

3.7 A mechanical system has the mathematical model

$$3\ddot{x}_1 + 5\dot{x}_1 - 5\dot{x}_2 + 10x_1 - 5x_2 = 0$$

$$4\ddot{x}_2 - 5\dot{x}_1 + 5\dot{x}_2 - 5x_1 + 12x_2 = f(t)$$

The input is $f(t)$ and the outputs are $x_1(t)$ and $x_2(t)$. Determine the transfer functions defined as $G_1(s) = \frac{X_1(s)}{F(s)}$ and $G_2(s) = \frac{X_2(s)}{F(s)}$.

3.8 An electrical system has the mathematical model

$$2\frac{di_1}{dt} + 10i_1 - 5i_2 + 20\int_0^t i_1\,dt - 20\int_0^t i_2\,dt = v(t)$$

$$\frac{di_2}{dt} - 5i_1 + 10i_2 - 20\int_0^t i_1\,dt + 30\int_0^t i_2\,dt = 0$$

The input is $v(t)$ and the outputs are $i_1(t)$ and $i_2(t)$. Determine the transfer functions defined as $G_1(s) = \frac{I_1(s)}{V(s)}$ and $G_2(s) = \frac{I_2(s)}{V(s)}$.

3.9 A three-tank liquid-level problem has the mathematical model

$$3\frac{dh_1}{dt} + 10h_1 - 4h_2 = q$$

$$4\frac{dh_2}{dt} - 4h_1 + 7h_2 - 3h_3 = 0$$

$$2\frac{dh_3}{dt} - 3h_2 + 10h_3 = 0$$

The input is $q(t)$ and the outputs are $h_1(t)$, $h_2(t)$, and $h_3(t)$. Determine the transfer functions defined as $G_1(s) = \frac{H_1(s)}{Q(s)}$, $G_2(s) = \frac{H_2(s)}{Q(s)}$, and $G_3(s) = \frac{H_3(s)}{Q(s)}$.

3.10 The mathematical model of a biological system is

$$2\dot{p}_1 + 4p_1 - 3p_2 = f(t)$$

$$5\dot{p}_2 - 3p_1 + 6p_2 = 0$$

The input is $f(t)$ and the outputs are $p_1(t)$ and $p_2(t)$. Determine the transfer functions defined as $G_1(s) = \frac{P_1(s)}{F(s)}$ and $G_2(s) = \frac{P_2(s)}{F(s)}$.

3.11 The mathematical model of a biological system is

$$2\dot{p}_1 + \dot{p}_2 + 4p_1 - 3p_2 = f(t)$$

$$\dot{p}_1 + 5\dot{p}_2 - 3p_1 + 6p_2 = 0$$

The input is $f(t)$ and the outputs are $p_1(t)$ and $p_2(t)$. Determine the transfer functions defined as $G_1(s) = \frac{P_1(s)}{F(s)}$ and $G_2(s) = \frac{P_2(s)}{F(s)}$.

3.12 A mechanical system has the mathematical model

$$3\ddot{x}_1 + 5\dot{x}_1 - 5\dot{x}_2 + 10x_1 - 5x_2 = f_1(t)$$

$$4\ddot{x}_2 - 5\dot{x}_1 + 5\dot{x}_2 - 5x_1 + 12x_2 = f_2(t)$$

The inputs are $f_1(t)$ and $f_2(t)$. Determine the transfer function matrix.

3.13 A mechanical system has the mathematical model

$$3\ddot{x}_1 + 2\ddot{x}_2 + 5\dot{x}_1 - 5\dot{x}_2 + 10x_1 - 5x_2 = f_1(t)$$

$$2\ddot{x}_1 + 4\ddot{x}_2 - 5\dot{x}_1 + 5\dot{x}_2 - 5x_1 + 12x_2 = f_2(t)$$

The inputs are $f_1(t)$ and $f_2(t)$. Determine the transfer function matrix.

3.14 An electrical system has the mathematical model

$$2\frac{di_1}{dt} + 10i_1 - 5i_2 + 20\int_0^t i_1\,dt - 20\int_0^t i_2\,dt = v_1(t)$$

$$\frac{di_2}{dt} - 5i_1 + 10i_2 - 20\int_0^t i_1\,dt + 30\int_0^t i_2\,dt = v_2(t)$$

The inputs are $v_1(t)$ and $v_2(t)$. Determine the transfer function matrix.

3.15 A three-tank liquid-level problem has the mathematical model

$$3\frac{dh_1}{dt} + 10h_1 - 4h_2 = q_1(t)$$

$$4\frac{dh_2}{dt} - 4h_1 + 7h_2 - 3h_3 = q_2(t)$$

$$2\frac{dh_3}{dt} - 3h_2 + 10h_3 = q_3(t)$$

The inputs are $q_1(t)$, $q_2(t)$, and $q_3(t)$. Determine the transfer function matrix.

3.16 The mathematical model of a biological system is

$$2\dot{p}_1 + \dot{p}_2 + 4p_1 - 3p_2 = f_1(t)$$

$$\dot{p}_1 + 5\dot{p}_2 - 3p_1 + 6p_2 = f_2(t)$$

The inputs are $f_1(t)$ and $f_2(t)$, and the outputs are $p_1(t)$ and $p_2(t)$. Determine the transfer function matrix.

3.17 A three-tank liquid-level problem has the mathematical model

$$3\frac{dh_1}{dt} + 10h_1 - 4h_2 = q_1(t)$$

$$4\frac{dh_2}{dt} - 4h_1 + 7h_2 - 3h_3 = 0$$

$$2\frac{dh_3}{dt} - 3h_2 + 10h_3 = q_2(t)$$

The inputs are $q_1(t)$ and $q_2(t)$. Determine the transfer function matrix.

3.18 A mechanical system has the mathematical model

$$3\ddot{x}_1 + 5\dot{x}_1 - 5\dot{x}_2 + 10x_1 - 5x_2 = f_1(t)$$
$$4\ddot{x}_2 - 5\dot{x}_1 + 5\dot{x}_2 - 5x_1 + 12x_2 - 3x_3 = f_2(t)$$
$$\ddot{x}_3 + 3\dot{x}_3 - 3x_2 + 3x_3 = 0$$

The inputs are $f_1(t)$ and $f_2(t)$. Determine the transfer function matrix.

3.19 An electrical system has the mathematical model

$$2\frac{di_1}{dt} + 10i_1 - 5i_2 + 20\int_0^t i_1\,dt - 20\int_0^t i_2\,dt = v_1(t)$$

$$\frac{di_2}{dt} - 5i_1 + 10i_2 - 3i_3 - 20\int_0^t i_1\,dt + 20\int_0^t i_2\,dt = 0$$

$$2\frac{di_3}{dt} - 3i_2 + 6i_3 + 20\int_0^t i_3\,dt = v_2(t)$$

The inputs are $v_1(t)$ and $v_2(t)$. Determine the transfer function matrix.

3.20 A mechanical system has the mathematical model

$$3\ddot{x}_1 + 5\dot{x}_1 - 5\dot{x}_2 + 10x_1 - 5x_2 = 3\dot{y}_1 + \dot{y}_2$$

$$4\ddot{x}_2 - 5\dot{x}_1 + 5\dot{x}_2 - 5x_1 + 12x_2 = 5y_1$$

The inputs are $y_1(t)$ and $y_2(t)$. Determine the transfer function matrix.

3.21 A mechanical system has the mathematical model

$$5\ddot{x} + 12\dot{x} + 100x = f(t)$$

Determine a state-space formulation for this system.

3.22 A mechanical system has the mathematical model

$$2\ddot{x} + 7\dot{x} + 50x = 7\dot{y} + 50y$$

Determine a state-space formulation for this system.

3.23 A mechanical system has the mathematical model

$$2\ddot{x} + 7\dot{x} + 50x = 3\dot{y}$$

Determine a state-space formulation for this system.

3.24 A system has the mathematical model

$$\dddot{x} + 4\ddot{x} + 10\dot{x} + 20x = f(t)$$

Determine a state-space formulation for this system.

3.25 A system has the mathematical model

$$\dddot{x} + 4\ddot{x} + 10\dot{x} + 20x = 4\dot{y} + 2y$$

Determine a state-space formulation for this system.

3.26 A system has the mathematical model

$$\dddot{x} + 4\ddot{x} + 10\dot{x} + 20x = \ddot{y} + 4\dot{y} + 2y$$

Determine a state-space formulation for this system.

3.27 An electrical system has the model

$$10\frac{di}{dt} + 100i + 5000\int_0^t i\,dt = v(t)$$

Determine a state-space formulation for this system.

3.28 An electrical system has the model

$$1 \times 10^{-6}\frac{di}{dt} + 4 \times 10^{-4}i + 0.03\int_0^t i\,dt = v(t)$$

Determine a state-space formulation for this system.

3.29 A mechanical system has the mathematical model

$$3\ddot{x}_1 + 5\dot{x}_1 - 5\dot{x}_2 + 10x_1 - 5x_2 = f_1(t)$$
$$4\ddot{x}_2 - 5\dot{x}_1 + 5\dot{x}_2 - 5x_1 + 12x_2 - 3x_3 = 0$$
$$\ddot{x}_3 + 3\dot{x}_3 - 3x_2 + 3x_3 = f_2(t)$$

Determine a state-space formulation for this system.

3.30 A mechanical system has the mathematical model

$$3\ddot{x}_1 + 5\dot{x}_1 - 5\dot{x}_2 + 10x_1 - 5x_2 = 3\dot{y}_1 + \dot{y}_2$$
$$4\ddot{x}_2 - 5\dot{x}_1 + 5\dot{x}_2 - 5x_1 + 12x_2 = 5y_1$$

Determine a state-space formulation for this system.

3.31 A two-tank liquid-level problem has the mathematical model

$$3\frac{dh_1}{dt} + 10h_1 - 4h_2 = q_1(t)$$
$$4\frac{dh_2}{dt} - 4h_1 + 7h_2 = q_2(t)$$

Determine a state-space formulation for this system.

3.32 Use the transfer function of Problem 3.1 to determine a state-space formulation for the system.

3.33 Use the transfer function of Problem 3.2 to determine a state-space formulation for the system.

3.34 Use the transfer function of Problem 3.3 to determine a state-space formulation for the system.

3.35 Use the transfer function of Problem 3.5 to determine a state-space formulation for the system.

3.36 Use the transfer function of Problem 3.6 to determine a state-space formulation for the system when the output is defined as $x_2(t)$.

3.37 Use the transfer function of Problem 3.7 to determine a state-space formulation for the system when the output is defined as $x_1(t)$.

3.38 Use the transfer function of Problem 3.7 to determine a state-space formulation for the system when the output is defined as $x_2(t)$.

3.39 Use the transfer function of Problem 3.8 to determine a state-space formulation for the system when the output is defined as $i_1(t)$.

3.40 Use the transfer function of Problem 3.8 to determine a state-space formulation for the system when the output is defined as $i_2(t)$.

3.41 Use the transfer function of Problem 3.11 to determine a state-space formulation for the system when the output is defined as $p_1(t)$.

3.42 Use the transfer function matrix of Problem 3.14 to determine a state-space formulation for the system.

3.43 Use the transfer function matrix of Problem 3.15 to determine a state-space formulation for the system.

3.44 Use the transfer function matrix of Problem 3.16 to determine a state-space formulation for the system.

3.45 Use the state-space formulation of Problem 3.21 to determine the transfer function for the system.

3.46 Use the state-space formulation of Problem 3.22 to determine the transfer function for the system.

3.47 Use the state-space formulation of Problem 3.23 to determine the transfer function for the system.

3.48 Use the state-space formulation of Problem 3.24 to determine the transfer function for the system.

3.49 Use the state-space formulation of Problem 3.25 to determine the transfer function for the system.

3.50 Use the state-space formulation of Problem 3.26 to determine the transfer function for the system.

3.51 Use the state-space formulation of Problem 3.27 to determine the transfer function for the system.

3.52 Use the state-space formulation of Problem 3.29 to determine the transfer function matrix for the system.

3.53 Use the state-space formulation of Problem 3.30 to determine the transfer function matrix for the system.

3.54 Use the state-space formulation of Problem 3.31 to determine the transfer function matrix for the system.

3.55 A machine tool has the mathematical model

$$10\ddot{x} + 100\dot{x} + 14{,}000x - 2000x(t - 0.3) = f(t)$$

Determine the transfer function for this model.

3.56 A chemical system has the mathematical model

$$4\frac{dc_1}{dt} + 3c_1 - 2c_2 = f(t)$$

$$4\frac{dc_2}{dt} - c_1 - c_1(t - 0.7) + 2c_2 = 0$$

Determine the transfer functions $G_1(s) = \frac{C_1(s)}{F(s)}$ and $G_2(s) = \frac{C_2(s)}{F(s)}$.

3.57 (a) Use MATLAB to define the transfer function of Example 3.16. (b) Given the transfer function of Example 3.16, use MATLAB to define a state-space formulation.

3.58 A state-space formulation is given in Equations (k) and (l) of Example 3.16. (a) Use MATLAB to define the state-space formulation. (b) Given the state-space formulation of Example 3.16, use MATLAB to determine the transfer function for the system.

4

Mechanical Systems

Learning Objectives

At the conclusion of this chapter, you should be able to:

- Determine the inertia properties of a rigid body undergoing planar motion
- Calculate the moments of inertia about a noncentroidal axis using the Parallel Axis Theorem
- Use the relative velocity and acceleration equations to calculate the velocity and acceleration for any particle
- Determine the kinematic relations of a rigid body undergoing planar motion
- Develop the force-displacement relationship in a spring
- Determine the equivalent stiffness of a combination of springs
- Apply the concept of viscous damping
- Draw a free-body diagram of a mechanical system at an arbitrary instant that illustrates the body forces (gravity) and the surface forces acting on the body
- Apply Newton's laws using a free-body diagram of a particle or a rigid body
- Derive the mathematical model of a mechanical system by applying Newton's laws to the free-body diagram of the system at an arbitrary instant
- Derive mathematical models for multidegree-of-freedom systems
- Use Lagrange's equations (the energy method) to derive mathematical models
- Calculate the transfer function from the mathematical model of a mechanical system
- Develop a state-space formulation for a mechanical system

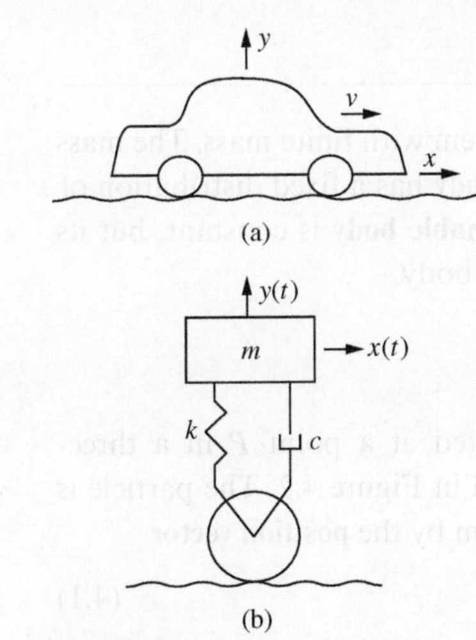

Figure 4.1
(a) The automobile as a complex mechanical system; (b) a simplified model of vehicle motion when it is traveling on a straight bumpy road.

Mechanical system components include inertia elements (elements with mass and kinetic energy), stiffness elements (flexible elements that store energy), and damping elements (elements that dissipate energy). External sources of force or motion provide input to the system. Dependent variables used in the modeling of mechanical systems are displacements of particles within the system or angular displacements. Application of conservation laws leads to differential equations whose solutions provide time-dependent descriptions of the motion of the system's components.

The automobile in Figure 4.1(a) is a familiar mechanical system. To simplify the modeling of the system, assume that when traveling on a straight bumpy road the body of the vehicle has displacements in two directions. Let x be the displacement, measured from a reference position, of the vehicle in the direction along the road and let y be the displacement, measured from an established equilibrium position, in the direction perpendicular to the average road surface. The vehicle's drive train provides an external force that propels the car along the road. The body of the automobile is assumed to be rigid and acts as the system's inertia element. The vehicle has a suspension system which protects the body from large accelerations caused by vertical changes in road contour. There are a variety of suspension elements, but many are modeled by a spring in parallel with a viscous damper. The flexible elements that store energy are the suspension system and the tires. Energy dissipation is provided in the vertical direction by the suspension system and in the horizontal direction by application of the brakes. A simplified model of the automobile is shown in Figure 4.1(b). The body of the vehicle and the suspension system serve as passive components. The road serves as an active component.

Implicit assumptions used throughout this chapter include

- the Earth is an inertial reference frame;
- particles are moving at speeds such that Newtonian mechanics apply;
- gravity is the only force field present;
- the acceleration due to gravity g is constant and equal to 9.81 m/s^2 or 32.2 ft/s^2;
- force-displacement relations for springs are linear;
- unless otherwise specified, viscous damping is the only dissipative mechanism considered (specifically all forms of aerodynamic drag, friction at pin supports, belt friction, pulley friction, and internal damping are assumed to be negligible);
- all system parameters are constant.

Explicit assumptions will be stated as needed. Assumptions such as the small angle assumption are often required to linearize nonlinear differential equations obtained through the mathematical modeling process.

4.1 Inertia Elements

An inertia element is a component of a mechanical system with finite mass. The mass of a **particle** is concentrated at a single point. A **rigid body** has a fixed distribution of mass about its center of gravity. The mass of a **deformable body** is constant, but its distribution may change relative to a fixed point on the body.

4.1.1 Particles

Consider a particle of mass m, instantaneously located at a point P in a three-dimensional Cartesian coordinate system, as illustrated in Figure 4.2. The particle is located relative to the origin of a fixed coordinate system by the **position vector**

$$\mathbf{r}_P = x\mathbf{i} + y\mathbf{j} + z\mathbf{k} \tag{4.1}$$

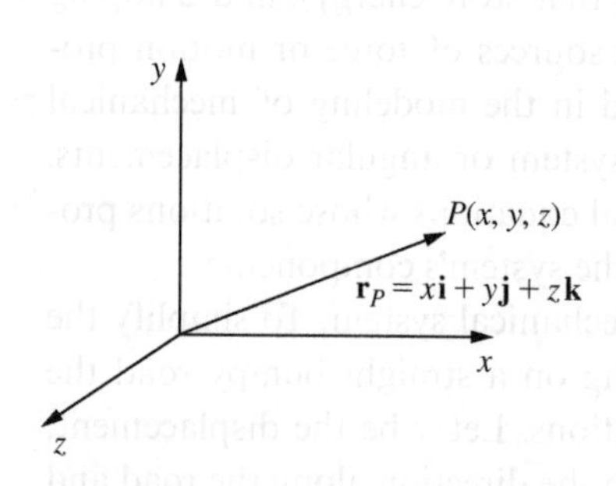

Figure 4.2
Position vector of a particle moving in a Cartesian system.

where x, y, and z are the coordinates of P in the Cartesian system, and **i**, **j**, and **k** are unit vectors parallel to the x, y, and z axes, respectively. As the particle moves, the location of P changes; thus x, y, and z and the position vector **r** are functions of time.

The **velocity vector** is defined as the time rate of change of the position vector

$$\mathbf{v} = \frac{d\mathbf{r}_P}{dt} = \frac{dx}{dt}\mathbf{i} + \frac{dy}{dt}\mathbf{j} + \frac{dz}{dt}\mathbf{k} \tag{4.2}$$

Using the shorthand dot notation described in Section 1.1, **v** is written as

$$\mathbf{v} = \dot{\mathbf{r}}_P = \dot{x}\mathbf{i} + \dot{y}\mathbf{j} + \dot{z}\mathbf{k} \tag{4.3}$$

The **acceleration vector** is defined as the time rate of change of the velocity vector

$$\mathbf{a} = \dot{\mathbf{v}} = \ddot{\mathbf{r}}_p \tag{4.4}$$

$$\mathbf{a} = \ddot{x}\mathbf{i} + \ddot{y}\mathbf{j} + \ddot{z}\mathbf{k} \tag{4.5}$$

The energy associated with the motion of a particle, its **kinetic energy**, is

$$T = \frac{1}{2}m\mathbf{v}\cdot\mathbf{v} \tag{4.6}$$

Using Equation (4.3) in Equation (4.6) leads to

$$T = \frac{1}{2}m(\dot{x}^2 + \dot{y}^2 + \dot{z}^2) \tag{4.7}$$

A particle's **linear momentum** is defined as

$$\mathbf{L} = m\mathbf{v} \tag{4.8}$$

4.1.2 Rigid Bodies

The mass of a rigid body is distributed throughout the body, but the distance between two particles on the body is fixed as the body moves in space. Consider particles A and B on the rigid body of Figure 4.3. Let $\mathbf{r}_A$ and $\mathbf{r}_B$ be the position vectors between the origin and particles A and B, respectively. Let $\mathbf{r}_{B/A}$ be the position vector of particle B relative to particle A. It is a vector between A and B directed toward B. As the body moves, the direction of $\mathbf{r}_{B/A}$ may change, but its magnitude is constant.

Application of the triangle law of vector addition gives

$$\mathbf{r}_B = \mathbf{r}_A + \mathbf{r}_{B/A} \tag{4.9}$$

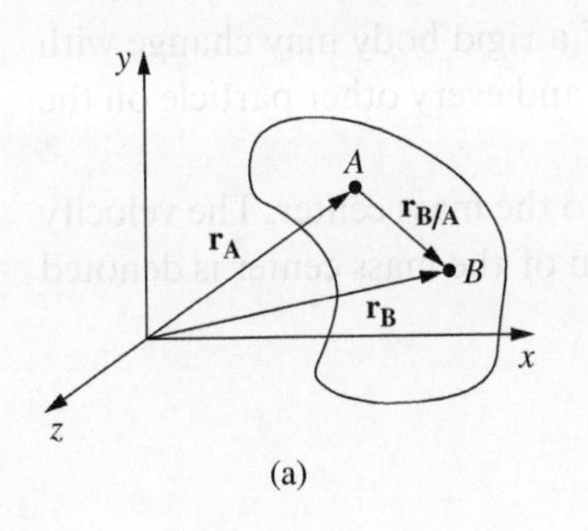

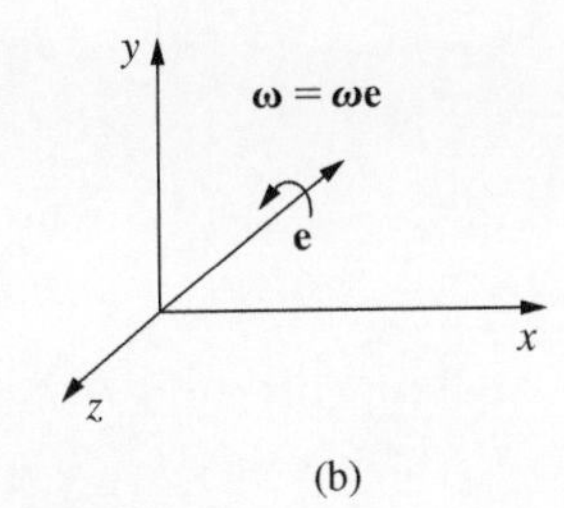

Figure 4.3

(a) As the rigid body moves, $\mathbf{r}_{B/A}$ may change direction, but its magnitude is constant. (b) The axis of rotation is defined by the unit vector **e**. The angular velocity vector is a vector about **e** which is positive if the motion is counterclockwise about **e**.

Differentiation of Equation (4.9) with respect to time leads to

$$\mathbf{v}_B = \mathbf{v}_A + \mathbf{v}_{B/A} \tag{4.10}$$

Rigid body motion is a superposition of translation and rotation. An observer fixed to the body at A only views particle B as rotating about an axis through A.

The **angular velocity** vector, illustrated in Figure 4.3(b), $\boldsymbol{\omega}$ for the body is

$$\boldsymbol{\omega} = \omega\mathbf{e} \tag{4.11}$$

where ω is the rate at which the rigid body is rotating about an axis defined by the unit vector **e**. The angular velocity is positive when the rotation is counterclockwise about the axis of rotation, viewed from the positive axis. The angular velocity vector is a property for the rigid body that may change with time; in general both ω and **e** may change with time. Equation (4.10) is written using the angular velocity vector as

$$\mathbf{v}_B = \mathbf{v}_A + \boldsymbol{\omega} \times \mathbf{r}_{B/A} \tag{4.12}$$

Differentiating Equation (4.12) with respect to time leads to the relative acceleration equation

$$\mathbf{a}_B = \mathbf{a}_A + \boldsymbol{\alpha} \times \mathbf{r}_{B/A} + \boldsymbol{\omega} \times (\boldsymbol{\omega} \times \mathbf{r}_{B/A}) \tag{4.13}$$

where the angular acceleration vector is

$$\boldsymbol{\alpha} = \frac{d\boldsymbol{\omega}}{dt} \tag{4.14}$$

A rigid body undergoes **planar motion** if the path of each point lies within a plane and the axis of rotation is perpendicular to the plane. An observer fixed at

A on a rigid body undergoing planar motion views particle B moving in a circular path whose center is at A. The relative velocity and acceleration between any two particles A and B on a rigid body undergoing planar motion are given by

$$\mathbf{v}_{B/A} = |\mathbf{r}_{B/A}|\omega\mathbf{i}_t \tag{4.15}$$

$$\mathbf{a}_{B/A} = |\mathbf{r}_{B/A}|\alpha\mathbf{i}_t - |\mathbf{r}_{B/A}|\omega^2\mathbf{i}_n \tag{4.16}$$

where $\mathbf{i}_t$ and $\mathbf{i}_n$ are unit vectors tangent and normal to the circular path, respectively. These unit vectors and the components of relative velocity and relative acceleration are illustrated in Figure 4.4.

The **center of mass** of a rigid body is defined as the point whose position vector is given by

$$\bar{\mathbf{r}} = \frac{1}{m}\int_m \mathbf{r}dm \tag{4.17}$$

or in component form

$$\bar{x} = \frac{1}{m}\int_m xdm \quad \bar{y} = \frac{1}{m}\int_m ydm \quad \bar{z} = \frac{1}{m}\int_m zdm \tag{4.18}$$

A bar above a quantity indicates that the quantity is evaluated at the mass center of the body. The position vector for the center of mass of a rigid body may change with time; however, the distance between the center of mass and every other particle on the rigid body is constant.

An overbar is used to denote quantities referenced to the mass center. The velocity of the mass center is denoted by $\bar{\mathbf{v}}$ while the acceleration of the mass center is denoted by $\bar{\mathbf{a}}$.

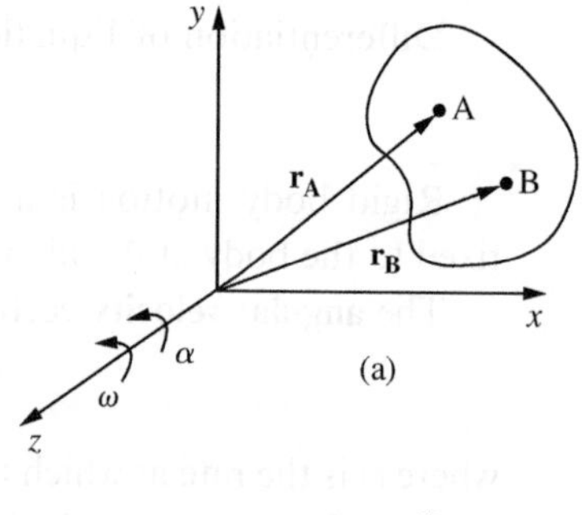

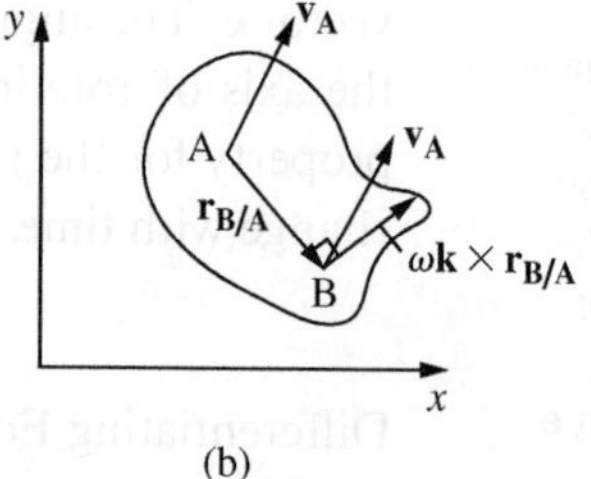

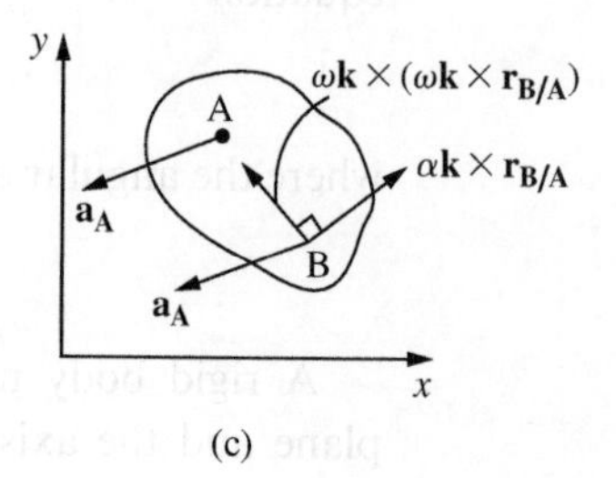

Figure 4.4
(a) The rigid body is undergoing planar motion. Particles A and B travel in the *x-y* plane. The body rotates about the *z* axis with an angular velocity ω and angular acceleration α. (b) Illustration of the relative velocity equation used to calculate the velocity of particle B. (c) Illustration of the relative acceleration equation used to calculate the acceleration of particle B.

The **moment of inertia** of a rigid body about an axis is a measure of how the body's mass is distributed about that axis. The larger the moment of inertia, the more mass is concentrated away from the axis. The definition of moment of inertia about an axis AA through a particle is

$$I_{AA} = \int_m r^2 dm \tag{4.19}$$

where r is the distance from the axis to a particle of mass dm. An **inertia tensor** can be developed at every point in the body

$$\mathbf{I} = \begin{pmatrix} I_{xx} & -I_{xy} & -I_{xz} \\ -I_{yx} & I_{yy} & -I_{yz} \\ -I_{zx} & -I_{zy} & I_{zz} \end{pmatrix} \tag{4.20}$$

The term tensor, in this context, is used for a second-order tensor which behaves like a 3×3 matrix. For example, the moment of inertia of a body about the x axis is

$$I_{xx} = \int_m (y^2 + z^2)dm \tag{4.21}$$

The off-diagonal elements of the inertia tensor are called **products of inertia**. For example,

$$I_{xy} = \int_m xydm \tag{4.22}$$

The x, y, and z axes used to define the inertia tensor are mutually perpendicular and form a right-handed coordinate system. The components of the inertia tensor depend on the orientation of these axes. A set of axes for which all products of inertia are zero is called a set of principal axes. An axis of symmetry is a principal axis.

If x and y are any set of orthogonal axes in a plane with an origin at a point O, then the sum of the moments of inertia about these axes is a constant called the **polar moment of inertia** about O,

$$J_O = I_{xx} + I_{yy} \tag{4.23}$$

The polar moment of inertia is a useful quantity for rigid bodies undergoing planar motion or in pure rotation about a fixed axis.

The components of the inertia tensor depend on the location of the particle for which the tensor is calculated. Centroidal moments of inertia and products of inertia are calculated for axes passing through the center of mass and are denoted by an overbar. The **Parallel Axis Theorem** relates moments of inertia calculated about a centroidal axis and a parallel axis passing through another particle. For moments of inertia about the x axis, the Parallel Axis Theorem is stated as

$$I_{xx} = \bar{I}_{xx} + md_x^2 \tag{4.24}$$

where $\bar{I}_{xx}$ is the moment of inertia about an x axis passing through the center of mass of a rigid body of mass m, and I_{xx} is the moment of inertia about an axis parallel to the x axis that passes through a particle such that the distance between the two parallel axes is d_x. Equation (4.24) shows that the moment of inertia is smallest through an axis passing through the center of mass. The corresponding form of the Parallel Axis Theorem for a product of inertia is

$$I_{xy} = \bar{I}_{xy} + m\bar{x}\,\bar{y} \tag{4.25}$$

where $\bar{x}$ and $\bar{y}$ are the x and y coordinates of the x-y axes relative to the centroidal axes. Moments of inertia of several rigid bodies about principal axes though the body's center of mass are given in Table 4.1.

Table 4.1 Moments of Inertia of Three-Dimensional Bodies

Body	General Shape	Centroidal Moments of Inertia
General shape		$\bar{I}_x = \int (y^2 + z^2)dm$ $\bar{I}_y = \int (x^2 + z^2)dm$ $\bar{I}_z = \int (x^2 + y^2)dm$
Slender rod		$\bar{I}_x \approx 0$ $\bar{I}_y = \frac{1}{12}mL^2$ $\bar{I}_z = \frac{1}{12}mL^2$
Thin disk		$\bar{I}_x = \frac{1}{2}mr^2$ $\bar{I}_y = \frac{1}{4}mr^2$ $\bar{I}_z = \frac{1}{4}mr^2$
Thin plate		$\bar{I}_x = \frac{1}{12}m(w^2 + h^2)$ $\bar{I}_y = \frac{1}{12}mw^2$ $\bar{I}_z = \frac{1}{12}mh^2$
Circular cylinder		$\bar{I}_x = \frac{1}{2}mr^2$ $\bar{I}_y = \frac{1}{12}m(3r^2 + L^2)$ $\bar{I}_z = \frac{1}{12}m(3r^2 + L^2)$
Sphere		$\bar{I}_x = \frac{2}{5}mr^2$ $\bar{I}_y = \frac{2}{5}mr^2$ $\bar{I}_z = \frac{2}{5}mr^2$

The mass center of a rigid body in planar motion travels in a plane, say the x-y plane. The body rotates only about the z axis. For bodies undergoing planar motion, only the moment of inertia about the z axis is important in mathematical modeling. Without loss of generality, the subscript zz is dropped, and thus $\bar{I}$ refers to the moment of inertia about the z axis through the center of mass. Then I_O refers to the moment of inertia about an axis parallel to the z axis and passing through O.

Example 4.1

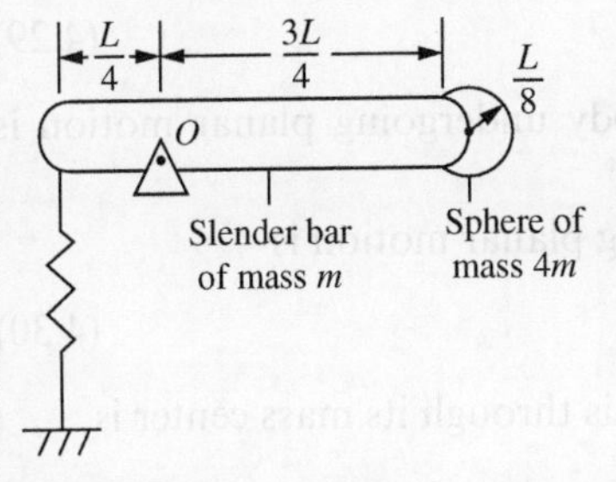

Figure 4.5

The system of Example 4.1.

A slender bar of mass m and length L is pinned as shown in Figure 4.5. A sphere of mass $4m$ and radius $L/8$ is attached to the end of the bar. The bar rotates about an axis through O, at the pin support. Determine the mass moment of inertia of the assembly about the axis of rotation.

Solution

The mass moment of inertia of the slender bar about its centroidal axis is obtained using Table 4.1 as

$$\bar{I}_b = \frac{1}{12}mL^2 \tag{a}$$

Application of the Parallel Axis Theorem between the centroidal axis and the axis of rotation, which is a distance $L/4$ from the centroidal axis, leads to

$$I_{O_b} = \bar{I}_b + m\left(\frac{L}{4}\right)^2$$

$$= \frac{7}{48}mL^2 \tag{b}$$

The mass moment of inertia of the sphere about its centroidal axis is determined using Table 4.1 as

$$\bar{I}_s = \frac{2}{5}(4m)\left(\frac{L}{8}\right)^2$$

$$= \frac{1}{40}mL^2 \tag{c}$$

The Parallel Axis Theorem is used to determine the mass moment of inertia of the sphere about an axis through O:

$$I_{O_s} = \frac{1}{40}mL^2 + (4m)\left(\frac{3}{4}L\right)^2$$

$$= \frac{37}{40}mL^2 \tag{d}$$

The moment of inertia of the assembly about an axis through O is

$$I_O = I_{Ob} + I_{Os}$$

$$= \frac{7}{48}mL^2 + \frac{37}{40}mL^2$$

$$= \frac{257}{240}mL^2 \tag{e}$$

The total kinetic energy of a rigid body is

$$T = \frac{1}{2}\int_m \mathbf{v}\cdot\mathbf{v}dm \tag{4.26}$$

For a rigid body undergoing planar motion, evaluation of Equation (4.26) leads to

$$T = \frac{1}{2}m\bar{v}^2 + \frac{1}{2}\bar{I}\omega^2 \tag{4.27}$$

where $\bar{v}$ is the magnitude of the velocity of the body's center of mass.

If the rigid body is rotating about a fixed axis at O a distance d from the center of mass, the velocity of the mass center is $\bar{v} = d\omega$, which when substituted into Equation (4.27) leads to

$$T = \frac{1}{2}(\bar{I} + md^2)\omega^2 \tag{4.28}$$

Use of the Parallel Axis Theorem, Equation (4.24), in Equation (4.28) leads to

$$T = \frac{1}{2}I_O\omega^2 \tag{4.29}$$

Equation (4.29) is applicable only when a rigid body undergoing planar motion is rotating about a fixed axis.

The linear momentum of a rigid body undergoing planar motion is

$$\mathbf{L} = m\bar{\mathbf{v}} \tag{4.30}$$

The angular momentum of a rigid body about an axis through its mass center is

$$\bar{\mathbf{H}} = \bar{I}\boldsymbol{\omega} \tag{4.31}$$

The angular momentum of a rigid body about an axis O is

$$\mathbf{H_O} = \bar{\mathbf{H}} + \mathbf{r_{O/G}} \times \mathbf{L} \tag{4.32}$$

If O is a fixed axis of rotation,

$$\mathbf{H_O} = I_O\boldsymbol{\omega} \tag{4.33}$$

Example 4.2

At the instant shown, the massless collar of Figure 4.6 has a velocity of 20 m/s to the right. Slender bar AB has mass 2.8 kg, while bar BC has mass 2.1 kg. Determine the kinetic energy of the system at this instant.

Solution

The law of sines is used to determine the angle that bar BC makes with the horizontal at this instant

$$\frac{\sin 30°}{\ell_{BC}} = \frac{\sin\theta}{\ell_{AB}} \tag{a}$$

leading to $\theta = 42.1°$. The relative velocity equation, Equation (4.12), is used on bar AB between A and B:

$$\mathbf{v_B} = \mathbf{v_A} + \boldsymbol{\omega}_{\mathbf{AB}} \times \mathbf{r_{B/A}}$$

$$\mathbf{v_B} = \mathbf{0} + \omega_{AB}\mathbf{k} \times 0.8(\cos 30°\mathbf{i} + \sin 30°\mathbf{j})$$

$$= -0.4\omega_{AB}\mathbf{i} + 0.693\omega_{AB}\mathbf{j} \tag{b}$$

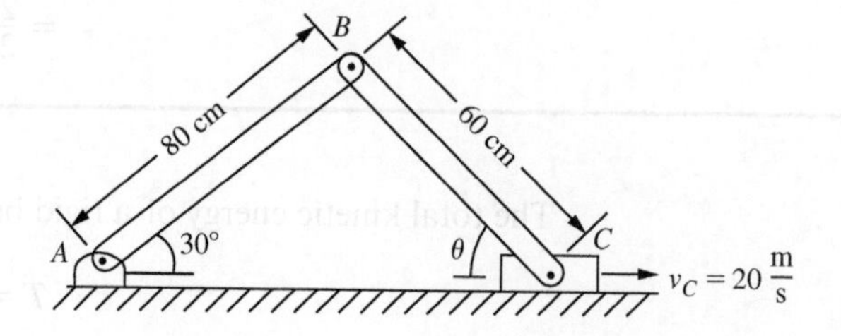

Figure 4.6

System of Example 4.2.

The relative velocity equation applied on bar BC between B and C is

$$\mathbf{v_C} = \mathbf{v_B} + \boldsymbol{\omega}_{\mathbf{BC}} \times \mathbf{r_{C/B}} \tag{c}$$

Substitution of Equation (b) into Equation (c) leads to

$$\begin{aligned}(20\mathbf{i}) &= -0.4\,\omega_{AB}\mathbf{i} + 0.693\,\omega_{AB}\mathbf{j} + \omega_{BC}\mathbf{k} \times 0.6(\cos 42.1°\,\mathbf{i} - \sin 42.1°\,\mathbf{j})\\ (20\mathbf{i}) &= (-0.4\,\omega_{AB} + 0.402\,\omega_{BC})\mathbf{i} + (0.693\,\omega_{AB} + 0.445\,\omega_{BC})\mathbf{j}\end{aligned} \tag{d}$$

The component forms of Equation (d) are

$$-0.4\,\omega_{AB} + 0.402\,\omega_{BC} = 20 \tag{e}$$

$$0.693\,\omega_{AB} + 0.445\,\omega_{BC} = 0 \tag{f}$$

The solution of Equations (e) and (f) is $\omega_{AB} = -19.5$ r/s and $\omega_{BC} = 30.4$ r/s. Since bar AB rotates about A, the velocity of its mass center is

$$\bar{v}_{AB} = \left|\frac{\ell_{AB}}{2}\omega_{AB}\right| \tag{g}$$

which is evaluated, yielding $\bar{v}_{AB} = 7.80$ m/s. The relative velocity equation, Equation (4.12), is used to determine the velocity of the mass center of bar BC:

$$\begin{aligned}\bar{\mathbf{v}}_{\mathbf{BC}} &= \omega_{AB}\mathbf{k} \times \ell_{AB}(\cos 30°\,\mathbf{i} + \sin 30°\,\mathbf{j}) + \omega_{BC}\mathbf{k} \times \frac{\ell_{BC}}{2}(\cos 42.1°\mathbf{i} - \sin 42.1°\mathbf{j})\\ &= (13.9\mathbf{i} - 6.74\mathbf{j})\text{ m/s}\end{aligned} \tag{h}$$

The magnitude of the velocity of the mass center of bar BC is $\bar{v}_{BC} = 15.45$ m/s.

The kinetic energy of the system is obtained using Equation (4.27) as

$$T = \frac{1}{2}m_{AB}\bar{v}_{AB}^2 + \frac{1}{2}\bar{I}_{AB}\omega_{AB}^2 + \frac{1}{2}m_{BC}\bar{v}_{BC}^2 + \frac{1}{2}\bar{I}_{BC}\omega_{BC}^2 \tag{i}$$

Substituting calculated values in Equation (i) gives

$$\begin{aligned}T = \frac{1}{2}\Big[&(2.8\text{ kg})(7.80\text{ m/s})^2 + \frac{1}{12}(2.8\text{ kg})(0.8\text{ m})^2(-19.5\text{ r/s})^2\\ &+ (2.1\text{ kg})(15.5\text{ m/s})^2 + \frac{1}{12}(2.1\text{ kg})(0.6\text{ m})^2(30.4\text{ r/s})^2\Big] = 394.8\text{ N·m}\end{aligned} \tag{j}$$

Since bar AB rotates about a fixed axis through A, Equation (4.29) may be used in lieu of Equation (4.27) to calculate its kinetic energy. However, since the axis of rotation for bar BC is not fixed, Equation (4.29) cannot be used to determine its kinetic energy.

4.1.3 Deformable Bodies

A deformable body changes its original shape when external forces or moments are applied. The elastic bar of Figure 4.7(a) (page 150) has a change in length when a longitudinal force is applied to its free end. The shaft of Figure 4.7(b) twists from its original shape when a torque is applied at its end. The beam of Figure 4.7(c) has a transverse deflection when a vertical load is applied. In each of theses cases the relative positions of particles on the body change when external forces or moments are applied. The distance between particles on a deformable body changes when forces are applied; the change in distance is dependent on the magnitude and direction of the forces and where they are applied.

Assumptions are often made to simplify the analysis of the motion of a deformable body. A common assumption used in the analysis of the bar of Figure 4.7(a) is that all particles in a plane perpendicular to the axis of the bar have the same displacement. Particles in a cross section of the shaft made by a plane perpendicular to the axis of

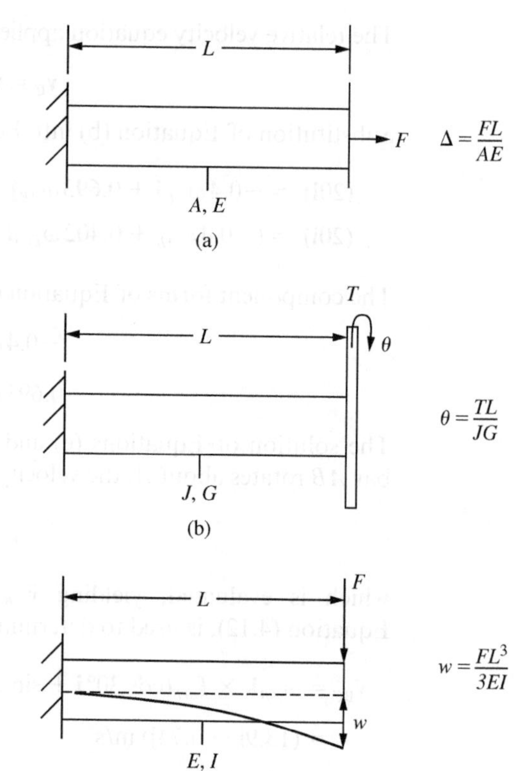

Figure 4.7

Examples of deformable bodies: (a) The elastic bar increases in length and decreases in cross section when an axial load is applied. (b) The elastic shaft twists when a torque is applied. (c) Application of a transverse load to the end of the beam leads to a deflected shape.

the shaft of Figure 4.7(b) are assumed to remain on the same radius as twisting occurs. Planar sections of the beam of Figure 4.7(c) are assumed to remain plane during deformation.

4.1.4 Degrees of Freedom

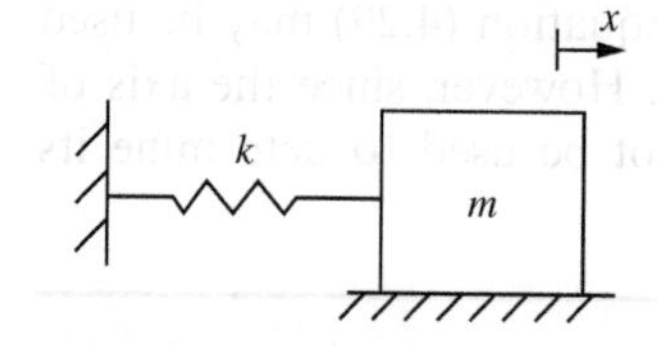

Figure 4.8

Since the block is constrained to move only in the x direction, the system has one degree of freedom.

The particle of Figure 4.2 may move independently in each of the three coordinate directions. However, its motion may be limited by constraints. For example, the particle in the system of Figure 4.8 is constrained to move along the surface and in the horizontal direction. The number of independent directions in which a particle may move is the number of **degrees of freedom** used in the analysis of the particle's motion. An unconstrained particle has three degrees of freedom. The number of degrees of freedom necessary for the analysis of the motion of a particle is

$$n_p = 3 - k \tag{4.34}$$

where k is the number of constraints.

The mass center of a rigid body may move independently in each of the three coordinate directions. The rigid body may also be free to rotate independently about each of the coordinate axes. Thus, a rigid body without any constraints has six degrees of freedom.

The mass center of a rigid body undergoing planar motion is constrained to move in a plane and may rotate only about an axis perpendicular to the plane. Thus a rigid body undergoing planar motion has at most three degrees of freedom. The system of Figure 4.9(a) has three degrees of freedom. The system of Figure 4.9(b) is constrained from side-to-side motion, but the displacement of its mass center is independent of the angular rotation of the bar and the system has two degrees of freedom. The pin support provides an additional constraint to the motion of the system of Figure 4.9(c), which only has one degree of freedom.

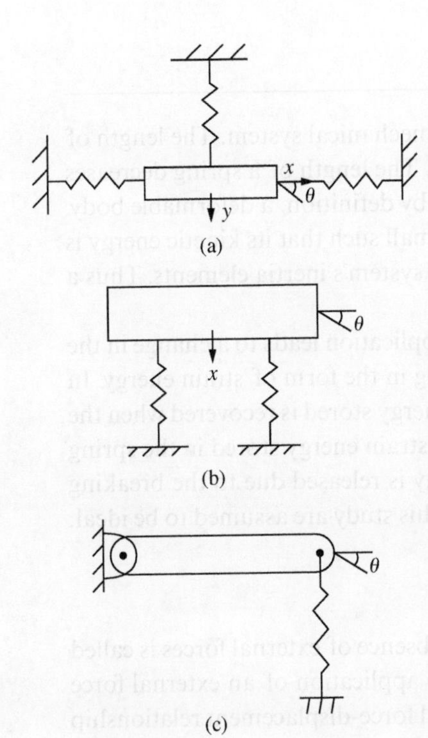

Figure 4.9

Rigid bodies undergoing planar motion with (a) three degrees of freedom, (b) two degrees of freedom, and (c) one degree of freedom.

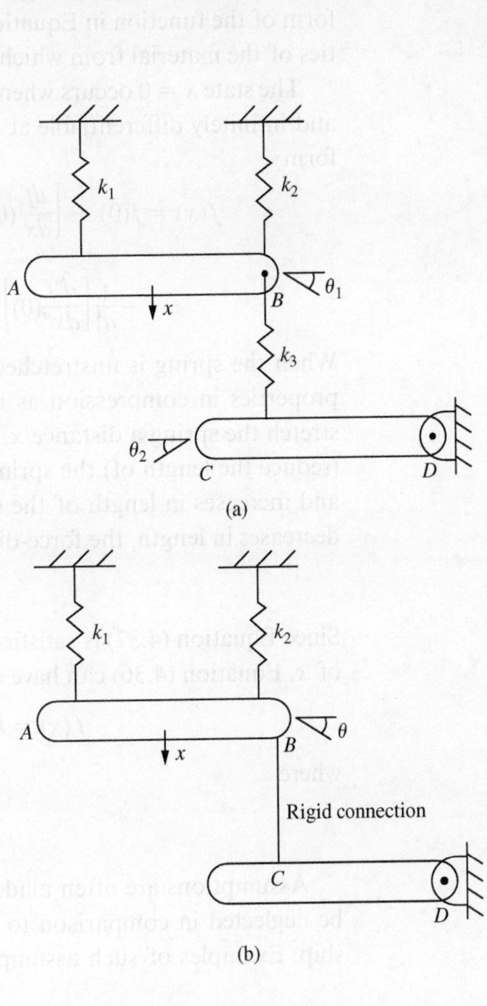

The number of degrees of freedom for a system composed of multiple particles or rigid bodies is the sum of the number of degrees of freedom for each body in the system minus the number of constraints added through the connections between the bodies. The system of Figure 4.10(a) has three degrees of freedom: (1) the vertical displacement of the mass center, bar AB, (2) the angular rotation of bar AB, and (3) the angular rotation of bar CD. However, the system of Figure 4.10(b) has only two degrees of freedom because the rigid connection between bar AB and bar CD adds a constraint: the displacement of particle B must equal the displacement of particle C. A system composed of a finite number of particles and rigid bodies is called a **discrete system**.

An n-degree-of-freedom discrete system requires n kinematically independent coordinates to describe the motion of the system. These coordinates are functions of time and serve as the dependent variables in the system modeling. Whereas the number of degrees of freedom necessary to model the system is a unique number, the choice of the corresponding dependent variables is not unique.

The displacement of any particle along the neutral axis of the beam of Figure 4.7(c) is independent of the displacement of every other particle on the neutral axis. Deformable bodies such as this beam have an infinite number of degrees of freedom. Such a system is called a **continuous system** or a **distributed parameter system**.

Two independent variables are necessary for modeling the motion of the beam of Figure 4.7(c). Let x be a spatial variable that measures the distance along the neutral axis of the beam from its left support when the beam is undeflected. The dependent variable, the transverse displacement, is a function of both x and time t. Thus, $w = w(x, t)$.

Figure 4.10

(a) Three-degree-of-freedom system composed of two rigid bodies; (b) Since the connection between bar AB and bar CD is rigid, the system has two degrees of freedom.

4.2 Springs

A spring is a flexible link between two particles in a mechanical system. The length of a spring increases when it is subject to a tensile force. The length of a spring decreases when it is subject to a compressive force. A spring is, by definition, a deformable body. In this book, the mass of a spring is assumed to be small such that its kinetic energy is negligible in comparison to the kinetic energy of the system's inertia elements. Thus a spring is assumed to be massless.

A force, applied to a spring, does work when its application leads to a change in the length of the spring and energy is stored in the spring in the form of strain energy. In an ideal system, the process is reversible and all the energy stored is recovered when the force is removed. The system is conservative and the strain energy stored in the spring is a form of potential energy. In a real system, energy is released due to the breaking of bonds and the motion of dislocations. Springs in this study are assumed to be ideal.

4.2.1 Force-Displacement Relations

Figure 4.11
Application of a tensile force F increases the length of a spring from its unstretched length.

The length of a spring in an equilibrium state in the absence of external forces is called its **unstretched length**. Since the spring is flexible, the application of an external force results in a change in length of the spring. The general force-displacement relationship for the spring shown schematically in Figure 4.11 is of the form

$$F = f(x) \tag{4.35}$$

where x is the change in length of the spring from its unstretched length. The specific form of the function in Equation (4.35) depends on the spring's geometry and properties of the material from which the spring is fabricated.

The state $x = 0$ occurs when the spring is unstretched. Assuming $f(x)$ is continuous and infinitely differentiable at $x = 0$, then it has a McLaurin series expansion of the form

$$f(x) = f(0) + \left[\frac{df}{dx}(0)\right] x + \frac{1}{2}\left[\frac{d^2f}{dx^2}(0)\right] x^2 + \frac{1}{6}\left[\frac{d^3f}{dx^3}(0)\right] x^3 + \cdots + \frac{1}{n!}\left[\frac{d^nf}{dx^n}(0)\right] x^n + \cdots \tag{4.36}$$

When the spring is unstretched, $F = 0$; thus $f(0) = 0$. Many springs have the same properties in compression as in tension. That is, if a tensile force F is required to stretch the spring a distance x, then a compressive force of F is required to compress (reduce the length of) the spring a distance x. Using positive values for tensile forces and increases in length of the spring and negative values for compressive forces and decreases in length, the force-displacement relation must satisfy

$$F(-x) = -F(x) \tag{4.37}$$

Since Equation (4.37) is satisfied by odd powers of x and not satisfied by even powers of x, Equation (4.36) can have only odd powers of x and reduces to

$$f(x) = k_1 x + k_3 x^3 + k_5 x^5 + \cdots \tag{4.38}$$

where

$$k_n = \frac{1}{n!}\frac{d^nf}{dx^n}(0) \tag{4.39}$$

Assumptions are often made that allow the nonlinear terms in Equation (4.38) to be neglected in comparison to $k_1 x$, thus linearizing the force-displacement relationship. Examples of such assumptions are that the spring is made of a linearly elastic material and stresses are below the yield stress or that the change in length of the spring is small. When linearized, Equation (4.38) is written as

$$F = kx \tag{4.40}$$

where

$$k = \frac{df}{dx}(0) \tag{4.41}$$

is called the spring stiffness which has dimensions of [F/L]. In SI units it is expressed in N/m or N/cm. In English units the spring stiffness is expressed in lb/ft or lb/in.

The value of x used in Equation (4.40) is the change in length of the spring, measured from its unstretched length. Consider the system of Figure 4.12(a). Assume x is positive. Then the spring of stiffness k is stretched and a tensile force exists in the spring. A free-body diagram of the spring is shown in Figure 4.12(b); equal and opposite forces labeled kx are drawn away from the axis of the spring. Now consider the free-body diagram of the block of mass m shown in Figure 4.12(c). The force from the spring on the block is equal to and opposite the force of the block on the spring. Thus, it is acting away from the block and labeled kx. If the force on the spring is compressive, then F is acting against the axis of the spring, as shown in Figure 4.12(d) and x is negative. The force from the spring on the block is acting against the block, as illustrated in Figure 4.12(e). Since x is negative, the force labeled kx on the diagram is actually acting away from the body. For the mass in Figure 4.13(a) (page 154), assuming x is positive to the right as shown, the force from the spring acting on the left of the block is acting away from the body on the free-body diagram of the block, as in Figure 4.13(b). The spring acting on the right of the block is compressed for a positive value of x; thus, its force is acting against the block.

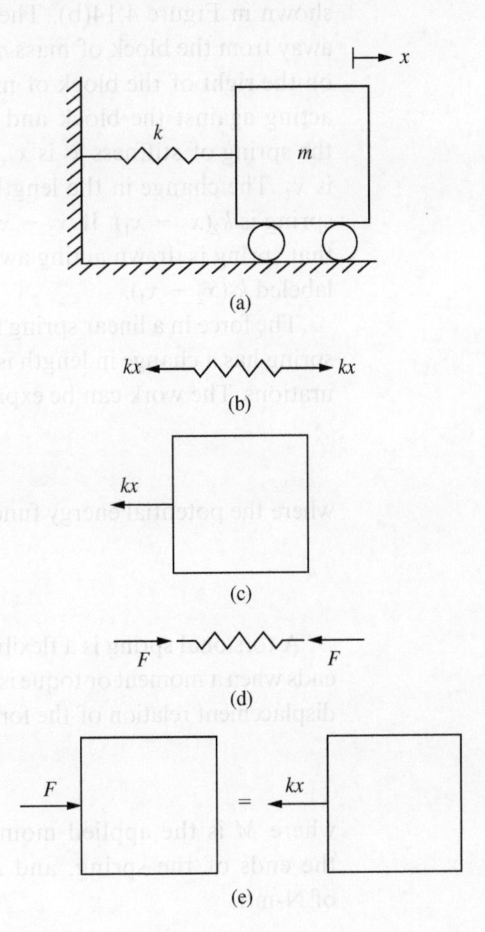

Figure 4.12
(a) Mass-spring system has a displacement x from equilibrium. (b) Free-body diagram of the spring when it has a change in length of x. The spring forces act away from the axis of the spring and in opposite directions from each other. (c) Newton's third law implies the force from the spring on the block is equal and opposite to the force developed in the spring. (d) For a negative value of x, the spring has decreased in length and is compressed. In this case, the spring forces are acting toward each other. (e) Newton's third law implies the force from the spring is acting against the block. However, x is negative and the spring force is labeled kx, acting away from the block.

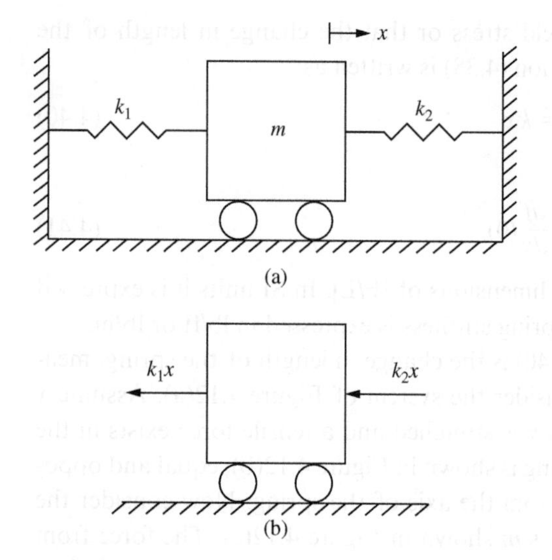

Figure 4.13

(a) System with a mass between two walls with a spring connecting the mass to each wall. The mass experiences a displacement x, measured positive to the right. (b) The free-body diagram of the mass is drawn consistent with the positive direction of x.

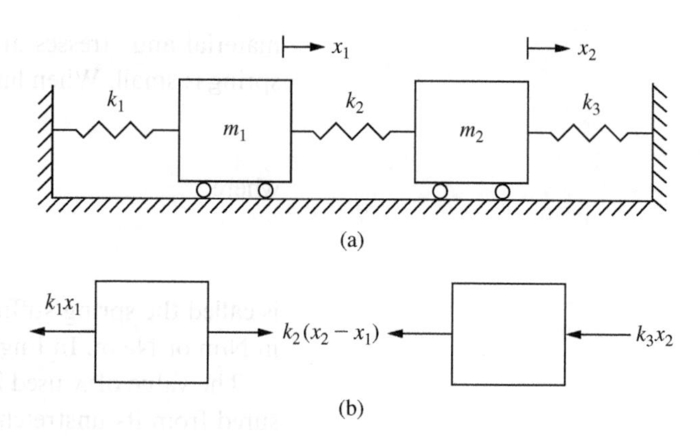

Figure 4.14

(a) System with two masses connected by springs. The displacements of the masses are independent and designated as x_1 and x_2, both positive to the right. (b) Free-body diagrams of the masses show the spring forces are drawn in a direction consistent with positive values of x_1 and x_2.

The result can be summarized as follows. When drawing the force from a spring on a free-body diagram, first assume a positive value for the dependent variable. If the spring is stretched for that positive value, the force from the spring on the body is drawn acting away from the body. If the spring is compressed, the force is drawn acting against the body.

Consider the two-degree-of-freedom system of Figure 4.14(a), where x_1 and x_2 represent the displacements of the blocks. Free-body diagrams of both blocks are shown in Figure 4.14(b). The force from the spring of stiffness k_1 is drawn acting away from the block of mass m_1 and is labeled $k_1 x_1$. The force from the spring acting on the right of the block of mass m_2 is drawn on a free-body diagram of that block acting against the block and is labeled $k_3 x_2$. The displacement of the left end of the spring of stiffness k_2 is x_1, while the displacement of the right end of the spring is x_2. The change in the length of the spring is $x_2 - x_1$. The force developed in the spring is $k_2(x_2 - x_1)$. If $x_2 - x_1$ is positive, the spring is stretched and the force from that spring is drawn acting away from both blocks on their free-body diagrams. It is labeled $k_2(x_2 - x_1)$.

The force in a linear spring is conservative; the work done by the spring force as the spring has a change in length is dependent on only the spring's initial and final configurations. The work can be expressed as a difference in potential energies

$$W_{1\to2} = V_1 - V_2 \tag{4.42}$$

where the potential energy function for a linear spring is

$$V = \frac{1}{2}kx^2 \tag{4.43}$$

A torsional spring is a flexible element that has an angular displacement between its ends when a moment or toque is applied. A linear torsional spring has a moment-angular displacement relation of the form

$$M = k_t\theta \tag{4.44}$$

where M is the applied moment, θ is the relative angular displacement between the ends of the spring, and k_t is the torsional stiffness, which has typical units of N-m/r.

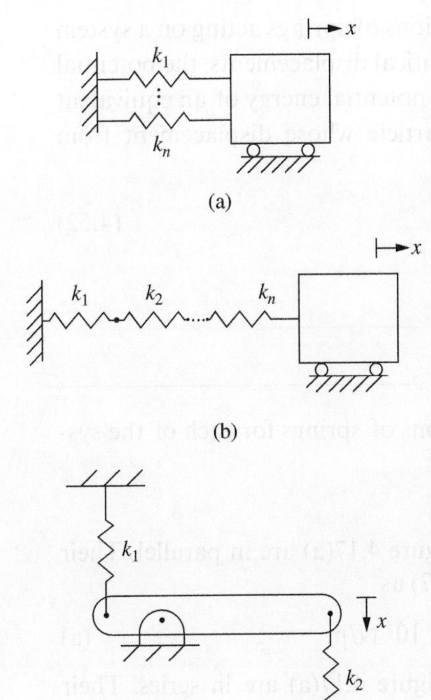

Figure 4.15

Combinations of springs: (a) springs in parallel; (b) springs in series; (c) springs that are neither in parallel nor in series.

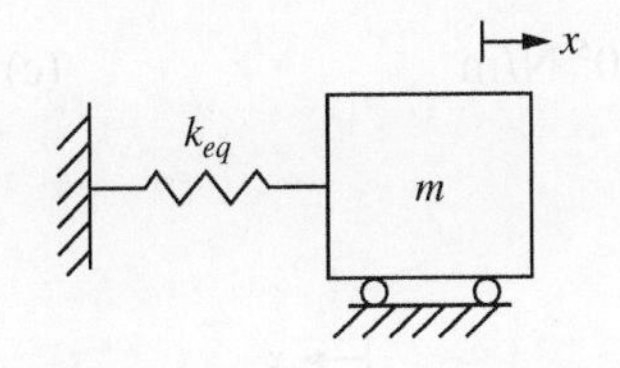

Figure 4.16

Equivalent model system for a system with a combination of springs.

4.2.2 Combinations of Springs

Each of the mechanical systems of Figure 4.15 contains a combination of linear springs. Modeling of these systems is simplified when the combination of springs is replaced by a single spring of an equivalent stiffness, as in Figure 4.16. The model system of Figure 4.16 is equivalent to one of the systems of Figure 4.15 if both systems have the same displacement when they are acted on by identical forces. When the mass in the model system is displaced a distance x from the system's equilibrium position, the force developed in the equivalent spring is

$$F = k_{eq}x \tag{4.45}$$

When the mass in the system of Figure 4.15(a) is displaced from equilibrium, the change in length of each of the springs in the combination is the same. The resultant force acting on the mass from the springs is the sum of the forces in the individual springs. These springs are said to be in **parallel**. A combination of springs is a parallel combination if all the springs have the same displacement and the resultant force from the springs in the combination is the sum of the individual spring forces.

Let x be the displacement from equilibrium of the block in Figure 4.15(a). Since each spring in a parallel combination of n springs has the same displacement, the resultant force acting on the block from this combination is

$$F = \left(\sum_{i=1}^{n} k_i\right)x \tag{4.46}$$

Equating Equation (4.46) with Equation (4.45) leads to the equivalent stiffness of a parallel combination of springs as

$$k_{eq} = \sum_{i=1}^{n} k_i \tag{4.47}$$

When the mass in the system of Figure 4.15(b) is displaced from equilibrium, the same force is developed in each spring in the combination. The change in length of each spring may be different, but the total change in length of the combination of springs, which is the sum of the changes in length of the individual springs, must equal the displacement of the mass. These springs are said to be in **series**. A combination of springs is in series if each spring has the same force and the total change in the length of the combination is the sum of the changes in the lengths of the individual springs.

If x is the displacement of the block of Figure 4.15(b), then the force developed in the ith spring in the combination is

$$F = k_i x_i \tag{4.48}$$

where x_i is the change in length of the ith spring in the series combination. The total change in length of the springs is

$$x = \sum_{i=1}^{n} x_i \tag{4.49}$$

Solving Equation (4.48) for x_i and then substituting into Equation (4.49) leads to

$$F = \left(\frac{1}{\sum_{i=1}^{n} \frac{1}{k_i}}\right)x \tag{4.50}$$

Comparing Equation (4.50) with Equation (4.45) leads to the equivalent stiffness of a series combination of springs as

$$k_{eq} = \frac{1}{\sum_{i=1}^{n} \frac{1}{k_i}} \tag{4.51}$$

Any combination of springs in a linear one-degree-of-freedom system can be replaced by a single spring of equivalent stiffness, even if they do not form a parallel combination or a series combination. Two combinations of springs acting on a system are equivalent if, when the two systems are given identical displacements, the potential energy of each of the combinations is the same. The potential energy of an equivalent spring connected to a mechanical system at a particle whose displacement from equilibrium is x is

$$V = \frac{1}{2}k_{eq}x^2 \tag{4.52}$$

Example 4.3

Determine the equivalent stiffness of the combinations of springs for each of the systems of Figure 4.17.

Solution

a. The springs acting on the left of the block of Figure 4.17(a) are in parallel. Their equivalent stiffness is calculated using Equation (4.47) as

$$k_{eq_a} = 2 \times 10^5 + 4 \times 10^5 = 6 \times 10^5 \text{ N/m} \tag{a}$$

The springs acting on the right of the block of Figure 4.17(a) are in series. Their equivalent stiffness is calculated using Equation (4.51) as

$$k_{eq_b} = \frac{1}{\frac{1}{2 \times 10^5} + \frac{1}{4 \times 10^5}} = 1.33 \times 10^5 \text{ N/m} \tag{b}$$

The results are summarized in Figure 4.18(a). When the block of Figure 4.18(a) is given a displacement x, the spring on the left of the block increases in length a distance x while the spring on the right decreases in length by x. The forces from the springs acting on the block are in the same direction and their resultant is their sum. Thus these springs are in parallel with an equivalent stiffness determined using Equation (4.47) of

$$k_{eq_c} = 6 \times 10^5 + 1.33 \times 10^5 = 7.33 \times 10^5 \text{ N/m} \tag{c}$$

The equivalent system is illustrated in Figure 4.18(b).

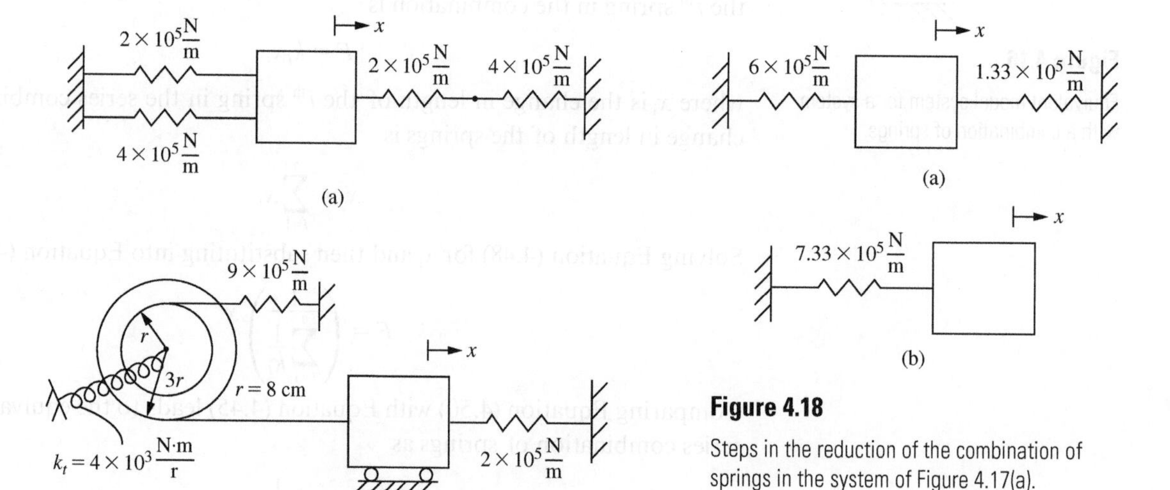

Figure 4.17
Systems for Example 4.3.

Figure 4.18
Steps in the reduction of the combination of springs in the system of Figure 4.17(a).

b. Let x be the displacement of the block of the system of Figure 4.17(b), let θ be the counterclockwise angular rotation of the disk, and let y be the increase in length of the spring connected to the inner radius of the disk. The total potential energy developed in the springs at an arbitrary instant is

$$V = \frac{1}{2}k_1x^2 + \frac{1}{2}k_2y^2 + \frac{1}{2}k_t\theta^2 \tag{d}$$

Assuming no slip between the cables and the disk, the coordinates are related by

$$x = 3r\theta \qquad y = r\theta = x/3 \tag{e}$$

Substitution of Equation (e) into Equation (d) leads to

$$V = \frac{1}{2}k_1x^2 + \frac{1}{2}k_2\left(\frac{x}{3}\right)^2 + \frac{1}{2}k_t\left(\frac{x}{3r}\right)^2$$

$$= \frac{1}{2}\left(k_1 + \frac{k_2}{9} + \frac{k_t}{9r^2}\right)x^2 \tag{f}$$

Comparison of Equation (f) with Equation (4.52) reveals

$$k_{eq} = k_1 + \frac{k_2}{9} + \frac{k_t}{9r^2} \tag{g}$$

Substitution of the given values into Equation (g) gives

$$k_{eq} = 2 \times 10^5\,\text{N/m} + \frac{1}{9}(9 \times 10^5\,\text{N/m}) + \frac{1}{9(0.08\,\text{m})^2}(4 \times 10^3\,\text{N·m/r})$$

$$= 3.69 \times 10^5\,\text{N/m} \tag{h}$$

4.2.3 Static Deflections

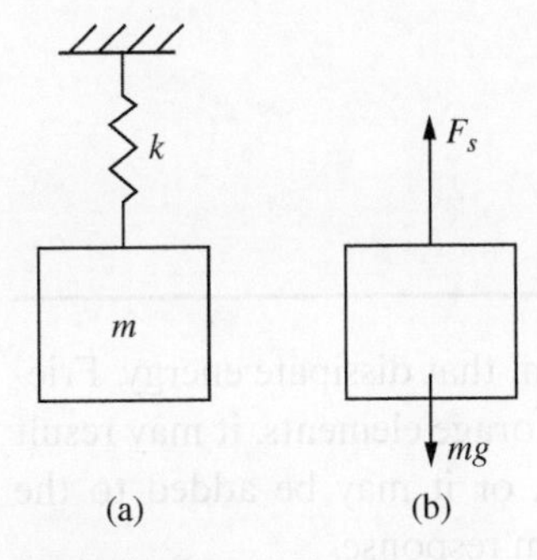

Figure 4.19

(a) When the system is in equilibrium position, a static force is developed in the spring to balance the gravity force of the block; (b) free-body diagram of the equilibrium position.

The system of Figure 4.19(a) is in static equilibrium. Application of the equilibrium condition that $\sum F = 0$ to the free-body diagram of Figure 4.19(b) leads to

$$mg - F_s = 0 \tag{4.53}$$

where F_s is the force developed in the spring to maintain equilibrium. Using the force-displacement relationship for a linear spring, Equation (4.40), the static spring force can be written as

$$F_s = k\Delta_s \tag{4.54}$$

The **static deflection** of the spring Δ_s is obtained by substituting Equation (4.54) into Equation (4.53) and rearranging, leading to

$$\Delta_s = \frac{mg}{k} \tag{4.55}$$

The static deflection of a spring in a system is the change in the length of the spring from its unstretched length when the system is in static equilibrium without the application of external forces. Since external forces are not present, a static spring force is often necessary to balance gravity forces in order to maintain equilibrium. If the dependent variables used in the system modeling are measured from the system's static equilibrium position, then a change in the dependent variable measures the change in the length of the spring from its equilibrium position, not from its unstretched length. If at some instant x is the downward displacement of the mass of Figure 4.19(a), measured from the system's static equilibrium position, then the force developed in the spring at this instant is

$$F_s = k(x + \Delta_s) \tag{4.56}$$

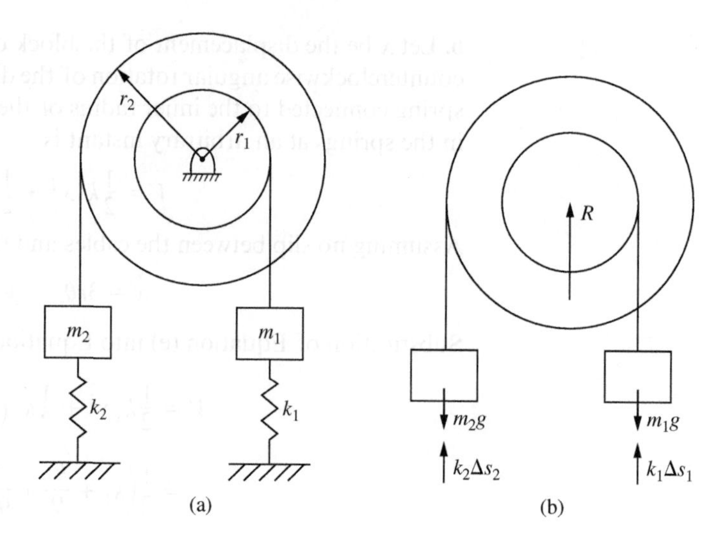

Figure 4.20

(a) Statically indeterminate system in which static deflections cannot be determined only by applying equilibrium conditions; (b) free-body diagram of the system in its equilibrium position.

The potential energy in the spring at this instant is

$$V = \frac{1}{2}k(x + \Delta_s)^2 \tag{4.57}$$

It is not possible to determine the static deflections solely from the equilibrium equations for a system that is statically indeterminate, such as the system of Figure 4.20(a). Application of the equilibrium condition, $\sum M_O = 0$, to the free-body diagram of Figure 4.20(b) leads to

$$m_1gr_1 - k_1\Delta_{s1}r_1 - m_2gr_2 + k_2\Delta_{s2}r_2 = 0 \tag{4.58}$$

A second equation, developed from geometric considerations, is necessary to solve for the static deflections. However, the development of an equilibrium condition like Equation (4.58) is sufficient for purposes of modeling the dynamics and response of a linear system.

4.3 Friction Elements

Friction elements are components of a mechanical system that dissipate energy. Friction may be inherent in the system's inertia and energy storage elements, it may result from contact between the system and its surroundings, or it may be added to the system through a system component to control the system response.

The dissipation of energy from a mechanical system is called **damping**, and devices that dissipate energy are called dampers. Damping mechanisms dissipate energy by converting stored energy into heat, which is then transferred from a mechanical system to its surroundings. Thus, as opposed to inertia and stiffness elements, friction elements do not store energy but dissipate energy.

4.3.1 Viscous Damping

Viscous damping refers to any form of damping in which the damping force is proportional to the velocity of a system component

$$F = cv \tag{4.59}$$

where c is the **damping coefficient**, which has dimensions of $[FTL^{-1}]$. Viscous damping leads to linear terms in differential equations governing the motion of a system.

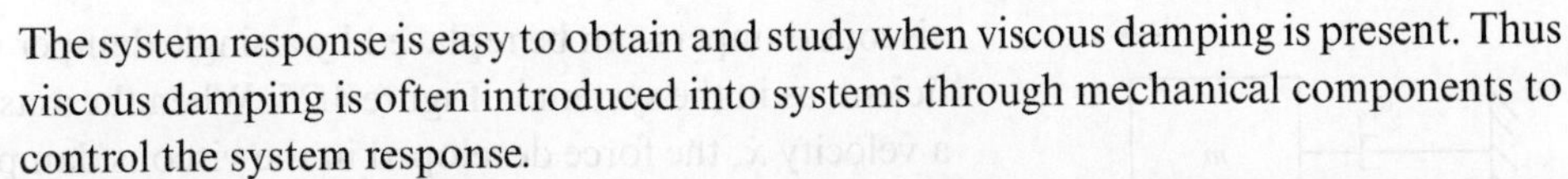

The system response is easy to obtain and study when viscous damping is present. Thus viscous damping is often introduced into systems through mechanical components to control the system response.

A **dashpot** is a mechanical device used to provide viscous damping. Dashpots use viscous fluids to develop forces proportional to the velocity of moving components. A common dashpot configuration, shown in Figure 4.21, is a piston moving in a cylinder of oil. As the piston moves in the cylinder with a velocity v, its motion is resisted by the fluid in the cylinder. If the piston is moving with a velocity v and the cylinder is stationary, then the force acting on the piston is of the form of Equation (4.59). If the cylinder has a velocity v_c, then the resisting force is proportional to the velocity of the piston relative to the cylinder

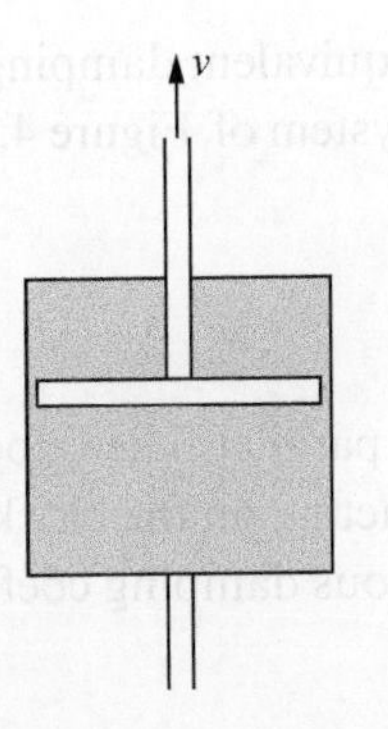

Figure 4.21

A piston-cylinder model of a viscous damper in which the piston moves in a cylinder filled with oil.

$$F = c(v - v_c) \tag{4.60}$$

A **torsional viscous damper** develops a resisting moment that is proportional to the angular velocity, ω, of a rotating component

$$M = c_t\omega \tag{4.61}$$

where c_t is the **torsional viscous damping coefficient**, which has dimensions of $[FTL]$ and typical units of N·m·s/r. The common torsional damper configuration illustrated in Figure 4.22 consists of a thin disk attached to the end of a shaft rotating at an angular speed ω. As the disk rotates in a dish of viscous oil, a shear stress distribution develops on both faces of the disk. The resultant of the shear stress distribution is a resisting moment that satisfies Equation (4.61).

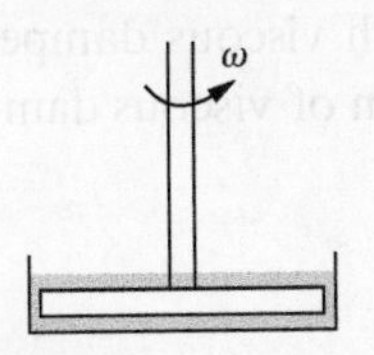

Figure 4.22

Model of a torsional viscous damper in which the disk rotates with angular velocity ω in a dish of viscous liquid.

Viscoelastic materials exhibit behaviors of both elastic materials and viscous fluids. Rubber is a viscoelastic material used in many mechanical systems. It is used, for example, in engine mounts or vibration isolators for both its elastic behavior as well as its damping properties. A common model for rubber, called a Kelvin model, is that of a spring in parallel with a viscous damper, as illustrated in Figure 4.23. The Kelvin model was developed to model the stress-strain behavior of the viscoelastic material, but it can also be used in modeling mechanical systems. The total force developed in a material modeled using a Kelvin model is

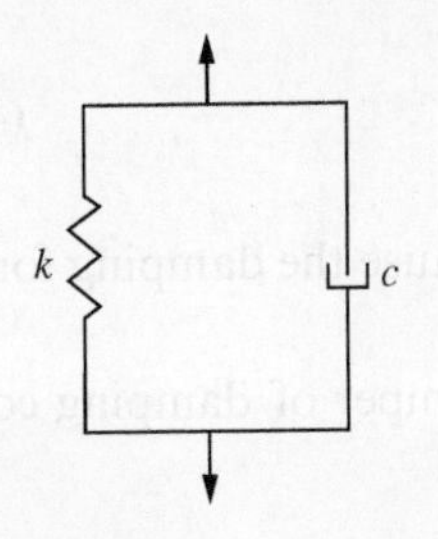

Figure 4.23

The Kelvin model for a viscoelastic material such as rubber consists of a spring in parallel with a viscous damper.

$$F = kx + c\dot{x} \tag{4.62}$$

Viscous dampers can be placed in combinations, as illustrated by the systems in Figure 4.24. For the purpose of modeling a mechanical system, a combination of

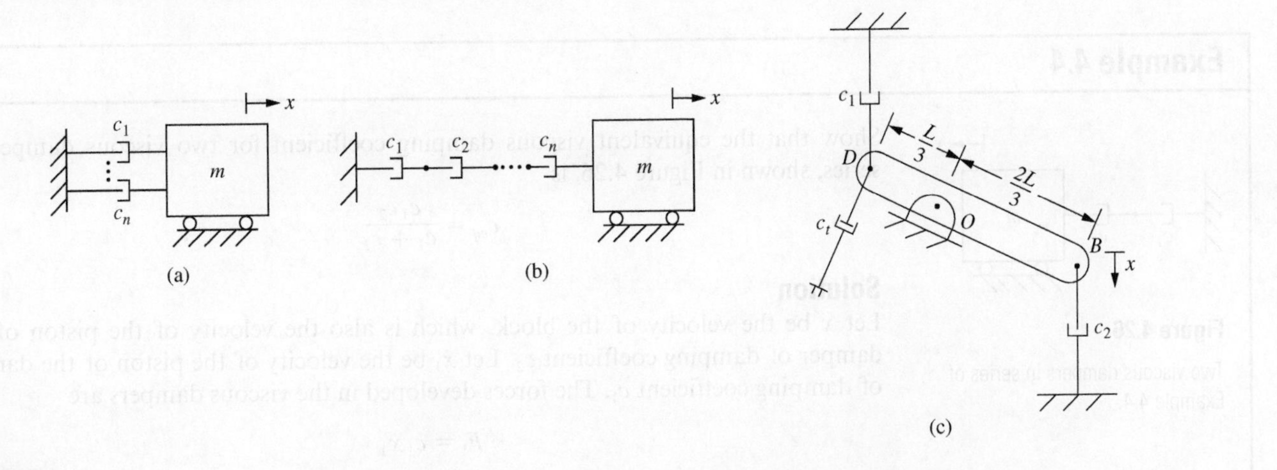

Figure 4.24

Combinations of viscous dampers: (a) dampers in parallel; (b) dampers in series; (c) dampers neither in parallel nor in series.

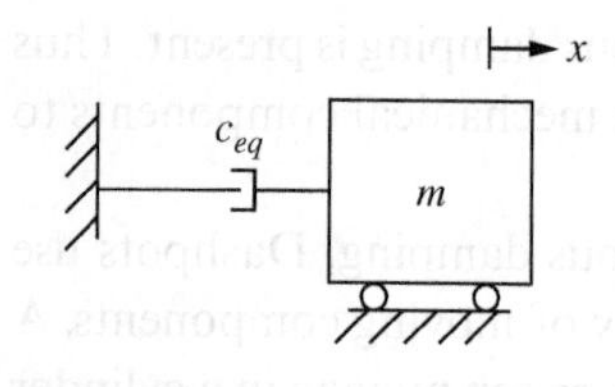

Figure 4.25
A combination of viscous dampers can, for modeling purposes, be replaced by a damper with the equivalent damping coefficient.

viscous dampers can be replaced by a single damper with an equivalent damping coefficient, as in the system in Figure 4.25. When the mass in the system of Figure 4.25 has a velocity $\dot{x}$, the force developed in the viscous damper is

$$F = c_{eq}\dot{x} \tag{4.63}$$

The viscous dampers of the system of Figure 4.24(a) are in parallel. The velocity of the piston in each damper is the same, and the resultant force acting on the block is the sum of the forces in the parallel dampers. The equivalent viscous damping coefficient for parallel viscous dampers is

$$c_{eq} = \sum_{i=1}^{n} c_i \tag{4.64}$$

The viscous dampers of the system of Figure 4.24(b) are in series. The force developed in each viscous damper is the same, and the velocity of the block is the sum of the relative velocities between the piston and cylinder of each viscous damper. The equivalent viscous damping coefficient for a series combination of viscous dampers is

$$c_{eq} = \frac{1}{\sum_{i=1}^{n} \frac{1}{c_i}} \tag{4.65}$$

The energy dissipated by a viscous damper is equal to the work done by the viscous damping force. If the piston of the viscous damper moves between two positions described by x_1 and x_2, then the work done by the viscous damping force in the system of Figure 4.25 is

$$W_{1\to 2} = -\int_{x_1}^{x_2} c_{eq}\dot{x}dx \tag{4.66}$$

The work done by the viscous damping force is negative because the damping force is always in the opposite direction of the motion.

The work done by the moment in a torsional viscous damper of damping coefficient c_t as it rotates between angles θ_1 and θ_2 is

$$W = -\int_{\theta_1}^{\theta_2} c_t\dot{\theta}d\theta \tag{4.67}$$

Example 4.4

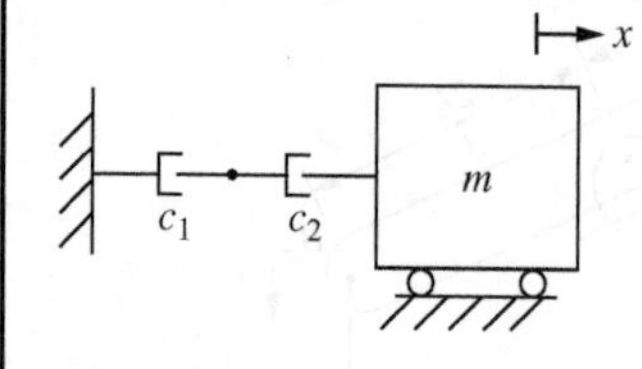

Figure 4.26

Two viscous dampers in series of Example 4.4.

Show that the equivalent viscous damping coefficient for two viscous dampers in series, shown in Figure 4.26, is

$$c_{eq} = \frac{c_1 c_2}{c_1 + c_2} \tag{a}$$

Solution

Let $\dot{x}$ be the velocity of the block, which is also the velocity of the piston of the damper of damping coefficient c_2. Let $\dot{x}_1$ be the velocity of the piston of the damper of damping coefficient c_1. The forces developed in the viscous dampers are

$$F_1 = c_1\dot{x}_1 \tag{b}$$

$$F_2 = c_2(\dot{x}_2 - \dot{x}_1) \tag{c}$$

The forces developed in viscous dampers in series are equal; thus Equations (b) and (c) lead to

$$c_1\dot{x}_1 = c_2(\dot{x}_2 - \dot{x}_1) \tag{d}$$

Equation (d) is rearranged to yield

$$\dot{x}_1 = \frac{c_2}{c_1 + c_2}\dot{x}_2 \tag{e}$$

Substitution of Equation (e) into Equation (b) leads to

$$F = \frac{c_1 c_2}{c_1 + c_2}\dot{x} \tag{f}$$

which, when compared with Equation (4.63), gives the desired result.

Example 4.5

Replace the combination of viscous dampers in the system of Figure 4.24(c) with a viscous damper of an equivalent viscous damping coefficient attached to the system at B.

Solution

Let x_D be the displacement of the left end of the bar, x_B the displacement of the right end of the bar, and θ the clockwise rotation of the bar, all measured from the system's equilibrium position. The work done by the viscous damping forces as the bar rotates between two positions is determined using Equations (4.66) and (4.67) as

$$W_{1\to 2} = -\int_{\theta_1}^{\theta_2} c_t\dot{\theta}d\theta - \int_{x_{D1}}^{x_{D2}} c_1\dot{x}_D dx_D - \int_{x_{B1}}^{x_{B2}} c_2\dot{x}_B dx_B \tag{a}$$

The equivalent viscous damper is to be placed at B; thus it is necessary to relate x_D and θ to x_B. Assuming small displacements, these relations are

$$x_B = -2x_D \tag{b}$$

$$\theta = \frac{3}{2L}x_B \tag{c}$$

Substituting Equations (b) and (c) into Equation (a) leads to

$$W_{1\to 2} = -\int_{x_{B1}}^{x_{B2}} c_t\left(\frac{3}{2L}\dot{x}_B\right)d\left(\frac{3}{2L}x_B\right) - \int_{x_{B1}}^{x_{B2}} c_1\left(-\frac{1}{2}\dot{x}_B\right)d\left(-\frac{1}{2}x_B\right) - \int_{x_{B1}}^{x_{B2}} c_2\dot{x}_B dx_B \tag{d}$$

Equation (d) is simplified to

$$W_{1\to 2} = -\int_{x_{B1}}^{x_{B2}} \left(\frac{9}{4L^2}c_t + \frac{1}{4}c_1 + c_2\right)dx_B \tag{e}$$

Comparison of Equation (e) with Equation (4.66) leads to

$$c_{eq} = \frac{9}{4L^2}c_t + \frac{1}{4}c_1 + c_2 \tag{f}$$

4.3.2 Coulomb Damping

The reaction force between two surfaces has two components: (1) a normal force, which is perpendicular to the plane of contact between the surfaces, and (2) a friction force, which lies within the plane of contact. Coulomb's friction law states that

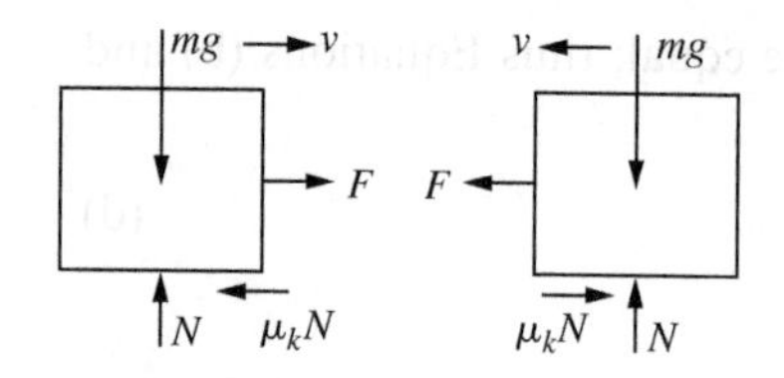

Figure 4.27

The friction force between two surfaces moving relative to one another is proportional to the normal force and opposes the direction of motion.

the magnitude of the friction force is directly proportional to the magnitude of the normal force. When there is no relative motion between the bodies, the constant of proportionality is called the **static coefficient of friction** μ_s. One body will move relative to the other when an externally applied force in the direction of the friction force exceeds the static friction force. When relative motion exists, the constant of proportionality between the normal force and the friction force decreases and is called the **kinetic coefficient of friction** μ_k. The friction force is normal to the plane of contact and opposite to the direction of the motion, as illustrated in Figure 4.27. The force required to impart the relative motion is greater than the force required to maintain the motion as $\mu_s \geq \mu_k$.

Coulomb's friction law for sliding contact between two bodies is

$$F = \mu_k N \tag{4.68}$$

The kinetic coefficient of friction is determined empirically. The sliding between two surfaces results in energy dissipation from the mechanical system in the form of heat transfer to the surroundings. The energy dissipated by friction between two surfaces in contact moving relative to each other is called **Coulomb damping**. The amount of energy dissipated through Coulomb damping is equal to the work done by the friction force

$$\Delta E = W \tag{4.69}$$

where ΔE is the change in energy of the mechanical system and W is the work done by the friction force. Since the friction force always opposes the direction of motion, its work is negative and the amount of energy stored in the system decreases. If the distance traveled by the point of application of the friction force is x, then the work done by the friction force is

$$W = -\mu_k N x \tag{4.70}$$

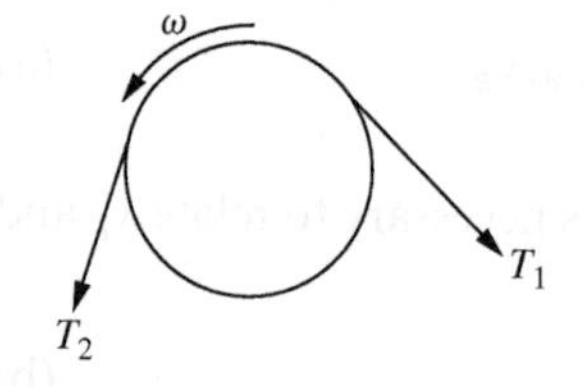

Figure 4.28

Belt friction occurs due to the sliding of a belt relative to a fixed drum. As a result of belt friction, $T_2 > T_1$ when the belt moves in the direction of T_2.

Coulomb damping also occurs when a belt or cable slides over a drum such as a rotor or a pulley, as illustrated in Figure 4.28. In the absence of friction, the tension in both sides of the cable is the same. The resultant of the friction as the belt slides over the drum is a moment about the center of the cable opposite the direction of the motion. The tensions are related by

$$T_2 = T_1 e^{\mu\beta} \tag{4.71}$$

where μ is the coefficient of belt friction and β is the angle of contact between the belt and the drum.

A friction force is developed between a surface and a rolling body, as illustrated in Figure 4.29. If the disk is rolling without sliding, then there is no relative velocity between the surface and its point of contact with the disk. Application of the relative velocity equation, Equation (4.12), between the mass center of the disk and the point of contact leads to

$$\bar{v} = r\omega \tag{4.72}$$

When the disk is rolling without sliding, the friction force developed is less than the maximum available friction force

$$F < \mu_k N \tag{4.73}$$

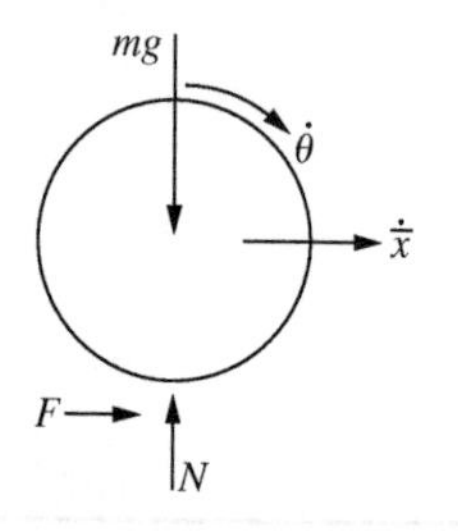

Figure 4.29

Illustration of rolling friction. If the disk rolls without sliding, then $\dot{x} = r\dot{\theta}$ and $F < \mu_k N$. If the disk rolls and slides, then $F = \mu_k N$ and no kinematic relation exists between $\dot{x}$ and θ.

Since there is no relative motion between the two bodies, the friction force does no work and no energy is dissipated by friction.

The disk will roll and slide when there is a relative velocity between the surface and its point of contact with the disk. In this case the friction force is its maximum value

$$F = \mu_k N \tag{4.74}$$

When sliding occurs, Equation (4.72) is not applicable, and kinematics alone cannot be used to determine a relation between the velocity of the center of the disk and the angular velocity of the disk.

Coulomb damping is present in all systems. However, when included in the modeling of mechanical systems, Coulomb damping leads to nonlinear terms in the differential equations governing the motion of the system. These terms result from the friction force having a constant absolute value, but always opposing the direction of motion. In contrast, the force in a viscous damper is proportional to the velocity and leads to linear terms in the governing differential equations.

Since Coulomb damping leads to nonlinear terms in differential equations, it is not included in the modeling of mechanical systems in this book. When viscous damping is added to a system, the effects of Coulomb damping are overshadowed by the effects of viscous damping. However, Coulomb damping is always present and leads to system behavior that is not predicted using viscous damping, such as cessation of motion.

4.3.3 Hysteretic Damping

In modeling a mechanical system with a linear spring, it is usually assumed that the spring's force-displacement relation, Equation (4.40), does not change with time. That is, at any time when a force F is developed in the spring, the spring has a change in length of x from its unstretched length, regardless of the history of the applied force. Equation (4.40) is assumed to apply while a force is being applied to the spring as well as when the spring is being unloaded.

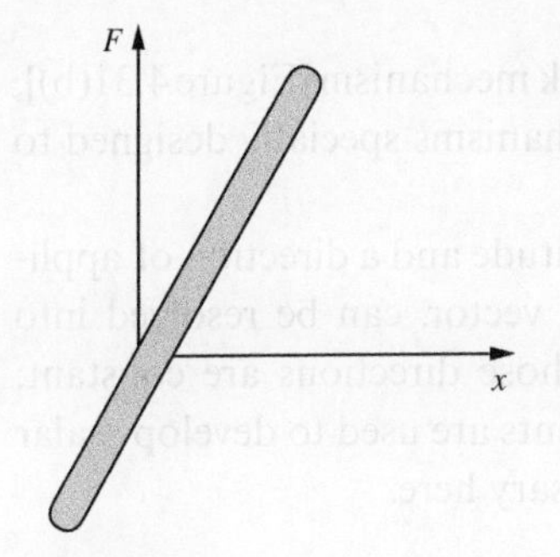

Figure 4.30

Force-displacement curve of a real engineering material during one cycle of completely reversed loading. The area enclosed by the hysteresis loop is the energy dissipated.

Springs are made of engineering materials such as steel, in which imperfections exist. As the spring is loaded, bonds break between molecules, releasing energy. These and other nonconservative effects due to real material behavior lead to the true force-displacement curve of Figure 4.30 for one cycle of completely reversed loading, which illustrates a **hysteresis loop**. The area enclosed by the hysteresis loop is the energy dissipated during the cycle due to nonideal material behavior. This dissipation of energy in a mechanical system is called **hysteretic damping**.

Hysteretic damping is present in all systems with elastic components, but it is quantified only empirically. Differential equations derived assuming hysteretic damping are often nonlinear. In some cases, the behavior of a system with hysteretic damping is similar to the behavior of a system with viscous damping. In these cases hysteretic damping is modeled using an equivalent viscous damping coefficient. The effects of hysteretic damping in systems with viscous damping are often overshadowed by the effects of viscous damping.

In this book it is assumed that all elastic materials are ideal and the effects of hysteretic damping are not included.

4.4 Mechanical System Input

External forces or externally imposed motions provide input for mechanical systems and provide energy to the system. The work done by these forces is nonconservative and is converted into mechanical energy. For example, a torque applied to a shaft with rotors and gears is developed by an electric motor, as discussed in Chapter 5. The operation of reciprocating machines produces forces arising from the rotation of unbalanced components that act as external excitation to the machine. Hydraulic servomotors, discussed in Chapter 6, act as actuators to deliver motion to mechanical

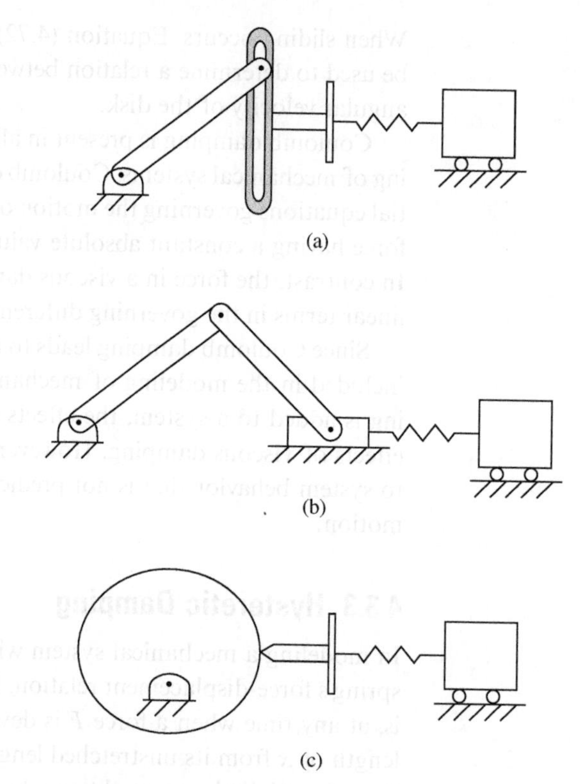

Figure 4.31

Mechanisms that impart a prescribed motion to a system: (a) Scotch yoke; (b) slider-crank; (c) cam and follower.

systems. The Scotch yoke [Figure 4.31(a)], the slider-crank mechanism [Figure 4.31(b)], and cam and follower systems [Figure 4.31(c)] are mechanisms specially designed to deliver specific motions to system components.

A force is a vector quantity in that it requires a magnitude and a direction of application for its complete specification. A force, like any vector, can be resolved into components. All problems in this book have forces whose directions are constant; thus, while the forces are still vectors, the force components are used to develop scalar equations. The full theory involving vectors is not necessary here.

4.4.1 External Forces and Torques

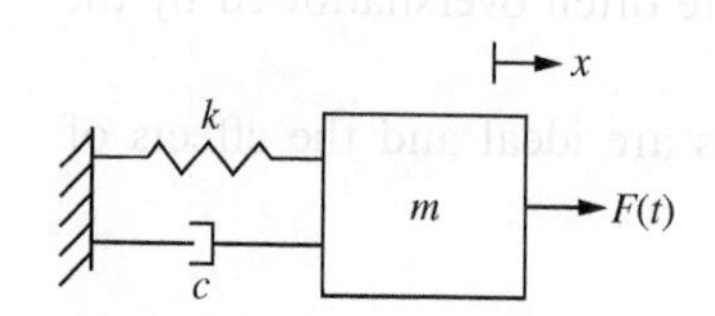

Figure 4.32

A force actuator provides a time-dependent force to a mechanical system.

Figure 4.32 illustrates a mechanical system subject to a time-dependent external force $F(t)$, which may be provided by an external actuator. The work done by the force as its point of application moves between two positions described by the coordinates x_1 and x_2 is

$$W_{1\to 2} = \int_{x_1}^{x_2} F(t)dx \tag{4.75}$$

An alternate expression for the work is obtained by noting that $dx = \dot{x}dt$, which when used in Equation (4.75) leads to

$$W_{1\to 2} = \int_{t_1}^{t_2} F(t)\dot{x}dt \tag{4.76}$$

The work done by the force on the system is converted into kinetic and potential (or stored) energy.

The work done by an applied moment or torque M as the component to which it is applied rotates between two positions described by θ_1 and θ_2 is

$$W_{1\to 2} = \int_{\theta_1}^{\theta_2} M(t)d\theta \tag{4.77}$$

Noting that $d\theta = \dot{\theta}dt$, Equation (4.77) can be rewritten as

$$W_{1\to 2} = \int_{\theta_1}^{\theta_2} M(t)\dot{\theta}dt \tag{4.78}$$

The power delivered to the system by an external force is

$$P = \frac{dW}{dt} \tag{4.79}$$

Use of Equation (4.76) in Equation (4.79) where the upper limit of integration is an arbitrary time leads to

$$P = F(t)\dot{x} \tag{4.80}$$

Using Equation (4.77), the power delivered by an externally applied moment is

$$P = M(t)\dot{\theta} \tag{4.81}$$

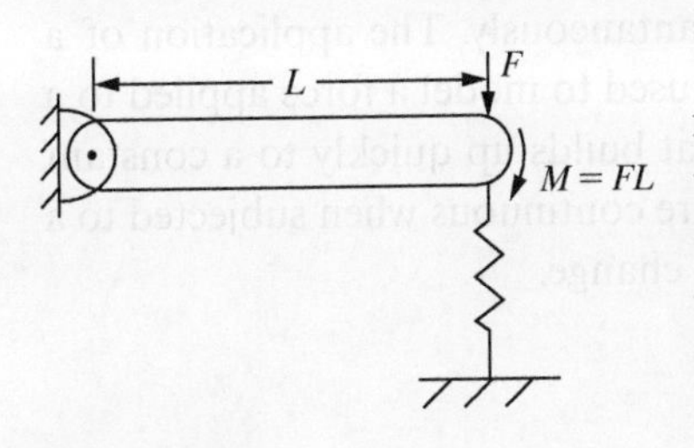

Figure 4.33

The power delivered to the system by the force F applied at the end of the bar is the same as the power delivered by a moment of magnitude FL.

For the purposes of modeling a mechanical system, two forces applied to a mechanical system are considered to be equivalent if they deliver the same power to the system at all times. For example, a force F applied to the end of the bar of Figure 4.33 delivers a power equal to $FL\dot{\theta}$, which is the same power as a clockwise moment of magnitude $M = FL$. Thus, for purposes of dynamic system modeling, the two are equivalent. The dynamic response of the system is the same whether the force or the moment is applied.

The assumption that the response of the system is the same whether the force F or the moment $M = FL$ is applied is true only for the dynamic response of the bar. The time-dependent reaction at the pin support depends upon whether the force or moment is applied. The same is true when other assumptions of equivalencies are used. The equivalency applies only to the motion of the system and not necessarily to constraint forces such as reactions.

4.4.2 Impulsive Forces

The **impulse** due to a force applied from time equal to zero to an arbitrary time t is defined as

$$I = \int_{0}^{t} F(t)dt \tag{4.82}$$

An **impulsive force** is a large force applied to a system over a short interval such that the response of the system is affected only by the total impulse imparted by the force, not by the time history of the force.

An impulsive force is often modeled as a force of very large but undetermined magnitude applied instantaneously such that the impulse applied by the force is finite. In this context, the mathematical representation of an impulsive force $F(t)$ is

$$F(t) = I\delta(t - t_0) \tag{4.83}$$

where I is the total impulse imparted due to the force, $\delta(t)$ is the unit impulse function as described in Section 1.7.1, and t_0 is the time at which the force is applied.

The application of an impulsive force to a system leads to an instantaneous change in the system's velocity. An impulse of magnitude I applied to a particle of mass m leads to an instantaneous change in the particle's velocity of

$$\Delta v = \frac{I}{m} \tag{4.84}$$

Although the velocity of the particle is discontinuous at the time of application of the impulse, its resulting displacement is continuous.

4.4.3 Step Forces

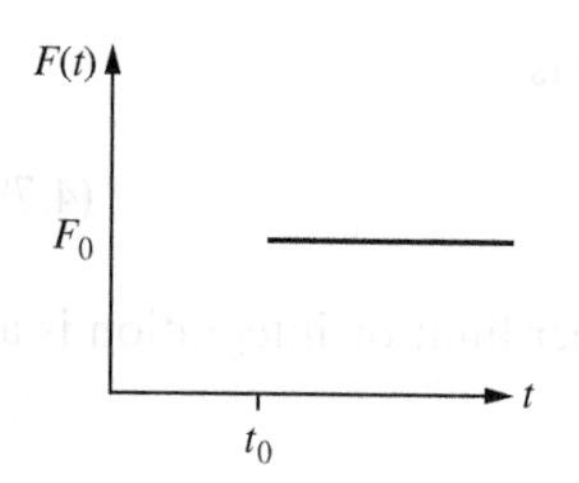

Figure 4.34

A step force applied to a system at time t_0 has the mathematical representation $F_0u(t-t_0)$.

A **step force** is a force that has a finite change at some time. A step force is discontinuous at the time of the step change. The mathematical representation of the step force illustrated in Figure 4.34 is

$$F(t) = F_0 u(t - t_0) \tag{4.85}$$

where $u(t)$ is the unit step function defined in Section 1.7.2 and t_0 is the time at which the step change occurs. Since $u(t)$ is zero when its argument is negative and has a value of 1 when its argument is positive, $F(t)$ as evaluated using Equation (4.85) has a value of 0 for $t < t_0$ and a value of F_0 for $t \geq t_0$.

It is physically impossible to apply a force instantaneously. The application of a force occurs over a finite time. A step force is often used to model a force applied to a particle at rest when suddenly subject to a force that builds up quickly to a constant value. The displacement and velocity of a particle are continuous when subjected to a step force, but its acceleration has an instantaneous change.

4.4.4 Periodic Forces

A force $F(t)$ is **periodic** if there is a value T such that

$$F(t + T) = F(t) \tag{4.86}$$

for all t. The value T is called the **period** of the force. Periodic forces are often developed during the continuous operation of machines. The most familiar examples of periodic functions are the trigonometric functions $\sin(\omega t)$ and $\cos(\omega t)$. The sine and cosine functions repeat when their arguments increase by 2π. The functions $\sin(\omega t)$ and $\cos(\omega t)$ are periodic of period

$$T = \frac{2\pi}{\omega} \tag{4.87}$$

where ω is the frequency of the force and has typical units of r/s.

Figure 4.35 provides examples of other periodic forces. The **fundamental frequency** of a periodic force is defined as

$$\omega = \frac{2\pi}{T} \tag{4.88}$$

Any periodic force has a **Fourier series representation**, which is an infinite series of trigonometric terms with frequencies that are integer multiples of the fundamental frequency.

A periodic force is often referred to as a **harmonic excitation**. A force that is represented as

$$F(t) = F_0 \sin(\omega t + \psi) \tag{4.89}$$

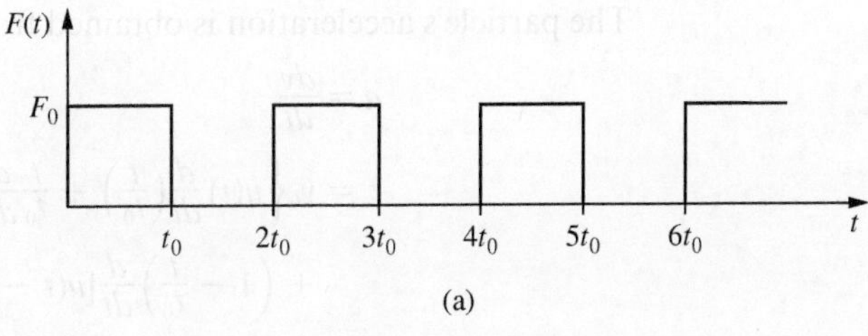

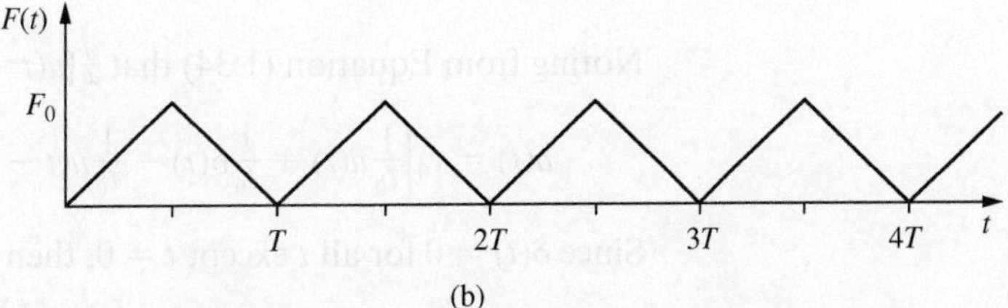

Figure 4.35

Examples of periodic functions: (a) square wave; (b) triangular wave.

is a single-frequency harmonic excitation with amplitude F_0, frequency ω, and phase angle ψ. The phase angle represents the shift from a purely sinusoidal excitation. A trigonometric identity can be used to write Equation (4.89) as

$$F(t) = F_0 \cos\psi \sin(\omega t) + F_0 \sin\psi \cos(\omega t) \tag{4.90}$$

4.4.5 Motion Input

Motion input occurs when a system particle is subject to a prescribed displacement, velocity, or acceleration. If one of these kinematic quantities is specified, the kinematic relations in Equations (4.3) and (4.4) may be used to determine the other two if necessary.

Example 4.6

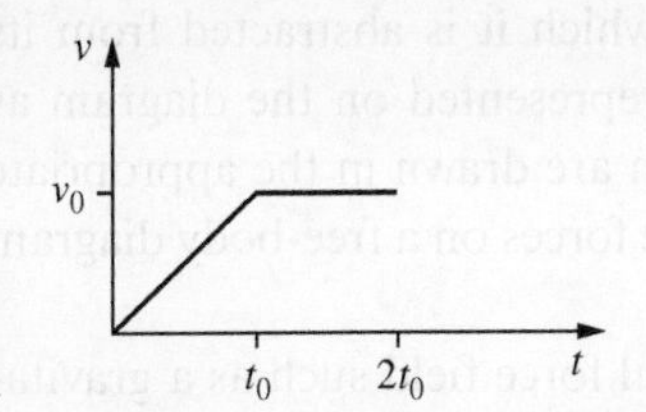

Figure 4.36

Velocity input for Example 4.6.

The velocity of a particle in a mechanical system is prescribed as in Figure 4.36. Determine the time-dependent displacement and acceleration of the particle.

Solution

The mathematical representation of the velocity, written using unit step functions, is

$$v(t) = v_0\left[\frac{t}{t_0}u(t) + \left(1 - \frac{t}{t_0}\right)u(t - t_0) - u(t - 2t_0)\right]$$

The particle displacement is obtained by

$$x(t) = \int_0^t v(\tau)d\tau$$

$$= v_0\int_0^t\left[\frac{\tau}{t_0}u(\tau) + \left(1 - \frac{\tau}{t_0}\right)u(\tau - t_0) - u(\tau - 2t_0)\right]d\tau \tag{a}$$

Equation (1.36) is used to evaluate the integrals in Equation (a), leading to

$$x(t) = v_0\left[u(t)\int_0^t \frac{\tau}{t_0}d\tau + u(t - t_0)\int_{t_0}^t\left(1 - \frac{\tau}{t_0}\right)d\tau - u(t - 2t_0)\int_{2t_0}^t d\tau\right]$$

$$= v_0\left[u(t)\frac{\tau^2}{2t_0}\bigg|_{\tau=0}^{\tau=t} + u(t - t_0)\left(\tau - \frac{\tau^2}{2t_0}\right)\bigg|_{\tau=t_0}^{\tau=t} - u(t - 2t_0)\tau\Big|_{\tau=2t_0}^{\tau=t}\right]$$

$$= v_0\left[\frac{t^2}{2t_0}u(t) + \left(t - \frac{t^2}{2t_0} - \frac{t_0}{2}\right)u(t - t_0) - (t - 2t_0)u(t - 2t_0)\right] \tag{b}$$

The particle's acceleration is obtained from

$$a = \frac{dv}{dt}$$

$$= v_0\left\{u(t)\frac{d}{dt}\left(\frac{t}{t_0}\right) + \frac{t}{t_0}\frac{d}{dt}[u(t)] + u(t - t_0)\frac{d}{dt}\left(1 - \frac{t}{t_0}\right)\right.$$

$$\left. + \left(1 - \frac{t}{t_0}\right)\frac{d}{dt}[u(t - t_0)] - \frac{d}{dt}[u(t - 2t_0)]\right\} \qquad \text{(c)}$$

Noting from Equation (1.34) that $\frac{d}{dt}[u(t - t_0)] = \delta(t - t_0)$, Equation (c) becomes

$$a(t) = v_0\left[\frac{1}{t_0}u(t) + \frac{t}{t_0}\delta(t) - \frac{1}{t_0}u(t - t_0) + \left(1 - \frac{t}{t_0}\right)\delta(t - t_0) - \delta(t - 2t_0)\right] \qquad \text{(d)}$$

Since $\delta(t) = 0$ for all t except $t = 0$, then $t\delta(t) = 0$ for all t. Similarly,

$$\left(1 - \frac{t}{t_0}\right)\delta(t - t_0) = 0$$

for all t. Thus Equation (d) becomes

$$a(t) = \frac{v_0}{t_0}[u(t) - u(t - t_0)] - v_0\delta(t - 2t_0) \qquad \text{(e)}$$

The presence of the unit impulse function in Equation (e) is a result of the impulse that must be applied to cause a discrete change in velocity at $t = 2t_0$.

4.5 Free-Body Diagrams

An essential step in the mathematical modeling of mechanical systems is the application of appropriate conservation laws to the entire system or its individual components. The application of a conservation law to a system component is facilitated by drawing a **free-body diagram (FBD)** of the component in which it is abstracted from its surroundings and the effects of the surroundings are represented on the diagram as forces and moments. The forces on a free-body diagram are drawn in the appropriate directions and labeled with appropriate magnitudes. The forces on a free-body diagram are of two types: body forces and surface forces.

A **body force** results from the presence of an external force field such as a gravitational field or an electromagnetic field. A body of mass m is subject to a force from the earth's gravitational field, called the **gravity force**, which is the weight of the body

$$W = mg \qquad (4.91)$$

where g is the acceleration due to gravity. On the surface of the earth

$$g = 32.2 \text{ ft/s}^2 \quad \text{or} \quad 9.81 \text{ m/s}^2 \qquad (4.92)$$

The gravity force is drawn on the free-body diagram with its point of application at the center of gravity and directed toward the center of the earth, usually the vertical direction. The point of application of the gravity force on a free-body diagram is at the center of mass. If the y coordinate in a Cartesian system is the vertical coordinate, the vector representation of the gravity force is

$$\mathbf{F} = -mg\mathbf{j} \qquad (4.93)$$

Surface forces show the effects of contact between the body and its surroundings. Examples of surface forces include reactions at supports, forces due to sliding friction, forces exerted by springs, and externally applied forces. Surface forces act only on the surface of a body, whereas a body force is applied at an internal point.

Mathematical modeling of a mechanical system requires the identification of time-dependent coordinates that describe the motion of the system. For modeling purposes, free-body diagrams are drawn at an arbitrary time and thus for arbitrary values of the dependent coordinates. Forces and moments are drawn on the free-body diagram in directions consistent with the chosen positive directions of the coordinates. The magnitudes of the forces are written in terms of the coordinates, where appropriate.

The **resultant** of the forces acting on a free-body diagram is the vector sum of the forces. The **resultant moment** about an axis is the sum of the moments about that axis of the forces acting on the free-body diagram.

Example 4.7

Draw free-body diagrams for each of the systems shown in Figure 4.37 at an arbitrary instant. Label the forces in terms of the indicated coordinate. All coordinates are measured from the system's static equilibrium position.

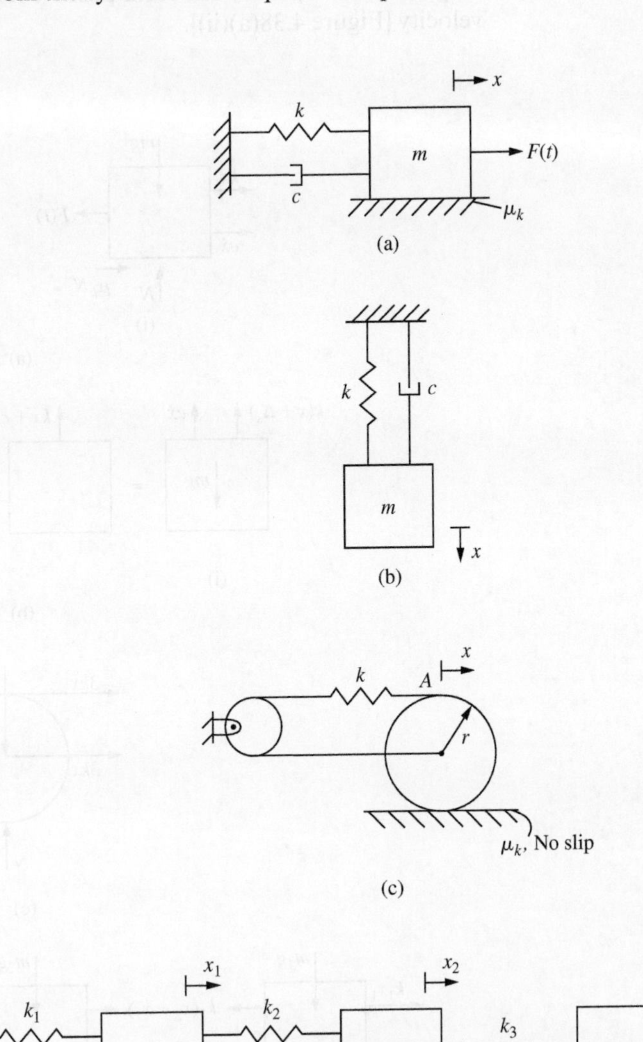

Figure 4.37

Systems for Example 4.7.

Solution

a. The free-body diagram for the system of Figure 4.37(a) is drawn in Figure 4.38(a). Gravity is the only body force. The surface forces include the force exerted by the spring on the block, the force from the viscous damper, the reaction at the surface, and the external force. When the block's displacement is positive, the spring is stretched and pulls on the block. When the block's displacement is negative, the spring is compressed and pushes against the block. Labeling the spring force as kx and drawing it so that it acts against the block is consistent with both positive and negative values of x.

At an arbitrary instant, the velocity of the block is $\dot{x}$ and thus the force acting on the block from the viscous damper is $c\dot{x}$. Since the viscous damping force always acts in the opposite direction of the velocity, it is drawn to the left on the free-body diagram, consistent with a positive velocity. This is also consistent with a negative velocity since the damping force is labeled $c\dot{x}$; when the velocity is negative, the damping force is actually acting to the right, again opposing the direction of motion.

The reaction force between the block and the surface has two components: (1) a force perpendicular to the surface, called the normal force, and (2) a force opposite the direction of motion, the friction force. However, unlike the spring force and viscous damping force, the direction of the force from Coulomb friction does not take care of itself when drawn and labeled on a free-body diagram. Two separate free-body diagrams are required, one for a positive $\dot{x}$ [Figure 4.38(a)(i)] and one for a negative velocity [Figure 4.38(a)(ii)].

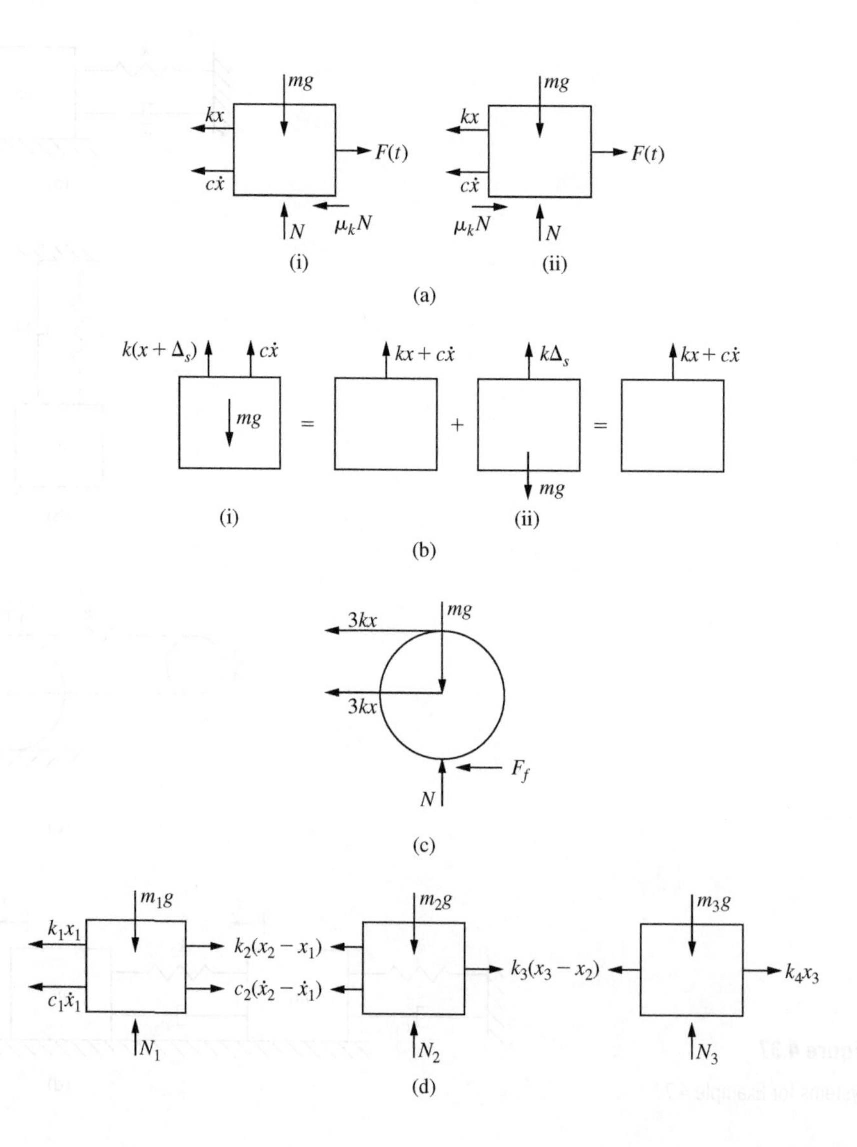

Figure 4.38

Free-body diagrams at an arbitrary instant for the systems of Example 4.7.

b. The free-body diagram for the system of Figure 4.37(b), drawn at an arbitrary instant, is shown in Figure 4.38(b)(i). The system is similar to that of Figure 4.37(a), except that the mass is suspended and its motion is in the vertical direction. As illustrated in Section 4.2.3, when the system is in equilibrium, the spring is stretched with a static deflection $\Delta_s = \frac{mg}{k}$. The dependent variable x is measured from the system's equilibrium position. Thus, at an arbitrary instant the change in length of the spring from its unstretched length is $x + \Delta_s$.

The resultant of the forces acting on the body is

$$\begin{aligned}\sum F &= -mg + c\dot{x} + k(x + \Delta_s)\\ &= -mg + c\dot{x} + kx + k\left(\frac{mg}{k}\right)\\ &= c\dot{x} + kx\end{aligned} \tag{a}$$

The canceling of the static spring force with the gravity force is illustrated in Figure 4.38(b)(ii). When the equilibrium condition is applied, the static spring force cancels the gravity force. Since the resultant of the static spring force and the gravity force is zero, the free-body diagram of the system could have been drawn without including either.

c. The force developed in the spring is its stiffness times its change in length. One end of the spring is attached to the mass center of the disk, which has a displacement x to the right. The other end of the spring is attached to point A. When the center of the disk has a displacement x, point A has rotated through an angle θ and the end of the spring attached to point A has increased in length by $x + r\theta$. Thus the total change in length of the spring is $2x + r\theta$. If the disk rolls without slip, then $x = r\theta$ and the change in length of the spring is $3x$. At any instant the disk has a single point of contact with the surface. The reaction at the surface has two components: a normal force and a friction force. Even when the disk rolls without slip, a friction force is developed but is less than the maximum available friction force. The free-body diagram for this system assuming no slip is drawn in Figure 4.38(c).

d. Modeling of this three-degree-of-freedom system requires free-body diagrams of each mass as drawn in Figure 4.38(d). The force developed in a spring is its stiffness times its change in length. Consider the spring of stiffness k_2, which links the particles of mass m_1 and m_2. The right end of the spring is attached to the block of mass m_2, which has a displacement x_2. The left end of the spring is attached to the block of mass m_1, which has a displacement x_1. The change in length of the spring is the difference between the displacements of its two ends, $x_2 - x_1$. If $x_2 > x_1$, then the spring's length has increased from its unstretched length and the spring is in tension. Thus if the spring force is labeled as $k_2(x_2 - x_1)$, it must be drawn directed away from each of the particles on their free-body diagrams.

Systems such as the one in Figure 4.37(b) include springs that are stretched or compressed when the system is in equilibrium. The dependent variable in Example 4.7(b) is measured from the system's equilibrium position. The spring force at an arbitrary instant is the sum of the static spring force ($k\Delta_{st}$) and the spring force developed due to the displacement from equilibrium (kx). The resultant of the static spring force and the gravity force is zero, and thus it is not necessary to include either when applying conservation laws to the system.

The result in the example can be generalized for any system. Consider a system that has springs that are at lengths different from their unstretched lengths when the system is in equilibrium. An equilibrium condition exists between the static spring forces and the gravity forces that cause these forces. The equilibrium condition is derived from requiring either that the resultant of the static forces and the gravity forces equals zero

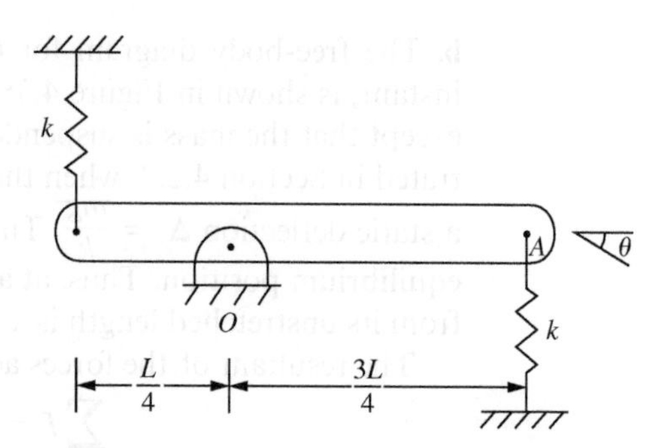

Figure 4.39
Static spring forces cancel with gravity when deriving the differential equation governing the motion of the system, but not when determining the reaction at O.

or that the resultant moment due to these forces is zero. If the system's dependent variables are measured from the system's equilibrium position, then, when the conservation laws are applied to the system at an arbitrary instant, the application of the equilibrium condition leads to the static spring forces and the gravity forces canceling from the resulting equation.

Therefore it is not necessary to include static spring forces or gravity forces that lead to static spring forces in free-body diagrams when mathematically modeling a linear system and the dependent variables are measured from the system's equilibrium position. These forces must be included if the system is nonlinear or the dependent variable is measured from a position other than the static equilibrium position. The forces must also be included if a different conservation law is to be applied than the one used to develop the equilibrium condition. For example, the equilibrium condition between the static spring force and gravity for the system of Figure 4.39 is obtained by summing moments about an axis through O. When the moment equation about an axis through O at an arbitrary instant is applied to derive the differential equation governing the motion of the system, the equilibrium condition appears and is equal to zero. Then the static spring forces cancel with gravity. However, these forces do not cancel when the summation of forces is applied at an arbitrary instant to determine the reaction force at O.

The coordinate θ is used to denote the clockwise angular displacement of the bar in Figure 4.39 from its equilibrium position. A free-body diagram of the bar, drawn at an arbitrary instant, is shown in Figure 4.40(a). If it is assumed that θ is small and the small angle approximation is used, then the gravity force in the bar will cancel with the static spring forces when summing moments about an axis through O. Thus, neither are included on the free-body diagram. The particle at the right end of the bar has a downward vertical displacement of $\frac{3L}{4}\sin\theta$ and a horizontal displacement to the left of $\frac{3L}{4}(1-\cos\theta)$. The use of the small angle approximation, Equation (1.15), leads to an approximate vertical displacement of $\frac{3L}{4}\theta$ and a horizontal displacement of $\frac{3L}{4}\theta^2$. Since θ is assumed to be small, $\theta^2 \ll \theta$. Thus, it is consistent with the small angle approximation that the horizontal displacement is neglected compared with the vertical. The displacement of the particle can be approximated as being vertical with a value of $\frac{3L}{4}\theta$.

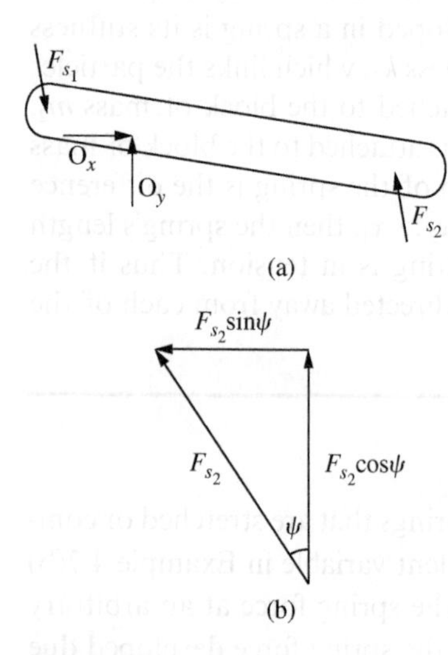

Figure 4.40
(a) Free-body diagram of system of Figure 4.39 drawn assuming the static spring forces cancel with the gravity force that causes them.
(b) Illustration of the geometry that allows the spring force to be replaced by a vertical force applied at the end of the bar if the small angle assumption is used.

The direction of the spring forces is along the axis of the springs and is consistent with the stretching or compression of the spring for a positive θ. The spring at the right end of the bar is compressed for a positive θ, and the spring force makes an angle of ψ with the vertical. The value of ψ is dependent on the value of θ such that if θ is small then ψ is also small. The spring force is resolved into a vertical component $F_s\cos\psi$ and a horizontal component $F_s\sin\psi$, as shown in Figure 4.40(b), where F_s is the magnitude of the spring force. Using the small angle approximation leads to $F_s = k(\frac{3L}{4}\theta)$. Then taking the moment of the spring force about a vertical axis through O leads to

$$M_O = F_H\left(\frac{3L}{4}\cos\theta\right) - F_V\left(\frac{3L}{4}\sin\theta\right) \tag{4.94}$$

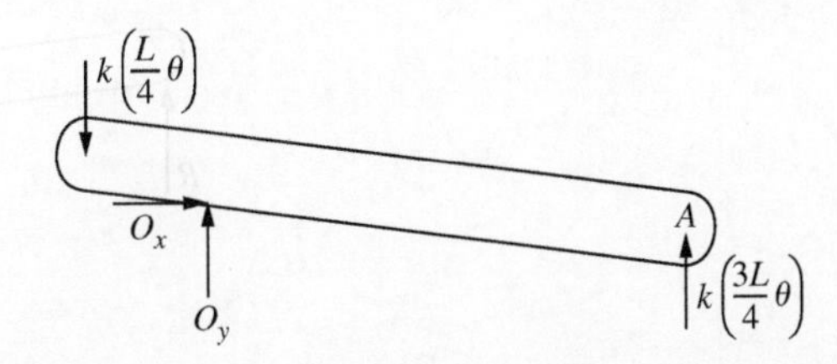

Figure 4.41

Free-body diagram of the system of Figure 4.39 using the small angle approximation.

Substituting for F_s and using the small angle approximation, Equation (1.15), in Equation (4.94) leads to

$$M_O = \frac{9kL^2}{16}\theta - \frac{9kL^2}{16}\psi\theta^2 \tag{4.95}$$

However, $\psi\theta^2 \ll \theta$ and Equation (4.95) is approximated as

$$M_O = \frac{9kL^2}{16}\theta \tag{4.96}$$

The small angle assumption is used in drawing the free-body diagram of Figure 4.41 in the following ways: (1) the displacement of particle A is approximated as vertical and equal to $\frac{3L}{4}\theta$, (2) the direction of the spring force is approximated as vertical, and (3) the moment of the spring force about a vertical axis through O is approximated by Equation (4.96). The same principles are used to draw the spring force acting at the left end of the bar as a vertical force of $k\frac{L}{4}\theta$.

Example 4.8

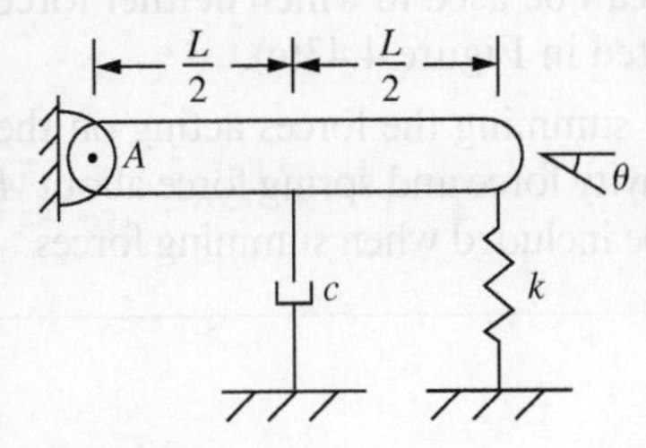

Figure 4.42

Systems for Example 4.8.

Consider the system of Figure 4.42. Assume small angles. **a.** Determine θ_{st}, the angular rotation of the bar, measured clockwise from the horizontal, when the system is in equilibrium. **b.** Draw a free-body diagram of the system at an arbitrary instant including gravity and the static forces in the spring. **c.** Show that the resultant moment about an axis through A is independent of the gravity force and θ_{st}. **d.** Redraw the free-body diagram not including the gravity force and the static force in the spring. **e.** What restrictions are placed on the use of this free-body diagram?

Solution

a. The free-body diagram of the static equilibrium position, assuming small angular displacements, is illustrated in Figure 4.43(a). Assuming the spring force is vertical is consistent with the small angle assumption. Summing moments, assuming clockwise positive, about the pin support leads to

$$\begin{aligned} \sum M_A &= 0 \\ mg(L/2) - kL\theta_{st}(L) &= 0 \\ \theta_{st} &= \frac{mg}{2kL} \end{aligned} \tag{a}$$

b. Let θ represent the clockwise angular displacement of the bar, measured from the system's equilibrium position. A free-body diagram of the system at an arbitrary instant, assuming small angular displacements, is illustrated in Figure 4.43(b). The bar is rotating about an axis fixed at A; thus the velocity of the center of the bar is $\frac{L}{2}\dot{\theta}$ and the force form the viscous damper is $c\frac{L}{2}\dot{\theta}$. Assuming small θ, the total change in length of the spring from its unstretched length is $L(\theta + \theta_{st})$.

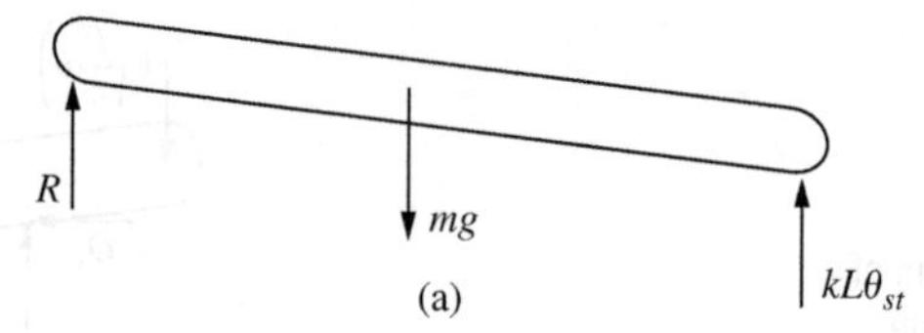

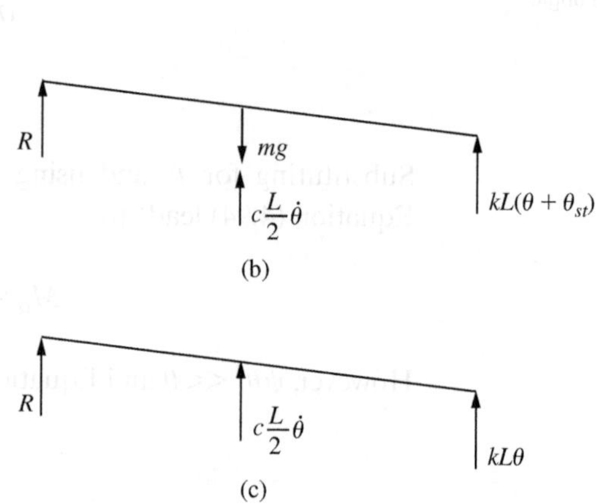

Figure 4.43

(a) Free-body diagram of static equilibrium position of system of Example 4.8; (b) free-body diagram at an arbitrary instant, including gravity and static spring force; (c) modified free-body diagram in which gravity and static spring force are not included because from the equilibrium condition the moment of the static spring force and the moment of the gravity force about an axis through A equal zero.

c. Summing moments, assuming positive clockwise, at an arbitrary instant about an axis at A gives

$$\sum M_A = mg(L/2) - c\frac{L}{2}\dot{\theta}(L/2) - kL(\theta + \theta_{st})(L) \tag{b}$$

Substitution of Equation (a) into Equation (b) leads to

$$\sum M_A = -c\frac{L^2}{4}\dot{\theta} - kL^2\theta \tag{c}$$

d. In view of Equation (c), the resultant moment about A is independent of the gravity force and the static spring force. Thus for the purposes of deriving a mathematical model for the system, a modified free-body diagram can be used in which neither force is included. This useful free-body diagram is illustrated in Figure 4.43(c).

e. The reaction at the pin support A is determined by summing the forces acting on the bar. However, whereas the resultant moment of the gravity force and spring force about A is zero, their resultant itself is not zero and thus must be included when summing forces.

4.6 Newton's Laws

Newton's three laws of motion provide a foundation for the modeling of mechanical systems. Newton's first law states that a body at rest or in motion with a constant velocity will continue to remain at rest or move with a constant velocity in the absence of external forces. Newton's first law is the basis of static analysis. Newton's third law, often stated as "every action has an equal and opposite reaction," is the basis on which surface forces are drawn on free-body diagrams. For example, consider the system of Figure 4.37(a). When the block of mass m is displaced a distance x to the right from equilibrium, the spring has an increase in length of x from its unstretched length. The change in length results in a force $F = kx$ pulling on the spring by the block. Newton's third law implies that the spring exerts a force $F = kx$ on the block in the opposite direction of the force from the block on the spring. This force is illustrated on the free-body diagram of Figure 4.38(a).

Newton's second law, the subject of this section, is the basic conservation law used in the mathematical modeling of mechanical systems.

4.6.1 Particles

Newton's second law applied to a particle states that the resultant force acting on a particle is equal to the time rate of change of linear momentum of the particle. If the mass of the particle is constant, then the time rate of change of linear momentum is equal to the mass of the particle times its acceleration. In this case, Newton's second law is written as

$$\sum \mathbf{F} = m\mathbf{a} \tag{4.97}$$

where $\sum \mathbf{F}$ is the sum of all external forces acting on the particle, which is the resultant of the external forces. The component forms of Newton's second law are

$$\sum F_x = m\ddot{x} \tag{4.98a}$$

$$\sum F_y = m\ddot{y} \tag{4.98b}$$

$$\sum F_z = m\ddot{z} \tag{4.98c}$$

Newton's second law is a basic law of nature. It cannot be derived from any law more basic and it can be proven only by empirical observation. The Principle of Work and Energy is obtained by integrating Newton's law spatially. The Principle of Impulse and Momentum is obtained by integrating Newton's law over time. While these principles are not independent of Newton's second law, they are often applied as an alternative to the direct application of it in the modeling of a mechanical system.

Mathematical modeling of a mechanical system requires the application of Newton's second law to the free-body diagram of a system component drawn at an arbitrary instant, for an arbitrary value of the chosen dependent variable. The external forces acting on the particle are labeled in terms of the dependent variable. The particle acceleration is often expressed in terms of time derivatives of the dependent variable. The application of Newton's second law to the free-body diagram leads to a differential equation whose solution is the time-dependent behavior of the system.

Example 4.9

A projectile of mass m is fired at $t = 0$ with velocity v_0 at an angle α with the horizontal. Determine a mathematical model for the path of the projectile to find the range of the projectile.

Solution

The procedure of Section 1.3. is followed in performing the modeling.

Step 1: The objective of the modeling is to determine the range of the projectile. In terms of systems, the inputs are the external forces acting on the projectile during flight, and the outputs are the coordinates of the projectile as functions of time.

Step 2: Explicit assumptions include (1) no air resistance, (2) no wind that can cause out-of-plane motion, and (3) a short distance projectile.

Step 3: There are no system components to quantify.

Step 4: The independent variable is the time after firing t. The dependent variable is the position vector of the projectile as it travels through the atmosphere. Assumption (2) implies that the motion occurs in a plane, say the x-y plane. The position vector is represented by the two dependent variables x and y.

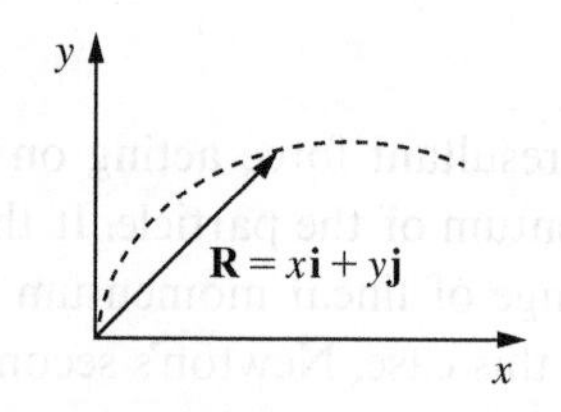

Figure 4.44

The position vector to the particle at an arbitrary instant is $\mathbf{R} = x\mathbf{i} + y\mathbf{j}$.

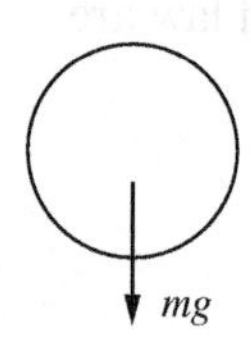

Figure 4.45

Free-body diagram of projectile at an arbitrary instant.

The particle's flight in the x-y plane is illustrated in Figure 4.44. Assumption (1) implies that the only force acting on the projectile is the force due to gravity. Thus, the mass of the projectile m and the acceleration due to gravity g are parameters. The initial velocity v_0 and the angle the initial velocity makes with the horizontal plane α are necessary to model the system. Since the range is required, an equation for the terrain $h(x)$ is necessary.

Step 5: A free-body diagram of the particle at an arbitrary instant is shown in Figure 4.45. It simply shows the particle at an arbitrary position in its flight with the body force due to gravity as the only force.

Step 6: Newton's laws are applied to the free-body diagram in the coordinate directions, yielding

$$\sum F_x = ma_x \Rightarrow 0 = m\ddot{x} \tag{a}$$

$$\sum F_y = ma_y \Rightarrow -mg = m\ddot{y} \tag{b}$$

Step 7: The particle is fired from the origin of the coordinate system at $t = 0$. Thus,

$$x(0) = 0 \tag{c}$$

$$y(0) = 0 \tag{d}$$

The particle is fired with an initial velocity v_0 that makes an angle α with the horizontal. Resolving the initial velocity into component form leads to

$$\dot{x}(0) = v_0 \cos\alpha \tag{e}$$

$$\dot{y}(0) = v_0 \sin\alpha \tag{f}$$

Thus, the mathematical model for this problem is Equation (a) subject to Equations (c) and (e) and Equation (b) subject to Equations (d) and (f).

Step 8: The differential equations, Equations (a) and (b), can be solved by applying indefinite integration twice to each equation, resulting in

$$x(t) = C_1 t + C_2 \tag{g}$$

$$y(t) = -\frac{1}{2}gt^2 + C_3 t + C_4 \tag{h}$$

In Equations (g) and (h), C_1, C_2, C_3, and C_4 are constants of integration that are determined by application of initial conditions. Doing so gives

$$x(0) = 0 \Rightarrow C_1(0) + C_2 = 0 \Rightarrow C_2 = 0 \tag{i}$$

$$\dot{x}(0) = v_0 \cos\alpha \Rightarrow C_1 = v_0 \cos\alpha \tag{j}$$

$$y(0) = 0 \Rightarrow -\frac{1}{2}g(0^2) + C_3(0) + C_4 = 0 \Rightarrow C_4 = 0 \tag{k}$$

$$\dot{y}(0) = v_0 \sin\alpha \Rightarrow -g(0) + C_3 = v_0 \sin\alpha \Rightarrow C_3 = v_0 \sin\alpha \tag{l}$$

Using Equations (i) and (j) in Equation (g) gives

$$x(t) = (v_0 \cos\alpha)t \tag{m}$$

Using Equations (k) and (l) in Equation (h) leads to

$$y(t) = -\frac{1}{2}gt^2 + (v_0 \sin\alpha)t \tag{n}$$

Step 9: Equations (m) and (n) are the result of the mathematical modeling of a projectile under the assumptions. Now the task is to use these equations to meet the objective. The equation for the trajectory of the projectile is obtained by solving Equation (m) for t in terms of x and substituting into Equation (n), resulting in

$$y = -\frac{1}{2}g\left(\frac{x}{v_0 \cos\alpha}\right)^2 + \frac{v_0 \sin\alpha}{v_0 \cos\alpha}x$$

$$= -\frac{g}{2(v_0 \cos\alpha)}x^2 + (\tan\alpha)x \tag{o}$$

The range of the projectile is the value of x where it hits the ground. If $y = h(x)$ is the equation for the terrain, then the range is determined from

$$h(x) = -\frac{g}{2(v_0 \cos\alpha)}x^2 + (\tan\alpha)x \tag{p}$$

If $h(x) = 0$, a flat terrain, then the range is determined as

$$x = \frac{2\,v_0^2 \cos^2\alpha}{\tan\alpha} \tag{q}$$

Step 10: The mathematical model should be validated by experiment. The projectile can be fired at different angles from the horizontal at different speeds and the range measured in each case.

Example 4.10

Use Newton's law to derive the differential equations governing the motion of **a.** the system of Figure 4.37(a) assuming $\mu_k = 0$ and **b.** the system of Figure 4.37(b).

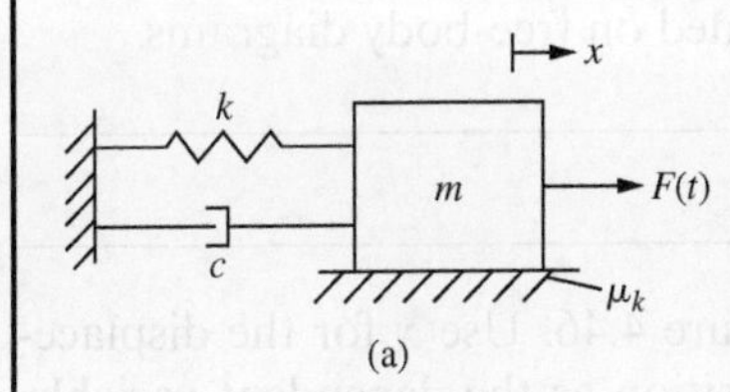

(b)

Figure 4.37

Systems for Examples 4.7 and 4.10. (Repeated)

Solution

The free-body diagrams for these systems drawn at an arbitrary instant are illustrated Figures 4.38(a) and 4.38(b).

a. The acceleration of the particle at an arbitrary instant is $\ddot{x}$. The component form of Newton's second law in the x direction, Equation (4.98a), is applied using the free-body diagram of Figure 4.38(a), redrawn below, with $\mu_k = 0$. Taking positive forces acting to the right leads to

$$-kx - c\dot{x} + F(t) = m\ddot{x} \tag{a}$$

Equation (a) is rearranged so that all the terms involving x are on the left-hand side, leading to

$$m\ddot{x} + c\dot{x} + kx = F(t) \tag{b}$$

Equation (b) is a second-order linear ordinary differential equation whose solution describes the motion of the system.

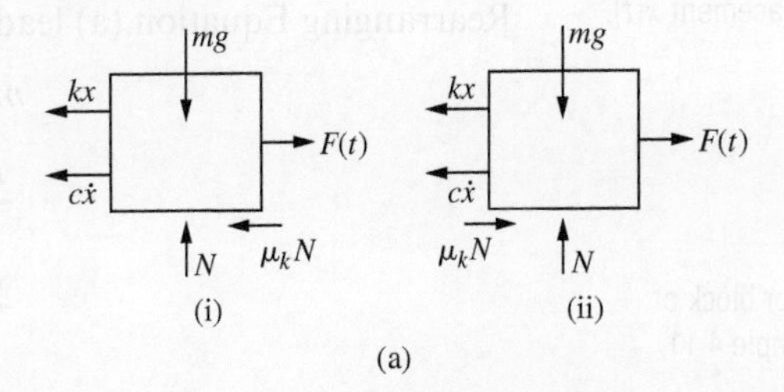

(b)

Figure 4.38

Free-body diagrams at an arbitrary instant for systems of Examples 4.7 and 4.10. (Repeated)

b. The application of Newton's second law in the x direction with the positive direction downward to the free-body diagram of Figure 4.38(b) leads to

$$-mg - k(x + \Delta s) - c\dot{x} + F(t) = m\ddot{x} \tag{c}$$

The equilibrium condition for this system is

$$mg + k\Delta_s = 0 \tag{d}$$

Using Equation (d) in Equation (c) and rearranging leads to

$$m\ddot{x} + c\dot{x} + kx = F(t) \tag{e}$$

Note that Equation (e) of Example 4.10 is identical to Equation (b) of Example 4.10.

The following comments are in agreement with the discussion in Section 4.5 regarding the use of the equilibrium condition in mathematical modeling of a linear system. The static equilibrium position of a system is identified. A force or moment balance to maintain this position, called the equilibrium condition, is developed. When the equilibrium condition is applied, and a differential equation governing the motion of the system at an arbitrary time is derived, the static spring forces and gravity cancel one another in the resulting differential equation. Thus, for purposes of deriving a mathematical model for a linear system, static spring forces and the gravity forces leading to static spring forces do not need to be included in the analysis. From this point in this book, these forces will not be included on free-body diagrams.

Example 4.11

Derive a mathematical model for the system of Figure 4.46. Use x for the displacement of the mass from the system's equilibrium position as the dependent variable. The input $z(t)$ is the prescribed motion of the movable base.

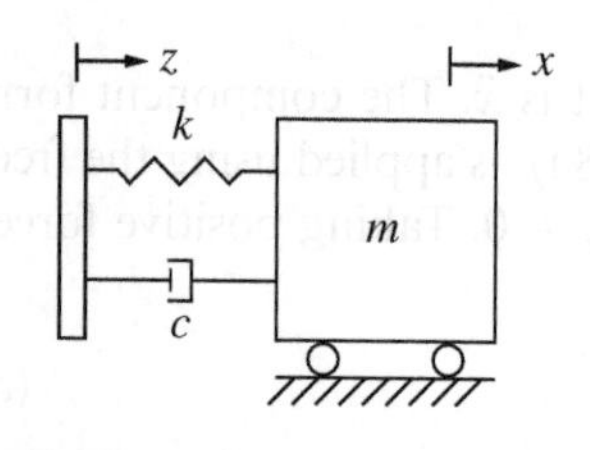

Figure 4.46

A mass is attached to a massless movable base by a spring in parallel with a viscous damper. The movable support has a prescribed displacement $z(t)$ which causes the mass to have a displacement $x(t)$.

Solution

A free-body diagram of the block is shown at an arbitrary instant in Figure 4.47. The left end of the spring has a displacement from equilibrium of z, while the right end of the spring has a displacement of x. The spring is compressed a total of $x - z$. The relative velocity between the two ends of the viscous damper is $\dot{x} - \dot{z}$, which if positive would mean the block is moving faster than its base. In this case, the force from the viscous damper on the block acts away from the block. Using Newton's second law on the free-body diagrams leads to

$$-c(\dot{x} - \dot{z}) - k(x - z) = m\ddot{x} \tag{a}$$

Rearranging Equation (a) leads to

$$m\ddot{x} + c\dot{x} + kx = c\dot{z} + kz \tag{b}$$

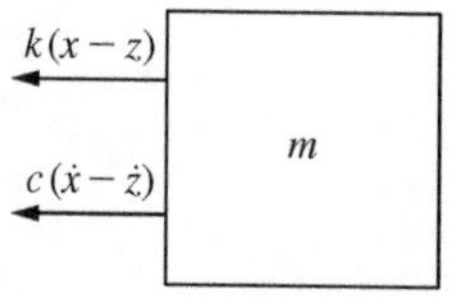

Figure 4.47

Free-body diagram for block of Figure 4.46 and Example 4.11.

4.6.2 Rigid Body Motion

Newton's law, as written in Equation (4.97), applies to a single particle. A rigid body is a system with an infinite number of particles. A form of Newton's law for a rigid body may be derived by first applying the law to an arbitrary particle on the rigid body of mass dm, as illustrated in Figure 4.48. The surface forces shown acting on a particle

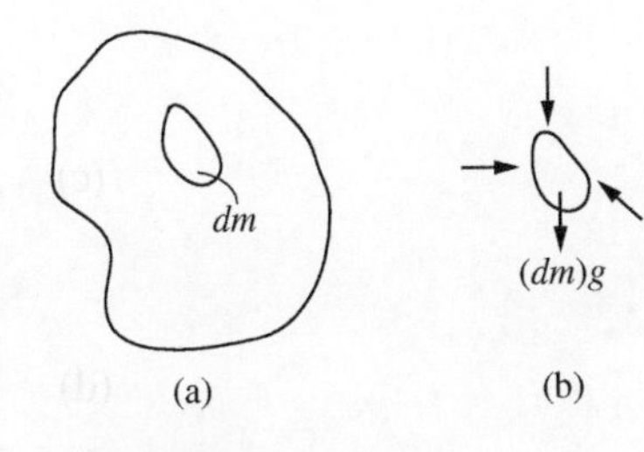

Figure 4.48

(a) A rigid body composed of an infinite number of particles of infinitesimal mass; (b) the free-body diagram of a particle of mass *dm* illustrates the internal forces between particles.

are internal forces from the surrounding particles. When equations for adjacent particles are added together, the internal forces cancel as a result of Newton's third law. When the equations for all particles comprising the rigid body are added together, the summation becomes an integration over the mass of the body, resulting in

$$\sum \mathbf{F} = m\bar{\mathbf{a}} \tag{4.99}$$

where $\bar{\mathbf{a}}$ is the acceleration of the mass center of the rigid body, m is its total mass, and $\sum \mathbf{F}$ is the resultant of the external forces as shown on a free-body diagram of the rigid body.

A general rigid-body motion is a combination of rotation and translation. The rotational motion of a rigid body leads to the necessity for applying a moment equation when deriving a mathematical model for the motion of the body. A moment equation for a rigid body is derived using a process similar to that used to derive the force equation. The result is

$$\sum \mathbf{M}_G = \dot{\mathbf{H}}_G \tag{4.100}$$

where $\sum \mathbf{M}_G$ is the resultant moment about the mass center of all external forces and moments and $\dot{\mathbf{H}}_G$ is the time rate of change of angular momentum of the rigid body about its mass center.

4.6.3 Pure Rotational Motion About a Fixed Axis of Rotation

The angular momentum of a rigid body undergoing planar motion about its mass center is obtained using Equation (4.31) as

$$H_G = \bar{I}\omega \tag{4.31}$$

where $\bar{I}$ is the body's mass moment of inertia about its centroidal axis and ω is the angular velocity about the axis of rotation. The angular momentum is in the direction of the axis of rotation. Equation (4.31) also applies for a symmetrical body undergoing pure rotation (no translation) about a fixed axis parallel to an axis of symmetry. The time rate of change of the angular momentum, determined using Equation (4.31), is

$$\dot{H}_G = \bar{I}\alpha \tag{4.101}$$

Thus, Equation (4.100) becomes

$$\sum M_G = \bar{I}\alpha \tag{4.102}$$

It is usually convenient to apply the moment equation about the axis of rotation. When the axis of rotation is not a centroidal axis, but parallel to a centroidal axis through a point O, the Parallel Axis Theorem is used to rewrite Equation (4.102) as

$$\sum M_O = I_O\alpha \tag{4.103}$$

where I_O is the moment of inertia of the rigid body about an axis through O.

Example 4.12

A rotor of moment of inertia I is rotating about its centroidal axis at a constant angular speed ω_0, as shown in Figure 4.49(a), when a constant braking torque T is applied. Determine a mathematical model for the system and determine the time required to stop the rotor.

Solution

Application of Equation (4.102) to the free-body diagram of Figure 4.49(b) gives

$$\sum M_G = I\alpha \tag{a}$$

$$-T = I\alpha \tag{b}$$

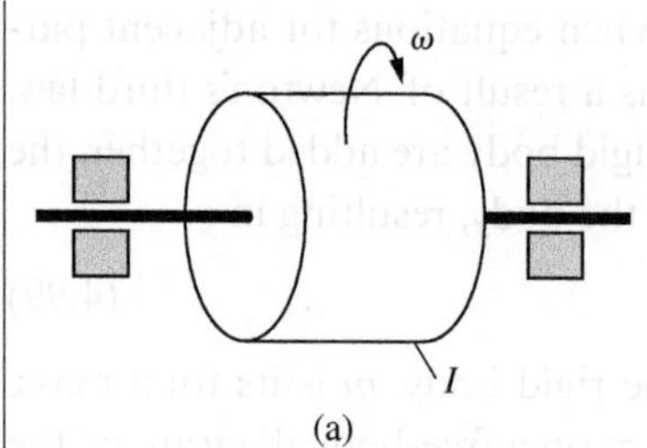

Figure 4.49

(a) The rotor has constant speed when a braking torque T is applied; (b) the free-body diagram of the rotor.

Noting that $\alpha = \dot{\omega}$ leads to

$$\dot{\omega} = -\frac{T}{I} \tag{c}$$

The torque is applied at $t = 0$ when

$$\omega(0) = \omega_0 \tag{d}$$

Integration of Equation (c) gives

$$\omega(t) = -\frac{T}{I}t + C \tag{e}$$

Application of the initial condition, Equation (d), leads to $C = \omega_0$ and

$$\omega(t) = -\frac{T}{I}t + \omega_0 \tag{f}$$

Setting $\omega = 0$ in Equation (f) to determine the time at which the rotor will stop leads to

$$t = \frac{I\omega_0}{T} \tag{g}$$

Example 4.13

A rotor of moment of inertia I is connected to a torsional spring of stiffness k_t and a torsional viscous damper of damping coefficient c_t, as shown in Figure 4.50(a). Let θ be the angular displacement of the rotor, measured from the system's equilibrium position. Derive a mathematical model for the system.

Solution

Application of Equation (4.102) to the free-body diagram of Figure 4.50(b)

$$\sum M_G = \bar{I}\alpha \tag{a}$$

$$M - k_t\theta - c_t\dot{\theta} = I\ddot{\theta} \tag{b}$$

Rearranging Equation (b) gives

$$I\ddot{\theta} + c_t\dot{\theta} + k_t\theta = M \tag{c}$$

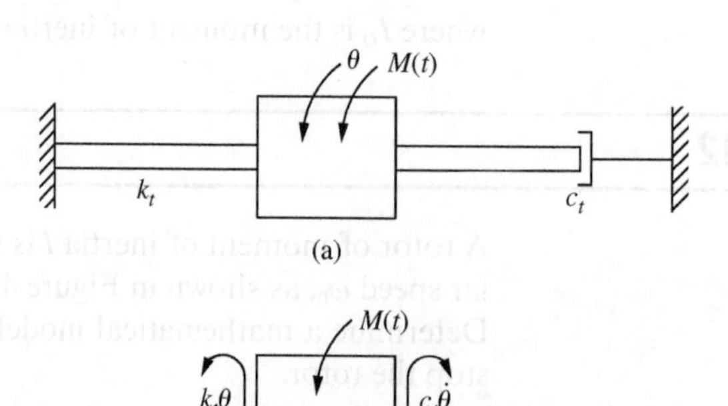

Figure 4.50

(a) The system of Example 4.13; (b) the free-body diagram of the system at an arbitrary instant.

4.6.4 Planar Motion of a Rigid Body

The applicable form of the force equation for a rigid body undergoing planar motion is Equation (4.99). In addition, because the axis of rotation is always perpendicular to the plane of motion, the moment Equation (4.102) also applies. Furthermore, if the axis of rotation is fixed, then Equation (4.103) applies. If the planar motion is constrained, then a kinematic relationship exists between $\bar{a}$ and α.

Example 4.14

Determine a mathematical model for the system of Figure 4.51(a). Let θ be the clockwise angular displacement of the bar, measured from the system's equilibrium position. Assume small θ.

Solution

The free-body diagram of the system at an arbitrary instant is shown in Figure 4.51(b). Use of the small angle assumption linearizes the system. In addition, θ is measured from the system's equilibrium position, so the static force in the spring and the gravity force of the bar cancel each other when the differential equation is derived. Thus these forces are not drawn on the free-body diagram. For a positive value of θ, the displacement of point B is $(2L/3)\theta$ downward and the velocity of point A is $(L/3)\dot{\theta}$ upward. The forces from the spring and viscous damper are drawn consistently on the free-body diagram.

The moment of inertia of a slender bar about an axis through its mass center is given in Table 4.1 as $\bar{I} = \frac{1}{12}mL^2$. Noting that the distance between the mass center and O is $L/6$, the Parallel Axis Theorem is used to determine I_O:

$$I_O = \frac{1}{12}mL^2 + m\left(\frac{L}{6}\right)^2 = \frac{1}{9}mL^2 \tag{a}$$

Applying Equation (4.103) to the free-body diagram of Figure 4.51(b) to sum moments about O leads to

$$\sum M_O = I_O\alpha \tag{b}$$

$$M(t) - \left(c\frac{L}{3}\dot{\theta}\right)\left(\frac{L}{3}\right) - \left(k\frac{2L}{3}\theta\right)\left(\frac{2L}{3}\right) = \frac{1}{9}mL^2\ddot{\theta} \tag{c}$$

Rearranging Equation (d) results in

$$\frac{1}{9}mL^2\ddot{\theta} + \frac{1}{9}cL^2\dot{\theta} + \frac{4}{9}kL^2\theta = M(t) \tag{d}$$

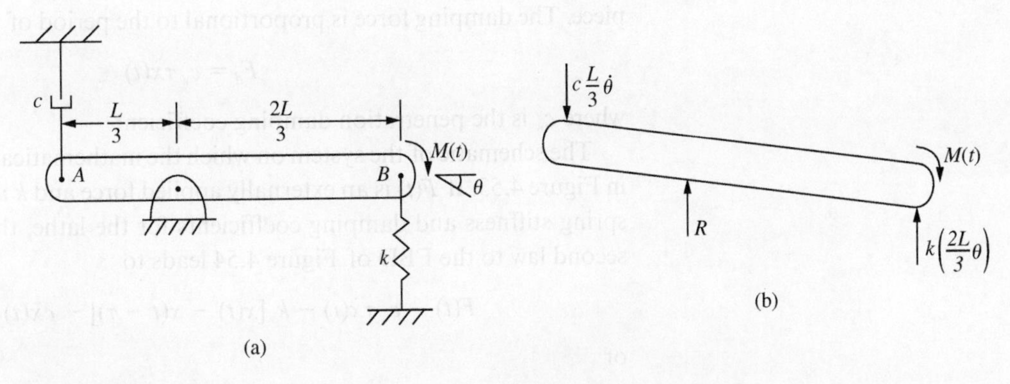

Figure 4.51
(a) The system of Example 4.14; (b) the free-body diagram at an arbitrary instant applying the small angle assumption.

Example 4.15

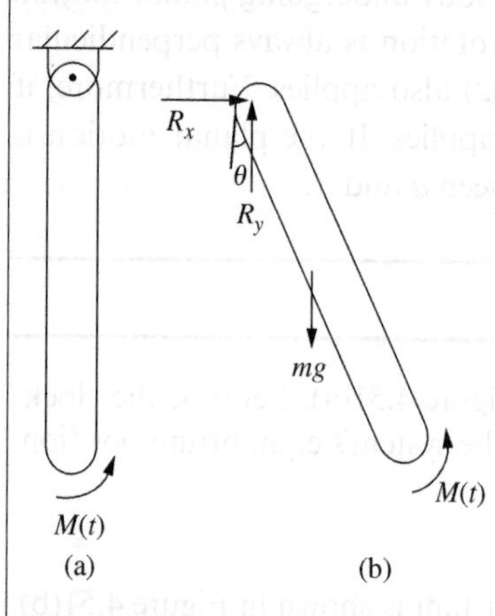

Figure 4.52

(a) The compound pendulum of Example 4.15; (b) the free-body diagram of the compound pendulum at an arbitrary instant.

Determine a linear mathematical model for the compound pendulum of Figure 4.52(a). Let θ be the counterclockwise angular rotation of the bar from the system's equilibrium position. Assume small θ.

Solution

The free-body diagram of the pendulum at an arbitrary instant is shown in Figure 4.52(b). The moment of inertia of the bar about O is obtained using Table 4.1 and the Parallel Axis Theorem as

$$I_O = \frac{1}{12}mL^2 + m\left(\frac{L}{2}\right)^2 = \frac{1}{3}mL^2 \tag{a}$$

Summing moments about O using Equation (4.103) gives

$$M(t) - mg\left(\frac{L}{2}\sin\theta\right) = \frac{1}{3}mL^2\ddot{\theta} \tag{b}$$

For small θ, $\sin\theta \approx \theta$. Using this approximation, Equation (b) is approximated by

$$\frac{1}{3}mL^2\ddot{\theta} + mg\frac{L}{2}\theta = M(t) \tag{c}$$

Mathematical modeling of machine tool vibrations often leads to differential equations with a time delay. Consider the cutting operation of a lathe, as illustrated in Figure 4.53. The workpiece rotates at a constant angular velocity ω. During the cutting operation, the machine tool provides a cutting force to the workpiece. The machine tool is modeled by a mass-spring-viscous damper system. Let $x(t)$ represent the penetration of the cutting tool into the workpiece. The workpiece resists the penetration and develops a cutting force that acts on the tool. The cutting force is modeled as a spring in parallel with a viscous damper, so that the total force provided to the tool is

$$F_c = F_s + F_d \tag{4.104}$$

The spring force is proportional to the instantaneous change in thickness of the workpiece, which is the difference between the current penetration of the tool and its penetration one revolution earlier:

$$F_s = k_w[x(t) - x(t-\tau)] \tag{4.105}$$

where k_w is the cutting stiffness and $\tau = 2\pi/\omega$ is the period of revolution of the workpiece. The damping force is proportional to the period of revolution of the workpiece:

$$F_d = c_w\tau\dot{x}(t) \tag{4.106}$$

where c_w is the penetration damping coefficient.

The schematic of the system on which the mathematical model is based is illustrated in Figure 4.53. If $F(t)$ is an externally applied force and k and c are naturally occurring spring stiffness and damping coefficients for the lathe, then application of Newton's second law to the FBD of Figure 4.54 leads to

$$F(t) - c_w\tau\dot{x}(t) - k_w[x(t) - x(t-\tau)] - c\dot{x}(t) - kx(t) = m\ddot{x}$$

or

$$m\ddot{x} + (c + c_w\tau)\dot{x} + (k + k_w)x - k_w x(t-\tau) = F(t) \tag{4.107}$$

Equation (4.107) is a time delay differential equation.

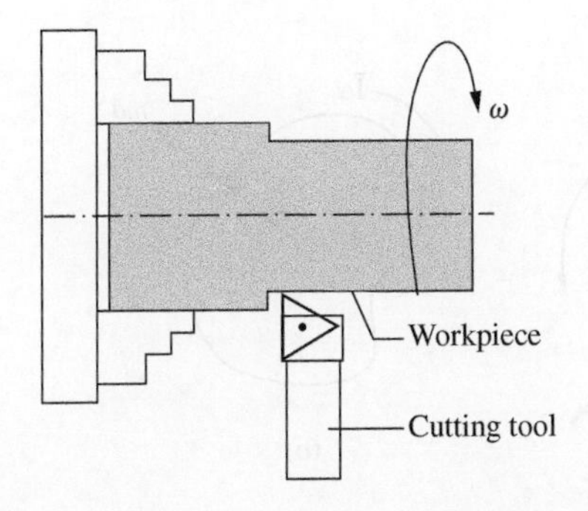

Figure 4.53

As the cutting tool penetrates the workpiece, which is rotating in the lathe at a constant speed ω, a cutting force is imparted to the tool from the workpiece.

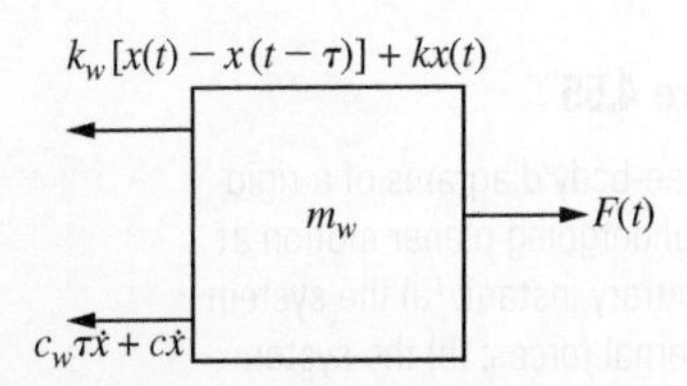

Figure 4.54

A cutting tool is modeled by a mass-spring-viscous damper system. An elastic force is developed which is proportional to the instantaneous thickness of the workpiece. The instantaneous thickness is the difference between the current penetration of the tool and the penetration one revolution earlier.

4.6.5 Three-Dimensional Motion of Rigid Bodies

The position vector of the mass center of a rigid body has three components, each of which may vary independently with time. A rigid body may have angular velocities about three independent axes of rotation. Thus a rigid body may have up to six degrees of freedom and it may be necessary to develop up to six equations to model the motion of the system. The component forms of Equation (4.99) can be used to develop three independent equations. The angular momentum equation, Equation (4.100), is used to develop Euler's equations, which are

$$\sum M_x = \bar{I}_x \dot{\omega}_x - (\bar{I}_y - \bar{I}_z)\omega_y \omega_z \tag{4.108a}$$

$$\sum M_y = \bar{I}_y \dot{\omega}_y - (\bar{I}_z - \bar{I}_x)\omega_z \omega_x \tag{4.108b}$$

$$\sum M_z = \bar{I}_z \dot{\omega}_z - (\bar{I}_x - \bar{I}_y)\omega_x \omega_y \tag{4.108c}$$

Euler's equations, as written in Equations (4.108), are for a set of principal axes located at the mass center of the rigid body.

4.6.6 D'Alembert's Principle

Particles

Newton's second law applied to a particle, Equation (4.97), can be rewritten as

$$\sum \mathbf{F} - m\mathbf{a} = 0 \tag{4.109}$$

Defining the inertia force by $m\mathbf{a}$, Equation (4.109) states that the sum of the external forces acting on the particle minus the inertia force is equal to zero. That is, the resultant of the external forces and the negative of the inertia force is the zero vector. This reformulation of Newton's second law converts the dynamics problem into a statics problem and is called D'Alembert's principle.

Rigid Bodies Undergoing Planar Motion

D'Alembert's principle for a particle is a trivial reformulation of Newton's second law. It is more substantial and useful when developed for a rigid body. For this case, the principles of equivalent force systems are used to develop an alternate formulation of the equations of motion.

The conservation laws that may be applied to rigid bodies undergoing planar motion are summarized by Equations (4.99) and (4.102). The application of these equations, which can be used to derive the differential equations governing the motion of any rigid body undergoing planar motion, is illustrated in the free-body diagrams

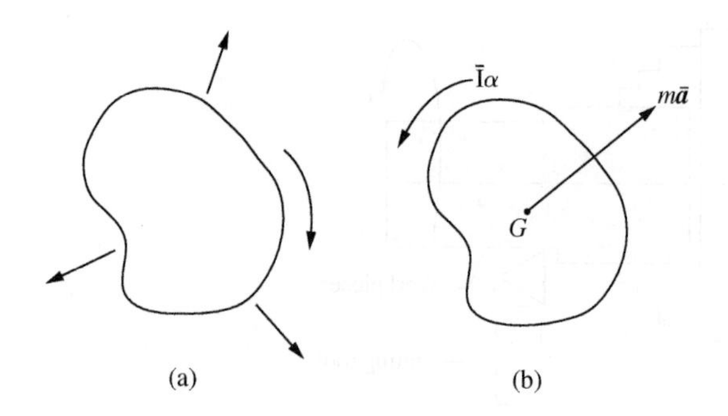

Figure 4.55

The free-body diagrams of a rigid body undergoing planar motion at an arbitrary instant: (a) the system of external forces; (b) the system of effective forces. D'Alembert's principle states that the system of external forces is equivalent to the system of effective forces.

of Figure 4.55. Figure 4.55(a) illustrates the system of external forces acting on a rigid body at an arbitrary instant. Figure 4.55(b) defines the system of effective forces, a force equal to $m\bar{a}$ that is applied at the center of mass of the body and a couple equal to $\bar{I}\alpha$. From Equations (4.99) and (4.102) it is clear that the resultant forces of the two systems are the same and the moment about the mass center for each of the systems is the same. Thus the system of external forces and the system of effective forces are equivalent force systems. This is a statement of **D'Alembert's principle**: at any instant of time, the system of external forces acting on a rigid body undergoing planar motion is equivalent to the body's system of effective forces. The system of effective forces is defined as a force equal to $m\bar{a}$ applied at the mass center and a couple equal to $\bar{I}\alpha$.

The properties of equivalent force systems lead to the following statements of D'Alembert's principle for planar motion:

$$\left(\sum \mathbf{F}\right)_{ext} = \left(\sum \mathbf{F}\right)_{eff} \tag{4.110}$$

$$\left(\sum M_A\right)_{ext} = \left(\sum M_A\right)_{eff} \tag{4.111}$$

The moments for Equation (4.111) are taken about an axis perpendicular to the plane of motion through A, where A is any point in the plane.

D'Alembert's principle provides an alternative to the direct application of Equations (4.99) and (4.102). If A is a fixed axis of rotation, then the Parallel Axis Theorem can be used to reconcile Equation (4.103) with Equation (4.111). Two problems in which the application of D'Alembert's principle is more convenient and provides more flexibility than the direct application of Equations (4.99) and (4.102) or (4.103) are illustrated in Figures 4.56 and 4.57. The axis of rotation for bar BC of the slider-crank mechanism of Figure 4.56(a) is not fixed; it changes as the bar rotates. Thus Equation (4.103) is not applicable. Free-body diagrams of external and effective forces for bar BC are shown in Figure 4.56(b). Principles of rigid-body kinematics must be applied to relate the acceleration of the mass center of the bar and its angular acceleration to the displacement of the collar. Note that gravity and the static deflection of the spring have not been included on the free-body diagram of the external forces because they cancel one another in the governing differential equation. D'Alembert's principle may be applied to this set of free-body diagrams.

The system of Figure 4.57(a) is composed of three bodies. The application of Newton's law requires drawing free-body diagrams and writing Equation (4.99) for each of the blocks and Equation (4.102) for the disk. In a drawing of the individual free-body diagrams, the tensions in the cables connecting the blocks to the disks are identified as unknown quantities. Algebra is used to eliminate the tensions from the equations to derive a single differential equation to solve for $x(t)$. Only one set of free-body diagrams is necessary for application of D'Alembert's principle, as illustrated in Figure 4.57(b). Kinematics is used to relate the accelerations of the various bodies. (This is also necessary when using the direct application of Newton's laws.) A single differential equation is derived by applying Equation (4.111) about the pin support of the disk.

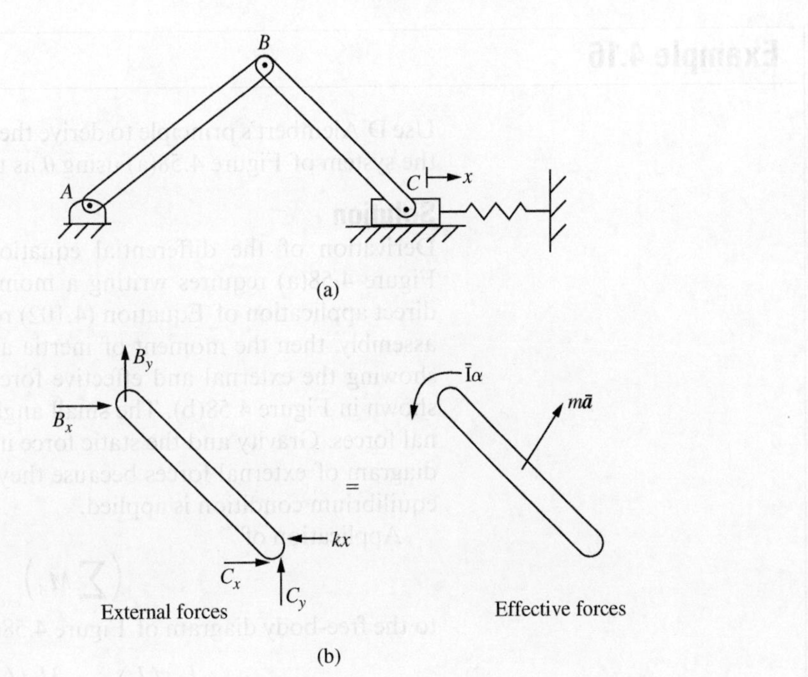

Figure 4.56

(a) The axis of rotation of bar BC changes as it rotates; (b) the free-body diagrams for the application of D'Alembert's principle to bar BC.

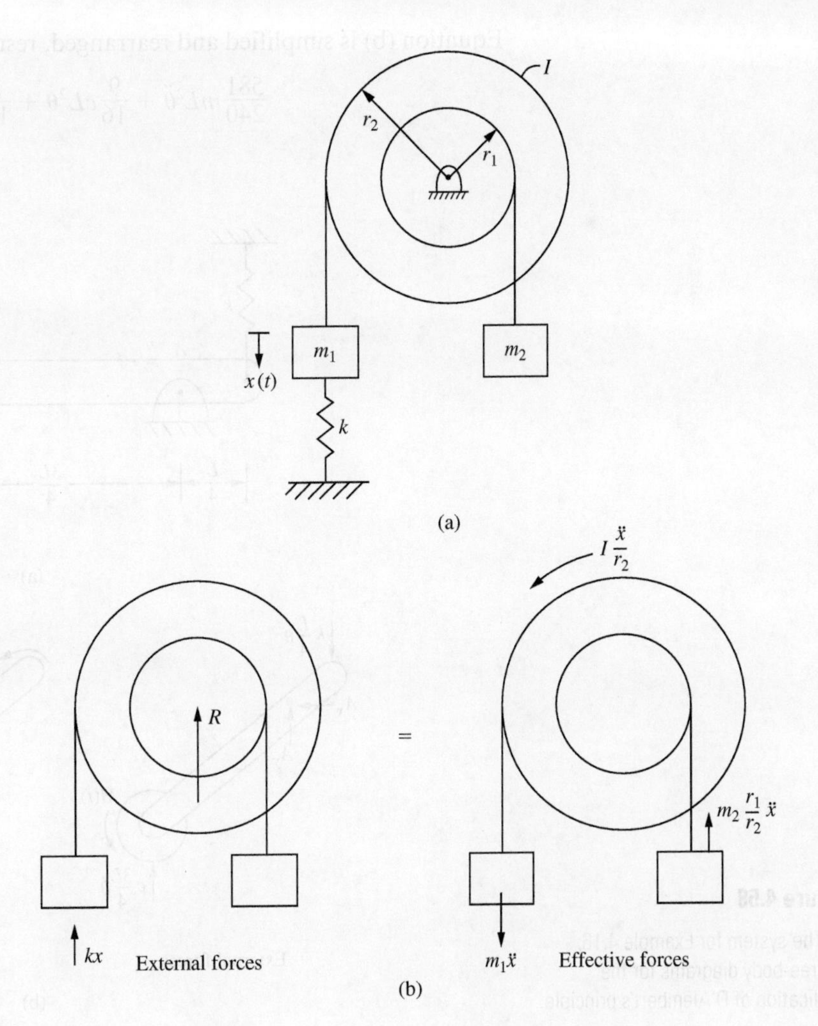

Figure 4.57

(a) Three free-body diagrams and the algebraic elimination of the tensions in the cables are necessary when directly applying Newton's law to derive the governing differential equation; (b) the free-body diagrams for the application of D'Alembert's principle.

Example 4.16

Use D'Alembert's principle to derive the differential equation governing the motion of the system of Figure 4.58(a) using θ as the dependent variable. Assume small θ.

Solution

Derivation of the differential equation governing the motion of the system of Figure 4.58(a) requires writing a moment equation about an axis through A. The direct application of Equation (4.102) requires determining the center of mass of the assembly, then the moment of inertia about the centroidal axis. Free-body diagrams showing the external and effective forces for the system at an arbitrary instant are shown in Figure 4.58(b). The small angle approximation is used in labeling the external forces. Gravity and the static force in the spring are not included on the free-body diagram of external forces because they cancel from the resulting equation when the equilibrium condition is applied.

Application of

$$\left(\sum M_A\right)_{ext} = \left(\sum M_A\right)_{eff} \tag{a}$$

to the free-body diagram of Figure 4.58(b) leads to

$$M(t) - k\frac{L}{4}\theta\left(\frac{L}{4}\right) - c\frac{3L}{4}\dot{\theta}\left(\frac{3L}{4}\right) = m\frac{L}{4}\ddot{\theta}\left(\frac{L}{4}\right) + 4m\frac{3L}{4}\ddot{\theta}\left(\frac{3L}{4}\right) + \frac{1}{12}mL^2\ddot{\theta} + \frac{2}{5}(4m)\left(\frac{L}{8}\right)^2\ddot{\theta} \tag{b}$$

Equation (b) is simplified and rearranged, resulting in

$$\frac{581}{240}mL^2\ddot{\theta} + \frac{9}{16}cL^2\dot{\theta} + \frac{1}{16}kL^2\theta = M(t) \tag{c}$$

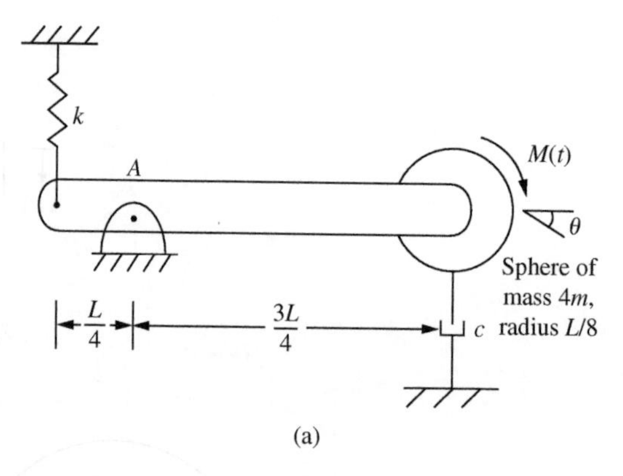

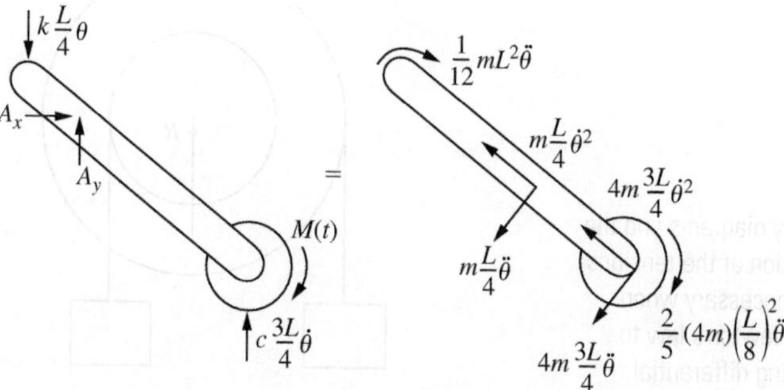

Figure 4.58

(a) The system for Example 4.16; (b) free-body diagrams for the application of D'Alembert's principle.

Example 4.17

Determine a mathematical model for the system of Figure 4.59(a) using x as the dependent variable.

Solution

Free-body diagrams of the system at an arbitrary instant are shown in Figure 4.59(b). If the block of mass m_1 moves down a distance x from the system's equilibrium position, then the disk rotates counterclockwise through an angle $\theta = x/r$, and the block of mass m_2 moves up a distance $3x$. Since x is measured from the system's equilibrium position, static spring forces and gravity forces are not included on the free-body diagrams.

D'Alembert's principle is applied in the form

$$\left(\sum M_A\right)_{ext} = \left(\sum M_A\right)_{eff} \tag{a}$$

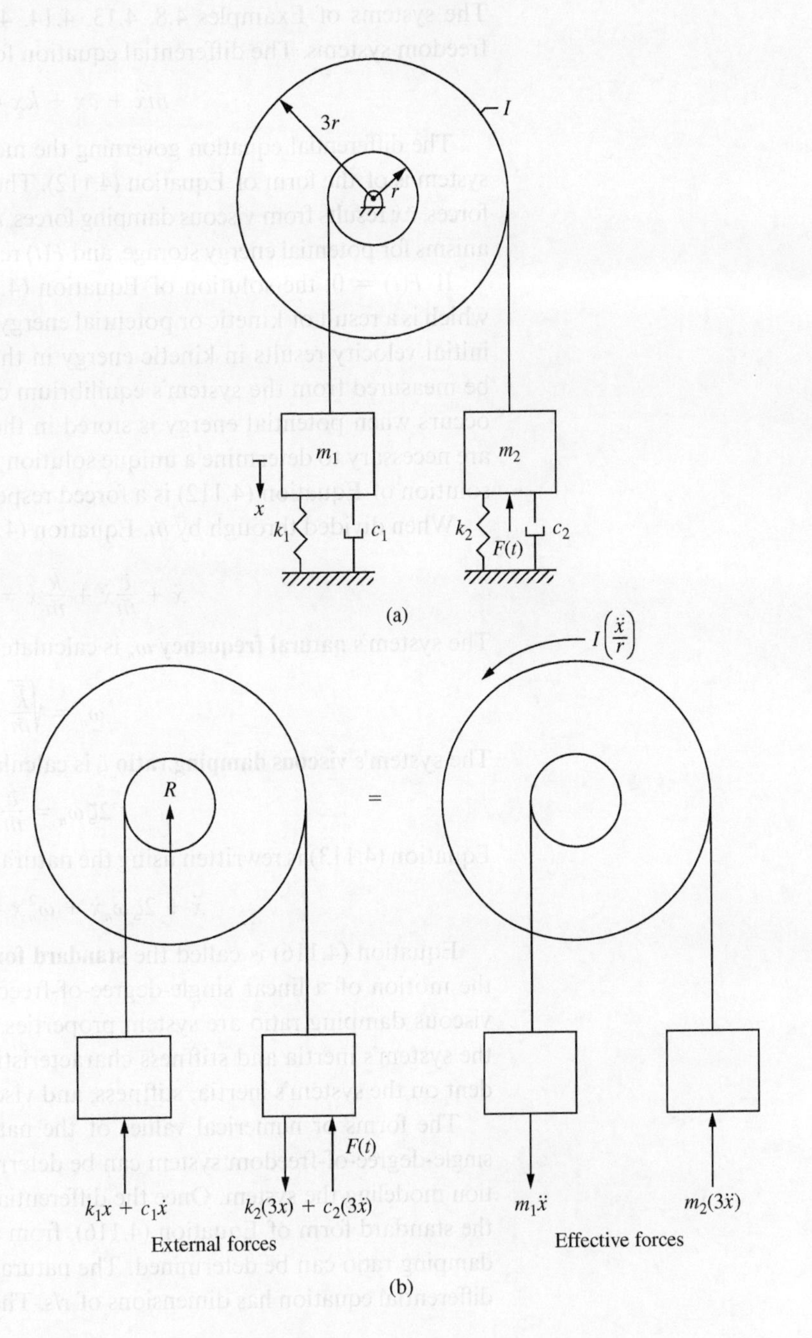

Figure 4.59
(a) The system of Example 4.17; (b) the free-body diagrams at an arbitary instant.

Application of Equation (a) to the free-body diagrams using counterclockwise moments as positive yields

$$F(t)(3r) - (k_1x + c_1\dot{x})r - (3k_2x + 3c_2\dot{x})(3r) = (m_1\ddot{x})r + I\left(\frac{\ddot{x}}{r}\right) + (3m_2\ddot{x})(3r) \qquad \text{(b)}$$

Rearrangement of Equation (b) leads to

$$\left(m_1r + 9m_2r + \frac{I}{r}\right)\ddot{x} + (c_1r + 9c_2r)\dot{x} + (k_1r + 9kr)x = 3rF(t) \qquad \text{(c)}$$

4.7 Single-Degree-of-Freedom Systems

Modeling of the motion of a single-degree-of-freedom (one-degree-of-freedom) system requires one time-dependent variable. The modeling process results in a differential equation whose solution provides the time-dependent behavior of the system. The systems of Examples 4.8, 4.13, 4.14, 4.15, 4.16, and 4.17 are single-degree-of-freedom systems. The differential equation for each of these systems is of the form

$$\tilde{m}\ddot{x} + \tilde{c}\dot{x} + \tilde{k}x = F(t) \qquad (4.112)$$

The differential equation governing the motion of any linear one-degree-of-freedom system is of the form of Equation (4.112). The term $\tilde{m}\ddot{x}$ results from the system's inertia forces, $\tilde{c}\dot{x}$ results from viscous damping forces, $\tilde{k}x$ results from forces in the system's mechanisms for potential energy storage, and $F(t)$ results from externally applied forces.

If $F(t) = 0$, the solution of Equation (4.112) is called the system's free response, which is a result of kinetic or potential energy present in the system at $t = 0$. A nonzero initial velocity results in kinetic energy in the system at $t = 0$. Since x is assumed to be measured from the system's equilibrium condition, a nonzero initial displacement occurs when potential energy is stored in the system at $t = 0$. The initial conditions are necessary to determine a unique solution of Equation (4.112). If $F(t) \neq 0$, then the solution of Equation (4.112) is a forced response.

When divided through by $\tilde{m}$, Equation (4.112) becomes

$$\ddot{x} + \frac{\tilde{c}}{\tilde{m}}\dot{x} + \frac{\tilde{k}}{\tilde{m}}x = \frac{1}{\tilde{m}}F(t) \qquad (4.113)$$

The system's **natural frequency** ω_n is calculated as

$$\omega_n = \sqrt{\frac{\tilde{k}}{\tilde{m}}} \qquad (4.114)$$

The system's **viscous damping ratio** ζ is calculated from

$$2\zeta\omega_n = \frac{\tilde{c}}{\tilde{m}} \qquad (4.115)$$

Equation (4.113) is rewritten using the natural frequency and viscous damping ratio as

$$\ddot{x} + 2\zeta\omega_n\dot{x} + \omega_n^2x = \frac{1}{\tilde{m}}F(t) \qquad (4.116)$$

Equation (4.116) is called the **standard form of the differential equation** governing the motion of a linear single-degree-of-freedom system. The natural frequency and viscous damping ratio are system properties. The natural frequency is dependent on the system's inertia and stiffness characteristics. The viscous damping ratio is dependent on the system's inertia, stiffness, and viscous damping characteristics.

The forms or numerical values of the natural frequency and damping ratio for a single-degree-of-freedom system can be determined directly from the differential equation modeling the system. Once the differential equation is derived, it can be written in the standard form of Equation (4.116), from which the natural frequency and viscous damping ratio can be determined. The natural frequency determined directly from the differential equation has dimensions of r/s. The viscous damping ratio is dimensionless.

Example 4.18

Numerical values for the parameters of the system of Example 4.14 are $m = 10$ kg, $k = 2 \times 10^5$ N/m, $c = 120$ N·s/m, and $L = 2.4$ m. The moment is a single-frequency harmonic excitation of amplitude 200 N·m, frequency 100 r/s, and phase 30°. **a.** Determine the system's natural frequency. **b.** Determine the system's viscous damping ratio. **c.** Write the differential equation governing the angular displacement of the bar in the standard form of Equation (4.115), using the numerical values obtained in parts c. and b.

Solution

Equation (d) of Example 4.14 is divided by $mL^2/9$, leading to

$$\ddot{\theta} + \frac{c}{m}\dot{\theta} + \frac{4k}{m}\theta = \frac{9}{mL^2}M(t) \tag{a}$$

a. The natural frequency is obtained by comparing Equation (a) with Equation (4.116) as

$$\omega_n = \sqrt{4k/m}$$

$$= \sqrt{\frac{4(2 \times 10^5\ \text{N/m})}{10\ \text{kg}}} = 282.8\ \text{r/s} \tag{b}$$

b. Comparison of Equation (a) with Equation (4.116) leads to

$$2\zeta\omega_n = \frac{c}{m} \tag{c}$$

Equation (c) is rearranged to give

$$\zeta = \frac{c}{2m\omega_n}$$

$$= \frac{120\ \text{N·s/m}}{2(10\ \text{kg})(282.8\ \text{r/s})} = 0.0212 \tag{d}$$

c. From the information given, the time-dependent form of $M(t)$ is

$$M(t) = 200\sin\left(100t + \frac{\pi}{6}\right)\ \text{N·m} \tag{e}$$

Substitution of Equations (b), (d), and (e) in Equations (a) and (4.116) leads to

$$\ddot{\theta} + 2(0.0212)(282.8)\dot{\theta} + (282.8)^2\theta = \frac{9}{(10)(2.4)^2}\left[200\sin\left(100t + \frac{\pi}{6}\right)\right] \tag{f}$$

$$\ddot{\theta} + 12\dot{\theta} + 80{,}000\theta = 31.25\sin\left(100t + \frac{\pi}{6}\right) \tag{g}$$

4.8 Multidegree-of-Freedom Systems

The application of Newton's law to the modeling of multidegree-of-freedom systems is similar to that of one-degree-of-freedom systems. Free-body diagrams are drawn of system components at an arbitrary instant and the appropriate forms of Newton's law applied. The differential equations for linear multidegree-of-freedom systems are summarized in a matrix form as

$$\mathbf{M}\ddot{\mathbf{x}} + \mathbf{C}\dot{\mathbf{x}} + \mathbf{K}\mathbf{x} = \mathbf{F} \tag{4.117}$$

For an n-degree-of-freedom system, $\mathbf{x}$ is a $n \times 1$ column vector whose elements are the chosen dependent variables, $\mathbf{M}$ is a $n \times n$ mass matrix, $\mathbf{C}$ is a $n \times n$ damping matrix, $\mathbf{K}$ is a $n \times n$ stiffness matrix, and $\mathbf{F}$ is a $n \times 1$ force vector. A dot above a vector represents differentiation of all elements of the vector with respect to time.

Example 4.19

Derive the differential equations modeling the three-degree-of-freedom system shown in Figure 4.60(a). Dependent variables are measured from the system's equilibrium position. Write the differential equations in matrix form.

Solution

Free-body diagrams of each mass are shown in Figure 4.60(b). The static spring forces cancel the gravity forces, and thus neither is included in the free-body diagrams. The differential equations are derived by applying Newton's second law to each of the blocks. Forces are summed assuming positive downward.

Block 1

$$\sum F = m_1\ddot{x}_1 \tag{a}$$

$$-k_1x_1 - c_1\dot{x}_1 + k_2(x_2 - x_1) + c_2(\dot{x}_2 - \dot{x}_1) = m_1\ddot{x}_1 \tag{b}$$

Block 2

$$\sum F = m_2\ddot{x}_2 \tag{c}$$

$$-k_2(x_2 - x_1) - c_2(\dot{x}_2 - \dot{x}_1) + k_3(x_3 - x_2) + c_3(\dot{x}_3 - \dot{x}_2) = m_2\ddot{x}_2 \tag{d}$$

Block 3

$$\sum F = m_3\ddot{x}_3 \tag{e}$$

$$-k_3(x_3 - x_2) - c_3(\dot{x}_3 - \dot{x}_2) + F(t) = m_3\ddot{x}_3 \tag{f}$$

Simplifying Equations (b), (d), and (f) leads to

$$m_1\ddot{x}_1 + (c_1 + c_2)\dot{x}_1 - c_2\dot{x}_2 + (k_1 + k_2)x_1 - k_2x_2 = 0 \tag{g}$$

$$m_2\ddot{x}_2 - c_2\dot{x}_1 + (c_2 + c_3)\dot{x}_2 - c_3\dot{x}_3 - k_2x_2 + (k_2 + k_3)x_2 - k_3x_3 = 0 \tag{h}$$

$$m_3\ddot{x}_3 - c_3\dot{x}_2 + c_3\dot{x}_3 - k_3x_2 + k_3x_3 = F(t) \tag{i}$$

Equations (g)–(i) are summarized in matrix form as

$$\begin{bmatrix} m_1 & 0 & 0 \\ 0 & m_2 & 0 \\ 0 & 0 & m_3 \end{bmatrix}\begin{bmatrix} \ddot{x}_1 \\ \ddot{x}_2 \\ \ddot{x}_3 \end{bmatrix} + \begin{bmatrix} c_1 + c_2 & -c_2 & 0 \\ -c_2 & c_2 + c_3 & -c_3 \\ 0 & -c_3 & c_3 \end{bmatrix}\begin{bmatrix} \dot{x}_1 \\ \dot{x}_2 \\ \dot{x}_3 \end{bmatrix}$$

$$+ \begin{bmatrix} k_1 + k_2 & -k_2 & 0 \\ -k_2 & k_2 + k_3 & -k_3 \\ 0 & -k_3 & k_3 \end{bmatrix}\begin{bmatrix} x_1 \\ x_2 \\ x_3 \end{bmatrix} = \begin{bmatrix} 0 \\ 0 \\ F(t) \end{bmatrix} \tag{j}$$

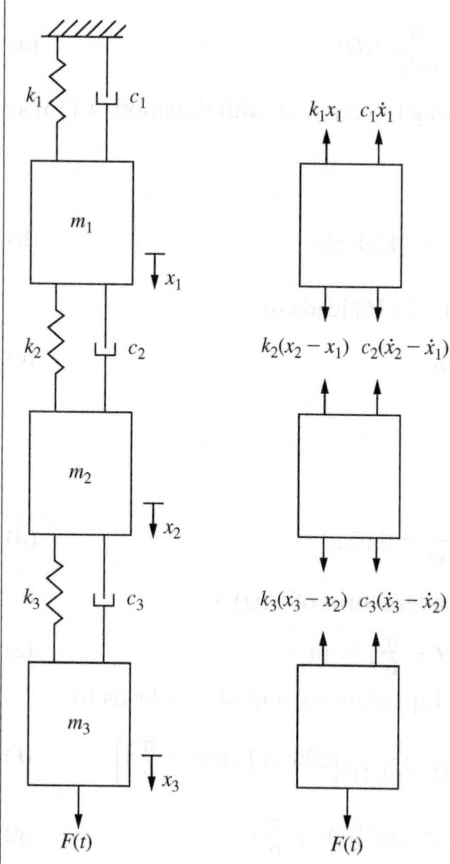

Figure 4.60

(a) The system of Example 4.19; (b) the free-body diagrams at an arbitrary instant.

4.9 Energy Methods

Energy methods provide a convenient alternative to the direct application of Newton's laws in the derivation of differential equations modeling the motion of a mechanical system and are usually preferred for the derivation of equations governing the motion of multidegree-of-freedom systems. The application of energy methods does not require the drawing of free-body diagrams and is often less confusing than the direct application of Newton's laws in that it is not necessary to identify the directions of spring forces to determine the potential energy of the system. Energy methods can be applied to a system of particles and rigid bodies.

4.9.1 Principle of Work and Energy

The Principle of Work and Energy is derived by taking the dot product of Newton's second law with a differential displacement vector and both sides of the resulting equation integrated between two points in the path of motion of the system, represented in the final equation by subscripts 1 and 2. The general form of the Principle of Work and Energy is

$$T_1 + W_{1\to 2} = T_2 \tag{4.118}$$

where T is the kinetic energy of the system and $W_{1\to 2}$ is the work done by all external forces as the system moves between position 1 and position 2.

The work done by a conservative force is independent of the path of motion and dependent only on the initial and final positions of the system. Gravity and linear spring forces are examples of conservative forces. Viscous damping, Coulomb damping, and externally applied forces are examples of nonconservative forces. For any conservative force, a potential energy function V can be defined so that its work is expressed as a difference in potential energies

$$(W_{1\to 2})_c = V_1 - V_2 \tag{4.119}$$

The potential energy of a system at a given instant is dependent on the system parameters and the instantaneous position of the system.

Most systems are subject to a combination of conservative and nonconservative forces. In such a case, the total work is represented as

$$W_{1\to 2} = V_1 - V_2 + (W_{1\to 2})_{NC} \tag{4.120}$$

where $(W_{1\to 2})_{NC}$ is the work done by all nonconservative forces. Using Equation (4.120), the Principle of Work and Energy, Equation (4.118) can be rewritten as

$$T_1 + V_1 + (W_{1\to 2})_{NC} = T_2 + V_2 \tag{4.121}$$

The total energy in a system is the sum of the kinetic and potential energies. If the system is conservative (that is, all forces are conservative), then Equation (4.121) shows that the total energy is a constant and the system satisfies the Principle of Conservation of Energy

$$T + V = C \tag{4.122}$$

4.9.2 Equivalent Systems

The Principle of Work and Energy is used to develop the equivalent systems method, which can be directly applied to derive the differential equation of a linear one-degree-of-freedom system. Let x be the chosen dependent variable for the system. The kinetic energy of a rigid body in planar motion is given by Equation (4.27). Since the system is linear and has only one degree of freedom, both v and ω are proportional to $\dot{x}$. Thus the kinetic energy of the rigid body is of the form

$$T = \frac{1}{2}m_{eq}\dot{x}^2 \tag{4.123}$$

where m_{eq} is a constant dependent on the inertia properties of the system. It is shown in Section 4.2 that a combination of springs can be replaced by a single spring of equivalent stiffness k_{eq} such that the potential energy of the combination is

$$V = \frac{1}{2}k_{eq}x^2 \tag{4.124}$$

It is shown in Section 4.3 that a combination of viscous dampers can be replaced by a single viscous damper of damping coefficient c_{eq} so that the work done by the viscous dampers is

$$W_{1\to 2} = -\int_{x_1}^{x_2} c_{eq}\dot{x}dx \tag{4.125}$$

Define position 1 as the system's initial position and position 2 as the position of the system at an arbitrary instant. Applying the Principle of Work and Energy, Equation (4.121), and using Equations (4.123), (4.124), and (4.125) leads to

$$T_1 + V_1 - \int_{x_1}^{x_2} c_{eq}\dot{x}dx = \frac{1}{2}m_{eq}\dot{x}^2 + \frac{1}{2}k_{eq}x^2 \tag{4.126}$$

Differentiating Equation (4.126) with respect to time, noting that $T_1 + V_1$ is a constant, gives

$$\frac{1}{2}m_{eq}\frac{d\dot{x}^2}{dt} + \frac{1}{2}k_{eq}\frac{dx^2}{dt} + \frac{d}{dt}\left(\int_{x_1}^{x_2} c_{eq}\dot{x}dx\right) = 0 \tag{4.127}$$

The following algebraic relations are noted:

$$\frac{d\dot{x}^2}{dt} = 2\dot{x}\frac{d\dot{x}}{dt} = 2\dot{x}\ddot{x} \tag{4.128a}$$

$$\frac{dx^2}{dt} = 2x\frac{dx}{dt} = 2x\dot{x} \tag{4.128b}$$

$$\frac{d}{dt}\left(\int_{x_1}^{x}\dot{x}dx\right) = \frac{d}{dt}\left(\int_{t_1}^{t}\dot{x}\frac{dx}{dt}dt\right) = \frac{d}{dt}\left(\int_{t_1}^{t}\dot{x}^2 dt\right) = \dot{x}^2 \tag{4.128c}$$

Using Equations (4.128) in Equation (4.127) leads to

$$\dot{x}(m_{eq}\ddot{x} + c_{eq}\dot{x} + k_{eq}x) = 0 \tag{4.129}$$

Equation (4.129) implies that either $\dot{x} = 0$ for all t or

$$m_{eq}\ddot{x} + c_{eq}\dot{x} + k_{eq}x = 0 \tag{4.130}$$

The corresponding form of Equation (4.130) for a rotational system is

$$I_{eq}\ddot{\theta} + c_{t,eq}\dot{\theta} + k_{t,eq}\theta = 0 \tag{4.131}$$

where

$$T = \frac{1}{2}I_{eq}\dot{\theta}^2 \tag{4.132}$$

$$V = \frac{1}{2}k_{t,eq}\theta^2 \tag{4.133}$$

$$W_{1\to 2} = -\int_{\theta_1}^{\theta} c_{t,eq}\dot{\theta}d\theta \tag{4.134}$$

Any linear one-degree-of-freedom system with inertia, stiffness, and/or viscous damping elements can be modeled using either Equation (4.130) or Equation (4.131). The coefficients m_{eq} and I_{eq} can be determined directly from the system's kinetic energy at an arbitrary instant, Equation (4.123) or Equation (4.132). The coefficients k_{eq} and $k_{t,eq}$ can be determined directly from the system's potential energy at an arbitrary instant, Equation (4.124) or Equation (4.133). If the system has viscous damping components, the coefficients c_{eq} and $c_{t,eq}$ can be determined from the work done by the viscous damping forces, Equation (4.125) or Equation (4.134).

If the system is subject to external forces, Equations (4.130) and (4.131) are modified as

$$m_{eq}\ddot{x} + c_{eq}\dot{x} + k_{eq}x = F_{eq}(t) \tag{4.135}$$

$$I_{eq}\ddot{\theta} + c_{t,eq}\dot{\theta} + k_{t,eq}\theta = M_{eq}(t) \tag{4.136}$$

where $F_{eq}(t)$ or $M_{eq}(t)$ is obtained by equating the power delivered by the external forces to the power requirement for an equivalent force.

As is the case when applying Newton's second law to model a mechanical system, the static forces developed in springs and the gravity forces leading to static deflections cancel one another if the system is linear and the dependent variable is measured from the system's equilibrium position. In determining the potential energy developed in a spring, the value of x used in Equation (4.43) is the change in length of the spring measured from the system's equilibrium position. The potential energy due to gravity forces that induce static deflections is not included in the potential energy calculation.

Example 4.20

Use the energy method to derive the differential equation for the system of Figure 4.61 using x, measured from the system's equilibrium position, as the dependent variable.

Solution

Let θ be the clockwise angular rotation of the pulley and y the displacement of the mass center of the disk, both measured from the system's equilibrium position. These are related to x by

$$\theta = \frac{x}{r_1} \tag{a}$$

$$y = \frac{r_2}{r_1}x \tag{b}$$

The total kinetic energy of the system is the sum of the kinetic energy in each of the bodies:

$$T = \frac{1}{2}m_1\dot{x}^2 + \frac{1}{2}I_p\dot{\theta}^2 + \frac{1}{2}m_2\dot{y}^2 + \frac{1}{2}\left(\frac{1}{2}m_2r_D^2\right)\omega^2 \tag{c}$$

where the moment of inertia of the disk is obtained using Table 4.1, r_D is the radius of the disk, and ω is its angular velocity. Since the disk rolls without slip,

$$\omega = \frac{\dot{y}}{r_D} \tag{d}$$

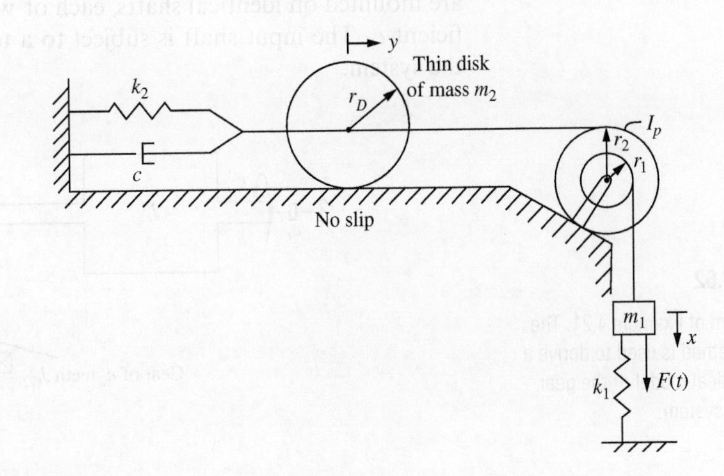

Figure 4.61

The system of Example 4.20 illustrating use of the energy method to derive the equation of motion for this one-degree-of-freedom system.

Substituting Equations (a), (b), and (d) into Equation (c) leads to

$$T = \frac{1}{2}\left(m_1 + \frac{I_p}{r_1^2} + \frac{3r_2^2}{2r_1^2}m_2\right)\dot{x}^2 \tag{e}$$

Comparison of Equation (e) with Equation (4.123) gives

$$m_{eq} = m_1 + \frac{I_p}{r_1^2} + \frac{3}{2}\left(\frac{r_2}{r_1}\right)^2 m_2 \tag{f}$$

The total potential energy of the system is

$$V = \frac{1}{2}k_1 x^2 + \frac{1}{2}k_2 y^2 = \frac{1}{2}\left(k_1 + \frac{r_2^2}{r_1^2}k_2\right)x^2 \tag{g}$$

Comparison of Equation (g) with Equation (4.124) leads to

$$k_{eq} = k_1 + \left(\frac{r_2}{r_1}\right)^2 k_2 \tag{h}$$

The work done by the viscous damper as the system moves between its initial position and an arbitrary position is

$$W_{1\to 2} = -\int_{y_1}^{y} c\dot{y}dy = -\int_{x_1}^{x} c\left(\frac{r_2}{r_1}\dot{x}\right)d\left(\frac{r_2}{r_1}x\right) = -\int_{x_1}^{x} c\left(\frac{r_2}{r_1}\right)^2 \dot{x}dx \tag{i}$$

Comparison of Equation (i) with Equation (4.125) leads to

$$c_{eq} = \left(\frac{r_2}{r_1}\right)^2 c \tag{j}$$

The power delivered by the force at an arbitrary instant is $P = F(t)\dot{x}$. Thus,

$$F_{eq}(t) = F(t) \tag{k}$$

The differential equation is obtained by using Equations (f), (h), (j), and (k) in Equation (4.135), leading to

$$\left[m_1 + \frac{I_p}{r_1^2} + \frac{3}{2}\left(\frac{r_2}{r_1}\right)^2 m_2\right]\ddot{x} + \left(\frac{r_2}{r_1}\right)^2 c\dot{x} + \left[k_1 + \left(\frac{r_1}{r_2}\right)^2 k_2\right]x = F(t) \tag{l}$$

Example 4.21

A gear reduction system is used to decrease the angular velocity from an input motor as illustrated in Figure 4.62. The angular velocity of the output rotor is ω. The rotors are mounted on identical shafts, each of which has a torsional viscous damping coefficient c_t. The input shaft is subject to a torque T. Derive a mathematical model for the system.

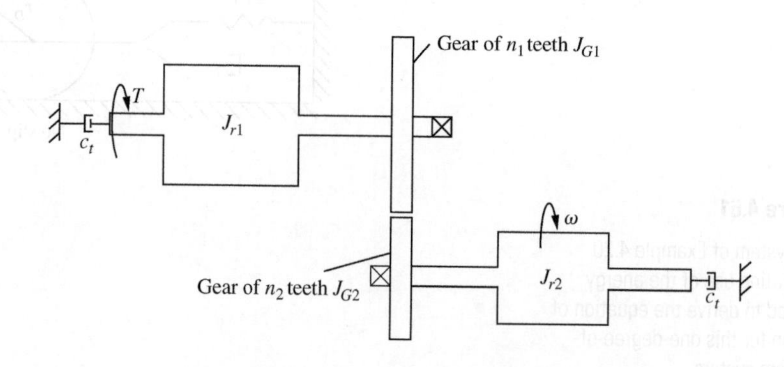

Figure 4.62

The system of Example 4.21. The energy method is used to derive a mathematical model of the gear reduction system.

Solution

Let ω_1 be the angular velocity of the input shaft. The meshing equation for the gears is

$$n_1\omega_1 = n_2\omega \tag{a}$$

$$\omega_1 = \frac{n_2}{n_1}\omega \tag{b}$$

The total kinetic energy of the system is

$$T = \frac{1}{2}J_{r1}\omega_1^2 + \frac{1}{2}J_{G1}\omega_1^2 + \frac{1}{2}J_{r2}\omega^2 + \frac{1}{2}J_{G2}\omega^2 \tag{c}$$

Using Equation (b) in Equation (c) leads to

$$T = \frac{1}{2}\left[(J_{r1} + J_{G1})\left(\frac{n_2}{n_1}\right)^2 + J_{r2} + J_{G2}\right]\omega^2 \tag{d}$$

Comparison of Equation (d) with Equation (4.132) leads to

$$I_{eq} = (J_{r1} + J_{G1})\left(\frac{n_2}{n_1}\right)^2 + J_{r2} + J_{G2} \tag{e}$$

The work done by the torsional viscous dampers is

$$W_{1\to 2} = -\int_{\theta_{1,0}}^{\theta_1} c_t\dot{\theta}_1 d\theta_1 - \int_{\theta_0}^{\theta} c_t\dot{\theta}d\theta = -\int_{\theta_0}^{\theta} c_t\left(\frac{n_2}{n_1}\dot{\theta}\right)d\left(\frac{n_2}{n_1}\theta\right) - \int_{\theta_0}^{\theta} c_t\dot{\theta}d\theta$$

$$W_{1\to 2} = -\int_{\theta_0}^{\theta}\left[1 + \left(\frac{n_2}{n_1}\right)^2\right]c_t\dot{\theta}d\theta \tag{f}$$

Comparison of Equation (f) to Equation (4.134) leads to

$$c_{t,eq} = \left[1 + \left(\frac{n_2}{n_1}\right)^2\right]c_t \tag{g}$$

The torque is applied to the input shaft. An equivalent torque applied to the output shaft is obtained by equating the power developed by the torque to the power developed by a torque if applied to the output shaft:

$$T\omega_1 = T_{eq}\omega$$

$$T_{eq} = \frac{\omega_1}{\omega}T$$

$$= \frac{n_2}{n_1}T \tag{h}$$

The differential equation modeling the system is obtained by substituting Equations (e), (g), and (h) into Equation (4.136), resulting in

$$\left[(J_{r1} + J_{G1})\left(\frac{n_2}{n_1}\right)^2 + J_{r2} + J_{G2}\right]\ddot{\theta} + \left[1 + \left(\frac{n_2}{n_1}\right)^2\right]c_t\dot{\theta} = \left(\frac{n_2}{n_1}\right)T \tag{i}$$

Noting that the angular velocity of the output shaft is $\omega = \dot{\theta}$, Equation (i) can be written as a differential equation for ω as

$$\left[(J_{r1} + J_{G1})\left(\frac{n_2}{n_1}\right)^2 + J_{r2} + J_{G2}\right]\dot{\omega} + \left[1 + \left(\frac{n_2}{n_1}\right)^2\right]c_t\omega = \left(\frac{n_2}{n_1}\right)T \tag{j}$$

4.9.3 Energy Storage

Equation (4.122) shows that in the absence of nonconservative forces the total energy (the sum of the kinetic energy and the potential energy) remains constant. A decrease in kinetic energy is compensated by an increase in potential energy and vice versa. The term "energy" in this context refers to stored energy or energy that can be converted from kinetic to potential or vice versa.

The kinetic energy of a system consisting of a single particle of mass m moving with a velocity $\dot{x}$ is $T = \frac{1}{2}m\dot{x}^2$. The initial velocity of the particle is determined by factors external to the system. However, the mass is constant. In this context, the mass can be viewed as a measure of the system's capacity for the storage of kinetic energy. Similarly m_{eq}, a system's equivalent mass, is the total capacity of the system to store kinetic energy.

The potential energy of a single spring change in length x is $V = \frac{1}{2}kx^2$. The initial displacement of the spring is determined by factors external to the system, but the stiffness is constant. The spring stiffness is a measure of its capacity to store potential energy. Similarly k_{eq}, a system's equivalent stiffness, is the total capacity of the system to store potential energy.

Equation (4.125) or Equation (4.134) show that the work done by the viscous damping forces is negative. Thus Equation (4.121) show that when viscous damping is present, the total energy is continually decreasing. A viscous damper does not store energy; it dissipates stored energy. The damping coefficient is a measure of the viscous damper's capacity to dissipate energy.

4.9.4 Lagrange's Equations for Multidegree-of-Freedom Systems

The energy method can be extended to multidegree-of-freedom systems by using a set of equations called Lagrange's equations. The development of Lagrange's equations and their use for nonconservative systems are beyond the scope of this book. Lagrange's equations can be used for both nonlinear and linear systems. However, only their use for linear systems is considered in this book.

Consider an n-degree-of-freedom system with dependent variables $x_1, x_2, \ldots, x_n$ chosen for modeling of the system. The kinetic energy of a linear system at an arbitrary time can be developed in the form of

$$T = T(\dot{x}_1, \dot{x}_2, \ldots, \dot{x}_n) \tag{4.137}$$

The potential energy of a linear system at an arbitrary instant can be developed in the form of

$$V = V(x_1, x_2, \ldots, x_n) \tag{4.138}$$

The lagrangian L is defined as the difference between the kinetic and potential energies

$$L = T - V \tag{4.139}$$

Thus the lagrangian is a function of the dependent variables and their first time derivatives.

Lagrange's equations for a conservative system are

$$\frac{d}{dt}\left(\frac{\partial L}{\partial \dot{x}_i}\right) - \frac{\partial L}{\partial x_i} = 0 \quad i = 1, 2, \ldots, n \tag{4.140}$$

The application of Lagrange's equations for $i = 1, 2, \ldots, n$ leads to n equations that become the differential equations governing the motion of the n-degree-of-freedom system.

Example 4.22

Use Lagrange's equations to derive the differential equations governing the motion of the two-degree-of-freedom system of Figure 4.63.

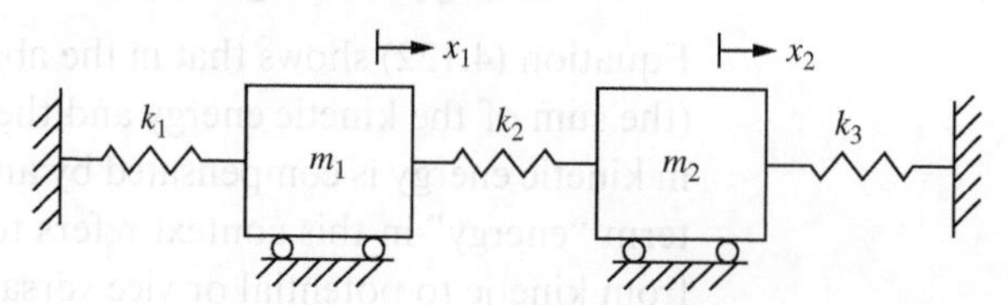

Figure 4.63
The two-degree-of-freedom system of Example 4.22.

Solution

The kinetic energy of the system at an arbitrary instant is

$$T = \frac{1}{2}m_1\dot{x}_1^2 + \frac{1}{2}m_2\dot{x}_2^2 \tag{a}$$

The potential energy of the system at an arbitrary instant is

$$V = \frac{1}{2}k_1x_1^2 + \frac{1}{2}k_2(x_2 - x_1)^2 + \frac{1}{2}k_3x_2^2 \tag{b}$$

Equations (a) and (b) are used in Equation (4.140) to determine the lagrangian

$$L = \frac{1}{2}m_1\dot{x}_1^2 + \frac{1}{2}m_2\dot{x}_2^2 - \left(\frac{1}{2}k_1x_1^2 + \frac{1}{2}k_2(x_2 - x_1)^2 + \frac{1}{2}k_3x_2^2\right) \tag{c}$$

Using Equation (c) in Lagrange's equations, Equation (4.140) leads to

$i = 1$:

$$\frac{d}{dt}\left(\frac{\partial L}{\partial \dot{x}_1}\right) - \frac{\partial L}{\partial x_1} = 0 \tag{d}$$

$$\frac{d}{dt}(m_1\dot{x}_1) - [-k_1x_1 - k_2(x_2 - x_1)(-1)] = 0 \tag{e}$$

$$m_1\ddot{x}_1 + (k_1 + k_2)x_1 - k_2x_2 = 0 \tag{f}$$

$i = 2$:

$$\frac{d}{dt}\left(\frac{\partial L}{\partial \dot{x}_2}\right) - \frac{\partial L}{\partial x_2} = 0 \tag{g}$$

$$\frac{d}{dt}(m_2\dot{x}_2) - [-k_2(x_2 - x_1) - k_3x_2] = 0 \tag{h}$$

$$m_2\ddot{x}_2 - k_2x_1 + (k_2 + k_3)x_2 = 0 \tag{i}$$

For a nonconservative system, Lagrange's equations are of the form

$$\frac{d}{dt}\left(\frac{\partial L}{\partial \dot{x}_i}\right) - \frac{\partial L}{\partial x_i} = Q_i \qquad i = 1, 2, \ldots, n \tag{4.141}$$

where Q_i, the generalized forces, are developed using the method of virtual work. Virtual displacements, $\delta x_1, \delta x_2, \ldots, \delta x_n$, are assumed for the dependent variables. The work done by the force is calculated assuming the system has moved through the virtual displacements. This is the virtual work δW. The virtual work can be written as

$$\delta W = \sum_{i=1}^{n} Q_i \delta x_i \tag{4.142}$$

Example 4.23

Derive a mathematical model for the system of Figure 4.64 using θ, x_1, and x_2 as dependent variables. Assume small θ.

Solution

The kinetic energy of the system at an arbitrary instant is

$$T = \frac{1}{2}m\left(\frac{L}{2}\dot{\theta}\right)^2 + \frac{1}{2}\left(\frac{1}{12}mL^2\right)\dot{\theta}^2 + \frac{1}{2}(2m)\dot{x}_1^2 + \frac{1}{2}(3m)\dot{x}_2^2$$

The potential energy of the system at an arbitrary instant is

$$V = \frac{1}{2}k(L\theta)^2 + \frac{1}{2}k\left(x_1 - \frac{3L}{4}\theta\right)^2 + \frac{1}{2}k(x_2 - x_1)^2$$

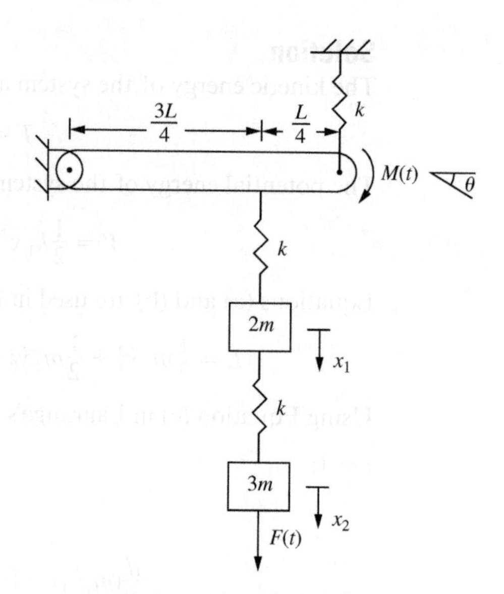

Figure 4.64

System of Example 4.23.

Assume virtual displacements $\delta\theta$, δx_1, and δx_2. The work done by the external forces as the system experiences these virtual displacements is

$$\delta W = M(t)\delta\theta + F(t)\delta x_2$$

Application of Lagrange's equations, Equation (4.141), gives

$$\frac{d}{dt}\left(\frac{\partial L}{\partial \dot{\theta}}\right) - \frac{\partial L}{\partial \theta} = M(t)$$

$$\frac{1}{2}\left(\frac{1}{3}mL^2\right)\frac{d}{dt}(\dot{\theta}^2) - \left[-k(2)\theta - k(2)\left(x_1 - \frac{3L}{4}\theta\right)\left(-\frac{3L}{4}\right)\right] = M(t)$$

$$\frac{1}{3}mL^2\ddot{\theta} + \frac{25}{16}kL^2\theta - \frac{3L}{4}kx_1 = M(t) \tag{a}$$

$$\frac{d}{dt}\left(\frac{\partial L}{\partial \dot{x}_1}\right) - \frac{\partial L}{\partial x_1} = 0$$

$$\frac{1}{2}(2m)\frac{d}{dt}(\dot{x}_1) - \left[-\frac{1}{2}k(2)\left(x_1 - \frac{3L}{4}\theta\right) - \frac{1}{2}k(2)(x_2 - x_1)(-1)\right] = 0$$

$$2m\ddot{x}_1 - \frac{3L}{4}k\theta + 2kx_1 - kx_2 = 0 \tag{b}$$

$$\frac{d}{dt}\left(\frac{\partial L}{\partial \dot{x}_2}\right) - \frac{\partial L}{\partial x_2} = F(t)$$

$$\frac{1}{2}(3m)\frac{d}{dt}(\dot{x}_2) - \left[-\frac{1}{2}k(2)(x_2 - x_1)\right] = F(t)$$

$$3m\ddot{x}_2 - kx_1 + kx_2 = F(t) \tag{c}$$

Summarizing Equations (a), (b), and (c) in matrix form leads to

$$\begin{bmatrix} \frac{1}{3}mL^2 & 0 & 0 \\ 0 & 2m & 0 \\ 0 & 0 & 3m \end{bmatrix}\begin{bmatrix} \ddot{\theta} \\ \ddot{x}_1 \\ \ddot{x}_2 \end{bmatrix} + \begin{bmatrix} \frac{25}{16}kL^2 & -\frac{3L}{4}k & 0 \\ -\frac{3L}{4}k & 2k & -k \\ 0 & -k & k \end{bmatrix}\begin{bmatrix} \theta \\ x_1 \\ x_2 \end{bmatrix} = \begin{bmatrix} M(t) \\ 0 \\ F(t) \end{bmatrix} \tag{d}$$

4.10 Transfer Functions for Mechanical Systems

4.10.1 One-Degree-of-Freedom Systems

A linear one-degree-of-freedom system composed of a mass, a spring, and a viscous damper with a force input has the mathematical model given by Equation (4.112). The tildes over the parameters are dropped for convenience. This leads to the differential equation governing a linear one-degree-of-freedom mechanical system as

$$m\ddot{x} + c\dot{x} + kx = F(t) \tag{4.143}$$

The system is defined with an input $F(t)$ and an output $x(t)$. Define $X(s) = \mathcal{L}\{x(t)\}$ and $F(s) = \mathcal{L}\{f(t)\}$. Taking the Laplace transform of Equation (4.143), assuming all initial conditions are zero, leads to

$$(ms^2 + cs + k)X(s) = F(s) \tag{4.144}$$

The transfer function, the ratio of the transform of the output to the transform of the input $G(s) = \frac{X(s)}{F(s)}$, is thus obtained from Equation (4.144) as

$$G(s) = \frac{1}{ms^2 + cs + k} \tag{4.145}$$

Dividing the numerator and denominator of Equation (4.145) by m leads to

$$G(s) = \frac{1/m}{s^2 + \frac{c}{m}s + \frac{k}{m}} \tag{4.146}$$

Rewriting Equation (4.146) using the definitions of natural frequency, Equation (4.114), and damping ratio, Equation (4.115), leads to

$$G(s) = \frac{1/m}{s^2 + 2\zeta\omega_n s + \omega_n^2} \tag{4.147}$$

The differential equation for a model linear system with motion input, as in Example 4.11 and Figure 4.46, is

$$m\ddot{x} + c\dot{x} + kx = c\dot{z} + kz \tag{4.148}$$

The output for the system is $x(t)$ and the input is $y(t)$. Let $X(s) = \mathcal{L}\{x(t)\}$ and $Z(s) = \mathcal{L}\{z(t)\}$. Taking the Laplace transform of Equation (4.148) with all initial conditions for $x(t)$ and $y(t)$ set equal to zero gives

$$(ms^2 + cs + k)X(s) = (cs + k)Z(s) \tag{4.149}$$

The transfer function defined by $G(s) = \frac{X(s)}{Z(s)}$ is

$$G(s) = \frac{cs + k}{ms^2 + cs + k} \tag{4.150}$$

Dividing the numerator and denominator of Equation (4.150) by m and using the definitions of natural frequency and damping ratio leads to

$$G(s) = \frac{2\zeta\omega_n s + \omega_n^2}{s^2 + 2\zeta\omega_n s + \omega_n^2} \tag{4.151}$$

The denominators of Equations (4.147) and (4.151) are both second-order polynomials. This indicates that there are two energy storage elements in each system: a spring, which stores potential energy, and a mass, which stores kinetic energy. The zeroes of the transfer function are determined by setting the denominator to zero. The quadratic formula is applied, leading to

$$s = -\omega_n\left(\zeta \pm \sqrt{\zeta^2 - 1}\right)$$

The transfer function can have real or complex poles, depending on the value of ζ. The implications of real or complex poles are considered in Chapter 7.

The transfer function of Equation (4.147) has no zeroes, as its numerator is a polynomial of zeroth-order. The transfer function given by Equation (4.151) has a zero of $s = -\frac{\omega_n}{2\zeta}$ because this is the value where its first-order numerator is zero.

Example 4.24

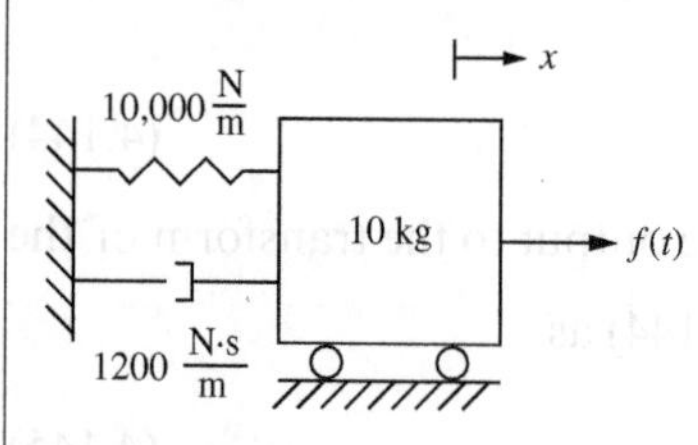

Figure 4.65

System of Example 4.24.

The components of the system of Figure 4.65 have the parameters: $m = 10$ kg, $c = 1200\ \text{N}\cdot\frac{\text{s}}{\text{m}}$, and $k = 10{,}000$ N/m. **a.** Specify the system's transfer function, $G(s) = \frac{X(s)}{F(s)}$. **b.** Use the transfer function to determine the system's natural frequency and damping ratio.

Solution

a. Substituting the parameters into the transfer function of Equation (4.146) leads to

$$G(s) = \frac{1}{10s^2 + 1200s + 10{,}000} \tag{a}$$

Writing the transfer function with a form where the leading coefficient of the denominator is one leads to

$$G(s) = \frac{0.1}{s^2 + 120s + 1000} \tag{b}$$

b. Comparing Equation (b) with Equation (4.147) gives

$$\omega_n^{\ 2} = 1000 \Rightarrow \omega_n = 31.6\ \frac{\text{rad}}{\text{s}} \tag{c}$$

$$2\zeta\omega_n = 120 \Rightarrow \zeta = \frac{120}{2(31.6)} = 1.90 \tag{d}$$

Example 4.25

Determine the transfer function for the system of Example 4.16 and Figure 4.58 when $\theta(t)$ is used as the output and $M(t)$ is the input. Write the transfer function as the ratio of two polynomials such that the leading coefficient of the polynomial in the denominator is 1.

Solution

The mathematical model for Figure 4.58 is Equation (c) of Example 4.16

$$\frac{581}{240}mL^2\ddot{\theta} + \frac{9}{16}cL^2\dot{\theta} + \frac{1}{16}kL^2\theta = M(t) \tag{a}$$

Let $\Theta(s) = \mathcal{L}\{\theta(t)\}$ and define $M(s) = \mathcal{L}\{M(t)\}$. Taking the Laplace transform of Equation (a), assuming all initial conditions are zero, gives

$$\left(\frac{581}{240}mL^2s^2 + \frac{9}{16}cL^2s + \frac{1}{16}kL^2\right)\Theta(s) = M(s) \tag{b}$$

Noting that $G(s) = \frac{\Theta(s)}{M(s)}$,

$$G(s) = \frac{1}{\frac{581}{240}mL^2s^2 + \frac{9}{16}cL^2s + \frac{1}{16}kL^2} \tag{c}$$

Dividing the numerator and denominator of the right side of Equation (c) by $\frac{581}{240}mL^2$ to make the leading coefficient of the denominator equal to one leads to

$$G(s) = \frac{\frac{240}{581mL^2}}{s^2 + \frac{135c}{581m}s + \frac{15k}{581m}} \tag{d}$$

Example 4.26

Determine the transfer function for the system of Example 4.21 and Figure 4.62 when $\omega(t)$ is viewed as the output and $T(t)$ is the input. Write the transfer function as the ratio of two polynomials such that the leading coefficient of the polynomial in the denominator is one.

Solution

The mathematical model for $\omega(t)$ is Equation (j) of Example 4.21. It is repeated here:

$$\left[(J_{r1} + J_{G1})\left(\frac{n_2}{n_1}\right)^2 + J_{r2} + J_{G2}\right]\dot{\omega} + \left[1 + \left(\frac{n_2}{n_1}\right)^2\right]c_t\omega = \left(\frac{n_2}{n_1}\right)T \tag{a}$$

Define $\Omega(s) = \mathscr{L}\{\omega(t)\}$ and $T(s) = \mathscr{L}\{T(t)\}$. Taking the Laplace transform of Equation (a), assuming all initial conditions are zero, leads to

$$\left\{\left[(J_{r1} + J_{G1})\left(\frac{n_2}{n_1}\right)^2 + J_{r2} + J_{G2}\right]s + \left[1 + \left(\frac{n_2}{n_1}\right)^2\right]\right\}\Omega(s) = \left(\frac{n_2}{n_1}\right)T(s) \tag{b}$$

The transfer function $G(s) = \frac{\Omega(s)}{T(s)}$ is determined from Equation (b) as

$$G(s) = \frac{\frac{n_2}{n_1}}{\left[(J_{r1} + J_{G1})\left(\frac{n_2}{n_1}\right)^2 + J_{r2} + J_{G2}\right]s + 1 + \left(\frac{n_2}{n_1}\right)^2} \tag{c}$$

Requiring the leading coefficient of the first-order polynomial in the denominator to be one leads to

$$G(s) = \frac{\dfrac{\frac{n_2}{n_1}}{(J_{r1} + J_{G1})\left(\frac{n_2}{n_1}\right)^2 + J_{r2} + J_{G2}}}{s + \dfrac{1 + \left(\frac{n_2}{n_1}\right)^2}{(J_{r1} + J_{G1})\left(\frac{n_2}{n_1}\right)^2 + J_{r2} + J_{G2}}} \tag{d}$$

The transfer function of Equation (d) is first-order because there is one independent energy storage device in the system; the rotors store kinetic energy.

Example 4.27

Determine the transfer function for the mathematical model of a machine tool given by Equation (4.107).

Solution

The differential equation governing machine tool vibrations is Equation (4.107):

$$m\ddot{x} + (c + c_w\tau)\dot{x} + (k + k_w)x - k_w x(t - \tau) = F(t) \tag{a}$$

The second shifting theorem implies that

$$\mathcal{L}\{x(t-\tau)\} = e^{-\tau s}X(s) \tag{b}$$

Taking the Laplace transform of the differential equation, assuming all initial conditions are zero, and using Equation (b) gives

$$ms^2X(s) + (c + c_w\tau)sX(s) + (k + k_w)X(s) - k_w e^{-\tau s}X(s) = F(s) \tag{c}$$

Rearranging Equation (c) leads to

$$X(s) = \frac{F(s)}{ms^2 + (c + c_w\tau)s + k + k_w - k_w e^{-\tau s}} \tag{d}$$

The transfer function is determined from Equation (d) as

$$G(s) = \frac{X(s)}{F(s)} = \frac{1}{ms^2 + (c + c_w\tau)s + k + k_w - k_w e^{-\tau s}} \tag{e}$$

4.10.2 Multidegree-of-Freedom Systems

The mathematical models for linear mechanical systems where more than one dependent variable necessary to model the system are summarized by Equation (4.117) and repeated here:

$$\mathbf{M}\ddot{\mathbf{x}} + \mathbf{C}\dot{\mathbf{x}} + \mathbf{K}\mathbf{x} = \mathbf{F} \tag{4.117}$$

The vectors $\mathbf{x}$ and $\mathbf{F}$ are $n \times 1$ column vectors. The input vector may be a different size than $\mathbf{F}$. For example, the three-degree-of-freedom system of Example 4.19 is subject to only a single force.

Consider an n-degree-of-freedom system subject to m forces. Define an $m \times 1$ column vector $\mathbf{q}$, a vector of inputs, and an $n \times m$ matrix $\mathbf{R}$ such that

$$\mathbf{F} = \mathbf{R}\mathbf{q} \tag{4.152}$$

Equation (4.117) becomes

$$\mathbf{M}\ddot{\mathbf{x}} + \mathbf{C}\dot{\mathbf{x}} + \mathbf{K}\mathbf{x} = \mathbf{R}\mathbf{q} \tag{4.153}$$

Define $\mathbf{X}(s)$ as the vector of Laplace transforms of the dependent variables and $\mathbf{Q}(s)$ as a vector of Laplace transforms of the outputs. Taking the Laplace transform of Equation (4.153), assuming all initial conditions are zero, leads to

$$(\mathbf{M}s^2 + \mathbf{C}s + \mathbf{K})\mathbf{X}(s) = \mathbf{R}\mathbf{Q}(s) \tag{4.154}$$

Equation (4.154) is the equation from which the transfer functions are determined for systems with force input. $G_{i,j}(s)$ for $i = 1, 2, \ldots, n$ and $j = 1, 2, \ldots, m$ is defined by

$$G_{i,j}(s) = \frac{X_i(s)}{Q_j(s)} \quad \text{with} \quad Q_k(s) = 0, \text{ for} \quad k \neq j \tag{4.155}$$

Equation (4.154) can be solved, leading to

$$\mathbf{X}(s) = (\mathbf{M}s^2 + \mathbf{C}s + \mathbf{K})^{-1}\mathbf{R}\mathbf{Q}(s) \tag{4.156}$$

Equation (4.156) can be written as

$$\mathbf{X}(s) = \mathbf{G}(s)\mathbf{Q}(s) \tag{4.157}$$

where $\mathbf{G}(s)$ is the matrix of transfer functions. If $m = n$, $\mathbf{G}(s)$ is a square matrix.

Example 4.28

Determine the matrix of transfer functions for the system of Figure 4.66. The blocks move on a frictionless surface.

Solution

The mathematical model for the system of Figure 4.66 is obtained by setting the summation of forces to the mass times acceleration of each block using free-body diagrams drawn at an arbitrary instant. This results in

$$10\ddot{x}_1 + 3000x_1 - 1000x_2 = F_1(t) \tag{a}$$

$$5\ddot{x}_2 - 1000x_1 + 1000x_2 = F_2(t) \tag{b}$$

Taking Laplace transforms of Equations (a) and (b), assuming all initial conditions are zero, leads to

$$(10s^2 + 3000)\,X_1(s) - 1000X_2(s) = F_1(s) \tag{c}$$

$$-1000X_1(s) + (5s^2 + 1000)\,X_2(s) = F_2(s) \tag{d}$$

Summarizing Equations (c) and (d) in matrix form yields

$$\begin{bmatrix} 10s^2 + 3000 & -1000 \\ -1000 & 5s^2 + 1000 \end{bmatrix} \begin{bmatrix} X_1(s) \\ X_2(s) \end{bmatrix} = \begin{bmatrix} F_1(s) \\ F_2(s) \end{bmatrix} \tag{e}$$

It is noted that

$$D(s) = \begin{vmatrix} 10s^2 + 3000 & -1000 \\ -1000 & 5s^2 + 1000 \end{vmatrix} = (10s^2 + 3000)(5s^2 + 1000) - (-1000)(-1000)$$

$$= 50s^4 + 25{,}000s^2 + 2 \times 10^6 \tag{f}$$

Then $G_{1,1}(s) = \dfrac{X_1(s)}{F_1(s)}$, where $X_1(s)$ is the solution of Equation (e) with $F_2(s) = 0$. Using Cramer's rule,

$$X_1(s) = \frac{1}{D(s)} \begin{vmatrix} F_1(s) & -1000 \\ 0 & 5s^2 + 1000 \end{vmatrix}$$

$$= \frac{5s^2 + 1000}{D(s)} F_1(s) \tag{g}$$

Equation (g) is used to determine

$$G_{1,1}(s) = \frac{5s^2 + 1000}{50s^4 + 25{,}000s^2 + 2 \times 10^6} \tag{h}$$

Then $G_{2,1}(s) = \dfrac{X_2(s)}{F_1(s)}$, where $X_2(s)$ is the solution of Equation (e) for $X_2(s)$ with $F_2(s) = 0$. Using Cramer's rule,

$$X_2(s) = \frac{1}{D(s)} \begin{vmatrix} 10s^2 + 3000 & F_1(s) \\ -1000 & 0 \end{vmatrix}$$

$$= \frac{1000}{D(s)} F_1(s) \tag{i}$$

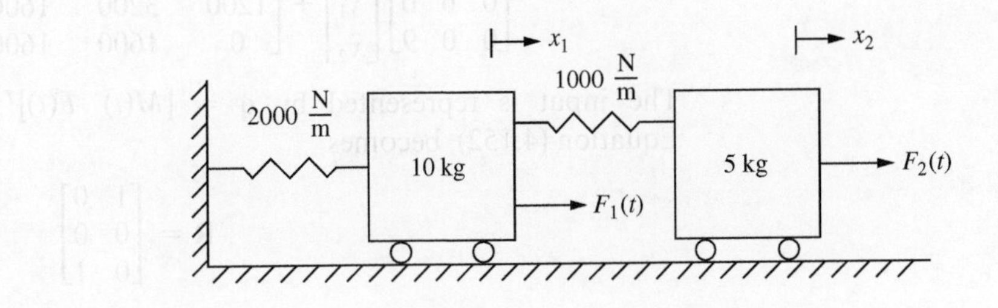

Figure 4.66

Two-degree-of-freedom mechanical system of Example 4.28.

The solution for $G_{2,1}(s)$

$$G_{2,1}(s) = \frac{1000}{50s^4 + 25{,}000s^2 + 2 \times 10^6}$$

The transfer function $G_{1,2}(s) = \frac{X_1(s)}{F_2(s)}$, where $X_1(s)$ is the solution of Equation (e) with $F_1(s) = 0$. Using Cramer's rule,

$$X_1(s) = \frac{1}{D(s)}\begin{vmatrix} 0 & -1000 \\ F_2(s) & 5s^2 + 1000 \end{vmatrix}$$

$$= \frac{1000}{D(s)} F_2(s) \tag{j}$$

The solution for $G_{1,2}(s)$ is

$$G_{1,2}(s) = \frac{1000}{50s^4 + 25{,}000s^2 + 2 \times 10^6} \tag{k}$$

Then $G_{2,2}(s) = \frac{X_2(s)}{F_2(s)}$, where $X_2(s)$ is the solution of Equation (e) with $F_1(s) = 0$. Using Cramer's rule,

$$X_2(s) = \frac{1}{D(s)}\begin{vmatrix} 10s^2 + 3000 & 0 \\ -1000 & F_2(s) \end{vmatrix}$$

$$= \frac{10s^2 + 3000}{D(s)} F_2(s) \tag{l}$$

The solution for $G_{2,2}(s)$ is

$$G_{2,2}(s) = \frac{10s^2 + 3000}{50s^4 + 25{,}000s^2 + 2 \times 10^6} \tag{m}$$

The transfer function matrix becomes

$$G(s) = \frac{1}{50s^4 + 25{,}000s^2 + 2 \times 10^6}\begin{bmatrix} 5s^2 + 1000 & 1000 \\ 1000 & 10s^2 + 3000 \end{bmatrix} \tag{n}$$

Example 4.29

Determine the transfer function matrix for the system of Figure 4.64 and Example 4.23. Use $L = 1$ m, $m = 3$ kg, and $k = 1600$ N/m.

Solution

Equation (d) of Example 4.23 governs the motion of the system and is repeated here:

$$\begin{bmatrix} \frac{1}{3}mL^2\ddot{\theta} & 0 & 0 \\ 0 & 2m & 0 \\ 0 & 0 & 3m \end{bmatrix}\begin{bmatrix} \ddot{\theta} \\ \ddot{x}_1 \\ \ddot{x}_2 \end{bmatrix} + \begin{bmatrix} \frac{25}{16}kL^2 & -\frac{3L}{4}k & 0 \\ -\frac{3L}{4}k & 2k & k \\ 0 & k & k \end{bmatrix}\begin{bmatrix} \theta \\ x_1 \\ x_2 \end{bmatrix} = \begin{bmatrix} M(t) \\ 0 \\ F(t) \end{bmatrix} \tag{a}$$

Substituting given values, Equation (a) becomes

$$\begin{bmatrix} 1 & 0 & 0 \\ 0 & 6 & 0 \\ 0 & 0 & 9 \end{bmatrix}\begin{bmatrix} \ddot{\theta} \\ \ddot{x}_1 \\ \ddot{x}_2 \end{bmatrix} + \begin{bmatrix} 2500 & -1200 & 0 \\ 1200 & 3200 & 1600 \\ 0 & 1600 & 1600 \end{bmatrix}\begin{bmatrix} \theta \\ x_1 \\ x_2 \end{bmatrix} = \begin{bmatrix} M(t) \\ 0 \\ F(t) \end{bmatrix} \tag{b}$$

The input is represented by $\mathbf{q} = [M(t) \;\; F(t)]^T$. The matrix $\mathbf{R}$, as defined by Equation (4.152), becomes

$$\mathbf{R} = \begin{bmatrix} 1 & 0 \\ 0 & 0 \\ 0 & 1 \end{bmatrix} \tag{c}$$

Equation (b) is rewritten as

$$\begin{bmatrix} 1 & 0 & 0 \\ 0 & 6 & 0 \\ 0 & 0 & 9 \end{bmatrix} \begin{bmatrix} \ddot{\theta} \\ \ddot{x}_1 \\ \ddot{x}_2 \end{bmatrix} + \begin{bmatrix} 2500 & -1200 & 0 \\ 1200 & 3200 & 1600 \\ 0 & 1600 & 1600 \end{bmatrix} \begin{bmatrix} \theta \\ x_1 \\ x_2 \end{bmatrix} = \begin{bmatrix} 1 & 0 \\ 0 & 0 \\ 0 & 1 \end{bmatrix} \begin{bmatrix} M(t) \\ F(t) \end{bmatrix} \tag{d}$$

Taking the Laplace transform of Equation (d), assuming all initial conditions are zero, leads to

$$\left\{ \begin{bmatrix} 1 & 0 & 0 \\ 0 & 6 & 0 \\ 0 & 0 & 9 \end{bmatrix} s^2 + \begin{bmatrix} 2500 & -1200 & 0 \\ 1200 & 3200 & 1600 \\ 0 & 1600 & 1600 \end{bmatrix} \right\} \begin{bmatrix} \Theta(s) \\ X_1(s) \\ X_2(s) \end{bmatrix} = \begin{bmatrix} 1 & 0 \\ 0 & 0 \\ 0 & 1 \end{bmatrix} \begin{bmatrix} M(s) \\ F(s) \end{bmatrix} \tag{e}$$

The solution to Equation (e) is

$$\begin{bmatrix} \Theta(s) \\ X_1(s) \\ X_2(s) \end{bmatrix} = \frac{1}{27s^6 + 8.856 \times 10^4 s^4 + 5.41 \times 10^7 s^2 + 3.99 \times 10^9} \times$$

$$\begin{bmatrix} 27s^4 + 2.71 \times 10^4 s^2 + 1.6 \times 10^6 & -8 \times 10^4 \\ 450s^2 + 8 \times 10^4 & -800s^2 - 2 \times 10^7 \\ -8 \times 10^4 & 3s^4 + 4.30 \times 10^4 s^2 + 4.50 \times 10^6 \end{bmatrix} \begin{bmatrix} M(s) \\ F(s) \end{bmatrix} \tag{f}$$

Equation (f) is written as $X(s) = G(s)Q(s)$. Thus from Equation (4.157), the matrix of transfer functions is

$$\begin{bmatrix} G_{1,1}(s) & G_{1,2}(s) \\ G_{2,1}(s) & G_{2,2}(s) \\ G_{3,1}(s) & G_{3,2}(s) \end{bmatrix} = \frac{1}{27s^6 + 8.856 \times 10^4 s^4 + 5.41 \times 10^7 s^2 + 3.99 \times 10^9} \times$$

$$\begin{bmatrix} 27s^4 + 2.71 \times 10^4 s^2 + 1.6 \times 10^6 & -8 \times 10^4 \\ 450s^2 + 8 \times 10^4 & -800s^2 - 2 \times 10^7 \\ -8 \times 10^4 & 3s^4 + 4.30 \times 10^4 s^2 + 4.50 \times 10^6 \end{bmatrix} \tag{g}$$

4.11 State-Space Formulation for Mechanical Systems

The number of independent energy storage elements corresponds to the number of states necessary for the state-space formulation of a mechanical system. The dependent variables defined for a mechanical system are $x_1, x_2, \ldots, x_n$, and the states are defined by $z_1, z_2, \ldots, z_m$. Typically, $m = 2n$.

4.11.1 One-Degree-of-Freedom Mechanical Systems with Force Input

Consider the one-degree-of-freedom linear mechanical system governed by a differential equation of the form of Equation (4.112). There are two energy storage elements in the system, so two states are expected. Define

$$z_1 = x \tag{4.158a}$$

$$z_2 = \dot{x} \tag{4.158b}$$

It is noted that Equations (4.158) implies that

$$\dot{z}_1 = z_2 \tag{4.159}$$

Substituting Equation (4.159) into Equation (4.112) leads to

$$m\dot{z}_2 + cz_2 + kz_1 = F(t) \tag{4.160}$$

Equation (4.160) is rearranged, yielding

$$\dot{z}_2 = -\frac{k}{m}z_1 - \frac{c}{m}z_2 + \frac{1}{m}F(t) \tag{4.161}$$

Equations (4.160) and (4.161) are the basis of the state-space model. Summarizing these in matrix form leads to

$$\begin{bmatrix} \dot{z}_1 \\ \dot{z}_2 \end{bmatrix} = \begin{bmatrix} 0 & 1 \\ -\frac{k}{m} & -\frac{c}{m} \end{bmatrix} \begin{bmatrix} z_1 \\ z_2 \end{bmatrix} + \begin{bmatrix} 0 \\ \frac{1}{m} \end{bmatrix} F(t) \tag{4.162}$$

The input to the system is the force $F(t)$. In terms of the notation of Section 3.4, there is only one input and the vector $\mathbf{u} = [F(t)]$. The output vector is defined as $\mathbf{x} = [x(t)]$. Relating $\mathbf{x}$ to $\mathbf{z}$ and $\mathbf{u}$ in the form of Equation (3.21) leads to

$$[x(t)] = [1 \quad 0]\begin{bmatrix} z_1 \\ z_2 \end{bmatrix} + [0]\,[F(t)] \tag{4.163}$$

Equations (4.162) and (4.163) form the state-space model for a linear one-degree-of-freedom mechanical system with force input. The matrices are from Equations (3.20) and (3.21) as

$$\text{State matrix } \mathbf{A} = \begin{bmatrix} 0 & 1 \\ -\frac{k}{m} & -\frac{c}{m} \end{bmatrix} \tag{4.164}$$

$$\text{Input matrix } \mathbf{B} = \begin{bmatrix} 0 \\ \frac{1}{m} \end{bmatrix} \tag{4.165}$$

$$\text{Output matrix } \mathbf{C} = [1 \quad 0] \tag{4.166}$$

$$\text{Transmission matrix } \mathbf{D} = [0] \tag{4.167}$$

Example 4.30

Derive a state-space model from the mathematical model for the system of Figure 4.58 and Example 4.16.

Solution

The mathematical model relating the output $\theta(t)$ to the input $M(t)$ is given by Equation (c) of Example 4.16 and is repeated here:

$$\frac{581}{240}mL^2\ddot{\theta} + \frac{9}{16}cL^2\dot{\theta} + \frac{1}{16}kL^2\theta = M(t) \tag{a}$$

Define $z_1 = \theta$ and $z_2 = \dot{\theta}$ and $\mathbf{u} = [M(t)]$. Then

$$\dot{z}_1 = z_2 \tag{b}$$

Substituting the state variables into Equation (a) yields

$$\frac{581}{240}mL^2\dot{z}_2 + \frac{9}{16}cL^2z_2 + \frac{1}{16}kL^2z_1 = u_1 \tag{c}$$

Rearrangement of Equation (c) gives

$$\dot{z}_2 = -\frac{15k}{581m}z_1 - \frac{135c}{581m}z_2 + \frac{240}{581mL^2}u_1 \tag{d}$$

Writing Equations (b) and (d) in matrix form,

$$\begin{bmatrix} \dot{z}_1 \\ \dot{z}_2 \end{bmatrix} = \begin{bmatrix} 0 & 1 \\ -\frac{15k}{581m} & -\frac{135c}{581m} \end{bmatrix} \begin{bmatrix} z_1 \\ z_2 \end{bmatrix} + \begin{bmatrix} 0 \\ \frac{240}{581mL^2} \end{bmatrix} [u_1] \tag{e}$$

The relation between the input and output may be written as

$$[\theta(t)] = [1 \;\; 0]\begin{bmatrix} z_1 \\ z_2 \end{bmatrix} + [0][u_1] \tag{f}$$

From Equations (e) and (f), the mathematical model in state-space notation is

$$\text{State matrix } \mathbf{A} = \begin{bmatrix} 0 & 1 \\ -\dfrac{15k}{581m} & -\dfrac{135c}{581m} \end{bmatrix} \tag{g}$$

$$\text{Input matrix } \mathbf{B} = \begin{bmatrix} 0 \\ \dfrac{240}{581mL^2} \end{bmatrix} \tag{h}$$

$$\text{Output matrix } \mathbf{C} = [1 \;\; 0] \tag{i}$$

$$\text{Transmission matrix } \mathbf{D} = [0] \tag{j}$$

4.11.2 One-Degree-of-Freedom Systems with Motion Input

The model system is a one-degree-of-freedom mass-spring-viscous damper system where the spring and viscous damper connect the mass to a support given the prescribed motion $y(t)$. The input is $y(t)$ and the output is $x(t)$. The mathematical model is

$$m\ddot{x} + c\dot{x} + kx = c\dot{z} + kz \tag{4.168}$$

The mathematical model has the derivative of the input in the model. As discussed in Section 3.5, the appropriate definitions of the state variables are

$$y_1 = x \tag{4.169}$$

$$y_2 = \dot{x} + \mu z \tag{4.170}$$

The procedure to determine a state-space formulation is the same as discussed in Example 3.9.

4.12 Further Examples

The examples in this section illustrate how the principles developed in previous sections are synthesized into the development of a mathematical model for a mechanical system. In addition to the implicit assumptions listed at the beginning of this chapter, the following assumptions are used throughout the examples:

- All springs are linear and massless.
- All viscous dampers are massless.
- Viscous damping is the only form of friction.
- Small displacements are assumed where appropriate to linearize the differential equations.
- All dependent variables are measured from the system's equilibrium position.
- Rigid bodies undergo planar motion.

Since all systems are linear and all coordinates are measured from the system's equilibrium position, the static forces developed in springs cancel the gravity forces that cause the static deflections. Thus neither are included in the free-body diagrams or the expressions of potential energy.

Two methods have been developed for the mathematical modeling of mechanical systems. The free-body diagram method requires the application of Newton's second law to a free-body diagram, drawn at an arbitrary instant, illustrating the external forces acting on the system. Problems involving rotational motion also require the application of a moment equation. For rigid bodies undergoing planar motion, an alternate formulation using D'Alembert's principle is developed in which the system of external forces is equivalent to the system of effective forces, defined as a force of $m\bar{a}$ applied at the mass center and a couple of $\bar{I}\alpha$.

The energy method is based on determining the equivalent mass, stiffness, and viscous damping coefficient for an equivalent mass spring-viscous damper model. The equivalent mass and stiffness are determined from expressions for the kinetic and potential energies written at an arbitrary instant. The equivalent viscous damping coefficient is obtained from an expression for the work done by viscous damping forces.

Example 4.31

Consider the system of Figure 4.67(a). The thin disk of mass m_1 and radius r is attached to a shaft of torsional stiffness k_t. **a.** Determine a mathematical model for the system using θ as the dependent variable. **b.** Determine the system's natural frequency and damping ratio using the given values.

Solution

a. Any of the methods can be used to model the system. The polar moment of inertia of the disk obtained from Table 4.1 is $J = (1/2)m_1r^2$. Assuming no slip between the disk and the cable, the velocity of a particle on the circumference of the disk is $r\dot{\theta}$ and the displacement of the suspended mass is $y = r\theta$.

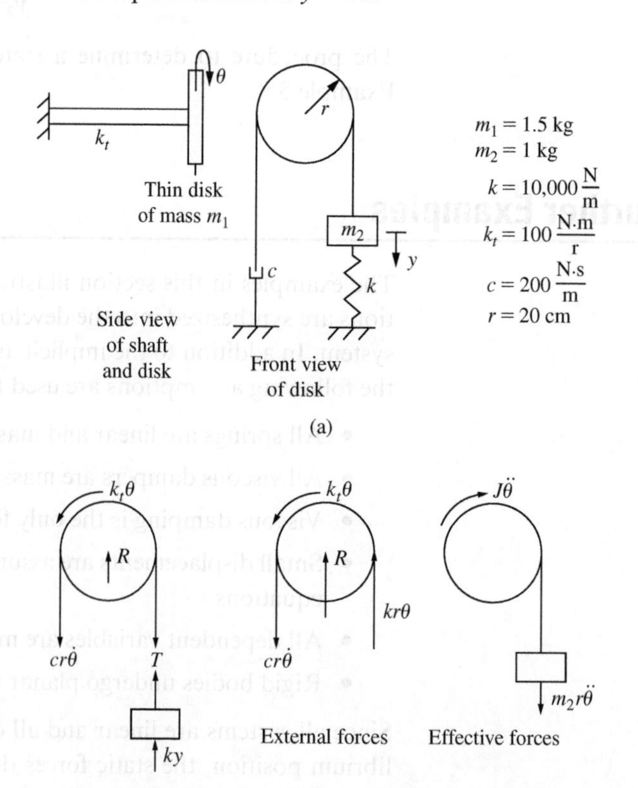

Figure 4.67

(a) The system of Example 4.31. The thin disk is attached to the end of a shaft. A mass is suspended from the disk, and a viscous damper is attached to the outer circumference of the disk; (b) the free-body diagrams of the disk and the suspended mass at an arbitrary instant; (c) the free-body diagrams of the disk and suspended mass at an arbitraty instant showing external and effective forces.

Direct application of Newton's second law: Free-body diagrams of the disk and the suspended mass are shown in Figure 4.67(b). Applying Newton's law to the suspended mass using positive forces downward

$$\sum F = m_2\ddot{y} \tag{a}$$

$$-T - ky = m_2\ddot{y} \tag{b}$$

$$T = -m_2\ddot{y} - ky \tag{c}$$

Since the disk is undergoing pure rotational motion, Equation (4.103) is used with positive moments clockwise

$$\sum M_G = J\ddot{\theta} \tag{d}$$

$$M(t) - k_t\theta - (cr\dot{\theta})r + Tr = \frac{1}{2}m_1 r^2\ddot{\theta} \tag{e}$$

Substituting Equations (c) and (e) and noting that $y = r\theta$ and rearranging leads to

$$\left(\frac{1}{2}m_1r^2 + m_2r^2\right)\ddot{\theta} + cr^2\dot{\theta} + (k_t + kr^2)\theta = M(t) \tag{f}$$

D'Alembert's principle: Free-body diagrams showing the external forces and the effective forces acting on the system composed of the disk and the suspended mass are shown in Figure 4.67(c). Application of D'Alembert's principle in the form of Equation (4.111) taking positive moments clockwise leads to

$$\left(\sum M_G\right)_{ext} = \left(\sum M_G\right)_{eff} \tag{g}$$

$$M(t) - k_t\theta - (cr\dot{\theta})r - (kr\theta)r = \frac{1}{2}m_1r^2\ddot{\theta} + (m_2r\ddot{\theta})r \tag{h}$$

Rearranging Equation (h) leads to Equation (f).

Energy method: The kinetic energy of the system at an arbitrary instant is

$$T = \frac{1}{2}I_D\dot{\theta}^2 + \frac{1}{2}m_2(r\dot{\theta})^2 = \frac{1}{2}\left(\frac{1}{2}m_1r^2 + m_2r^2\right)\dot{\theta}^2 \tag{i}$$

Comparison of Equation (i) with Equation (4.132) leads to

$$I_{eq} = \frac{1}{2}m_1r^2 + m_2r^2 \tag{j}$$

The potential energy of the system at an arbitrary instant is

$$V = \frac{1}{2}k_t\theta^2 + \frac{1}{2}k(r\theta)^2 = \frac{1}{2}(k_t + kr^2)\theta^2 \tag{k}$$

Comparison of Equation (k) with Equation (4.133) leads to

$$k_{t,eq} = k_t + kr^2 \tag{l}$$

The work done by the viscous damper between two arbitrary positions is

$$W_{1\to 2} = -\int_{\theta_1}^{\theta_2} c(r\dot{\theta})d(r\theta) = -\int_{\theta_1}^{\theta_2} cr^2\dot{\theta}d\theta \tag{m}$$

Comparison of Equation (m) and Equation (4.134) leads to

$$c_{t,eq} = cr^2 \tag{n}$$

The power delivered by the external moment is

$$P = M(t)\dot{\theta} \tag{o}$$

Comparison of Equation (o) with the power delivered by an equivalent moment leads to

$$M_{eq}(t) = M(t) \tag{p}$$

When Equations (j), (l), and (n) are used in Equation (4.136), Equation (f) is obtained.

b. Equation (f) is put in the standard form of Equation (4.116) by dividing by its leading coefficient, yielding

$$\ddot{\theta} + \frac{c}{\frac{1}{2}m_1 + m_2}\dot{\theta} + \frac{k_t + kr^2}{\left(\frac{1}{2}m_1 + m_2\right)r^2}\theta = \frac{M(t)}{\left(\frac{1}{2}m_1 + m_2\right)r^2} \tag{q}$$

Comparison of Equation (q) with Equation (4.116) yields the natural frequency

$$\omega_n = \sqrt{\frac{k_t + kr^2}{\left(\frac{1}{2}m_1 + m_2\right)r^2}}$$

$$= \sqrt{\frac{100\text{ N·m/r} + (10{,}000\text{ N/m})(0.2\text{ m})^2}{\left[\frac{1}{2}(1.5\text{ kg}) + 1\text{ kg}\right](0.2\text{ m})^2}} = 84.5\text{ r/s} \tag{r}$$

and viscous damping ratio

$$\zeta = \frac{c}{2\omega_n\left(\frac{1}{2}m_1 + m_2\right)}$$

$$= \frac{200\text{ N·s/m}}{2(84.5\text{ r/s})\left[\frac{1}{2}(1.5\text{ kg}) + 1\text{ kg}\right]} = 0.676 \tag{s}$$

Example 4.32

Figure 4.68(a) illustrates a simplified model of a vehicle suspension system. The vehicle of mass m is attached to an element that is modeled as a spring of stiffness k_1 in parallel with a viscous damper of damping coefficient c. This element is in series with an element that is modeled as a spring of stiffness k_2. The other end of this spring is subject to a time-dependent displacement $z(t)$ caused by the terrain over which the vehicle is traveling. Determine a mathematical model for the motion of the vehicle.

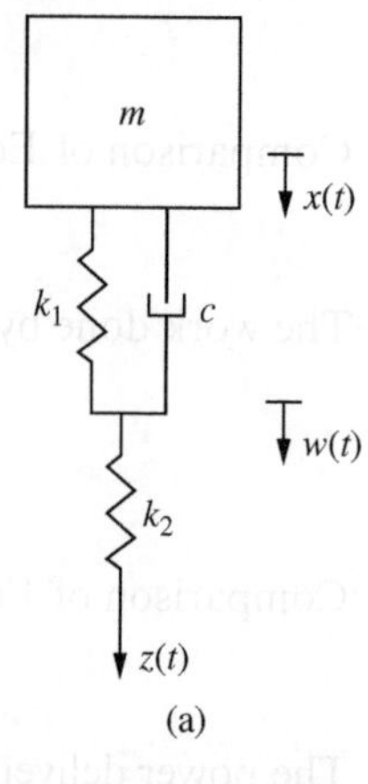

$k_1(w - x)$ $c(\dot{w} - \dot{x})$

$k_1(w - x)$ $c(\dot{w} - \dot{x})$

$k_2(z - w)$

(b)

Figure 4.68
(a) The system of Example 4.32; (b) the free-body diagrams at an arbitrary instant.

Solution

Let $x(t)$ represent the displacement of the vehicle's body from the system's equilibrium position. Let $w(t)$ represent the displacement of the junction between the two suspension elements. Free-body diagrams of the body and the junction at an arbitrary time are shown in Figures 4.68(b).

Applying Newton's second law to the vehicle body

$$\sum F = m\ddot{x} \tag{a}$$

$$k_1(w - x) + c(\dot{w} - \dot{x}) = m\ddot{x} \tag{b}$$

$$m\ddot{x} + c\dot{x} - c\dot{w} + k_1 x - k_1 w = 0 \tag{c}$$

Application of Newton's second law to the junction, which is massless, leads to

$$k_2(z - w) - k_1(w - x) - c(\dot{w} - \dot{x}) = 0 \tag{d}$$

$$-c\dot{x} + c\dot{w} - k_1 x + (k_1 + k_2)w = k_2 z \tag{e}$$

Equations (c) and (e) provide a mathematical model for the system and can be summarized in matrix form as

$$\begin{bmatrix} m & 0 \\ 0 & 0 \end{bmatrix}\begin{bmatrix} \ddot{x} \\ \ddot{w} \end{bmatrix} + \begin{bmatrix} c & -c \\ -c & c \end{bmatrix}\begin{bmatrix} \dot{x} \\ \dot{w} \end{bmatrix} + \begin{bmatrix} k_1 & -k_1 \\ -k_1 & k_1 + k_2 \end{bmatrix}\begin{bmatrix} x \\ w \end{bmatrix} = \begin{bmatrix} 0 \\ k_2 z \end{bmatrix} \tag{f}$$

A single equation to solve for $x(t)$ can be obtained by eliminating $w(t)$ using Equations (b) and (d). The result is

$$\frac{cm}{k_2}\dddot{x} + m\left(1 + \frac{k_1}{k_2}\right)\ddot{x} + c\dot{x} + k_1 x = c\dot{z} + k_1 z \tag{g}$$

Example 4.33

Determine **a.** the transfer function model and **b.** a state-space model for the system of Example 4.32.

Solution

The input to this system is $z(t)$. The output is $x(t)$.

a. The transfer function is $G(x) = \frac{X(s)}{Z(s)}$. Taking the Laplace transform of Equation (f) of Example 4.32, assuming all initial conditions are zero, leads to

$$\left\{\begin{bmatrix} m & 0 \\ 0 & 0 \end{bmatrix}s^2 + \begin{bmatrix} c & -c \\ -c & c \end{bmatrix}s + \begin{bmatrix} k_1 & -k_1 \\ -k_1 & k_1 + k_2 \end{bmatrix}\right\}\begin{bmatrix} X(s) \\ W(s) \end{bmatrix} = \begin{bmatrix} 0 \\ k_2 Z(s) \end{bmatrix} \tag{a}$$

Solving for $X(s)$ using Cramer's rule gives

$$X(s) = \frac{\begin{vmatrix} 0 & -cs - k_1 \\ k_2 Z(s) & k_1 + k_2 \end{vmatrix}}{\begin{vmatrix} ms^2 + cs + k_1 & -cs - k_1 \\ -cs - k_1 & cs + k_1 + k_2 \end{vmatrix}}$$

$$= \frac{cs + k_1}{mcs^3 + m(k_1 + k_2)s^2 + ck_2 s + k_2 k_1}k_2 Z(s) \tag{b}$$

Using Equation (b), it is found that

$$G(s) = \frac{k_2(cs + k_1)}{mcs^3 + m(k_1 + k_2)s^2 + ck_2 s + k_2 k_1} \tag{c}$$

Dividing the numerator and denominator of the right-hand side of Equation (c) by mc in order to make the leading coefficient of the denominator 1 gives

$$G(s) = \frac{\frac{k_2}{mc}(cs + k_1)}{s^3 + \frac{k_1 + k_2}{c}s^2 + \frac{k_2}{m}s + \frac{k_2 k_1}{mc}} \tag{d}$$

It is noted that determining the transfer function from Equation (g) of Example 4.32 directly leads to Equation (d).

b. Define the states as $y_1 = x(t)$, $y_2 = \dot{x}(t)$, $y_3 = w(t)$. The input is $u_1 = z(t)$. From the definition of the states,

$$\dot{y}_1 = y_2 \tag{e}$$

Substituting into the mathematical model, Equation (f) of Example 4.32 yields

$$m\dot{y}_2 + cy_2 - c\dot{y}_3 + k_1 y_1 - k_1 y_3 = 0 \tag{f}$$

$$-cy_2 + c\dot{y}_3 - k_1 y_1 + (k_1 + k_2)y_3 = k_2 u_1 \tag{g}$$

Equations (g) and (h) are solved for $\dot{y}_2$ and $\dot{y}_3$, leading to

$$\dot{y}_2 = -\frac{k_2}{m}y_3 + \frac{k_2}{m}u_1 \tag{h}$$

$$\dot{y}_3 = \frac{k_1}{c}y_1 + y_2 - \frac{k_1 + k_2}{c}y_3 + \frac{k_2}{m}u_1 \tag{i}$$

Equations (e), (h), and (i) are summarized in matrix form as

$$\begin{bmatrix} \dot{y}_1 \\ \dot{y}_2 \\ \dot{y}_3 \end{bmatrix} = \begin{bmatrix} 0 & 1 & 0 \\ 0 & 0 & -\frac{k_2}{m} \\ \frac{k_1}{c} & 1 & -\frac{k_1 + k_2}{c} \end{bmatrix} \begin{bmatrix} y_1 \\ y_2 \\ y_3 \end{bmatrix} + \begin{bmatrix} 0 \\ \frac{k_2}{m} \\ \frac{k_2}{m} \end{bmatrix} [u_1] \tag{j}$$

The relation between the state variables and the input and output is

$$[x(t)] = [1 \quad 0 \quad 0] \begin{bmatrix} y_1 \\ y_2 \\ y_3 \end{bmatrix} + [0][u_1] \tag{k}$$

The matrices involved in the state-space formulation of Equation (j) are

$$\text{State matrix } \mathbf{A} = \begin{bmatrix} 0 & 1 & 0 \\ 0 & 0 & -\frac{k_2}{m} \\ \frac{k_1}{c} & 1 & -\frac{k_1 + k_2}{c} \end{bmatrix} \tag{l}$$

$$\text{Input matrix } \mathbf{B} = \begin{bmatrix} 0 \\ \frac{k_2}{m} \\ \frac{k_2}{m} \end{bmatrix} \tag{m}$$

$$\text{Output matrix } \mathbf{C} = [1 \quad 0 \quad 0] \tag{n}$$

$$\text{Transmission matrix } \mathbf{D} = [0] \tag{o}$$

Example 4.34

A centrifuge, illustrated schematically in Figure 4.69, of total mass m operates at a constant rotational speed ω in r/s. The centrifuge is mounted on a foundation that is modeled as a spring of stiffness k in parallel with a viscous damper of damping

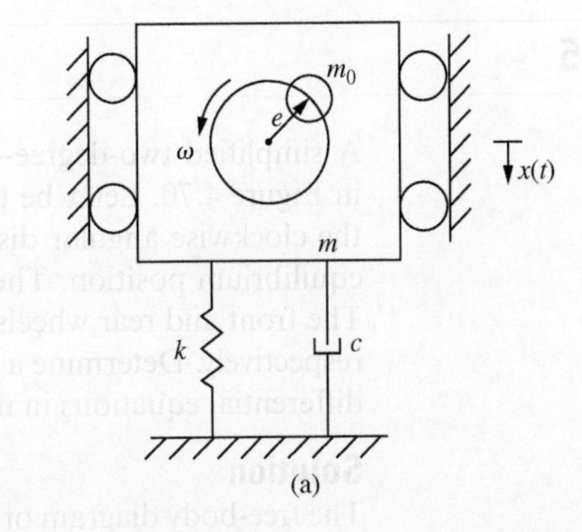

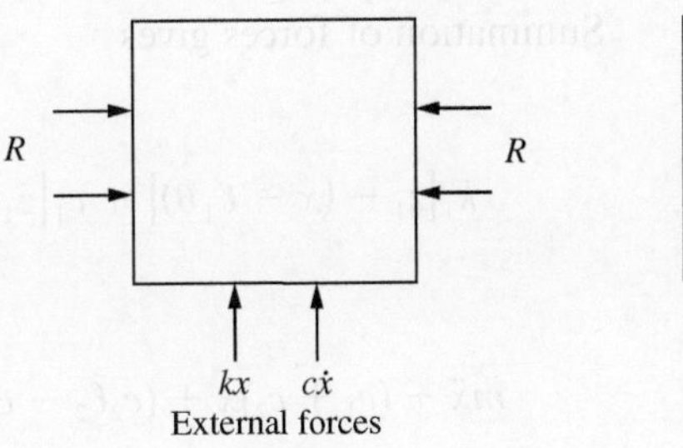

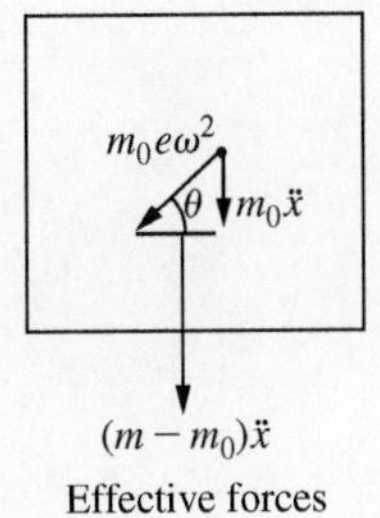

(b)

Figure 4.69

(a) The centrifuge with a rotating mass m_0 and a distance e from the axis of rotation; (b) the free-body diagrams of the centrifuge, and the rotating mass at an arbitrary instant.

coefficient c. The centrifuge has a rotational mass m_0 whose center of gravity is a distance e from the axis of rotation. Determine a mathematical model for the motion of the centrifuge.

Solution

Let $x(t)$ represent the displacement of the centrifuge from its equilibrium position. The matter in the centrifuge has a motion relative to the centrifuge. The acceleration of the mass center of the rotating component is determined using the relative acceleration equation, Equation (4.13), as

$$\mathbf{a}_p = (-e\omega^2 \cos\theta)\mathbf{i} + (\dot{x} + e\omega^2 \sin\theta)\mathbf{j} \tag{a}$$

where θ is the angle between a line drawn from the mass center of the centrifuge to the instantaneous position of the mass center of the matter and the horizontal. By definition of the angular velocity,

$$\omega = \frac{d\theta}{dt} \tag{b}$$

Since ω is constant and assuming $\theta = 0$ at $t = 0$, Equation (b) is integrated to yield

$$\theta = \omega t \tag{c}$$

Free-body diagrams of the centrifuge at an arbitrary instant are given in Figure 4.69(b). The external forces acting on the centrifuge are the spring force and viscous damping force. The effective forces include the inertia force of the machine and the inertia force of the matter. Application of D'Alembert's principle in the form of

$$\sum F_{ext} = \sum F_{eff} \tag{d}$$

in the vertical direction gives

$$-kx - c\dot{x} = (m - m_0)\ddot{x} + m_0(\ddot{x} + e\omega^2 \sin\omega t) \tag{e}$$

Equation (e) is rearranged to

$$m\ddot{x} + c\dot{x} + kx = -m_0 e\omega^2 \sin\omega t \tag{f}$$

The input to the system is $z(t) = \sin\omega t$ and the output is $x(t)$.

Example 4.35

A simplified two-degree-of-freedom model of a vehicle suspension system is shown in Figure 4.70. Let x be the displacement of G, the mass center of the vehicle, and θ the clockwise angular displacement of the vehicle, both measured from the system's equilibrium position. The vehicle has a mass m and centroidal moment of inertia $\bar{I}$. The front and rear wheels of the vehicle have vertical displacements of $z_1(t)$ and $z_2(t)$, respectively. Determine a mathematical model for the motion of the vehicle. Write the differential equations in matrix form.

Solution

The free-body diagram of the system at an arbitrary instant is shown in Figure 4.70(b). Summation of forces gives

$$\sum F = m\ddot{x} \tag{a}$$

$$k_1[z_1 - (x - \ell_1\theta)] + c_1[\dot{z}_1 - (\dot{x} - \ell_1\dot{\theta})] + k_2[z_2 - (x + \ell_2\theta)] + c_2[\dot{z}_2 - (\dot{x} + \ell_2\dot{\theta})] = m\ddot{x} \tag{b}$$

$$m\ddot{x} + (c_1 + c_2)\dot{x} + (c_2\ell_2 - c_1\ell_1)\dot{\theta} + (k_1 + k_2)x + (k_2\ell_2 - k_1\ell_1)\theta = k_1z_1 + k_2z_2 + c_1\dot{z}_1 + c_2\dot{z}_2 \tag{c}$$

Summing moments about the mass center gives

$$\sum M_G = \bar{I}\ddot{\theta} \tag{d}$$

$$-k_1[z_1 - (x - \ell_1\theta)]\ell_1 - c_1[\dot{z}_1 - (\dot{x} - \ell_1\dot{\theta})]\ell_1 + k_2[z_2 - (x + \ell_2\theta)]\ell_2 + c_2[\dot{z}_2 - (\dot{x} + \ell_2\dot{\theta})]\ell_2 = \bar{I}\ddot{\theta} \tag{e}$$

$$\bar{I}\ddot{\theta} + (c_2\ell_2 - c_1\ell_1)\dot{x} + (c_1\ell_1^2 + c_2\ell_2^2)\dot{\theta} + (k_2\ell_2 - k_1\ell_1)x + (k_2\ell_2^2 + k_1\ell_1^2)\theta = -k_1\ell_1z_1 + k_2\ell_2z_2 - c_1\ell_1\dot{z}_1 + c_2\ell_2\dot{z}_2 \tag{f}$$

Equations (c) and (f) are summarized in matrix form as

$$\begin{bmatrix} m & 0 \\ 0 & \bar{I} \end{bmatrix}\begin{bmatrix} \ddot{x} \\ \ddot{\theta} \end{bmatrix} + \begin{bmatrix} c_1 + c_2 & c_2\ell_2 - c_1\ell_1 \\ c_2\ell_2 - c_1\ell_1 & c_1\ell_1^2 + c_2\ell_2^2 \end{bmatrix}\begin{bmatrix} \dot{x} \\ \dot{\theta} \end{bmatrix} + \begin{bmatrix} k_1 + k_2 & k_2\ell_2 - k_1\ell_1 \\ k_2\ell_2 - k_1\ell_1 & k_1\ell_1^2 + k_2\ell_2^2 \end{bmatrix}\begin{bmatrix} x \\ \theta \end{bmatrix} = \begin{bmatrix} k_1z_1 + k_2z_2 + c_1\dot{z}_1 + c_2\dot{z}_2 \\ -k_1\ell_1z_1 + k_2\ell_2z_2 - c_1\ell_1\dot{z}_1 + c_2\ell_2\dot{z}_2 \end{bmatrix} \tag{g}$$

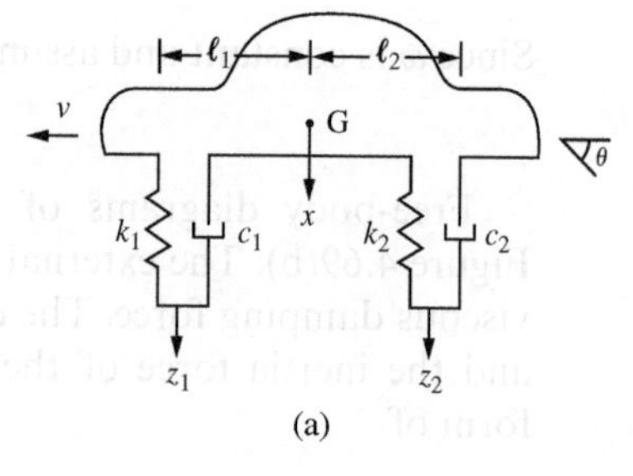

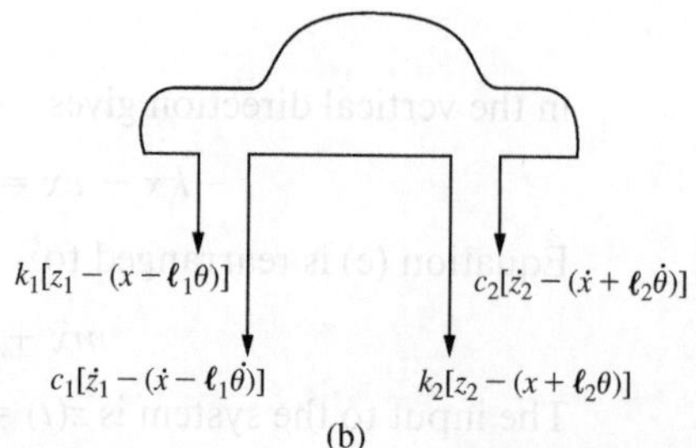

Figure 4.70

(a) The two-degree-of-freedom model of a vehicle suspension system; (b) the free-body diagram at an arbitrary instant.

Example 4.36

Determine a state-space model for the system of Example 4.35. Use $m = 400$ kg, $\bar{I} = 200\ \text{kg}\cdot\text{m}^2$, $k_1 = k_2 = 1 \times 10^5 \frac{\text{N}}{\text{m}}$, $c_1 = c_2 = 10{,}000 \frac{\text{N}\cdot\text{s}}{\text{m}}$, $\ell_1 = 1$ m, and $\ell_2 = 1.5$ m.

Solution

The inputs to this system are the displacements of the wheels, z_1 and z_2. The outputs are x and θ. Substituting the given values into Equation (g) of Example 4.35 leads to

$$\begin{bmatrix} 400 & 0 \\ 0 & 200 \end{bmatrix}\begin{bmatrix} \ddot{x} \\ \ddot{\theta} \end{bmatrix} + \begin{bmatrix} 20{,}000 & 5000 \\ 5000 & 25{,}000 \end{bmatrix}\begin{bmatrix} \dot{x} \\ \dot{\theta} \end{bmatrix} + \begin{bmatrix} 2 \times 10^5 & 50{,}000 \\ 50{,}000 & 2.5 \times 10^5 \end{bmatrix}\begin{bmatrix} x \\ \theta \end{bmatrix}$$

$$= \begin{bmatrix} 10{,}000\dot{z}_1 + 10{,}000\dot{z}_2 + 1 \times 10^5 z_1 + 1 \times 10^5 z_1 \\ -10{,}000\dot{z}_1 + 15{,}000\dot{z}_2 - 1 \times 10^5 z_1 + 1.5 \times 10^5 z_1 \end{bmatrix} \tag{a}$$

Since the derivatives of the input appear in the mathematical model, the standard definition of state variables does not suffice. Define

$$y_1 = x \tag{b}$$

$$y_2 = \theta \tag{c}$$

$$y_3 = \dot{x} + \mu_1 z_1 + \mu_2 z_2 \tag{d}$$

$$y_4 = \dot{\theta} + \nu_1 z_1 + \nu_2 z_2 \tag{e}$$

Substitution of Equations (b)–(e) into Equation (a) leads to

$$\begin{aligned} &400(\dot{y}_3 - \mu_1\dot{z}_1 - \mu_2\dot{z}_2) + 20{,}000(y_3 - \mu_1 z_1 - \mu_2 z_2) + 5000(y_4 - \nu_1 z_1 - \nu_2 z_2) \\ &+ 2 \times 10^5 y_1 + 50{,}000 y_2 = 10{,}000\dot{z}_1 + 10{,}000\dot{z}_2 + 1 \times 10^5 z_1 + 1 \times 10^5 z_2 \end{aligned} \tag{f}$$

$$\begin{aligned} &200(\dot{y}_4 - \nu_1\dot{z}_1 - \nu_2\dot{z}_2) + 5000(y_3 - \mu_1 z_1 - \mu_2 z_2) + 25{,}000(y_4 - \nu_1 z_1 - \nu_2 z_2) + 50{,}000 y_1 \\ &+ 2.5 \times 10^5 y_2 = -10{,}000\dot{z}_1 + 15{,}000 \times 10^4 \dot{z}_2 - 1 \times 10^5 z_1 + 1.5 \times 10^5 z_2 \end{aligned} \tag{g}$$

The values of μ_1, μ_2, ν_1, and ν_2 are chosen to eliminate the derivatives of the input from Equations (f) and (g), leading to

$$\mu_1 = -25,\ \mu_2 = -25,\ \nu_1 = 50,\ \nu_2 = -75 \tag{h}$$

Substituting Equation (h) into Equations (d) and (e) and rearranging leads to

$$\dot{y}_1 = y_3 + 25z_1 + 25z_2 \tag{i}$$

$$\dot{y}_2 = y_4 - 50z_1 + 75z_2 \tag{j}$$

Substituting Equation (h) into Equations (f) and (g) and rearranging gives

$$\dot{y}_3 = -625y_1 - 125y_2 - 50y_3 - 12.5y_4 - 62.5z_1 - 1937.5z_2 \tag{k}$$

$$\dot{y}_4 = -250y_1 - 1250y_2 - 25y_3 - 12.5y_4 - 875z_1 - 4625.5z_2 \tag{l}$$

Expressing Equations (i)–(l) in matrix form leads to

$$\begin{bmatrix} \dot{y}_1 \\ \dot{y}_2 \\ \dot{y}_3 \\ \dot{y}_4 \end{bmatrix} = \begin{bmatrix} 0 & 0 & 1 & 0 \\ 0 & 0 & 0 & 1 \\ -625 & -125 & -50 & -12.5 \\ -250 & -1250 & -25 & -12.5 \end{bmatrix}\begin{bmatrix} y_1 \\ y_2 \\ y_3 \\ y_4 \end{bmatrix} + \begin{bmatrix} 25 & 25 \\ -50 & 75 \\ -62.5 & -1937.5 \\ -875 & 4625.5 \end{bmatrix}\begin{bmatrix} z_1 \\ z_2 \end{bmatrix} \tag{m}$$

The relation among the output, the state variables, and the input is

$$\begin{bmatrix} x \\ \theta \end{bmatrix} = \begin{bmatrix} 1 & 0 & 0 & 0 \\ 0 & 1 & 0 & 0 \end{bmatrix} \begin{bmatrix} y_1 \\ y_2 \\ y_3 \\ y_4 \end{bmatrix} + \begin{bmatrix} 0 & 0 \\ 0 & 0 \end{bmatrix} \begin{bmatrix} z_1 \\ z_2 \end{bmatrix} \tag{n}$$

From Equations (m) and (n),

$$\text{State matrix } \mathbf{A} = \begin{bmatrix} 0 & 0 & 1 & 0 \\ 0 & 0 & 0 & 1 \\ -625 & -125 & -50 & -12.5 \\ -250 & -1250 & -25 & -12.5 \end{bmatrix} \tag{o}$$

$$\text{Input matrix } \mathbf{B} = \begin{bmatrix} 25 & 25 \\ -50 & 75 \\ -62.5 & -1937.5 \\ -875 & -4625.5 \end{bmatrix} \tag{p}$$

$$\text{Output matrix } \mathbf{C} = \begin{bmatrix} 1 & 0 & 0 & 0 \\ 0 & 1 & 0 & 0 \end{bmatrix} \tag{q}$$

$$\text{Transmission matrix } \mathbf{D} = \begin{bmatrix} 0 & 0 \\ 0 & 0 \end{bmatrix} \tag{r}$$

Example 4.37

Derive the mathematical model for the system of Figure 4.71(a) using θ, the clockwise angular rotation of the bar, as the dependent variable; assume small θ. The bar is massless with a particle attached at each end. A time-dependent force is applied to the particle of mass $2m$.

Solution

The Parallel Axis Theorem is used to determine the mass moment of inertia of the system about O. The moment of inertia of a particle about an axis through the particle is zero. Thus

$$I_O = (2m)(\ell)^2 + m(2\ell)^2 = 6m\ell^2 \tag{a}$$

A free-body diagram of the system, drawn at an arbitrary instant, is shown in Figure 4.71(b). The forces are labeled assuming small θ. Since O is a fixed axis of rotation, the appropriate conservation law is $\sum M_O = I_O\ddot{\theta}$. Assuming moments are taken as positive clockwise, application of the moment equation leads to

$$F(t)(\ell) - (k\ell\theta)\ell - (2c\ell)(2\ell) = 6m\ell^2\ddot{\theta} \tag{b}$$

Equation (b) is rearranged to yield

$$6m\ell^2\ddot{\theta} + 4c\ell^2\dot{\theta} + k\ell^2\theta = \ell F(t) \tag{c}$$

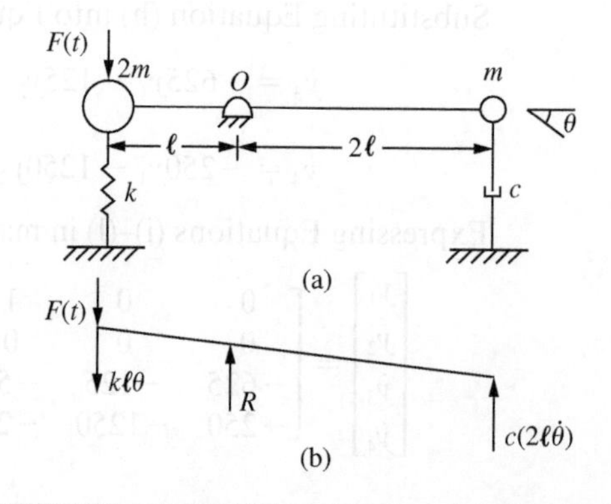

Figure 4.71
(a) The system of Example 4.37; (b) the free-body diagram of system at arbitrary instant.

Example 4.38

Consider the system of Figure 4.72(a). **a.** Determine a mathematical model for the system. **b.** Determine the transfer function with $F(t)$ as the input and $x(t)$ as the output. **c.** Determine a state-space formulation using $x(t)$ as the output.

Solution

a. Let $x(t)$ represent the displacement of the mass from equilibrium. Let $z(t)$ represent the displacement of the joint between the spring and the viscous damper. The force developed in the viscous damper at an arbitrary instant is $c(\dot{x} - \dot{z})$ and is directed away from the mass. Free-body diagrams of the mass and the joint at an arbitrary instant are shown in Figures 4.72(b) and 4.72(c). Summing the forces on the mass leads to

$$-c(\dot{x} - \dot{z}) + F(t) = m\ddot{x} \tag{a}$$

Summing the forces at the joint at an arbitrary instant leads to

$$c(\dot{x} - \dot{z}) - kz = 0 \tag{b}$$

Equations (a) and (b) form the mathematical model for the system.

b. Taking the Laplace transform of Equations (a) and (b) using the properties of linearity of the transform and transform of derivatives, assuming all initial conditions are zero, leads to

$$\begin{bmatrix} ms^2 + cs & -cs \\ -cs & cs + k \end{bmatrix} \begin{bmatrix} X(s) \\ Z(s) \end{bmatrix} = \begin{bmatrix} F(s) \\ 0 \end{bmatrix} \tag{c}$$

Equation (c) is solved, yielding

$$X(s) = \frac{F(s)(cs + k)}{s(mcs^2 + mks + ck)} \tag{d}$$

The transfer function is determined from Equation (d) as

$$G(s) = \frac{X(s)}{F(s)} = \frac{cs + k}{s(mcs^2 + mks + ck)} \tag{e}$$

c. Define the state variables as

$$y_1 = x \tag{f}$$

$$y_2 = \dot{x} \tag{g}$$

$$y_3 = x - z \tag{h}$$

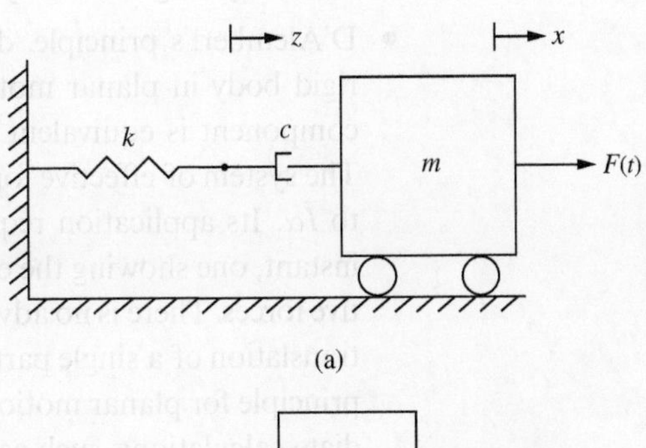

Figure 4.72

(a) System of Example 4.38; (b) free-body diagram of block at an arbitrary instant; (c) free-body diagram of joint between viscous damper and spring at an arbitrary instant.

From Equations (f) and (g), it is obvious that

$$\dot{y}_1 = y_2 \tag{i}$$

Using the state variables in Equations (a) and (b),

$$-c\dot{y}_3 + F(t) = m\dot{y}_2 \tag{j}$$

and

$$c\dot{y}_3 - k(y_3 - y_1) = 0 \tag{k}$$

Writing Equations (i), (j), and (k) in matrix form leads to

$$\begin{bmatrix} 1 & 0 & 0 \\ 0 & m & c \\ 0 & 0 & c \end{bmatrix} \begin{bmatrix} \dot{y}_1 \\ \dot{y}_2 \\ \dot{y}_3 \end{bmatrix} = \begin{bmatrix} 0 & 1 & 0 \\ 0 & 0 & 0 \\ -k & 0 & k \end{bmatrix} \begin{bmatrix} y_1 \\ y_2 \\ y_3 \end{bmatrix} + \begin{bmatrix} 0 \\ F(t) \\ 0 \end{bmatrix} \tag{l}$$

Equation (l) is multiplied by the inverse of the matrix to the left of the vector of derivatives of the state variables, leading to

$$\begin{bmatrix} \dot{y}_1 \\ \dot{y}_2 \\ \dot{y}_3 \end{bmatrix} = \begin{bmatrix} 0 & 1 & 0 \\ k/m & 0 & -k/m \\ -k/c & 0 & k/c \end{bmatrix} \begin{bmatrix} y_1 \\ y_2 \\ y_3 \end{bmatrix} + \begin{bmatrix} 0 \\ 1/m \\ 0 \end{bmatrix} [F(t)] \tag{m}$$

The relation between the output and the state variables is

$$[x(t)] = [1 \quad 0 \quad 0] \begin{bmatrix} y_1 \\ y_2 \\ y_3 \end{bmatrix} + [0][F(t)] \tag{n}$$

4.13 Summary

4.13.1 Modeling Methods

This chapter focuses on the mathematical modeling of mechanical systems. Three methods have been developed and can be applied to a one-degree-of-freedom linear system.

- The direct application of Newton's second law to free-body diagrams of system components is the basic method. Appropriate forms of Newton's second law can be written for particles and rigid bodies, and the law can be specialized for rigid bodies undergoing planar motion. The method is fundamental and easy to remember, and it can always be applied. Its application requires drawing correct free-body diagrams of system components at an arbitrary instant.
- D'Alembert's principle, developed from Newton's second law, states that for a rigid body in planar motion, the system of external forces acting on a system component is equivalent to the system of effective forces for that component. The system of effective forces is defined as a force equal to $m\bar{a}$ and a couple equal to $\bar{I}\alpha$. Its application requires drawing two free-body diagrams at an arbitrary instant, one showing the external forces and the other showing the system's effective forces. There is no advantage to the use of D'Alembert's principle for the pure translation of a single particle or rotation about a fixed axis. Use of D'Alembert's principle for planar motion of rigid bodies often eliminates the need for intermediate calculations, such as the determination of mass centers and the calculation of intermediate forces, necessary when applying Newton's second law.
- The energy method or the equivalent systems method is based on the model equation derived for the motion of a linear one-degree-of-freedom system. The coefficients in the model equation are determined using the kinetic and potential energies of the system as well as the work done by viscous damping forces. An

advantage of using the energy method is that its application does not require drawing free-body diagrams, and thus mistakes associated with drawing incorrect free-body diagrams (such as drawing forces in the wrong direction) are eliminated.

A method that can be applied in the mathematical modeling of multidegree-of-freedom systems is the application of Lagrange's equations. This can be developed using calculus of variations and energy methods. Lagrange's equations can be applied to linear and nonlinear systems. Application to a system requires the development of kinetic and potential energy at an arbitrary instant in terms of the dependent variables and their time derivatives. The method of virtual work is necessary if nonconservative forces are to be modeled.

4.13.2 Chapter Highlights

Other important points developed in this chapter are as follows:

- Passive components of a mechanical system include inertia elements, which store kinetic energy; springs, which store potential energy; and viscous dampers, which dissipate energy.
- Active elements include externally applied forces and elements that provide motion input.
- A particle's mass is concentrated at a single point. The mass of a rigid body is distributed about its mass center. The distribution of mass is measured by the moment of inertia.
- Springs are flexible links in mechanical systems that store energy when stretched or compressed. A linear spring has a linear force-displacement relation.
- A combination of springs can be replaced by a single spring of an equivalent stiffness, which can be determined so that the equivalent spring has the same potential energy at any instant as the total potential energy of the combination.
- A viscous damper has a force proportional to the velocity of the particle to which it is attached. A viscous damper is useful because it leads to a linear term in the differential equation obtained through system modeling.
- A free-body diagram (FBD) is a diagram of a body at an arbitrary instant, abstracted from its surroundings and showing the effect of the surroundings in the form of forces. Both surface forces and body forces are illustrated and labeled consistently with chosen dependent variables.
- Newton's second law is the basic law of nature that forms the basis for the mathematical modeling of mechanical systems.
- A moment equation, as well as an appropriate form of Newton's second law, is necessary for the modeling of rigid bodies undergoing planar motion.
- D'Alembert's principle, an alternate formulation of Newton's second law, is useful for modeling systems with rigid bodies undergoing planar motion.
- An energy method uses a spatially integrated form of Newton's second law (the Principle of Work and Energy) to provide an alternate method for the mathematical modeling of mechanical systems.
- The differential equation governing the motion of a one-degree-of-freedom linear system can be written in a standard form in terms of two system parameters: the natural frequency and the damping ratio.
- Lagrange's equations provide an alternative to Newton's law for deriving the mathematical model. They are mostly used for multidegree-of-freedom systems.
- Transfer functions are derived by taking the Laplace transform of the mathematical model, assuming all initial conditions are zero.
- Transfer functions for one-degree-of-freedom systems are second-order.

- State variables are associated with some form of energy.
- State variables for systems with the derivative of the input in the mathematical model involve the input.

4.13.3 Important Equations

Chapter 4 contains the following important equations:

- Definition of velocity and acceleration

$$\mathbf{v} = \frac{d\mathbf{r}_P}{dt} = \frac{dx}{dt}\mathbf{i} + \frac{dy}{dt}\mathbf{j} + \frac{dz}{dt}\mathbf{k} \tag{4.2}$$

$$\mathbf{a} = \dot{\mathbf{v}} = \ddot{\mathbf{r}}_p \tag{4.4}$$

- Kinetic energy of particle and rigid body under planar motion

$$\text{particle } T = \frac{1}{2}m\mathbf{v}\cdot\mathbf{v} \tag{4.6}$$

$$\text{rigid body } T = \frac{1}{2}m\bar{v}^2 + \frac{1}{2}\bar{I}\omega^2 \tag{4.27}$$

- Parallel Axis Theorem

$$I_{xx} = \bar{I}_{xx} + md_x^2 \tag{4.24}$$

- Linear force-displacement relation in a spring

$$F = kx \tag{4.40}$$

- Potential energy function for a spring

$$V = \frac{1}{2}kx^2 \tag{4.43}$$

- Equivalent stiffness of spring combinations

$$\text{parallel } k_{eq} = \sum_{i=1}^{n} k_i \tag{4.47}$$

$$\text{series } k_{eq} = \frac{1}{\sum_{i=1}^{n}\frac{1}{k_i}} \tag{4.51}$$

- Force-velocity relation for a viscous damper

$$F = cv \tag{4.59}$$

- Newton's second law for a particle

$$\sum \mathbf{F} = m\mathbf{a} \tag{4.97}$$

- Newton's second law for a rigid body undergoing planar motion

$$\sum \mathbf{F} = m\bar{\mathbf{a}} \tag{4.99}$$

$$\sum M_G = \bar{I}\alpha \tag{4.102}$$

- Moment equation for rotation about a fixed axis

$$\sum M_O = I_O\alpha \tag{4.103}$$

- D'Alembert's principle

$$\left(\sum \mathbf{F}\right)_{ext} = \left(\sum \mathbf{F}\right)_{eff} \tag{4.110}$$

$$\left(\sum M_A\right)_{ext} = \left(\sum M_A\right)_{eff} \tag{4.111}$$

- Standard form of differential equation for one-degree-of-freedom system

$$\ddot{x} + 2\zeta\omega_n\dot{x} + \omega_n^2 x = \frac{1}{m}F(t) \tag{4.116}$$

- Principle of Work and Energy

$$T_1 + V_1 + (W_{1\to 2})_{NC} = T_2 + V_2 \tag{4.121}$$

- Equivalent systems

$$T = \frac{1}{2} m_{eq} \dot{x}^2 \tag{4.123}$$

$$V = \frac{1}{2} k_{eq} x^2 \tag{4.124}$$

$$W_{1\to 2} = -\int_{x_1}^{x_2} c_{eq} \dot{x} dx \tag{4.125}$$

- Lagrange's equations

$$\frac{d}{dt}\left(\frac{\partial L}{\partial \dot{x}_i}\right) - \frac{\partial L}{\partial x_i} = 0 \tag{4.140}$$

- Transfer function for a one-degree-of-freedom system with force input

$$G(s) = \frac{1/m}{s^2 + 2\zeta\omega_n s + \omega_n^2} \tag{4.147}$$

- Transfer function for a one-degree-of-freedom system with motion input

$$G(s) = \frac{2\zeta\omega_n s + \omega_n^2}{s^2 + 2\zeta\omega_n s + \omega_n^2} \tag{4.151}$$

- Transfer functions for multidegree-of-freedom systems

$$\mathbf{X}(s) = (\mathbf{M}s^2 + \mathbf{C}s + \mathbf{K})^{-1}\mathbf{RQ}(s) \tag{4.156}$$

$$\mathbf{X}(s) = \mathbf{G}(s)\mathbf{Q}(s) \tag{4.157}$$

Short Answer Problems

Short Answer Problems SA4.1–SA4.10 are true or false. If the statement is false, rewrite it to make it true.

SA4.1 Given the moment of inertia of an object about an axis through the centroid of the object, the Parallel Axis Theorem may be used to determine the moment of inertia about an axis parallel to the centroidal axis.

SA4.2 The moment of inertia is largest when calculated about a centroidal axis.

SA4.3 The kinetic energy of a rigid body is always $\frac{1}{2}m\bar{v}^2$, where $\bar{v}$ is the speed of the body's center of mass.

SA4.4 Every particle on a rigid body undergoing pure rotational motion moves in parallel planes.

SA4.5 Springs in parallel all have the same change in length.

SA4.6 Springs in series are characterized as having the same force developed in each spring.

SA4.7 For a linear system, when the dependent variable is measured from the system's equilibrium position, any static forces present in the spring cancel the gravity force that causes them when the mathematical model is derived.

SA4.8 The force from a spring acting on a body is an example of a surface force on an FBD of the body.

SA4.9 Lagrange's equations result from the summation of forces and moments on an FBD.

SA4.10 D'Alembert's principle applied to a particle is $\sum\mathbf{F} - m\mathbf{a} = \mathbf{0}$.

Short Answer Problems SA4.11–SA4.29 require a short answer.

SA4.11 For a rigid body, axis A is parallel to an axis B that passes through the body's centroid. State the Parallel Axis Theorem for axes A and B.

SA4.12 What is the definition of planar motion of a rigid body?

SA4.13 Under what conditions is the kinetic energy of a rigid body undergoing planar motion equal to $\frac{1}{2}I_A\omega^2$?

SA4.14 Under what condition is the equation $\sum\mathbf{M}_G = \dot{\mathbf{H}}_G$ valid, where $\mathbf{H}_G$ is the angular momentum about the mass center?

SA4.15 Explain how the small angle assumption is used to linearize differential equations a priori.

SA4.16 Fill in the blanks: Two springs are in parallel. Both springs have the same ______________, but the forces from the springs on an FBD ______________.

SA4.17 Fill in the blanks: Two springs are in series. Both springs have the same ________________, but their ______________________________ add together.

SA4.18 Explain D'Alembert's principle and how it can be used to derive a mathematical model for a mechanical system.

SA4.19 Explain the use of Lagrange's equations.

SA4.20 Why are static deflections not included on FBDs of linear mechanical systems?

SA4.21 Draw an FBD to be used in the mathematical modeling of the mechanical system shown in Figure SA4.21. The input to the system is $f(t)$ and the output is $x(t)$. Label all forces at an arbitrary instant.

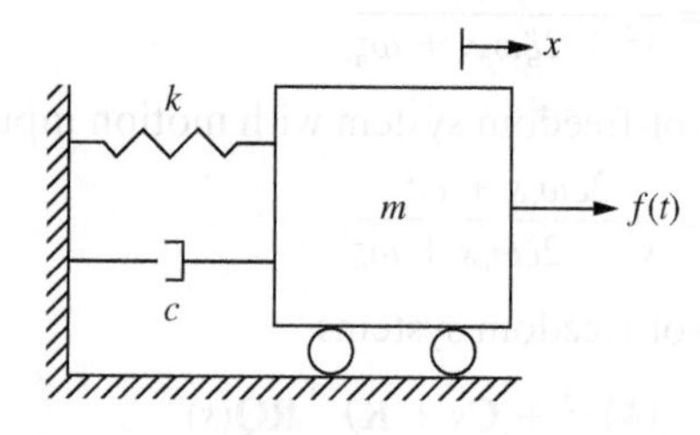

Figure SA4.21

SA4.22 Draw an FBD to be used in the mathematical modeling of the mechanical system shown in Figure SA4.22. The input to the system is $f(t)$ and the output is $x(t)$. Label all forces at an arbitrary instant.

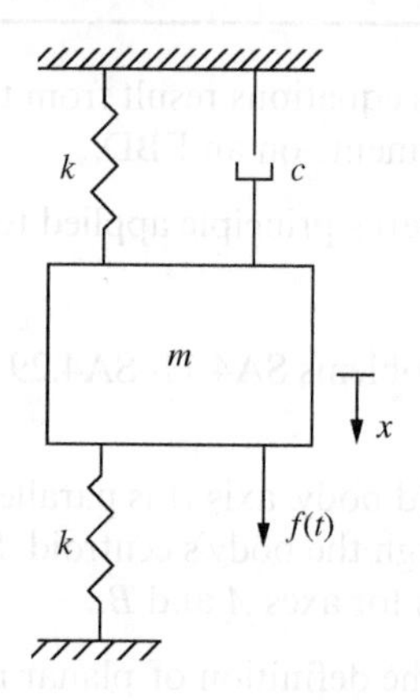

Figure SA4.22

SA4.23 Draw an FBD to be used in the mathematical modeling of the mechanical system shown in Figure SA4.23. The input to the system is $z(t)$ and the output is $x(t)$. Label all forces at an arbitrary instant.

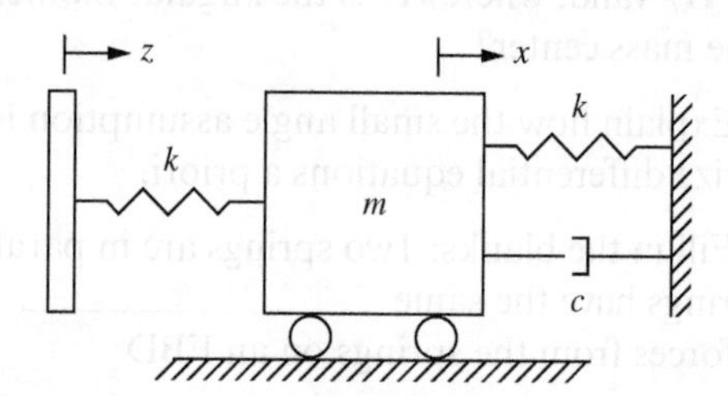

Figure SA4.23

SA4.24 Draw an FBD to be used in the mathematical modeling of the mechanical system shown in Figure SA4.24. The input to the system is $z(t)$ and the output is $x(t)$. Label all forces at an arbitrary instant.

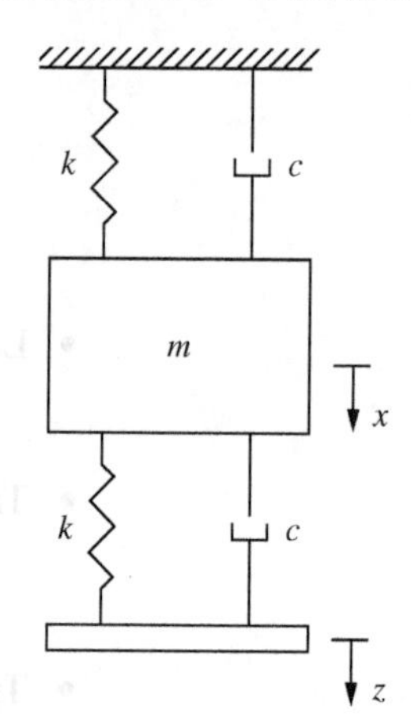

Figure SA4.24

SA4.25 Draw an FBD to be used in the mathematical modeling of the mechanical system shown in Figure SA4.25. The input to the system is $M(t)$ and the output is $\theta(t)$. Label all forces at an arbitrary instant. Assume a slender bar of mass m.

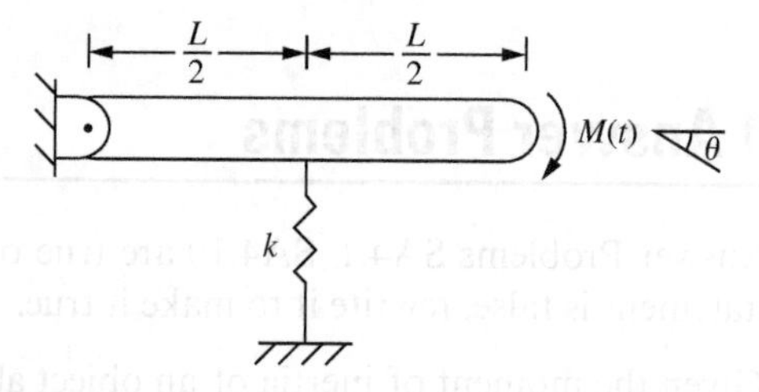

Figure SA4.25

SA4.26 Draw an FBD to be used in the mathematical modeling of the mechanical system shown in Figure SA4.26. The input to the system is $f(t)$ and the output is $x(t)$. Label all forces at an arbitrary instant. Assume a slender bar of mass m.

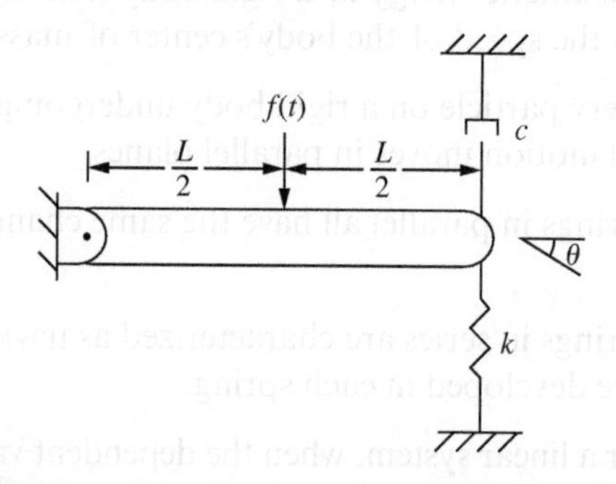

Figure SA4.26

SA4.27 Draw an FBD to be used in the mathematical modeling of the mechanical system shown in Figure SA4.27. The input to the system is $f(t)$ and the output is $x(t)$. Label all forces at an arbitrary instant. Assume a slender bar of mass m.

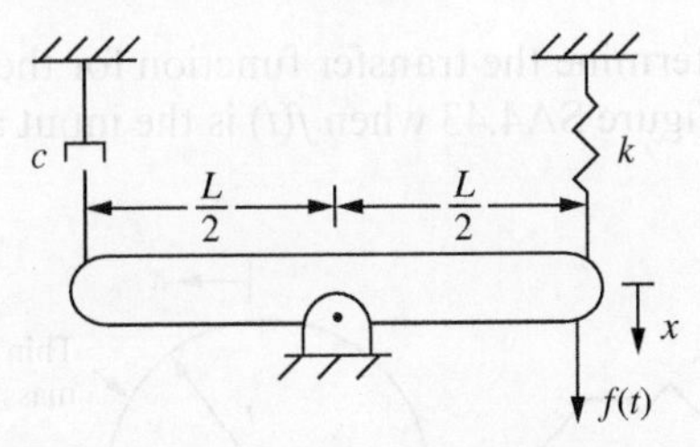

Figure SA4.27

SA4.28 Draw an FBD to be used in the mathematical modeling of the mechanical system shown in Figure SA4.28. The input to the system is $f(t)$ and the outputs are $x_1(t)$ and $x_2(t)$. Label all forces at an arbitrary instant.

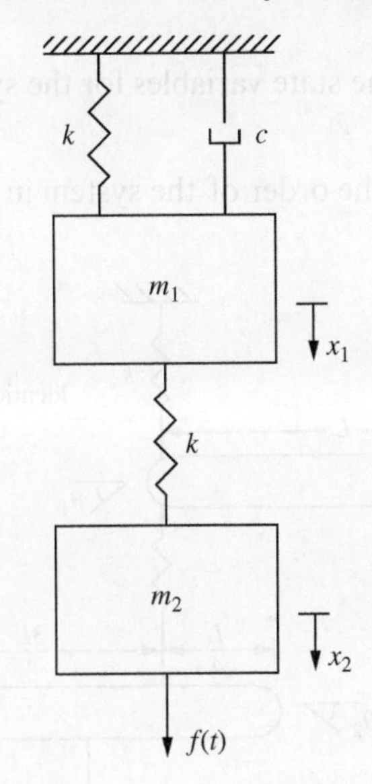

Figure SA4.28

SA4.29 Draw an FBD to be used in the mathematical modeling of the mechanical system shown in Figure SA4.29. The inputs to the system are $M(t)$ and $f(t)$ and the outputs are $\theta(t)$ and $x(t)$. Label all forces at an arbitrary instant.

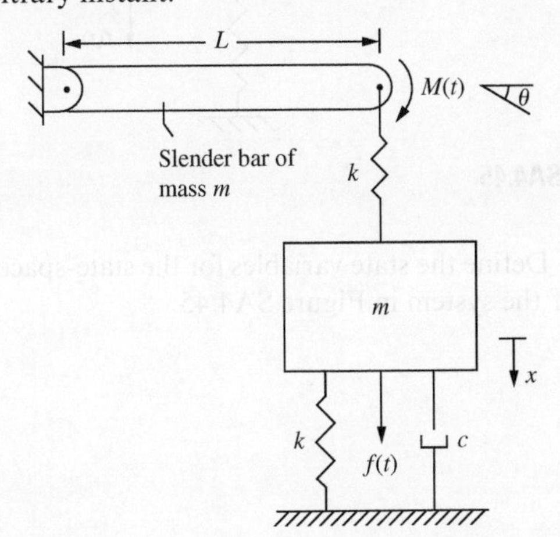

Figure SA4.29

Short Answer Problems SA4.30–SA4.40 require a short calculation.

SA4.30 What is the equivalent stiffness of a spring of stiffness k_1 placed in series with a spring of stiffness k_2?

SA4.31 What is the equivalent stiffness of a spring of stiffness k_1 placed in parallel with a spring of stiffness k_2?

SA4.32 What is the equivalent stiffness of the combination of springs in Figure SA4.32?

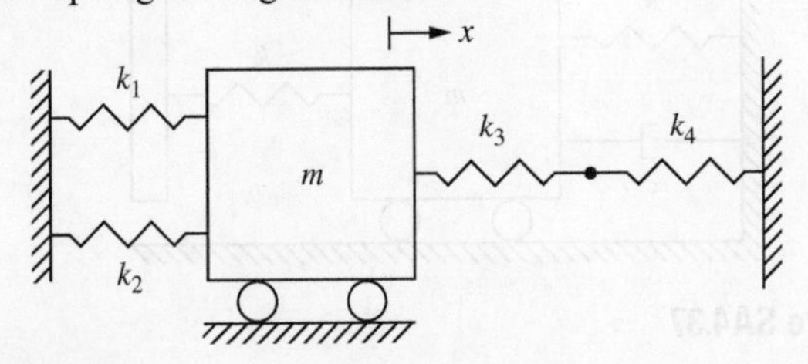

Figure SA4.32

SA4.33 What is the equivalent stiffness of the combination of springs in Figure SA4.33?

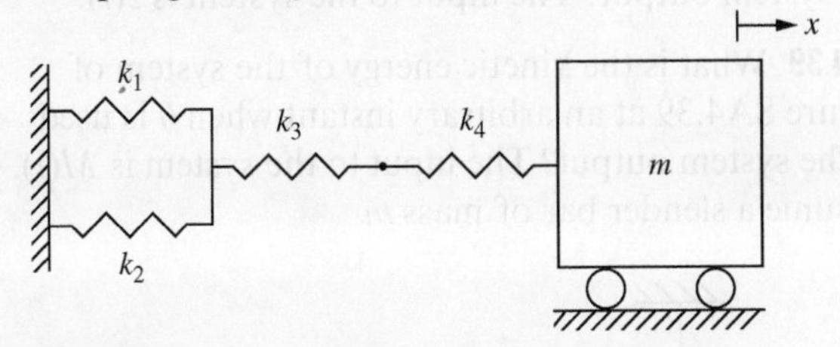

Figure SA4.33

SA4.34 What is the equivalent torsional stiffness of the combination of springs in Figure SA4.34 when θ is used as the system output?

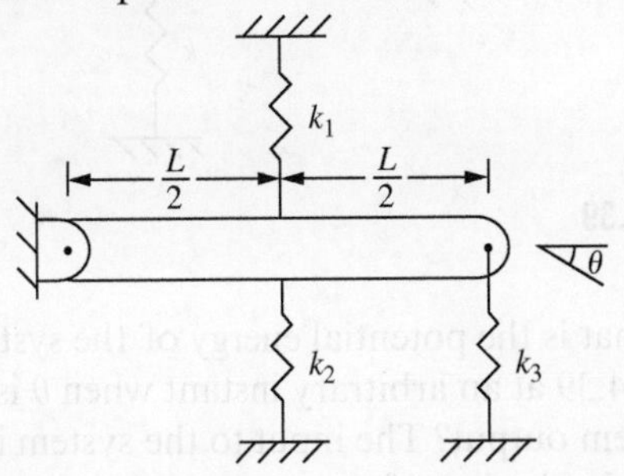

Figure SA4.34

SA4.35 What is the kinetic energy of the system of Figure SA4.35 at an arbitrary instant when x is used as the system output? The input to the system is $f(t)$.

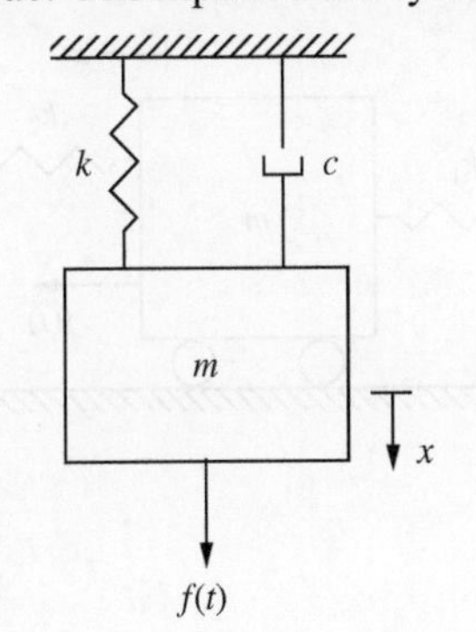

Figure SA4.35

SA4.36 What is the potential energy of the system of Figure SA4.35 at an arbitrary instant when x is used as the system output? The input to the system is $f(t)$.

SA4.37 What is the kinetic energy of the system of Figure SA4.37 at an arbitrary instant when x is used as the system output? The input to the system is $z(t)$.

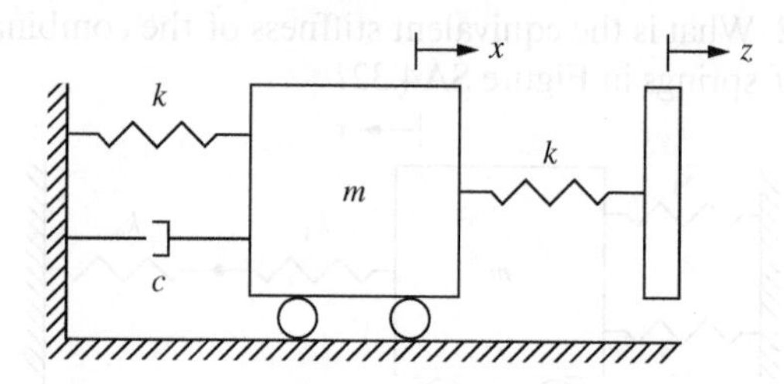

Figure SA4.37

SA4.38 What is the potential energy of the system of Figure SA4.37 at an arbitrary instant when x is used as the system output? The input to the system is $z(t)$.

SA4.39 What is the kinetic energy of the system of Figure SA4.39 at an arbitrary instant when θ is used as the system output? The input to the system is $M(t)$. Assume a slender bar of mass m.

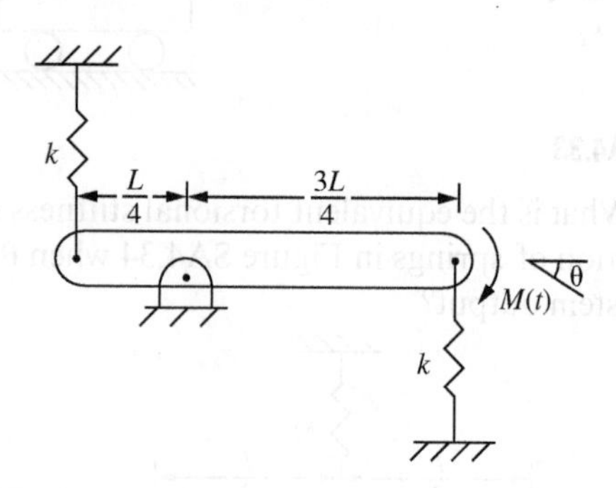

Figure SA4.39

SA4.40 What is the potential energy of the system of Figure SA4.39 at an arbitrary instant when θ is used as the system output? The input to the system is $M(t)$. Assume a slender bar of mass m.

SA4.41 Determine the transfer function for the system shown in Figure SA4.41 when $f(t)$ is the input and $x(t)$ is the output.

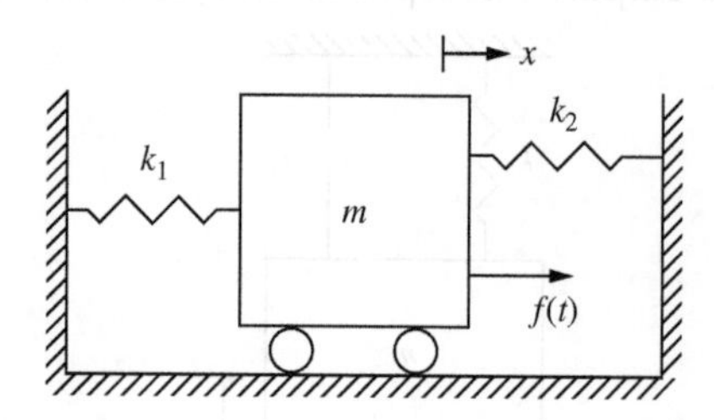

Figure SA4.41

SA4.42 Define the state variables for the system of Figure SA4.41.

SA4.43 Determine the transfer function for the system shown in Figure SA4.43 when $f(t)$ is the input and $x(t)$ is the output.

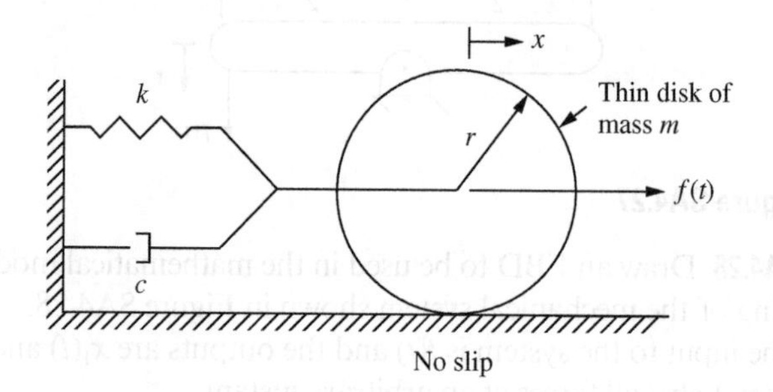

Figure SA4.43

SA4.44 Define the state variables for the system of Figure SA4.43.

SA4.45 What is the order of the system in Figure SA4.45?

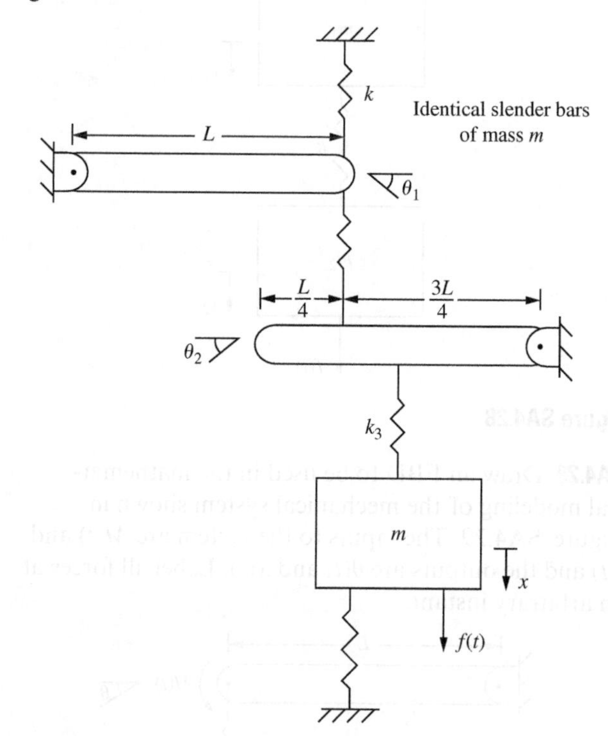

Figure SA4.45

SA4.46 Define the state variables for the state-space modeling of the system in Figure SA4.45.

SA4.47 What is the order of the system in Figure SA4.47?

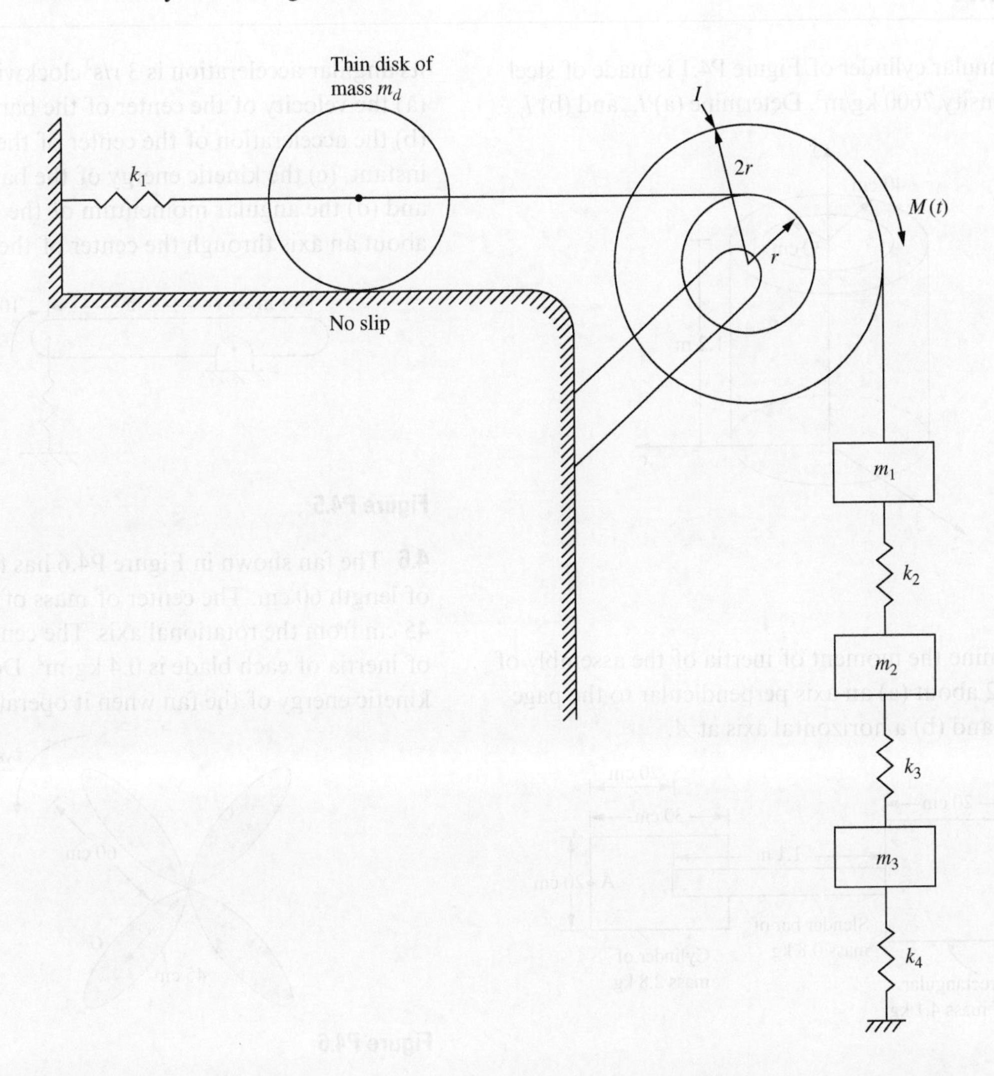

Figure SA4.47

SA4.48 Define the state variables for the state-space modeling of the system in Figure SA4.47.

SA4.49 The transfer function $G(s) = \dfrac{\Omega(s)}{M(s)}$ for the angular velocity of a shaft is $G(s) = \dfrac{0.5}{s + 0.1}$. What is the transfer function for the angular position of the shaft $G(s) = \dfrac{\Theta(s)}{M(s)}$?

SA4.50 Determine a state-space model for the angular velocity of the shaft of Short Answer Problem 4.49.

SA4.51 Determine a state-space model for the angular position of the shaft of Short Answer Problem 4.49.

SA4.52 The transfer function for the displacement of a specific mechanical system with force input is $G(s) = \dfrac{1}{2s^2 + 5s + 25}$. What is the transfer function for the velocity?

SA4.53 The transfer function for the displacement x of a mechanical system with force input is $G(s) = \dfrac{1}{2s^2 + 5s + 25}$. The transmitted force between the system and its foundation is $F_T = 5\dot{x} + 25x$. Determine the transfer function for the transmitted force.

SA4.54 What are the output matrix and the transmission matrix for the system of Short Answer Problem 4.53 when the transmitted force is used as the output?

Problems

4.1 The annular cylinder of Figure P4.1 is made of steel of mass density 7600 kg/m^3. Determine (a) I_{xx} and (b) I_{yy}.

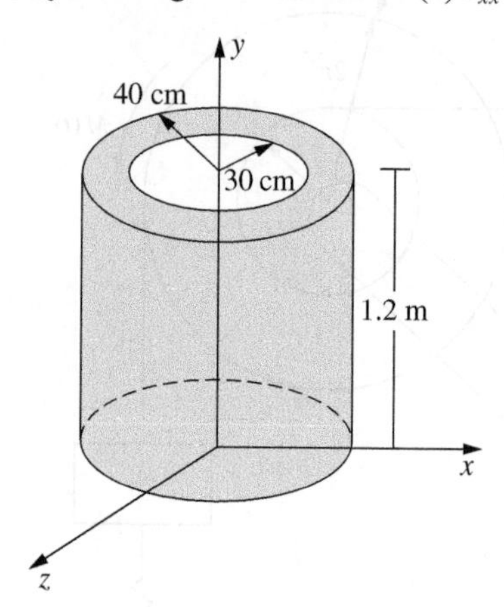

Figure P4.1

4.2 Determine the moment of inertia of the assembly of Figure P4.2 about (a) an axis perpendicular to the page through A and (b) a horizontal axis at A.

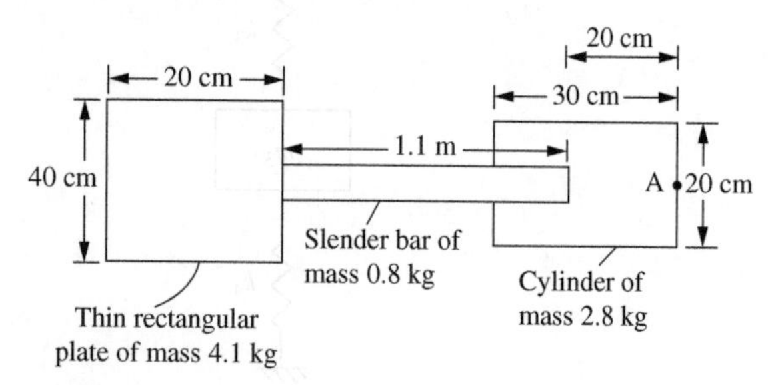

Figure P4.2

4.3 Determine the moment of inertia of the assembly of Figure P4.3 about an axis perpendicular to the page through O.

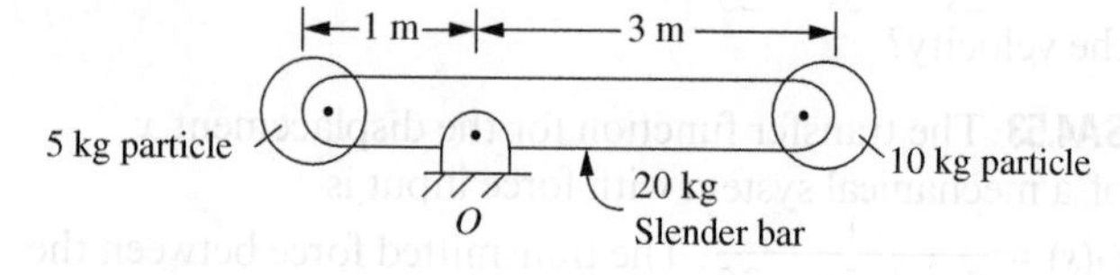

Figure P4.3

4.4 A 2-kg particle has a time-dependent velocity of

$$\mathbf{v}(t) = -2\sin(3t)\mathbf{i} + 4\cos(3t)\mathbf{j} + 2\mathbf{k} \text{ m/s}$$

The particle is at the origin of the coordinate system at $t = 0$. Determine (a) the position vector of the particle, (b) the acceleration of the particle, (c) the kinetic energy of the particle at $t = 1$ s, and (d) the linear momentum of the particle at $t = 1$ s.

4.5 At the instant shown, the angular velocity of the 6-kg bar of Figure P4.5 is 10 r/s counterclockwise and its angular acceleration is 3 r/s^2 clockwise. Determine (a) the velocity of the center of the bar at this instant, (b) the acceleration of the center of the bar at this instant, (c) the kinetic energy of the bar at this instant, and (d) the angular momentum of the bar at this instant about an axis through the center of the bar.

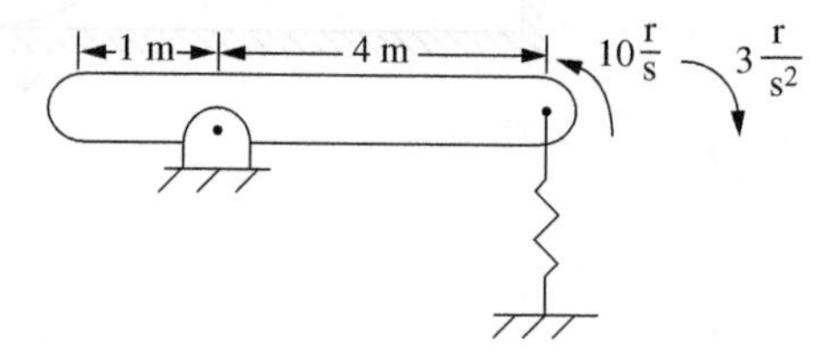

Figure P4.5

4.6 The fan shown in Figure P4.6 has four 1.2-kg blades of length 60 cm. The center of mass of each blade is 45 cm from the rotational axis. The centroidal moment of inertia of each blade is 0.4 kg·m^2. Determine the kinetic energy of the fan when it operates at 250 rpm.

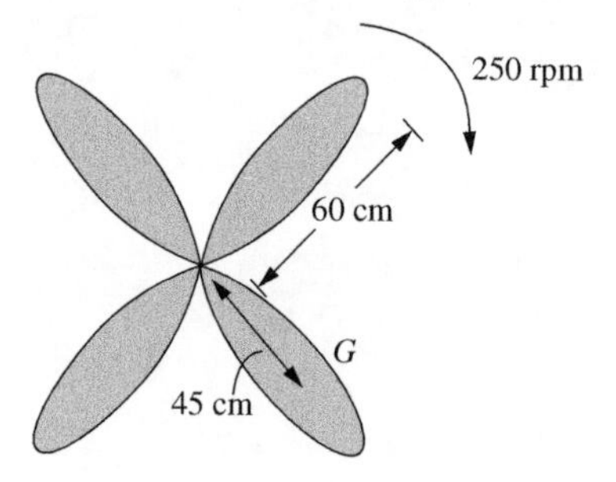

Figure P4.6

4.7 A schematic representation of a small single-cylinder engine is shown in Figure P4.7. The 0.5-kg crank is of length 10 cm and rotates about its mass center O at a constant speed of 250 rpm clockwise. Its moment of inertia is $I_O = 0.4$ kg·m^2. The 0.4-kg connecting rod is a slender rod of length 30 cm. A 0.3-kg piston, constrained to move vertically, is attached to the connecting rod. Determine (a) the kinetic energy of the engine (crank, connecting rod, and piston) at the instant shown and (b) the acceleration of the piston at this instant.

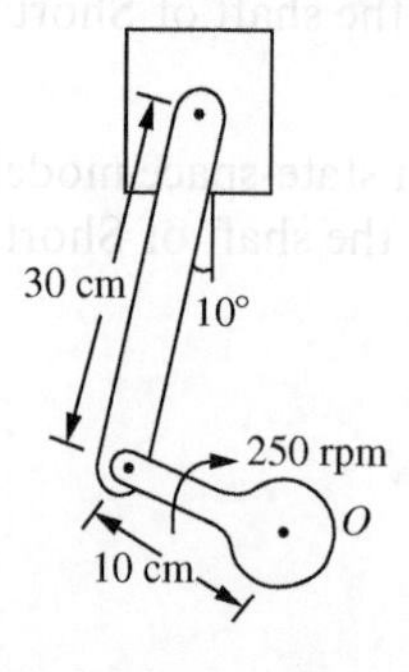

Figure P4.7

4.8 Determine the kinetic energy of the system of Figure P4.8 in terms of (a) $\dot{x}$, the velocity of the block, and (b) $\dot{\theta}$, the clockwise angular velocity of the disk.

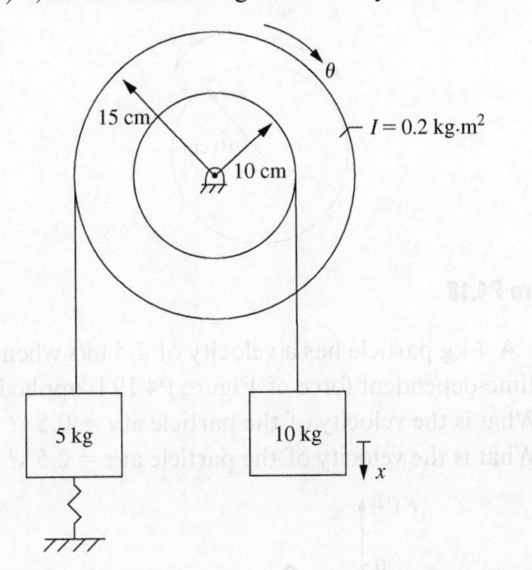

Figure P4.8

4.9 Determine the kinetic energy of the rotor and gear system of Figure P4.9 when shaft AB rotates at 250 rpm.

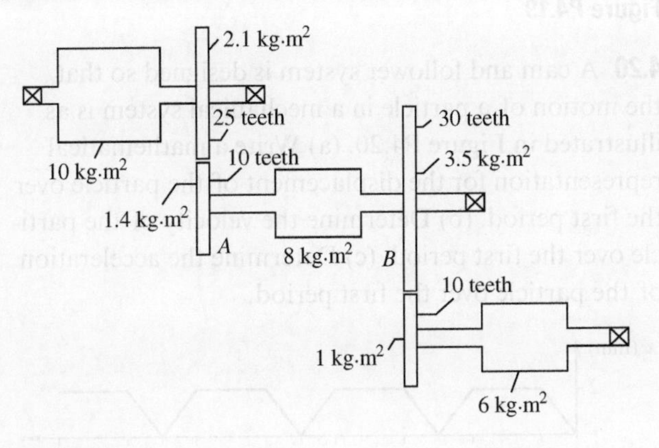

Figure P4.9

4.10 Determine the equivalent stiffness of the springs in the system of Figure P4.10.

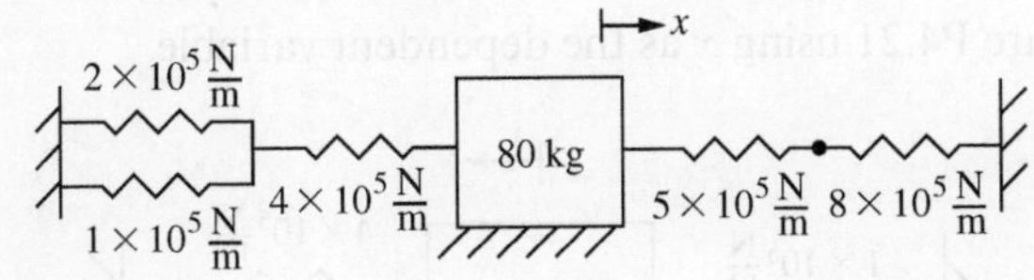

Figure P4.10

4.11 Determine the equivalent torsional stiffness of the springs in the system of Figure P4.11.

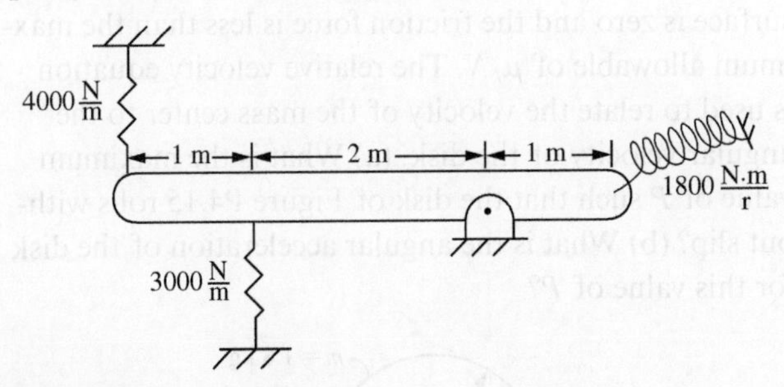

Figure P4.11

4.12 Determine the equivalent stiffness of a spring placed at A for the system of springs in the system of Figure P4.12.

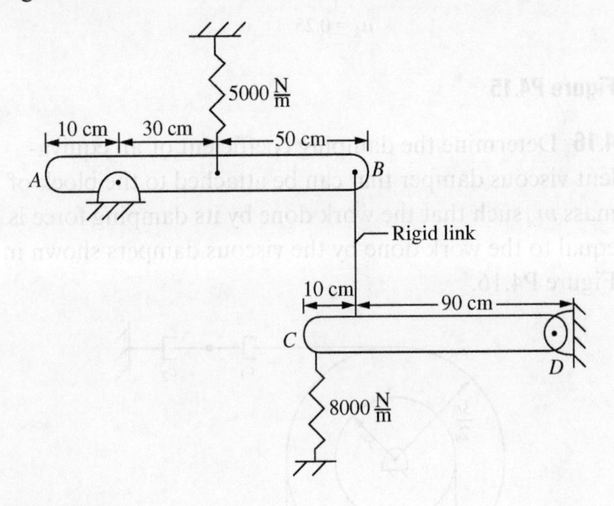

Figure P4.12

4.13 Using a potential energy argument, show that the equivalent stiffness of two springs in series is

$$k_{eq} = \frac{k_1 k_2}{k_1 + k_2} \tag{a}$$

4.14 A model of a viscoelastic mechanical component is shown in Figure P4.14. Let $x(t)$ represent the displacement of the end of the component. Determine a relation between force developed in the component and x.

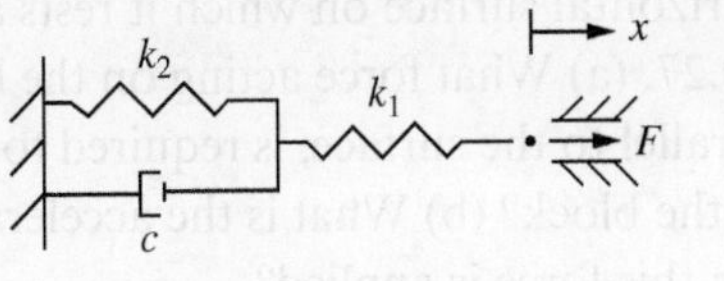

Figure P4.14

4.15 When a disk rolls on a surface without slip, the velocity of the point of contact of the disk with the surface is zero and the friction force is less than the maximum allowable of $\mu_k N$. The relative velocity equation is used to relate the velocity of the mass center to the angular velocity of the disk. (a) What is the maximum value of P such that the disk of Figure P4.15 rolls without slip? (b) What is the angular acceleration of the disk for this value of P?

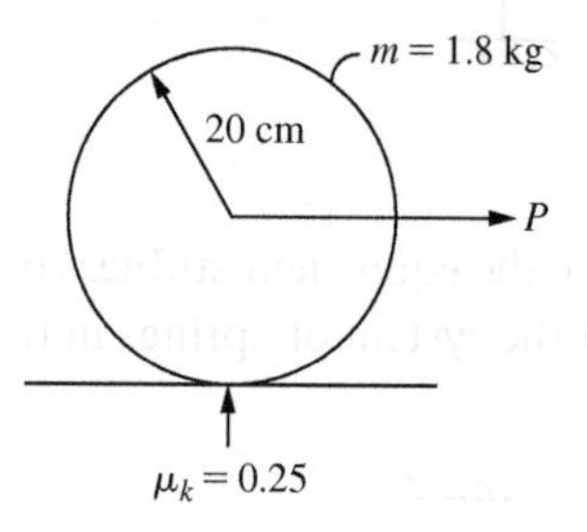

Figure P4.15

4.16 Determine the damping coefficient of an equivalent viscous damper that can be attached to the block of mass m_A such that the work done by its damping force is equal to the work done by the viscous dampers shown in Figure P4.16.

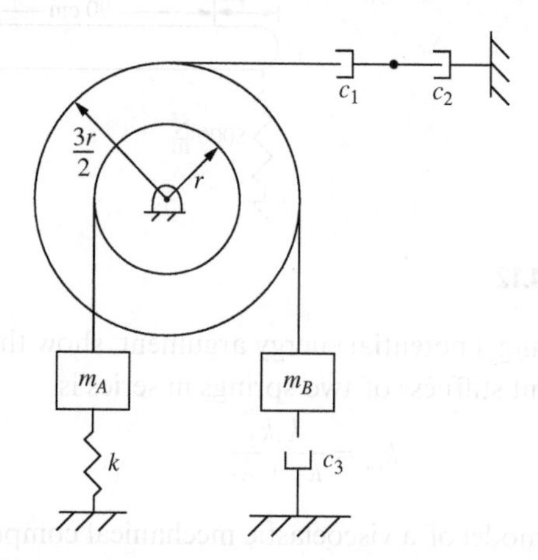

Figure P4.16

4.17 The coefficients of friction between a 45-kg block and the horizontal surface on which it rests are $\mu_s = 0.30$ and $\mu_k = 0.27$. (a) What force acting on the block, applied parallel to the surface, is required to initiate motion of the block? (b) What is the acceleration of the block when this force is applied?

4.18 The disk of Figure P4.18 has a mass of 0.75 kg and a radius of 10 cm. It is rotating at an angular velocity of 1200 rpm when a constant torque T is applied to bring the disk to rest. (a) What is the minimum value of T needed to stop the disk in 1 s? (b) What is the minimum value of T needed to stop the disk within 100 revolutions? (c) Redo parts (a) and (b) if the disk has a frictional moment of 10 N·m.

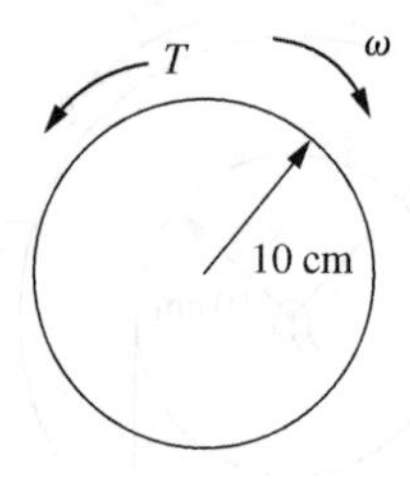

Figure P4.18

4.19 A 4-kg particle has a velocity of 2.5 m/s when the time-dependent force of Figure P4.19 is applied. (a) What is the velocity of the particle at $t = 0.5$ s? (b) What is the velocity of the particle at $t = 2.5$ s?

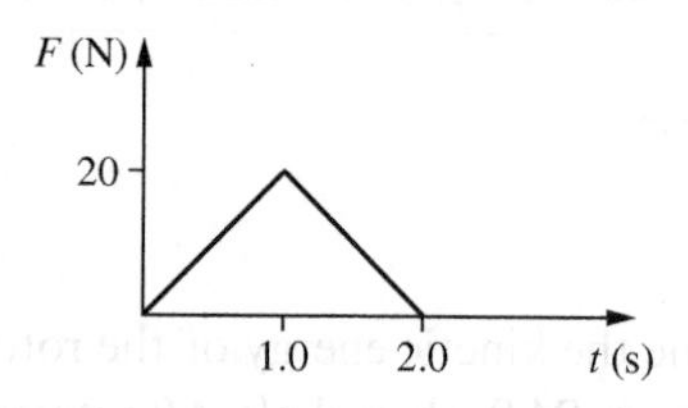

Figure P4.19

4.20 A cam and follower system is designed so that the motion of a particle in a mechanical system is as illustrated in Figure P4.20. (a) Write a mathematical representation for the displacement of the particle over the first period. (b) Determine the velocity of the particle over the first period. (c) Determine the acceleration of the particle over the first period.

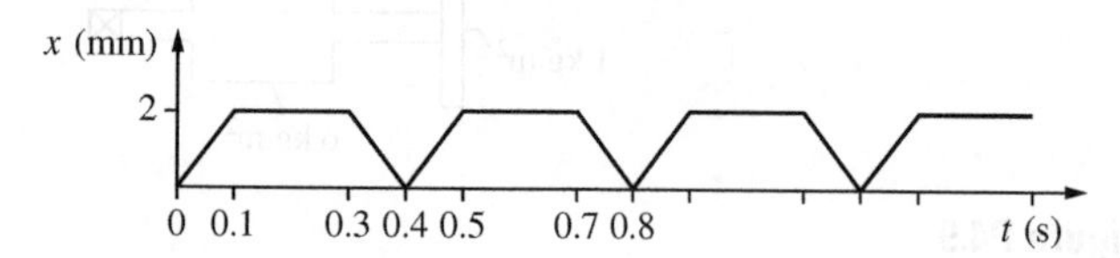

Figure P4.20

4.21 Derive a mathematical model for the system of Figure P4.21 using x as the dependent variable.

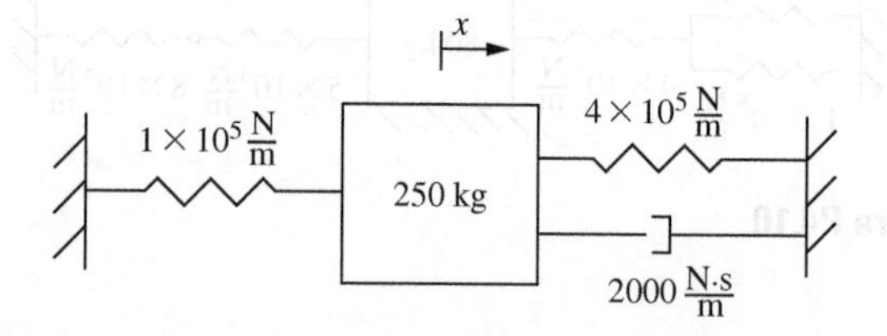

Figure P4.21

4.22 A 2000-kg vehicle is traveling along a steep road at a speed of 30 m/s, as shown in Figure P4.22, when the driver spots a stalled car 300 m from the vehicle. The driver immediately applies the brakes, resulting in a constant braking force of F_B. Let $x(t)$ represent the vehicle's displacement measured from when the driver applies the brakes. (a) Derive a mathematical model for $x(t)$. (b) Use the model to determine the minimum value of F_B such that the vehicle stops before it hits the stalled car.

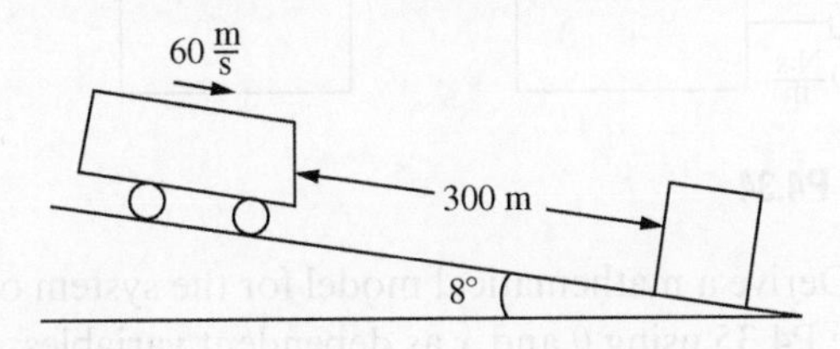

Figure P4.22

4.23 A worker is using a torque wrench to remove a bolt from the flange of a pressure vessel. The worker applies a torque of 500 N·m with the wrench. As it turns, a resisting moment of 480 N·m is applied to the bolt, which has a moment of inertia of 0.035 kg·m^2. It takes 50 revolutions to remove the bolt. How long will it take the worker to remove the bolt assuming the applied torque and the resisting torque are constant?

4.24 A 10-kg projectile is fired at a speed of 800 m/s at an angle of 30° with the horizontal. The projectile will land 100 m below the elevation of firing. (a) How long will the projectile be in motion? (b) What is the horizontal range of the projectile? (c) What is the maximum altitude attained by the projectile?

4.25 A 0.2-kg projectile is fired at a speed of 600 m/s at an angle of 40° with the horizontal. The projectile is fired into the wind, which is estimated to provide a constant horizontal force of 5 N opposing motion of the projectile. Let $x(t)$ represent the horizontal displacement of the projectile and $y(t)$ represent the vertical displacement of the projectile, both measured from the firing location. (a) Derive a mathematical model for $x(t)$ and $y(t)$. (b) What is the maximum altitude attained by the projectile? (c) What is the range of the projectile?

4.26 Derive a mathematical model for the system of Figure P4.26 using x as the dependent variable.

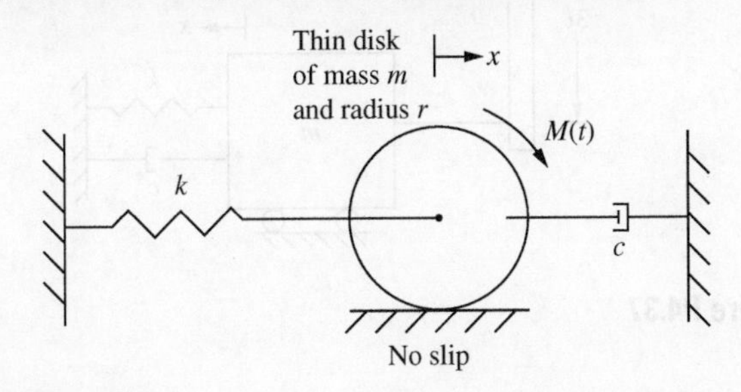

Figure P4.26

4.27 Derive a mathematical model for the system of Figure P4.27 where $y(t)$, the input to the system, is the prescribed displacement of the base of the system and $x(t)$ is the system output.

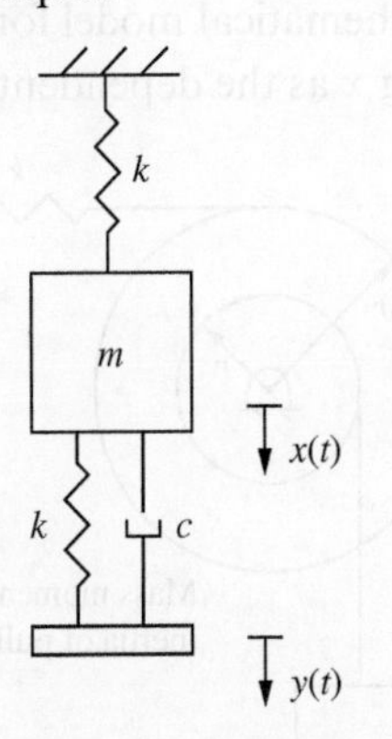

Figure P4.27

4.28 Derive a mathematical model for the system of Figure P4.28 using x as the dependent variable.

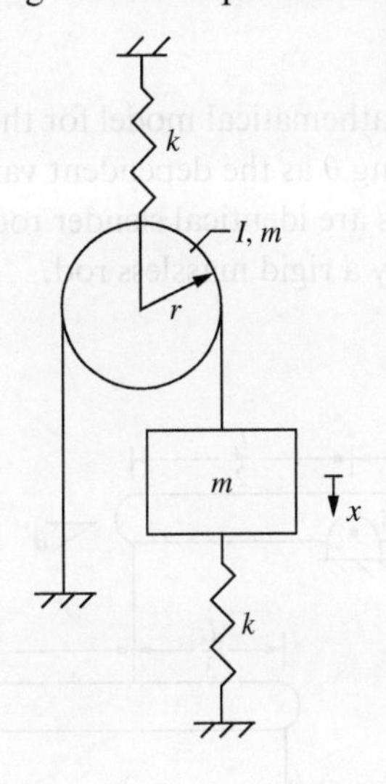

Figure P4.28

4.29 Derive a mathematical model for the system of Figure P4.29 using θ, the angular displacement of the bar from the system's equilibrium position, as the dependent variable. Assume the rod is rigid and massless and assume small θ.

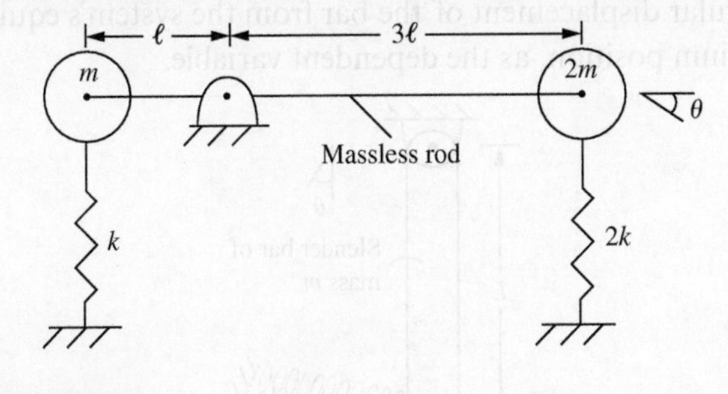

Figure P4.29

4.30 Derive a mathematical model for the system of Figure P4.29 and Problem 4.29 assuming the bar is a rigid slender bar of mass m.

4.31 Derive a mathematical model for the system of Figure P4.31 using x as the dependent variable.

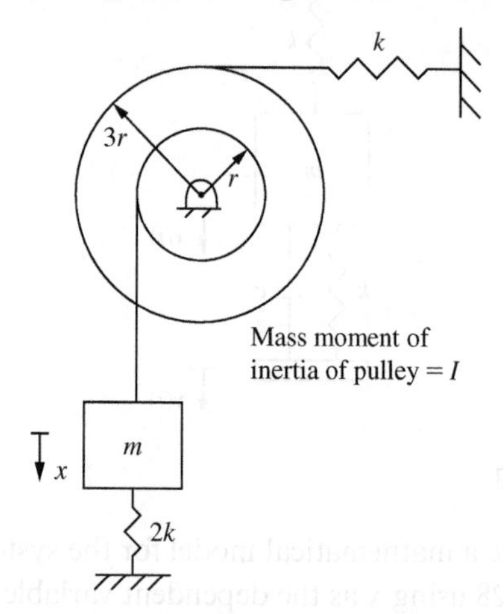

Figure P4.31

4.32 Derive a mathematical model for the system of Figure P4.32 using θ as the dependent variable assuming small θ. The bars are identical slender rods of mass m and connected by a rigid massless rod.

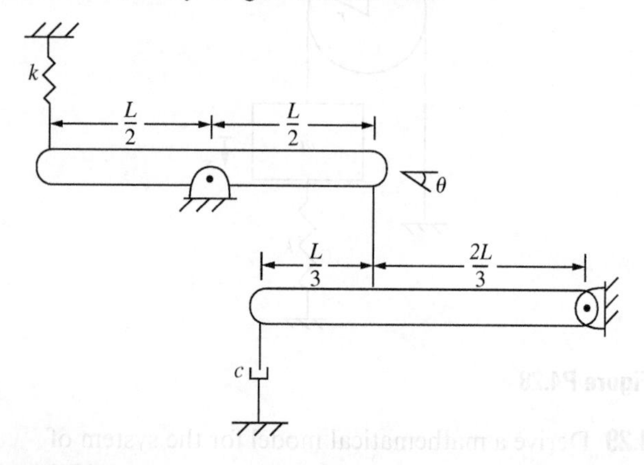

Figure P4.32

4.33 Derive a mathematical model for the compound pendulum of Figure P4.33. Use θ, the counterclockwise angular displacement of the bar from the system's equilibrium position, as the dependent variable.

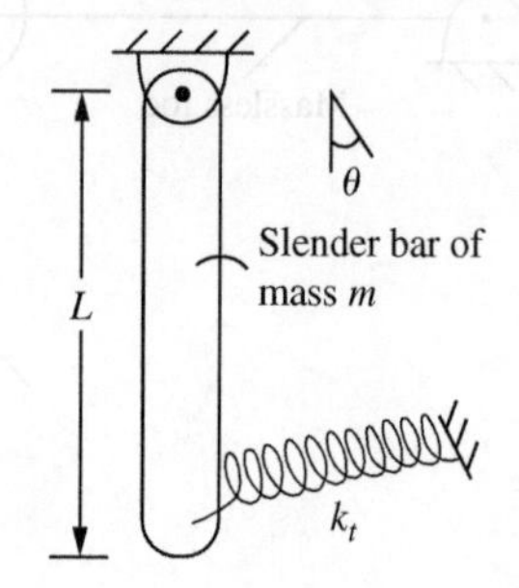

Figure P4.33

4.34 Derive a mathematical model for the system of Figure P4.34 using x_1 and x_2 as dependent variables.

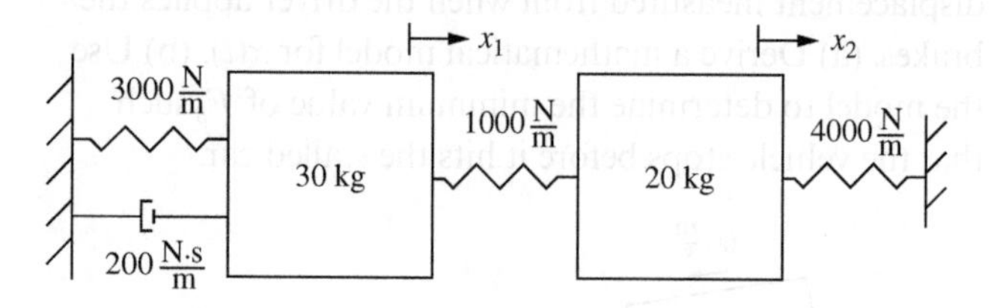

Figure P4.34

4.35 Derive a mathematical model for the system of Figure P4.35 using θ and x as dependent variables.

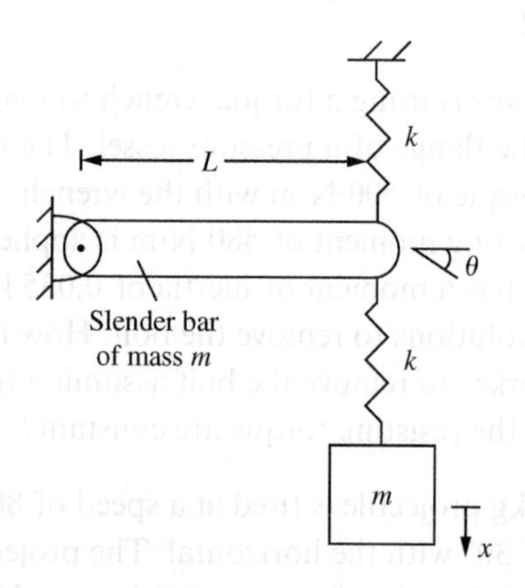

Figure P4.35

4.36 Derive a mathematical model for the system of Figure P4.36 using x_1, x_2, and x_3 as dependent variables.

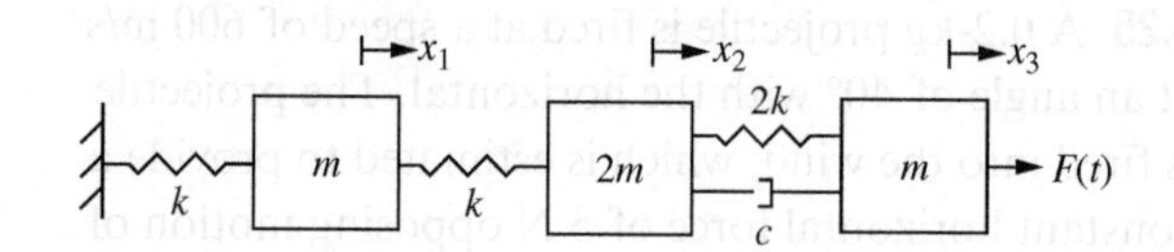

Figure P4.36

4.37 Derive a mathematical model for the system of Figure P4.37 using x as the dependent variable.

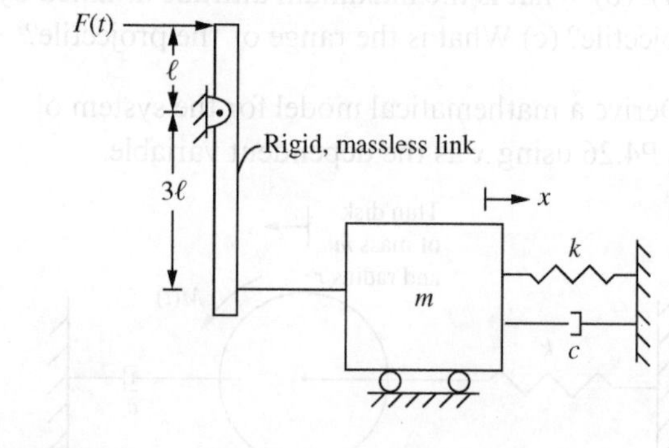

Figure P4.37

4.38 Derive a mathematical model for the system of Figure P4.38 using θ as the dependent variable, noting that $y(t)$ is a prescribed displacement.

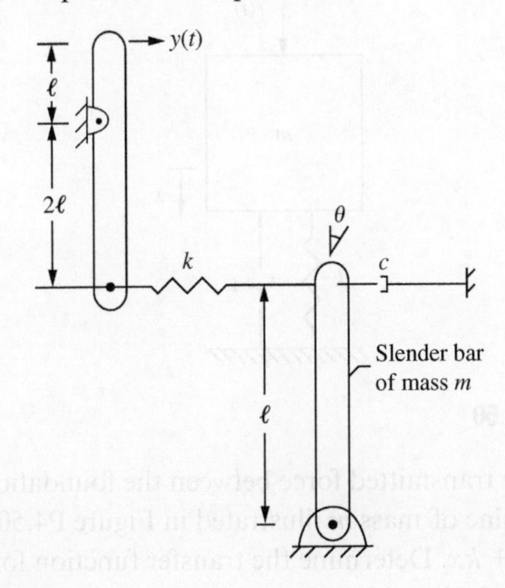

Figure P4.38

4.39 The single-cylinder engine of Problem 4.7 and Figure P4.7 is mounted in a housing, as illustrated in Figure P4.39, such that the total mass of the engine and the housing is M. The housing is mounted on an elastic foundation to allow vertical motion, but it is constrained from horizontal or rotational motion. The properties of the crank, connecting rod, and piston are those given in Problem 4.7. (a) Determine the displacement, velocity, and acceleration of the piston as a function of time if the crank rotates at a constant speed of 250 rpm. Assume that the crank, connecting rod, and piston are aligned at $t = 0$. (b) Derive a mathematical model for the motion of the engine in its housing and on the elastic foundation.

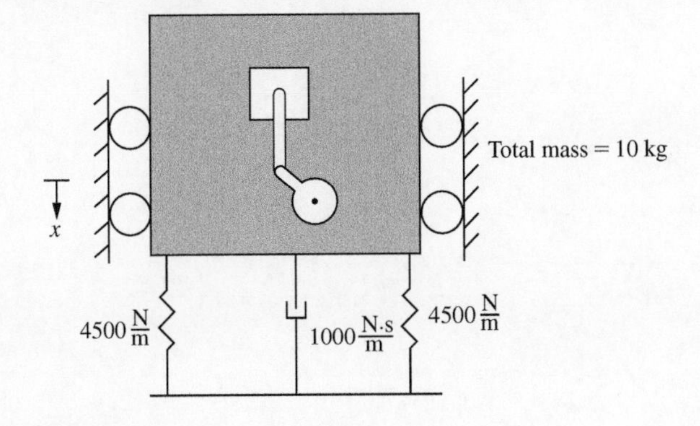

Figure P4.39

4.40 Determine the transfer function defined as $G(s) = \dfrac{X(s)}{M(s)}$ for the system of Figure P4.40. The uniform thin disk has a radius of 30 cm and a mass of 0.7 kg. It rolls without slip.

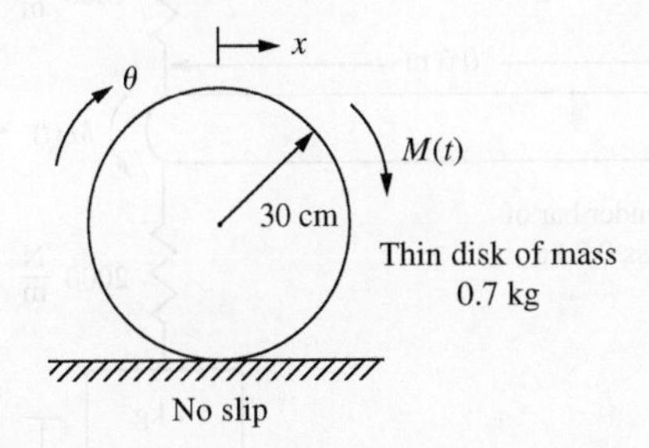

Figure P4.40

4.41 Determine the transfer function defined as $G(s) = \dfrac{\Theta(s)}{M(s)}$ for the system of Figure P4.40, where $\theta(t)$ is the angular displacement of the disk from when the disk is in equilibrium. The uniform thin disk has a radius of 30 cm and a mass of 0.7 kg. It rolls without slip.

4.42 Develop a state-space model for the system of Figure P4.40 where $M(t)$ is the input and $x(t)$, the displacement of the mass center of the disk, is the output.

4.43 Develop a state-space model for the system of Figure P4.40 where $M(t)$ is the input and $\theta(t)$ is the output.

4.44 Determine the transfer function $G(s) = \dfrac{X_2(s)}{F_1(s)}$ for the system of Figure P4.44.

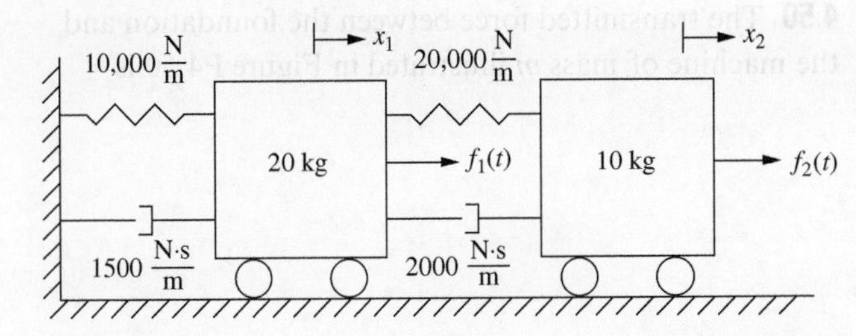

Figure P4.44

4.45 Determine the transfer function matrix $\mathbf{G}(s)$ for the system of Figure P4.44.

4.46 Determine a state-space model for the system of Figure P4.44 when the outputs are defined as x_1 and x_2.

4.47 Determine the transfer function $G(s) = \frac{X_2(s)}{M(s)}$ for the system of Figure P4.47.

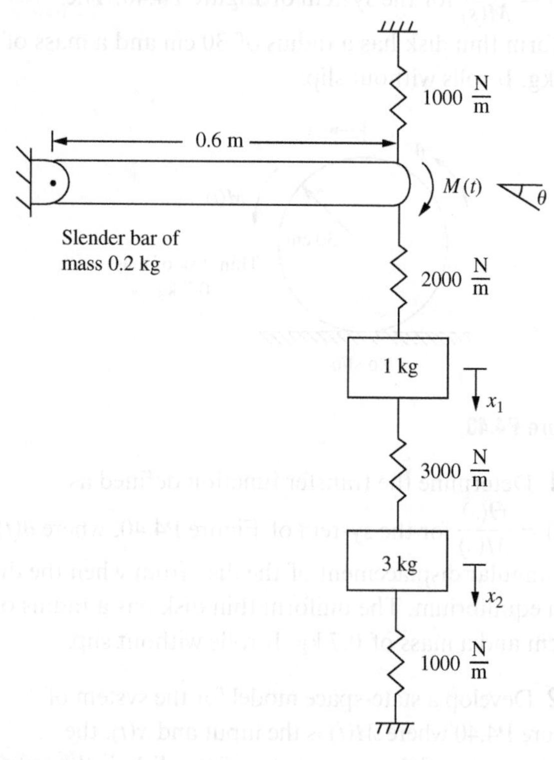

Figure P4.47

4.48 Determine the transfer function vector $\mathbf{G}(s)$ for the system of Figure P4.47.

4.49 Determine a state-space model for the system of Figure P4.47 when the outputs are defined as x_1 and x_2.

4.50 The transmitted force between the foundation and the machine of mass m illustrated in Figure P4.50 is $F_T = c\dot{x} + kx$. Determine a state-space model for the system if the input is $F(t)$ and the output is F_T.

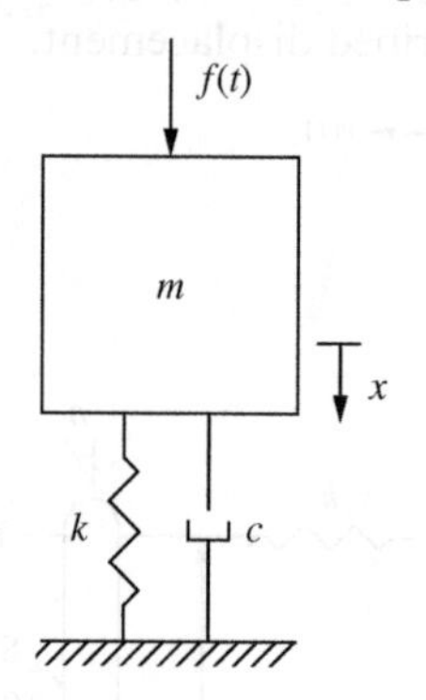

Figure P4.50

4.51 The transmitted force between the foundation and the machine of mass m illustrated in Figure P4.50 is $F_T = c\dot{x} + kx$. Determine the transfer function for the system if the input is $F(t)$ and the output is F_T.

4.52 The transfer function for a system with motion input $y(t)$ is $G(s) = \frac{X(s)}{Y(s)} = \frac{3s + 15}{s^2 + 3s + 15}$. The relative motion between the input and output is defined by $w(t) = x(t) - y(t)$. Determine the transfer function for the relative motion defined by $\frac{W(s)}{Y(s)}$.

4.53 Determine a state-space model for the system of Problem 4.52 when the output of the system is $w(t)$.

4.54 One measure of the performance of the suspension system of Example 4.35 and Example 4.36 is the acceleration of the mass center of the vehicle, $a = \ddot{x}_3$. Determine the transfer function $G(s) = \frac{A(s)}{Z_1(s)}$.

5

Electrical Systems

Learning Objectives

At the conclusion of this chapter, you should be able to:

- Understand the definitions of charge, current, electric potential, electric field, and electric power
- Understand the basic circuit components: the resistor, the capacitor, the inductor, and the source
- Understand, quantify, and apply Kirchhoff's Current Law and Kirchhoff's Voltage Law to circuits
- Understand the concepts of parallel and series resistors, capacitors, and inductors, and know how these are used in circuit reduction
- Understand the mathematical modeling of circuits, and apply KCL and KVL to single and multiloop circuits involving resistors, capacitors, and inductors
- Understand the analogies between electrical systems and mechanical systems for electrical systems with a voltage source and for circuits with a current source; also know how these analogies can be extended to multiloop circuits
- Understand the ideal operational amplifier and how it is used in electrical controllers
- Understand the coupling field in an electromechanical system, know how to mathematically model electromechanical systems, and apply the modeling to dc servomotors
- Know how to obtain a transfer function from the mathematical model of an electrical or electromechanical system
- Know how to obtain a state-space model from the mathematical model of an electrical or electromechanical system

Electrical systems involve the motion of charged particles through electric and magnetic fields established in system components. Electrical systems are often used to provide power for mechanical and thermal systems. For example, an electrical system may be used to provide power for a household appliance such as a toaster or for the engine of an automobile. Such systems are often used in conjunction with mechanical or thermal systems for the purpose of power generation. In electromechanical systems such as motors, robots, transducers, and loudspeakers, the electrical and mechanical systems are coupled and interact with one another.

Many electrical devices can be modeled using electric circuits, which are schematically represented by circuit diagrams. A circuit diagram illustrates the arrangement of the device's components and provides a format for the application of conservation laws. A circuit consists of **active elements** (elements that add energy to the circuit) and **passive elements** (elements that dissipate or store energy).

The mathematical modeling of electrical systems is introduced in this chapter. The modeling follows the process outlined in Section 1.3.

1. The system to be modeled is an electrical device that is represented by a circuit diagram.
2. Implicit assumptions used in this study of modeling of electrical systems include the following:
   - Circuit components are two-terminal elements.
   - Circuit components obey linear laws.
   - Circuit components are lumped and represented on a circuit diagram at specific locations on the circuit.
   - A circuit is not affected by external electromagnetic fields.
3. The components of an electric circuit are discussed in Section 5.2.
4. The dependent variables in circuit analysis are current and voltage, as introduced in Section 5.1. The parameters of circuit components are discussed in Section 5.2.
5. A circuit diagram is drawn, illustrating the circuit components and the independent variables used.
6. The applicable physical laws, Kirchhoff's laws, are discussed in Section 5.3. Their application is illustrated in Sections 5.4–5.6.
7. Appropriate initial conditions are specified.
8. Mathematical analysis is discussed in Chapters 2 and 3.
9. Analysis of system response is discussed in Chapters 7 and 8.

This chapter leads to an understanding of the fundamental principles of circuit analysis in order to understand the modeling of electromechanical systems and the fundamentals of electronic controllers.

Details of the application of steps 1–7 are presented in Sections 5.1–5.5. The process is applied to passive circuits used to model devices such as filters and power generation systems as well as active circuits containing operational amplifiers. Op-amp circuits that can be used as electronic controllers are disussed in Section 5.6. Analogies between electrical and mechanical systems are presented using energy methods. Electromechanical devices in which the motion of a mechanical system is coupled with an electromagnetic field are disussed in Section 5.8. Transfer functions for electrical systems are discussed in Section 5.9, and a state-space formulation of the mathematical model is considered in Section 5.10.

5.1 Charge, Current, Voltage, and Power

Atoms consist of electrons orbiting a nucleus, which contains neutrons and protons. Forces between the particles in the nucleus and the electrons keep the electrons in orbit. These forces result due to the electric **charge** of the atomic particles. An electron is negatively charged, a proton is positively charged, and a neutron is neutrally charged. A repulsive force is developed between charged particles if the particles have charges of the same sign; the force is attractive if the particles have charges of opposite sign. An object with an excess of electrons has a net negative charge, while an object with a deficiency of electrons has a net positive charge. The unit of charge is the coulomb (C), and one electron has a charge of -1.602×10^{-19} C. The net charge of an object is the number of excess electrons times the charge of an electron.

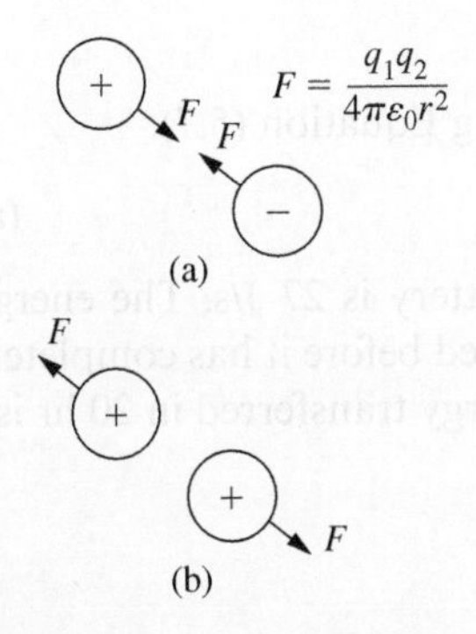

Figure 5.1

Forces between charged particles of (a) opposite sign and (b) like sign.

The magnitude of the force between two particles of charge q_1 and q_2 is given by Coulomb's Law

$$F = \frac{q_1 q_2}{4\pi\varepsilon_0 r^2} \tag{5.1}$$

where ε_0 is the permittivity of free space, which has a numerical value of 8.85×10^{-12} $C^2/N{\cdot}m^2$. The line of action of the force is between the two particles. If the particles have charges of opposite signs, the force between the particles is directed between the particles, as shown in Figure 5.1(a). If the particles have charges of the same sign, the force from one particle on the other is directed away form the particle, as shown in Figure 5.1(b).

An **electric field** develops in the vicinity of a charged particle. The magnitude of the electric field at a given point due to the presence of a nearby charged particle is the force per unit charge of a force that would be developed at that point if a particle of unit charge was placed at that point. Electric field is a vector quantity. The total electric field at a point in space is the sum of the electric fields due to all charged particles.

Electric current is the flow of charged particles. One ampere (A) of current results from 1 C of charge moving through a surface in a second:

$$1 \text{ ampere (A)} = \frac{1 \text{ coulomb (C)}}{\text{second (s)}} \tag{5.2}$$

The algebraic symbol for electric current is i. Electric current is considered to be positive when it is in the direction of the motion of positively charged particles, or opposite the direction of negatively charged particles. The mathematical relationship between current and charge at a location in space is

$$i = \frac{dq}{dt} \tag{5.3}$$

Electric potential is the work required to move a 1-C charge against an electric field. The unit of electric potential, called the volt (V), is defined by

$$1 \text{ volt (V)} = \frac{1 \text{ N·m}}{1 \text{ C}} = 1\frac{\text{joule (J)}}{\text{C}} \tag{5.4}$$

Electric potential is also called **voltage** and is represented by the algebraic symbol v.

The existence of an electric potential indicates the ability to do work to move particles against an electric field, thus inducing current. Electric potential is analogous to the potential energy due to gravity, which represents the ability to move a mass against a gravitational field, thus initiating kinetic energy.

Electric power is the rate at which energy is transferred from an electric source. The basic unit of electric power is the watt (W), defined by

$$1 \text{ watt (W)} = 1 \text{ J/s} \tag{5.5}$$

Equation (5.5) can be rewritten as

$$\begin{aligned} 1 \text{ W} &= 1 \text{ (J/C)(C/s)} \\ &= (1 \text{ V})(1 \text{ A}) \end{aligned} \tag{5.6}$$

Thus from Equation (5.2) and Equation (5.5), electric power is the product of current and voltage, or

$$P = vi \tag{5.7}$$

Example 5.1

When a 9-V battery is placed in a device such that the current is 3 A, it completely discharges in 20 hours (hr). What is the energy stored in the battery?

Solution

The power delivered during the discharge is calculated using Equation (5.7):

$$P = (9\text{ V})(3\text{ A}) = 27\text{ W} = 27\text{ J/s} \tag{a}$$

Thus, the rate at which energy is transferred from the battery is 27 J/s. The energy stored in the battery is the total amount of energy transferred before it has completely discharged. Assuming the current is constant, the total energy transferred in 20 hr is

$$\begin{aligned} E &= Pt \\ &= (27\text{ W})(20\text{ hr}) \\ &= (27\text{ J/s})(20\text{ hr})(3600\text{ s/hr}) \\ &= 1.944\text{ MJ} \end{aligned} \tag{b}$$

5.2 Circuit Components

Active circuit components provide a source of energy to a circuit, while a passive circuit component stores or dissipates energy. Passive circuit components used in electrical and electromechanical systems that are modeled in this book, and illustrated in Table 5.1, are resistors, capacitors, and inductors. Active circuit components, illustrated in Table 5.2, include voltage sources, current sources, and operational amplifiers. A circuit in which all sources provide constant input is called a direct current (dc) circuit. A circuit in which a source provides input that varies with time is called an alternating current (ac) circuit.

5.2.1 Resistors

A **resistor** is a circuit component that dissipates energy by converting electrical energy to heat. There is a decrease in voltage, a voltage drop, across the resistor in the direction of positive current. In general, the voltage drop is a function of the current, $v = f(i)$. A linear resistor is one in which **Ohm's law**, which states that the voltage drop across a linear resistor is proportional to the current flowing through the resistor, is satisfied:

$$v = iR \tag{5.8}$$

Table 5.1 Passive Circuit Elements

Element	Schematic	Voltage-Current Relations	Power	Energy
Resistor	R	$v = iR$ $i = \frac{v}{R}$	$P = i^2R = \frac{v^2}{R}$	Dissipated $E = \int_0^t i^2Rdt$
Capacitor	C	$i = C\frac{dv}{dt}$ $v = \frac{1}{C}\int_0^t i\,dt + v(0)$	$P = Cv\frac{dv}{dt}$	Stored in electric field $E = \frac{1}{2}Cv^2$
Inductor	L	$v = L\frac{di}{dt}$ $i = \frac{1}{L}\int_0^t vdt + i(0)$	$P = Li\frac{di}{dt}$	Stored in magnetic field $E = \frac{1}{2}Li^2$

Table 5.2 Active Circuit Elements

Element	Schematic	Comments
dc voltage source		Voltage increases continuously from − to +.
Time-varying voltage source		Voltage increases in the direction of current from − to +.
Current source		Current flows in direction of arrow.
Operational amplifier		Voltages are referenced to ground, connected to external dc source. Output is proportional to difference of inputs.

The constant of proportionality R is called the resistance. The unit of resistance, the ohm (Ω), is defined as

$$1 \text{ ohm } (\Omega) = \frac{1 \text{ V}}{1 \text{ A}} \tag{5.9}$$

The voltage drop in a resistor occurs in the direction of the current; the voltage is greater at the end of the resistor where the current enters.

Many components of electrical devices dissipate energy into heat. After use, the cord of an electric appliance is often warm as a result of energy dissipation within the wires. Batteries have internal resistances that dissipate energy when connected in a circuit. The resistance of an electrical component is dependent on the material the component is made from. The resistance of a wire of length L and cross-sectional area A is calculated by

$$R = \frac{\rho L}{A} \tag{5.10}$$

where ρ is the resistivity, with units of Ω·m, of the material from which the wire is made. Conductivity σ is the reciprocal of resistivity:

$$\sigma = \frac{1}{\rho} \tag{5.11}$$

Metals are good conductors of electricity and have high conductivity and low resistivity. Insulating materials have high resistivity and low conductivity.

The power dissipated by a resistor is the product of the current times the voltage drop across the resistor. Ohm's law, Equation (5.8), shows that the dissipated power can be written as either

$$P = i^2 R \tag{5.12}$$

or

$$P = \frac{v^2}{R} \tag{5.13}$$

Example 5.2

The voltage drop between two ends of a wire is measured as 3 V. The heat transferred from the wire in 1 s is measured as 90 J. Determine **a.** the resistance of the wire and **b.** the current in the wire.

Solution

The power dissipated in the wire is equal to the heat transferred from the wire in 1 s, $P = 90 \text{ J}/1 \text{ s} = 90 \text{ W}$.

a. Solving Equation (5.13) for R leads to

$$R = \frac{v^2}{P}$$

$$= \frac{(3 \text{ V})^2}{90 \text{ W}} = 0.1 \ \Omega \tag{a}$$

b. The current in the wire is calculated using Ohm's law, Equation (5.8), as

$$i = \frac{v}{R}$$

$$= \frac{3 \text{ V}}{0.1 \ \Omega} = 30 \text{ A} \tag{b}$$

5.2.2 Capacitors

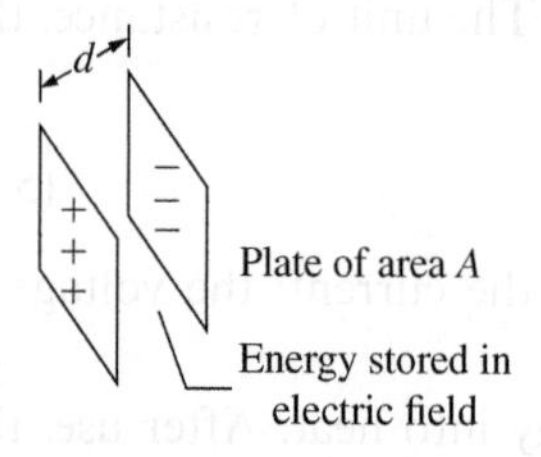

Figure 5.2
Parallel plate capacitor, $C = K\frac{\varepsilon_0 A}{d}$. An electric field is developed between the plates in which energy is stored.

A **capacitor** is a device that stores energy in the presence of an electric field. A capacitor develops a charge when it is subject to a potential difference. A **parallel plate capacitor**, illustrated in Figure 5.2, consists of two parallel plates made of a material that conducts electricity. The thin gap between the plates is of a nonconducting, or dielectric, material. When the plates of the capacitor are charged, one with a positive charge and one with a negative charge, an electric field develops in the dielectric material and stores energy.

When a capacitor is connected to a battery, the battery does work to charge the plates of the capacitor until the potential difference across them is equal to the battery voltage. When the voltage source is removed, the capacitor remains charged until discharged by equalizing the charges on the parallel plates.

The charge stored in the capacitor is proportional to the voltage:

$$q = Cv \tag{5.14}$$

The constant of proportionality in Equation (5.14) is called the **capacitance**, whose units are expressed in **farads** defined by

$$1 \text{ farad (F)} = 1 \text{ C/V} = 1 \text{ A·s/V} \tag{5.15}$$

The capacitance of a parallel plate capacitor of plate area A whose plates are separated by a distance d is

$$C = K\varepsilon_0\frac{A}{d} \tag{5.16}$$

where K is the dielectric constant of the material in the gap separating the plates and $\varepsilon_0 = 8.85 \times 10^{-12} \text{ C}^2/(\text{N·m}^2)$ is the permittivity of free space.

Differentiation of Equation (5.14) with respect to time leads to

$$\frac{dq}{dt} = C\frac{dv}{dt} \tag{5.17}$$

Current is the time rate of change of charge; thus using Equation (5.3) in Equation (5.17) leads to

$$i = C\frac{dv}{dt} \tag{5.18}$$

Integration of Equation (5.18) from $t = 0$ to an arbitrary time t leads to

$$\int_0^t i\,dt = C\int_0^t \frac{dv}{dt}dt$$

$$= C[v(t) - v(0)] \tag{5.19}$$

where $i(0) = 0$ and $v(0)$ is the potential in the capacitor at $t = 0$. (Remember a capacitor may store energy even when current does not flow through the capacitor.) Rearranging Equation (5.19) to solve for $v(t)$ leads to

$$v(t) = \frac{1}{C}\int_0^t i\,dt + v(0) \tag{5.20}$$

The power for the capacitor is obtained using Equations (5.7) and (5.18) as

$$\begin{aligned} P &= vi \\ &= v\left(C\frac{dv}{dt}\right) \end{aligned} \tag{5.21}$$

The energy stored in a capacitor between $t = 0$ and an arbitrary time t is

$$\begin{aligned} E &= \int_0^t P dt \\ &= \int_0^t Cv\frac{dv}{dt}\,dt \\ &= \int_0^t C\frac{d}{dt}\left(\frac{1}{2}v^2\right)dt \\ &= \frac{1}{2}C\left[v(t)^2 - v(0)^2\right] \end{aligned} \tag{5.22}$$

Equation (5.22) gives the energy stored in the capacitor since $t = 0$. The energy stored in the capacitor at $t = 0$ is $\frac{1}{2}Cv(0)^2$ and the total energy stored in the capacitor is

$$E = \frac{1}{2}Cv^2 \tag{5.23}$$

Example 5.3

A capacitor of capacitance 40 μF initially has a charge of 200 μC when it is subject to an alternating current

$$i(t) = 0.5\cos(100t)\text{ A} \tag{a}$$

Determine **a.** $v(t)$, **b.** $P(t)$, and **c.** $E(t)$, the energy stored since $t = 0$.

Solution

The initial voltage in the capacitor is obtained using Equation (5.14) as

$$\begin{aligned} v(0) &= \frac{q(0)}{C} \\ &= \frac{200\ \mu\text{C}}{40\ \mu\text{F}} \\ &= 5\text{ V} \end{aligned} \tag{b}$$

a. The voltage in the capacitor is obtained using Equation (5.20) as

$$\begin{aligned} v &= \frac{1}{40\ \mu\text{F}}\int_0^t [0.5\cos(100t)\text{ A}]\,dt + 5 \\ &= \frac{0.5}{(40\times 10^{-6})(100)}\sin(100t)\Big|_0^t + 5 \\ &= [125\sin(100t) + 5]\text{ V} \end{aligned} \tag{c}$$

b. The power in the capacitor is determined using Equation (5.7) as

$$\begin{aligned} P &= vi \\ &= \{[125 \sin(100t) + 5] \text{ V}\}[0.5 \cos(100t) \text{ A}] \\ &= [62.5 \sin(100t) \cos(100t) + 2.5 \cos(100t)] \text{ W} \\ &= [31.25 \sin(200t) + 2.5 \cos(100t)] \text{ W} \end{aligned} \quad \text{(d)}$$

c. The energy stored in the capacitor is obtained using Equation (5.22)

$$\begin{aligned} E &= \frac{1}{2}(40 \ \mu\text{F})\{[125 \sin(100t) + 5]^2 \text{ V}^2 - (5 \text{ V})^2\} \\ &= (20 \times 10^{-6}) [15{,}625 \sin^2(100t) + 1250 \sin(100t)] \text{ J} \\ &= [0.3125 \sin^2(100t) + 0.0250 \sin(100t)] \text{ J} \\ &= \left\{(0.3125) \frac{1}{2}[1 - \cos(200t)] + 0.0250 \sin(100t)\right\} \text{ J} \\ &= [156.25 - 156.25 \cos(200t) + 25 \sin(100t)] \text{ mJ} \end{aligned} \quad \text{(e)}$$

Equation (e) shows that, for an alternating source, the energy stored in the capacitor increases while the voltage is increasing but is released when the voltage begins to decrease.

5.2.3 Inductors

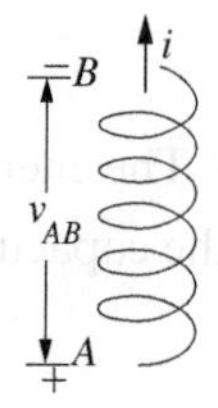

Figure 5.3

When current moves through a coil of wire, a magnetic field is induced in which energy is stored.

When a coil of wire is placed in a moving magnetic field, current is induced in the wire. The moving magnetic field leads to a potential difference across the wire, which is referred to as an electromotive force (emf). The electromotive force induces current in the wire. Similarly, when current moves through a coil of wire, as illustrated in Figure 5.3, a magnetic field is induced and energy is stored in the magnetic field. **Induction** is the process by which a current is developed in a moving magnetic field or by which a magnetic field is developed in a coil of wire due to the motion of charged particles.

An **inductor** stores energy in the presence of a magnetic field. Most inductors are coils of wire. When current flows through the wire, work is done to establish a magnetic field and energy is stored in the magnetic field. When the current source is removed, the magnetic field vanishes and the stored energy is released.

The magnetic flux ϕ developed in an inductor is proportional to the current such that $d\phi/di = \alpha$, a constant. **Faraday's law** states that the electric potential is proportional to the rate of change of the magnetic flux,

$$v = N\frac{d\phi}{dt} \quad (5.24)$$

where N is the number of coils in the inductor. Using the chain rule for derivatives in Equation (5.24) leads to

$$\begin{aligned} v &= N\frac{d\phi}{dt}\frac{di}{dt} \\ &= L\frac{di}{dt} \end{aligned} \quad (5.25)$$

where $L = N\frac{d\phi}{di}$ is a constant called the **inductance**, which is measured in henrys (H):

$$1 \text{ henry (H)} = 1\frac{\text{V}}{\text{A/s}} = 1\frac{\text{V}\cdot\text{s}}{\text{A}} \quad (5.26)$$

The unit of V·s, called a weber (Wb), is a unit in which magnetic flux is measured. Thus, Equation (5.26) can also be written as

$$1 \text{ H} = 1 \text{ Wb/A} \quad (5.27)$$

Integration of Equation (5.25) leads to

$$i(t) = \frac{1}{L}\int_0^t v(t)dt + i(0) \tag{5.28}$$

The power of the inductor is obtained using Equations (5.7) and (5.25), leading to

$$P = Li\frac{di}{dt} \tag{5.29}$$

The energy stored in the inductor is

$$E = \int_0^t Pdt$$

$$= \frac{1}{2}L\{[i(t)]^2 - [i(0)]^2\} \tag{5.30}$$

When two inductors are near one another in the same circuit or in circuits of close proximity, the inductance of one inductor is affected by the magnetic field of the other. Consider the inductors of Figure 5.4, where a current i_1 is flowing through the inductor of inductance L_1. Therefore, a magnetic field is established in the vicinity. The magnetic field then induces a current i_2 in the inductor of inductance L_2. This current causes a change in the properties of the magnetic field and changes the current flowing through the inductor of inductance L_1. A mutual inductance is established between the inductors. The two inductors have a property called mutual inductance M. The mutual inductance is measured in henrys, and its sign is determined by the directions of the magnetic fluxes of the magnetic fields developed in the inductors. The relation for the potential difference across each inductor is

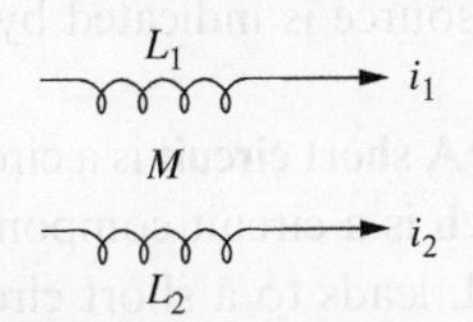

Figure 5.4

Two inductors are in close proximity to one another. When a current i_1 flows through the first inductor, a magnetic field is established surrounding the inductor. This magnetic field induces a current i_2 flowing through the second inductor.

$$v_1 = L_1\frac{di_1}{dt} + M\frac{di_2}{dt} \tag{5.31a}$$

$$v_2 = M\frac{di_1}{dt} + L_2\frac{di_2}{dt} \tag{5.31b}$$

Example 5.4

The current through a 4.2-mH inductor is

$$i(t) = 15(1 - e^{-1250t})\text{ A} \tag{a}$$

Determine **a.** the voltage in the inductor, **b.** the power in the inductor, and **c.** the energy stored in the inductor.

Solution

a. The voltage in the inductor is determined using Equation (5.25) as

$$v(t) = (4.2\text{ mH})\frac{d}{dt}\{15(1 - e^{-1250t})\text{ A}\}$$

$$= (4.2 \times 10^{-3})(15)(1250)e^{-1250t}\text{ V}$$

$$= 78.75e^{-1250t}\text{ V} \tag{b}$$

b. The power in the inductor is obtained using Equation (5.7) as

$$P = (78.75e^{-1250t}\text{ V})[15(1 - e^{-1250t})\text{ A}]$$

$$= 1.18(e^{-1250t} - e^{-2500t})\text{ kW} \tag{c}$$

c. The energy stored in the inductor is calculated using Equation (5.30) as

$$E = \frac{1}{2}(4.2 \text{ mH}) [15(1 - e^{-1250t}) \text{ A}]^2$$
$$= 472.5(1 - 2e^{-1250t} + e^{-2500t}) \text{ mJ} \qquad \text{(d)}$$

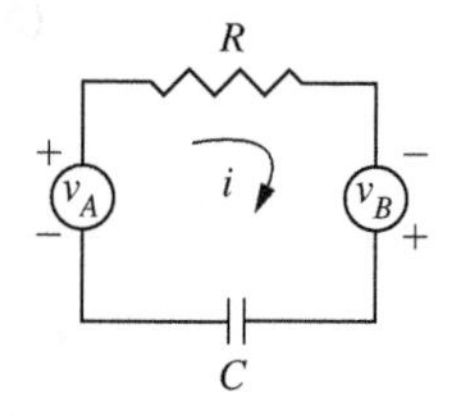

Figure 5.5

Circuit diagram for a device in which energy is provided by a direct current source v_A, is dissipated by a resistor, is absorbed by a source v_B, and is stored by a capacitor.

5.2.4 Voltage and Current Sources

A **voltage source** is an external source of electric potential. The electric potential increases across a voltage source if the current is flowing from the negative (−) terminal of the source to the positive (+) terminal. If the current flows from the positive to the negative terminal, the voltage decreases across the source. This source is absorbing power. If, in the circuit of Figure 5.5, the current is flowing in the clockwise direction, v_A is a time-varying voltage source, energy is absorbed and converted to heat by the resistor, a constant energy is absorbed by v_B, and energy is stored by the electric field developed by the capacitor.

A **current source** is an element that leads to a specific current flowing through a circuit. The direction of the current provided by a current source is indicated by an arrow on the source.

An **open circuit** is a circuit in which current is not flowing. A **short circuit** is a circuit component across which there is no voltage change. A **switch** is a circuit component that, when open, leads to an open circuit and, when closed, leads to a short circuit across the switch, as illustrated in Figure 5.6.

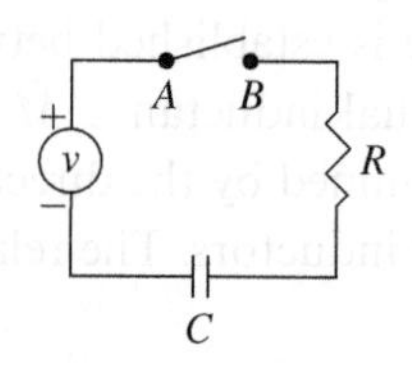

Figure 5.6

When the switch between A and B is open, no current flows through the circuit and the circuit is an open circuit. When the switch is closed, there is no voltage drop between A and B and the circuit is a short circuit.

5.2.5 Operational Amplifiers

An **amplifier** is a device that magnifies a signal. An **operational amplifier** is a device with a positive terminal and a negative terminal and is connected to dc power sources. The output of the amplifier is proportional to the difference of the input voltages. Operational amplifiers are discussed in more detail in Section 5.7.

5.2.6 Electric Circuits and Mechanical Systems

Direct analogies between mechanical systems and electric circuits are developed in detail in Section 5.6. It is possible from the preceding discussion to draw similarities between the two.

The passive elements in the mechanical systems of Chapter 4 are inertia elements, stiffness elements, and damping elements. The active elements in mechanical systems are the force and motion input. The passive elements of a circuit are inductors, capacitors, and resistors. The active elements in a circuit are voltage and current sources. In both types of systems, the passive elements store or dissipate energy and the active elements provide energy to the system. In a mechanical system, kinetic energy is stored by inertia elements, potential energy is stored by stiffness elements, and energy is dissipated by damping elements. In an electric circuit, energy is stored in magnetic fields by inductors, is stored in electric fields by capacitors, and is dissipated by resistors.

5.3 Kirchhoff's Laws

A circuit diagram illustrates how components in an electric circuit are connected to one another. An example circuit diagram is illustrated in Figure 5.7. The junction between three or more wires in a circuit is called a **node**. The nodes in the circuit of Figure 5.7 are at B and E. A closed path in a circuit is called a **loop**. Two loops are independent if one loop is not enclosed in the second loop. Independent loops in the circuit of Figure 5.7 are

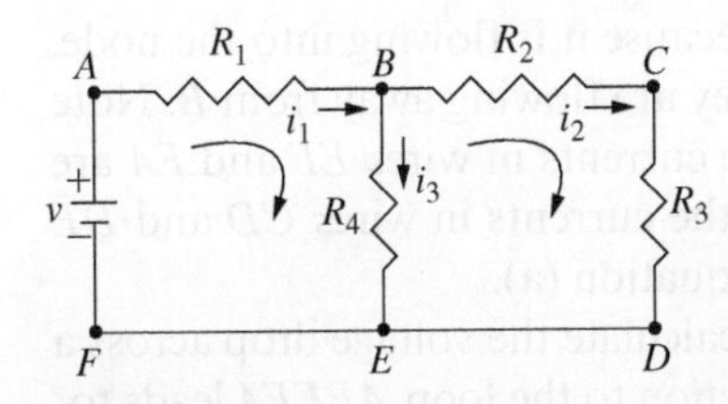

Figure 5.7

Circuit with nodes at B and E and loops $ABEFA$ and $BCDEB$.

$ABEFA$ and $BCDEB$. The closed path $ACDFA$ is also a loop, but it is not independent of $ABEFA$ and $BCDEB$. The circuit in Figure 5.7 is an example of a two-loop circuit.

The motion of charged particles is governed by a set of equations called **Maxwell's equations**, which provide the basis for all electromagnetic theory. A discussion of the full set of Maxwell's equations is beyond the scope of this study; however, under certain assumptions, Maxwell's equations can be reduced to a simpler form that can be applied to model circuits. In a manner analogous to the development of the Principle of Work and Energy from Newton's second law, principles can be derived from Maxwell's equations that can be applied to circuits in which the properties are constant (lumped) across a circuit component in lieu of the full set of Maxwell's equations.

Circuits are idealized models of electrical devices consisting of discrete components and lumped properties. The application of Maxwell's equations is achieved by applying two principles:

1. *Conservation of Charge:* Charge does not accumulate at the junction between circuit components. Thus the rate of change of charge entering the junction is equal to the rate of change of charge leaving the junction.
2. *Conservation of Energy:* Energy is conserved around any loop in a circuit. The energy supplied by the sources is dissipated by resistors and stored by capacitors and inductors.

The laws of Conservation of Charge and Conservation of Energy are formulated into **Kirchhoff's circuit laws**.

- **Kirchhoff's Current Law (KCL)** states that the sum of currents at a node is zero. The sign of a current is determined by the direction of the current in the wire. A current flowing into the node is taken as positive, and a current flowing away from the node is taken as negative.
- **Kirchhoff's Voltage Law (KVL)** states that the total change in voltage around any loop in a circuit must be zero. A voltage change across a circuit component is positive if the current flow through the component leads to an increase in voltage in the circuit. A current flowing through a voltage source from the negative terminal to the positive terminal of the source leads to a voltage increase. A voltage change across a circuit component is negative if the component draws voltage from the circuit. Voltage changes across a resistor that dissipates energy into heat and across a capacitor that stores energy in an electric field are negative. KVL can be applied in either the clockwise or the counterclockwise direction around a loop. If the direction of application of KVL is opposite that of the assumed direction of a current, the current is taken as negative.

The node law KCL, as stated, applies to the junction between three or more wires. However, Conservation of Charge also occurs at the junction of two wires. A trivial application of this principle is that the currents in two wires that meet at a simple junction (only two wires) have the same value and are in the same direction.

Example 5.5

Apply Kirchhoff's circuit laws to the two-loop circuit of Figure 5.7.

Solution

Let i_1 be the current in wire AB flowing from A to B. Let i_2 be the current in wire BC flowing from B to C. Let i_3 be the current in wire BE flowing from B to E. Application of KCL at node B yields

$$i_1 - i_2 - i_3 = 0 \tag{a}$$

Note that in Equation (a) the sign on i_1 is positive because it is flowing into the node, whereas the signs on i_2 and i_3 are negative because they are flowing away from B. Note that, because of the simple junctions at A and F, the currents in wires EF and FA are i_1, and because of the simple junctions at C and D, the currents in wires CD and DE are i_2. Application of KCL at node E also leads to Equation (a).

Noting that Ohm's law, Equation (5.8), is used to calculate the voltage drop across a resistor, the application of KVL in the clockwise direction to the loop $ABEFA$ leads to

$$v - R_1 i_1 - R_4 i_3 = 0 \tag{b}$$

The application of KVL in the clockwise direction to loop $BCDEB$ leads to

$$-R_2 i_2 - R_3 i_2 + R_4 i_3 = 0 \tag{c}$$

The application of KVL in the clockwise direction to loop $BCDEB$ requires the determination of the voltage change across the resistor of resistance R_4 going from E to B, which is opposite the assumed direction of i_3. Thus the current through the resistor is taken to be $-i_3$.

The currents in the circuit are obtained by the simultaneous solution of Equations (a)–(c). There are many methods for the solution of simultaneous equations. The application of Cramer's rule (see Appendix B) to a matrix formulation of the system of equations is used here. Equations (a)–(c) are summarized in matrix form as

$$\begin{bmatrix} 1 & -1 & -1 \\ R_1 & 0 & R_4 \\ 0 & -(R_2 + R_3) & R_4 \end{bmatrix} \begin{bmatrix} i_1 \\ i_2 \\ i_3 \end{bmatrix} = \begin{bmatrix} 0 \\ v \\ 0 \end{bmatrix} \tag{d}$$

The application of Cramer's rule to Equation (d) to solve for i_1 leads to

$$i_1 = \frac{\begin{vmatrix} 0 & -1 & -1 \\ v & 0 & R_4 \\ 0 & -(R_2 + R_3) & R_4 \end{vmatrix}}{\begin{vmatrix} 1 & -1 & -1 \\ R_1 & 0 & R_4 \\ 0 & -(R_2 + R_3) & R_4 \end{vmatrix}}$$

$$= \frac{v(R_2 + R_3) + v(R_4)}{R_1(R_2 + R_3) + R_4(R_2 + R_3) + R_1 R_4}$$

$$= \frac{R_2 + R_3 + R_4}{(R_2 + R_3)(R_1 + R_4) + R_1 R_4} v \tag{e}$$

Cramer's rule is used to solve for i_2 and i_3, leading to

$$i_2 = \frac{R_4}{(R_2 + R_3)(R_1 + R_4) + R_1 R_4} v \tag{f}$$

$$i_3 = \frac{R_2 + R_3}{(R_2 + R_3)(R_1 + R_4) + R_1 R_4} v \tag{g}$$

5.4 Circuit Reduction

Components of electric circuits, like components of mechanical systems, can be placed in series or in parallel. As is the case for mechanical systems, series and parallel combinations of like components may be replaced by a single equivalent component. The criteria for the equivalence of two sets of electrical system components is that, when the same current is passed through each set, the same voltage change occurs.

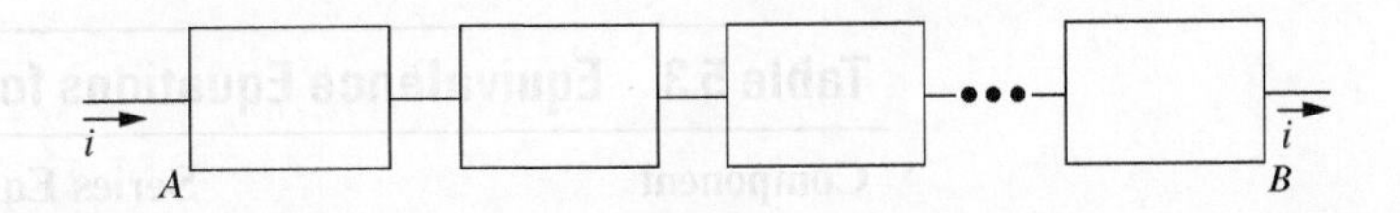

Figure 5.8

Circuit components in series: The same current passes through each component, and the voltage change from A to B is the sum of the voltage changes in the components.

5.4.1 Combinations of Circuit Components

Circuit components placed in series, as illustrated in Figure 5.8, have the same current flowing through each component and the total voltage change across the series combination is the sum of the voltage changes. Thus if there are n series components in Figure 5.8,

$$i_1 = i_2 = \cdots = i_n \tag{5.32}$$

$$v_{AB} = v_1 + v_2 + \cdots + v_n \tag{5.33}$$

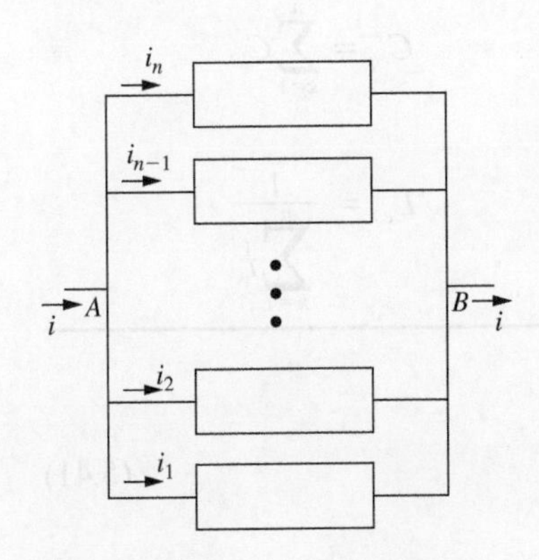

Figure 5.9

Circuit components in parallel: The voltage change is the same across each component, and the current is the sum of the currents in the components.

Circuit components placed in parallel are illustrated in Figure 5.9. Applying KCL to each of the nodes in the parallel combination, beginning with the node at A, leads to

$$i = i_1 + i_2 + \cdots + i_n \tag{5.34}$$

The voltage change from A to B is the same across each component

$$v_{AB} = v_1 = v_2 = \cdots = v_n \tag{5.35}$$

5.4.2 Series Combinations

Figure 5.10(a) illustrates a series combination of n resistors. It is desired to replace the combination by a single resistor of resistance R_{eq}, as illustrated in Figure 5.10(b). The voltage change across the resistor of Figure 5.10(b) when a current i flows through the component is

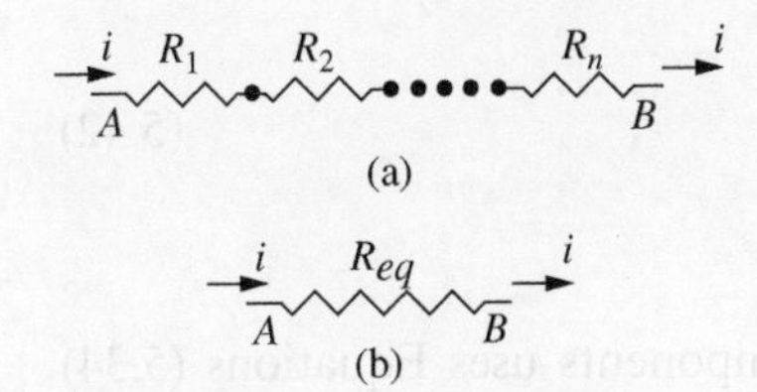

Figure 5.10

(a) n resistors in series; (b) the combination is replaced by a single resistor of equivalent resistance which is the sum of the individual resistances.

$$v = R_{eq} i \tag{5.36}$$

The total voltage change across the series combination of Figure 5.10(a) is

$$v = R_1 i + R_2 i + \cdots + R_n i = \left(\sum_{k=1}^{n} R_k\right) i \tag{5.37}$$

The comparison of Equations (5.36) and (5.37) leads to

$$R_{eq} = \sum_{k=1}^{n} R_k \tag{5.38}$$

Equation (5.38) shows that the equivalent resistance of a series combination of resistors is the sum of the individual resistances. Equation (5.38) is similar to the equation for the equivalent stiffness of a parallel combination of springs.

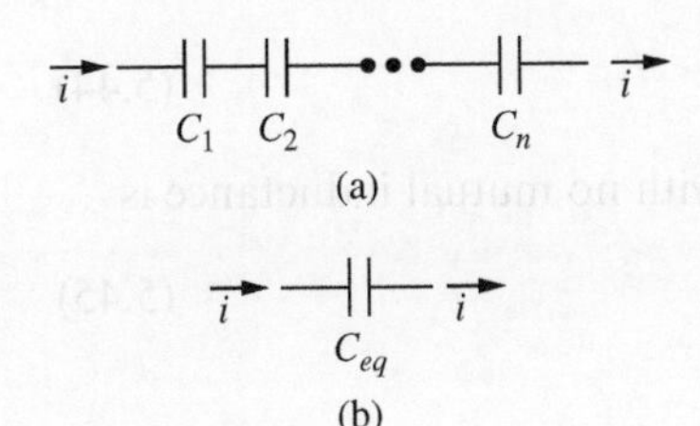

Figure 5.11

(a) n capacitors in series; (b) the equivalent capacitance is given by Equation (5.41).

A series combination of n capacitors is illustrated in Figure 5.11(a). An equivalent system with a single capacitor of capacitance C_{eq} is illustrated in Figure 5.11(b). The voltage change across the capacitor of Figure 5.11(b), assuming the capacitor is uncharged at $t = 0$, is

$$v_{AB} = \frac{1}{C_{eq}} \int_0^t i\,dt \tag{5.39}$$

The voltage change across the series combination of Figure 5.11(a), assuming all capacitors are uncharged at $t = 0$, is

$$v_{AB} = \frac{1}{C_1}\int_0^t i\,dt + \frac{1}{C_2}\int_0^t i\,dt + \cdots + \frac{1}{C_n}\int_0^t i\,dt \tag{5.40}$$

Table 5.3 Equivalence Equations for Circuit Components

Component	Series Equivalence	Parallel Equivalence
Resistor	$R_{eq} = \sum_{k=1}^{n} R_k$	$R_{eq} = \dfrac{1}{\sum_{k=1}^{n} \frac{1}{R_k}}$
Capacitor	$C_{eq} = \dfrac{1}{\sum_{k=1}^{n} \frac{1}{C_k}}$	$C_{eq} = \sum_{k=1}^{n} C_k$
Inductor	$L_{eq} = \sum_{k=1}^{n} L_k$	$L_{eq} = \dfrac{1}{\sum_{k=1}^{n} \frac{1}{L_k}}$

Comparison of Equations (5.39) and (5.40) leads to

$$C_{eq} = \frac{1}{\sum_{k=1}^{n} \frac{1}{C_k}} \tag{5.41}$$

Equation (5.41) shows that the equivalent capacitance of a series combination of capacitors is the reciprocal of the sum of the reciprocals of the capacitances of the members of the series combination. Equation (5.41) is similar to the equation for the equivalent stiffness of springs in parallel.

It is easy to show that the equivalent inductance of a combination of n inductors placed in series with no mutual inductance is

$$L_{eq} = \sum_{k=1}^{n} L_k \tag{5.42}$$

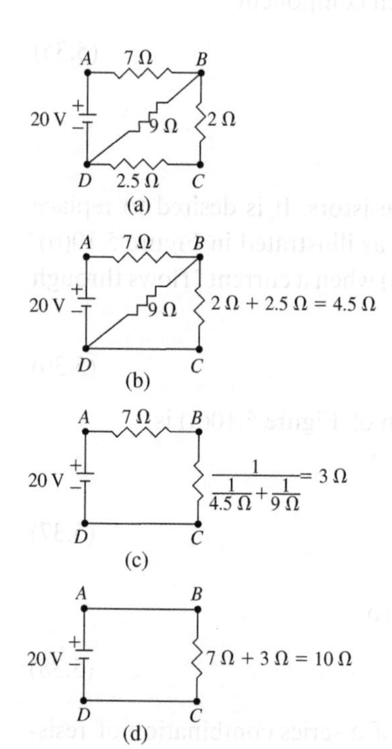

Figure 5.12

(a) The resistive circuit of Example 5.6; (b) resistors from B to C and C to D in series; (c) resistors from B to C and B to D in parallel; (d) resistors from A to B and B to C in series.

5.4.3 Parallel Combinations

The analysis of parallel combinations of circuit components uses Equations (5.34) and (5.35). Their applications to circuit components to determine equivalent values of resistors and inductors are similar to those of springs in series. The determination of an equivalent capacitance is similar to that of springs in parallel.

The equivalent resistance for n resistors in parallel is

$$R_{eq} = \frac{1}{\sum_{k=1}^{n} \frac{1}{R_k}} \tag{5.43}$$

The equivalent capacitance for n capacitors in parallel is

$$C_{eq} = \sum_{k=1}^{n} C_k \tag{5.44}$$

The equivalent inductance for n inductors in parallel with no mutual inductance is

$$L_{eq} = \frac{1}{\sum_{k=1}^{n} \frac{1}{L_k}} \tag{5.45}$$

The equations for equivalent components for combinations of circuit elements are summarized in Table 5.3.

Example 5.6

Determine the power delivered by the 20-V source to the resistive circuit of Figure 5.12(a).

Solution

For purposes of analysis, the combination of resistors can be replaced by a single resistor of resistance R_{eq}. The steps in the circuit reduction are shown in Figures 5.12(b), (c), and (d). The same current flows through the resistors between B and C and C and D, and the total voltage drop between B and D is the sum of the voltage drops across the resistors. Thus, these resistors are in series. Using KCL, the current entering node B is divided into the current flowing from B to C and the current flowing from B to D. Using KVL, the voltage drops across the resistors between B and D and across the equivalent resistor between B and C are equal. Thus these resistors are in parallel. Finally, this equivalent resistor is in series with the resistor between A and B. The equivalent resistance of the circuit is $R_{eq} = 10\ \Omega$. The power delivered to the circuit is obtained using Equation (5.13) as

$$\begin{aligned} P &= \frac{v^2}{R_{eq}} \\ &= \frac{(20\text{ V})^2}{10\ \Omega} \\ &= 40\text{ W} \end{aligned} \tag{a}$$

Example 5.7

Consider the capacitive circuit of Figure 5.13. **a.** Replace the capacitors with a single capacitor of equivalent capacitance. **b.** What are the charges on each capacitor?

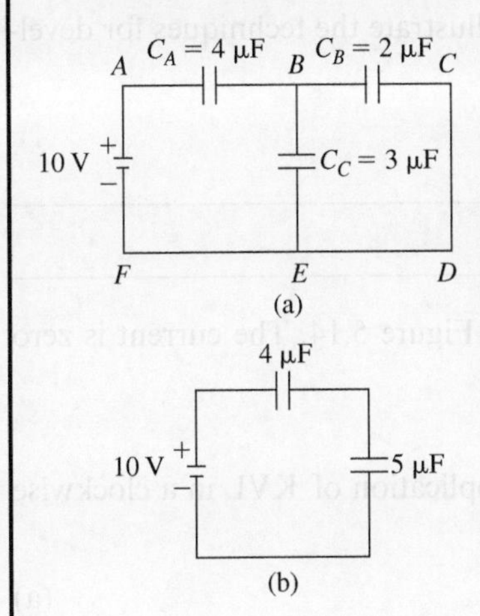

Figure 5.13

(a) The capacitive circuit of Example 5.7; (b) capacitors from B to C and E to B are in parallel and replaced by an equivalent capacitor; (c) the remaining two capacitors are in series. They are replaced by a single capacitor with an equivalent capacitance of 20/9 μF.

Solution

a. The 2-μF capacitor and the 3-μF capacitor have the same voltage change (from KVL) and thus are in parallel. Using Equation (5.44), they can be replaced by a single capacitor of $C_{eq1} = 5\ \mu$F, as illustrated in Figure 5.13(b). The total voltage drop across the capacitors in Figure 5.13(b) is the sum of the individual voltage drops. The current through each capacitor is the same. Thus, these capacitors are in series. Using Equation (5.41), these series capacitors can be replaced by a single capacitor of equivalent capacitance of

$$C_{eq} = \frac{1}{\dfrac{1}{4\ \mu\text{F}} + \dfrac{1}{5\ \mu\text{F}}} = \frac{20}{9}\ \mu\text{F} \tag{a}$$

The equivalent capacitance is illustrated in Figure 5.13(c).

b. Let v_A, v_B, and v_C represent the potential in the capacitors and let q_A, q_B, and q_C represent the charges on each capacitor. From Conservation of Charge at node B,

$$q_A - q_B - q_C = 0 \tag{b}$$

Application of KVL around loop $ABEFA$ gives

$$10 - v_A - v_C = 0 \tag{c}$$

Since capacitors B and C are in parallel,

$$v_B - v_C = 0 \tag{d}$$

From Equation (5.14) the charge on each capacitor is $q_A = (4\ \mu\text{F})v_A$, $q_B = (2\ \mu\text{F})v_B$, and $q_C = (3\ \mu\text{F})v_C$. Substituting into Equation (b) and summarizing Equations (b)–(d) in matrix form leads to

$$\begin{bmatrix} 4 & -2 & -3 \\ 1 & 0 & 1 \\ 0 & 1 & -1 \end{bmatrix} \begin{bmatrix} v_A \\ v_B \\ v_C \end{bmatrix} = \begin{bmatrix} 0 \\ 10 \\ 0 \end{bmatrix} \tag{e}$$

Simultaneous solution of Equation (e) leads to $v_A = 50/9$ V, $v_B = v_C = 40/9$ V. The charges on the capacitors are

$$q_A = C_A v_A = (4\ \mu\text{F})\ (50/9\ \text{V}) = 22.2\ \mu\text{C}$$
$$q_B = C_B v_B = (2\ \mu\text{F})\ (40/9\ \text{V}) = 8.89\ \mu\text{C}$$
$$q_C = C_C v_C = (3\ \mu\text{F})\ (40/9\ \text{V}) = 13.3\ \mu\text{C} \tag{f}$$

5.5 Mathematical Modeling of Electric Circuits

Capacitors and inductors are circuit components that store energy. The voltage change across an inductor is proportional to the rate of change of current, while the current through a capacitor is proportional to the rate of change of voltage across the capacitor. Thus the presence of these components in an electric circuit leads to time-dependent current and voltage changes. Time-dependent sources also lead to time-dependent currents and time-dependent voltage changes in a circuit.

The system response of an electric circuit is described by the time-dependent currents and voltage changes, which are determined through mathematical modeling of the circuit. Kirchhoff's laws, both KCL and KVL, are applied to the circuit diagram, resulting in differential equations or integrodifferential equations whose solution provides the system response. The following examples illustrate the techniques for developing mathematical models for electric circuits.

Example 5.8

Derive a mathematical model for the *LR* circuit of Figure 5.14. The current is zero when the voltage source is connected at $t = 0$.

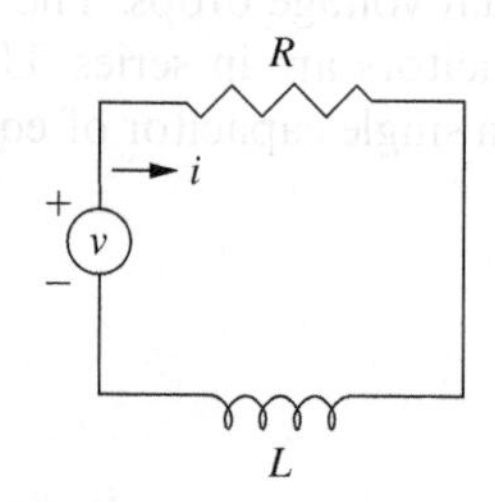

Figure 5.14
The *LR* circuit of Example 5.8.

Solution

Let $i(t)$ represent the current through the circuit. Application of KVL in a clockwise direction around the circuit leads to

$$v(t) - Ri - L\frac{di}{dt} = 0 \tag{a}$$

$$L\frac{di}{dt} + Ri = v(t) \tag{b}$$

Equation (b) is a first-order differential equation whose solution is the time-dependent current through the circuit. Since the source is connected at $t = 0$, the appropriate initial condition is $i(0) = 0$. Equation (b) is a mathematical model for the circuit with input $v(t)$ and output $i(t)$.

Example 5.9

Derive a mathematical model for the *RC* circuit of Figure 5.15. The switch is open and then closed at $t = 0$. The capacitor is uncharged before the switch is closed. **a.** Use the current through the circuit as the system output. **b.** Use the voltage change across the capacitor as the system output.

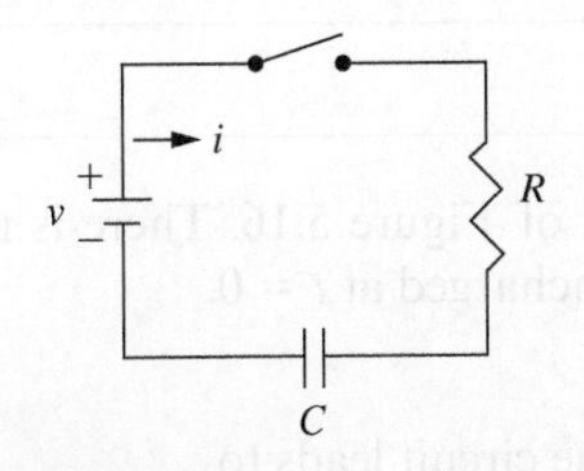

Figure 5.15

The *RC* circuit of Example 5.9: Current flows through the circuit after the switch is closed.

Solution

a. Let $i(t)$ represent the current through the circuit. Since the circuit is open at $t = 0$, the source is not connected until $t = 0$. The appropriate mathematical representation of the voltage source is

$$v(t) = vu(t) \tag{a}$$

where $u(t)$ is the unit step function defined in Equation (1.33). The application of KVL in a clockwise fashion around the circuit leads to

$$vu(t) - Ri - \frac{1}{C}\int_0^t i\, dt = 0$$

$$Ri + \frac{1}{C}\int_0^t i\, dt = vu(t) \tag{b}$$

Equation (b) is an integrodifferential equation whose solution is the time-dependent current in the circuit.

Equation (b) can be converted into a first-order differential equation through differentiation with respect to time. Recalling that $du(t)/dt = \delta(t)$, where $\delta(t)$ is the unit impulse function, differentiation of Equation (b) leads to

$$R\frac{di}{dt} + \frac{1}{C}i = v\delta(t) \tag{c}$$

Solution of Equation (c) requires the application of the initial condition $i(0) = 0$.

The solution of Equation (b) does not require the application of an initial condition. Substitution of $t = 0$ into Equation (b) leads to $i(0) = 0$. However, the evaluation of Equation (b) at a time $t = 0^+$ shows that $i(0^+) = v/R$, and thus the solution of Equation (b) leads to a current that is discontinuous at $t = 0$. The discontinuous current is a result of the assumption that when the switch is closed, the voltage source is instantaneously connected to the circuit with full potential being delivered. The solution of Equation (c), subject to the initial condition $i(0) = 0$, also leads to the prediction of a discontinuous current.

b. Let $v_c(t)$ represent the voltage change across the capacitor. A trivial application of KCL shows that the current through the resistor is equal to the current through the capacitor

$$i_r = i_C \tag{d}$$

Applying KVL in a clockwise fashion around the circuit gives

$$vu(t) - v_r - v_c = 0 \tag{e}$$

Using Ohm's law, Equation (5.8), in Equation (e) and rearranging leads to

$$i_r = \frac{vu(t) - v_c}{R} \tag{f}$$

The current through the capacitor is obtained using Equation (5.18) as

$$i_C = C\frac{dv_C}{dt} \tag{g}$$

Substituting Equations (f) and (g) into Equation (d) gives

$$\frac{vu(t) - v_C}{R} = C\frac{dv_C}{dt}$$

$$C\frac{dv_C}{dt} + \frac{1}{R}v_C = \frac{v}{R}u(t) \tag{h}$$

Example 5.10

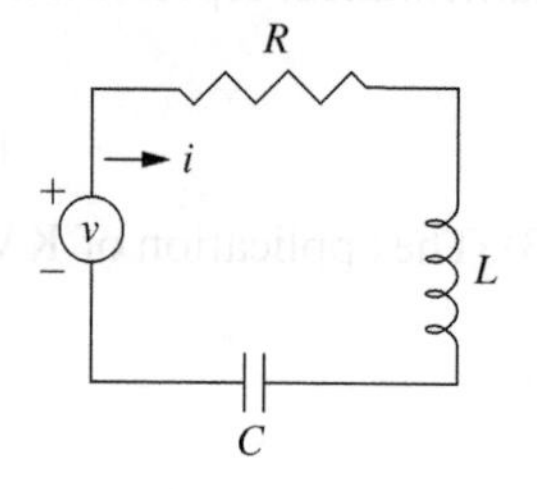

Figure 5.16

The series *LRC* circuit of Example 5.10.

Derive a mathematical model for the series *LRC* circuit of Figure 5.16. There is no current flowing in the circuit at $t = 0$. The capacitor is uncharged at $t = 0$.

Solution

Application of KVL in the clockwise direction around the circuit leads to

$$v(t) - Ri - L\frac{di}{dt} - \frac{1}{C}\int_0^t i\,dt = 0 \quad \text{(a)}$$

$$L\frac{di}{dt} + Ri + \frac{1}{C}\int_0^t i\,dt = v(t) \quad \text{(b)}$$

Equation (b) is an integrodifferential equation whose solution, subject to $i(0) = 0$, is the time-dependent current through the circuit. The input is $v(t)$ and the output is $i(t)$.

Equation (b) can be rewritten as a second-order differential equation through differentiation with respect to time

$$L\frac{d^2i}{dt^2} + R\frac{di}{dt} + \frac{1}{C}i = \frac{dv}{dt} \quad \text{(c)}$$

If Equation (c) is used in lieu of Equation (b), a value of $di/dt(0)$ is necessary. Evaluation of Equation (b) at $t = 0$ leads to

$$L\frac{di}{dt}(0) = v(0) \quad \text{(d)}$$

Example 5.11

Obtain a mathematical model for the three-loop circuit of Figure 5.17. Use i_1, i_2, and i_3 as dependent variables.

Solution

The variables i_4 and i_5, as indicated on Figure 5.17, are introduced as intermediate variables that will not appear in the mathematical model. Application of KCL at nodes *B* and *C* leads to

$$i_1 - i_2 - i_4 = 0 \quad \text{(a)}$$

$$i_2 - i_3 - i_5 = 0 \quad \text{(b)}$$

The application of KVL in the clockwise direction around loop *ABGHA* and using Equation (a) leads to

$$v(t) - L\frac{di_1}{dt} - \frac{1}{C}\int_0^t i_4\,dt - Ri_1 = 0 \quad \text{(c)}$$

$$L\frac{di_1}{dt} + Ri_1 + \frac{1}{C}\int_0^t (i_1 - i_2)\,dt = v(t) \quad \text{(d)}$$

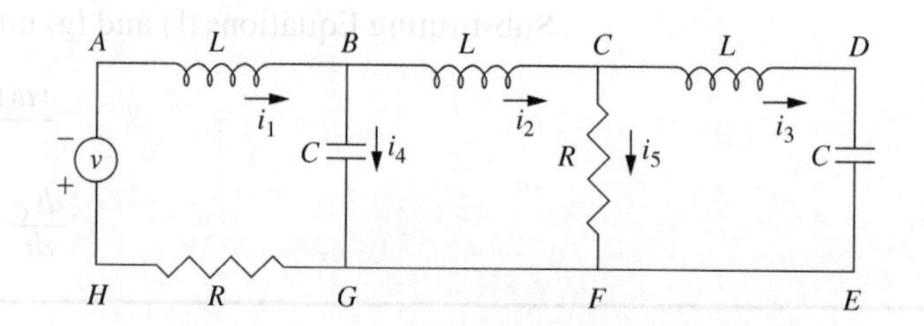

Figure 5.17

The three-loop circuit of Example 5.11.

Application of KVL in the clockwise direction around loop *BCFGB* and using Equations (a) and (b) leads to

$$-L\frac{di_2}{dt} - Ri_5 + \frac{1}{C}\int_0^t i_4\,dt = 0 \tag{e}$$

$$L\frac{di_2}{dt} + R(i_2 - i_3) - \frac{1}{C}\int_0^t (i_1 - i_2)dt = 0 \tag{f}$$

Application of KVL in the clockwise direction around loop *CDEFC* and using Equation (b) leads to

$$-L\frac{di_3}{dt} - \frac{1}{C}\int_0^t i_3\,dt + Ri_5 = 0 \tag{g}$$

$$L\frac{di_3}{dt} - R(i_2 - i_3) + \frac{1}{C}\int_0^t i_3\,dt = 0 \tag{h}$$

Equations (d), (f), and (h) are a mathematical model for the circuit with input $v(t)$ and outputs $i_1(t), i_2(t)$, and $i_3(t)$.

In each of the previous examples, nodal currents (currents entering or leaving nodes) were used as dependent variables. In some problems, especially those with current sources, it is more convenient to use voltages as dependent variables. The dependent variable chosen for Example 5.12 is the voltage drop across a circuit component, while the dependent variables in Example 5.13 are the voltages at nodes in the circuit.

Example 5.12

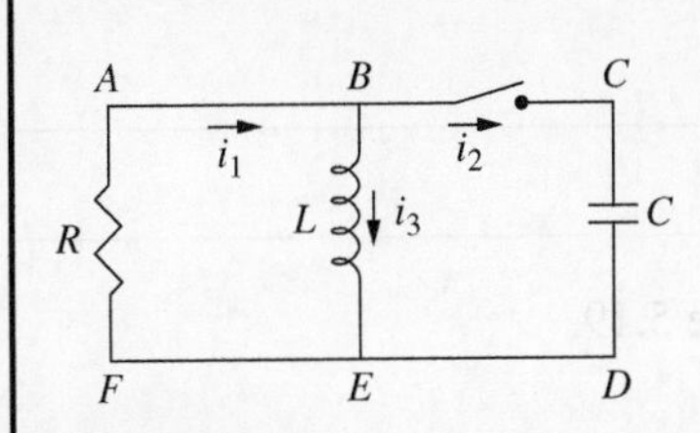

Figure 5.18

The parallel *LRC* circuit of Example 5.12.

Derive a mathematical model for the parallel *LRC* circuit of Figure 5.18. The capacitor has a charge q at $t = 0$, when the switch is closed, allowing the capacitor to discharge.

Solution

Define v_R as the voltage change across the resistor, v_C as the voltage change across the capacitor, and v_L as the voltage change across the inductor. The application of KVL to loop *ABEFA* in the clockwise direction leads to

$$v_R + v_L = 0 \tag{a}$$

The application of KVL to loop *BCDEB* in the clockwise direction leads to

$$v_C - v_L = 0 \tag{b}$$

From Equations (a) and (b),

$$v_L = v_C = -v_R \tag{c}$$

The voltage change across the resistor is used as the dependent variable in the modeling of the parallel *LRC* circuit.

Application of KCL to the node at *B* using the definitions of currents shown in Figure 5.18 leads to

$$i_1 - i_2 - i_3 = 0 \tag{d}$$

The current-voltage laws for the circuit components [Equations (5.8), (5.18), and (5.28)] are used to give

$$i_1 = \frac{v_R}{R} \tag{e}$$

$$i_2 = C\frac{dv_C}{dt} = -C\frac{dv_R}{dt} \tag{f}$$

$$i_3 = \frac{1}{L}\int_0^t v_L\,dt = -\frac{1}{L}\int_0^t v_R\,dt \tag{g}$$

Substitution of Equations (e)–(g) into Equation (d) leads to

$$\frac{v_R}{R} + C\frac{dv_R}{dt} + \frac{1}{L}\int_0^t v_R dt = 0 \tag{h}$$

Equation (h) must be supplemented by an initial condition. The initial voltage in the capacitor due to the initial charge is obtained using Equation (5.14)

$$v_C(0) = \frac{q}{C} \tag{i}$$

Equations (c) and (i) are used to obtain

$$v_R(0) = -\frac{q}{C} \tag{j}$$

Equation (h) is an integrodifferential equation without any input. It must satisfy Equation (j), which makes this an initial value problem.

Equation (h) can be rewritten as a second-order differential equation through differentiation with respect to time:

$$C\frac{d^2v_R}{dt^2} + \frac{1}{R}\frac{dv_R}{dt} + \frac{1}{L}v_R = 0 \tag{k}$$

Since Equation (k) is a second-order differential equation, two initial conditions must be specified to determine its response. One initial condition is provided by Equation (j). The second is obtained by evaluating Equation (h) at $t = 0$,

$$\frac{1}{R}v_R(0) + C\frac{dv_R}{dt}(0) = 0 \tag{l}$$

$$\frac{dv_R}{dt}(0) = -\frac{q}{RC^2} \tag{m}$$

Example 5.13

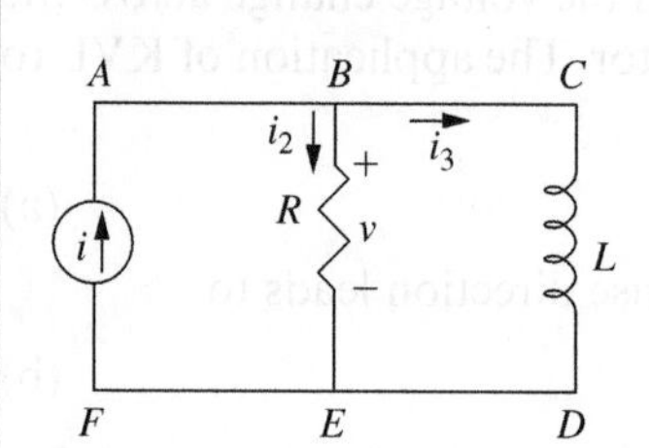

Figure 5.19

Circuit of Example 5.13.

Derive a mathematical model for the circuit of Figure 5.19.

Solution

Application of KCL at node B gives

$$i(t) - i_2 - i_3 = 0 \tag{a}$$

Let v be the voltage drop across the resistor. Then

$$i_2 = \frac{v}{R} \tag{b}$$

KVL applied around loop $BCDEB$ reveals that the voltage drop across the inductor is the same as that across the resistor. Thus

$$i_3 = \frac{1}{L}\int_0^t v(t)dt \tag{c}$$

Substitution of Equations (b) and (c) into Equation (a) leads to

$$\frac{v}{R} + \frac{1}{L}\int_0^t vdt = i(t) \tag{d}$$

The mathematical model for this circuit is the integrodifferential equation, Equation (d). The input is $i(t)$, the current provided by the source. The output is $v(t)$, which is the potential difference across the resistor. Since the resistor is in parallel with the inductor, $v(t)$ is also the potential difference across the inductor.

5.6 Mechanical Systems Analogies

5.6.1 Energy Principles

The second-order differential equation derived for the series *LRC* circuit in Example 5.10, Equation (c), has the same form as the differential equation for a mass-spring-viscous damper system. Thus, it is reasonable to assume that the two systems are somehow analogous. The analogy between electrical and mechanical systems is best examined using energy methods.

It is shown in Section 4.9 that the kinetic and potential energies of a linear one-degree-of-freedom system have quadratic forms given by Equations (4.123) $\left(\frac{1}{2}m_{eq}\dot{x}^2\right)$ and (4.124) $\left(\frac{1}{2}k_{eq}x^2\right)$, respectively. The work done by the viscous damping forces has the form of Equation (4.125). The principle of work and energy is used to show that any linear one-degree-of-freedom mechanical system has a differential equation of the form

$$m_{eq}\ddot{x} + c_{eq}\dot{x} + k_{eq}x = F_{eq}(t) \tag{4.135}$$

Energies stored in electric circuit components also have quadratic forms. The energy stored in a capacitor in the presence of an electric field is $\frac{1}{2}Cv^2$. The energy stored in an inductor in the presence of a magnetic field is $\frac{1}{2}Li^2$. The energy lost across a resistor is of the form $\int Ri^2\,dt$, which is similar to the expression for energy dissipated by a viscous damper when written as $\int c_{eq}\dot{x}^2 dt$. The similarities of the quadratic forms for stored energy and dissipated energy also suggest that making analogies between mechanical and electrical systems is valid.

5.6.2 Single-Loop Circuits with Voltage Sources

Consider a one-loop circuit with resistors, inductors, capacitors, and a voltage source. The analysis in Section 5.4 shows that the resistors can be replaced by a single resistor of equivalent resistance R_{eq}, the capacitors can be replaced by a single capacitor of equivalent capacitance C_{eq}, and the inductors can be replaced by a single inductor of equivalent inductance L_{eq}. Since energy is conserved in the circuit, the energy provided by the source E_s is equal to the energy dissipated by the resistors E_R, plus the energy stored by the inductors E_L, plus the energy stored by the capacitors E_C:

$$E_s = E_R + E_C + E_L \tag{5.46}$$

Differentiation of Equation (5.46) with respect to time leads to an equation for power in the circuit:

$$P_s = P_R + P_C + P_L \tag{5.47}$$

The power from the source is $P_s = vi$ and the expressions for power from the circuit components are given in Equations (5.12), (5.21), and (5.29). Substitution into Equation (5.47) leads to

$$vi = i^2R_{eq} + C_{eq}v\frac{dv}{dt} + L_{eq}i\frac{di}{dt} \tag{5.48}$$

Using Equation (5.19), the relation between current and voltage change across a capacitor, in Equation (5.48) leads to

$$vi = i^2R_{eq} + \frac{1}{C_{eq}}i\int_0^t i\,dt + L_{eq}i\frac{di}{dt} \tag{5.49}$$

Factoring i from both sides of Equation (5.49) and rejecting the possibility that $i(t) = 0$ for all t leads to

$$L_{eq}\frac{di}{dt} + R_{eq}i + \frac{1}{C_{eq}}\int_0^t i\,dt = v(t) \tag{5.50}$$

Differentiation of Equation (5.50) with respect to time gives

$$L_{eq}\frac{d^2i}{dt^2} + R_{eq}\frac{di}{dt} + \frac{1}{C_{eq}}i = \frac{dv}{dt} \tag{5.51}$$

Equation (5.51) is of the same form as Equation (4.135), which governs the motion of a one-degree-of-freedom mechanical system. Both equations can be derived using energy principles, and their systems' behavior can be explained using energy principles.

When the mechanical system is subject to an external force resulting in motion, the spring stores potential energy and the inertia elements store kinetic energy. The viscous damper dissipates the energy. At any time, the energy stored in the system is the sum of the kinetic energy and the potential energy.

When the series *LRC* circuit is connected to a voltage source that results in current flow, the inductor stores energy in a magnetic field and the capacitor stores energy in an electric field. The resistor dissipates the energy. At any time, the energy stored in the system is the sum of the energy stored in the inductor and the energy stored in the capacitor.

It is clear there is a direct analogy between the mechanical system and the electrical system. The resistor is analogous to the viscous damper, the capacitor is analogous to the spring, and the inductor is analogous to the mass. The preceding discussion shows the analogy between the parameters of a mechanical system and the parameters of a series *LRC* circuit.

However, the analogy between the dependent variables is not quite as clear. An alternative formulation of Equation (5.50) is to use electric charge as a dependent variable rather than current. Recalling $q = di/dt$, where q is electric charge, Equation (5.50) can be written as

$$L\frac{d^2q}{dt^2} + R\frac{dq}{dt} + \frac{1}{C}q = v(t) \tag{5.52}$$

Comparing Equation (5.52) with Equation (4.135), it is clear that there is a direct analogy between the external force in a mechanical system and the voltage source for a series *LRC* circuit when electric charge is analogous to displacement. Note that the power delivered to a mechanical system by an external force is $F\dot{x}$ and the power delivered to an electric circuit by an external voltage source is $vi = v\dot{q}$. Thus the direct physical analogies are between displacement and electric charge and between force and voltage. However, since the dependent variable used in circuit analysis is usually current rather than electric charge, a mathematical analogy is used between current and displacement and force and rate of change of voltage. This analogy is illustrated in Figure 5.20.

5.6.3 Single-Loop Circuits with Current Sources

Now consider a circuit in which the external energy is supplied by a current source, such as the parallel *LRC* circuit of Figure 5.21(a). The differential equation for this circuit in the absence of the current source, using KVL and KCL, is derived in Example 5.12. Using a similar analysis, note that if KVL is applied to each of the loops, it is clear that the voltage change across each of the circuit components is equal. Application of KCL at the two nodes leads to

$$i_s = i_R + i_L + i_C \tag{5.53}$$

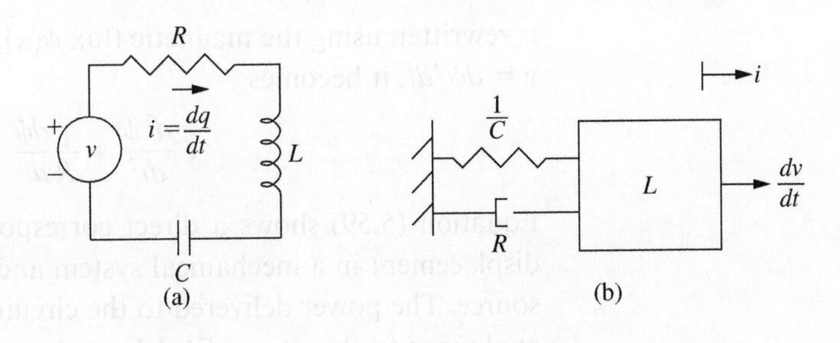

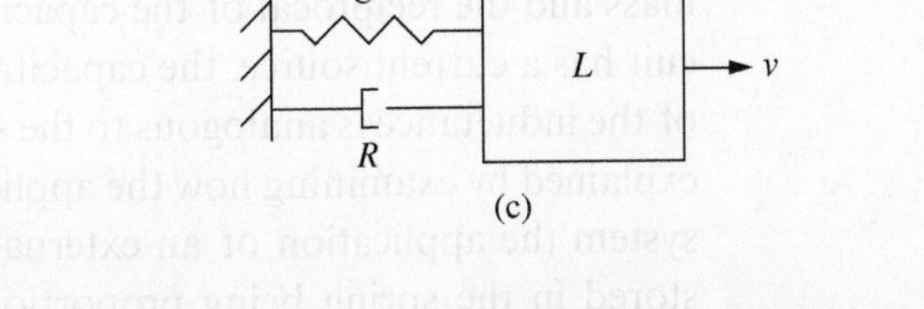

Figure 5.20

(a) A series *LRC* circuit; (b) an equivalent one-degree-of-freedom mechanical system using the analogy between current and displacement; (c) the equivalent mechanical system using the analogy between charge and displacement.

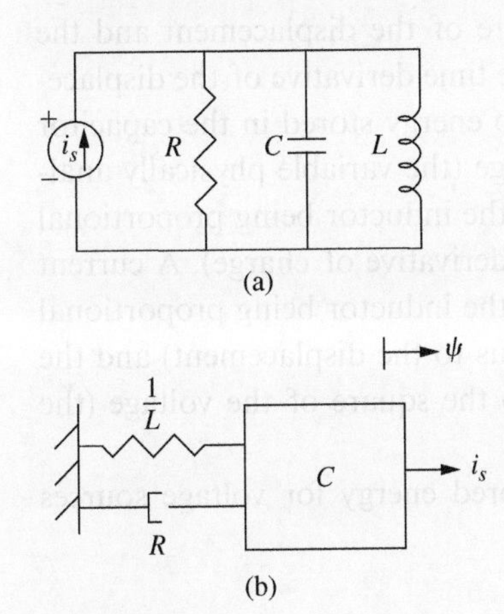

Figure 5.21

(a) A parallel *LRC* circuit; (b) the equivalent mechanical system using the analogy between magnetic flux and displacement.

Multiplying Equation (5.53) by v, the voltage across any of the system components, leads to

$$vi_s = vi_R + vi_L + vi_C \tag{5.54}$$

Equation (5.54) represents a balance of power, or a time-differentiated form of Conservation of Energy. The left-hand side of Equation (5.54) is the power delivered to the circuit by the current source. The right-hand side is the sum of the power in the resistor, the capacitor, and the inductor. When these terms are replaced by Equations (5.13), (5.21), and (5.29), respectively, Equation (5.54) can be rewritten as

$$vi_s = \frac{v^2}{R} + Li_L\frac{di_L}{dt} + Cv\frac{dv}{dt} \tag{5.55}$$

Equation (5.28) is used in Equation (5.55) to substitute for the current in the inductor, leading to

$$Cv\frac{dv}{dt} + \frac{v^2}{R} + \frac{1}{L}v\int_0^t v\,dt = vi_s \tag{5.56}$$

Factoring v from both sides of Equation (5.56) and noting that $v(t) \neq 0$ for all t leads to

$$C\frac{dv}{dt} + \frac{1}{R}v + \frac{1}{L}\int_0^t v\,dt = i_s(t) \tag{5.57}$$

Differentiation of Equation (5.57) with respect to time gives

$$C\frac{d^2v}{dt^2} + \frac{1}{R}\frac{dv}{dt} + \frac{1}{L}v = \frac{di_s}{dt} \tag{5.58}$$

Equation (5.58) is also analogous to Equation (4.135). Energy principles are used to derive both equations. If a current source is applied to a parallel *LRC* circuit, leading to the development of electric potential in the system, then energy is stored in the capacitor and the inductor, and energy is dissipated by the resistor. The total stored energy is the sum of the energy in the capacitor and the energy in the inductor.

The mechanical-electrical analogy for a circuit with a current source is that the mass m_{eq} corresponds to the capacitance C, the equivalent viscous damping coefficient corresponds to the reciprocal of the resistance $1/R$, and the equivalent stiffness corresponds to the reciprocal of the inductance $1/L$. However, as in the *LRC* circuit with the voltage source, there is no direct physical correspondence between the dependent variables of displacement and electric potential. When Equation (5.57) is rewritten using the magnetic flux $\psi(x)$, which is related to the electric potential by $v = d\psi/dt$, it becomes

$$C\frac{d^2\psi}{dt^2} + \frac{1}{R}\frac{d\psi}{dt} + \frac{1}{L}\psi = i_s(t) \tag{5.59}$$

Equation (5.59) shows a direct correspondence between the magnetic flux and the displacement in a mechanical system and between the external force and the current source. The power delivered to the circuit by the external source is $vi_s = \dot{\psi} i_s$, which is analogous to the power $F\dot{x}$ delivered to the mechanical system by the external force. This analogy is illustrated in Figure 5.21(b).

Note that when the circuit has a voltage source, the inductance is analogous to the mass and the reciprocal of the capacitance is analogous to the stiffness; when the circuit has a current source, the capacitance is analogous to the mass and the reciprocal of the inductance is analogous to the stiffness. This switch in analogous components is explained by examining how the applied load leads to stored energy. In the mechanical system the application of an external force leads to a displacement with the energy stored in the spring being proportional to the square of the displacement and the kinetic energy being proportional to the square of the time derivative of the displacement. A voltage source in the electric circuit leads to energy stored in the capacitor being proportional to the square of the electric charge (the variable physically analogous to the displacement) and the energy stored in the inductor being proportional to the square of the current (the square of the time derivative of charge). A current source in an electric circuit leads to energy stored in the inductor being proportional to the magnetic flux (the variable physically analogous to the displacement) and the energy stored in the capacitor being proportional to the square of the voltage (the square of the time derivative of the magnetic flux).

The mechanical-electrical analogies including stored energy for voltage sources and current sources are summarized in Table 5.4.

Table 5.4 Summary of Mechanical-Electrical Analogies

Mechanical System	Circuit with Voltage Source	Circuit with Current Source
System parameters		
Mass (m)	L	C
Stiffness (k)	$\frac{1}{C}$	$\frac{1}{L}$
Viscous damper (c)	R	$\frac{1}{R}$
Physical analogy		
Displacement (x)	q	ψ
Force (F)	v	i
Velocity ($\dot{x}$)	i	v
Energy and power		
Kinetic energy $\left(\frac{1}{2}m\dot{x}^2\right)$	$\frac{1}{2}Li^2$	$\frac{1}{2}Cv^2$
Potential energy $\left(\frac{1}{2}kx^2\right)$	$\frac{1}{2}\frac{q^2}{C}$	$\frac{1}{2}\frac{\psi^2}{L}$
Dissipated energy $\int c\dot{x}dx$	$\int Ridq$	$\int \frac{1}{R}vd\psi$
Power ($F\dot{x}$)	$v\dot{q} = vi$	$i\dot{\psi} = iv$

5.6.4 Multiple-Loop Circuits

Multiple-loop circuits have direct analogies to multidegree-of-freedom mechanical systems. When the circuit has a voltage source, an inductor is analogous to a mass, a resistor is analogous to a viscous damper, and a capacitor is analogous to a spring. The circuit components common to multiple loops provide the coupling between the loops, and their analogous components provide the coupling between the masses. Series components in parallel in a multiloop circuit are analogous to parallel components in a mechanical system. Parallel components in a multiloop circuit are analogous to series components in a mechanical system.

Example 5.14

Draw the analogous mechanical system for the circuit of Figure 5.22(a). Write the differential equations for the mathematical model for both the mechanical system and the electrical system.

Solution

The analogous mechanical system is shown in Figure 5.22(b). The differential equations for the circuit derived by applying KVL and KCL are

$$L_1\frac{di_1}{dt} + R_1 i_1 + R_2(i_1 - i_2) + \frac{1}{C}\int_0^t (i_1 - i_2)dt = v(t) \quad \text{(a)}$$

$$L_2\frac{di_2}{dt} + R_3 i_2 + R_2(i_2 - i_1) + \frac{1}{C}\int_0^t (i_2 - i_1)dt = 0 \quad \text{(b)}$$

The differential equations for the analogous system are

$$m_1\ddot{x}_1 + (c_1 + c_2)\dot{x}_1 - c_2\dot{x}_2 + kx_1 - kx_2 = F(t) \quad \text{(c)}$$

$$m_2\ddot{x}_2 - c_2\dot{x}_1 + (c_2 + c_3)\dot{x}_2 - kx_1 + kx_2 = 0 \quad \text{(d)}$$

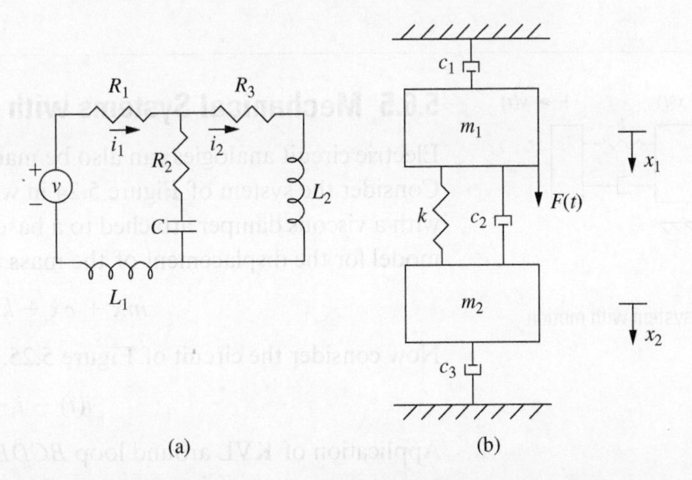

Figure 5.22
(a) The electric circuit of Example 5.14; (b) the analogous mechanical system.

Example 5.15

Draw the analogous electric circuit for the mechanical system of Figure 5.23(a). Write the differential equations for the mathematical model for both the mechanical system and the electric circuit.

Solution

The analogous electric circuit is shown in Figure 5.23(b). The differential equations for the mechanical system are derived by applying Newton's second law to free-body

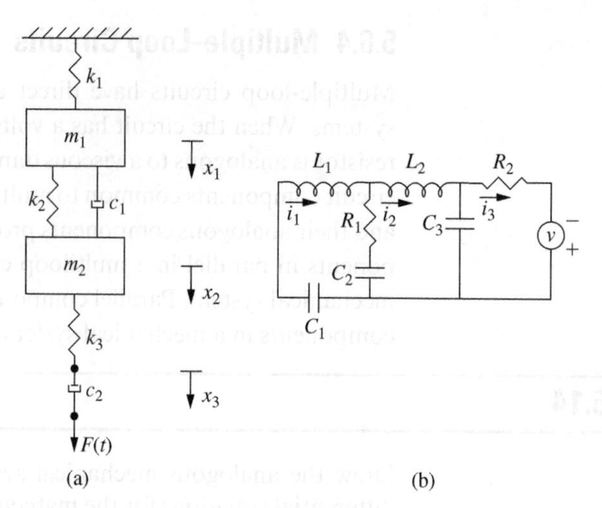

Figure 5.23

(a) The mechanical system of Example 5.15; (b) the analogous electric circuit.

diagrams of the masses and the massless connection whose displacement is x_3. The differential equations are

$$m_1\ddot{x}_1 + c_1\dot{x}_1 - c_1\dot{x}_2 + (k_1 + k_2)x_1 - k_2x_2 = 0 \tag{a}$$

$$m_2\ddot{x}_2 - c_1\dot{x}_1 + c_1\dot{x}_2 - k_2x_1 + (k_2 + k_3)x_2 - k_3x_3 = 0 \tag{b}$$

$$c_2\dot{x}_3 - k_3x_2 + k_3x_3 = F(t) \tag{c}$$

The differential equations of the analogous electrical system are

$$L_1\frac{di_1}{dt} + R_1(i_1 - i_2) + \frac{1}{C_1}\int_0^t i_1 dt + \frac{1}{C_2}\int_0^t (i_1 - i_2)dt = 0 \tag{d}$$

$$L_2\frac{di_2}{dt} - R_1(i_1 - i_2) - \frac{1}{C_2}\int_0^t (i_1 - i_2)dt + \frac{1}{C_3}\int_0^t (i_2 - i_3)dt = 0 \tag{e}$$

$$R_2 i_3 - \frac{1}{C_3}\int_0^t (i_2 - i_3)dt = v(t) \tag{f}$$

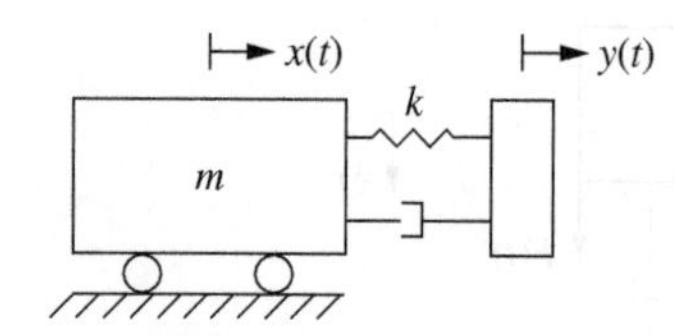

Figure 5.24

A mechanical system with motion input.

5.6.5 Mechanical Systems with Motion Input

Electric circuit analogies can also be made for mechanical systems with motion input. Consider the system of Figure 5.24 in which a mass is attached to a spring in parallel with a viscous damper attached to a base with a prescribed motion. The mathematical model for the displacement of the mass is

$$m\ddot{x} + c\dot{x} + kx = c\dot{y} + ky \tag{5.60}$$

Now consider the circuit of Figure 5.25. Application of KCL at node B gives

$$i(t) - i_1 - i_2 = 0 \tag{5.61}$$

Application of KVL around loop $BCDEB$ leads to

$$L\frac{di_2}{dt} - Ri_1 - \frac{1}{C}\int_0^t i_1 dt = 0 \tag{5.62}$$

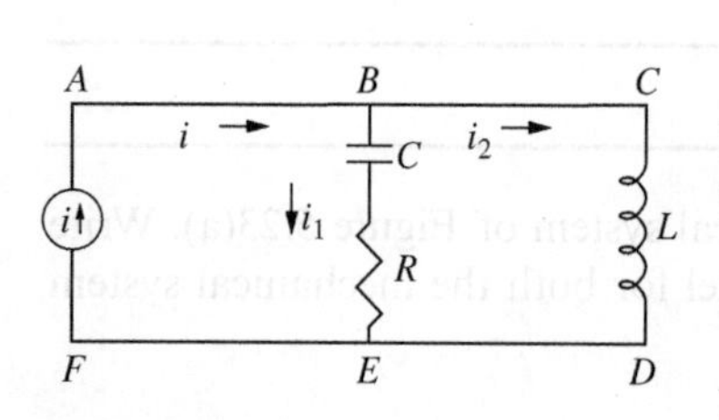

Figure 5.25

The electric circuit analogy for a mechanical system with motion input.

Equation (5.61) is rearranged as $i_1 = i(t) - i_2$, which when substituted into Equation (5.62) leads to

$$L\frac{di_2}{dt} + Ri_2 + \frac{1}{C}\int_0^t i_2 dt = Ri + \frac{1}{C}\int_0^t i\, dt \tag{5.63}$$

Using the relationships between current and charge, $i_2 = dq_2/dt$ and $i = dq/dt$, in Equation (5.63) leads to

$$L\frac{d^2q_2}{dt^2} + R\frac{dq_2}{dt} + \frac{1}{C}q_2 = R\frac{dq}{dt} + \frac{1}{C}q \tag{5.64}$$

Comparison of Equation (5.64) with Equation (5.60) illustrates the analogy. The displacement of the inertia element in the mechanical system is analogous to the charge in the inductor, and the base displacement is analogous to the charge provided by the current source. The inertia element is analogous to the inductor, the viscous damper is analogous to the resistor, and the spring is analogous to the capacitor. These component analogies are the same as in the analogy between a mechanical system with force input and an electric circuit with a voltage source.

In general, the electric circuit analogy for a mechanical system with motion input is that of a current source in parallel with circuit components analogous to those of the mechanical system to which the motion input is prescribed. The analogy corresponding to the remainder of the mechanical system is placed in parallel with this combination. The component analogy is the same as that used for the analogy between a mechanical system with a force input and an electric circuit with a voltage source.

Example 5.16

Draw the electric circuit analogy to the suspension system model of Figure 5.26(a).

Solution

The motion input for the suspension system is provided to the combination of the spring of stiffness k_2 in parallel with the viscous damper. This combination is in series with the spring of stiffness k_1. The analogous electric circuit has a current source in parallel with a capacitor and resistor. The analogy for the remainder of the mechanical system is that of a capacitor added in parallel and then an inductor added in parallel, as illustrated in Figure 5.26(b).

The differential equations modeling the mechanical system are

$$m\ddot{x}_1 + k_1(x_1 - x_2) = 0 \tag{a}$$

$$k_1(x_1 - x_2) + c(\dot{z} - \dot{x}_2) + k_2(z - x_2) = 0 \tag{b}$$

The differential equations modeling the electric circuit obtained through the application of KCL at the nodes and KVL around the loops are

$$L\frac{di_1}{dt} + \frac{1}{C_1}\int_0^t (i_1 - i_2)dt = 0 \tag{c}$$

$$\frac{1}{C_2}\int_0^t (i - i_2)dt + R(i - i_2) + \frac{1}{C_1}\int_0^t (i_1 - i_2)dt = 0 \tag{d}$$

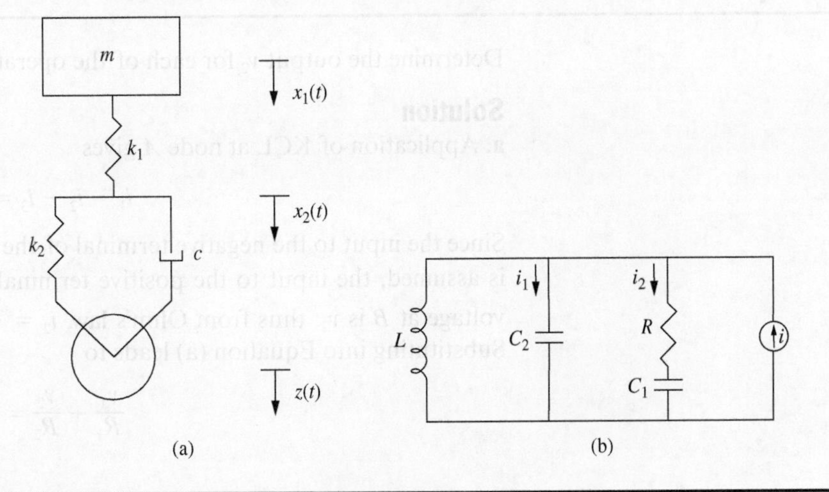

Figure 5.26

(a) The suspension system mode of Example 5.16; (b) the analogous electric circuit.

5.7 Operational Amplifiers

An operational amplifier, schematically illustrated in Figure 5.27, is a device used in circuits that are used for control systems and sensors. The operational amplifier has two input terminals, one positive and one negative. An input voltage passes through a resistor of very high resistance such that the current through the amplifier is

$$i_a = \frac{v^+ - v^-}{R_a} \tag{5.65}$$

Since the resistance is very high, the current is small. The operational amplifier contains a dependent source with a voltage

$$v_0 = K_a(v^+ - v^-) \tag{5.66}$$

where K_a is the amplifier gain. If $|v^+ - v^-|$ exceeds that maximum voltage provided by an external power supply, then the amplifier is saturated and its voltage output is this maximum. The amplifier may have a second resistor with very low resistance such that unless saturated, the output voltage from the amplifier is given by Equation (5.66).

An ideal operational amplifier has $R_a = \infty$, $R_o = 0$, $K_a = \infty$, and $v^+ = v^-$. The mathematical modeling of circuits with operational amplifiers presented assumes ideal unsaturated amplifiers.

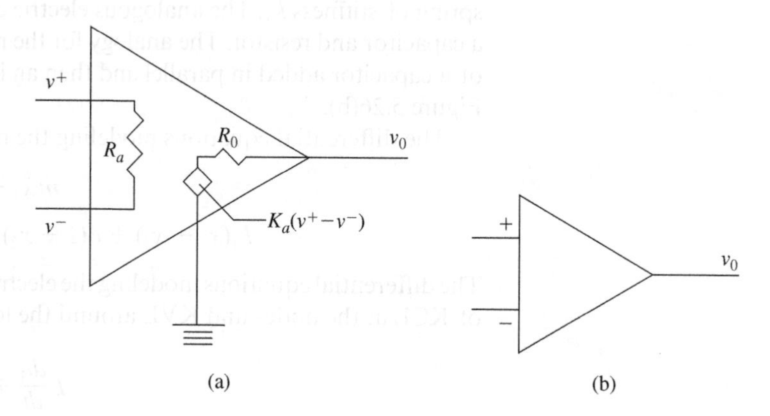

Figure 5.27
(a) The output from an operational amplifier is $K_a(v^+ - v^-)$; (b) an ideal operational amplifier has no current and $v^+ = v^-$ and $K_a = \infty$.

Example 5.17

Determine the output v_2 for each of the operational amplifier circuits of Figure 5.28.

Solution

a. Application of KCL at node A gives

$$i_1 - i_2 - i_3 = 0 \tag{a}$$

Since the input to the negative terminal of the amplifier is zero and an ideal amplifier is assumed, the input to the positive terminal v_A must also be zero and $i_3 = 0$. The voltage at B is v_2; thus from Ohm's law, $i_2 = \frac{v_A - v_2}{R_2} = \frac{-v_2}{R_2}$ and $i_1 = \frac{v_1 - v_A}{R_1} = \frac{v_1}{R_1}$. Substituting into Equation (a) leads to

$$\frac{v_1}{R_1} + \frac{v_2}{R_2} = 0 \tag{b}$$

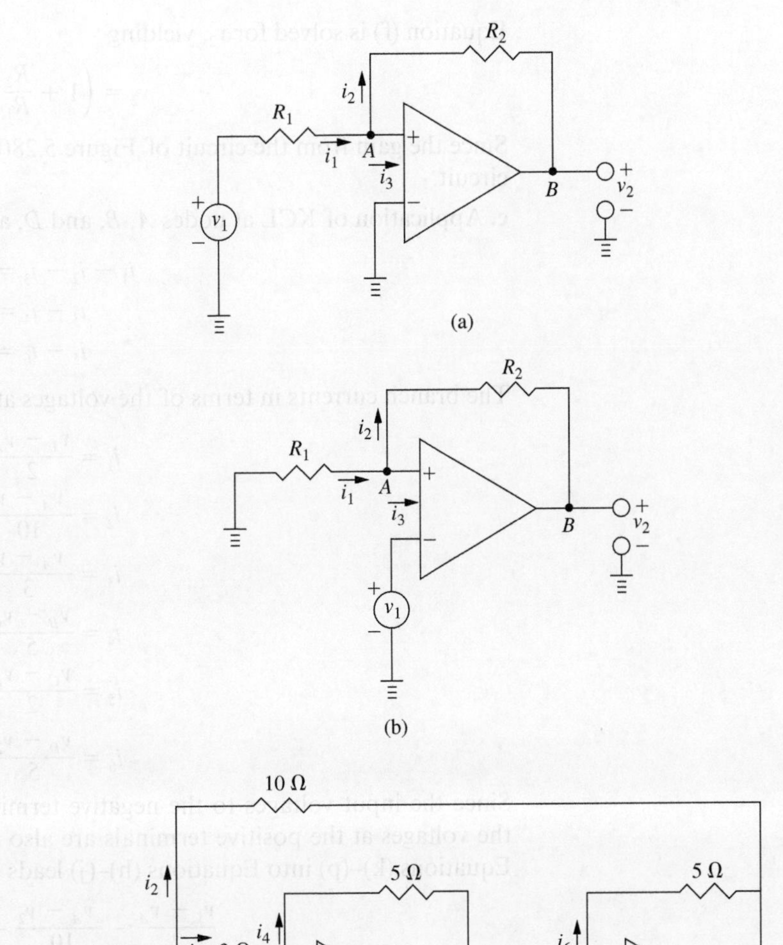

Figure 5.28

Circuits for Example 5.17.

Equation (b) is rearranged, leading to

$$v_2 = -\frac{R_2}{R_1} v_1 \tag{c}$$

The output from the circuit of Figure 5.28(a) is proportional to the input, but with the opposite sign. This circuit is called an inverting circuit.

b. Application of KCL at node A leads to Equation (a) of part a. The input to the positive terminal of the operational amplifier is v_1. Assuming an ideal amplifier, $i_3 = 0$ and $v_A = v_1$. The currents are then identified as

$$i_1 = \frac{-v_A}{R_1} = \frac{-v_1}{R_1} \tag{d}$$

$$i_2 = \frac{v_A - v_B}{R_2} = \frac{v_1 - v_2}{R_2} \tag{e}$$

Substitution of Equations (d) and (e) into Equation (a) leads to

$$\frac{-v_1}{R_1} - \frac{v_1 - v_2}{R_2} = 0 \tag{f}$$

Equation (f) is solved for v_2, yielding

$$v_2 = \left(1 + \frac{R_2}{R_1}\right) v_1 \tag{g}$$

Since the gain from the circuit of Figure 5.28(b) is positive, it is called a noninverting circuit.

c. Application of KCL at nodes A, B, and D, assuming ideal amplifiers, leads to

$$i_1 - i_2 - i_3 = 0 \tag{h}$$

$$i_3 - i_4 = 0 \tag{i}$$

$$i_5 - i_6 = 0 \tag{j}$$

The branch currents in terms of the voltages at the nodes are

$$i_1 = \frac{v_1 - v_A}{2} \tag{k}$$

$$i_2 = \frac{v_A - v_2}{10} \tag{l}$$

$$i_3 = \frac{v_A - v_B}{3} \tag{m}$$

$$i_4 = \frac{v_B - v_C}{5} \tag{n}$$

$$i_5 = \frac{v_C - v_D}{2} \tag{o}$$

$$i_6 = \frac{v_D - v_2}{5} \tag{p}$$

Since the input voltages to the negative terminals of each of the amplifiers is zero, the voltages at the positive terminals are also zero, $v_B = 0$, and $v_D = 0$. Substituting Equations (k)–(p) into Equations (h)–(j) leads to

$$\frac{v_1 - v_A}{2} - \frac{v_A - v_2}{10} - \frac{v_A}{3} = 0 \tag{q}$$

$$\frac{v_A}{3} - \frac{-v_C}{5} = 0 \tag{r}$$

$$\frac{v_C}{2} - \frac{-v_2}{5} = 0 \tag{s}$$

Equations (q), (r), and (s) are solved simultaneously, leading to

$$v_2 = \frac{125}{11} v_1 \tag{t}$$

An alternate solution to part c. is to note that both operational amplifiers in the system act as inverting amplifiers. Using the results of part a., $v_C = -\frac{5}{3}v_A$ and $v_2 = -\frac{5}{2}v_C$. These are used in conjunction with Equations (h), (k), (l), and (m) to arrive at the same solution.

Consider the operational amplifier circuit of Figure 5.29, in which a capacitor is used to bridge the amplifier. Application of KCL at node A and recalling that the relation between current and voltage drop across a capacitor may be written as $i = C\frac{dv}{dt}$ leads to

$$\frac{v_1}{R} + C\frac{dv_2}{dt} = 0 \tag{5.67}$$

Integrating both sides of Equation (5.67) with respect to time and solving for v_2 leads to

$$v_2 = -\frac{1}{RC}\int_0^t v_1 dt \tag{5.68}$$

The output voltage from this circuit is proportional to the integral of the input. Thus the circuit of Figure 5.29 is an integrating circuit and, in conjuction with an inverting circuit, may be used as an electronic integral controller.

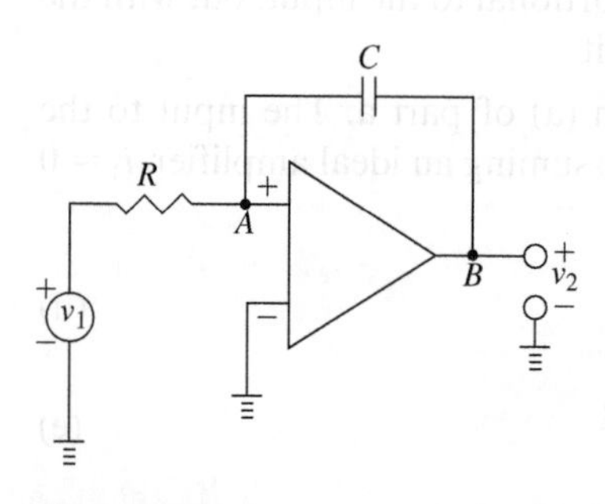

Figure 5.29

An operational amplifier circuit with a capacitor used to integrate input voltage.

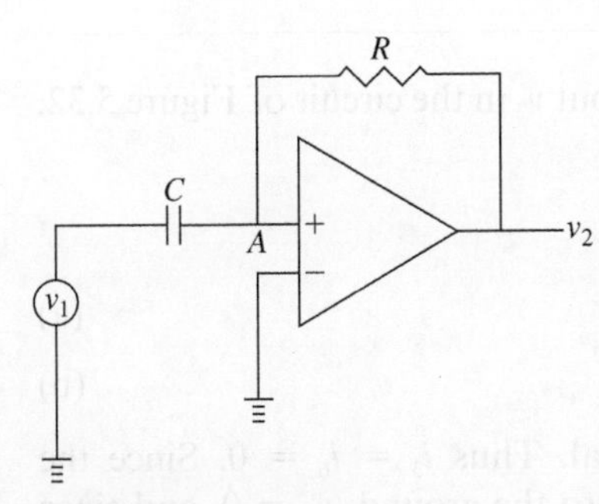

Figure 5.30

An operational amplifier circuit with a capacitor where the output is proportional to the derivative of the input.

Consider the amplifier circuit of Figure 5.30 in which a capacitor is placed such that its output voltage feeds into the amplifier. Application of KVL across the capacitor leads to $v_1 - v_A = \frac{1}{C}\int_0^v i_1 dt$. Since that amplifier is ideal, $v_A = 0$ and $i_1 = C\frac{dv_1}{dt}$. Substituting into KCL written at node A leads to

$$C\frac{dv_1}{dt} + \frac{v_2}{R} = 0 \tag{5.69}$$

Rearranging Equation (5.67) gives

$$v_2 = -RC\frac{dv_1}{dt} \tag{5.70}$$

The amplifier shown in Figure 5.30 may be used as a differential controller, since the output is proportional to the derivative of the input.

Operational amplifier circuits with capacitors or inductors are used in the design of electronic control systems. Analog computers were designed using operational amplifiers, resistors, capacitors, and indicators to simulate the solution of differential equations.

Example 5.18

Determine the relation between the input v_1 and output v_2 in the circuit of Figure 5.31.

Solution

Application of KCL at node A gives

$$i_1 - i_2 - i_3 = 0 \tag{a}$$

Since the positive terminal of the operational amplifier is connected to the ground, the voltage at A is zero, and since the amplifier is assumed to be ideal, $i_3 = 0$ and thus

$$i_1 = \frac{v_1}{R_1} \tag{b}$$

The current in the parallel combination of the inductor and the resistor is the sum of the currents in each component. The voltage change across each component is $v_A - v_2 = -v_2$. Using Equation (5.7) for the current passing through the resistor and Equation (5.28) for the current passing through the inductor gives

$$i_2 = -\frac{v_2}{R_2} - \frac{1}{L}\int_0^t v_2 dt \tag{c}$$

Substitution of Equations (b) and (c) into Equation (a) leads to

$$\frac{v_1}{R_1} + \frac{v_2}{R_2} + \frac{1}{L}\int_0^t v_2 dt = 0 \tag{d}$$

Equation (d) can be rewritten as

$$v_2 + \frac{R_2}{L}\int_0^t v_2 dt = -\frac{R_2}{R_1}v_1 \tag{e}$$

The output from the circuit is the solution to the integral equation, Equation (e).

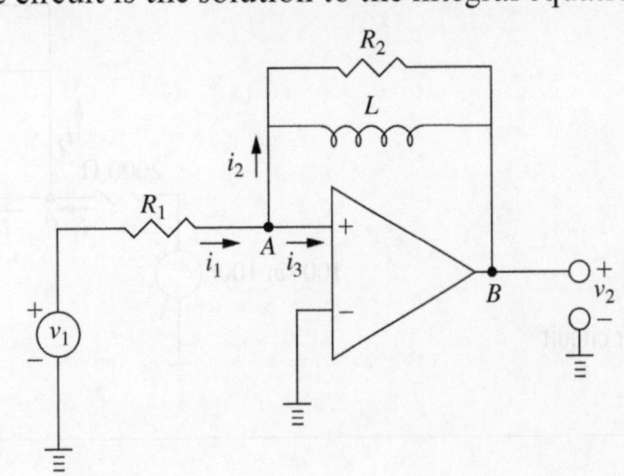

Figure 5.31

The operational amplifier circuit of Example 5.18 simulates the solution of the integral equation.

Example 5.19

Determine the relation between the input v_1 and output v_2 in the circuit of Figure 5.32.

Solution

Application of KCL at nodes A and C gives

$$i_1 - i_2 - i_3 = 0 \tag{a}$$

$$i_4 - i_5 - i_6 = 0 \tag{b}$$

The operational amplifiers are assumed to be ideal. Thus $i_3 = i_6 = 0$. Since the positive terminal of the first amplifier is connected to the ground, $v_A = 0$, and since the positive terminal of the second amplifier is connected to the external source, $v_C = 100\sin(80t)$. Applying the current and voltage drop laws across the circuit elements, the other currents become

$$i_1 = \frac{100\sin(100t)}{2000} \tag{c}$$

$$i_2 = -\frac{v_B}{4000} - 30 \times 10^{-6}\frac{dv_B}{dt} \tag{d}$$

$$i_4 = \frac{v_B - 100\sin(80t)}{6000} \tag{e}$$

$$i_5 = -30 \times 10^{-6}\frac{d}{dt}(v_2 - 100\sin 80t) \tag{f}$$

Substitution of Equations (c)–(f) in Equations (a) and (b) leads to

$$\frac{100\sin(100t)}{2000} + \frac{v_B}{4000} + 30 \times 10^{-6}\frac{dv_B}{dt} = 0 \tag{g}$$

$$\frac{v_B - 100\sin(80t)}{6000} + 30 \times 10^{-6}\frac{d}{dt}[v_2 - 100\sin(80t)] = 0 \tag{h}$$

Equation (h) is used to solve for v_B, which is then substituted into Equation (g), leading to

$$\frac{100\sin(100t)}{2000} + \frac{100\sin(80t)}{4000} - \frac{6000}{4000}(30 \times 10^{-6})\frac{d}{dt}[v_2 - 100\sin(80t)]$$
$$+(30 \times 10^{-4})\frac{d}{dt}[\sin(80t)] - (30 \times 10^{-6})^2(6000)\frac{d^2}{dt^2}[v_2 - 100\sin(80t)] \tag{i}$$

Simplification of Equation (i) leads to

$$\frac{d^2v_2}{dt^2} + 8.33\frac{dv_2}{dt} = 1.11 \times 10^5\cos(80t) - 6.34 \times 10^5\sin(80t) + 9.26 \times 10^3\sin(100t) \tag{j}$$

The output from the circuit is the solution of the differential equation, Equation (j).

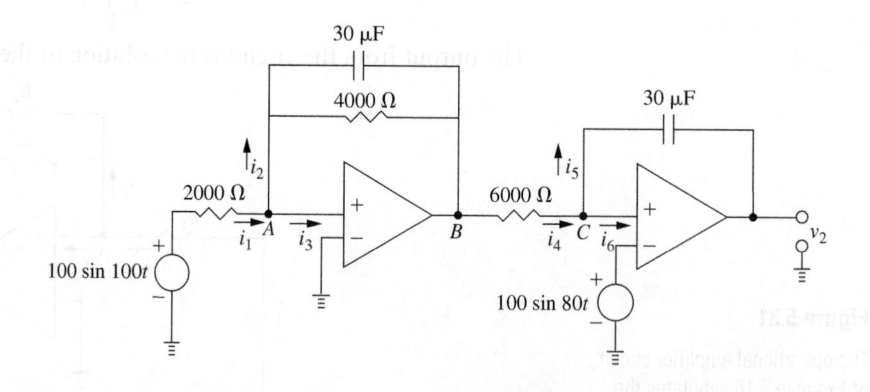

Figure 5.32

The operational amplifier circuit of Example 5.19.

5.8 Electromechanical Systems

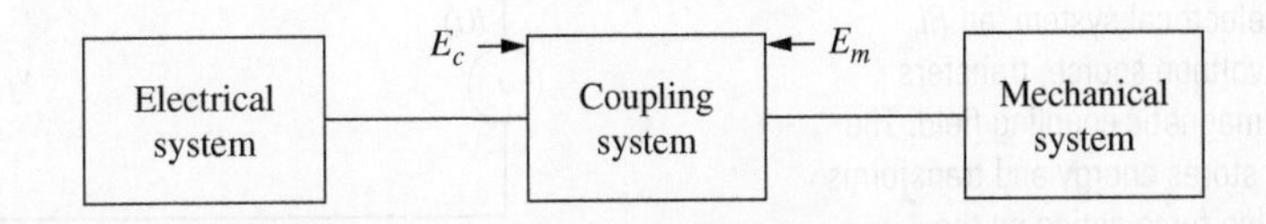

Figure 5.33

Schematic of an electromechanical system: Energy is transferred from the electrical and mechanical systems and stored in the coupling field.

An **electromechanical system** is a system in which mechanical and electrical systems interact such that energy is transferred from one system to another. The energy transfer occurs in a **coupling field**, also called a **transducer**, as illustrated schematically in Figure 5.33. The coupling field may be an electrostatic or electromagnetic field. Whereas the general theory is the same regardless of which field is used, motors and many electromechanical energy conversion devices use electromagnetic fields. A brief review of the necessary concepts of magnetic fields is presented first, followed by an outline of the general theory of electromechanical systems. The theory is applied to loudspeakers, dc servomotors, and the developing field of microelectromechanical systems (MEMS).

5.8.1 Magnetic Fields

A magnetic field surrounds an electric charge in motion. A force acting perpendicular to the direction of motion acts on the charge. The force is given by Lorenz's law as

$$\mathbf{F} = q(\mathbf{V} \times \mathbf{B}) \tag{5.71}$$

where q is the charge of the particle, $\mathbf{V}$ is its velocity, and $\mathbf{B}$ is the flux density of the magnetic field. When the charge is moving through a conductor perpendicular to the magnetic field, as shown in Figure 5.34, Lorenz's law shows that the magnitude of the force is $F = Bi\ell$, where i is the current and ℓ is the length of the conductor. A voltage increase across the conductor of $v = B\ell V$ is induced.

This type of conductor may act as a generator or a motor. If it acts as a generator, a mechanical system must deliver power equal to $P = FV = Bi\ell V$, and the power generated by the conductor is $P = vi = Bi\ell V$. If the conductor acts as a motor, it moves freely, being acted on and accelerated by the magnetic force. The conductor produces an increase in voltage, often called an electromotive force, to balance the magnetic force.

The simple descriptions of a generator and a motor can be generalized and made more complex by considering a time-varying magnetic field and a conductor consisting of a wire of N coils. In this case the voltage is given by Faraday's law:

$$v(t) = \frac{d\lambda}{dt} \tag{5.72}$$

where

$$\lambda = N\phi(t) \tag{5.73}$$

is called the flux linkage and $\phi(t)$ is the time-varying flux of the magnetic field (flux density equals flux/area).

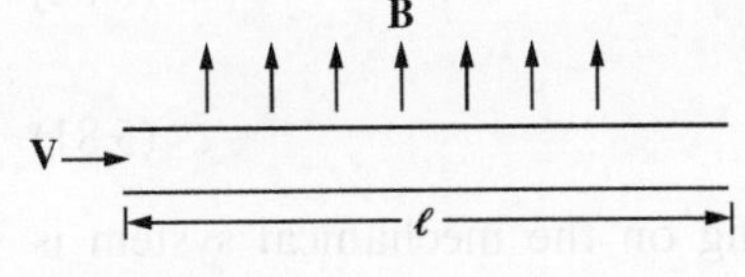

Figure 5.34

Current moving through a conductor of length ℓ with a velocity perpendicular to a magnetic field of flux density **B** has a magnetic force $F = Bi\ell$ and leads to an increase in voltage $v = B\ell V$ across the conductor.

Faraday's law is used to write the inductance of the conductor, defined as in Equation (5.25), in terms of the flux linkage as

$$L = \frac{\lambda}{i} \tag{5.74}$$

5.8.2 General Theory

A model of an electromechanical system is illustrated in Figure 5.33. The coupling field stores energy provided by the electrical system and the mechanical system. Assuming

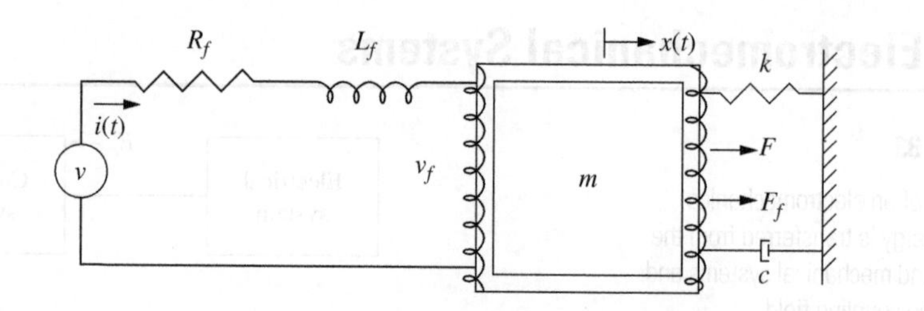

Figure 5.35

The schematic of an electromechanical actuator. The electrical system, an *RL* circuit with a voltage source, transfers energy to the magnetic coupling field. The coupling field stores energy and transforms it into a coupling force acting on the mechanical system.

the coupling field is conservative, that is, no energy is dissipated in the coupling field, the application of Conservation of Energy to the coupling field leads to

$$E_c = E_e + E_m \tag{5.75}$$

where E_c is the energy stored in the coupling field, E_e is the energy transferred by the electric system, and E_m is the energy transferred by the mechanical system. Differentiating Equation (5.75) with respect to time leads to a power form of the conservation law.

A simple electromechanical actuator is shown in Figure 5.35. The electrical system is a circuit with a voltage source, an inductor, and a resistor. The coupling field is a magnetic circuit connected to a mass-spring-viscous damper system. The motion of the mechanical system induces a voltage drop v_f in the coupling, and the coupling field produces a force F_f which acts on the mechanical system. Let $i(t)$ represent the current developed in the circuit and $x(t)$ represent the displacement of the mechanical system from its equilibrium position. The power developed in the transducer due to the induced voltage drop is $v_f i$, while the power used by the mechanical system is $F_f \dot{x}$. Thus the total power in the transducer is

$$\frac{dE_c}{dt} = v_f i - F_f \dot{x} \tag{5.76}$$

A differential form of Equation (5.76) is

$$dE_c = v_f i \, dt - F_f dx \tag{5.77}$$

Using the definition of flux linkage, Equation (5.72), Equation (5.77) becomes

$$\begin{aligned} dE_c &= i\frac{d\lambda}{dt}dt - F_f dx \\ &= i d\lambda - F_f dx \end{aligned} \tag{5.78}$$

The energy stored in the transducer is a function of the flux linkage and the displacement, $E_c = E_c(\lambda, x)$. Since E_c is a function of two variables, its total differential is

$$dE_c = \frac{\partial E_c}{\partial \lambda} d\lambda + \frac{\partial E_c}{\partial x} dx \tag{5.79}$$

Comparison of Equations (5.78) and (5.79) leads to

$$i = \frac{\partial E_c}{\partial \lambda} \tag{5.80}$$

$$F_f = -\frac{\partial E_c}{\partial x} \tag{5.81}$$

The force generated in the coupling field and acting on the mechanical system is determined using Equation (5.81).

Example 5.20

Consider the electromechanical actuator of Figure 5.35. The mechanical system is subject to an external force $F(t)$ and the electrical system has a voltage source $v(t)$. Let x

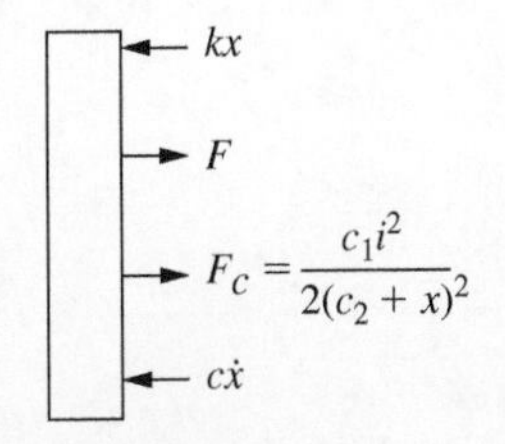

Figure 5.36

The free-body diagram of the mechanical system of Example 5.20.

be the displacement of the mechanical system from where the spring is unstretched. It can be shown that the energy stored in the transducer is

$$E_c = \frac{1}{2} L i^2 \tag{a}$$

and that the inductance varies with x according to

$$L = \frac{c_1}{c_2 + x} \tag{b}$$

Derive a mathematical model for the current in the circuit and the displacement of the mass from equilibrium.

Solution

Combining Equations (a) and (b) gives

$$E_c = \frac{1}{2}\left(\frac{c_1}{c_2 + x}\right) i^2 \tag{c}$$

Equations (a) and (5.74) are combined to give

$$E_c = \frac{1}{2}\frac{\lambda^2}{L} \tag{d}$$

Using Equation (d), note that

$$\frac{\partial E_c}{\partial \lambda} = \frac{\lambda}{L} = i \tag{e}$$

Thus Equation (5.80) is satisfied.

The coupling force applied to the mechanical system is determined from Equations (5.81) and (c) as

$$F_c = -\frac{\partial E_c}{\partial x} = \frac{c_1 i^2}{2(c_2 + x)^2} \tag{f}$$

Application of Newton's law to the free-body diagram of Figure 5.36 gives

$$\sum F = ma$$

$$F(t) + F_c - kx + c\dot{x} = m\ddot{x}$$

$$m\ddot{x} + c\dot{x} + kx - \frac{c_1 i^2}{2(c_2 + x)^2} = F(t) \tag{g}$$

Application of Kirchhoff's Voltage Law to the electrical system gives

$$v = Ri + L\frac{di}{dt} + v_f \tag{h}$$

Example 5.21

Loudspeakers are electromechanical systems in which an electric circuit is used to produce a force that moves an object, usually a cone, in a direction to amplify sound waves. A voltage provided to the voice coil results in a coupling force that moves the cone, thus radiating acoustic waves. An electromechanical model for the loudspeaker of Figure 5.37(a) is illustrated in Figure 5.37(b). The coupling force acting on the cone is $K_1 i_2$, while the induced voltage in the voice coil (also called the back emf or back electromotive force) is $K_2\dot{x}$. Derive a mathematical model for the response of the loudspeaker.

Solution

Let x be the displacement of the cone from its equilibrium position. The application of Newton's law to a free-body diagram of the cone leads to

$$K_1 i_2 - kx - c\dot{x} = m\ddot{x}$$

$$m\ddot{x} + c\dot{x} + kx = K_1 i_2 \tag{a}$$

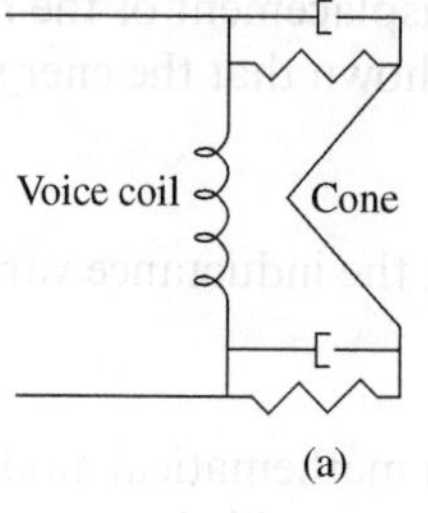

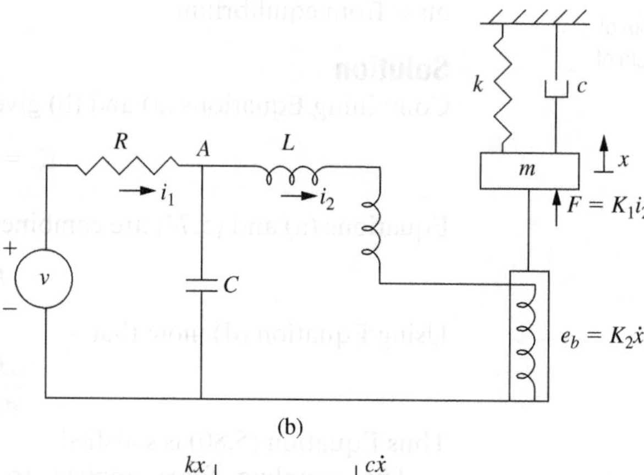

Figure 5.37
(a) A voltage is supplied to the voice coil of a loudspeaker, resulting in a coupling force that involves the loudspeaker cone; (b) the typical electric circuit of a loudspeaker; (c) the free-body diagram of the speaker cone.

The application of KCL to node A shows that the current through the capacitor is $i_1 - i_2$. Subsequent application of KVL to each of the loops gives

$$Ri_1 + \frac{1}{C}\int_0^t (i_1 - i_2)dt = v \tag{b}$$

$$L\frac{di_2}{dt} - \frac{1}{C}\int_0^t (i_1 - i_2)dt = -K_2\dot{x} \tag{c}$$

Equations (a), (b), and (c) provide the mathematical model for the loudspeaker. The input to the system is $v(t)$, while the outputs are $x(t)$, $i_1(t)$, and $i_2(t)$.

5.8.3 Direct Current (dc) Servomotors

Direct current (dc) servomotors consist of a field circuit and an armature circuit. Figures 5.38 and 5.39 show separately controlled dc motors. Other possible configurations of the two circuits are shunt connection, series connection, and compound connection. For the separately connected configurations, let i_a be the current in the armature circuit and let i_f be the current in the field circuit. The armature circuit contains a motor of moment of inertia J, which rotates with an angular displacement $\theta(t)$.

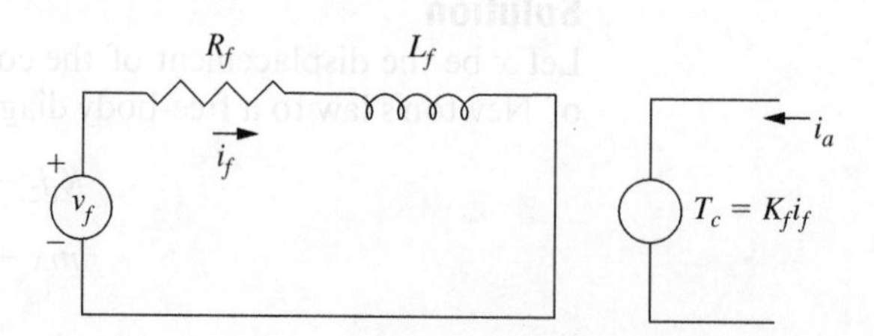

Figure 5.38

A field-controlled dc motor with the armature current constant.

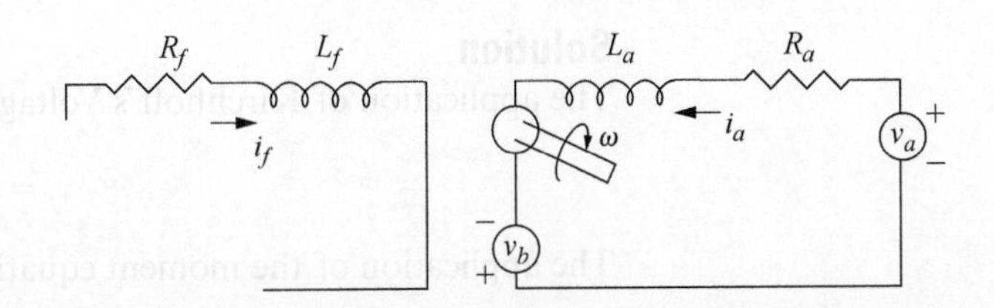

Figure 5.39

The armature-controlled dc motor has a constant field current. Back emf is induced by the rotation of the motor.

The technique for developing mathematical models of dc servomotors is as described in Section 5.8.2 except that the coupling field leads to a coupling torque applied to the armature and the torque is related to the coupling energy by

$$T_c = -\frac{\partial E_c}{\partial \theta} \tag{5.82}$$

Equation (5.82) is used to show that

$$T_c = M_{af} i_a i_f \tag{5.83}$$

where M_{af} is the mutual inductance between the inductors in the armature circuit and the field circuit.

A field-controlled dc motor is one in which the armature current is constant and the field current varies with time. In this case, Equation (5.83) is written as

$$\begin{aligned} T_c &= (M_{af} i_a) i_f \\ &= K_f i_f \end{aligned} \tag{5.84a}$$

In an armature-controlled motor, the field current is constant while the armature current varies with time. In this case, Equation (5.83) is written as

$$\begin{aligned} T_c &= (M_{af} i_f) i_a \\ &= K_a i_a \end{aligned} \tag{5.84b}$$

In an armature-controlled motor, the rotation of the motor induces a voltage in the armature circuit, commonly called the back electromotive force v_b, which is proportional to the angular speed of the shaft

$$v_b = M_{af} i_f \omega = k_f \omega \tag{5.85}$$

Example 5.22

Derive a mathematical model for the field-controlled dc motor of Figure 5.40. The motor of moment of inertia J is on a shaft with friction in its bearings with an equivalent torsional damping coefficient c_t. The shaft is subject to an external torque load T. The field circuit has a voltage source v_f, a resistance R_f, and an inductance L_f.

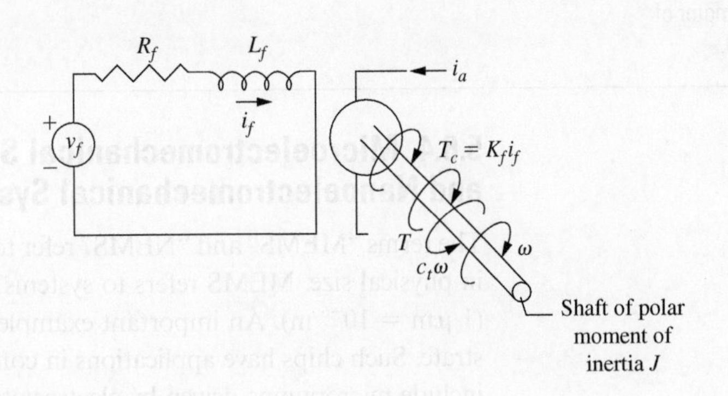

Figure 5.40

The field-controlled dc motor of Example 5.22. The shaft is mounted on bearings with torsional damping coefficient c_t and subject to an external load T.

Solution

The application of Kirchhoff's Voltage Law to the field circuit leads to

$$v_f = R_f i_f + L_f \frac{di_f}{dt} \tag{a}$$

The application of the moment equation to the motor leads to

$$\sum M = J\alpha$$

$$T_c + T - c_t\omega = J\frac{d\omega}{dt}$$

$$J\frac{d\omega}{dt} + c_t\omega - K_f i_f = T \tag{b}$$

Equations (a) and (b) constitute a mathematical model of the motor.

Example 5.23

Derive a mathematical model for the armature-controlled dc servomotor of Figure 5.41. The motor of moment of inertia J is on a shaft with friction in its bearings with an equivalent torsional damping coefficient c_t. The shaft is subject to an external torque T. The armature circuit has a voltage source v_a, a resistance R_a, and an inductance L_a.

Solution

The rotation of the motor induces a voltage in the armature circuit given by $K_a\omega$ Equation (5.84b). Application of Kirchhoff's Voltage Law to the armature circuit leads to

$$v_a = R_a i_a + L_a \frac{di_a}{dt} - v_b$$

$$L_a \frac{di_a}{dt} + R_a i_a - K_a\omega = v_a \tag{a}$$

Application of the moment equation to the motor gives

$$\sum M = J\alpha$$

$$T_c + T - c_t\omega = J\frac{d\omega}{dt}$$

$$J\frac{d\omega}{dt} + c_t\omega - K_a i_a = T \tag{b}$$

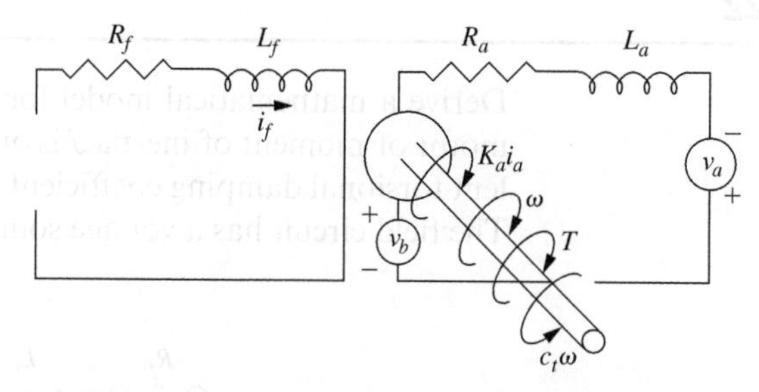

Figure 5.41

The dc servomotor of Example 5.23.

5.8.4 Microelectromechanical Systems (MEMSs) and Nanoelectromechanical Systems (NEMSs)

The terms "MEMS" and "NEMS" refer to electromechanical systems that are very small in physical size. MEMS refers to systems that operate on approximately the microscale ($1\ \mu\text{m} = 10^{-6}$ m). An important example of a MEMS device is a chip on a silica substrate. Such chips have applications in computer technology. Other MEMS applications include micropumps driven by electrostatic or magnetic forces, micromirrors, and magnetic levitation devices. The principles used to model systems at the macroscale generally apply to MEMS systems. However, since displacements and motions are on a smaller scale, forces that are often ignored in macroscale modeling assume more importance in MEMS modeling. Various forms of friction, for example, are often ignored or assumed insignificant when modeling at the macroscale. However, the effect of energy dissipation caused by friction is often significant in MEMS systems. For this reason, care is often taken to minimize the existence of unwanted nonconservative forces in MEMS systems.

NEMS operate on approximately the nanoscale (1 nm $= 10^{-9}$ m). The radius of one carbon atom is approximately 0.34 nm. Thus, modeling of NEMS is approaching modeling at the atomistic level at which the continuum assumption breaks down. Investigators have suggested that the limit at which the continuum model is valid is 50 nm. Modeling of devices with length scales below 50 nm requires applications of the principles of molecular dynamics. The approach followed here applies only to systems where continuum models are adequate. Assumptions used at the macroscale and even at the microscale break down at the nanoscale. For example, van der Waals forces, the repulsive and attractive forces between atoms, are often important in NEMS modeling.

The mathematical models derived for most MEMS and NEMS are nonlinear. The linearizing assumptions previously used and discussed in Chapter 1 are often not applicable. Small displacement assumptions are inherent in MEMS and NEMS modeling. Linear models obtained through perturbing about a steady state produce results of only limited applicability.

Example 5.24

A schematic of a MEMS sensor, such as an accelerometer, is shown in Figure 5.42. A plate of mass m and area A is attached to a piezoelectric material whose effect is modeled as that of a spring of stiffness k in parallel with a viscous damper of damping coefficient c. The bottom of the plate is subject to an applied voltage $v(t)$. The plate is parallel to a stationary, grounded plate. The total distance between the two fixed supports is ℓ. Derive a mathematical model for $x(t)$, the distance from the charged plate to the upper fixed support.

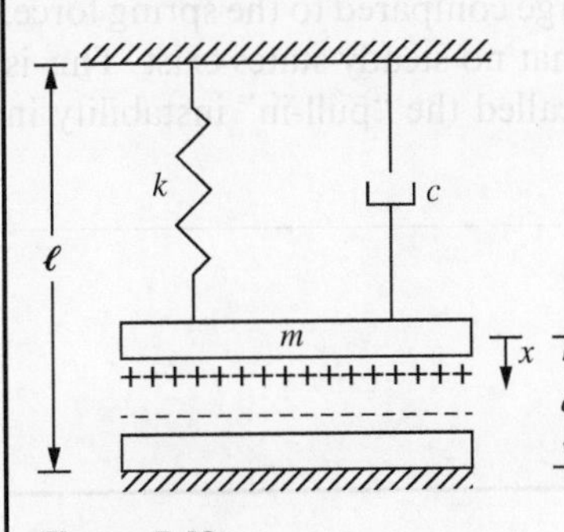

Figure 5.42

The MEMS device of Example 5.24 consists of a mass suspended from a spring and viscous damper in parallel with an applied voltage to its bottom surface. As the plate moves, the capacitance between the plates varies and exerts an electrostatic coupling force on the plate.

Solution

The two plates, separated at time t by a distance $d = \ell - x$, serve as a parallel plate capacitor. Assuming the gap between the plates has a dielectric constant of one, the capacitance is obtained using Equation (5.16) as

$$C = \varepsilon_0 \frac{A}{\ell - x} \tag{a}$$

The energy stored in the capacitor is obtained using Equation (5.23) as

$$E = \frac{1}{2}Cv^2 = \frac{\varepsilon_0 A v^2}{2(\ell - x)} \tag{b}$$

The system is an example of an electromechanical system discussed earlier in this section with an electrical system supplying the voltage, the mass-spring-viscous damper as the mechanical system, and the capacitor as the coupling field. The force acting on the mechanical system from the coupling is obtained by applying Equation (5.81) to Equation (b), resulting in

$$\begin{aligned} F_c &= -\frac{\varepsilon_0 A v^2}{2}\frac{\partial}{\partial x}\left[\frac{1}{\ell - x}\right] \\ &= \frac{\varepsilon_0 A v^2}{2(\ell - x)^2} \end{aligned} \tag{c}$$

Let x_0 be the value of x when the system is in static equilibrium in the absence of an applied voltage and let ℓ_0 be the unstretched length of the spring. A force balance on the free-body diagram of this equilibrium position leads to

$$k(x_0 - \ell) = mg \tag{d}$$

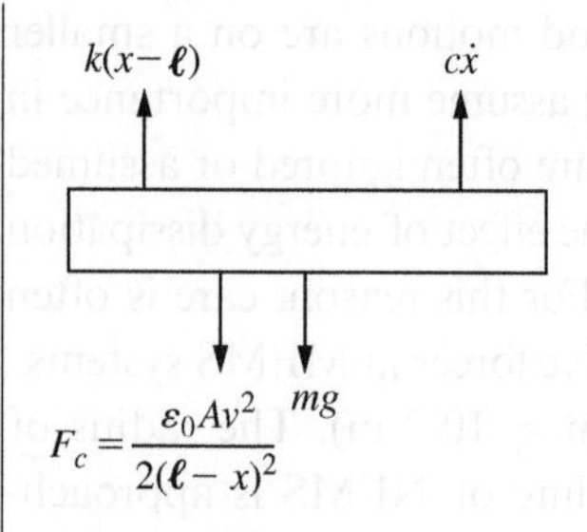

Figure 5.43

The free-body diagram of the upper plate of Example 5.24, drawn at an arbitrary instant.

The application of Newton's second law to the free-body diagram of the charged plate at an arbitrary instant, shown in Figure 5.43, leads to

$$\sum F = ma$$

$$mg - k(x - \ell) - c\dot{x} + \frac{\varepsilon_0 A v^2}{2(\ell - x)^2} = m\ddot{x} \tag{e}$$

Equation (d) is used in Equation (e), leading to

$$m\ddot{x} + c\dot{x} + kx = kx_0 + \frac{\varepsilon_0 A v^2}{2(\ell - x)^2} \tag{f}$$

Equation (f) of Example 5.24 is a nonlinear differential equation. Since the dependent variable is not measured from the system's equilibrium position, the static spring force appears explicitly in the equation. However, if the dependent variable were measured from the static equilibrium position, the gravity force would explicitly appear in the differential equation. The statement put forth in Chapter 4 that the static spring force and gravity force cancel in the differential equations was under the caveats that the system is linear and that the dependent variable is measured from the system's equilibrium position; neither applies in this problem.

If the voltage source is a dc source, a steady state may occur after some time in which $\dot{x} = \ddot{x} = 0$. An equation for the steady-state length x_s is determined using Equation (*f*):

$$kx_s = kx_0 + \frac{\varepsilon_0 A v^2}{2\left(\ell - x_s\right)^2} \tag{g}$$

The number of real roots with $x_s < \ell$ depends on the values of the parameters. Multiple, physically possible roots exist for certain values of the parameters. In such cases, only one steady state is stable. If the electrostatic force is large compared to the spring force, Equation (g) may have no real solutions, indicating that no steady states exist. This is the origin of an instability problematic for MEMS called the "pull-in" instability in which the top plate is pulled into the lower plate.

5.9 Transfer Functions for Electrical Systems

The mathematical model for the current in a single-loop series *LRC* circuit *i* given the input potential $v(t)$ is the integrodifferential equation, Equation (b) of Example 5.10:

$$L\frac{di}{dt} + Ri + \frac{1}{C}\int_0^t i\,dt = v \tag{5.86}$$

Let $I(s) = \mathscr{L}\{i(t)\}$ and $V(s) = \mathscr{L}\{v(t)\}$. The transfer function for this system is defined as $G(s) = \frac{I(s)}{V(s)}$. Taking the Laplace transform of Equation (5.86), assuming all initial conditions are zero, leads to

$$\left(Ls + R + \frac{1}{Cs}\right) I(s) = V(s) \tag{5.87}$$

Since the transfer function is a ratio of two polynomials, the numerator and denominator of the right-hand side of Equation (5.87) are multiplied by s, leading to

$$\left(Ls^2 + Rs + \frac{1}{C}\right) I(s) = sV(s) \tag{5.88}$$

The transfer function for a single-loop series *LRC* circuit is determined from Equation (5.87) as

$$G(s) = \frac{s}{Ls^2 + Rs + \frac{1}{C}} \tag{5.89}$$

The transfer function for a series *LRC* circuit, Equation (5.89), is rewritten with a leading coefficient of 1 in the denominator as

$$G(s) = \frac{s/L}{s^2 + \frac{R}{L}s + \frac{1}{CL}} \tag{5.90}$$

The transfer function specified in Equation (5.90) is second-order because the circuit has two independent storage elements: an inductor that stores energy in a magnetic field and a capacitor that stores energy in an electric field.

The charge in the circuit is related to the current by Equation (5.3) as

$$i = \frac{dq}{dt} \tag{5.91}$$

If $Q(s) = \mathcal{L}\{q(t)\}$, then from taking the Laplace transform of Equation (5.91),

$$I(s) = sQ(s) \tag{5.92}$$

Define the transfer function $G_q(s) = \frac{Q(s)}{V(s)}$. Then from Equation (5.92),

$$G_q(s) = \frac{1}{s}G(s) \tag{5.93}$$

which from Equation (5.90) becomes

$$G_q(s) = \frac{1/L}{s^2 + \frac{R}{L}s + \frac{1}{CL}} \tag{5.94}$$

It is noted that the transfer function for the charge in a series *LRC* circuit is of the same form as the transfer function for the displacement in a mass-spring-viscous damper mechanical system. This reaffirms the electrical and mechanical systems analogy in which charge in an electric circuit is analogous to displacement in a mechanical system.

Example 5.25

Determine the transfer function for the operational amplifier circuit of Figure 5.31 and Example 5.18.

Solution

The input to the circuit is $v_1(t)$ and the output is $v_2(t)$. The transfer function is $G(s) = \frac{V_2(s)}{V_1(s)}$. The mathematical model for the circuit is specified in Equation (e) of Example 5.18:

$$v_2(t) + \frac{R_2}{L}\int_0^t v_2\,dt = -\frac{R_2}{R_1}v_1 \tag{a}$$

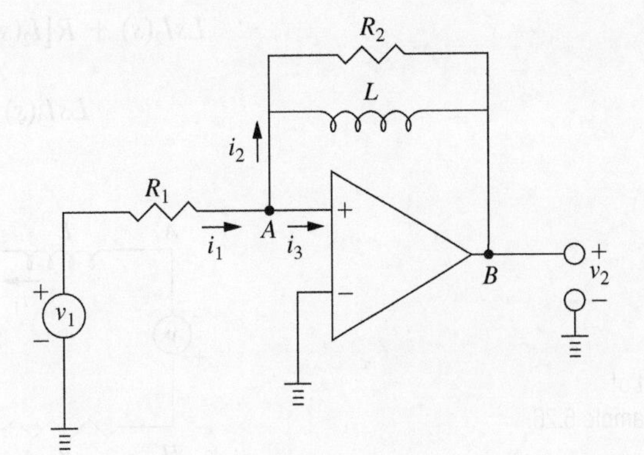

Figure 5.31

The operational amplifier circuit of Example 5.18 and Example 5.25 simulates the solution of the integral equation. (Repeated)

Taking the Laplace transform of Equation (a), using the property of linearity of the transform, leads to

$$V_2(s) + \frac{R_2}{L}\mathcal{L}\left\{\int_0^t v_2\,dt\right\} = -\frac{R_2}{R_1}V_1(s) \tag{b}$$

Using the property of the transform of an integral in Equation (b) gives

$$\left(1 + \frac{R_2}{Ls}\right)V_2(s) = -\frac{R_2}{R_1}V_1(s) \tag{c}$$

Multiplying Equation (c) by s and rearranging results in

$$G(s) = \frac{V_2(s)}{V_1(s)} = \frac{-\dfrac{R_2}{R_1}s}{s + \dfrac{R_2}{L}} \tag{d}$$

Equation (d) represents the transfer function of a first-order system, since there is one energy storage device, the inductor, in the system.

Example 5.26

Determine the transfer function vector for the three-loop circuit of Figure 5.17 and Example 5.11.

Solution

The mathematical model for the circuit is specified in Equations (d), (f), and (h) of Example 5.11. They are

$$L\frac{di_1}{dt} + Ri_1 + \frac{1}{C}\int_0^t (i_1 - i_2)\,dt = v(t) \tag{a}$$

$$L\frac{di_2}{dt} + R(i_2 - i_3) - \frac{1}{C}\int_0^t (i_1 - i_2)\,dt = 0 \tag{b}$$

$$L\frac{di_3}{dt} - R(i_2 - i_3) + \frac{1}{C}\int_0^t i_3 dt = 0 \tag{c}$$

The input is $v(t)$ and the outputs are i_1, i_2, and i_3. The matrix of transfer functions has three rows and one column. It is really a column vector. Taking the Laplace transform of Equations (a), (b), and (c) using the properties of linearity of the transform, the transforms of the first derivative and transform of an integral, and assuming all initial conditions are zero leads to

$$LsI_1(s) + RI_1(s) + \frac{1}{Cs}[I_1(s) - I_2(s)] = V(s) \tag{d}$$

$$LsI_2(s) + R[I_2(s) - I_3(s)] - \frac{1}{Cs}[I_1(s) - I_2(s)] = 0 \tag{e}$$

$$LsI_3(s) - R[I_2(s) - I_3(s)] + \frac{1}{Cs}I_3(s) = 0 \tag{f}$$

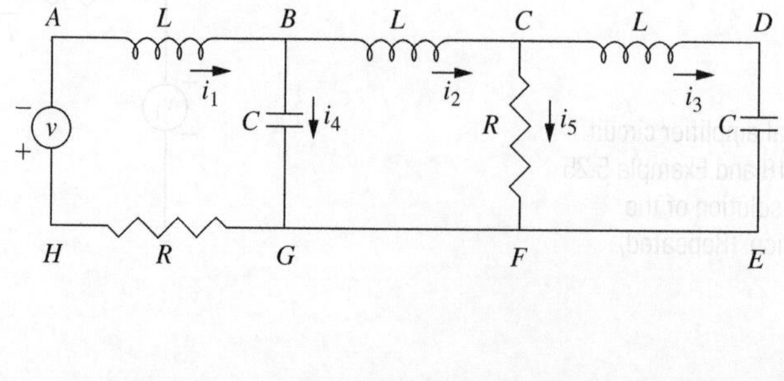

A L B L C L D
i_1 i_2 i_3
v C i_4 R i_5 C
H R G F E

Figure 5.17

The three-loop circuit of Example 5.11 and Example 5.26. (Repeated)

Writing Equations (d), (e), and (f) in matrix form gives

$$\begin{bmatrix} Ls + R + \frac{1}{Cs} & -\frac{1}{Cs} & 0 \\ -\frac{1}{Cs} & Ls + R + \frac{1}{Cs} & -R \\ 0 & -R & Ls + R + \frac{1}{Cs} \end{bmatrix} \begin{bmatrix} I_1(s) \\ I_2(s) \\ I_3(s) \end{bmatrix} = \begin{bmatrix} V(s) \\ 0 \\ 0 \end{bmatrix} \quad \text{(g)}$$

Solving the three equations represented by Equation (g) simultaneously leads to

$$\begin{bmatrix} I_1(s) \\ I_2(s) \\ I_3(s) \end{bmatrix} = \frac{s}{Ls^3 + 2RLs^2 + \frac{2L}{C}s + \frac{2R}{C}} \begin{bmatrix} \frac{1}{Cs}V(s) \\ V(s) \\ RV(s) \end{bmatrix} \quad \text{(h)}$$

The vector of transfer functions is determined from Equation (h) as

$$\begin{bmatrix} G_1(s) \\ G_2(s) \\ G_3(s) \end{bmatrix} = \frac{s}{Ls^3 + 2RLs^2 + \frac{2L}{C}s + \frac{2R}{C}} \begin{bmatrix} \frac{1}{Cs} \\ 1 \\ R \end{bmatrix} \quad \text{(i)}$$

Example 5.27

Determine the transfer function for the ideal amplifier circuit of Figure 5.32 and Example 5.19. Assume the input voltages are arbitrary. Replace 100 sin 100t by v_1 and assume there is no input to the second amplifier.

Solution

The input to the circuit is $v_1(t)$ and the output is $v_2(t)$. The transfer function is $G(s) = \frac{V_3(s)}{V_1(s)}$. The problem can be solved by performing the algebra in the transform domain rather than in the time domain. Applying KCL at A and C leads to

$$i_1 - i_2 - i_3 = 0 \quad \text{(a)}$$

and

$$i_4 - i_5 - i_6 = 0 \quad \text{(b)}$$

Assuming the amplifiers are ideal implies that $i_3 = i_6 = 0$. Taking the Laplace transform of Equations (a) and (b) leads to

$$I_1(s) = I_2(s) \quad \text{(c)}$$

and

$$I_4(s) = I_5(s) \quad \text{(d)}$$

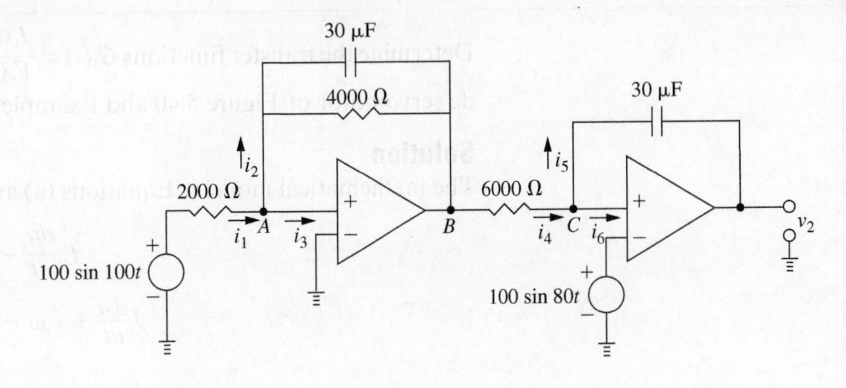

Figure 5.32

The operational amplifier circuit of Example 5.19 and Example 5.27. (Repeated)

Using the voltage and current laws across the circuit component gives

$$v_{i1}(t) = \frac{1}{C_1}\int_0^t i_1\,dt \Rightarrow i_1 = C_1\frac{dv_1}{dt} \tag{e}$$

$$i_2 = -\frac{v_B}{R_1} - C_2\frac{dv_B}{dt} \tag{f}$$

$$i_4 = \frac{v_2}{R_2} \tag{g}$$

$$v_2 = R_3 i_5 + \frac{1}{C_3}\int_0^t i_5\,dt \tag{h}$$

Taking the Laplace transform of Equations (e)–(h) leads to

$$I_1(s) = C_1 s V_1(s) \tag{i}$$

$$I_2(s) = -\frac{V_2(s)}{R_1} - C_2 s V_B(s) \tag{j}$$

$$I_4(s) = \frac{V_B(s)}{R_2} \tag{k}$$

$$V_3(s) = -R_3 I_5(s) - \frac{1}{C_3 s} I_5(s) \tag{l}$$

Equations (i) and (j) are substituted into Equation (c) to solve for $V_2(s)$ in terms of $V_1(s)$. Equations (k) and (l) are substituted into Equation (d), thus relating $V_2(s)$ to $V_B(s)$. The results are

$$V_B(s) = \frac{C_1 s}{C_2 s + \frac{1}{R_1}} V_1(s) \tag{m}$$

and

$$V_2(s) = -\left(\frac{R_3}{R_2}s + \frac{1}{C_3 R_2}\right) V_B(s) \tag{n}$$

Equations (m) and (n) are combined, leading to

$$V_3(s) = \left(\frac{R_3}{R_2}s + \frac{1}{C_3 R_2}\right)\left(\frac{C_1 s}{C_2 s + \frac{1}{R_1}}\right) V_1(s) \tag{o}$$

The transfer function is determined from Equation (o) as

$$G(s) = \frac{V_2(s)}{V_1(s)} = \frac{R_1 C_1 s(R_3 C_3 s + 1)}{C_3 R_2(R_1 C_2 s + 1)} \tag{p}$$

Example 5.28

Determine the transfer functions $G_1(s) = \frac{I_f(s)}{V_f(s)}$ and $G_2(s) = \frac{\Omega(s)}{V_f(s)}$ for the field-controlled dc servomotor of Figure 5.40 and Example 5.22.

Solution

The mathematical model is Equations (a) and (b) of Example 5.22:

$$L_f\frac{di_f}{dt} + R_f i_f = v_f \tag{a}$$

$$J\frac{d\omega}{dt} + c_t\omega - K_f i_f = T \tag{b}$$

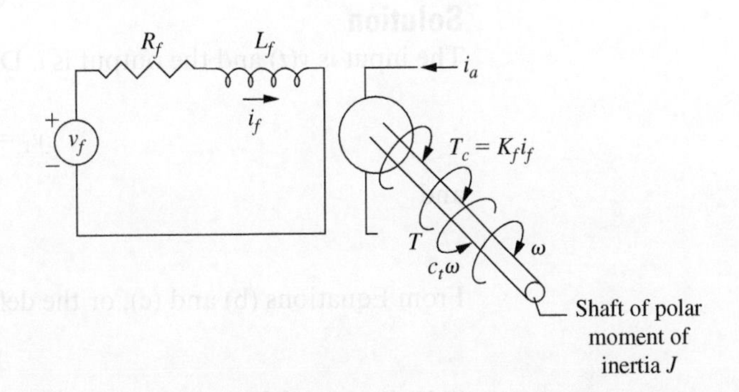

Figure 5.40

The field-controlled dc motor of Example 5.22 and Example 5.28. The shaft is mounted on bearings with torsional damping coefficient c_t and subject to a load T. (Repeated)

where the inputs are v_f, the potential source in the field circuit, and T, the applied torque. Taking the Laplace transforms of Equations (a) and (b) and requiring that all initial conditions are zero leads to

$$L_f s I_f(s) + R_f I_f(s) = V_f(s) \tag{c}$$

$$Js\Omega(s) + c_t\Omega(s) - K_f I_f = T(s) \tag{d}$$

The transfer functions defined are determined with $T(s) = 0$. Using Equation (c),

$$G_1(s) = \frac{I_f(s)}{V_f(s)} = \frac{1}{L_f s + R_f} \tag{e}$$

Using Equations (d) and (e),

$$G_2(s) = \frac{\Omega(s)}{V_f(s)} = \frac{\Omega(s)}{I_f(s)}\frac{I_f(s)}{V_f(s)} = \left(\frac{1}{Js + c_t}\right)\left(\frac{1}{L_f s + R_f}\right) = \frac{1}{(Js + c_t)(L_f s + R_f)} \tag{f}$$

The transforms of Equations (e) and (f) are of different orders and have different denominators even though they are part of the same system. The only energy source that affects the current in the field circuit is the inductor. Both the inductor and the kinetic energy stored in the shaft affect the angular velocity.

5.10 State-Space Modeling of Electric Circuits

State variables for an electric circuit are determined by considering the analogy between mechanical systems and electrical systems. For a mass, spring, and viscous damper system, the state variables may be chosen as $y_1 = x$, the displacement of the mass, and $y_2 = \dot{x}$, its velocity. Referring to Table 5.4, comparing a mechanical system with a displacement x and an electrical system with a voltage source, it is clear that x is analogous to q, the charge, and $\dot{x}$ is analogous to the current i. Thus, one choice to define state variables for a series LRC circuit with current input is to let $y_1 = q$ and $y_2 = i$.

Example 5.29

The mathematical model for a series LRC circuit is

$$L\frac{di}{dt} + Ri + \frac{1}{C}\int_0^t i\,dt = v \tag{a}$$

Determine a state-space model for the series LRC circuit.

Solution

The input is $v(t)$ and the output is i. Define the state variables as

$$y_1 = q = \int_0^t i\,dt \tag{b}$$

and

$$y_2 = i \tag{c}$$

From Equations (b) and (c), or the definition of current,

$$y_2 = \dot{y}_1 \tag{d}$$

Substitution of Equations (b) and (c) into Equation (a) leads to

$$L\dot{y}_2 + Ry_2 + \frac{1}{C}y_1 = v \tag{e}$$

Rearranging Equation (e) results in

$$\dot{y}_2 = -\frac{1}{LC}y_1 - \frac{R}{L}y_2 + \frac{1}{L}v \tag{f}$$

Equations (d) and (f) form a state-space model for the series *LRC* circuit. Formulating these in matrix form gives

$$\begin{bmatrix} \dot{y}_1 \\ \dot{y}_2 \end{bmatrix} = \begin{bmatrix} 0 & 1 \\ -\frac{1}{LC} & -\frac{R}{L} \end{bmatrix} \begin{bmatrix} y_1 \\ y_2 \end{bmatrix} + \begin{bmatrix} 0 \\ \frac{1}{L} \end{bmatrix} [v] \tag{g}$$

The relationship among the output and the state variables and the input is

$$[i] = [0 \quad 1]\begin{bmatrix} y_1 \\ y_2 \end{bmatrix} + [0][v] \tag{h}$$

From Equations (g) and (h), it is determined that

$$\text{State matrix: } \mathbf{A} = \begin{bmatrix} 0 & 1 \\ -\frac{1}{LC} & -\frac{R}{L} \end{bmatrix}$$

$$\text{Input matrix: } \mathbf{B} = \begin{bmatrix} 0 \\ \frac{1}{L} \end{bmatrix}$$

$$\text{Output matrix: } \mathbf{C} = [0 \quad 1]$$

$$\text{Transmission matrix: } \mathbf{D} = [0]$$

Example 5.30

The mathematical formulation for the three-loop circuit of Figure 5.17 and Examples 5.11 and 5.26 is

$$L\frac{di_1}{dt} + Ri_1 + \frac{1}{C}\int_0^t (i_1 - i_2)dt = v(t) \tag{a}$$

$$L\frac{di_2}{dt} + R(i_2 - i_3) - \frac{1}{C}\int_0^t (i_1 - i_2)dt = 0 \tag{b}$$

$$L\frac{di_3}{dt} - R(i_2 - i_3) + \frac{1}{C}\int_0^t i_3 dt = 0 \tag{c}$$

Determine a state-space formulation for this circuit.

Solution

The input is $v(t)$ and the outputs are i_1, i_2, and i_3. Define as the state variables

$$y_1 = q_1, y_2 = q_2, y_3 = q_3 \tag{d}$$

$$y_4 = i_1, y_5 = i_2, y_6 = i_3 \tag{e}$$

where $q_j = \frac{1}{C}\int_0^t i_j dt$ is the charge in the wire and i_j is the current. From the definitions of the state variables,

$$\dot{y}_1 = y_4,\ \dot{y}_2 = y_5,\ \dot{y}_3 = y_6 \tag{f}$$

Substituting Equations (d) and (e) into Equations (a), (b), and (c) and rearranging leads to

$$\dot{y}_4 = -\frac{1}{CL}y_1 + \frac{1}{CL}y_2 - \frac{R}{L}y_4 + \frac{1}{L}v(t) \tag{g}$$

$$\dot{y}_5 = \frac{1}{CL}y_1 - \frac{1}{CL}y_2 - \frac{R}{L}y_5 + \frac{R}{L}y_6 \tag{h}$$

$$\dot{y}_6 = \frac{1}{CL}y_3 + \frac{R}{L}y_5 - \frac{R}{L}y_6 \tag{i}$$

Equations (g), (h), and (i) are summarized in matrix form as

$$\begin{bmatrix} \dot{y}_1 \\ \dot{y}_2 \\ \dot{y}_3 \\ \dot{y}_4 \\ \dot{y}_5 \\ \dot{y}_6 \end{bmatrix} = \begin{bmatrix} 0 & 0 & 0 & 1 & 0 & 0 \\ 0 & 0 & 0 & 0 & 1 & 0 \\ 0 & 0 & 0 & 0 & 0 & 1 \\ -\frac{1}{CL} & \frac{1}{CL} & 0 & -\frac{R}{L} & 0 & 0 \\ \frac{1}{CL} & -\frac{1}{CL} & 0 & 0 & -\frac{R}{L} & \frac{R}{L} \\ 0 & 0 & -\frac{1}{CL} & 0 & \frac{R}{L} & -\frac{R}{L} \end{bmatrix} \begin{bmatrix} y_1 \\ y_2 \\ y_3 \\ y_4 \\ y_5 \\ y_6 \end{bmatrix} + \begin{bmatrix} 0 \\ 0 \\ 0 \\ \frac{1}{L} \\ 0 \\ 0 \end{bmatrix} [v] \tag{j}$$

The relationship among the output and the state variables and the input is

$$\begin{bmatrix} i_1 \\ i_2 \\ i_3 \end{bmatrix} = \begin{bmatrix} 0 & 0 & 0 & 1 & 0 & 0 \\ 0 & 0 & 0 & 0 & 1 & 0 \\ 0 & 0 & 0 & 0 & 0 & 1 \end{bmatrix} \begin{bmatrix} y_1 \\ y_2 \\ y_3 \\ y_4 \\ y_5 \\ y_6 \end{bmatrix} + \begin{bmatrix} 0 \\ 0 \\ 0 \end{bmatrix} [v] \tag{k}$$

The state matrix and input matrix are defined in Equation (j). The output matrix and transmission matrix are defined in Equation (k).

Example 5.31

The mathematical model for the circuit of Example 5.16 and Figure 5.26(b) is Equations (c) and (d) of Example 5.16.

$$L\frac{di_1}{dt} + \frac{1}{C_1}\int_0^t (i_1 - i_2)\,dt = 0 \tag{a}$$

$$\frac{1}{C_2}\int_0^t (i - i_2)dt + R(i - i_2) + \frac{1}{C_1}\int_0^t (i_1 - i_2)dt = 0 \tag{b}$$

Determine a state-space model for the circuit.

Solution

The input is $i(t)$ and the output is i_1. There are three energy storage elements, so three state variables are expected. If the state variables are chosen as the charge of each wire

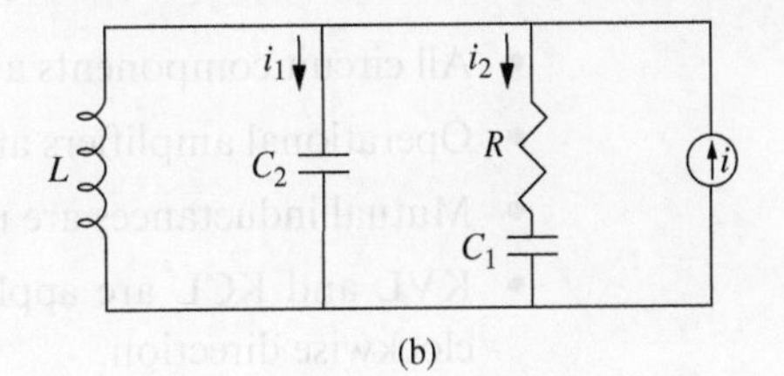

Figure 5.26

(b) The analogous electric circuit of Example 5.16 is used in Example 5.31. (Repeated) and the current through the inductor, then the derivative of the input appears in the mathematical model. The state variables are amended such that

$$y_1 = i_1 \tag{c}$$

$$y_2 = q_1 \tag{d}$$

$$y_3 = q_2 + \mu q \tag{e}$$

where

$$q = \int_0^t i\,dt \tag{f}$$

is the charge provided by the source. From Equations (d) and (e),

$$\dot{y}_2 = y_1 \tag{g}$$

Using the state variables of Equations (c), (d), and (e) in the mathematical model leads to

$$L\dot{y}_1 + \frac{1}{C_1}y_1 - \frac{1}{C_1}(y_3 - \mu q) = 0 \tag{h}$$

$$-\frac{1}{C_2}(y_3 - \mu q) - R(\dot{y}_3 - \mu\dot{q}) + \frac{1}{C_1}(y_2 - y_3 - \mu q) = -R\dot{q} - \frac{1}{C_1}q \tag{i}$$

Choose $\mu = -1$ to eliminate the $\dot{q}$ terms from Equation (i). Then Equations (h) and (i) can be rewritten as

$$\dot{y}_1 = -\frac{1}{LC_1}y_1 + \frac{1}{LC_1}y_3 + \frac{1}{L}q \tag{j}$$

$$\dot{y}_3 = -\frac{1}{R}\left(\frac{1}{C_1} + \frac{1}{C_2}\right)y_3 + \frac{1}{RC_2}y_2 + \frac{1}{RC_2}q \tag{k}$$

Equations (f), (j), and (k) can be summarized in matrix form as

$$\begin{bmatrix} \dot{y}_1 \\ \dot{y}_2 \\ \dot{y}_3 \end{bmatrix} = \begin{bmatrix} \frac{1}{LC_1} & 0 & \frac{1}{LC_1} \\ 1 & 0 & 0 \\ 0 & -\frac{1}{RC_2} & -\frac{1}{R}\left(\frac{1}{C_1} + \frac{1}{C_2}\right) \end{bmatrix} \begin{bmatrix} y_1 \\ y_2 \\ y_3 \end{bmatrix} + \begin{bmatrix} \frac{1}{L} \\ 0 \\ \frac{1}{RC_2} \end{bmatrix} [q] \tag{l}$$

The relationship among the output variables and the state variables and the input variable is

$$[i_1] = [1 \quad 0 \quad 0] \begin{bmatrix} y_1 \\ y_2 \\ y_3 \end{bmatrix} + \begin{bmatrix} 0 \\ 0 \end{bmatrix} [q] \tag{m}$$

The state matrix and input matrix are defined in Equation (l). The output matrix and transmission matrix are defined in Equation (m).

5.11 Further Examples

The below assumptions are used in the following examples:

- All circuit components are linear and have lumped properties.
- Operational amplifiers are ideal and unsaturated.
- Mutual inductances are negligible unless otherwise specified.
- KVL and KCL are applicable for all circuits, with KVL always applied in a clockwise direction.

Example 5.32

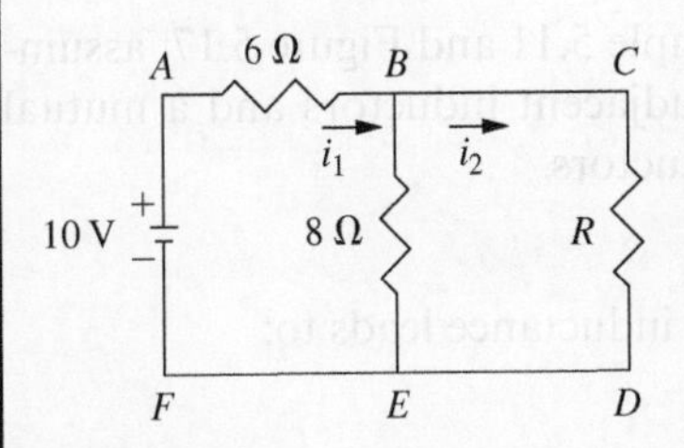

Figure 5.44

The circuit for Example 5.32.

Determine the value of R in the circuit of Figure 5.44 such that the power through that resistor is a maximum.

Solution

Let i_1 be the current in the 6-Ω resistor and i_2 be the current in the resistor of resistance R. The application of KCL at node B shows that the current through the 8-Ω resistor is $i_1 - i_2$. Application of KVL to the two loops in the circuit leads to

$$10 - 6i_1 - 8(i_1 - i_2) = 0 \tag{a}$$

$$-Ri_2 + 8(i_1 - i_2) = 0 \tag{b}$$

Equations (a) and (b) are rearranged to

$$-14i_1 + 8i_2 = -10 \tag{c}$$

$$8i_1 - (8 + R)i_2 = 0 \tag{d}$$

Equations (c) and (d) are solved simultaneously, leading to

$$i_1 = \frac{5(8 + \mathrm{R})}{24 + 7R} \tag{e}$$

$$i_2 = \frac{40}{24 + 7R} \tag{f}$$

The power through the resistor is

$$P = i_2^2 R$$

$$= \frac{1600R}{(24 + 7R)^2} \tag{g}$$

The value of R that leads to maximum power is obtained by setting $dP/dR = 0$ and solving for R. To this end, using the quotient rule for differentiation,

$$\frac{dP}{dR} = \frac{1600(24 + 7R)^2 - (1600R)(2)(24 + 7R)(7)}{(24 + 7R)^4}$$

$$= \frac{1600(24 + 7R)(24 - 7R)}{(24 + 7R)^4} \tag{h}$$

Setting $dP/dR = 0$ leads to $R = 24/7\ \Omega$. The maximum power is calculated as

$$P_{\max} = \frac{1600(24/7)}{[24 + 7(24/7)]^2}\ \mathrm{W}$$

$$= 2.38\ \mathrm{W} \tag{i}$$

Example 5.33

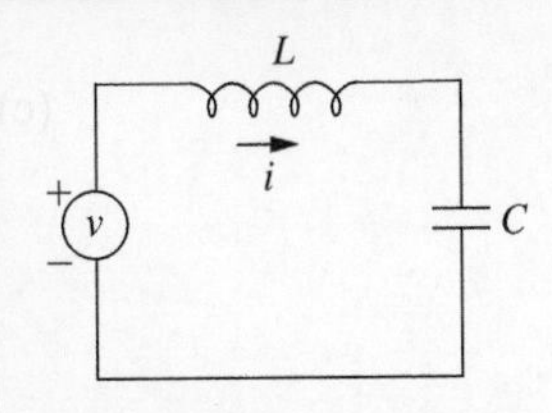

Figure 5.45

The *LC* circuit of Example 5.33.

Derive a mathematical model for the *LC* circuit of Figure 5.45.

Solution

Define $i(t)$ as the current flowing through the circuit. Application of KVL in the clockwise direction leads to

$$v(t) - L\frac{di}{dt} - \frac{1}{C}\int_0^t i\,dt = 0 \tag{a}$$

Rearranging Equation (a) leads to

$$L\frac{di}{dt} + \frac{1}{C}\int_0^t i\,dt = v(t) \tag{b}$$

Example 5.34

Derive a mathematical model for the circuit of Example 5.11 and Figure 5.17, assuming mutual inductance of M between each of the adjacent inductors and a mutual inductance of zero between the two nonadjacent inductors.

Solution

Application of KVL for each loop assuming mutual inductance leads to:

Loop *ABGHA*

$$v(t) - \left(L\frac{di_1}{dt} + M\frac{di_2}{dt}\right) - \frac{1}{C}\int_0^t (i_1 - i_2)dt - Ri_1 = 0 \tag{a}$$

$$L\frac{di_1}{dt} + M\frac{di_2}{dt} + Ri_1 + \frac{1}{C}\int_0^t i_1 dt - \frac{1}{C}\int_0^t i_2 dt = v(t) \tag{b}$$

Loop *BCFGB*

$$-\left(L\frac{di_2}{dt} + M\frac{di_1}{dt} + M\frac{di_3}{dt}\right) - R(i_2 - i_3) + \frac{1}{C}\int_0^t (i_1 - i_2)dt = 0 \tag{c}$$

$$M\frac{di_1}{dt} + L\frac{di_2}{dt} + M\frac{di_3}{dt} + Ri_2 - Ri_3 - \frac{1}{C}\int_0^t i_1 dt + \frac{1}{C}\int_0^t i_2 dt = 0 \tag{d}$$

Loop *CDEFC*

$$-\left(L\frac{di_3}{dt} + M\frac{di_2}{dt}\right) - \frac{1}{C}\int_0^t i_3 dt + R(i_2 - i_3) = 0 \tag{e}$$

$$M\frac{di_2}{dt} + L\frac{di_3}{dt} - Ri_2 + Ri_3 + \frac{1}{C}\int_0^t i_3 dt = 0 \tag{f}$$

The circuit is modeled using Equations (b), (d), and (f).

Example 5.35

Derive a mathematical model for the circuit of Figure 5.46.

Solution

The application of KCL at node B leads to

$$i_1 - i_2 - i_3 = 0 \tag{a}$$

Applying KVL around loop *ABEFA* leads to

$$v - Ri_1 - \frac{1}{C}\int_0^t i_3 dt = 0 \tag{b}$$

Applying KVL to loop *BCDEB* leads to

$$-L\frac{di_2}{dt} - Ri_2 + \frac{1}{C}\int_0^t i_3 dt = 0 \tag{c}$$

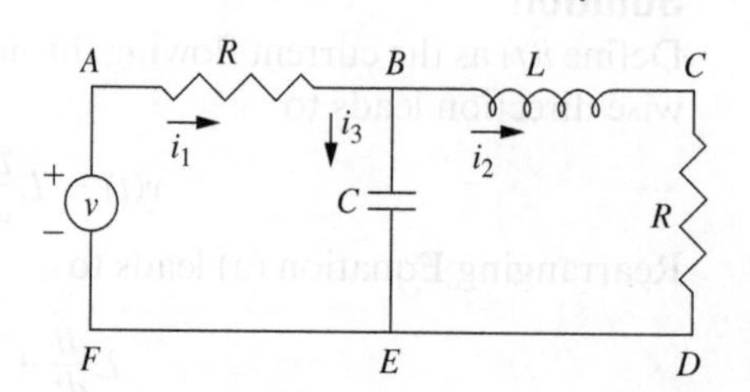

Figure 5.46
The circuit for Example 5.35.

Substituting Equation (a) into Equations (b) and (c) and rearranging leads to

$$Ri_1 + \frac{1}{C}\int_0^t i_1 dt - \frac{1}{C}\int_0^t i_2 dt = v \tag{d}$$

$$-\frac{1}{C}\int_0^t i_1 dt + L\frac{di_2}{dt} + Ri_2 + \frac{1}{C}\int_0^t i_2 dt = 0 \tag{e}$$

Example 5.36

Determine the output voltage from the operational amplifier circuit of Figure 5.47. Assume both amplifiers are ideal.

Solution

Since both amplifiers are ideal, the current through each amplifier is zero and the voltage at each terminal is the same. Note that both amplifiers are inverting amplifiers as in Example 5.17a. The output voltage from an inverting amplifier is the negative of the ratio of the resistances times the input voltage. Thus

$$v_B = -\frac{R_2}{R_1}v_1 \tag{a}$$

and

$$v_2 = -\frac{R_3}{R_3}v_B$$

$$= \frac{R_2}{R_1}v_1 \tag{b}$$

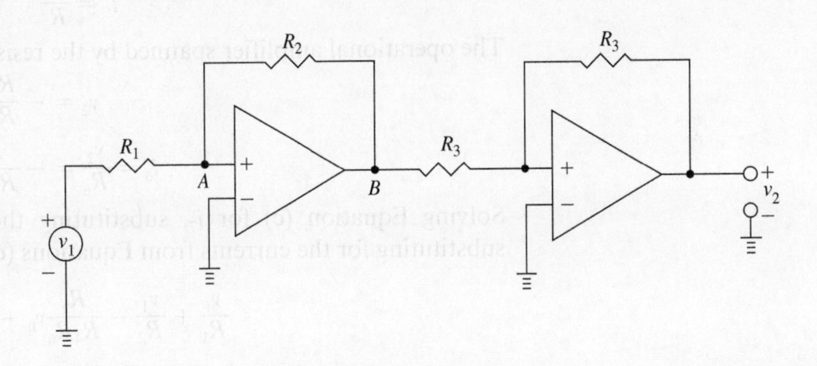

Figure 5.47

The two inverting operational amplifiers in series of Example 5.36.

Example 5.37

Determine the relationship between the output voltage $v_0(t)$ and the input voltage $v_i(t)$ for the operational amplifier circuit of Figure 5.48.

Solution

Defining currents as illustrated in Figure 5.48, the application of KCL at nodes A, B, and C leads to

$$i_1 + i_7 - i_2 = 0 \tag{a}$$

$$i_3 - i_4 = 0 \tag{b}$$

$$i_5 + i_6 - i_7 = 0 \tag{c}$$

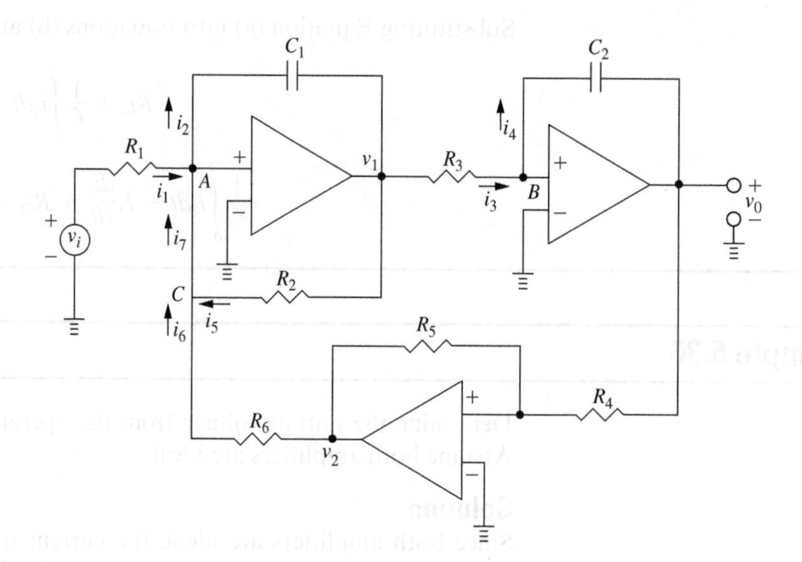

Figure 5.48

The operational amplifier circuit of Example 5.37.

The currents are given by

$$i_1 = \frac{v_1}{R_1} \tag{d}$$

$$i_2 = -C_1\frac{dv_1}{dt} \tag{e}$$

$$i_3 = \frac{v_1}{R_3} \tag{f}$$

$$i_4 = -C_2\frac{dv_O}{dt} \tag{g}$$

$$i_5 = \frac{v_1}{R_2} \tag{h}$$

The operational amplifier spanned by the resistor R_5 is an inverting amplifier. Thus

$$v_2 = -\frac{R_5}{R_4}v_O \tag{i}$$

$$i_6 = \frac{v_2}{R_6} = -\frac{R_5}{R_4R_6}v_O \tag{j}$$

Solving Equation (c) for i_7, substituting the result into Equation (a), and then substituting for the currents from Equations (d)–(i) leads to

$$\frac{v_i}{R_1} + \frac{v_1}{R_2} - \frac{R_5}{R_4R_6}v_O + C_1\frac{dv_1}{dt} = 0 \tag{k}$$

Substitution of Equations (f) and (g) into Equation (b) leads to

$$\frac{v_1}{R_3} + C_2\frac{dv_O}{dt} = 0 \tag{l}$$

Equation (l) is rearranged to give

$$v_1 = -R_3C_2\frac{dv_O}{dt} \tag{m}$$

Substitution of Equation (m) into Equation (k) gives

$$\frac{v_i}{R_1} - \frac{R_3C_2}{R_2}\frac{dv_O}{dt} - \frac{R_5}{R_4R_6}v_O - C_1C_2R_3\frac{d^2v_O}{dt^2} = 0 \tag{n}$$

Rearrangement of Equation (m) leads to

$$C_1C_2R_3\frac{d^2v_O}{dt^2} + \frac{R_3C_2}{R_2}\frac{dv_O}{dt} + \frac{R_5}{R_4R_6}v_O = \frac{v_i}{R_1} \tag{o}$$

The time-dependent output from the operational amplifier circuit is the solution to the differential equation, Equation (o).

Example 5.38

Derive a mathematical model for the generator of Figure 5.49. The rotation of the shaft provides a voltage v_b to the circuit.

Solution

Application of KCL at node B leads to

$$i_1 - i_2 - i_3 = 0 \tag{a}$$

Application of KVL around loop $ABEFA$ gives

$$-L\frac{di_1}{dt} - R_1 i_1 - R_2 i_3 - v_b = 0 \tag{b}$$

Application of KVL around loop $BCDEB$ yields

$$-R_3 i_2 + v_b + R_2 i_3 = 0 \tag{c}$$

Substituting Equation (a) into Equations (b) and (c) and rearranging leads to

$$L\frac{di_1}{dt} + (R_1 + R_2)i_1 - R_2 i_2 + v_b = 0 \tag{d}$$

$$-R_2 i_1 + (R_2 + R_3)i_2 = v_b \tag{e}$$

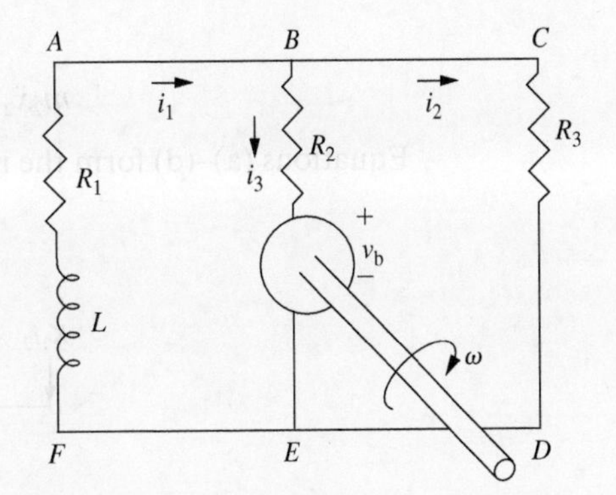

Figure 5.49

The generator circuit of Example 5.38.

Example 5.39

Derive a mathematical model for the electromechanical system of Figure 5.50. The magnetic coupling field leads to a force $K_1 i_2$ applied to the end of the lever, and the motion of the plunger through the magnetic field leads to an induced electromotive force of $K_2\dot{y}$, where $\dot{y}$ is the velocity of the plunger.

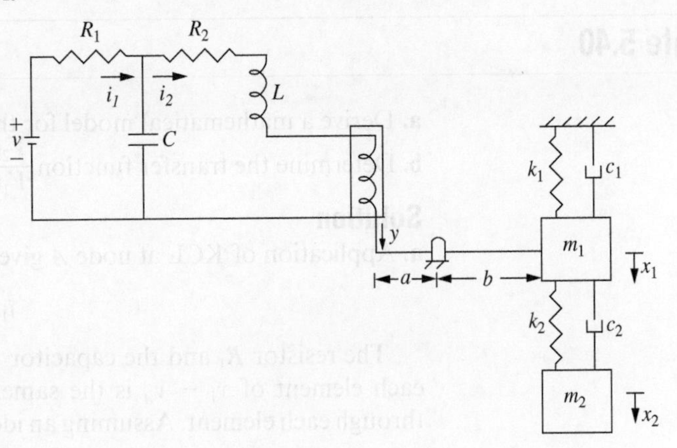

Figure 5.50

The system of Example 5.39.

Solution

Application of KVL around the loops of the circuit, noting that $y = \frac{b}{a}x_1$, leads to

$$R_1 i_1 + \frac{1}{C}\int_0^t i_1 dt - \frac{1}{C}\int_0^t i_2 dt = v \tag{a}$$

$$-\frac{1}{C}\int_0^t i_1 dt + L\frac{di_2}{dt} + R_2 i_2 + \frac{1}{C}\int_0^t i_2 dt - K_2\frac{b}{a}\dot{x}_1 = 0 \tag{b}$$

Summing moments on the free-body diagram of Figure 5.51(a) assuming small displacements gives

$$\sum M_O = I_O \alpha$$

$$-K_1 i_2 a - (k_1 x_1 + c_1 \dot{x}_1)b + [k_2(x_2 - x_1) + c_2(\dot{x}_2 - \dot{x}_1)]b = m_1 b^2\left(\frac{\ddot{x}_1}{b}\right)$$

$$m_1\ddot{x}_1 + (c_1 + c_2)\, b^2\dot{x}_1 - c_2 b^2\dot{x}_2 + (k_1 + k_2)\, b^2 x_1 - k_2 b^2 x_2 + K_1 ab i_2 = 0 \tag{c}$$

Summing forces on the free-body diagram of Figure 5.51(b) gives

$$\sum F = m_2\ddot{x}_2$$

$$-k_2(x_2 - x_1) - c_2(\dot{x}_2 - \dot{x}_1) = m_2\ddot{x}_2$$

$$m_2\ddot{x}_2 - c_2\dot{x}_1 + c_2\dot{x}_2 - k_2 x_1 + k_2 x_2 = 0 \tag{d}$$

Equations (a)–(d) form the mathematical model for this system.

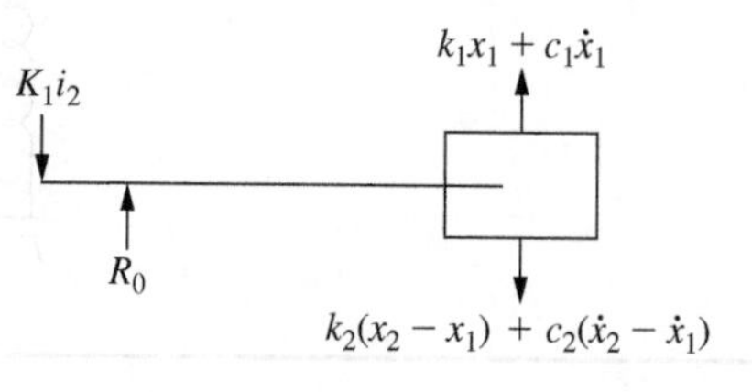

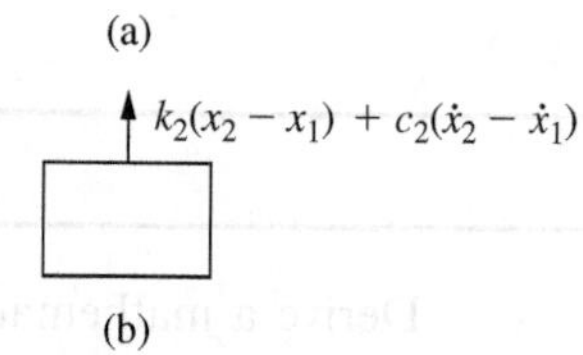

Figure 5.51

The free-body diagrams for system of Example 5.39.

Example 5.40

a. Derive a mathematical model for the operational amplifier circuit of Figure 5.52.

b. Determine the transfer function $\frac{V_2(s)}{V_1(s)}$.

Solution

a. Application of KCL at node A gives

$$i_1 - i_2 = 0 \tag{a}$$

The resistor R_1 and the capacitor C_1 are in parallel. Thus the voltage drop across each element of $v_1 - v_A$ is the same, and the current i_1 is the sum of the currents through each element. Assuming an ideal operational amplifier $v_A = 0$, the total current

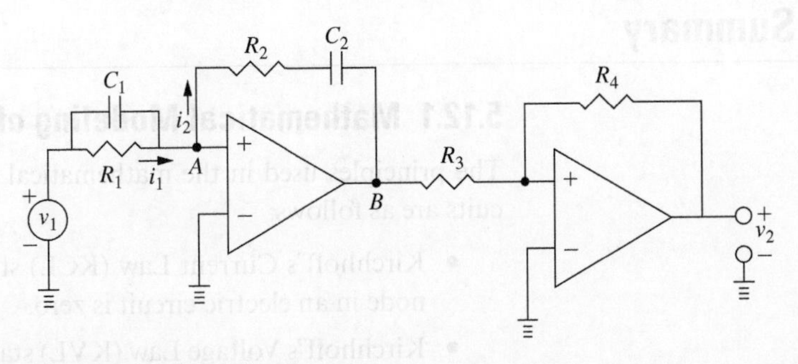

Figure 5.52
The operational amplifier circuit of Example 5.40.

through the parallel combination is

$$i_1 = \frac{v_1}{R_1} + C_1\frac{dv_1}{dt} \tag{b}$$

The resistor R_2 and the capacitor C_2 are in series. Thus the current through each element i_2 is the same and the total voltage drop $-v_B$ is the sum of the voltage drops across each element. Thus

$$-v_B = R_2 i_2 + \frac{1}{C_2}\int_0^t i_2\, dt \tag{c}$$

Use of Equations (a) and (b) in Equation (c) leads to

$$-v_B = R_2\left(\frac{v_1}{R_1} + C_1\frac{dv_1}{dt}\right) + \frac{1}{C_2}\int_0^t \left(\frac{v_1}{R_1} + C_1\frac{dv_1}{dt}\right)dt$$

$$= C_1 R_2\frac{dv_1}{dt} + \left(\frac{R_2}{R_1} + \frac{C_1}{C_2}\right)v_1 + \frac{1}{R_1 C_2}\int_0^t v_1\, dt \tag{d}$$

The second operational amplifier acts as an inverter such that

$$v_2 = -\frac{R_4}{R_3}v_B \tag{e}$$

Substitution of Equation (d) into Equation (e) gives

$$v_2 = \frac{C_1 R_2 R_4}{R_3}\frac{dv_1}{dt} + \left(\frac{R_2 R_4}{R_1 R_3} + \frac{C_1 R_4}{C_2 R_3}\right)v_1 + \frac{R_4}{C_2 R_1 R_3}\int_0^t v_1\, dt \tag{f}$$

It is shown in Chapter 9 that this operational amplifier circuit functions as a proportional plus integral plus derivative controller.

b. Taking the Laplace transforms of Equations (a), (b), (c), and (e) leads to

$$I_1(s) = I_2(s) \tag{g}$$

$$I_1(s) = \left(\frac{1}{R_1} + C_1 s\right)V_1(s) \tag{h}$$

$$V_B(s) = -\left(R_2 + \frac{1}{C_2 s}\right)I_2(s) \tag{i}$$

and

$$V_2(s) = -\frac{R_4}{R_3}V_B(s) \tag{j}$$

Combining Equations (g)–(j) leads to

$$V_2(s) = \frac{R_4}{R_3}\left(C_1 R_2 s + \frac{R_2}{R_1} + \frac{C_2}{C_1} + \frac{1}{C_2 R_1 s}\right)V_1(s) \tag{k}$$

Equation (k) is rearranged, leading to a transfer function of

$$G(s) = \frac{R_4}{C_2 R_1 R_3}\left[\frac{C_1 R_2 s^2 + \left(\frac{R_2}{R_1} + \frac{C_2}{C_1}\right)s + 1}{s}\right] \tag{l}$$

5.12 Summary

5.12.1 Mathematical Modeling of Electrical Systems

The principles used in the mathematical modeling of lumped parameter electric circuits are as follows:

- Kirchhoff's Current Law (KCL) states that the sum of the currents entering a node in an electric circuit is zero.
- Kirchhoff's Voltage Law (KVL) states that the sum of the voltage drops around any closed loop in a circuit is zero.
- Use of KVL and KCL on lumped parameter circuits leads to differential and integrodifferential equations whose independent variable is time and whose dependent variables are mesh currents or voltage drops across circuit components.

5.12.2 Other Chapter Highlights

- Electric current is a result of the motion of charged particles.
- Electric potential (voltage) is the work required to move a charged particle through an electric field.
- Passive circuit components are capacitors that store energy in an electric field, inductors that store energy in an electric field, and resistors that dissipate energy.
- Active circuit components are voltage sources, current sources, and operational amplifiers.
- Kirchhoff's laws are statements of Conservation of Charge and Conservation of Energy.
- Circuit components are in series if they have the same current and the total voltage drop across the combination is the sum of the individual voltage drops.
- Circuit components are in parallel if each component has the same voltage drop and the total current through the combination is the sum of the currents through individual components.
- Components in series or parallel can be replaced by a single component with an equivalent parameter.
- The differential equations modeling circuits are analogous to the differential equations governing mechanical systems. The direct analogies for circuits with voltage sources are that resistors are analogous to viscous dampers, capacitors are analogous to springs, inductors are analogous to inertia elements, charge is analogous to displacement, and current is analogous to velocity.
- Operational amplifiers have external voltage sources. An ideal amplifier has no current and the voltage at each terminal is the same.
- Electromechanical systems have an electrical system and a mechanical system coupled through a coupling field. The coupling field provides an electrostatic or electromagnetic force that acts on the mechanical system.
- Field-controlled dc servomotors have a coupling torque proportional to the current in the field circuit.
- Armature-controlled dc servomotors have a coupling torque proportional to the current in the armature circuit and an induced back emf.

- Transfer functions for an electrical system are developed using the methods of Chapter 3.
- A state-space formulation is developed from the mathematical model using the methods of Chapter 3.
- Charge and its time derivatives are usually chosen as state variables when a voltage source is the input.

5.12.3 Important Equations

- Electric power

$$P = vi \tag{5.7}$$

- Ohm's law

$$v = iR \tag{5.8}$$

- Voltage-current relation in a capacitor

$$i = C\frac{dv}{dt} \tag{5.18}$$

- Voltage-current relation in an inductor

$$v = L\frac{di}{dt} \tag{5.25}$$

- Series LRC circuit

$$L_{eq}\frac{di}{dt} + R_{eq}i + \frac{1}{C_{eq}}\int_0^t i\,dt = v(t) \tag{5.50}$$

- Electrostatic or electromagnetic force in coupling field

$$F_f = -\frac{\partial E_c}{\partial x} \tag{5.81}$$

- Transfer function for a series LRC circuit

$$G(s) = \frac{s/L}{s^2 + \frac{R}{L}s + \frac{1}{CL}} \tag{5.90}$$

Short Answer Problems

State whether the statements in Short Answer Problems SA5.1–SA5.10 are true or false. If false, rewrite the statement to make it true.

SA5.1 Resistivity is a material property.

SA5.2 Resistance is a material property.

SA5.3 Inductance is a material property.

SA5.4 The mutual inductance between two inductors is measured in henrys.

SA5.5 The current in the wire between points A and B enters the wire at A. The potential is greater at B.

SA5.6 A magnetic field exists between the plates of a parallel plate capacitor.

SA5.7 In a circuit, current is necessary to store energy in a capacitor.

SA5.8 In a circuit, current is necessary to store energy in an inductor.

SA5.9 In a circuit, a statement of conservation of charge leads to Kirchhoff's Current Law.

SA5.10 The voltage change across two circuit elements in parallel is the same.

Short Answer Problems SA5.11–SA5.23 require a short answer.

SA5.11 What are the dimensions of watts using the mass-length-time-temperature-current system?

SA5.12 What are the dimensions of watts using the force-length-time-temperature-current system?

SA5.13 What are the dimensions of farads using the mass-length-time-temperature-current system?

SA5.14 What are the dimensions of webers using the force-length-time-temperature-current system?

SA5.15 What are the dimensions of resistivity using the mass-length-time-temperature-current system?

SA5.16 Particles *A*, *B*, and *C* are colinear. Particle *A* has a negative charge, while particles *B* and *C* have a positive charge. What is the direction of the force between (a) *A* and *B*, (b) *A* and *C*, and (c) *B* and *C*?

SA5.17 Particle *C* is twice as far from particle *A* as from particle *B*. The force between particles *A* and *C* is what fraction or multiple of the force between *A* and *B*?

SA5.18 Define in words the term "voltage."

SA5.19 Define in words an electric field.

SA5.20 What do *n* capacitors in series have in common?

SA5.21 What do *n* resistors in series have in common?

SA5.22 The operation amplifier circuit of Figure SA5.22 serves as what type of controller?

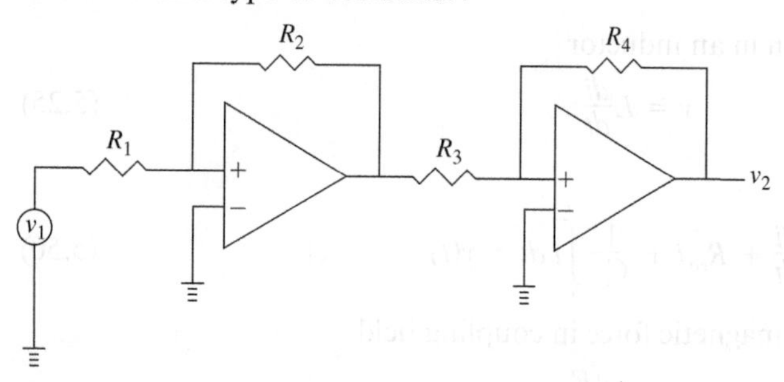

Figure SA5.22

SA5.23 The operation amplifier circuit of Figure SA5.23 serves as what type of controller?

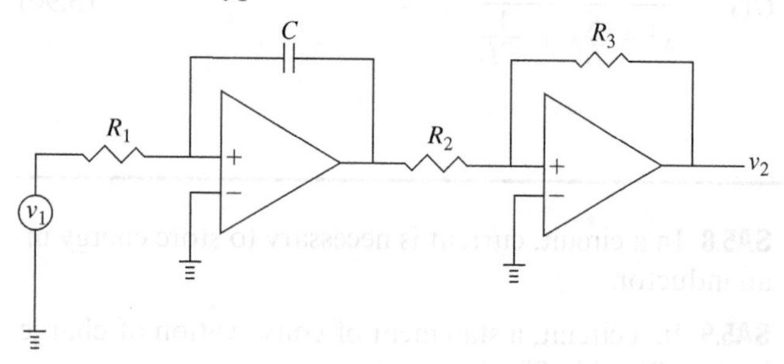

Figure SA5.23

Short Answer Problems SA5.24–SA5.54 require a numerical answer.

SA5.24 What is the charge on a particle that has an excess of 5 electrons?

SA5.25 Particle *A* has an excess of 5 electrons. Particle *B* has a deficiency of 10 electrons. The particles are a distance of 3 μm from each other. What is the force of attraction between the particles?

SA5.26 What is the electric field at *O* in Figure SA5.26 due to the presence of particles *A* and *B*? What is the direction of the electric field at *O*?

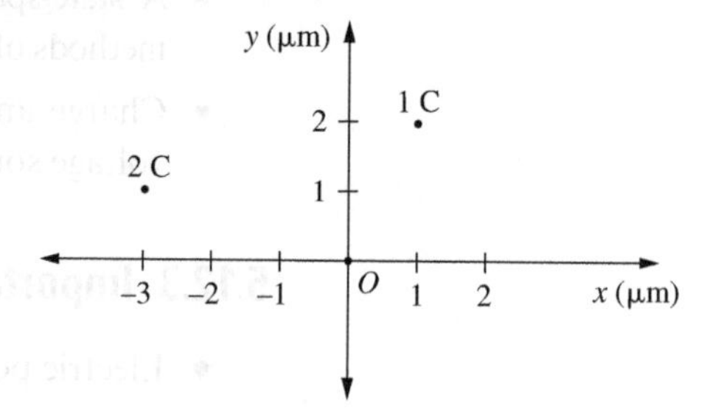

Figure SA5.26

SA5.27 Particles that have a total charge of 2.4×10^{-5} C move by point *A* in 1.2×10^{-3} s. What is the current at *A*?

SA5.28 There exists an electric potential of 3 V at point *A*. How much work is required to move a charged particle of 1.2 C through the electric field at *A*?

SA5.29 A battery has a potential of 5 V and a current of 10 A. How much power does the battery deliver?

SA5.30 A wire has a resistance of 10 Ω. A potential of 20 V exists at the entrance to the wire. The current passing through the wire is 1.5 A. What is the potential at the exit of the wire?

SA5.31 What is the power dissipated by the wire described in Problem SA5.30?

SA5.32 What is the capacitance of a parallel plate capacitor that has a plate of area 1.6×10^{-6} m². The plates are 3.2×10^{-3} m apart.

SA5.33 Three wires meet at a junction. The currents in two of the wires are 2.3 A leaving the junction and 2.1 A entering the junction. What is the current in the third wire? Is the current entering the junction or leaving the junction?

SA5.34 The potential provided to the circuit of Figure SA5.34 is 10 V. What is the current in the circuit?

Figure SA5.34

SA5.35 Three resistors are in series. The individual resistances are 100 Ω, 80 Ω, and 150 Ω. What is their equivalent resistance?

SA5.36 Three resistors are in parallel. The individual resistances are 100 Ω, 80 Ω, and 150 Ω. What is their equivalent resistance?

SA5.37 Three capacitors are in series. Their individual capacitances are 20 μF, 40 μF, and 100 μF. What is the equivalent capacitance?

SA5.38 Three capacitors are in parallel. Their individual capacitances are 20 μF, 40 μF, and 100 μF. What is the equivalent capacitance?

SA5.39 What is the order of the transfer function $G(s) = \frac{I(s)}{V(s)}$ for the circuit of Figure SA5.39?

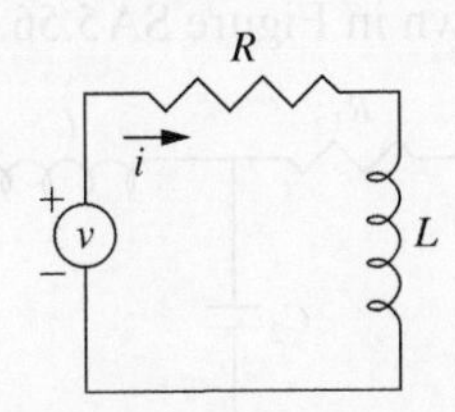

Figure SA5.39

SA5.40 What is the order of the transfer function $G(s) = \frac{I(s)}{V(s)}$ for the circuit of Figure SA5.40?

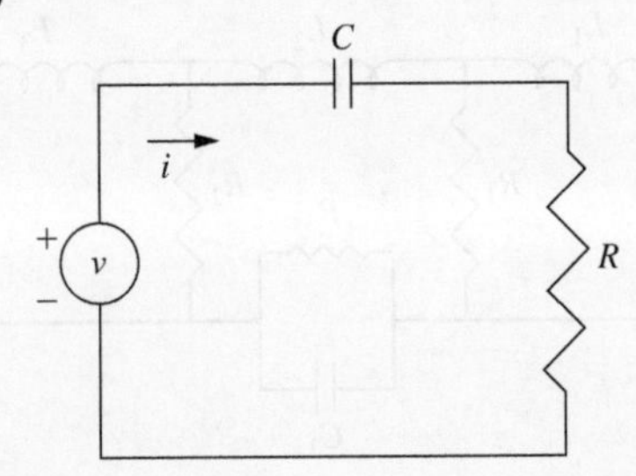

Figure SA5.40

SA5.41 What is the order of the transfer function $G(s) = \frac{I_2(s)}{V(s)}$ for the circuit of Figure SA5.41?

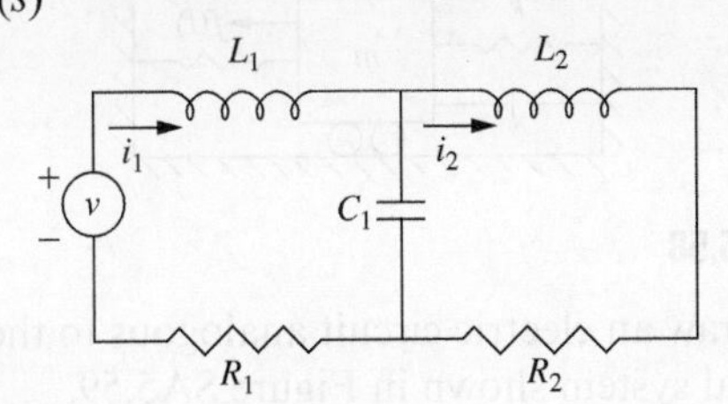

Figure SA5.41

SA5.42 What is the order of the transfer function $G(s) = \frac{V(s)}{I(s)}$ for the circuit of Figure SA5.42 where $v(t) = v_B - v_E$?

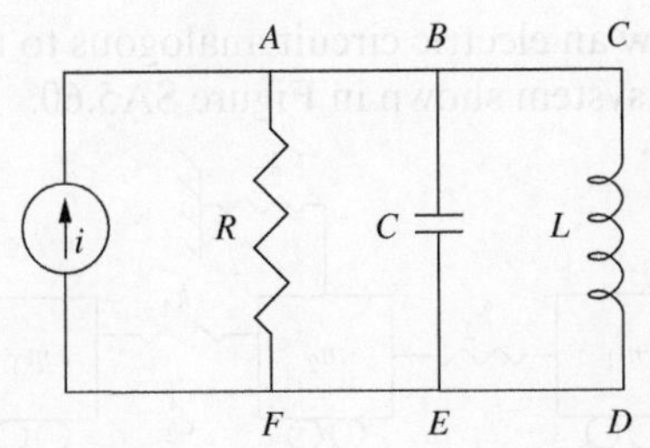

Figure SA5.42

SA5.43 What is the order of the transfer function $G(s) = \frac{I_3(s)}{V_2(s)}$ for the circuit of Figure SA5.43?

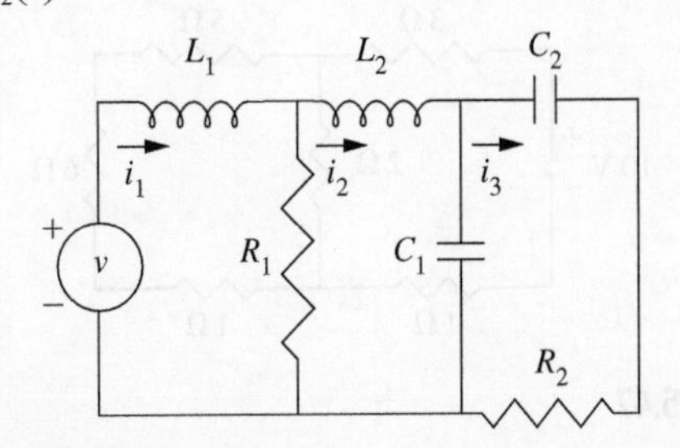

Figure SA5.43

SA5.44 In a state-space formulation for the mathematical model of the circuit of Figure SA5.44, how many state variables are required?

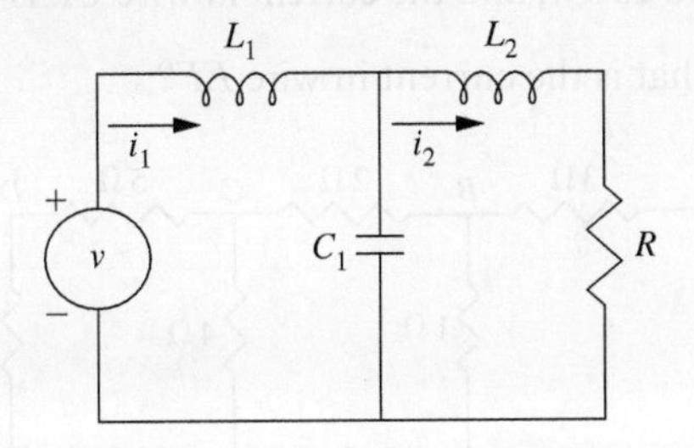

Figure SA5.44

SA5.45 In a state-space formulation for the mathematical model of the circuit of Figure SA5.45, how many state variables are required?

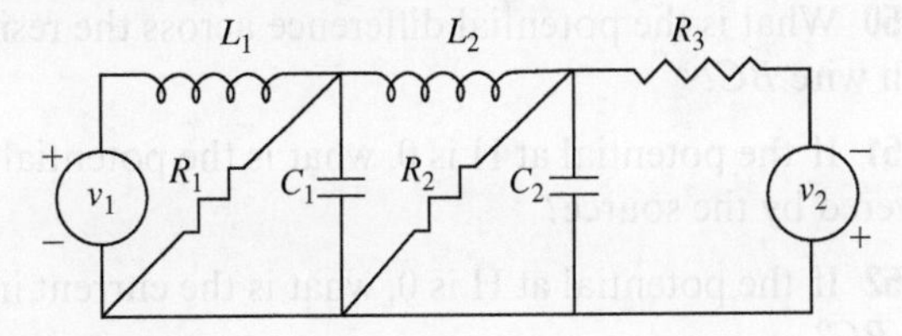

Figure SA5.45

SA5.46 In a state-space formulation for the mathematical model of the circuit of Figure SA5.46, how many state variables are required?

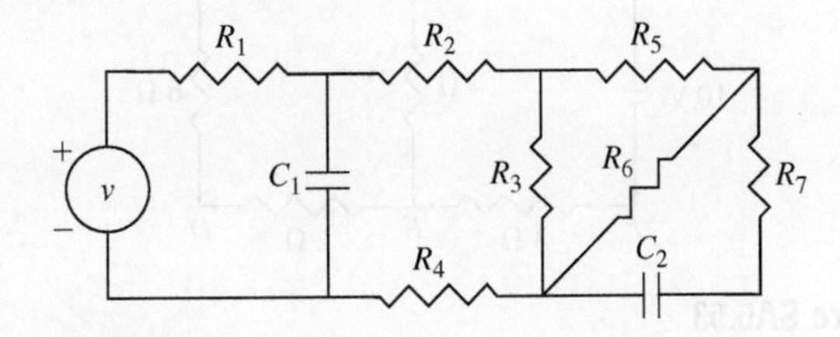

Figure SA5.46

SA5.47 What are the currents in each of the wires in the circuit shown in Figure SA5.47?

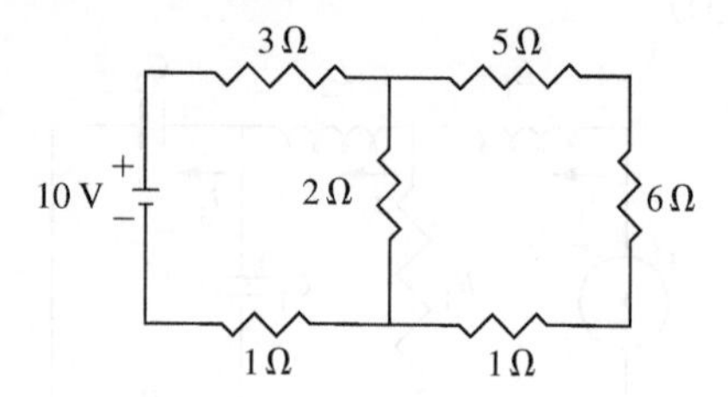

Figure SA5.47

Short Answer Problems SA5.48–SA5.52 refer to Figure SA5.48. The potential difference between D and E is measured as 3 V, and the current in wire CF is 2 A.

SA5.48 What is the current in wire EF?

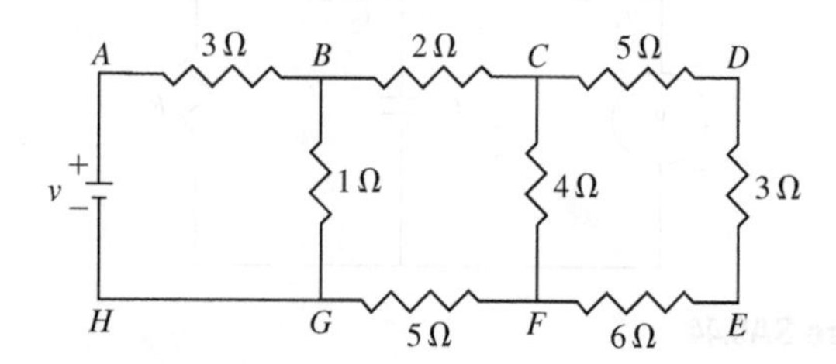

Figure SA5.48

SA5.49 What is the current in wire FG?

SA5.50 What is the potential difference across the resistor in wire BC?

SA5.51 If the potential at H is 0, what is the potential delivered by the source?

SA5.52 If the potential at H is 0, what is the current in wire BG?

SA5.53 What are the potentials at each of the nodes in the circuit of Figure SA5.53?

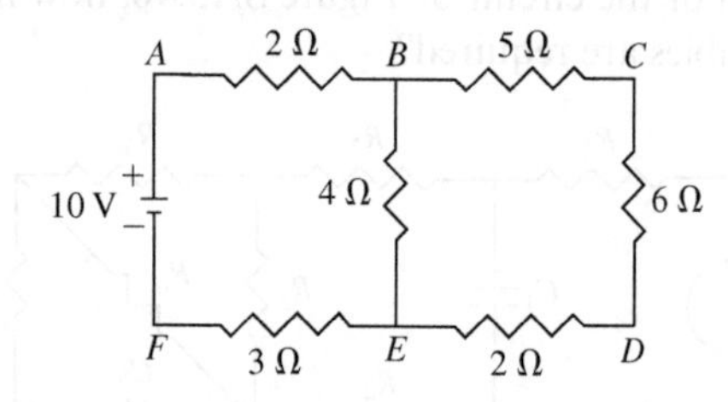

Figure SA5.53

SA5.54 What are the currents in each of the wires in the circuit shown in Figure SA5.53?

SA5.55 Draw a mechanical system analogous to the electric circuit shown in Figure SA5.55.

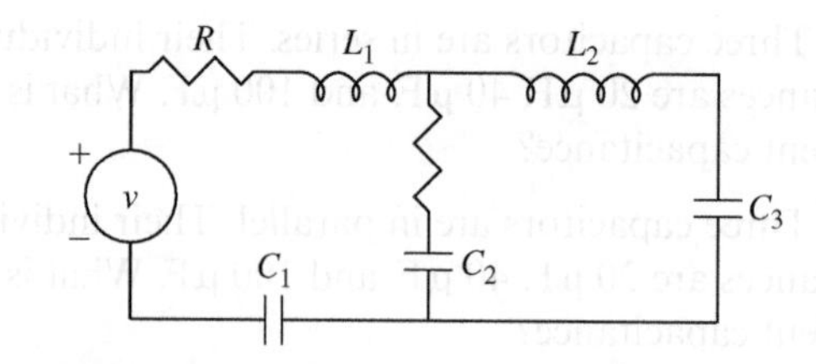

Figure SA5.55

SA5.56 Draw a mechanical system analogous to the electric circuit shown in Figure SA5.56.

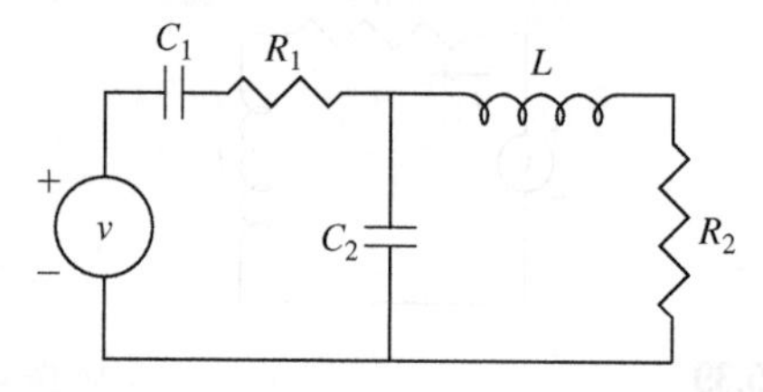

Figure SA5.56

SA5.57 Draw a mechanical system analogous to the electric circuit shown in Figure SA5.57.

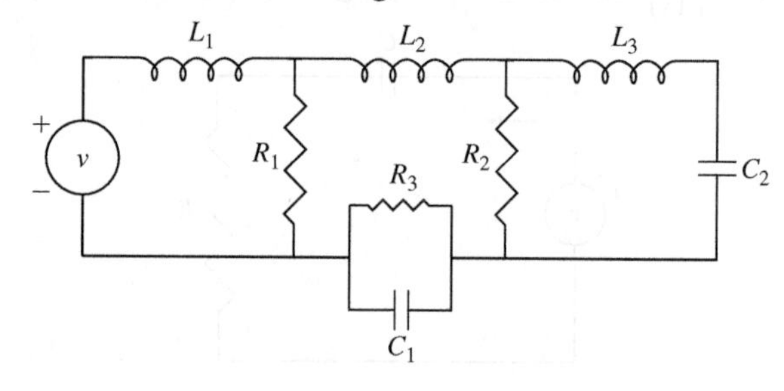

Figure SA5.57

SA5.58 Draw an electric circuit analogous to the mechanical system shown in Figure SA5.58.

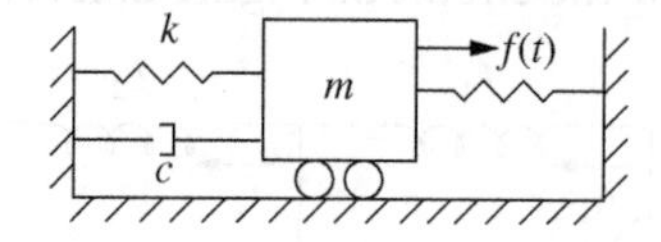

Figure SA5.58

SA5.59 Draw an electric circuit analogous to the mechanical system shown in Figure SA5.59.

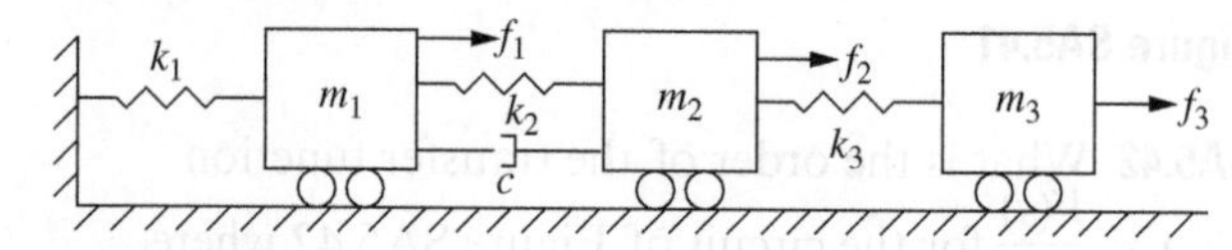

Figure SA5.59

SA5.60 Draw an electric circuit analogous to the mechanical system shown in Figure SA5.60.

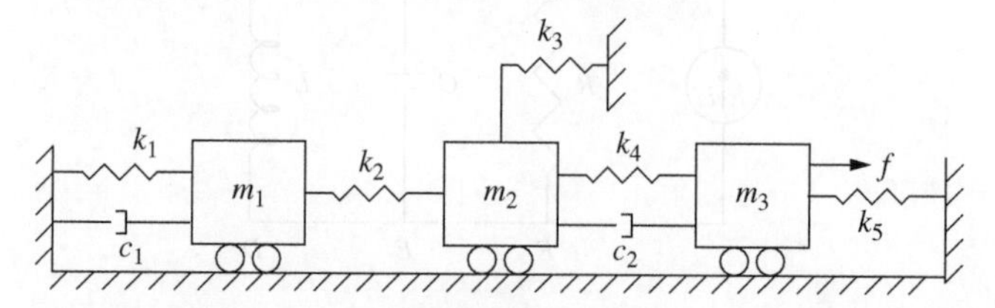

Figure SA5.60

SA5.61 Draw an electric circuit analogous to the mechanical system shown in Figure SA5.61.

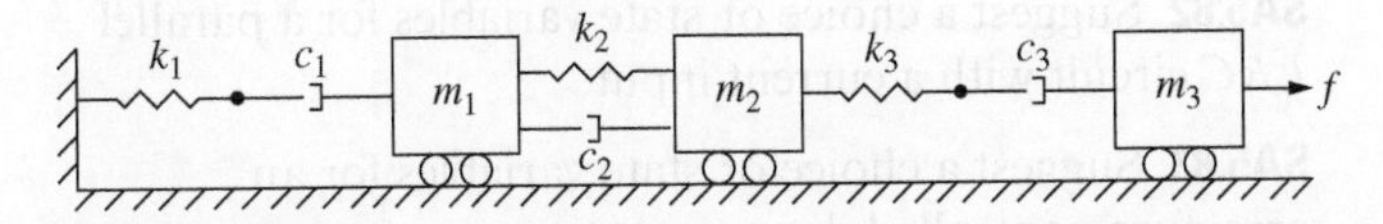

Figure SA5.61

SA5.62 What is the mathematical model for the current in the circuit of Figure SA5.62?

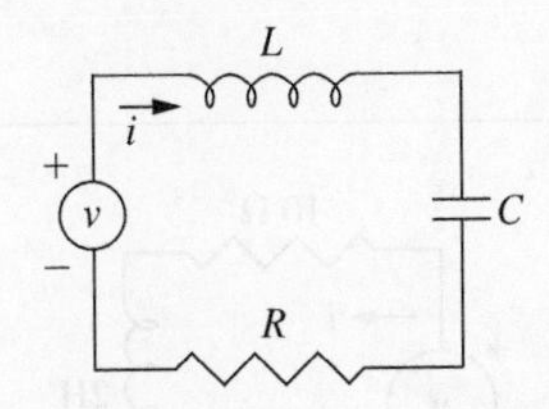

Figure SA5.62

SA5.63 What is the mathematical model for the current in the circuit of Figure SA5.63?

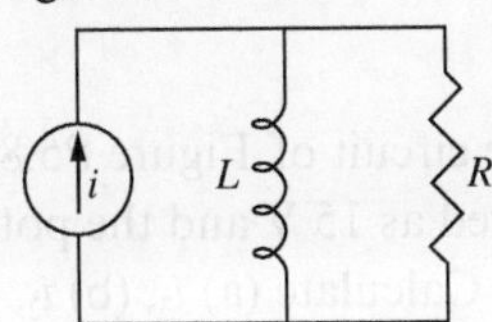

Figure SA5.63

All operational amplifiers in Short Answer Problems SA5.64–SA5.73 are ideal.

SA5.64 What is the potential at node A in the amplifier circuit in Figure SA5.64?

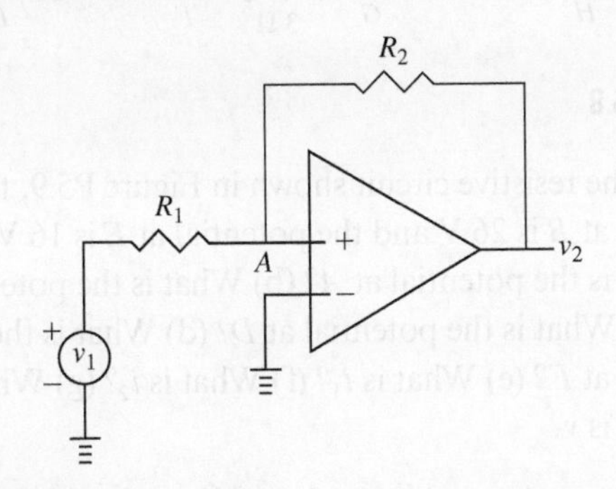

Figure SA5.64

SA5.65 What is the current entering node A from the amplifier in Figure SA5.64?

SA5.66 What is the current entering node A from the resistor of resistance R_1 in Figure SA5.64?

SA5.67 What is v_2 in the amplifier circuit in Figure SA5.64?

SA5.68 What is the potential at node A in the amplifier circuit in Figure SA5.68?

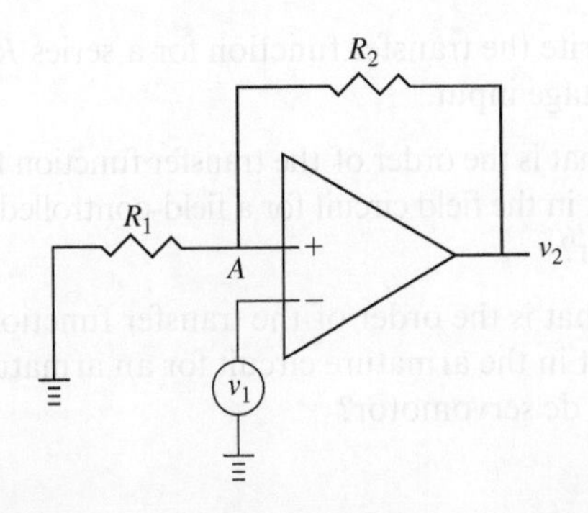

Figure SA5.68

SA5.69 What is the current entering node A from the amplifier in Figure SA5.68?

SA5.70 What is the current entering node A from the resistor of resistance R_1 in Figure SA5.68?

SA5.71 What is v_2 in the amplifier circuit in Figure SA5.68?

SA5.72 What is v_0 in the amplifier circuit of Figure SA5.72?

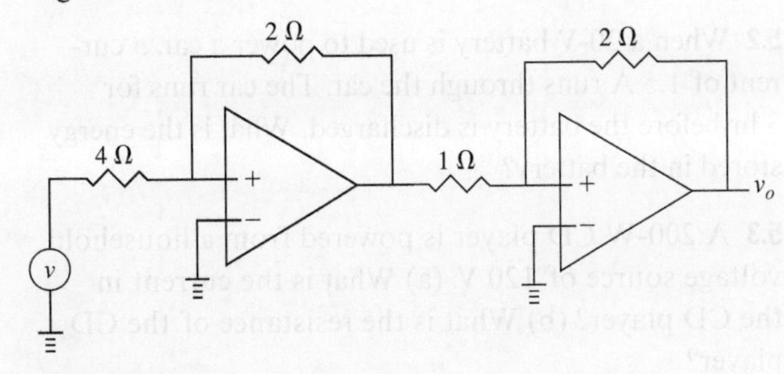

Figure SA5.72

SA5.73 What is v_0 in the amplifier circuit of Figure SA5.73?

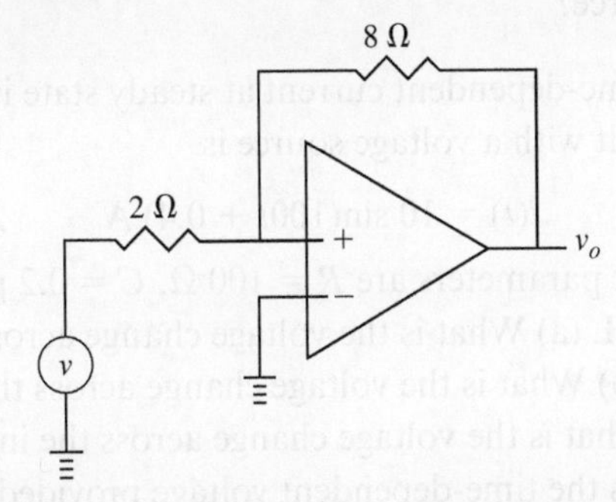

Figure SA5.73

SA5.74 Write the mathematical model for a series RC circuit with a voltage input.

SA5.75 Write the mathematical model for a series LRC circuit with a voltage input.

SA5.76 Write the mathematical model for a parallel LRC circuit with a current input.

SA5.77 Write the transfer function for a parallel LRC circuit with a current input.

SA5.78 Write the transfer function for a series *RC* circuit with a voltage input.

SA5.79 What is the order of the transfer function for the current in the field circuit for a field-controlled dc servomotor?

SA5.80 What is the order of the transfer function for the current in the armature circuit for an armature-controlled dc servomotor?

SA5.81 Suggest a choice of state variables for a series *LRC* circuit with a voltage input.

SA5.82 Suggest a choice of state variables for a parallel *LRC* circuit with a current input.

SA5.83 Suggest a choice of state variables for an armature-controlled dc servomotor.

Problems

5.1 (a) What is the required separation distance for a parallel plate capacitor of plate area 0.5 m² such that its capacitance is 0.05 μF when glass ($K = 4.6$) is used as a dielectric? (b) What is the charge on the capacitor when it is connected to a 12-V battery? (c) What is the energy stored in the capacitor when it is connected to a 12-V battery?

5.2 When a 20-V battery is used to power a car, a current of 1.5 A runs through the car. The car runs for 3 hr before the battery is discharged. What is the energy stored in the battery?

5.3 A 200-W CD player is powered from a household voltage source of 120 V. (a) What is the current in the CD player? (b) What is the resistance of the CD player?

5.4 What is the required resistance of a heating coil that generates 1500 kJ of heat per hour when connected to a 120-V source?

5.5 The time-dependent current at steady state in a series *RLC* circuit with a voltage source is

$$i(t) = 10 \sin(100t + 0.4) \text{ A}$$

The circuit parameters are $R = 100\ \Omega$, $C = 0.2\ \mu\text{F}$, and $L = 0.25$ H. (a) What is the voltage change across the resistor? (b) What is the voltage change across the capacitor? (c) What is the voltage change across the inductor? (d) What is the time-dependent voltage provided by the source?

5.6 For the circuit described in Problem 5.5, calculate and plot as a function of time (a) the energy stored in the capacitor, (b) the energy stored by the inductor, and (c) the energy dissipated by the resistor.

5.7 The current through the circuit of Figure P5.7 is $i(t) = 10 \sin 100t$ mA. At a time when the current is equal to 10 mA, determine (a) the power dissipated by the resistor, (b) the energy stored in the capacitor, and (c) the energy stored in the inductor.

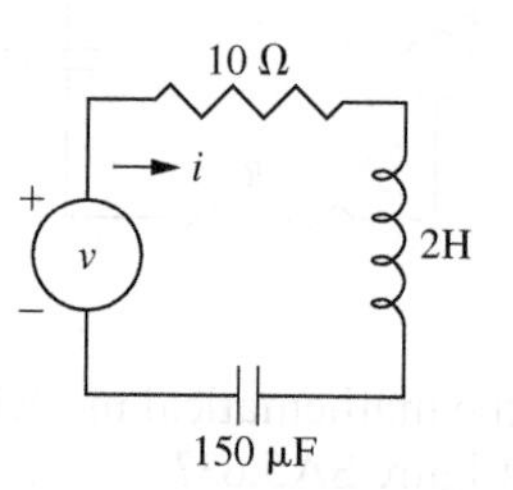

Figure P5.7

5.8 In the resistive circuit of Figure P5.8, the potential at *C* is measured as 15 V and the potential at *D* is measured as 11 V. Calculate (a) i_3, (b) i_2, (c) i_1, (d) v, and (e) P, the power delivered by the circuit.

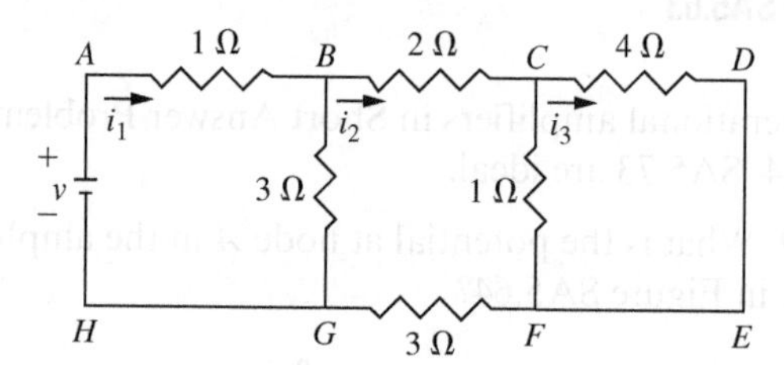

Figure P5.8

5.9 For the resistive circuit shown in Figure P5.9, the potential at *B* is 26 V and the potential at *E* is 16 V. (a) What is the potential at *A*? (b) What is the potential at *C*? (c) What is the potential at *D*? (d) What is the potential at *F*? (e) What is i_1? (f) What is i_2? (g) What is i_3? (h) What is v?

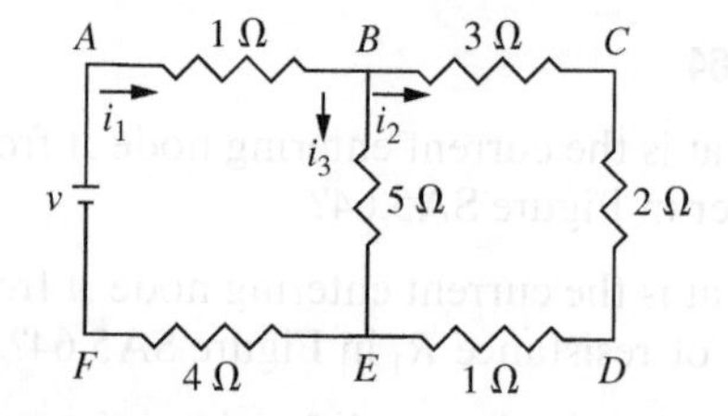

Figure P5.9

5.10 A resistive circuit is illustrated in Figure P5.10. The potential at point *A* is measured as 28 V, and the potential at point *B* is measured as 25 V. Fill out the table accompanying the figure.

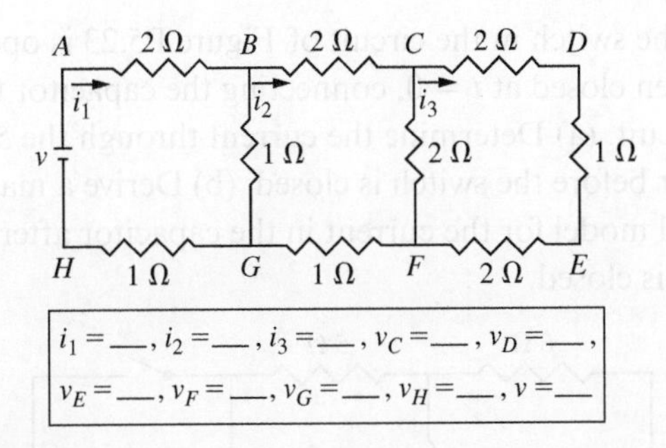

Figure P5.10

5.11 A resistive circuit is illustrated in Figure P5.11. The potential at point A is measured as 12 V, and the potential at point F is measured as 6 V. Fill out the table accompanying the figure.

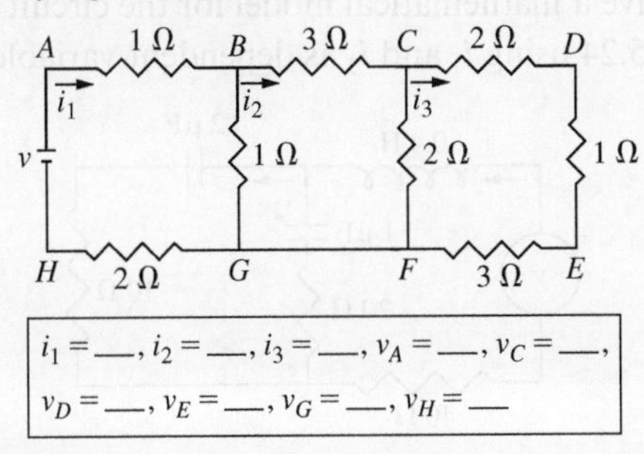

Figure P5.11

5.12 (a) Determine the current through each resistor in the resistive circuit of Figure P5.12. (b) Determine the voltage drop across each resistor. (c) Determine the power in each resistor.

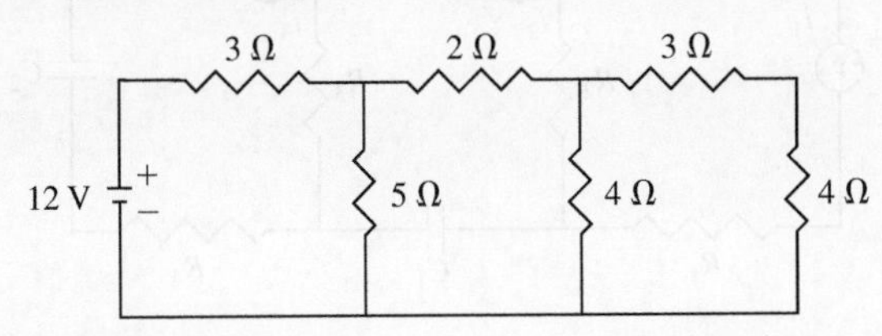

Figure P5.12

5.13 (a) Replace the resistors in the circuit of Figure P5.12 by a single resistor of an equivalent resistance. (b) Determine the power delivered to the circuit.

5.14 Replace the resistors in the circuit of Figure P5.14 by a single resistor of an equivalent resistance.

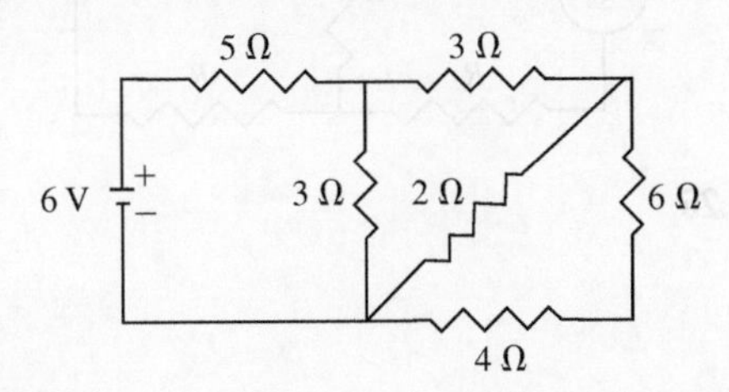

Figure P5.14

5.15 A 10-V battery serves as input to the circuit of Figure P5.15. The output from the circuit is to be the voltage across the 11-Ω resistor. What is the voltage measured across the resistor?

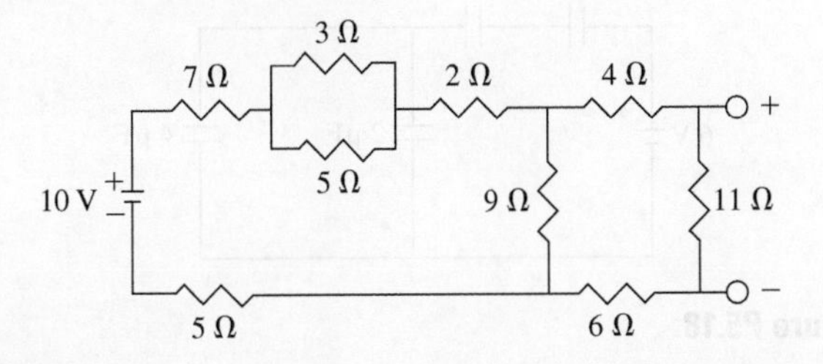

Figure P5.15

5.16 A Wheatstone bridge is a resistive circuit that is used in strain gauge transducers. A typical Wheatstone bridge circuit is shown in Figure P5.16. The resistor whose resistance is labeled as R_4 is the strain gauge. (a) Determine the output voltage v_o. (b) Derive a condition between the resistances such that the output voltage is zero. (c) Explain how this circuit serves as a strain gauge, noting that the strain in a wire is the ratio of its change in length to its original length.

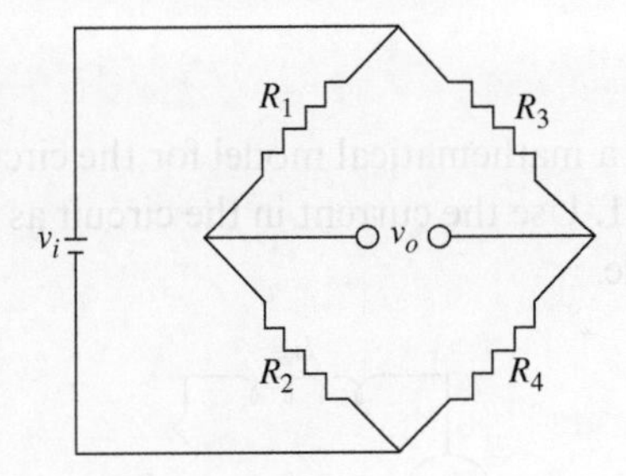

Figure P5.16

5.17 A circuit for a potentiometer is shown in Figure P5.17. The output is the voltage change measured across the resistor of resistance R_2. Determine this voltage in terms of R_1, R_2, and v_1.

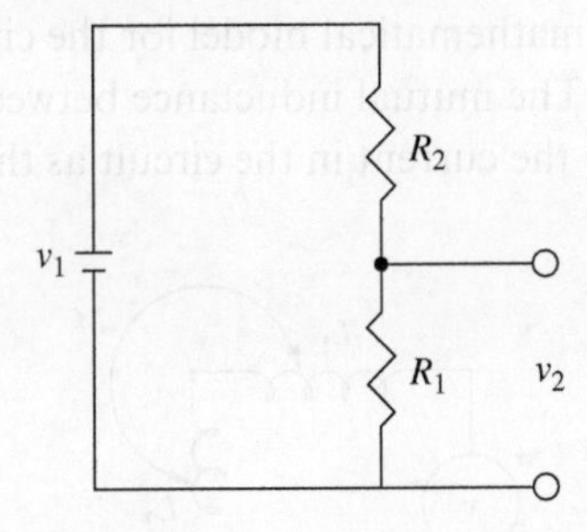

Figure P5.17

5.18 (a) Determine the voltage in each of the capacitors in the circuit of Figure P5.18. (b) Determine the charge on each of the capacitors.

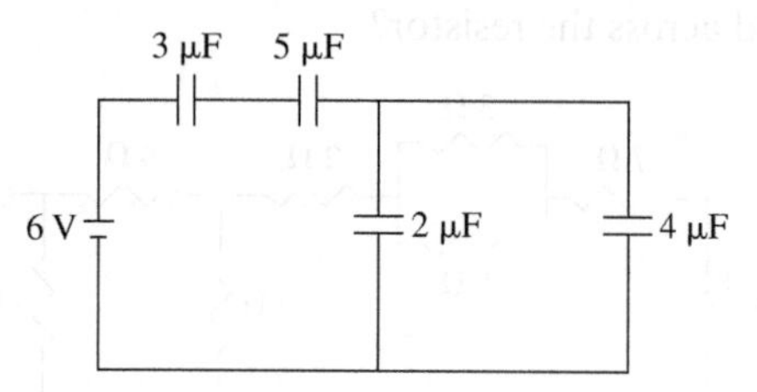

Figure P5.18

5.19 Replace the capacitors in the circuit of Figure P5.18 with a single capacitor of an equivalent capacitance.

5.20 Derive a mathematical model for the *RC* circuit of Figure P5.20. Use the voltage drop across the resistor as the dependent variable.

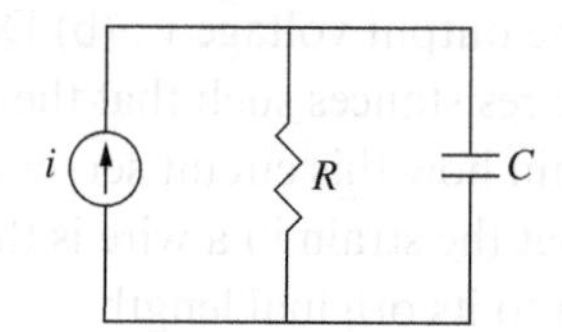

Figure P5.20

5.21 Derive a mathematical model for the circuit of Figure P5.21. Use the current in the circuit as the dependent variable.

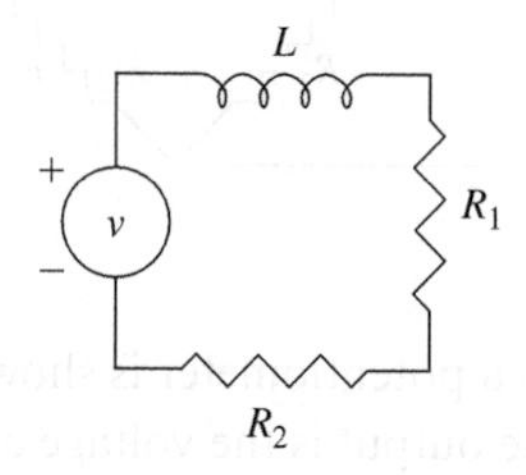

Figure P5.21

5.22 Derive a mathematical model for the circuit of Figure P5.22. The mutual inductance between the inductors is M. Use the current in the circuit as the dependent variable.

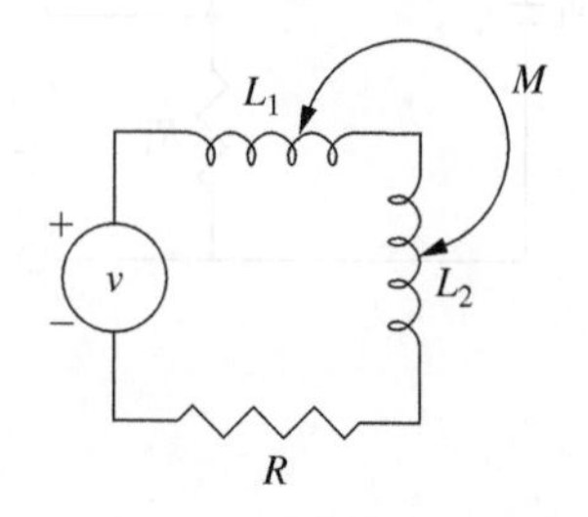

Figure P5.22

5.23 The switch in the circuit of Figure P5.23 is open and then closed at $t = 0$, connecting the capacitor to the circuit. (a) Determine the current through the 5-Ω resistor before the switch is closed. (b) Derive a mathematical model for the current in the capacitor after the switch is closed.

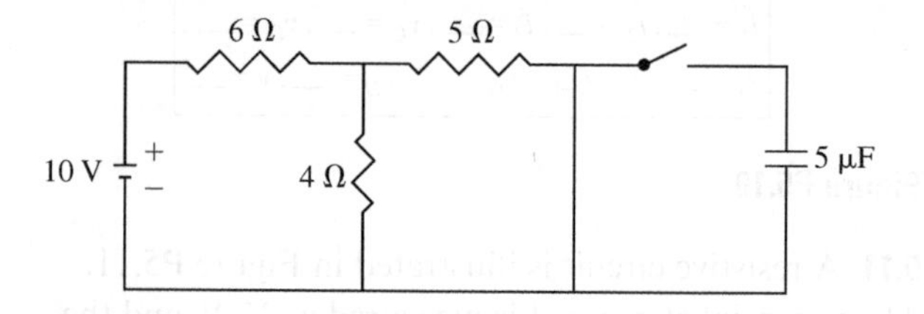

Figure P5.23

5.24 Derive a mathematical model for the circuit of Figure P5.24 using i_1 and i_2 as dependent variables.

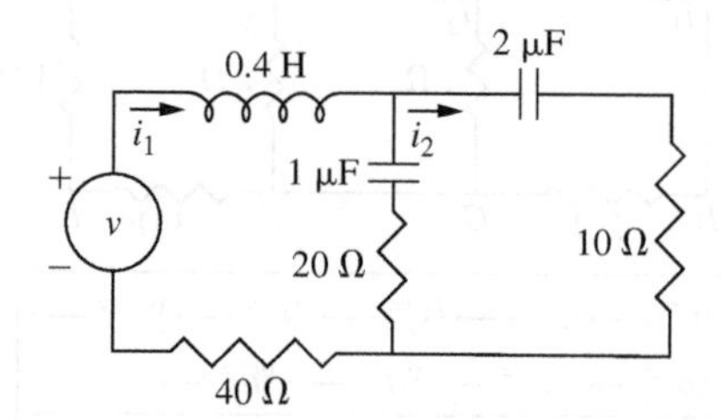

Figure P5.24

5.25 Derive a mathematical model for the circuit of Figure P5.25 using i_1, i_2, and i_3 as dependent variables.

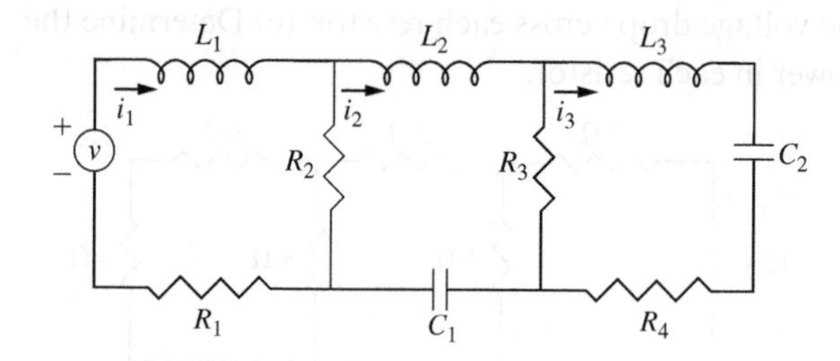

Figure P5.25

5.26 Derive a mathematical model for the circuit of Figure P5.26 using i_1 and i_2 as dependent variables.

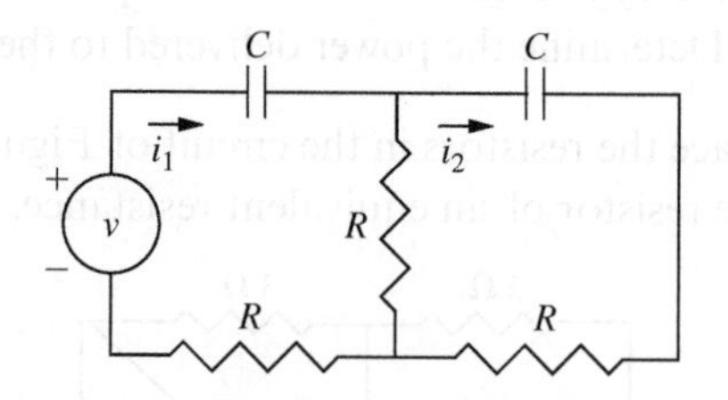

Figure P5.26

5.27 Derive a mathematical model for the circuit of Figure P5.27 using i_1 and i_2 as dependent variables.

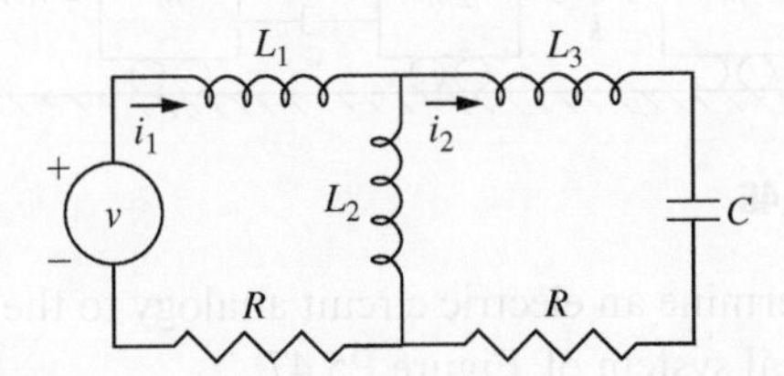

Figure P5.27

5.28 Derive a mathematical model for the circuit of Figure P5.28. Use the current through the inductor as the dependent variable.

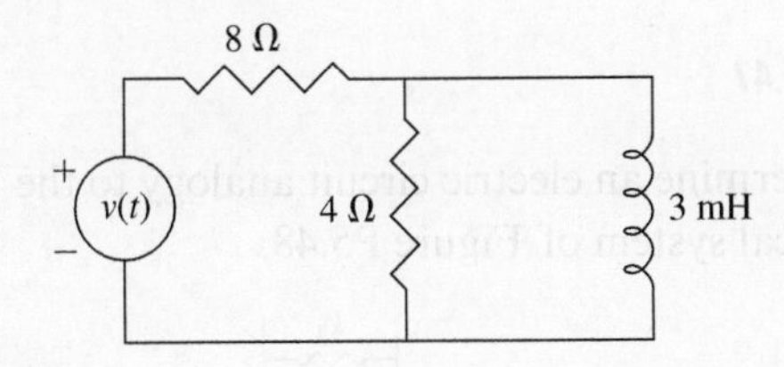

Figure P5.28

5.29 Derive a mathematical model for the circuit of Figure P5.29. Use the current through the inductor as the dependent variable.

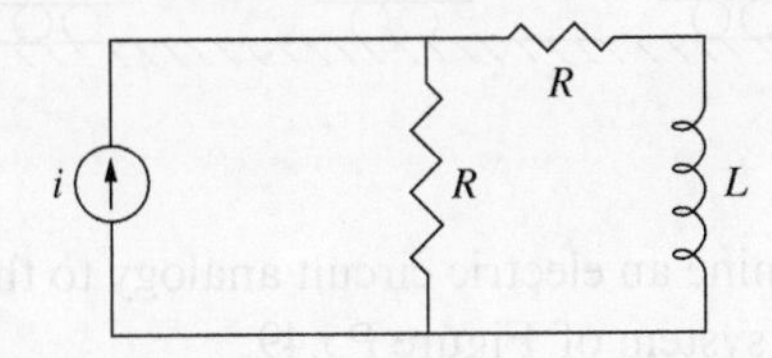

Figure P5.29

5.30 Derive a mathematical model for the circuit of Figure P5.30. Use i_1, i_2, and i_3 as the dependent variables.

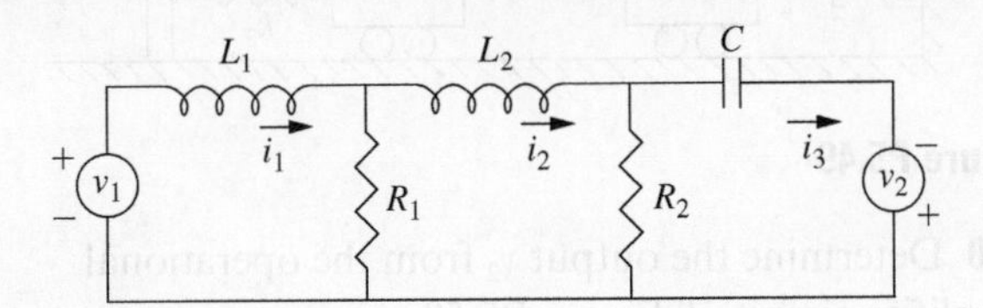

Figure P5.30

5.31 The capacitor C_1 in the circuit of Figure P5.31 has an initial charge q while the capacitor C_2 is uncharged. Derive a mathematical model for the current in the circuit after the switch is closed at $t = 0$. Include the appropriate initial condition.

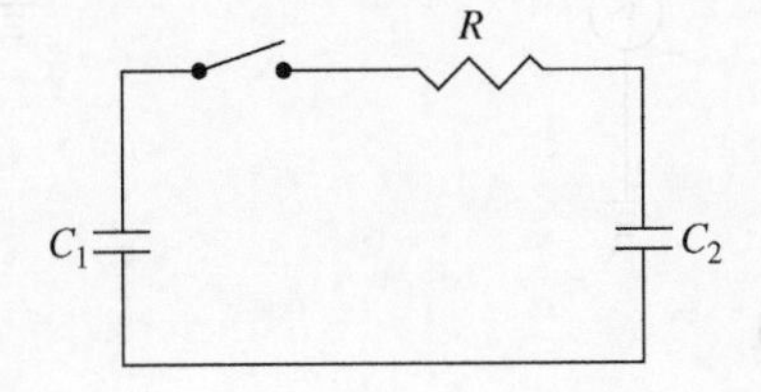

Figure P5.31

5.32 Derive a mathematical model for the circuit of Figure P5.32. Use the currents through the resistors as the dependent variables.

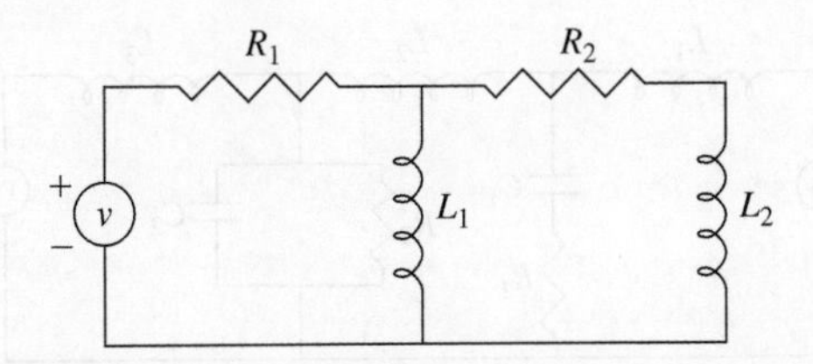

Figure P5.32

5.33 Derive a mathematical model for the circuit of Figure P5.33. Use the voltage across the 3-kΩ resistor and the current through the inductor as the dependent variables.

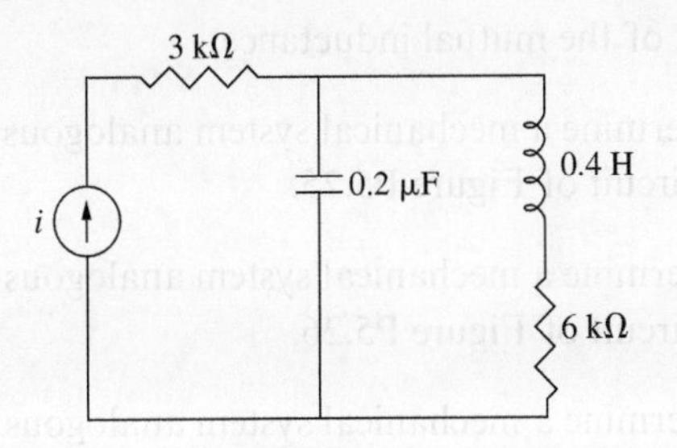

Figure P5.33

5.34 Derive a mathematical model for the circuit of Figure P5.34. Use i_1 and i_2 as the dependent variables.

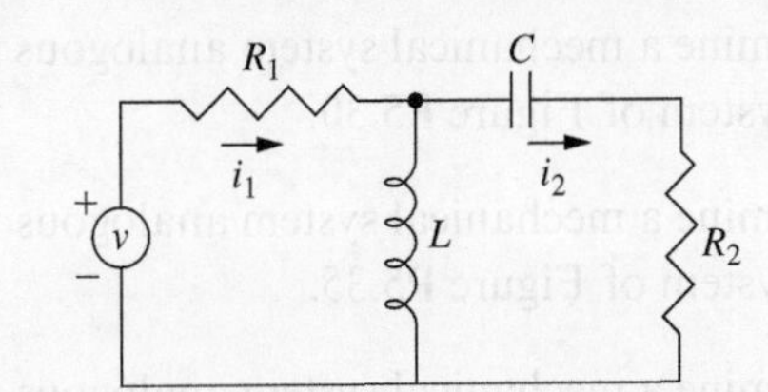

Figure P5.34

5.35 Derive a mathematical model for the circuit of Figure P5.35. Use the currents through the inductors as dependent variables.

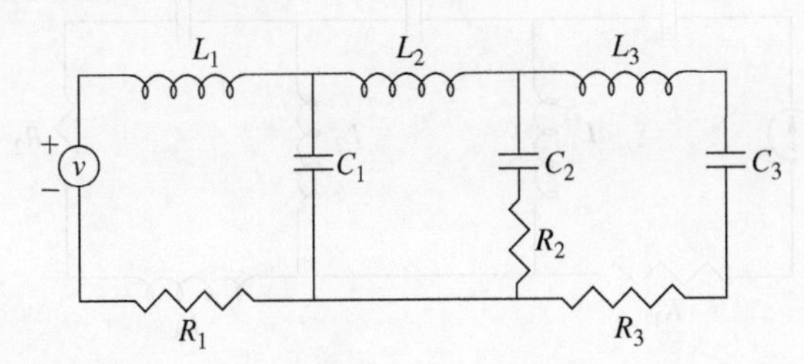

Figure P5.35

5.36 Derive a mathematical model for the circuit of Figure P5.36. Use the currents through the inductors as dependent variables.

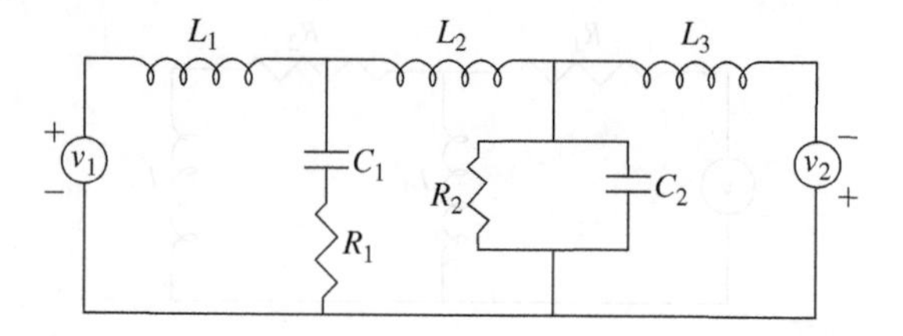

Figure P5.36

5.37 The mutual inductance between two adjacent inductors in the circuit of Figure P5.36 is M; the mutual inductance between nonadjacent inductors is zero. Derive a mathematical model for the circuit, including the effect of the mutual inductance.

5.38 Determine a mechanical system analogous to the electric circuit of Figure P5.25.

5.39 Determine a mechanical system analogous to the electric circuit of Figure P5.26.

5.40 Determine a mechanical system analogous to the electric circuit of Figure P5.27.

5.41 Determine a mechanical system analogous to the electric circuit of Figure P5.28.

5.42 Determine a mechanical system analogous to the electrical system of Figure P5.30.

5.43 Determine a mechanical system analogous to the electrical system of Figure P5.35.

5.44 Determine a mechanical system analogous to the electrical system of Figure P5.36.

5.45 Determine a mechanical system analogous to the electrical system of Figure P5.45.

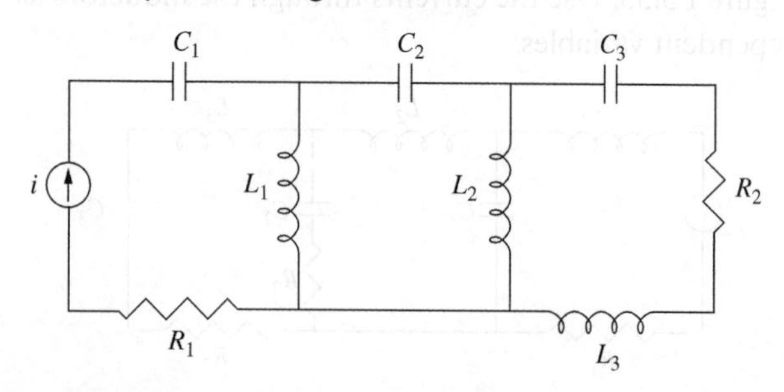

Figure P5.45

5.46 Determine an electric circuit analogy to the mechanical system of Figure P5.46.

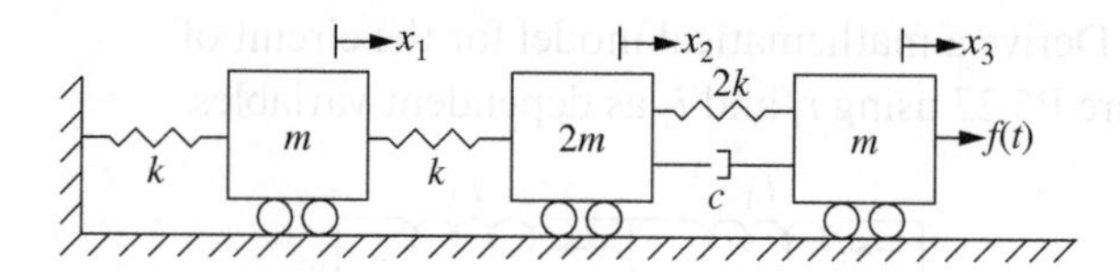

Figure P5.46

5.47 Determine an electric circuit analogy to the mechanical system of Figure P5.47.

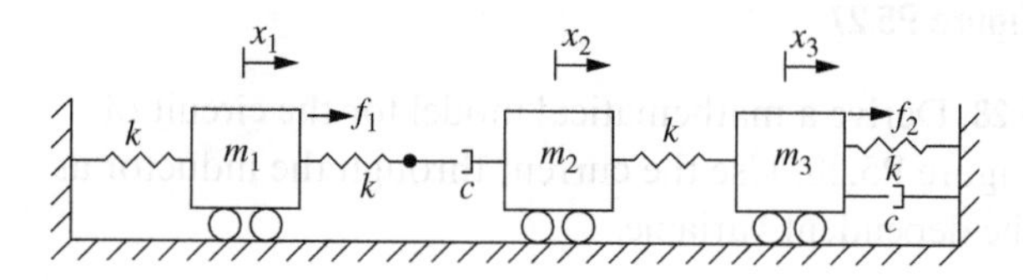

Figure P5.47

5.48 Determine an electric circuit analogy to the mechanical system of Figure P5.48.

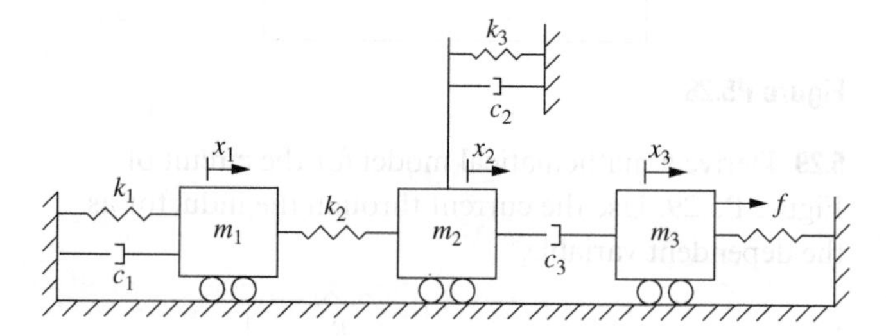

Figure P5.48

5.49 Determine an electric circuit analogy to the mechanical system of Figure P5.49.

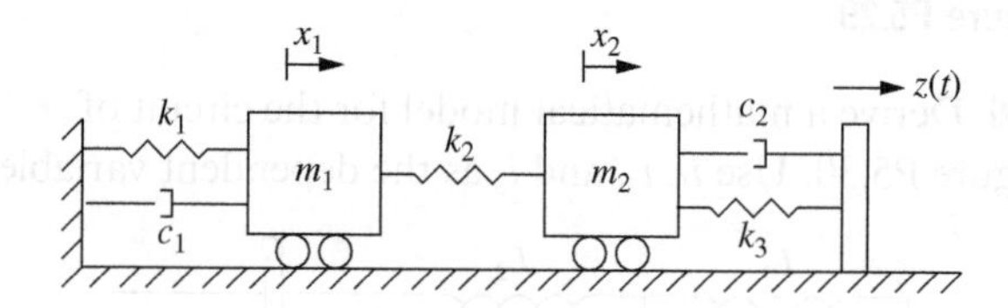

Figure P5.49

5.50 Determine the output v_2 from the operational amplifier circuit of Figure P5.50.

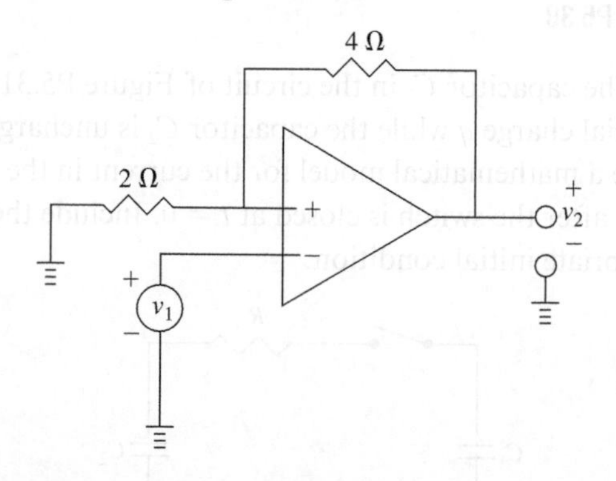

Figure P5.50

5.51 Determine the output v_2 from the operational amplifier circuit of Figure P5.51.

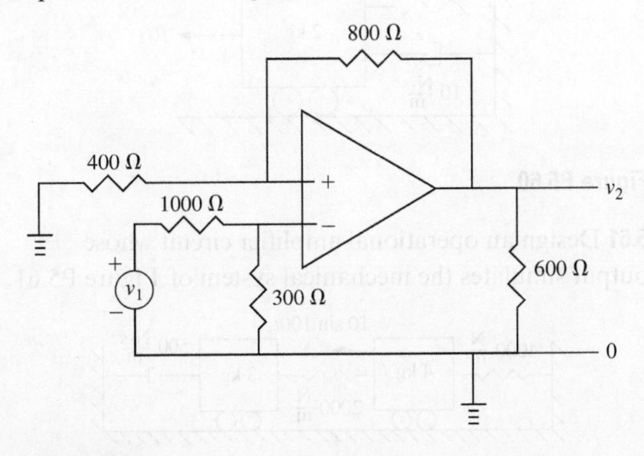

Figure P5.51

5.52 Determine the output v_3 from the operational amplifier circuit of Figure P5.52.

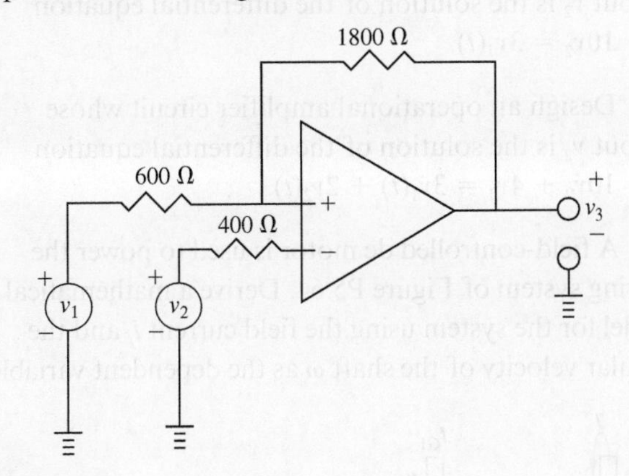

Figure P5.52

5.53 Determine the output v_2 from the operational amplifier circuit of Figure P5.53.

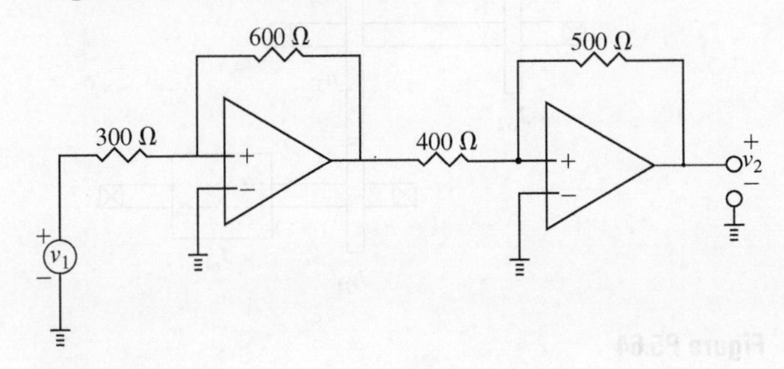

Figure P5.53

5.54 Determine the output v_2 from the operational amplifier circuit of Figure P5.54.

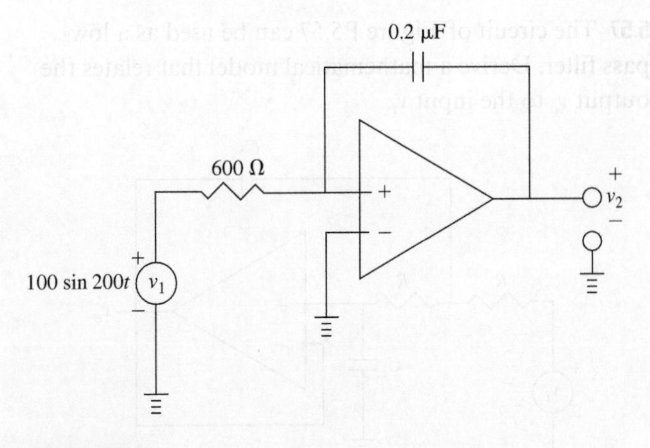

Figure P5.54

5.55 Determine the output v_2 from the operational amplifier circuit of Figure P5.55.

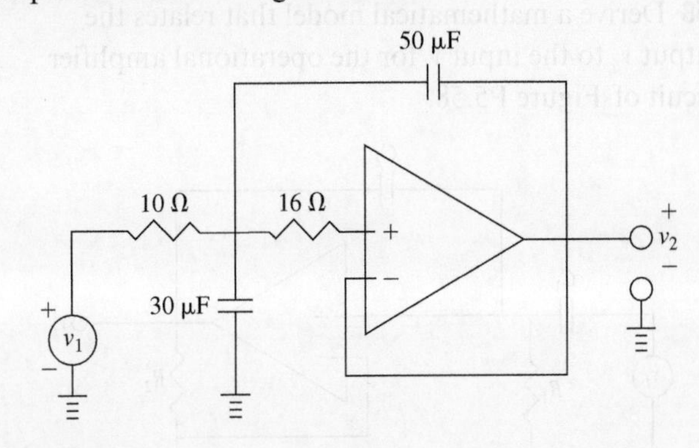

Figure P5.55

5.56 The circuit of Figure 5.56 is used in a capacitive proximity probe. Derive a mathematical model that relates the output v_o to the input v_i.

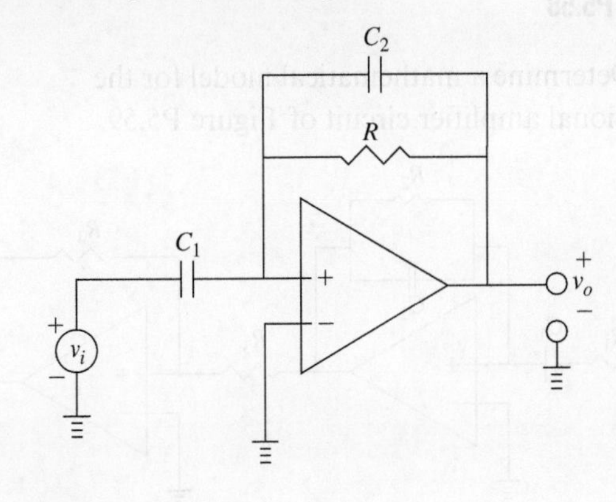

Figure P5.56

5.57 The circuit of Figure P5.57 can be used as a low-pass filter. Derive a mathematical model that relates the output v_o to the input v_i.

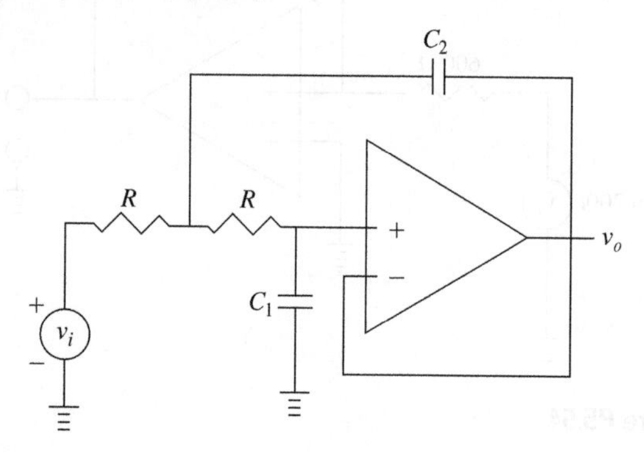

Figure P5.57

5.58 Derive a mathematical model that relates the output v_o to the input v_i for the operational amplifier circuit of Figure P5.58.

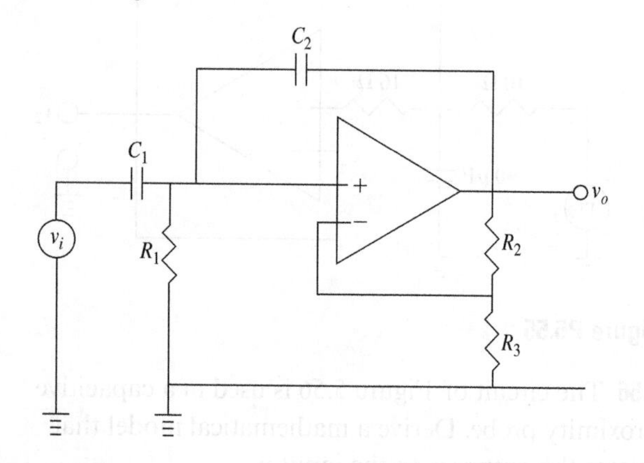

Figure P5.58

5.59 Determine a mathematical model for the operational amplifier circuit of Figure P5.59.

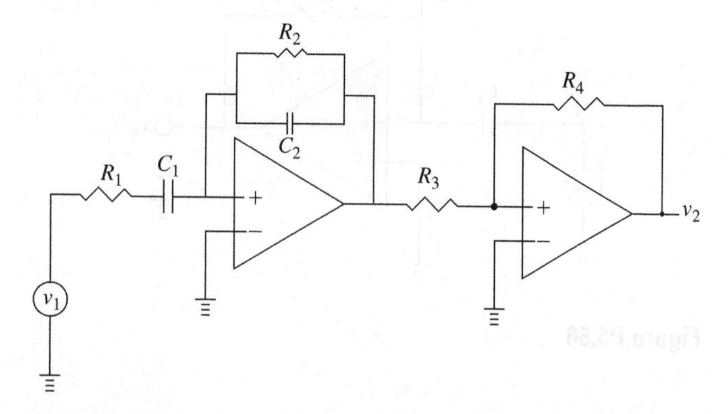

Figure P5.59

5.60 Design an operational amplifier circuit whose output v_2 simulates the response of the mechanical system of Figure P5.60.

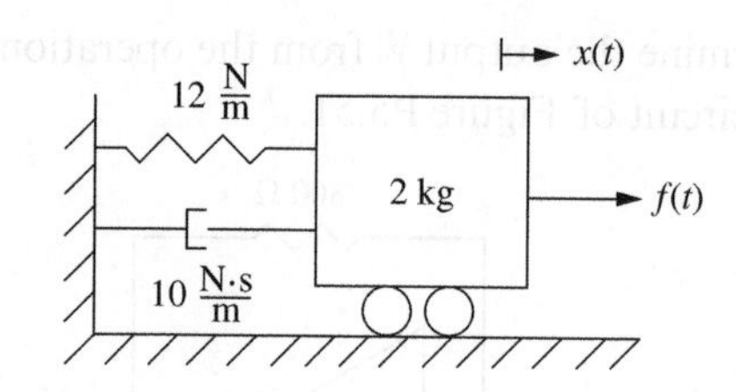

Figure P5.60

5.61 Design an operational amplifier circuit whose output simulates the mechanical system of Figure P5.61.

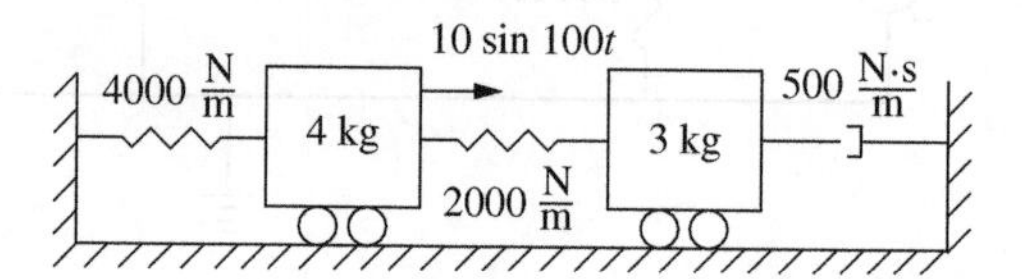

Figure P5.61

5.62 Design an operational amplifier circuit whose output v_2 is the solution of the differential equation $\dot{v}_2 + 10v_2 = 3v_1(t)$.

5.63 Design an operational amplifier circuit whose output v_2 is the solution of the differential equation $\ddot{v}_2 + 10\dot{v}_2 + 4v_1 = 3\dot{v}_1(t) + 2v_2(t)$.

5.64 A field-controlled dc motor is used to power the gearing system of Figure P5.64. Derive a mathematical model for the system using the field current i_f and the angular velocity of the shaft ω as the dependent variables.

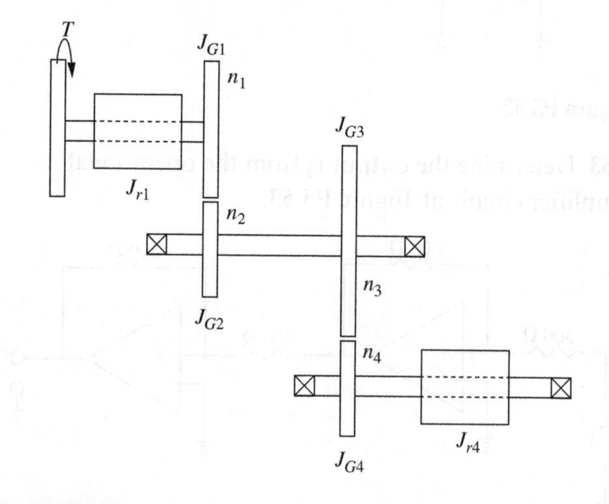

Figure P5.64

5.65 Repeat Problem 5.64 when an armature-controlled dc motor is used, with the armature current i_a and the angular velocity of the shaft ω as dependent variables.

5.66 Derive a mathematical model for the system of Problem 5.64 and Figure P5.64 when the shafts are supported by identical bearings, each with a torsional damping coefficient c_t and a torsional stiffness k_t.

5.67 Derive a mathematical model for the electromechanical system of Figure P5.67. Use i_1, i_2, x_1, and x_2 as dependent variables. The motion of the bar through the magnetic field of the inductor leads to a force F_c acting on the horizontal bar as well as a back emf e_b in the circuit.

5.68 Derive Equation (5.43).

5.69 Derive Equation (5.44).

5.70 Derive Equation (5.45).

5.71 Determine the transfer function matrix for the system of Problem 5.25.

5.72 Determine the transfer function $G(s) = \frac{I_2(s)}{V(s)}$ for the system of Problem 5.27.

5.73 Determine the transfer function $G(s) = \frac{I_2(s)}{V(s)}$ for the system of Problem 5.30.

5.74 Determine the transfer functions for the system of Problem 5.33.

5.75 Determine the transfer function $G(s) = \frac{I_2(s)}{V(s)}$ for the system of Problem 5.35.

5.76 Determine the transfer function $G(s) = \frac{V_2(s)}{V_1(s)}$ for the system of Problem 5.55.

5.77 Determine the transfer function $G(s) = \frac{V_o(s)}{V_i(s)}$ for the system of Problem 5.56.

5.78 Determine the transfer function $G(s) = \frac{V_o(s)}{V_i(s)}$ for the system of Problem 5.57.

5.79 Determine the transfer function $G(s) = \frac{V_o(s)}{V_i(s)}$ for the system of Problem 5.58.

5.80 Determine the transfer function $G(s) = \frac{V_2(s)}{V_1(s)}$ for the system of Problem 5.59.

5.81 Determine the transfer function $G(s) = \frac{X(s)}{V(s)}$ for the loudspeaker of Example 5.21.

5.82 Derive the transfer function $G(s) = \frac{I_2(s)}{V_b(s)}$ for the generator of Example 5.38.

5.83 Determine a state-space model from the mathematical model of the system of Problem 5.25.

5.84 Determine a state-space model from the mathematical model of the system of Problem 5.27.

5.85 Determine a state-space model from the mathematical model of the system of Problem 5.30.

5.86 Determine a state-space model from the mathematical model of the system of Problem 5.33.

5.87 Determine a state-space model from the mathematical model of the system of Problem 5.56.

5.88 Determine a state-space model from the mathematical model of the system of Problem 5.57.

5.89 Determine a state-space model from the mathematical model of the system of Problem 5.59.

5.90 Determine a state-space model for the loudspeaker of Example 5.21 when x is used as the output.

5.91 Determine a state-space model for the generator of Example 5.38 when the angular velocity of the shaft is used as the output.

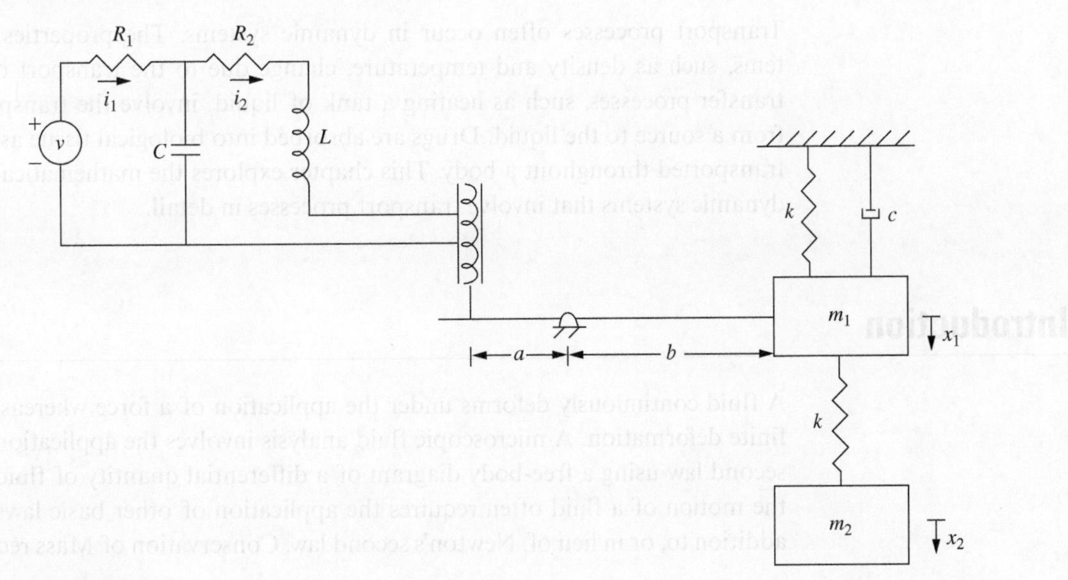

Figure P5.67

6

Fluid, Thermal, and Chemical Systems

Learning Objectives

At the conclusion of this chapter, you should be able to:

- List the assumptions in the mathematical modeling of transport systems
- Apply the control volume forms of Conservation of Mass and Conservation of Energy to problems involving transport processes
- Model problems that involve pipes by calculating the pipe resistance based on the head loss through the pipe
- Derive mathematical models for liquid-level systems
- Derive mathematical models for systems that include hydraulic servomotors
- Derive mathematical models for pneumatic systems
- Understand the concepts of thermal resistance and thermal capacitance
- Derive mathematical models for thermal systems
- Derive mathematical models for systems that include continuous stirred tank reactors (CSTR)
- Derive mathematical models for drug delivery systems

Transport processes often occur in dynamic systems. The properties of fluid systems, such as density and temperature, change due to the transport of mass. Heat transfer processes, such as heating a tank of liquid, involve the transport of energy from a source to the liquid. Drugs are absorbed into biological tissue as the drugs are transported throughout a body. This chapter explores the mathematical modeling of dynamic systems that involve transport processes in detail.

6.1 Introduction

A fluid continuously deforms under the application of a force whereas a solid has a finite deformation. A microscopic fluid analysis involves the application of Newton's second law using a free-body diagram of a differential quantity of fluid. Analysis of the motion of a fluid often requires the application of other basic laws of nature in addition to, or in lieu of, Newton's second law. Conservation of Mass requires that the mass of a system of fluid particles is constant. Conservation of Energy, a statement of the First Law of Thermodynamics, when applied includes forms of energy other than kinetic and potential. Dependent variables in fluid flow problems may include velocity components and fluid pressure.

The motion of a fluid is very complex and depends on many parameters. Many simplifying assumptions are used to develop models of fluid systems. An incompressible fluid is one in which the density is approximately constant for a wide variety of fluid conditions. Most liquids, such as water and oils, are assumed to be incompressible fluids. The density of a compressible fluid changes as other properties, such as pressure and temperature, change. An ideal gas such as air is an example of a compressible fluid.

Pipe flows and liquid-level problems are studied in Sections 6.3 and 6.4. Hydraulic systems, which use liquids, and pneumatic systems, which use gases, often serve as actuators for mechanical systems and are studied in Section 6.5.

A thermal system is defined as a system in which heat transfer occurs. The heat transfer may occur through conduction in a solid or through convection in a fluid. Heat transfer leads to a change in temperature, a thermodynamic potential, which is a dependent variable in thermal systems. Thermal systems are studied in Section 6.6.

A chemical system is any system in which a chemical reaction occurs such that one or more components are changed into other components. A Conservation of Mass equation that includes the effect of the chemical reaction is necessary for every component in the system. Component concentrations are typically used as dependent variables in chemical systems. Chemical reactions often require heat to occur or give off heat while occurring. Thus many chemical systems are also thermal systems. Many chemical systems also have fluid flow. Chemical systems involving the continuous stirred tank reactor (CSTR) are studied in Section 6.7.

The principles of transport processes also apply to many other physical systems. A biological systems example is presented in Section 6.8.

The analyses of fluid, thermal, and chemical systems in this chapter use **lumped parameter assumptions**; that is, all dependent variables and all properties are constant across all spatial coordinates; time is the only independent variable. Fluid properties (such as density and viscosity), thermal properties (such as thermal conductivity and specific heat), and chemical properties (such as rate of reaction) are all assumed to be constant in space but may vary with time. The lumped parameter assumption allows systems to be modeled using ordinary differential equations.

Consider a tank containing a liquid. The flow rate Q into the tank is equal to the flow rate out. Thus the volume of liquid in the tank is constant. For a tank of uniform area, the height of the column of liquid in the tank, called the liquid level, is constant. The system operates at a steady state in which all properties are constant. Perturbations of properties of the fluid entering the tank lead to dynamic responses of the properties of the fluid in the tank. A change in flow rate affects the liquid level in the tank; a change in inlet stream temperature causes a time-dependent change in temperature in the tank; and a change in the concentrations of the components of the incoming stream leads to changes in component concentrations in the tank. Whereas the properties of the incoming stream may be assumed to change instantaneously, the properties in the tank are continuous with time. Mathematical modeling of this system leads to ordinary differential equations whose solutions provide the time-dependent response of the dependent variables. Often perturbations in properties lead to a new steady state, but it is of interest to know how long it takes to reach the new steady state and the dynamic response until the steady state is reached.

In view of all this, most dependent variables are broken down into the original steady-state value plus a perturbation from the steady state. For example, if H represents the level of liquid in a tank that is changing with time due to a time dependent flow rate Q, then

$$H(t) = h_s + h(t) \tag{6.1a}$$

$$Q(t) = q_s + q(t) \tag{6.1b}$$

where $H(t)$ and $Q(t)$ are the total values of the quantities, h_s and q_s are the steady-state values, and $h(t)$ and $q(t)$ are the time-dependent perturbations from steady state. Often the differential equations for the total quantities are nonlinear. The equations are linearized, as shown in Section 1.6, by assuming that the perturbation quantities are small compared to the steady-state quantities and by using binominal or Taylor series expansions of the nonlinear terms about the steady state.

Lumped parameter modeling of fluid, thermal, and chemical systems requires a method of analysis different from that of mechanical and electrical systems. Instead of tracking the motion of a fixed set of particles, the changes occurring in a defined region in space, called a control volume, are tracked.

6.2 Control Volume Analysis

The term "system," as used so far in this book, refers to a set of integrated components designed to achieve a certain objective. A system converts an input into a desired output. An alternate usage of the word "system" is applied for the mathematical modeling of fluid, thermal, and chemical problems. This alternate meaning is introduced to provide a framework for such analysis, but it retains its former definition throughout the remainder of the book.

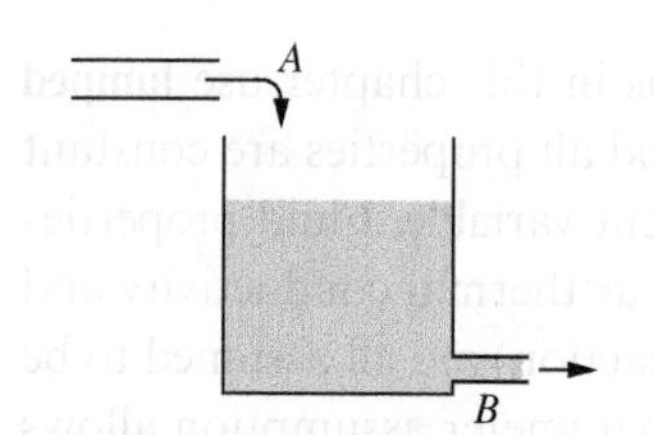

Figure 6.1
It is more practical to use a control volume analysis for the liquid in the tank rather than a system analysis.

A **system**, in the context just described, is defined as a fixed set of matter. A system does not gain or lose mass. Using this definition, it is clear that the principles developed in Chapter 4 are applied to systems of particles including rigid bodies. The members of a system of particles are tracked in space with time. However, it is impractical to track every particle in a system of matter in a liquid or gas state. Consider, for example, the tank of Figure 6.1. Liquid enters the tank through the pipe at A and exits through the pipe at B. Instead of tracking specific fluid particles after leaving the tank, it is of interest to analyze the properties of the liquid remaining in the tank. The properties of the liquid in the tank that may change with time include the fluid level, the temperature of the fluid in the tank, and the concentrations of various species in the tank.

Consequently, a **control volume**, a region in space with defined boundaries, is defined and used for fluid, thermal, and chemical analysis in lieu of a system. An essential difference between a system and a control volume is that matter may cross the boundaries of a control volume, whereas the matter of a system is fixed. The tank and the liquid in the tank of Figure 6.1 is an example of a control volume. Mass crosses the boundary of the control volume as fluid enters at A and leaves at B.

A control volume is a region with defined boundaries; however, it is not necessary for it to be fixed in space or have fixed boundaries. A rocket is an example of an accelerating control volume where mass leaves the control volume as exhaust gases. A helium balloon, while it is being filled, is an example of a control volume whose boundaries are defined, but vary with time.

Systems that include fluid flow, heat transfer, or chemical reactions are usually best analyzed by using a control volume analysis. The basic laws of Conservation of Mass, Conservation of Energy, and Conservation of Momentum are modified from the conservation laws used for system analysis to account for mass and energy crossing the boundaries of a control volume.

6.2.1 Conservation of Mass

The control volume form of **Conservation of Mass** is summarized as follows:

$$\begin{pmatrix}\text{Rate at which}\\ \text{mass enters a}\\ \text{control volume}\end{pmatrix} - \begin{pmatrix}\text{Rate at which}\\ \text{mass leaves a}\\ \text{control volume}\end{pmatrix} = \begin{pmatrix}\text{Rate at which mass}\\ \text{accumulates in a}\\ \text{control volume}\end{pmatrix} \tag{6.2}$$

Mass enters and leaves a control volume through its boundaries. The rate at which mass crosses the boundary of a control volume is called the **mass flow rate** $\dot{m}$, which has dimensions of $[M][T^{-1}]$. Letting m be the total mass of the control volume at any instant, Equation (6.2) is written as

$$\dot{m}_{in} - \dot{m}_{out} = \frac{dm}{dt} \tag{6.3}$$

The total mass in the control volume at any instant is

$$m = \int_V \rho\, dV \tag{6.4}$$

where ρ is the mass density of the fluid in the control volume. The mass flow rate can be written as

$$\dot{m} = \rho Q \tag{6.5}$$

where Q is the **volumetric flow rate,** which has dimensions of $[L^3][T^{-1}]$. Using Equations (6.4) and (6.5) in Equation (6.3) yields

$$\rho_{in} Q_{in} - \rho_{out} Q_{out} = \frac{d}{dt}\left(\int_V \rho\, dV\right) \tag{6.6}$$

If the fluid entering the control volume is of a constant species, the fluid in the control volume is homogeneous, and if the fluid is **incompressible**, the density is constant throughout the fluid and does not vary with time and $\rho = \rho_{in} = \rho_{out}$. For homogeneous, incompressible fluids, Equation (6.6) reduces to

$$Q_{in} - Q_{out} = \frac{dV}{dt} \tag{6.7}$$

where V is the total volume of fluid in the control volume.

Most liquids are incompressible and all gases are compressible. The Conservation of Mass equation is often referred to as the **continuity equation**. Thus Equation (6.7) is the control volume form of the continuity equation for an incompressible fluid. Equation (6.6) is the appropriate form of the continuity equation for a compressible fluid.

Example 6.1

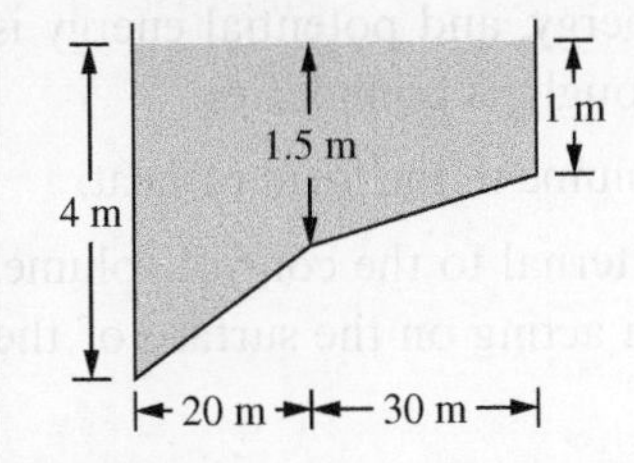

Figure 6.2
Swimming pool depth profile for Example 6.1.

A swimming pool of width 15 m has the depth profile shown in Figure 6.2. The pool, initially empty, is to be filled until the depth in the shallow end is 1 m. The pool is to be filled using a 2-cm-diameter hose with a fluid velocity of 5 m/s. How long will it take to fill the pool?

Solution

Take the control volume as the pool and the water in the pool. The amount of water in the control volume increases as water enters through the hose. The flow rate of the fluid entering through the hose is

$$\begin{aligned} Q &= vA \\ &= (5\text{ m/s})\left[\frac{\pi}{4}(0.02\text{ m})^2\right] = 1.57 \times 10^{-3}\text{ m}^3/\text{s} \end{aligned} \tag{a}$$

Conservation of Mass in the form of Equation (6.7) is applied to give

$$1.57 \times 10^{-3} = \frac{dV}{dt} \quad \text{(b)}$$

The volume of the pool when filled is

$$V = \left[(1.5 \text{ m})(50 \text{ m}) - (0.5 \text{ m})(30 \text{ m}) + \frac{1}{2}(2.5 \text{ m})(20 \text{ m})\right](15 \text{ m}) = 1385 \text{ m}^2 \quad \text{(c)}$$

Equation (b) is integrated to find the time required to fill the pool:

$$\int_0^t 1.57 \times 10^{-3}\, dt = \int_0^{1385} dV$$

$$1.57 \times 10^{-3} t = 1385$$

$$t = 8.83 \times 10^5 \text{ s} = 245.5 \text{ hr} \quad \text{(d)}$$

6.2.2 Energy Equation

The total energy in a control volume is

$$E = U + T + V + E_{other} \quad (6.8)$$

where U is the **internal energy** of all particles in the control volume, T is the total kinetic energy, V is the potential energy due to gravity, and E_{other} represents other forms of energy such as energy stored in mechanical or electric components and energy associated with chemical or nuclear reactions. The total energy can be obtained by integrating the **specific energy**, the energy per unit mass e, over the control volume:

$$E = \int_m e\, dm = \int_V e\rho\, dV \quad (6.9)$$

The internal energy is the energy associated with the random motion of molecules and thus is a function of temperature. The specific internal energy is given by

$$u = c_v T \quad (6.10)$$

where c_v is the specific heat at constant volume. The specific kinetic energy is $v^2/2$ and the specific potential energy is gz, where z is the elevation above a defined datum. Using these definitions, Equation (6.9) becomes

$$E = \int_V \left(u + \frac{v^2}{2} + gz + e_{other}\right)\rho\, dV \quad (6.11)$$

Energy is added to or subtracted from a control volume in four ways:

- Energy in the form of internal energy, kinetic energy, and potential energy is transported into or out of the control volume through its boundaries.
- Energy is transferred into or out of the control volume in the form of heat.
- Work is done on the control volume by forces external to the control volume. These include pressure forces and viscous friction acting on the surface of the control volume.
- Energy is transferred out of the control volume due to work done by the components of the control volume on components external to the control volume. The work done during this energy transfer is referred to as **shaft work**. Examples of shaft work are energy required to drive a turbine and work done on the control volume by a pump.

The control volume form of the energy equation is summarized as

$$\begin{pmatrix}\text{Rate at which}\\ \text{energy is transported}\\ \text{into the control}\\ \text{volume through its}\\ \text{boundaries}\end{pmatrix} - \begin{pmatrix}\text{Rate at which}\\ \text{energy is transported}\\ \text{out of the control}\\ \text{volume through}\\ \text{its boundaries}\end{pmatrix} + \begin{pmatrix}\text{Rate at}\\ \text{which heat}\\ \text{is transferred}\\ \text{into the control}\\ \text{volume}\end{pmatrix}$$

$$+ \begin{pmatrix}\text{Rate at which}\\ \text{work is done}\\ \text{on the control}\\ \text{volume by pressure}\\ \text{and friction}\end{pmatrix} - \begin{pmatrix}\text{Rate at which}\\ \text{the control}\\ \text{volume does}\\ \text{work on its}\\ \text{surroundings}\end{pmatrix} = \begin{pmatrix}\text{Rate at}\\ \text{which energy}\\ \text{accumulates}\\ \text{within the}\\ \text{control volume}\end{pmatrix} \tag{6.12}$$

For a control volume with one inlet and one outlet, Equation (6.12) can be written as

$$\dot{m}_{in}\left(u + \frac{v^2}{2} + gz\right)_{in} - \dot{m}_{out}\left(u + \frac{v^2}{2} + gz\right)_{out} + \dot{Q} + \dot{W}_p - \dot{W}_f - \dot{W}_s$$
$$= \frac{d}{dt}\left[\int_V \left(u + \frac{v^2}{2} + gz + e_{other}\right)\rho dV\right] \tag{6.13}$$

The rate of heat transfer $\dot{Q}$ is positive when heat is transferred into the control volume and negative when heat is transferred from the control volume. The rate of shaft work $\dot{W}_s$ is positive when the control volume does work on its surroundings and negative when the surroundings do work on the control volume. The work done by friction forces $\dot{W}_f$ is always negative. The rate of work done by pressure forces is given by

$$\dot{W}_p = \left(\dot{m}\frac{p}{\rho}\right)_{in} - \left(\dot{m}\frac{p}{\rho}\right)_{out} \tag{6.14}$$

Using Equation (6.14) in Equation (6.13) leads to

$$\dot{m}_{in}\left(u + \frac{p}{\rho} + \frac{v^2}{2} + gz\right)_{in} - \dot{m}_{out}\left(u + \frac{p}{\rho} + \frac{v^2}{2} + gz\right)_{out} + \dot{Q} - \dot{W}_s - \dot{W}_f$$
$$= \frac{d}{dt}\left[\int_V \left(u + \frac{v^2}{2} + gz\right)\rho dV\right] \tag{6.15}$$

Equation (6.15) is the most general form of the energy equation for a control volume with one inlet and one outlet. Assumptions may be used to reduce the equation to simpler forms.

6.2.3 Bernoulli's Equation

The control volume form of the momentum equation is useful for problems involving accelerating control volumes such as rockets or for determining external forces acting on the control volume such as reactions at joints in pipes. However, the control volume form of the momentum equation is not useful for problems studied in this book and is thus not presented. Instead an equation that can be derived from the control volume momentum equation is presented: Bernoulli's equation.

Recall that for a mechanical system the Principle of Work and Energy is derived from Newton's second law by integrating the dot product of Newton's second law with a differential displacement vector between two points in the path of motion. Bernoulli's equation is derived in a similar manner. A **streamline** in a flow is a line tangent to the velocity vector. For a steady flow, the streamline is the same as a pathline, the path traveled by a fluid particle. The control volume form of the momentum equation can be applied to a control volume enclosing a streamline. For a frictionless flow of an incompressible fluid, the following equation results

$$\frac{p_1}{\rho} + \frac{v_1^2}{2} + gz_1 = \frac{p_2}{\rho} + \frac{v_2^2}{2} + gz_2 \tag{6.16}$$

where 1 and 2 refer to two points along a streamline in a flow. Equation (6.16) is referred to as **Bernoulli's equation.**

Under the assumptions of steady frictionless incompressible flow with no shaft work, the control volume energy equation, Equation (6.15), reduces to

$$\dot{m}_{in}\left(u + \frac{p}{\rho} + \frac{v^2}{2} + gz\right)_{in} - \dot{m}_{out}\left(u + \frac{p}{\rho} + \frac{v^2}{2} + gz\right)_{out} + \dot{Q} = 0 \tag{6.17}$$

For steady flow there is no mass accumulation within the control volume and Equation (6.3) reduces to $\dot{m}_{in} = \dot{m}_{out} = \dot{m}$. Dividing Equation (6.17) by $\dot{m}$ and defining $q = \dot{Q}/\dot{m}$, Equation (6.17) reduces to

$$\left(u + \frac{p}{\rho} + \frac{v^2}{2} + gz\right)_{in} - \left(u + \frac{p}{\rho} + \frac{v^2}{2} + gz\right)_{out} + q = 0 \tag{6.18}$$

Defining 1 as the inlet and 2 as the outlet, Equation (6.18) reduces to Bernoulli's equation [Equation (6.16)] if

$$q = u_2 - u_1 \tag{6.19}$$

Equation (6.19) implies that heat transferred into the control volume leads to an increase of internal energy. Thus, Bernoulli's equation reconciles with the control volume form of the energy equation under the assumptions of steady incompressible flow with no friction and no shaft work. In such problems the energy equation and Bernoulli's equation, which is derived from Conservation of Momentum, are identical. The application of only Conservation of Mass and Bernoulli's equation is necessary when modeling such systems. Bernoulli's equation is analogous to the Principle of Work and Energy considered in Section 4.9 for mechanical systems.

Bernoulli's equation may be applied only along a streamline in the flow. The subscripts 1 and 2 in Equations (6.16) and (6.19) refer to two points on the same streamline. Since Bernoulli's equation involves the term $(p_2 - p_1)/\rho$, **gauge pressure**, the difference between absolute pressure and atmospheric pressure, may be used in expressing the pressure terms. Gauge pressure is zero when the flow is exposed to the open atmosphere.

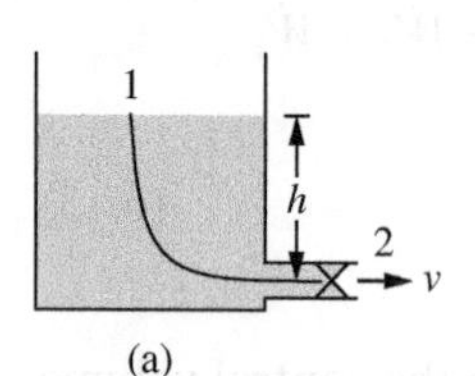

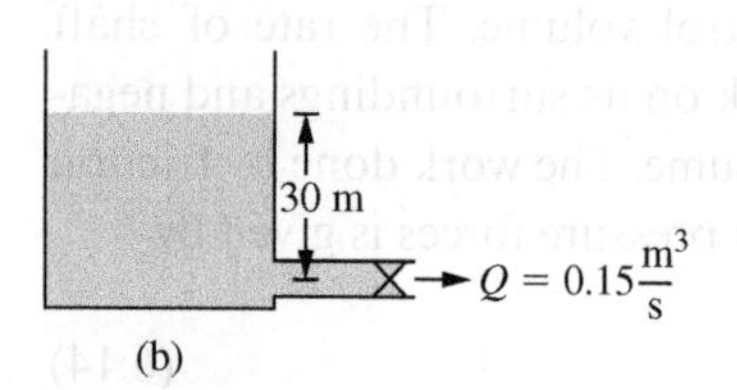

Figure 6.3

(a) The reservoir of Example 6.2. A streamline is taken from the free surface of the reservoir to the tank outlet for application of Bernoulli's equation. (b) The reservoir of Example 6.3. Head losses occur due to the valve and pipe.

Example 6.2

The reservoir of Figure 6.3(a) has a valve that, when open, allows water to flow out of the tank. Use Bernoulli's equation to determine the velocity of the fluid leaving the tank.

Solution

The application of Bernoulli's equation requires the flow to be steady, which is a reasonable assumption provided the reservoir is large and the change in liquid level is negligible. Bernoulli's equation is applied along a streamline from 1, the free surface of the liquid, to 2, the exit of the reservoir. Define $z = 0$ at the surface of the reservoir, thus $z_2 = -h$. The pressure at the free surface and at the exit and the pressure at the exit are both atmospheric. Since the reservoir is large, the velocity of the fluid on the free surface is negligible, $v_1 = 0$. Application of Bernoulli's equation, Equation (6.16), applied between the free surface of the tank where $p_1 = 0$, $v_1 = 0$, and $z_1 = 0$ and the exit of the reservoir where $p_2 = 0$ and $z_2 = -h$, leads to

$$0 = \frac{v_2^2}{2} - gh \tag{a}$$

Equation (a) is solved for v, leading to

$$v_2 = \sqrt{2gh} \tag{b}$$

Dividing Equation (6.16) by g leads to

$$h_1 = h_2 \tag{6.20}$$

where the **head** is defined by

$$h = \frac{p}{\rho g} + \frac{v^2}{2g} + z \tag{6.21}$$

The head, which has dimensions of $[L]$, is a property of the flow. Equation (6.21), the head form of Bernoulli's equation, shows that the head is constant in a flow with no losses.

In a static case, $v = 0$ and Equation (6.21) reduces to

$$\frac{p_2 - p_1}{\rho g} = z_1 - z_2 \tag{6.22}$$

Equation (6.22) is called the **manometer equation** because it is the basis of the use of a manometer to measure static pressure.

A head in a flow is a measure of the total energy in the flow. A reduction in the head in a flow is called a **head loss**. If a head loss h_ℓ occurs between points 1 and 2 in a flow, the head equation is modified as

$$h_1 - h_\ell = h_2 \tag{6.23}$$

Equation (6.23) is called the **extended Bernoulli's equation**, that is, extended to take into account energy losses. The extended equation is really a head form of the energy equation. Head losses occur in a real flow due to friction, an obstacle in the flow, or a change in the flow. Head losses are often expressed in the form

$$h_\ell = K\frac{v^2}{2g} \tag{6.24}$$

where K is a loss coefficient.

Example 6.3

Fluid exits the reservoir of Figure 6.3(b) when the valve is open. Energy is lost due to friction in the pipe and changes in the flow pattern in the valve. The fluid exits through a pipe of diameter 10 cm at a flow rate of 0.15 m³/s. Determine the loss coefficient for the pipe and valve system.

Solution

The application of the extended Bernoulli's equation between the free surface of the reservoir and the exit of the pipe using Equation (6.24) for the head loss gives

$$\frac{p_1}{\rho g} + \frac{v_1^2}{2g} + z_1 = K\frac{v_2^2}{2g} + \frac{p_2}{\rho g} + \frac{v_2^2}{2g} + z_2 \tag{a}$$

Both the free surface and the exit of the pipe are open to the atmosphere; thus $p_1 = p_2 = 0$. The reservoir is assumed to be large, thus $v_1 \approx 0$. Defining the datum at the level of the exit of the pipe leads to $z_1 = 30$ m and $z_2 = 0$ m. Using this information in Equation (a) leads to

$$z_1 = (K + 1)\frac{v_2^2}{2g} \tag{b}$$

The velocity of the fluid in the pipe is calculated as

$$v_2 = \frac{Q}{A} = \frac{0.15 \text{ m}^3/\text{s}}{\frac{\pi}{4}(0.1 \text{ m})^2} = 19.1 \text{ m/s} \tag{c}$$

Solving Equation (b) for K and substituting known values leads to

$$K = \frac{2gz_1}{v_2^2} - 1 = \frac{2(9.81 \text{ m/s}^2)(30 \text{ m})}{(19.1 \text{ m/s})^2} - 1 = 0.613 \tag{d}$$

6.3 Pipe Flow

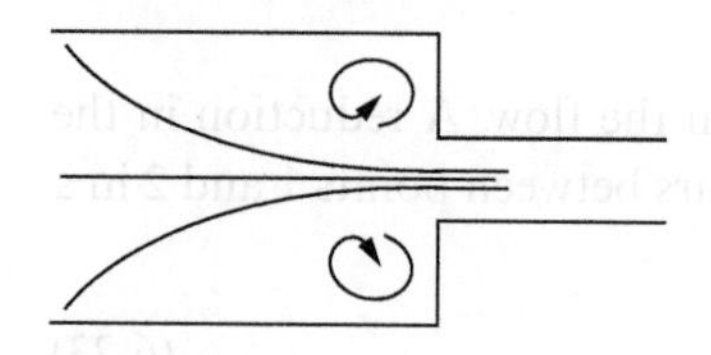

Figure 6.4

The sudden contraction causes fluid to be entrained. The energy of the flow entering the contraction is less than the energy of the flow leaving the contraction.

Most pipe flows may be analyzed using the head form of the extended Bernoulli's equation, Equation (6.23). Losses occur in a pipe due to friction and changes in flow pattern caused by elbows, bends, fittings, valves, and nozzles. Thus the head at the exit of the pipe is equal to the head at the pipe inlet minus the head losses. If the pipe is of constant area and at a constant elevation, the head loss leads to a decrease in pressure along the pipe.

Streamlines for a fluid entering a sudden contraction are illustrated in Figure 6.4. The fluid near the contraction is entrained in a recirculating pattern, leading to a loss of kinetic energy.

6.3.1 Losses

Head losses in pipes due to friction are called **major losses**, while losses due to valves, expansions and contractions, bends, joints, and other changes in the flow are called **minor losses**. All losses are expressed in the form of Equation (6.24). Minor loss coefficients are highly dependent on geometry and vary greatly. Loss coefficients for pipe elbows range between 0.2 and 1.5, loss coefficients for contractions range up to 0.5, and loss coefficients for valves vary greatly depending on whether they are open or partially closed.

The total head loss in a piping system is

$$h_\ell = h_f + h_m \tag{6.25}$$

where h_f is the head loss due to friction and h_m is the sum of all minor head losses. The frictional head loss is dependent on the geometry of the pipe as well as on flow properties. The frictional head loss is written as

$$h_f = f\frac{L}{D}\frac{v^2}{2g} \tag{6.26}$$

where L is the length of the pipe, D is its diameter, and f is called the friction factor. It has been shown that the friction factor is a function of two dimensionless parameters

$$f = f\left(\mathrm{Re}, \frac{\varepsilon}{D}\right) \tag{6.27}$$

where Re is the **Reynolds number**

$$\mathrm{Re} = \frac{\rho v D}{\mu} \tag{6.28}$$

and ε/D is the ratio of the pipe's roughness to its diameter. In Equation (6.28), ρ is the mass density of the fluid and μ is its dynamic viscosity. The Reynolds number represents the ratio of the flow's inertia forces to its friction forces.

A pipe flow is **laminar** if the flow moves along smooth parallel streamlines and the velocity of the fluid is constant at any point in the pipe along a streamline. A pipe flow is **turbulent** if there are random fluctuations in the velocity profile and mixing of the fluid occurs at a macroscopic level. The Reynolds number provides a basis for predicting whether a pipe flow is laminar or turbulent. Laminar flow occurs in pipes for $\mathrm{Re} < 2300$ and turbulent flow occurs in pipes for $\mathrm{Re} > 4000$. For $2300 < \mathrm{Re} < 4000$, the flow is transitional, fluctuating between laminar and turbulent.

For laminar flow, analytical methods can be used to show that the friction factor is

$$f = \frac{64}{\mathrm{Re}} \tag{6.29}$$

Noting that the velocity of a fluid in a circular pipe is related to volumetric flow rate by

$$v = \frac{4Q}{\pi D^2} \tag{6.30}$$

the substitution of Equations (6.28), (6.29), and (6.30) into Equation (6.26) leads to

$$h_f = \frac{128 \mu L}{\rho g \pi D^4} Q \tag{6.31}$$

Thus for laminar flow, the frictional head loss in a pipe is proportional to the flow rate.

Empirical methods are used to quantify Equation (6.27) for turbulent flow. The **Colebrook equation** provides a quantitative fit to empirical data for a pressure drop in a pipe:

$$\frac{1}{\sqrt{f}} = -2.0 \log\left(\frac{\varepsilon/D}{3.7} + \frac{2.51}{\mathrm{Re}\sqrt{f}}\right) \tag{6.32}$$

It is not possible to solve Equation (6.32) in closed form to determine f given values of Re and ε/D. The friction factor is obtained by solving the Colebrook equation using numerical methods or by referring to the **Moody diagram** (Figure 6.5), a chart that provides curves of f vs. Re for different values of ε/D. When the fluid velocity is unknown, the Reynolds number is not known and the Moody diagram cannot be used to directly determine the friction factor. In such cases, iterative methods are used to determine the velocity and the friction factor.

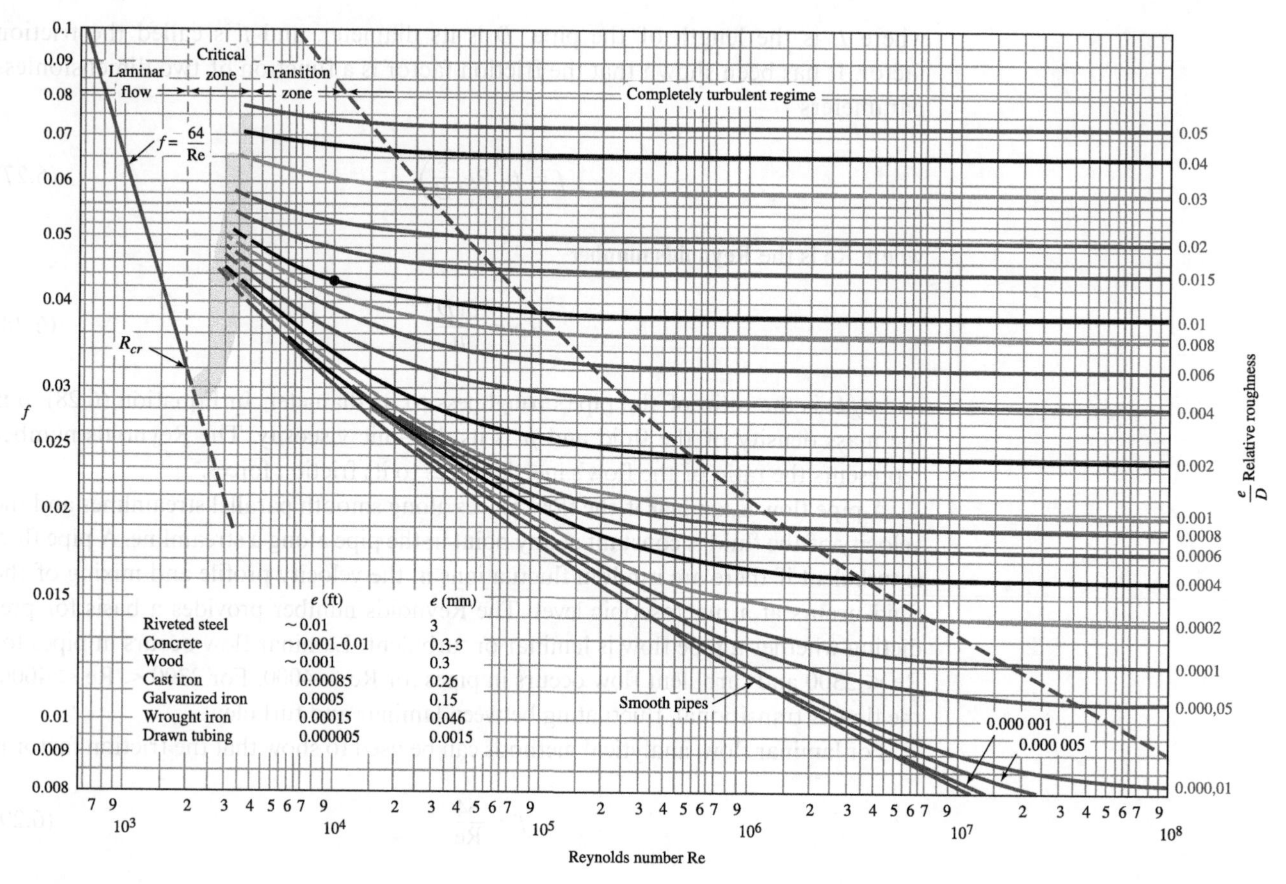

Figure 6.5

Moody diagram.

Source: L. F. Moody from *Trans. ASME*, Vol. 66, 1944. Reprinted by permission.

Example 6.4

Water at 10°C ($\rho = 1000$ kg/m^3, $\mu = 1.31 \times 10^{-3}$ N·s/m^2) enters the 600-m-long pipe of Figure 6.6 at A, where the gauge pressure is 900 kPa. The fluid exits into the atmosphere at B, where the pressure is atmospheric. Pipe AB is made of commercial steel with a roughness of $\varepsilon = 0.046$ mm and a diameter of 10 cm. The total of the minor loss coefficients is 12.4. Determine the flow rate through the pipe.

Solution

The application of the extended Bernoulli's equation between A and B gives

$$\frac{p_A}{\rho g} + \frac{v_A^2}{2g} + z_A = h_\ell + \frac{p_B}{\rho g} + \frac{v_B^2}{2g} + z_B \tag{a}$$

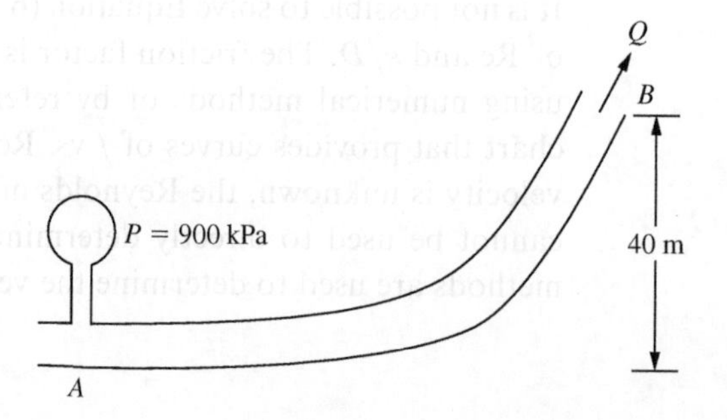

Figure 6.6

Pipe of Examples 6.4 and 6.5.

The pipe is of constant diameter; thus Conservation of Mass implies $v_A = v_B$. The fluid exits into the atmosphere at B; thus the gauge pressure at B is zero. The datum is taken at the level of the pipe at A: $z_A = 0$ m, $z_B = 40$ m. The total head loss is the sum of the minor losses and the major losses

$$\begin{aligned} h_\ell &= h_f + h_m \\ &= f\frac{L}{D}\frac{v^2}{2g} + \sum K\frac{v^2}{2g} \\ &= \left(f\frac{600\text{ m}}{0.1\text{ m}} + 12.4\right)\frac{v^2}{2g} \\ &= (6000f + 12.4)\frac{v^2}{2g} \end{aligned} \tag{b}$$

The flow rate is related to the velocity by

$$\begin{aligned} v &= \frac{Q}{A} \\ &= \frac{4Q}{\pi D^2} \\ &= \frac{4Q}{\pi(0.1\text{ m})^2} = 1.27 \times 10^2\, Q \end{aligned} \tag{c}$$

Substituting into Equation (a) leads to

$$\frac{900 \times 10^3\text{ N/m}^2}{(1000\text{ kg/m}^3)(9.81\text{ m/s}^2)} = (6000f + 12.4)\frac{(1.27 \times 10^2\, Q)^2}{2(9.81\text{ m/s}^2)} + 40 \tag{d}$$

Equation (d) simplifies to

$$6.29 \times 10^{-2} = (6000f + 12.4)\,Q^2 \tag{e}$$

The Reynolds number for the flow is calculated in terms of the flow rate as

$$\begin{aligned} \text{Re} &= \frac{\rho v D}{\mu} \\ &= \frac{(1000\text{ kg/m}^3)(1.27 \times 10^2\, Q)(0.1\text{ m})}{1.31 \times 10^{-3}\text{ N}\cdot\text{s/m}^2} \\ &= 9.69 \times 10^6\, Q \end{aligned} \tag{f}$$

The roughness ratio for the pipe is

$$\frac{\varepsilon}{D} = \frac{4.6 \times 10^{-5}\text{ m}}{0.1\text{ m}} = 4.6 \times 10^{-4} \tag{g}$$

An iterative procedure is used to determine the flow rate using Equations (e) and (f) and the curve on the Moody diagram corresponding to the roughness ratio of Equation (g). A value of Q is guessed. Equation (f) is used to determine the Reynolds number. The Moody diagram is used to determine the friction factor. Using this value of f, Equation (e) is used to determine a new value of Q. The iteration continues until convergence is achieved. The calculations for the iteration are given in Table 6.1. The process converges in three iterations to a flow rate of 2.23×10^{-2} m³/s.

Table 6.1 Iteration to Determine Flow Rate for Example 6.4

Q (m³/s)	Re [Equation (f)]	f (Moody Diagram)	Q (m³/s) [Equation (e)]
1.0	9.69×10^6	0.0165	2.38×10^{-2}
2.38×10^{-2}	2.30×10^5	0.0185	2.26×10^{-2}
2.26×10^{-2}	2.19×10^5	0.019	2.23×10^{-2}

The **resistance** of a pipe is defined as

$$R = \frac{dh_\ell}{dQ} \tag{6.33}$$

The resistance of a pipe with laminar flow and no minor losses is determined using Equation (6.31) as

$$R = \frac{128\mu L}{\rho g \pi D^4} \tag{6.34}$$

The resistance of a pipe with turbulent flow can be determined by developing a graph of h_f as a function of Q. Note that for a pipe of constant diameter with minor losses, Equation (6.25) can be written in terms of flow rate as

$$h_\ell = \left(f\frac{L}{D} + \sum K\right)\frac{8}{\pi^2 D^4 g}Q^2 \tag{6.35}$$

Examination of the Moody diagram shows that for large Reynolds numbers the friction factor f has little dependence on Re, varying mostly with ε/D. In this case, Equation (6.35) can be written as

$$Q = B\sqrt{h_\ell} \tag{6.36}$$

where the constant of proportionality is

$$B = \left[\left(f\frac{L}{D} + \sum K\right)\frac{8}{\pi^2 D^4 g}\right]^{-1/2} \tag{6.37}$$

Using Equation (6.37), the pipe's resistance is determined as

$$R = 2\frac{Q}{B^2} = 2\frac{h_\ell}{Q} \tag{6.38}$$

Equation (6.38) shows that the resistance of a pipe with turbulent flow is a function of flow rate. Given a curve for head loss vs. flow rate, the resistance can be determined as the slope of the curve for a specified flow rate. A curve of frictional head loss vs. flow rate for a specified pipe could be developed using the Colebrook equation and the Moody diagram. However, in a practical situation where minor losses may be significant, the curve of head loss for the piping system vs. flow rate can be determined empirically.

Example 6.5

Determine the resistance of the pipe of Example 6.4.

Solution

Using Equations (b) and (c) of Example 6.4, the head loss is related to the flow rate as

$$h_\ell = (6000f + 12.4)\frac{(1.27 \times 10^2 Q)^2}{2(9.81)}$$
$$= 8.22 \times 10^2(6000f + 12.4)Q^2 \tag{a}$$

Using the calculated values of $f = 0.019$ and $Q = 2.23 \times 10^{-2}$ m^3/s in Equation (b) leads to $h_\ell = 51.7$ m. The pipe's resistance at this flow rate is obtained using Equation (6.38) as

$$R = 2\frac{h_\ell}{Q}$$
$$= 2\frac{51.7 \text{ m}}{2.23 \times 10^{-2} \text{ m}^3/\text{s}} = 4.64 \times 10^3 \text{ s/m}^2 \tag{b}$$

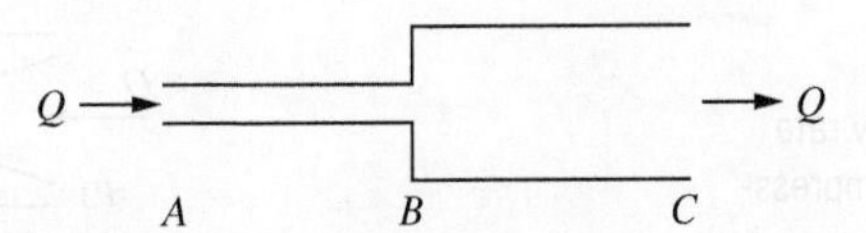

Figure 6.7

Pipes *AB* and *BC* are in series. The flow rate through each pipe is the same, and the head loss from *A* to *C* is the head loss from *A* to *B* plus the head loss from *B* to *C*.

Pipes are similar to resistors in electric circuits in that they dissipate energy, which is converted into heat. Similar to resistors, pipe systems can be classified as being in series or parallel. Two pipes in series are illustrated in Figure 6.7. These pipes of different diameters are joined directly together such that use of the continuity equation implies that the flow rate in each pipe is the same. The total head loss across the system is the sum of the head losses in the two pipes. Thus pipes in series are characterized as having the same flow rate but with the total head loss as the sum of the head losses.

An equivalent resistance for pipes with turbulent flow is defined by

$$R_{eq} = 2\frac{h_{\ell,tot}}{Q} \tag{6.39}$$

Referring to Figure 6.7 for the two pipes in series,

$$h_{\ell,tot} = h_{\ell,AB} + h_{\ell,BC} \tag{6.40}$$

Using Equation (6.38) in Equation (6.40) leads to

$$h_{\ell,tot} = \frac{1}{2}R_{AB}Q + \frac{1}{2}R_{BC}Q \tag{6.41}$$

Substitution of Equation (6.41) in Equation (6.39) leads to

$$R_{eq} = R_{AB} + R_{BC} \tag{6.42}$$

Thus the equivalent resistance of pipes in series is the sum of the resistances of the pipes.

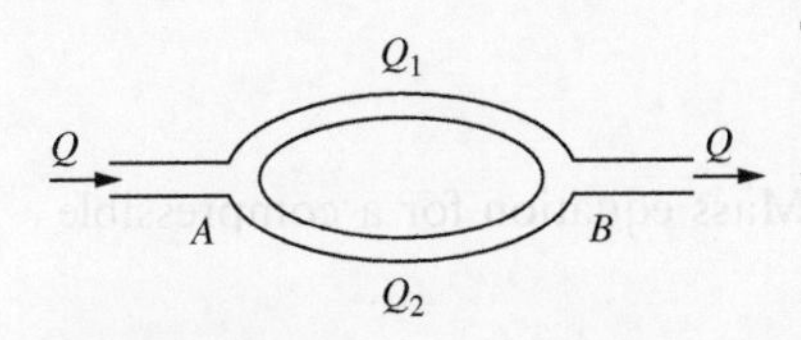

Figure 6.8

The two parallel pipes have the same head loss, and the total flow rate is the sum of the flow rates in the pipes.

The application of the continuity equation to the parallel pipes of Figure 6.8 shows that

$$Q = Q_1 + Q_2 \tag{6.43}$$

The application of the extended Bernoulli's equation between *A* and *B* leads to

$$\frac{p_A}{\rho} + \frac{v_A^2}{2g} + z_A = h_\ell + \frac{p_B}{\rho} + \frac{v_B^2}{2g} + z_B \tag{6.44}$$

However, Equation (6.44) can be applied along a streamline through either pipe, which implies that the head loss in each pipe is the same. Thus parallel pipes are characterized as having the same head loss, and the total flow rate is the sum of the flow rates through the pipes.

Using a method similar to that used to derive the equivalent resistance of resistors in parallel, it is determined that the equivalent resistance of n pipes in parallel is

$$R_{eq} = \frac{1}{\sum_{i=1}^{n}\frac{1}{R_i}} \tag{6.45}$$

6.3.2 Orifices

Consider the flow of an incompressible fluid in a pipe with an orifice, as illustrated in Figure 6.9. The application of Bernoulli's equation between the upstream position 1 and the downstream position 2, assuming no change in elevation between positions, leads to

$$\frac{p_1}{\rho} + \frac{v_1^2}{2} = \frac{p_2}{\rho} + \frac{v_2^2}{2} \tag{6.46}$$

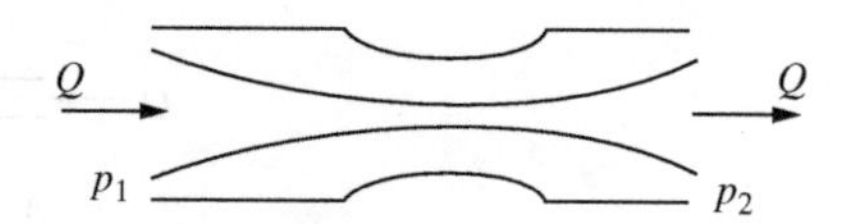

Figure 6.9

The relationship between flow rate and pressure drop for an incompressible flow through an orifice is given in Equation (6.49).

Since the flow is incompressible, the density and thus the volumetric flow rate Q are constant. Conservation of Mass is expressed as

$$Q = v_1 A_1 = v_2 A_2 \tag{6.47}$$

Equation (6.47) is used to express the velocities in terms of the flow rate which are then substituted into Equation (6.46), leading to

$$Q = A_2 \sqrt{\frac{2(p_1 - p_2)}{\rho[1 - (A_2/A_1)^2]}} \tag{6.48}$$

Equation (6.48) is for an ideal flow without losses. Losses are taken into account by the introduction of a loss coefficient C_d such that for a real flow

$$Q = \frac{C_d A_2}{\sqrt{1 - (A_2/A_1)^2}} \sqrt{\frac{2(p_1 - p_2)}{\rho}} \tag{6.49}$$

The flow coefficient C_Q is defined as

$$C_Q = \frac{C_d}{\sqrt{1 - (A_2/A_1)^2}} \tag{6.50}$$

such that Equation (6.49) can be written as

$$Q = C_Q A_2 \sqrt{\frac{2(p_1 - p_2)}{\rho}} \tag{6.51}$$

6.3.3 Compressible Flows

The control volume form of the Conservation of Mass equation for a compressible flow in a pipe is

$$\rho_1 V_1 A_1 = \rho_2 V_2 A_2 \tag{6.52}$$

Since the density is not constant, the volumetric flow rate is not constant. If the velocity of the fluid is greater than the speed of sound, the flow is said to be supersonic and a normal shock may form in a duct or nozzle. If the velocity of the fluid is less than the speed of sound, the flow is subsonic. Only subsonic flows are considered here.

Consider the system of Figure 6.10. The system is at a steady state when the upstream pressure is increased, forcing air through the pipe into the reservoir. The mass of air in the reservoir increases, leading to an increase in pressure. Compressible flow in a constant-area pipe with friction is called Fanno flow, which has complicated relationships between the pressure change across the length of the pipe and the mass flow rate. For a small increase in upstream pressure, the flow is subsonic. The mass flow rate into the reservoir is roughly proportional to the square root of the difference between the upstream pressure p_1 and the pressure in the reservoir p_2. This relationship is analogous to the relationship for volumetric flow rate and head loss for an incompressible flow, Equation (6.36). Thus the concept of resistance is defined for subsonic compressible flows in a fashion similar to that of liquid flow in pipes. Specifically, the resistance is defined by

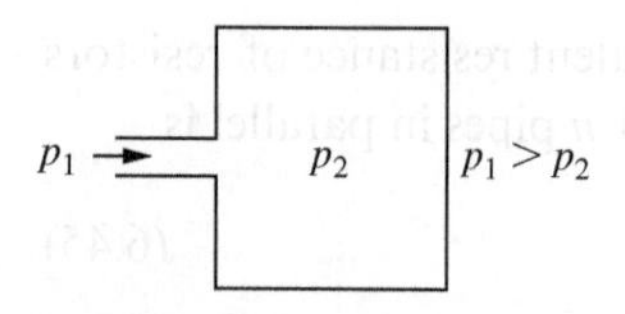

Figure 6.10

Compressible flow through a pipe into a reservoir. As the upstream pressure increases, more mass is forced into the reservoir, increasing the pressure of the fluid in the reservoir. The resistance of the pipe is defined by Equation (6.53).

$$R = \frac{d}{d\dot{m}}(p_1 - p_2) \tag{6.53}$$

The resistance, defined by Equation (6.53), is not constant. For flow of a liquid in a pipe, the resistance is calculated at the steady state, which linearizes the mathematical modeling of the system. However, for the system considered in Figure 6.10, the steady state corresponds to a mass flow rate of zero. Using Equation (6.51) at $\dot{m} = 0$ leads to significant error in the modeling. Thus an average value of R is used. The linearizing approximation is

$$\dot{m} = \frac{p_1 - p_2}{R} \tag{6.54}$$

where R is determined by developing a curve for $p_1 - p_2$ as a function of $\dot{m}$ (at the exit of the pipe) and using Equation (6.53) at an average value of $\dot{m}$.

6.4 Modeling of Liquid-Level Systems

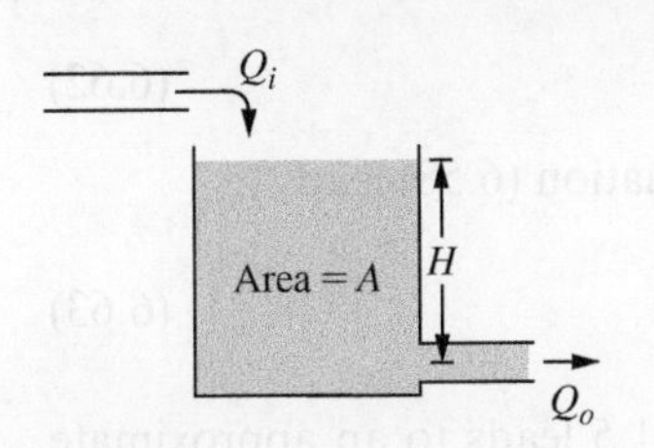

Figure 6.11

The reservoir used as an example for modeling liquid-level systems. The system is operating at a steady state. A change in inlet flow rate Q_i leads to perturbations in the liquid-level H and the outlet flow rate Q_o.

The level of the liquid in the tank of Figure 6.11 is constant when the flow rate of the inlet stream is equal to the flow rate through the pipe. When this steady state is disturbed, for example by a sudden increase or decrease in the inlet flow rate, the level of the liquid in the tank varies with time. Mathematical modeling of the change in liquid level requires application of the control volume form of Conservation of Mass as well as the pipe flow equations of Section 6.3. Such problems are referred to as liquid-level problems.

Liquid enters a tank of area A at a volumetric flow rate Q_i. Liquid leaves the tank through a pipe of length L, diameter D, and roughness ε. The liquid is at a constant temperature and has a dynamic viscosity μ and a mass density ρ. Let $H(t)$ be the liquid level in the tank and Q_o be the volumetric flow rate of the fluid leaving the pipe. A control volume consisting of the tank and the pipe is selected. Application of Conservation of Mass, in the form of Equation (6.7), to the control volume leads to

$$Q_i - Q_o = \frac{d}{dt}(AH) \tag{6.55}$$

Consider a streamline from the entrance of the pipe to its exit. Since the area of the pipe is constant, the velocity of the fluid at the inlet is the same as the velocity of the fluid at the outlet. The pressure head at the inlet of the pipe is H. The application of the extended Bernoulli's equation [Equation (6.23)] along this streamline leads to

$$H + \frac{v^2}{2g} - h_\ell = \frac{v^2}{2g}$$
$$h_\ell = H \tag{6.56}$$

Thus the head loss due to friction in the pipe is equal to the level of the liquid in the tank.

Assuming turbulent flow, the flow rate at the outlet is given by Equation (6.36), which, using Equation (6.56), leads to

$$Q_o = B\sqrt{H} \tag{6.57}$$

Substitution of Equation (6.57) into Equation (6.55) leads to

$$A\frac{dH}{dt} + B\sqrt{H} = Q_i \tag{6.58}$$

At $t = 0$, the system is at a steady state with an inlet flow rate of $q_{i,s}$ and a liquid level of h_s. The steady state implies $dh_s/dt = 0$ and thus Conservation of Mass requires that the outlet flow rate is equal to the inlet flow rate. Thus from Equation (6.57),

$$q_{o,s} = q_{i,s} = B\sqrt{h_s} \tag{6.59}$$

Suppose the inlet flow rate changes at $t = 0$ and is given by $Q_i(t)$. The time-dependent level of the liquid in the tank is obtained by solving Equation (6.58) subject to the initial condition, $H(0) = h_s$. However, Equation (6.58) is a nonlinear differential equation for which a solution is difficult to obtain. If the expected change in liquid level is small compared to h_s, then Equation (6.58) may be linearized using the methods of Section 1.6. Indeed, Equation (6.58) is the equation used in Section 1.6 to illustrate the method of linearization through the introduction of perturbation variables.

The steady state is disturbed due to a perturbation in the inlet flow rate defined such that

$$Q_i = q_{i,s} + q_i(t) \tag{6.60}$$

The perturbation in flow rate induces a time-dependent perturbation in the liquid level $h(t)$, defined such that

$$H(t) = h_s + h(t) \tag{6.61}$$

The outlet flow rate is obtained from Equation (6.57) as

$$Q_o(t) = B\sqrt{h_s + h(t)} \tag{6.62}$$

Substituting Equations (6.60), (6.61), and (6.62) into Equation (6.58) leads to

$$q_{i,s} + q_i(t) - B\sqrt{h_s + h(t)} = A\frac{dh}{dt} \tag{6.63}$$

Linearization of Equation (6.63) described in Example 1.5 leads to an approximate equation of

$$A\frac{dh}{dt} + \frac{B}{2\sqrt{h_s}}h = q_i(t) \tag{6.64}$$

The resistance of a pipe with turbulent flow, where the relationship between the flow rate and head loss is given by Equation (6.36), is developed in Equation (6.38), which leads to $R = \sqrt{h_s}/B$, where R is the pipe resistance calculated for the original steady state. Using this result in Equation (6.64) gives

$$A\frac{dh}{dt} + \frac{1}{R}h = q_i(t) \tag{6.65}$$

Equation (6.65) is analogous to the differential equation for the voltage change across the capacitor of an RC circuit, Equation (h) of Example 5.9. The head in the tank is analogous to the potential across the capacitor. The friction in the pipe dissipates energy into heat like a resistor. A loss of head occurs due to friction as flow passes through a pipe. The resistance of the pipe is the ratio of the change in liquid level to the change in flow rate. This is analogous to a resistor in an electric circuit. Potential is lost as current passes through a resistor. A definition of electrical resistance for a linear resistor is the ratio of the change in current to the change in potential. Note, however, that the dimensions of flow resistance and electrical resistance are not the same.

The tank stores potential energy through gravity. The total potential energy of the liquid in the tank is the weight of the liquid in the tank multiplied by the distance from the mass center of the liquid to a reference plane, or the total energy stored in the tank is

$$V = \frac{1}{2}\rho g A H^2 \tag{6.66}$$

The total energy stored in the electric field of a capacitor is $E = \frac{1}{2}Cv^2$, where the capacitance could be defined as the ratio of the change in the charge on the plates of a capacitor to the change in potential difference across the capacitor. The analogous definition of capacitance for the liquid-level system is the ratio of the change in volume of stored liquid to the head, leading to the capacitance of the tank being simply equal to its area.

The technique used to derive Equation (6.65) can be applied to any liquid-level problem, including problems involving multiple tanks. The control volume form of Conservation of Mass is applied to each tank. The extended Bernoulli's equation is used to relate the head loss in the pipes to the fluid levels. Perturbation variables, measured from the steady state, are introduced. Conservation of Mass equations are written using perturbation variables as dependent variables. The equations are linearized as in Section 1.6.

The resistance used in Equation (6.65) is evaluated at the original steady-state conditions. This is a result of the linearization process. The resistance is a function of flow rate.

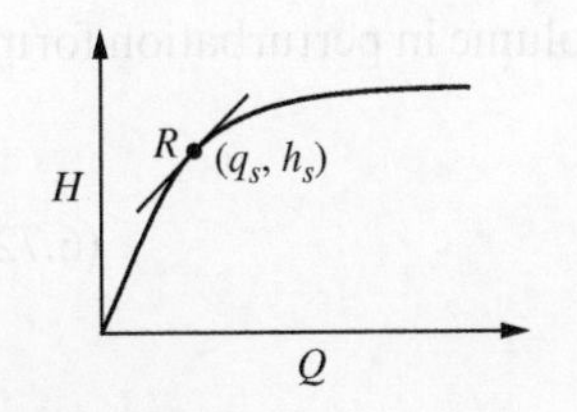

Figure 6.12
The resistance of a pipe is a function of flow rate through the pipe.

An alternate derivation of equations governing liquid-level problems involves the definition of resistance at the steady state, defined using Equation (6.33):

$$R = \left.\frac{dH}{dQ_o}\right|_{Q_o = q_{o,s}} \tag{6.67}$$

This use of the resistance calculated for the original steady-state conditions is a result of the linearization process. The resistance changes with flow rate, as illustrated in Figure 6.12. At a given flow rate the resistance is the slope of the tangent to the H vs. Q curve. The a priori use of resistance at the steady state leads to an a priori alternate method for linearizing the differential equation; that is, the use of the linearizing assumption in deriving the differential equation leads to the formulation of a linear differential equation. Using the definition of the derivative, and since $h(t)$ and $q_o(t)$ are small perturbations from the steady state, the resistance is approximated as

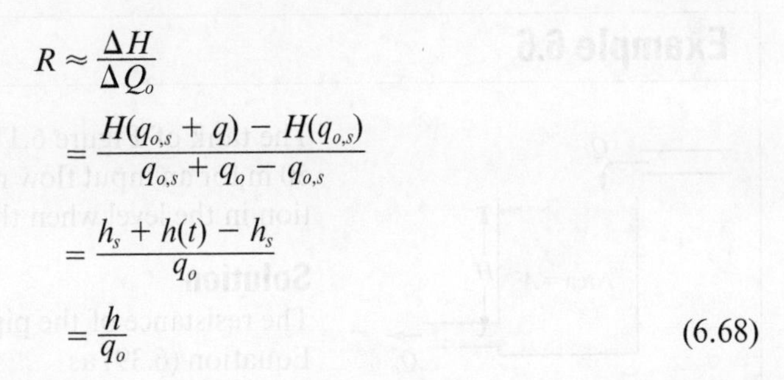

$$\begin{aligned} R &\approx \frac{\Delta H}{\Delta Q_o} \\ &= \frac{H(q_{o,s} + q) - H(q_{o,s})}{q_{o,s} + q_o - q_{o,s}} \\ &= \frac{h_s + h(t) - h_s}{q_o} \\ &= \frac{h}{q_o} \end{aligned} \tag{6.68}$$

Equation (6.68) can be rewritten as

$$q_o = \frac{h}{R} \tag{6.69}$$

The output flow rate is then written as

$$Q_o = q_{o,s} + \frac{h}{R} \tag{6.70}$$

Using Equations (6.57) and (6.70), the Conservation of Mass equation, Equation (6.55), can be written as

$$q_{i,s} + q_i(t) - q_{o,s} - \frac{h}{R} = A\frac{dh}{dt} \tag{6.71}$$

Since $q_{i,s} = q_{o,s}$ is the Conservation of Mass equation for the steady state, Equation (6.71) reduces to Equation (6.65).

The linearization using this method occurs in the approximation of the resistance by Equation (6.68). In general, the linear approximation for the resistance is the change in head loss due to the perturbations divided by the perturbation of the flow rate in the pipe. This, in turn, leads to a linear approximation for the perturbation in flow rate in a pipe as the change in head loss across the pipe divided by its resistance at steady state, Equation (6.69). Direct use of Equation (6.69) in the Conservation of Mass equation applied to a tank provides an alternate method for the mathematical modeling of liquid-level problems, which avoids the necessity for formal linearization of the resulting equations.

The modeling of multitank liquid-level systems can be performed by drawing a control volume around each of the tanks and their outlet pipes. The control volume form of Conservation of Mass is applied to each control volume in perturbation form. That is, for tank j,

$$A_j \frac{dh_j}{dt} = q_{j,i} - q_{j,o} \tag{6.72}$$

The inlet perturbation in flow rate is

$$q_{j,i} = \frac{h_j - h_{j-1}}{R_j} + q_j(t) \tag{6.73}$$

The outlet perturbation in flow rate is

$$q_{j+1,o} = \frac{h_{j+1} - h_j}{R_{j+1}} \tag{6.74}$$

In Equations (6.73) and (6.74), R_j is the resistance of pipe j at steady state and $q_j(t)$ is the flow rate of fluid flowing into tank j from an external source.

Example 6.6

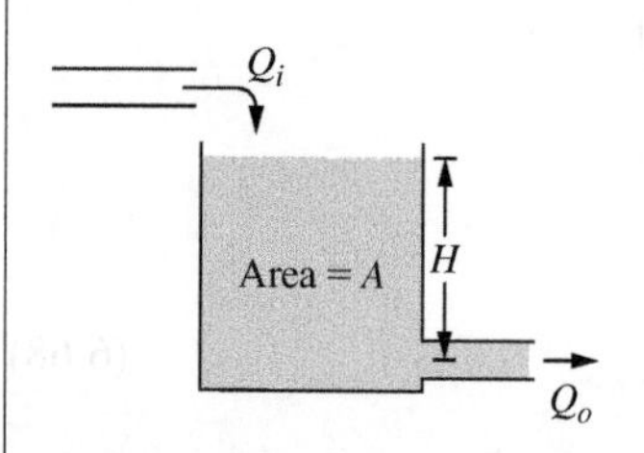

Figure 6.11

System for Example 6.6. (Repeated)

The tank of Figure 6.11 has a cross-sectional area of 900 m^2. The steady-state level is 50 m for an input flow rate of 2.5 m^3/s. Derive a mathematical model for the perturbation in the level when the input flow rate is suddenly changed to 2.45 m^3/s.

Solution

The resistance of the pipe at steady state, assuming turbulent flow, is determined using Equation (6.39) as

$$\begin{aligned} R &= 2\frac{h_s}{q_{o,s}} \\ &= 2\frac{50 \text{ m}}{2.50 \text{ m}^3/\text{s}} \\ &= 40 \text{ s/m}^2 \end{aligned} \tag{a}$$

The perturbation in the input flow rate is written using the unit step function as

$$q_i(t) = -0.05u(t) \text{ m}^3/\text{s} \tag{b}$$

Using Equations (a) and (b) in Equation (6.65) yields

$$900\frac{dh}{dt} + \frac{1}{40}h = -0.05u(t) \tag{c}$$

Example 6.7

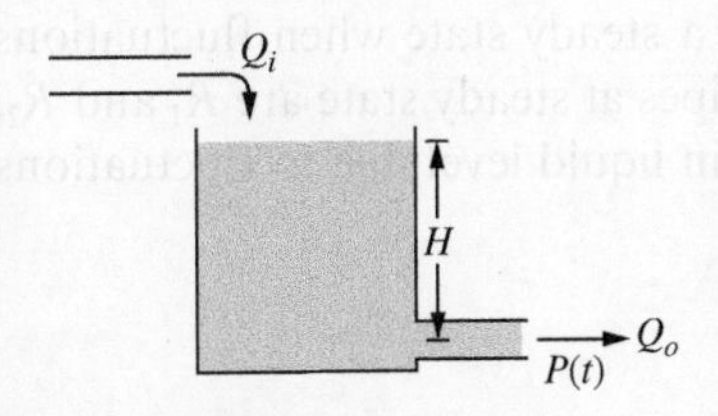

Figure 6.13

The system of Example 6.7. Pressure fluctuations at the tank's exit lead to perturbation in liquid level from the steady state.

The tank of Figure 6.13 is operating at a steady state when the pressure at the exit of the pipe begins to fluctuate according to $P = p_s + p(t)$. The input flow rate remains constant. Derive a mathematical model for the perturbation of the liquid level in the tank.

Solution

The steady state is defined by $H = h_s$, $Q_i = q_{i,s}$, $Q_o = q_{o,s}$, $dh_s/dt = 0$, and $Q_i = Q_o$. The perturbations in liquid-level and outlet flow rate are defined by

$$Q_o = q_{o,s} + q_o(t) \tag{a}$$

$$H = h_s + h(t) \tag{b}$$

The application of Conservation of Mass to the tank leads to Equation (6.71), which applied to this system becomes

$$q_{i,s} - Q_o = A\frac{dh}{dt} \tag{c}$$

The application of the extended Bernoulli's equation along a streamline in the pipe gives

$$H = h_\ell + \frac{p(t)}{\rho g} \tag{d}$$

The outlet flow rate is related to head loss through Equation (6.57), leading to

$$H = \frac{Q_o^2}{B^2} + \frac{p}{\rho g} \tag{e}$$

Equation (e) is rearranged as

$$Q_o = B\sqrt{H - \frac{p}{\rho g}}$$

$$= B\sqrt{h_s + h - \frac{p}{\rho g}} \tag{f}$$

Substituting Equation (f) into Equation (c) leads to

$$q_{i,s} - B\sqrt{h_s + h - \frac{p}{\rho g}} = A\frac{dh}{dt} \tag{g}$$

Assuming $|h - \frac{p}{\rho g}| \ll h_s$, Equation (g) is linearized using the methods of Section 1.6. h_s is factored out of the square root term. The binomial expansion is applied to the square root term and the result is truncated after the first-order term. The result is

$$q_{i,s} - B\sqrt{h_s}\left(1 + \frac{1}{2}\frac{h - \frac{p}{\rho g}}{h_s}\right) = A\frac{dh}{dt} \tag{h}$$

Applying the steady-state condition [Equation (6.59)] to Equation (h) leads to

$$\frac{B}{2\sqrt{h_s}}\left(\frac{p}{\rho g} - h\right) = A\frac{dh}{dt} \tag{i}$$

The resistance of the pipe at the steady state is obtained using Equations (6.36) and (6.38) as

$$R = \frac{2\sqrt{h_s}}{B} \tag{j}$$

Using Equation (j) in Equation (i) leads to

$$A\frac{dh}{dt} + \frac{1}{R}h = \frac{p}{\rho g R} \tag{k}$$

Example 6.8

The two-tank system of Figure 6.14 is operating at a steady state when fluctuations in the inlet flow rates occur. The resistances of the pipes at steady state are R_1 and R_2. Derive a mathematical model for the perturbations in liquid level due to fluctuations in the inlet flow rates.

Solution

The following steady-state variables are defined:

Liquid level in tank 1	h_{s1}
Liquid level in tank 2	h_{s2}
Inlet flow rate to tank 1	q_{si1}
Inlet flow rate to tank 2	q_{si2}
Outlet flow rate from pipe 2	$q_{so} = q_{si1} + q_{si2}$

Perturbation variables are introduced:

$$H_1(t) = h_{s1} + h_1(t) \tag{a}$$

$$H_2(t) = h_{s2} + h_2(t) \tag{b}$$

$$Q_{i1}(t) = q_{si1} + q_{i1}(t) \tag{c}$$

$$Q_{i2}(t) = q_{si2} + q_{i2}(t) \tag{d}$$

$$Q_o(t) = q_{so} + q_o(t) \tag{e}$$

Define

$$Q_{p1}(t) = q_{sp1} + q_{p1}(t) \tag{f}$$

as the flow rate in pipe 1, connecting the two tanks.

The application of Conservation of Mass to each tank leads to

$$Q_{i1} - Q_{p1} = A_1 \frac{dH_1}{dt} \tag{g}$$

$$Q_{i2} + Q_{p1} - Q_o = A_2 \frac{dH_2}{dt} \tag{h}$$

The steady-state conditions are

$$q_{si1} - q_{sp1} = 0 \tag{i}$$

$$q_{sp1} + q_{si2} - q_{so} = 0 \tag{j}$$

The application of the extended Bernoulli's equation between the free surface of tank 1 and the exit of the pipe connecting the two tanks, noting that the pressure head at the exit of the pipe is P and assuming the kinetic energy of the flow is lost as the pipe exits into the second tank, leads to

$$H_1 = h_{\ell 1} + H_2 \tag{k}$$

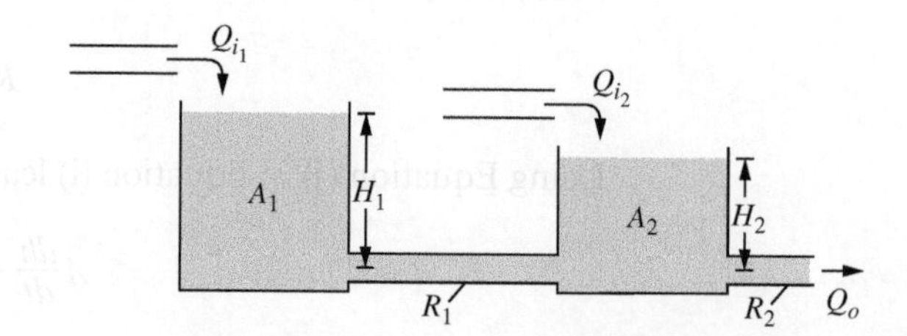

Figure 6.14
Two-tank system of Example 6.8.

The application of the extended Bernoulli's equation between the free surface of tank 2 and the outlet of the second pipe leads to

$$H_2 = h_{\ell 2} \tag{l}$$

Equations (k) and (l) can be rewritten as

$$h_1 = \Delta h_{\ell 1} + h_2 \tag{m}$$

$$h_2 = \Delta h_{\ell 2} \tag{n}$$

where $\Delta h_{\ell 1}$ and $\Delta h_{\ell 2}$ are the changes in head loss from the steady state. The linearized resistances, defined as in Equation (6.68), are related to the flow rate perturbations by

$$q_{p1} = \frac{\Delta h_{\ell 1}}{R_1} \tag{o}$$

$$q_o = \frac{\Delta h_{\ell 2}}{R_2} \tag{p}$$

Using Equations (o) and (p) in Equations (m) and (n) leads to

$$q_{p1} = \frac{1}{R_1}(h_1 - h_2) \tag{q}$$

$$q_o = \frac{1}{R_2}h_2 \tag{r}$$

Substitution of Equations (a)–(f) and (q) and (r) in Equations (g) and (h) leads to

$$q_{si1} + q_{i1} - q_{sp1} - \frac{1}{R_1}(h_1 - h_2) = A_1\frac{dh_1}{dt} \tag{s}$$

$$q_{si2} + q_{i2} + q_{sp1} + \frac{1}{R_1}(h_1 - h_2) - q_{so} - \frac{1}{R_2}h_2 = A_2\frac{dh_2}{dt} \tag{t}$$

Eliminating the steady-state terms from Equations (s) and (t) and rearranging leads to

$$A_1\frac{dh_1}{dt} + \frac{1}{R_1}h_1 - \frac{1}{R_1}h_2 = q_{i1}(t) \tag{u}$$

$$A_2\frac{dh_2}{dt} - \frac{1}{R_1}h_1 + \left(\frac{1}{R_1} + \frac{1}{R_2}\right)h_2 = q_{i2}(t) \tag{v}$$

Equations (u) and (v) can be derived by defining two control volumes, one encompassing each tank and outlet pipe. Conservation of Mass is applied to each control volume using perturbation variables. The equations are pre-linearized by using the assumption that the resistances of the pipes are obtained using steady-state values and applying Equations (6.73) and (6.74).

6.5 Pneumatic and Hydraulic Systems

Many dynamic systems have **pneumatic** components, which involve the flow of a compressed gas (usually air) thorough a passage or use air to supply power. Many dynamic systems have **hydraulic** components, which involve the flow of an incompressible liquid through a passage or use liquid to supply power. Pneumatic and hydraulic systems are used in motion control devices.

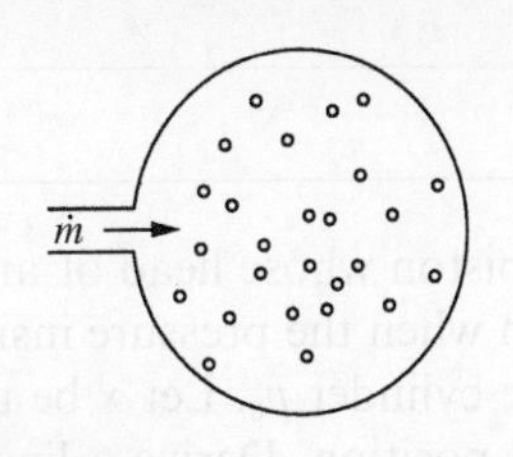

Figure 6.15

When mass flows into the pressure vessel, the density and pressure of the gas increase.

6.5.1 Pneumatic Systems

Consider the pressure vessel of Figure 6.15. When the valve between the pressure vessel and the supply pipe is open, gas flows into the pressure vessel at a mass flow rate $\dot{m}$.

The application of the control volume form of Conservation of Mass to a control volume consisting of the pressure vessel leads to

$$\dot{m} = \frac{dm_v}{dt} \tag{6.75}$$

where m_v is the mass of the gas in the vessel. As gas flows into the pressure vessel, the total mass in the vessel increases. If the volume of the vessel is constant, then as the mass increases the density and pressure of the gas in the pressure vessel also increase. An **isothermal** process occurs at a constant temperature. Under certain conditions, the temperature may also change, affecting the pressure.

The relationship between density, pressure, and temperature in a gas is called an **equation of state**. The equation of state for an ideal gas, such as air, helium, or nitrogen, is called the ideal gas law and is written as

$$p = \rho R_g T \tag{6.76}$$

where ρ is the density of the gas, p is the absolute pressure of the gas, and T is the absolute temperature of the gas. Equation (6.76) is valid if the gas in question has a low density. The constant of proportionality R_g is called the gas constant, which for an ideal gas is defined by

$$R_g = \frac{\overline{R}}{M} \tag{6.77}$$

where M is the molecular mass of the gas and $\overline{R}$ is the universal gas constant:

$$\begin{aligned} \overline{R} &= 8314.3 \text{ J/(kg·mol·K)} \\ &= 1545.3 \text{ ft·lb/(lbm·mol·R)} \end{aligned} \tag{6.78}$$

For air, often used in pneumatic systems, $M = 28.9/\text{mol}$ and the gas constant is

$$R_g = 287.0 \text{ J/(kg·K)} = 53.35 \text{ ft·lb/(lbm·R)} \tag{6.79}$$

A process is **isentropic** (constant entropy) if it is reversible and adiabatic. Throughout an isentropic process, the pressure and density of an ideal gas are related by

$$\frac{p}{\rho^{\gamma}} = \text{constant} \tag{6.80}$$

where $\gamma = c_p/c_v$, the specific heat ratio. For air, $\gamma = 1.4$.

Many processes are neither isothermal nor isentropic, but instead polytropic, in which the pressure and density are related throughout the process by

$$\frac{p}{\rho^{n}} = \text{constant} \tag{6.81}$$

where n is the polytropic exponent. It is noted that isothermal processes ($n = 1$) and isentropic processes ($n = \gamma$) are special cases of polytropic processes.

Example 6.9

The mass m in Figure 6.16(a) is connected to the end of a piston whose head of area A is in a sealed cylinder of air. The system is in equilibrium when the pressure inside the cylinder is the same pressure as the air surrounding the cylinder p_0. Let x be the displacement of the particle from the system's equilibrium position. Derive a linear mathematical model for the system, assuming the expansion and contraction of the air in the cylinder are isentropic processes.

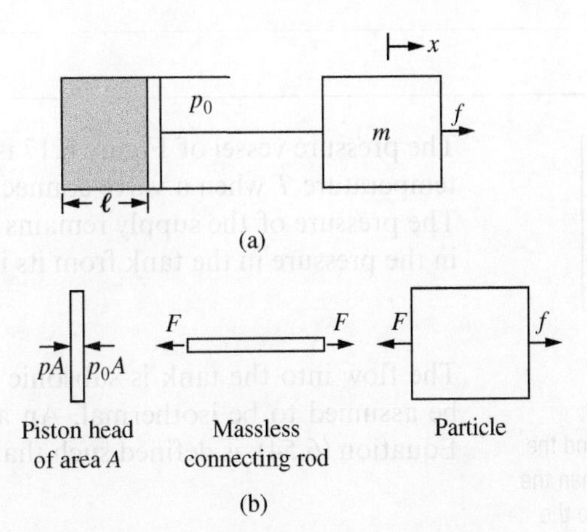

Figure 6.16
(a) The system of Example 6.9, in which air expands and contracts in a sealed cylinder as a piston moves in the cylinder; (b) the free-body diagrams at an arbitrary instant.

Solution

The application of Newton's second law to the free-body diagram of the particle drawn at an arbitrary instant [Figure 6.16(b)] leads to

$$-F + f(t) = m\ddot{x} \tag{a}$$

where F is the resultant force on the piston. Assuming the connecting rod and the piston head are massless,

$$F = p_0 A - pA \tag{b}$$

Since the cylinder is sealed, the mass of air contained in the cylinder is a constant

$$m_a = \rho_0 \ell A \tag{c}$$

where ℓ is the length of the column of air when the system is in equilibrium. When an external force $f(t)$ is applied to the piston and the piston is displaced a distance x from its equilibrium position, the length of the column of air is increased by x, leading to

$$m_a = \rho(\ell + x)A \tag{d}$$

Equations (c) and (d) are used to determine the density as

$$\rho = \rho_0\left(1 + \frac{x}{\ell}\right)^{-1} \tag{e}$$

The process is assumed to be isentropic. Thus Equations (6.80) and (e) lead to

$$p = p_0\left(1 + \frac{x}{\ell}\right)^{-\gamma} \tag{f}$$

Substitution of Equations (b) and (f) into Equation (a) leads to

$$m\ddot{x} - p_0 A\left(1 + \frac{x}{\ell}\right)^{-\gamma} = -p_0 A + f(t) \tag{g}$$

Equation (g) is linearized by applying the binomial expansion

$$\left(1 + \frac{x}{\ell}\right)^{-\gamma} = 1 - \gamma\frac{x}{\ell} + \frac{\gamma(\gamma + 1)}{2}\left(\frac{x}{\ell}\right)^2 + \cdots \tag{h}$$

Using Equation (h) in Equation (g) and truncating after the linear term leads to

$$m\ddot{x} + \frac{\gamma p_0 A}{\ell}x = f(t) \tag{i}$$

Equation (i) provides a linear mathematical model for the system in which the air in the piston and cylinder acts as a spring of stiffness $k = \gamma p_0 A/\ell$.

Example 6.10

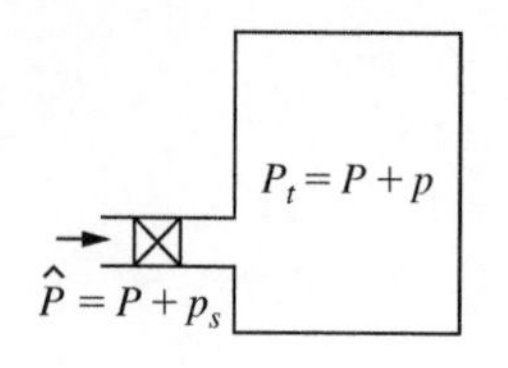

Figure 6.17

The system of Example 6.10: When the valve is opened and the supply pressure is greater than the tank pressure, gas flows into the pressure vessel.

The pressure vessel of Figure 6.17 is of volume V and contains air at a pressure P and temperature T when a valve connecting the vessel to an external air supply is opened. The pressure of the supply remains constant at $\hat{P} = P + p_s$. Let p be the perturbation in the pressure in the tank from its initial value defined such that

$$P_t = P + p \tag{a}$$

The flow into the tank is subsonic at a very low Mach number and the process can be assumed to be isothermal. An average resistance for the pipe and valve R, as in Equation (6.54), is defined such that

$$R = \frac{p_s - p}{\dot{m}} \tag{b}$$

Derive a mathematical model for $p(t)$.

Solution

The application of the ideal gas law to the air in the tank before the valve is opened leads to

$$P = \rho_0 R_a T$$
$$T = \frac{P}{\rho_0 R_a} \tag{c}$$

The application of the control volume form of Conservation of Mass to the pressure vessel results in Equation (6.75). The mass of the air in the vessel at any time is

$$m_v = \rho V \tag{d}$$

The application of the ideal gas law at an arbitrary time during the process leads to

$$\rho = \frac{P_t}{R_a T} \tag{e}$$

Since the process is isothermal, Equation (c) is used in Equation (e), leading to

$$\rho = \rho_0 \frac{P_t}{P} = \rho_0\left(1 + \frac{p}{P}\right) \tag{f}$$

Use of Equations (b), (d), and (f) in Equation (6.75) gives

$$\frac{d}{dt}\left[\rho_0 V\left(1 + \frac{p}{P}\right)\right] = \frac{p_s - p}{R} \tag{g}$$

Noting from Equation (c) that $\rho_0/P = 1/R_a T$, Equation (g) is simplified to

$$\frac{V}{R_a T}\frac{dp}{dt} + \frac{p}{R} = \frac{p_s}{R} \tag{h}$$

Equation (h) is a first-order differential equation analogous to those for an RC circuit or single-tank liquid-level system. The resistance of a pneumatic system is defined in general as the ratio of the change in pressure difference to the change in mass flow rate. The pressure difference is a nonlinear function of the mass flow rate. Thus the resistance, which is the slope of the line tangent to the $p - p_s$ vs. $\dot{m}$ curve, varies with mass flow rate.

The capacitance of a pressure vessel is defined as the ratio of the change in the mass in the pressure vessel to the change in pressure:

$$C = \frac{dm}{dp}$$
$$= \frac{d}{dp}(\rho V)$$
$$= V\frac{d\rho}{dp} \tag{6.82}$$

The application of Equation (6.82) to a pressure vessel being filled during an isothermal process at a temperature T with an ideal gas of gas constant R_g leads to

$$C = \frac{V}{R_g T} \tag{6.83}$$

6.5.2 Hydraulic Systems

Hydraulic power is used in machine tools, precision control systems, and a variety of applications in which high power and high precision are required.

The pressure variation in an incompressible fluid varies linearly with depth into the liquid. If the pressure at a certain depth into a liquid of density ρ is p_0, then the pressure at a depth h further into the liquid is

$$p = p_0 + \rho g h \tag{6.84}$$

The pressure in a static liquid is constant at a given depth, not varying across the cross section of its container. These concepts are used to design hydraulic systems such as lifts that provide great mechanical advantage.

Example 6.11

A hydraulic lift is illustrated in Figure 6.18. At $t = 0$, a mass m rests on the surface of the right tank while the left tank is subject to a pressure such that the system is in equilibrium with both sides at the same level. Compressed air is then fed into the left tank such that its free surface pressure increases at a constant rate $\dot{p}_1$, which causes the mass to rise. Let $y(t)$ be the upward displacement of the mass from its initial position. Derive a mathematical model for $y(t)$, neglecting the inertia of the liquid.

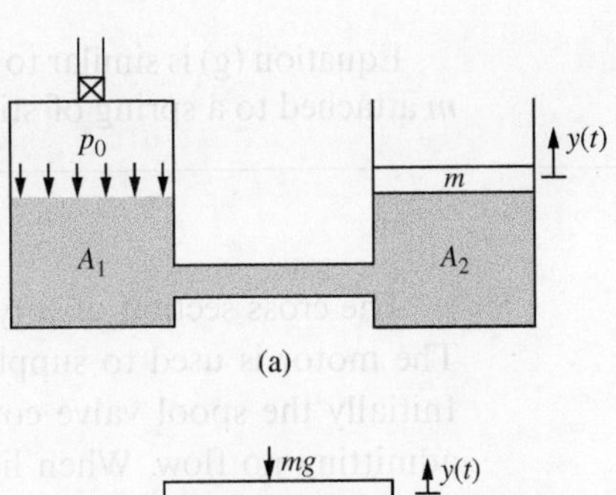

Figure 6.18
(a) The hydraulic lift of Example 6.11; (b) the free-body diagram of mass at an arbitrary instant.

Solution

Since the liquid in both columns is at the same level at $t = 0$, the pressures on both surfaces are equal to

$$p_0 = \frac{mg}{A_2} \tag{a}$$

If the rate of change of pressure on the left surface is constant, then

$$p_1 = p_0 + \dot{p}_1 t \tag{b}$$

The increase in pressure on the free surface of the left column leads to an increase in pressure on the right column, which in turn causes the mass to rise. Let y be the upward displacement of the mass from its initial position, and let F_2 be the force acting on the mass due to the pressure. The application of Newton's law to the free-body diagram of Figure 6.18(b) leads to

$$F_2 = m\ddot{y} + mg \tag{c}$$

The pressure on the surface of the right column is

$$\begin{aligned} p_2 &= \frac{F_2}{A_2} \\ &= \frac{m\ddot{y}}{A_2} + \frac{mg}{A_2} \\ &= \frac{m\ddot{y}}{A_2} + p_0 \end{aligned} \tag{d}$$

Let $x(t)$ be the downward displacement of the left column of liquid. Since the liquid is incompressible, the decrease in volume on the left side equals the increase in volume on the right side:

$$A_1 x = A_2 y$$

$$x = \frac{A_1}{A_2} y \tag{e}$$

The difference in liquid level between both sides is $y + x$. The pressure varies linearly with depth into the liquid. Thus

$$p_1 = p_2 + \rho g(y + x) \tag{f}$$

Substitution of Equations (b), (d), and (e) into Equation (f) leads to

$$p_0 + \dot{p}_1 t = \frac{m\ddot{y}}{A_2} + p_0 + \rho g\left(1 + \frac{A_1}{A_2}\right)y$$

$$m\ddot{y} + \rho g(A_1 + A_2)y = \dot{p}_1 A_2 t \tag{g}$$

Equation (g) is similar to the equation for a mechanical system composed of a mass m attached to a spring of stiffness $\rho g(A_1 + A_2)$ acted on by an external force $\dot{p}_1 A_2 t$.

The cross section of a typical hydraulic servomotor is illustrated in Figure 6.19. The motor is used to supply power to a mechanical system, often called the load. Initially the spool valve covers the opening to the cylinder containing the piston, admitting no flow. When liquid is supplied to the servomotor, the change in pressure forces the spool valve to open, allowing flow into the cylinder. The pressure difference created across the head of the piston causes it to move and supply power to the load.

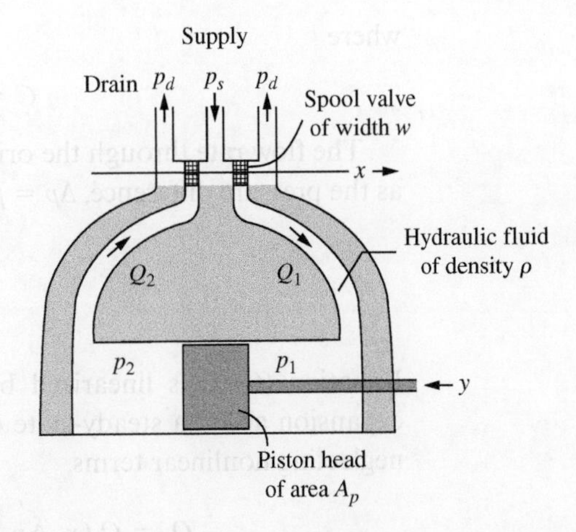

Figure 6.19
Cross section of a hydraulic servomotor.

Liquid is supplied to the servomotor at a pressure p_s. The flow drains into a line of pressure p_d. The flow from the drain may be filtered and circulated to resupply the servomotor. The area of the face of the piston is A_p. The spool valve is of width w and is assumed to be critically lapped such that any displacement of the spool valve allows flow into the cylinder. Let $x(t)$ represent the displacement of the spool valve and let $y(t)$ represent the displacement of the piston head.

Consider a control volume as defined in Figure 6.20, consisting of the fluid in the orifice flowing from the valve to the cylinder and the fluid to the right of the piston head. Let Q_1 be the flow rate through the spool valve into the cylinder. The rate at which mass accumulates in the control volume is $\rho A_p \dot{y}$. The application of Conservation of Mass to this control volume leads to

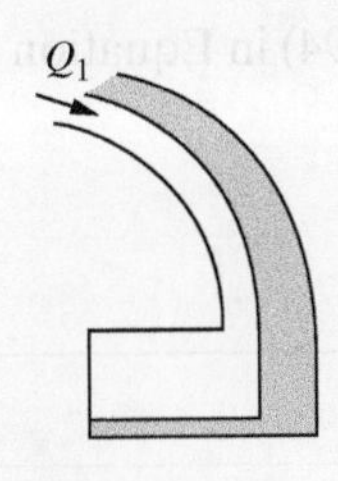

Figure 6.20
A control volume used for the analysis of the hydraulic servomotor.

$$\rho Q_1 = \rho A_p \frac{dy}{dt} \tag{6.85}$$

The flow through the spool valve is the same as the flow through an orifice. Application of Equation (6.51), the relation between the pressure drop across an orifice and the volumetric flow rate through an orifice, to the spool valve leads to

$$Q_1 = C_Q A \sqrt{\frac{2(p_s - p_1)}{\rho}} \tag{6.86}$$

where C_Q is the flow coefficient and A, the area of the valve opening, is

$$A = wx \tag{6.87}$$

Substitution of Equations (6.86) and (6.87) into Equation (6.85) leads to

$$C_Q wx \sqrt{\frac{2(p_s - p_1)}{\rho}} = A_p \frac{dy}{dt} \tag{6.88}$$

The application of Conservation of Mass to a control volume encasing the entire servomotor and piston leads to $Q_1 = Q_2$. Assuming the discharge coefficients and areas of both spool valves are the same, the pressure drop across both valves must also be the same. Thus

$$p_2 - p_d = p_s - p_1 \tag{6.89}$$

A steady-state condition occurs when the pressure on each side of the piston is equal and constant. In this case, Equation (6.88) becomes

$$\hat{C}x = A_p \dot{y} \tag{6.90}$$

where

$$\hat{C} = C_Q w\sqrt{\frac{2(p_s - p_1)}{\rho}} \tag{6.91}$$

The flow rate through the orifice is a function of the length of the opening as well as the pressure difference, $\Delta p = p_s - p_1$:

$$\begin{aligned} Q_1 &= Q(x, \Delta p) \\ &= C_d w\sqrt{\frac{2}{\rho}}\, x\sqrt{\Delta p} \end{aligned} \tag{6.92}$$

Equation (6.92) is linearized by expanding Q_1 using a two-variable Taylor series expansion about a steady-state operating point, defined by x_s, and Δp_s. To this end, neglecting nonlinear terms,

$$\begin{aligned} Q_1 &= Q_1(x_s, \Delta p_s) + x\frac{\partial Q_1}{\partial x}(x_s, \Delta p_s) + \Delta p\frac{\partial Q_1}{\partial \Delta p}(x_s, \Delta p_s) \\ &= C_d w\sqrt{\frac{2}{\rho}}\left(x_s\sqrt{\Delta p_s} + \sqrt{\Delta p_s}\, x + \frac{1}{2}\frac{x_s}{\sqrt{\Delta p_s}}\Delta p\right) \end{aligned} \tag{6.93}$$

Equation (6.93) can be written as

$$Q_1 = Q_s + Q_x x + Q_p \Delta p \tag{6.94}$$

A linear model for the relation between the pressure difference across the orifice and the displacement of the piston is obtained by using Equation (6.94) in Equation (6.88), leading to

$$Q_s + Q_x x + Q_p \Delta p = A_p \dot{y} \tag{6.95}$$

Example 6.12

The piston of the hydraulic servomotor of Figure 6.21 is connected to a mass-spring system. Derive a mathematical model for $y(t)$, the displacement of the mass from the system's equilibrium position.

Solution

Since the piston is connected to the mass, its displacement is also $y(t)$. The force acting on the piston is

$$F_p = (p_2 - p_1)A_p \tag{a}$$

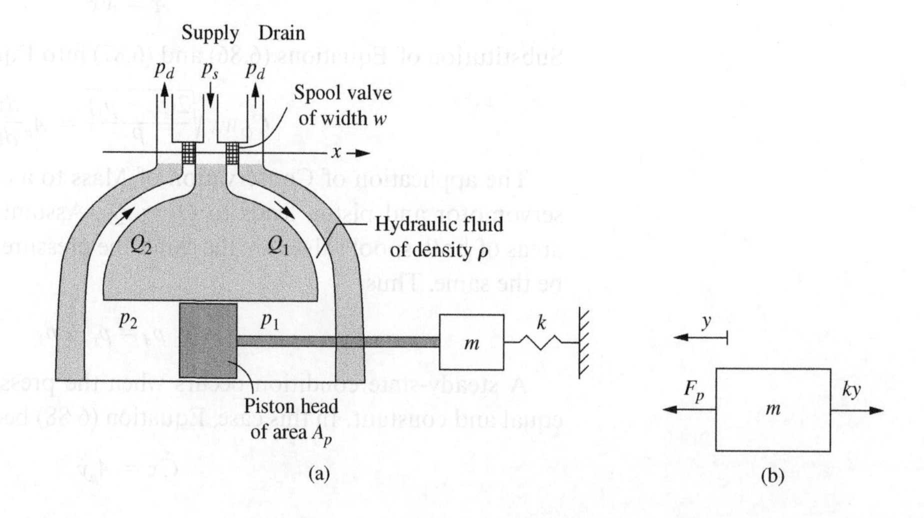

Figure 6.21
(a) The hydraulic servomotor of Example 6.12 is used to provide displacement to a mass-spring system. (b) Free-body diagram of mass attached to piston.

The application of Newton's second law to a free-body diagram of the load (Figure 6.21b) gives

$$F_p - ky = m\ddot{y} \tag{b}$$

The total pressure drop across the servomotor is $p_s - p_d$, which is the sum of the pressure drops across the orifices and across the head of the piston

$$p_s - p_d = p_s - p_1 + p_1 - p_2 + p_2 - p_d \tag{c}$$

Since the flow rates through the orifices, given by Equation (6.86), must be equal, $\Delta p = p_s - p_1 = p_d - p_2$, which when used in Equation (c) leads to

$$\Delta p = \frac{1}{2}[p_s - p_d - (p_2 - p_1)] \tag{d}$$

Substitution of Equation (b) into Equation (d) gives

$$\Delta p = \frac{1}{2}\left[p_s - p_d - \frac{1}{A_p}(m\ddot{y} + ky)\right] \tag{e}$$

Using Equation (e) in Equation (6.95) leads to

$$Q_s + Q_x x + Q_p \frac{1}{2}\left[p_s - p_d - \frac{1}{A_p}(m\ddot{y} + ky)\right] = A_p\dot{y} \tag{f}$$

Since y is measured from equilibrium, $Q_s = 0$, Equation (f) reduces to

$$m\ddot{y} + \frac{2A_p^2}{Q_p}\dot{y} + ky = \frac{2Q_x}{Q_p}x + A_p(p_d - p_s) \tag{g}$$

Example 6.13

A walking beam is connected to the hydraulic servomotor, as shown in Figure 6.22. Let $z(t)$ be the input to the system and $y(t)$ be the output from the piston. Ignoring the load, derive a mathematical model for the system.

Solution

Equation (6.90) defines the relation between the displacement of the spool valve and the displacement of the piston. Assume that the walking beam is vertical when the system is in equilibrium. The geometry of the walking beam is illustrated in Figure 6.23 for arbitrary values of z and y. Applying the principles of similar triangles leads to

$$\frac{z + y}{a + b} = \frac{x + y}{b} \tag{a}$$

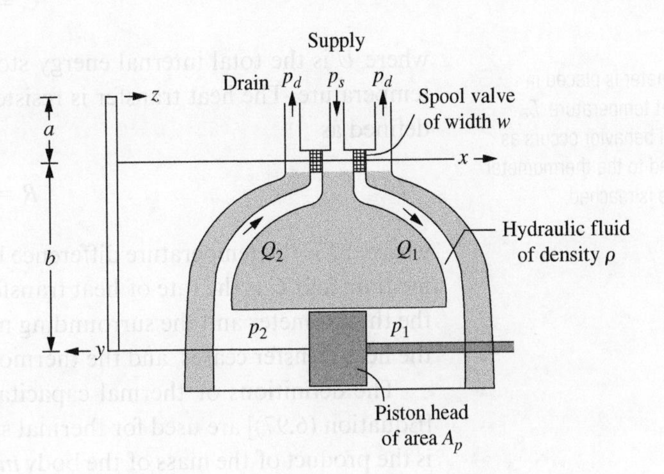

Figure 6.22
The hydraulic servomotor with walking beam of Example 6.13.

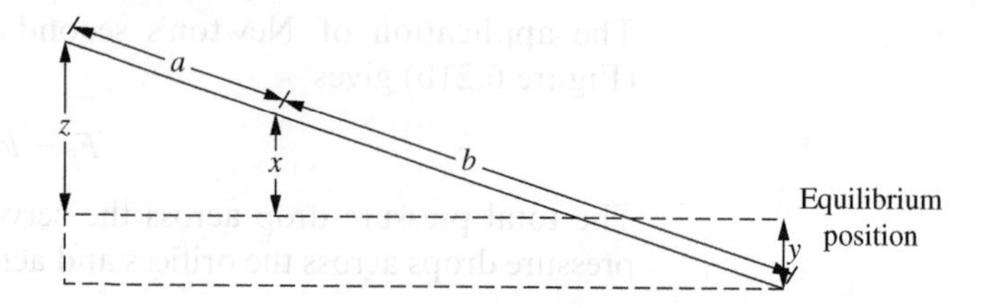

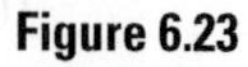
Figure 6.23

The geometry of the walking beam of Example 6.13.

Equation (a) is solved for x as

$$x = \frac{b}{a+b}z - \frac{a}{a+b}y \tag{b}$$

Substitution of Equation (b) into Equation (6.90) leads to

$$\frac{\hat{C}}{a+b}(bz - ay) = A_p\dot{y}$$

$$A_p(a+b)\dot{y} + \hat{C}ay = \hat{C}bz \tag{c}$$

Hydraulic servomotors, such as the one in Example 6.13, serve as actuators for control systems. The input to the servomotor is $z(t)$, and its output is $y(t)$. A relation such as Equation (c) of Example 6.13 is derived between the input and the output. This relation leads to the development of a transfer function between the input and output. When $b \gg a$, the servomotor of Example 6.13 acts as an integral controller.

6.6 Thermal Systems

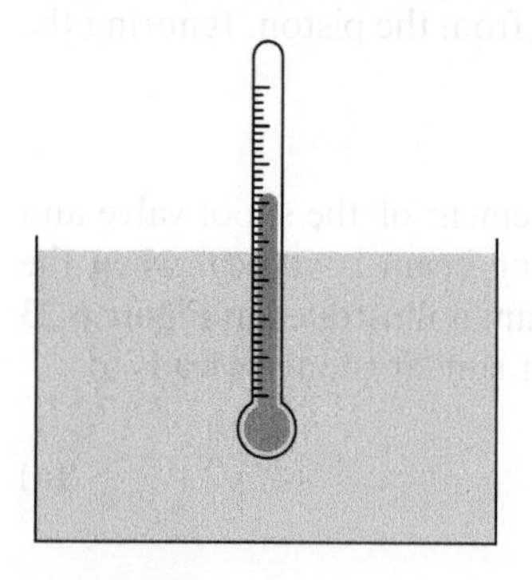

Figure 6.24

When a thermometer is placed in a bath of ambient temperature T_∞, transient thermal behavior occurs as heat is transferred to the thermometer until steady state is reached.

A thermal system is a system in which heat transfer occurs. A thermometer is a classic example of a dynamic thermal system. When the thermometer of Figure 6.24, initially in a steady state at a uniform temperature, is suddenly placed in a medium at a higher temperature, heat is transferred to the thermometer. A transient response occurs as the temperature of the thermometer gradually approaches the temperature of the surrounding medium. The thermometer has a capacity for storing internal energy that accumulates due to the transfer of heat from the surrounding medium. The capacitance of the thermometer is defined as

$$C = \frac{dU}{dT} \tag{6.96}$$

where U is the total internal energy stored in the thermometer and T is the absolute temperature. The heat transfer is resisted from the medium with a thermal resistance defined as

$$R = \frac{\Delta T}{\dot{Q}} \tag{6.97}$$

where ΔT is the temperature difference between the thermometer and the surrounding medium and $\dot{Q}$ is the rate of heat transfer between the two. When the temperatures of the thermometer and the surrounding medium are equal, thermal equilibrium occurs, the heat transfer ceases, and the thermometer stops storing energy.

The definitions of thermal capacitance [Equation (6.96)] and thermal resistance [Equation (6.97)] are used for thermal systems in general. The total internal energy U is the product of the mass of the body m and the specific internal energy u. Recall from

Equation (6.10) that the specific internal energy is equal to the specific heat c_v times the temperature T. The application of Equation (6.96) leads to the thermal capacitance of a body as

$$C = mc_v \tag{6.98}$$

The transient behavior of the thermometer of Figure 6.24 is modeled by applying Conservation of Energy to a control volume containing the thermometer. Energy, in the form of heat transfer, is transferred into the control volume through its boundary. Conservation of Energy reduces to the rate of change of internal energy is equal to the rate at which heat is transferred to the thermometer:

$$\frac{dU}{dt} = \dot{Q} \tag{6.99}$$

Assuming constant specific heat and using the definition of thermal resistance [Equation (6.97)], Equation (6.99) becomes

$$mc_v \frac{dT}{dt} = -\frac{1}{R}(T - T_\infty) \tag{6.100}$$

where T_∞ is the temperature of the surrounding medium. Equation (6.100) can be rearranged as

$$mc_v \frac{dT}{dt} + \frac{1}{R}T = \frac{1}{R}T_\infty \tag{6.101}$$

Equation (6.101) is a first-order differential equation similar to that obtained for an *RC* circuit, a single-tank liquid-level problem, and a linear pneumatic system.

Heat transfer occurs through three modes: conduction, convection, and radiation. Conduction heat transfer occurs when a body is not at a uniform temperature. Molecular motion results due to the difference in internal energy between higher temperature regions and lower temperature regions. Conduction heat transfer occurs between these molecules. The rate of conduction heat transfer per unit area is given by Fourier's conduction law as

$$\frac{\dot{Q}}{A} = -k\frac{\partial T}{\partial n} \tag{6.102}$$

where k is the thermal conductivity of the material and n is the direction normal to the area through which heat is transferred. For one-dimensional heat transfer in the x direction, Equation (6.102) becomes

$$\frac{\dot{Q}}{A} = -k\frac{dT}{dx} \tag{6.103}$$

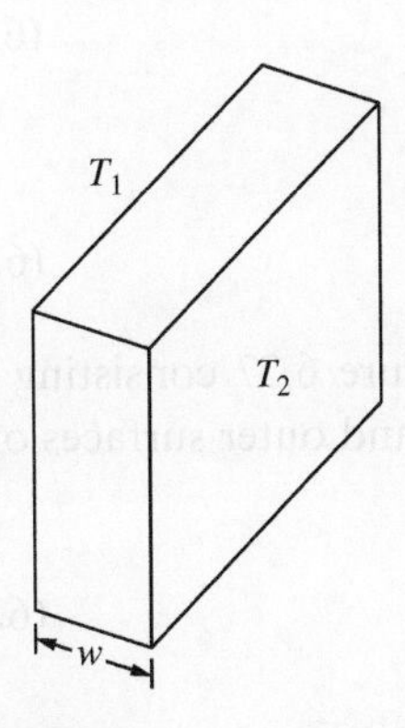

Figure 6.25

One-dimensional conduction heat transfer occurs through the wall of surface area A and width w when exterior surfaces are at different temperatures.

Consider a region of width w and surface area A as illustrated in Figure 6.25. The temperatures on the exterior surfaces of the region are T_1 and T_2. In the steady state, the temperature gradient across the wall is a constant $\frac{dT}{dx} = \frac{T_2 - T_1}{w}$, in which case Equation (6.103) becomes

$$\dot{Q} = \frac{kA}{w}(T_1 - T_2) \tag{6.104}$$

Comparison of Equation (6.104) with Equation (6.97) shows that the resistance due to conduction is

$$R = \frac{w}{kA} \tag{6.105}$$

A similar analysis is used to show that the resistance due to conduction in an annular cylinder of length L, inner radius r_i, and outer radius r_o, which is made of a material of thermal conductivity k, is

$$R = \frac{1}{2\pi kL}\ln\left(\frac{r_o}{r_i}\right) \tag{6.106}$$

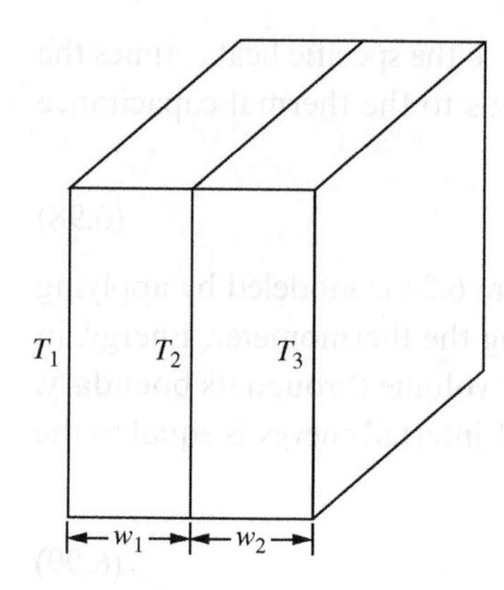

Figure 6.26
The equivalent thermal resistance of two regions in series is the sum of the resistances.

Conduction heat transfer occurs within the two adjacent regions illustrated in Figure 6.26. In the absence of other heat sources, the rate of heat transfer across each region is the same and the total temperature change is the sum of the temperature changes across each region. Thus the temperature gradient across the entire region can be written as

$$\begin{aligned} \Delta T &= \Delta T_1 + \Delta T_2 \\ &= R_1\dot{Q} + R_2\dot{Q} \\ &= R_{eq}\dot{Q} \end{aligned} \tag{6.107}$$

Equation (6.107) shows that the equivalent resistance is the sum of their individual resistances. This is the same relation as two resistors in series in an electric circuit. Thus, regions where the rate of heat transfer is the same across the entire region and the temperature change is the sum of the temperature changes across individual regions are said to be in series.

If the temperature change is the same in both regions but the rate of heat transfer is the sum of the rates of heat transfer, then the two regions are in parallel. The region has an equivalent resistance equal to the reciprocal of the sum of the reciprocals of the individual regions. This is analogous to the equation for the equivalent resistance of resistors in parallel in an electric circuit.

Convection heat transfer occurs between a body and its surrounding medium when the body and the medium have different temperatures. While convection occurs between a moving fluid and a body, the following discussion is limited to bodies at rest. Fluid particles adjacent to a solid body must be at the same temperature as the body. Since the temperature of the surrounding medium is different from that of the body, a thin region near the surface of the body must exist where a temperature gradient exists. This region is called a thermal boundary layer. The rate of heat transfer per unit area due to convection between a body whose surface is at a temperature T and the surrounding medium at a temperature T_∞ due to the existence of the thermal boundary layer is given by Newton's law of cooling as

$$\frac{\dot{Q}}{A} = h(T - T_\infty) \tag{6.108}$$

where h is called the film coefficient or heat transfer coefficient. Equation (6.108) can be rewritten as

$$\dot{Q} = \frac{1}{R_c}(T - T_\infty) \tag{6.109}$$

where the convective resistance is

$$R_c = \frac{1}{hA} \tag{6.110}$$

The equivalent thermal resistance of the system of Figure 6.27 consisting of a multilayer wall with convection occurring at both the inner and outer surfaces of the wall is

$$R_{eq} = \frac{1}{h_i A} + \sum_j \frac{w_j}{k_j A} + \frac{1}{h_o A} \tag{6.111}$$

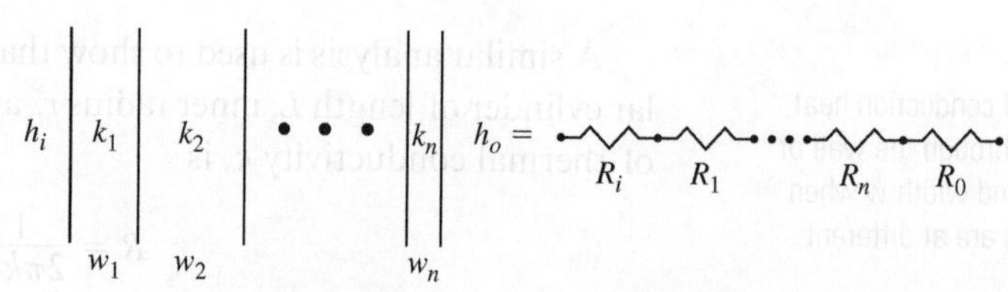

h_i k_1 k_2 • • • k_n h_o = R_i R_1 R_n R_o
w_1 w_2 w_n

Figure 6.27
The equivalent thermal resistance of a region due to convection and conduction is modeled using resistors in series.

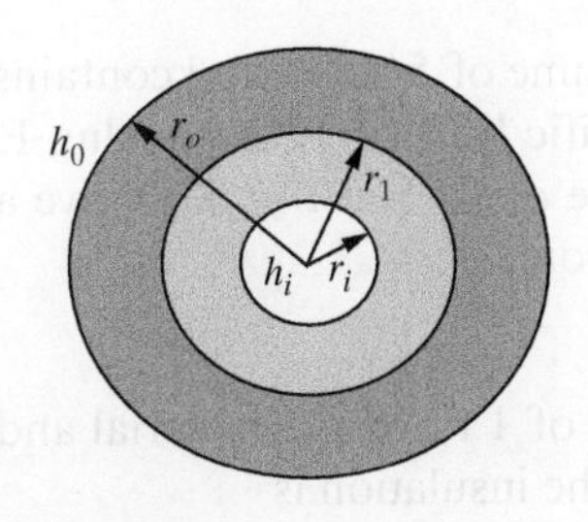

Figure 6.28

Resistance in an annular cylinder due to conduction and convection.

Equation (6.111) may be modified for the multilayer cylinder of Figure 6.28 using Equation (6.106) for the resistance for a layer of the cylinder and calculating the area over which convection occurs at the inner and outer radii of the cylinder

$$R = \frac{1}{h_i 2\pi r_i L} + \sum_j \frac{1}{2\pi k_j L} \ln\left(\frac{r_j}{r_{j-1}}\right) + \frac{1}{h_o 2\pi r_o L} \tag{6.112}$$

The **overall heat transfer coefficient** U_o is defined between two media of temperatures T_1 and T_2 over an area A:

$$\dot{Q} = U_o A(T_2 - T_1) \tag{6.113}$$

For a rectangular region, Equation (6.105) yields

$$U_o = \frac{1}{R_{eq} A} \tag{6.114}$$

When the external surface area of a cylinder, $A = 2\pi r_o L$, is used to define the overall heat transfer coefficient,

$$U_o = \frac{1}{\frac{r_o}{r_i h_i} + r_o\left[\sum_j \frac{1}{k_j}\ln\left(\frac{r_j}{r_{j-1}}\right)\right] + \frac{1}{h_o}} \tag{6.115}$$

Radiation heat transfer occurs between two surfaces at different temperatures. A black body is one that is nonreflective and opaque. The rate of radiation heat transfer per unit area between a surface of a black body at temperature T_1 and a surface of a black body at a lower temperature T_2 is

$$\frac{\dot{Q}}{A} = \sigma\mathfrak{F}(T_1^4 - T_2^4) \tag{6.116}$$

where σ is the Stefan-Boltzmann constant and $\mathfrak{F}$ is a shape factor that depends on how one surface is viewed from the other. Radiation heat transfer is significant in the transient response of a thermal system only when one surface is at a much higher temperature. The modeling of a transient thermal system with radiation heat transfer is inherently nonlinear due to the presence of the fourth power of temperatures in Equation (6.116).

Example 6.14

The exterior face of a building is made of 2-in.-thick brick with a thermal conductivity of 0.74 Btu/(hr·ft·F). The wall, which, as illustrated in Figure 6.29, has a surface area of 450 ft^2, contains a cavity with insulation having an R value of 12, and its interior surface is made of 0.5-in.-thick gypsum, which has a thermal conductivity of 0.24 Btu/(hr·ft·F). The film coefficient between the ambient and the brick is 5.3 Btu/(hr·ft^2·F). The film coefficient between the gypsum and the interior air is 2.8 Btu/(hr·ft^2·F). **a.** Determine the resistance and overall heat transfer coefficient between the ambient and the interior surfaces of the wall. **b.** The ambient temperature in the summer varies over the course of a day as

$$T_\infty = 80 + 10\sin\left(\frac{\pi t}{12}\right)\text{F} \tag{a}$$

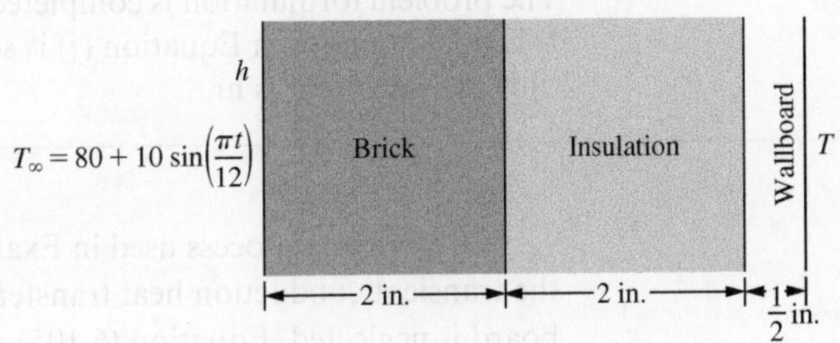

Figure 6.29

The system of Example 6.14: The exterior wall of a building is exposed to a periodic ambient temperature.

where t is measured in hours. The interior room has a volume of 5400 ft^3 and contains air with a density of 2.38×10^{-3} slug/ft^3 and has a specific heat of 7.72 Btu/slug·F. The room is not heated or cooled and is at a temperature of 74 F at $t = 0$. Derive a mathematical model for the temperature in the interior room.

Solution

a. The R value of the insulation as given is the resistance of 1 ft^2 of the material and has implied units of hr·ft^2·F/Btu. Thus the resistance of the insulation is

$$R_i = \frac{R}{A} = \frac{12 \text{ hr}\cdot\text{ft}^2\cdot\text{F/Btu}}{450 \text{ ft}^2} = 0.0267 \text{ hr}\cdot\text{F/Btu} \tag{b}$$

The resistances of the brick and gypsum are obtained using Equation (6.105),

$$R_b = \frac{(2 \text{ in.})(1 \text{ ft}/12 \text{ in.})}{(0.74 \text{ Btu/hr}\cdot\text{ft}\cdot\text{F})(450 \text{ ft}^2)} = 5.01 \times 10^{-4} \text{ hr}\cdot\text{F/Btu} \tag{c}$$

$$R_g = \frac{(0.5 \text{ in.})(1 \text{ ft}/12 \text{ in.})}{(0.24 \text{ Btu/hr}\cdot\text{ft}\cdot\text{F})(450 \text{ ft}^2)} = 3.86 \times 10^{-4} \text{ hr}\cdot\text{F/Btu} \tag{d}$$

The convective resistances are obtained using Equation (6.110),

$$R_{ci} = \frac{1}{(5.3 \text{ Btu/hr}\cdot\text{ft}^2\cdot\text{F})(450 \text{ ft}^2)} = 4.19 \times 10^{-4} \text{ hr}\cdot\text{F/Btu} \tag{e}$$

$$R_{co} = \frac{1}{(2.8 \text{ Btu/hr}\cdot\text{ft}^2\cdot\text{F})(450 \text{ ft}^2)} = 7.94 \times 10^{-4} \text{ hr}\cdot\text{F/Btu} \tag{f}$$

The equivalent resistance of the wall is

$$\begin{aligned} R_{eq} &= R_{co} + R_b + R_i + R_g + R_{ci} \\ &= (4.19 \times 10^{-4} + 5.01 \times 10^{-4} + 2.67 \times 10^{-2} + 3.86 \times 10^{-4} \\ &\quad + 7.94 \times 10^{-4}) \text{ hr}\cdot\text{F/Btu} \\ &= 2.88 \times 10^{-2} \text{ hr}\cdot\text{F/Btu} \end{aligned} \tag{g}$$

The overall heat transfer coefficient is calculated from Equation (6.114) as

$$U = \frac{1}{R_{eq}A} = \frac{1}{(2.88 \times 10^{-2} \text{ hr}\cdot\text{F/Btu})(450 \text{ ft}^2)} = 7.72 \times 10^{-2} \text{ Btu/(hr}\cdot\text{ft}^2\cdot\text{F)} \tag{h}$$

b. The interior room is assumed to be at a uniform temperature at any time. The effect of temperature change on density is neglected, and the density used is that at the initial temperature. The application of Conservation of Energy leads to Equation (6.101). To apply this equation, it is noted that

$$mc_v = (2.38 \times 10^{-3} \text{ slug/ft}^3)(5400 \text{ ft}^3)[7.72 \text{ Btu/(slug}\cdot\text{F)}] = 99.2 \text{ Btu/F} \tag{i}$$

The mathematical model for this problem is obtained by applying Equation (6.101), which becomes

$$99.2\frac{dT}{dt} + \frac{1}{2.88 \times 10^{-2}}T = \frac{1}{2.88 \times 10^{-2}}\left[80 + 10 \sin\left(\frac{\pi t}{12}\right)\right] \tag{j}$$

The problem formulation is completed by specifying the initial condition, $T(0) = 74$ F. It is noted that when Equation (j) is solved subject to the initial condition, the unit of time measurement is hr.

The modeling process used in Example 6.14 assumes lumped temperatures; that is, the transient conduction heat transfer through the brick, the insulation, and the wall-board is neglected. Equation (6.105) defining the resistance of a rectangular region is derived assuming a steady state and a constant temperature gradient. In a transient situation, the temperature gradient is nonuniform through the region; that is, $\dot{Q}$ varies with x, a spatial coordinate measured into the wall from its exterior, as well as time. The modeling used in Example 6.14 neglects this variation and assumes an average value of the temperature gradient resulting in a lumped approximation for the temperature gradient. Such an assumption is reasonable for slowly varying changes in temperature and for thin regions.

Thermal energy can be added to a fluid in a control volume through an external heater, which transfers heat to the fluid at a constant rate. Application of the control volume form of the energy equation to a control volume with heat added and with one inlet and one outlet leads to

$$\frac{d}{dt}\int \rho u dV = \dot{Q} + \dot{m}_{in}u_{in} - \dot{m}_{out}u_{out} \tag{6.117}$$

where u is the specific internal energy. Assuming the density and internal energy are constant throughout the control volume and the mass of the control volume does not change, Equation (6.117) is rewritten as

$$m\frac{du}{dt} = \dot{Q} + \dot{m}_{in}u_{in} - \dot{m}_{out}u_{out} \tag{6.118}$$

The internal energy is related to the temperature by Equation (6.10): $u = c_v T$. The specific heat for most liquids and gases varies with temperature. However, the assumption is often made that the specific heat is a constant or that, if the variation with temperature is included, when the equation is linearized, the value of the specific heat at the steady-state condition is used. If the specific heat is assumed to be constant, then Equation (6.118) becomes

$$mc_v\frac{dT}{dt} = \dot{Q} + \dot{m}_{in}c_v T_{in} - \dot{m}_{out}c_v T_{out} \tag{6.119}$$

Example 6.15

Water enters a tank of volume 40 m^3 at a flow rate of 0.5 kg/s at a temperature of 50°C. A heater delivers 4000 J/s to the water in the tank. **a.** Assuming a constant specific heat of $c_v = 4.19 \times 10^3$ J/(kg·C), what is the temperature of the outlet flow? **b.** The rate at which heat is added is suddenly changed to 4500 J/(kg·C). Derive a mathematical model for the temperature of the water in the tank as a function of time. The density of water is 1000 kg/m^3.

Solution

The application of Conservation of Mass to the tank, assuming no accumulation of water, shows that $\dot{m}_{in} = \dot{m}_{out} = \dot{m}$. It is assumed that the heat is added such that the temperature in the tank is the same as the temperature of the water at the outlet, the specific heat is constant with temperature, the density of the water is constant, and no phase changes occur. Using these assumptions, Equation (6.119) reduces to

$$\rho V c_v \frac{dT}{dt} = \dot{Q} + \dot{m}c_v(T_{in} - T) \tag{a}$$

a. In the steady state, the temperature is constant and Equation (a) reduces to

$$\dot{Q}_s + \dot{m}c_v(T_{in,s} - T_s) = 0 \tag{b}$$

Equation (b) is solved for the outlet temperature, leading to

$$T_s = T_{in,s} + \frac{\dot{Q}_s}{\dot{m}c_v}$$
$$= 50°\text{C} + \frac{4000 \text{ J/s}}{(0.5 \text{ kg/s})(4.19 \times 10^3 \text{ J/kg·C})} = 51.9°\text{C} \tag{c}$$

b. The rate of heat transfer into the tank is written as the heat transfer at the initial steady state plus a perturbation

$$\dot{Q} = \dot{Q}_s + \dot{q} \tag{d}$$

where from the given information

$$\dot{q} = 500\, u(t) \tag{e}$$

The perturbation in the rate of heat transfer leads to a perturbation, $\theta(t)$, of temperature in the tank and the temperature in the outlet stream such that

$$T(t) = T_s + \theta(t) \tag{f}$$

Substitution of Equations (d) and (f) into Equation (a) leads to

$$\rho c_v V\frac{d}{dt}(T_s + \theta) = \dot{Q}_s + \dot{q} + \dot{m}c_v(T_{in,s} - T_s - \theta) \tag{g}$$

Noting that $dT_s/dt = 0$ and using Equation (b) in Equation (g) leads to

$$\rho Vc_v\frac{d\theta}{dt} = \dot{q} - \dot{m}c_v\theta$$
$$\rho Vc_v\frac{d\theta}{dt} + \dot{m}c_v\theta = \dot{q} \tag{h}$$

Substituting given values leads to

$$1.68 \times 10^8\frac{d\theta}{dt} + 2.095 \times 10^3\theta = 500\, u(t) \tag{i}$$

The differential equation derived to model the change in temperature of the outlet stream in Example 6.15, Equation (i), is a first-order differential equation similar to that derived to describe the transient behavior of an *RC* circuit and the dynamic response of the liquid level in a tank due to a perturbation in the inlet flow rate. Equation (i) is of the form

$$C\frac{d\theta}{dt} + \frac{1}{R}\theta = \dot{q}(t) \tag{6.120}$$

Analogous to the *RC* circuit, the capacitance for the system is defined as $C = \rho Vc_v$ and the resistance is defined as $R = 1/(\dot{m}c_v)$.

6.7 Chemical Systems

A chemical system is a system in which a chemical reaction occurs by which one component is converted into another component. The specific chemical system modeled in this study is the continuous stirred tank reactor (CSTR), illustrated in Figure 6.30. The input stream to the CSTR shown has two components, which include the reactant *A* and the product of the reaction *B*. The concentrations of the components in the input stream are known. The output stream includes the same components but at different concentrations than the input stream. The assumption that the reactor is a CSTR implies that the reactions occur continuously and that the composition of the mixture in the reactor is the same as the composition of the outlet stream.

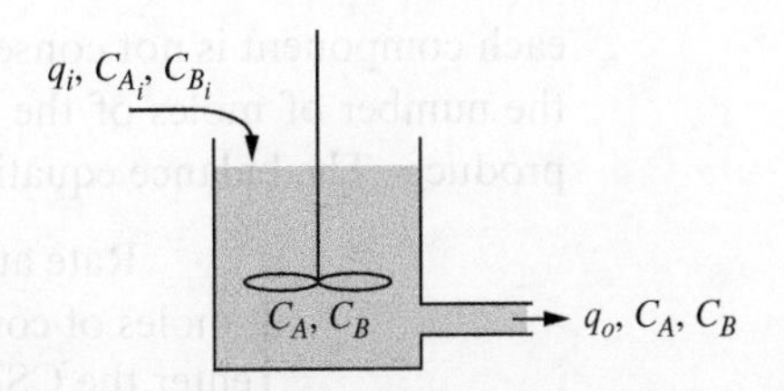

Figure 6.30
A chemical reaction occurs in a continuous stirred tank reactor (CSTR), in which component A is converted to component B.

The amount of a component in a mixture is quantified by its concentration, defined as the number of moles of the component per unit volume of the mixture. If n_A is the number of moles of component A in a mixture of volume V, the concentration of component A is

$$C_A = \frac{n_A}{V} \tag{6.121}$$

The chemical reaction that converts component A into component B is represented as

$$A \xrightarrow{k} B \tag{6.122}$$

where k is the rate at which the reaction occurs. The one-way arrow indicates that the reaction is irreversible. The rate at which the concentration of the reactant changes in a first-order reaction is

$$\frac{dC_A}{dt} = -kC_A \tag{6.123}$$

while for the product B,

$$\frac{dC_B}{dt} = kC_A \tag{6.124}$$

The negative sign in Equation (6.123) indicates that moles of component A are being consumed by the reaction, while the positive sign in Equation (6.124) indicates that the reaction is producing moles of component B.

The rate of reaction has temperature dependence according to the Arrhenius law

$$k = \alpha e^{-E/(\bar{R}T)} \tag{6.125}$$

where α is a constant, $\bar{R}$ is the gas constant, and E is the activation energy of the reaction. An isothermal reaction occurs at a constant temperature and its rate of reaction is constant. Endothermic reactions require heat to be added to affect the reaction, while exothermic reactions generate heat. Such reactions are not isothermal and their rate of reaction changes as the temperature changes according to Equation (6.125). The heat of reaction λ is defined as the energy required per mole of reactant, where λ is positive for an endothermic reaction. The rate at which heat is generated in a CSTR of constant volume V due to a reaction is obtained using Equations (6.121) and (6.123) as

$$\begin{aligned} \dot{Q} &= \lambda \frac{dn_A}{dt} \\ &= \lambda \frac{d}{dt}(VC_A) \\ &= -\lambda V k C_A \end{aligned} \tag{6.126}$$

Modeling of a CSTR requires the application of the control volume forms of Conservation of Mass and Conservation of Energy to the reactor. While the overall Conservation of Mass equation is applicable, an equation balancing the number of moles of each component in the reactor can also be written. The number of moles of each component is not conserved because the chemical reaction leads to a decrease in the number of moles of the reactants and an increase in the number of moles of the products. The balance equation for the number of moles of component A is

$$\begin{pmatrix}\text{Rate at which}\\ \text{moles of component } A\\ \text{enter the CSTR through}\\ \text{the inlet stream}\end{pmatrix} - \begin{pmatrix}\text{Rate at which the}\\ \text{moles of component } A\\ \text{exit the CSTR through}\\ \text{the outlet stream}\end{pmatrix}$$

$$+ \begin{pmatrix}\text{Rate at which the}\\ \text{moles of component } A\\ \text{are produced through}\\ \text{the chemical reaction}\end{pmatrix} = \begin{pmatrix}\text{Rate at which the}\\ \text{moles of component } A\\ \text{accumulate in}\\ \text{the CSTR}\end{pmatrix} \tag{6.127}$$

An equation of the form of Equation (6.127) can be written for each component in the reactor. Thus for a reactor with n components, n equations of the form of Equation (6.127) may be written. When the n equations are combined, they lead to the overall Conservation of Mass equation.

Example 6.16

An inlet stream for a CSTR of constant volume V contains components A and B at concentrations C_{Ai} and C_{Bi}, respectively. The volumetric flow rate into and out of the tank is q. An isothermal first-order irreversible reaction converts A into B within the tank, with a constant rate of reaction k. Derive a mathematical model for the concentrations of A and B in the outlet stream.

Solution

An assumption used in the modeling of a CSTR is that the concentrations of the components in the outlet stream are the same as those in the reactor. The rate at which moles of component A enter the tank is qC_{Ai} (volume/time $\times$ moles/volume), and the rate at which moles of component A leave the tank is qC_A. The rate at which the chemical reaction changes the concentration of component A is

$$\left(\frac{dC_A}{dt}\right)_{reaction} = -kC_A \tag{a}$$

Thus the rate of change in the number of moles of component A due to the chemical reaction is

$$\left(\frac{dn_A}{dt}\right)_{reaction} = V\left(\frac{dC_A}{dt}\right)_{reaction} = -VkC_A \tag{b}$$

Thus the application of Equation (6.127) for component A leads to

$$\frac{d}{dt}(VC_A) = qC_{Ai} - qC_A - VkC_A \tag{c}$$

Equation (6.127) is applied to component B in a similar fashion, with the exception that

$$\left(\frac{dn_B}{dt}\right)_{reaction} = VkC_A \tag{d}$$

Thus

$$\frac{d}{dt}(VC_B) = qC_{iB} - qC_B + VkC_A \tag{e}$$

Equations (d) and (e) constitute a mathematical model for this CSTR. If V is constant, these equations become

$$V\frac{dC_A}{dt} + (q + kV)C_A = qC_{Ai} \tag{f}$$

$$V\frac{dC_B}{dt} - kVC_A + qC_B = qC_{Bi} \tag{g}$$

Example 6.17

An irreversible reaction of the form of Equation (6.122) occurs in the CSTR of Figure 6.31. The reaction is exothermic with a heat of reaction λ. A cooler removes heat from the tank at a constant rate $\dot{Q}$. **a.** Derive a mathematical model for the system if the volume of the tank V is constant, the input and output flow rates are q, the temperature of the inlet stream is T_i, and the concentrations of the inlet stream are C_{Ai} and C_{Bi}. **b.** Develop a set of equations defining the system's steady state. **c.** The system is at steady state when the cooling rate decreases to $\dot{Q}_1$. Derive a linearized model for the perturbations in concentration and temperature that occur due to the change in the rate of cooling.

Solution

a. The equations for the rate of change of concentration for a constant volume CSTR are derived as Equations (f) and (g) in Example 6.16. Since the reaction gives off heat, the system is not isothermal and the temperature in the reactor changes with time. Thus the Arrhenius law, Equation (6.125), is used in these equations for the rate of reaction, resulting in

$$V\frac{dC_A}{dt} + (q + \alpha Ve^{-E/(RT)})C_A = qC_{Ai} \tag{a}$$

$$V\frac{dC_B}{dt} - \alpha Ve^{-E/(RT)}C_A + qC_B = qC_{Bi} \tag{b}$$

The application of the energy equation, Equation (6.15), to a control volume enclosing the reactor leads to

$$\rho qc_v T_i - \rho qc_v T - \dot{Q} + \lambda V\alpha e^{-E/(RT)}C_A = \rho Vc_v\frac{dT}{dt} \tag{c}$$

where Equation (6.126) is used for the rate at which heat is given off by the reaction. Equations (a), (b), and (c) constitute a mathematical model for the system with dependent variables T, C_A, and C_B.

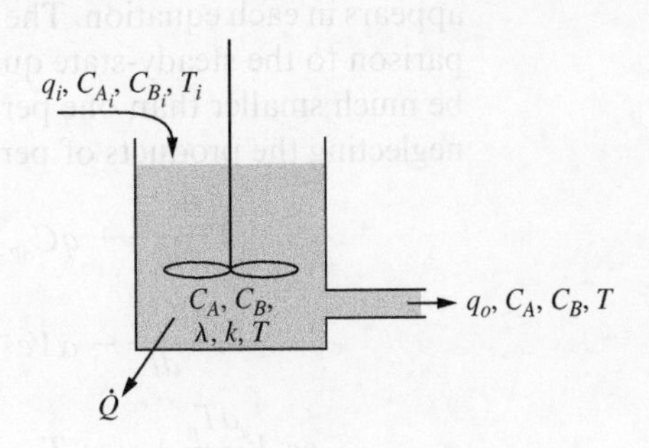

Figure 6.31
The CSTR of Example 6.17: The reaction is exothermic and a cooler is used to remove heat generated by the reaction.

b. The steady state is defined by setting $dC_A/dt = 0$, $dC_B/dt = 0$, and $dT/dt = 0$. The nonlinear algebraic equations defining the steady-state variables C_{As}, C_{Bs}, and T_s are obtained from Equations (a)–(c) as

$$(q + \alpha V e^{-E/(RT_s)}) C_{As} = q C_{Ai} \tag{d}$$

$$-q C_{as} + \alpha V e^{-E/(RT_s)} C_{Bs} = q C_{Bi} \tag{e}$$

$$\rho q c_v T_i - \rho q c_v T_s - \dot{Q} + \lambda \alpha V e^{-E/(RT_s)} C_{As} = 0 \tag{f}$$

c. When the system is at the steady state described by Equations (d)–(f) and the rate of cooling decreases to $\dot{Q}_1$, transient behavior results. Perturbations of concentrations and temperature from the steady state are introduced by

$$C_A = C_{As} + C_{Ap} \tag{g}$$

$$C_B = C_{Bs} + C_{Bp} \tag{h}$$

$$T = T_s + T_p \tag{i}$$

Substitution of Equations (g)–(i) in Equations (a)–(c) leads to

$$V\frac{d}{dt}(C_{As} + C_{Ap}) + (q + \alpha V e^{-E/[R(T_s+T_p)]})(C_{As} + C_{Ap}) = q C_{Ai} \tag{j}$$

$$V\frac{d}{dt}(C_{Bs} + C_{Bp}) - \alpha V e^{-E/[R(T_s+T_p)]}(C_{As} + C_{Ap}) + q(C_{Bs} + C_{Bp}) = q C_{Bi} \tag{k}$$

$$\rho V c_v \frac{d}{dt}(T_s + T_p) + \rho q c_v (T_s + T_p) - \lambda \alpha V e^{-E/[R(T_s+T_p)]}(C_{As} + C_{Ap}) = \rho q c_v T_i - \dot{Q}_1 \tag{l}$$

Setting the time derivatives of the steady-state terms to zero and using the linearization of the Arrhenius law obtained in Example 1.6 leads to

$$V\frac{dC_{Ap}}{dt} + qC_{As} + qC_{Ap} + \alpha V e^{-E/(RT_s)}\left(1 + \frac{E}{RT_s^2}T_p\right)(C_{As} + C_{Ap}) = qC_{Ai} \tag{m}$$

$$V\frac{dC_{Bp}}{dt} - \alpha V e^{-E/(RT_s)}\left(1 + \frac{E}{RT_s^2}T_p\right)(C_{As} + C_{Ap}) + qC_{Bs} + qC_{Bp} = qC_{Bi} \tag{n}$$

$$\rho V c_v \frac{dT_p}{dt} + \rho q c_v T_s + \rho q c_v T_p - \lambda \alpha V e^{-E/(RT_s)}\left(1 + \frac{E}{RT_s^2}T_p\right)(C_{As} + C_{Ap}) = \rho c_v T_i - (\dot{Q}_1 - \dot{Q}) - \dot{Q} \tag{o}$$

Equations (d)–(f) are used in Equations (m)–(o) to subtract the steady state, leading to

$$V\frac{dC_{Ap}}{dt} + qC_{Ap} + \alpha V e^{-E/(RT_s)}\left(C_{Ap} + \frac{E}{RT_s^2}T_p C_{As} + \frac{E}{RT_s^2}T_p C_{Ap}\right) = 0 \tag{p}$$

$$V\frac{dC_{Bp}}{dt} - \alpha V e^{-E/(RT_s)}\left(C_{Ap} + \frac{E}{RT_s^2}T_p C_{As} + \frac{E}{RT_s^2}T_p C_{Ap}\right) + qC_{Bp} = 0 \tag{q}$$

$$\rho c_v V \frac{dT_p}{dt} + \rho q c_v T_p - \lambda \alpha V e^{-E/(RT_s)}\left(C_{Ap} + \frac{E}{RT_s^2}T_p C_{As} + \frac{E}{RT_s^2}T_p C_{Ap}\right) = \dot{Q} - \dot{Q}_1 \tag{r}$$

Equations (p)–(r) are nonlinear due to the presence of the product $T_p C_{Ap}$, which appears in each equation. The perturbation quantities are assumed to be small in comparison to the steady-state quantities. The product of two perturbation terms should be much smaller than one perturbation term. Equations (p) and (q) are linearized by neglecting the products of perturbations, resulting in

$$V\frac{dC_{Ap}}{dt} + qC_{Ap} + \alpha V e^{-E/(RT_s)}\left(C_{Ap} + \frac{E}{RT_s^2}T_p C_{As}\right) = 0 \tag{s}$$

$$V\frac{dC_{Bp}}{dt} - \alpha V e^{-E/(RT_s)}\left(C_{Ap} + \frac{E}{RT_s^2}T_p C_{As}\right) + qC_{Bp} = 0 \tag{t}$$

$$\rho c_v V \frac{dT_p}{dt} + \rho q c_v T_p - \lambda \alpha V e^{-E/(RT_s)}\left(C_{Ap} + \frac{E}{RT_s^2}T_p C_{As}\right) = \dot{Q} - \dot{Q}_1 \tag{u}$$

6.8 Biological Systems

Dynamic behavior occurs in many biological systems. Biological fluids are transported through the body, muscles and tissue move within the body, and body parts move relative to joints. Biological systems are modeled, much like chemical systems, by writing equations that express the conservation of species. The principles of chemical reaction kinetics may be applied to model processes involving the absorption and elimination of drugs in the body. The process of drug infusion is modeled in the following examples.

Example 6.18

A drug is infused into the plasma through a controlled release mechanism such as a patch. The concentration of the drug in the plasma is defined as

$$C = \frac{m_p}{V} \tag{a}$$

where m_p is the mass of the drug in the plasma and V is the volume of the plasma. The drug is eliminated from the plasma due to normal biological processes at a rate k_e. The rate of infusion is I. Derive a mathematical model for the concentration of the drug in the plasma assuming that all of the drug is infused into the plasma and none is infused into tissue.

Solution

Consider a control volume consisting of the plasma. The appropriate form of Conservation of Mass applied to the drug in the control volume is

$$\text{Rate of accumulation} = \text{Rate of infusion} - \text{Rate of elimination} \tag{b}$$

Using the defined parameters in Equation (b) leads to

$$\frac{d}{dt}(VC) = I - k_e(VC) \tag{c}$$

$$V\frac{dC}{dt} + k_e VC = I \tag{d}$$

Example 6.19

When a drug is infused into the plasma, as in Example 6.18, it is then transported by the plasma and distributed to the tissue in the body. This is a complex process, but a simple model uses only the concentration of the drug in the plasma C_p and the concentration of the drug in the tissue C_t as dependent variables. The concentrations are defined as

$$C_p = \frac{m_p}{V_p} \tag{a}$$

$$C_t = \frac{m_t}{V_t} \tag{b}$$

where m_p is the mass of the drug in the plasma, V_p is the volume of the plasma, m_t is the mass of the drug in the tissue, and V_t is the volume of the tissue. Through a transport process, the drug is transported from the plasma to the tissue at a rate k_1. Simultaneously, the drug is transported from the tissue to the plasma at a rate k_2. Normal biological processes cause the drug to be eliminated from the plasma at a rate of elimination k_e. The drug is infused into the plasma at a rate $I(t)$. Derive a mathematical model for the concentrations of the drug in the plasma and tissue. The mathematical model derived is called a two-compartment model for this pharmokinetic process.

Solution

Conservation of Mass for the drug is applied to both the plasma and the tissue. The appropriate forms are

$$\begin{pmatrix}\text{Rate of}\\ \text{accumulation of the}\\ \text{drug in the plasma}\end{pmatrix} = \begin{pmatrix}\text{Rate of}\\ \text{infusion}\end{pmatrix} - \begin{pmatrix}\text{Rate of}\\ \text{elimination}\end{pmatrix} - \begin{pmatrix}\text{Rate of transport}\\ \text{from the plasma}\\ \text{to the tissue}\end{pmatrix} + \begin{pmatrix}\text{Rate of transport}\\ \text{from the tissue}\\ \text{to the plasma}\end{pmatrix} \tag{c}$$

$$\begin{pmatrix}\text{Rate of}\\ \text{accumulation of the}\\ \text{drug in the tissue}\end{pmatrix} = \begin{pmatrix}\text{Rate of transport}\\ \text{from the plasma}\\ \text{to the tissue}\end{pmatrix} - \begin{pmatrix}\text{Rate of transport}\\ \text{from the tissue}\\ \text{to the plasma}\end{pmatrix} \tag{d}$$

Use of the defined variables and parameters in Equations (c) and (d) leads to

$$V_p\frac{dC_p}{dt} = I(t) - k_e V_p C_p - k_1 V_p C_p + k_2 V_t C_t \tag{e}$$

$$V_t\frac{dC_t}{dt} = k_1 V_p C_p - k_2 V_t C_t \tag{f}$$

Equations (e) and (f) are rearranged to

$$V_p\frac{dC_p}{dt} + (k_e + k_1)V_p C_p - k_2 V_t C_t = I(t) \tag{g}$$

$$V_t\frac{dC_t}{dt} - k_1 V_p C_p + k_2 V_t C_t = 0 \tag{h}$$

6.9 Transfer Functions for Transport Systems

The determination of transfer functions for systems with transport processes is very similar to that for mechanical and electrical systems. The Laplace transform of the mathematical model is taken assuming that all initial conditions are zero. The transfer function for an SISO (single input and single output) system is defined as the ratio of the Laplace transform of the system output to the Laplace transform of the system input. The order of the system, which is equivalent to the order of the polynomial in the denominator of the transform, is equal to the number of independent energy storage elements in the system. In liquid-level systems, the tanks store potential energy. Since perturbations of the liquid levels are used as the dependent variables, changes in kinetic energies are usually very small and neglected. The energy in systems in which heat transfer occurs is internal energy due to the change in temperature: $u = c_v T$. A reaction in a CSTR system is endothermic when it uses energy from the surroundings to cause the reaction. It is exothermic if it gives off heat and uses stored energy from the heat of reaction. In either case, the enthalpy changes. Each compartment in a drug delivery model serves as an energy storage element.

A matrix of transfer functions is defined for a MIMO (multiple input and multiple output) system. A liquid-level problem involving n tanks with flow rate perturbations in m tanks has a transfer matrix of n rows and m columns ($n \times m$).

Example 6.20

a. Determine transfer function for the linearized single-tank liquid-level problem.
b. Determine the matrix of transfer functions for the two-tank liquid-level problem of Example 6.8.

Solution

a. The linearized mathematical model for the single-tank liquid-level problem is Equation (6.65), repeated here:

$$A\frac{dh}{dt} + \frac{1}{R}h = q_i(t) \tag{a}$$

Letting $H(s) = \mathcal{L}\{h(t)\}$ and $Q_i(s) = \mathcal{L}\{q_i(t)\}$ and taking the Laplace transform of Equation (a), assuming $h(0) = 0$, leads to

$$AsH(s) + \frac{1}{R}H(s) = Q_i(s) \tag{b}$$

Rearranging Equation (b) and solving for $G(s) = \dfrac{H(s)}{Q_i(s)}$ yields

$$G(s) = \frac{1}{As + \frac{1}{R}} \tag{c}$$

b. The linearized mathematical model for the two-tank liquid-level problem is given by Equations (u) and (v) of Example 6.8:

$$A_1\frac{dh_1}{dt} + \frac{1}{R_1}h_1 - \frac{1}{R_1}h_2 = q_{i1} \tag{d}$$

$$A_2\frac{dh_2}{dt} - \frac{1}{R_1}h_1 + \left(\frac{1}{R_1} + \frac{1}{R_2}\right)h_2 = q_{i2} \tag{e}$$

Taking the Laplace transforms of Equations (d) and (e), assuming $h_1(0) = h_2(0)$, leads to

$$\begin{bmatrix} A_1 s + \frac{1}{R_1} & -\frac{1}{R_1} \\ -\frac{1}{R_1} & A_2 s + \frac{1}{R_1} + \frac{1}{R_2} \end{bmatrix} \begin{bmatrix} H_1(s) \\ H_2(s) \end{bmatrix} = \begin{bmatrix} Q_{i1}(s) \\ Q_{i2}(s) \end{bmatrix} \tag{f}$$

The determinant of the coefficient matrix is

$$\begin{aligned} D(s) &= \left(A_1 s + \frac{1}{R_1}\right)\left(A_2 s + \frac{1}{R_1} + \frac{1}{R_2}\right) - \left(-\frac{1}{R_1}\right)\left(-\frac{1}{R_1}\right) \\ &= A_1 A_2 s^2 + \left(\frac{A_1}{R_1} + \frac{A_1}{R_2} + \frac{A_2}{R_1}\right)s + \frac{1}{R_1 R_2} \end{aligned} \tag{g}$$

For example, the transfer function $G_{1,2}(s) = \dfrac{H_1(s)}{Q_{i2}(s)}$, assuming $Q_{i1}(s) = 0$. The transfer matrix is determined using Equation (f) as

$$\mathbf{G}(s) = \frac{1}{D(s)} \begin{bmatrix} A_2 s + \frac{1}{R_1} + \frac{1}{R_2} & \frac{1}{R_1} \\ \frac{1}{R_1} & A_1 s + \frac{1}{R_1} \end{bmatrix} \tag{h}$$

Example 6.21

Determine the transfer function for the thermal problem of Example 6.14 if the outside temperature is $T_\infty(t)$.

Solution

The mathematical model for the indoor temperature is given as Equation (j) of Example 6.14 with an arbitrary function for the outside temperature, which becomes

$$99.2\frac{dT}{dt} + 34.7T = 34.7\,T_\infty(t) \tag{a}$$

Let $\theta_\infty(t)$ be the perturbation in the outdoor temperature from steady state. This induces a perturbation in temperature $\theta(t)$. The mathematical model for $\theta(t)$ is obtained from Equation (a) as

$$99.2\frac{d\theta}{dt} + 34.7\theta = 34.7\,\theta_\infty(t) \tag{b}$$

Taking the Laplace transform of Equation (b), assuming $\theta(0) = 0$, leads to

$$99.2s\Theta(s) + 34.7\Theta(s) = 34.7\Theta_\infty(s) \tag{c}$$

The transfer function is determined from Equation (c) as

$$G(s) = \frac{\Theta(s)}{\Theta_\infty(s)} = \frac{0.35}{s + 0.35} \tag{d}$$

6.10 State-Space Modeling of Transport Systems

State-space modeling of transport systems is very similar to state-space modeling of mechanical or electrical systems. Since the mathematical models of transport systems often involve first-order differential equations, the state variables are often chosen as the dependent variables in the differential equations. The state variables are often related to the energy present in the systems.

The order of a system is equal to the number of independent energy storage elements in the system. For an nth-order system, n state variables are chosen and the mathematical model is reformulated as n first-order differential equations. The differential equations for a linear system are summarized in matrix form. From this matrix form, a state matrix and an input matrix are defined. The relation between the state variables and the output is written in matrix form as well, and the output matrix and the transmission matrix are defined.

Example 6.22

Determine a state-space model for the linearized mathematical model of the system of Example 6.15.

Solution

Equation (h) of Example 6.15 provides the mathematical model:

$$\rho V c_v\frac{d\theta}{dt} + \dot{m}\,c_v\theta = \dot{q} \tag{a}$$

The input for this system is $\dot{q}$. The output is $\theta(t)$. The mathematical model of the system, given by Equation (a), is that of a first-order system. Define a state variable by

$$y_1 = \theta \tag{b}$$

Rewriting Equation (a) using the state variable,

$$\dot{y}_1 = -\frac{\dot{m}}{\rho V} y_1 + \frac{1}{\rho V c_v} \dot{q} \tag{c}$$

The output variable is related to the state variable and the output by

$$\theta = y_1 + 0\dot{q} \tag{d}$$

From Equation (c), the state matrix and the input matrix are

$$\mathbf{A} = \left[-\frac{\dot{m}}{\rho V}\right] \qquad \mathbf{B} = \left[\frac{1}{\rho V c_v}\right] \tag{e}$$

Using Equation (d), the output matrix and the transmission matrix for this model become

$$\mathbf{C} = [1] \qquad \mathbf{D} = [0] \tag{f}$$

Example 6.23

Determine a state-space model for the mathematical model of the CSTR in Example 6.16.

Solution

The mathematical model is given in Equations (f) and (g) of Example 6.16:

$$V\frac{dC_A}{dt} + (q + kV)C_A = qC_{Ai} \tag{a}$$

$$V\frac{dC_B}{dt} - kVC_A + qC_B = qC_{Bi} \tag{b}$$

Define the state variables as

$$y_1 = C_A \tag{c}$$

$$y_2 = C_B \tag{d}$$

Rewriting Equations (c) and (d) in terms of the state variables,

$$\dot{y}_1 = -\left(\frac{q}{V} + k\right)y_1 + \frac{q}{V}C_{Ai} \tag{e}$$

$$\dot{y}_2 = k y_1 - \frac{q}{V}y_2 + \frac{q}{V}C_{Bi} \tag{f}$$

Rewriting Equations (e) and (f) in matrix form leads to

$$\begin{bmatrix}\dot{y}_1\\ \dot{y}_2\end{bmatrix} = \begin{bmatrix}-\left(\frac{q}{V}+k\right) & 0\\ k & -\frac{q}{V}\end{bmatrix}\begin{bmatrix}y_1\\ y_2\end{bmatrix} + \begin{bmatrix}\frac{q}{V} & 0\\ 0 & \frac{q}{V}\end{bmatrix}\begin{bmatrix}C_{Ai}\\ C_{Bi}\end{bmatrix} \tag{g}$$

The state matrix and the input matrix are determined using Equation (g). The relation between the output and the state variables and the input is

$$\begin{bmatrix}C_A\\ C_B\end{bmatrix} = \begin{bmatrix}1 & 0\\ 0 & 1\end{bmatrix}\begin{bmatrix}y_1\\ y_2\end{bmatrix} + \begin{bmatrix}0 & 0\\ 0 & 0\end{bmatrix}\begin{bmatrix}C_{Ai}\\ C_{Bi}\end{bmatrix} \tag{h}$$

The output matrix and the transmission matrix are determined from Equation (h).

6.11 Further Examples

The following assumptions are used in the examples in this section:

- The applicable control volume forms of Conservation of Mass and Conservation of Energy are given by Equations (6.2) and (6.12), respectively.
- Liquids are incompressible.
- There is no shaft work acting on any control volume.
- The acceleration effects of liquids are negligible in large tanks.
- The equations developed using major and minor losses in pipes conveying liquids are linearized using Equation (6.68) and defining the resistance at the steady-state operating condition.
- The equations developed for the pressure drop in pipes conveying gases are linearized using Equation (6.54) and defining the resistance as determined for an average value of the mass flow rate.
- The variation of specific heat with temperature is negligible.
- The ideal gas law applies for all gases.
- Equation (6.94) provides a linearization for the pressure drop across an orifice in a hydraulic servomotor, and Equation (6.90) defines the servomotor's operating condition.
- Thermal properties are lumped spatially by defining the thermal resistance, as in Equations (6.105) and (6.106).
- Tanks in thermal and chemical systems are well stirred such that the properties of the exit stream are those in the tank.
- All chemical reactions are reversible and of the first order, and the rate of reaction is related to temperature by the Arrhenius law.
- Chemical reactions do not lead to phase changes.

Example 6.24

The two-tank system of Figure 6.32 is operating at a steady state as shown. **a.** Derive a mathematical model for the changes in liquid levels that occur when the flow rates are suddenly changed to $Q_1 = 1.2\ \text{m}^3/\text{s}$ and $Q_2 = 1.2\ \text{m}^3/\text{s}$. **b.** Predict the liquid levels in the tank that result when the system reaches steady state at these flow rates.

Solution

a. The notation used in the solution is the same as that in Example 6.8. The flow rate through the pipe at steady state is q_{s1}. The head loss through this pipe is $h_{s2} - h_{s1}$. The application of Conservation of Mass to the entire two-tank system reveals that the flow rate through the outlet pipe is $q_{s1} + q_{s2}$. The head loss through the exit pipe is h_{s2}.

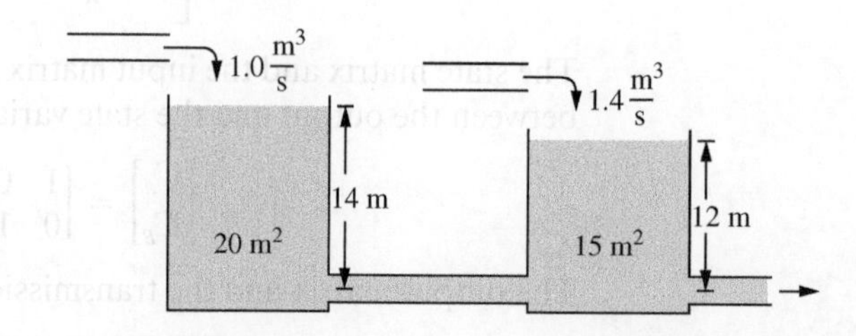

Figure 6.32 Two-tank liquid level system of Example 6.24 shown at steady state. The inlet flow rate into both tanks is suddenly changed to 1.2 m³/s.

Assuming turbulent flow in the pipes, their resistances at steady state are calculated using Equation (6.38):

$$R_1 = 2\frac{h_{s1} - h_{s2}}{q_{s1}} = 2\frac{(14\text{ m}) - (12\text{ m})}{1.0\text{ m}^3/\text{s}} = 4.0\text{ s/m}^2 \tag{a}$$

$$R_2 = 2\frac{h_{s2}}{q_{s1} + q_{s2}} = 2\frac{12\text{ m}}{(1.0\text{ m}^3/\text{s}) + (1.4\text{ m}^3/\text{s})} = 10.0\text{ s/m}^2 \tag{b}$$

The perturbations in the flow rates are

$$q_1(t) = 0.2u(t) \tag{c}$$

$$q_2(t) = -0.2u(t) \tag{d}$$

The modeling of this system is the same as that of the system of Example 6.8. Substitution of the given and calculated values into Equations (u) and (v) of Example 6.8 leads to

$$20\frac{dh_1}{dt} + 0.25h_1 - 0.25h_2 = 0.2u(t) \tag{e}$$

$$15\frac{dh_2}{dt} - 0.25h_1 + 0.35h_2 = -0.2u(t) \tag{f}$$

b. Steady state occurs when $dh_1/dt = dh_2/dt = 0$, in which case Equations (e) and (f) become

$$0.25h_1 - 0.25h_2 = 0.2 \tag{g}$$

$$-0.25h_1 + 0.35h_2 = -0.2 \tag{h}$$

Simultaneous solution of Equations (g) and (h) yields $h_1 = 0.8$ m and $h_2 = 0$ m. Thus the new steady-state levels are $H_1 = h_{s1} + h_1 = 14.8$ m and $H_2 = h_{s2} + h_2 = 12$ m.

Example 6.25

Figure 6.33 on the next page shows, at steady state, a series of n cascading tanks. **a.** Derive a mathematical model to describe the change in liquid levels from steady state when the inlet flow rate into the first tank has a perturbation from steady state of $q_1(t)$. Define R_i as the resistance in the pipe between tanks i and $i + 1$. **b.** Determine the transfer function $G_i(s) = \dfrac{H_i(s)}{Q(s)}$.

Solution

a. Let

$$H_i(t) = h_{is} + h_i(t) \tag{a}$$

represent the liquid level in tank i, $i = 1, 2, \ldots, n$. Let

$$Q_i(t) = q_{is} + q_i(t) \tag{b}$$

represent the flow rate into tank i, $i = 1, 2, \ldots, n$, and $Q_{n+1}(t)$ the flow rate out of the system. Using Equation (6.69) the perturbation flow rates and liquid levels are related through the resistances by

$$q_{i+1} = \frac{h_i}{R_i} \qquad i = 1, 2, \ldots, n \tag{c}$$

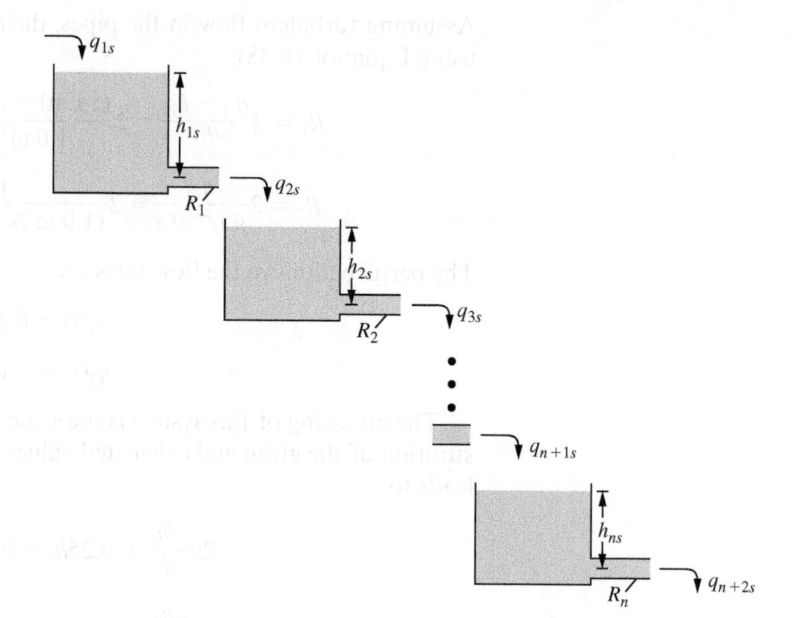

Figure 6.33

A series of n tanks, shown at steady state, of Example 6.25.

The application of the control volume form of Conservation of Mass applied to tank i using perturbation variables and linearization of the resulting equation leads to

$$A_i \frac{dh_i}{dt} = q_i - q_{i+1} \qquad i = 1, 2, \ldots, n \tag{d}$$

Using Equation (c) in Equation (d) for $i = 1$ leads to

$$A_1 \frac{dh_1}{dt} + \frac{1}{R_1} h_1 = q_1(t) \tag{e}$$

Substitution of Equation (c) into Equation (d) for $i = 2, 3, \ldots, n$ yields

$$A_i \frac{dh_i}{dt} - \frac{1}{R_{i-1}} h_{i-1} + \frac{1}{R_i} h_i = 0 \tag{f}$$

Equations (e) and (f) provide the mathematical model for the system.

b. Taking the Laplace transform of Equation (e) leads to

$$G_1(s) = \frac{1}{A_1 s + \frac{1}{R_1}} \tag{g}$$

Taking the Laplace transform of Equation (f) with $i = 2$ leads to

$$H_2(s) = \frac{\frac{1}{R_1}}{A_2 s + \frac{1}{R_2}} H_1(s) \tag{h}$$

Note that

$$G_2(s) = \frac{H_2(s)}{Q(s)} = \left[\frac{H_2(s)}{H_1(s)}\right]\left[\frac{H_1(s)}{Q(s)}\right] = \left(\frac{\frac{1}{R_1}}{A_2 s + \frac{1}{R_2}}\right)\left(\frac{1}{A_1 s + \frac{1}{R_1}}\right) \tag{i}$$

Taking the Laplace transform of Equation (f) leads to

$$H_i(s) = \frac{\frac{1}{R_{i-1}}}{A_i s + \frac{1}{R_i}} H_{i-1}(s) \tag{j}$$

Writing an equation like Equation (i) gives

$$G_i(s) = \frac{H_i(s)}{Q(s)} = \left[\frac{H_i(s)}{H_{i-1}(s)}\right]\left[\frac{H_{i-1}(s)}{H_{i-2}(s)}\right]\left[\frac{H_{i-2}(s)}{H_{i-3}(s)}\right]\cdots\left[\frac{H_1(s)}{Q(s)}\right] \tag{k}$$

By induction, the transfer function is written as

$$G_i(s) = \frac{\left(\frac{1}{R_{i-1}}\right)\left(\frac{1}{R_{i-2}}\right)\cdots\left(\frac{1}{R_1}\right)}{\left(A_is + \frac{1}{R_i}\right)\left(A_{i-1}s + \frac{1}{R_{i-1}}\right)\left(A_{i-2}s + \frac{1}{R_{i-2}}\right)\cdots\left(A_1s + \frac{1}{R_1}\right)} \tag{l}$$

Example 6.26

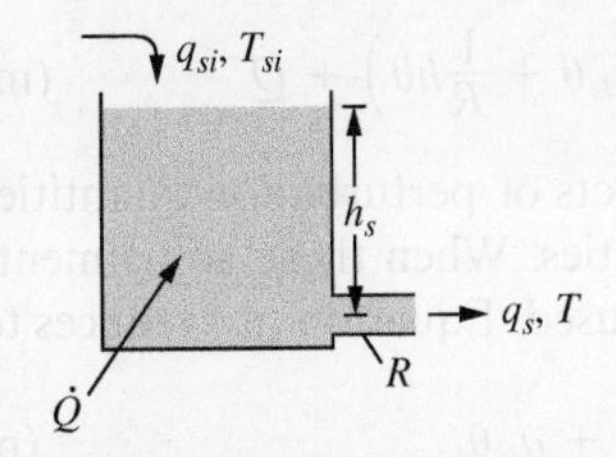

Figure 6.34

The steady-state operation of the heated tank for Example 6.26.

Liquid enters the tank of Figure 6.34 at a flow rate q_{si} and temperature T_{si}. The liquid is heated in the tank at a rate $\dot{Q}$ such that its temperature is T_s. The liquid is thoroughly heated such that the temperature of the outlet stream is also T_s. The steady state is disturbed when the flow rate and temperature of the inlet stream are suddenly changed such that

$$Q_i = q_{si} + q_i(t) \tag{a}$$

$$T_i = T_{si} + \theta_i(t) \tag{b}$$

Derive a linear mathematical model to describe the change in liquid level in the tank and the temperature of the outlet stream.

Solution

Perturbations in the liquid level, the temperature of the liquid in the tank, and the outlet flow rate are introduced such that

$$H = h_s + h(t) \tag{c}$$

$$T(t) = T_s + \theta(t) \tag{d}$$

$$Q_o = q_s + q_o(t) \tag{e}$$

Assuming turbulent flow in the pipe, the resistance at steady state is determined using Equation (6.38) as

$$R = 2\frac{h_s}{q_{si}} \tag{f}$$

A linear model for the liquid level is obtained by using Equation (6.69) to relate the perturbation in the outlet flow to the perturbation in the liquid level via the resistance determined at steady state:

$$q_o(t) = \frac{h(t)}{R} \tag{g}$$

The application of Conservation of Mass to the tank, assuming the density of the fluid remains constant, and linearization of the resulting equation leads to

$$A\frac{dh}{dt} = q_i(t) - q_o(t) \tag{h}$$

Substitution of Equation (g) into Equation (h) leads to

$$A\frac{dh}{dt} + \frac{1}{R}h = q_i(t) \tag{i}$$

Application of the control volume form of Conservation of Energy at the initial steady state gives

$$\rho q_{is} c_v T_{is} - \rho c_v q_{is} T_s + \dot{Q} = 0 \tag{j}$$

Application of the control volume form of Conservation of Energy to the tank at an arbitrary time gives

$$\frac{d}{dt}(\rho A H c_v T) = \rho Q_i c_v T_i - \rho c_v Q_o T + \dot{Q} \tag{k}$$

Substitution of Equations (a)–(e) in Equation (k) leads to

$$\rho A c_v \frac{d}{dt}[(h_s + h)(T_s + \theta)] = \rho c_v (q_{is} + q_i)(T_{is} + \theta_i) - \rho c_v (q_{is} + q_o)(T_s + \theta) + \dot{Q} \tag{l}$$

The expansion of products in Equation (l), using the product rule for differentiation, setting derivatives of steady-state variables to zero, and using Equation (g) lead to

$$\rho A c_v \left[h_s \frac{d\theta}{dt} + T_s \frac{dh}{dt} + \frac{d}{dt}(h\theta)\right] = \rho c_v (q_{is} T_{is} + q_i T_{is} + q_{is}\theta_i + q_i \theta_i)$$
$$-\rho c_v \left(q_{is} T_s + \frac{1}{R} h T_s + q_{is}\theta + \frac{1}{R} h\theta\right) + \dot{Q} \tag{m}$$

The nonlinear terms in Equation (m) arising from products of perturbation quantities are negligible in comparison to the perturbation quantities. When these adjustments are made and the steady-state condition [Equation (j)] is used, Equation (m) reduces to

$$A h_s \frac{d\theta}{dt} + A T_s \frac{dh}{dt} + \frac{T_s}{R} h + q_{is}\theta = T_{is} q_i + q_{is}\theta_i \tag{n}$$

Equations (i) and (n) form the linear model for this system.

Example 6.27

The system of Figure 6.35 contains two tanks with compressed air. The tanks of volume V_1 and V_2 are connected by a hose of average resistance R_2. The upstream tank is connected to a supply tank through a hose of resistance R_1. Initially, the valve between the supply and the tanks is closed and both tanks are in equilibrium at equal pressure P. The valve is suddenly opened, allowing air to flow into the tanks. The upstream change in pressure is p_s. Derive a mathematical model for pressure in each of the tanks assuming the process is isothermal at a temperature T.

Solution

Perturbations in pressure in each of the tanks are defined such that

$$P_{t1} = P + p_1 \tag{a}$$

$$P_{t2} = P + p_2 \tag{b}$$

Let $\dot{m}_1$ be the mass flow rate into the first tank downstream of the supply and $\dot{m}_2$ the mass flow rate into the second tank. Equation (6.54) is used to relate the mass flow rates and pressures through the resistances by

$$R_1 = \frac{p_s - p_1}{\dot{m}_1} \tag{c}$$

$$R_2 = \frac{p_1 - p_2}{\dot{m}_2} \tag{d}$$

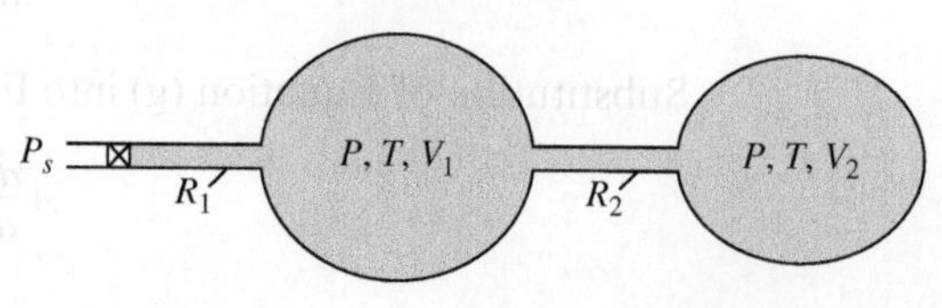

Figure 6.35

Two tanks of compressed air are at the same pressure when the upstream valve is opened.

The application of the control volume form of Conservation of Mass to each tank leads to

$$\dot{m}_1 - \dot{m}_2 = \frac{dm_1}{dt} \tag{e}$$

$$\dot{m}_2 = \frac{dm_2}{dt} \tag{f}$$

The total mass of air in each tank is

$$m_1 = \rho_1 V_1 \tag{g}$$

$$m_2 = \rho_2 V_2 \tag{h}$$

The initial pressure and temperature in both tanks are the same. Thus the ideal gas law implies that the initial density in each tank is the same: $\rho_0 = P/R_aT$. As in Example 6.10, the process is isothermal, which implies that the densities are related to the pressures by

$$\rho_1 = \frac{P}{R_aT}\left(1 + \frac{p_1}{P}\right) \tag{i}$$

$$\rho_2 = \frac{P}{R_aT}\left(1 + \frac{p_2}{P}\right) \tag{j}$$

Use of Equations (c), (d), (g), (h), (i), and (j) in Equations (e) and (f) leads to

$$\frac{d}{dt}\left[\frac{PV_1}{R_aT}\left(1 + \frac{p_1}{P}\right)\right] = \frac{p_s - p_1}{R_1} - \frac{p_1 - p_2}{R_2} \tag{k}$$

$$\frac{d}{dt}\left[\frac{PV_2}{R_aT}\left(1 + \frac{p_2}{P}\right)\right] = \frac{p_1 - p_2}{R_2} \tag{l}$$

Simplification of Equations (k) and (l) leads to

$$\frac{V_1}{R_aT}\frac{dp_1}{dt} + \left(\frac{1}{R_1} + \frac{1}{R_2}\right)p_1 - \frac{1}{R_2}p_2 = \frac{1}{R_1}p_s \tag{m}$$

$$\frac{V_2}{R_aT}\frac{dp_2}{dt} - \frac{1}{R_2}p_1 + \frac{1}{R_2}p_2 = 0 \tag{n}$$

Example 6.28

A walking beam is attached to a hydraulic servomotor, as illustrated in Figure 6.36. One end of the walking beam is subject to the input $z(t)$, while the other end is attached to a viscous damper of damping coefficient c and a linear spring of stiffness k.

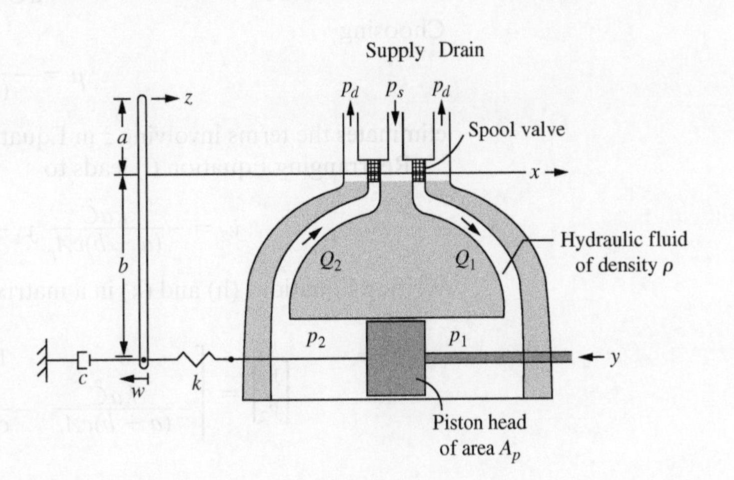

Figure 6.36 The hydraulic servomotor and walking beam system of Example 6.28.

The spring is also attached to the piston, whose displacement is $y(t)$.

a. Derive a mathematical model of the system relating $y(t)$ to $z(t)$.

b. Determine a state-space model for this system.

Solution

a. Let $w(t)$ be the displacement at the end of the walking beam that is attached to the spring and viscous damper. The application of similar triangles to relate the displacements of points on the walking beam, as in Example 6.13, leads to

$$\frac{z+w}{a+b} = \frac{x+w}{b}$$

$$w = \frac{b}{a}z - \frac{a+b}{a}x \tag{a}$$

For arbitrary values of z, y, and w the force developed in the viscous damper is $c\dot{w}$, while the force developed in the spring is $k(w - y)$. Since the walking beam has no mass, the forces must sum to zero

$$c\dot{w} + k(w - y) = 0 \tag{b}$$

The application of Conservation of Mass to the hydraulic servomotor leads to the equation defining the operation of the servomotor, Equation (6.90)

$$\hat{C}x = A_p\dot{y} \tag{c}$$

where $\hat{C}$ is defined in Equation (6.91). Substitution of Equation (c) into Equation (a) leads to

$$w = \frac{b}{a}z - \frac{(a+b)A_p}{a\hat{C}}\dot{y} \tag{d}$$

Substitution of Equation (d) into Equation (b) leads to

$$\frac{(a+b)cA_p}{a\hat{C}}\ddot{y} + \frac{(a+b)k}{a\hat{C}}\dot{y} + ky = \frac{cb}{a}\dot{z} + \frac{kb}{a}z \tag{e}$$

b. Since the derivative of the input appears explicitly in the mathematical model the state variables are introduced as in Section 3.5 as

$$y_1 = y \tag{f}$$

$$y_2 = \dot{y} + \mu z \tag{g}$$

Equations (f) and (g) imply

$$\dot{y}_1 = y_2 - \mu z \tag{h}$$

Substituting Equations (f) and (g) into Equation (e) yields

$$\frac{(a+b)cA_p}{a\hat{C}}(\dot{y}_2 - \mu\dot{z}) + \frac{(a+b)k}{a\hat{C}}(y_2 - \mu z) + ky_1 = \frac{bc}{a}\dot{z} + \frac{bk}{a}z \tag{i}$$

Choosing

$$\mu = -\frac{b\hat{C}}{(a+b)A_p} \tag{j}$$

eliminates the terms involving $\dot{z}$ in Equation (i).

Rearranging Equation (i) leads to

$$\dot{y}_2 = -\frac{ka\hat{C}}{(a+b)cA_p}y_1 - \frac{k}{cA_p}y_2 + \frac{bk\hat{C}}{(a+b)cA_p}z \tag{k}$$

Writing Equations (h) and (k) in a matrix form using Equation (j) gives

$$\begin{bmatrix}\dot{y}_1\\ \dot{y}_2\end{bmatrix} = \begin{bmatrix}0 & 1\\ -\frac{ka\hat{C}}{(a+b)cA_p} & -\frac{k}{cA_p}\end{bmatrix}\begin{bmatrix}y_1\\ y_2\end{bmatrix} + \begin{bmatrix}\frac{b\hat{C}}{(a+b)A_p}\\ \frac{bk\hat{C}}{(a+b)cA_p}\end{bmatrix}[z] \tag{l}$$

The relation between the output and the state variables is

$$[y] = [1 \quad 0]\begin{bmatrix} y_1 \\ y_2 \end{bmatrix} + [0]\,[z] \tag{m}$$

A state-space model for this system is given by Equations (l) and (m).

Example 6.29

The hydraulic servomotor of Figure 6.37, under appropriate conditions, acts as a controller. Derive the transfer function $G(s) = X(s)/Y(s)$.

Solution

For convenience introduce the intermediate variables $z(t)$ as the displacement of the point on the walking beam where the spring and viscous damper are attached, let $w(t)$ represent the point on the lever of length ℓ where the spring is attached, and let $v(t)$ represent the displacement of the rod connected to the valve. Assuming small displacements and using similar triangles leads to

$$w(t) = \frac{\ell}{a}x(t) \tag{a}$$

Geometry is also used to obtain

$$v(t) = \frac{y(t) - z(t)}{2} \tag{b}$$

The flow equation for the servomotor is obtained using Equation (6.90) as

$$\hat{C}v = A\dot{x} \tag{c}$$

where $\hat{C}$ is the flow coefficient of the motor and A is the area of the piston head. A force balance on the walking beam leads to

$$k(w - z) = c\dot{z} \tag{d}$$

Taking the Laplace transform of Equations (a)–(d) leads to

$$W(s) = \frac{\ell}{a}X(s) \tag{e}$$

$$V(s) = \frac{1}{2}[Y(s) - Z(s)] \tag{f}$$

$$\hat{C}V(s) = AsX(s) \tag{g}$$

$$k[W(s) - Z(s)] = csZ(s) \tag{h}$$

Equations (e)–(h) are combined, yielding

$$G(s) = \frac{\frac{c}{k}s + 1}{\frac{2cA}{\hat{C}k}s^2 + \frac{2A}{\hat{C}}s + \frac{\ell}{a}} \tag{i}$$

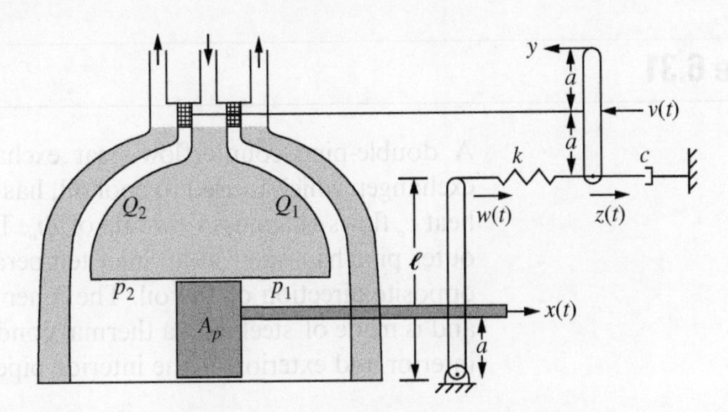

Figure 6.37
Hydraulic servomotor system of Example 6.29.

Example 6.30

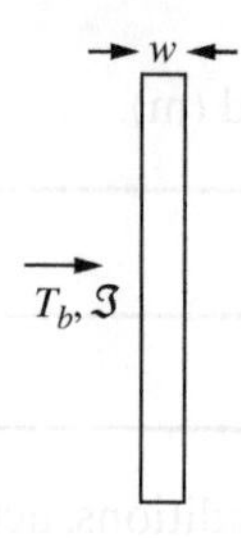

Figure 6.38

The system of Example 6.30: The wall of thickness w and surface area A receives radiation from a body whose temperature is T_b with a shape factor $\Im$.

The exterior surface of a wall of surface area A, illustrated in Figure 6.38, receives radiation heat transfer from a body whose temperature is T_b. The shape factor between the body and the wall is $\Im$. The wall has a small thickness w and is made of a material of specific heat c_v. Assuming no other modes of heat transfer and a uniform wall temperature T, at any time **a.** derive a mathematical model for the temperature in the wall, and **b.** derive a linear model for the temperature in the wall when the temperature of the radiating body is perturbed by $\theta_b(t)$ from steady state.

Solution

a. The total heat transfer to the wall due to radiation is given by the Stefan-Boltzmann equation as

$$\dot{Q}_w = \sigma \Im A(T_b^4 - T_w^4) \tag{a}$$

The application of Conservation of Energy to the wall leads to

$$\rho c_v A w \frac{dT_w}{dt} = \sigma \Im A(T_b^4 - T_w^4) \tag{b}$$

b. When the system is in steady state, the temperature of the wall equals the temperature of the radiating body $T_w = T_b = T_s$. Let θ_o be perturbation in the wall temperature due to the perturbation in the temperature of the radiating body θ_b

$$T_b = T_s + \theta_b \tag{c}$$

$$T_w = T_s + \theta_w \tag{d}$$

Substitution of Equations (c) and (d) in Equation (b) leads to

$$\rho c_v w \frac{d\theta_w}{dt} = \sigma \Im \left[(T_s + \theta_b)^4 - (T_s + \theta_w)^4\right] \tag{e}$$

Equation (e) is linearized using a binomial expansion truncated after the linear term as illustrated in Section 1.6:

$$\begin{aligned}(T_s + \theta_w)^4 &= T_s^4 \left(1 + \frac{\theta_w}{T_s}\right)^4 \\ &= T_s^4 \left(1 + 4\frac{\theta_w}{T_s} + \cdots\right) \\ &= T_s^4 + 4T_s^3\theta_w + \cdots \end{aligned} \tag{f}$$

Substitution of Equation (f) into Equation (e) and using a similar expansion for the temperature from the radiating body leads to

$$\rho c_v w \frac{d\theta_w}{dt} + 4\sigma \Im T_s^3 \theta_w = 4\sigma \Im T_s^3 \theta_b \tag{g}$$

Example 6.31

A double-pipe counterflow heat exchanger is illustrated in Figure 6.39. The heat exchanger, which is used to cool oil, has a central pipe through which the oil of specific heat c_o flows at a mass flow rate of $\dot{m}_o$. The temperature of the oil at the inlet is T_{oi}. The outer pipe has water at an inlet temperature T_{wi} and mass flow rate $\dot{m}_w$ flowing in the opposite direction of the oil. The inner pipe has an inner radius r_i and outer radius r_o and is made of steel with a thermal conductivity k. The heat transfer coefficients on the interior and exterior of the interior pipe are h_i and h_o, respectively. The outer pipe has

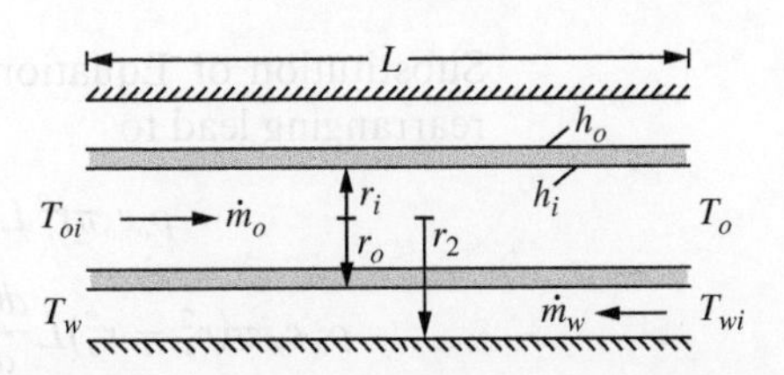

Figure 6.39

The double-pipe counter flow heat exchanger of Example 6.31 is used to cool oil using water as a cooling liquid.

an outer radius r_2 and is perfectly insulated, allowing no heat transfer from the water. The total length of the heat exchanger is L. **a.** Assuming lumped parameters, derive a mathematical model for the temperatures of the oil and water in the heat exchanger and define the steady state. **b.** Derive a model for the transient changes in temperatures if the temperature of the inlet stream of water is suddenly increased by θ_w.

Solution

a. Heat transfer occurs though the inner pipe between the water and the oil due to both conduction and convection. The total resistance is obtained using Equation (6.112) as

$$R = \frac{1}{2\pi r_i L h_i} + \frac{1}{2\pi L k}\ln\left(\frac{r_o}{r_i}\right) + \frac{1}{2\pi r_o L h_o} \tag{a}$$

Thus the heat transfer between the oil and the water is

$$\dot{Q} = \frac{1}{R}(T_o - T_i) \tag{b}$$

An alternative to Equation (b) involving the overall heat transfer coeffcient, as defined in Equation (6.115), is often used in heat exchanger analysis. Using this equation with

$$U = \frac{1}{\dfrac{r_o}{r_i h_i} + \dfrac{r_o}{k}\ln\left(\dfrac{r_o}{r_i}\right) + \dfrac{1}{h_o}} \tag{c}$$

leads to an alternative form of Equation (b) as

$$\dot{Q} = 2\pi r_o LU(T_o - T_w) \tag{d}$$

A lumped model is used, which neglects the variation of temperature across the length of the heat exchanger. Consistent with this model is an assumption that the heat transfer occurs as soon as the fluid enters the heat exchanger and that the temperature of the fluid flowing through the heat exchanger is the same as the temperature of the fluid at the outlet.

The application of Conservation of Energy to a control volume consisting of the oil in the heat exchanger gives

$$\rho_o c_o \pi r_i^2 L \frac{dT_o}{dt} = c_o \dot{m}_o (T_{oi} - T_o) - 2\pi r_o LU(T_o - T_w) \tag{e}$$

The application of Conservation of Energy to a control volume consisting of the water in the heat exchanger gives

$$\rho_w c_w \pi (r_2^2 - r_o^2) L \frac{dT_w}{dt} = c_w \dot{m}_w (T_{wi} - T_w) + 2\pi r_o LU(T_o - T_w) \tag{f}$$

The steady state is defined by $dT_o/dt = dT_w/dt = 0$. Adding Equations (e) and (f) together for the steady state leads to

$$c_o \dot{m}_o (T_{oi} - T_{os}) = c_w \dot{m}_w (T_{ws} - T_{wis}) \tag{g}$$

b. Define perturbations in temperature from steady state by

$$T_w = T_{ws} + \theta_w \tag{h}$$

$$T_o = T_{os} + \theta_o \tag{i}$$

Transient behavior is induced by a perturbation in the inlet temperature of the water such that

$$T_{wi} = T_{wis} + \theta_{wi} \tag{j}$$

Substitution of Equations (h)–(j) in Equations (e) and (f), using Equation (g), and rearranging lead to

$$\rho_o c_o \pi r_i^2 L \frac{d\theta_o}{dt} + (2\pi r_o LU + c_o \dot{m}_o)\theta_o - 2\pi r_o UL\theta_w = c_o \dot{m}_o \theta_{oi} \quad \text{(k)}$$

$$\rho_w c_w \pi (r_2^2 - r_o^2) L \frac{d\theta_w}{dt} - 2\pi r_o LU\theta_o + (2\pi r_o LU + c_w \dot{m}_w)\theta_w = c_w \dot{m}_w \theta_{wi} \quad \text{(l)}$$

Example 6.32

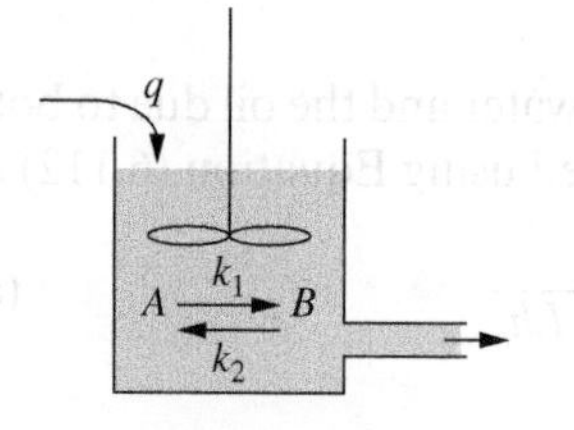

Figure 6.40

The CSTR of Example 6.32.

The CSTR of Figure 6.40 has a two-way reaction represented by

$$A \underset{k_2}{\overset{k_1}{\rightleftarrows}} B \quad \text{(a)}$$

Both reactions are first order. Derive a mathematical model for the concentrations of components A and B in the reactor assuming both reactions are isothermal.

Solution

The two-way reaction implies that moles of A are converted to moles of B with a rate of reaction k_1 and that moles of B are converted to moles of A with a rate of reaction k_2. Thus the rate of change of the number of moles of A in the reactor due to the reactions is

$$\left(\frac{dn_A}{dt}\right)_{reaction} = -k_1 V C_A + k_2 V C_B \quad \text{(b)}$$

Similarly, the rate of change of the number of moles of B in the reactor due to the reactions is

$$\left(\frac{dn_B}{dt}\right)_{reaction} = k_1 V C_A - k_2 V C_B \quad \text{(c)}$$

The mathematical model is obtained by writing component material balances for both components, as in Example 6.16. Equations (a) and (b) are used in these material balances, resulting in

$$V\frac{dC_A}{dt} + (q + k_1 V)C_A - k_2 V C_B = qC_{Ai} \quad \text{(d)}$$

$$V\frac{dC_B}{dt} - k_1 V C_A + (q + k_2 V)C_B = qC_{Bi} \quad \text{(e)}$$

Example 6.33

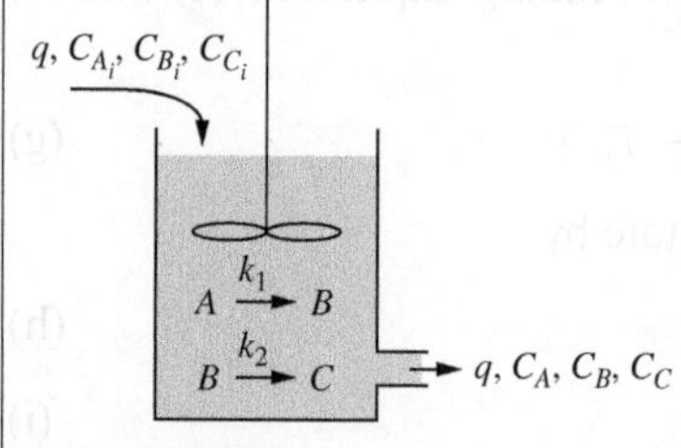

Figure 6.41

The CSTR of Example 6.33, in which two consecutive reactions occur.

The CSTR of Figure 6.41 is of constant volume V. Two consecutive reactions occur according to

$$A \xrightarrow{k_1} B \quad \text{(a)}$$

$$B \xrightarrow{k_2} C \quad \text{(b)}$$

where both reactions are first order. **a.** Derive a mathematical model that describes the component concentrations assuming the reactions are isothermal. **b.** Derive a mathematical model for the component concentrations assuming that the reactions are exothermic with heats of reaction λ_1 and λ_2 and that the tank is cooled with a rate of heat transfer $\dot{Q}$.

Solution

a. The rates of change of the number of moles of each component in the reactor due to the reactions are

$$\frac{d}{dt}(VC_A) = -k_1 VC_A \tag{c}$$

$$\frac{d}{dt}(VC_B) = k_1 VC_A - k_2 VC_B \tag{d}$$

$$\frac{d}{dt}(VC_C) = k_2 VC_B \tag{e}$$

The application of the equation to balance the number of moles of each component in the reactor [Equation (6.127)] gives

$$qC_{Ai} - qC_A - k_1 VC_A = \frac{d}{dt}(VC_A) \tag{f}$$

$$qC_{Bi} - qC_B + k_1 VC_A - k_2 VC_B = \frac{d}{dt}(VC_B) \tag{g}$$

$$qC_{Ci} - qC_C + k_2 VC_B = \frac{d}{dt}(VC_C) \tag{h}$$

Rearranging Equations (f)–(h), noting that V is constant, leads to

$$V\frac{dC_A}{dt} + (q + k_1 V)\, C_A = qC_{Ai} \tag{i}$$

$$V\frac{dC_B}{dt} - k_1 VC_A + (q + k_2 V)\, C_B = qC_{Bi} \tag{j}$$

$$V\frac{dC_C}{dt} - k_2 VC_B + qC_C = qC_{Ci} \tag{k}$$

b. The rates of reaction are assumed to follow the Arrhenius law such that

$$k_1 = \alpha e^{-E_1/(RT)} \tag{l}$$

$$k_2 = \alpha e^{-E_2/(RT)} \tag{m}$$

When the reactions are nonisothermal, the temperature in the tank changes and becomes a fourth dependent variable. The component balance equations are the same as those for the nonisothermal case [Equations (f)–(h)], except that the rates of reaction are given by Equations (l) and (m). This substitution results in a nonlinear set of equations.

The fourth independent equation is obtained through the application of the control volume form of Conservation of Energy to the reactor. In the application of the energy equation, it is assumed that the density and specific heat of the inlet stream are the same as those in the outlet stream. The resulting energy balance is

$$\rho q c_v T_i - \rho q c_v T - \dot{Q} + \lambda_1 V\alpha e^{-E_1/(RT)} C_A + \lambda_2 V\alpha e^{-E_2/(RT)} C_B = \rho V c_v \frac{dT}{dt} \tag{n}$$

Equations (f)–(h) and (n) are nonlinear and may be linearized by assuming perturbations from a steady state as in Example 6.17.

Example 6.34

Consider the two-stage reactor of Figure 6.42. The first-order reaction

$$A \xrightarrow{k_1} B \tag{a}$$

occurs in the first reactor, which is of area A_1. In addition to Reaction (a), a second first-order reaction

$$B \xrightarrow{k_2} C \tag{b}$$

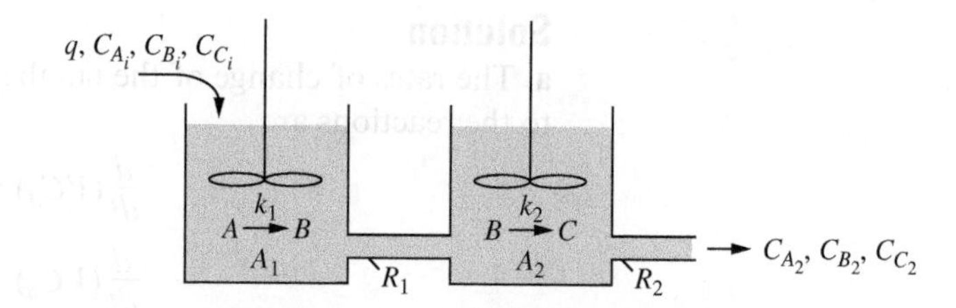

Figure 6.42

The two-stage reactor of Example 6.34.

takes place in the second reactor, which is of area A_2. Assume the system is isothermal. **a.** Derive a mathematical model for the concentrations of the components in each reactor, assuming the volumes remain constant. **b.** Derive a linear mathematical model for the system when the inlet flow rate is perturbed from steady state by q_p. The hydraulic resistances of the pipes at steady state are R_1 and R_2. **c.** Note that it takes a finite amount of time for the mixture to travel along the pipe connecting the reactors. Thus the concentrations of the components in the inlet stream to the second reactor are those that were transported from the first reactor a finite time before. Modify the mathematical model of part b. to account for this delay.

Solution

a. Since the volumes of both reactors remain constant, the flow rate through the system q is a constant. However, for generality and for use of the modeling in part b., let q_1, q_2, and q_3 be the volumetric flow rates into the first reactor, from the first reactor to the second reactor, and from the second reactor, respectively. The application of the component balance equation, Equation (6.127), to each component in the first reactor gives

$$q_1C_{Ai} - q_2C_{A1} - k_1V_1C_{A1} = \frac{d}{dt}(V_1C_{A1}) \tag{c}$$

$$q_1C_{Bi} - q_2C_{B1} + k_1V_1C_{A1} = \frac{d}{dt}(V_1C_{B1}) \tag{d}$$

$$q_1C_{Ci} - q_2C_{C1} = \frac{d}{dt}(V_1C_{C1}) \tag{e}$$

The concentrations of the components leaving the first reactor are the concentrations entering the second reactor. The application of the component balance equations to each component in the second reactor leads to

$$q_2C_{A1} - q_3C_{A2} - k_1V_2C_{A2} = \frac{d}{dt}(V_2C_{A2}) \tag{f}$$

$$q_2C_{B1} - q_3C_{B2} + k_1V_2C_{A2} - k_2V_2C_{B2} = \frac{d}{dt}(V_2C_{B2}) \tag{g}$$

$$q_2C_{C1} - q_3C_{C2} + k_2V_2C_{B2} = \frac{d}{dt}(V_2C_{C2}) \tag{h}$$

The mathematical model is Equations (c)-(h).

b. The system is operating at steady state when the inlet flow rate of q_s is perturbed by q_p. The steady state is defined by setting the time derivatives to zero in Equations (c)–(h), leading to

$$q_sC_{Ai} - q_sC_{A1s} - k_1V_{1s}C_{A1s} = 0 \tag{i}$$

$$q_sC_{Bi} - q_sC_{B1s} + k_1V_{1s}C_{A1s} = 0 \tag{j}$$

$$q_sC_{Ci} - q_sC_{C1s} = 0 \tag{k}$$

$$q_sC_{A1s} - q_sC_{A2s} - k_1V_{2s}C_{A2s} = 0 \tag{l}$$

$$q_sC_{B1s} - q_sC_{B2s} + k_1V_{2s}C_{A2s} - k_2V_{2s}C_{B2s} = 0 \tag{m}$$

$$q_sC_{C1s} - q_sC_{C2s} + k_2V_{2s}C_{C2s} = 0 \tag{n}$$

Perturbations from steady state are defined for the concentrations and volumes such that for a variable z, the perturbation z_p is defined by $z = z_s + z_p$. Equations (c)–(h), as written, are still valid. However, when the volumes of the tanks are variable, they are nonlinear due to the products of concentration and volume. Consider a representative term

$$V_1 C_{A1} = (V_{1s} + V_{1p})(C_{A1s} + C_{A1p})$$

$$= V_{1s}C_{1s} + V_{1s}C_{A1p} + V_{1p}C_{A1s} + V_{1p}C_{A1p} \tag{o}$$

The first term on the right-hand side of Equation (o) is the product of two steady-state terms which cancel from the model due to the steady-state conditions, Equations (i)-(n). The second and third terms on the right-hand side of Equation (o) are linear terms which remain in the model. The last term on the right-hand side is nonlinear due to the product of two perturbation variables. The nonlinear term is neglected when writing the mathematical model, as it is small compared to a single perturbation term, such as the second and third terms.

The application of the control volume form of Conservation of Mass to each tank leads to

$$\frac{dV_{1p}}{dt} = q_{1p} - q_{2p} \tag{p}$$

$$\frac{dV_{2p}}{dt} = q_{2p} - q_{3p} \tag{q}$$

Noting that the liquid level in a tank is equal to the volume of the tank divided by the area, the flow rates are related to the hydraulic resistances and volumes by

$$q_{2p} = \frac{V_{2p}/A_2 - V_{1p}/A_1}{R_1} \tag{r}$$

$$q_{3p} = \frac{V_{2p}/A_2}{R_2} \tag{s}$$

Substitution of Equations (r) and (s) into Equations (p) and (q) leads to

$$\frac{dV_{1p}}{dt} = \frac{1}{R_1 A_2} V_{2p} - \frac{1}{R_1 A_1} V_{1p} + q_{1p} \tag{t}$$

$$\frac{dV_{2p}}{dt} = \frac{1}{R_1 A_1} V_{1p} - \left(\frac{1}{R_1} + \frac{1}{R_2}\right) \frac{V_{2p}}{A_2} \tag{u}$$

The use of approximations similar to that made with Equation (o) in Equations (c)–(h) and simplifying by using Equations (i)–(n) lead to

$$V_{1s}\frac{dC_{A1p}}{dt} + C_{A1s}\frac{dV_{1p}}{dt} + (q_s + k_1 V_{1s})\,C_{A1p} + C_{A1s}q_{2p} + k_1 C_{A1s} V_{1p} = q_p C_{Ai} \tag{v}$$

$$V_{1s}\frac{dC_{B1p}}{dt} + C_{B1s}\frac{dV_{1p}}{dt} - k_1 V_{1s} C_{A1p} - k_1 C_{A1s} V_{1p} + (q_s + k_1 V_{1s})\,C_{B1p} + C_{B1s}q_{2p} - k_1 C_{B1s} V_{1p} = q_p C_{Bi} \tag{w}$$

$$V_{1s}\frac{dC_{C1p}}{dt} + C_{C1s}\frac{dV_{1p}}{dt} + q_s C_{C1p} + C_{C1s}q_{2p} = q_p C_{Ci} \tag{x}$$

$$V_{2s}\frac{dC_{A2p}}{dt} + C_{A2s}\frac{dV_{2p}}{dt} + (q_s + k_1 V_{2s})\,C_{A2p} + C_{A2s}q_{3p} + k_1 C_{A2s} V_{2s} - q_s C_{A1p} - C_{A1s}q_{2p} = 0 \tag{y}$$

$$V_{2s}\frac{dC_{B2}}{dt} + C_{B2s}\frac{dV_{2p}}{dt} + (q_s + k_2 V_{2s})\,C_{B2s} - k_1 V_{2s} C_{A2p} - k_1 C_{A2s} V_{2p} - k_2 C_{B2s} V_{2p} - C_{B1s}q_{2p} - q_s C_{B1p} + C_{B2s}q_{3p} = 0 \tag{z}$$

$$V_{2s}\frac{dC_{C2p}}{dt} + C_{C2s}\frac{dV_{2p}}{dt} + q_sC_{C2p} + C_{C2s}q_{3p} - q_sC_{C1p} - C_{C1s}q_{2p}$$
$$- k_2V_{2s}C_{B2p} - k_2C_{B2s}V_{2p} = 0 \qquad \text{(aa)}$$

Equations (v)–(aa) provide a mathematical model for the perturbed system. Note that Equations (p) and (q) may be used to further simplify the system.

c. The perturbed variables are functions of time. Assuming no reactions occur in the pipe connecting the two reactors, the concentrations of the stream fed into the second reactor at time t are not the concentrations in the stream leaving the reactor at time t, but rather the concentrations leaving the first reactor a time t_d earlier. The delay time t_d is calculated as

$$t_d = \frac{L}{q_2A_p} \qquad \text{(bb)}$$

where A_p is the area of the pipe. The concentrations entering the second reactor at time t are the concentrations leaving the first reactor at time $t - t_d$. For example, C_{A1} in Equation (f) should be replaced by

$$C_{A1}(t - t_d) = C_{A1s} + C_{A1p}(t - t_d) \qquad \text{(cc)}$$

Linearization requires that q_2 in Equation (bb) be approximated by q_{2s}.

Example 6.35

The double CSTR system of Figure 6.43 is operating at a steady state when the concentration of component A in the inlet stream is suddenly changed by C_{Aip}. The flow rate and concentration of all other components in the inlet stream are unchanged. All reactions are first order, the system is isothermal, and the volume of components in each reactor is constant. The time required for the fluid particles to travel between the tanks is τ. **a.** Determine a mathematical model for the perturbations in concentrations. **b.** Determine the transfer functions for the concentrations in each tank. The input to the system is C_{Aip}.

Solution

a. Conservation of mass along with the rate of reaction equations are applied to the species in the tanks, leading to the following mathematical model for the perturbations in concentrations of A, B, and C in each reactor:

$$V_1\frac{dC_{A1p}}{dt} + (q + k_1V_1)C_{A1p} = C_{Aip} \qquad \text{(a)}$$

$$V_1\frac{dC_{B1p}}{dt} - k_1V_1C_{A1p} + qC_{B1p} = 0 \qquad \text{(b)}$$

$$V_1\frac{dC_{C1p}}{dt} + qC_{C1p} = 0 \qquad \text{(c)}$$

$$V_2\frac{dC_{A2p}}{dt} - qC_{A1p}(t - \tau) + (q + k_1V_2)C_{A2p} = 0 \qquad \text{(d)}$$

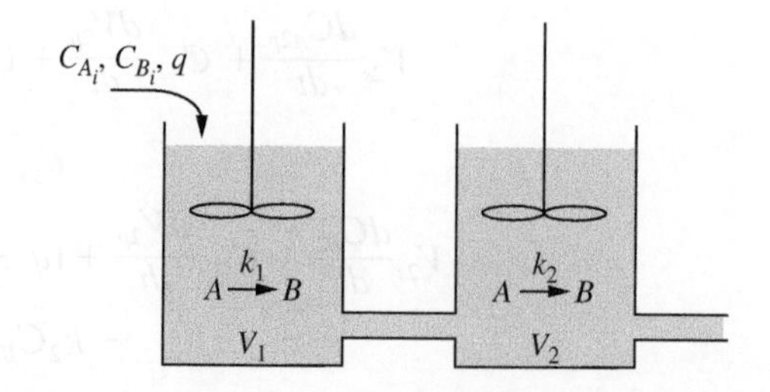

Figure 6.43

The double CSTR system of Example 6.35.

$$V_2\frac{dC_{B2p}}{dt} - qC_{B1p}(t-\tau) - k_1V_2C_{A2p} + (q + k_2V_2)C_{B2p} = 0 \qquad \text{(e)}$$

$$V_2\frac{dC_{C2p}}{dt} - k_1V_2C_{B2p} + qC_{C2p} = 0 \qquad \text{(f)}$$

b. Taking the Laplace transforms of Equation (a) and setting all initial conditions to zero leads to

$$C_{A1p}(s) = \frac{C_{Aip}(s)}{s + \frac{q}{V_1} + k} \Rightarrow G_{A1p}(s) = \frac{1}{s + \frac{q}{V_1} + k} \qquad \text{(g)}$$

Taking the Laplace transform of Equation (b), setting all initial conditions to zero, and using Equation (g) gives

$$C_{B1p}(s) = \frac{k_1C_{A1p}(s)}{s + \frac{q}{V_1}} = \frac{k_1C_{Aip}(s)}{\left(s + \frac{q}{V_1}\right)\left(s + \frac{q}{V_1} + k_1\right)} \Rightarrow G_{B1p}(s) = \frac{k_1}{\left(s + \frac{q}{V_1}\right)\left(s + \frac{q}{V_1} + k_1\right)} \qquad \text{(h)}$$

Because Equation (c) does not contain any input, taking the Laplace transform of Equation (c) leads to

$$G_{C1p}(s) = 0 \qquad \text{(i)}$$

Taking the Laplace transform of Equation (d) and using Equation (g) leads to

$$C_{A2p}(s) = \frac{\frac{q}{V_2}e^{-\tau s}C_{Aip}(s)}{\left(s + \frac{q}{V_1} + k_1\right)\left(s + \frac{q}{V_2} + k_1\right)} \Rightarrow G_{A2p}(s) = \frac{\frac{q}{V_2}e^{-\tau s}}{\left(s + \frac{q}{V_1} + k_1\right)\left(s + \frac{q}{V_2} + k_1\right)} \qquad \text{(j)}$$

Taking the Laplace transform of Equation (e) and using Equations (h) and (j) leads to

$$C_{B2p}(s) = \frac{\frac{q}{V_2}e^{-\tau s}C_{B1}(s)}{s + \frac{q}{V_2} + k_2} + \frac{k_1C_{A2p}(s)}{s + \frac{q}{V_2} + k_2}$$

$$= \frac{\frac{q}{V_2}e^{-\tau s}C_{Aip}(s)}{\left(s + \frac{q}{V_1}\right)\left(s + \frac{q}{V_1} + k_1\right)\left(s + \frac{q}{V_2} + k_2\right)} + \frac{k_1\frac{q}{V_2}e^{-\tau s}C_{Aip}(s)}{\left(s + \frac{q}{V_1} + k_1\right)\left(s + \frac{q}{V_2} + k_1\right)\left(s + \frac{q}{V_2} + k_2\right)}$$

$$\Rightarrow C_{B2p}(s) = \frac{\frac{q}{V_2}e^{-\tau s}\left[s + \frac{q}{V_2} + k_1 + k_1\left(s + \frac{q}{V_1}\right)\right]}{\left(s + \frac{q}{V_1}\right)\left(s + \frac{q}{V_1} + k_1\right)\left(s + \frac{q}{V_2} + k_2\right)\left(s + \frac{q}{V_2} + k_1\right)} \qquad \text{(k)}$$

Taking the Laplace transform of Equation (f) and using Equation (k) leads to

$$G_{C2p}(s) = \frac{k_1\frac{q}{V_2}e^{-\tau s}\left[s + \frac{q}{V_2} + k_1 + k_1\left(s + \frac{q}{V_1}\right)\right]}{\left(s + \frac{q}{V_1}\right)\left(s + \frac{q}{V_1} + k_1\right)\left(s + \frac{q}{V_2} + k_2\right)\left(s + \frac{q}{V_2} + k_1\right)\left(s + \frac{q}{V_2}\right)} \qquad \text{(l)}$$

6.12 Summary

6.12.1 Mathematical Modeling of Transport Systems

The development of mathematical models for the dynamic behavior of fluid, thermal, chemical, and biological systems requires the modeling of transport processes. Macroscopic modeling of transport processes for lumped parameter systems requires the application of conservation laws to a distinct control volume.

The general principles for the mathematical modeling of transport processes for lumped parameter systems are:

- A control volume, a region with defined boundaries through which mass can enter and leave, is identified.
- The control volume form of Conservation of Mass is applied, as stated in Equation (6.2).
- The control volume form of Conservation of Energy is applied, as stated in Equation (6.12).
- A mathematical model is obtained through the application of basic conservation laws to the control volume at an arbitrary instant.

6.12.2 Chapter Highlights

Some important concepts covered in Chapter 6 are:

- Bernoulli's equation is derived from the momentum equation and applies along a streamline in a flow. Under appropriate assumptions it can be reconciled with the control volume form of Conservation of Energy.
- Head is a measure of the total kinetic and potential energy and the potential of the pressure to do work.
- Minor head losses in piping systems occur due to changes in flow geometry at locations such as valves, elbows, and joints.
- Major head losses that occur due to friction are dependent on the Reynolds number and relative roughness ratio.
- The extended head form of Bernoulli's equation, which includes head losses, is used to analyze flows in pipes.
- A pipe's resistance is the rate of change of head loss with flow rate.
- Liquid-level problems deal with the determination of transient perturbations of liquid level in a tank due to perturbation in a parameter such as flow rate.
- Liquid-level problems are linearized by using the system's resistances at steady state.
- Many pneumatic problems involve a polytropic process. Isothermal and isentropic processes are special cases of a polytropic process.
- A flow equation is developed for hydraulic servomotors that relates the rate of change of position of the piston to the opening of the spool valve.
- Heat transfer occurs in thermal systems by conduction, convection, or radiation.
- An equivalent thermal resistance can be determined for combined heat transfer by conduction and convection.
- A conservation of species equation, along the lines of Equation (6.127), can be developed for each species component in a CSTR.
- Nonisothermal reactions in CSTRs require the application of Conservation of Energy as well as Conservation of Species.
- Mathematical models for biological processes such as drug infusion are developed using conservation laws.

6.12.3 Important Equations

The following equations are intended to reinforce the important points in this chapter. When using any of these equations, it is necessary to understand its inherent assumptions as well as the meaning of all parameters. This information is found in the text near the equation.

- Control volume form of Conservation of Mass

$$\dot{m}_{in} - \dot{m}_{out} = \frac{dm}{dt} \tag{6.3}$$

- Control volume form of Conservation of Energy

$$\dot{m}_{in}\left(u + \frac{p}{\rho} + \frac{v^2}{2} + gz\right)_{in} - \dot{m}_{out}\left(u + \frac{p}{\rho} + \frac{v^2}{2} + gz\right)_{out} + \dot{Q}$$

$$-\dot{W}_s - \dot{W}_f = \frac{d}{dt}\left[\int_V \left(u + \frac{v^2}{2} + gz\right)\rho dV\right] \tag{6.15}$$

- Bernoulli's equation

$$\frac{p_1}{\rho} + \frac{v_1^2}{2} + gz_1 = \frac{p_2}{\rho} + \frac{v_2^2}{2} + gz_2 \tag{6.16}$$

- Definition of head

$$h = \frac{p}{\rho g} + \frac{v^2}{2g} + z \tag{6.21}$$

- Extended Bernoulli's equation

$$h_1 - h_\ell = h_2 \tag{6.23}$$

- Friction head loss

$$h_\ell = f\frac{L}{D}\frac{v^2}{2g} \tag{6.26}$$

- Pipe resistance for turbulent flow

$$R = 2\frac{h_\ell}{Q} \tag{6.38}$$

- Orifice equation

$$Q = C_Q A_2\sqrt{\frac{2(p_1 - p_2)}{\rho}} \tag{6.51}$$

- Resistance for compressible flows

$$\dot{m} = \frac{p_1 - p_2}{R} \tag{6.54}$$

- Mathematical model of liquid-level systems

$$A\frac{dh}{dt} + \frac{1}{R}h = q_i(t) \tag{6.65}$$

- Flow equation for hydraulic servomotors

$$\hat{C}x = A_p\dot{y} \tag{6.90}$$

- Thermal resistance for a rectangular wall

$$R_{eq} = \frac{1}{h_i A} + \sum \frac{w_j}{k_j A} + \frac{1}{h_o A} \tag{6.111}$$

- Thermal resistance for a cylinder

$$R = \frac{1}{h_i 2\pi r_i L} + \sum_j \frac{1}{2\pi k_j L}\ln\left(\frac{r_j}{r_{j-1}}\right) + \frac{1}{h_o 2\pi r_o L} \tag{6.112}$$

- Rate of reaction

$$\frac{dC_A}{dt} = -kC_A \tag{6.123}$$

- Arrhenius law

$$k = \alpha e^{-E/(\overline{R}T)} \tag{6.125}$$

Short Answer Problems

SA6.1 Explain the lumped parameter assumption.

SA6.2 Explain the difference between a control volume and a system.

SA6.3 Give an example of a control volume that changes shape with time.

SA6.4 State in words the control volume form of Conservation of Mass.

SA6.5 Explain the term "streamline."

SA6.6 Explain the difference between laminar flow and turbulent flow.

SA6.7 Define the term "head."

SA6.8 What are the dimensions of head loss?

SA6.9 What ratio forms the Reynolds number?

SA6.10 Describe how the linearization assumption is used in liquid-level problems.

SA6.11 A vessel contains an incompressible liquid of mass density 1000 kg/m^3. If fluid enters the vessel at a constant flow rate of 3.0 m^3/s, what is the mass flow rate of the entering stream of fluid?

SA6.12 A vessel of cross-sectional area 0.27 m^2 contains an incompressible liquid of mass density 1000 kg/m^3. If fluid enters the vessel through the only inlet at a constant flow rate of 3.0 m^3/s, what is the rate of change of the height of the fluid?

SA6.13 A vessel of cross-sectional area 0.27 m^2 contains an incompressible liquid of mass density 1000 kg/m^3. The initial height of the fluid in the vessel is 0.78 m. If fluid enters the vessel through the only inlet at a constant flow rate of 3.0 m^3/s, how long will it take for the height of the fluid in the vessel to reach 1.0 m?

SA6.14 Two pipes are in parallel with a total flow rate of 30 m^3/s. The flow rate through one pipe is 21 m^3/s. What is the flow rate through the second pipe?

SA6.15 Two pipes are in series. The flow rate through the first pipe is 30 m^3/s. What is the flow rate through the second pipe?

SA6.16 Two pipes are in parallel. The head loss across one pipe is 250 mm. What is the head loss across the second pipe?

SA6.17 Two pipes are in series with a total head loss across the system of 650 mm. The head loss across one pipe is 250 mm. What is the head loss across the second pipe?

SA6.18 Water at 20°C is flowing through a pipe of diameter 20 mm at a speed of 14 m/s. What is the Reynolds number for the flow?

SA6.19 Water at 20°C is flowing through a smooth pipe with a Reynolds number of 6×10^5. What is the friction factor for this flow?

SA6.20 Water at 20°C is flowing through a smooth pipe with a Reynolds number of 640. What is the friction factor for this flow?

SA6.21 Water at 20°C is flowing through a pipe of diameter 300 mm at a Reynolds number of 1200. The head loss across the pipe is 1.6 m. What is the resistance of the pipe?

SA6.22 Water at 20°C is flowing through a pipe of diameter 100 mm with a Reynolds number of 6.5×10^4. The head loss across the pipe is 4.1 m. What is the resistance of the pipe?

SA6.23 Air is flowing through a pipe at 20°C. The mass flow rate at the exit of the pipe is 20 kg/s. The pressure drop across the pipe is 200 Pa. What is the resistance of the pipe?

SA6.24 Water exits a reservoir at a depth of 14 m and flows through a pipe at a flow rate of 28 m^3/s. The water exits into the atmosphere. What is the resistance of the pipe?

SA6.25 The two tanks of Figure SA6.25 are shown at a steady state. What is the resistance of the pipe connecting the two tanks?

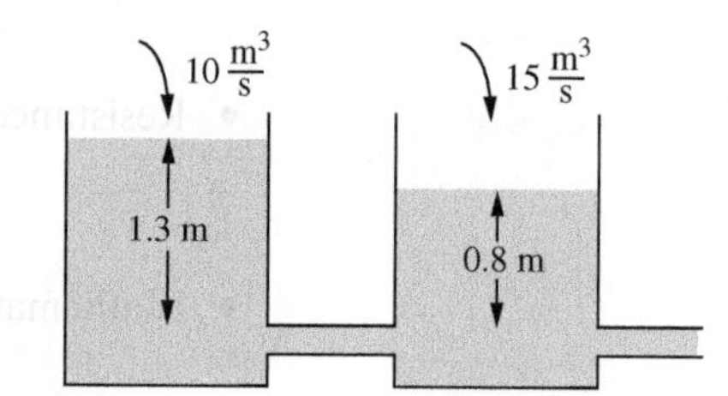

Figure SA6.25

SA6.26 The two tanks of Figure SA6.25 are shown at a steady state. What is the resistance of the pipe that exits from the second tank?

SA6.27 Write the mathematical model for the system of Figure SA6.27 if h is the perturbation in the liquid level due to a perturbation in the flow rate of q.

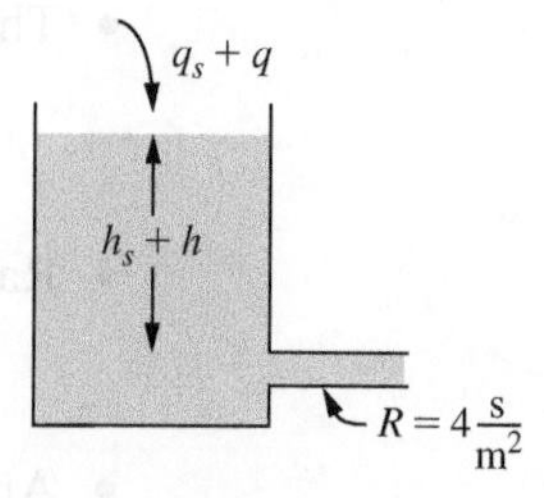

Figure SA6.27

SA6.28 For the equation $\hat{C}x = A_p\dot{y}$, what is (a) x, (b) y, (c) $\hat{C}$, and (d) A_p?

SA6.29 What is the capacitance of a pressure vessel being filled during an isothermal process?

SA6.30 What is the value of n in the equation $\frac{p}{\rho^n} = C$ for an isentropic process?

SA6.31 A wall of thickness 2 in. is made of brick with a surface area of 150 ft^2. What is the thermal resistance of the wall?

SA6.32 The wall of Question SA6.31 is adjacent to the ambient. The heat transfer coefficient between the ambient and the wall is 4 Btu/(hr·ft^2·F). What is the equivalent resistance of the wall and the thermal boundary layer between the wall and the ambient?

SA6.33 An annular cylinder of length 30 cm is made of steel and has an inner radius of 1 cm and an outer radius of 3 cm. What is the thermal resistance of the cylinder?

SA6.34 How is thermal resistance defined?

SA6.35 What is Fourier's conduction law for heat transfer in one dimension?

SA6.36 A wall that has a thermal resistance of R, a total mass of m, and a specific heat of c_v is subject to a time-dependent temperature T_∞ at its exterior surface. What is the mathematical model for the temperature at the interior surface due to a change in T_∞? List all assumptions.

SA6.37 What assumptions are made in modeling a CSTR?

SA6.38 What is meant by an exothermic reaction?

SA6.39 What is the Arrhenius law?

SA6.40 What is meant by the term "heat of reaction"?

SA6.41 What is the order of the transfer function for a CSTR system with an isothermal first-order irreversible reaction? What are the energy sources?

SA6.42 What is the order of the transfer function of a three-tank liquid-level problem?

SA6.43 What is the order of the transfer function of the two-compartment model for drug delivery?

SA6.44 What is the transfer function for a single-tank liquid-level problem?

SA6.45 How many state variables are required for state-space modeling of the hydraulic servomotor of Example 6.13?

SA6.46 How many state variables are required for state-space modeling of the heat exchanger of Example 6.31?

Problems

6.1 The water clock of Figure P6.1 is of conical shape. The water exits through a 3-mm diameter hole. To what height should the clock be filled if the water is to drain in exactly 1 hr?

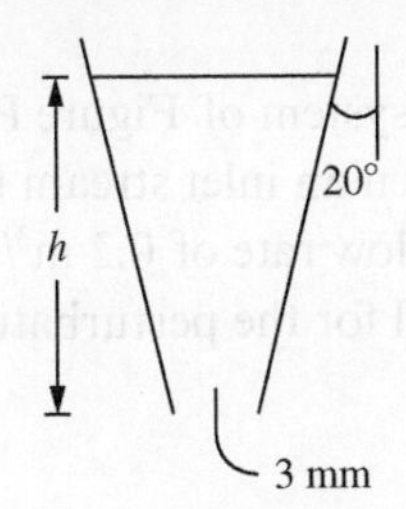

Figure P6.1

6.2 A spherical balloon is being filled with helium ($\rho = 0.169$ kg/m^3) at a constant mass flow rate of 0.1 kg/s. The balloon is considered to be filled when it reaches a diameter of 20 cm. How long does it take to fill the balloon?

6.3 Water is contained in the large reservoir of Figure P6.3 under surface pressure of 2.0 kPa. Determine ℓ, the range of the water jet.

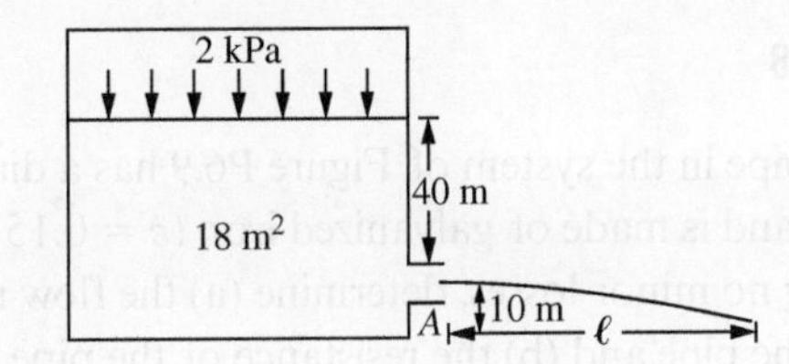

Figure P6.3

6.4 The reservoir of Figure P6.4 drains into a 20-cm diameter pipe of length 26 m. What is the velocity of the water leaving the pipe if the pressure on the surface of the reservoir is atmospheric? Assume a friction factor of 0.003.

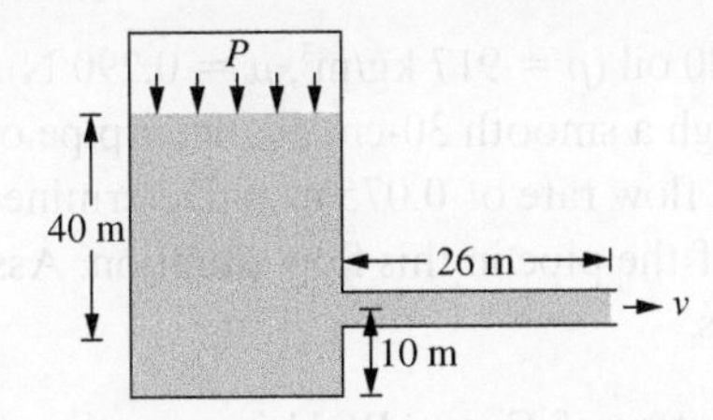

Figure P6.4

6.5 Repeat Problem 6.4 if the gauge pressure on the surface of the reservoir is 50 kPa.

6.6 Repeat Problem 6.4 if the pipe has a relative roughness ratio of 0.002.

6.7 What is the required diameter of a 10-m smooth pipe conveying water from a pressure head of 18 m that will deliver a flow rate of 0.2 m^3/s?

6.8 For purposes of modeling, consider a toy water rocket to be a cylinder of diameter D and length L, as illustrated in Figure P6.8. The rocket initially contains air at atmospheric pressure when it is filled to a height h with water. (a) What is the pressure on the surface of the water after the rocket is filled and inverted? (b) After the rocket is released, water exits through a hole of diameter d. Let $z(t)$ be the instantaneous height of the water in the rocket. Determine the pressure on the surface of the water as a function of z. (c) Apply Bernoulli's equation between the surface of the water and the exit to approximate the velocity of the water leaving the rocket. (d) Apply the control volume form of Conservation of Mass to the rocket to derive a mathematical model for $z(t)$.

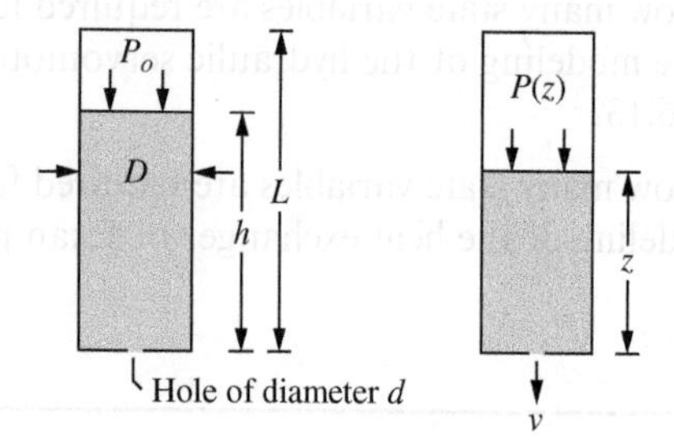

Figure P6.8

6.9 The pipe in the system of Figure P6.9 has a diameter of 30 cm and is made of galvanized iron ($\varepsilon = 0.15$ mm). Assuming no minor losses, determine (a) the flow rate through the pipe and (b) the resistance of the pipe at this flow rate.

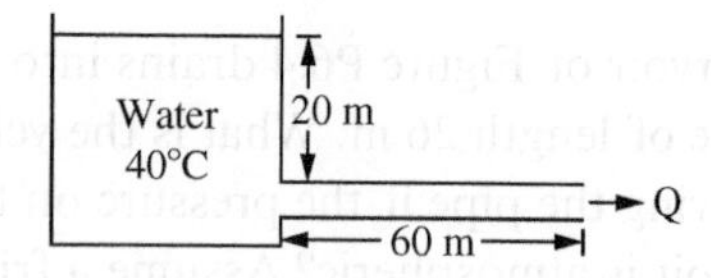

Figure P6.9

6.10 SAE 30 oil ($\rho = 917$ kg/m^3, $\mu = 0.290$ N·s/m^2) flows through a smooth 30-cm diameter pipe of length 30 m with a flow rate of 0.075 m^3/s. Determine the resistance of the pipe in this flow situation. Assume no minor losses.

6.11 The system of Figure P6.11 is operating at a steady state. Determine the resistances in pipes A and B.

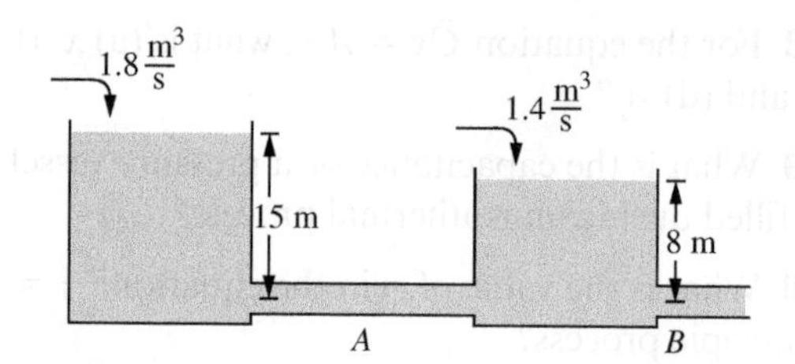

Figure P6.11

6.12 Determine the head loss between A and B and the flow rate in each of the pipes in the system of Figure P6.12.

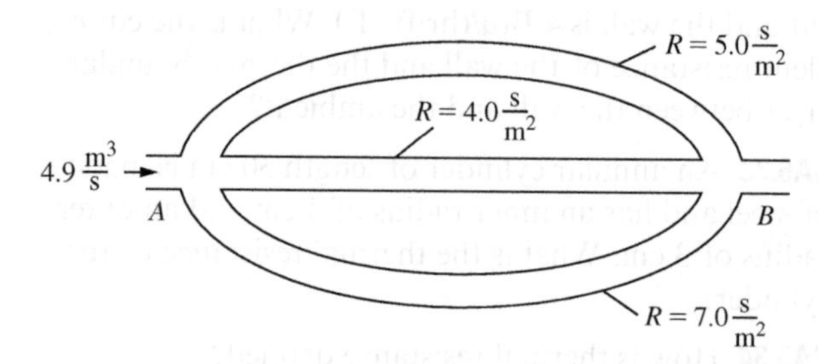

Figure P6.12

6.13 The system of Figure P6.13 is operating at a steady state when the input flow rate is suddenly decreased to 2.2 m^3/s. Derive a linearized mathematical model for the resulting perturbation of the liquid level in the tank.

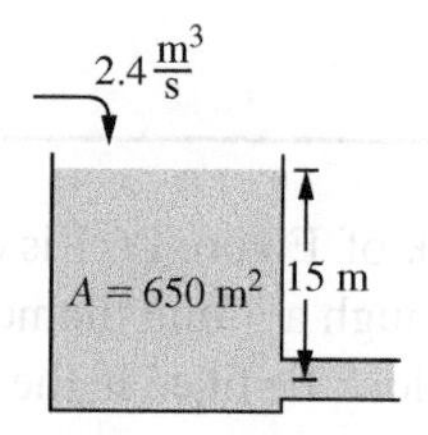

Figure P6.13

6.14 The two-tank system of Figure P6.14 is operating at a steady state when an inlet stream begins to flow into tank B with a flow rate of 0.2 m^3/s. Derive a mathematical model for the perturbations in the liquid levels in the tanks.

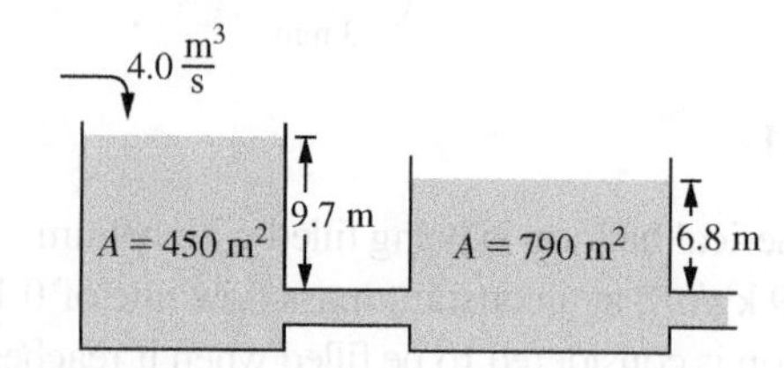

Figure P6.14

6.15 The two-tank system of Figure P6.15 is operating at a steady state when the inlet flow rate is increased to 4.6 m^3/s. Derive a mathematical model for the perturbations in liquid levels in the tanks. Both tanks are circular in cross section.

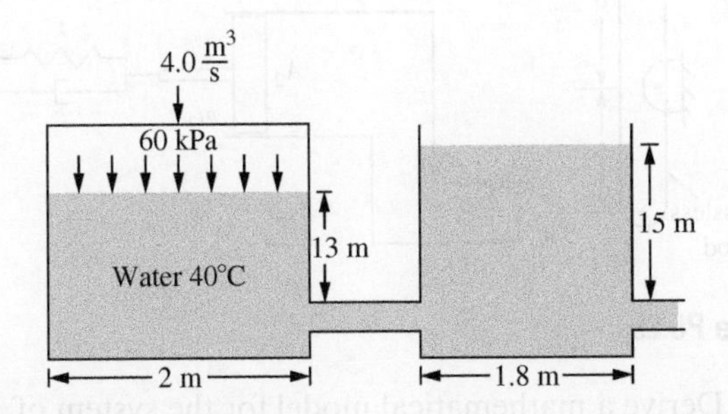

Figure P6.15

6.16 The system of Figure P6.16 is operating at a steady state when the inlet flow rate has a perturbation given by $q_p(t)$. Derive a linear mathematical model for the perturbations in liquid levels in the tanks. The tanks are connected by horizontal pipes.

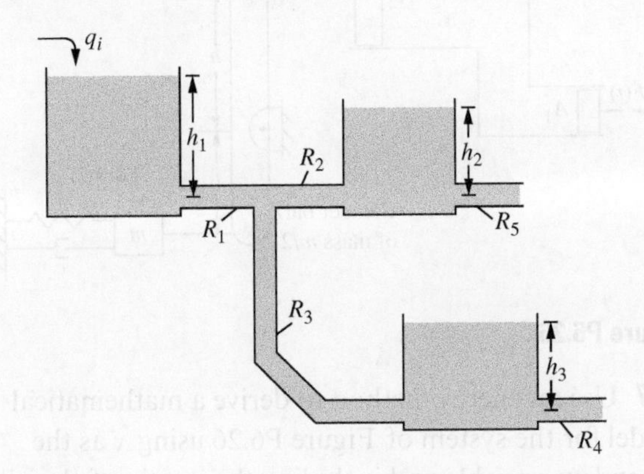

Figure P6.16

6.17 Derive a mathematical model for the perturbation in the liquid level in the conical tank of Figure P6.17 when it is subject to a perturbation in the inlet flow rate. The pipe that drains the tank from its bottom has resistance R. List all assumptions.

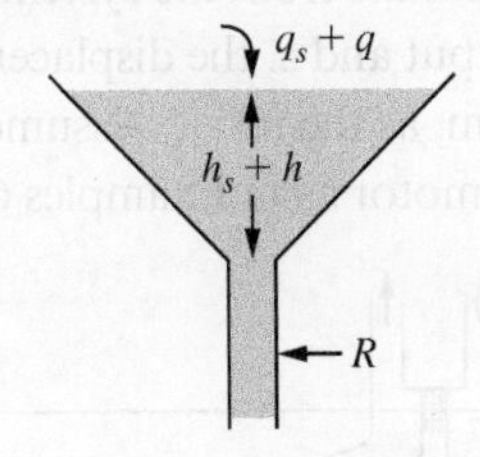

Figure P6.17

6.18 Derive a mathematical model for the perturbation in the liquid level in the diameter D spherical tank of Figure P6.18 when it is subject to a perturbation in the inlet flow rate. The pipe that drains the tank from its bottom has resistance R. List all assumptions.

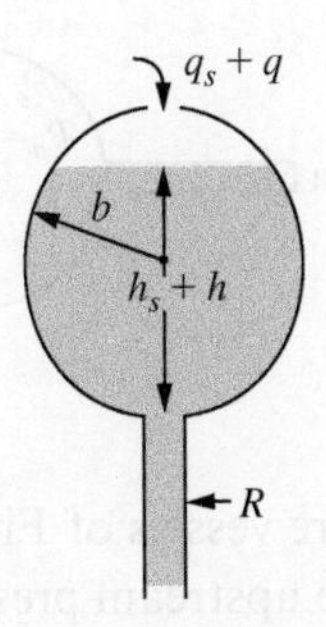

Figure P6.18

6.19 A cylindrical tank of diameter D and length L is shown in Figure P6.19. The tank is at a steady state at a level H when the flow rate into the tank is perturbed such that it becomes $Q + q$. The pipe that is connected at the bottom of the tank has resistance R. Derive a mathematical model for the perturbation in the liquid level h.

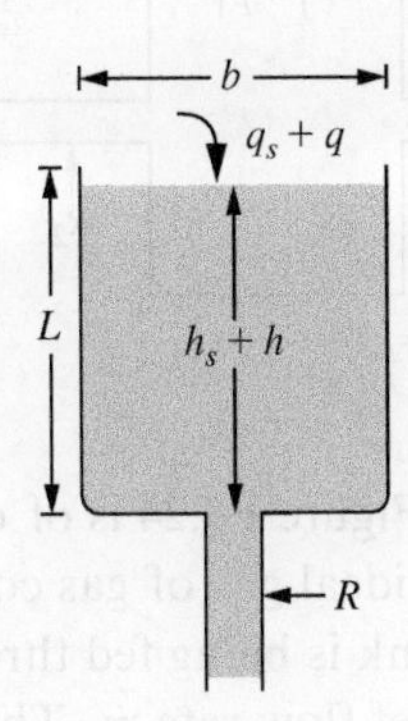

Figure P6.19

6.20 Derive a mathematical model for the system of Example 6.10 using temperature as the dependent variable and assuming the process is isentropic.

6.21 Derive a mathematical model for the system of Example 6.10 assuming the process is isothermal and using density as the dependent variable.

6.22 The tank of Figure P6.22 is spherical with a diameter of 2 m. It contains helium ($R_{He} = 2077\ m^2/(s^2 \cdot K)$, $\gamma = 1.66$) at a pressure of 2.5 kPa and a temperature of 60°C. At $t = 0$, the valve is open, allowing helium to flow into the tank from an upstream source at a pressure of 10 kPa. Previous measurements have shown that the initial mass flow rate into the tank from the connecting pipe is 0.6 kg/s when the pressure in the tank is 2.5 kPa and the supply pressure is 8 kPa. Derive a linear mathematical model for the process assuming it is isothermal and using the pressure in the tank as the dependent variable.

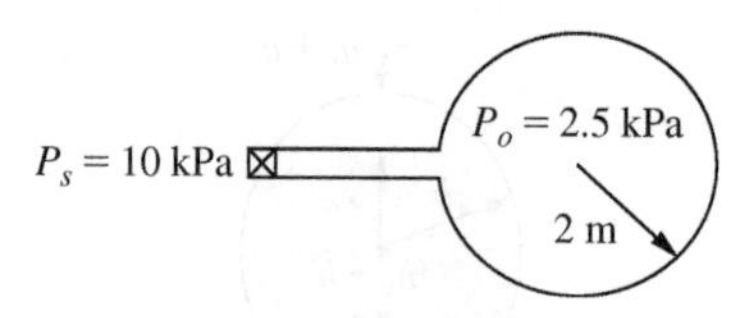

Figure P6.22

6.23 The two pressure vessels of Figure P6.23 are at a steady state when the upstream pressure P is perturbed such that is becomes $P + p$. This leads to perturbations of pressure in the tanks p_1 and p_2. The pipe from the supply to the first tank has resistance R_1, and the pipe connecting the two tanks has resistance R_2. Derive a linearized mathematical model for the perturbations in pressures in the tanks using the perturbation in the upstream pressure as the input.

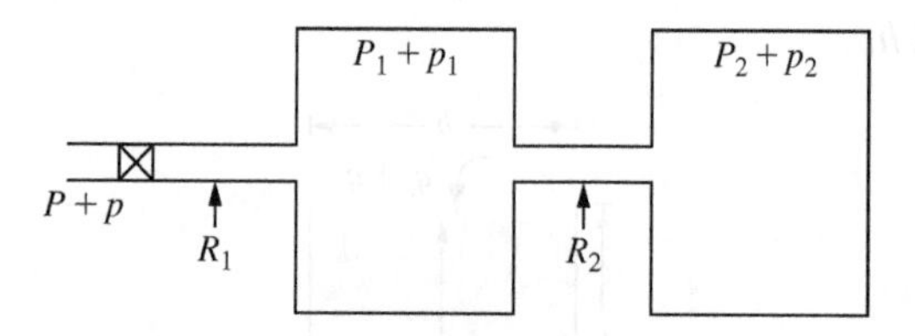

Figure P6.23

6.24 The tank of Figure P6.24 is of constant volume V and contains an ideal gas of gas constant R_g and viscosity μ. The tank is being fed through a tube with a constant inlet flow rate $\dot{m}_i$. The tank is being bled through a pipe of length L and diameter D into a large reservoir at an exit pressure p_e. Assume the process is isothermal at a temperature T and the flow through the pipe is laminar such that the friction factor is $f = 64/\text{Re}$. The gas in the tank is initially at a pressure p_o. Derive a mathematical model for the process using the pressure in the tank as the dependent variable.

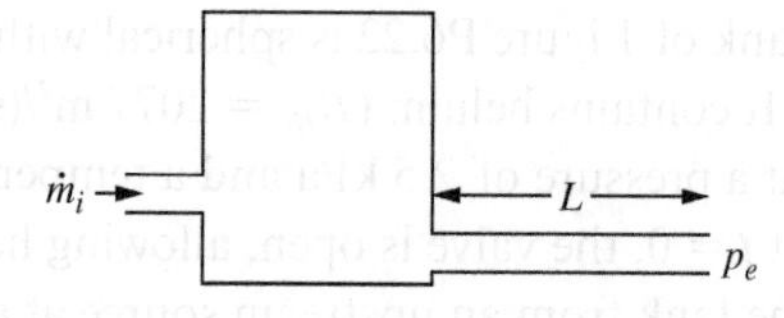

Figure P6.24

6.25 Derive a mathematical model for the system of Figure P6.25 using x, the displacement of the piston from the system's equilibrium position, as the dependent variable and $F(t)$ as the input to the system. The hydraulic system lies in a horizontal plane.

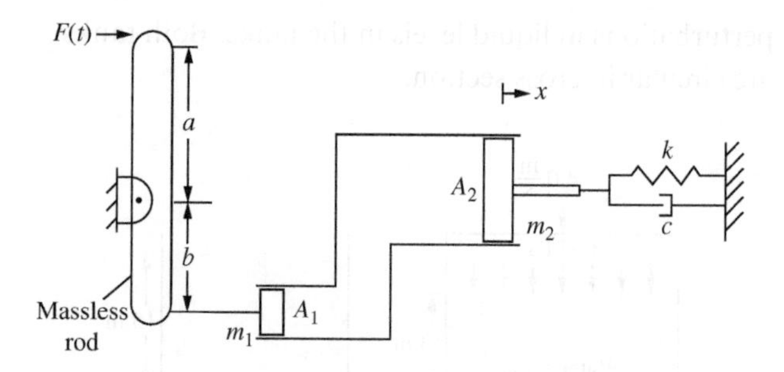

Figure P6.25

6.26 Derive a mathematical model for the system of Figure P6.26 using x, the displacement of the block of mass m from the system's equilibrium position, and $F(t)$, the force applied to the smaller piston, as input to the system. Ignore the mass of the pistons. The hydraulic system lies in a horizontal plane.

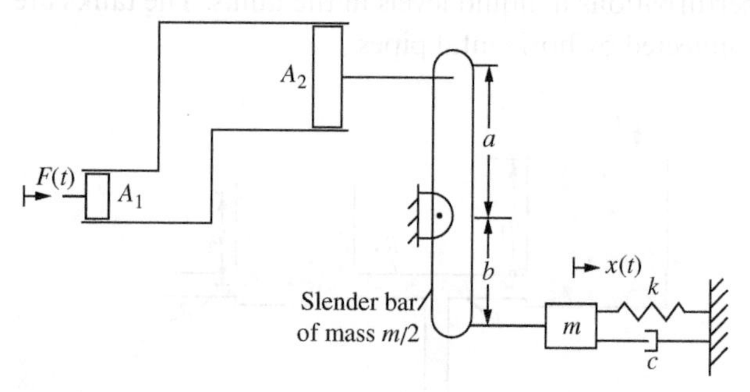

Figure P6.26

6.27 Use an energy method to derive a mathematical model for the system of Figure P6.26 using x as the dependent variable and including the inertia of the pistons, but neglecting the inertia of the fluid. The pistons are of mass m_1 and m_2, respectively.

6.28 The hydraulic servomotor of Figure P6.28 is connected to a walking beam as in Example 6.13. However, the end of the piston is connected to a mechanism as shown. Derive a mathematical model for the system using x, the displacement of the mass from the system's equilibrium position, as the output and z, the displacement of the end of the walking beam, as the input. Assume parameters of the hydraulic servomotor as in Examples 6.12 and 6.13.

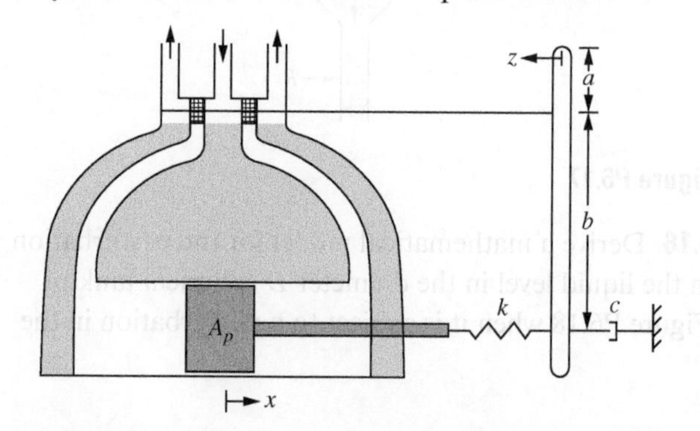

Figure P6.28

6.29 Derive the mathematical model between the input $z(t)$ and the output $x(t)$ for the hydraulic servomotor of Figure P6.29. Assume the appropriate flow equation is $\hat{C}x_A = A_P\dot{x}$, where x_A is the displacement of the spool valve.

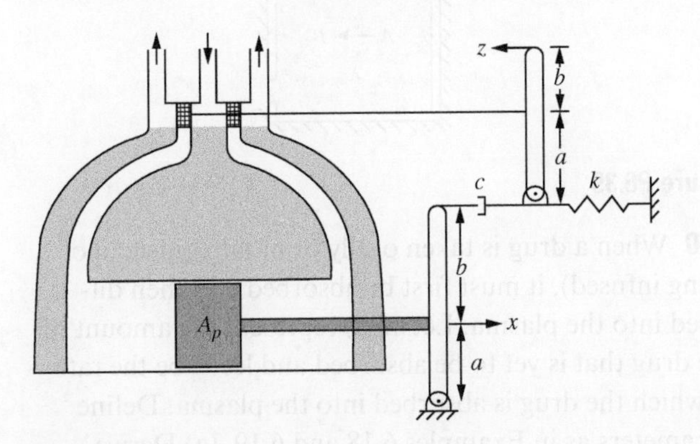

Figure P6.29

6.30 Derive the mathematical model relating the applied force $F(t)$ to the displacement of the piston $x(t)$ for the hydraulic servomotor of Figure P6.30.

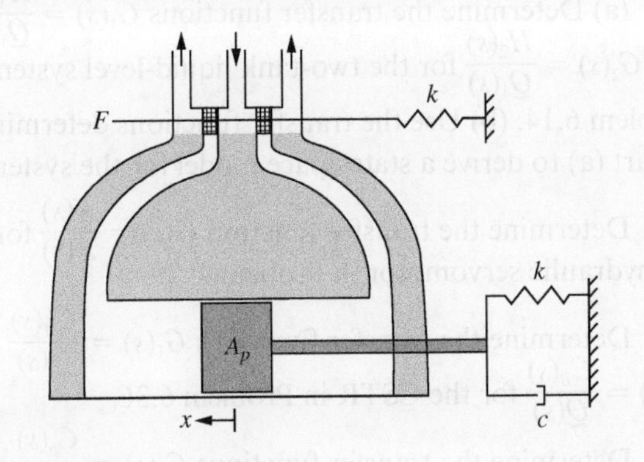

Figure P6.30

6.31 For the system of Figure P6.31, derive the mathematical model using $z(t)$ as the input and $x(t)$ as the output.

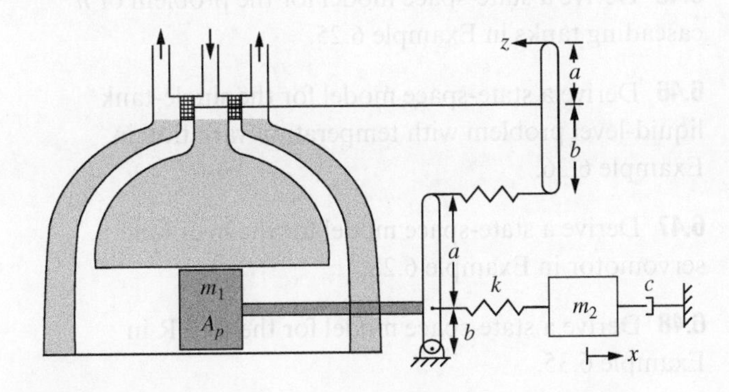

Figure P6.31

6.32 The thin plate of Figure P6.32 is at the ambient temperature T_0 when it begins to slide with constant velocity V on a surface. The plate is made of a material of mass density ρ and specific heat c_v. The plate is of uniform thickness ℓ and has a rectangular surface of width w and depth d. The coefficient of kinetic friction between the plate and surface is μ, and the heat transfer coefficient with the ambient is h. Derive a mathematical model for the temperature in the plate as it slides across the surface.

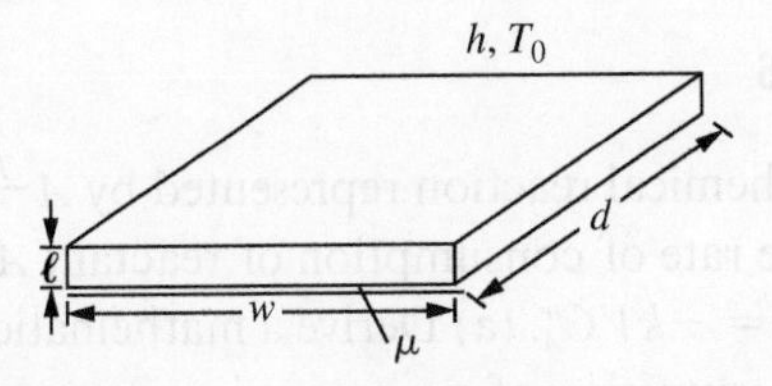

Figure P6.32

6.33 An aquarium houses a 600-m³ tank of seawater. The aquatic life requires that the temperature be maintained near 30°C. A filtration and recirculation system maintains the tank at a constant volume. After filtration, the entering stream has a flow rate of 1.8 m³/s and a temperature of 20°C. Seawater has a density of 1.03×10^3 kg/m³ and a specific heat of 4.0×10^3 J/kg·C. (a) What is the rate at which heat must be added to the tank to maintain the temperature of the tank at 30°C? (b) The system is operating at a steady state when the heater stops working. Derive a mathematical model for the change in temperature from the steady state, assuming constant specific heat.

6.34 The aquarium in Problem 6.33 is operating at a steady state with heat added such that the temperature is maintained at 30°C when the filtration system stops working. The water is recirculated without cooling or filtration. Derive a mathematical model for the temperature in the tank after the filtration system stops, assuming that it takes 15 s to recirculate the flow through the system.

6.35 The aquarium tank of Problem 6.33 is operating at steady state when the flow rate suddenly changes to 1.6 m³/s. The heater continues to provide heat to the tank at the same rate. Derive a mathematical model for the change in temperature in the tank. Assume the volume of the tank and the temperature of the inlet stream are unchanged.

6.36 The CSTR of Figure P6.36 has a first-order reaction represented by

$$2A \xrightarrow{k} B \tag{a}$$

In the reaction described by Equation (a), two moles of component A are used to form one mole of component B.

Derive a mathematical model for the concentrations of components A and B in the reactor.

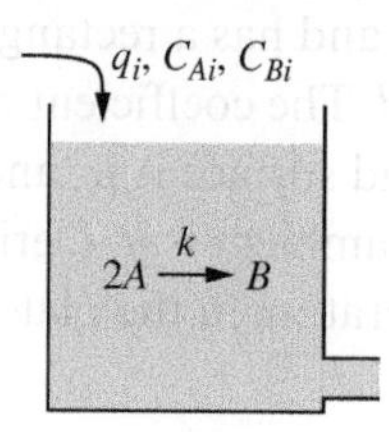

Figure P6.36

6.37 If a chemical reaction represented by $A \xrightarrow{k} B$ is of order n, the rate of consumption of reactant A is given by $dC_A/dt = -kVC_A^n$. (a) Derive a mathematical model for the concentration of a reactant in a constant-volume CSTR with an isothermal reaction of order 1.5. (b) The reactor is operating at a steady state when a sudden perturbation in the concentration of the reactant in the input stream occurs such that $C_{Ai} = C_{Ai,s} + C_{Ai,p}$, where $C_{Ai,s}$ is the previous steady-state concentration of A in the input stream and $C_{Ai,p}$ is its perturbation. Derive a linear model for the perturbation in the reactant concentration in the tank.

6.38 The reaction in a CSTR is represented by

$$A + B \xrightarrow{k} C \tag{a}$$

in which components A and B are the reactants and component C is the product. The rate of change of the number of moles of reactant A due to the reaction is

$$\left(\frac{dn_A}{dt}\right)_{reaction} = -kV^2C_AC_B \tag{b}$$

Develop a mathematical model for the concentrations of components A, B, and C in the reactor assuming the reaction is isothermal.

6.39 A first-order reaction given by $A \xrightarrow{k} B$ occurs in a CSTR. The reaction is exothermic with a heat of reaction of λ. The reactor is cooled by surrounding it with a jacket of cooling water, as illustrated in Figure P6.39. Water flows into the jacket at a rate q_w at a temperature T_{wi}. A constant volume of water is maintained in the jacket. Let R_w be the thermal resistance in the water and R_1 and R_2 be the thermal resistances from the interior and exterior walls of the jacket. Let R_c be the convective resistance between the jacket and the component mixture in the reactor. Assume the exterior surface of the jacket is insulated. Derive a mathematical model for the process using the concentrations of the reactant and product C_A and C_B, the temperature in the reactor T, and the temperature of the water in the jacket T_w as dependent variables.

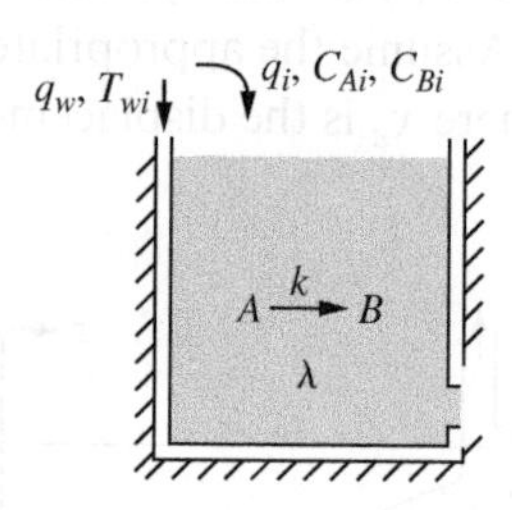

Figure P6.39

6.40 When a drug is taken orally or nasally (instead of being infused), it must first be absorbed and then diffused into the plasma. Let $m_a(t)$ represent the amount of the drug that is yet to be absorbed and let k_a be the rate at which the drug is absorbed into the plasma. Define parameters as in Examples 6.18 and 6.19. (a) Derive a mathematical model for m_a and C_p assuming the one-compartment model of Example 6.18. (b) Derive a mathematical model for m_a, C_p, and C_t assuming the two-compartment model of Example 6.19.

6.41 (a) Determine the transfer functions $G_1(s) = \frac{H_1(s)}{Q_1(s)}$ and $G_2(s) = \frac{H_2(s)}{Q_1(s)}$ for the two-tank liquid-level system of Problem 6.14. (b) Use the transfer functions determined in part (a) to derive a state-space model for the system.

6.42 Determine the transfer function $G(s) = \frac{X(s)}{Z(s)}$ for the hydraulic servomotor in Problem 6.29.

6.43 Determine the transfer functions $G_1(s) = \frac{C_A(s)}{Q(s)}$ and $G_2(s) = \frac{C_B(s)}{Q(s)}$ for the CSTR in Problem 6.36.

6.44 Determine the transfer functions $G_1(s) = \frac{C_p(s)}{I(s)}$ and $G_2(s) = \frac{C_t(s)}{I(s)}$ for the two-compartment model of drug delivery in Example 6.19.

6.45 Derive a state-space model for the problem of n cascading tanks in Example 6.25.

6.46 Derive a state-space model for the single-tank liquid-level problem with temperature variation in Example 6.26.

6.47 Derive a state-space model for the hydraulic servomotor in Example 6.28.

6.48 Derive a state-space model for the CSTR in Example 6.35.

7 Transient Analysis and Time Domain Response

Learning Objectives

At the conclusion of this chapter, you should be able to:

- Identify the poles, zeros, and residues of a transfer function
- Understand the definitions of free response, impulsive response, step response, ramp response, and general transient response of a system
- Discuss the initial continuity of the impulsive response of a system
- Determine the stability of a system from knowing its poles
- Calculate the final value of the step response of a system and determine other parameters of the step response
- Identify the time constant for a first-order system and understand its role in determining the responses of the system
- Identify the natural frequency and damping ratio for a second-order system and understand their role in determining the responses of the system
- Calculate the free, impulsive, and step responses for first-order and second-order systems
- Understand that the response of a higher-order system can be written as a linear combination of the responses of first-order and second-order systems
- Understand how to determine the responses of a system with time delay

7.1 System Response Using Transfer Functions

This chapter focuses on the transient response of systems given the mathematical model for the system and the system's transfer function.

7.1.1 Transfer Functions

Recall that the transfer function for a SISO system with input $f(t)$ and output $x(t)$ is defined by the ratio of the Laplace transform of the output $x(s)$ to the Laplace transform of the input $F(s)$:

$$G(s) = \frac{X(s)}{F(s)} \tag{7.1}$$

Equation (7.1) may be rearranged, leading to

$$X(s) = G(s)\,F(s) \tag{7.2}$$

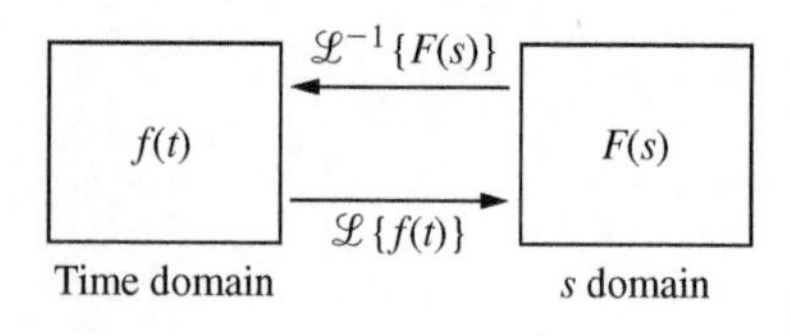

Figure 7.1

A function is naturally in the time domain. The function is in the *s* domain after taking the Laplace transform of the function in the time domain.

Applying the inverse Laplace transform to both sides of Equation (7.2) gives

$$x(t) = \mathcal{L}^{-1}\{G(s)F(s)\} \tag{7.3}$$

The response of the system in the time domain is given by Equation (7.3). The response of the system in the transform or "s" domain is given in Equation (7.2). Figure 7.1 shows the relationship between the time domain and the s domain.

This chapter uses the relationships given in Equations (7.2) and (7.3) to determine the time-dependent response of a system. Both a system's short-term behavior and its long-term behavior are investigated.

A system's **transient response** is its response due to changes in input. A transient response may occur when a nonzero initial condition is imposed on a system. A transient response may also occur due to a perturbation from a steady-state operating condition.

7.1.2 System Order

Transfer functions are usually written as the ratio of two polynomials:

$$G(s) = \frac{N(s)}{D(s)} \tag{7.4}$$

The order of the numerator $N(s)$ is taken to be n, while the order of the denominator $D(s)$ is taken to be m. Additionally, and without loss of generality, the leading coefficient of the denominator is taken to be 1. The order of the transfer function is the order of $D(s)$, which is m. The order is also equal to the number of independent energy storage elements in the system. Then $N(s)$ and $D(s)$ have the forms

$$N(s) = b_n s^n + b_{n-1}s^{n-1} + b_{n-2}s^{n-2} + \cdots + b_1 s + b_0 \tag{7.5}$$

$$D(s) = s^m + a_{m-1}s^{m-1} + a_{m-2}s^{m-2} + \cdots + a_1 s + a_0 \tag{7.6}$$

Transfer functions for systems with a time delay provide exceptions to the form of Equation (7.4). These transfer functions involve functions of the form $e^{-\tau s}$, where τ is the time delay.

The **poles** of the system are the roots of $D(s)$, the values of s such that $D(s) = 0$. The poles of a dynamic system may be real, purely imaginary, complex, or some combination of these possibilities. Since the coefficients of $D(s)$ are all real, complex poles must occur in complex conjugate pairs.

The **zeros** of a transfer function are the roots of $N(s)$. Not all transfer functions have zeros. For example, the transfer function of a mechanical system with force input has no zeros.

The **residues** of the transfer function are the numerical values in the numerator when the transfer function is written as a linear combination of first-order transfer functions. The residues may be complex when the poles of the transfer function are complex.

Example 7.1

Consider the transfer function

$$G(s) = \frac{2s + 3}{(s + 2)(s^2 + 6s + 25)} \tag{a}$$

Determine **a.** the poles of the transfer function, **b.** the zeros of the transfer function, and **c.** the residues of the transfer function.

Solution

a. The poles of the transfer function are the values of s for which the denominator is zero. The transfer function is third order, so there are three poles. Setting the denominator to zero leads to

$$s_p = -2, -3 \pm 4j \tag{b}$$

b. The zeros are the values of s for which the numerator of the transfer function is zero. This is clearly

$$s_z = -\frac{3}{2} \tag{c}$$

c. A partial fraction decomposition for Equation (a) gives

$$G(s) = \frac{A_1}{s+2} + \frac{A_2}{s+3-4j} + \frac{A_3}{s+3+4j} \tag{d}$$

where the residues A_1, A_2, and A_3 are determined using the residue theorem, Equation (2.38), as

$$A_1 = \lim_{s\to-2}(s+2)\left[\frac{2s+3}{(s+2)(s^2+6s+25)}\right] = \frac{2(-2)+3}{(-2)^2+6(-2)+25} = \frac{-1}{17} \tag{e}$$

$$A_2 = \lim_{s\to-3+4j}(s+3-4j)\left[\frac{2s+3}{(s+2)(s+3-4j)(s+3+4j)}\right]$$

$$= \frac{2(-3+4j)+3}{(-3+4j+2)(-3+4j+3+4j)} = \frac{-3+8j}{32+8j} = -0.0294 + 0.257j \tag{f}$$

$$A_2 = \lim_{s\to-3-4j}(s+3+4j)\left[\frac{2s+3}{(s+2)(s+3-4j)(s+3+4j)}\right]$$

$$= \frac{2(-3-4j)+3}{(-3-4j+2)(-3-4j+3-4j)} = \frac{-3-8j}{-32+8j} = -0.0294 - 0.257j \tag{g}$$

The appropriate partial fraction decomposition is

$$G(s) = \frac{-0.0588}{s+2} + \frac{-0.0294+0.257j}{s+3-4j} + \frac{-0.0294-0.257j}{s+3+4j} \tag{h}$$

The residues are

$$r = -0.0588, -0.0294 \pm 0.257j \tag{i}$$

7.1.3 Benchmark Problems

For this study, five benchmark problems are considered. Their transfer functions are referenced to illustrate the concepts.

1. A single-tank liquid-level problem has as input the perturbation in flow rate $q(t)$ and as output the perturbation of the height of a column of liquid $h(t)$. The transfer function for such a system is first order:

$$G(s) = \frac{1/A}{s + \frac{1}{RA}} \tag{7.7}$$

where A is the cross-sectional area of the tank and R is the resistance of the pipe (see Figure 7.2).

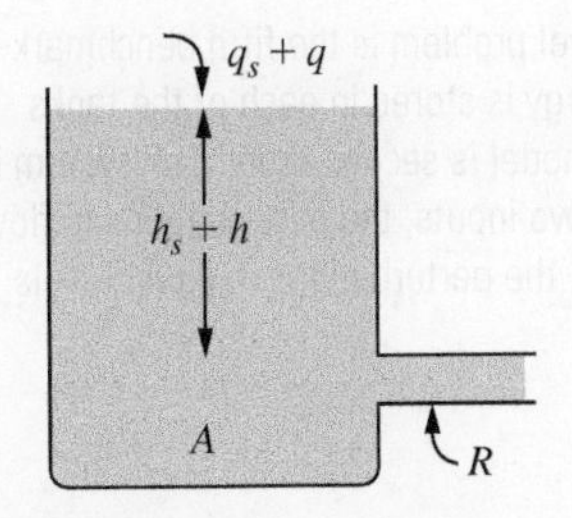

Figure 7.2

The single tank liquid level problem is the first benchmark problem. It has one energy storage element and its mathematical model is first order.

2. A mechanical mass, spring, and viscous-damper system is subject to a force input $f(t)$. The output is $x(t)$, the displacement of the mass from equilibrium. The transfer function is second order:

$$G(s) = \frac{1/m}{s^2 + \frac{c}{m}s + \frac{k}{m}} \tag{7.8}$$

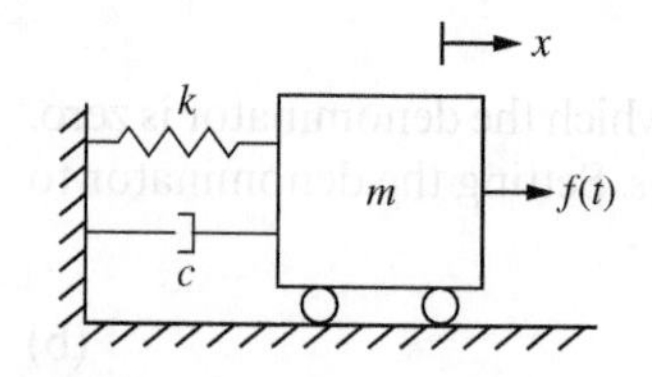

Figure 7.3

A mass, spring, and viscous-damper problem with force input is the second benchmark problem. Potential energy is stored by the spring, while kinetic energy is stored by the mass. Its mathematical model is second order.

In Equation (7.8), m is the mass of the block, c is the viscous damping coefficient, and k is the spring stiffness, as illustrated in Figure 7.3.

3. A mechanical mass, spring, and viscous-damper system is subject to motion input $y(t)$. The output is $x(t)$, the displacement of the mass from equilibrium. The transfer function is second order:

$$G(s) = \frac{\frac{c}{m}s + \frac{k}{m}}{s^2 + \frac{c}{m}s + \frac{k}{m}} \tag{7.9}$$

The system is illustrated in Figure 7.4.

4. The series LRC circuit has input voltage $v(t)$ and output current $i(t)$. The transfer function is second order:

$$G(s) = \frac{s/L}{s^2 + \frac{R}{L}s + \frac{1}{LC}} \tag{7.10}$$

The circuit is illustrated in Figure 7.5, where L is the inductance of the inductor, R is the resistance of the resistor, and C is the capacitance of the capacitor.

5. A two-tank liquid-level problem has inputs of the perturbations in the flow rates into the tanks, q_{i1} and q_{i2}, and outputs of the perturbations in the liquid levels, h_1 and h_2. The system has a matrix of transfer functions

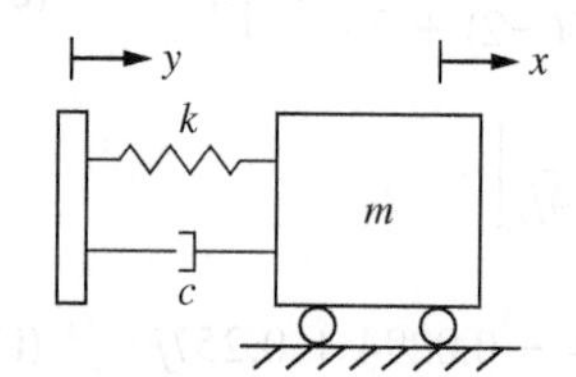

Figure 7.4

A mass, spring, and viscous-damper problem with motion input is the third benchmark problem. Its mathematical model is second order but contains a derivative of the input.

$$\mathbf{G}(s) = \frac{1/(A_1 A_2)}{D(s)}\begin{bmatrix} A_2 s + \frac{1}{R_1} + \frac{1}{R_2} & \frac{1}{R_1} \\ \frac{1}{R_1} & A_1 s + \frac{1}{R_1} \end{bmatrix} \tag{7.11}$$

where

$$D(s) = s^2 + \left(\frac{1}{A_2 R_1} + \frac{1}{A_2 R_2} + \frac{1}{A_1 R_1}\right)s + \frac{1}{A_1 A_2 R_1 R_2} \tag{7.12}$$

In Equations (7.11) and (7.12), A_1 and A_2 are the areas of the two tanks, and R_1 and R_2 are the resistances of the two pipes, as illustrated in Figure 7.6.

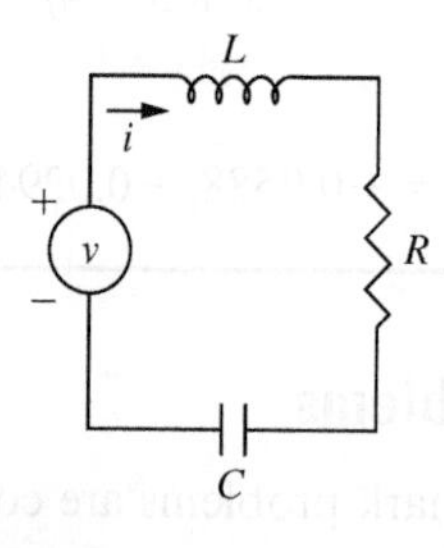

Figure 7.5

The series *LRC* circuit is the fourth benchmark problem. Energy is stored in an electric field by the capacitor, while energy is stored in a magnetic field by the inductor. Its mathematical model is second order.

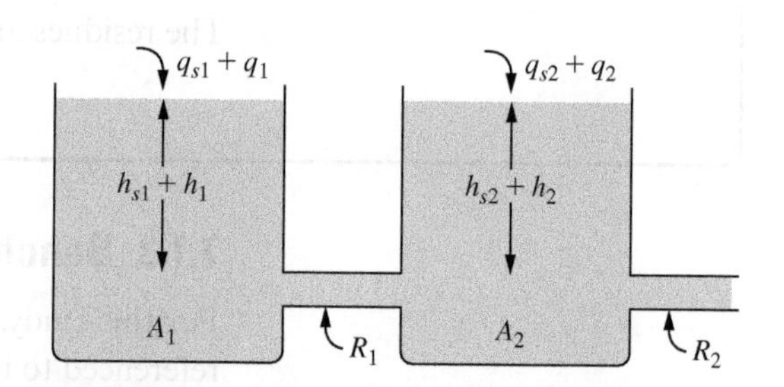

Figure 7.6

The two-tank liquid-level problem is the fifth benchmark problem. Potential energy is stored in each of the tanks and its mathematical model is second order. The system is a MIMO system with two inputs, the perturbations in flow rates, and two outputs, the perturbations in liquid levels.

7.2 Transient Response Specification

The transient response of a dynamic system depends on the input. The types of transient response and the parameters defining transient response are considered in this section.

7.2.1 Free Response

The free response occurs due to nonzero initial conditions in the absence of any additional input. The free response of a system is determined by applying the Laplace transform method to the mathematical model of the system, as discussed in Chapter 2.

7.2.2 Impulsive Response

Noting that $\{\mathcal{L}\{\delta(t)\} = 1$, it is clear from Equation (7.2) that the transfer function is equal to the Laplace transform of the system response when the input is the unit impulse function. Thus the system response due to an input of a unit impulse, called the **impulsive response**, is

$$x_i(t) = \mathcal{L}^{-1}\{G(s)\} \tag{7.13}$$

The impulsive response of a system may be discontinuous. The principle of impulse and momentum shows that the application of a unit impulse to a mechanical system leads to a discontinuity in velocity. However, displacement, usually used as the output in a mathematical model, is continuous. Recall, in the analogy between electrical systems powered by voltage sources and mechanical systems, that the velocity in a mechanical system is analogous to the current and that displacement is analogous to electric charge. Continuing with this analogy, an impulsive voltage leads to a discontinuity in current but a continuous charge. In the analogy between a mechanical system and a circuit with a current source, velocity is analogous to voltage and displacement is analogous to magnetic flux. Thus if a circuit is subject to an impulsive current, the voltage is discontinuous but the magnetic flux is continuous.

The value of the impulsive response immediately after the application of the impulse is obtained using the initial value theorem (IVT), property 15 of Table 2.2:

$$x_i(0) = \lim_{s\to\infty} sG(s) \tag{7.14}$$

Suppose $G(s) = N(s)/D(s)$ is the ratio of two polynomials that are given by Equations (7.5) and (7.6). The following can be deduced from Equation (7.14), where n and m are the order of $N(s)$ and $D(s)$, respectively:

- The response is continuous at $t = 0$, $x_i(0) = 0$, if the order of $D(s)$ is at least two greater than the order of $N(s)$; that is, $m \geq n + 2$.
- The response is discontinuous but finite at $t = 0$, $x_i(0) = C$, where C is a nonzero constant, if the order of $D(s)$ is one greater than the order of $N(s)$; that is, $m = n + 1$. The constant is obtained by taking the limit of $sG(s)$ as s approaches infinity. The result is b_n.
- The response is unbounded at $t = 0$, $x_i(0) = \infty$ if the order of $D(s)$ is less than or equal to the order of $N(s)$; that is, $m \leq n$.

All first-order systems have, at best, a discontinuous impulsive response. For a first-order system, $m = 1$. If $n = 0$, then $m = n + 1$. The first benchmark problem, the single-tank liquid-level problem, has an impulsive response that is discontinuous due to an impulsive flow rate. However, an impulsive flow rate means that the tank is subject to a very large amount of liquid in a very short time, thus increasing the height almost instantaneously.

The mechanical system in the second benchmark problem has a continuous impulsive response as $m = 2$ and $n = 0$. This reaffirms that all the information about a dynamic system is contained in the system's transfer function. The impulsive force input causes a discontinuous velocity and a continuous displacement. Noting that $v = \dot{x}$, the transfer function for the velocity is

$$G_v(s) = sG(s)$$

$$= \frac{s/m}{s^2 + \frac{c}{m}s + \frac{k}{m}} \tag{7.15}$$

In this case, $m = 2$, $n = 1$, and $m = n + 1$. This implies that the velocity is discontinuous due to a unit impulse.

The transfer function for the mechanical system in the third benchmark problem, Equation (7.9), has $m = 2$, $n = 1$, and $m = n + 1$. This means that the impulsive response of a mechanical system due to motion input if there is an impulsive motion is discontinuous. However, an impulsive motion input means that the base is suddenly subject to a very large motion, which cannot happen.

The fourth benchmark problem is the electrical system that was alluded to earlier. The transfer function has $m = 2$, $n = 1$, and $m = n + 1$, and hence the system has a discontinuous impulsive response.

The fifth benchmark problem has the matrix of transfer functions given by Equation (7.11). Each transfer function in the matrix is second order. The transfer functions $G_{1,1}(s)$ and $G_{2,2}(s)$ have a numerator with $n = 1$, and the impulsive response of each tank is discontinuous when that tank is subject to an impulsive flow rate. However, $G_{1,2}(s)$ and $G_{2,1}(s)$ have a numerator with $n = 0$, which implies that if one tank is subject to an impulsive flow rate, the perturbation in liquid level in the other tank is continuous.

7.2.3 Step Response

The **step response** of a system is its response when the input is the unit-step function $u(t)$. Noting that $\mathcal{L}\{u(t)\} = 1/s$, Equation (7.2) shows that the transform of the step response is

$$X_s(s) = \frac{1}{s}G(s) \tag{7.16}$$

The step responses of many systems have a definite limit for large t; that is, they approach a final value. The application of the final value theorem (FVT), property 14 of Table 2.2, to Equation (7.16) shows that the final value of the step response, when it exists, is

$$\begin{aligned} x_{s,f} &= \lim_{t\to\infty} x_s(t) \\ &= \lim_{s\to 0} sX_s(s) \\ &= \lim_{s\to 0} s\left(\frac{1}{s}G(s)\right) \\ &= G(0) \end{aligned} \tag{7.17}$$

Parameters defining the transient response of a dynamic system are often defined in terms of its step response. Since, for a stable system (see Section 7.3), the step response has a defined final value, certain benchmark times and parameters are defined.

The **2 percent settling time** is the time t_s or $t_{2\%}$ such that the step response is permanently within 2 percent of its final value, $G(0)$. The 2 percent settling time is used as a benchmark in the transient response of dynamic systems, but settling times can be defined for other percentages. For example, $t_{4\%}$ is the time it takes for the step response to be permanently within 4 percent of its final value.

The **10–90 percent rise time** is the time t_r it takes for the response to initially grow from 10 percent of its final value to 90 percent of its final value. The selection of

10 percent and 90 percent to define the settling time provides benchmark values of the rise time, but certainly a rise time can be defined using different percentages.

If, during the step response, $x(t)$ exceeds $G(0)$, then the system is said to have **overshoot**. In such cases, the **percentage overshoot** is defined by

$$\eta_o = 100\left(\frac{x_p - G(0)}{G(0)}\right) \tag{7.18}$$

where x_p is the maximum, or peak, value of $x(t)$. The time at which x attains x_p is called the **peak time** t_p.

Figure 7.7 shows possible step responses of systems. Figure 7.7(a) shows a system where there is no overshoot in the step response, or $\eta_o = 0$. In such a case, the peak response is the final value $G(0)$, but it is never reached, so $t_p = \infty$. The 2 percent settling time is defined as the time it takes the step response to reach 98 percent of $G(0)$. The 10–90 percent rise time is shown on the graph.

Figure 7.7(b) illustrates the step response of a system for which the peak response is reached. Often the step response oscillates about the final value. Then the 2 percent settling time is defined as the last time the step response is either 0.98 $G(0)$ or 1.02 $G(0)$. The 10–90 percent rise time, the peak time, and the overshoot are noted on the graph.

Equations (7.13) and (7.16) show that

$$X_i(s) = sX_s(s) \tag{7.19}$$

The application of the property of transform of derivative leads to

$$\mathscr{L}\left\{\frac{dx_s}{dt}\right\} = sX_s(s) - x_s(0) \tag{7.20}$$

Substitution of Equation (7.19) into Equation (7.20) leads to

$$\mathscr{L}\left\{\frac{dx_s}{dt}\right\} = X_i(s) - x_s(0) \tag{7.21}$$

Inversion of both sides of Equation (7.21) leads to

$$x_i(t) = \frac{dx_{rs}}{dt} + x_s(0)\delta(t) \tag{7.22}$$

Thus $x_i(t) = dx_s/dt$ except possibly at $t = 0$.

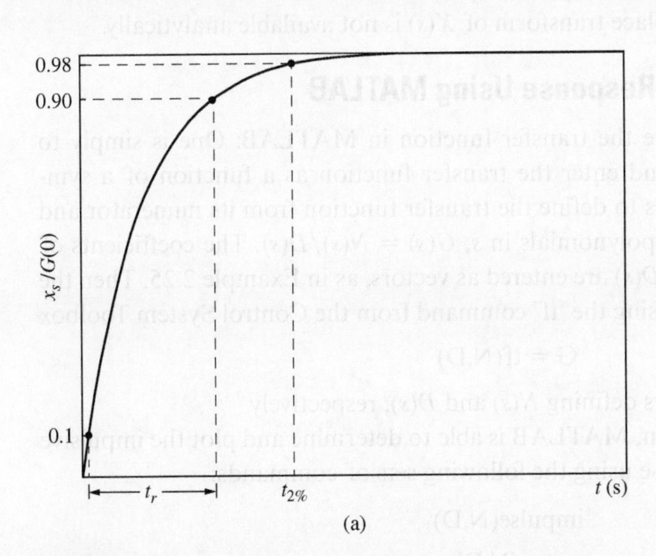

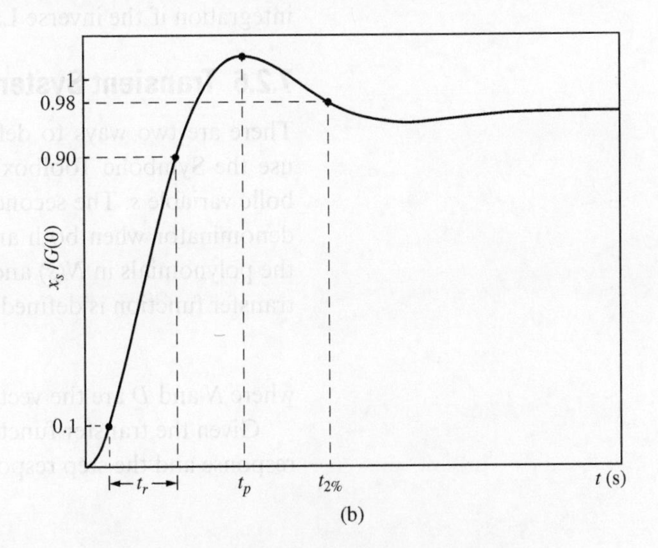

Figure 7.7

Typical step responses: (a) system with no overshoot has $t_p = \infty$; the 10–90 percent rise time and the 2 percent settling time are illustrated; (b) system with overshoot; the peak time, the 10–90 percent rise time, and the 2 percent settling time are illustrated.

7.2.4 Ramp Response

The ramp response of a system, useful in control system design, is the response due to a unit ramp input, $r(t) = t$. Since $\mathscr{L}\{t\} = 1/s^2$, the transform of the ramp response of a dynamic system is

$$X_r(s) = \frac{1}{s^2}G(s) \tag{7.23}$$

Use of Equation (7.16) in Equation (7.23) leads to

$$X_r(s) = \frac{1}{s}X_s(s) \tag{7.24}$$

Equation (7.24) implies that the ramp response of a system is the step response of a system whose transfer function is $X_s(s)$, the transform of the step response of the system. A procedure similar to the procedure used to derive Equation (7.22) shows that

$$\frac{dx_r}{dt} = x_s(t) \tag{7.25}$$

In deriving Equation (7.25), it is assumed that $x_r(0) = 0$, which is expected since the ramp input is continuous at $t = 0$.

7.2.5 Convolution Integral

The transient response to an arbitrary input $F(t)$ is obtained by applying Equation (7.1) in the form of Equation (7.2), $X(s) = G(s)\,F(s)$. Equation (7.13) shows that the transfer function is the transform of the system's impulsive response. Property 13 of Table 2.2, the convolution property, states that the inverse transform of a product of transforms is the convolution of the inverse transforms. Applying the convolution property to Equation (7.2) yields

$$x(t) = x_i(t) * f(t) \tag{7.26}$$

or

$$x(t) = \int_0^t f(\tau)\,x_i(t - \tau)d\tau \tag{7.27}$$

Equation (7.27) is called the **convolution integral** solution for the response of a system. In the application of the integral, it is noted that $f(t)$ is the system input and $x_i(t)$ is the response due to a unit impulse. Equation (7.27) can be used for numerical integration if the inverse Laplace transform of $X(s)$ is not available analytically.

7.2.6 Transient System Response Using MATLAB

There are two ways to define the transfer function in MATLAB. One is simply to use the Symbolic Toolbox and enter the transfer function as a function of a symbolic variable s. The second is to define the transfer function from its numerator and denominator when both are polynomials in s, $G(s) = N(s)/D(s)$. The coefficients of the polynomials in $N(s)$ and $D(s)$ are entered as vectors, as in Example 2.25. Then the transfer function is defined using the 'tf' command from the Control System Toolbox

```
G = tf(N,D)
```

where N and D are the vectors defining $N(s)$ and $D(s)$, respectively.

Given the transfer function, MATLAB is able to determine and plot the impulsive response and the step response using the following sets of commands:

```
impulse(N,D)
step(N,D)
```

or

```
impulse(tf)
step(tf)
```

The information about the settling time, rise time, peak time, and percent overshoot can be obtained using MATLAB with the command

stepinfo(G)

The MATLAB commands tf, impulse, step, and stepinfo are available in the Control Systems Toolbox. The commands impulse and step can be modified by adding a final value of the time, tfinal, at which the calculations are to be stopped. When this parameter is added, the commands become

step(N,D,tfinal)
impulse(N,D,tfinal)

There is no MATLAB command for the ramp response. The ramp response is obtained by defining a new transfer function using

D1=[D, 0]
G1=tf(N,D1)

Then take the step response of G1. A value of tfinal should be specified small enough so that the details are visible near $t = 0$:

step(G1,tfinal)

Example 7.2

A system has a transfer function of

$$G(s) = \frac{s+3}{s^2 + 6s + 20} \tag{a}$$

a. Determine the initial value of the response when the system is subjected to a unit impulse input. **b.** Determine the final value of the response when the system is subjected to a unit step input. **c.** Define the transfer function using MATLAB, and use the transfer function to determine the system's impulsive response, its step response, and its ramp response. **d.** Use the command stepinfo from the Control Systems Toolbox in MATLAB to approximate the 2 percent settling time, the 10–90 percent rise time, the percent overshoot, and the peak time.

Solution

a. The initial value of the response when the system is subject to a unit impulse is obtained using Equation (7.14)

$$x_1(0) = \lim_{s\to\infty} s\frac{s+3}{s^2+6s+20}$$

$$= \lim_{s\to\infty} \frac{s^2+3s}{s^2+6s+20}$$

$$= \lim_{s\to\infty} \frac{1+\frac{3}{s}}{1+\frac{6}{s}+\frac{20}{s^2}} = 1 \tag{b}$$

b. The final value of the response when the system is subject to a unit step input is obtained using Equation (7.17) as

$$x_f = G(0) = \frac{3}{20} = 0.15 \tag{c}$$

c. The MATLAB work session is shown in Figure 7.8. The resulting plots are shown in Figure 7.9. Note that to determine the ramp response, a transfer function G1(s) is defined, which has the same numerator as G(s) but its denominator is the denominator of G(s) multiplied by s. The ramp response of the system whose transfer function is G(s) is the step response of the system whose transfer function is G1(s).

d. Figure 7.10 shows the output from using the MATLAB command stepinfo(G). It provides the settling time, rise time, peak time, and percent overshoot.

```
>> % Defining numerator and denominator of transfer function
>> N=[1 3]

N =

      1 3

>> D=[1 6 20]

D =

1 6 20

>> % Defining transfer function
>> G=tf(N,D)

Transfer function:
G=
         s + 3
    --------------
    s^2 + 6 s + 20

>> % Impulsive response
>> impulse(G)
>> %Step response
>> step(G)
>> % Defining transfer function for ramp response
>> % Numerator is the same, but denominator is multiplied by s
>> D1=[1 6 20 0]

D1 =

    1  6  20  0
>> G1=tf(N,D1)

Transfer function:
G1=
        s + 3
    -------------------
    s^3 + 6 s^2 + 20 s

>> step(G1)
>>
```

Figure 7.8

The MATLAB work session for the solution of Example 7.2.

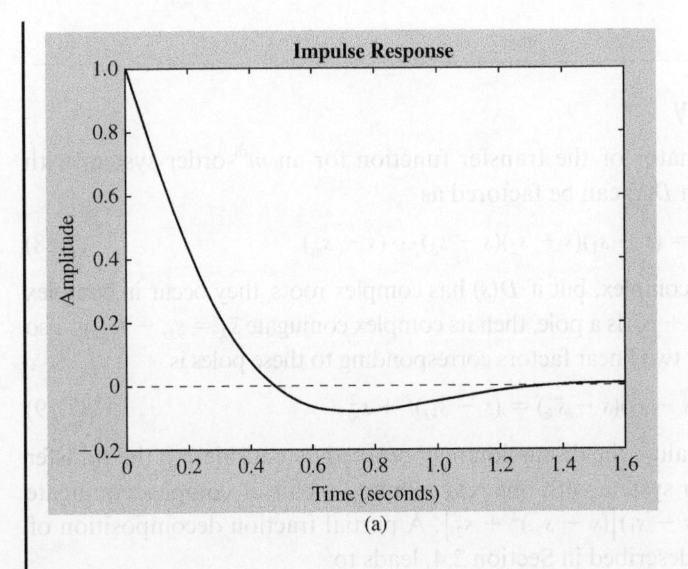

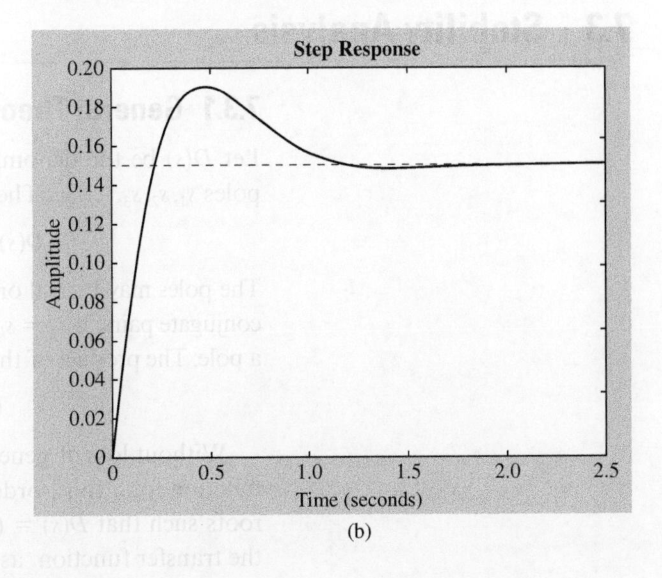

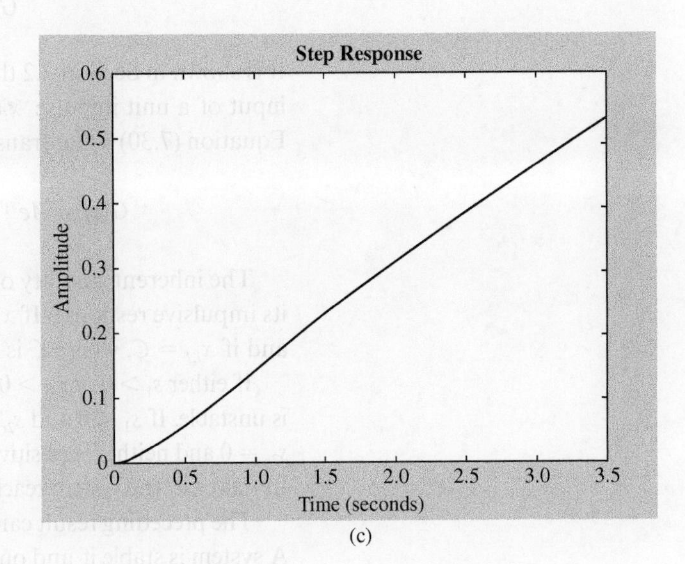

Figure 7.9

The MATLAB output for Example 7.2: (a) the impulsive response; (b) the step response; (c) the ramp response.

```
        RiseTime: 0.1710
   TransientTime: 1.0677
    SettlingTime: 1.0677
     SettlingMin: 0.1410
     SettlingMax: 0.1900
       Overshoot: 26.6987
      Undershoot: 0
            Peak: 0.1900
        PeakTime: 0.4759
```

Figure 7.10

MATLAB output when the command stepinfo(G) is used for the transfer function of Example 7.2.

7.3 Stability Analysis

7.3.1 General Theory

Let $D(s)$ be the denominator of the transfer function for an mth-order system with poles $s_1, s_2, s_3, \ldots, s_m$. Then $D(s)$ can be factored as

$$D(s) = (s - s_1)(s - s_2)(s - s_3)\cdots(s - s_m) \tag{7.28}$$

The poles may be real or complex, but if $D(s)$ has complex roots, they occur in complex conjugate pairs. If $s_k = s_{kr} + js_{kj}$ is a pole, then its complex conjugate $\bar{s}_k = s_{kr} - js_{kj}$ is also a pole. The product of the two linear factors corresponding to these poles is

$$(s - s_k)(s - \bar{s}_k) = (s - s_{kr})^2 + s_{kj}^2 \tag{7.29}$$

Without loss of generality, the discussion may proceed by considering the transfer function for a third-order system with one real root and a pair of complex conjugate roots such that $D(s) = (s - s_1)\left[(s - s_{2r})^2 + s_{2j}^2\right]$. A partial fraction decomposition of the transfer function, as described in Section 2.4, leads to

$$G(s) = \frac{A}{s - s_1} + \frac{Bs + C}{(s - s_{2r})^2 + s_{2j}^2} \tag{7.30}$$

It is shown in Section 7.2 that $x_i(t) = \mathcal{L}^{-1}\{G(s)\}$ is the response of the system due to the input of a unit impulse. $x_i(t)$ is obtained for this system by inverting the transforms in Equation (7.30) using transform pairs 1, 12, and 13 from Table 2.1, resulting in

$$G(t) = Ae^{s_1 t} + e^{s_{2r} t}\left[B\cos(s_{2j}t) + \frac{C + Bs_{2r}}{s_{2j}}\sin(s_{2j}t)\right] \tag{7.31}$$

The inherent stability of a system is determined by $x_{i,f} = \lim_{t\to\infty} x_i(t)$, the final value of its impulsive response. If $x_{i,f} = 0$, the system is stable; if $x_{i,f} = \infty$, the system is unstable; and if $x_{i,f} = C$, where C is a nonzero constant, the system is neutrally stable.

If either $s_1 > 0$ or $s_{2r} > 0$, then $x_i(t)$ is unbounded as t increases. In this case, the system is unstable. If $s_1 < 0$ and $s_{2r} < 0$, then $\lim_{t\to\infty} x_i(t)$ and the system is stable. If either $s_1 = 0$ or $s_{2r} = 0$ and neither is positive, then $x_i(t)$ is bounded but does not approach 0 as t increases. In this case, the system reaches a nonzero steady state and the system is neutrally stable.

The preceding result can be generalized for any system with the following statements. A system is stable if and only if all poles of its transfer function have negative real parts. If any pole has a positive real part, then the system is unstable. If any pole has a real part of zero and all other poles have nonpositive real parts, then the system is neutrally stable.

Example 7.3

The mathematical model for the motion of the system of Figure 7.11, an inverted compound pendulum with springs and viscous dampers subject to an applied moment, is

$$\frac{1}{3}m(a^2 - ab + b^2)\ddot{\theta} + 2cb^2\dot{\theta} + \left[2ka^2 - mg\left(\frac{a-b}{2}\right)\right]\theta = M(t) \tag{a}$$

Determine a stability criterion for this system involving k, a, b, m, and c.

Solution

Taking the Laplace transform of Equation (a) leads to

$$\Theta(s) = \frac{M(s)}{\frac{1}{3}m(a^2 - ab + b^2)s^2 + 2cbs + 2ka^2 - mg\left(\frac{a-b}{2}\right)} \tag{b}$$

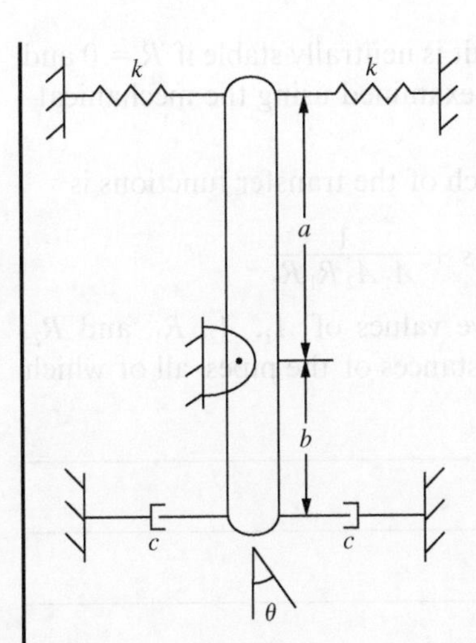

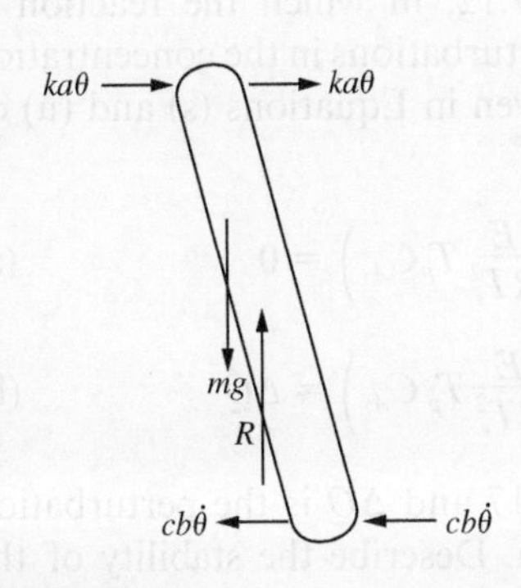

Figure 7.11

System is unstable if $ka^2 < mg\left(\frac{a-b}{4}\right)$ and neutrally stable if $c = 0$. Otherwise the system is stable.

The transfer function $G(s) = \Theta(s)/M(s)$ for this system is determined from Equation (a) as

$$G(s) = \frac{1}{\frac{1}{3}m(a^2 - ab + b^2)s^2 + 2cbs + 2ka^2 - mg\left(\frac{a-b}{2}\right)} \tag{c}$$

The poles are the values of s such that the denominator of $D(s)$ is zero:

$$\frac{1}{3}m(a^2 - ab + b^2)s^2 + 2cbs + 2ka^2 - mg\left(\frac{a-b}{2}\right) = 0 \tag{d}$$

The quadratic formula is used to determine the poles as

$$s_1 = \frac{3}{m(a^2 - ab + b^2)}(-cb - \sqrt{D}) \tag{e}$$

$$s_2 = \frac{3}{m(a^2 - ab + b^2)}(-cb + \sqrt{D}) \tag{f}$$

where

$$D = c^2b^2 - \frac{1}{3}(a^2 - ab + b^2)\left[2ka^2 - mg\left(\frac{a-b}{2}\right)\right] \tag{g}$$

If $D < 0$, then both poles are complex with negative real parts and thus the system is stable. If $D > 0$, then both poles are real and s_1 is negative while the sign of s_2 depends on the value of D. The system is stable if $s_2 < 0$, which requires

$$-cb + \sqrt{c^2b^2 - \frac{1}{3}(a^2 - ab + b^2)\left[2ka^2 - mg\left(\frac{a-b}{2}\right)\right]} < 0 \tag{h}$$

Algebraic manipulation of Equation (h) leads to a stability criterion of

$$2ka^2 > mg\left(\frac{a-b}{2}\right) \tag{i}$$

If the system is undamped, $c = 0$, then both poles are purely imaginary and the system is neutrally stable.

The stability criterion is a statement that the moment of the spring forces about the pin support is greater than the moment of the gravity force. The potential energy of the system at an arbitrary instant, assuming small displacements, is $V = \frac{1}{2}\{2ka^2 - mg\,[(a-b)/2]\}\,\theta^2$. Thus the stability criterion is also the condition under which the potential energy is positive.

Example 7.4

a.–e. Determine the stability of each of the benchmark problems of Section 7.1.3.

Solution

a. The single-tank liquid-level problem is always stable, as the pole of its transfer function is $-\frac{1}{RA}$. Both the cross-sectional area of tank A and the resistance in the pipe must be positive.

b. The poles for the mechanical system with force input are

$$s_{1,2} = \frac{1}{2}\left[-\frac{c}{m} \pm \sqrt{\left(\frac{c}{m}\right)^2 - 4\frac{k}{m}}\right] \tag{a}$$

If $c = 0$, the system is neutrally stable as the poles become $\pm j\sqrt{\frac{k}{m}}$, where both k and m must be positive. In this case, both poles have a real part of zero. If $c > 0$, the system is stable. If $\left(\frac{c}{m}\right)^2 \leq 4\frac{k}{m}$, this is clear since the square root produces imaginary terms and the real part of $-\frac{c}{2m}$ is negative. If $\left(\frac{c}{m}\right)^2 > 4\frac{k}{m}$, the square root produces real terms that are less than $\frac{c}{m}$. Hence the larger root is always negative. The mass, spring, and viscous-damper system is never unstable.

c. The mechanical system with motion input has a transfer function with the same $D(s)$ as the mechanical system with force input. Hence the stability of these systems is the same. The system is neutrally stable if $c = 0$ and stable if $c > 0$. The numerator of the transfer function does not affect stability.

d. Following the logic of part b., the series *LRC* circuit is neutrally stable if $R = 0$ and stable otherwise. The stability of this system can be examined using the mechanical–electrical system analogy.

e. Repeating Equation (7.12), the denominator of each of the transfer functions is

$$D(s) = s^2 + \left(\frac{1}{A_2 R_1} + \frac{1}{A_2 R_2} + \frac{1}{A_1 R_1}\right)s + \frac{1}{A_1 A_2 R_1 R_2}$$

The polynomial has only negative roots for positive values of A_1, A_2, R_1, and R_2. These parameters are the areas of the tanks and resistances of the pipes, all of which must be positive. Hence the system is always stable.

Example 7.5

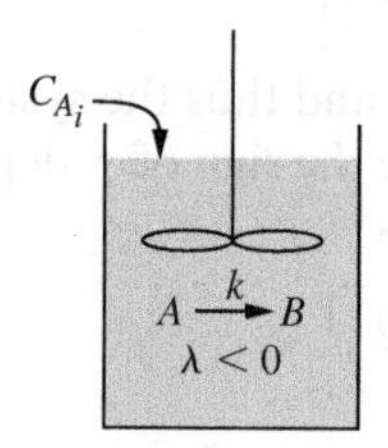

Figure 7.12

The CSTR with an exothermic reaction is unstable when the heat of the reaction is greater than a critical value.

Consider the CSTR of Example 6.17 and Figure 7.12, in which the reaction is exothermic. The linearized equations describing the perturbations in the concentration of reactant A, C_{Ap}, and reactor temperature T_p are given in Equations (s) and (u) of Example 6.17 as

$$V\frac{dC_{Ap}}{dt} + qC_{Ap} + \alpha Ve^{-E/(RT_s)}\left(C_{Ap} + \frac{E}{RT_s^2}T_pC_{As}\right) = 0 \tag{a}$$

$$\rho Vc_p\frac{dT_p}{dt} + \rho qc_pT_p - \lambda\alpha Ve^{-E/(RT_s)}\left(C_{Ap} + \frac{E}{RT_s^2}T_pC_{As}\right) = \Delta\dot{Q} \tag{b}$$

where the parameters are as described in Example 6.17 and $\Delta\dot{Q}$ is the perturbation of the rate at which heat is removed from the system. Describe the stability of the system in terms of system parameters. The concentration of the reactant, C_{Bp}, does not appear in Equation (a) or Equation (b) and has no effect on the stability.

Solution

The transfer function $G(s) = \mathcal{L}\{T_p\}/\mathcal{L}(\Delta\dot{Q})$ is determined as

$$G(s) = \frac{Vs + q + b}{\rho c_pV^2s^2 + [(2q + b)\rho c_pV - \lambda\mu bV]s + \rho c_pq^2 + b\rho c_pq - \lambda b\mu q} \tag{c}$$

where

$$b = \alpha Ve^{-E/(RT_s)} \tag{d}$$

$$\mu = \frac{EC_{As}}{RT_s^2} \tag{e}$$

The quadratic formula is applied to determine the poles of the transfer function as

$$s_1 = -\frac{q + b}{V} \tag{f}$$

$$s_2 = \frac{\lambda\mu b - q\rho c_p}{\rho c_pV} \tag{g}$$

The pole s_2 is positive and the system unstable when

$$\lambda > \frac{q\rho c_p}{\mu b} \tag{h}$$

The parameter λ is the rate at which heat is released from the reaction. If λ exceeds the critical value determined by Equation (h), the temperature increases without bound. However, Equations (a) and (b) are obtained from the linearization of nonlinear differential equations. In the process, the temperature will increase until the linearizing assumptions are no longer valid and the system's inherent nonlinearity takes over. The nonlinearity will lead to a saturation of the temperature and the concentration. While the temperature will remain bounded, it will be at a high value and the reactor will not function as desired.

7.3.2 Routh's Method

Consider an mth-order system with j pairs of complex conjugate poles and $m - 2j$ real poles. In this case, Equation (7.28) can be written as

$$D(s) = \left\{ \prod_{k=1}^{j} \left[(s - s_{kr})^2 + s_{kj}^2 \right] \right\} \left[\prod_{k=1}^{m-2j} (s - s_k) \right] \tag{7.32}$$

When the products in Equation (7.32) are expanded, the result is an mth-order polynomial of the form

$$D(s) = s^m + a_1 s^{m-1} + a_2 s^{m-2} + \cdots + a_k s^{m-k} + \cdots + a_{m-1}s + a_m \tag{7.33}$$

The coefficients $a_1, a_2, \ldots, a_m$ are sums of products of the negatives of the real parts of the poles and the square of the imaginary parts. If the system is stable, then the real parts of all poles are negative. Thus all coefficients are sums of positive numbers and are thus all positive.

This shows that a necessary condition for stability is that all the coefficients of $D(s)$, the polynomial in the denominator of the transfer function, are positive. Thus a necessary condition for the stability of a second-order system whose mathematical model for its free response is $\ddot{x} + 2\zeta\omega_n\dot{x} + \omega_n^2 x = 0$ is that both ζ and ω_n^2 are positive numbers. If $\omega_n^2 < 0$, then the potential energy of a mechanical system, as measured from the system's equilibrium position, could be negative and the system would be unstable, as is possible in Example 7.3. If the damping ratio is negative (negative damping), then instead of dissipating energy the damping mechanism is actually adding energy to the system, leading to an unstable system. If the damping ratio of a second-order system is zero (undamped), then $D(s) = s^2 + \omega_n^2$; thus $a_1 = 0$ and the system is neutrally stable.

While the requirement that all coefficients in the polynomial of the denominator of the transfer function be positive is a necessary condition for stability, it is not a sufficient condition. If the necessary condition for stability is met, then the application of Routh's method, presented without derivation, determines whether the system is stable without solving for the poles. Routh's method is an algebraic method that determines the number of poles with positive real parts.

To apply Routh's method, develop a matrix of numbers formed from the coefficients of $D(s)$. The matrix has $m + 1$ rows. If m is even, the matrix has $k = (m/2) + 1$ columns. If m is odd, the matrix has $k = (m + 1)/2$ columns. The matrix for an even value of m is shown in Equation (7.34).

$$\mathbf{B} = \begin{bmatrix} 1 & a_2 & a_4 & \cdots & a_{m-2} & a_m \\ a_1 & a_3 & a_5 & \cdots & a_{m-1} & 0 \\ b_{3,1} & b_{3,2} & b_{3,3} & \cdots & b_{3,m-1} & 0 \\ b_{4,1} & b_{4,2} & b_{4,3} & \cdots & b_{4,m-1} & 0 \\ \vdots & \vdots & \vdots & \cdots & \vdots & \vdots \\ b_{m+1,1} & b_{m+1,2} & b_{m+1,3} & \cdots & b_{m+1,m-1} & 0 \end{bmatrix} \tag{7.34}$$

where

$$b_{i,j} = \frac{b_{i-1,1}\, b_{i-2,j+1} - b_{i-2,1}\, b_{i-1,j+1}}{b_{i-1,1}} \qquad i = 3, \ldots, m + 1; j = 1, \ldots, k - 1 \tag{7.35}$$

The matrix for an odd value of m is

$$\mathbf{B} = \begin{bmatrix} 1 & a_2 & a_4 & \cdots & a_{m-3} & a_{m-1} \\ a_1 & a_3 & a_5 & \cdots & a_{m-2} & a_m \\ b_{3,1} & b_{3,2} & b_{3,3} & \cdots & b_{3,m-1} & 0 \\ b_{4,1} & b_{4,2} & b_{4,3} & \cdots & b_{4,m-1} & 0 \\ \vdots & \vdots & \vdots & \cdots & \vdots & \vdots \\ b_{m+1,1} & b_{m+1,2} & b_{m+1,3} & \cdots & b_{m+1,m-1} & 0 \end{bmatrix} \tag{7.36}$$

where Equation (7.35) is used to determine the entries after the second row. Routh's method specifies that the number of poles with positive real parts is equal to the number of sign changes in the elements of the first column of **B**. Thus if all elements of the first column of **B** are positive, then there are no sign changes in the column, the system has no poles with positive real parts, and it is stable. If any element of the first column of B is negative, then there is at least one pole with a positive real part and the system is unstable.

A special case occurs when an element of the first row column of **B** is zero. This occurs when the system has either a pair of purely imaginary roots or a repeated root. To determine which case occurs, replace the zero in the first column by an arbitrarily small positive value, say δ, and continue to calculate the remainder of the elements of **B** according to Equation (7.35). If the next element in the first column is zero, the system has a pair of purely imaginary roots and is neutrally stable. If the next element in the first column is negative, the system has a repeated positive root and is unstable.

Example 7.6

Use Routh's method to determine a stability criterion for a third-order system with $D(s) = s^3 + a_1 s^2 + a_2 s + a_3$.

Solution

For $m = 3$, **B** is a 4×2 matrix, initially set up as

$$\mathbf{B} = \begin{bmatrix} 1 & a_2 \\ a_1 & a_3 \\ b_{3,1} & 0 \\ b_{4,1} & 0 \end{bmatrix} \tag{a}$$

Using Equation (7.35),

$$b_{3,1} = \frac{b_{2,1} b_{1,2} - b_{1,1} b_{2,2}}{b_{2,1}} = \frac{a_1 a_2 - (1) a_3}{a_1} \tag{b}$$

$$b_{4,1} = \frac{b_{3,1} b_{2,2} - b_{2,1} b_{3,2}}{b_{3,1}} = \frac{\dfrac{a_1 a_2 - a_3}{a_1} a_3 - a_1 (0)}{\dfrac{a_1 a_2 - a_3}{a_1}} = a_3 \tag{c}$$

Thus the first column of **B** is

$$\begin{bmatrix} 1 \\ a_1 \\ \dfrac{a_1 a_2 - a_3}{a_1} \\ a_3 \end{bmatrix} \tag{d}$$

All elements of the column are positive and the system is stable if

$$a_1 a_2 > a_3 \tag{e}$$

Example 7.7

Use Routh's method to determine the stability of a system with the transfer function $G(s) = (s + 2)/(s^4 + 5.5s^3 + 8s^2 + 0.5s + 1)$.

Solution

The matrix **B** for this transfer function is determined as

$$\mathbf{B} = \begin{bmatrix} 1 & 8 & 1 \\ 5.5 & 0.5 & 0 \\ 7.910 & 1 & 0 \\ -0.1953 & 0 & 0 \\ 1 & 0 & 0 \end{bmatrix} \tag{a}$$

The first column of **B** has two sign changes. Thus $G(s)$ has two poles with positive real parts and the system is unstable.

7.3.3 Relative Stability

Routh's method, as described, determines the absolute stability of a system. In some cases it is also necessary to determine the relative stability of a system, that is, how close the system is to being unstable. A system is stable when all its poles have negative real parts. For a stable system, a generalization of Equation (7.31) is used to determine the system's impulsive response. The dominant pole in a stable system is the pole whose real part has the smallest absolute value. The term in the impulsive response corresponding to this pole has the slowest rate of decay and is dominant in the solution for large t. In subsequent sections it is shown how properties of transient response such as settling time and peak time depend on the dominant pole. A design consideration in a control system may be whether the settling time is less than a given value. If the dominant pole is too close to the imaginary axis, the system may not satisfy a design constraint.

To address these concerns, the concept of relative stability is used. A system is stable relative to the line $s = -p$ if all poles lie to the left of $s = -p$. To determine the relative stability, consider $D(s - p)$. Equation (7.28) is used to show that if s_k is a root of $D(s)$, then $s_k + p$ is a root of $D(s - p)$. Thus the stability of the system relative to $s = -p$ may be determined by applying Routh's method to $D(s - p)$.

Example 7.8

A control system for a guidance system is being designed such that its transfer function is

$$G(s) = \frac{1}{s^3 + 10s^2 + (K + 3)s + 50} \tag{a}$$

where K is a design parameter in the control system. **a.** Determine the values of K for which the system is absolutely stable. **b.** Determine the values of K for which the system is stable relative to $s = -5$. **c.** Determine the values of K for which the system is stable relative to $s = -1$.

Solution

a. Equation (e) of Example 7.6 is a stability criterion, derived using Routh's method, for third-order systems. The application of Equation (e) to the transfer function of Equation (a) leads to

$$\begin{aligned} 10(K + 3) &> 50 \\ K &> 2 \end{aligned} \tag{b}$$

Thus the system is absolutely stable if $K > 2$.

b. The stability of the system relative to $s = -5$ is determined by applying Routh's method to $D(s-5)$. To this end,

$$\begin{aligned} D(s-5) &= (s-5)^3 + 10(s-5)^2 + (K+3)(s-5) + 50 \\ &= s^3 - 5s^2 + (K-22)s + 160 - 5K \end{aligned} \tag{c}$$

A necessary requirement for stability is that all coefficients in the polynomial are positive. Since $D(s-5)$ has a negative coefficient, independent of the value of K, there are no values of K such that the system is stable relative to $s = -5$.

c. The stability of the system relative to $s = -1$ is determined by applying Routh's method to $D(s-1)$. To this end

$$\begin{aligned} D(s-1) &= (s-1)^3 + 10(s-1)^2 + (K+3)(s-1) + 50 \\ &= s^3 + 7s^2 + (K-14)s + 56 - K \end{aligned} \tag{d}$$

Application of Equation (e) of Example 7.6 to Equation (d) shows that the system is stable relative to $s = -1$ when

$$\begin{aligned} 7(K-14) &> 56 - K \\ K &> 19.25 \end{aligned} \tag{e}$$

7.4 First-Order Systems

The standard form of the differential equation governing the behavior of a linear first-order system is

$$T\frac{dx}{dt} + x = f(t) \tag{7.37}$$

The parameter T is called the system's **time constant**. First-order systems studied in Chapters 4–6 are summarized in Table 7.1. Specification of an initial condition of the form

$$x(0) = x_0 \tag{7.38}$$

is necessary to determine a unique solution to Equation (7.37).

Taking the Laplace transform of Equation (7.37) using Equation (7.38) leads to

$$X(s) = \frac{F(s) + x_0 T}{T\left(s + \frac{1}{T}\right)} \tag{7.39}$$

The transfer function for a first-order system is determined from Equation (7.39) as

$$G(s) = \frac{1/T}{s + \frac{1}{T}} \tag{7.40}$$

A first-order system has one pole

$$s_1 = -\frac{1}{T} \tag{7.41}$$

Thus a first-order system is stable if $T > 0$.

7.4.1 Free Response

The free response of a first-order system is obtained by inverting Equation (7.39) with $F(s) = 0$, leading to

$$x(t) = x_0 e^{-\frac{t}{T}} \tag{7.42}$$

Table 7.1 First-Order Systems

Name	Schematic	Differential Equation	Time Constant
Spring and viscous-damper		$c\dot{x} + kx = F(t)$	$\frac{c}{k}$
LR Circuit		$L\frac{di}{dt} + Ri = v(t)$	$\frac{L}{R}$
RC Circuit		$C\frac{dv}{dt} + \frac{1}{R}v = i(t)$	RC
Liquid-level problem		$A\frac{dh}{dt} + \frac{1}{R}h = q$	AR
Lumped thermal system		$\rho cV\frac{d\theta}{dt} + hA\theta = hA\theta_\infty$	$\frac{\rho cV}{hA}$

Table 7.2 Free Response of First-Order System, $x(t) = x_0 e^{-t/T}$

$x(t)/x_0$	t/T
0.1	2.30
0.2	1.61
0.3	1.20
0.4	0.916
0.5	0.693
0.6	0.511
0.7	0.357
0.8	0.223
0.9	0.105

The free response of a first-order system, illustrated in Figure 7.13, decays exponentially and approaches zero asymptotically. The free-response of a linear first-order system depends on one parameter, the time constant. Table 7.2 illustrates the number of time constants it takes the response to decay to a fraction of its initial value.

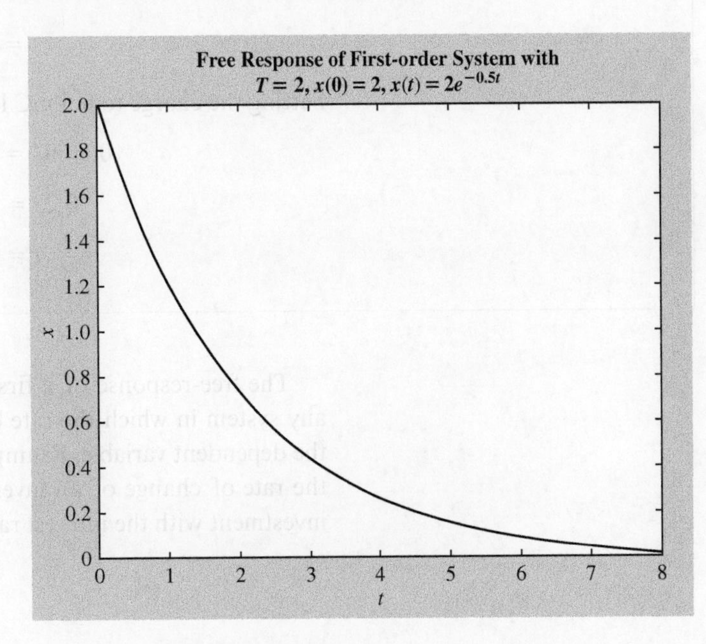

Figure 7.13

The free response of a first-order system.

Example 7.9

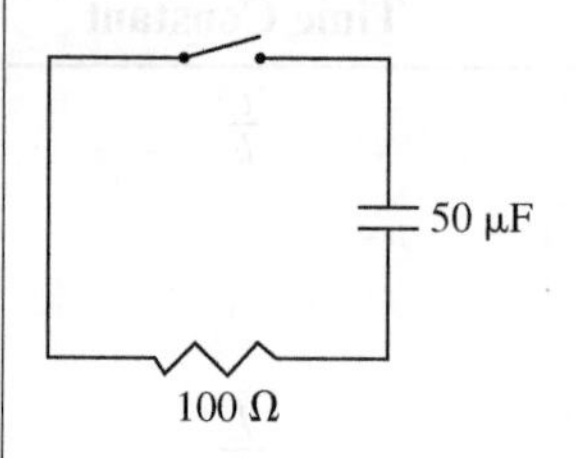

Figure 7.14

The *RC* circuit of Example 7.9.

The capacitor in the *RC* circuit of Figure 7.14 has an initial charge of 2 mC. The switch is initially open and then closed at $t = 0$, allowing the capacitor to discharge. **a.** Determine the power delivered by the capacitor as a function of time. **b.** How long does it take for the charge to drop to 0.5 mC?

Solution

a. The differential equation for the voltage in the circuit is

$$C\frac{dv}{dt} + \frac{1}{R}v = 0 \tag{a}$$

$$RC\frac{dv}{dt} + v = 0 \tag{b}$$

From Equation (b), the time constant for this first-order system is determined as

$$T = RC = (100\ \Omega)(50 \times 10^{-6}\ \text{F}) = 5 \times 10^{-4}\text{s} \tag{c}$$

The initial voltage in the capacitor is

$$v(0) = \frac{q(0)}{C} = \frac{2 \times 10^{-3}\ \text{C}}{50 \times 10^{-6}\ \text{F}} = 40\ \text{V} \tag{d}$$

Equations (c) and (d) are used in Equation (7.42), leading to

$$v(t) = 40e^{-200t}\ \text{V} \tag{e}$$

The power in the capacitor is determined using Equation (5.21) as

$$\begin{aligned} P &= Cv\frac{dv}{dt} \\ &= (50 \times 10^{-6}\ \text{F})(40e^{-200t}\ \text{V})(-8000e^{-200t}\ \text{V/s}) \\ &= -16e^{-400t}\ \text{W} \end{aligned} \tag{f}$$

b. The charge in the capacitor is

$$\begin{aligned} q &= Cv \\ &= (50 \times 10^{-6}\ \text{F})(40e^{-200t}\ \text{V}) \\ &= 2e^{-200t}\ \text{mC} \end{aligned} \tag{g}$$

Setting the charge to 0.5 mC leads to

$$\begin{aligned} 0.5\ \text{mC} &= 2e^{-200t}\ \text{mC} \\ 0.25 &= e^{-200t} \\ t &= -\frac{\ln(0.25)}{200} = 6.93 \times 10^{-3}\ \text{s} \end{aligned} \tag{h}$$

The free-response of a first-order system is a model for the dynamic response of any system in which the rate of change of the dependent variable is proportional to the dependent variable. Examples include continuously compounded interest in which the rate of change of an investment $m(t)$ is proportional to the current value of the investment with the interest rate i at the constant of proportionality

$$\frac{dm}{dt} = im \tag{7.43}$$

The solution of Equation (7.43) is $m(t) = m_0 e^{it}$, where m_0 is the amount initially invested. This is an example of an unstable first-order system. Theoretically the investment grows without bound over time. One would not put money in a bank if the amount deposited decayed with time.

Example 7.10

The rate of decay of a radioactive isotope is proportional to the instantaneous amount of the isotope. The half-life of an isotope is defined as the time it takes for a quantity of the isotope to decay to half of its initial mass. The isotope ^{14}C, used in dating artifacts, has a half-life of 5700 years. How long will it take 20 g of the isotope to decay to 18 g?

Solution

Since the rate of change of the mass of the isotope is proportional to the amount of the isotope $m(t)$,

$$\frac{dm}{dt} = -rm \tag{a}$$

where r is the rate of decay. The rate of change is negative because the isotope is losing mass. Equation (a) can be rewritten in the form of Equation (7.37) as

$$\frac{1}{r}\frac{dm}{dt} + m = 0 \tag{b}$$

Comparing Equation (b) with Equation (7.37), it is clear that the time constant for the system is equal to the reciprocal of the rate of decay. The solution of Equation (b) is obtained using Equation (7.42) as

$$m(t) = m_0 e^{-rt} \tag{c}$$

Let t_h be the half-life of an isotope. Evaluation of Equation (c) at $t = t_h$ leads to

$$m(t_h) = \frac{1}{2}m_0 = m_0 e^{-rt_h} \tag{d}$$

Equation (d) is used to solve for r as

$$r = \frac{1}{t_h}\ln(2) \tag{e}$$

Substitution of Equation (e) into Equation (c) leads to

$$m(t) = m_0 e^{-\ln(2)\frac{t}{t_h}} \tag{f}$$

Application of Equation (f) to the data given for the ^{14}C isotope leads to

$$18\text{ g} = (20\text{ g})\, e^{-\ln(2)\frac{t}{5700\text{ yr}}} \tag{g}$$

The solution of Equation (g) leads to $t = 866$ yr.

7.4.2 Impulsive Response

The response of a first-order system when $f(t) = I\delta(t)$ is found by inverting $IG(s)$, which is obtained using Equation (7.40) as

$$x_i(t) = \frac{I}{T}e^{-\frac{t}{T}}u(t) \tag{7.44}$$

Evaluation of Equation (7.44) at $t = 0$ leads to $x(0) = I/T$, which is in conflict with the applied initial condition $x(0) = 0$. This illustrates the necessity of using the unit step function in the response given by Equation (7.44). It is clear that in this context, $u(0) = 0$ for $t = 0$ and $u(t) = 1$ for $t > 0$. Differentiating Equation (7.44) with respect to time leads to

$$\dot{x}_i(t) = -\frac{I}{T^2}e^{-\frac{t}{T}}u(t) + \frac{I}{T}\delta(t) \tag{7.45}$$

Equation (7.45) shows that the derivative of the dependent variable is infinite at $t = 0$, leading to a discontinuity in the dependent variable at $t = 0$. It is shown in Section 7.2 that the response of a first-order system with an impulsive input is discontinuous at $t = 0$.

The discontinuity in the dependent variable can be understood by considering the first-order mechanical system in Table 7.1. The application of the principle of impulse and momentum to a mechanical system reveals that application of an impulse leads to a finite change in linear momentum. The linear momentum of a system is mass times velocity. Since a first-order system has no inertia element and hence no mass, a finite change in linear momentum occurs only if the velocity is infinite. A mechanical system without an inertia element is an idealized system in which the inertia of the spring and viscous damper is neglected. In a real system, the mass of these elements absorbs the impulse, leading to a finite change in velocity and a continuous displacement.

Example 7.11

The bumper of Figure 7.15(a) is modeled as a spring in parallel with a viscous damper. Determine the response of the system predicted by this model when it is struck by a body that imparts an impulse of 10 N·s to the bumper.

Solution

The differential equation governing the displacement of the bumper is

$$2000\frac{dx}{dt} + 100x = 10\delta(t)$$
$$20\frac{dx}{dt} + x = 0.1\delta(t) \tag{a}$$

The time constant for the system is

$$T = \frac{2000 \text{ N·s/m}}{100 \text{ N/m}} = 20 \text{ s} \tag{b}$$

The system response is obtained using Equation (7.44) as

$$x(t) = \frac{0.1 \text{ m·s}}{20 \text{ s}}e^{-0.05t}u(t)$$
$$= 0.005e^{-0.05t}u(t) \text{ m} \tag{c}$$

Equation (c) is plotted in Figure 7.15(b). The initial displacement of 5 cm is really the displacement after a small but finite time. After 1 min, the displacement is approximately 0.4 mm.

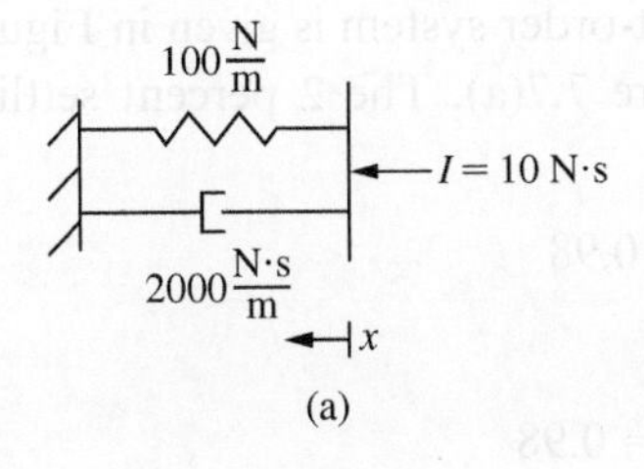

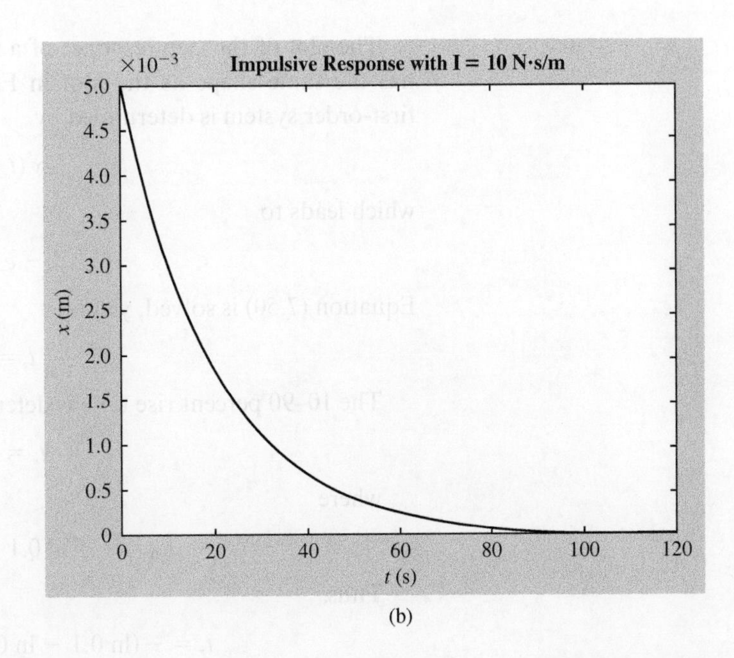

Figure 7.15

(a) The bumper, modeled by a spring in parallel with a viscous damper, is subject to an impulse when struck by a vehicle. (b) Response of bumper due to impact which leads to an impulse of magnitude 10 N·s/m.

7.4.3 Step Response

If $f(t) = u(t)$ and $x_0 = 0$, Equation (7.39) becomes, with $x_0 = 0$,

$$X_s(s) = \frac{\frac{1}{T}}{s\left(s + \frac{1}{T}\right)} \tag{7.46}$$

Equation (7.46) is inverted using transform pair 17 of Table 2.1, leading to

$$x_s(t) = \left(1 - e^{-\frac{t}{T}}\right)u(t) \tag{7.47}$$

The response given by Equation (7.47) and illustrated in Figure 7.16 shows that a first-order system subject to a step input approaches a new steady state. It is shown in Section 7.2, through application of the final value theorem, that $\lim_{t\to\infty} x_s(t) = G(0)$. For a first-order system whose Laplace transform is given by Equation (7.46),

$$\lim_{t\to\infty} x_s(t) = G(0) = 1 \tag{7.48}$$

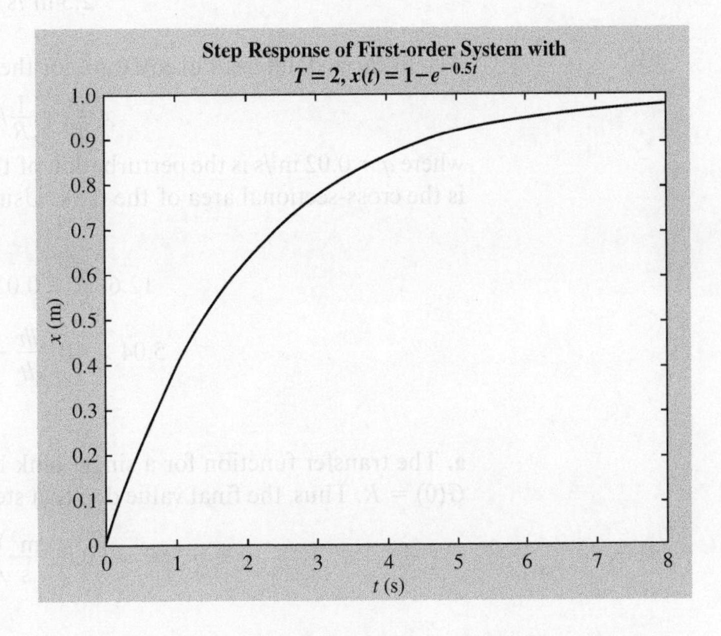

Figure 7.16

The step response of a first-order system.

The plot of the step response of a first-order system is given in Figure 7.16, which has the same shape as the plot in Figure 7.7(a). The 2 percent settling time for a first-order system is determined by

$$x_s(t_s) = 0.98 \tag{7.49}$$

which leads to

$$1 - e^{-\frac{t_s}{T}} = 0.98 \tag{7.50}$$

Equation (7.50) is solved, yielding

$$t_s = 3.92T \tag{7.51}$$

The 10–90 percent rise time is determined from

$$t_r = t_{90} - t_{10} \tag{7.52}$$

where

$$t_{90} = -T \ln 0.1 \qquad t_{10} = -T \ln 0.9 \tag{7.53}$$

Thus,

$$t_r = -(\ln 0.1 - \ln 0.9)T = T \ln 9 = 2.20T \tag{7.54}$$

Example 7.12

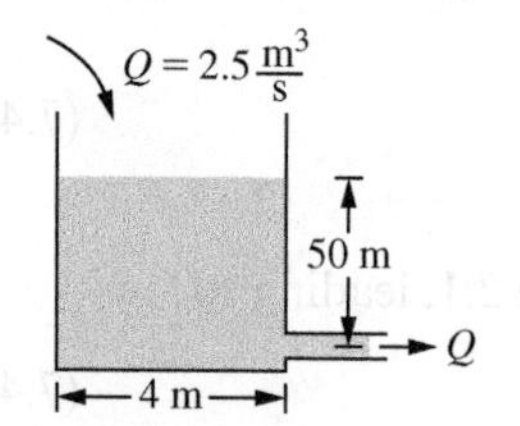

Figure 7.17
The steady-state operation of the tank of Example 7.12.

The circular tank of Figure 7.17 has a steady-state level of 50 m for a flow rate of 2.5 m^3/s. The flow rate is suddenly increased to 2.52 m^3/s. Determine **a.** the level of the liquid when the system reaches steady state, **b.** the time-dependent response of the system, **c.** the 2 percent settling time for the system, and **d.** the 10–90 percent rise time.

Solution

Assuming turbulent flow of the liquid in the pipe, its resistance at the initial steady state is determined using Equation (6.38) as

$$R = 2\frac{H}{Q}$$

$$= 2\frac{50 \text{ m}}{2.5 \text{ m}^3/\text{s}} = 40 \text{ s/m}^2 \tag{a}$$

The linearized differential equation for the perturbation in liquid level is

$$A\frac{dh}{dt} + \frac{1}{R}h = qu(t) \tag{b}$$

where $q = 0.02$ m^3/s is the perturbation of the flow rate and $A = (\pi/4)(4 \text{ m})^2 = 12.6 \text{ m}^2$ is the cross-sectional area of the tank. Using these values, Equation (b) becomes

$$12.6\frac{dh}{dt} + 0.025h = 0.02u(t)$$

$$5.04 \times 10^2\frac{dh}{dt} + h = 0.8u(t) \tag{c}$$

a. The transfer function for a single-tank liquid-level problem, Equation (7.7), yields $G(0) = R$. Thus, the final value due to a step input of magnitude q is

$$h_f = qR = \left(0.02\frac{\text{m}^3}{\text{s}}\right)\left(40\frac{\text{s}}{\text{m}^2}\right) = 0.8 \text{ m} \tag{d}$$

b. The time constant for the system is $T = RA = (40\ \frac{\text{s}}{\text{m}^2})(12.6\ \text{m}^2) = 504$ s. The system response is obtained using Equation (7.47) as

$$h(t) = 0.8(1 - e^{-1.99 \times 10^{-3}t})u(t) \tag{e}$$

c. The 2 percent settling time is calculated using Equation (7.51) as

$$t_s = 3.92T = 3.92(5.04 \times 10^2\ \text{s}) = 1.98 \times 10^3\ \text{s} \tag{f}$$

d. The 10–90 percent rise time is obtained using Equation (7.54):

$$t_r = T \ln 9 = (504\ \text{s}) \ln 9 = 1.11 \times 10^3 \text{s} \tag{g}$$

7.4.4 Ramp Response

The ramp response is obtained by inverting

$$X_r(s) = \frac{\frac{1}{T}}{s^2\left(s + \frac{1}{T}\right)} \tag{7.55}$$

Inversion of Equation (7.55) leads to

$$x_r(t) = t - T + Te^{-\frac{t}{T}} \tag{7.56}$$

7.4.5 General Response

The response of a first-order system due to an input $f(\tau)$ is obtained using the convolution integral, Equation (7.27), and the impulsive response of the system due to a unit impulse, Equation (7.44), with $I = 1$ N·s:

$$\begin{aligned} x(t) &= \int_0^t f(\tau)x_i(t-\tau)d\tau \\ &= \frac{1}{T}\int_0^t f(\tau)e^{-(t-\tau)/T}d\tau \\ &= \frac{e^{-t/T}}{T}\int_0^t f(\tau)e^{\tau/T}d\tau \end{aligned} \tag{7.57}$$

Example 7.13

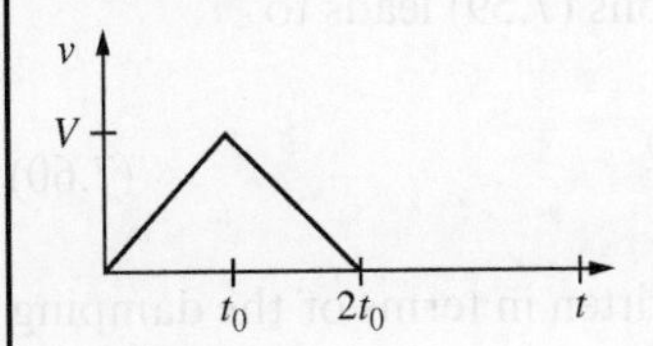

Figure 7.18

Triangular pulse voltage input of Example 7.13.

Determine the current through an LR circuit when it is subjected to the triangular voltage input of Figure 7.18.

Solution

The mathematical model for the current in an LR circuit is

$$L\frac{di}{dt} + Ri = v(t) \tag{a}$$

The transfer function for the circuit is

$$G(s) = \frac{I(s)}{V(s)} = \frac{1/L}{s + R/L} \tag{b}$$

The inverse of the transfer function is the impulsive response

$$i_i(t) = \mathcal{L}^{-1}\left\{\frac{1/L}{s + R/L}\right\} = \frac{1}{L}e^{-(R/L)t} \tag{c}$$

The mathematical function describing the voltage input of Figure 7.18 is

$$v(t) = \frac{V}{t_0}[tu(t) - 2(t - t_0)u(t - t_0) + (t - 2t_0)u(t - 2t_0)] \quad \text{(d)}$$

Applying Equation (7.57), the response due to the triangular voltage input is

$$v(t) = \int_0^t v(\tau)\, i_i(t - \tau)d\tau$$

$$= \int_0^t \left\{\frac{V}{t_0}[\tau u(\tau) - 2(\tau - t_0)u(\tau - t_0) + (\tau - 2t_0)u(\tau - 2t_0)]\right\}\left\{\frac{1}{L}e^{-\left(\frac{R}{L}\right)(t-\tau)}\right\}d\tau$$

$$= \frac{V}{Lt_0}e^{-\left(\frac{R}{L}\right)t}\int_0^t [\tau u(\tau) - 2(\tau - t_0)u(\tau - t_0) + (\tau - 2t_0)u(\tau - 2t_0)]e^{\left(\frac{R}{L}\right)\tau}d\tau \quad \text{(e)}$$

Evaluation of the integral in Equation (e) gives

$$v(t) = \frac{V}{Rt_0}e^{-\left(\frac{R}{L}\right)t}\left\{\left[te^{\left(\frac{R}{L}\right)t} + \frac{L}{R}\left(e^{\left(\frac{R}{L}\right)t} - 1\right)\right]u(t) - 2\left[te^{\left(\frac{R}{L}\right)t} - t_0e^{\left(\frac{R}{L}\right)t_0} - t_0e^{\left(\frac{R}{L}\right)t} + 1\right.\right.$$

$$\left.\left. + \frac{L}{R}\left(e^{\left(\frac{R}{L}\right)t} - e^{\left(\frac{R}{L}\right)t_0}\right)\right]u(t - t_0) + \left[te^{\left(\frac{R}{L}\right)t} - 2t_0e^{\left(\frac{R}{L}\right)2t_0} - 2t_0e^{\left(\frac{R}{L}\right)t} + 1 + \frac{L}{R}\left(e^{\left(\frac{R}{L}\right)t} - e^{\left(\frac{R}{L}\right)2t_0}\right)\right]u(t - 2t_0)\right\} \quad \text{(f)}$$

7.5 Second-Order Systems

The standard form of the differential equation for the response of a second-order mechanical system determined in Equation (4.116) is

$$\ddot{x} + 2\zeta\omega_n\dot{x} + \omega_n^2 x = \frac{1}{m}f(t) \quad (7.58)$$

The response of a first-order system is dependent on one system parameter: the time constant. The response of a second-order system is dependent on two system parameters: the damping ratio ζ and the natural frequency ω_n. Table 7.3 illustrates second-order systems encountered in Chapters 4–6 and the algebraic forms of the system parameters.

Initial conditions required to determine the response of a second-order system are

$$x(0) = x_0 \quad (7.59a)$$

$$\dot{x}(0) = \dot{x}_0 \quad (7.59b)$$

Taking the Laplace transform of Equation (7.58) using appropriate transform properties and applying the initial conditions of Equations (7.59) leads to

$$X(s) = \frac{\frac{F(s)}{m} + (s + 2\zeta\omega_n)x_0 + \dot{x}_0}{s^2 + 2\zeta\omega_n s + \omega_n^2} \quad (7.60)$$

The denominator of a second-order system can be written in terms of the damping ratio and the natural frequency as

$$D(s) = s^2 + 2\zeta\omega_n s + \omega_n^2 \quad (7.61)$$

The standard form of the transfer function for a second-order system is

$$G(s) = \frac{As + B}{s^2 + 2\zeta\omega_n s + \omega_n^2} \quad (7.62)$$

For a mechanical system with force input, $A = 0$ and $B = 1/m$. For a series LRC circuit, $A = 1/L$ and $B = 0$. For a mechanical system with motion input, $A = 2\zeta\omega_n$ and

Table 7.3 Second-Order Systems

Name	Schematic	Differential Equation	Natural Frequency	Damping Ratio
Mass, spring, and viscous damper	k, c, m, x, $f(t)$	$m\ddot{x} + c\dot{x} + kx = f(t)$	$\sqrt{\frac{k}{m}}$	$\frac{c}{2\sqrt{mk}}$
Mass, spring, and viscous damper with motion input	z, x, k, c, m	$m\ddot{x} + c\dot{x} + kx = c\dot{z} + kz$	$\frac{k}{m}$	$\frac{c}{2\sqrt{mk}}$
Series LRC circuit	L, v, i, R, C	$L\frac{di}{dt} + Ri + \frac{1}{C}\int_0^t i\,dt = v(t)$	$\frac{1}{\sqrt{LC}}$	$\frac{R}{2}\sqrt{\frac{C}{L}}$
Parallel LRC circuit	i, L, R, C	$C\frac{dv}{dt} + \frac{v}{R} + \frac{1}{L}\int_0^t v\,dt = i(t)$	$\frac{1}{\sqrt{LC}}$	$\frac{1}{2R}\sqrt{\frac{L}{C}}$

$B = \omega_n^2$. Second-order systems include systems whose mathematical modeling results in two first-order differential equations. In such cases, it is usually more convenient to develop transfer functions from the two first-order equations rather than algebraically combine the equations into a second-order differential equation. The system's natural frequency and damping ratio are then determined from the transfer function.

Example 7.14

Determine the natural frequency and damping ratio of benchmark problems of Section 7.1.3: **a.** 2, **b.** 3, **c.** 4, and **d.** 5.

Solution

a. Comparing the standard form of the denominator for a second-order system to the denominator of the transfer function of benchmark problem 2, Equation (7.8), leads to

$$\omega_n^2 = \frac{k}{m} \Rightarrow \omega_n = \sqrt{\frac{k}{m}} \tag{a}$$

and

$$2\zeta\omega_n = \frac{c}{m} \Rightarrow \zeta = \frac{c}{2m\omega_n} = \frac{c}{2\sqrt{mk}} \tag{b}$$

b. The transfer function for benchmark problem 3 has the same denominator as benchmark problem 2. Hence the natural frequency and damping ratio are given by Equations (a) and (b), respectively. The numerator of the transfer function does not affect these parameters.

c. Comparing the standard form of the denominator for a second-order system to the denominator of the transfer function of benchmark problem 4, Equation (7.10), leads to

$$\omega_n^2 = \frac{1}{LC} \Rightarrow \omega_n = \sqrt{\frac{1}{LC}} \tag{c}$$

and

$$2\zeta\omega_n = \frac{R}{L} \Rightarrow \zeta = \frac{R}{2L\omega_n} = \frac{1}{2R}\sqrt{\frac{C}{L}} \tag{d}$$

d. Comparing the standard form of the denominator for a second-order system to the denominator of the transfer function of benchmark problem 5, Equation (7.12), leads to

$$\omega_n^2 = \frac{1}{A_1 A_2 R_1 R_2} \Rightarrow \omega_n = \sqrt{\frac{1}{A_1 A_2 R_1 R_2}} \tag{e}$$

and

$$2\zeta\omega_n = \frac{1}{A_2 R_1} + \frac{1}{A_2 R_2} + \frac{1}{A_1 R_1} \Rightarrow$$

$$\zeta = \frac{1}{2}\sqrt{\frac{A_1 R_2}{A_2 R_1} + \frac{A_1 R_1}{A_2 R_2} + \frac{A_2 R_2}{A_1 R_1}} \tag{f}$$

7.5.1 Free Response

The free response of a system is a transient response that occurs due to nonzero initial conditions. Nonzero initial conditions are the result of an energy input. When the mechanical system in Table 7.3 is displaced from its equilibrium position, potential energy is stored in the spring. When the inertia element has a nonzero initial velocity, the system has an initial kinetic energy. The total energy in the system at $t = 0$ is

$$E = \frac{1}{2}kx_0^2 + \frac{1}{2}m\dot{x}_0^2 \tag{7.63}$$

If the capacitor in the series LRC circuit is initially charged, then energy is stored in an electric field and the system has a nonzero value of $q(0)$. If current is flowing in the circuit at $t = 0$, then energy is stored in the inductor, resulting in a nonzero value of $dq/dt(0)$.

$$E = \frac{1}{2}L\left[\frac{dq}{dt}(0)\right]^2 + \frac{1}{2}\frac{1}{C}[q(0)]^2 \tag{7.64}$$

The poles of the transfer function for a second-order system are the roots of $D(s)$,

$$s_1 = -\zeta\omega_n - \omega_n\sqrt{\zeta^2 - 1} \tag{7.65a}$$

$$s_2 = -\zeta\omega_n + \omega_n\sqrt{\zeta^2 - 1} \tag{7.65b}$$

The mathematical form of the free response depends on the value of the damping ratio ζ. Four cases are considered.

Case 1: $\zeta = 0$ (Undamped Response) For an undamped system, $\zeta = 0$, the poles of the transfer function are purely imaginary, $s_{1,2} = \pm j\omega_n$, and Equation (7.60) becomes, with $F(s) = 0$,

$$X(s) = \frac{x_0 s + \dot{x}_0}{s^2 + \omega_n^2} \tag{7.66}$$

Equation (7.66) is inverted using transform pairs (2) and (3) from Table 2.1, yielding

$$x(t) = x_0\cos(\omega_n t) + \frac{\dot{x}_0}{\omega_n}\sin(\omega_n t) \tag{7.67}$$

An alternate, but equivalent, representation of the undamped response is

$$x(t) = A\sin(\omega_n t + \phi) \tag{7.68}$$

where the **amplitude** A and the **phase** ϕ are given by

$$A = \sqrt{x_0^2 + (\dot{x}_0/\omega_n)^2} \tag{7.69}$$

$$\phi = \tan^{-1}(\omega_n x_0/\dot{x}_0) \tag{7.70}$$

The free response of a second-order undamped system, illustrated in Figure 7.19, is periodic of period

$$T = \frac{2\pi}{\omega_n} \tag{7.71}$$

One period has occurred when the system returns to the state defined by the initial conditions. This is equivalent to the system executing one **cycle**. Thus the period represents the time it takes the system to execute one cycle, and the reciprocal of the period, called the **frequency**, is the number of cycles executed per unit of time. Using seconds as the unit of time, the frequency is measured in cycles per second [hertz (Hz)]. The frequency in hertz is determined as

$$f = \frac{\omega_n}{2\pi} \tag{7.72}$$

After execution of one cycle, the argument of the sine function increases by 2π radians. Thus 1 cycle $= 2\pi$ radians and from Equation (7.72), it is clear that ω_n is the frequency in radians per second (r/s). The system parameter ω_n is called the system's **natural frequency** because it is the frequency of the system response in the absence of viscous damping and external excitation. The natural frequency is the frequency of the response due to nonzero initial conditions. The natural frequency is dependent on system parameters and has no dependence on the initial conditions.

During the free response of an undamped system, energy is neither added to the system by an external force nor dissipated from the system. Thus, energy is conserved. The amplitude of the system response being constant is a statement of Conservation of Energy. The amplitude of an undamped system is proportional to the energy stored in the system. Equations (7.63) and (7.68) are used to show that for a mechanical system

$$A = \sqrt{\frac{2E}{k}} \tag{7.73}$$

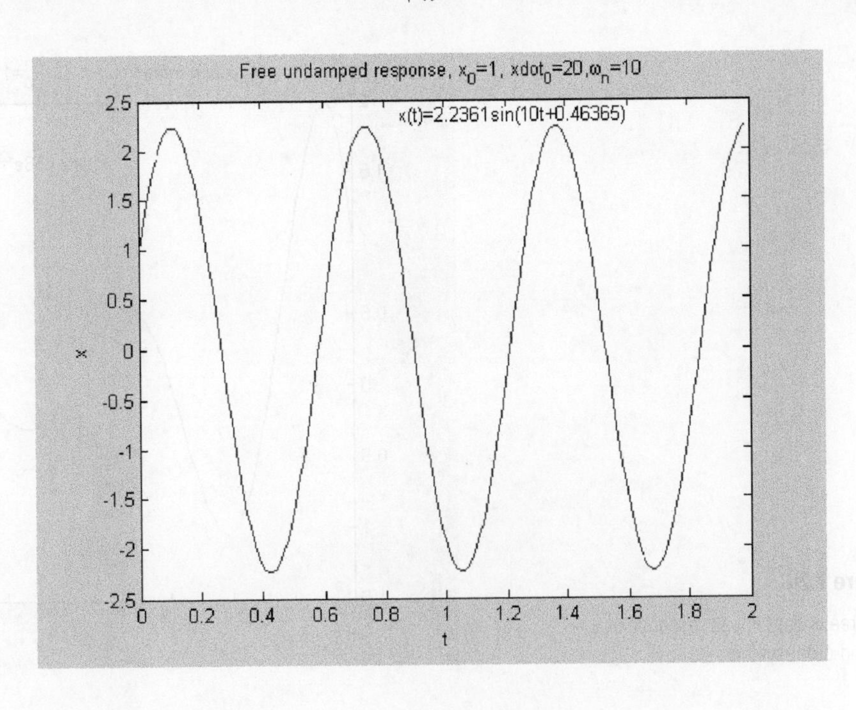

Figure 7.19

The free undamped response of a second-order system.

while for a series LRC circuit

$$A = \sqrt{2CE} \tag{7.74}$$

The phase represents the difference between the system response and a purely sinusoidal response. If $x_0 = 0$, then $\phi = 0$. If $\dot{x}_0 = 0$ and $x_0 > 0$, then $\phi = \pi/2$. If $\dot{x}_0 = 0$ and $x_0 < 0$, then $\phi = -\pi/2$. If both initial conditions are nonzero and of the same sign, then the phase is positive, $0 < \phi < \pi/2$. If both initial conditions are nonzero and of opposite signs, then the phase is negative, $-\pi/2 < \phi < 0$. If the phase is positive, the response is said to lead a sinusoidal response. If the phase is negative, the response is said to lag a sinusoidal response.

Case 2: $0 < \zeta < 1$ (Underdamped Response) For $0 < \zeta < 1$, the poles of the transfer function are complex conjugates with negative real parts. The inversion of Equation (7.60) with $F(s) = 0$ is achieved by completing the square in $D(s)$

$$X(s) = \frac{(s + \zeta\omega_n)x_0 + \zeta\omega_n x_0 + \dot{x}_0}{(s + \zeta\omega_n)^2 + \omega_d^2} \tag{7.75}$$

where

$$\omega_d = \omega_n\sqrt{1 - \zeta^2} \tag{7.76}$$

Equation (7.75) is inverted using Table 2.1, leading to

$$x(t) = A_d e^{-\zeta\omega_n t}\sin(\omega_d t + \phi_d) \tag{7.77}$$

where the amplitude and phase for the damped system are

$$A_d = \sqrt{x_0^2 + \left(\frac{\dot{x}_0 + \zeta\omega_n x_0}{\omega_d}\right)^2} \tag{7.78}$$

$$\phi_d = \tan^{-1}\left(\frac{x_0\omega_d}{\dot{x}_0 + \zeta\omega_n x_0}\right) \tag{7.79}$$

The free response of an underdamped system is illustrated in Figure 7.20. The response is not periodic, but cyclic with an exponentially decaying amplitude. The time

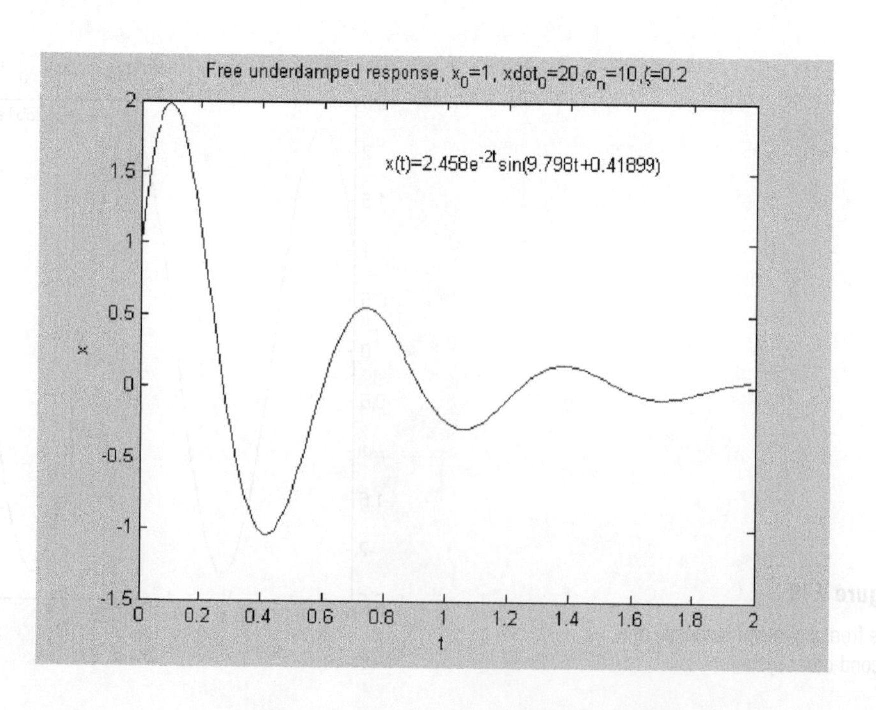

Figure 7.20

The free underdamped response of a second-order system.

between peaks on the response curve constitutes one cycle. The time to execute one cycle is the **damped period**

$$T_d = \frac{2\pi}{\omega_d} \tag{7.80}$$

and ω_d is called the **damped natural frequency**.

An underdamped system has a component that dissipates energy, such as a viscous damper in a mechanical system or a resistor in an electric circuit. The resistance provided by the dissipative mechanism slows the system down. The larger the resistance is, the longer it takes the system to execute one cycle and the smaller the damped frequency. For an underdamped system, the energy dissipated during a cycle is less than the energy stored in the system. The energy dissipated over one cycle is a constant fraction of the energy stored in the system at the beginning of the cycle.

Example 7.15

A simplified model of a vehicle suspension system is illustrated in Figure 7.21. For parameter values of $m = 200$ kg, $k = 5 \times 10^5$ N·m, and $c = 1.5 \times 10^4$ N·s/m, determine **a.** the damping ratio for the system, **b.** the system's damped period, and **c.** the system response when the vehicle experiences a sudden change in road contour that leads to the initial conditions $x(0) = 0.001$ m, $\dot{x}(0) = 0$.

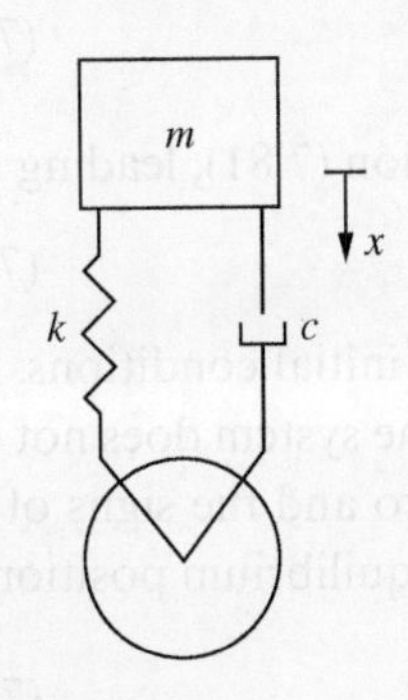

Figure 7.21

The simplified one-degree-of-freedom model for the vehicle suspension system in Example 7.15.

Solution

The differential equation governing the free response of the vehicle after it encounters a sudden change in road contour is

$$m\ddot{x} + c\dot{x} + kx = 0 \tag{a}$$

which is put in the standard form of Equation (7.58) with $f(t) = 0$ by dividing by m, leading to

$$\ddot{x} + \frac{c}{m}\dot{x} + \frac{k}{m}x = 0 \tag{b}$$

The natural frequency of the system is determined from Equation (b) as

$$\omega_n = \sqrt{\frac{k}{m}}$$

$$= \sqrt{\frac{5 \times 10^5 \text{ N/m}}{200 \text{ kg}}} = 50 \text{ r/s} \tag{c}$$

a. The damping ratio is obtained from Equations (b) and (c) as

$$\zeta = \frac{c}{2m\omega_n}$$

$$= \frac{1.5 \times 10^4 \text{ N·s/m}}{2(200 \text{ kg})(50 \text{ r/s})} = 0.75 \tag{d}$$

b. The damped natural frequency is calculated using Equation (7.76) as

$$\omega_d = \omega_n\sqrt{1 - \zeta^2}$$

$$= 50 \text{ r/s}\sqrt{1 - (0.75)^2} = 33.1 \text{ r/s} \tag{e}$$

The damped period is calculated from Equation (7.80) as

$$T_d = \frac{2\pi}{\omega_d}$$

$$= \frac{2\pi}{33.1 \text{ r/s}} = 0.190 \text{ s} \tag{f}$$

c. For the given initial conditions, the amplitude and phase are calculated using Equations (7.78) and (7.79) as

$$A_d = \sqrt{(0.001)^2 + \left[\frac{0.75(50)(0.001)}{33.1}\right]^2} = 1.51 \times 10^{-3}\ \text{m} \tag{g}$$

$$\phi_d = \tan^{-1}\left[\frac{(0.001)(33.1)}{(0.75)(50)(0.001)}\right] = 0.723\ \text{r} \tag{h}$$

The system response is calculated using Equation (7.77) as

$$x(t) = 1.51 \times 10^{-3} e^{-37.5t} \sin(33.1t + 0.723)\ \text{m} \tag{i}$$

Case 3: $\zeta = 1$ (Critically Damped System) For $\zeta = 1$, the poles of the transfer function for a second-order system are equal, $s_{1,2} = -\omega_n$. In this case, Equation (7.60) with $F(s) = 0$ becomes

$$X(s) = \frac{x_0}{s + \omega_n} + \frac{\dot{x}_0 + \omega_n x_0}{(s + \omega_n)^2} \tag{7.81}$$

Transform pairs 1 and 11 of Table 2.1 are used to invert Equation (7.81), leading to

$$x(t) = x_0 e^{-\omega_n t} + (\dot{x}_0 + \omega_n x_0) t e^{-\omega_n t} \tag{7.82}$$

Equation (7.82) is plotted in Figure 7.22 for several sets of initial conditions. The response of a critically damped system decays exponentially. The system does not execute one full cycle of motion. If neither initial condition is zero and the signs of the initial conditions are opposite, then the system overshoots its equilibrium position if

$$\frac{x_0}{\dot{x}_0 + \omega_n x_0} < 0 \tag{7.83}$$

Case 4: $\zeta > 1$ (Overdamped System) For $\zeta > 1$, the poles of the transfer function are real and negative:

$$s_1 = -\omega_n\left(\zeta + \sqrt{\zeta^2 - 1}\right) \tag{7.84a}$$

$$s_2 = -\omega_n\left(\zeta - \sqrt{\zeta^2 - 1}\right) \tag{7.84b}$$

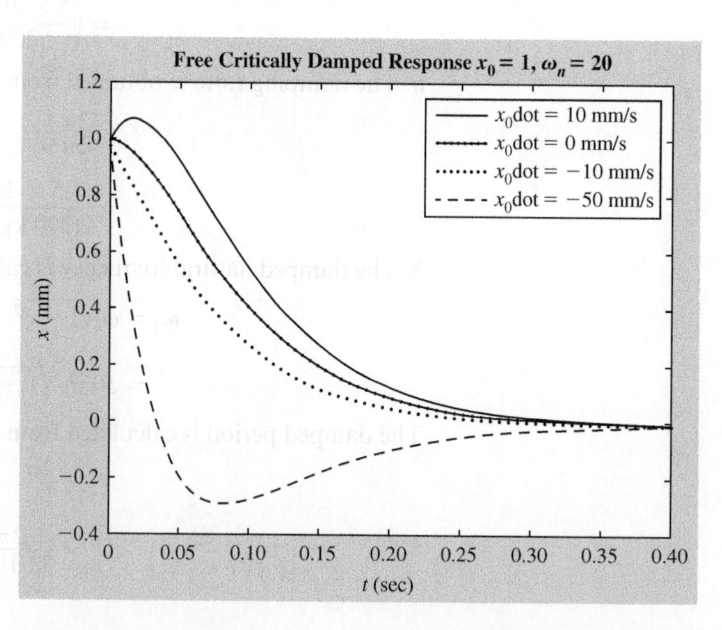

Figure 7.22

The free response of a critically damped system for a fixed initial displacement but various values of initial velocity.

Equation (7.60) with $F(s) = 0$ becomes

$$X(s) = \frac{(s + \zeta\omega_n)x_0 + \dot{x}_0}{(s - s_1)(s - s_2)} \tag{7.85}$$

Use of transform pairs 16 and 18 from Table 2.1 in the inversion of the transform of Equation (7.85) yields

$$x(t) = \frac{1}{2\sqrt{\zeta^2 - 1}}\left\{\left[x_0\left(-\zeta + \sqrt{\zeta^2 - 1}\right) - \frac{\dot{x}_0}{\omega_n}\right]e^{s_1 t} + \left[x_0\left(\zeta + \sqrt{\zeta^2 - 1}\right) + \frac{\dot{x}_0}{\omega_n}\right]e^{s_2 t}\right\} \tag{7.86}$$

The free response of an overdamped system, as illustrated in Figure 7.23 for several sets of initial conditions, decays exponentially. An overdamped system takes longer to approach equilibrium than a critically damped system.

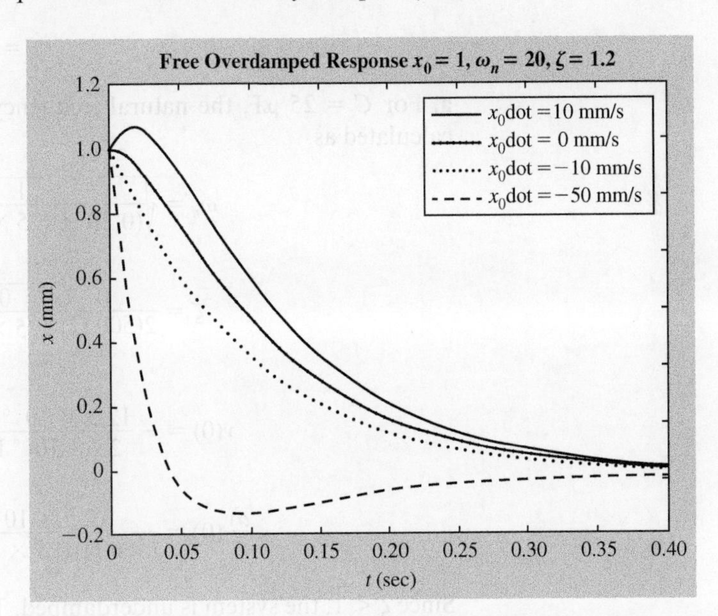

Figure 7.23

The free response of an overdamped system for a fixed initial displacement but various values of initial velocity.

Example 7.16

The parallel LRC circuit of Example 5.12 and Figure 5.18 has resistance $R = 500\ \Omega$ and inductance $L = 0.25$ H. The capacitor has an initial charge of $q = 12.5\ \mu$C. Determine the voltage after the switch is closed at $t = 0$ if **a.** $C = 25\ \mu$F, **b.** $C = 0.25\ \mu$F, and **c.** $C = 0.16\ \mu$F.

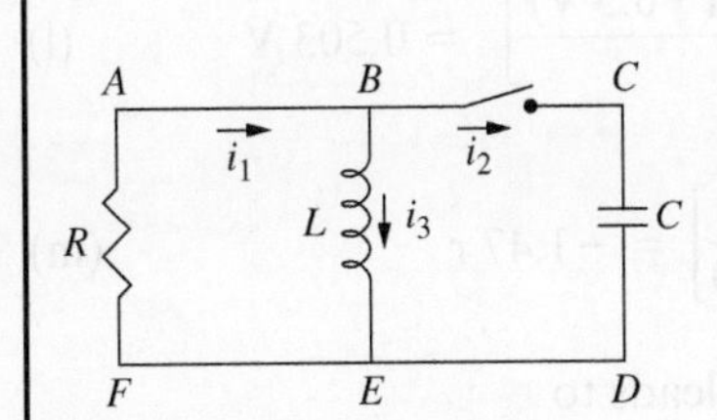

Figure 5.18

The parallel *LRC* circuit of Example 5.12 and Example 7.16. (Repeated)

Solution

It is shown in Example 5.12 that the voltage change across each of the circuit elements is the same for the parallel LRC circuit. The differential equation governing the voltage change across the resistor is obtained in Equation (k) of Example 5.12 as

$$C\frac{d^2v}{dt^2} + \frac{1}{R}\frac{dv}{dt} + \frac{1}{L}v = 0 \tag{a}$$

The initial conditions corresponding to an initial charge of q on the capacitor are given as Equations (j) and (m) of Example 5.12 as

$$v(0) = -\frac{q}{C} \tag{b}$$

$$\frac{dv}{dt}(0) = \frac{q}{RC^2} \tag{c}$$

The natural frequency and damping ratio for this second-order system are determined from Equation (a) as

$$\omega_n = \sqrt{\frac{1}{LC}} \tag{d}$$

$$\zeta = \frac{1}{2RC\omega_n}$$

$$= \frac{1}{2R}\sqrt{\frac{L}{C}} \tag{e}$$

a. For $C = 25\ \mu\text{F}$, the natural frequency, damping ratio, and initial conditions are calculated as

$$\omega_n = \sqrt{\frac{1}{(0.25\text{ H})(25 \times 10^{-6}\text{ F})}} = 400\text{ r/s} \tag{f}$$

$$\zeta = \frac{1}{2(500\ \Omega)}\sqrt{\frac{0.25\text{ H}}{25 \times 10^{-6}\text{ F}}} = 0.1 \tag{g}$$

$$v(0) = -\frac{12.5 \times 10^{-6}\text{ C}}{25 \times 10^{-6}\text{ F}} = -0.5\text{ V} \tag{h}$$

$$\frac{dv}{dt}(0) = \frac{12.5 \times 10^{-6}\text{ C}}{(500\ \Omega)(25 \times 10^{-6}\text{ F})^2} = 40\text{ V/s} \tag{i}$$

Since $\zeta < 1$, the system is underdamped. The damped natural frequency is

$$\omega_d = 400\text{ r/s}\sqrt{1 - (0.1)^2} = 398.0\text{ r/s} \tag{j}$$

The response is determined using Equation (7.77) as

$$v(t) = A_d e^{-40t}\sin(398t + \phi_d) \tag{k}$$

where the initial conditions are used in Equations (7.78) and (7.79) to obtain

$$A_d = \sqrt{(-0.5\ V)^2 + \left[\frac{(40\text{ V/s}) + (0.1)(400\text{ r/s})(-0.5\text{ V})}{398.0\text{ r/s}}\right]^2} = 0.503\text{ V} \tag{l}$$

$$\phi_d = \tan^{-1}\left[\frac{(-0.5\text{ V})(398.0\text{ r/s})}{(40\text{ V/s}) + (0.1)(400\text{ r/s})(-0.5\text{ V})}\right] = -1.47\text{ r} \tag{m}$$

Substituting Equations (l) and (m) into Equation (k) leads to

$$v(t) = 0.503e^{-40t}\sin(398.0t - 1.47)\text{ V} \tag{n}$$

b. For $C = 0.25\ \mu\text{F}$, the natural frequency, damping ratio, and initial conditions are calculated using Equations (d), (e), (b), and (c) as $\omega_n = 4000$ r/s, $\zeta = 1$, $v(0) = -50$ V, and $dv/dt(0) = 4 \times 10^5$ V/s. The damping ratio is one; thus the system is critically damped. The response is determined using Equation (7.82) as

$$v(t) = (-50 \text{ V})e^{-4000t} + [4 \times 10^5 \text{ V/s} + (4000 \text{ r/s})(-50 \text{ V})]te^{-4000t}$$

$$= (2 \times 10^5 t - 50)e^{-4000t} \text{ V} \quad \text{(o)}$$

c. For $C = 0.16$ μF, the natural frequency, damping ratio, and initial conditions are determined using Equations (d), (e), (b), and (c) as $\omega_n = 5000$ r/s, $\zeta = 1.25$, $v(0) = -78.1$ V, and $dv/dt(0) = 9.77 \times 10^5$ V/s. Since the damping ratio is greater than one, the system is overdamped. Equations (7.84a) and (7.84b) yield

$$s_1 = -5000\left[1.25 + \sqrt{(1.25)^2 - 1}\right] = -10{,}000 \quad \text{(p)}$$

$$s_2 = -5000\left[1.25 - \sqrt{(1.25)^2 - 1}\right] = -2500 \quad \text{(q)}$$

The response of the system is obtained using Equation (7.86) as

$$v(t) = \frac{1}{2(0.75)}\left\{\left[(-78.1 \text{ V})(-0.5) - \frac{9.77 \times 10^5 \text{ V/s}}{5000 \text{ r/s}}\right]e^{-10{,}000t} + \left[(-78.1 \text{ V})(2) + \frac{9.77 \times 10^5 \text{ V/s}}{5000 \text{ r/s}}\right]e^{-2500t}\right\}$$

$$= (26.16e^{-2500t} - 104.2e^{-10{,}000t}) \text{ V} \quad \text{(r)}$$

7.5.2 Impulsive Response

An impulsive force applied to a mechanical system leads to an instantaneous discrete change in velocity. An impulsive voltage applied to a parallel *LRC* circuit leads to an instantaneous discrete change in the current through the inductor. An impulsive current applied to a series *LRC* circuit leads to an instantaneous discrete change in voltage in the capacitor.

These changes can be observed by examining the differential equations with an impulsive input. Consider for example the integrodifferential equation for the series *LRC* circuit when subject to an impulsive voltage

$$L\frac{di}{dt} + Ri + \frac{1}{C}\int_0^t i\,dt = I_V\delta(t) \quad (7.87)$$

If there is no current in the circuit when the impulse is applied at $t = 0$, then evaluation of Equation (7.87) at $t = 0$ leads to

$$L\frac{di}{dt}(0) = I_V\delta(0) \quad (7.88)$$

Thus $di/dt(0)$ is infinite and the current is discontinuous at $t = 0$. The current at $t = 0^+$ is I_V/L. The impulsive response of this second-order system is similar to that of a circuit with $i(0) = I_V/L$. The mathematical forms of the responses are identical for $t > 0$.

In a similar fashion, it is reasoned that the impulsive response of a second-order mechanical system is equal to the free response due to a nonzero initial velocity equal to $1/m$ and multiplied by $u(t)$. Impulsive responses of second-order systems are catalogued in Table 7.4 for systems with $A = 1$ and $B = 0$ and for systems with $A = 0$ and $B = 1$. The impulsive response for nonzero values of both A and B is obtained by forming a linear combination of the appropriate responses.

Table 7.4 Impulsive Response of Second-Order Systems

Transfer Function	Condition	$x(t)$
$\frac{1}{s^2+\omega_n^2}$	Undamped	$\frac{1}{\omega_n}\sin\omega_n t$
$\frac{1}{s^2+2\zeta\omega_n s+\omega_n^2}$	Underdamped $\omega_d = \omega_n\sqrt{1-\zeta^2}$	$\frac{1}{\omega_d}e^{-\zeta\omega_n t}\sin(\omega_d t)$
$\frac{1}{s^2+2\zeta\omega_n s+\omega_n^2}$	Critically damped	$te^{-\omega_n t}$
$\frac{1}{s^2+2\zeta\omega_n s+\omega_n^2}$	Overdamped $s_1 = -\omega_n(\zeta+\sqrt{\zeta^2-1})$ $s_2 = -\omega_n(\zeta-\sqrt{\zeta^2-1})$	$\frac{1}{s_2-s_1}(e^{s_2 t}-e^{s_1 t})$
$\frac{s}{s^2+\omega_n^2}$	Undamped	$\cos\omega_n t$
$\frac{s}{s^2+2\zeta\omega_n s+\omega_n^2}$	Underdamped $\omega_d = \omega_n\sqrt{1-\zeta^2}$	$e^{-\zeta\omega t}\left[\cos(\omega_d t)+\frac{\zeta\omega_n}{\omega_d}\sin(\omega_d t)\right]$
$\frac{s}{s^2+2\zeta\omega_n s+\omega_n^2}$	Critically damped	$e^{-\omega_n t}(1+\omega_n t)$
$\frac{s}{s^2+2\zeta\omega_n s+\omega_n^2}$	Overdamped $s_1 = -\omega_n(\zeta+\sqrt{\zeta^2-1})$ $s_2 = -\omega_n(\zeta-\sqrt{\zeta^2-1})$	$\frac{1}{s_2-s_1}(s_2 e^{s_1 t}-s_1 e^{s_2 t})$

Example 7.17

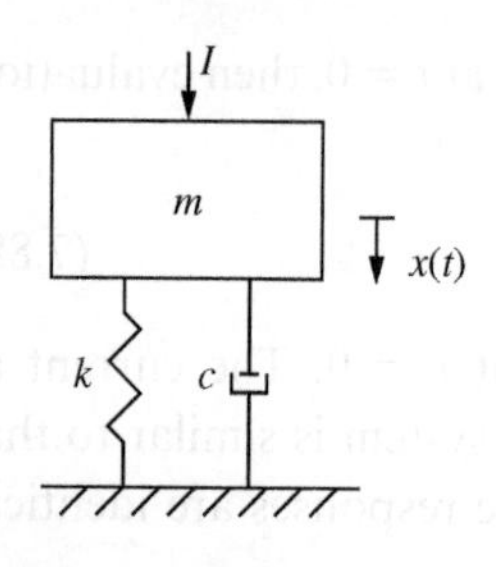

Figure 7.24
The schematic representation of the stamping machine of Example 7.17.

A stamping machine, illustrated schematically in Figure 7.24, is of mass 200 kg and rests on an isolator that is modeled as a spring of stiffness 3×10^5 N/m in parallel with a viscous damper of damping coefficient 5000 N·s/m. During its operation, the machine is subject to an impulse of magnitude 20 N·s. **a.** Determine the response of the machine after the application of the impulse. **b.** Determine the maximum displacement of the machine after the application of the impulse and the time at which the maximum displacement occurs. **c.** Determine the maximum value of the force developed in the isolator, which is calculated by

$$F = c\dot{x} + kx \tag{a}$$

d. Determine the isolator efficiency, which is defined by

$$E = \frac{I^2}{2mF_{\max}x_{\max}} \tag{b}$$

Solution

The mathematical model for the system is the same as benchmark problem 2 of Section 7.1.3, whose transfer function is given by Equation (7.8). The natural frequency and damping ratio are determined using Table 7.3 as

$$\omega_n = \sqrt{k/m}$$

$$= \sqrt{\frac{3 \times 10^5\ \text{N/m}}{200\ \text{kg}}} = 38.7\ \text{r/s} \tag{c}$$

$$\zeta = \frac{c}{2m\omega_n}$$

$$= \frac{5000\ \text{N}\cdot\text{s/m}}{2(200\ \text{kg})(38.7\ \text{r/s})} = 0.323 \tag{d}$$

The damping ratio is $\zeta < 1$, so the system is underdamped with a damped natural frequency of

$$\omega_d = \omega_n\sqrt{1-\zeta^2}$$

$$= 38.7\ \text{r/s}\sqrt{1-(0.323)^2} = 36.6\ \text{r/s} \tag{e}$$

a. The impulsive response of an underdamped system is obtained using Table 7.4 for a transfer function of the form of Equation (7.8) as

$$x(t) = \frac{I}{m\omega_d}e^{-\zeta\omega_n t}\sin(\omega_d t) \tag{f}$$

Substitution of the given and calculated values in Equation (f) leads to

$$x(t) = \frac{20\ \text{N}\cdot\text{s}}{(200\ \text{kg})(36.6\ \text{r/s})}e^{-(0.323)(38.7)t}\sin(36.6t)$$

$$= 2.73e^{-12.5t}\sin(36.6t)\ \text{mm} \tag{g}$$

b. The maximum displacement occurs the first time the velocity is zero. From Equation (g), the velocity is determined as

$$\dot{x}(t) = e^{-12.5t}[-34.1\sin(36.6t) + 99.9\cos(36.6t)]\ \text{mm/s} \tag{h}$$

The time t_p at which the maximum displacement occurs is obtained by setting $\dot{x}$ to zero, which leads to

$$-34.1\sin(36.6t_p) + 99.9\cos(36.6t_p) = 0$$

$$\tan(36.6t_p) = 99.9/34.1 = 2.91$$

$$t_p = \frac{1}{36.6}\tan^{-1}(2.91) = 3.39 \times 10^{-2}\ \text{s} \tag{i}$$

The maximum displacement is

$$x_{\max} = x(t_p)$$

$$= 2.73e^{-12.5(3.39\times10^{-2})}\sin[36.6(3.39\times10^{-2})] = 1.69\ \text{mm} \tag{j}$$

c. The force developed in the isolator is

$$F = c\dot{x} + kx$$

$$= 5000\ \text{N}\cdot\text{s/m}\{e^{-12.5t}[-34.1\sin(36.6t) + 99.9\cos(36.6t)]\ \text{mm/s}\}$$

$$+ 3\times10^5\ \text{N/m}\ [2.73e^{-12.5t}\sin(36.6t)\ \text{mm}]$$

$$= e^{-12.5t}[648.5\sin(36.6t) + 499.5\cos(36.6t)]\ \text{N} \tag{k}$$

The time t_f at which the maximum force occurs is obtained by setting $\dot{F}$ to zero, leading to

$$0 = -12.5e^{-12.5t_f}[648.5\sin(36.6t_f) + 499.5\cos(36.6t_f)]$$
$$+ e^{-12.5t_f}[(648.5)(36.6)\cos(36.6t_f) - (499.5)(36.6)\sin(36.6t_f)]$$
$$= e^{-12.5t_f}[-2.64\times10^4\sin(36.6t_f) + 1.75\times10^4\cos(36.6t_f)]$$
$$t_f = \frac{1}{36.6}\tan^{-1}\left(\frac{1.75\times10^4}{2.64\times10^4}\right) = 1.60\times10^{-2}\text{ s} \tag{l}$$

The maximum force is

$$F_{\max} = F(t_f) = e^{-12.5(1.60\times10^{-2})}\{648.5\sin[(36.6)(1.60\times10^{-2})]$$
$$+ 499.5\cos[(36.6)(1.60\times10^{-2})]\}$$
$$= 634.2\text{ N} \tag{m}$$

d. The isolator efficiency is calculated as

$$E = \frac{I^2}{2mF_{\max}x_{\max}}$$
$$= \frac{(20\text{ N/s})^2}{2(200\text{ kg})(634.2\text{ N})(1.69\times10^{-3}\text{ m})} = 0.933 \tag{n}$$

7.5.3 Step Response

The mathematical form of the step response of a second-order system depends on whether the system is undamped, underdamped, critically damped, or overdamped. The step response is obtained by taking the inverse transform of Equation (7.16). Consider a mechanical system with force input, such as benchmark problem 2 of Section 7.1.3. The transfer function is given in Equation (7.9) and its step response becomes

$$x_s(t) = \mathcal{L}^{-1}\left\{\frac{1/m}{s\left(s^2 + \frac{c}{m}s + \frac{k}{m}\right)}\right\} = \mathcal{L}^{-1}\left\{\frac{1/m}{s(s^2 + 2\zeta\omega_n s + \omega_n^2)}\right\} \tag{7.89}$$

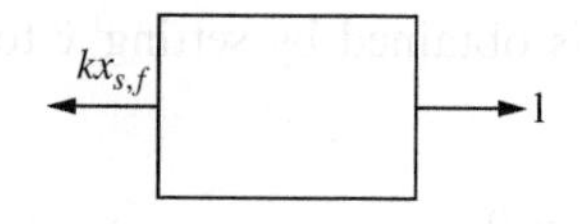

Figure 7.25

Free-body diagram of mass in a mass-spring-viscous damper system due to a step input at its final value.

The final value of the step response is $G(0) = \frac{1}{k}$. A free-body diagram of the system when the system is at its final value due to a unit step input is shown in Figure 7.25. At its final value the velocity is zero; thus there is no viscous damping force. The acceleration at the system's final value is also zero, so application of Newton's law gives

$$\sum F = 0 \Rightarrow -kx_{s,f} + 1 = 0 \Rightarrow x_{s,f} = \frac{1}{k} \tag{7.90}$$

From the transfer function, the final value of the step response is

$$G(0) = \frac{1}{k} \tag{7.91}$$

which confirms that the transfer function contains all the information regarding a dynamic system.

A partial fraction decomposition of Equation (7.89) with $m = 1$ is performed, resulting in

$$X_s(s) = \frac{1}{\omega_n^2}\left(\frac{1}{s} - \frac{s + 2\zeta\omega_n}{s^2 + 2\zeta\omega_n s + \omega_n^2}\right) \tag{7.92}$$

The value of ζ affects the inverse of the transform of Equation (7.92) in much the same way it affects the free response. The forms of the responses are given in Table 7.5 for undamped ($\zeta = 0$), underdamped ($0 < \zeta < 1$), critically damped ($\zeta = 1$), and overdamped ($\zeta > 0$) systems. Also given in Table 7.5 are the responses of systems with s in the numerator, much like the series LRC circuit. The step responses are illustrated in Figure 7.26.

Table 7.5 Step Response of Second-Order Systems

Transfer Function	Condition	$x(t)$
$\frac{1}{s^2+\omega_n^2}$	Undamped	$\frac{1}{\omega_n^2}(1-\cos\omega_n t)$
$\frac{1}{s^2+2\zeta\omega_n s+\omega_n^2}$	Underdamped $\omega_d = -\omega_n\left(\sqrt{1-\zeta^2}\right)$	$\frac{1}{\omega_n^2}\left[1-e^{-\zeta\omega_n t}\cos(\omega_d t)-\frac{\zeta\omega_n}{\omega_d}e^{-\zeta\omega_n t}\sin(\omega_d t)\right]$
$\frac{1}{s^2+2\zeta\omega_n s+\omega_n^2}$	Critically damped	$\frac{1}{\omega_n^2}[1-e^{-\omega_n t}-\omega_n te^{-\omega_n t}]$
$\frac{1}{s^2+2\zeta\omega_n s+\omega_n^2}$	Overdamped $s_1 = -\omega_n\left(\zeta+\sqrt{\zeta^2-1}\right)$ $s_2 = -\omega_n\left(\zeta-\sqrt{\zeta^2-1}\right)$	$\frac{1}{s_1 s_2(s_1-s_2)}(s_1-s_2+s_2e^{s_1 t}-s_1e^{s_2 t})$
$\frac{s}{s^2+\omega_n^2}$	Undamped	$\frac{1}{\omega_n}\sin\omega_n t$
$\frac{s}{s^2+2\zeta\omega_n s+\omega_n^2}$	Underdamped $\omega_d = \omega_n\sqrt{1-\zeta^2}$	$\frac{1}{\omega_d}e^{-\zeta\omega_n t}\sin\omega_d t$
$\frac{s}{s^2+2\zeta\omega_n s+\omega_n^2}$	Critically damped	$te^{-\omega_n t}$
$\frac{s}{s^2+2\zeta\omega_n s+\omega_n^2}$	Overdamped $s_1 = -\omega_n\left(\zeta+\sqrt{\zeta^2-1}\right)$ $s_2 = -\omega_n\left(\zeta-\sqrt{\zeta^2-1}\right)$	$\frac{1}{s_2-s_1}(e^{s_2 t}-e^{s_1 t})$

The step response for a transfer function of the form of Equation (7.62) can be obtained by taking the sum of the appropriate value of A times the step response for a transfer function with s in the numerator and the appropriate value of B times the step response for a transfer function with a 1 in the numerator.

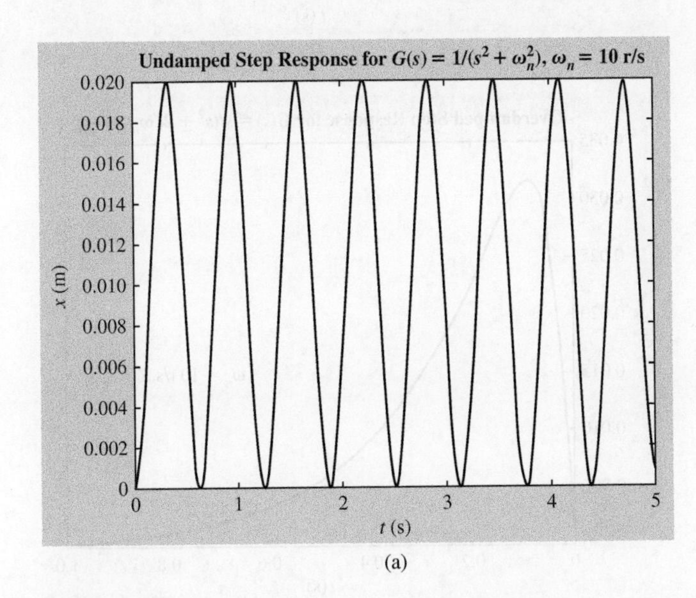

(a)

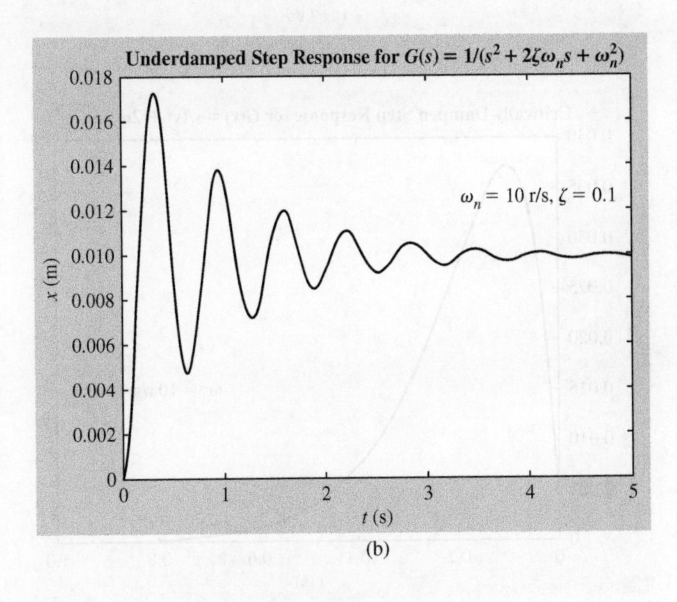

(b)

Figure 7.26

The responses of second-order systems due to unit step input: (a) The undamped mechanical system with $\omega_n = 10$ r/s; (b) The underdamped mechanical system with $\omega_n = 10$ r/s and $\zeta = 0.1$; (c) the critically damped mechanical system with $\omega_n = 10$ r/s; (d) the overdamped mechanical system with $\omega_n = 10$ r/s and $\zeta = 1.3$; (e) the undamped series *LRC* circuit with $\omega_n = 10$ r/s; (f) the underdamped series *LRC* circuit with $\omega_n = 10$ r/s and $\zeta = 0.1$; (g) the critically damped series *LRC* circuit with $\omega_n = 10$ r/s; (h) the overdamped series *LRC* circuit with $\omega_n = 10$ r/s and $\zeta = 1.3$.

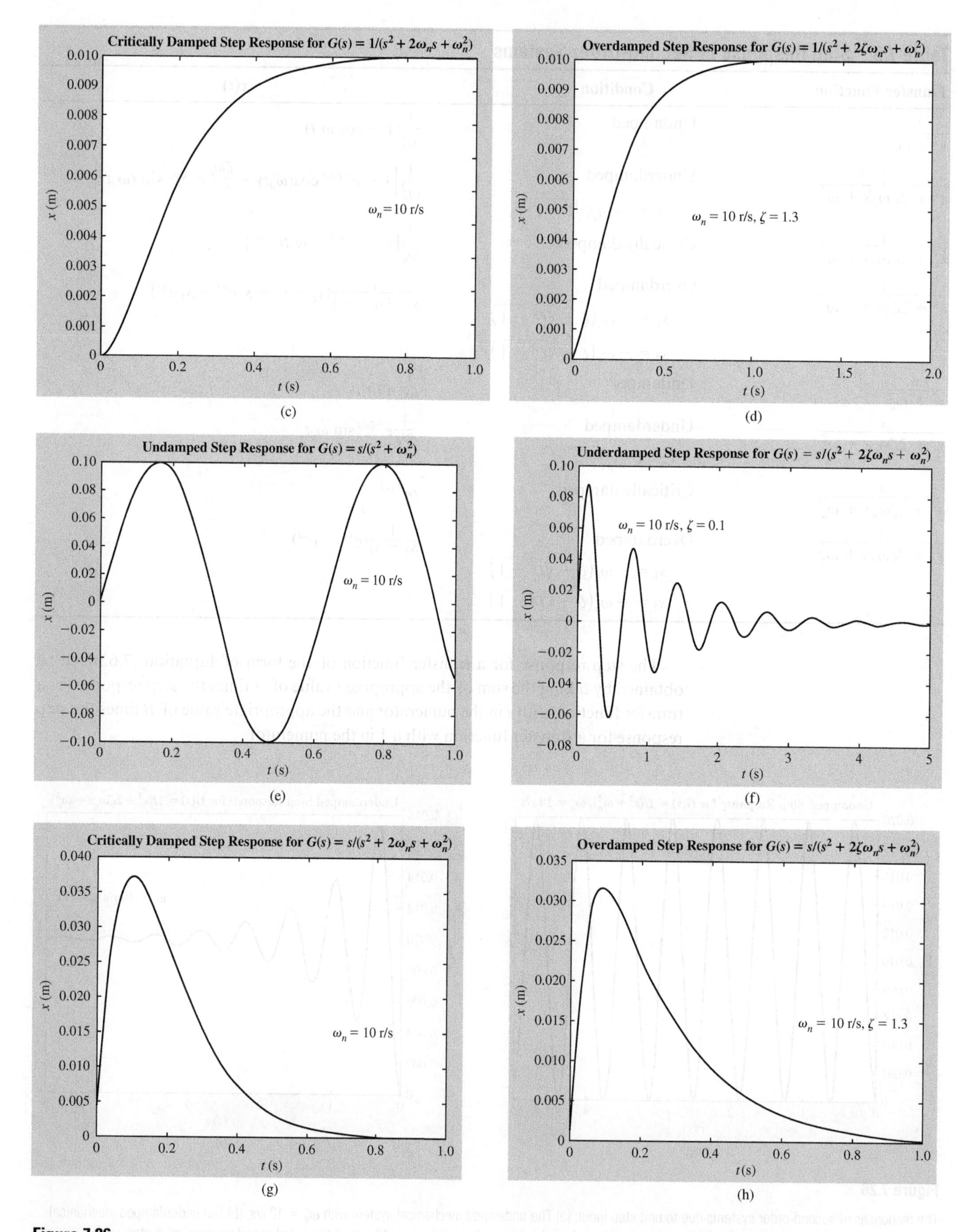

Figure 7.26
(Continued)

The responses of undamped, underdamped, critically damped, and overdamped systems due to a step input are catalogued in Table 7.5 for the cases $A = 0$ and $B = 1$ and $A = 1$ and $B = 0$. The step response for a transfer function of the form of Equation (7.62) is obtained by taking the value of A times the response of a system with s in the numerator plus the value of B times the response of a system with a one in the numerator. The step responses are illustrated in Figure 7.26.

The response specifications of settling time, rise time, overshoot, and peak time are determined from the system response and depend on the damping ratio and natural frequency. Since the final value of a system with $B = 0$ is zero, the definitions of rise time percent overshoot do not apply to the response.

Since an undamped second-order system is neutrally stable, the definitions of 10–90 percent rise time and 2 percent settling time do not apply. For an underdamped system, the exact 2 percent settling time may have to be determined numerically. However, noting that for large t, $x(t)$ is oscillating about $G(0)$ with a decaying amplitude. For the purpose of approximating the 2 percent settling time, the oscillatory behavior can be neglected. The step response for a system with a transfer function with only a 1 in the numerator is approximated by

$$\lim_{t\to\infty} x_s(t) = \frac{1}{\omega_n^2}(1 - e^{-\zeta\omega_n t}) \tag{7.93}$$

Approximating the 2 percent settling time from Equation (7.93) leads to

$$t_s = \frac{3.92}{\zeta\omega_n} \tag{7.94}$$

If the transfer function has a numerator of the form $As + B$, then the settling time has to be calculated using the definition.

For an overdamped system under the same conditions, the term $e^{s_2 t}$ is much larger than $e^{s_1 t}$ in the range of the settling time and thus the latter term is neglected. Then the 2 percent settling time is approximated from

$$\begin{aligned} 0.02 &= \frac{s_1}{s_1 - s_2} e^{s_2 t_s} \\ &= \frac{\zeta + \sqrt{\zeta^2 - 1}}{2\sqrt{\zeta^2 - 1}} e^{s_2 t_s} \end{aligned} \tag{7.95}$$

For a specific value of ζ, Equation (7.95) provides a good approximation for the 2 percent settling time. The larger the value of ζ, the better the approximation.

The denominator of a second-order overdamped system is reducible, meaning it can be factored with real factors, $D(s) = (s - s_1)(s - s_2)$. A partial fraction decomposition of the transform of the step response for an overdamped system is of the form

$$X_s(s) = \frac{C}{s} + \frac{D_1}{s - s_1} + \frac{D_2}{s - s_2} \tag{7.96}$$

Thus, an overdamped system can be treated as a linear combination of two first-order systems with time constants:

$$\tau_1 = \frac{1}{s_1} \quad \tau_2 = \frac{1}{s_2} \tag{7.97}$$

Example 7.18

A system has a second-order transfer function of the form

$$G(s) = \frac{2s + 3}{s^2 + 4s + 100} \tag{a}$$

a. Determine the final value of the system's step response. **b.** Determine the step response of the system. **c.** Approximate the system's 2 percent settling time. **d.** Find the percent overshoot of the step response. **e.** Use MATLAB to determine the plot of the step response and the parameters. Compare the parameters with the answers in parts c. and d.

Solution

The parameters of the second-order system are determined from Equation (a):

$$\omega_n = \sqrt{100} = 10 \text{ r/s} \tag{b}$$

$$\zeta = \frac{4}{2\omega_n} = \frac{4}{2(10)} = 0.2 \tag{c}$$

The damped natural frequency is obtained from Equation (7.76) as

$$\omega_d = 10\sqrt{1 - 0.2^2} = 9.80 \text{ r/s} \tag{d}$$

a. The final value of the system's step response is determined from Equation (a) and using Equation (7.17) as

$$x_{s,f} = G(0) = 0.03 \tag{e}$$

b. The system is underdamped. The step response is obtained by using a linear combination of the second and sixth entries in Table 7.5:

$$\begin{aligned} x_s(t) &= 2\left(\frac{1}{9.80}e^{-0.2(10)t}\sin 9.80t\right) + 3\left[\frac{1}{10^2}\left(1 - e^{-0.2(10)t}\cos 9.80t - \frac{(0.2)(10)}{9.80}e^{-0.2(10)t}\sin 9.80t\right)\right] \\ &= 0.03 - 0.03e^{-2t}\cos 9.80t + 0.198e^{-2t}\sin 9.80t \end{aligned} \tag{f}$$

c. Since the numerator of $G(s)$ is of the form $As + B$, the 2 percent settling time cannot be approximated using Equation (7.94). Instead, set

$$1.02(0.03) = 0.03 - 0.03e^{-2t_s}\cos 9.80t_s + 0.198e^{-2t_s}\sin 9.80t_s \tag{g}$$

Equation (g) is solved numerically, leading to $t_s = 2.81$. Note that in this case, an approximation to the settling time can be calculated by neglecting the $0.03e^{-2t_s}\cos 9.80t_s$ term in comparison to the $0.198e^{-2t_s}\sin 9.80t_s$ term. Equation (g) can be solved analytically, resulting in $t_s = 2.90$.

d. The overshoot is determined by setting $\dot{x}_s = 0$, leading to

$$0 = 2.004e^{-2t_p}\cos 9.80t_p - 0.102e^{-2t_p}\sin 9.80t_p \tag{h}$$

Equation (h) is solved for the peak time of 0.155 s. Substituting the peak time into Equation (f) leads to a peak value of 0.174. The overshoot is

$$\eta = 100\left(\frac{0.174 - 0.03}{0.03}\right) = 483\%$$

e. The MATLAB code to plot and obtain information about the step response of this system is given in Figure 7.27(a). The MATLAB plot is shown in Figure 7.27(b), and

```
% Step response of Example 7.18
%
clear
% Numerator and denominator of transfer function
N=[2 3];
D=[1 4 100);
% Defining transfer function from numerator and
denominator
G=tf(N,D)
% Plotting step response of system
step(G)
% Obtaining information about step response
stepinfo(G)
% End
```

(a)

Figure 7.27

(a) MATLAB code to obtain and plot information about the step response for Example 7.18; (b) Step response; (c) Output from MATLAB code.

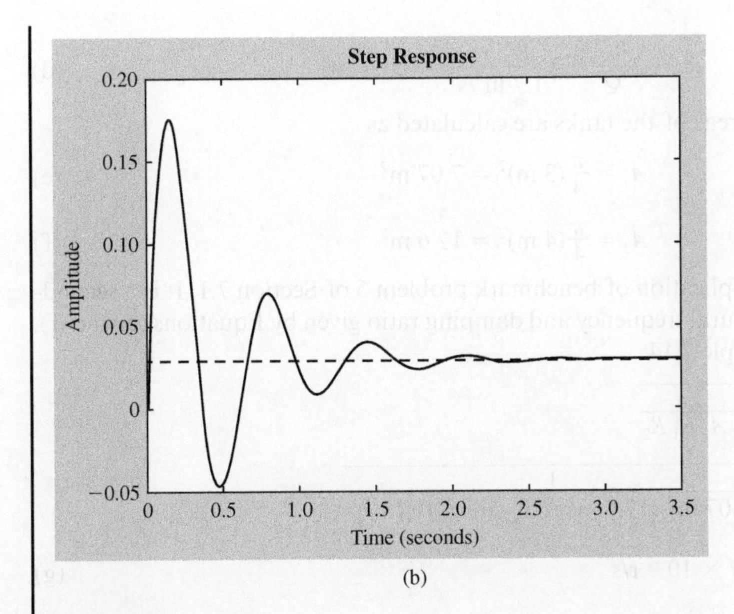

(b)

```
2s+3
---------------
s^2 + 4s + 100

Continuous-time transfer function.

ans=

struct with fields:

      RiseTime: 0.0125
 TransientTime: 2.1150
  SettlingTime: 2.8033
   SettlingMin: -0.0455
   SettlingMax: 0.1736
     Overshoot: 478.7237
    Undershoot: 151.8084
          Peak: 0.1736
      PeakTime: 0.1612
```

(c)

Figure 7.27

(Continued)

the output from the code is given in Figure 7.27(c). The settling time and the overshoot compare well with the calculated values. The undershoot is the percentage with respect to the final value that the step response drops below zero.

Example 7.19

The steady-state operation of a two-tank system is illustrated in Figure 7.28. Both tanks are circular in cross section. The input flow rate has a step increase of 0.2 m^3/s. Determine **a.** the steady-state liquid levels in each tank after the increase in flow rate, **b.** the time-dependent response of each tank, **c.** the 2 percent settling time, and **d.** the 10–90 percent rise time. **e.** Use MATLAB to plot the system response.

Solution

The application of Conservation of Mass to the tanks in the steady state shows that the flow rates through each of the pipes is equal to the input flow rate. Thus the head losses in the pipes at the original steady state are

$$h_{\ell 1} = H_2 - H_1 = 5.1 \text{ m} \tag{a}$$

$$h_{\ell 2} = H_2 = 8.5 \text{ m} \tag{b}$$

The resistances of the pipes at steady state are calculated using Equation (6.38) as

$$R_1 = 2\frac{h_{\ell 1}}{Q} = 2\frac{5.1 \text{ m}}{1.7 \text{ m}^3/\text{s}} = 6.0 \text{ s/m}^2 \tag{c}$$

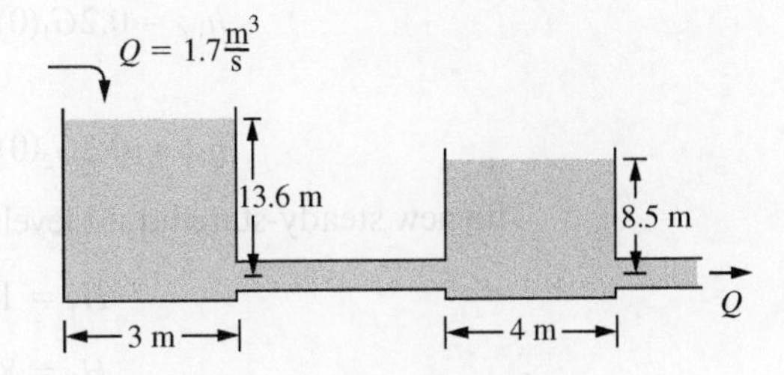

Figure 7.28

The steady-state operation of the two-tank liquid-level system of Example 7.19.

$$R_2 = 2\frac{h_{\ell 2}}{Q} = 2\frac{8.5 \text{ m}}{1.7 \text{ m}^3/\text{s}} = 10.0 \text{ s/m}^2 \tag{d}$$

The cross-sectional areas of the tanks are calculated as

$$A_1 = \frac{\pi}{4}(3 \text{ m})^2 = 7.07 \text{ m}^2 \tag{e}$$

$$A_2 = \frac{\pi}{4}(4 \text{ m})^2 = 12.6 \text{ m}^2 \tag{f}$$

This problem is an application of benchmark problem 5 of Section 7.1. It is a second-order system with natural frequency and damping ratio given by Equations (e) and (f), respectively, of Example 7.14:

$$\omega_n = \sqrt{\frac{1}{A_1 A_2 R_1 R_2}}$$

$$= \sqrt{\frac{1}{(7.07 \text{ m}^2)(12.6 \text{ m}^2)(6.0 \text{ s/m}^2)(10 \text{ s/m}^2)}}$$

$$= 1.37 \times 10^{-2} \text{ r/s} \tag{g}$$

$$\zeta = \frac{1}{2\omega_n}\sqrt{\frac{1}{A_2 R_1} + \frac{1}{A_2 R_2} + \frac{1}{A_1 R_1}}$$

$$= \frac{1}{2(1.37 \times 10^{-2} \text{ r/s})}$$

$$\times \sqrt{\frac{1}{12.6 \text{ m}^2(6.0 \text{ s/m}^2)} + \frac{1}{12.6 \text{ m}^2(10 \text{ s/m}^2)} + \frac{1}{7.07 \text{ m}^2(6.0 \text{ s/m}^2)}}$$

$$= 7.75 \tag{h}$$

The denominator of the transfer function is

$$D(s) = s^2 + 2(7.75)(0.0137)s + (0.0137)^2$$

$$= s^2 + 0.212s + 1.88 \times 10^{-4} \tag{i}$$

The transfer functions are obtained using Equation (7.11) with Equation (i) used for $D(s)$. Only the first column of the matrix in Equation (7.11) is used, as this system only has input into the first tank.

$$G_1(s) = \frac{1}{D(s)}\frac{1}{A_1 A_2}\left(A_2 s + \frac{1}{R_1} + \frac{1}{R_2}\right) = \frac{\frac{1}{(12.6)(7.07)}\left(12.6s + \frac{1}{6.0} + \frac{1}{10.0}\right)}{s^2 + 0.212s + 1.88 \times 10^{-4}}$$

$$= \frac{(0.141s + 0.003)}{s^2 + 0.212s + 1.88 \times 10^{-4}} \tag{j}$$

and

$$G_2(s) = \frac{\frac{1}{R_1}}{A_1 A_2 D(s)} = \frac{\left(\frac{1}{6}\right)\frac{1}{(12.6)(7.07)}}{s^2 + 0.212s + 1.88 \times 10^{-4}} = \frac{0.00187}{s^2 + 0.212s + 1.88 \times 10^{-4}} \tag{k}$$

a. The input to the system is $q_1(t) = 0.2u(t)$. The final values are

$$h_{1,f} = 0.2G_1(0) = (0.2)\left(\frac{0.003}{1.88 \times 10^{-4}}\right) = 3.18 \text{ m} \tag{l}$$

$$h_{2,f} = 0.2G_2(0) = (0.2)\left(\frac{0.00187}{1.88 \times 10^{-4}}\right) = 2.0 \text{ m} \tag{m}$$

The new steady-state liquid levels are

$$H_1 = 13.6 \text{ m} + 3.18 \text{ m} = 16.78 \text{ m} \tag{n}$$

$$H_2 = 8.5 \text{ m} + 2.0 \text{ m} = 10.50 \text{ m} \tag{o}$$

b. Since the system is overdamped, the poles of the transfer function are

$$s_1 = -\omega_n\left(\zeta + \sqrt{\zeta^2 - 1}\right)$$
$$= -1.37 \times 10^{-2}\left(7.75 + \sqrt{(7.75)^2 - 1}\right) = -0.211 \quad \text{(p)}$$

$$s_2 = -\omega_n\left(\zeta - \sqrt{\zeta^2 - 1}\right)$$
$$= -1.37 \times 10^{-2}\left(7.75 - \sqrt{(7.75)^2 - 1}\right) = -8.88 \times 10^{-4} \quad \text{(q)}$$

The responses of the system are 0.2 times the step response obtained from the transfer functions, Equations (j) and (k), and Table 7.5. These are

$$h_1(t) = 0.2\Bigg[\frac{0.141}{-8.88 \times 10^{-4} - (-0.211)}\left(e^{-8.88\times10^{-4}t} - e^{-0.211t}\right)$$
$$+ \frac{0.003}{(-0.211)(-8.88 \times 10^{-4})(-0.211 - (-8.88 \times 10^{-4}))}$$
$$\left(-0.211 - (-8.88 \times 10^{-4}) - 8.88 \times 10^{-4}e^{-0.211t} + 0.211\,e^{-8.88\times10^{-4}t}\right)\Bigg]$$
$$= 3.18u(t) - 0.0622\,e^{-0.211t} - 3.13\,e^{-8.88\times10^{-4}t} \quad \text{(r)}$$

$$h_2(t) = 0.2(0.00187)\Bigg[\frac{0.003}{(-0.211)(-8.88 \times 10^{-4})(-0.211 - (-8.88 \times 10^{-4}))}$$
$$\left(-0.211 - (-8.88 \times 10^{-4}) - 8.88 \times 10^{-4}e^{-0.211t} + 0.211e^{-8.88\times10^{-4}t}\right)\Bigg]$$
$$= 1.99u(t) + 0.00844e^{-0.211t} - 2.00\,e^{-8.88\times10^{-4}t} \quad \text{(s)}$$

c. The 2 percent settling time is the time that it takes for both tanks to reach 98 percent of their new steady-state values:

$$0.98(3.18) = 3.18 - 0.0622\,e^{-0.211t_s} - 3.13\,e^{-8.91\times10^{-4}t_s} \quad \text{(t)}$$

The solution of Equation (t) is $t_s = 4.37 \times 10^3$ s.

d. In a manner similar to the determination of the 2 percent settling time, the time required to rise to 10 percent of the system's final value is 100 s and the time required to rise to 90 percent of the system's final value is 2570 s. Thus the rise time is $t_r = 2570\text{ s} - 100\text{ s} = 2470\text{ s}$.

e. MATLAB is used to develop plots for the time-dependent responses of the liquid level perturbations, as shown in Figure 7.29.

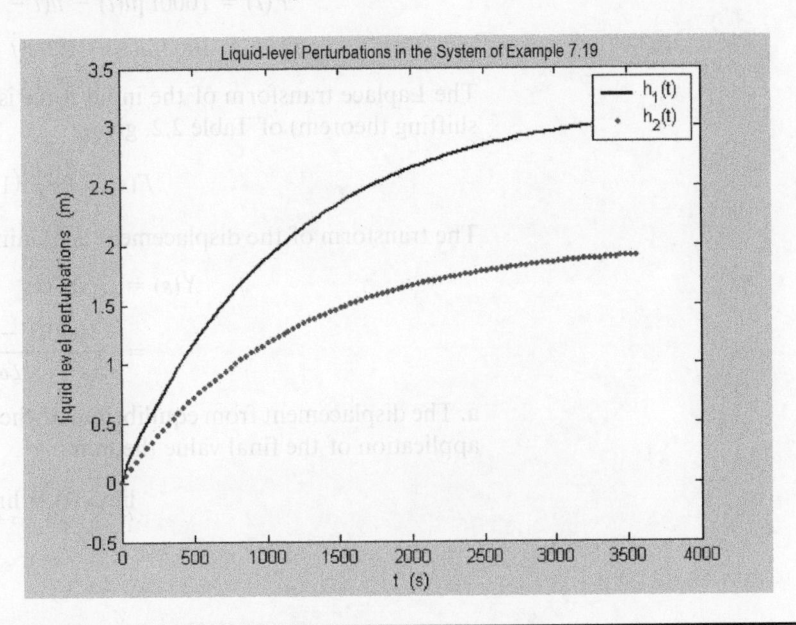

Figure 7.29

Liquid-level perturbations for both tanks in the two-tank liquid-level problem of Example 7.19.

7.5.4 General Transient Response

The transient response of a second-order system due to any input can be determined using either the Laplace transform method or the convolution integral solution, which is derived using the Laplace transform method.

Example 7.20

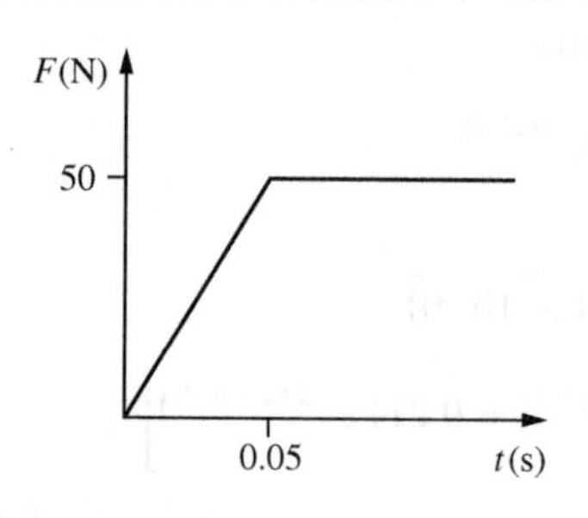

Figure 7.30

The input force for Example 7.20.

The time-dependent force applied to a one-degree-of-freedom mass, spring, and viscous-damper system is illustrated in Figure 7.30. The system parameters are $m = 1.0$ kg, $k = 1 \times 10^4$ N/m, and $c = 600$ N·s/m. Determine **a.** the steady-state displacement of the system from its equilibrium position and **b.** the time-dependent response of the system by inversion of the Laplace transform. **c.** Set up the convolution integral that can be used as an alternative to find the solution.

Solution

The transfer function for the system is that of benchmark problem 2 of Section 7.1.3

$$G(s) = \frac{1}{m(s^2 + 2\zeta\omega_n s + \omega_n^2)} \tag{a}$$

where the natural frequency and damping ratio are determined using Table 7.3 as

$$\omega_n = \sqrt{\frac{k}{m}}$$

$$= \sqrt{\frac{1 \times 10^4 \text{ N/m}}{1 \text{ kg}}} = 100 \text{ r/s} \tag{b}$$

$$\zeta = \frac{c}{2m\omega_n}$$

$$= \frac{600 \text{ N·s/m}}{2(10 \text{ kg})(100 \text{ r/s})} = 0.3 \tag{c}$$

The system is underdamped and its damped natural frequency is calculated using Equation (7.76) as

$$\omega_d = (100 \text{ r/s})\sqrt{1 - (0.3)^2} = 95.4 \text{ r/s} \tag{d}$$

The input force is written using unit-step functions as

$$F(t) = 1000t\,[u(t) - u(t - 0.05)] + 50u(t - 0.05)$$

$$= 1000tu(t) - 1000(t - 0.05)u(t - 0.05) \tag{e}$$

The Laplace transform of the input force is determined using Property 6 (the second shifting theorem) of Table 2.2, giving

$$F(s) = \frac{1000}{s^2}(1 - e^{-0.05s}) \tag{f}$$

The transform of the displacement is obtained using Equations (a) and (f) as

$$X(s) = G(s)F(s)$$

$$= \frac{1000(1 - e^{-0.05s})}{s^2(s^2 + 2\zeta\omega_n s + \omega_n^2)} \tag{g}$$

a. The displacement from equilibrium of the mass in the steady state is obtained by the application of the final value theorem:

$$\lim_{t\to\infty} x(t) = \lim_{s\to 0} sX(s)$$

$$= \lim_{s\to 0} \frac{1000(1 - e^{-0.05s})}{s(s^2 + 2\zeta\omega_n s + \omega_n^2)}$$

$$= \lim_{s\to 0} \frac{1000(1 - e^{-0.05s})}{s\omega_n^2} \tag{h}$$

The limit in Equation (h) is indeterminate but can be evaluated using L'Hospital's rule:

$$\lim_{t\to\infty} x(t) = \frac{1000(0.05)}{\omega_n^2} = 5 \text{ mm} \tag{i}$$

b. A partial fraction decomposition of Equation (g) leads to

$$X(s) = \frac{1000}{\omega_n^4}\left(\frac{\omega_n^2}{s^2} - \frac{2\zeta\omega_n}{s} + \frac{2\zeta\omega_n s}{s^2 + 2\zeta\omega_n s + \omega_n^2}\right)(1 - e^{-0.05s}) \tag{j}$$

Inversion of the transform in Equation (j) leads to

$$\begin{aligned} x(t) = \frac{1000}{\omega_n^4}\Bigg\{\Bigg[&\omega_n^2 t - 2\zeta\omega_n + 2\zeta\omega_n e^{-\zeta\omega_n t}\cos(\omega_d t) \\ &- 2\frac{\zeta^2\omega_n^2}{\omega_d}e^{-\zeta\omega_n t}\sin(\omega_d t)\Bigg]u(t) - \Bigg[\omega_n^2(t - 0.05) - 2\zeta\omega_n \\ &+ 2\zeta\omega_n e^{-\zeta\omega_n(t-0.05)}\cos[(\omega_d(t - 0.05)] \\ &- 2\frac{\zeta^2\omega_n^2}{\omega_d}e^{-\zeta\omega_n(t-0.05)}\sin[\omega_d(t - 0.05)]\Bigg]u(t - 0.05)\Bigg\} \end{aligned} \tag{k}$$

Substitution of numerical values leads to

$$\begin{aligned} x(t) = \{&[0.1t - 6\times 10^{-4} + 6\times 10^{-4}e^{-30t}\cos(95.4t) - 1.89\times 10^{-4}e^{-30t}\sin(95.4t)]\} \\ &-[0.1(t - 0.05) - 6\times 10^{-4} + 6\times 10^{-4}e^{-30(t-0.05)}\cos[95.4(t - 0.05)] \\ &-1.89\times 10^{-4}e^{-30(t-0.05)}\sin[95.4(t - 0.05)]]u(t - 0.05) \end{aligned} \tag{l}$$

The script of a MATLAB program, Example7_20.m, which symbolically determines the system response from Equation (g) and then plots the response, is shown in Figure 7.31(a). The symbolic form of the response obtained from the execution of Example7_20.m, shown in Figure 7.31(b), agrees with Equation (l). The response is plotted in Figure 7.31(c).

c. An alternate method to the Laplace transform method is to apply the convolution integral. The impulsive response for this underdamped system is determined using Table 7.4 as

$$x_i(t) = \frac{1}{95.4}e^{-30t}\sin 95.4t \tag{m}$$

The convolution integral is used to determine the response of the system as

$$x(t) = \frac{1}{95.4}\int_0^t F(\tau)e^{-30(t-\tau)}\sin[95.4(t - \tau)]d\tau \tag{n}$$

Equation (n) is written for $t < 0.05$ s as

$$x(t) = \frac{1}{95.4}\int_0^t 1000\tau e^{-30(t-\tau)}\sin[95.4(t - \tau)]d\tau \tag{o}$$

and for $t > 0.05$ s as

$$x(t) = \frac{1}{95.4}\int_0^{0.05} 1000\tau e^{-30(t-\tau)}\sin[95.4(t-\tau)]d\tau + \frac{1}{95.4}\int_{0.05}^{t} 1000 e^{-30(t-\tau)}\sin[95.4(t-\tau)]d\tau \tag{p}$$

```
% Example 7.20
%
% Declaring symbolic variables
%
syms s t
%
% Defining parameters
%
zeta=0.3;
wn=100;
%
% Defining transform, Equation (g) of Example 7.20
%
X=1000*(1-exp(-0.05*s))/(s^2*(s^2+2*zeta*wn*s+wn^2));
%
% Taking inverse Laplace transform
%
x=ilaplace(X);
%
% Simplifying symbolic form of response
%
x1=simplify(x);
x2=vpa(x1,3)
%
% Plotting response
%
t=linspace(0,0.2,501);
x3=subs (x2, t);
plot(t,x3)
%
% Annotating plot
%
xlabel('t (s)')
ylabel('x (m)')
title('Response of System of Example 7.20')
%
% End of Example 7.20
```

(a)

```
x2 =

0.1*t+6.0e-4*exp(-30.0*t)*(cos(95.4*t)-1.43*sin(95.4*t))
-1000.0*heaviside(1.0*t-0.05)*(1.0e-4*t+6.0e-7*exp(1.5
-30.0*t)*(cos(95.4*t-4.77)-1.43*sin(95.4*t-4.77))
-5.6e-6)-6.0e-4
```

(b)

Figure 7.31

(a) Script of the file Example7_20.m; (b) partial output from the execution of the file; (c) the plot of system response.

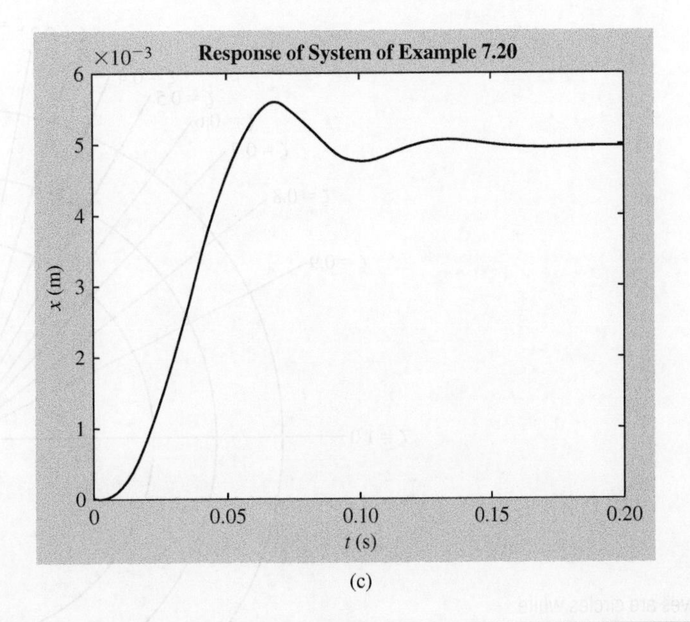

Figure 7.31
(Continued)

7.6 Higher-Order Systems

The transfer function, $G(s)$, of an nth-order system has n poles: s_i, $i = 1, 2, \ldots, n$. If k poles $s_1, s_2, \ldots, s_k$ are real, then $n - k$ poles are complex and occur in conjugate pairs. It is assumed that the system is stable or neutrally stable and thus all poles have non-positive real parts.

The time constant of a first-order system is the negative of the reciprocal of the pole of its transfer function. By analogy, the time constants for a higher-order system can be defined as the negative reciprocals of its real poles,

$$T_i = -\frac{1}{s_i} \quad i = 1, 2, \ldots, k \tag{7.98}$$

The complex conjugate poles of a second-order system are represented in terms of a natural frequency ω_n and a damping ratio ζ. By analogy, natural frequencies and damping ratios for higher-order systems are defined according to

$$s_\ell, \overline{s}_\ell = \omega_\ell\left(-\zeta_\ell \pm j\sqrt{1-\zeta_\ell^2}\right) \quad \ell = 1, 2, \ldots, \frac{1}{2}(n-k) \tag{7.99}$$

It is noted that $(s - s_\ell)(s - \overline{s}_\ell) = s^2 + 2\zeta_\ell\omega_\ell s + \omega_\ell^2(1 - \zeta_\ell^2)$.

The general form of a transfer function for an nth-order system is

$$G(s) = \frac{N(s)}{\left(s+\frac{1}{T_1}\right)\cdots\left(s+\frac{1}{T_k}\right)\left[s^2+2\zeta_1\omega_1 s+\omega_1^2(1-\zeta_1^2)\right]\cdots\left[s^2+2\zeta_p\omega_p s+\omega_p^2\left(1-\zeta_p^2\right)\right]} \tag{7.100}$$

where $p = \frac{1}{2}(n - k)$. A partial fraction decomposition of Equation (7.100) is of the form

$$G(s) = \sum_{i=1}^{k}\frac{A_i}{s-\frac{1}{T_i}} + \sum_{\ell=1}^{p}\frac{B_\ell s + C_\ell}{s^2 + 2\zeta_\ell\omega_\ell s + \omega_\ell^2(1-\zeta_\ell^2)} \tag{7.101}$$

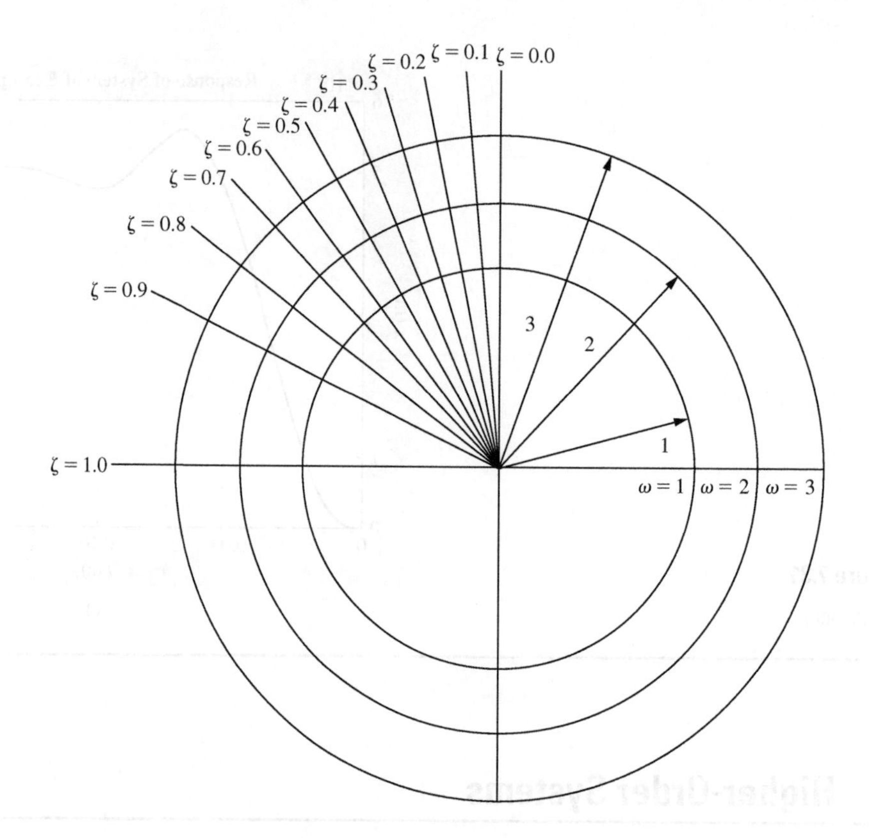

Figure 7.32

Constant ω curves are circles while constant ζ curves for $0 \leq \zeta \leq 1$ are lines in the second quadrant of the imaginary plane. The distance from the origin to a point on these curves is ω.

For a system input $F(s)$, the system output is

$$X(s) = \sum_{i=1}^{k} \frac{A_i F(s)}{s - \frac{1}{T_i}} + \sum_{\ell=1}^{p} \frac{(B_\ell s + C_\ell)F(s)}{s^2 + 2\zeta_\ell \omega_\ell s + \omega_\ell^2(1 - \zeta_\ell^2)} \tag{7.102}$$

Equation (7.102) shows that the response of a higher-order system whose initial conditions are all zero is the sum of responses of first- and second-order systems.

An nth-order stable system has poles such that $\text{Re}(s_1) \geq \text{Re}(s_2) \geq \cdots \geq \text{Re}(s_n)$. If s_1 is real, then it is the dominant pole because $e^{s_1 t}$ decays slower than the responses due to the other poles. The time constant corresponding to the dominant pole is $T_1 = \frac{1}{|s_1|}$.

A pair of complex poles, s_k and s_{k+1}, have a real part of $-\zeta_k \omega_k$ and their imaginary parts are $\pm \omega_k \sqrt{1 - \zeta_k^2}$. Plotted in the complex s plane, $\zeta_k = 0$ corresponds to the imaginary axis and $\zeta_k = 1$ corresponds to the negative real axis. The point $s_k \Rightarrow \left(-\zeta_k \omega_k, \omega_k \sqrt{1 - \zeta_k^2}\right)$ is in the third quadrant of the plane, while $s_{k+1} \Rightarrow \left(-\zeta_k \omega_k, -\omega_k \sqrt{1 - \zeta_k^2}\right)$ lies in the fourth quadrant. The distance from this point to the origin (0, 0) is $\sqrt{(\zeta_k \omega_k)^2 + \omega_k^2(1 - \zeta_k^2)} = \omega_k$. For a fixed ω_k, s_k is a function of ζ_k, where $0 < \zeta_k < 1$. Then the envelope of the plots of s_k is a semicircle, as illustrated in Figure 7.32. The distance from the origin in the s plane is important for understanding the system response. The smallest distance corresponds to the pole with the smallest ω_k.

Example 7.21

Determine the step response of the system of Figure 7.33(a). For the system of Figure 7.33(a), **a.** determine the impulsive response by inverting the transform and **b.** determine the plot of the step response using MATLAB.

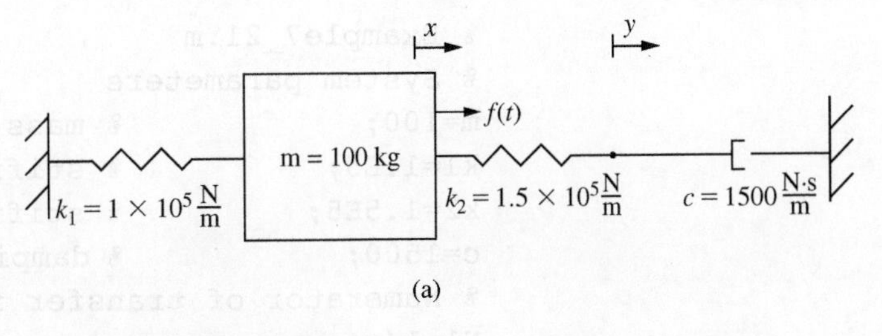

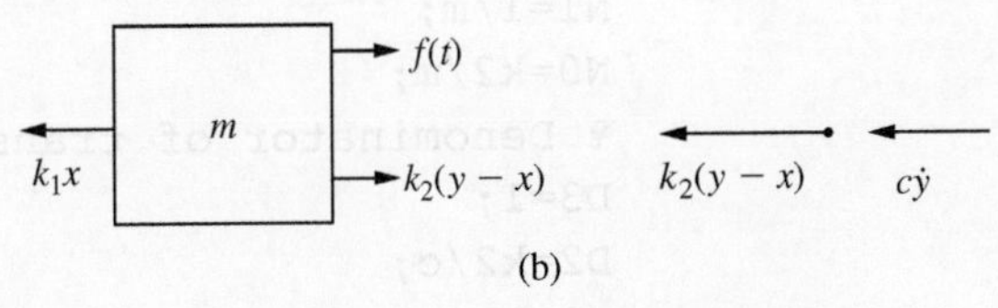

Figure 7.33

(a) The system of Example 7.21; (b) the free-body diagrams of the block and the joint between the viscous damper and the spring of stiffness k_2 at an arbitrary instant.

Solution

The application of Newton's law to the free-body diagrams of Figure 7.33(b) leads to

$$m\ddot{x} + (k_1 + k_2)x - k_2 y = f(t) \tag{a}$$

$$-k_2 x + k_2 y + c\dot{y} = 0 \tag{b}$$

Taking the Laplace transforms of Equations (a) and (b) leads to

$$(ms^2 + k_1 + k_2)X(s) - k_2 Y(s) = F(s) \tag{c}$$

$$-k_2 X(s) + (cs + k_2)Y(s) = 0 \tag{d}$$

Simultaneous solution of Equations (c) and (d) leads to

$$X(s) = \frac{(cs + k_2)F(s)}{mc\left(s^3 + \frac{k_2}{c}s^2 + \frac{k_1 + k_2}{m}s + \frac{k_1 k_2}{mc}\right)} \tag{e}$$

Substitution of given values leads to the transfer function

$$G(s) = \frac{X(s)}{F(s)} = \frac{0.01s + 1500}{s^3 + 100s^2 + 2500s + 100{,}000} \tag{f}$$

a. The MATLAB program used to calculate the poles and residues of the transfer function and plot the step response is given in Figure 7.34(a). The output for the poles and residues is given in Figure 7.34(b). They lead to the partial fraction decomposition of the transfer function

$$G(s) = \frac{2.143}{s + 84.18} + \frac{-0.107 - 0.234j}{s + 7.79 - 33.5j} + \frac{-0.107 + 0.234j}{s + 7.79 + 33.5j}$$

$$= \frac{2.143}{s + 84.18} + \frac{-0.107(s + 7.79) + 15.68}{(s + 7.79)^2 + (33.5)^2} \tag{g}$$

The inverse transform of the transfer function is the impulsive response of

$$x_i(t) = 2.143e^{-84.18t} + e^{-7.79t}(-0.107 \cos 33.5t + 0.465 \sin 33.5t) \tag{h}$$

b. The MATLAB plot of the step response is given in Figure 7.34(c).

```
% Example7_21.m
% System parameters
m=100;                    % mass in kg
k1=1.E5;                  % stiffness in N/m
k2=1.5E5;                 % stiffness in N/m
c=1500;                   % damping coefficient in N-s/m
% Numerator of transfer function
N1=1/m;
N0=k2/m;
% Denominator of transfer function
D3=1;
D2=k2/c;
D1=(k1+k2)/m;
D0=k1*k2/(m*c);
% Transfer function
N=[N1 N0];
D=[D3 D2 D1 D0];
% Poles of transfer fucntion
[r,p,k]=residue(N,D)
% Step response
step(N,D)
title('Step response for System of Example 7.21')
ylabel('x (m)')
xlabel('t')
%
% End of Example7_21.m
```

(a)

```
r =

   0.2143 + 0.0000i
  -0.1072 - 0.2451i
  -0.1072 + 0.2451i


p =

 -84.4178 + 0.0000i
  -7.7911 +33.5244i
  -7.7911 -33.5244i


k =

     []
```

(b)

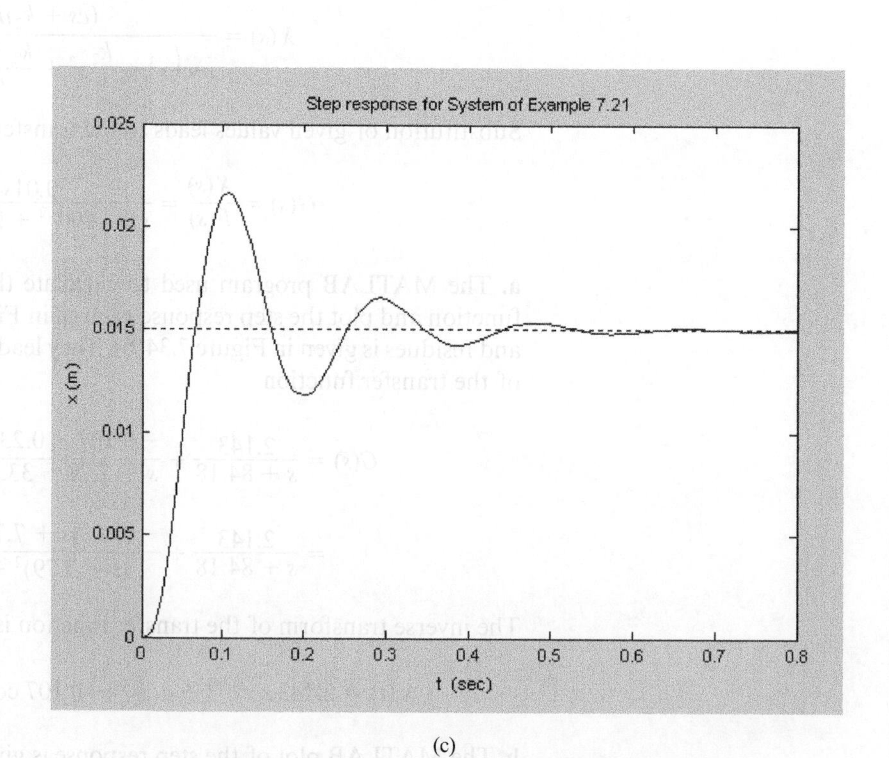

(c)

Figure 7.34

(a) Script of Example7_21.m, which determines the poles of the transfer function for Example 7.21 and plots the step response; (b) the output from the execution of Example7_21.m; (c) the step response plotted using Example7_21.m.

Example 7.22

Consider a cascade of n CSTRs, illustrated in Figure 7.35, in which the reactant A is fed into the first tank. Reactant A is consumed in the ith reactor of volume V_i at a rate k_i. The tanks all have constant volume and all reactions are isothermal. **a.** Determine the transient behavior of the reactant in each reactor when the concentration in the feed stream suddenly changes by ΔC_{Ai}. **b.** Determine the final concentration of reactant A in each tank.

Solution

a. The molecular balance equation for component A in the first reactor is

$$V_1 \frac{dC_{A1}}{dt} + (q + k_1 V_1) = q\Delta C_{Ai} \tag{a}$$

The molecular balance equation for component A in reactor p for $p = 2, 3, \ldots$ is

$$V_p \frac{dC_{Ap}}{dt} - qC_{Ap-1} + (q + k_p V_p)C_{Ap} = 0 \tag{b}$$

where C_{Ap} is the perturbation in the concentration of component A from the steady state due to the perturbation in the inlet stream.

Taking the Laplace transform of Equations (a) and (b) and defining $\bar{C}_{Ap}$ as the Laplace transfer of C_{Ap} leads to

$$(V_1 s + q + k_1 V_1)\bar{C}_{A1} = q\Delta C_{Ai} \tag{c}$$

$$(V_p s + q + k_p V_p)\bar{C}_{Ap} - q\bar{C}_{Ap-1} = 0 \qquad p = 2, 3, \ldots \tag{d}$$

The first transfer function is determined using Equation (c) as

$$G_1(s) = \frac{q}{V_1\left(s + \frac{q}{V_1} + k_1\right)} \tag{e}$$

Equation (d) is rearranged to

$$G_p(s) = \frac{q}{V_p\left(s + \frac{q}{V_p} + k_p\right)} G_{p-1}(s) \tag{f}$$

Mathematical induction, using Equations (e) and (f), is used to show

$$G_p(s) = \frac{q^p}{V_1 V_2 \ldots V_p \left(s + \frac{q}{V_1} + k_1\right)\left(s + \frac{q}{V_2} + k_2\right)\ldots\left(s + \frac{q}{V_p} + k_p\right)} \tag{g}$$

Thus the transfer function $G_p(s)$ is p^{th} order and has p real poles

$$s_k = -\left(\frac{q}{V_k} + k_k\right) \quad k = 1, 2, \ldots, p \tag{h}$$

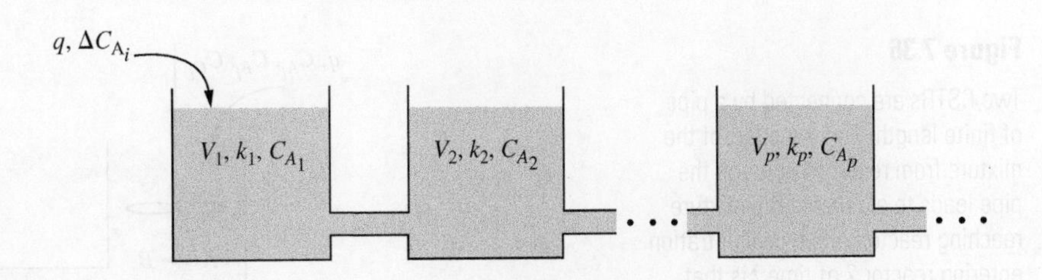

Figure 7.35

The cascade of CSTRs for Example 7.22.

The time constants for the reactor are

$$T_k = \frac{V_k}{q + k_k V_k} \quad k = 1, 2, \ldots, p \tag{i}$$

A partial fraction decomposition of Equation (g) leads to

$$G_p(s) = \sum_{k=1}^{p} \frac{A_{k,p}}{s + \frac{1}{T_k}} \tag{j}$$

where

$$A_{k,p} = \frac{q^p}{\left(\prod_{i=1}^{p} V_i\right)\left[\prod_{\substack{i=1 \\ i \neq k}}^{p}\left(\frac{q}{V_p} + k_p - \frac{q}{V_k} - k_k\right)\right]} \tag{k}$$

The response due to a step change in inlet concentration, $F(s) = \Delta C_{Ai}/s$, is

$$C_p(t) = \Delta C_{Ai} \sum_{k=1}^{p} A_{k,p}\left(1 - e^{-\frac{t}{T_k}}\right) \tag{l}$$

b. The final value theorem can be applied to Equation (g) to determine the final value for the perturbation in the concentration of component A in each reactor:

$$C_{p,f} = \Delta C_{Ai} \lim_{s \to 0} G(s)$$

$$= \Delta C_{Ai} \frac{q^p}{\prod_{k=1}^{p}(q + k_k V_k)} \tag{m}$$

7.7 Systems with Time Delay

A dynamic system experiences a **time delay** when its response at time t is explicitly affected by the system's response at a previous time, $t - \tau$ for a fixed value of τ. If $x(t)$ is a dependent variable in a system with a time delay of τ, then the mathematical model for the system may include a term of the form $x(t - \tau)$. It is necessary to employ a modified version of the second shifting theorem to determine the transfer function for the system; if $x(t) = 0$ for $t < 0$, then

$$\mathscr{L}\{x(t - \tau)\} = e^{-\tau s} X(s) \tag{7.103}$$

The pipe connecting the two CSTRs of Figure 7.36 is of finite length L. If the average velocity of the flow through the pipe is v, then the time required for fluid particles to travel between the two reactors is $\tau = L/v$. The properties of the mixture entering the second reactor at time t are the properties of the mixture leaving the first reactor at time $t - \tau$.

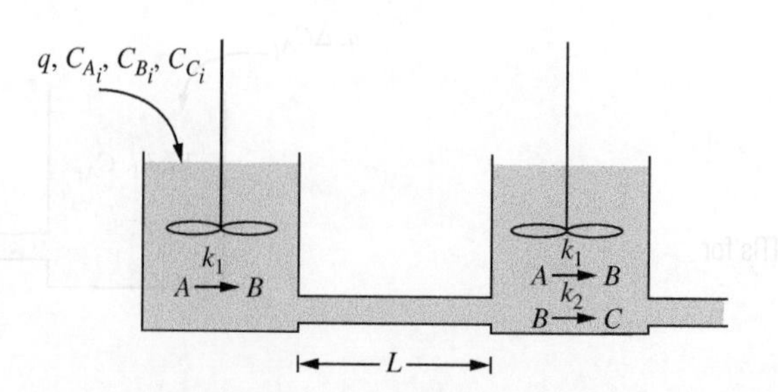

Figure 7.36

Two CSTRs are connected by a pipe of finite length. The transport of the mixture from reactor 1 through the pipe leads to a delay in the mixture reaching reactor 2. The concentration entering reactor 2 at time t is that leaving reactor 1 at time $t - L/v$.

Example 7.23

The double CSTR system of Figure 7.36 is operating at steady state when the concentration of component A in the inlet stream is suddenly changed by C_{Aip}. The flow rate and concentrations of all other components in the inlet stream are unchanged. All reactions are first order, the system is isothermal, and the volume of the mixtures in each reactor is constant. Determine the response of the perturbation of concentration for component A in each reactor if $V_1 = 1.25 \times 10^{-3}$ m^3, $V_2 = 2.08 \times 10^{-3}$ m^3, $q = 2.4 \times 10^{-6}$ m^3/s, $k_1 = 2.05 \times 10^{-3}$ s^{-1}, $\tau = 10.0$ s, and the inlet concentration of component A has a step increase of 0.2 mol/L at $t = 0$.

Solution

Conservation equations are applied to each component in each tank. Each concentration is written as the sum of the steady-state concentration and a perturbation in concentration. The steady state is subtracted, leading to the mathematical model for the perturbations in concentration:

$$V_1 \frac{dC_{A1p}}{dt} + (q + k_1 V_1) C_{A1p}(t) = qC_{Aip} \tag{a}$$

$$V_1 \frac{dC_{B1p}}{dt} - k_1 V_1 C_{A1p}(t) + q\, C_{B1p}(t) = 0 \tag{b}$$

$$V_1 \frac{dC_{C1p}}{dt} + q\, C_{C1p}(t) = 0 \tag{c}$$

$$V_2 \frac{dC_{A2p}}{dt} - q\, C_{A1p}(t-\tau) + (q + k_1 V_2) C_{A2p}(t) = 0 \tag{d}$$

$$V_2 \frac{dC_{B2p}}{dt} - q\, C_{B1p}(t-\tau) - k_1 V_2 C_{A2p}(t) + (q + k_2 V_2) C_{B2p}(t) = 0 \tag{e}$$

$$V_2 \frac{dC_{C2p}}{dt} - k_2 V_2 C_{Bp2}(t) + q\, C_{C2p}(t) = 0 \tag{f}$$

The transfer functions corresponding to C_{A1p} and C_{A2p} are given by Equations (g) and (j) of Example 6.35 and are repeated here:

$$G_{A1p}(s) = \frac{1}{s + \frac{q}{V_1} + k} \tag{g}$$

$$G_{A2p}(s) = \frac{\frac{q}{V_2} e^{-\tau s}}{\left(s + \frac{q}{V_1} + k_1\right)\left(s + \frac{q}{V_2} + k_1\right)} \tag{h}$$

Noting that $C_{Aip}(s) = 0.2/s$, the substitution of numerical values into Equations (g) and (h) leads to

$$C_{A1p}(s) = \frac{3.84 \times 10^{-4}}{s(s + 3.97 \times 10^{-3})} \tag{i}$$

$$C_{A2p}(s) = \frac{8.51 \times 10^{-10} e^{-10.0s}}{s(s + 3.97 \times 10^{-3})(s + 3.20 \times 10^{-3})} \tag{j}$$

Partial fraction decomposition and inversion of the transforms leads to

$$C_{A1p}(t) = 9.67 \times 10^{-2}(1 - e^{-3.97\times10^{-3}t})u(t) \tag{k}$$

$$C_{A2p}(t) = \big(6.67 \times 10^{-7} + 2.79 \times 10^{-4} e^{3.97\times10^{-3}(t-10.0)} - 5.32 \times 10^{-3} e^{3.20\times10^{-3}(t-10.0)}\big)u(t - 10.0) \tag{l}$$

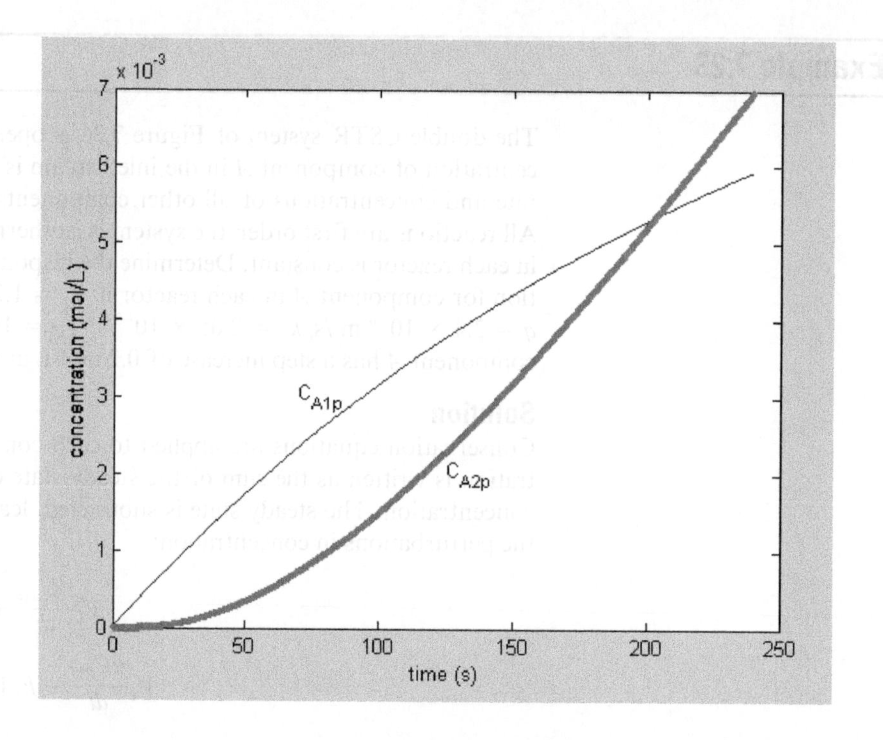

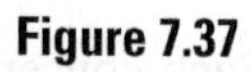
Figure 7.37
The perturbation concentrations of component A in each reactor of Example 7.23. The perturbation concentration in reactor B is zero until $t = 10$ s due to the dead time, during which the mixture from reactor A is transported to reactor B.

The concentration perturbations are plotted in Figure 7.37. The perturbation in the second reactor is zero until $t = 0.5$ s. The time between the perturbation in the inlet stream and the response of the second reactor is called dead time.

Systems with a delay time often have the exponential appear in the denominator of the transfer function. For example, consider the machine tool of Example 4.27 whose transfer function is

$$G(s) = \frac{X(s)}{F(s)} = \frac{1}{ms^2 + (c + c_w\tau)s + k + k_w - k_w e^{-\tau s}} \tag{7.104}$$

The poles of Equation (7.104) are given by setting

$$ms^2 + (c + c_w\tau)s + k + k_w - k_w e^{-\tau s} = 0 \tag{7.105}$$

One common method of approximating this denominator is to use a Pade approximation for $e^{-\tau s}$. A Pade approximation of order n for a continuous function is a ratio of polynomials, each of order n, such that the McLaurin series for the function is equal to the McLaurin series of the Pade approximation through the first $2n$ terms. To this end, the Pade approximations of $e^{-\tau s}$ are

$$e^{-\tau s} = \frac{e^{-\frac{\tau}{2}s}}{e^{\frac{\tau}{2}s}} = \frac{1 - \frac{\tau}{2}s}{1 + \frac{\tau}{2}s} \qquad n = 1 \tag{7.106}$$

$$e^{-\tau s} = \frac{1 - \frac{\tau}{2}s + \frac{\tau^2}{12}s^2}{1 + \frac{\tau}{2}s + \frac{\tau^2}{12}s^2} \qquad n = 2 \tag{7.107}$$

$$e^{-\tau s} = \frac{1 - \frac{\tau}{2}s + \frac{\tau^2}{10}s^2 - \frac{\tau^3}{120}s^3}{1 + \frac{\tau}{2}s + \frac{\tau^2}{10}s^2 + \frac{\tau^3}{120}s^3} \qquad n = 3 \tag{7.108}$$

The Pade approximation is substituted in place of $e^{-\tau s}$ in the transfer function. For the transfer function of Equation (7.104) when the Pade approximation for $n = 1$ is substituted,

$$G(s) = \frac{1}{ms^2 + (c + c_w\tau)s + k + k_w - k_w\frac{1 - \frac{\tau}{2}s}{1 + \frac{\tau}{2}s}} \tag{7.109}$$

Equation (7.109) is rearranged by multiplying the numerator and denominator by $\left(1 + \frac{\tau}{2}s\right)$, resulting in

$$G(s) = \frac{1 + \frac{\tau}{2}s}{m\frac{\tau}{2}s^3 + \left[m + (c + c_w\tau)\frac{\tau}{2}\right]s^2 + \left(c + c_w\tau + k\frac{\tau}{2}\right)s + k + k_w} \tag{7.110}$$

Equation (7.110) is an approximation to the transfer function of the system using a first-order Pade approximation. The approximate transfer function is third order with three poles. It also has a zero at $s = -\frac{2}{\tau}$. This approximation may be used to approximate the responses of the system. Higher-order Pade approximations provide increasingly better approximations, but the transfer function is of a higher order.

Example 7.24

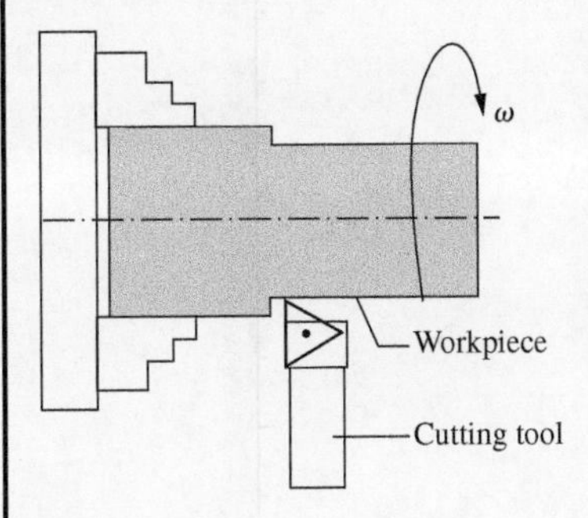

Figure 4.53

As the cutting tool penetrates the workpiece, which is rotating in the lathe at a constant speed ω, a cutting force is imparted to the tool from the workpiece. (Repeated)

The machine tool of Figure 4.53 and Example 4.27 has parameters $m = 1$ kg, $c = 100\,\frac{\text{N}\cdot\text{s}}{\text{m}}$, $c_w = 40\,\frac{\text{N}}{\text{m}}$, $k = 10{,}000\,\frac{\text{N}}{\text{m}}$, $k_w = 4000\,\frac{\text{N}}{\text{m}}$, and $\tau = 0.2\ \text{s}^{-1}$. Approximate the step response for the advancement of the tool using MATLAB and using a Pade approximation of order **a.** 1, **b.** 2, and **c.** 3.

Solution

Substituting the given parameters into Equation (7.104) leads to

$$G(s) = \frac{1}{s^2 + 120s + 14{,}000 - 4000e^{-0.2s}} \tag{a}$$

a. Using the first-order Pade approximation for $e^{-0.5s}$ in Equation (a) leads to

$$G(s) = \frac{1}{s^2 + 120s + 14{,}000 - 4000\dfrac{1 - 0.1s}{1 + 0.1s}} \tag{b}$$

Upon rearrangement, Equation (b) becomes

$$G(s) = \frac{1 + 0.1s}{0.1s^3 + 13s^2 + 1520s + 10{,}000} \tag{c}$$

The poles of the transfer function represented by Equation (c) are -0.0697, $-0.615 + 1.028j$, and $-0.615 - 1.028j$. The plot of the step response generated by MATLAB is shown in Figure 7.38(a), while the `stepinfo` results are given in Figure 7.39(a).

b. Substituting the Pade approximation for $n = 2$, Equation (7.107), into Equation (a) leads to

$$G(s) = \frac{1}{s^2 + 120s + 14{,}000 - 4000\left(\dfrac{1 - 0.1s + 0.00333s^2}{1 + 0.1s + 0.00333s^2}\right)} \tag{d}$$

Multiplying the numerator and denominator of Equation (d) by $1 + 0.1s + 0.00333s^2$ gives

$$G(s) = \frac{1 + 0.1s + 0.00333s^2}{0.00333s^4 + 0.5s^3 + 43.33s^2 + 1920s + 10{,}000} \tag{e}$$

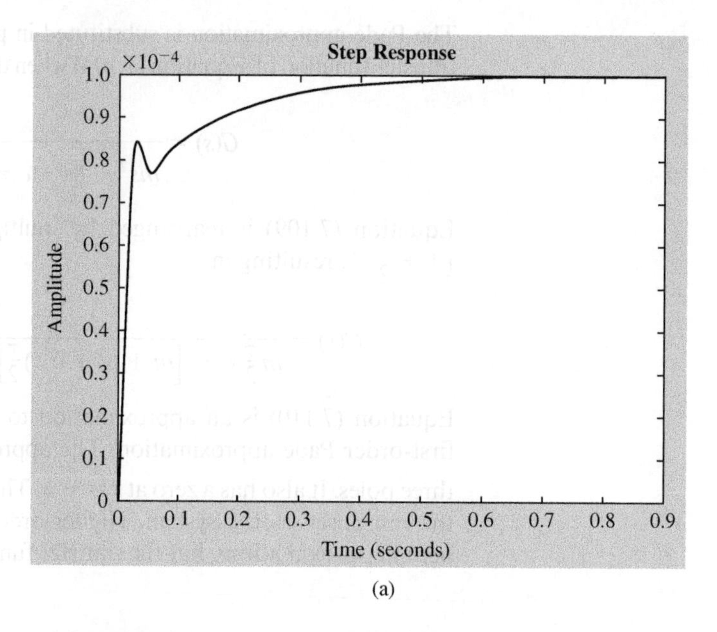

(a)

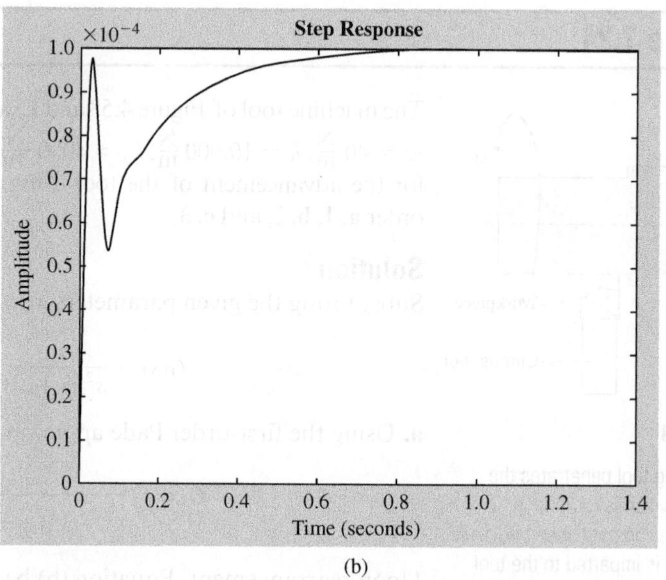

(b)

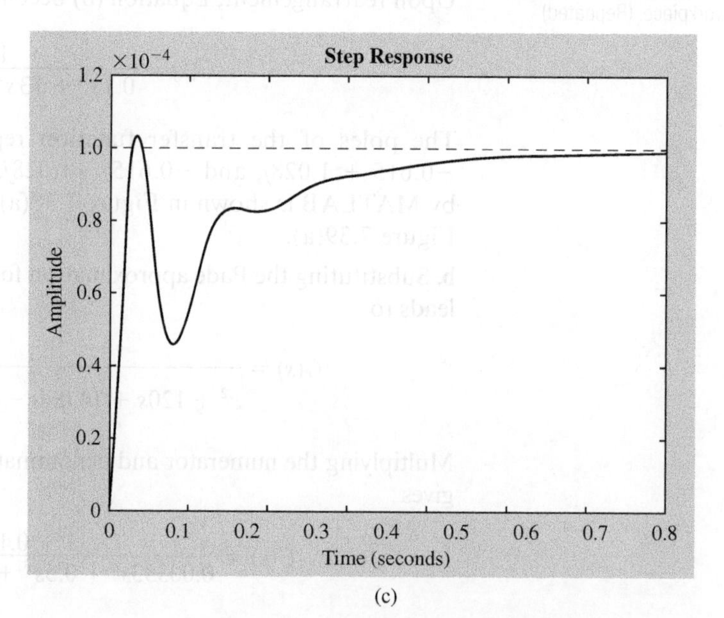

(c)

Figure 7.38

MATLAB generated plots of the step responses of Example 7.24 with a Pade approximation for the exponential term in the denominator of the transfer function. The actual transfer function is of infinte order, but a Pade approximation of order n gives a denominator of order $n + 2$. (a) $n = 1$; (b) $n = 2$; (c) $n = 3$.

```
  RiseTime: 0.1622
TransientTime: 0.3980
SettlingTime: 0.3980
SettlingMin: 9.0036e-05
SettlingMax: 9.9978e-05
 Overshoot: 0
 Undershoot: 0
     Peak: 9.9978e-05
  PeakTime: 1.0466
```
(a)

```
  RiseTime: 0.0193
TransientTime: 0.5720
SettlingTime: 0.5720
SettlingMin: 5.2904e-05
SettlingMax: 9.9989e-05
 Overshoot: 0
 Undershoot: 0
     Peak: 9.9989e-05
  PeakTime: 1.4482
```
(b)

```
  RiseTime: 0.0189
TransientTime: 0.5613
SettlingTime: 0.5613
SettlingMin: 4.6099e-05
SettlingMax: 1.0305e-04
 Overshoot: 3.0474
 Undershoot: 0
     Peak: 1.0305e-04
  PeakTime: 0.0343
```
(c)

Figure 7.39

MATLAB generated output from the stepinfo command for Example 7.24 with a Pade approximation for $e^{-0.2s}$ of order (a) $n = 1$; (b) $n = 2$; (c) $n = 3$.

The poles of the transfer function of Equation (e) are calculated as -5.96, -77.44, $-35.87 \pm 75.31j$. The plot of the step response generated by MATLAB is shown in Figure 7.38(b) while the stepinfo results are given in Figure 7.39(b).

c. Substituting the Pade approximation for $n = 3$, Equation (7.108), into Equation (a) leads to

$$G(s) = \frac{1}{s^2 + 120s + 14{,}000 - 4000\left(\frac{1 - 0.1s + 0.004s^2 - 6.67 \times 10^{-5}s^3}{1 + 0.1s + 0.004s^2 + 6.67 \times 10^{-5}s^3}\right)} \tag{f}$$

Multiplying the numerator and denominator of Equation (f) by $1 + 0.1s + 0.004s^2 + 6.67 \times 10^{-5}s^3$ gives

$$G(s) = \frac{1 + 0.1s + 0.004s^2 + 6.67 \times 10^{-5}s^3}{6.67 \times 10^{-5}s^5 + 0.012s^4 + 1.247s^3 + 53s^2 + 1920s + 10{,}000} \tag{g}$$

The poles of the transfer function of Equation (f) are calculated as -6.09, $-17.16 \pm 45.23j$, $-69.76 \pm 75.12j$. The plot of the step response generated by MATLAB is shown in Figure 7.38(c) while the stepinfo results are given in Figure 7.39(c).

The transfer function, Equation (a), actually has an infinite number of poles. The greater the value of n, the better the approximation to the step response and the poles with the smallest absolute value of the real part, which is demonstrated in Figures 7.38 and 7.39.

7.8 Further Examples

The examples in this section develop the transient response for systems whose models are developed in examples in Chapters 4–6.

Example 7.25

The differential equation derived in Equation (i) of Example 6.15 for the perturbation in the temperature of water being heated in a tank when the rate at which heat is added to the tank suddenly changes is

$$1.68 \times 10^8 \frac{d\theta}{dt} + 2.10 \times 10^3 \theta = 500u(t) \tag{a}$$

a. Determine the time constant for the system.
b. Determine the response of the system.
c. Determine the increase in temperature when the system reaches its new steady state.
d. Determine the 8 percent settling time for the system.

Solution

a. Equation (a) is rewritten in the form of Equation (7.37) as

$$8 \times 10^4 \frac{d\theta}{dt} + \theta = 0.238u(t) \tag{b}$$

from which the time constant is determined as $T = 80{,}000$ s.

b. Application of Equation (7.47), the response of a first-order system due to a unit step input, to this system leads to

$$\theta(t) = 0.238(1 - e^{-t/80{,}000}) \tag{c}$$

c. From Equation (c), it is clear that $\lim_{t\to\infty} \theta(t) = 0.238$. Thus the increase in the rate at which heat is transferred to the tank leads to an increase in the steady-state temperature of the water of $\theta_s = 0.238°\text{C}$.

d. The 8 percent settling time is the time t_s such that

$$\theta(t_s) = 0.92\,\theta_s \tag{d}$$

Substitution of Equation (c) into Equation (d) leads to

$$0.238(1 - e^{-t_s/80{,}000}) = (0.92)(0.238) \tag{e}$$

The solution of Equation (e) is $t_s = 2.01 \times 10^5$ s.

Example 7.26

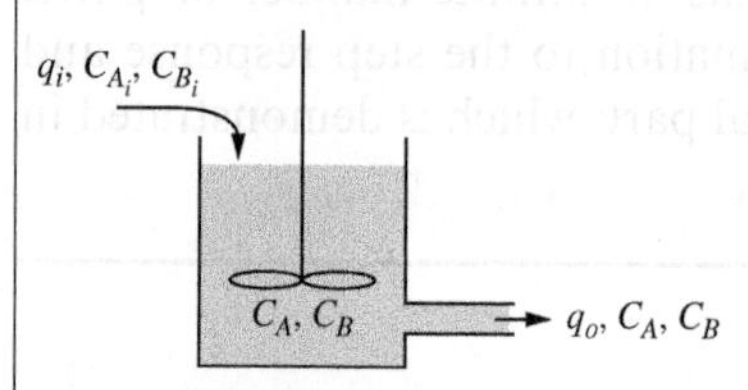

Figure 6.30

A chemical reaction occurs in a continuous stirred tank reactor (CSTR), in which component A is converted to component B. (Repeated)

The CSTR of Example 6.16 and Figure 6.30 has a constant volume of 1.32×10^{-3} m^3 with an inlet flow rate $q = 1.5 \times 10^{-6}$ m^3/s. Initially, neither the tank nor the inlet stream contains any reactant A. The initial concentration of the product B in the inlet stream and tank is 0.12 mol/L. At $t = 0$, a step change occurs such that the concentrations of the reactant A and the product B in the inlet stream are 0.25 mol/L and 0.12 mol/L, respectively. The rate of the first-order reaction that occurs in the CSTR is 1.8×10^{-3} s^{-1}. Assume the process is isothermal. **a.** Determine the time constant for the CSTR for the reactant A. **b.** Determine $C_A(t)$. **c.** Determine $G_B(s)$. **d.** Determine $C_B(t)$.

Solution

The mathematical model derived for the concentrations of the reactants in Example 6.16 is

$$V\frac{dC_A}{dt} + (q + kV)\,C_A = qC_{Ai} \tag{a}$$

$$V\frac{dC_B}{dt} - kVC_A + qC_B = qC_{Bi} \tag{b}$$

The initial conditions given in the problem statement are $C_A(0) = 0$, $C_B(0) = 0.12$ mol/L.

a. Equation (a) is rewritten in the standard from of the equation for a first-order system as

$$\frac{V}{q + kV}\frac{dC_A}{dt} + C_A = \frac{q}{q + kV}C_{Ai} \tag{c}$$

The time constant is obtained by comparing Equation (c) with the standard form of the Equation for a first-order system, Equation (7.37), leading to Equation (c) as

$$T = \frac{V}{q + kV}$$

$$= \frac{1.32 \times 10^{-3}\ \text{m}^3}{(1.5 \times 10^{-6}\ \text{m}^3/\text{s}) + (1.8 \times 10^{-3}\ \text{s}^{-1})(1.32 \times 10^{-3}\ \text{m}^3)} = 3.41 \times 10^2\ \text{s} \tag{d}$$

b. The input to the system is $C_{Ai}(t) = 0.25u(t)$ mol/L. Substituting numerical values into Equation (c) leads to

$$3.41 \times 10^2 \frac{dC_A}{dt} + C_A = 0.387u(t) \tag{e}$$

Equation (e) is that of a first-order system with an input given by 0.387 times the unit step function. Comparing with Equation (7.40), it is determined that the transfer function for the system is

$$G_A(s) = \frac{1/T}{s + 1/T} = \frac{2.94 \times 10^{-3}}{s + 2.94 \times 10^{-3}} \tag{f}$$

The response is 0.387 times the step response of a first-order system. Using Equation (7.47), this becomes

$$C_A(t) = 0.387(1 - e^{2.94 \times 10^{-3}t})\ \text{mol/L} \tag{g}$$

c. The problem for $C_B(t)$ is an initial value problem with $C_B(0) = 0.12$ mol/L. The problem also has input $C_A(t)$ in the form of a unit step function. The transfer function for $C_B(t)$ is obtained as

$$G_B(s) = \frac{kVq}{(Vs + q)(Vs + q + kV)} \tag{h}$$

d. The transfer function for $C_B(t)$ is that of a second-order system. However, the system has real poles of $-\frac{q}{V}$ and $-\left(k + \frac{q}{V}\right)$. Thus, the system is overdamped. Using Table 7.5, the step response of $C_B(t)$ due to a unit step input is

$$C_B(t) = \frac{kq}{\left(\frac{q}{V}\right)\left(k + \frac{q}{V}\right)(k)}\left[k - \left(k + \frac{q}{V}\right)e^{-\frac{q}{V}t} + \frac{q}{V}e^{-\left(k + \frac{q}{V}\right)t}\right] \tag{i}$$

The response for $C_B(t)$ is

$$C_B(t) = C_{Bi} + 0.25C_B(t)$$

$$= 0.12 + 0.25\left(0.613 - e^{-1.14 \times 10^{-3}t} + e^{-2.94 \times 10^{-3}t}\right)\frac{\text{mol}}{\text{L}} \tag{j}$$

Example 7.27

The hydraulic servomotor of Example 6.13 and Figure 6.22 contains oil of density $\rho = 900$ kg/m³. The discharge coefficient through the orifice is $C_Q = 0.65$, the width of the spool valve is $w = 30$ mm, the area of the piston is $A_p = 14$ cm², and the dimensions of the segments of the walking beam are $a = 2$ cm and $b = 12$ cm. The pressure difference is $p_s - p_1 = 400$ kPa. **a.** Determine the time constant for the system. **b.** Determine $y(t)$ when the end of the beam is given a sudden displacement $z(t) = 0.003u(t)$ m.

Solution

The mathematical mode derived for the system, Equation (c) of Example 6.13, is

$$A_p(a + b)\dot{y} + \hat{C}ay = \hat{C}bz \tag{a}$$

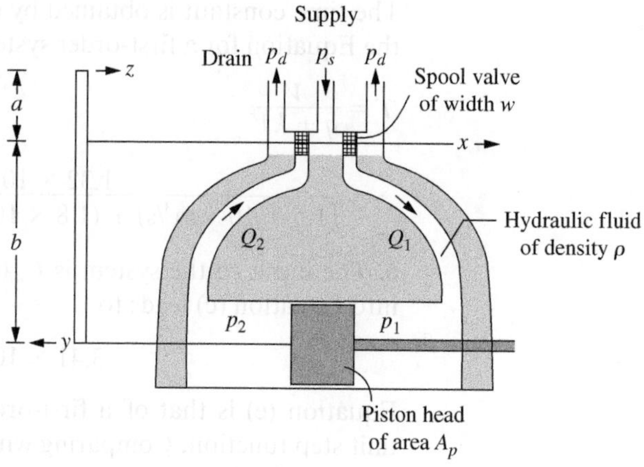

Figure 6.22
The hydraulic servomotor with walking beam of Example 6.13 and Example 7.27. (Repeated)

where

$$\hat{C} = C_Q w \sqrt{\frac{2(p_s - p_1)}{\rho}}$$

$$= (0.65)(0.030 \text{ m}) \sqrt{\frac{2(4.0 \times 10^5 \text{ N/m}^2)}{900 \text{ kg/m}^3}} = 0.581 \text{ m}^2/\text{s} \tag{b}$$

Equation (a) is rewritten in the standard form of a first-order equation as

$$\frac{A_p}{\hat{C}}\left(1 + \frac{b}{a}\right)\dot{y} + y = \frac{b}{a}z(t) \tag{c}$$

a. The system's time constant is obtained from Equation (c) as

$$T = \frac{A_p}{\hat{C}}\left(1 + \frac{b}{a}\right)$$

$$= \frac{(14 \text{ cm}^2)(1 \text{ m}/100 \text{ cm})^2}{0.581 \text{ m}^2/\text{s}}\left(1 + \frac{12 \text{ cm}}{2 \text{ cm}}\right) = 1.69 \times 10^{-2} \text{s} \tag{d}$$

b. Substitution of numerical values into Equation (c) leads to

$$1.69 \times 10^{-2}\dot{y} + y = 0.018u(t) \tag{e}$$

The solution of Equation (e) is obtained using Equation (7.47) as

$$y(t) = 0.018\left(1 - e^{-\frac{t}{1.69 \times 10^{-2}}}\right) \text{ m} \tag{f}$$

Example 7.28

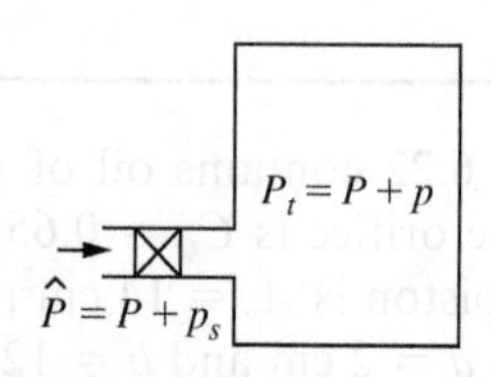

Figure 6.17
The system of Example 6.10 and Example 7.28: when the valve is opened and the supply pressure is greater than the tank pressure, gas flows into the pressure vessel. (Repeated)

The pressure vessel of Example 6.10 and Figure 6.17 has a volume $V = 1.2 \text{ m}^3$ and contains air at a pressure $P = 100$ kPa and temperature $T_a = 25°\text{C}$. An upstream valve is open from a supply of pressure $\hat{P} = 500$ kPa. The pipe between the supply and the pipe has an average resistance $R = 1 \times 10^4 \text{ m}^{-1}\text{s}^{-1}$. How long will it take the pressure in the pressure vessel to reach 450 kPa? Assume the flow through the connector is subsonic and the process is isothermal.

Solution

The differential equation of the mathematical model for the perturbation pressure in the pressure vessel is derived in Example 6.10 as

$$\frac{V}{R_a T_a}\frac{dp}{dt} + \frac{p}{R} = \frac{p_s}{R} \tag{a}$$

where $p_s = 400$ kPa is the difference from the initial pressure in the tank. Equation (a) is rewritten in the standard form of the differential equation for a first-order system as

$$\frac{VR}{R_a T_a}\frac{dp}{dt} + p = p_s \tag{b}$$

The time constant is determined from Equation (b) as

$$T = \frac{VR}{R_a T_a}$$

$$= \frac{(1.2 \text{ m}^3)(1 \times 10^4 \text{ m}^{-1}\text{s}^{-1})}{(287 \text{ J·kg/K})(298 \text{ K})} = 0.140 \text{ s} \tag{c}$$

Substitution of numerical values into Equation (b) leads to

$$0.140\frac{dp}{dt} + p = 400u(t) \tag{d}$$

The solution of Equation (d) subject to $p(0) = 0$ is obtained using Equation (7.47) as

$$p(t) = 400(1 - e^{-t/0.140}) \text{ kPa} \tag{e}$$

The pressure in the pressure vessel reaches 450 kPa when the perturbation pressure is 350 kPa. Thus the time required to reach this pressure is obtained by solving

$$350 = 400(1 - e^{-t/0.140}) \tag{f}$$

The solution of Equation (f) is $t = 0.291$ s.

Example 7.29

Consider the RC circuit of Figure 7.40. Determine the current in the circuit and the voltage drop across the capacitor due to an impulsive source.

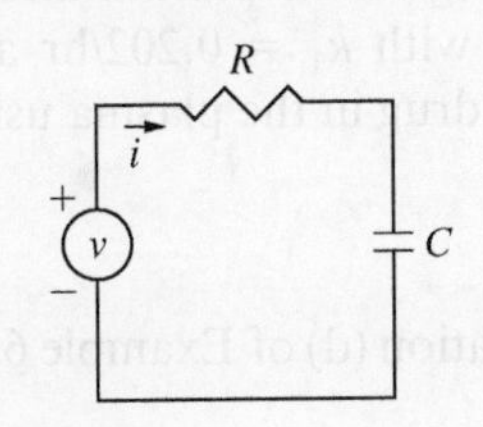

Figure 7.40

The *RC* circuit of Example 7.29.

Solution

Application of KVL around the loop leads to

$$Ri + \frac{1}{C}\int_0 i dt = v(t) \tag{a}$$

The transfer function for the circuit is

$$G(s) = \frac{I(s)}{V(s)} = \frac{\frac{1}{R}s}{s + \frac{1}{RC}}$$

$$= \frac{1}{R}\left(1 - \frac{1}{RC}\frac{1}{s + \frac{1}{RC}}\right) \tag{b}$$

The impulsive response is determined by inverting the transfer function, resulting in

$$i(t) = \frac{1}{R}\left[\delta(t) - \frac{1}{RC}e^{-t/(RC)}u(t)\right] \tag{c}$$

Equation (c) shows that an impulsive voltage applied to an RC circuit leads to an impulsive current. The current is 0 before the impulse is applied and jumps to $-1/R^2C$ after application of the impulse. The voltage across the capacitor has a discrete jump at $t = 0$ to $1/RC$. The impulsive response is the response for the current in the circuit assuming the voltage is $\delta(t)$. The step response assumes a voltage input of $u(t)$. Thus the values of the charge after the impulse and the initial voltage across the capacitor are per unit input volt.

The voltage drop across the capacitor is

$$v_C(t) = \frac{1}{C}\int_0^t i(t)dt \tag{d}$$

Taking the Laplace transform of Equation (d) gives

$$V_C(s) = \frac{1}{Cs}I(s) = \frac{1}{Cs}\left(\frac{\frac{1}{R}s}{s + \frac{1}{RC}}\right)V(s) \tag{e}$$

The transfer function for v_C is obtained from Equation (e) as

$$\frac{V_C(s)}{V(s)} = \frac{1}{Cs}\left(\frac{\frac{1}{R}s}{s + \frac{1}{RC}}\right) = \frac{1}{RC}\frac{1}{s + \frac{1}{RC}} \tag{f}$$

The impulsive response for $v_C(t)$ is the inverse of its transfer function, Equation (f),

$$v_C(t) = \frac{1}{RC}e^{-\frac{t}{RC}} \tag{g}$$

Example 7.30

Consider the pharmokinetic problem of Examples 6.18 and 6.19. A drug is infused into the plasma at a constant rate of 0.302 mg/hr until it is expended after 12 hr. The rate of elimination is 0.105/hr. Assume the available volume of plasma is 40 L. **a.** Using the one-compartment model of Example 6.18, determine the concentration of the drug in the plasma as a function of time. **b.** Use the two-compartment model of Example 6.19 to determine the concentration of the drug in the plasma and in the tissue, assuming the available volume of tissue is 4 L with $k_1 = 0.202$/hr and $k_2 = 0.133$/hr. **c.** Plot and compare the concentration of the drug in the plasma using the two models.

Solution

a. The rate of infusion is $I(t) = 0.302[1 - u(t - 12)]$ mg/hr. Equation (d) of Example 6.18 is rearranged as

$$\frac{1}{k_e}\frac{dC}{dt} + C = \frac{I(t)}{k_e V} \tag{a}$$

The time constant for the first-order system is $T = 1/k_e = 9.52$ hr. Substitution of values into Equation (a) leads to

$$9.52\frac{dC}{dt} + C = 0.0719[1 - u(t - 12)] \tag{b}$$

where t is measured in hours and the concentration is in milligrams per liter. The response of the system during the time that the drug is being infused is that of a first-order system with a step input. After the drug is expended, the response of the system is that of a first-order system with a nonzero initial condition and decays, and it asymptotically approaches zero over a long time. Taking the Laplace transform of Equation (b) leads to

$$C(s) = \frac{0.0719(1 - e^{-12s})}{s(9.52s + 1)} = \frac{7.553 \times 10^{-3}(1 - e^{-12s})}{(s + 0.105)s} \tag{c}$$

Inversion of Equation (c) leads to

$$C(t) = 7.19 \times 10^{-2}[(1 - e^{-0.105t})u(t) - (1 - e^{-0.105(t-12)})u(t - 12)] \text{ mg/L} \tag{d}$$

b. The transfer functions for the system are determined using Equations (g) and (h) of Example 6.19 as

$$G_1(s) = \frac{C_p(s)}{I(s)} = \frac{s + k_2}{V_p[s^2 + (k_e + k_1 + k_2)s + k_e k_2]} \tag{e}$$

$$G_2(s) = \frac{C_t(s)}{I(s)} = \frac{k_1}{V_t[s^2 + (k_e + k_1 + k_2)s + k_e k_2]} \tag{f}$$

The two-compartment model is a second-order system of natural frequency

$$\omega_n = \sqrt{k_e k_2} = \sqrt{(0.105/\text{hr})(0.133/\text{hr})} = 0.118 \text{ r/hr} \tag{g}$$

and damping ratio

$$\zeta = \frac{k_e + k_1 + k_2}{2\omega_n} = \frac{(0.105/\text{hr}) + (0.202/\text{hr}) + (0.133/\text{hr})}{2(0.118/\text{hr})} = 1.86 \tag{h}$$

The poles of the transfer function are

$$s_{1,2} = \omega_n\left(-\zeta \pm \sqrt{\zeta^2 - 1}\right) = -0.405, -0.0344 \tag{i}$$

Substitution of the given and calculated values into Equations (e) and (f) leads to

$$G_1(s) = \frac{0.025(s + 0.133)}{(s + 0.405)(s + 0.0344)} \tag{j}$$

$$G_2(s) = \frac{0.0333}{(s + 0.405)(s + 0.133)} \tag{k}$$

Noting that $I(s) = 0.302(1 - e^{-12s})/s$ and performing partial fraction decompositions on Equations (j) and (k) lead to

$$C_p(s) = \left(\frac{0.0721}{s} - \frac{0.0137}{s + 0.405} - \frac{0.0589}{s + 0.0344}\right)(1 - e^{-12s}) \tag{l}$$

$$C_t(s) = \left(\frac{0.722}{s} + \frac{0.0670}{s + 0.405} - \frac{0.788}{s + 0.0344}\right)(1 - e^{-12s}) \tag{m}$$

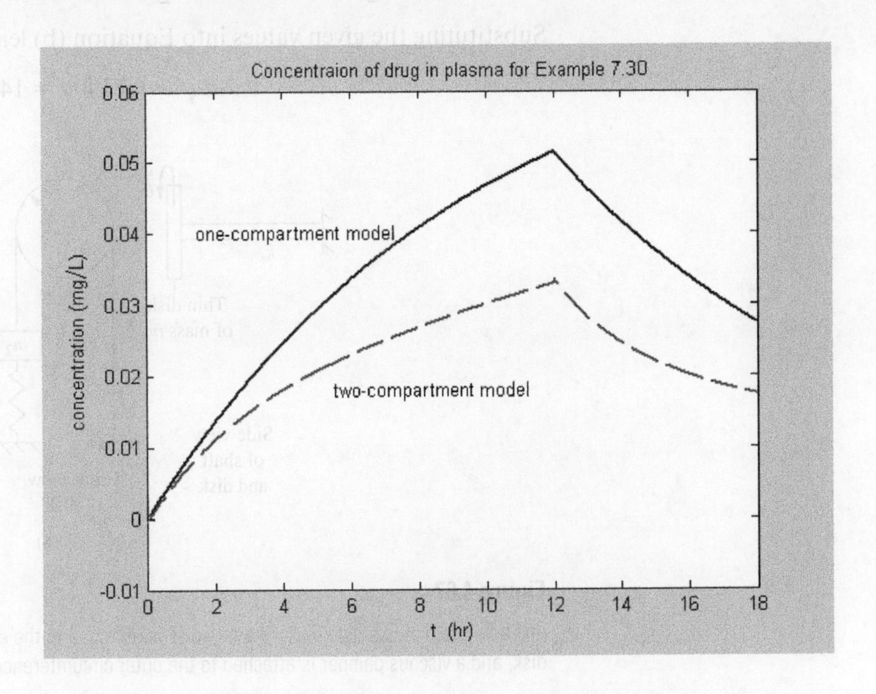

Figure 7.41

The time-dependent response for the concentration of a drug in plasma during and after infusion.

Equations (l) and (m) are inverted, leading to

$$C_p(t) = (0.0721 - 0.0137e^{-0.405t} - 0.0589e^{-0.0344t})u(t)$$
$$- (0.0721 - 0.0137e^{-0.405(t-12)} - 0.0589e^{-0.0344(t-12)})u(t-12) \quad \text{(n)}$$

$$C_t(t) = (0.722 + 0.0670e^{-0.405t} - 0.788e^{-0.0344t})u(t)$$
$$- (0.722 + 0.0670e^{-0.405(t-12)} - 0.788e^{-0.0344(t-12)})u(t-12) \quad \text{(o)}$$

c. Equations (d) and (n) are plotted using MATLAB on the same graph in Figure 7.41.

Example 7.31

Consider the second-order system of Example 4.31 and Figure 4.67. The mathematical model for the system is given in Equation (f) of Example 4.31 as

$$\left(\frac{1}{2}m_1 r^2 + m_2 r^2\right)\ddot{\theta} + cr^2\dot{\theta} + (k_t + kr^2)\theta = M(t) \quad \text{(a)}$$

Determine the response of the system when **a.** $M(t) = 0$ and the disk has an initial angular rotation of 0.05 r and an initial angular velocity of zero, **b.** the disk is in equilibrium and subject to an angular impulse of 2.8 N·m·s, and **c.** the disk is in equilibrium and then subject to a constant moment of 14 N·m.

Solution

The mathematical model, Equation (a), is put in standard form by dividing through by the coefficient of the $\ddot{\theta}$ term, leading to Equation (q) of Example 4.31

$$\ddot{\theta} + \frac{cr^2}{\frac{1}{2}m_1 r^2 + m_2 r^2}\dot{\theta} + \frac{k_t + kr^2}{\frac{1}{2}m_1 r^2 + m_2 r^2}\theta = \frac{1}{\frac{1}{2}m_1 r^2 + m_2 r^2}M(t) \quad \text{(b)}$$

Substituting the given values into Equation (b) leads to

$$\ddot{\theta} + 114\dot{\theta} + 7140\theta = 14.3M(t) \quad \text{(c)}$$

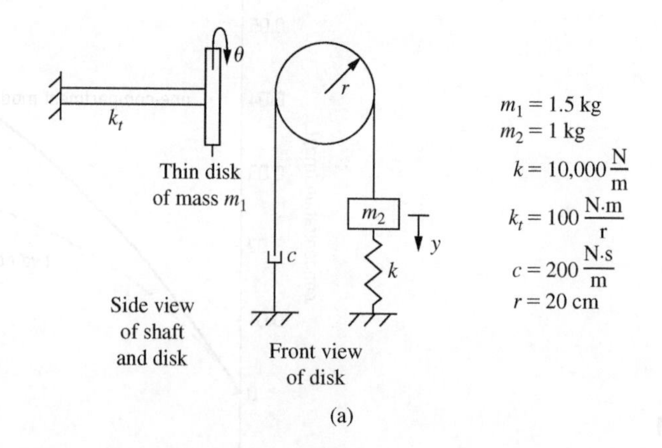

Figure 4.67

(a) The system of Example 4.31. The thin disk is attached to the end of a shaft. A mass is suspended from the disk, and a viscous damper is attached to the outer circumference of the disk. (Repeated)

The natural frequency and damping ratio are calculated by comparing Equation (c) with Equation (7.58), leading to

$$\omega_n = \sqrt{7140} = 84.5\frac{\text{r}}{\text{s}} \tag{d}$$

$$2\zeta\omega_n = 114 \Rightarrow \zeta = \frac{114}{2(84.5)} = 0.676 \tag{e}$$

Since $\zeta < 1$, the system is underdamped and has a damped natural frequency given by Equation (7.76) of

$$\omega_d = \omega_n\sqrt{1-\zeta^2} = 62.3\frac{\text{r}}{\text{s}} \tag{f}$$

a. With $M(t) = 0$, the system has a free response given by Equation (7.77) subject to the initial conditions of $\theta(0) = 0.05$ r and $\dot{\theta}(0) = 0$. The amplitude and phase angle are given by Equations (7.78) and (7.79) respectively, leading to

$$A_d = 0.05\sqrt{1 + \left[\frac{(0.676)(84.5)}{62.3}\right]^2} = 0.068 \text{ r} \tag{g}$$

$$\phi_d = \tan^{-1}\left[\frac{62.3}{(0.676)(84.5)}\right] = 0.83 \text{ r} \tag{h}$$

Substituting Equations (f)-(h) in Equation (7.77) gives

$$\begin{aligned}\theta(t) &= 0.068\, e^{-(0.676)(84.5)t}\sin(62.3t + 0.83) \text{ r} \\ &= 0.068e^{-57.1t}\sin(62.3t + 0.83) \text{ r}\end{aligned} \tag{i}$$

b. The system's transfer function is obtained from Equation (c) as

$$G(s) = \frac{1}{s^2 + 114s + 7140} \tag{j}$$

With $M(t) = 2.8\delta(t)$, the response is obtained by using the second entry in Table 7.4 as

$$\begin{aligned}\theta(t) &= 14.3(2.8)\frac{1}{62.3}e^{-57.1t}\sin(62.3t) \\ &= 0.64\, e^{-57.1t}\sin(62.3t) \text{ r}\end{aligned} \tag{k}$$

With $M(t) = 14u(t)$, the response is obtained by using the second entry in Table 7.5 to give

$$\theta(t) = 14.3(14)\frac{1}{84.5^2}\left\{1 - e^{-57.1t}\left[\cos 62.3t + \frac{(0.676)(84.5)}{62.3}\sin 62.3t\right]\right\} \tag{l}$$

Example 7.32

A dynamometer test is run on an armature-controlled dc servomotor, as modeled in Example 5.23. During the test, the dynamometer applies a constant armature voltage, and the output torque and shaft speed are measured. For a certain motor, when a dynamometer test is run with $v_a = 50$ V, the torque curve, the plot of torque vs. speed as shown in Figure 7.42, is approximately linear. The armature resistance is 100 Ω and the armature inductance is 0.15 H. **a.** Use the torque curve to determine the values of K_a and K_b. **b.** The motor is used to run a shaft with $J = 1.8$ kg·m^2 and

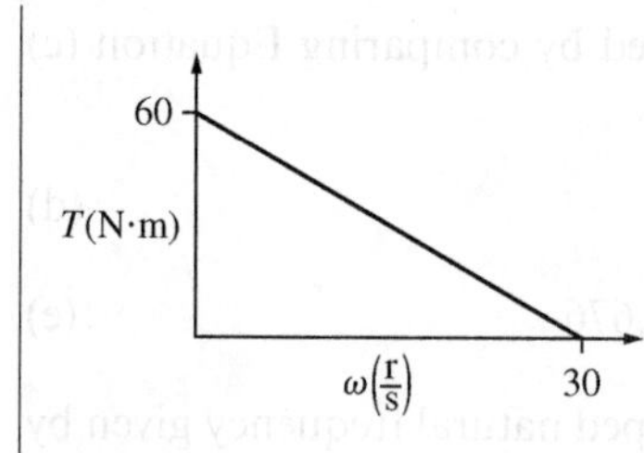

Figure 7.42

The torque curve for the motor of Example 7.32.

$c_t = 2.15$ N·m·s/r. Determine the transient response of the armature current and the shaft speed when a 100-V source is connected to the circuit. **c.** How long will it take for the speed to become within 2 percent of its eventual steady-state value?

Solution

a. The system is operating at steady state for the dynamometer test. Equation (a) of Example 5.23, evaluated at steady state, becomes

$$R_a i_a + K_b \omega = v_a \tag{a}$$

Equation (5.84b) is used to relate the armature current to the torque $i_a = T/K_a$, which when used in Equation (a) leads to

$$\frac{R_a}{K_a}T + K_b \omega = v_a \tag{b}$$

Equation (b) is rearranged in the form of the equation of a line as

$$T = -\frac{K_a K_b}{R_a}\omega + \frac{K_a}{R_a}v_a \tag{c}$$

The slope of the torque curve, determined from Figure 7.42, is -2.0 N·m·s/r and the intercept with the torque axis is 60 N·m. Thus Equation (c) implies

$$\frac{K_a}{R_a}v_a = 60 \text{ N·m}$$

$$K_a = \frac{(60 \text{ N·m})(100\ \Omega)}{50 \text{ V}} = 120 \text{ N·m/A} \tag{d}$$

and

$$-\frac{K_a K_b}{R_a} = -2 \text{ N·m·s/r}$$

$$K_b = \frac{(2 \text{ N·m·s/r})(100\ \Omega)}{120 \text{ N·m/A}} = 1.67 \text{ V·s/r} \tag{e}$$

b. Substitution of given and calculated values into Equations (a) and (b) of Example 5.23 leads to

$$0.15\frac{di_a}{dt} + 100i_a + 1.67\omega = v_a(t) \tag{f}$$

$$1.8\frac{d\omega}{dt} + 2.15\omega - 120 i_a = 0 \tag{g}$$

Transfer functions are obtained using Equations (f) and (g) as

$$G_1(s) = \frac{I(s)}{V_a(s)} = \frac{6.672 + 7.96}{s^2 + 666.7s + 1.54 \times 10^3} \tag{h}$$

$$G_2(s) = \frac{\Omega(s)}{V_a(s)} = \frac{444.4}{s^2 + 666.7s + 1.54 \times 10^3} \tag{i}$$

The response is that of a second-order system of natural frequency $\omega_n = \sqrt{1.54 \times 10^3} = 39.4$ r/s and damping ratio $\zeta = 666.7/[2(39.4)] = 8.46$. The input is $v_a(t) = 100u(t)$. Hence the responses are 100 times the step responses. Table 7.5 is used for an overdamped second-order system, leading to

$$i_a(t) = (0.506 - 1.005e^{-664.3t} + 1.52e^{-2.34t})u(t) \tag{j}$$

$$\omega(t) = (28.3 + 0.101e^{-664.3t} - 28.4e^{-2.34t})u(t) \tag{k}$$

The eventual steady-state speed is 28.3 r/s. The two percent settling time is the time it takes to reach 98 percent of the final speed. It is obtained by solving

$$0.98(28.3) = 28.35 - 0.101e^{-664.3t_s} - 28.4e^{-2.34t_s} \tag{l}$$

The solution of Equation (l) is

$$t_s = 1.67 \text{ s} \tag{m}$$

Example 7.33

Design the operational amplifier circuit of Example 5.37 and Figure 5.48 such that it simulates a critically damped second-order system with a natural frequency of 250 r/s.

Solution

Equation (o) of Example 5.37 is the differential equation simulated by the operational amplifier circuit,

$$C_1C_2R_3\frac{d^{23}v_0}{dt^2} + \frac{R_3C_2}{R_2}\frac{dv_0}{dt} + \frac{R_5}{R_4R_6}v_0 = \frac{v_i}{R_1} \tag{a}$$

Dividing Equation (a) by $C_1C_2R_3$ leads to

$$\frac{d^2v_0}{dt^2} + \frac{1}{C_1R_2}\frac{dv_0}{dt} + \frac{R_5}{R_3R_4R_6C_1C_2}v_0 = \frac{1}{R_1R_3C_1C_2}v_i \tag{b}$$

The transfer function for the circuit is obtained from Equation (b) as

$$G(s) = \frac{V_0(s)}{V_i(s)} = \frac{\frac{1}{R_1R_3C_1C_2}}{s^2 + \frac{1}{C_1R_2}s + \frac{R_5}{R_3R_4R_6C_1C_2}} \tag{c}$$

Equation (c) is the standard form of the transfer function for a second-order system. System parameters are identified by comparing Equation (c) with the

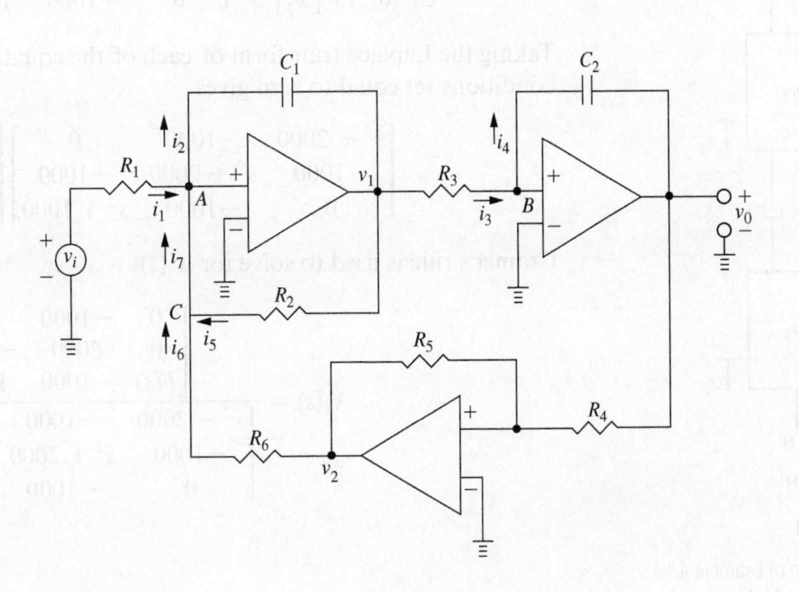

Figure 5.48

The operational amplifier circuit of Example 5.37 and Example 7.33. (Repeated)

standard form of the denominator of the transfer function for a second-order system, Equation (7.62). Through this comparison, the natural frequency and damping ratio are

$$\omega_n = \sqrt{\frac{R_5}{R_3 R_4 R_6 C_1 C_2}} \tag{d}$$

$$\zeta = \frac{1}{2\omega_n C_1 R_2} \tag{e}$$

Since there are seven circuit parameters in Equations (d) and (e) and only two conditions to satisfy, the design is not unique. Arbitrarily choose $R_4 = R_5$, $C_1 = C_2 = 1\ \mu\text{F}$. Substituting into Equation (d) and setting the natural frequency to 250 r/s leads to

$$\sqrt{\frac{1 \times 10^{12}}{R_3 R_6}} = 250 \tag{f}$$

Choosing $R_3 = R_6$ in Equation (f) gives $R_3 = R_6 = 4\ \text{k}\Omega$.

A second-order system is critically damped when its damping ratio is one. Thus from Equation (e),

$$\frac{1}{2(250\ \text{r/s})(1 \times 10^{-6}\ \text{F})R_2} = 1$$

$$R_2 = 2\ \text{k}\Omega \tag{g}$$

Example 7.34

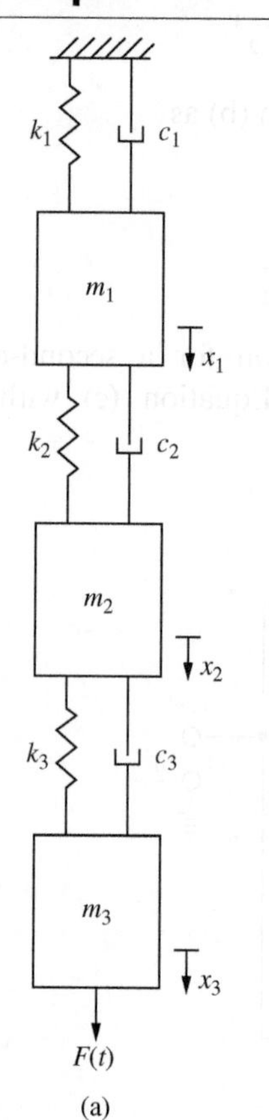

Figure 4.60

(a) The system of Example 4.19 and Example 7.34. (Repeated)

a. Determine the natural frequencies of the three-degree-of-freedom system of Example 4.19 and Figure 4.60 when $m_1 = m_2 = m_3 = 1$ kg and $k_1 = k_2 = k_3 = 10{,}000$ N/m.

b. Use MATLAB to determine the step response for $x_1(t)$, for the undamped system.

Solution

a. Substitution of given values into Equation (j) of Example 4.19 with $c_1 = c_2 = c_3 = 0$ leads to

$$\begin{bmatrix} 1 & 0 & 0 \\ 0 & 1 & 0 \\ 0 & 0 & 1 \end{bmatrix}\begin{bmatrix} \ddot{x}_1 \\ \ddot{x}_2 \\ \ddot{x}_3 \end{bmatrix} + \begin{bmatrix} 2000 & -1000 & 0 \\ -1000 & 2000 & -1000 \\ 0 & -1000 & 1000 \end{bmatrix}\begin{bmatrix} x_1 \\ x_2 \\ x_3 \end{bmatrix} = \begin{bmatrix} 0 \\ 0 \\ F(t) \end{bmatrix} \tag{a}$$

Taking the Laplace transform of each of the equations of Equation (a) with all initial conditions set equal to zero gives

$$\begin{bmatrix} s^2 + 2000 & -1000 & 0 \\ -1000 & s^2 + 2000 & -1000 \\ 0 & -1000 & s^2 + 1000 \end{bmatrix}\begin{bmatrix} X_1(s) \\ X_2(s) \\ X_3(s) \end{bmatrix} = \begin{bmatrix} 0 \\ 0 \\ F_1(s) \end{bmatrix} \tag{b}$$

Cramer's rule is used to solve for $X_1(s)$:

$$X_1(s) = \frac{\begin{bmatrix} 0 & -1000 & 0 \\ 0 & 2000 & -1000 \\ F(s) & -1000 & 1000 \end{bmatrix}}{\begin{bmatrix} s^2 + 2000 & -1000 & 0 \\ -1000 & s^2 + 2000 & -1000 \\ 0 & -1000 & s^2 + 1000 \end{bmatrix}} \tag{c}$$

from which a transfer function is obtained as

$$G_1(s) = \frac{X_1(s)}{F(s)} = \frac{1 \times 10^6 F(s)}{s^6 + 5 \times 10^3 s^4 + 6 \times 10^6 s^2 + 1 \times 10^9} \qquad \text{(d)}$$

The poles of the transfer function are obtained using MATLAB as $s = \pm j57.0$, $\pm j39.4$, and $\pm j14.1$. Thus the natural frequencies are $\omega_1 = 14.1$ r/s, $\omega_2 = 39.4$ r/s, and $\omega_3 = 57.0$ r/s.

b. The MATLAB work session to determine the poles of the transfer function and the step response for $x_1(t)$, as well as the step response, are shown in Figure 7.43.

```
>> N=[1E6];
>> D=[1 0 5E3 0 6E6 0 1E9];
>> [r,p,k]=residue(N,D)
r =
      0 - 0.0017i
      0 + 0.0017i
      0 + 0.0055i
      0 - 0.0055i
      0 - 0.0086i
      0 + 0.0086i

p =
      0 + 56.9823i
      0 - 56.9823i
      0 + 39.4330i
      0 - 39.4330i
      0 + 14.0735i
      0 - 14.0735i
k =
      []
>> step(N,D)
>>
```

(a)

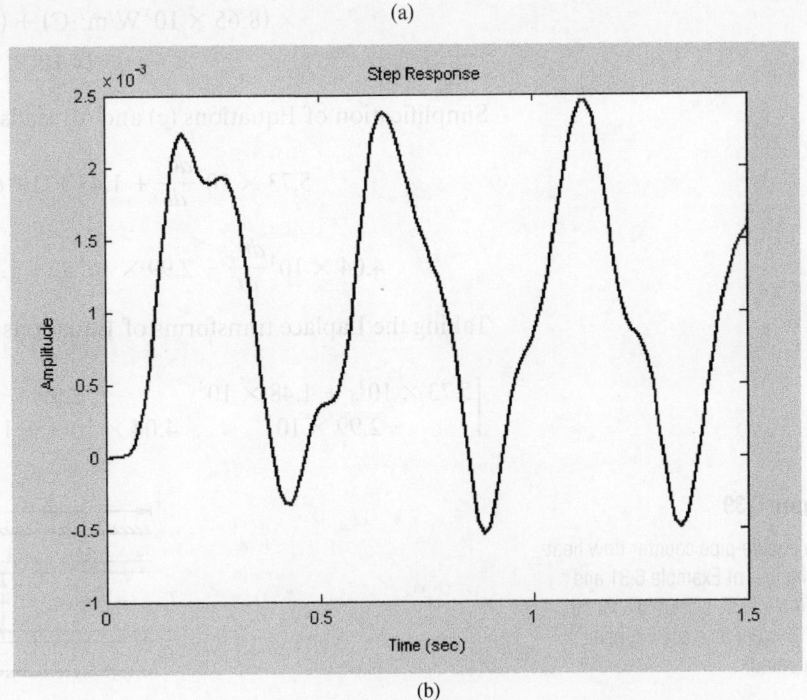

(b)

Figure 7.43

(a) The MATLAB work session used to determine the natural frequencies of the system of Example 7.34 and (b) the step response for $x_1(t)$.

Example 7.35

The heat exchanger of Example 6.31 and Figure 6.39 uses water at 20°C to cool oil at 300°C. The properties of the heat exchanger and the flow are $\rho_w = 9.60 \times 10^2$ kg/m³, $\rho_o = 7.50 \times 10^2$ kg/m³, $c_w = 4.19 \times 10^3$ J/(kg·C), $c_o = 2.4 \times 10^3$ J/(kg·C), $U = 8.65 \times 10^2$ W/(m²·C), $\dot{m}_w = 2.51$ kg/s, $\dot{m}_o = 4.85$ kg/s, $r_i = 2$ cm, $r_o = 2.2$ cm, $r_2 = 4.2$ cm, and $L = 2.5$ m. **a.** Determine the steady-state outlet temperatures of the oil and water. **b.** Determine the transient perturbations in outlet temperatures when the temperature of the water is suddenly increased to 30°C. Assume the properties of both fluids remain constant.

Solution

a. Substituting the given values into Equations (e) and (f) of Example 6.31 with the time derivatives set to zero leads to the steady-state equations:

$$\begin{aligned}(2.43 \times 10^3\ \text{J/kg·C})(4.85\ \text{kg/s})(300\ \text{C} - T_{os}) &= 2\pi(0.022\ \text{m}) \\ &\times (2.5\ \text{m})(8.65 \times 10^2\ \text{W/m}^2\text{· C})(T_{os} - T_{ws})\end{aligned} \quad \text{(a)}$$

$$\begin{aligned}(4.19 \times 10^3\ \text{J/kg·C})(2.51\ \text{kg/s})(20\ \text{C} - T_{ws}) &= -2\pi(0.022\ \text{m}) \\ &\times (2.5\ \text{m})(8.65 \times 10^2\ \text{W/m}^2\text{· C})(T_{os} - T_{ws})\end{aligned} \quad \text{(b)}$$

Equations (a) and (b) are solved, yielding $T_{os} = 293.3$°C, $T_{ws} = 27.5$°C.

b. Substitution of given values into Equations (k) and (l) of Example 6.31 gives

$$\begin{aligned}(7.50 \times 10^2\text{kg/m}^3)(2.43 \times 10^3\text{J/kg·C})\pi(0.02\ \text{m})^2(2.5\ \text{m})\frac{d\theta_o}{dt} \\ +[2\pi(0.022\ \text{m})(2.5\ \text{m})(8.65 \times 10^3\text{W/m}^2\text{·C}) \\ +(2.43 \times 10^3\text{J/kg·C})(4.85\ \text{kg/s})]\,\theta_o - 2\pi(0.022\ \text{m}) \\ \times (8.65 \times 10^3\text{W/m}^2\text{·C})(2.5\ \text{m})\,\theta_w = 0\end{aligned} \quad \text{(c)}$$

$$\begin{aligned}(9.60 \times 10^2\text{kg/m}^3)(4.19 \times 10^3\text{J/kg·C})\pi[(0.042\ \text{m})^2 - (0.022\ \text{m})^2](2.5\ \text{m})\frac{d\theta_w}{dt} \\ - 2\pi(0.022\ \text{m})(8.65 \times 10^3\text{W/m}^2\text{·C})(2.5\ \text{m})\,\theta_0 + [2\pi(0.022\ \text{m})(2.5\ \text{m}) \\ \times (8.65 \times 10^3\text{W/m}^2\text{·C}) + (4.19 \times 10^3\text{J/kg·C})(2.51\ \text{kg/s})]\,\theta_w \\ = (4.19 \times 10^3\text{J/kg·C})(2.51\ \text{kg/s})(10\text{°C})u(t)\end{aligned} \quad \text{(d)}$$

Simplification of Equations (c) and (d) leads to

$$5.73 \times 10^3\frac{d\theta_o}{dt} + 1.48 \times 10^4\theta_o - 2.99 \times 10^3\theta_w = 0 \quad \text{(e)}$$

$$4.04 \times 10^4\frac{d\theta_w}{dt} - 2.99 \times 10^3\theta_o + 1.35 \times 10^4\theta_w = 1.05 \times 10^5u(t) \quad \text{(f)}$$

Taking the Laplace transforms of Equations (e) and (f) leads to

$$\begin{bmatrix}5.73 \times 10^3s + 1.48 \times 10^4 & -2.99 \times 10^3 \\ -2.99 \times 10^3 & 4.04 \times 10^4s + 1.35 \times 10^4\end{bmatrix}\begin{bmatrix}\Theta_o(s) \\ \Theta_w(s)\end{bmatrix} = \begin{bmatrix}0 \\ \dfrac{1.05 \times 10^5}{s}\end{bmatrix} \quad \text{(g)}$$

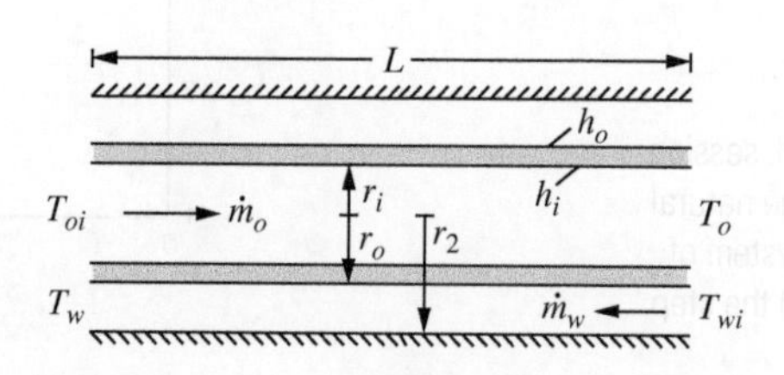

Figure 6.39

The double-pipe counter flow heat exchanger of Example 6.31 and Example 7.35 is used to cool oil using water as a cooling liquid. (Repeated)

The system of Equation (g) is solved simultaneously, leading to

$$\Theta_o(s) = \frac{1.36}{s(s^2 + 2.92s + 0.827)} \quad \text{(h)}$$

$$\Theta_w(s) = \frac{2.61s + 6.73}{s(s^2 + 2.92s + 0.827)} \quad \text{(i)}$$

Use of MATLAB to invert the transforms leads to

$$\theta_o(t) = (1.64 + 0.222e^{-2.63t} - 1.87e^{-0.314t})u(t) \quad \text{(j)}$$

$$\theta_w(t) = (8.14 + 0.246e^{-0.263t} - 8.11e^{-0.314t})u(t) \quad \text{(k)}$$

The eventual steady-state temperatures are T_o = 293.3 C + 1.64 C = 295.0 C and T_w = 27.5 C + 8.14 C = 35.6 C.

Example 7.36

After the water leaves the heat exchanger of Example 6.31, it is cooled to the inlet temperature and recirculated through the heat exchanger and used for the inlet stream. The heat exchanger system of Example 6.31 is operating at steady state under the conditions described in Example 7.35 when the system cooling the water breaks. The water is recirculated through the system at the temperature it leaves the heat exchanger but takes 5 s to again reach the heat exchanger inlet. Determine the transforms of the perturbation in temperatures due to the breakdown of the cooling system. Determine the eventual steady-state temperatures.

Solution

After the cooling system stops functioning, the inlet water temperature is that of the exit water temperature, delayed by $\tau = 5$ s, $T_{wi} = T_w(t-\tau)$. From the definitions of the temperature perturbations of Equations (h)–(j) of Example 6.31,

$$\begin{aligned}\theta_{wi} &= T_{wi} - T_{wis} \\ &= T_w(t-\tau) - T_{wis} \\ &= T_{ws} + \theta_w(t-\tau)\,T_{wis} \\ &= (T_{ws} - T_{wis}) + \theta_w(t-\tau)\end{aligned} \quad \text{(a)}$$

Substituting Equation (a) into Equation (l) of Example 6.31 leads to

$$\begin{aligned}\rho_w c_w \pi(r_2^2 - r_o^2)L\frac{d\theta_w}{dt} &- 2\pi r_o LU\theta_o + (2\pi r_o LU + c_w\dot{m}_w)\,\theta_w \\ &= c_w\dot{m}_w[T_{ws} - T_{wis} + \theta_w(t-\tau)]\end{aligned} \quad \text{(b)}$$

The substitution of the given values of Example 7.35 into Equation (k) of Example 6.31 and Equation (a) of this example leads to

$$5.73 \times 10^3\frac{d\theta_o}{dt} + 1.48 \times 10^4\theta_o - 2.99 \times 10^3\theta_w = 0 \quad \text{(c)}$$

$$4.04 \times 10^4\frac{d\theta_w}{dt} - 2.99 \times 10^3\theta_o + 1.35 \times 10^4\theta_w = 1.05 \times 10^4[7.5 + \theta_w(t-5)] \quad \text{(d)}$$

Taking the Laplace transform of Equations (c) and (d) using Equation (7.103) leads to

$$\begin{bmatrix} 5.73 \times 10^3 s + 1.48 \times 10^4 & -2.99 \times 10^3 \\ -2.99 \times 10^3 & 4.04 \times 10^4 s + 1.35 \times 10^4 - 1.05 \times 10^4 e^{-5s} \end{bmatrix} \times \begin{bmatrix} \Theta_o(s) \\ \Theta_w(s) \end{bmatrix} = \begin{bmatrix} 0 \\ \dfrac{7.88 \times 10^4}{s} \end{bmatrix} \quad \text{(e)}$$

Simultaneous solution of the equations represented by Equation (e) leads to

$$\Theta_o(s) = \frac{0.949}{s(s^2 + 2.92s + 0.827 - 0.260se^{-5s} - 0.671e^{-5s})} \quad \text{(f)}$$

$$\Theta_w(s) = \frac{1.82s + 4.70}{s(s^2 + 2.92s + 0.827 - 0.260se^{-5s} - 0.671e^{-5s})} \quad \text{(g)}$$

Using a second-order Pade approximation for e^{-5s} in Equations (f) and (g) leads to

$$\Theta_o(s) = \frac{0.949}{s\left(s^2 + 2.92s + 0.827 - (0.260s + 0.671)\dfrac{1 - 2.5s + 2.5s^2}{1 + 2.5s + 2.5s^2}\right)} \quad \text{(h)}$$

$$\Theta_w(s) = \frac{1.82s + 4.70}{s\left(s^2 + 2.92s + 0.827 - (0.260s + 0.671)\dfrac{1 - 2.5s + 2.5s^2}{1 + 2.5s + 2.5s^2}\right)} \quad \text{(i)}$$

Multiplying the numerator and denominator of Equations (h) and (i) by $1 + 2.5s + 2.5s^2$ leads to

$$\Theta_o(s) = \frac{0.949(1 + 2.5s + 2.5s^2)}{s(2.5s^4 + 9.15s^3 + 9.34s^2 + 6.405s + 0.156)} \quad \text{(j)}$$

$$\Theta_w(s) = \frac{4.55s^3 + 16.35s^2 + 13.62s + 4.72}{s(2.5s^4 + 9.15s^3 + 9.34s^2 + 6.405s + 0.156)} \quad \text{(k)}$$

The time-dependent responses are plotted in Figure 7.44. The poles of the transform are $s = 0, -0.0253, -2.598$, and $-5.184 \pm 0.826j$. The transforms are made up of three linear factors and an irreducible quadratic factor. The frequency of the sinusoidal term is 0.908 r/s. However, the damping rate for the quadratic factor is higher than the linear factors, so the response is close to that of a first-order response. The final value theorem is applied to determine the eventual steady-state temperatures of the oil and water: $\theta_{fo} \lim_{s\to 0} s\Theta_o(s) = 6.09$, $\theta_{fw} = \lim_{s\to 0} s\Theta_w(s) = 30.1$.

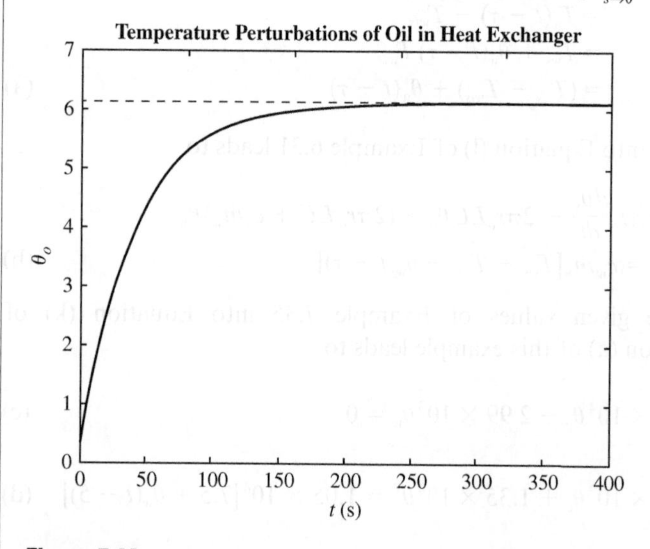

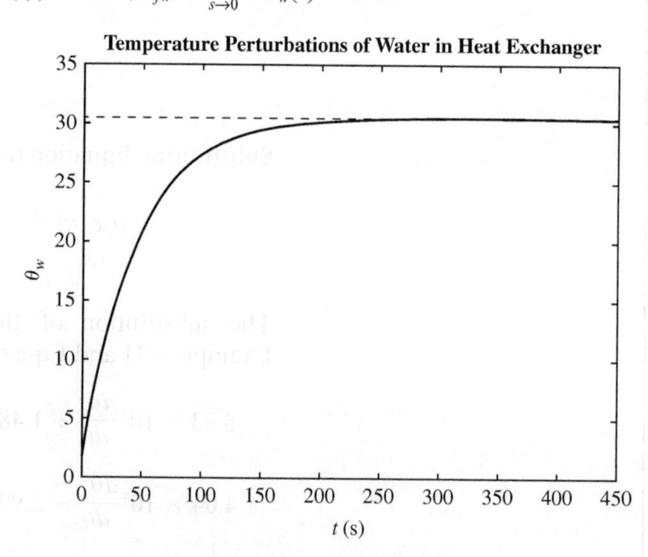

Figure 7.44

Temperature perturbations in (a) oil and (b) water for heat exchanger with a time delay in recirculating water.

7.9 Summary

7.9.1 Chapter Highlights

- Chapter 7 is the first part of the study of system response: the transient response of dynamic systems.
- The chapter uses the transfer function to derive and study transient response.
- The free response is the system response due to nonzero initial conditions and in the absence of additional system input.
- The impulsive response is the system response due to the input of a unit impulse.
- The impulsive response is obtained as the inverse of the transfer function.
- The impulsive response is continuous at $t = 0$ if the order of $N(s)$ is at least two less than the order of $D(s)$. Impulsive responses of first-order systems are always discontinuous.
- The step response is the system response due to the input of a unit step function.
- Transient response specifications, defined for the step response, include final value, settling time, rise time, percent overshoot, and peak time.
- The ramp response is the system response due to the input of a unit ramp function.
- A system is stable if and only if all its poles have negative real parts.
- If one or more poles have a real part of zero and all other poles have negative real parts, then the system is neutrally stable.
- If any pole has a positive real part, then the system is unstable.
- A necessary, but not sufficient, condition for stability is if all coefficients of $D(s)$ are positive.
- Routh's criterion is used to determine stability from $D(s)$.
- Routh's criterion may be used to determine the relative stability with respect to $s = -p$ by examining the absolute stability of $D(s - p)$.
- The transient response of a first-order system is dependent on one parameter, the time constant.
- The transient response of a stable first-order system exponentially approaches its final value.
- The transient response of a second-order system is dependent on two system parameters: the natural frequency ω_n and the damping ratio ζ.
- The free response of an undamped system ($\zeta = 0$) is periodic.
- The free response of an underdamped system ($0 < \zeta < 1$) is cyclic, but not periodic and decays exponentially.
- The free-response of a critically damped system ($\zeta = 1$) decays exponentially.
- The free response of an overdamped system ($\zeta > 1$) decays exponentially, but at a slower rate than the response of a critically damped system.
- The impulsive and step responses of second-order systems are given in Tables 7.4 and 7.5, respectively.
- The transient response of a higher-order system is a linear combination of first-order and second-order system responses.
- The transfer function for systems with time delay includes exponential terms.
- Pade approximations are used to approximate the exponential terms that appear in the transfer functions for systems with time delays.
- MATLAB can be used to determine and plot the impulsive, step, and ramp responses, given the system's transfer function.
- MATLAB's symbolic capabilities can be used to determine the transfer functions for multidegree-of-freedom systems from the impedance matrix.

7.9.2 Important Equations

- Impulsive response

$$x_i(t) = \mathcal{L}^{-1}\{G(s)\} \tag{7.13}$$

- Initial value of the impulsive response

$$x_i(0) = \lim_{s\to\infty} sG(s) \tag{7.14}$$

- Transform of step response

$$X_s(s) = \frac{1}{s}G(s) \tag{7.16}$$

- Final value after step input

$$x_{s,f} = G(0) \tag{7.17}$$

- Convolution integral

$$x(t) = \int_0^t f(\tau)\, x_i(t-\tau)d\tau \tag{7.27}$$

- General form of the differential equation for a first-order system

$$T\frac{dx}{dt} + x = f(t) \tag{7.37}$$

- Transfer function for a first-order system

$$G(s) = \frac{1/T}{s + \frac{1}{T}} \tag{7.40}$$

- Free response of a first-order system

$$x(t) = x_0 e^{-\frac{t}{T}} \tag{7.42}$$

- Impulsive response of a first-order system

$$x_i(t) = \frac{1}{T}e^{-\frac{t}{T}}u(t) \tag{7.44}$$

- Step response of a first-order system

$$x_s(t) = (1 - e^{-\frac{t}{T}})u(t) \tag{7.47}$$

- Standard form of a differential equation for a second-order system

$$\ddot{x} + 2\zeta\omega_n\dot{x} + \omega_n^2 x = \frac{1}{m}f(t) \tag{7.58}$$

- Standard form of a transfer function for a second-order system

$$G(s) = \frac{As + B}{s^2 + 2\zeta\omega_n s + \omega_n^2} \tag{7.62}$$

- Poles of a transfer function of a second-order system

$$s_1 = -\zeta\omega_n - \omega_n\sqrt{\zeta^2 - 1} \tag{7.65a}$$

$$s_2 = -\zeta\omega_n + \omega_n\sqrt{\zeta^2 - 1} \tag{7.65b}$$

- Undamped free response of a second-order system

$$x(t) = A\sin(\omega_n t + \phi) \tag{7.68}$$

- Underdamped free response of a second-order system

$$x(t) = A_d e^{-\zeta\omega_n t}\sin(\omega_d t + \phi_d) \tag{7.77}$$

- Critically damped free response of a second-order system

$$x(t) = x_0 e^{-\omega_n t} + (\dot{x}_0 + \omega_n x_0)te^{-\omega_n t} \tag{7.82}$$

- Overdamped free response of a second-order system

$$x(t) = \frac{1}{2\sqrt{\zeta^2 - 1}}\left\{\left[x_0\left(-\zeta + \sqrt{\zeta^2 - 1}\right) - \frac{\dot{x}_0}{\omega_n}\right]e^{s_1 t} + \left[x_0\left(\zeta + \sqrt{\zeta^2 - 1}\right) + \frac{\dot{x}_0}{\omega_n}\right]e^{s_2 t}\right\} \tag{7.86}$$

Short Answer Problems

SA7.1 A transfer function has a numerator of order 3 and a denominator of order 5. Is the impulsive response for this system continuous?

SA7.2 A transfer function has a numerator of order 2 and a denominator of order 3. Is the impulsive response for this system continuous?

SA7.3 A transfer function has a numerator of order 4 and a denominator of order 4. Is the impulsive response for this system continuous?

SA7.4 A transfer function has a numerator of order 0 and a denominator of order 2. Is the impulsive response for this system continuous?

SA7.5 The transfer function for a system is

$$G(s) = \frac{3s + 15}{s^2 + 4s + 60}$$

Is the impulsive response continuous? If not, what is the initial value of the system's impulsive response?

SA7.6 The transfer function for a system is

$$G(s) = \frac{2s^2 + 6s + 19}{s^2 + 12s + 150}$$

Is the impulsive response continuous? If not, what is the initial value of the system's impulsive response?

SA7.7 The transfer function for a system is

$$G(s) = \frac{3s + 15}{s^2 + 4s + 60}$$

What is the final value of the system's step response?

SA7.8 Does the final value of the step response of the system whose transfer function is

$$G(s) = \frac{3s + 15}{s^3 + 2s^2 + 16s + 60}$$

exist? If so, what is it?

SA7.9 Does the final value of the step response of the system whose transfer function is

$$G(s) = \frac{5}{s^3 + 2s^2 + 16s + 30}$$

exist? If so, what is it?

SA7.10 Does the final value of the step response of the system whose transfer function is

$$G(s) = \frac{4s^2 + 12s + 29}{s^3 + 3s^2 + 12}$$

exist? If so, what is it?

For Problems SA7.11 through SA7.20, knowing that the system has the given poles, choose one of the below

a. The system is unstable.

b. The system is neutrally stable.

c. The system is stable.

d. There is not enough information to determine the system's stability.

SA7.11 A third-order system has poles of -2, -3, and -5.

SA7.12 A fifth-order system has poles of -5, $-3 + 2j$, and $-5 - 12j$.

SA7.13 A fifth-order system has poles of -4, $2j$, and $-3 + 4j$.

SA7.14 A fifth-order system has poles of -2, $3 + 2j$, and $-5 + 12j$.

SA7.15 A fourth-order system has poles of -2 and $-3 + 2j$.

SA7.16 A seventh-order system has poles of -3, $2j$, $-4 - 10j$, and $5 + 12j$.

SA7.17 A second-order system has a pole of $-3 + 2j$.

SA7.18 A fifth-order system has poles of 2, $3 + 2j$, and $5 + 12j$.

SA7.19 A fifth-order system has poles of -3 and $2j$.

SA7.20 A fifth-order system has poles of 2 and $2j$.

SA7.21 Determine the stability of the system which has the third-order transfer function

$$G(s) = \frac{2s + 3}{s^3 + 3s^2 + 7s + 25}$$

SA7.22 What is the time constant for a system whose transfer function is $G(s) = \frac{3}{s + 0.5}$?

SA7.23 What is the time constant for a system whose transfer function is $G(s) = \frac{1}{2s + 5}$?

SA7.24 What is the time constant for a system whose transfer function is $G(s) = \frac{3}{3s + 0.15}$?

SA7.25 What is the 2 percent settling time for a system whose transfer function is $G(s) = \frac{3}{s + 1}$?

SA7.26 What is the 2 percent settling time for a system whose transfer function is $G(s) = \frac{2}{s + 7}$?

SA7.27 What is the 10–90 percent rise time for a system whose transfer function is $G(s) = \frac{3}{s + 1}$?

SA7.28 What is the 10–90 percent rise time for a system whose transfer function is $G(s) = \frac{2}{s + 8}$?

SA7.29 What is the percent overshoot for a system whose transfer function is $G(s) = \frac{3}{s + 1}$?

SA7.30 What is the peak time for a system whose transfer function is $G(s) = \frac{3}{s + 1}$?

SA7.31 An *RC* circuit with a current source has a resistance of 100 Ω and a capacitance of 400 μF. What is the time constant for the system?

SA7.32 A single-tank liquid-level problem consists of a circular tank of radius 30 cm and a pipe of resistance $2 \frac{\text{s}}{\text{m}^2}$. What is the 2 percent settling time of the system?

SA7.33 A single-tank liquid-level problem consists of a circular tank of radius 30 cm and a pipe of resistance $2 \frac{\text{s}}{\text{m}^2}$. What is the final value of the perturbation in liquid level if the flow rate into the tank is suddenly changed from $15.0 \frac{\text{m}^3}{\text{s}}$ to $16.5 \frac{\text{m}^3}{\text{s}}$?

SA7.34 A spring of stiffness 200 N/m is in parallel with a viscous damper of damping coefficient $15 \frac{\text{N}\cdot\text{s}}{\text{m}}$. The system is subject to a force of 10 N that is applied suddenly. What is the final value of the displacement?

SA7.35 A spring of stiffness 200 N/m is in parallel with a viscous damper of damping coefficient $15 \frac{\text{N}\cdot\text{s}}{\text{m}}$. The system is subject to a force of 10 N that is applied suddenly. How long does it take for the response to reach 0.04 m?

SA7.36 A bumper is modeled by a spring of stiffness 100 N/m in parallel with a viscous damper of damping coefficient $400 \frac{\text{N}\cdot\text{s}}{\text{m}}$. The bumper is struck by an object that imparts an impulse of 20 N·s. How long does it take for the bumper to return to within 0.001 m of equilibrium?

SA7.37 If \$100,000 is deposited in a bank that offers an interest rate of 2.5% that is compounded continuously, how much money is available for withdrawal after 3 years?

SA7.38 If \$100,000 is deposited in a bank that offers an interest rate of 3.5% that is compounded continuously, how long will it take for the sum to grow to \$120,000?

SA7.39 If the birth rate of a species is 4% and the death rate is 7%, how long does it take for a population of 1,000,000 to decrease to 300,000? Assume a first-order model.

SA7.40 How do you find the ramp response using MATLAB?

Problems SA7.41–SA7.51 refer to the convolution integral solution written as

$$x(t) = \int_0^t F(\tau)\, x_i(t - \tau)\, d\tau$$

SA7.41 What is the meaning of $x_i(t)$?

SA7.42 What is $x_i(t)$ corresponding to a transfer function of $G(s) = \frac{3}{s + 1}$?

SA7.43 What is $x_i(t)$ corresponding to a transfer function of $G(s) = \frac{1}{s^2 + 4s + 49}$?

SA7.44 Set up, but do not evaluate, the convolution integral solution if $F(t) = e^{-2t}$ and $G(s) = \frac{3}{s + 3}$.

SA7.45 Set up, but do not evaluate, the convolution integral solution if $F(t) = e^{-t}$ and $G(s) = \frac{3}{s + 1}$.

SA7.46 Set up, but do not evaluate, the convolution integral solution at $t = 1.2$ s if $F(t)$ is as given in Figure SA7.46 and $G(s) = \frac{3}{s + 1}$.

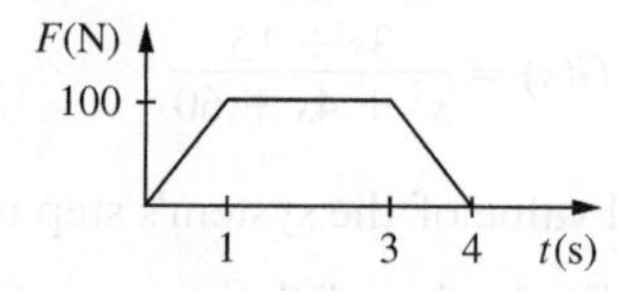

Figure SA7.46

SA7.47 Set up, but do not evaluate, the convolution integral solution at $t = 5$ s if $F(t)$ is as given in Figure SA7.46 and $G(s) = \frac{3}{s+1}$.

SA7.48 Set up, but do not evaluate, the convolution integral solution if $F(t) = e^{-t}$ for a critically damped second-order system with only a 1 in the numerator of its transfer function.

SA7.49 Set up, but do not evaluate, the convolution integral solution if $F(t) = e^{-t}$ for an overdamped second-order system with only a 1 in the numerator of its transfer function for $\zeta = 1.5$.

SA7.50 Set up, but do not evaluate, the convolution integral solution if $F(t) = e^{-t}$ for a critically damped second-order system with only s in the numerator of its transfer function.

SA7.51 Set up, but do not evaluate, the convolution integral solution if $F(t) = e^{-t}$ for an overdamped second-order system with only s in the numerator of its transfer function for $\zeta = 1.5$.

SA7.52 Determine the natural frequency and damping ratio for a system with a transfer function of $G(s) = \frac{1}{s^2 + 4s + 49}$.

SA7.53 Determine the natural frequency and damping ratio for a system with a transfer function of

$$G(s) = \frac{1}{s^2 + 28s + 49}.$$

SA7.54 Determine the damped natural frequency for a system with a transfer function of $G(s) = \frac{1}{s^2 + 9s + 81}$.

SA7.55 Determine the impulsive response of a second-order mechanical system with mass 1 kg, damping ratio 0.4, and natural frequency 10 r/s.

SA7.56 Determine the impulsive response of a second-order mechanical system with mass 1 kg, damping ratio 1, and natural frequency 10 r/s.

SA7.57 Determine the impulsive response of a second-order mechanical system with mass 1 kg, damping ratio 1.5, and natural frequency 10 r/s.

SA7.58 Determine the impulsive response of a series LRC circuit with $L = 1$ H, a natural frequency of 10 r/s, and a damping ratio of 0.5.

SA7.59 Determine the impulsive response of a series LRC circuit with $L = 1$ H, a natural frequency of 10 r/s, and a damping ratio of 1.

SA7.60 Determine the impulsive response of a series LRC circuit with $L = 1$ H, a natural frequency of 10 r/s, and a damping ratio of 1.5.

SA7.61 Determine the impulsive response for the system whose transfer function is

$$G(s) = \frac{2s + 3}{s^2 + 8s + 80}$$

SA7.62 Determine the impulsive response for the system whose transfer function is

$$G(s) = \frac{2s + 3}{s^2 + 14s + 49}$$

SA7.63 Determine the impulsive response for the system whose transfer function is

$$G(s) = \frac{2s + 3}{s^2 + 20s + 80}$$

SA7.64 Determine the step response of a second-order mechanical system with mass 1 kg, damping ratio 0.4, and natural frequency 10 r/s.

SA7.65 Determine the step response of a second-order mechanical system with mass 1 kg, damping ratio 1, and natural frequency 10 r/s.

SA7.66 Determine the step response of a second-order mechanical system with mass 1 kg, damping ratio 1.5, and natural frequency 10 r/s.

SA7.67 Determine the step response of a series LRC circuit with $L = 1$ H, a natural frequency of 10 r/s, and a damping ratio of 0.5.

SA7.68 Determine the step response of a series LRC circuit with $L = 1$ H, a natural frequency of 10 r/s, and a damping ratio of 1.

SA7.69 Determine the step response of a series LRC circuit with $L = 1$ H, a natural frequency of 10 r/s, and a damping ratio of 1.5.

SA7.70 Determine the step response for the system whose transfer function is

$$G(s) = \frac{2s + 3}{s^2 + 8s + 80}$$

SA7.71 Determine the step response for the system whose transfer function is

$$G(s) = \frac{2s + 3}{s^2 + 14s + 49}$$

SA7.72 Determine the step response for the system whose transfer function is

$$G(s) = \frac{2s + 3}{s^2 + 20s + 80}$$

SA7.73 Construct the transfer function for a circuit whose poles are $-2, -5, -1 \pm 3j$, and $-2 \pm 2j$ and whose zeros are $-3, -4$, and $-5 \pm 4j$.

SA7.74 A system has a transfer function of

$$G(s) = \frac{1}{s^2 + 10s + 89 - 12e^{-s}}$$

Use a first-order Pade approximation for e^{-s} to write the transfer function as a ratio of two polynomials.

Problems

7.1 The transfer function of a 3rd-order system is

$$G(s) = \frac{s + 4}{s^3 + 8s^2 + 15s + 20}$$

Use Routh's stability criterion to determine if the system is stable.

7.2 The transfer function for a 4th-order system is

$$G(s) = \frac{3s^2 + 10s + 120}{s^4 + 12s^3 + 20s^2 + 50s + 100}$$

Use Routh's criterion to determine if the system is stable.

7.3 The transfer function for a 5th-order system is

$$G(s) = \frac{s^3 + 5s^2 + 14s + 150}{s^5 + 2s^4 + 10s^3 + 20s^2 + 60s + 100}$$

Use Routh's criterion to determine if the system is stable.

7.4 The transfer function for a 3rd-order system is

$$G(s) = \frac{3s^2 + 10s + 120}{s^3 + 3s^2 + 50s + 100}$$

Use Routh's criterion to determine if the system is stable relative to $s = -2$.

7.5 The transfer function for a 4th-order system is

$$G(s) = \frac{s^2 + 100s + 120}{s^4 + 15s^3 + 30s^2 + 100s + 800}$$

Use Routh's criterion to determine if the system is stable relative to $s = -1$.

7.6 The transfer function of a 2nd-order system in a feedback loop with an integral controller is

$$G(s) = \frac{K_i(4s + 25)}{s^3 + 5s^2 + (10 + 4K_i)s + 25K_i}$$

(a) Determine a criterion for K_i, the integral gain, such that the system is stable. (b) Use MATLAB to draw impulsive responses of the system for several values of K_i to confirm the result of part (a).

7.7 The transfer function of a 3rd-order system is

$$G(s) = \frac{2s + 4}{s^3 + 4s^2 + 5s + 3}$$

Use Routh's criteria to determine (a) the absolute stability of the system and (b) the stability of the system relative to $s = -1$.

7.8 The transfer function for a 5th-order system is

$$G(s) = \frac{s^3 + 150s + 1200}{s^5 + 10s^4 + 20s^3 + 40s^2 + 100s + 500}$$

Use Routh's criterion to determine if the system is stable.

7.9 The closed-loop transfer function for a plant placed in a feedback control loop with a proportional controller is

$$H(s) = \frac{KG(s)}{1 + KG(s)}$$

where K is the proportional gain and $G(s)$ is the transfer function of the plant. The transfer function of a fourth-order plant is

$$G(s) = \frac{s + 2}{s(s - 2)(s^2 + 6s + 5)}$$

For what values of K is the closed-loop system stable?

7.10 The transfer function for a third-order plant is

$$G(s) = \frac{2s + 5}{(s + 3)(s^2 + 2s + 10)}$$

For what values of K is the plant closed loop stable (see Problem 7.9)?

7.11 The transfer function of a third-order plant is

$$G(s) = \frac{s^2 + 6s + 25}{(s + 4)(s^2 + 10s + 106)}$$

For what values of K is the plant closed loop stable relative to $s = -3$ (see Problem 7.9)?

7.12 The two-tank liquid-level system of Example 7.19 is placed in a feedback control loop with a proportional controller as discussed in Problem 7.9. For what values of the proportional gain K is the closed-loop system stable?

7.13 Consider the circuit of Figure P7.13. (a) Determine the time constant for the system. (b) Determine the response of the system when $v(t) = 10u(t)$ V.

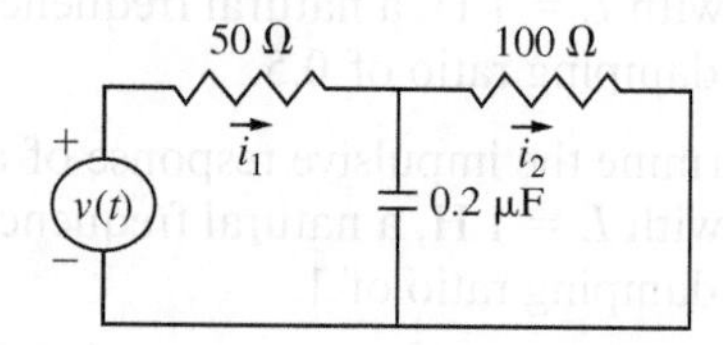

Figure P7.13

7.14 The capacitor in the system of Figure P7.14 has an initial charge of 1.3×10^{-3} C. The switch is closed at $t = 0$, allowing the capacitor to discharge. How long after the switch is closed does it take for the charge on the capacitor to drop to 5×10^{-6} C?

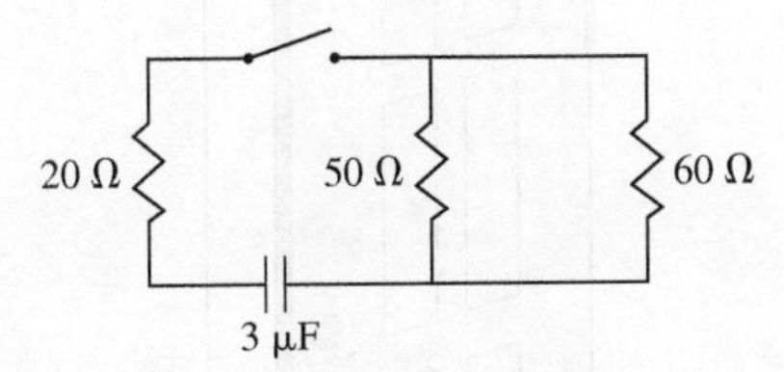

Figure P7.14

7.15 Consider the RL circuit of Figure P7.15 when the input current is a unit impulse. Determine (a) the voltage drop across the inductor and (b) the current through the inductor.

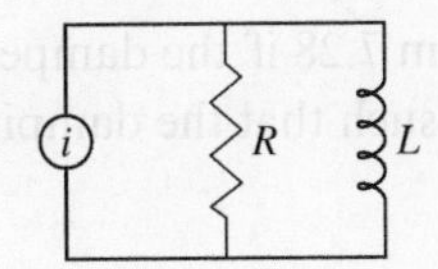

Figure P7.15

7.16 A first-order system has the transfer function

$$G(s) = \frac{2}{4s + 15}$$

Determine (a) the system's step response; (b) the system's impulsive response; (c) the system's ramp response; (d) the system's 2 percent settling time; (e) the system's 10–90 percent rise time; (f) the system's final value; (g) the system's peak time; and (h) the percent overshoot.

7.17 A first-order plant has a transfer function of

$$G(s) = \frac{s}{s + 17}$$

Determine (a) the time constant for the system; (b) the system's impulsive response; (c) the system's step response; (d) the system's final value; and (e) the system's peak time.

7.18 A first-order plant has a transfer function of

$$G(s) = \frac{s + 4}{s + 8}$$

Determine (a) the time constant for the system; (b) the system's impulsive response; (c) the system's step response; (d) the system's final value; (e) the system's peak time; and (f) the percent overshoot.

7.19 Reconsider the pharmokinetic system of Example 7.30. (a) Determine the concentration in the plasma as a function of time using the one compartment model when the drug is infused again 2 hr after being expended. (b) Set up a general model for the plasma concentration when this process of infusion is continually repeated; that is, the drug is infused at a constant rate for 12 hr, no drug is infused for 2 hr, then the drug is infused at a constant rate for the next 12 hr and so on.

7.20 The system of Figure P7.20 is at steady state when the exit pressure is suddenly changed from atmospheric pressure to a gauge pressure of 2.5 kPa. The differential equation obtained in Equation (k) of Example 6.7 for the perturbation in liquid level due to a perturbation in pressure is

$$A\frac{dh}{dt} + \frac{1}{R}h = \frac{p}{\rho g R}$$

Determine (a) the level of the liquid when a new steady state is reached; (b) the time-dependent response of the system; and (c) the 4 percent settling time.

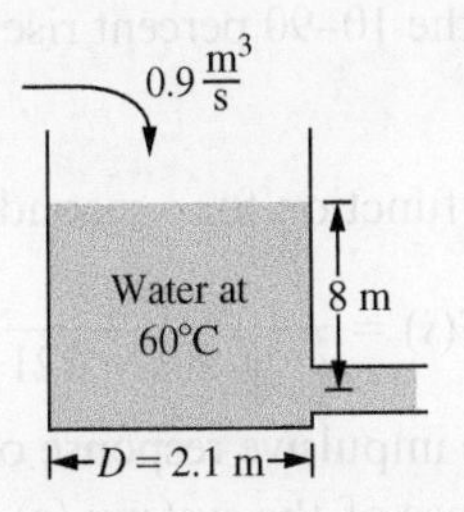

Figure P7.20

7.21 A 20-ft steel 32-Btu/(hr·ft·F) pipe has an inner diameter of 12 in. and an outer diameter of 14 in. A 0.5-in. layer of insulation of R value 15 covers the pipe. The film coefficients at the inner and outer surfaces of the system are 2.1 Btu/(hr·ft²·F) and 3.2 Btu/(hr·ft²·F), respectively. The pipe is at a uniform temperature of 70 F when the external temperature is suddenly changed to 68 F. (a) What is the time constant for the system? (b) Determine the time-dependent temperature in the interior of the pipe. (c) How long does it take for the temperature in the interior of the pipe to reach 68.3 F?

7.22 A tank of volume 20 m³ initially contains air at a temperature of 40 C and a pressure of 12 kPa. An upstream valve is opened, allowing air from a supply of pressure 20 kPa to flow into the tank. The resistance between the supply and the tank is $16.01(\text{m}\cdot\text{s})^{-1}$. The process is isothermal. (a) What is the time constant for the system? (b) Determine the pressure in the tank as a function of time. (c) How long does it take for the pressure in the tank to reach 18 kPa?

7.23 The transfer function for a second-order system is

$$G(s) = \frac{2s + 3}{4s^2 + 10s + 45}$$

Determine the natural frequency and damping ratio for the system.

7.24 The transfer function for a second-order system is

$$G(s) = \frac{3s + 5}{s^2 + 20s + 500}$$

Determine (a) the impulsive response of the system; (b) the step response of the system; (c) the 2 percent settling time; (d) the 10–90 percent rise time; (e) the percent overshoot of the step response; and (f) the peak time of the step response.

7.25 The transfer function for a second-order system is

$$G(s) = \frac{1}{s^2 + 20s + 50}$$

Determine (a) the impulsive response of the system; (b) the step response of the system; (c) the 2 percent settling time; (d) the 10–90 percent rise time; and (e) the peak time.

7.26 The transfer function for a second-order system is

$$G(s) = \frac{1}{s^2 + 20s + 221}$$

Determine (a) the impulsive response of the system; (b) the step response of the system; (c) the 2 percent settling time, (d) the 10–90 percent rise time; and (e) the peak time.

7.27 For the system for Figure P7.27, determine (a) its free response when the block is displaced 2 mm from equilibrium and then released; (b) its impulsive response; (c) its step response; and (d) its ramp response.

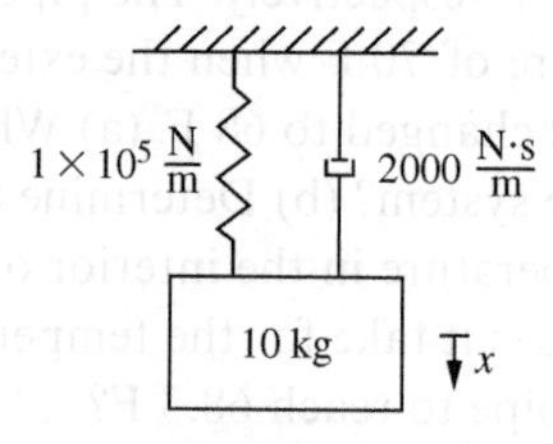

Figure P7.27

7.28 The door of Figure P7.28 is free to rotate about an axis through its hinges. The door has a moment of inertia about an axis through its hinges of 25.2 kg·m². Each hinge has a torsional stiffness of 15 N·m/r. A door damper provides torsional damping such that the system is critically damped. (a) If the door is held open at an angle of 50° and then released, how long does it take for the door to close to an angle of 5° where it automatically latches shut? (b) If the door is initially closed, what initial angular velocity must be imparted to it to cause it to open to an angle of 70°?

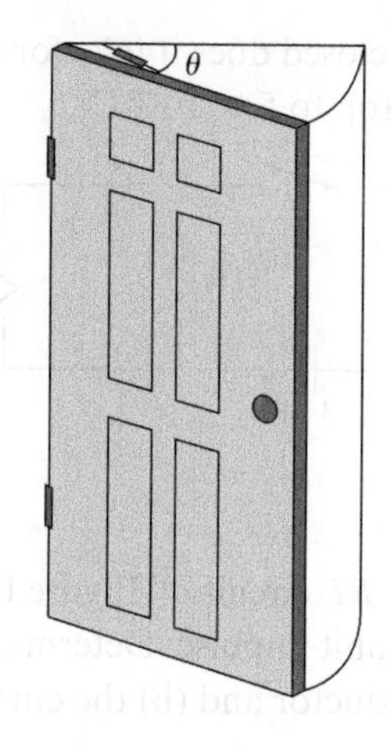

Figure P7.28

7.29 Repeat Problem 7.28 if the damper provides torsional damping such that the damping ratio of the system is 1.15.

7.30 Figure P7.30 illustrates a simplified model of a vehicle suspension system. Let $y(t)$ be the displacement of the wheel as it traverses the road contour, and $x(t)$ be the displacement of the vehicle. (a) Show that the mathematical model for $x(t)$ is

$$m\ddot{x} + c\dot{x} + kx = c\dot{y} + ky$$

(b) Consider a vehicle with $m = 500$ kg, $k = 3.2 \times 10^5$ N/m, and $c = 10{,}000$ N·s/m. The vehicle is traveling with a constant horizontal velocity of 80 km/hr. Determine the response of the vehicle after it encounters a sudden dip of 12.8 mm in the road.

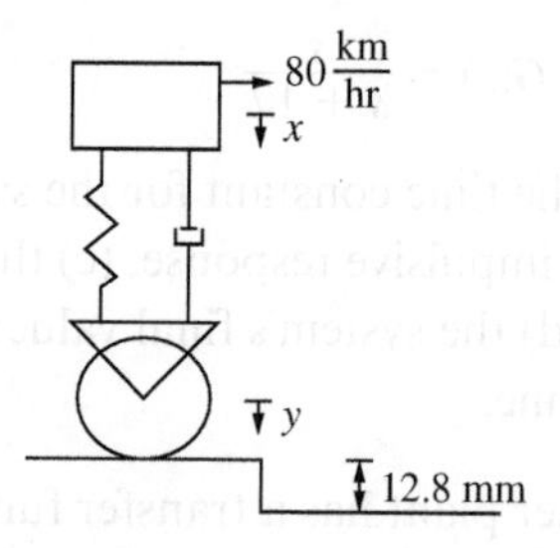

Figure P7.30

7.31 The capacitor in the circuit of Figure P7.31 has an initial charge q_0. The switch is open and then closed at $t = 0$. Determine the resulting current through the inductor if $L = 0.2$ H, $C = 0.4$ μF, $q_0 = 3.1$ μC, and (a) $R = 1000\ \Omega$, (b) $R = 2000\ \Omega$, and (c) $R = 3000\ \Omega$.

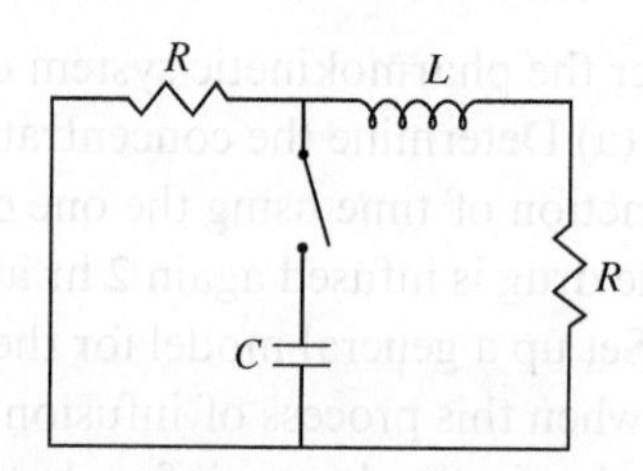

Figure P7.31

7.32 The switch in the circuit of Figure P7.32 is open and then closed at $t = 0$. (a) Determine the transfer function $G(s) = V_2(s)/V_1(s)$. (b) Determine the response $v_2(t)$ if $L = 0.25$ H, $R = 3.6$ kΩ, $C = 0.15$ μH, and $v_1 = 12$ V.

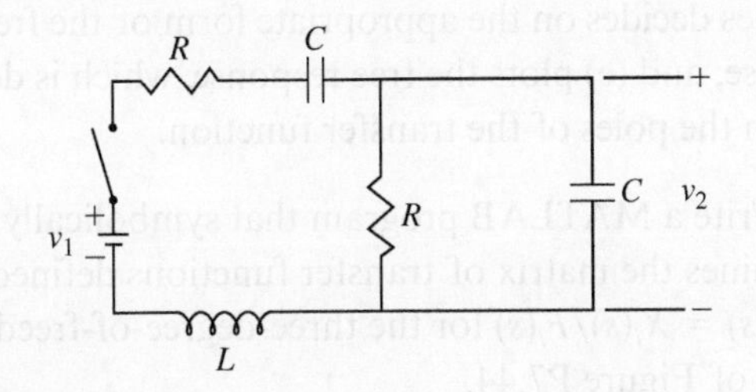

Figure P7.32

7.33 Determine $v_2(t)$ when $v_1(t) = 20u(t)$ V for the operational amplifier circuit of Figure P7.33.

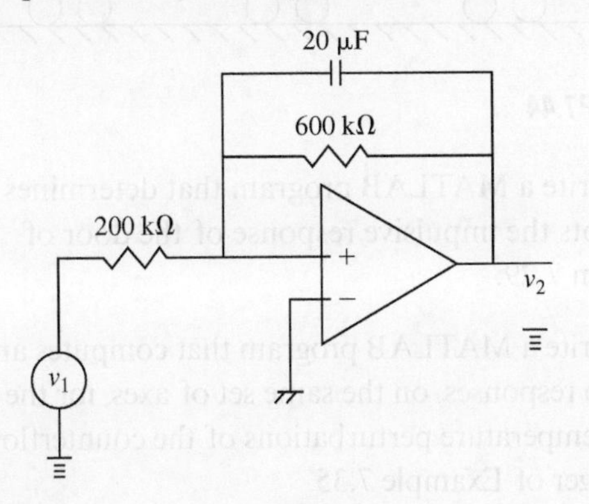

Figure P7.33

7.34 Determine the liquid-level perturbations $h_1(t)$ and $h_2(t)$ for the system of Problem 6.14.

7.35 Determine the liquid-level perturbations in each of the tanks of Problem 6.16 if the steady-state levels are $h_{1s} = 12$ m, $h_{2s} = 15$ m, and $h_{3s} = 12$ m. The steady-state inlet flow rate is 2.25 m³/s and the flow rate suddenly increases to 2.31 m³/s. The tanks are all circular with a diameter of 6.5 m.

7.36 Determine the natural frequencies of the two-degree-of-freedom mechanical system of Figure P7.36.

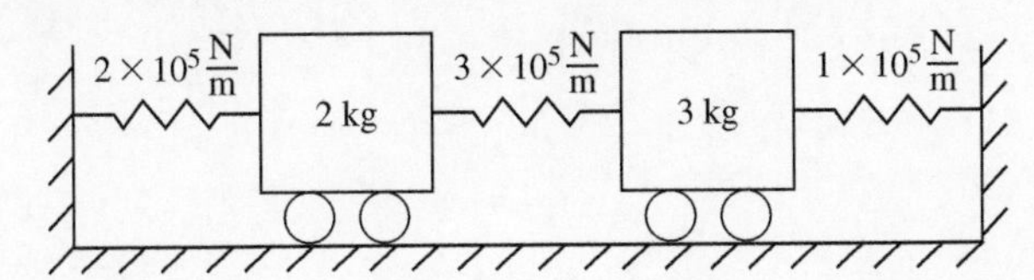

Figure P7.36

7.37 The machine of Figure P7.37 is mounted on a viscoelastic foundation that is modeled as a spring in parallel with a spring and viscous damper in series. This model of the isolator is called a Maxwell model. Let $x(t)$ represent the response of the machine when subject to a force $F(t)$. (a) Determine the transfer function $G_1(s) = X(s)/F(s)$. (b) The force transmitted to the foundation through the isolator is $F_T = k_1 x + c\dot{z}$, where $z(t)$ is the displacement of the joint between the spring and the viscous damper. Determine the transfer function $G_2(s) = F_T(s)/F(s)$. (c) Determine the impulsive response for $x(t)$ and $F_T(t)$.

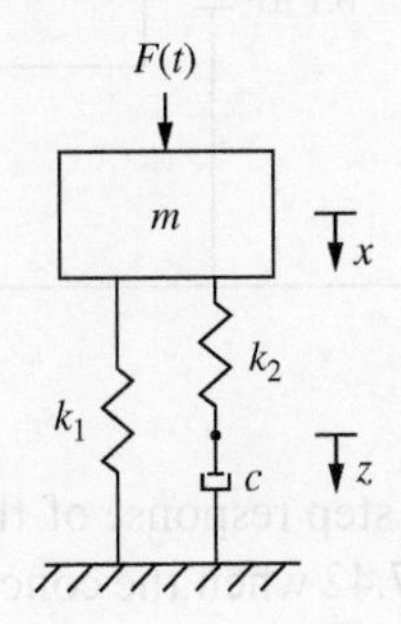

Figure P7.37

7.38 Determine and plot, for the system of Figure P7.38, its response $\theta(t)$ (a) when $F(t) = 10\delta(t)$ and (b) when $F(t) = 100u(t)$.

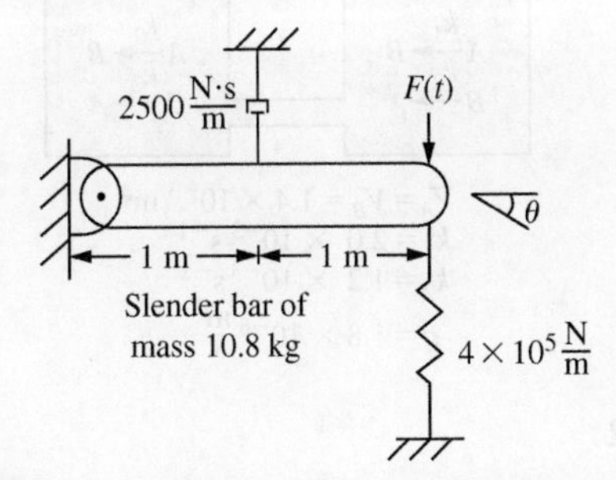

Figure P7.38

7.39 Determine and plot, for the system of Figure P7.39, its response $i(t)$ (a) when $v(t) = 10\delta(t)$, (b) when $v(t) = 10u(t)$, and (c) when $v(t) = 0.5t$.

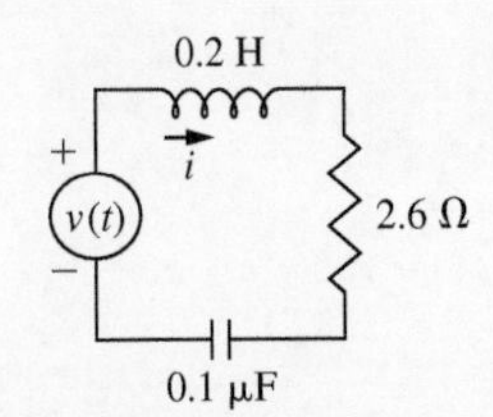

Figure P7.39

7.40 Determine the step response for v_2 in the circuit of Figure P7.40.

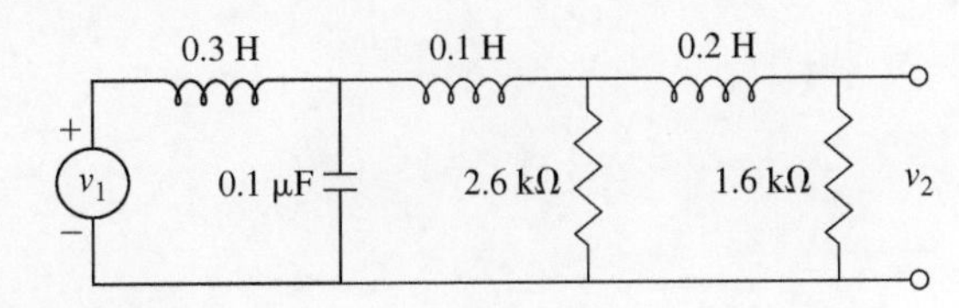

Figure P7.40

7.41 Determine the step response of the operational amplifier circuit of Figure P7.41.

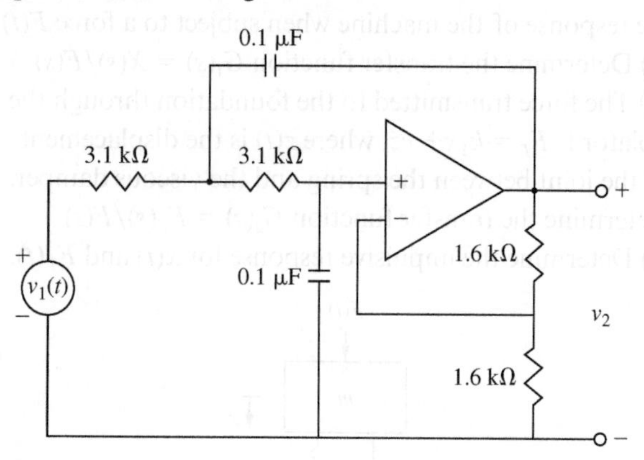

Figure P7.41

7.42 Determine the step response of the two-CSTR system of Figure P7.42 when the concentration of the reactant A in the inlet stream suddenly changes. The mixture takes 2.5 s to travel between the reactors.

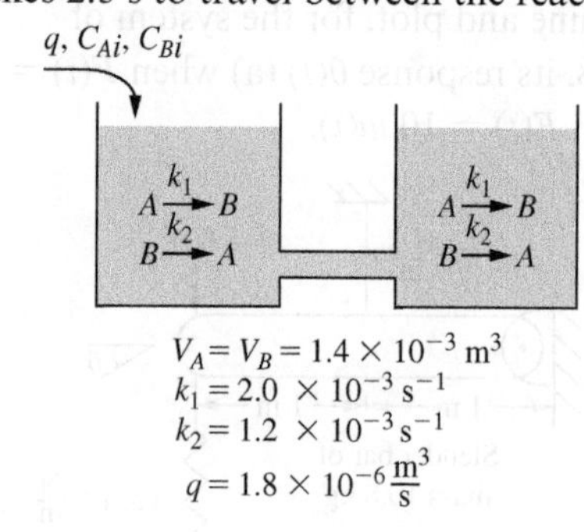

Figure P7.42

7.43 Consider the system of Problem 7.32. Write a MATLAB program that does all the following: (a) inputs values of all parameters; (b) specifies the transfer function from the numerator and denominator; (c) determines the poles of the transfer function; (d) from the poles decides on the appropriate form of the free response; and (e) plots the free response, which is dependent on the poles of the transfer function.

7.44 Write a MATLAB program that symbolically determines the matrix of transfer functions defined by $G_{i,j}(s) = X_i(s)/F_j(s)$ for the three-degree-of-freedom system of Figure P7.44.

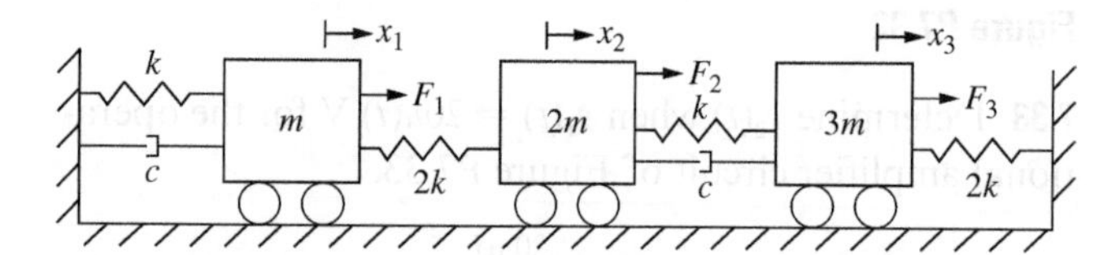

Figure P7.44

7.45 Write a MATLAB program that determines and plots the impulsive response of the door of Problem 7.29.

7.46 Write a MATLAB program that computes and plots the step responses, on the same set of axes, for the oil and water temperature perturbations of the counterflow heat exchanger of Example 7.35.

8 Frequency Response

Learning Objectives

At the conclusion of this chapter, you should be able to:

- Recognize that resonance, which is characterized by an unbounded growth of the response, occurs in an undamped system when the excitation frequency coincides with the natural frequency
- Recognize that beating, which is characterized by a continual buildup and decay in amplitude, occurs for an undamped system when the excitation frequency is close, but not equal, to the natural frequency
- Determine the sinusoidal transfer function, $G(j\omega)$, for a system and know how to represent it in polar form
- Use the sinusoidal transfer function to determine the steady-state amplitude
- Understand the concept of frequency response
- Use the sinusoidal transfer function to draw frequency response diagrams
- Construct Bode diagrams from the sinusoidal transfer function
- Construct Nyquist diagrams from the real and imaginary parts of the sinusoidal transfer function
- Use MATLAB to generate Bode diagrams and Nyquist plots
- Determine the response due to sinusoidal input and draw frequency response curves, Bode diagrams, and Nyquist diagrams for a first-order system
- Determine the response due to sinusoidal input and draw frequency response curves, Bode diagrams, and Nyquist diagrams for a second-order system, including mechanical systems with force input and with motion input
- Determine if a circuit may be used as a low-pass, high-pass, band-pass, or band-reject filter
- Understand the use of vibration isolators
- Understand the design of vibration absorbers
- Analyze systems with a general periodic input
- Determine the response due to sinusoidal input and draw frequency response curves, Bode diagrams, and Nyquist diagrams for a higher-order system

The general form of a single-frequency harmonic input is

$$F(t) = F_0 \sin(\omega t + \psi) \tag{8.1}$$

where ω is the frequency of the input, F_0 is its amplitude, and ψ is its phase. The parameters are defined such that $\omega > 0$, $F_0 > 0$, and $-\pi < \psi \leq \pi$. If $\psi = 0$ or $\psi = \pi$, the input is purely sinusoidal. If $\psi = \pi/2$, the input is a cosine function of frequency ω. For other values of ψ, the input is a linear combination of the sine and cosine functions.

The total response of a dynamic system due to a harmonic input is the sum of a transient response, which is independent of the frequency of the input, and a response that depends on the frequency of the input. For a stable system, as shown in Chapter 7, the transient response decays exponentially and eventually becomes insignificant compared to the response that is specific to the input. When this occurs, the system has reached steady state and the remaining response is called the **steady-state response.**

The steady-state response of a linear system with dependent variable $x(t)$, due to a sinusoidal input of the form of Equation (8.1), is shown to be of the form

$$x(t) = X\sin(\omega t + \psi + \phi) \tag{8.2}$$

where X is called the **steady-state amplitude** and ϕ is called the **steady-state phase**. The steady-state amplitude and steady-state phase are functions of the input frequency, the input amplitude, and system parameters, but are independent of the initial conditions and input phase. Thus, without loss of generality, the initial conditions and input phase are taken to be zero when studying steady-state response.

The **frequency response** is the study of the dependence of the steady-state response on the input frequency. The frequency response is studied in the **time domain** by determining the steady-state response of the system as a function of time, as well as the **frequency domain** (also called the **s domain**) by studying the Laplace transform of the response. The frequency domain response is often studied using graphical tools.

8.1 Undamped Second-Order Systems

The response of an undamped second-order or higher system due to a harmonic input is a special case in studying the frequency response and is treated separately. The general form of the differential equation for a second-order system with sinusoidal input is

$$\ddot{x} + \omega_n^2 x = \frac{1}{m}F_0\sin(\omega t) \tag{8.3}$$

where m is the mass of a mechanical mass-spring system.

Taking the Laplace transform of Equation (8.3) assuming both initial conditions are zero gives $(s^2 + \omega_n^2)X(s) = \dfrac{F_0\omega}{m(s^2 + \omega^2)}$. Solving for $X(s)$ leads to

$$X(s) = \frac{F_0\omega}{m(s^2 + \omega^2)(s^2 + \omega_n^2)} \tag{8.4}$$

A partial fraction decomposition of Equation (8.4) using entry 7 of Table 2.3 is obtained:

$$X(s) = \frac{F_0\omega}{m(\omega_n^2 - \omega^2)}\left[\frac{1}{s^2 + \omega^2} - \frac{1}{s^2 + \omega_n^2}\right] \tag{8.5}$$

For $\omega \neq \omega_n$, inversion of Equation (8.5) leads to

$$x(t) = \frac{F_0}{m(\omega_n^2 - \omega^2)}\left[\sin(\omega t) - \frac{\omega}{\omega_n}\sin(\omega_n t)\right] \tag{8.6}$$

The $\sin(\omega_n t)$ term in Equation (8.6) represents the transient response, whereas the $\sin(\omega t)$ term is the part of the response specific to the input. An undamped system model is often used to simplify the system analysis, but all real systems have some form of energy dissipation. Damping leads to the decay of the transient solution, leaving only the steady-state response after a long time. This response is

$$x(t) = \frac{F_0}{m(\omega_n^2 - \omega^2)}\sin(\omega t) \tag{8.7}$$

Equation (8.7) is of the form of Equation (8.2) with

$$X = \left|\frac{F_0}{m(\omega_n^2 - \omega^2)}\right| \tag{8.8}$$

$$\phi = \begin{cases} 0 & \omega < \omega_n \\ \pi & \omega > \omega_n \end{cases} \tag{8.9}$$

The steady-state amplitude is defined as positive. Then the response is in phase with the input when the input frequency is less than the natural frequency and 180° out of phase with the input when the input frequency is greater than the natural frequency.

A special case occurs when the input frequency coincides with the natural frequency. In this case, Equation (8.4) is written as

$$X(s) = \frac{F_0\omega_n}{m(s^2 + \omega_n^2)^2} \tag{8.10}$$

Equation (8.10) can be rewritten as

$$X(s) = \frac{F_0}{2m\omega_n}\left[\frac{1}{s^2 + \omega_n^2} - \frac{s^2 - \omega_n^2}{(s^2 + \omega_n^2)^2}\right] \tag{8.11}$$

Inversion of Equation (8.11) is performed using transform pairs 2 and 15 of Table 2.1, leading to

$$x(t) = \frac{F_0}{2m\omega_n^2}[\sin(\omega_n t) - \omega_n t\cos(\omega_n t)] \tag{8.12}$$

Equation (8.12) illustrates that when the input frequency for an undamped system coincides with its natural frequency, the system response grows without bound. This condition is called **resonance**. When resonance occurs, the system response grows without bound until either the system fails or the response becomes so large that basic assumptions made to linearize the system are no longer valid and nonlinear behavior occurs.

When an undamped second-order system has an initial energy input, its resulting free response is periodic with a frequency equal to the natural frequency. Since an undamped system has no dissipative mechanism, the response sustains itself without additional energy input. When the system is subject to a sinusoidal input, the response is a linear combination of a sinusoidal term at the natural frequency and a sinusoidal term at the input frequency. Some of the energy initiates the free response while much of the input energy leads to a response at the input frequency. However, when the input frequency coincides with the natural frequency, the energy input is not necessary to sustain the response. Thus energy continues to be stored in the system, leading to a system response that grows without bound.

When the input frequency is close to, but not exactly equal to, the natural frequency, the frequency ratio $r = \omega/\omega_n$ can be written as $r = 1 + \varepsilon$ where ε is a small number. In this case a trigonometric identity can be used with Equation (8.6) to yield

$$\begin{aligned} x(t) &= \frac{F_0}{m(\omega_n^2 - \omega^2)}[\sin(\omega t) - (1+\varepsilon)\sin(\omega_n t)] \\ &= \frac{F_0}{m(\omega_n^2 - \omega^2)}\left\{\sin\left[\left(\frac{\omega - \omega_n}{2}\right)t\right]\cos\left[\left(\frac{\omega + \omega_n}{2}\right)t\right] - \varepsilon\sin(\omega_n t)\right\} \\ &\approx \frac{F_0}{m(\omega_n^2 - \omega^2)}\sin\left[\left(\frac{\omega - \omega_n}{2}\right)t\right]\cos\left[\left(\frac{\omega + \omega_n}{2}\right)t\right] \end{aligned} \tag{8.13}$$

The response given by Equation (8.13) is characterized by a slow buildup and decrease in amplitude. The response is periodic of period $T = 4\pi/(\omega + \omega_n)$ but is cyclic of period $T_b = 2\pi/|\omega - \omega_n|$. This type of response is called **beating**, and T_b is the **period of beating**.

Example 8.1

A 40-kg machine is mounted on springs of equivalent stiffness 1×10^5 N/m. During operation it is subject to a harmonic force $F(t) = 100 \sin(\omega t)$ N. Determine and plot the system response for **a.** $\omega = 25$ r/s, **b.** $\omega = 50$ r/s, **c.** $\omega = 52$ r/s, and **d.** $\omega = 100$ r/s.

Solution

The system is modeled by a differential equation of the form of Equation (8.3) with $F_0 = 100$ N and

$$\begin{aligned} \omega_n &= \sqrt{\frac{k}{m}} \\ &= \sqrt{\frac{1 \times 10^5 \text{ N/m}}{40 \text{ kg}}} = 50 \text{ r/s} \end{aligned} \tag{a}$$

a. For $\omega = 25$ r/s, the response is given by Equation (8.6). The substitution of given values into Equation (8.6) leads to

$$\begin{aligned} x(t) &= \frac{100 \text{ N}}{40 \text{ kg}[(50 \text{ r/s})^2 - (25 \text{ r/s})^2]}\left[\sin(25t) - \frac{25 \text{ r/s}}{50 \text{ r/s}}\sin(50t)\right] \\ &= 1.33[\sin(25t) - 0.5\sin(50t)] \text{ mm} \end{aligned} \tag{b}$$

The effects of both the input frequency and the natural frequency are evident in the response, which is plotted in Figure 8.1(a). The response is periodic of period $T = 2\pi/25$ s.

b. The input frequency $\omega = 50$ r/s coincides with the system's natural frequency leading to resonance. The substitution of the given values into Equation (8.12) gives

$$\begin{aligned} x(t) &= \frac{100 \text{ N}}{2(40 \text{ kg})(50 \text{ r/s})^2}[\sin(50t) - 50t\cos(50t)] \\ &= 0.5[\sin(50t) - 50t\cos(50t)] \text{ mm} \end{aligned} \tag{c}$$

The response when the input frequency coincides with the natural frequency is illustrated in Figure 8.1(b). The response is cyclic but has a continual increase in amplitude illustrating the phenomenon of resonance.

c. The input frequency $\omega = 52$ r/s is close, but not exactly equal, to the natural frequency. Thus the response is closely approximated by Equation (8.13). Substituting the given values into Equation (8.13) leads to

$$x(t) = \frac{100 \text{ N}}{(40 \text{ kg})[(50 \text{ r/s})^2 - (52 \text{ r/s})^2]}\left[\sin\left(\frac{52-50}{2}t\right)\cos\left(\frac{52+50}{2}t\right)\right]$$
$$= -12.3 \sin(t)\cos(51t) \text{ mm} \quad \text{(d)}$$

Equation (d) is plotted in Figure 8.1(c), which illustrates the phenomenon of beating where there is a continual buildup and decay of amplitude. The period of the response is $T = (4\pi)/(\omega + \omega_n) = 0.123$ s. The period of beating is $T_b = (2\pi)/|\omega - \omega_n| = \pi$ s.

d. The input frequency $\omega = 100$ r/s is greater than the natural frequency. The substitution of values into Equation (8.6) leads to

$$x(t) = \frac{100 \text{ N}}{40 \text{ kg } [(50 \text{ r/s})^2 - (100 \text{ r/s})^2]}\left[\sin(100t) - \frac{100 \text{ r/s}}{50 \text{ r/s}}\sin(50t)\right]$$
$$= -0.333\,[\sin(100t) - 2\sin(50t)] \text{ mm} \quad \text{(e)}$$

Equation (e) is plotted in Figure 8.1(d).

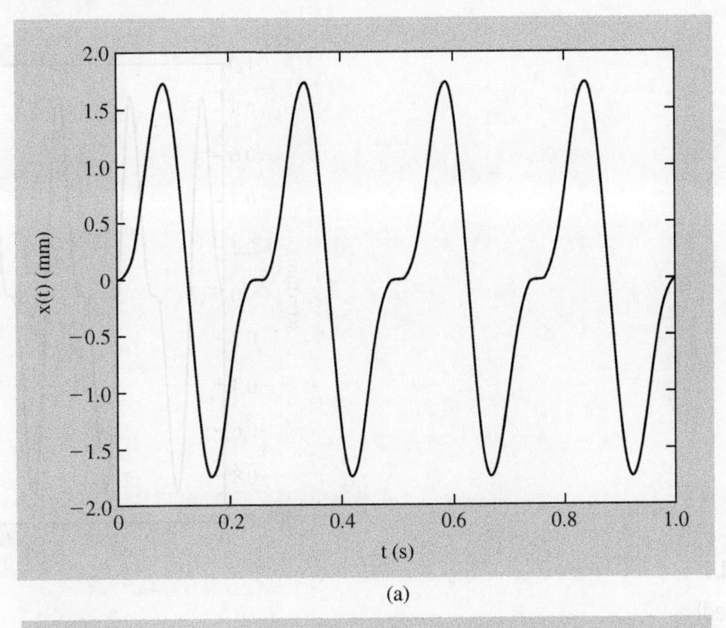

(a)

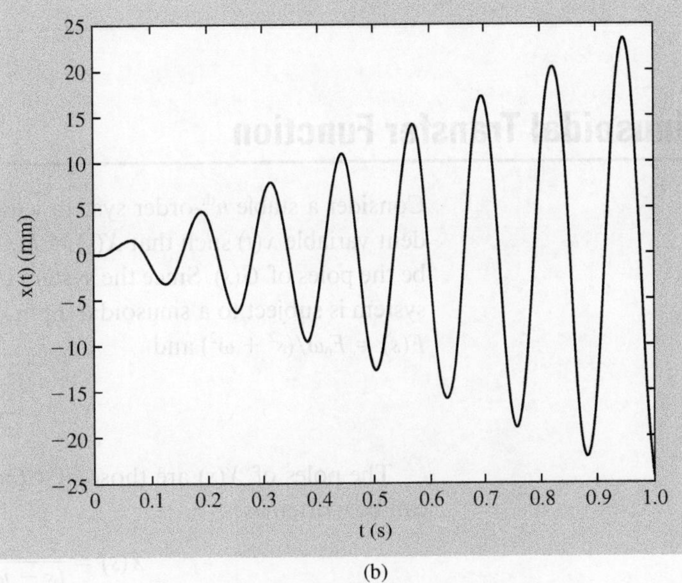

(b)

Figure 8.1

(a) The response of the undamped system of Example 8.1 with $\omega = 25$ r/s. The response is periodic of period $T = 2\pi/25$ s. (b) The response of the undamped system of Example 8.1 when the input frequency coincides with the natural frequency. The response illustrates the phenomenon of resonance characterized by unbounded amplitude growth. (c) The response of the undamped system of Example 8.1 when the input frequency is near, but not exactly equal to, the natural frequency. The beating phenomenon illustrated features a continual buildup and decrease in amplitude. (d) The response of the undamped system of Example 8.1 when the input frequency is twice the natural frequency.

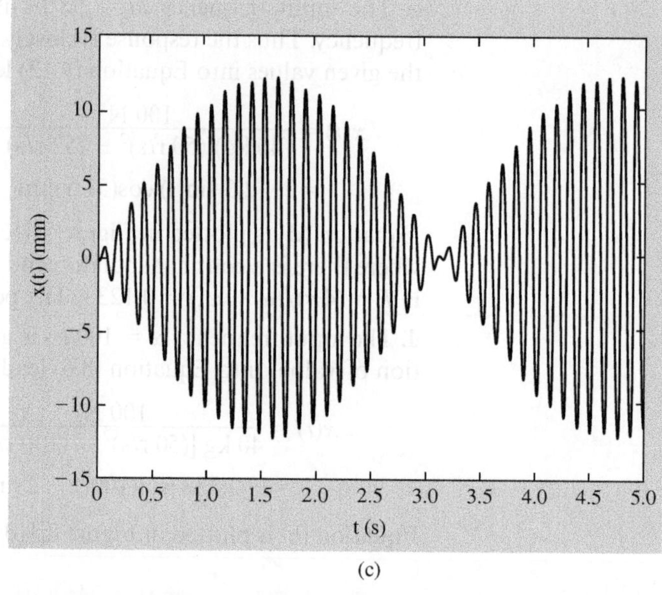

(c)

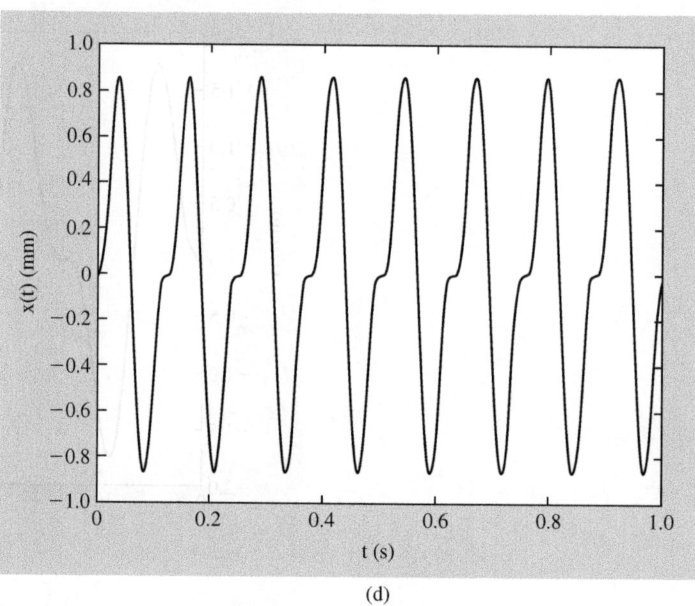

(d)

Figure 8.1
(Continued)

8.2 Sinusoidal Transfer Function

Consider a stable n^{th}-order system with a transfer function $G(s)$ defined for a dependent variable $x(t)$ such that $X(s) = F(s)G(s)$ for any system input $F(t)$. Let $s_1, s_2, \ldots, s_n$ be the poles of $G(s)$. Since the system is stable, all poles have negative real parts. If the system is subject to a sinusoidal input of the form of Equation (8.1) with $\psi = 0$, then $F(s) = F_0\omega/(s^2 + \omega^2)$ and

$$X(s) = \frac{F_0\omega}{s^2 + \omega^2}G(s) \tag{8.14}$$

The poles of $X(s)$ are those of $G(s)$ as well as $j\omega$ and $-j\omega$. Thus Equation (8.14) can be written as

$$X(s) = \frac{F_0\omega}{(s - j\omega)(s + j\omega)}G(s) \tag{8.15}$$

which has a partial fraction decomposition of the form

$$X(s) = \frac{B_1}{s - j\omega} + \frac{B_2}{s + j\omega} + \sum_{k=1}^{n} \frac{A_k}{s - s_k} \tag{8.16}$$

Inversion of Equation (8.16) leads to

$$x(t) = B_1 e^{j\omega t} + B_2 e^{-j\omega t} + \sum_{k=1}^{n} A_k e^{s_k t} \tag{8.17}$$

Since the system is stable and all poles of the transfer function have negative real parts, all terms in the summation in Equation (8.17) decay exponentially. After some period of time these terms will be insignificant compared to the first two terms, which constitute the steady-state response:

$$x_s(t) = B_1 e^{j\omega t} + B_2 e^{-j\omega t} \tag{8.18}$$

The coefficients B_1 and B_2 are the residues of linear factors and are determined using the residue theorem developed in Section 2.4.2 as

$$B_1 = (s - j\omega)X(s)|_{s = j\omega} \tag{8.19}$$

$$B_2 = (s + j\omega)X(s)|_{s = -j\omega} \tag{8.20}$$

Use of Equation (8.15) in Equations (8.19) and (8.20) leads to

$$B_1 = \frac{F_0 \omega}{2j\omega} G(j\omega)$$

$$= -j\frac{F_0}{2} G(j\omega) \tag{8.21}$$

$$B_2 = \frac{F_0 \omega}{-2j\omega} G(-j\omega)$$

$$= j\frac{F_0}{2} G(-j\omega) \tag{8.22}$$

Substitution of Equations (8.21) and (8.22) in Equation (8.18) leads to

$$x_s(t) = j\frac{F_0}{2}[G(-j\omega)e^{-j\omega t} - G(j\omega)e^{j\omega t}] \tag{8.23}$$

$G(j\omega)$ can be written as

$$G(j\omega) = \text{Re}[G(j\omega)] + j\,\text{Im}[G(j\omega)] \tag{8.24}$$

or in a polar form

$$G(j\omega) = |G(j\omega)|\,e^{j\phi} \tag{8.25}$$

where the magnitude of $G(j\omega)$ is

$$|G(j\omega)| = \sqrt{\{\text{Re}[G(j\omega)]\}^2 + \{\text{Im}[G(j\omega)]\}^2} \tag{8.26}$$

and its phase is

$$\phi = \tan^{-1}\left\{\frac{\text{Im}[G(j\omega)]}{\text{Re}[G(j\omega)]}\right\} \tag{8.27}$$

$G(-j\omega)$ is the complex conjugate of $G(j\omega)$. Thus from Equations (8.24) and (8.25),

$$G(-j\omega) = \text{Re}[G(j\omega)] - j\text{Im}[G(j\omega)] \tag{8.28}$$

and

$$G(-j\omega) = |G(j\omega)|\,e^{-j\phi} \tag{8.29}$$

Using Equations (8.25) and (8.27) in Equation (8.23) leads to

$$x_s(t) = j\frac{F_0}{2}|G(j\omega)|[e^{-j(\omega t+\phi)} - e^{j(\omega t+\phi)}] \tag{8.30}$$

Equation (8.30) is equivalent to

$$x_s(t) = F_0|G(j\omega)|\sin(\omega t + \phi) \tag{8.31}$$

Equation (8.31) is of the form of Equation (8.2) where the steady-state amplitude is $F_0|G(j\omega)|$ and the phase is the same as the phase of $G(j\omega)$.

Thus the steady-state response of a stable system due to a sinusoidal input is obtained from the system's transfer function. The steady-state amplitude of the system is the amplitude of the input times the magnitude of the transfer function evaluated for $s = j\omega$, and the phase in the steady state is the phase of the same transfer function.

$G(j\omega)$ is referred to as the **sinusoidal transfer function** because it is used to determine the steady-state response of a system due to a sinusoidal input. It is also used to determine the frequency response of a system, which is the variation of the steady-state amplitude and phase with ω.

Example 8.2

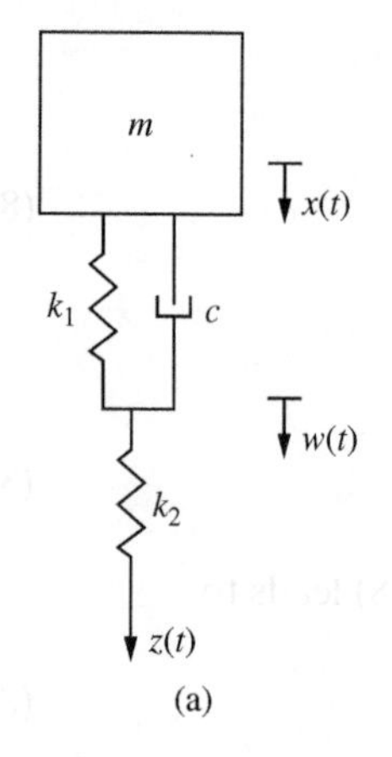

Figure 4.68

(a) The system of Example 4.32; and Example 8.2. (Repeated)

A simplified model of a vehicle suspension system including elasticity of the wheel in Example 4.32 is shown in Figure 4.68(a). Consider a system with parameters $m = 500$ kg, $c = 17{,}800 \frac{\text{N}\cdot\text{s}}{\text{m}}$, $k_1 = 2.45 \times 10^6$ N/m, and $k_2 = 1.47 \times 10^6$ N/m. The transfer function for such a model of the suspension system is obtained using the mathematical model, Equation (g) of Example 4.32, as the mathematical model

$$G(s) = \frac{X(s)}{Z(s)} = \frac{1.75 \times 10^4 s + 2.45 \times 10^6}{5.95s^3 + 1.33 \times 10^3 s^2 + 1.75 \times 10^4 s + 2.45 \times 10^6} \tag{a}$$

Determine the steady-state response of the system when

$$z(t) = 0.01\sin(100t) \tag{b}$$

Solution

The steady-state response is of the form of Equation (8.31) where ϕ is determined using Equation (8.27), $F_0 = 0.01$, and $\omega = 100$. From Equation (a)

$$\begin{aligned} G(j100) &= \frac{1.75 \times 10^4(100j) + 2.45 \times 10^6}{5.95(100j)^3 + 1.33 \times 10^3(100j)^2 + 1.75 \times 10^4(100j) + 2.45 \times 10^6} \\ &= \frac{2.45 \times 10^6 + 1.75 \times 10^6 j}{-5.95 \times 10^6 j - 1.33 \times 10^7 + 1.75 \times 10^6 j + 2.45 \times 10^6} \\ &= \frac{2.45 \times 10^6 + 1.75 \times 10^6 j}{-1.09 \times 10^7 - 4.20 \times 10^6 j} \end{aligned} \tag{c}$$

Equation (c) is written in the form of Equation (8.24) by multiplying the numerator and denominator by the complex conjugate of the denominator. To this end

$$\begin{aligned} &G(j100) \\ &= -\frac{2.45 \times 10^6 + 1.75 \times 10^6 j}{1.09 \times 10^7 + 4.20 \times 10^6 j}\,\frac{1.09 \times 10^7 - 4.20 \times 10^6 j}{1.09 \times 10^7 - 4.20 \times 10^6 j} \\ &= -\frac{[(2.45 \times 10^6)(1.09 \times 10^7) - (1.75 \times 10^6)(-4.20 \times 10^6)] + j[(1.75 \times 10^6)(1.09 \times 10^7) + (2.455 \times 10^6)(-4.20 \times 10^6)]}{(1.09 \times 10^7)^2 + (4.20 \times 10^6)^2} \\ &= -0.251 - 0.064j \end{aligned} \tag{d}$$

The magnitude and phase of $G(j100)$ are calculated as

$$|G(j100)| = \sqrt{(-0.251)^2 + (-0.064)^2} = 0.259 \quad \text{(e)}$$

$$\phi = \tan^{-1}\left(\frac{-0.064}{-0.251}\right) = 194.4° \quad \text{(f)}$$

Since both the numerator and denominator of the argument of the inverse tangent in Equation (f) are negative, the appropriate phase is in the third quadrant. The steady-state response of the vehicle is

$$x_s(t) = (0.01)(0.259)\sin(100t + 194.4°)$$
$$= 2.59 \times 10^{-3}\sin(100t + 194.4°) \text{ m} \quad \text{(g)}$$

MATLAB is useful for performing the tedious computations required in numerical evaluation of the sinusoidal transfer function. If z is a complex number, then the commands

A = abs(z)
B = angle(z)

respectively evaluate the magnitude and phase of z.

Example 8.3

Write a MATLAB program that determines and plots on the same graph the sinusoidal responses of the system of Example 8.2 for three different input frequencies.

Solution

Example 8_3.m, whose script is shown in Figure 8.2(a), does the following:

- Defines the transfer function of Equation (a) of Example 8.2 using the symbolic variable s.
- Inputs the amplitude of the input displacement and three values for input frequency.
- Evaluates the sinusoidal transfer function for each input frequency.
- Determines the steady-state amplitude and steady-state phase for each input frequency.
- Calculates and plots the steady-state response for each input frequency.

```
% Example 8.3
% Numerical evaluation of sinusoidal transfer function
%
% Defining symbolic variable
syms s
% Transfer function of Example 8.2 in terms of s
G=(1.75E4*s+2.45E6)/(5.95*s^3+1.33E3*s^2+1.75E4*s+2.45E6);
%
% Input amplitude and input frequencies
%
Y = input('Input amplitude of displacement in m ');
disp('Input three frequencies in r/s for which steady-state response is to be evaluated')
for k=1:3
  str = ['omega(',num2str(k),')= '];
  w(k) = input(str);
```

(a)

Figure 8.2

(a) MATLAB script for Example8_3.m; (b) the output from the execution of the script; (c) the steady-state responses for the three values of input frequency

```
  % Evaluate sinusoidal transfer function
    A(k) = subs(G,j*w(k));
  % Determine amplitude and phase
    phi(k) = angle(A(k));
    Q(k) = Y*abs(A(k));
  % Print amplitude and phase
    str1 = ['Steady state amplitude=',num2str(Q(k))),' m'];
    str2 = ['Steady-state phase=',num2str(eval(phi(k))),' rad'];
    disp(str1);
    disp(str2);
end
% Determine time scale for plotting
% The value of dt is chosen such that the graph will illustrate the
% response over at least three periods for each value of omega
C = min(w);
Tmax = 2*pi/C;
dt = 3*Tmax/100;
for i=1:101
  t(i)=(i-1)*dt;
  for k=1:3
    x(k,i)=Q(k)*sin(w(k)*t(i)+phi(k));
  end
end
%
% Plot steady-state responses
%
plot(t,x(1,:),'-',t,x(2,:),'-.',t,x(3,:),'--');
xlabel('t (s)')
ylabel('x (m)')
title('Steady-state response for Example 8.3')
str1=['\omega=',num2str(w(1)),' rad/s'];
str2=['\omega=',num2str(w(2)),' rad/s'];
str3=['\omega=',num2str(w(3)),' rad/s'];
legend(str1,str2,str3)
%
% End of Example8_3.m
```

(a) (Continued)

```
>>Example 8_3
Input amplitude of displacement in m .001
Input three frequencies in r/s for which steady-state response is to be evaluated
omega(1)= 100
Steady state amplitude=0.00025878 m
Steady-state phase=-2.8907 rad
omega(2)= 120
Steady state amplitude=0.0001735 m
Steady-state phase=-2.8885 rad
omega(3)= 150
Steady state amplitude=0.00011031 m
Steady-state phase=-2.8877 rad
>>
```

(b)

Figure 8.2

(Continued)

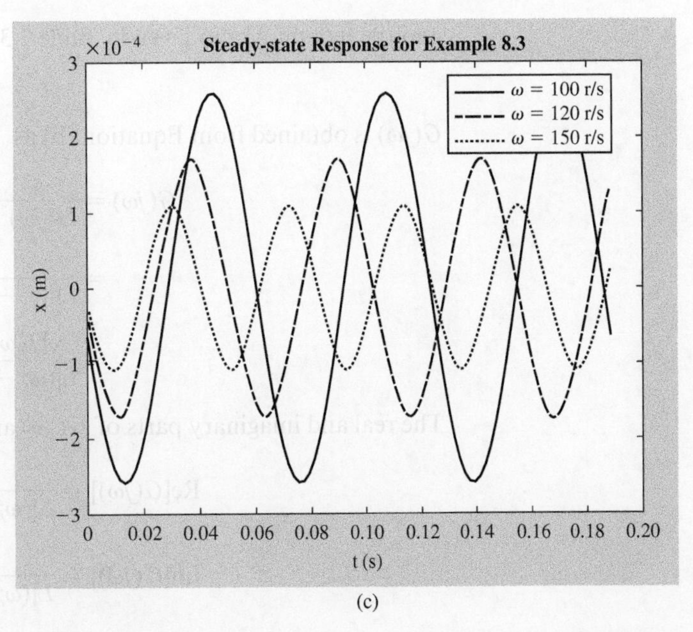

(c)

Figure 8.2
(Continued)

The output from the execution of Example8_3.m is shown in Figure 8.2(b), and the steady-state response for the input frequency is illustrated in Figure 8.2(c). The following is noted regarding the program:

- MATLAB allows use of either i or j to represent the complex number $\sqrt{-1}$. Note, however, that the output presents complex numbers using i. It is possible, as is done in this program, to redefine either i or j to represent another variable. In Example 8_3.m, i is used as a counter.
- The command 'subs(G,j*w(k))' numerically evaluates the symbolic function G at a value of s=j*w(k).
- The command 'C=min(w)' determines the minimum value of the elements of w. This command is used to determine the minimum value of the input frequency, which corresponds to the maximum value of the period of the input. The time scale for the plots is adjusted such that the steady-state response is plotted for at least three periods for each input frequency.

Example 8.4

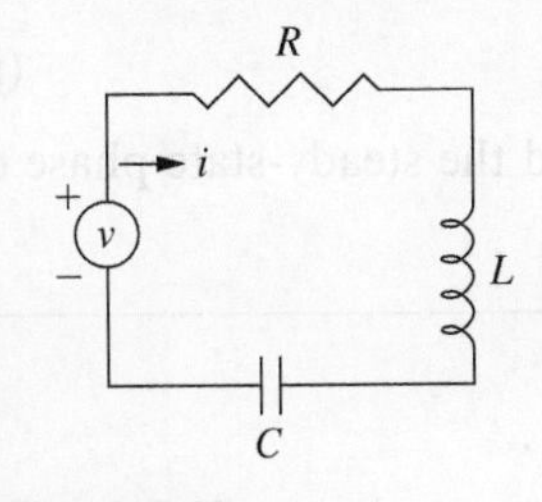

Figure 5.16

The series *LRC* circuit of Example 5.10 and Example 8.4. (Repeated)

Determine the frequency response of the series *LRC* circuit of Example 5.10 and Figure 5.16. This is also benchmark problem 4 of Section 7.1.3.

Solution

The differential equation modeling a series *LRC* circuit is given by Equation (b) of Example 5.10 and in Equation (5.86). This leads to the transfer function defined as $G(s) = \dfrac{I(s)}{V(s)}$ as in Equation (5.90), which is

$$G(s) = \frac{s/L}{s^2 + \frac{R}{L}s + \frac{1}{CL}} \tag{a}$$

Equation (a) can be written in the form of Equation (7.62)

$$G(s) = \frac{As + B}{s^2 + 2\zeta\omega_n s + \omega_n^2} \tag{b}$$

with $A = \dfrac{1}{L}$, $B = 0$, the natural frequency determined in Table 7.3 is

$$\omega_n = \frac{1}{\sqrt{LC}} \tag{c}$$

and the damping ratio given in Table 7.3 is

$$\zeta = \frac{R}{2}\sqrt{\frac{C}{L}} \tag{d}$$

$G(j\omega)$ is obtained from Equation (b) as

$$\begin{aligned} G(j\omega) &= \frac{j\omega}{L[(j\omega)^2 + 2\zeta\omega_n(j\omega) + \omega_n^2]} \\ &= \frac{j\omega}{L[(\omega_n^2 - \omega^2) + j2\zeta\omega\omega_n]} \\ &= \frac{\omega[2\zeta\omega\omega_n + j(\omega_n^2 - \omega^2)]}{L[(\omega_n^2 - \omega^2)^2 + (2\zeta\omega\omega_n)^2]} \end{aligned} \tag{e}$$

The real and imaginary parts of $G(j\omega)$ are determined from Equation (e) as

$$\text{Re}[G(j\omega)] = \frac{2\zeta\omega^2\omega_n}{L[(\omega_n^2 - \omega^2)^2 + (2\zeta\omega\omega_n)^2]} \tag{f}$$

$$\text{Im}[G(j\omega)] = \frac{\omega(\omega_n^2 - \omega^2)}{L[(\omega_n^2 - \omega^2)^2 + (2\zeta\omega\omega_n)^2]} \tag{g}$$

The steady-state amplitude and phase, due to a sinusoidal input of $V_0 \sin(\omega t)$, are determined using Equations (f) and (g) as

$$\begin{aligned} I &= V_0|G(j\omega)| \\ &= V_0\sqrt{\left\{\frac{2\zeta\omega^2\omega_n}{L[(\omega_n^2 - \omega^2)^2 + (2\zeta\omega\omega_n)^2]}\right\}^2 + \left\{\frac{\omega(\omega_n^2 - \omega^2)}{L[(\omega_n^2 - \omega^2)^2 + (2\zeta\omega\omega_n)^2]}\right\}^2} \\ &= \frac{V_0\omega}{L}\frac{1}{\sqrt{(\omega_n^2 - \omega^2)^2 + (2\zeta\omega\omega_n)^2}} \end{aligned} \tag{h}$$

$$\begin{aligned} \phi &= \tan^{-1}\left\{\frac{\text{Im}[G(j\omega)]}{\text{Re}[G(j\omega)]}\right\} \\ &= \tan^{-1}\left\{\frac{\dfrac{\omega(\omega_n^2 - \omega^2)}{L[(\omega_n^2 - \omega^2)^2 + (2\zeta\omega\omega_n)^2]}}{\dfrac{2\zeta\omega^2\omega_n}{L[(\omega_n^2 - \omega^2)^2 + (2\zeta\omega\omega_n)^2]}}\right\} \\ &= \tan^{-1}\left(\frac{\omega_n^2 - \omega^2}{2\zeta\omega\omega_n}\right) \end{aligned} \tag{i}$$

The steady-state response of the LRC circuit is of the form of Equation (8.1)

$$i(t) = I\sin(\omega_n t + \phi) \tag{j}$$

where the steady-state amplitude I is as in Equation (h) and the steady-state phase ϕ is given in Equation (i).

8.3 Frequency Response

The frequency response for a system is the variation of the system's steady-state amplitude and steady-state phase with frequency. The frequency response is often presented graphically. The most basic graphical representations are frequency response curves, which are plots of the steady-state amplitude versus the input frequency and the phase angle versus the input frequency. Frequency response curves are often presented as a family of curves on a common set of coordinate axes, with each curve representing the frequency response for a specific value of a system parameter such as the damping ratio. It is usually desirable to nondimensionalize the relationship between the steady-state amplitude and the input frequency. The vertical axis is used for this nondimensional dependent parameter. The horizontal axis is a nondimensional frequency. If the system is first order, the nondimensional parameter may be the time constant times the frequency $\omega\tau$. If the system is second order, the nondimensional parameter may be the ratio of the frequency and the natural frequency. This is called the frequency ratio r:

$$r = \frac{\omega}{\omega_n} \tag{8.32}$$

The behavior of the system is noted at low frequencies (near $\omega = 0$), at high frequencies (as $\omega \to \infty$), and near the value of the nondimensional frequency near 1. The maximum of the amplitude is noted, as well as the value of the frequency parameter for which it occurs.

The relationship between the steady-state amplitude of the current in a series *LRC* circuit and the frequency of a sinusoidal voltage source is determined as Equation (h) in Example 8.4:

$$I = \frac{V_0\omega}{L}\sqrt{\frac{1}{(\omega_n^2 - \omega^2)^2 + (2\zeta\omega\omega_n)^2}} \tag{8.33}$$

Equation (8.33) is rearranged by multiplying both sides by $L\omega_n/V_0$, leading to

$$\begin{aligned}\frac{L\omega_n I}{V_0} &= \omega\omega_n\sqrt{\frac{1}{(\omega_n^2 - \omega^2)^2 + (2\zeta\omega\omega_n)^2}} \\ &= \frac{\omega}{\omega_n}\sqrt{\frac{1}{\left[1 - \left(\frac{\omega}{\omega_n}\right)^2\right]^2 + \left(2\zeta\frac{\omega}{\omega_n}\right)^2}}\end{aligned} \tag{8.34}$$

Both sides of Equation (8.34) are nondimensional and the equation is written using the frequency ratio as

$$M = \frac{L\omega_n I}{V_0} = \frac{r}{\sqrt{(1 - r^2)^2 + (2\zeta r)^2}} \tag{8.35}$$

The nondimensional parameter M is called the magnification factor. It expresses how much the frequency increases or decreases the steady-state amplitude beyond that resulting from a step input. A family of curves illustrating the behavior of $L\omega_n I/V_0$ vs. r for several values of ζ is presented in Figure 8.3.

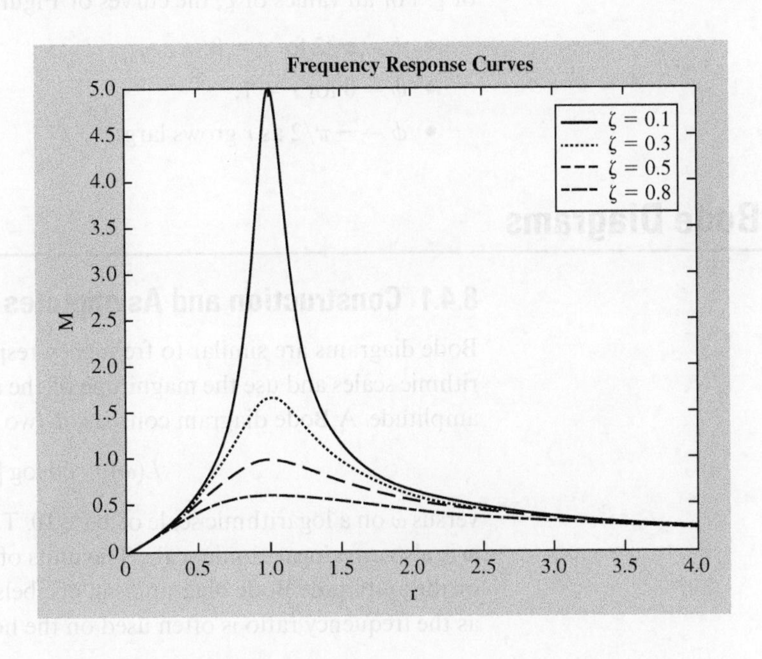

Figure 8.3
The frequency response curves for a series *LRC* circuit.

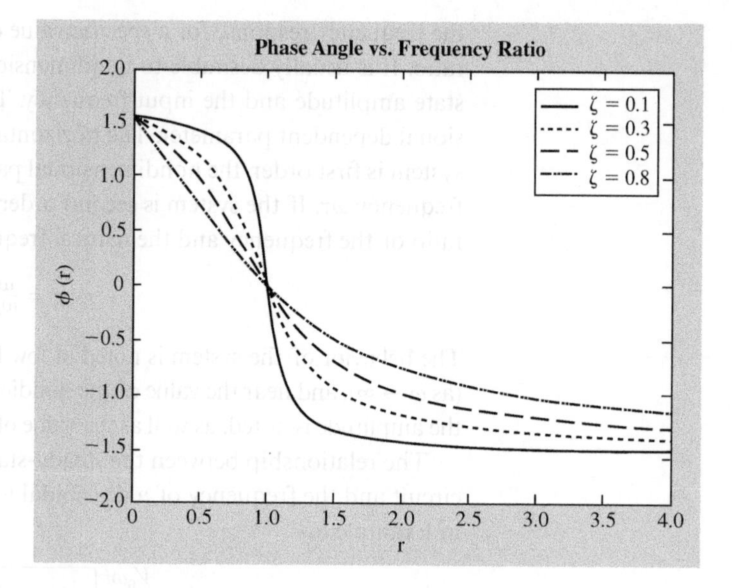

Figure 8.4
Phase angle vs. frequency ratio for an *LRC* circuit.

From Figure 8.3 and Equation (8.35), it is ascertained that

- $M(0) = 0$ for all ζ;
- $\lim_{r\to\infty} M = 0$ and behaves as $1/r$ for large r;
- M is a maximum near $r = 1$ and decreases as r increases from there. The value of r where the steady-state amplitude is a maximum is determined by setting $\frac{dM}{dr} = 0$ and solving for r. The corresponding value of the maximum is obtained by substituting that value into Equation (8.35).

The phase angle for the series *LRC* circuit is determined as Equation (i) in Example 8.4 as

$$\phi = \tan^{-1}\left(\frac{\omega_n^2 - \omega^2}{2\zeta\omega\omega_n}\right) \tag{8.36}$$

Equation (8.36) is rewritten using the frequency ratio as

$$\phi = \tan^{-1}\left(\frac{1 - r^2}{2\zeta r}\right) \tag{8.37}$$

Equation (8.37) is illustrated in Figure 8.4, which shows ϕ vs. r for several values of ζ. For all values of ζ, the curves of Figure 8.4 show

- $\phi = \pi/2$ for $r = 0$;
- $\phi = 0$ for $r = 1$;
- $\phi \to -\pi/2$ as r grows large.

8.4 Bode Diagrams

8.4.1 Construction and Asymptotes

Bode diagrams are similar to frequency response curves, but they are drawn on logarithmic scales and use the magnitude of the sinusoidal transfer function instead of the amplitude. A Bode diagram consists of two plots. One is a diagram plotting

$$L(\omega) = 20 \log |G(j\omega)| \tag{8.38}$$

versus ω on a logarithmic scale of base 10. The other is a diagram of ϕ versus ω, where ω is also on a logarithmic scale. The units of $L(\omega)$ are decibels (dB). The vertical scale of the amplitude Bode diagram is in decibels, while a nondimensional parameter such as the frequency ratio is often used on the horizontal axis.

The logarithm used in defining $L(\omega)$ and in calculating $\log \omega$ is a base 10 logarithm. When drawing the amplitude part of a Bode diagram, one can take advantage of the properties of logarithms such as $\log(ab) = \log a + \log b$, $\log \frac{a}{b} = \log a - \log b$, and $\log a^b = b \log a$ for nonnegative values of a and b.

The conversion of the sinusoidal transfer function into decibels using Equation (8.38) allows an easy comparison of frequency responses for different systems. A numerical value of the sinusoidal transfer function of one translates into zero decibels. A change in the order of magnitude in the sinusoidal transfer function (factor of 10) translates into a change of 20 dB. A tenfold increase in the sinusoidal transfer function leads to a 20-dB increase in $L(\omega)$, while a reduction in the sinusoidal transfer function by a factor of 10 leads to a 20-dB decrease in $L(\omega)$. A range of a factor of 10 in the frequency is called a decade, which is a unit change on the horizontal logarithmic scale. Since the horizontal scale is logarithmic, a Bode diagram can illustrate the frequency response over a wide range of frequencies. However, unlike frequency response curves, the Bode diagram cannot show the response for $\omega = 0$.

Bode diagrams for the series *LRC* circuit, determined using Equations (h) and (i) of Example 8.4, are illustrated in Figure 8.5 for a natural frequency of $\omega_n = 1$.

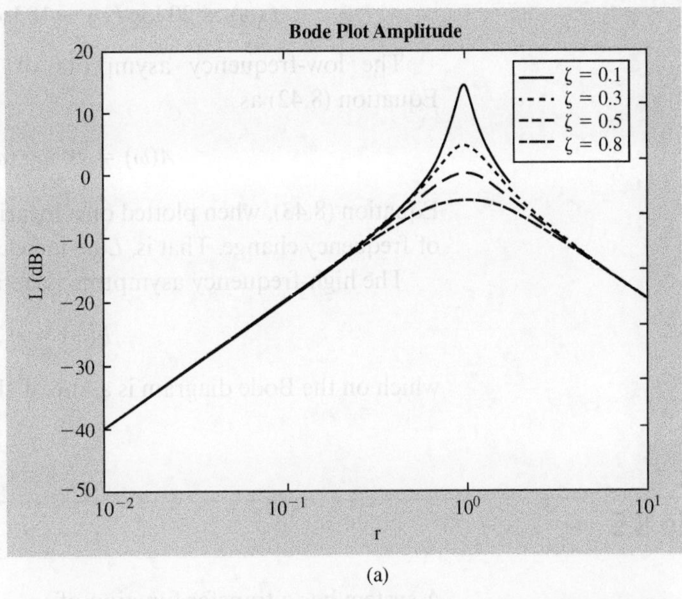

(a)

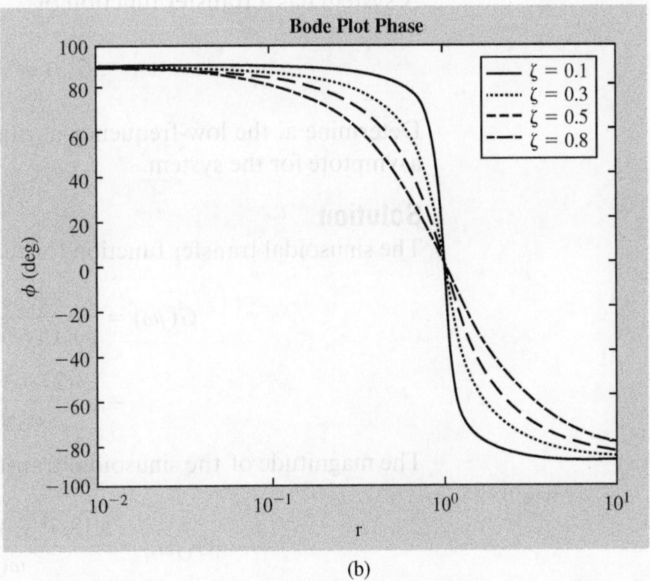

(b)

Figure 8.5

(a) The Bode plot of $|G(j\omega)|$ for the series *LRC* circuit; (b) $L(\omega) = 20 \log$ phase part of the Bode diagram for the series *LRC* circuit.

8.4.2 Asymptotes

Low-frequency asymptotes and high-frequency asymptotes of the Bode diagram are important in control system analysis. As their name implies, the low-and high-frequency asymptotes are the lines to which Equation (8.38) are asymptotic as $\omega \to 0$ and $\omega \to \infty$, respectively.

The low-frequency asymptote is defined as the line

$$A(\omega) = \lim_{\omega \to 0} L(\omega) \tag{8.39}$$

The high-frequency asymptote is defined as the line

$$B(\omega) = \lim_{\omega \to \infty} L(\omega) \tag{8.40}$$

Using Equation (h) of Example 8.4 for the series LRC circuit

$$L(\omega) = 20 \log\left[\frac{\omega}{\sqrt{(\omega_n^2 - \omega^2)^2 + (2\zeta\omega\omega_n)^2}}\right] \tag{8.41}$$

Properties of logarithms are used to rewrite Equation (8.41) as

$$L(\omega) = 20 \log(\omega) - 10 \log[(\omega_n^2 - \omega^2)^2 + (2\zeta\omega\omega_n)^2] \tag{8.42}$$

The low-frequency asymptote of the series LRC circuit is obtained from Equation (8.42) as

$$A(\omega) = 20 \log(\omega) - 40 \log(\omega_n) \tag{8.43}$$

Equation (8.43), when plotted on a logarithmic scale, is a line of slope 20 dB per decade of frequency change. That is, $L(\omega)$ increases by 20 dB as ω increases by a factor of 10.

The high-frequency asymptote is obtained using Equation (8.42) as

$$B(\omega) = -20 \log(\omega) \tag{8.44}$$

which on the Bode diagram is a line of slope -20 dB per decade of frequency change.

Example 8.5

A system has a transfer function of

$$G(s) = \frac{3(s+5)}{s(s^2 + 4s + 13)} \tag{a}$$

Determine **a.** the low-frequency asymptote for the system and **b.** the high-frequency asymptote for the system.

Solution

The sinusoidal transfer function for the system is

$$\begin{aligned} G(j\omega) &= \frac{3(j\omega + 5)}{j\omega(13 - \omega^2 + 4\omega j)} \\ &= \frac{-(21\omega + 3\omega^2) + (3\omega^2 - 195)j}{\omega(\omega^4 - 10\omega^2 + 169)} \end{aligned} \tag{b}$$

The magnitude of the sinusoidal transfer function is

$$|G(j\omega)| = \frac{\sqrt{18\omega^4 + 186\omega^3 - 1170\omega^2 + 38{,}025}}{\omega(\omega^4 - 10\omega^2 + 169)} \tag{c}$$

Using Equation (8.38),

$$L(\omega) = 20 \log \frac{\sqrt{18\omega^4 + 186\omega^3 - 1170\omega^2 + 38{,}025}}{\omega(\omega^4 - 10\omega^2 + 169)} \tag{d}$$

Using the properties of logarithms, Equation (d) is rewritten as

$$L(\omega) = $$
$$10 \log (18\omega^4 + 186\omega^3 - 1170\omega^2 + 38{,}025) - 20 \log \omega - 20 \log (\omega^4 - 10\omega^2 + 169) \tag{e}$$

a. The low-frequency asymptote is obtained using Equation (8.39):

$$\begin{aligned} A(\omega) = \lim_{\omega \to 0} L(\omega) &= 10 \log 38{,}025 - 20 \log \omega - 20 \log 169 \\ &= 1.24 - 20 \log \omega \end{aligned} \tag{f}$$

b. The high-frequency asymptote is obtained using Equation (8.40):

$$\begin{aligned} B(\omega) = \lim_{\omega \to \infty} L(\omega) &= 10 \log 18\omega^4 - 20 \log \omega - 20 \log \omega^4 \\ &= 10 \log 18 + 40 \log \omega - 20 \log \omega - 80 \log \omega \\ &= 12.6 - 60 \log \omega \end{aligned} \tag{g}$$

8.4.3 Products of Transfer Functions

Consider a system in which the transfer function $G(s)$ can be written as

$$G(s) = G_1(s)\, G_2(s) \tag{8.45}$$

The sinusoidal transfer function for the system is

$$\begin{aligned} G(j\omega) &= G_1(j\omega)\, G_2(j\omega) \\ &= |G_1(j\omega)|\, e^{j\phi_1} |G_2(j\omega)|\, e^{j\phi_2} \\ &= |G_1(j\omega)||G_2(j\omega)|\, e^{j(\phi_1 + \phi_2)} \end{aligned} \tag{8.46}$$

Thus

$$|G(j\omega)| = |G_1(j\omega)||G_2(j\omega)| \tag{8.47}$$

and defining ϕ as the phase of the response,

$$\phi = \phi_1 + \phi_2 \tag{8.48}$$

From Equation (8.48), it is clear that the phase angle component of the Bode plot of a transfer function given by Equation (8.45) can be obtained by summing the phases for the individual transfer functions.

Using Equation (8.38) and the property of logarithms of products,

$$\begin{aligned} L(\omega) &= 20 \log [|G(j\omega)|] \\ &= 20 \log [|G_1(j\omega)||G_2(j\omega)|] \\ &= 20 \log [|G_1(j\omega)|] + 20 \log [|G_2(j\omega)|] \\ &= L_1(\omega) + L_2(\omega) \end{aligned} \tag{8.49}$$

Thus the amplitude component of the Bode plot can be obtained by simply adding the amplitude components of the Bode plots for the individual transfer functions.

8.4.4 Bode Diagrams for Common Transfer Functions

Bode diagrams for some basic transfer functions can be drawn and catalogued. The property of products of transfer functions can be used to draw Bode diagrams for more complicated transfer functions. Some of the transfer functions considered are those corresponding to controllers discussed in Chapter 9.

The transfer function of an actuator that provides proportional control is of the form $G(s) = K$, where K is a constant called the gain. For this transfer function, $|G(j\omega)| = K$ and $\phi = 0$. The Bode diagram for proportional gain is constructed from

$$L(\omega) = 20 \log (K) \tag{8.50a}$$

$$\phi = 0° \tag{8.50b}$$

The Bode diagram for proportional gain is a horizontal line, as illustrated in Figure 8.6. The phase portion of the Bode diagram is a horizontal line of $\phi = 0$.

The transfer function of a differential controller (a differentiator) is $G(s) = s$. It can be shown that $G(j\omega) = j\omega$, $|G(j\omega)| = \omega$, and $\phi = 90°$. The Bode diagram for a differentiator is constructed from

$$L(\omega) = 20 \log (\omega) \tag{8.51a}$$

$$\phi = 90° \tag{8.51b}$$

The Bode diagram for a differentiator is a line of slope 20 dB/decade that passes through $L = 0$ when $\omega = 1$, as illustrated in Figure 8.7. The phase portion of the Bode diagram is a horizontal line of $\phi = 90°$.

The transfer function $G(s) = 1/s$ is the transfer function provided by an integral controller (an integrator). It can be shown that $G(j\omega) = -j(1/\omega)$, $|G(j\omega)| = 1/\omega$, and $\phi = -90°$. The Bode diagram for an integrator is constructed from

$$L(\omega) = -20 \log (\omega) \tag{8.52a}$$

$$\phi = -90° \tag{8.52b}$$

The Bode diagram for an integrator is a line of slope −20 dB/decade that passes through $L = 0$ when $\omega = 1$, as illustrated in Figure 8.8. The phase portion of the Bode diagram is a horizontal line of $\phi = -90°$.

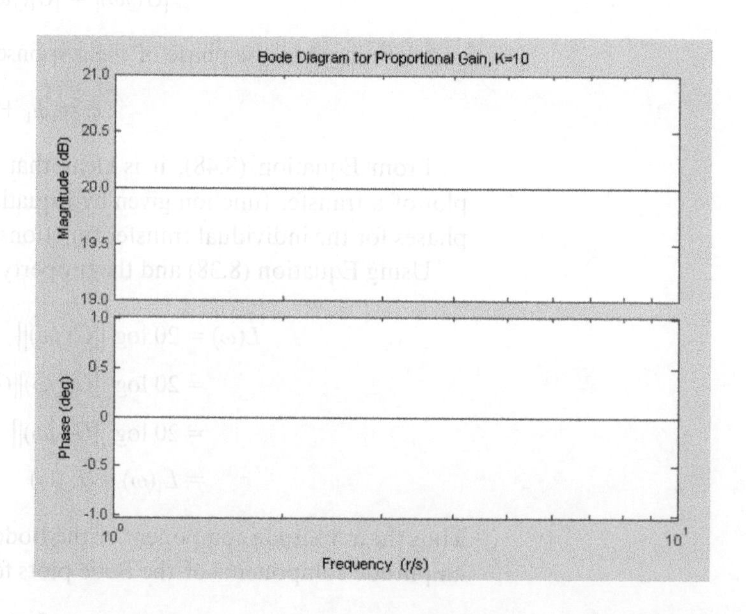

Figure 8.6

The Bode diagram for the transfer function corresponding to the proportional control with $K = 10$.

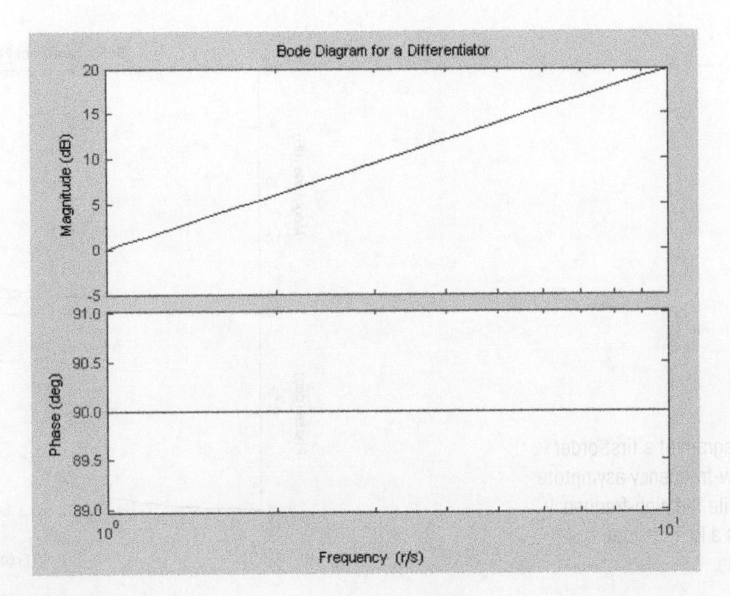

Figure 8.7

The Bode diagram for a differentiator. The amplitude part is a line with a slope of 20 dB/decade. The phase portion is the horizontal line $\phi = \pi/2$.

The numerator of transfer functions often includes a term of the form $G(s) = 1 + \tau s$, which is often referred to as a first-order lead. The sinusoidal transfer function for a first-order lead is $G(j\omega) = 1 + j\tau\omega$, which gives $|G(j\omega)| = \sqrt{1 + \tau^2\omega^2}$, $\phi = \tan^{-1}(\omega\tau)$. The transfer function $1 + \tau s$ is called a first-order lead because it is of order one and provides a positive phase component when used in series with another transfer function. The Bode diagram for a first-order lead is constructed from

$$L(\omega) = 10 \log (1 + \omega^2 \tau^2) \tag{8.53a}$$

$$\phi = \tan^{-1}(\omega\tau) \tag{8.53b}$$

The low-frequency asymptote for a first-order lead is the horizontal line $L = 0$, while the high-frequency asymptote is $L = 20 \log (\tau) + 20 \log (\omega)$, which is a line of slope 20 dB/decade and $L = 20 \log (\tau)$ when $\omega = 1$. The Bode diagram for a first-order lead is illustrated in Figure 8.9.

The transfer function corresponding to the denominator of a first-order system, called a first-order lag, is $G(s) = 1/(1 + \tau s)$. The corresponding sinusoidal transfer

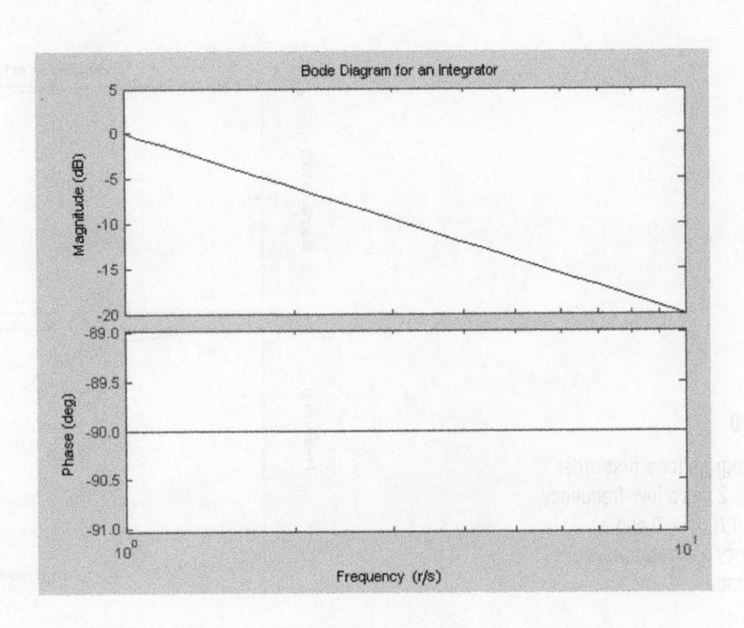

Figure 8.8

The amplitude part of a Bode diagram for an integrator is a line of slope −20 dB/decade. The phase part of a Bode diagram is a horizontal line $\phi = -\pi/2$.

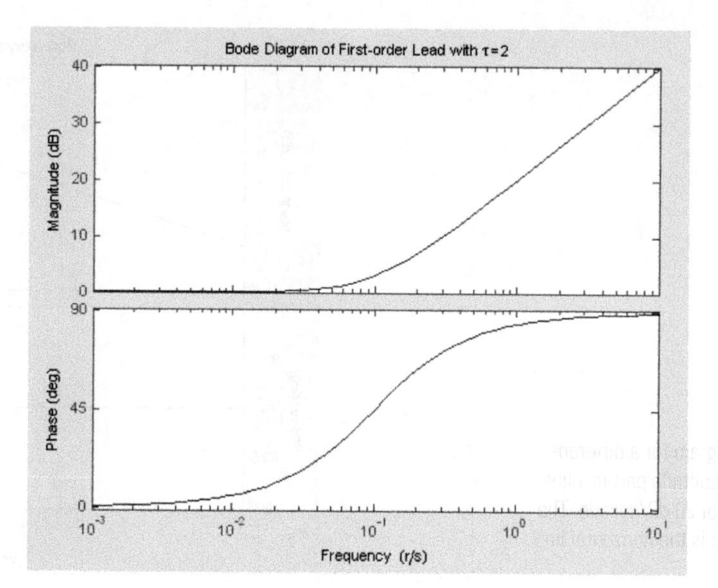

Figure 8.9

The Bode diagram of a first-order lead. The low-frequency asymptote is $L = 0$, while the high-frequency asymptote is a line of slope 20 dB/decade.

function is $G(j\omega) = (1 - j\omega\tau)/(1 + \omega^2\tau^2)$, which leads to $|G(j\omega)| = \dfrac{1}{\sqrt{1 + \omega^2\tau^2}}$ and $\phi = \tan^{-1}(-\omega\tau)$. The transfer function $1/(1 + \tau s)$ is called a first-order lag because it is of order one and provides a negative phase component when used in series with another transfer function. The Bode diagram for a first-order lag is constructed from

$$L(\omega) = -10 \log (1 + \omega^2\tau^2) \tag{8.54a}$$

$$\phi = \tan^{-1}(-\omega\tau) \tag{8.54b}$$

The low-frequency asymptote is the line $L = 0$, while the high-frequency asymptote is $L = -20 \log (\tau) - 20 \log (\omega)$, which on the Bode diagram is a line of slope -20 dB/decade that passes through $L = -20 \log (\tau)$ when $\omega = 1$. It is noted that the phase for a first-order lag is always negative while the phase for a first-order lead is always positive; thus these transfer functions are aptly named. The Bode diagram for a first-order lag is shown in Figure 8.10.

The transfer function corresponding to a second-order numerator with complex zeros is of the form $G(s) = s^2 + 2\alpha\beta s + \beta^2$. This transfer function corresponds

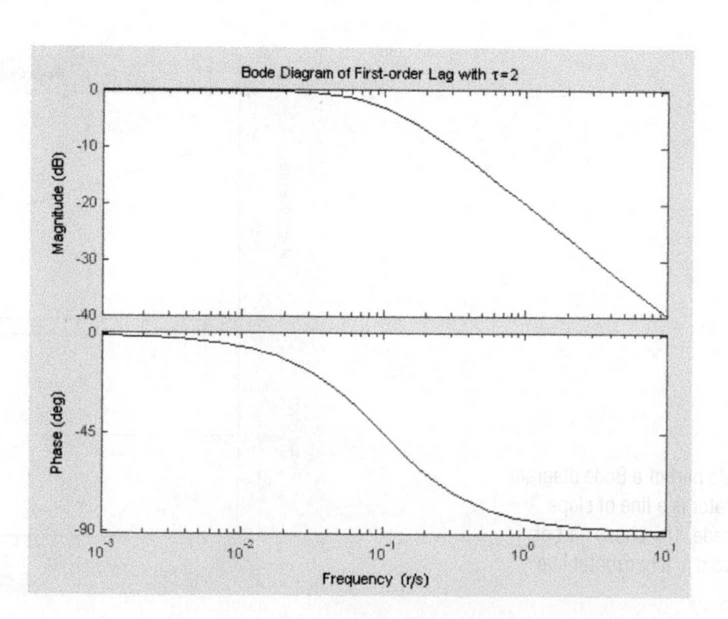

Figure 8.10

The Bode diagram for a first-order lag with $\tau = 2$ has a low-frequency asymptote of $L(\omega) = 0$ and a high-frequency asymptote of slope -20 dB/decade.

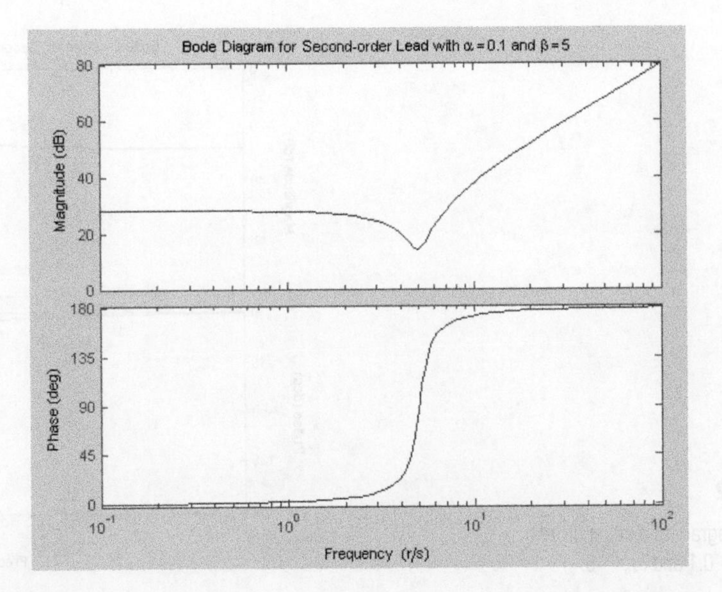

Figure 8.11

The Bode diagram for a second-order lead with $\alpha = 0.1$ and $\beta = 5$ has a low-frequency asymptote of $40 \log(5) = 27.99$. Its high-frequency asymptote is a line of slope 40 dB/decade.

to a second-order lead. The corresponding sinusoidal transfer function is $G(j\omega) = \beta^2 - \omega^2 + j2\alpha\beta\omega$, which leads to $|G(j\omega)| = \sqrt{(\beta^2 - \omega^2)^2 + (2\alpha\beta\omega)^2}$ and $\phi = \tan^{-1}[2\alpha\beta\omega/(\beta^2 - \omega^2)]$. The Bode diagram for a second-order lead is constructed from

$$L(\omega) = 10 \log [(\beta^2 - \omega^2)^2 + (2\alpha\beta\omega)^2] \tag{8.55a}$$

$$\phi = \tan^{-1}\left(\frac{2\alpha\beta\omega}{\beta^2 - \omega^2}\right) \tag{8.55b}$$

The Bode diagram for a second-order lead is illustrated in Figure 8.11. The low-frequency asymptote is $L = 40 \log (\beta)$, while the high-frequency asymptote is $L = 40 \log (\omega)$.

The transfer function corresponding to the denominator of a second-order system with complex poles (an underdamped system) is of the general form $G(s) = 1/(s^2 + 2\zeta\omega_n s + \omega_n^2)$. This system is also called a second-order lag. Its sinusoidal transfer function is $G(j\omega) = (\omega_n^2 - \omega^2 - j2\zeta\omega\omega_n)/[(\omega_n^2 - \omega^2)^2 + (2\zeta\omega\omega_n)^2]$, which leads to $|G(j\omega)| = 1/\sqrt{(\omega_n^2 - \omega^2)^2 + (2\zeta\omega\omega_n)^2}$ and $\phi = \tan^{-1}[-2\zeta\omega\omega_n/(\omega_n^2 - \omega^2)]$. The Bode diagram for a second-order underdamped system is constructed from

$$L(\omega) = -10 \log [(\omega_n^2 - \omega^2)^2 + (2\zeta\omega\omega_n)^2] \tag{8.56a}$$

$$\phi = \tan^{-1}\left(-\frac{2\zeta\omega\omega_n}{\omega_n^2 - \omega^2}\right) \tag{8.56b}$$

The low-frequency asymptote for an underdamped second-order system is $L = -40 \log (\omega_n)$, while its high-frequency asymptote is $L = -40 \log (\omega)$. The Bode diagram for an underdamped second-order system is shown in Figure 8.12.

The transfer function for a time delay is $G(s) = e^{-\tau s}$. The corresponding sinusoidal transfer function is $G(j\omega) = e^{-j\tau\omega} = \cos(\omega\tau) - j\sin(\omega\tau)$, which leads to $|G(j\omega)| = 1$ and $\phi = -\omega\tau$. The Bode diagram for a time delay is constructed from

$$L(\omega) = 0 \tag{8.57a}$$

$$\phi = -\omega\tau \tag{8.57b}$$

The amplitude portion of the Bode diagram for a time delay, as illustrated in Figure 8.13(a), is a horizontal line. The phase part of a Bode diagram for a time delay, illustrated in Figure 8.13(b), has a low frequency asymptote of $-\log(\tau)$ which is 0 for $\tau = 1$ and approaches $-\pi$ radians for large ω.

Figure 8.12

The Bode diagram of a second-order lag with $\zeta = 0.1$ and $\omega_n = 5$.

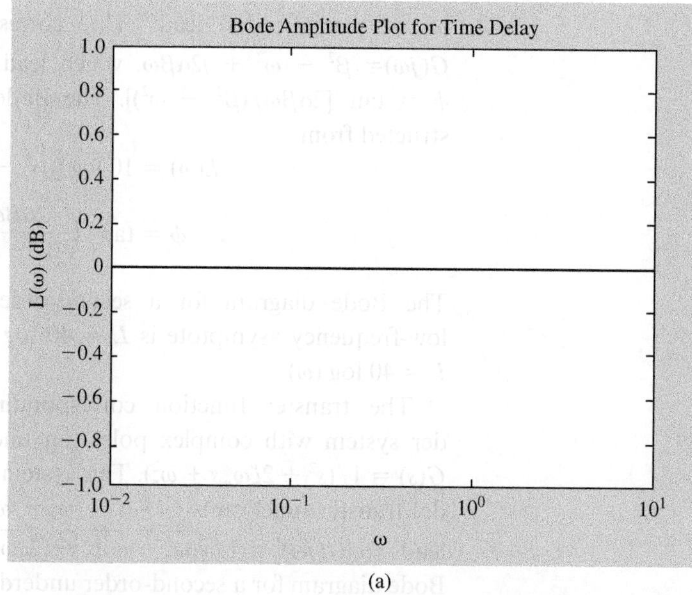

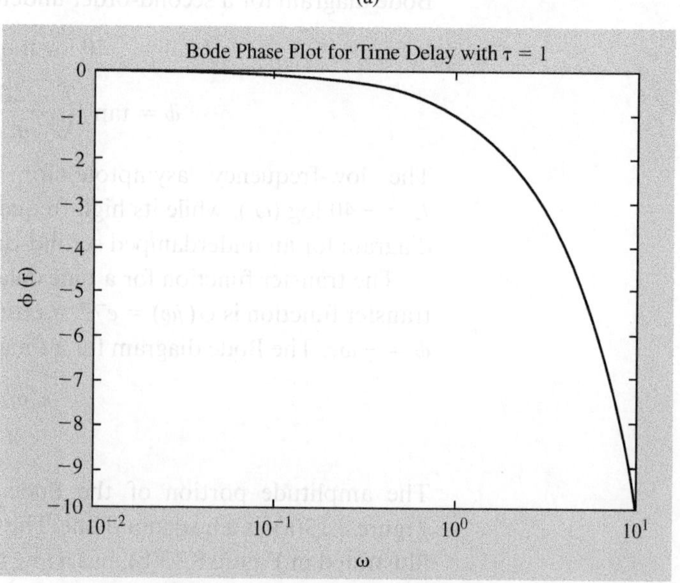

Figure 8.13

The Bode plot for the time delay transfer function with $\tau = 1$. (a) Amplitude and (b) phase.

Bode diagrams for many transfer functions may be constructed using the superposition formulas of Equations (8.48) and (8.49) and the forms of $L(\omega)$ and ϕ for the transfer functions of Equations (8.50)–(8.57). It is noted that if $G(s) = N(s)/D(s)$, where $N(s)$ and $D(s)$ are polynomials with real coefficients, then $N(s)$ and $D(s)$ can be factored using only linear and quadratic factors. Then $G(s)$ can be written as the product of first-order and second-order lead and lag transfer functions.

Example 8.6

Consider the series *LRC* circuit when the parameters are prescribed such that $\zeta = 1.2$, $\omega_n = 100$ r/s, and $L = 1$ H. Develop the Bode plot for this system and discuss its low- and high-frequency asymptotes.

Solution

Benchmark problem 4 of Section 7.1.3 is the series LRC circuit. Its transfer function is given by Equation (7.9), which is subsequently written in the form of Equation (7.62) with $A = 1$ (since $L = 1$ H) and $B = 0$. Since $\zeta > 1$, the system is overdamped and the real poles of the transfer function are calculated, using Equations (7.84a) and (7.84b), as

$$s_1 = -\omega_n\left(\zeta + \sqrt{\zeta^2 - 1}\right) = -186.3 \tag{a}$$

$$s_2 = -\omega_n\left(\zeta - \sqrt{\zeta^2 - 1}\right) = -53.7 \tag{b}$$

For the overdamped system, the transfer function can be written as

$$G(s) = \frac{s}{(s + s_1)(s + s_2)} \tag{c}$$

Equation (c) can be rewritten as the product of the transfer functions just defined by dividing the numerator and denominator by $s_1 s_2$, leading to

$$G(s) = \frac{\frac{1}{s_1 s_2} s}{(1 + \tau_1 s)(1 + \tau_2 s)} \tag{d}$$

where

$$\tau_1 = -\frac{1}{s_1} = 5.37 \times 10^{-3} \tag{e}$$

$$\tau_2 = -\frac{1}{s_2} = 1.86 \times 10^{-2} \tag{f}$$

Equation (d) can be written as

$$G(s) = G_1(s)\,G_2(s)\,G_3(s)\,G_4(s) \tag{g}$$

where

$$G_1(s) = \frac{1}{s_1 s_2} = 1 \times 10^{-4} \tag{h}$$

$$G_2(s) = s \tag{i}$$

$$G_3(s) = \frac{1}{1 + \tau_1 s} \tag{j}$$

$$G_4(s) = \frac{1}{1 + \tau_2 s} \tag{k}$$

Thus, the transfer function for the system is composed of the transfer functions for a proportional gain, a differentiator, and two first-order lags. Its Bode diagram is constructed using the superposition formulas of Equations (8.48) and (8.49), as well as Equations (8.50), (8.51), and (8.54).

$$L(\omega) = 20 \log (1 \times 10^{-4}) + 20 \log (\omega) - 10 \log (1 + 2.88 \times 10^{-5}\omega^2) - 10 \log (1 + 3.47 \times 10^{-4}\omega^2) \tag{l}$$

$$\phi = 0° + 90° + \tan^{-1}(-5.37 \times 10^{-3}\omega) + \tan^{-1}(1.86 \times 10^{-2}\omega) \tag{m}$$

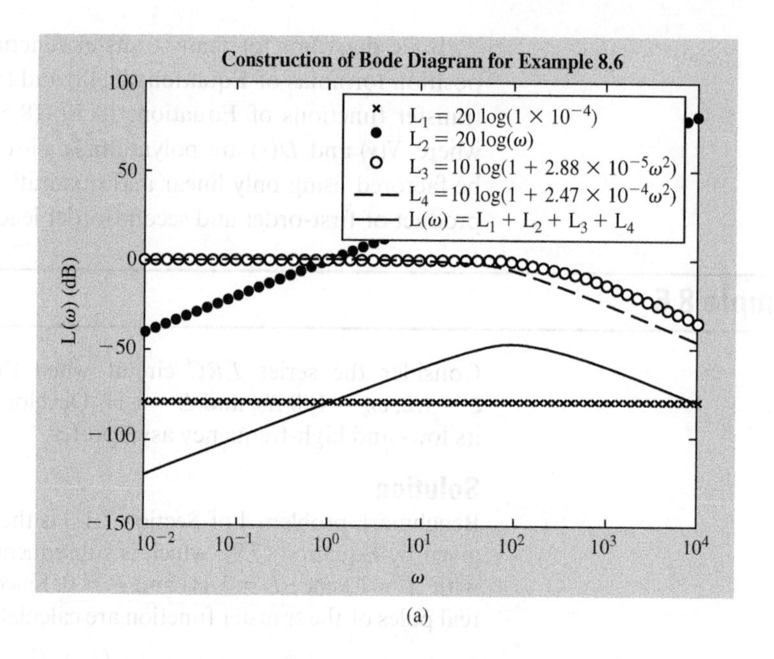

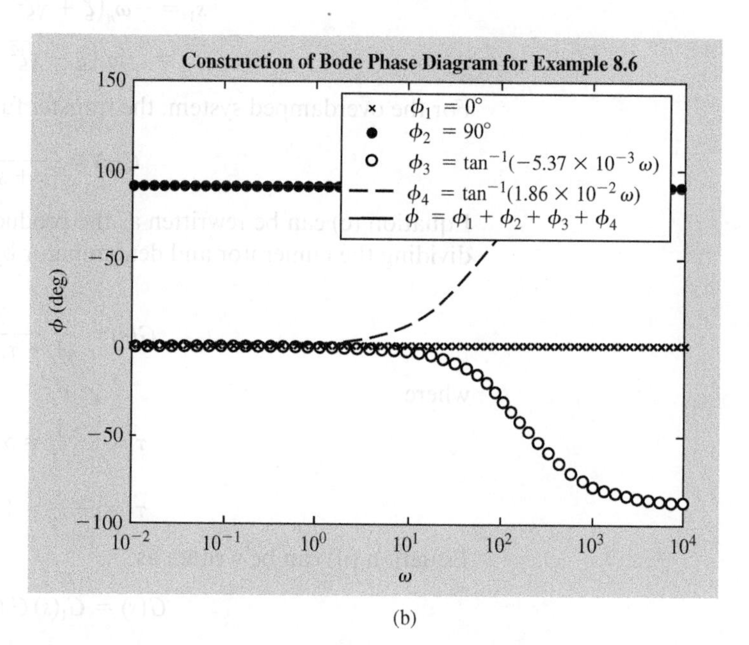

Figure 8.14

The construction of the Bode diagram using the properties for Bode diagrams from products of transfer functions: (a) The amplitude plot; (b) the phase plot.

The development of the Bode diagram for this overdamped system is illustrated in Figure 8.14. The low-frequency asymptote is obtained by using Equation (8.39) on Equation (l), resulting in

$$A(\omega) = -80 + 20 \log(\omega) \tag{n}$$

The high-frequency asymptote is obtained by using Equation (8.40) on Equation (l), leading to

$$\begin{aligned} B(\omega) &= -80 + 20 \log(\omega) - 10 \log(2.88 \times 10^{-5}\omega^2) - 10 \log(3.47 \times 10^{-4}\omega^2) \\ &= -80 + 20 \log(\omega) - 20 \log(\omega) + 45.41 - 20 \log(\omega) + 34.59 \\ &= -20 \log(\omega) \end{aligned} \tag{o}$$

8.4.5 Bode Diagram Parameters

Several parameters are defined to describe the frequency response as presented on the Bode diagram.

The **corner frequency**, also called the breakpoint frequency, is the frequency at which the low-frequency asymptote intersects the high-frequency asymptote. For example, the corner frequency for a first-order lag is obtained by requiring $0 = -20 \log(\tau) - 20 \log(\omega)$, which leads to a corner frequency of $\omega = 1/\tau$.

The **cutoff frequency** ω_c is the frequency at which

$$L(0) - L(\omega_c) = 3 \text{ dB} \tag{8.58}$$

For a first-order lag, $L(0) = 0$ and the cutoff frequency is obtained from $10 \log(1 + \omega_c^2\tau^2) = 3$, leading to $\omega_c = 0.9976/\tau$.

The **bandwidth** is the range of ω over which $L(\omega)$ is greater than 3 dB below L_{max}, the maximum value of L. The bandwidth for a first-order lag is from $\omega = 0$ to $\omega = \omega_C = 0.9976/\tau$. The bandwidth for a second-order lag is dependent on the damping ratio. If L_{max} is less than 3 dB greater than $L(0)$, then the bandwidth is between 0 and ω_c. If L_{max} is more than 3 dB greater than $L(0)$, the bandwidth is between two values of $\omega_c - \omega_l$, where $L(0) - L(\omega_l) = 3$ dB. That is, ω_l is the lower value where the frequency is 3 dB below L_{max}. This is the case if $\zeta > 0.384$.

If all the poles and zeros of a transfer function have negative real parts, the transfer function is a **minimum phase transfer function**. If any of the poles or zeros of a transfer function has a nonnegative real part, the transfer function is a **nonminimum phase transfer function**. As its name implies, the phase angle for a minimum phase transfer function has a defined minimum.

Example 8.7

A system has a transfer function

$$G(s) = \frac{(s^2 + 10s + 61)}{(s + 5)(s^2 + 12s + 100)} \tag{a}$$

Determine **a.** the Bode diagram for this system, **b.** the corner frequency, and **c.** the system's bandwidth.

Solution

Equation (a) is rewritten as

$$G(s) = \frac{0.2(s^2 + 10s + 61)}{\left(\frac{1}{5}s + 1\right)(s^2 + 4s + 100)} \tag{b}$$

The transfer function is the product of four transfer functions: one gain (0.2), one second-order lead $(s^2 + 10s + 61)$, one first-order lag $\left(\dfrac{1}{\frac{1}{5}s + 1}\right)$, and one second-order lag $\left(\dfrac{1}{s^2 + 12s + 100}\right)$. Using the known forms of $L(\omega)$, Equation (8.50a), Equation (8.54a), Equation (8.55a), and Equation (8.56a), and using the superposition formula for $L(\omega)$, Equation (8.49), leads to

$$L(\omega) = 20 \log 0.2 + 10 \log[(3721 - \omega^2)^2 + (100\omega)^2]$$
$$-10 \log\left(1 + \frac{1}{25}\omega^2\right) - 10 \log[(10{,}000 - \omega^2)^2 + (16\omega)^2] \tag{c}$$

The low-frequency asymptote is

$$\lim_{\omega \to 0} L(\omega) = 20 \log 0.3 + 20 \log 3721 - 20 \log 10{,}000 = -19.0 \text{ dB} \tag{d}$$

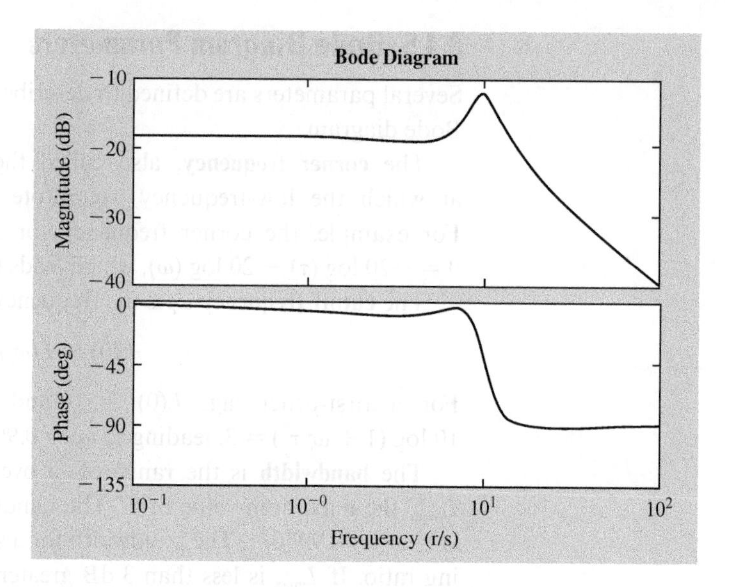

Figure 8.15
Bode diagram for transfer function of Example 8.7.

The high-frequency asymptote is

$$\lim_{\omega \to \infty} L(\omega) = 20 \log 0.2 + 40 \log \omega - 20 \log \omega + 20 \log 5 - 40 \log \omega = -20 \log \omega \tag{e}$$

a. The Bode diagram is plotted in Figure 8.15.

b. The corner frequency is obtained by setting the high-frequency asymptote equal to the low-frequency asymptote:

$$-19.0 = -20 \log \omega \Rightarrow \omega = 8.90 \text{ r/s} \tag{f}$$

c. The maximum value of L, obtained from the Bode diagram, is

$$L_{max} = -12.7 \text{ dB} \tag{g}$$

The bandwidth for this system is the frequencies that satisfy

$$L(\omega) > -15.7 \text{ dB} \tag{h}$$

The solution to Equation (h) is

$$7.7 \text{ r/s} < \omega < 12.3 \text{ r/s} \tag{i}$$

8.5 Nyquist Diagrams

A Nyquist diagram is a plot of $\text{Re}[G(j\omega)]$ vs. $\text{Im}[G(j\omega)]$ with $\text{Re}[G(j\omega)]$ as the abscissa and $\text{Im}[G(j\omega)]$ as the ordinate. A Nyquist diagram is equivalent to a plot of $|G(j\omega)|$ vs. ϕ in polar coordinates. Since $\text{Re}[G(j\omega)]$ and $\text{Im}[G(j\omega)]$ have parametric representations in terms of the frequency ω, a Nyquist diagram is developed by varying ω from $-\infty < \omega < \infty$. Key points on a Nyquist diagram are the points corresponding to $\omega = 0$ and $\omega \to \pm\infty$, as well as points where the diagram intercepts the axes.

The real and imaginary parts of the transfer function for the series *LRC* circuit were developed in Equations (f) and (g) of Example 8.4 and are repeated here with $L = 1$ H:

$$\text{Re}[G(j\omega)] = \frac{2\zeta\omega^2\omega_n}{[(\omega_n^2 - \omega^2)^2 + (2\zeta\omega\omega_n)^2]} \tag{8.59a}$$

$$\text{Im}[G(j\omega)] = \frac{\omega(\omega_n^2 - \omega^2)}{[(\omega_n^2 - \omega^2)^2 + (2\zeta\omega\omega_n)^2]} \tag{8.59b}$$

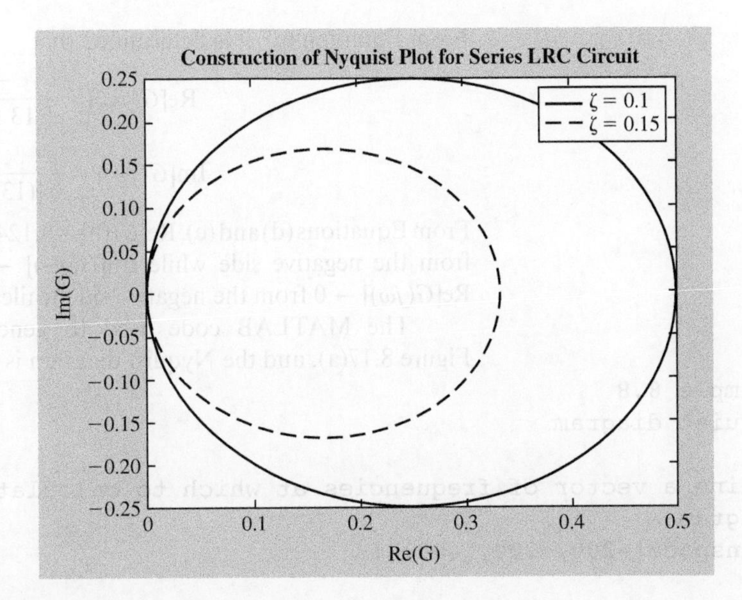

Figure 8.16

The Nyquist diagram for the series *LRC* circuit with $\omega_n = 10$.

Equations (8.59) are used to determine the following regarding the Nyquist diagram for the series *LRC* circuit:

- When $\omega = 0$, $\text{Re}[G(0)] = 0$ and $\text{Im}[G(0)] = 0$.
- As $\omega \to \pm\infty$, $\text{Re}[G(j\omega)] \to 0$ and $\text{Im}[G(j\omega)] \to 0$.
- $\text{Im}[G(j\omega)] = 0$ only for $\omega = 0$ and $\omega = \pm\omega_n$. When $\omega = \pm\omega_n$, $\text{Re}[G(j\omega_n)] = \pm\frac{1}{2\zeta\omega_n}$.
- $\text{Re}[G(j\omega)] = 0$ only for $\omega = 0$.
- $\text{Re}[G(j\omega)] > 0$ for all ω, while $\text{Im}[G(j\omega)] > 0$ for $\omega < \omega_n$ and $\text{Im}[G(j\omega)] < 0$ for $\omega > \omega_n$.

These statements suggest that the Nyquist plot for the transfer function for the series *LRC* circuit should be a closed curve (since it approaches the origin for both large and small ω) existing in the first and fourth quadrants of the *G* plane. It crosses the real axis only at 0 and $1/(2\zeta\omega_n)$. The Nyquist plot is tangent to the imaginary axis at the origin. These observations are confirmed in Figure 8.16, which shows the Nyquist plot for the transfer function for a series *LRC* circuit for $\omega_n = 10$ r/s and for several values of ζ.

Example 8.8

Draw the Nyquist diagram for a system whose transfer function is

$$G(s) = \frac{3(s + 5)}{s(s^2 + 4s + 13)} \tag{a}$$

Solution

The sinusoidal transfer function developed using Equation (a) is

$$G(j\omega) = \frac{3(j\omega + 5)}{j\omega(13 - \omega^2 + 4j\omega)} \tag{b}$$

The numerator and denominator of Equation (b) are multiplied by the complex conjugate of the denominator. The result is written as

$$G(j\omega) = \frac{-21\omega - 3\omega^3 - j(195 - 3\omega^2)}{\omega[(13 - \omega^2)^2 + 16\omega^2]} \tag{c}$$

From Equation (c), it is determined that

$$\mathrm{Re}[G(j\omega)] = \frac{-(21 + 3\omega^2)}{[(13 - \omega^2)^2 + 16\omega^2]} \tag{d}$$

$$\mathrm{Im}[G(j\omega)] = \frac{3\omega^2 - 195}{\omega[(13 - \omega^2)^2 + 16\omega^2]} \tag{e}$$

From Equations (d) and (e), $\mathrm{Re}[G(0)] = 0.124$ and $\mathrm{Im}[G(0)] = \infty$. As $\omega \to \infty$, $\mathrm{Re}[G(j\omega)] \to 0$ from the negative side while $\mathrm{Im}[G(j\omega)] \to 0$ from the positive side. As $\omega \to -\infty$, $\mathrm{Re}[G(j\omega)] \to 0$ from the negative side while $\mathrm{Im}[G(j\omega)] \to 0$ from the negative side.

The MATLAB code used to generate the Nyquist diagram is shown in Figure 8.17(a), and the Nyquist diagram is shown in Figure 8.17(b).

```
% Example 8.8
% Nyquist diagram
%
% Define a vector of frequencies at which to calculate point on Nyquist
% diagram
om=linspace(-200, 200, 4001);
%
% R=real part of sinusoidal transfer function
% I=imaginary part of sinusoidal transfer function
%
R=-(21+3.*om.^2}./(169-10.*om.^2+om.^4);
I=(3.*om.^2-125)./(169-10.*om.^2+om.^4)./om;
%
% Nyquist diagram is a plot of I (vetrical axis) versus R (horizontal axis)
%
figure
plot(R,I)
xlabel('Re[G(j\omega)]')
ylabel('Im[G(j\omega)]')
title('Nyquist diagram for G(s)=3(s+5)/(s(s^2+4s+13))')
%
% End of Example 8.8
```

(a)

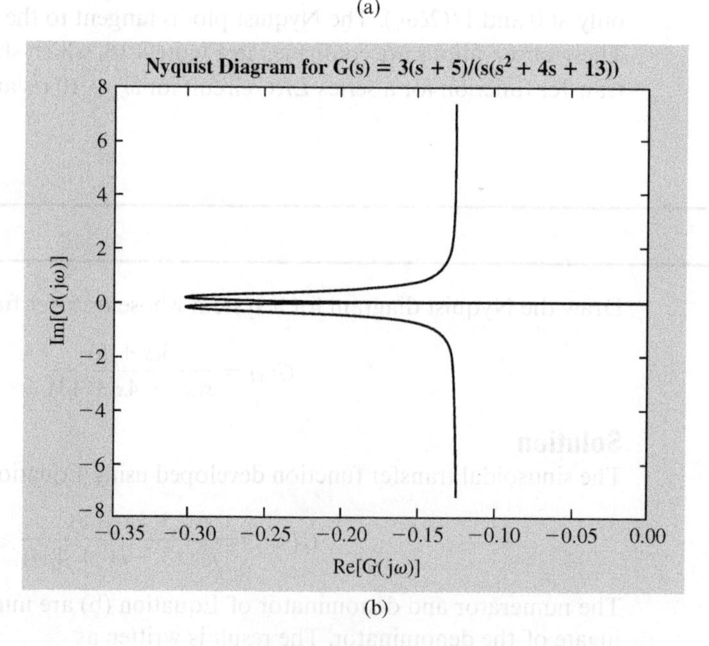

(b)

Figure 8.17

(a) MATLAB code to generate the Nyquist diagram for Example 8.8; (b) the Nyquist diagram.

8.6 Use of MATLAB to Develop Bode Plots and Nyquist Diagrams

MATLAB's Control System Toolbox has direct commands for developing Bode plots and Nyquist diagrams from the definition of the transfer function. If N is a vector containing the coefficients in the polynomial in the numerator of the transfer function, and D is a vector containing the coefficients in the polynomial in the denominator of the transfer function, the commands from the Control System Toolbox

```
bode(N,D)
nyquist(N,D)
```

plot on the screen the system's Bode plot and Nyquist diagram, respectively.

The default range for the Bode diagram is $-1 < \log(\omega) < 2$, while the Nyquist diagram is drawn for $-\infty < \omega < \infty$. The user may define specific points to use in developing either diagram. This allows, for example, the Bode diagram to be focused on a smaller range of frequencies or the Nyquist diagram to be drawn using only positive values of ω. The user supplies a vector specifying the values of ω to be used, say

```
omega=0.001:0.001:10
```

The commands

```
bode(N,D,omega)
nyquist(N,D,omega)
```

draw the Bode plot and Nyquist diagram using points calculated corresponding to only the values specified by the vector omega.

The preceding commands draw plots on the screen without saving any data used to generate the plots. It is often convenient, for comparison purposes, to draw several plots on the same set of axes. The command

```
[G,phi] = bode(N,D,omega)
```

calculates the magnitude and phase of the sinusoidal transfer function for the defined values in omega and stores them in the vectors G and phi, respectively. If the argument corresponding to omega is not specified, the values are calculated for the default values used by MATLAB in developing a Bode plot. The vector of magnitudes G can be converted to decibels using Equation (8.41).

The command

```
[R,I] = nyquist(N,D,omega)
```

calculates the real and imaginary parts of the sinusoidal transfer function for the specified values of omega and stores them in the vectors R and I, respectively.

Example 8.9

A proportional controller of proportional gain K is being designed for use in a feedback control system. When implemented, the transfer function for the system is

$$G(s) = \frac{Ks}{s^2 + (K+2)s + 25} \tag{a}$$

Bode diagrams and Nyquist plots both play an important role in control system design. Thus it is desired to draw these plots for different values of K. Write a MATLAB program that (i) inputs three values of K for which the system is to be analyzed, (ii) draws the Bode plots for each value of K on the same axes, and (iii) draws the Nyquist diagram for each value of K on the same axes.

```
% Example 8.9
clear
% Bode plots and Nyquist diagrams
%
% Input values of K
disp('Input three values of proportional gain')
for i=1:3
  str=['K(',num2str(i),')= '];
  K(i)=input(str);
end
%
% Set range of omegas for Bode plot
omegab=logspace(-3,3,201);
% Set range of omegas for Nyquist diagram
omegan=linspace(0,2000,2001);
for i=1:3
% Define transfer function
  N=[K(i) 0];
  D=[1 K(i)+2 25];
% Determine vectors for Bode plots
  [G(:,i),phi(:,i)]=bode(N,D,omegab);
% Determine vectors for Nyquist plots
  [R(:,i),I(:,i)]=nyquist(N,D,omegan);
end
% Plot Nyquist diagrams
figure
plot(R(:,1),I(:,1),'-',R(:,2),I(:,2),'-.',R(:,3),I(:,3),'--')
xlabel('Re')
ylabel('Im')
title('Nyquist Diagrams for G(s) = Ks/[s^2+(K + (K +2)s+25]')
str1=['K=',num2str(K(1))];
str2=['K=',num2str(K(2))];
str3=['K=',num2str(K(3))];
legend(str1,str2,str3)
figure
% Plot amplitude portion of Bode plots
semilogx(omegab,20*log10(G(:,1)),'-',omegab,20*log10(G(:,2)),'-.',omegab...
  ,20*log10(G(:,3)),'--')
xlabel('\omega')
ylabel('L(\omega) (dB)')
title('Amplitude Part of Bode Diagram for G(s) = Ks/[s^2 + (K + 2)s + 25]')
legend(str1,str2,str3)
figure
% Plot phase portion of Bode plots
semilogx(omegab,phi(:,1),'-',omegab,phi(:,2),'-.',omegab,phi(:,3),'--')
xlabel('\omega')
ylabel('\phi (degrees)')
title('Phase Part of Bode Diagram for G(s) = Ks/[s^2 + (K + 2)s + 25]')
legend(str1,str2,str3)
%
% End of Example8_9.m
```

(a)

Figure 8.18

(a) The script of Example8_9.m; (b) the Nyquist diagrams generated from the execution of the program; (c) the amplitude portion of the Bode diagram generated from the execution of the program; (d) the phase portion of the Bode diagram generated from the execution of the program.

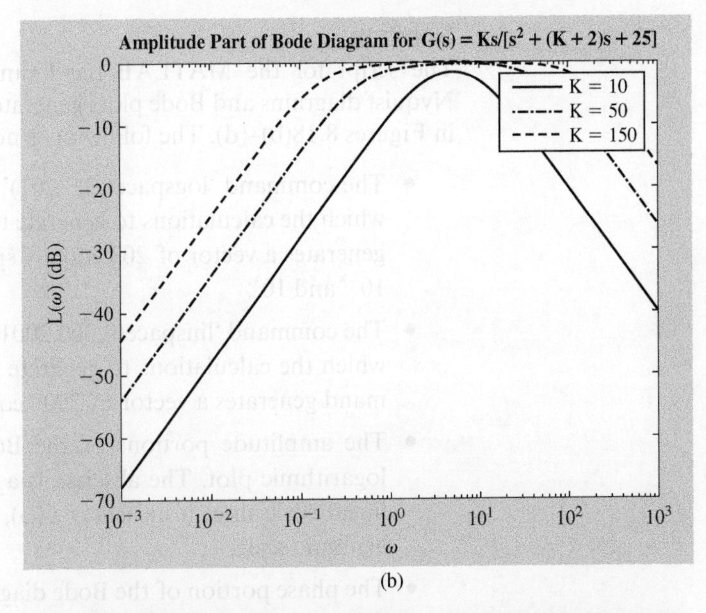

(b)

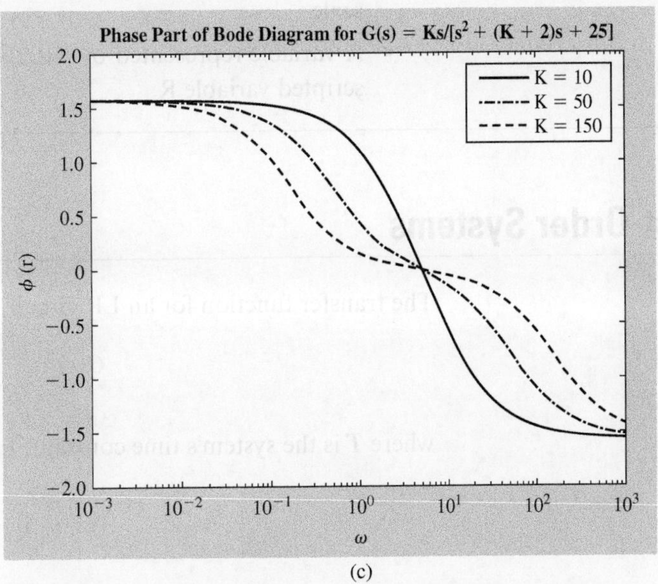

(c)

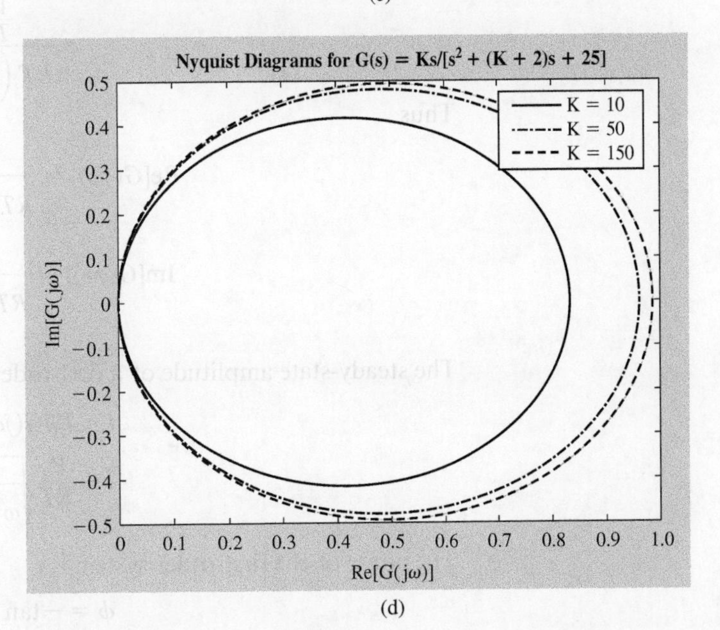

(d)

Figure 8.18
(Continued)

Solution

The script for the MATLAB file Example8_9.m is given in Figure 8.18(a). The Nyquist diagrams and Bode plots generated from the execution of this script are given in Figures 8.18(b)–(d). The following is noted about the script:

- The command 'logspace(-3,3,201)' is used to generate the vector of values for which the calculations to generate the Bode diagrams are made. This command generates a vector of 200 equally spaced values on a logarithmic scale between 10^{-3} and 10^3.
- The command 'linspace(0,200,2001)' is used to generate the vector of values for which the calculations to generate the Nyquist diagrams are made. This command generates a vector of 2000 equally spaced values between 0 and 200.
- The amplitude portions of the Bode diagrams are developed using a semilogarithmic plot. The abscissa is a logarithmic scale whereas the ordinate is a linear scale since it measures $L(\omega)$, which has already been converted to a logarithmic scale.
- The phase portion of the Bode diagrams are developed using a semilogarithmic scale.
- A variable represented by R(:,2) refers to the second column of a double subscripted variable R.

8.7 First-Order Systems

The transfer function for an LR circuit, which is a first-order system, is of the form

$$G(s) = \frac{1}{RT\left(s + \frac{1}{T}\right)} \tag{8.60}$$

where T is the system's time constant. Evaluation of $G(j\omega)$ leads to

$$G(j\omega) = \frac{1}{RT\left(j\omega + \frac{1}{T}\right)}$$

$$= \frac{\frac{1}{T} - j\omega}{RT\left(\omega^2 + \frac{1}{T^2}\right)} \tag{8.61}$$

Thus

$$\mathrm{Re}[G(j\omega)] = \frac{1}{RT^2\left(\omega^2 + \frac{1}{T^2}\right)} \tag{8.62a}$$

$$\mathrm{Im}[G(j\omega)] = \frac{-\omega}{RT\left(\omega^2 + \frac{1}{T^2}\right)} \tag{8.62b}$$

The steady-state amplitude of a first-order LR circuit is

$$I = V_0|G(j\omega)|$$

$$= \frac{V_0}{RT}\sqrt{\frac{1}{\omega^2 + \frac{1}{T^2}}} \tag{8.63}$$

The phase of the first-order system is

$$\phi = -\tan^{-1}(\omega T) \tag{8.64}$$

Defining the nondimensional frequency $\beta = \omega T$, Equations (8.63) and (8.64) can be written in nondimensional form as

$$M = \frac{RI}{V_0} = \sqrt{\frac{1}{\beta^2 + 1}} \tag{8.65}$$

$$\phi = -\tan^{-1}(\beta) \tag{8.66}$$

Frequency response curves for the first-order system are illustrated in Figure 8.19, which represents the frequency response curves for all first-order systems. The following is noted from these curves:

- As ω increases, the steady-state amplitude decreases.
- $M = 1/\sqrt{2}$ for $\beta = 1$.

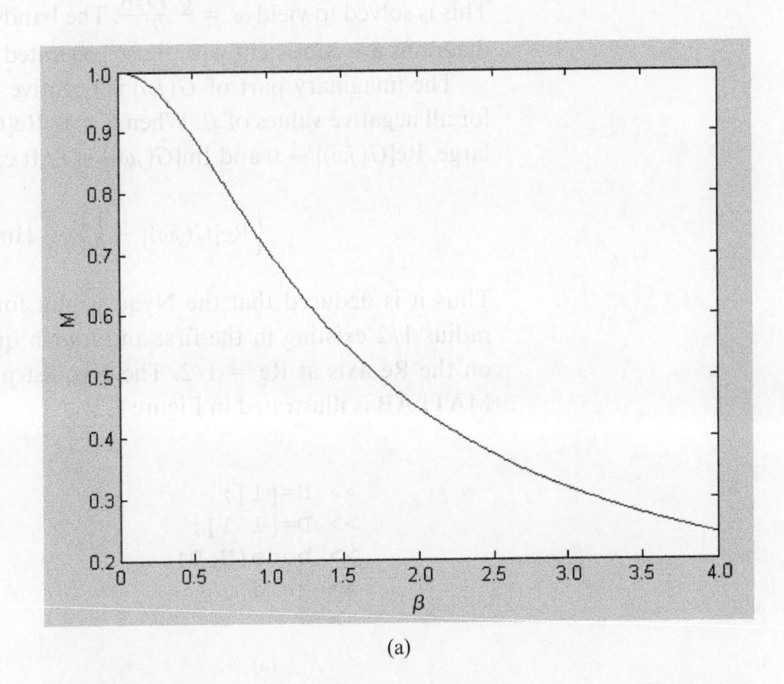

(a)

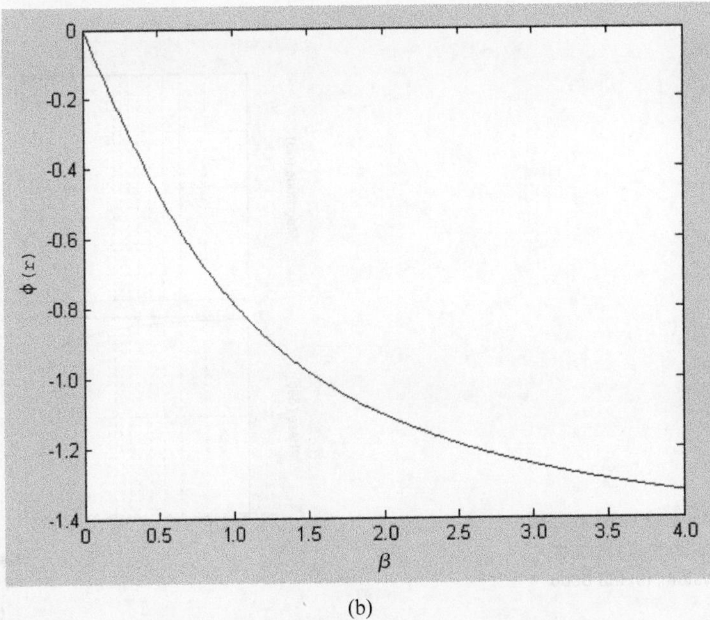

(b)

Figure 8.19

The frequency response curves for a first-order system whose transfer function is given by Equation (8.60): (a) The steady-state amplitude; (b) the phase angle.

- $\lim_{\beta \to \infty} M = 0$, but M behaves as $1/\beta$ for large β.
- ϕ is small and negative for small β.
- $\phi = -\pi/4$ for $\beta = 1$.
- $\lim_{\beta \to \infty} \phi = -\frac{\pi}{2}$.

Bode plots for a first-order system are shown in Figure 8.20(b). The script of the MATLAB work session used to generate these diagrams is given in Figure 8.20(a). The low-frequency asymptote is horizontal at log $(M) = 0$. The high-frequency asymptote is a line with a slope of -20 dB. The point of intersection of the two asymptotes, the corner frequency, is at $\beta = 1$, corresponding to $\omega = 1/T$. The cutoff frequency is determined using Equation (8.58), giving

$$L(0) - L(\omega_c) = 3 \text{ dB} = 0 - [-10 \log (1 + \omega^2 T^2)]$$

This is solved to yield $\omega_c = \frac{0.9976}{T}$. The bandwidth is from $\omega = 0$ to $\omega = \omega_c$. The Bode diagrams are consistent with those generated for a first-order lag in Section 8.4.4.

The imaginary part of $G(j\omega)$ is negative for all positive values of β and positive for all negative values of β. When $\beta = 0$, $\text{Re}[G(j\omega)] = 1$ and $\text{Im}[G(j\omega)] = 0$. As β gets large, $\text{Re}[G(j\omega)] \to 0$ and $\text{Im}[G(j\omega)] \to 0$. It can be shown algebraically that

$$\left\{ \text{Re}[G(j\omega)] - \frac{1}{2} \right\}^2 + \{\text{Im}[G(j\omega)]\}^2 = \frac{1}{4} \tag{8.67}$$

Thus it is deduced that the Nyquist plot for a first-order system is a semicircle of radius 1/2 existing in the first and fourth quadrants of the G plane with its center on the Re axis at Re $= 1/2$. The Nyquist plot of a first-order system drawn using MATLAB is illustrated in Figure 8.21.

```
>> N=[1];
>> D=[1 1];
>> bode(N,D)
>> grid
>>
```

(a)

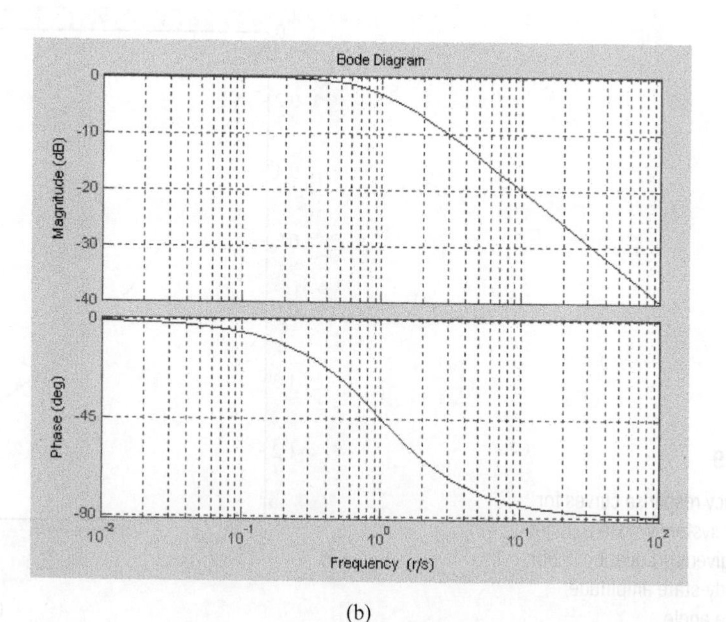

(b)

Figure 8.20

(a) The MATLAB work session used to generate the Bode diagram of a first-order system; (b) the Bode diagram.

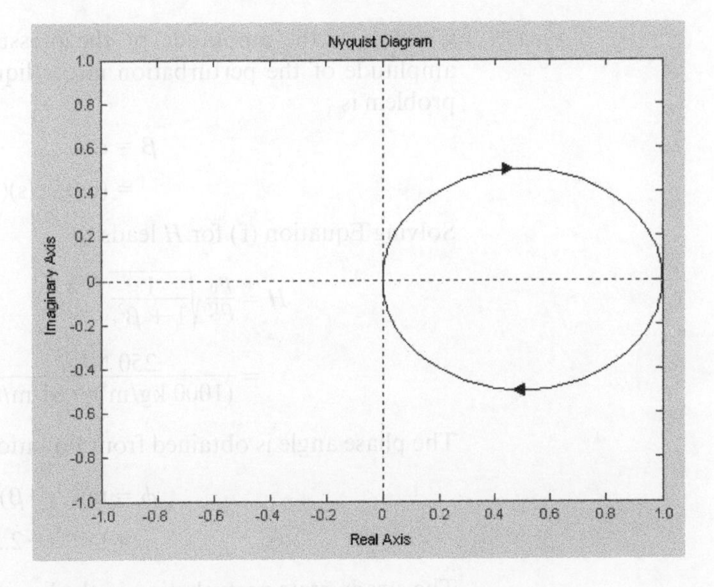

Figure 8.21

The Nyquist diagram for a first-order system.

Example 8.10

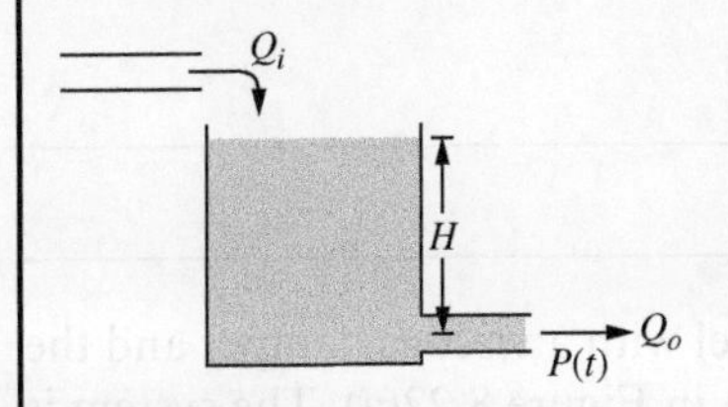

Figure 6.13

The system of Example 6.7 and Example 8.10. Pressure fluctuations at the tank's exit lead to perturbation in liquid level from the steady state. (Repeated)

The differential equation for the perturbation in liquid level $h(t)$ in a tank of area A due to a perturbation in exit pressure $p(t)$ is illustrated in Figure 6.13 and derived as Equation (k) in Example 6.7 as

$$A\frac{dh}{dt} + \frac{1}{R}h = \frac{p(t)}{\rho g R} \tag{a}$$

where R is the resistance in the exit pipe and ρ is the mass density of the liquid. Determine the steady-state response for $h(t)$ for a circular tank of diameter 4 m filled with water ($\rho = 1000$ kg/m^3) when the pipe resistance is 4.0 s/m^2 and the perturbation in exit pressure is

$$p(t) = 250 \sin(0.05t) \text{ Pa} \tag{b}$$

where t is measured in seconds.

Solution

Equation (a) is rewritten as

$$RA\frac{dh}{dt} + h = \frac{p(t)}{\rho g} \tag{c}$$

The time constant for the system is determined from Equation (c) as

$$\begin{aligned} T &= RA \\ &= \left(4.0\frac{\text{s}}{\text{m}^2}\right)\left[\frac{\pi}{4}(4\text{ m})^2\right] = 50.3\text{ s} \end{aligned} \tag{d}$$

The transfer function $G(s) = H(s)/P(s)$ is determined from Equation (c) as

$$G(s) = \frac{1}{\rho g T\left(s + \frac{1}{T}\right)} \tag{e}$$

The transfer function of Equation (e) is the same as the transfer function in Equation (8.60), but with ρg replacing R. The analogous form of Equation (8.65) is

$$\frac{\rho g H}{p_0} = \sqrt{\frac{1}{1 + \beta^2}} \tag{f}$$

where p_0 is the amplitude of the pressure perturbation and H is the steady-state amplitude of the perturbation in the liquid level. The frequency parameter for this problem is

$$\beta = \omega T$$
$$= (0.05 \text{ r/s})(50.3 \text{ s}) = 2.52 \tag{g}$$

Solving Equation (f) for H leads to

$$H = \frac{p_0}{\rho g}\sqrt{\frac{1}{1+\beta^2}}$$
$$= \frac{250 \text{ N/m}^2}{(1000 \text{ kg/m}^3)(9.81 \text{ m/s}^2)}\sqrt{\frac{1}{1+(2.2)^2}} = 10.6 \text{ mm} \tag{h}$$

The phase angle is obtained from Equation (8.66) as

$$\phi = \tan^{-1}(-\beta)$$
$$= \tan^{-1}(-2.52) = -1.19 \text{ r} \tag{i}$$

The steady-state perturbation in the liquid level is

$$h(t) = H\sin(\omega t + \phi)$$
$$= 10.6\sin(0.05t - 1.19) \text{ mm} \tag{j}$$

Example 8.11

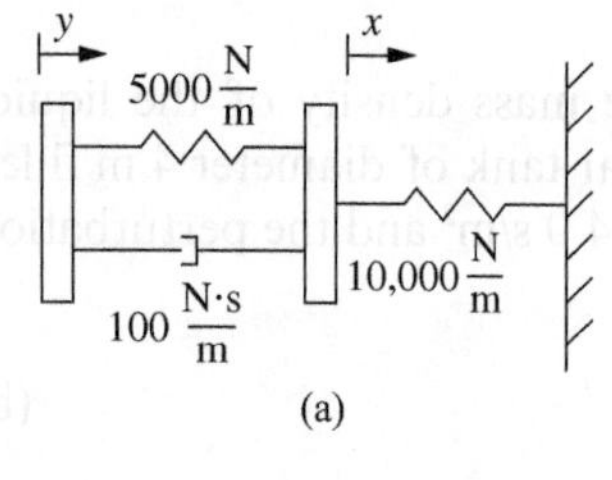

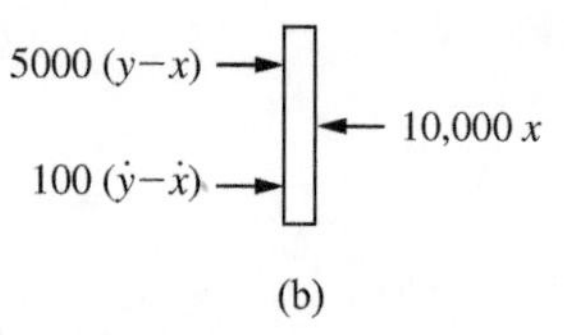

Figure 8.22

(a) Massless system with springs and a viscous damper; (b) free-body diagrams of the massless joint at an arbitrary instant.

A massless mechanical system with a spring in parallel with a viscous damper and the combination in parallel with another spring is shown in Figure 8.22(a). The system is given a prescribed base input of

$$y(t) = 0.002\sin\omega t \text{ m} \tag{a}$$

Let x be the displacement of the massless board and z be the displacement of the board relative to the base. **a.** Determine the steady-state displacement of the board for $\omega = 5$ r/s. **b.** Determine the steady-state displacement of the board relative to the base for $\omega = 5$ r/s. **c.** Draw the amplitude part of the Bode diagram for $x(t)$. **d.** What is the bandwidth for $x(t)$?

Solution

Applying Newton's second law to the free-body diagram of the massless block illustrated in Figure 8.22(b), leads to

$$100(\dot{y} - \dot{x}) + 5000(y - x) - 10{,}000x = 0 \tag{b}$$

Rearranging Equation (b) leads to

$$100\dot{x} + 15{,}000x = 100\dot{y} + 5000y \tag{c}$$

The transfer function is determined from Equation (c) as

$$G(s) = \frac{X(s)}{Y(s)} = \frac{100s + 5000}{100s + 15{,}000} \tag{d}$$

The transfer function is a ratio of a first-order lead over a first-order lag. Putting Equation (d) in this form gives

$$G(s) = \frac{1}{3}\left(\frac{0.2s + 1}{0.0667s + 1}\right) \tag{e}$$

Equation (e) is a transfer function with a gain of $\frac{1}{3}$, a first-order lead of time constant 0.2 s, and a first-order lag of time constant 0.667 s.

The mathematical model for the relative displacement of the board relative to the base,

$$z(t) = y(t) - x(t) \tag{f}$$

is

$$100\dot{z} + 5000z = 10{,}000x \tag{g}$$

The transfer function for the mathematical model of Equation (g) is

$$G_z(s) = \frac{100s + 5000}{10{,}000} = \frac{1}{2}(0.2s + 1) \tag{h}$$

The transfer function for the relative displacement is that of a gain of $\frac{1}{2}$ and a first-order lead with a time constant of 0.2 s.

a. The transfer function for $x(t)$ is written in the form of Equation (8.46) but a product of three functions. The magnitude of the sinusoidal transfer function is given by

$$|G(j\omega)| = \frac{1}{3}[1 + (0.2\omega)^2]^{1/2}\frac{1}{[1 + (0.0667\omega)^2]^{1/2}} \tag{i}$$

Evaluating Equation (i) for $\omega = 5$ r/s leads to $|G(5j)| = 0.447$. The phase of the sinusoidal transfer function is

$$\phi(\omega) = \phi\left(\frac{1}{3}\right) + \phi(0.2\omega j + 1) + \phi\left(\frac{1}{0.0667\omega j + 1}\right) \tag{j}$$

Evaluating Equation (i) for $\omega = 5$ r/s leads to $\phi(5) = 0.463$. Thus, the steady-state response is obtained using Equation (8.31) as

$$x(t) = 0.002(0.447)\sin(5t + 0.463) = 8.94 \times 10^{-4}\sin(5t + 0.463) \tag{k}$$

b. The transfer function for $z(t)$ is written in the form of Equation (8.46). The magnitude of the sinusoidal transfer function is given by

$$|G_z(j\omega)| = \frac{1}{2}[1 + (0.2\omega)^2]^{1/2} \tag{l}$$

Evaluating Equation (l) for $\omega = 5$ r/s leads to $|G_z(5j)| = 0.707$. The phase of the sinusoidal transfer function is

$$\phi(\omega) = \phi\left(\frac{1}{2}\right) + \phi(0.2\omega j + 1) \tag{m}$$

Evaluating Equation (m) for $\omega = 5$ r/s leads to $\phi(5) = 0.785$. Thus, the steady-state response is given by Equation (8.31) as

$$x(t) = 0.002(0.707)\sin(5t + 0.785) = 0.00141\sin(5t + 0.785) \tag{n}$$

c. The Bode diagram for x is formed from the superposition of three Bode diagrams: one for the gain, one for the first-order lead, and one for the first-order lag. The low-frequency asymptote is $-20 \log(3) = 9.53$ dB. The high-frequency asymptote is $20 \log(30) = 29.5$ dB. Construction of the amplitude part of the Bode diagram is shown in Figure 8.23.

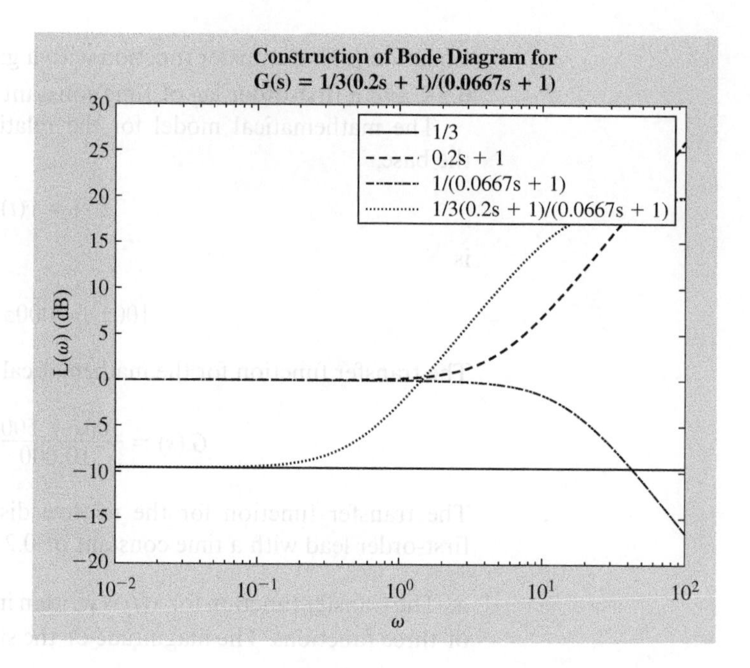

Figure 8.23

Amplitude part of Bode diagram for displacement of mass of Example 8.11 is the sum of the Bode diagrams for a gain, a first-order lead, and a first-order lag.

d. Note that for x, the maximum of $L(\omega)$ is 29.5 dB, the high-frequency asymptote. Then the bandwidth is the range of frequencies between a frequency that is 3 dB less than the maximum and infinity. The low frequency in this range is obtained by requiring that

$$29.5 - L(\omega) = 3 \tag{o}$$

$$29.5 - \left\{20\log\frac{1}{3} + 10\log\left[1 + (0.2\omega)^2\right] - 10\log\left[1 + (0.0667\omega)^2\right]\right\} = 3 \tag{p}$$

Equation (p) is solved, yielding 14.79 r/s. Thus, the bandwidth is

$$14.79 \text{ r/s} < \omega < \infty \tag{q}$$

8.8 Second-Order Systems

The frequency response for the series LRC circuit was used to illustrate the development of frequency response diagrams, Bode diagrams, and Nyquist diagrams. The frequency response for other important second-order systems is examined in this section.

8.8.1 One-Degree-of-Freedom Mechanical System

The transfer function for a second-order mechanical system with force input is given in Equation (7.8) and is written in terms of natural frequency and damping ratio as

$$G(s) = \frac{1}{m(s^2 + 2\zeta\omega_n s + \omega_n^2)} \tag{8.68}$$

The steady-state response of the system due to a sinusoidal input of the form $F(t) = F_0 \sin(\omega t)$ is obtained through the evaluation of $G(j\omega)$, which leads to

$$G(j\omega) = \frac{(\omega_n^2 - \omega^2) - j(2\zeta\omega\omega_n)}{m[(\omega_n^2 - \omega^2)^2 + (2\zeta\omega\omega_n)^2]} \tag{8.69}$$

The steady-state response of the second-order system is of the form of Equation (8.2), where

$$X = F_0|G(j\omega)|$$
$$= \frac{F_0}{m}\sqrt{\frac{1}{(\omega_n^2 - \omega^2)^2 + (2\zeta\omega\omega_n)^2}} \tag{8.70}$$

and

$$\phi = \tan^{-1}\left(\frac{2\zeta\omega\omega_n}{\omega - \omega_n^2}\right) \tag{8.71}$$

Nondimensional forms of Equations (8.70) and (8.71) are obtained in terms of the frequency ratio $r = \omega/\omega_n$ as

$$M = \frac{m\omega_n^2 X}{F_0} = \frac{1}{\sqrt{(1 - r^2)^2 + (2\zeta r)^2}} \tag{8.72}$$

$$\phi = \tan^{-1}\left(\frac{2\zeta r}{r^2 - 1}\right) \tag{8.73}$$

Frequency response curves for the second-order system are shown in Figure 8.24. Analysis of these curves, as well as Equations (8.72) and (8.73), shows the following:

- For $\zeta > 1/\sqrt{2}$ the steady-state amplitude decreases with increasing frequency, approaching zero for large r.
- For $\zeta < 1/\sqrt{2}$ the steady-state amplitude increases with increasing r until it reaches a maximum and then decreases, approaching zero for large r.
- For $\zeta = 1/\sqrt{2}$ the steady-state amplitude has its maximum of $F_0 = (m\omega_n^2)$ at $r = 0$ and decreases with increasing r, approaching zero for large r.
- The maximum value of X over all values of r occurs for the value of r for which $dX/dr = 0$. The maximum value occurs for $r = \sqrt{1 - 2\zeta^2}$ and is equal to

$$X_{\max} = \frac{F_0}{2m\omega_n^2\zeta\sqrt{1 - \zeta^2}} \tag{8.74}$$

- For a fixed r, the larger the value of ζ, the smaller the value of X.
- The phase is near zero for r near zero and is near π for large r.

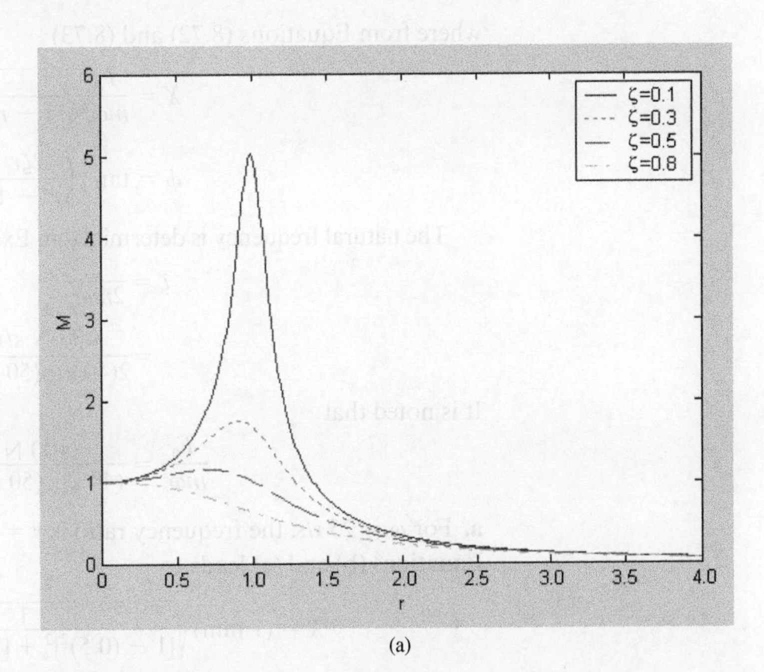

Figure 8.24

The frequency response curves for a mass, spring, and viscous damper system: (a) The steady-state amplitude vs. frequency ratio; (b) the phase angle vs. frequency ratio.

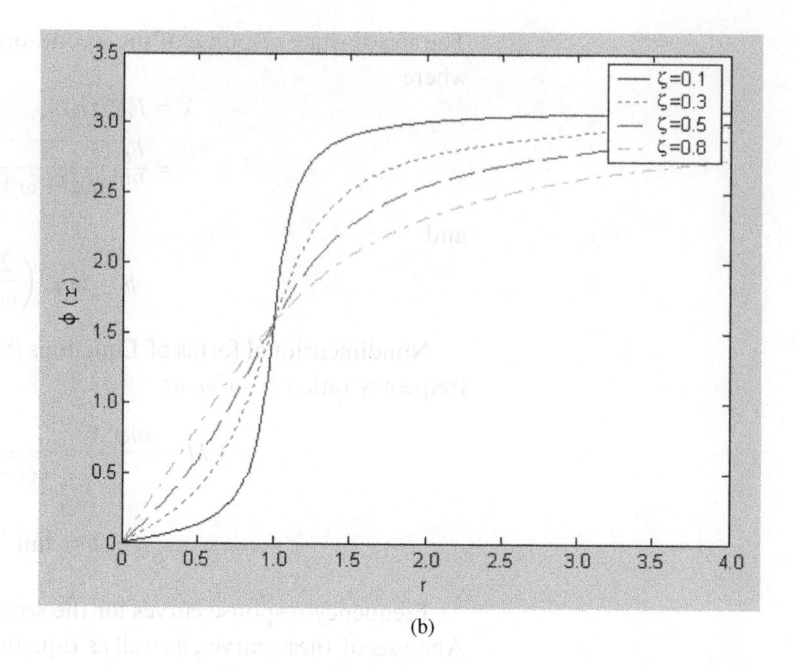

Figure 8.24
(Continued)

(b)

Example 8.12

Determine the steady-state response of the machine of Example 8.1 if the 40-kg machine is mounted on a spring of stiffness 1×10^5 N/m in parallel with a viscous damper of damping coefficient 400 N·s/m when it is subject to an input force of $F(t) = 100 \sin(\omega t)$ N for **a.** $\omega = 25$ r/s, **b.** $\omega = 50$ r/s, and **c.** $\omega = 100$ r/s. **d.** In addition, draw its Bode diagram, specifying its low-frequency and high-frequency asymptotes, corner frequency, and cutoff frequency. **e.** Draw the Nyquist diagram for a mass, spring, and viscous damper system for several values of ζ.

Solution

The steady-state response of the system for an input frequency ω is given by Equation (8.12) as

$$x(t) = X \sin(\omega t + \phi) \tag{a}$$

where from Equations (8.72) and (8.73)

$$X = \frac{F_0}{m\omega_n^2}\sqrt{\frac{1}{(1 - r^2)^2 + (2\zeta r)^2}} \tag{b}$$

$$\phi = \tan^{-1}\left(\frac{2\zeta r}{r^2 - 1}\right) \tag{c}$$

The natural frequency is determined in Example 8.1 as 50 r/s. The damping ratio is

$$\zeta = \frac{c}{2m\omega_n} = \frac{400 \text{ N·s/m}}{2(40 \text{ kg})(50 \text{ r/s})} = 0.1 \tag{d}$$

It is noted that

$$\frac{F_0}{m\omega_n^2} = \frac{100 \text{ N}}{(40 \text{ kg})(50 \text{ r/s})^2} = 1 \text{ mm} \tag{e}$$

a. For $\omega = 25$ r/s, the frequency ratio is $r = (25 \text{ r/s})/(50 \text{ r/s}) = 0.5$. The evaluation of Equations (b) and (c) leads to

$$X = (1 \text{ mm})\sqrt{\frac{1}{[1 - (0.5)^2]^2 + [2(0.1)(0.5)]^2}} = 1.32 \text{ mm} \tag{f}$$

$$\phi = \tan^{-1}\left(\frac{2(0.1)(0.5)}{(0.5)^2 - 1}\right) = -0.133 \text{ r} \quad \text{(g)}$$

The steady-state response for an input frequency of 25 r/s is

$$x(t) = 1.32 \sin(25t - 0.133) \text{ mm} \quad \text{(h)}$$

b. For $\omega = 50$ r/s, the frequency ratio is $r = (50 \text{ r/s})/(50 \text{ r/s}) = 1$. The evaluation of Equations (b) and (e) leads to

$$X = (1 \text{ mm})\sqrt{\frac{1}{[1 - (1)^2]^2 + [2(0.1)(1)]^2}} = 5 \text{ mm} \quad \text{(i)}$$

$$\phi = \tan^{-1}\left(\frac{2(0.1)(1)}{1 - (1)^2}\right) = -\frac{\pi}{2} \text{ r} \quad \text{(j)}$$

The steady-state response for an input frequency of 50 r/s is

$$x(t) = 5 \sin\left(50t - \frac{\pi}{2}\right) \text{ mm} \quad \text{(k)}$$

c. For $\omega = 100$ r/s, the frequency ratio is $r = (100 \text{ r/s})/(50 \text{ r/s}) = 2$. The evaluation of Equations (b) and (e) leads to

$$X = (1 \text{ mm})\sqrt{\frac{1}{[1 - (2)^2]^2 + [2(0.1)(2)]^2}} = 0.333 \text{ mm} \quad \text{(l)}$$

$$\phi = \tan^{-1}\left(\frac{2(0.1)(2)}{(2)^2 - 1}\right) = 0.133 \text{ r} \quad \text{(m)}$$

The steady-state response for an input frequency of 100 r/s is

$$x(t) = 0.333 \sin(100t + 0.133) \text{ mm} \quad \text{(n)}$$

d. Bode diagrams corresponding to the second-order system of Example 8.12 with $m = 1$ kg are shown in Figure 8.25. The Bode diagram is that of the second-order underdamped systems given in Equations (8.56). Its low-frequency asymptote is

$$L = -20 \log (\omega_n^2) = -68.0 \text{ dB} \quad \text{(o)}$$

while its high-frequency asymptote is

$$L = -40 \log (\omega) \quad \text{(p)}$$

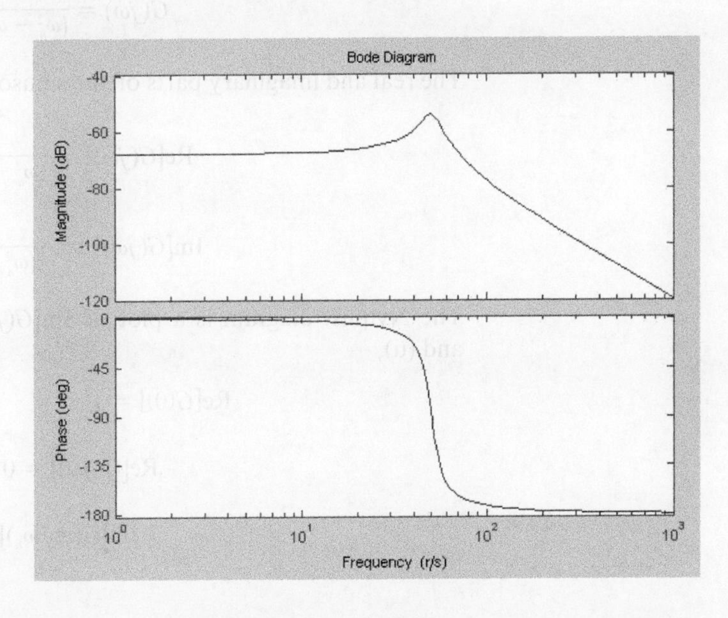

Figure 8.25

The Bode diagram for second-order mechanical system of Example 8.12 for $m = 1$ kg.

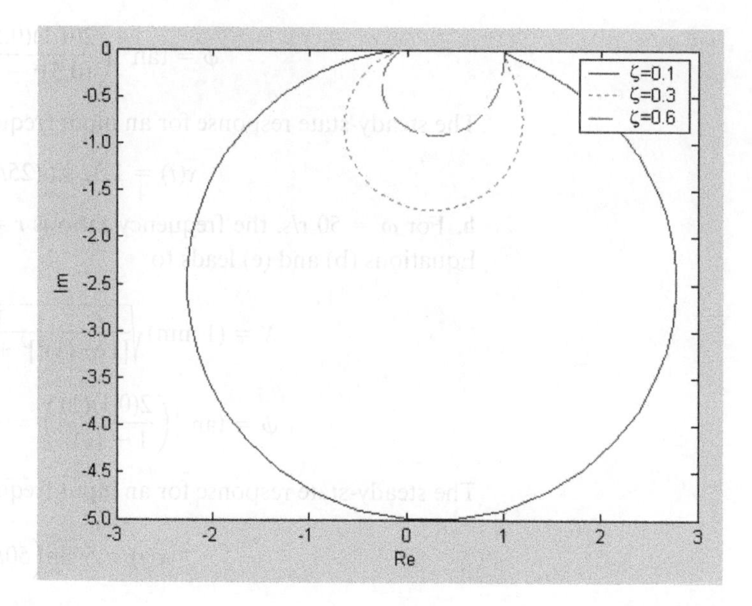

Figure 8.26

The Nyquist plots for a second-order mechanical system, for $m = 1$ kg.

The corner frequency is obtained by equating Equations (o) and (p), leading to

$$
\begin{aligned}
-40 \log(\omega) &= -68.0 \\
\omega &= 50 \text{ r/s}
\end{aligned} \tag{q}
$$

It is noted that the system's corner frequency is equal to its natural frequency, which is true for all transfer functions for a second-order lag system.

The system's cutoff frequency is the frequency when $L(\omega)$ is equal to 3 dB less than $L(0)$. To this end,

$$
\begin{aligned}
-68.0 + 10 \log [(50^2 - \omega_c^2)^2 + [2(0.1)(50)\omega_c]^2] &= 3 \\
10 \log [(2500 - \omega_c^2)^2 + 100\omega_c^2] &= 71.0 \\
\omega_c^4 - 4900\omega_c^2 + 6.25 \times 10^6 &= 1.25 \times 10^7
\end{aligned} \tag{r}
$$

Equation (r) is solved, leading to its only real and positive solution of $\omega_c = 77.1$ r/s.

e. The sinusoidal transfer function for the system is given by Equation (8.69). With $m = 1$ kg, this becomes

$$G(j\omega) = \frac{(\omega_n^2 - \omega^2) - j(2\zeta\omega\omega_n)}{(\omega_n^2 - \omega^2)^2 + (2\zeta\omega\omega_n)^2} \tag{s}$$

The real and imaginary parts of the sinusoidal transfer function are

$$\text{Re}[G(j\omega)] = \frac{\omega_n^2 - \omega^2}{(\omega_n^2 - \omega^2)^2 + (2\zeta\omega\omega_n)^2} \tag{t}$$

$$\text{Im}[G(j\omega)] = -\frac{2\zeta\omega\omega_n}{(\omega_n^2 - \omega^2)^2 + (2\zeta\omega\omega_n)^2} \tag{u}$$

The Nyquist diagram is a plot of $\text{Im}[G(j\omega)]$ versus $\text{Re}[G(j\omega)]$. From Equations (t) and (u),

$$\text{Re}[G(0)] = 1 \qquad \text{Im}[G(0)] = 0 \tag{v}$$

$$\text{Re}[G(j\omega)] = 0 \Rightarrow \omega = \pm\omega_n \tag{w}$$

$$\text{Im}[G(\pm j\omega_n)] = -\frac{1}{2\zeta\omega_n^2} \tag{x}$$

It is noted from Equation (u) that $\mathrm{Im}[G(j\omega)] < 0$ for all ω while $\mathrm{Re}[G(j\omega)] > 0$ for $\omega < \omega_n$ and $\mathrm{Re}[G(j\omega)] < 0$ for $\omega > \omega_n$. Also, from Equations (t) and (u),

$$\lim_{\omega \to \pm\infty} \mathrm{Re}[G(j\omega)] = 0 \tag{y}$$

$$\lim_{\omega \to \pm\infty} \mathrm{Im}[G(j\omega)] = 0 \tag{z}$$

Based upon Equations (v)-(z), the Nyquist diagram can be developed. The Nyquist diagram is drawn for several values of ζ in Figure 8.26.

8.8.2 Motion Input

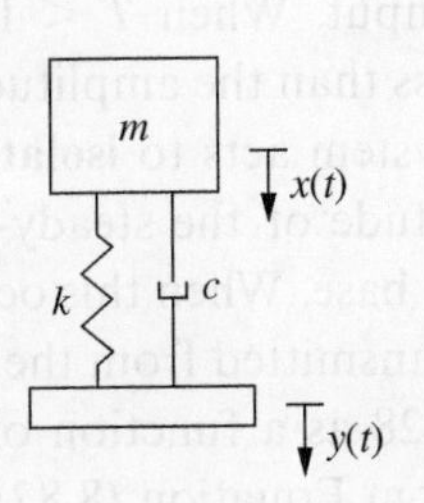

Figure 8.27

The mass-spring-viscous damper system with a prescribed motion of support.

The transfer function that is determined from the mathematical model of the mass-spring-viscous damper system connected to a movable base, shown in Figure 8.27, is that of Equation (7.9), which is written in the form of Equation (7.62) with $A = 2\zeta\omega_n$ and $B = \omega_n^2$:

$$G_x(s) = \frac{X(s)}{Y(s)} = \frac{2\zeta\omega_n s + \omega_n^2}{s^2 + 2\zeta\omega_n s + \omega_n^2} \tag{8.75}$$

The acceleration of the mass is $a(t) = \ddot{x}(t)$. Its transfer function is determined using Equation (8.75):

$$G_a(s) = \frac{A(s)}{Y(s)} = \frac{s^2(2\zeta\omega_n s + \omega_n^2)}{s^2 + 2\zeta\omega_n s + \omega_n^2} \tag{8.76}$$

The displacement of the mass relative to the base is

$$z(t) = x(t) - y(t) \tag{8.77}$$

The transfer function for the relative displacement is determined as

$$G_z(s) = \frac{Z(s)}{Y(s)} = \frac{s^2}{s^2 + 2\zeta\omega_n s + \omega_n^2} \tag{8.78}$$

The magnitudes of the sinusoidal transfer functions are obtained by evaluating Equations (8.75), (8.76), and (8.78) at $s = j\omega$ and then taking the magnitude of the resulting complex expression, leading to

$$|G_x(j\omega)| = \sqrt{\frac{\omega_n^4 + (2\zeta\omega\omega_n)^2}{(\omega_n^2 - \omega^2)^2 + (2\zeta\omega\omega_n)^2}} \tag{8.79}$$

$$|G_a(j\omega)| = \omega^2\sqrt{\frac{\omega_n^4 + (2\zeta\omega\omega_n)^2}{(\omega_n^2 - \omega^2)^2 + (2\zeta\omega\omega_n)^2}} \tag{8.80}$$

$$|G_z(j\omega)| = \frac{\omega^2}{\sqrt{(\omega_n^2 - \omega^2)^2 + (2\zeta\omega\omega_n)^2}} \tag{8.81}$$

Equations (8.79), (8.80), and (8.81) are written in terms of the frequency ratio $r = \omega/\omega_n$ as

$$T = |G_x(j\omega)| = \sqrt{\frac{1 + (2\zeta r)^2}{(1 - r^2)^2 + (2\zeta r)^2}} \tag{8.82}$$

$$\omega_n^2 \Gamma = |G_a(j\omega)| = \omega_n^2 r^2 \sqrt{\frac{1 + (2\zeta r)^2}{(1 - r^2)^2 + (2\zeta r)^2}} \tag{8.83}$$

$$\Lambda = |G_z(j\omega)| = \frac{r^2}{\sqrt{(1 - r^2)^2 + (2\zeta r)^2}} \tag{8.84}$$

The steady-state amplitudes due to a sinusoidal motion input of the form $y(t) = Y\sin(\omega t)$ are

$$X = TY \tag{8.85}$$

$$A = \omega_n^2 \Gamma Y \tag{8.86}$$

$$Z = \Lambda Y \tag{8.87}$$

The ratio $T = X/Y$ is called the transmissibility because it represents the ratio of the amplitude of the output to the amplitude of the input. When $T < 1$, the amplitude of the steady-state displacement of the mass is less than the amplitude of the support. In this case, the spring and viscous damper system acts to **isolate** the mass from the motion of its base. When $T > 1$, the amplitude of the steady-state displacement of the mass is larger than the amplitude of the base. When this occurs, the spring and viscous damper act to **amplify** the motion transmitted from the base to the mass. The transmissibility is illustrated in Figure 8.28 as a function of the frequency ratio for several values of the damping ratio. From Equation (8.82) and Figure 8.28, it is noted that

- $T = 0$ when $r = 0$ for all values of ζ.
- $T = 1$ when $r = \sqrt{2}$ for all values of ζ.
- $T > 1$ when $r < \sqrt{2}$ for all values of ζ.
- $T < 1$ when $r > \sqrt{2}$ for all values of ζ.
- T approaches zero for large r for all values of ζ.
- For $r > \sqrt{2}$, T is smaller for smaller values of ζ.

Thus it is clear that $r > \sqrt{2}$ is the **range of isolation** and $r < \sqrt{2}$ is the **range of amplification.**

For a specific system, Equation (8.86) can be used along with Equation (8.83) to determine the steady-state amplitude of the acceleration of the mass. Figure 8.29(a)

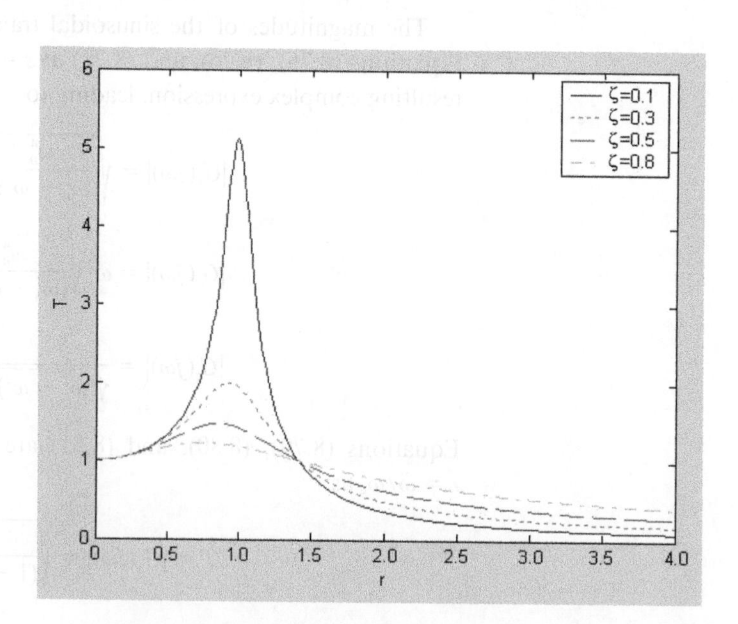

Figure 8.28

The transmissibility ratio as a function of the frequency ratio for several values of the damping ratio.

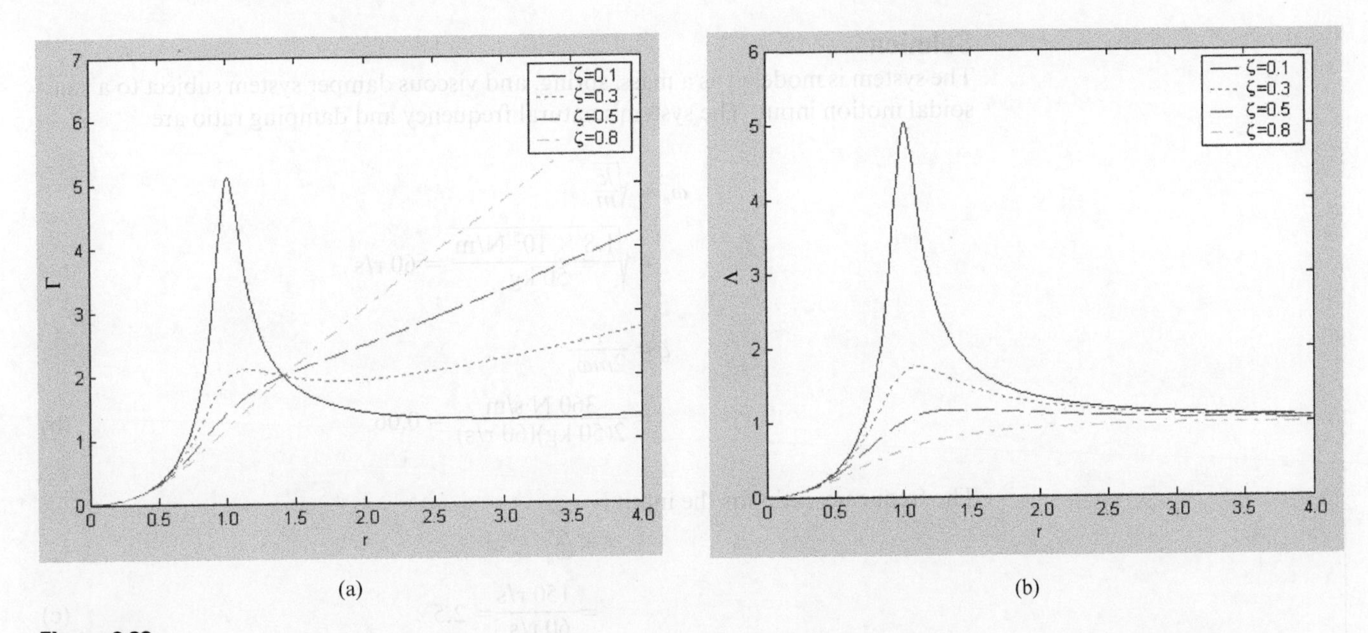

Figure 8.29

(a) Γ as a function of the frequency ratio for several values of the damping ratio. (b) Λ as a function of the frequency ratio for several values of the damping ratio.

illustrates the variation of Γ with the frequency ratio for several values of the damping ratio. The following is obtained from Figure 8.29(a) and Equation (8.83):

- $\Gamma = 0$ when $r = 0$ for all values of ζ.
- As $r \to \infty, \Gamma \to 2\zeta r$.
- $\Gamma = 2$ when $r = \sqrt{2}$ for all values of ζ.
- For $\zeta < \sqrt{2}/4$, the function Γ has a relative maximum for some value of $r < \sqrt{2}$ and a relative minimum for some value of $r > \sqrt{2}$.
- For $\zeta > \sqrt{2}/4$, the value of Γ increases with increasing r, achieving no relative maximum.

The steady-state amplitude of the displacement of the mass relative to the support is given by Equation (8.87), where Λ is defined in Equation (8.84). The variation of Λ with r is illustrated in Figure 8.29(b), which shows that:

- $\Lambda = 0$ when $r = 0$ for all values of ζ.
- $\Lambda \to 1$ as $r \to \infty$ for all values of ζ.
- For $\zeta < 1/\sqrt{2}$, Λ has a maximum of $1/\left(2\zeta\sqrt{1-\zeta^2}\right)$ corresponding to a value of $r = 1/\sqrt{1-2\zeta^2}$.
- For $\zeta > 1/\sqrt{2}$, Λ increases with increasing r, never attaining its maximum.

Example 8.13

A 50-kg machine is mounted on an isolator of stiffness 1.8×10^5 N/m and damping coefficient 360 N·s/m. The floor on which the isolator is mounted has a sinusoidal displacement of amplitude 0.2 mm and frequency 150 r/s. **a.** Determine the steady-state amplitude of the displacement of the machine. **b.** Determine the steady-state amplitude of the machine's acceleration. **c.** Determine the steady-state amplitude of the displacement of the machine relative to the floor.

Solution

The system is modeled as a mass, spring, and viscous damper system subject to a sinusoidal motion input. The system's natural frequency and damping ratio are

$$\omega_n = \sqrt{\frac{k}{m}}$$

$$= \sqrt{\frac{1.8 \times 10^5 \text{ N/m}}{50 \text{ kg}}} = 60 \text{ r/s} \quad \text{(a)}$$

$$\zeta = \frac{c}{2m\omega_n}$$

$$= \frac{360 \text{ N·s/m}}{2(50 \text{ kg})(60 \text{ r/s})} = 0.06 \quad \text{(b)}$$

The frequency ratio for the input is

$$r = \frac{\omega}{\omega_n}$$

$$= \frac{150 \text{ r/s}}{60 \text{ r/s}} = 2.5 \quad \text{(c)}$$

a. The steady-state amplitude of the machine's displacement is obtained using Equations (8.82) and (8.85) as

$$X = (Y)(T)$$

$$= (0.2 \text{ mm})\sqrt{\frac{1 + [2(0.06)(2.5)]^2}{[1 - (2.5)^2]^2 + [2(0.06)(2.5)]^2}} = 0.0397 \text{ mm} \quad \text{(d)}$$

b. The steady-state amplitude of the machine's acceleration is

$$A = \omega^2 X$$

$$= (150 \text{ r/s})^2(0.0397 \text{ mm}) = 0.893 \text{ m/s}^2 \quad \text{(e)}$$

c. The steady-state amplitude of the displacement of the machine relative to the floor is obtained using Equations (8.84) and (8.87) as

$$Z = (Y)(\Lambda)$$

$$= (0.2 \text{ mm})\frac{(2.5)^2}{\sqrt{[1 - (2.5)^2]^2 + [2(0.06)(2.5)]^2}} = 0.238 \text{ mm} \quad \text{(f)}$$

8.8.3 Filters

Electric circuits can be designed to act as filters. The input to such a circuit is a sinusoidal voltage source of constant amplitude. The output from the filter is a sinusoidal voltage at some amplitude. A filter amplifies the voltage from sources of some frequencies and greatly reduces the amplitude from sources at other frequencies. A **low-pass filter** amplifies low frequencies and filters out higher frequencies. A **high-pass filter** amplifies higher frequencies and filters out lower frequencies. A **band-pass filter** amplifies frequencies in a certain frequency range and filters out frequencies outside the designated range. A **band-reject filter** filters out frequencies in a certain range and amplifies frequencies outside the range. The range of frequencies amplified by a circuit and the range of frequencies filtered out by a circuit are determined from a frequency response analysis of the circuit.

Example 8.14

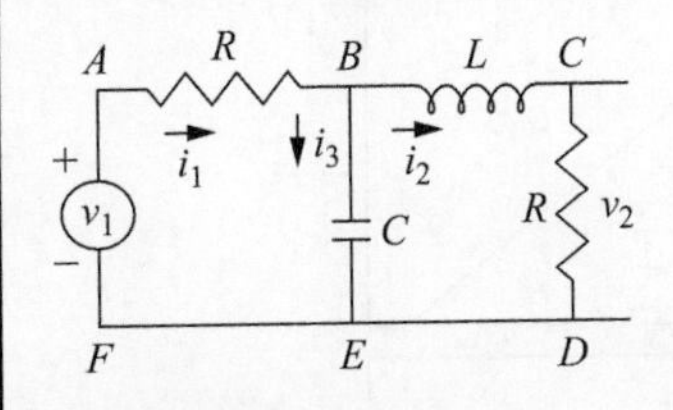

Figure 8.30

A second-order Butterworth filter.

The output from the circuit of Figure 8.30 is $v_2(t)$, the voltage across terminals at the ends of the resistor. Determine the steady-state response for $v_2(t)$ when $R = \sqrt{L/C}$. Discuss why this circuit is called a second-order low-pass filter (called a Butterworth filter).

Solution

Application of KCL at node B and application of KVL around loops $ABEFA$ and $BCDEB$ lead to a mathematical model for the circuit:

$$R_1 i_1 + \frac{1}{C}\int_0^t (i_1 - i_2)dt = v_1(t) \tag{a}$$

$$L\frac{di_2}{dt} + Ri_2 - \frac{1}{C}\int_0^t (i_1 - i_2)dt = 0 \tag{b}$$

Equations (a) and (b) are used to determine the transfer function for $i_2(t)$:

$$G_2(s) = \frac{I_2(s)}{V_1(s)} = \frac{1/(LRC)}{s^2 + \left(\frac{R}{L} + \frac{1}{RC}\right)s + \frac{2}{LC}} \tag{c}$$

The voltage change from C to D is

$$v_2(t) = Ri_2 \tag{d}$$

Equations (c) and (d) are used to determine the transfer function defined as $G(s) = \frac{V_2(s)}{V_1(s)}$:

$$G(s) = \frac{1}{LC\left[s^2 + \left(\frac{R}{L} + \frac{1}{RC}\right)s + \frac{2}{LC}\right]} \tag{e}$$

The denominator of Equation (e) is compared to the denominator of a second-order transfer function, Equation (7.61), and the natural frequency is determined as

$$\omega_n = \sqrt{\frac{2}{LC}} \tag{f}$$

and the damping ratio is determined as

$$\zeta = \frac{1}{2\omega_n}\left(\frac{R}{L} + \frac{1}{RC}\right)$$

$$= \frac{1}{2\sqrt{2}}\left(R\sqrt{\frac{C}{L}} + \frac{1}{R}\sqrt{\frac{L}{C}}\right) \tag{g}$$

For $R = \sqrt{L/C}$, Equation (g) evaluates to a damping ratio of $\zeta = 1/\sqrt{2}$ and the transfer function given by Equation (e) becomes

$$G(s) = \frac{\omega_n^2}{2(s^2 + \sqrt{2}\omega_n s + \omega_n^2)} \tag{h}$$

where the natural frequency is given by Equation (f).

The steady-state response of the output voltage due to a sinusoidal input voltage of the form $v_1(t) = V_1 \sin(\omega t)$ is

$$v_2(t) = V_2 \sin(\omega t + \phi) \tag{i}$$

where

$$V_2 = V_1 |G(j\omega)| \tag{j}$$

and ϕ is the phase of $G(j\omega)$. Substituting $s = j\omega$ in Equation (h) leads to

$$G(j\omega) = \frac{\omega_n^2}{2}\left[\frac{(\omega_n^2 - \omega^2) - j\sqrt{2}\omega\omega_n}{(\omega_n^2 - \omega^2)^2 + 2\omega_n^2\omega^2}\right] \tag{k}$$

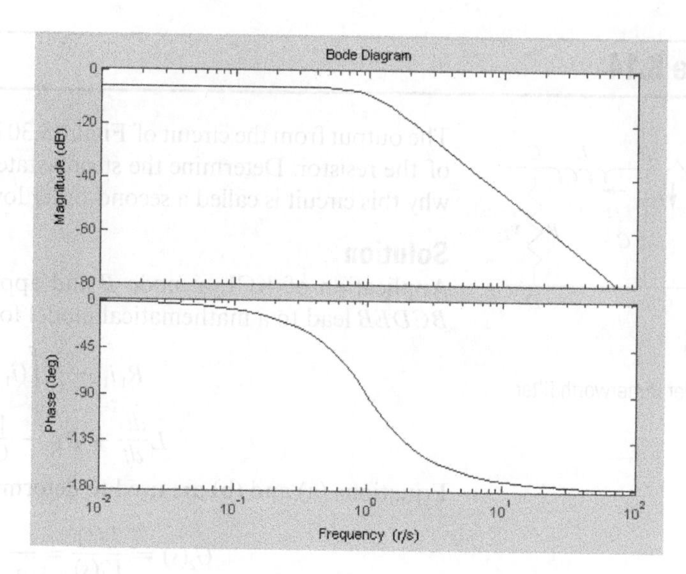

Figure 8.31

The Bode diagram for a second-order Butterworth filter.

Equation (k) is used to obtain

$$V_2 = \frac{V_1}{2\sqrt{1 + \left(\frac{\omega}{\omega_n}\right)^4}} \quad \text{(l)}$$

$$\phi = \tan^{-1}\left(\frac{-\sqrt{2}\frac{\omega}{\omega_n}}{1 - \left(\frac{\omega}{\omega_n^2}\right)}\right) \quad \text{(m)}$$

The Bode diagram for the transfer function of Equation (h) is given in Figure 8.31. Equation (l) and Figure 8.31 both show that the steady-state amplitude of the output for small frequencies is near the amplitude of the steady-state input, while the steady-state amplitude of the response corresponding to high-frequency input is very small. Thus this circuit filters out high-frequency input and passes on low-frequency input; hence the name low-pass filter. The circuit is a second-order filter because its transfer function is that of a second-order system.

8.9 Higher-Order Systems

The same tools are used to develop the steady-state response of higher-order systems (systems greater than second order) as are used for first-and second-order systems. The sinusoidal transfer function is used to determine the steady-state amplitudes and phase angles. Frequency response curves, Bode plots, and Nyquist plots illustrate qualitative and quantitative features of the frequency response.

A higher-order system is usually modeled using more than one dependent variable. Examples of higher-order systems include multidegree-of-freedom mechanical systems and multiloop electric circuits with capacitors or inductors.

If the system has multiple outputs, then a transfer function matrix **G**(s) is defined. The individual transfer function $G_{i,j}(s)$ is the impulsive response of the output defined as x_i due to the force $f_j(t)$. Consider the transfer function for a system without a time delay. The transfer function of this system, $G_{i,j}(s)$, has a denominator that is the product of irreducible second-order polynomials and first-order polynomials. The second-order polynomials each can be written in terms of a damping ratio and a natural frequency. The first-order polynomials can each be written in terms of a time constant.

Consider a system with n inputs and m outputs. If an input is $f_j(t)$, which has a Laplace transform $F_j(s)$, then the Laplace transform of an output x_i due solely to the input $f_j(t)$ may be written as

$$X_i(s) = F_j(s)\, G_{i,j}(s) \tag{8.88}$$

for $i = 1, 2, \ldots, m$. If n independent inputs are given, then

$$X_i(s) = \sum_{j=1}^{n} F_j(s)\, G_{i,j}(s) \tag{8.89}$$

If $f_j(t)$ is a harmonic input of the form of Equation (8.1), that is

$$f_j(t) = F_{0,j}\sin(\omega_j t + \psi_j) \tag{8.90}$$

The steady-state response is obtained using Equation (8.31) as

$$x_{s,i}(t) = F_{0,j}\left|G_{i,j}(j\omega_j)\right|\sin(\omega_j t + \psi_j + \phi_j) \tag{8.91}$$

where $\left|G_{i,j}(j\omega_j)\right|$ is the magnitude of the sinusoidal transfer function developed from $G_{i,j}(s)$ and ϕ_j is the phase of the transfer function. The general form of the response is obtained by taking the inverse Laplace transform of Equation (8.89) and recognizing that one term in the summation is the right-hand side of Equation (8.91).

$$x_{s,i}(t) = \sum_{j=1}^{n} F_{0,j}\left|G_{i,j}(j\omega_j)\right|\sin(\omega_j t + \psi_j + \phi_j) \tag{8.92}$$

The algebra required to determine system response for a higher-order system is complex and tedious. Higher-order systems are now often modeled using the state-space methods of Chapter 12.

Section 7.6 presents a general discussion of the free response of a higher-order system. The discussion of the frequency response of higher-order systems is limited to the specific applications of dynamic vibration absorbers for mechanical systems and higher-order filters.

8.9.1 Dynamic Vibration Absorbers

Consider the two-degree-of-freedom system of Figure 8.32. The matrix form of the differential equations governing the response of the system is

$$\begin{bmatrix} m_1 & 0 \\ 0 & m_2 \end{bmatrix}\begin{bmatrix} \ddot{x}_1 \\ \ddot{x}_2 \end{bmatrix} + \begin{bmatrix} k_1 + k_2 & -k_2 \\ -k_2 & k_2 \end{bmatrix}\begin{bmatrix} x_1 \\ x_2 \end{bmatrix} = \begin{bmatrix} F_0 \sin(\omega t) \\ 0 \end{bmatrix} \tag{8.93}$$

The transfer functions for the system are obtained as

$$G_1(s) = \frac{X_1(s)}{F(s)} = \frac{m_2 s^2 + k_2}{(m_1 s^2 + k_1 + k_2)(m_2 s^2 + k_2) - (k_2)^2} \tag{8.94}$$

$$G_2(s) = \frac{X_2(s)}{F(s)} = \frac{k_2}{(m_1 s^2 + k_1 + k_2)(m_2 s^2 + k_2) - (k_2)^2} \tag{8.95}$$

The steady-state amplitudes are obtained as the real parts of the sinusoidal transfer functions. After some algebra,

$$X_1 = F_0\left|G_1(j\omega)\right| = \frac{F_0(k_2 - m_2\omega^2)}{m_1 m_2\omega^4 - [(k_1 + k_2)m_2 + k_2 m_1]\omega^2 + k_1 k_2} \tag{8.96}$$

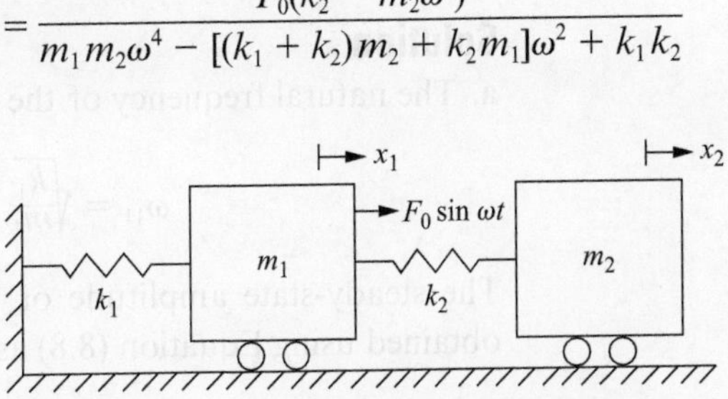

Figure 8.32

A two-degree-of-freedom mechanical system. If $k_2/m_2 = \omega^2$, then the steady-state amplitude $X_1 = 0$ and the mass m_2 acts as a vibration absorber.

$$X_2 = F_0|G_2(j\omega)|$$
$$= \frac{F_0 k_2}{m_1 m_2 \omega^4 - [(k_1 + k_2)m_2 + k_2 m_1]\omega^2 + k_1 k_2} \quad (8.97)$$

Equation (8.96) shows that if

$$\frac{k_2}{m_2} = \omega^2 \quad (8.98)$$

then $X_1 = 0$; the steady-state response of the particle of mass m_1 is zero.

Equation (8.98) is the basis of the theory on which an undamped dynamic vibration absorber operates. The primary system is composed of the particle of mass m_1 attached to a spring of stiffness k_1 and acted on by a harmonic force of magnitude F_0 at a frequency ω. A large steady-state response occurs when the primary system, by itself, is subject to an input with a frequency ω close to the natural frequency, $\omega_{11} = \sqrt{k_1/m_1}$. The auxiliary system, or the absorber, is composed of the particle of mass m_2 that is connected to the primary system through a spring of stiffness k_2. The resulting system has two degrees of freedom. The amplitudes of the steady-state responses of the primary system and the auxiliary system are given by Equations (8.96) and (8.97), respectively. Equation (8.96) shows that when the absorber is designed (or tuned) such that Equation (8.98) is satisfied, the amplitude of the primary system is zero.

The term "absorber" is something of a misnomer. The system is effective because the natural frequencies of the two-degree-of-freedom system are away from the natural frequency of the primary system. The natural frequencies are obtained by determining the poles of the transfer functions of Equations (8.94) and (8.95). The results are

$$\omega_{1,2} = \left[\frac{(k_1 + k_2)m_2 + k_2 m_1 \pm \sqrt{[(k_1 + k_2)m_2 + k_2 m_1]^2 - 4 m_1 m_2 k_1 k_2}}{2 m_1 m_2}\right]^{1/2} \quad (8.99)$$

The absorber is said to be tuned to the frequency $\omega_{22} = \sqrt{k_2/m_2}$. When the primary system is subject to a harmonic input at the tuned frequency, the steady-state response of the primary system is zero. The application of Newton's law to the free-body diagram of the absorber under these conditions shows that the steady-state amplitude of the absorber mass is

$$X_2 = -\frac{F_0}{k_2} \quad (8.100)$$

Example 8.15

A 500-kg machine is mounted on a foundation of stiffness 7.2×10^6 N/m. During operation, it is subject to a harmonic force of magnitude 5000 N at a frequency of 118 r/s. **a.** Determine the steady-state amplitude of the machine under these conditions. **b.** Specify the mass and stiffness of a dynamic vibration absorber that is tuned to 118 r/s and when added to the primary system has a steady-state amplitude of 5 mm. **c.** Plot the frequency response of the primary system both with and without the absorber.

Solution

a. The natural frequency of the primary system is

$$\omega_{11} = \sqrt{\frac{k_1}{m_1}} = \sqrt{\frac{7.2 \times 10^6 \text{ N/m}}{500 \text{ kg}}} = 120 \text{ r/s} \quad (a)$$

The steady-state amplitude of the primary system before an absorber is added is obtained using Equation (8.8) as

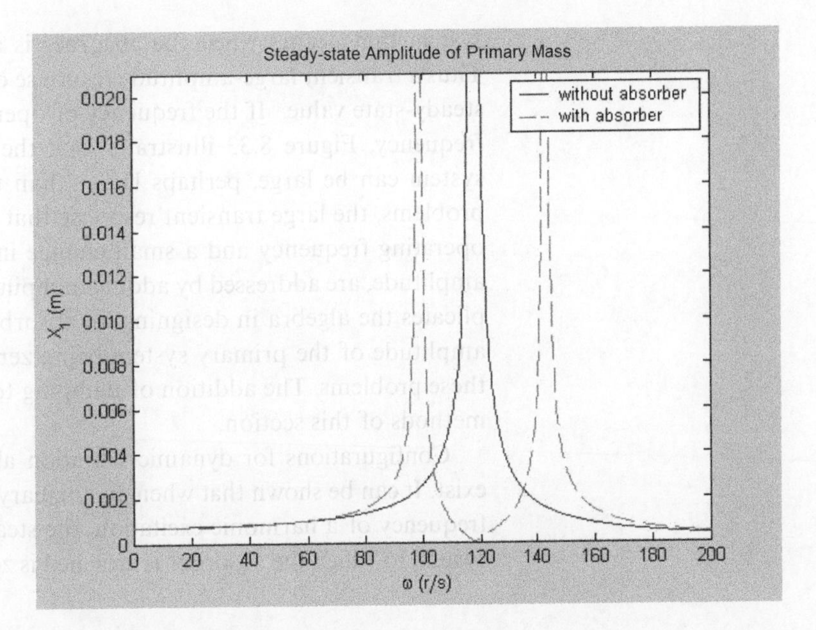

Figure 8.33

Contrast of the steady-state amplitude of a machine operating near resonance with and without the addition of a vibration absorber.

$$X_1 = \left|\frac{F_0}{m(\omega_1^2 - \omega^2)}\right|$$

$$= \left|\frac{5000 \text{ N}}{(500 \text{ kg})\,[(120 \text{ r/s})^2 - (118 \text{ r/s})^2]}\right| = 2.10 \text{ cm} \tag{b}$$

b. If the absorber is designed such that it is tuned to 118 r/s and the input frequency is 118 r/s, then the steady-state amplitude of the primary system is zero and the steady-state amplitude of the absorber is given by Equation (8.100). Requiring the steady-state amplitude of the absorber to be less than 5 mm leads to

$$\frac{F_0}{k_2} < 5 \text{ mm}$$

$$k_2 > \frac{5000 \text{ N}}{0.005 \text{ m}} = 1 \times 10^6 \text{ N/m} \tag{c}$$

Since the tuning frequency is fixed and equal to the excitation frequency, the absorber with the minimum mass is obtained by choosing the smallest possible stiffness. Tuning the absorber to 118 r/s requires

$$\frac{k_2}{m_2} = (118 \text{ r/s})^2$$

$$m_2 = \frac{1 \times 10^6 \text{ N/m}}{(118 \text{ r/s})^2} = 71.8 \text{ kg} \tag{d}$$

c. The substitution of the given and calculated values into Equation (8.96) leads to

$$X_1 = \frac{1.39 \times 10^5 - 10\omega^2}{\omega^4 - 3.03 \times 10^4\omega^2 + 2.01 \times 10^8} \tag{e}$$

Equations (b) and (e) are plotted on the same graph in Figure 8.33.

Figure 8.33 shows that when the absorber is tuned to the input frequency, the steady-state amplitude of the primary system is zero. The amplitude is large when the input frequency is away from the tuned frequency of the absorber. Figure 8.33 also illustrates that the lowest natural frequency of the two-degree-of-freedom system that occurs when the absorber is added is less than the tuned frequency. Thus a transient large-amplitude response occurs as the input frequency builds to its steady-state value. If the frequency of operation is slightly different from the tuned frequency, Figure 8.33 illustrates that the steady-state amplitude of the primary system can be large, perhaps larger than the system without the absorber. These problems, the large transient response that occurs as the frequency builds up to the operating frequency and a small change in operating frequency leading to a large amplitude, are addressed by adding damping to the absorber. Damping greatly complicates the algebra in designing an absorber and does not lead to the steady-state amplitude of the primary system being zero for any frequency, but it does address these problems. The addition of damping to the absorber can be analyzed using the methods of this section.

Configurations for dynamic vibration absorbers other than that of Figure 8.32 exist. It can be shown that when an auxiliary mass-spring system is tuned to the input frequency of a harmonic excitation, the steady-state amplitude of the particle in the system to which the absorber is attached is zero.

8.9.2 Higher-Order Filters

Passive filters that use only resistors, capacitors, and inductors do not have external energy sources. Active filters contain resistors and capacitors and amplifiers. Filters are often referred to by order, which is equal to the order of their transfer function. Higher-order filters provide greater design flexibility in achieving the required objectives.

The response of a filter is dependent on its transfer function. The polynomial in the numerator of the transfer function affects the frequency response, especially for small and large frequencies. For an nth-order filter, the numerator of its transfer function can be written as

$$N(s) = a_n s^n + a_{n-1} s^{n-1} + \cdots + a_1 s + a_0 \tag{8.101}$$

The following can be deduced about the frequency response from the numerator of the transfer function. Let $V(\omega)$ represent the frequency response of the filter, the amplitude of the steady-state output as a function of the input frequency:

- $V(0) = 0$ if $a_0 = 0$ and $V(0) = C$, a nonzero constant, when $a_0 \neq 0$.
- $\lim_{\omega\to\infty} V(\omega) = 0$ if $a_n = 0$ and $\lim_{\omega\to\infty} V(\omega) = C$, a nonzero constant, when $a_n \neq 0$. In this case, the order of the denominator is greater than the order of the numerator.
- When $a_n = 0$, the rate at which $V(\omega)$ approaches zero depends on the highest power of s with a nonzero coefficient. $V(\omega)$ approaches zero fastest when $N(s) = a_0$.

It is noted then that the low-frequency asymptote on the Bode diagram is a constant if $a_0 \neq 0$ and the high-frequency asymptote is a constant if $a_n \neq 0$. In order for a filter to pass through signals at low frequencies, it must have a low-frequency asymptote that is a constant. Similarly, in order for a filter to pass through signals at high frequencies, it must have a high-frequency asymptote that is a constant. These trends are used to specify the necessary conditions for the transfer function for the different types of filters:

- A low-pass filter, which allows low-frequency signals to pass but filters high-frequency signals, requires $V(0) = C$ and $\lim_{\omega\to\infty} V(\omega) = 0$; thus its transfer function must have $a_0 \neq 0$ and $a_n = 0$.
- A high-pass filter, which allows high-frequency signals to pass but filters low-frequency signals, must have $V(0) = 0$ and $\lim_{\omega\to\infty} V(\omega) = C$; thus its transfer function must have $a_0 = 0$ and $a_n \neq 0$.

- A band-pass filter, which allows signals in a band of frequencies to pass but filters signals outside the band, should have $V(0) = 0$ and $\lim_{\omega\to\infty} V(\omega) = 0$; thus its transfer function must have $a_0 = 0$ and $a_n = 0$.
- A band-reject filter, which filters signals in a bandwidth and allows signals outside that bandwidth to pass, should have $V(0) = C$ and $\lim_{\omega\to\infty} V(\omega) = C$; thus its transfer function must have $a_0 \neq 0$ and $a_n \neq 0$.

The potential use of a circuit as a filter is determined by the numerator of its transfer function. Properties of the filter such as bandwidth are determined by the location of the poles of the transfer function. The values of component parameters may be adjusted to place poles at the desired locations.

Example 8.16

Consider the circuit of Figure 8.34. **a.** Derive the transfer function $G(s) = V_2(s)/V_1(s)$. **b.** State the potential use of the circuit as a filter. **c.** Determine the poles of the transfer function. **d.** Draw the Bode plots and frequency response curves for this circuit.

Solution

a. The application of KCL at nodes A and B and KVL around each of the loops, using the currents illustrated in Figure 8.34, leads to

$$v_1(t) - i_1 - \sqrt{2}\frac{di_1}{dt} - \sqrt{2}\int_0^t i_1\,dt - \frac{1}{\sqrt{2}}\frac{d}{dt}(i_1 - i_2) = 0 \tag{a}$$

$$-\frac{1}{\sqrt{2}}\int_0^t (i_2 - i_3)dt + \frac{1}{\sqrt{2}}\frac{d}{dt}(i_1 - i_2) = 0 \tag{b}$$

$$-i_3 + \frac{1}{\sqrt{2}}\int_0^t (i_2 - i_3)dt = 0 \tag{c}$$

Transforming Equations (a), (b), and (c) into the Laplace domain leads to

$$\left(1 + \sqrt{2}s + \frac{1}{\sqrt{2}}s + \frac{\sqrt{2}}{s}\right)I_1(s) - \frac{1}{\sqrt{2}}sI_2(s) = V_1(s) \tag{d}$$

$$-\frac{1}{\sqrt{2}}sI_1(s) + \left(\frac{1}{\sqrt{2}}s + \frac{1}{\sqrt{2}s}\right)I_2(s) - \frac{1}{\sqrt{2}s}I_3(s) = 0 \tag{e}$$

$$-\frac{1}{\sqrt{2}s}I_2(s) + \left(1 + \frac{1}{\sqrt{2}s}\right)I_3(s) = 0 \tag{f}$$

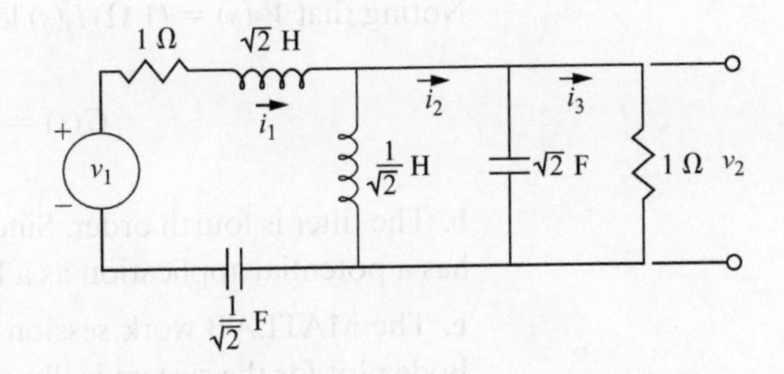

Figure 8.34

The circuit of Example 8.16 is an example of a band-bass filter.

```
>> N=[0.5 0 0];
>> D=[1 2^0.5 3 2^0.5 1];
>> [r,p,k]=residue(N,D)
r =
    0.0536 - 0.2454i
    0.0536 + 0.2454i
   -0.0536 + 0.1082i
   -0.0536 - 0.1082i
p =
   -0.4776 + 1.3612i
   -0.4776 - 1.3612i
   -0.2295 + 0.6541i
   -0.2295 - 0.6541i
k =
    []
>> bode(N,D)
>>
```

(a)

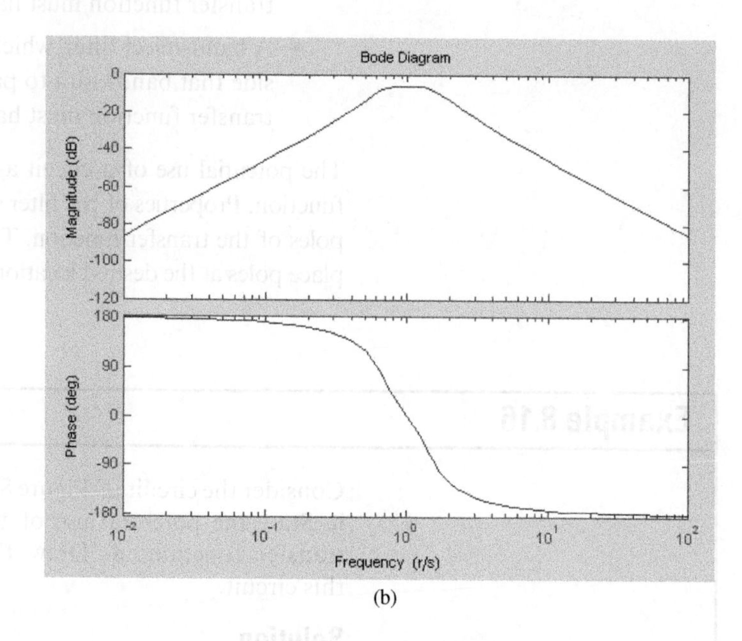

(b)

Figure 8.35

(a) The MATLAB work session used to determine the poles of the transfer function and develop the Bode plot of the filter of Example 8.16; (b) the Bode diagram shows that the circuit may be used as a band-pass filter.

Equations (d), (e), and (f) are solved using Cramer's rule, leading to

$$I_3(s) = \frac{\begin{vmatrix} 1+\sqrt{2}s+\frac{1}{\sqrt{2}}s+\frac{\sqrt{2}}{s} & -\frac{1}{\sqrt{2}}s & V_1(s) \\ -\frac{1}{\sqrt{2}}s & \frac{1}{\sqrt{2}}s+\frac{1}{\sqrt{2}s} & 0 \\ 0 & -\frac{1}{\sqrt{2}s} & 0 \end{vmatrix}}{\begin{vmatrix} 1+\sqrt{2}s+\frac{1}{\sqrt{2}}s+\frac{\sqrt{2}}{s} & -\frac{1}{\sqrt{2}}s & 0 \\ -\frac{1}{\sqrt{2}}s & \frac{1}{\sqrt{2}}s+\frac{1}{\sqrt{2}s} & -\frac{1}{\sqrt{2}s} \\ 0 & -\frac{1}{\sqrt{2}s} & 1+\frac{1}{\sqrt{2}s} \end{vmatrix}}$$

$$= \frac{0.5s^2}{s^4+\sqrt{2}s^3+3s^2+\sqrt{2}s+1}V_1(s) \tag{g}$$

Noting that $V_2(s) = (1\ \Omega)I_3(s)$ leads to

$$G(s) = \frac{0.5s^2}{s^4+\sqrt{2}s^3+3s^2+\sqrt{2}s+1} \tag{h}$$

b. The filter is fourth order. Since the numerator has both $a_4 = 0$ and $a_0 = 0$, the filter has a potential application as a band-pass filter.

c. The MATLAB work session used to determine the poles of $G(s)$ and to draw the Bode plot for the system is illustrated in Figure 8.35.

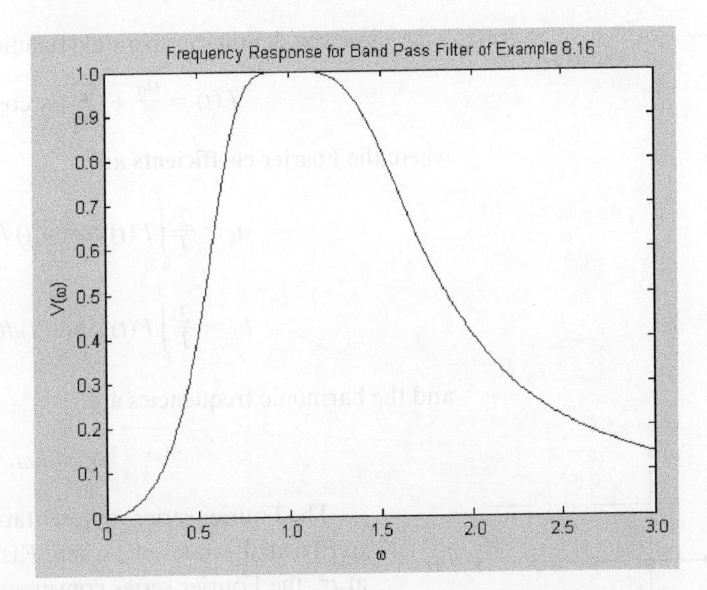

Figure 8.36

The flat nature of the frequency response curve near $\omega = 1$ illustrates the use of the circuit of Example 8.16 as a band-pass filter.

d. The sinusoidal transfer function is determined from Equation (h) as

$$G(j\omega) = \frac{-0.5\omega^2(1 - 3\omega^2 + \omega^4) + j0.5\sqrt{2}\omega^2(\omega - \omega^3)}{(1 - 3\omega^2 + \omega^4)^2 + 2(\omega - \omega^3)^2} \tag{i}$$

The steady-state amplitude is determined using Equation (i) as

$$V_1|G(j\omega)| = \frac{V_1\omega^2}{\sqrt{(1 - 3\omega^2 + \omega^4)^2 + 2(\omega - \omega^3)^2}} \tag{j}$$

The frequency response is illustrated in Figure 8.36. Both the Bode diagram and the frequency response curves show the nature of this band-pass filter.

8.10 Response Due to Periodic Input

A system input $F(t)$ is periodic if there exists a value of T, called the period, such that $F(t + T) = F(t)$ for all t. The fundamental frequency of a periodic input is defined as

$$\omega_1 = \frac{2\pi}{T} \tag{8.102}$$

Examples of periodic inputs are given in Figure 8.37.

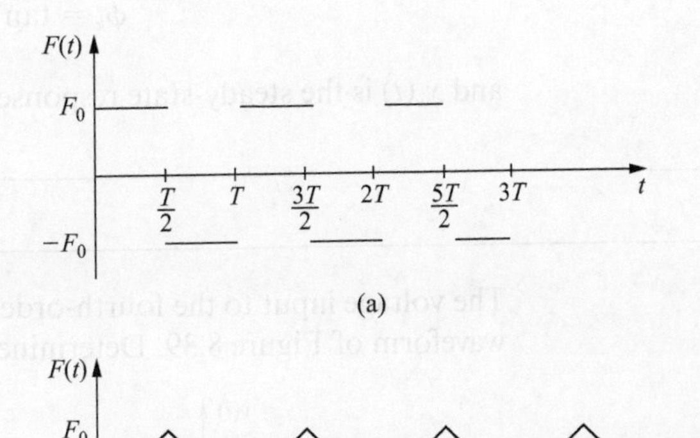

Figure 8.37

Example of periodic functions: (a) square wave; (b) triangular wave.

A piecewise continuous periodic function has a **Fourier series** representation given by

$$F(t) = \frac{a_0}{2} + \sum_{k=1}^{\infty}[a_k \cos(\omega_k t) + b_k \sin(\omega_k t)] \tag{8.103}$$

where the **Fourier coefficients** are

$$a_k = \frac{2}{T}\int_0^T F(t)\cos(\omega_k t)dt \qquad k = 0, 1, 2, \ldots \tag{8.104}$$

$$b_k = \frac{2}{T}\int_0^T F(t)\sin(\omega_k t)dt \qquad k = 1, 2, \ldots \tag{8.105}$$

and the harmonic frequencies are

$$\omega_k = k\omega_1 = \frac{2\pi k}{T} \tag{8.106}$$

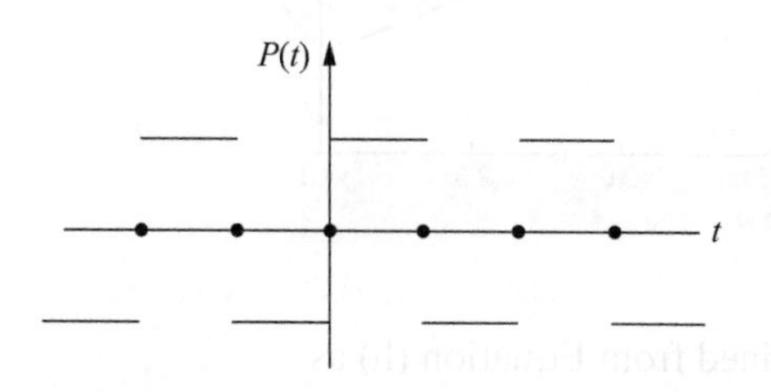

Figure 8.38

The convergence of the Fourier series representation for a square wave.

The Fourier series representation of Equation (8.103) converges pointwise to $F(t)$ at all values of t where F is continuous. If $F(t)$ has a jump discontinuity at t^*, the Fourier series converges to $\frac{1}{2}[F(t^{*+}) + F(t^{*-})]$. Figure 8.38 illustrates the convergence of the Fourier series representation of the square wave input of Figure 8.37(a).

The Laplace transform of a periodic function can be obtained using its Fourier series representation

$$\mathcal{L}\{F(t)\} = \frac{1}{2}\mathcal{L}\{a_0\} + \mathcal{L}\left\{\sum_{k=1}^{\infty}[a_k\cos(\omega_k t) + b_k\sin(\omega_k t)]\right\} \tag{8.107}$$

Since the Fourier series converges pointwise, the property of linearity of the transform may be applied, leading to

$$F(s) = \frac{a_0}{2s} + \sum_{k=1}^{\infty}\left(a_k\frac{s}{s^2 + \omega_k^2} + b_k\frac{\omega_k}{s^2 + \omega_k^2}\right) \tag{8.108}$$

If a periodic input is applied to a system whose transfer function is $G(s)$ and whose output is $x(t)$, then

$$X(s) = \left[\frac{a_0}{2s} + \sum_{k=1}^{\infty}\left(a_k\frac{s}{s^2 + \omega_k^2} + b_k\frac{\omega_k}{s^2 + \omega_k^2}\right)\right]G(s) \tag{8.109}$$

A procedure similar to that used in Section 8.2 to derive the steady-state response due to a sinusoidal input is applied to derive the steady-state response of the system as

$$x(t) = \frac{a_0}{2} + \sum_{k=1}^{\infty}|G(j\omega_k)|[a_k \cos(\omega t + \phi_k) + b_k \sin(\omega t + \phi_k)] \tag{8.110}$$

where

$$\phi_k = \tan^{-1}\left\{\frac{\mathrm{Im}[G(j\omega_k)]}{\mathrm{Re}[G(j\omega_k)]}\right\} \tag{8.111}$$

and $x_u(t)$ is the steady-state response of the system due to a unit step input.

Example 8.17

The voltage input to the fourth-order band-pass filter of Example 8.16 is the periodic waveform of Figure 8.39. Determine the output voltage from the filter.

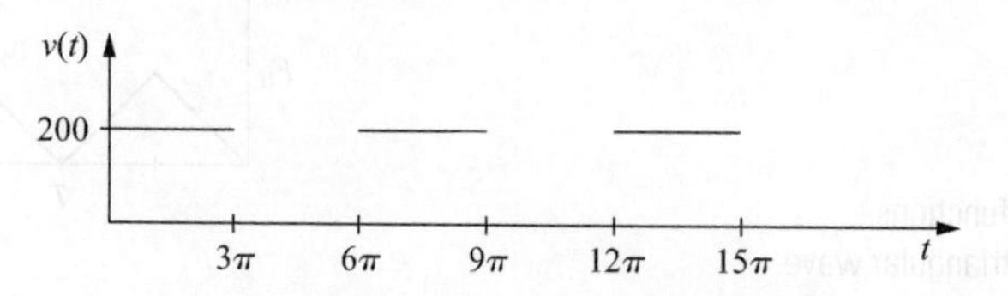

Figure 8.39

The periodic input for Example 8.17.

Solution

The voltage signal of Figure 8.39 is periodic of period $T = 6\pi$ s. The definition of the voltage over one period is

$$v_1(t) = \begin{cases} 200 \text{ V} & 0 < t < 3\pi \text{ s} \\ 0 & 3\pi \text{ s} < t < 6\pi \text{ s} \end{cases} \tag{a}$$

The fundamental frequency of the voltage is

$$\omega_1 = \frac{2\pi}{T} = \frac{1}{3} \tag{b}$$

The Fourier coefficients are calculated using Equations (8.104) and (8.105) as

$$\begin{aligned} a_0 &= \frac{2}{6\pi}\int_0^{6\pi} F(t)dt \\ &= \frac{2}{6\pi}\int_0^{3\pi}(200)dt = 200 \end{aligned} \tag{c}$$

$$\begin{aligned} a_k &= \frac{2}{6\pi}\int_0^{3\pi} 200 \cos\frac{k}{3}t\, dt \\ &= \frac{200}{k\pi}\sin\frac{k}{3}t\Big|_{t=0}^{t=3\pi} = 0 \end{aligned} \tag{d}$$

$$\begin{aligned} b_k &= \frac{2}{6\pi}\int_0^{3\pi} 200 \sin\frac{k}{3}t\, dt \\ &= -\frac{200}{k\pi}\cos\frac{k}{3}t\Big|_{t=0}^{t=3\pi} \\ &= \frac{200}{k\pi}(1 - \cos k\pi) \\ &= \begin{cases} 0 & k = 2, 4, 6, \ldots \\ \dfrac{400}{k\pi} & k = 1, 3, 5, \ldots \end{cases} \end{aligned} \tag{e}$$

Thus the Fourier series representation for $F(t)$ is obtained by using Equations (c), (d), and (e) in Equation (8.103), leading to

$$F(t) = 100 + \sum_{k=1,2,3}^{\infty}\frac{400}{k\pi}\sin\left(\frac{k}{3}t\right) \tag{f}$$

The sinusoidal transfer function obtained in Equation (j) of Example 8.16 leads to

$$|G(j\omega_k)| = \frac{\omega_k^2}{\sqrt{(1 - 3\omega_k^2 + \omega_k^4)^2 + 2(\omega_k - \omega_k^3)^2}} \tag{g}$$

$$\phi_k = \tan^{-1}\left(\frac{\sqrt{2}(\omega_k - \omega_k^3)}{1 - 3\omega_k^2 + \omega_k^4}\right) \tag{h}$$

The final value theorem is used to determine the steady-state response of the system due to a unit step input:

$$\begin{aligned} v_f &= \lim_{s\to 0} G(s) \\ &= \lim_{s\to 0}\frac{0.5s^2}{s^4 + \sqrt{2}s^3 + 3s^2 + \sqrt{2}s + 1} = 0 \end{aligned} \tag{i}$$

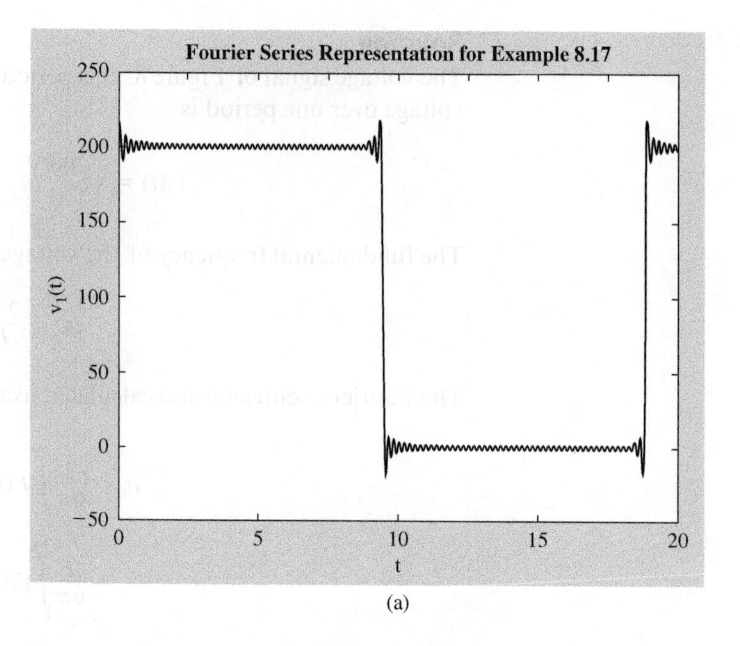

(a)

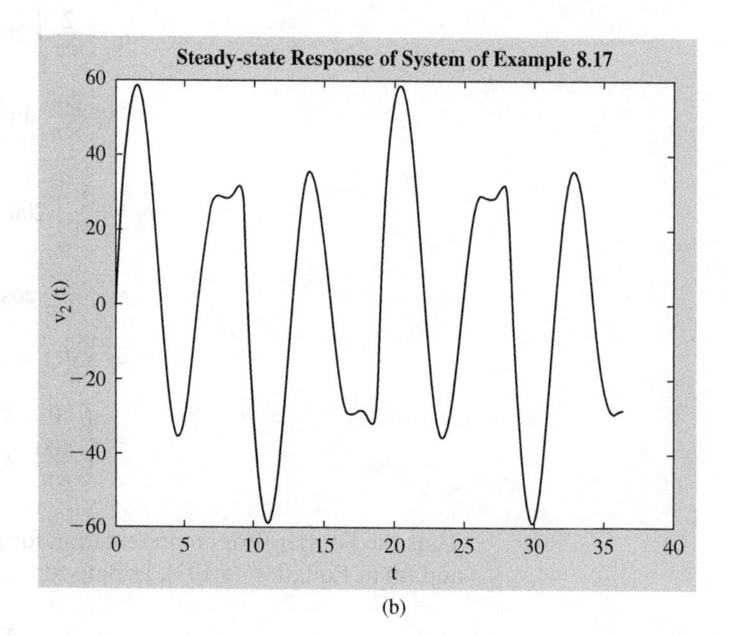

(b)

Figure 8.40

(a) The Fourier series representation of the voltage input of Figure 8.39; (b) the steady-state voltage across the terminals of the filter for the periodic input of Figure 8.39.

The steady-state response is obtained using Equation (8.109) as

$$v_2(t) = \frac{400}{\pi} \sum_{k=1,3,5}^{\infty} \frac{|G(j\omega_k)|}{k} \sin\left(\frac{k}{3}t + \phi\right) \tag{j}$$

Plots of the Fourier series representation for $F(t)$, evaluated using an upper limit on the summation of 50 terms and of the steady-state response, both obtained using MATLAB, are shown in Figure 8.40.

8.11 Further Examples

Aspects of the frequency response for several examples considered in Chapters 4–6 are considered in the following problems. MATLAB is used to draw Bode diagrams and Nyquist diagrams as well as to aid in calculations.

Example 8.18

An LR circuit with $L = 0.25$ H and $R = 1$ kΩ has a voltage source $v(t) = 100 \sin(1000t)$ V. Determine its steady-state response.

Solution

The time constant for the first-order circuit is $T = L/R = 0.25\text{ H}/1000\ \Omega = 2.5 \times 10^{-4}$ s. The transfer function for an LR circuit, when written in terms of its time constant, is the same as the transfer function for an RC circuit, Equation (8.60). The steady-state amplitude is determined using Equation (8.63) as

$$I = \frac{V_0}{RT}\sqrt{\frac{1}{\omega^2 + \frac{1}{T^2}}}$$

$$= \frac{100\text{ V}}{(1000\ \Omega)(2.5 \times 10^{-4})\text{s}}\sqrt{\frac{1}{(1000\text{ r/s})^2 + \frac{1}{(2.5 \times 10^{-4}\text{ s})^2}}}$$

$$= 9.70\text{ mA} \tag{a}$$

The phase is determined using Equation (8.84) as

$$\phi = -\tan^{-1}(\omega t) = -\tan^{-1}[(1000\text{ r/s})(2.5 \times 10^{-4}\text{ s})] = -0.245\text{ r} \tag{b}$$

Thus the steady-state response is

$$i(t) = 9.70 \sin(1000t - 0.245)\text{ mA} \tag{c}$$

Example 8.19

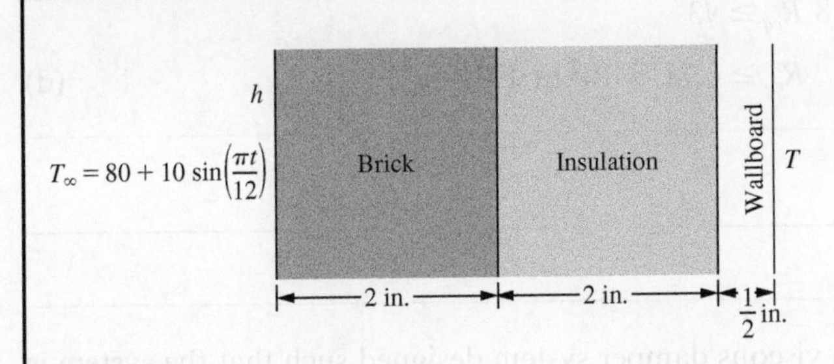

Figure 6.29
The system of Example 6.14: The exterior wall of a building is exposed to a periodic ambient temperature. (Repeated)

Consider the thermal system of Example 6.14 and Figure 6.29. **a.** Determine the maximum and minimum temperatures in the room during the day. **b.** At what time does the temperature in the room reach its maximum if the outside temperature is maximum at 3 p.m.?

Solution

a. Defining $\theta = T - 80$ F in Equation (j) of Example 6.14 and rearranging leads to

$$2.86\frac{d\theta}{dt} + \theta = 10\sin\left(\frac{\pi t}{12}\right) \tag{a}$$

where t is measured in hours. The time constant for this first-order system is $T = 2.86$ hr and the input frequency is $\pi/12$ r/s. The nondimensional frequency is $\beta = \omega T = 0.748$. The steady-state response is of the form $\theta(t) = \Theta \sin[(\pi/12)t + \phi]$, where $\Theta = (10\text{ F})\sqrt{1/(1 + \beta^2)} = 8.01$ F and $\phi = -\tan(0.748) = -0.642$ r. Thus the steady-state response for the room temperature is

$$\theta(t) = 8.01 \sin\left(\frac{\pi}{12}t - 0.642\right)\text{ F} \tag{b}$$

The maximum and minimum room temperatures are $T_{max} = 80\text{ F} + 8.01\text{ F} = 88.0\text{ F}$ and $T_{min} = 80\text{ F} - 8.01\text{ F} = 72.0\text{ F}$.

b. Since the maximum temperature occurs at 3 p.m., this must correspond to a value of t such that $\pi t/12 = \pi/2$ or $t = 6$ hr. The maximum room temperature occurs when $(\pi/12)t - 0.642 = \pi/2$ or $t = 8.45$ hr. Thus the maximum temperature in the room occurs 2.45 hr after 3 p.m. or at 5:27 p.m. This is obviously not a very accurate result. However, the model assumes all properties are lumped and the temperature in the room is uniform. A more accurate model considers a variation in temperature across the room at any instant.

Example 8.20

In Example 8.19, it is shown that the maximum temperature in the room of Example 6.14 is 88.0 F. What is the minimum value of R_{eq} such that the maximum temperature in the room is 85 F assuming all other parameters are as given in Example 6.14?

Solution

In terms of the notation of Example 8.19, for the maximum temperature of the room to be at most 85 F, the steady-state amplitude must be limited to 5 F, $\Theta \leq 5$ F, which leads to

$$5\text{ F} \leq (10\text{ F})\sqrt{\frac{1}{1+\beta^2}} \tag{a}$$

Equation (a) is rearranged to

$$\beta \geq \sqrt{3} \tag{b}$$

Equation (j) of Example 6.14 can be written in terms of the equivalent resistance as

$$99.2\, R_{eq}\frac{d\theta}{dt} + \theta = 10\sin\left(\frac{\pi t}{12}\right) \tag{c}$$

The time constant and the input frequency are determined from Equation (c) as $T = 99.2R_{eq}$ hr and $\omega = \pi/12$ r/hr, respectively. Note that $\beta = \omega T = 99.2\pi R_{eq}/12 = 25.8R_{eq}$. The substitution of this result into Equation (b) gives

$$25.8\, R_{eq} \geq \sqrt{3}$$

$$R_{eq} \geq 6.71 \times 10^2 \text{ hr·F/Btu} \tag{d}$$

Example 8.21

Consider a mass, spring, and viscous damper system designed such that the system is critically damped with a natural frequency of 1 r/s. **a.** Qualitatively discuss the Bode diagram. Determine its low-frequency and high-frequency asymptotes. Discuss the behavior of the phase. **b.** Determine the features of the system's Nyquist diagram. Determine its low-and high-frequency limits and its intercepts. **c.** Use MATLAB to plot the Bode diagram and the Nyquist diagrams.

Solution

The transfer function for a critically damped mechanical system with a natural frequency of 1 r/s is

$$G(s) = \frac{1}{(s+1)^2} \tag{a}$$

The sinusoidal transfer function for the system is obtained as

$$G(j\omega) = \frac{1 - \omega^2 - j2\omega}{(\omega^2 + 1)^2} \tag{b}$$

The real and imaginary parts of the transfer function and its magnitude are

$$\text{Re}[G(j\omega)] = \frac{1 - \omega^2}{(\omega^2 + 1)^2} \tag{c}$$

$$\text{Im}[G(j\omega)] = \frac{-2\omega}{(\omega^2 + 1)^2} \tag{d}$$

$$|G(j\omega)| = \frac{1}{1 + \omega^2} \tag{e}$$

The phase for the system is

$$\phi = -\tan^{-1}\left(\frac{-2\omega}{1 - \omega^2}\right) \tag{f}$$

a. The Bode diagram is a plot of

$$L(\omega) = 20 \log |G(j\omega)| = 20 \log \frac{1}{1 + \omega^2} = -20 \log (1 + \omega^2) \tag{g}$$

versus log ω. The low-frequency asymptote is obtained as

$$L(0) = -20 \log (1) = 0 \text{ dB} \tag{h}$$

The high-frequency asymptote is obtained as

$$\lim_{\omega \to \infty} L(\omega) = \lim_{\omega \to \infty} -20 \log (100 + \omega^2) = -20 \log \omega^2 = -40 \log \omega \tag{i}$$

Thus the low-frequency asymptote is a horizontal line $L(\omega) = 0$ dB while the high-frequency asymptote is a line of slope -40 dB.

From Equation (f) it is noted that the phase is zero for $\omega = 0$ and decreases with increasing ω. When $\omega = 1$ r/s, $\phi = -\pi/2$. For large ω, $\tan \phi \to 0$, but both the numerator and denominator are negative, thus $\phi = -\pi$ for large ω.

b. The Nyquist diagram is a plot of $\text{Re}|G(j\omega)|$, given by Equation (c) on the horizontal axis, vs. $\text{Im}|G(j\omega)|$, given by Equation (d) on the vertical axis. The diagram is plotted with ω as a parameter for $-\infty < \omega < \infty$. Equations (c) and (d) lead to the following:

- The Nyquist diagram intercepts the horizontal axis only for $\omega = 0$ when $\text{Re}[G(j\omega)] = 1$.
- The Nyquist diagram intercepts the vertical axis only for $\omega = \pm 1$. When $\omega = 1$, $\text{Im}[G(j\omega)] = -0.5$. When $\omega = -1$, $\text{Im}\,[G(j\omega)] = 0.5$.
- As $\omega \to \pm\infty$, $\text{Re}|G(j\omega)| \to 0$ and $\text{Im}|G(j\omega)| \to 0$.
- As ω increases from 0, Equation (d) shows that $\text{Im}|G(j\omega)|$ grows negatively while $\text{Re}|G(j\omega)|$ decreases from 1. This indicates that the Nyquist plot is traversed clockwise as ω increases.

c. The MATLAB-generated Bode diagrams and Nyquist plot for this system are illustrated in Figures 8.41 and 8.42.

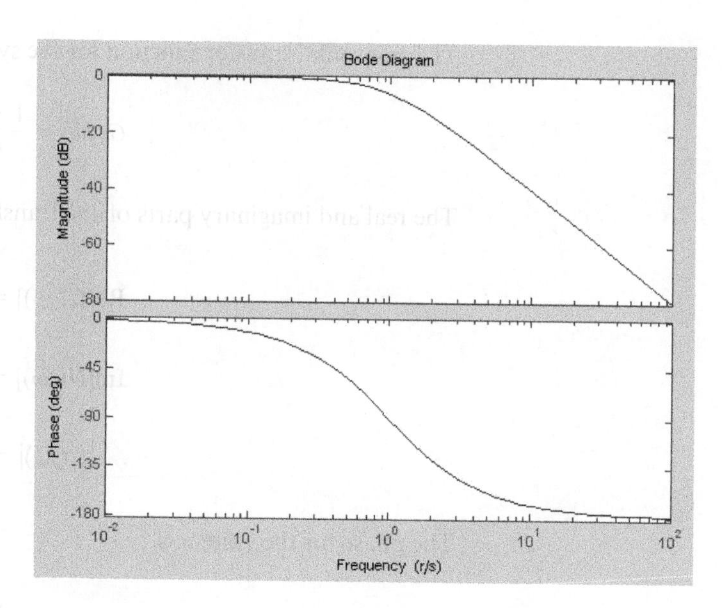

Figure 8.41

The Bode diagram for the system of Example 8.21. The low-frequency asymptote of 0 dB and the high-frequency asymptote of $-40 \log \omega$ are clearly illustrated.

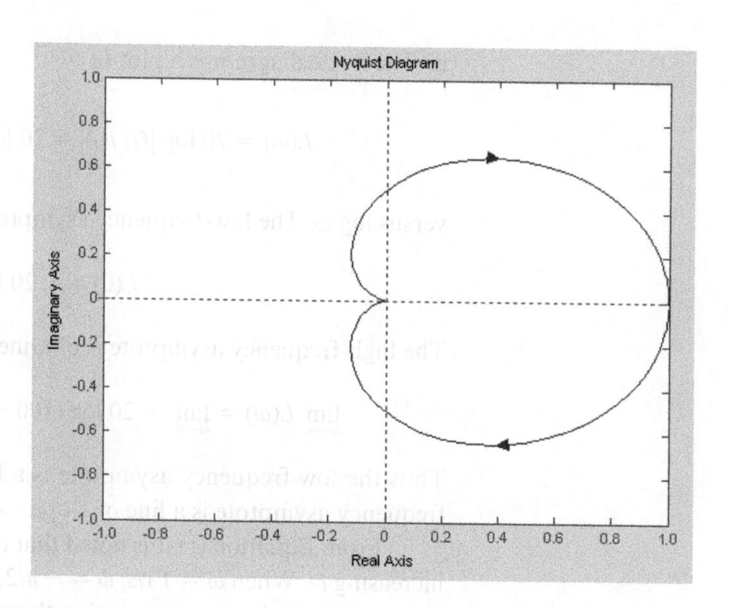

Figure 8.42

The Nyquist diagram for the system of Example 8.21.

Example 8.22

The centrifuge of Example 4.34 and Figure 4.69 has parameters $m = 99.5$ kg, $k = 1 \times 10^6$ N/m, $c = 2000$ N·s/m, $m_0 = 0.5$ kg, and $e = 0.1$ m. **a.** Determine the steady-state response for $\omega = 200$ r/s. **b.** Determine the frequency response for the system.

Solution

a. The substitution of the given values into Equation (f) of Example 4.34 leads to

$$100\ddot{x} + 2000\dot{x} + 1 \times 10^6 x = 0.05\omega^2 \sin(\omega t) \tag{a}$$

b. The natural frequency and damping ratio for the system are obtained as $\omega_n = \sqrt{1 \times 10^6/100} = 100\,\text{r/s}$ and $\zeta = (2000\ \text{N·s/m})/[2(100\ \text{kg})(100\ \text{r/s})] = 0.1$.

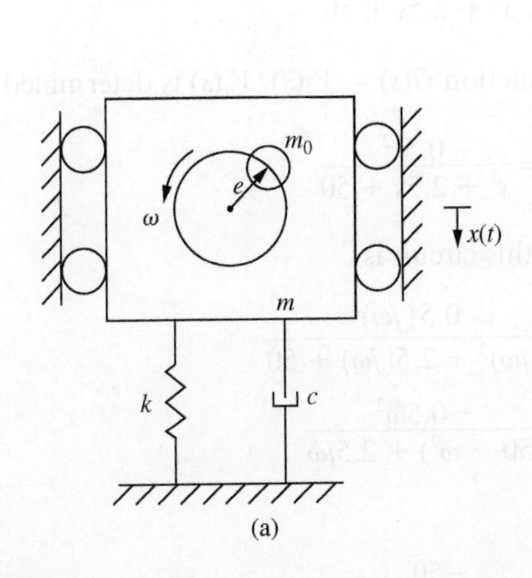

Figure 4.69

(a) The centrifuge with a rotating mass m_0 and a distance e from the axis of rotation and Example 8.22. (Repeated)

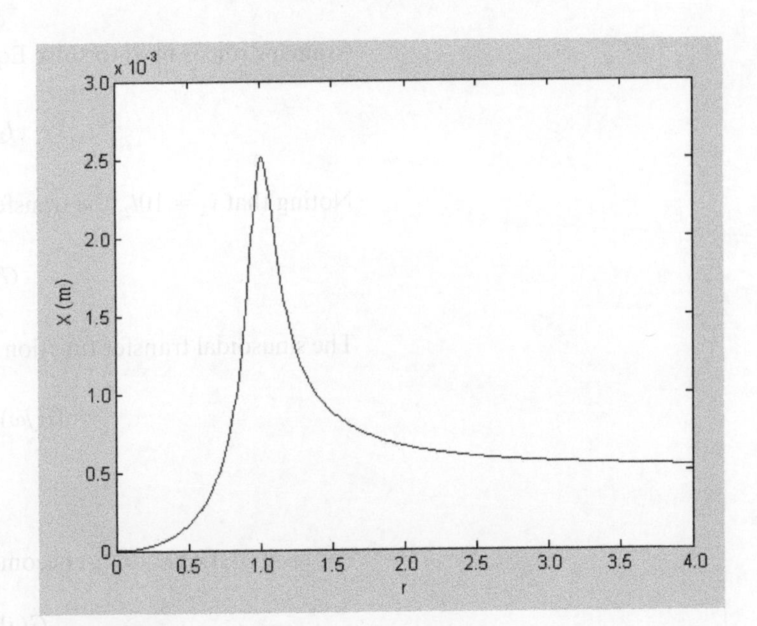

Figure 8.43

The frequency response of the centrifuge of Example 8.22.

The steady-state amplitude and phase for the system are obtained using Equations (8.70) and (8.71) with $F_0 = 0.05\omega^2$. The application of Equation (8.70) leads to

$$X = \frac{0.05\omega^2}{100\omega_n^2} \frac{1}{\sqrt{(1 - r^2)^2 + (0.2r)^2}} \tag{b}$$

Noting that $r = \omega/\omega_n$, Equation (b) is rewritten as

$$X = \frac{5 \times 10^{-4} r^2}{\sqrt{(1 - r^2)^2 + (0.2r)^2}} \tag{c}$$

Equation (c) is plotted in Figure 8.43.

Example 8.23

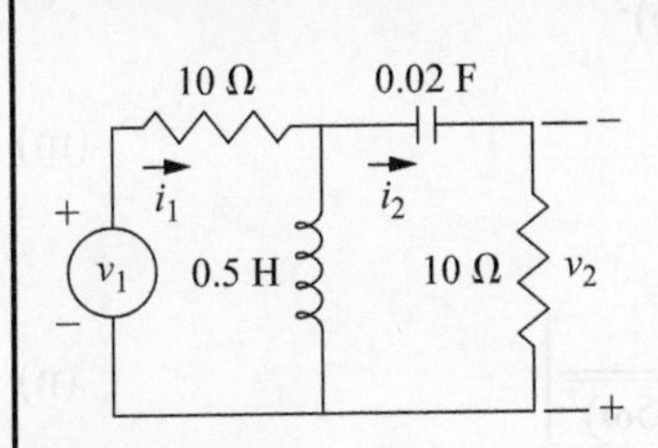

Figure 8.44

The circuit of Example 8.23.

For the circuit of Figure 8.44, **a.** determine the steady-state response for $v_2(t)$ when $v_1(t) = 100\sin(10t)$ V; **b.** draw the Bode plot for the system and comment on the use of this circuit as a filter; and **c.** qualitatively discuss and then plot the Nyquist diagram for this system.

Solution

a. Let $i_1(t)$ and $i_2(t)$ be the nodal currents. The application of KVL to each loop leads to

$$0.5\frac{di_1}{dt} \quad 0.5\frac{di_2}{dt} + 10i_1 = v_1(t) \tag{a}$$

$$-0.5\frac{di_1}{dt} + 0.5\frac{di_2}{dt} + 10i_2 + 50\int_0^t i_2 dt = 0 \tag{b}$$

Taking the Laplace transforms of Equations (a) and (b) leads to

$$(0.5s + 10)I_1(s) - 0.5sI_2(s) = V_1(s) \tag{c}$$

$$-0.5sI_1(s) + \left(0.5s + 10 + \frac{50}{s}\right)I_2(s) = 0 \tag{d}$$

Cramer's rule is used to solve Equations (c) and (d) for $I_2(s)$, leading to

$$I_2(s) = \frac{0.05s^2 V_1(s)}{s^2 + 2.5s + 50} \tag{e}$$

Noting that $v_2 = 10i_2$, the transfer function $G(s) = V_2(s)/V_1(s)$ is determined as

$$G(s) = \frac{0.5s^2}{s^2 + 2.5s + 50} \tag{f}$$

The sinusoidal transfer function for this circuit is

$$\begin{aligned} G(j\omega) &= \frac{0.5(j\omega)^2}{(j\omega)^2 + 2.5(j\omega) + 50} \\ &= \frac{-0.5\omega^2}{(50 - \omega^2) + 2.5j\omega} \end{aligned} \tag{g}$$

For $\omega = 10$, Equation (g) becomes

$$\begin{aligned} G(j10) &= \frac{-50}{-50 + 25j} \\ &= \frac{-50(-50 - 25j)}{3125} \\ &= 0.8 + 0.4j \end{aligned} \tag{h}$$

It is apparent from Equation (h) that $|G(j10)| = 0.894$ and $\phi = \tan^{-1}(0.4/0.8) = 0.464$ r. Thus the steady-state response is

$$v_2(t) = 89.4 \sin(10t + 0.464) \text{ V} \tag{i}$$

b. Equation (g) is used to show that

$$\text{Re}[G(j\omega)] = \frac{-0.5\omega^2(50 - \omega^2)}{(50 - \omega^2)^2 + (2.5\omega)^2} \tag{j}$$

$$\text{Im}[G(j\omega)] = \frac{-1.25\omega^3}{(50 - \omega^2)^2 + (2.5\omega)^2} \tag{k}$$

The magnitude and phase of the sinusoidal transfer function are obtained by using Equations (8.26) and (8.27) in conjunction with Equation (e), yielding

$$|G(j\omega)| = \frac{0.5\omega^2}{\sqrt{(50 - \omega^2)^2 + (2.5\omega)^2}} \tag{l}$$

$$\phi = \tan^{-1}\left(\frac{2.5\omega}{50 - \omega^2}\right) \tag{m}$$

Equation (8.38) is used with Equation (l) to obtain

$$L(\omega) = 20 \log \left[\frac{0.5\omega^2}{\sqrt{(50 - \omega^2)^2 + (2.5\omega)^2}}\right] \tag{n}$$

The low-frequency asymptote is $20 \log (0.5\omega^2/50) = 20 \log (0.01\omega^2) = -40 + 40 \log (\omega)$. The high-frequency asymptote on the Bode diagram is $20 \log (0.5) = -6.02$. The MATLAB-generated Bode diagram for the circuit, shown in Figure 8.45, confirms these asymptotes.

The denominator of the transfer function is second order. Since the numerator is also second order, but with the coefficient of its constant term equal to zero, the circuit could be used as a high-pass filter. This is confirmed by the Bode diagram.

c. In drawing the Nyquist plot, it is noted that $\text{Re}[G(j\omega)] = 0$ for $\omega = 0, \pm\sqrt{50}$. Thus the Nyquist plot intercepts the imaginary axis at $\text{Im}[G(0)] = 0$, $\text{Im}[G(j\sqrt{50})] = -1.41$,

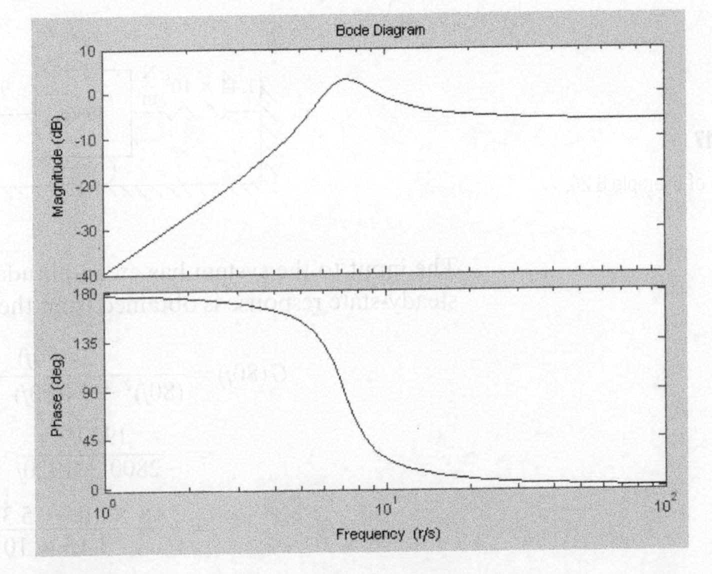

Figure 8.45

The Bode diagram for the circuit of Example 8.23. This diagram shows that the circuit can be used as a high-pass filter.

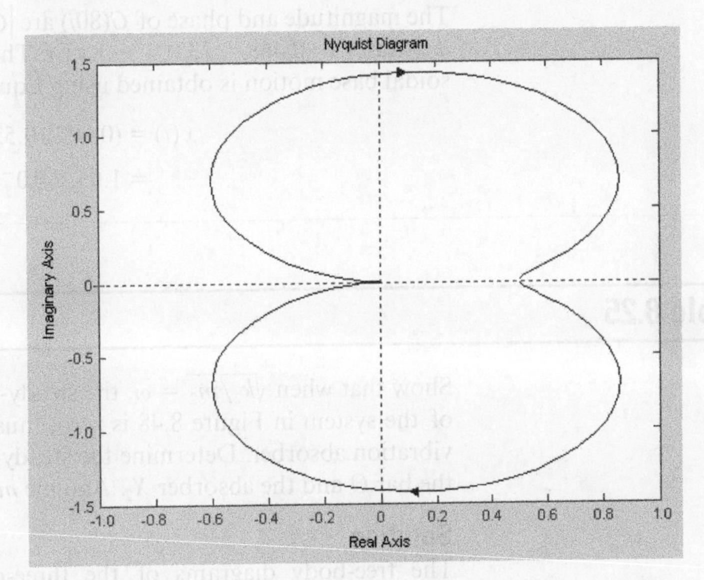

Figure 8.46

The Nyquist diagram for the circuit of Example 8.23.

and $\mathrm{Im}[G(-j\sqrt{50})] = 1.41$. It is further noted that $\mathrm{Im}[G(j\omega)] = 0$ only for $\omega = 0$ and as $\omega \to \pm\infty$. Thus the Nyquist plot intercepts the real axis only at $\mathrm{Re}[G(j\omega)] = 0$ and $\lim_{\omega \to \pm\infty} \mathrm{Re}[G(j\omega)] = 0.5$. The MATLAB-generated Nyquist diagram of Figure 8.46 confirms these points.

Example 8.24

Determine the steady-state response of the system of Figure 8.47.

Solution

The differential equation governing the motion of the system is

$$40\ddot{x} + 960\dot{x} + 1.44 \times 10^5 x = 960\dot{y}(t) \tag{a}$$

where $\dot{y}(t)$ is the prescribed displacement of the base to which the viscous damper is attached. The transfer function $G(s) = X(s)/Y(s)$ is obtained from Equation (a) as

$$G(s)\frac{24s}{s^2 + 24s + 3600} \tag{b}$$

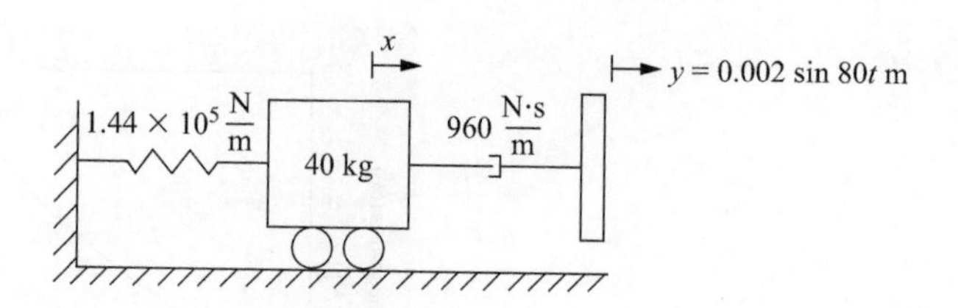

Figure 8.47

The system of Example 8.24.

The input to the system has an amplitude of 0.002 m and a frequency of 80 r/s. The steady-state response is obtained from the sinusoidal transfer function:

$$G(80j) = \frac{24(80j)}{(80j)^2 + 24(60j) + 3600}$$

$$= \frac{1920j}{-2800 + 1920j} \cdot \frac{1920j(-2800 + 1920j)}{-2800 + 1920j}$$

$$= \frac{2.68 \times 10^6 - 5.38 \times 10^6}{1.15 \times 10^7} = 0.233 - 0.468j \tag{c}$$

The magnitude and phase of $G(80j)$ are $|G(80j)| = \sqrt{(0.233)^2 + (-0.468)^2} = 0.523$ and $\phi = \tan^{-1}(-0.468/0.233) = -1.11$ r. Thus the steady-state response due to the sinusoidal base motion is obtained using Equation (8.31) as

$$x(t) = (0.002)(0.523)\sin(80t - 1.11)$$

$$= 1.05 \times 10^{-3}\sin(80t - 1.11)\,\text{m} \tag{d}$$

Example 8.25

Show that when $\sqrt{k_2/m_2} = \omega$, the steady-state amplitude of the middle of the bar X_1 of the system in Figure 8.48 is zero; thus the auxiliary mass-spring system acts as a vibration absorber. Determine the steady-state amplitudes of the angular rotation of the bar Θ and the absorber X_2. Assume $m_2 = 5$ kg in all calculations.

Solution

The free-body diagrams of the three-degree-of-freedom system at an arbitrary instant, assuming small θ, are shown in Figure 8.49. The small angle assumption also implies that gravity cancels with static spring forces and thus neither are shown on the free-body diagrams. The application of conservation laws to the free-body diagrams leads to

$$\begin{bmatrix} m_1 & 0 & 0 \\ 0 & I & 0 \\ 0 & 0 & m_2 \end{bmatrix}\begin{bmatrix} \ddot{x}_1 \\ \ddot{\theta} \\ \ddot{x}_2 \end{bmatrix} + \begin{bmatrix} 2k_1 & 0 & -k_2 \\ 0 & 2k_1\ell^2 & 0 \\ -k_2 & 0 & k_2 \end{bmatrix}\begin{bmatrix} x_1 \\ \theta \\ x_2 \end{bmatrix} = \begin{bmatrix} F(t) \\ F(t)\ell \\ 0 \end{bmatrix} \tag{a}$$

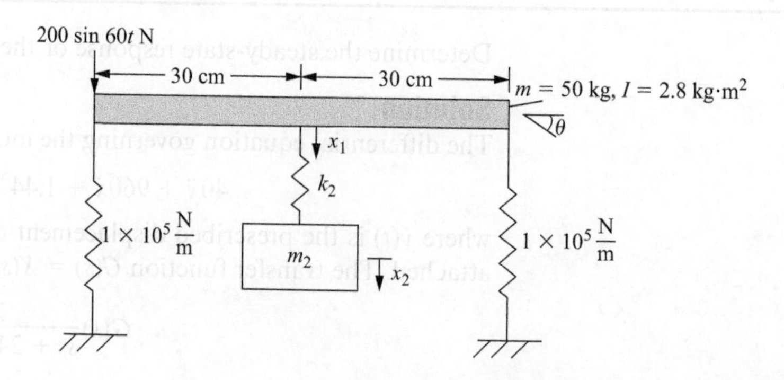

Figure 8.48

The system of Example 8.25.

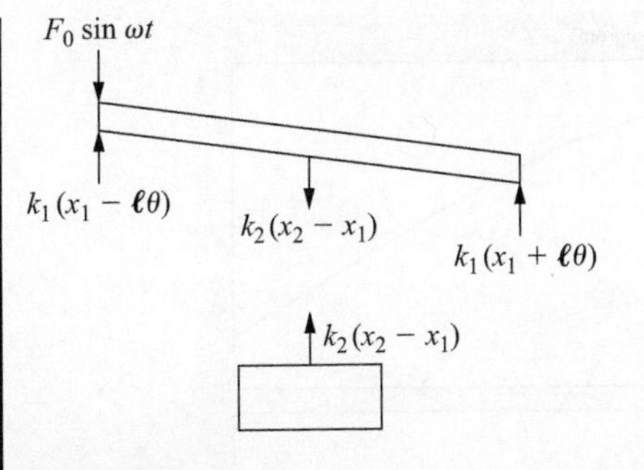

Figure 8.49

The free-body diagrams of a primary system and auxilary mass at an arbitrary instant.

Taking the Laplace transform of Equation (a), substituting in given numerical values, and assuming $m_2 = 5$ kg and $k_2 = (5 \text{ kg})(60 \text{ r/s})^2 = 1.8 \times 10^4$ N/m leads to

$$\begin{bmatrix} 50s^2 + 2 \times 10^5 & 0 & -1.8 \times 10^4 \\ 0 & 2.8s^2 + 7.2 \times 10^4 & 0 \\ -1.8 \times 10^4 & 0 & 5s^2 + 1.8 \times 10^4 \end{bmatrix} \begin{bmatrix} X_1 \\ \Theta \\ X_2 \end{bmatrix} = \begin{bmatrix} 1 \\ 0.3 \\ 0 \end{bmatrix} F(s) \tag{b}$$

Cramer's rule is used to determine the system's transfer functions as

$$G_1(s) = \frac{X_1(s)}{F(s)} = \frac{5s^2 + 1.8 \times 10^4}{250s^4 + 1.90 \times 10^6 s^2 + 3.28 \times 10^9} \tag{c}$$

$$G_2(s) = \frac{\Theta(s)}{F(s)} = \frac{0.3}{2.8s^2 + 7.2 \times 10^4} \tag{d}$$

$$G_3(s) = \frac{X_2(s)}{F(s)} = \frac{1.8 \times 10^4}{250s^4 + 1.90 \times 10^6 s^2 + 3.28 \times 10^9} \tag{e}$$

The steady-state amplitudes are obtained from the sinusoidal transfer functions as

$$X_1 = 200|G_1(60j)| = \left| \frac{200[5(60j)^2 + 1.8 \times 10^4]}{250(60j)^4 + 1.96 \times 10^6 (60j)^2 + 3.28 \times 10^9} \right| = 0 \tag{f}$$

$$\Theta = 200|G_2(60j)| = \left| \frac{200(0.3)}{2.8(60j)^2 + 7.2 \times 10^4} \right| = 9.70 \times 10^{-4} \text{ r} \tag{g}$$

$$X_2 = 200|G_3(60j)| = \left| \frac{200(1.8 \times 10^4)}{250(60j)^4 + 1.90 \times 10^6 (60j)^2 + 3.28 \times 10^9} \right| = 11.3 \text{ mm} \tag{h}$$

Example 8.26

Contrast the Bode plots and Nyquist diagrams for the concentration perturbation C_{A2p} of the system of Example 7.23 for the case where there is no time delay and for the case where the system has a time delay of 10 s.

Solution

The appropriate transfer function is determined from Equation (k) of Example 7.23, which leads to:

$$G(s) = \frac{4.43 \times 10^{-7} e^{-10s}}{(s + 3.97 \times 10^{-3})(s + 3.20 \times 10^{-3})} \tag{a}$$

The sinusoidal transfer function obtained for the system without the time delay is

$$G(j\omega) = \frac{4.43 \times 10^{-7}(1.27 \times 10^{-5} - \omega^2 - j7.17 \times 10^{-3}\,\omega)}{(1.27 \times 10^{-5} - \omega^2)^2 + (7.17 \times 10^{-3}\,\omega)^2} \tag{b}$$

The real and imaginary parts of the transfer function are

$$\text{Re}[G(j\omega)] = \frac{4.43 \times 10^{-7}(1.27 \times 10^{-5} - \omega^2)}{(1.27 \times 10^{-5} - \omega^2)^2 + (7.17 \times 10^{-3}\,\omega)^2} \tag{c}$$

$$\text{Im}[G(j\omega)] = \frac{3.18 \times 10^9\,\omega}{(1.27 \times 10^{-5} - \omega^2)^2 + (7.17 \times 10^{-3}\,\omega)^2} \tag{d}$$

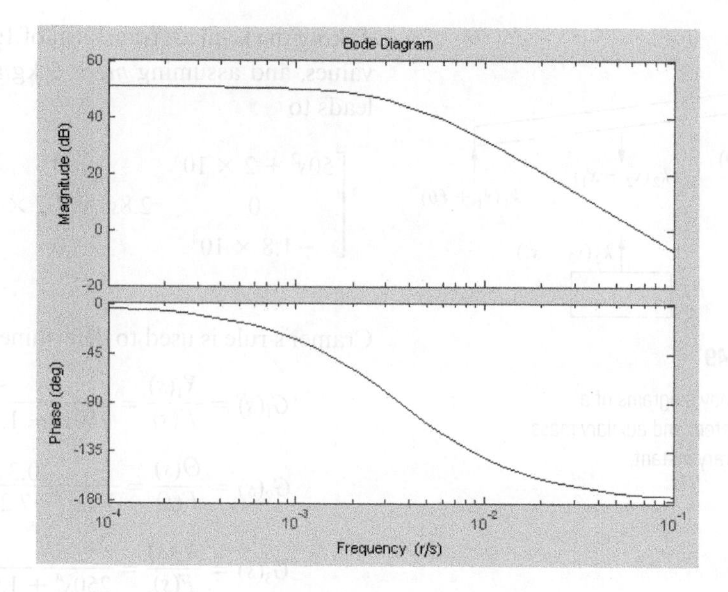

Figure 8.50

The Bode diagram for the system of Example 8.26 without time delay.

The magnitude and phase of the sinusoidal transfer function are

$$|G(j\omega)| = \frac{4.43 \times 10^{-7}}{\sqrt{(1.27 \times 10^{-5} - \omega^2)^2 + (7.17 \times 10^{-3}\,\omega)^2}} \tag{e}$$

$$\phi = \tan^{-1}\left(\frac{-7.17 \times 10^{-3}\,\omega}{1.27 \times 10^{-5}\,\omega^2}\right) \tag{f}$$

The MATLAB-generated Bode diagram and Nyquist plots for the system without the time delay are given in Figures 8.50 and 8.51, respectively.

The sinusoidal transfer function for the system with the time delay is

$$G(j\omega) = \frac{4.43 \times 10^{-7} e^{-j10\omega}(1.27 \times 10^{-5} - \omega^2 - j7.17 \times 10^{-3}\,\omega)}{(1.27 \times 10^{-5} - \omega^2)^2 + (7.17 \times 10^{-3}\,\omega)^2} \tag{g}$$

Noting that $e^{-j10\omega} = \cos(10\omega) - j\sin(10\omega)$, Equation (g) can be rewritten as

$$\mathrm{Re}[G(j\omega)] = \frac{4.43 \times 10^{-7}[\cos(10\omega)(1.27 \times 10^{-5} - \omega^2) + 7.17 \times 10^{-3}\,\omega \sin(10\omega)]}{(1.27 \times 10^{-5} - \omega^2)^2 + (7.17 \times 10^{-3}\,\omega)^2} \tag{h}$$

$$\mathrm{Im}[G(j\omega)] = \frac{4.43 \times 10^{-7}[7.17 \times 10^{-3}\,\omega \cos(10\omega) + (1.27 \times 10^{-5} - \omega^2)\sin(10\omega)]}{(1.27 \times 10^{-5} - \omega^2)^2 + (7.17 \times 10^{-3}\,\omega)^2} \tag{i}$$

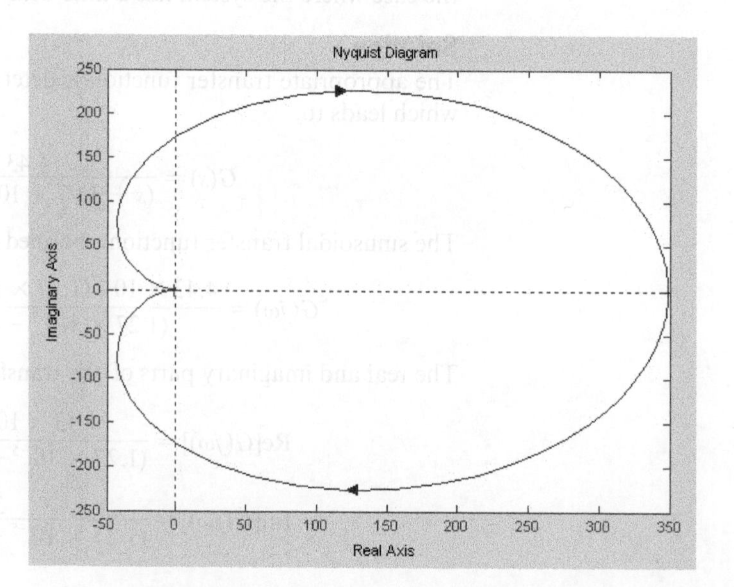

Figure 8.51

The Nyquist diagram for the system of Example 8.26 without time delay.

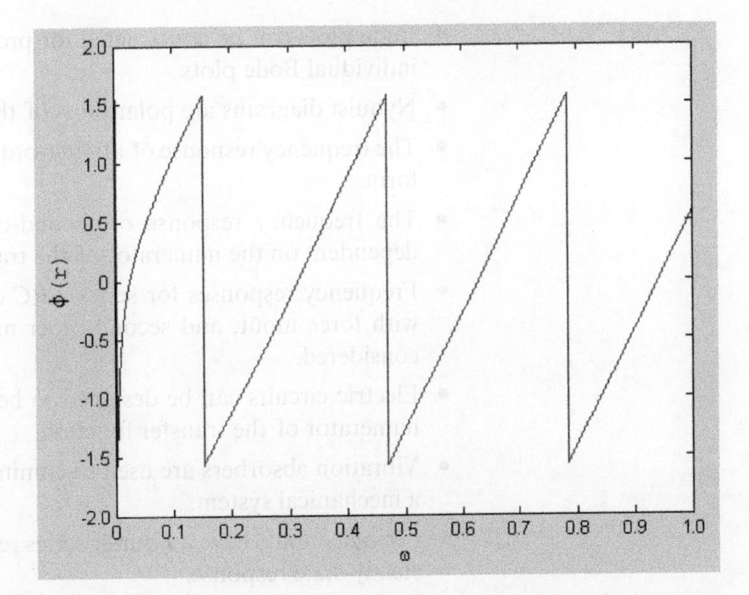

Figure 8.52

The phase portion of plot of the Bode diagram with time delay.

The magnitude of the sinusoidal transfer function is not affected by the time delay and is given by Equation (e). The phase is affected by the time delay and is determined as

$$\phi = \tan^{-1}\left(\frac{7.17 \times 10^{-3}\,\omega\cos(10\omega) + (1.27 \times 10^{-5} - \omega^2)\sin(10\omega)}{\cos(10\omega)(1.27 \times 10^{-5} - \omega^2) + 7.17 \times 10^{-3}\omega\sin(10\omega)}\right) \tag{j}$$

Figure 8.52 illustrates the steady-state phase as a function of the frequency. Note that the denominator of the argument of the inverse tangent function of Equation (j) is zero when

$$\tan(10\omega) = -\frac{1.27 \times 10^{-5} - \omega^2}{7.17 \times 10^{-3}\,\omega} \tag{k}$$

Equation (k) has an infinite number of solution ($\omega = \pi/20, 3\pi/20, 5\pi/20, \ldots$), each of which leads to $\phi = \pi/2$.

The distance from the origin to a point on the Nyquist diagram is the magnitude of the sinusoidal transfer function for the appropriate value of ω. Since the magnitude of the sinusoidal transfer function is unaffected by the time delay, the Nyquist diagram is also unaffected.

8.12 Summary

8.12.1 Chapter Highlights

- Chapter 8 studies the steady-state responses due to a harmonic input. The study is conducted in the frequency or transfer domain.
- The undamped second-order system is a special case. When the input frequency coincides with the natural frequency, a resonance condition occurs. Resonance is characterized by a response that grows without bound.
- The sinusoidal transfer function is developed from the transfer function. The steady-state response is developed using the sinusoidal transfer function.
- The steady-state response is characterized by an amplitude and a phase.
- A system's frequency response describes how its amplitude and phase change with input frequency.
- Bode plots show the frequency response using logarithmic scales.

- Bode plots can be constructed for products of transfer functions by adding the individual Bode plots.
- Nyquist diagrams are polar plots of the sinusoidal transfer function.
- The frequency response of all first-order systems is the same in a nondimensional form.
- The frequency response of second-order and higher-order systems is highly dependent on the numerator of the transfer function.
- Frequency responses for series *LRC* circuits, second-order mechanical systems with force input, and second-order mechanical systems with motion input are considered.
- Electric circuits can be designed to be filters. The type of filter depends on the numerator of the transfer function.
- Vibration absorbers are used to eliminate steady-state vibrations of a particle in a mechanical system.
- Periodic inputs have a Fourier series representation, which is used to develop the steady-state response.
- The steady-state response due to a general periodic input is an infinite series of harmonic terms whose frequencies are integer multiples of the fundamental frequency.

8.12.2 Important Equations

- General form of periodic input

$$F(t) = F_0 \sin(\omega t + \psi) \tag{8.1}$$

- General form of the steady-state response

$$x(t) = X \sin(\omega t + \psi + \phi) \tag{8.2}$$

- Resonance for undamped mechanical systems

$$x(t) = \frac{F_0}{2m\omega_n^2}[\sin(\omega_n t) - \omega_n t \cos(\omega_n t)] \tag{8.12}$$

- Steady-state response using the sinusoidal transfer function

$$\phi = \tan^{-1}\left\{\frac{\text{Im}[G(j\omega)]}{\text{Re}[G(j\omega)]}\right\} \tag{8.27}$$

$$x_s(t) = F_0|G(j\omega)|\sin(\omega t + \phi) \tag{8.31}$$

- Frequency response of a series *LRC* circuit

$$M = \frac{L\omega_n I}{V_0} = \frac{r}{\sqrt{(1 - r^2)^2 + (2\zeta r)^2}} \tag{8.35}$$

- Bode diagram

$$L(\omega) = 20 \log |G(j\omega)| \tag{8.38}$$

- Frequency response for first-order systems

$$M = \sqrt{\frac{1}{\beta^2 + 1}} \tag{8.65}$$

$$\phi = -\tan^{-1}(\beta) \tag{8.66}$$

- Frequency response for second-order mechanical systems with force input

$$\frac{m\omega_n^2 X}{F_0} = \frac{1}{\sqrt{(1-r^2)^2 + (2\zeta r)^2}} \tag{8.72}$$

$$\phi = \tan^{-1}\left(\frac{2\zeta r}{r^2 - 1}\right) \tag{8.73}$$

- Frequency response for mechanical systems with motion input

$$T = \sqrt{\frac{1 + (2\zeta r)^2}{(1-r^2) + (2\zeta r)^2}} \tag{8.82}$$

$$\omega_n^2 \Gamma = \omega_n^2 r^2 \sqrt{\frac{1 + (2\zeta r)^2}{(1-r^2) + (2\zeta r)^2}} \tag{8.83}$$

$$\Lambda = \frac{r^2}{\sqrt{(1-r^2)^2 + (2\zeta r)^2}} \tag{8.84}$$

- Fourier series representation of periodic function

$$F(t) = \frac{a_0}{2} + \sum_{k=1}^{\infty}[a_k \cos(\omega_k t) + b_k \sin(\omega_k t)] \tag{8.103}$$

- Fourier coefficients

$$a_k = \frac{2}{T}\int_0^T F(t)\cos(\omega_k t)\, dt \tag{8.104}$$

$$b_k = \frac{2}{T}\int_0^T F(t)\sin(\omega_k t)\, dt \tag{8.105}$$

Short Answer Problems

Indicate whether the statements in Problems SA8.1–SA8.15 are true or false. If a statement is false, rewrite the statement so that it is true.

SA8.1 Resonance occurs only in a neutrally stable system.

SA8.2 Beating is characterized by a continual buildup and decay of amplitude.

SA8.3 The sinusoidal transfer function for a system is $G(j\omega)$, where $G(s)$ is the system's transfer function and ω is the natural frequency of the system.

SA8.4 Frequency response is the variation of amplitude and phase with frequency.

SA8.5 The complex conjugate of the sinusoidal transfer function $G(j\omega)$ is $G(-j\omega)$.

SA8.6 If a system with transfer function $G(s)$ is subject to an input of the form $A\sin\omega t$, then its steady-state response is $x(t) = X\sin\omega t$, where $X = A|G(j\omega)|$.

SA8.7 The frequency response of a series *LRC* circuit is such that $\lim_{\omega\to\infty} I = 0$, where I is the steady-state amplitude of the current through the circuit.

SA8.8 The frequency response of a series *LRC* circuit is such that $\lim_{\omega\to\infty} \phi = 0$, where ϕ is the steady-state phase of the current through the circuit.

SA8.9 The amplitude part of the Bode diagram is a plot of $20\log|G(-j\omega)|$ versus $\log\omega$.

SA8.10 $\log\omega$ is measured in decibels.

SA8.11 If $G(s) = G_1(s)G_2(s)$, then $L(\omega) = L_1(\omega)L_2(\omega)$, where $L_i(\omega) = 20\log|G_i(j\omega)|$.

SA8.12 The Bode diagram for the transfer function of a gain is a horizontal line.

SA8.13 The low-frequency asymptote on the Bode diagram for a first-order lag is the horizontal line $L = 0$.

SA8.14 The phase of a first-order lead is always negative.

SA8.15 The Nyquist diagram is drawn using ω as a parameter.

SA8.16 What are the standard units of measurement of $L(\omega)$?

SA8.17 What is the steady-state response of a system with transfer function $G(s)$ due to an input of the form $F \sin \omega t$?

SA8.18 Define the corner frequency on the Bode diagram.

SA8.19 Explain the concept of bandwidth.

SA8.20 What is the phase angle of a gain?

SA8.21 A system has a transfer function of $G(s) = \frac{1}{2s+3}$. What is the sinusoidal transfer function?

SA8.22 A system has a transfer function of $G(s) = \frac{1}{2s+3}$. What is $L(\omega)$?

SA8.23 A system has a transfer function of $G(s) = \frac{1}{2s+3}$. What is the phase of the steady-state response for $\omega = 10$ r/s?

SA8.24 A system has a transfer function of $G(s) = \frac{4}{s^2 + 14s + 50}$. What is the steady-state phase when $\omega = 8$ r/s?

SA8.25 A system has a transfer function of $G(s) = \frac{0.25}{s^2 + 12s + 49}$. For what frequency will the steady-state phase be $\pi/2$ r?

SA8.26 A system has a transfer function of $G(s) = \frac{0.25}{s^2 + 12s + 49}$. What is $\lim_{\omega \to \infty} \phi$?

SA8.27 A system has a transfer function of $G(s) = \frac{2(s+10)}{s+15}$. What is the system's gain K?

SA8.28 Write the transfer function of a system that has a gain of 4, a first-order lead with a time constant of 2 s, and an integrator.

SA8.29 Write the transfer function for a system that has a gain of 0.2, a first-order lag with a time constant of 3 s, a second-order lag with a natural frequency of 10 r/s, and a damping ratio of 0.5.

SA8.30 What is the bandwidth of a system with a gain of 1.5 and a first-order lag with a time constant of 5 s?

SA8.31 What is the bandwidth of a system with a second-order lag with natural frequency 5 r/s and damping ratio 0.2?

SA8.32 A system that has a sinusoidal transfer function evaluated for $\omega = 5$ r/s of $G(5j) = 2 - 3j$ has an input of $0.3 \sin 5t$. What is the steady-state amplitude of the system?

SA8.33 A system that has a sinusoidal transfer function evaluated for $\omega = 5$ r/s of $G(5j) = 2 - 3j$ has an input of $0.3 \sin 5t$. What is the system's steady-state phase at 5 r/s?

SA8.34 A system that has a sinusoidal transfer function evaluated for $\omega = 5$ r/s of $G(5j) = 2 - 3j$ has an input of $0.3 \sin 5t$. What is the steady-state response of the system at 5 r/s?

SA8.35 A system that has a sinusoidal transfer function evaluated for $\omega = 5$ r/s of $G(5j) = -2 - 3j$ has an input of $0.3 \sin 5t$. What is the steady-state amplitude of the system?

SA8.36 A system that has a sinusoidal transfer function evaluated for $\omega = 5$ r/s of $G(5j) = -2 - 3j$ has an input of $0.3 \sin 5t$. What is the system's steady-state phase at 5 r/s?

SA8.37 A system that has a sinusoidal transfer function evaluated for $\omega = 5$ r/s of $G(5j) = -2 - 3j$ has an input of $0.3 \sin 5t$. What is the steady-state response of the system at 5 r/s?

SA8.38 A first-order system has a transfer function of $G(s) = \frac{1}{2s+3}$. What is the gain of the system?

SA8.39 A first-order system has a transfer function of $G(s) = \frac{1}{2s+3}$. What is the low-frequency asymptote for the system?

SA8.40 A first-order system has a transfer function of $G(s) = \frac{1}{2s+3}$. What is the high-frequency asymptote for the system?

SA8.41 A first-order system has a transfer function of $G(s) = \frac{1}{2s+3}$. What is corner frequency for the system?

SA8.42 A first-order system has a transfer function of $G(s) = \frac{1}{2s+3}$. What is the system's bandwidth?

SA8.43 A system has a transfer function of $G(s) = \frac{4s+1}{2s+3}$. What is the low-frequency asymptote for the system?

SA8.44 A system has a transfer function of $G(s) = \frac{4s+1}{2s+3}$. What is the high-frequency asymptote for the system?

SA8.45 A system has a transfer function of $G(s) = \frac{4s+1}{2s+3}$. What is the corner frequency for the system?

SA8.46 A system has the transfer function $G(s) = \frac{s}{s^2 + 20s + 200}$. What is the low-frequency asymptote for the system?

SA8.47 A system has the transfer function $G(s) = \frac{s}{s^2 + 20s + 200}$. What is the high-frequency asymptote for the system?

SA8.48 A system has the transfer function $G(s) = \frac{s}{s^2 + 20s + 200}$. What is the corner frequency for the system?

SA8.49 The sinusoidal transfer function for a system is $G(j\omega) = \frac{100 - 20\omega^2 + j(49\omega - \omega^3)}{625 - 14\omega^2 + \omega^4}$. What point on the Nyquist diagram corresponds to a frequency of 2 r/s?

SA8.50 The sinusoidal transfer function for a system is $G(j\omega) = \frac{100 - 20\omega^2 + j(49\omega - \omega^3)}{625 - 14\omega^2 + \omega^4}$. What is the point on the Nyquist diagram when $\omega \to \infty$?

SA8.51 The sinusoidal transfer function for a system is $G(j\omega) = \dfrac{100 - 20\omega^2 + j(49\omega - \omega^3)}{625 - 14\omega^2 + \omega^4}$. What are the real axis intercepts on the Nyquist diagram?

SA8.52 The sinusoidal transfer function for a system is $G(j\omega) = \dfrac{100 - 20\omega^2 + j(49\omega - \omega^3)}{625 - 14\omega^2 + \omega^4}$. What are the imaginary axis intercepts on the Nyquist diagram?

SA8.53 The system of Figure SA8.53 consists of a spring in parallel with a viscous damper that is attached to a massless board. The board is subject to a force of $100 \sin \omega t$ N. What is the maximum steady-state amplitude of the system over all values of ω?

Figure SA8.53

SA8.54 Given the system of Figure SA8.53, what is the value of ω such that the steady-state amplitude is $\sqrt{2}/2$ times its maximum value?

SA8.55 What is the steady-state phase of the system in Problem SA8.54 when the steady-state amplitude is $\sqrt{2}/2$ times its maximum value?

SA8.56 A mass, spring, and viscous damper system has a mass of 1 kg, a natural frequency of 100 r/s, and a damping ratio of 0.2. It is subject to a sinusoidal force of magnitude 20,000 N. What is the frequency at which the maximum steady-state amplitude is attained?

SA8.57 A mass, spring, and viscous damper system has a mass of 1 kg, a natural frequency of 100 r/s, and a damping ratio of 0.2. It is subject to a sinusoidal force of magnitude 20,000 N. What is the maximum value of the steady-state amplitude over all frequencies of excitation that the mass will achieve?

SA8.58 A mass, spring, and viscous damper system has a mass of 1 kg, a natural frequency of 100 r/s, and a damping ratio of 0.2. What is the mass's steady-state amplitude when it is subject to a force of 20,000 sin $50t$?

SA8.59 A mass, spring, and viscous damper system has a mass of 1 kg, a natural frequency of 100 r/s, and a damping ratio of 0.2. What is the mass's steady-state phase when it is subject to a force of 20,000 sin $50t$?

SA8.60 A mass, spring, and viscous damper system has a mass of 1 kg, a natural frequency of 100 r/s, and a damping ratio of 0.8. It is subject to a sinusoidal force of magnitude 20,000 N. What is the frequency at which the maximum steady-state amplitude is attained?

SA8.61 A mass, spring, and viscous damper system has a mass of 1 kg, a natural frequency of 100 r/s, and a damping ratio of 0.8. It is subject to a sinusoidal force of magnitude 20,000 N. What is the maximum value of the steady-state amplitude over all frequencies of excitation that the mass will achieve?

SA8.62 A 2-kg mass is connected to a spring in parallel with a viscous damper. The other end of this parallel combination is connected to a movable base. The system has a natural frequency of 20 r/s and a damping ratio of 0.25. What is the transfer function for the absolute displacement of the mass?

SA8.63 A 2-kg mass is connected to a spring in parallel with a viscous damper. The other end of this parallel combination is connected to a movable base. The system has a natural frequency of 20 r/s and a damping ratio of 0.25. What is the steady-state amplitude of the absolute displacement of the mass?

SA8.64 A 2-kg mass is connected to a spring in parallel with a viscous damper. The other end of this parallel combination is connected to a movable base. The system has a natural frequency of 20 r/s and a damping ratio of 0.25. What is the steady-state amplitude of the acceleration of the mass?

SA8.65 A 2-kg mass is connected to a spring in parallel with a viscous damper. The other end of this parallel combination is connected to a movable base. The system has a natural frequency of 20 r/s and a damping ratio of 0.25. What is the steady-state amplitude of the displacement of the mass relative to the base?

SA8.66 A 2-kg mass is connected to a spring in parallel with a viscous damper. The other end of this parallel combination is connected to a movable base. The system has a natural frequency of 20 r/s and a damping ratio of 0.25. When the base has a motion given by $y(t) = 0.003 \sin 35t$, is the acceleration of the mass less than the acceleration of its base?

SA8.67 A 2-kg mass is connected to a spring in parallel with a viscous damper. The other end of this parallel combination is connected to a movable base. The system has a natural frequency of 20 r/s and a damping ratio of 0.25. When the base has a motion given by $y(t) = 0.003 \sin 25t$, is the acceleration of the mass less than the acceleration of its base?

SA8.68 Two isolators are available to protect a machine from large accelerations generated by its base. They have the same stiffness and lead to a natural frequency in the range of isolation, but one has a damping ratio of 0.04 and the other has a damping ratio of 0.1. Which isolator offers better protection to the machine?

SA8.69 What type of filter might an electric circuit with a transfer function of $G(s) = \dfrac{2s + 3}{s^2 + 10s + 50}$ make?

SA8.70 What type of filter might an electric circuit with a transfer function of $G(s) = \dfrac{2s^3 + 3s}{s^3 + 6s^2 + 10s + 50}$ make?

SA8.71 What type of filter might an electric circuit with a transfer function of $G(s) = \dfrac{s^2 + 3}{s^2 + 10s + 50}$ make?

SA8.72 What type of filter might an electric circuit with a transfer function of $G(s) = \dfrac{s^2}{s^3 + 6s^2 + 10s + 50}$ make?

SA8.73 Sketch the Bode diagram for a low-pass filter with a bandwidth of 200 r/s.

SA8.74 Sketch the Bode diagram for a high-pass filter with a bandwidth of 50 r/s to infinity.

SA8.75 Sketch the Bode diagram for a band-pass filter with a bandwidth from 1000 r/s to 10,000 r/s.

SA8.76 Sketch the Bode diagram for a band-reject filter with a reject bandwidth from 1000 r/s to 10,000 r/s.

SA8.77 A primary system with a natural frequency of 100 r/s operates at $\omega = 105$ r/s. To eliminate large steady-state vibrations, a vibration absorber is to be added. What is the required stiffness of the absorber if its mass is to be 5 kg?

SA8.78 A primary system with a natural frequency of 100 r/s operates at $\omega = 95$ r/s. To eliminate large steady-state vibrations, a vibration absorber is to be added. What is the required mass of the absorber if its stiffness is to be 3×10^4 N/m?

SA8.79 A primary system with a natural frequency of 100 r/s operates at $\omega = 105$ r/s. To eliminate large steady-state vibrations, a vibration absorber is to be added. The force is of magnitude 10,000 N. Design an absorber with minimum mass such that the steady-state amplitude of the absorber is to be no greater than 1 cm.

SA8.80 What are the natural frequencies of the system of Problem SA8.79 with the absorber in place?

Problems

8.1 Determine, for the series RL circuit of Figure P8.1, (a) the sinusoidal transfer function when $\omega = 800$ r/s and (b) the steady-state current through the circuit when $\omega = 800$ r/s.

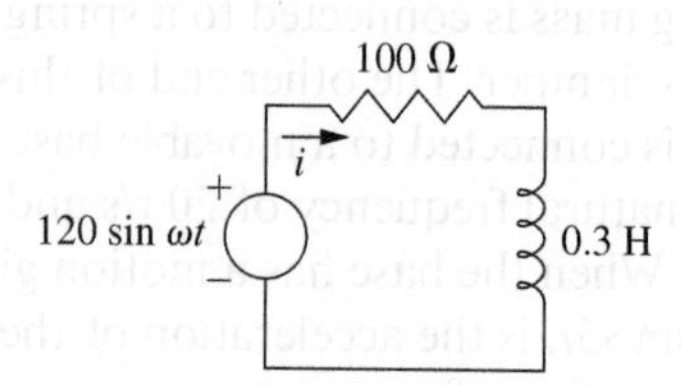

Figure P8.1

8.2 Determine, for the two-mesh circuit of Figure P8.2, (a) the sinusoidal transfer function $G(s) = I(s)/V_1(s)$ when the frequency of the sinusoidal voltage source is 2.1 kHz, and (b) the steady-state current through the 10-Ω resistor when the frequency of the sinusoidal voltage source is 2.1 kHz.

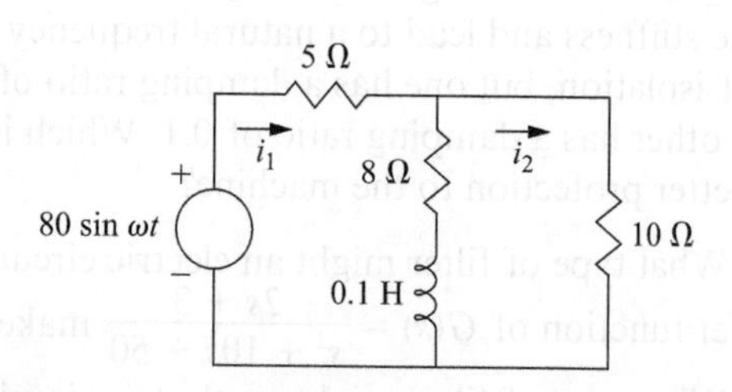

Figure P8.2

8.3 Determine the steady-state response of the spring-viscous damper system of Figure P8.3.

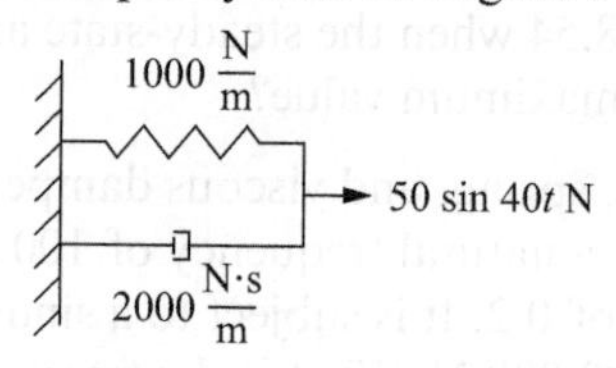

Figure P8.3

8.4 Determine the steady-state response of the system of Figure P8.4 when $\omega = 80$ r/s.

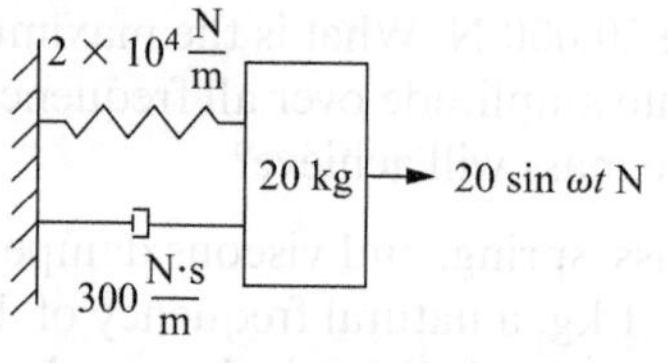

$2 \times 10^4 \frac{\text{N}}{\text{m}}$

20 kg

20 sin ωt N

$300 \frac{\text{N·s}}{\text{m}}$

Figure P8.4

8.5 Determine the steady-state response for the absolute displacement of the system of Figure P8.5.

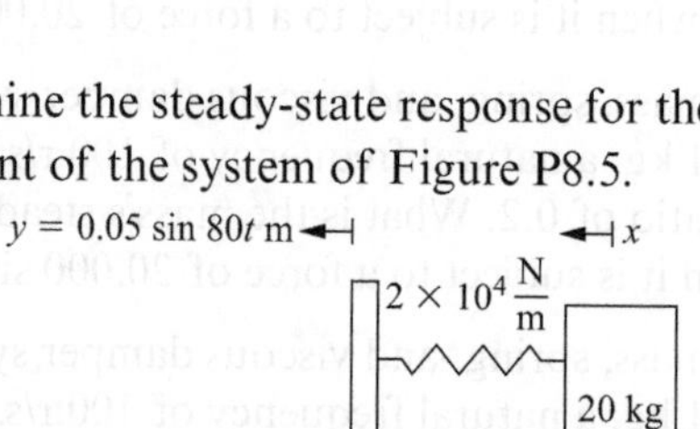

$y = 0.05 \sin 80t$ m

x

$2 \times 10^4 \frac{\text{N}}{\text{m}}$

20 kg

$300 \frac{\text{N·s}}{\text{m}}$

Figure P8.5

8.6 The thin disk of Example 4.17 and Figure 4.59(a) is subject to a counter-clockwise time-dependent moment of the form $M(t) = 20.5 \sin(200t)$ N·m instead of $F(t)$. Determine the steady-state amplitude of angular oscillation of the disk using the following parameters: $m_1 = 10$ kg, $m_2 = 15$ kg, $r = 20$ cm, $I = 0.1$ kg·m^2, $k_1 = k_2 = 1 \times 10^4$ N/m, and $c_1 = c_2 = 2.5 \times 10^3$ N·s/m.

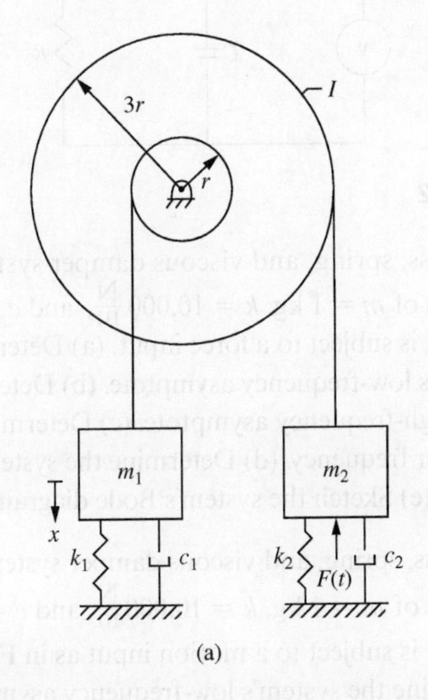

Figure 4.59

(a) The system of Example 4.17 and Problem 8.6. (Repeated)

8.7 Determine the steady-state current through the inductor in the circuit of Figure P8.7.

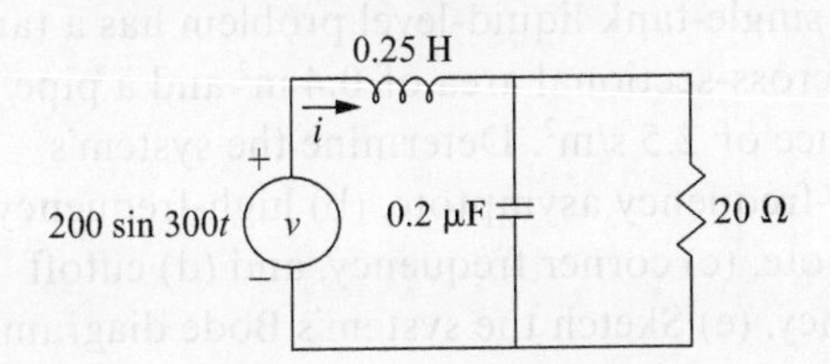

Figure P8.7

8.8 The two-tank system of Figure P8.8 is in the steady-state shown when the exit pressure is 2.5 kPa. Determine the steady-state liquid levels in the tanks of the system of Figure P8.8 when the exit pressure varies sinusoidally according to $p_e(t) = 2.5 + 0.6 \sin(50t)$ kPa.

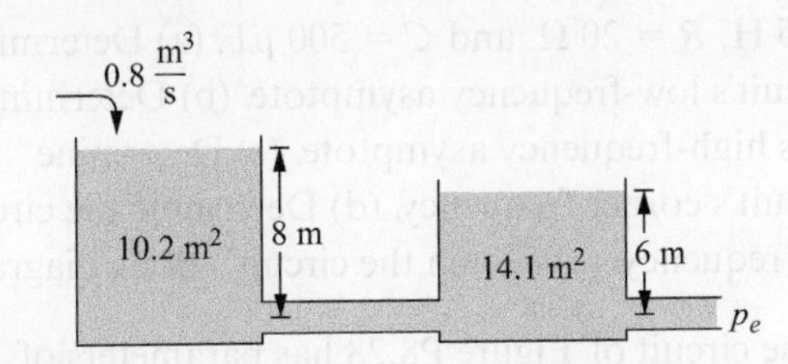

Figure P8.8

8.9 Determine the steady-state response for the acceleration of the mass in the system of Figure P8.9.

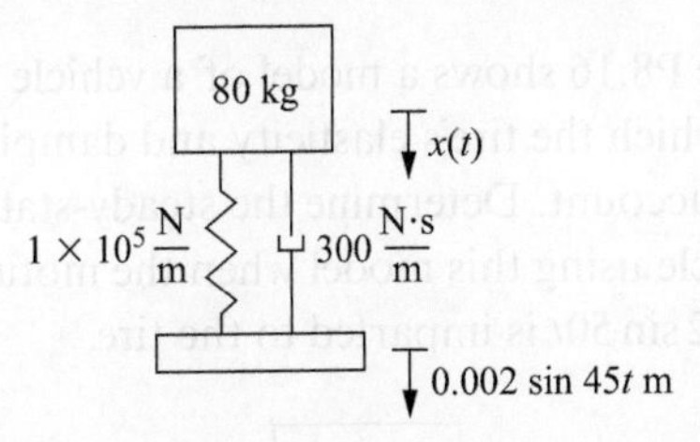

Figure P8.9

8.10 Determine the steady-state response for the relative displacement between the mass and the base for the system of Figure P8.9.

8.11 Determine the steady-state response of the absolute displacement of the mass for the system of Figure P8.9.

8.12 Determine the steady-state response for i_1 and i_2 of the circuit in Figure P8.12 if $v(t) = 10 \sin 1000t$.

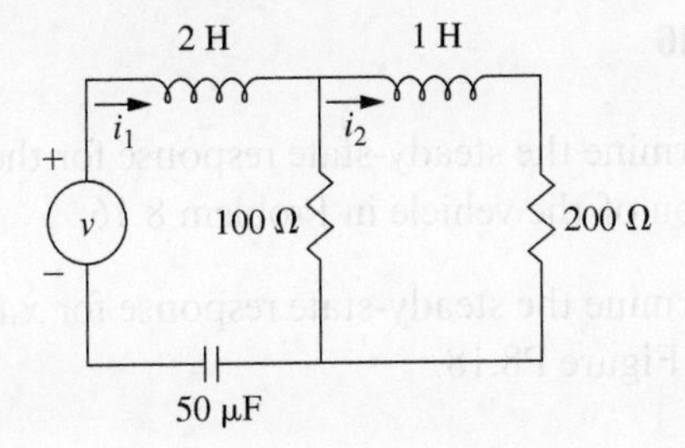

Figure P8.12

8.13 Determine the steady-state response for i_1, i_2, and i_3 of the circuit in Figure P8.13 if $v(t) = 10 \sin 100t$.

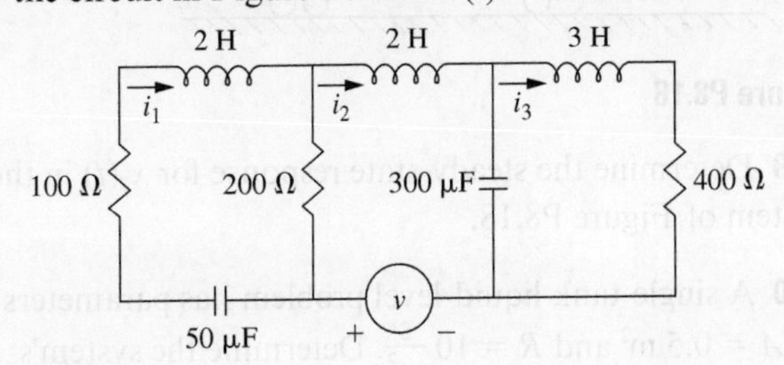

Figure P8.13

8.14 Determine the steady-state amplitude of the slender bar in the system in Figure P8.14 if $M(t) = 10 \sin 40t$.

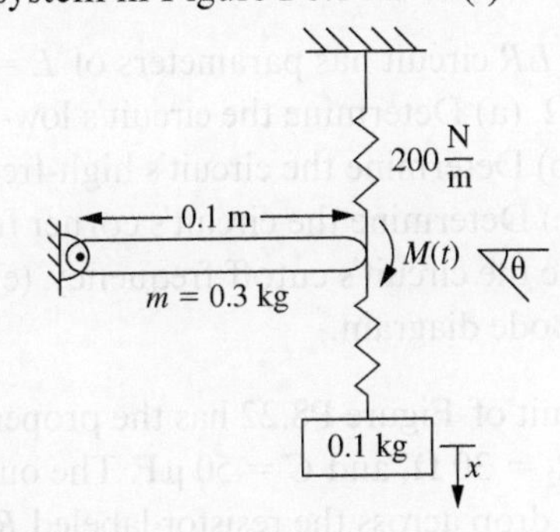

Figure P8.14

8.15 Determine the steady-state response of the 0.1-kg mass in the system in Figure P8.14.

8.16 Figure P8.16 shows a model of a vehicle suspension system in which the tire's elasticity and damping are taken into account. Determine the steady-state response of the vehicle using this model when the motion $y(t) = 0.002 \sin 50t$ is imparted to the tire.

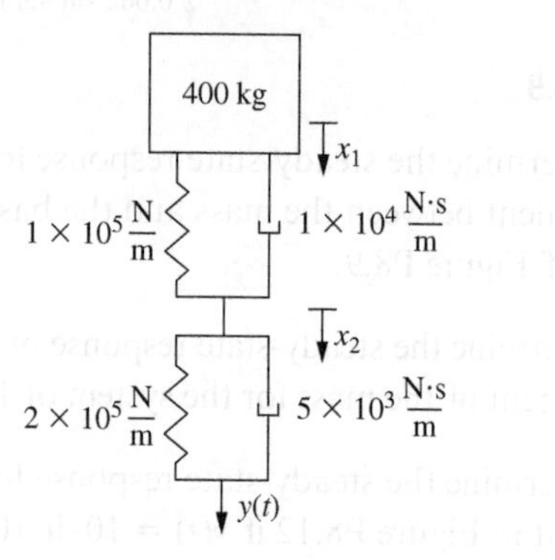

Figure P8.16

8.17 Determine the steady-state response for the vertical acceleration of the vehicle in Problem 8.16.

8.18 Determine the steady-state response for $x_1(t)$ for the system of Figure P8.18.

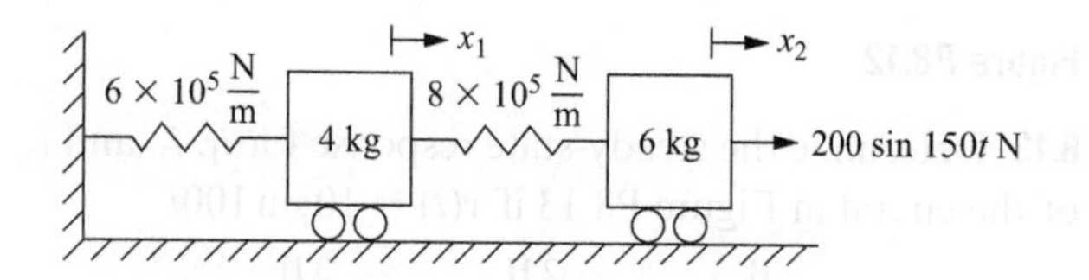

Figure P8.18

8.19 Determine the steady-state response for $x_2(t)$ in the system of Figure P8.18.

8.20 A single-tank liquid-level problem has parameters of $A = 0.5 \text{ m}^2$ and $R = 10 \frac{\text{s}}{\text{m}^2}$. Determine the system's (a) low-frequency asymptote, (b) high-frequency asymptote, (c) corner frequency, and (d) cutoff frequency. (e) Sketch the system's Bode diagram.

8.21 A series LR circuit has parameters of $L = 1.3$ H and $R = 20\ \Omega$. (a) Determine the circuit's low-frequency asymptote. (b) Determine the circuit's high-frequency asymptote. (c) Determine the circuit's corner frequency. (d) Determine the circuit's cutoff frequency. (e) Sketch the circuit's Bode diagram.

8.22 The circuit of Figure P8.22 has the properties $R_1 = 10\ \Omega$, $R_2 = 30\ \Omega$, and $C = 50\ \mu$F. The output is the voltage drop across the resistor labeled R_2. (a) Determine the circuit's low-frequency asymptote. (b) Determine the circuit's high-frequency asymptote. (c) Determine the circuit's corner frequency. (d) Determine the circuit's cutoff frequency. (e) Sketch the circuit's Bode diagram.

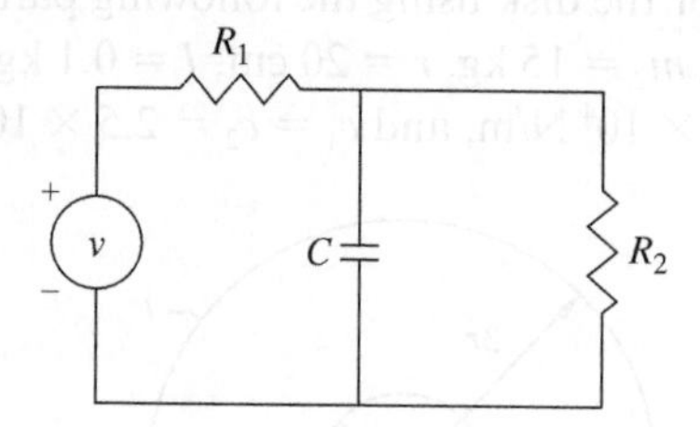

Figure P8.22

8.23 A mass, spring, and viscous damper system has parameters of $m = 1$ kg, $k = 10{,}000 \frac{\text{N}}{\text{m}}$, and $c = 200 \frac{\text{N}\cdot\text{s}}{\text{m}}$. The system is subject to a force input. (a) Determine the system's low-frequency asymptote. (b) Determine the system's high-frequency asymptote. (c) Determine the system's corner frequency. (d) Determine the system's cutoff frequency. (e) Sketch the system's Bode diagram.

8.24 A mass, spring, and viscous damper system has parameters of $m = 1$ kg, $k = 10{,}000 \frac{\text{N}}{\text{m}}$, and $c = 200 \frac{\text{N}\cdot\text{s}}{\text{m}}$. The system is subject to a motion input as in Figure P8.5. (a) Determine the system's low-frequency asymptote. (b) Determine the system's high-frequency asymptote. (c) Determine the system's corner frequency. (d) Determine the system's cutoff frequency. (e) Sketch the system's Bode diagram.

8.25 A single-tank liquid-level problem has a tank with a cross-sectional area of 0.4 m^2 and a pipe of resistance of 2.5 s/m^2. Determine the system's (a) low-frequency asymptote, (b) high-frequency asymptote, (c) corner frequency, and (d) cutoff frequency. (e) Sketch the system's Bode diagram.

8.26 A series LRC circuit has parameters of $L = 0.5$ H, $R = 20\ \Omega$, and $C = 500\ \mu$F. (a) Determine the circuit's low-frequency asymptote. (b) Determine the circuit's high-frequency asymptote. (c) Determine the circuit's corner frequency. (d) Determine the circuit's cutoff frequency. (e) Sketch the circuit's Bode diagram.

8.27 A parallel LRC circuit has parameters of $L = 0.5$ H, $R = 20\ \Omega$, and $C = 500\ \mu$F. (a) Determine the circuit's low-frequency asymptote. (b) Determine the circuit's high-frequency asymptote. (c) Determine the circuit's corner frequency. (d) Determine the circuit's cutoff frequency. (e) Sketch the circuit's Bode diagram.

8.28 The circuit of Figure P8.28 has parameters of $L = 0.5$ H, $R_1 = 10\ \Omega$, $R_2 = 5\ \Omega$, $C_1 = 100\ \mu$F, and

$C_2 = 200$ μF. The output is the voltage drop across the resistor labeled R_2. (a) Determine the circuit's low-frequency asymptote. (b) Determine the circuit's high-frequency asymptote. (c) Determine the circuit's corner frequency. (d) Determine the circuit's cutoff frequency. (e) Sketch the circuit's Bode diagram.

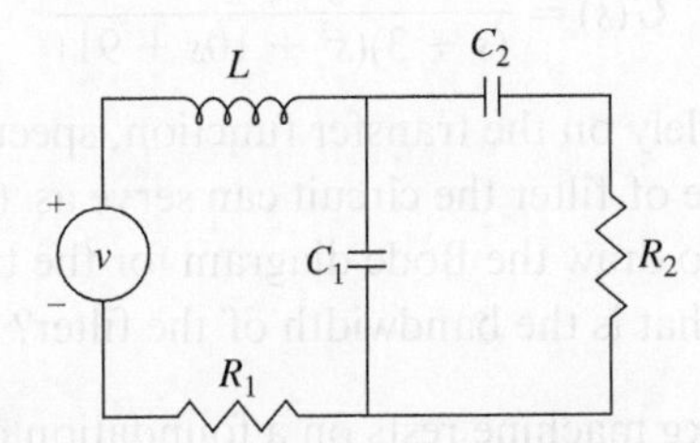

Figure P8.28

8.29 The circuit of Figure P8.29 has parameters of $L_1 = 0.5$ H, $L_2 = 0.1$ H, $R_1 = 1\ \Omega$, $R_2 = 2\ \Omega$, $R_3 = 5\ \Omega$, and $C = 200$ μF. The output is the voltage drop across the resistor labeled R_2. (a) Determine the circuit's low-frequency asymptote. (b) Determine the circuit's high-frequency asymptote. (c) Determine the circuit's corner frequency. (d) Determine the circuit's cutoff frequency. (e) Sketch the circuit's Bode diagram.

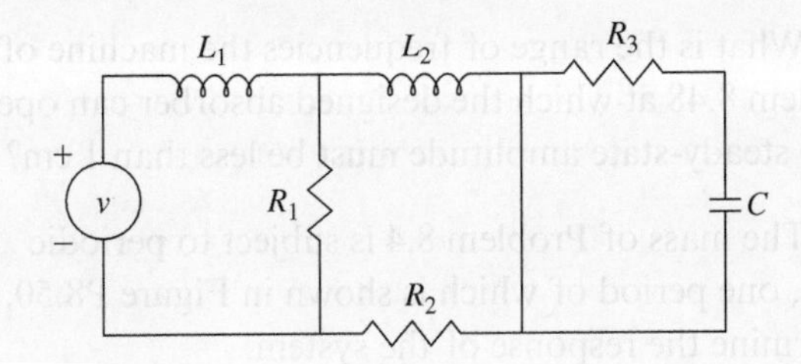

Figure P8.29

8.30 Draw a Bode diagram for a series *LRC* circuit with $\zeta = 1.2$ and $\omega_n = 500$ r/s, and discuss its low-frequency and high-frequency asymptotes.

8.31 Draw a Bode diagram for an undamped mass and spring system of natural frequency 100 r/s.

8.32 Draw a Bode diagram for the vehicle of Problem 7.30 when it traverses a road whose contour is sinusoidal, as illustrated in Figure P8.32. Note that the sinusoidal road contour provides a sinusoidal input to the suspension system in which the frequency is proportional to vehicle speed.

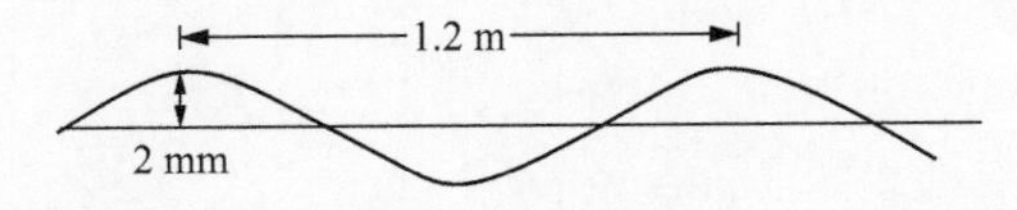

Figure P8.32

8.33 The transfer function for a dynamic system is

$$G(s) = \frac{s+3}{s^2 + 10s + 50}$$

(a) Determine the low-frequency and high-frequency asymptotes for its Bode diagram. (b) Qualitatively discuss the system's Nyquist diagram by determining (i) the points corresponding to $\omega = 0, \pm\infty$, (ii) its axes intercepts, and (iii) the quadrants in which the Nyquist diagram exists. (c) Use MATLAB to draw the Bode plot and Nyquist diagram for the system.

8.34 Repeat Problem 8.33 with a system transfer function of

$$G(s) = \frac{s^2 + 2s + 4}{s^3 + 5s^2 + 6s + 10}$$

8.35 Use superposition and the Bode diagrams of Section 8.4.4 to sketch the Bode diagram for the transfer function

$$G(s) = \frac{s+2}{(s+3)(s^2 + 6s + 45)}$$

8.36 Use superposition and the Bode diagrams of Section 8.4.4 to sketch the Bode diagram for the transfer function

$$G(s) = \frac{(s+2)(s^2 + 14s + 58)}{s(s+1)(s^2 + 2s + 40)}$$

8.37 Use superposition and the Bode diagrams of Section 8.4.4 to sketch the Bode diagram for the transfer function

$$G(s) = \frac{3s(s^2 + 16s + 100)}{(s+4)(s^2 + 8s + 80)}$$

8.38 Draw a Bode diagram for the time delay system whose transfer function is

$$G(s) = \frac{(2s+3)e^{-3s}}{s(s+6)(s+3)(s+2)}$$

Determine the low- and high-frequency asymptotes.

8.39 Draw a Bode diagram for a system whose transfer function is

$$G(s) = \frac{s^2 + 3s + 4}{s(s^2 + 5s + 12)}$$

8.40 Without plotting the Nyquist diagram, determine its (a) coordinates for $\omega = 0$, (b) coordinates as $\omega \to \infty$, (c) coordinates as $\omega \to -\infty$, (d) real axis intercepts, and (e) imaginary axis intercepts for a system with a transfer function for a single-tank liquid-level problem with $A = 0.5\ \text{m}^2$ and $R = 10\ \frac{\text{s}}{\text{m}^2}$.

8.41 Without plotting the Nyquist diagram, determine its (a) coordinates for $\omega = 0$, (b) coordinates as $\omega \to \infty$, (c) coordinates as $\omega \to -\infty$, (d) real axis intercepts, and (e) imaginary axis intercepts for the transfer function of a series *LRC* circuit with $L = 0.5$ H, $R = 20\ \Omega$, and $C = 500\ \mu$F.

8.42 Without plotting the Nyquist diagram, determine its (a) coordinates for $\omega = 0$, (b) coordinates as $\omega \to \infty$, (c) coordinates as $\omega \to -\infty$, (d) real axis intercepts, and (e) imaginary axis intercepts for a system with a transfer function of

$$G(s) = \frac{s + 2}{s^2 + 6s + 45}$$

8.43 Without plotting the Nyquist diagram, determine its (a) coordinates for $\omega = 0$, (b) coordinates as $\omega \to \infty$, (c) coordinates as $\omega \to -\infty$, (d) real axis intercepts, and (e) imaginary axis intercepts for a system with a transfer function of

$$G(s) = \frac{s + 2}{(s + 3)(s^2 + 6s + 45)}$$

8.44 The transfer function for a circuit being proposed as a filter is

$$G(s) = \frac{s^3 + 2s + 6}{s^3 + 10s^2 + 20s + 100}$$

(a) From the transfer function, discuss the type of filter for which this circuit may be suitable. (b) Use MATLAB to draw the Bode diagram for this transfer function and use the Bode diagram to discuss its suitability for use as a filter.

8.45 An electric circuit has a transfer function of

$$G(s) = \frac{s + 2}{s^2 + 10s + 91}$$

(a) Based solely on the transfer function, speculate on what type of filter the circuit can serve as. (b) Use MATLAB to draw the Bode diagram for the transfer function. What is the bandwidth of the filter?

8.46 An electric circuit has a transfer function of

$$G(s) = \frac{s^3 + 25s}{(s + 6)(s^2 + 12s + 85)}$$

(a) Based solely on the transfer function, speculate on what type of filter the circuit can serve as. (b) Use MATLAB to draw the Bode diagram for the transfer function. What is the bandwidth of the filter?

8.47 An electric circuit has a transfer function of

$$G(s) = \frac{s^3 + 2}{(s + 3)(s^2 + 10s + 91)}$$

(a) Based solely on the transfer function, speculate on what type of filter the circuit can serve as. (b) Use MATLAB to draw the Bode diagram for the transfer function. What is the bandwidth of the filter?

8.48 A 500-kg machine rests on a foundation of stiffness 5×10^6 N/m. During operation it is subject to a harmonic force of magnitude 150 N at a frequency of 110 r/s. (a) What is the steady-state response of the machine assuming the system is undamped? (b) What is the minimum stiffness of a dynamic vibration absorber that could be added to the machine to eliminate steady-state vibrations when the machine operates at 100 r/s if the steady-state amplitude of the absorber is to be limited to 1 cm? (c) What is the steady-state amplitude of the machine with the absorber of part (b) in place when a 10-kg component is removed from the machine?

8.49 What is the range of frequencies the machine of Problem 8.48 at which the designed absorber can operate if the steady-state amplitude must be less than 1 cm?

8.50 The mass of Problem 8.4 is subject to periodic input, one period of which is shown in Figure P8.50. Determine the response of the system.

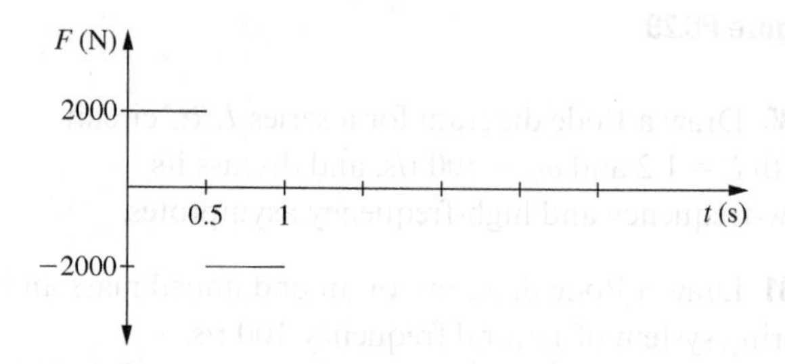

Figure P8.50

9

Feedback Control Systems

Learning Objectives

At the conclusion of this chapter, you should be able to:

- Draw a block diagram model of a system using its transfer function
- Determine closed-loop transfer functions for systems with a feedback control loop
- Use Simulink to model a system from its block diagram model
- Differentiate between P, PD, PI, and PID control
- Specify when a lag, lead, or lead-lag compensator may be used
- Calculate the error and offset for a system when subject to a step input
- Determine the type of a system and the static error constants for that system
- Identify the types of controllers, such as operational amplifier systems, mechanical systems, hydraulic servomotors, and pneumatic circuits
- Determine the changes in properties when a controller is used with a first-order plant
- Determine the changes in properties when a controller is used with a second-order plant
- Determine the closed-loop transfer functions for parallel feedback loops

The previous chapters have focused on the modeling and response of dynamic systems. The transient response of a first-order system is characterized by the system's time constant. The transient response of a second-order system is characterized by the system's natural frequency and damping ratio. The steady-state response due to a harmonic input is characterized by the steady-state amplitude and phase, and it is graphically represented by frequency response curves, Bode diagrams, and Nyquist plots.

It is often necessary to modify the transient response of a system by decreasing its rise time or settling time or by eliminating oscillations of an underdamped system. This can be done by using an actuator, a device that acts on the system to modify its behavior. Actuators themselves are dynamic systems with transfer functions. An actuator can be a mechanical system, an electrical system, a hydraulic system, or a combination of systems.

An actuator, by itself, is generally not sufficient to modify a dynamic system to achieve a desired response. Often actuators are placed in combination with dynamic systems in a feedback control loop. The purpose of this chapter is to introduce feedback control systems and to provide examples of their analysis and how they can be used to achieve the desired response of a dynamic system.

The chapter begins with a survey of block diagrams that provides graphical models of feedback control systems in terms of transfer functions of system components. Common control strategies are introduced and their effects on first- and second-order systems are examined. This chapter is simply an introduction to feedback control systems, which in itself is an exhaustive topic.

9.1 Block Diagrams

A **block diagram** provides a schematic representation of system components and their input and output. A block diagram shows the path of a signal through a system and how the signal is modified. The modifications occur through actuators and functions. Block diagrams are drawn in the transform domain or the *s* domain. The transfer function for a system component $G(s)$ is represented by a block, as illustrated in Figure 9.1. The input signal to the system component is represented by its transfer function $A(s)$ and its output signal has a transfer function $C(s)$. Recall from using Equation (3.9) that the output is written as the product of the transfer function and the input is written as

$$C(s) = G(s)A(s) \tag{9.1}$$

9.1.1 Block Diagram Algebra

Block diagrams show the relationship between the transfer functions of system components. Two systems, represented by block diagrams, are taken to be equivalent if, given the same input, they have the same output. Block diagram algebra is a tool used to simplify the block diagram of a complex system by replacing it with a simpler but equivalent block diagram in which the equivalent system has a transfer function obtained using Equation (9.1). The transfer functions on the lines between the blocks are the transfer functions of the signal as it passes between components.

Figure 9.2 illustrates two components in **series** where the output from the first component is fed as input to the second component. Thus the output from the second series component is

$$C(s) = G_2(s)[G_1(s)A(s)] = [G_1(s)\,G_2(s)]A(s) \tag{9.2}$$

Equation (9.2) shows that the equivalent transfer function for system components in series is the product of the transfer functions of the individual components.

A **branch point** in a block diagram illustrates output from one component used as input for multiple system components. The output from the left-most block of Figure 9.3 is fed as input to both components out of the branch point which have transfer functions $G_1(s)$ and $G_2(s)$.

A **summing point**, also called a summing junction, in a block diagram represents a junction where two or more signals are added to or subtracted from one another. The output from the summing point illustrated in Figure 9.4 is the signal whose transfer function is $G_1(s) - G_2(s)$.

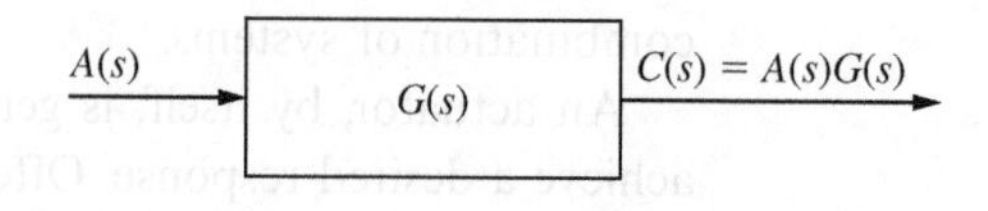

Figure 9.1

The representation of a system component in a block diagram using its transfer function.

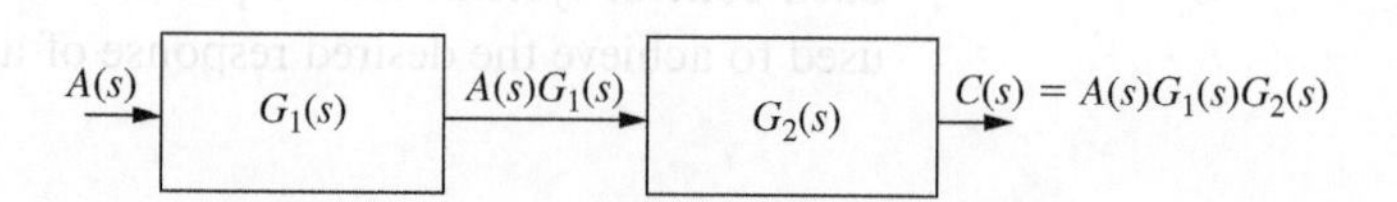

Figure 9.2

Series components.

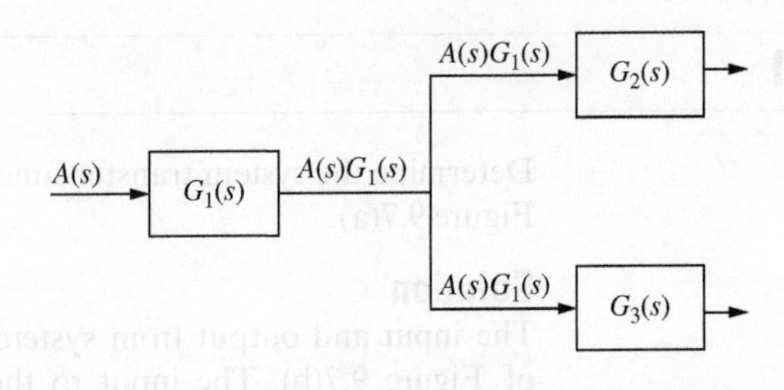

Figure 9.3

A branch point in a block diagram.

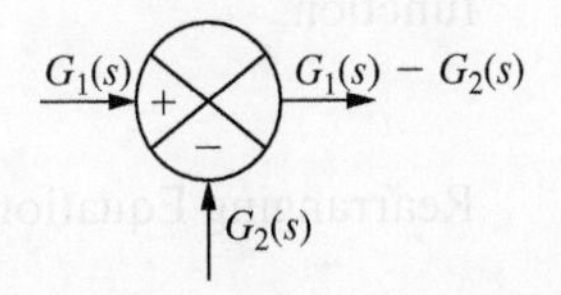

Figure 9.4

A summing point or summing junction in a block diagram adds or subtracts signals. The signals are represented by transfer functions on the block diagram.

Parallel system components are illustrated in Figure 9.5, where both components have the same input fed from the branch point and their outputs are summed at the summing point. If the input to the components is $A(s)$, then the output from the parallel combination is

$$C(s) = G_1(s)A(s) + G_2(s)A(s)$$
$$= [G_1(s) + G_2(s)]A(s) \tag{9.3}$$

Equation (9.3) shows that two parallel system components can be replaced by a single component whose transfer function is the sum of the component transfer functions.

In the **feedback loop** of Figure 9.6, the branch point sends the output to a summing point where it is subtracted from the loop's input. Let $A(s)$ be the input to the feedback loop, let $C(s)$ be its output, and let $G(s)$ be the transfer function of the system component. The input to the component is $A(s) - C(s)$. Thus the output from the component is

$$C(s) = [A(s) - C(s)]G(s) \tag{9.4}$$

Rearranging Equation (9.4) leads to

$$H(s) = \frac{C(s)}{A(s)} = \frac{G(s)}{1 + G(s)} \tag{9.5}$$

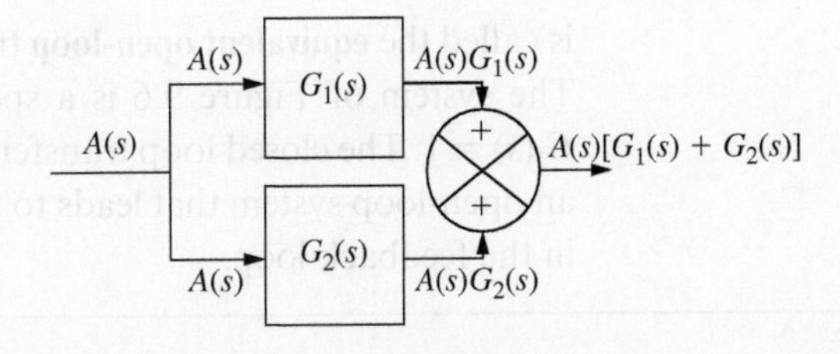

Figure 9.5

Parallel components.

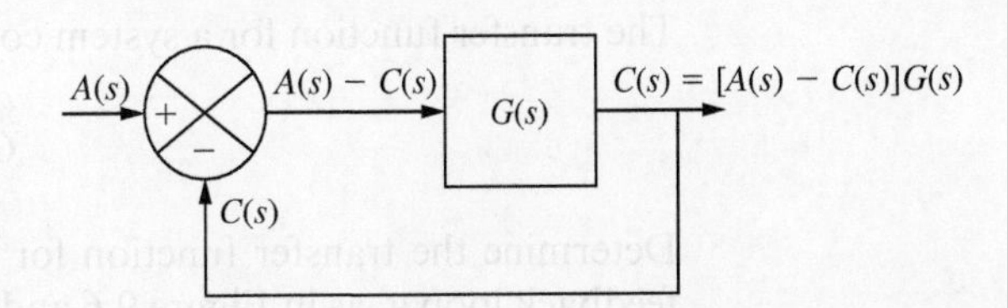

Figure 9.6

A system with closed-loop feedback.

Example 9.1

Determine the system transfer function $H(s) = C(s)/A(s)$ for the feedback system of Figure 9.7(a).

Solution

The input and output from system components is illustrated on the block diagram of Figure 9.7(b). The input to the component whose transfer function is $G_1(s)$ is $A(s) - G_2(s)C(s)$ while its output is $C(s)$. Thus using the definition of a transfer function

$$C(s) = G_1(s)[A(s) - G_2(s)C(s)] \tag{a}$$

Rearranging Equation (a) leads to

$$H(s) = \frac{C(s)}{A(s)} = \frac{G_1(s)}{1 + G_1(s)G_2(s)} \tag{b}$$

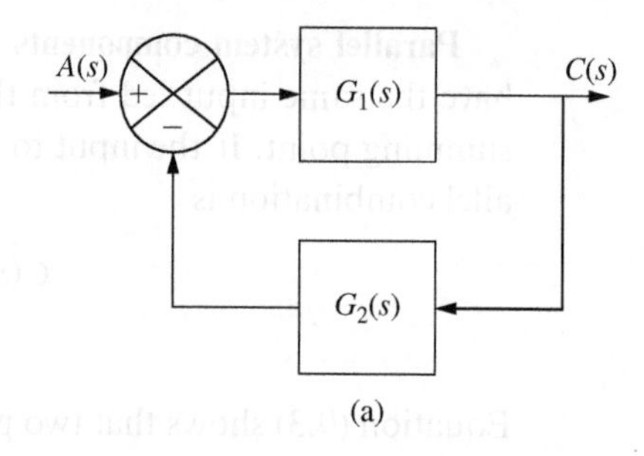

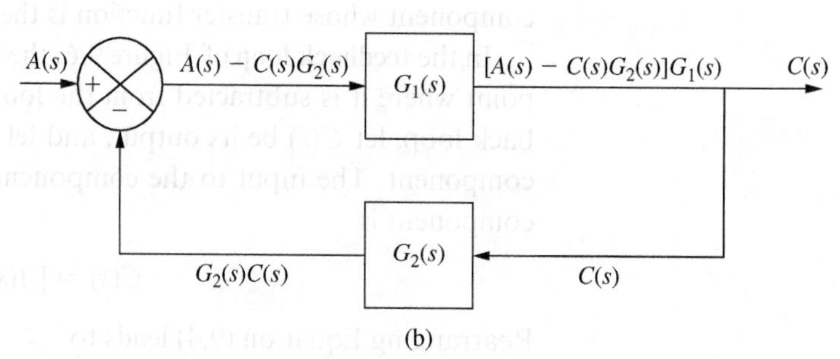

Figure 9.7

(a) The system of Example 9.1; (b) a block diagram illustrating input and output for each system component.

The transfer function

$$H(s) = \frac{G_1(s)}{1 + G_1(s)G_2(s)} \tag{9.6}$$

is called the **equivalent open-loop transfer function** or the closed loop transfer function. The system of Figure 9.6 is a special case of the system of Figure 9.7(a) in which $G_2(s) = 1$. The closed loop transfer function is the transfer function of a system used in an open-loop system that leads to the same system response as the system when placed in the feedback loop.

Example 9.2

The transfer function for a system component is

$$G(s) = \frac{1}{s + 2} \tag{a}$$

Determine the transfer function for the system when this component is placed in a feedback loop **a.** as in Figure 9.6 and **b.** as in Figure 9.7(a) with $G_2(s) = 0.5$.

Solution

a. The transfer function of the system when the component is placed in a feedback loop as in Figure 9.6 is obtained by applying Equation (9.5):

$$\frac{C(s)}{A(s)} = \frac{\frac{1}{s+2}}{1 + \frac{1}{s+2}} \tag{b}$$

Using a common denominator for the denominator in Equation (a) yields

$$\frac{C(s)}{A(s)} = \frac{\frac{1}{s+2}}{\frac{s+2+1}{s+2}} \tag{c}$$

Multiplying the right-hand side of Equation (b) by $\frac{s+2}{s+2}$ leads to

$$\frac{C(s)}{A(s)} = \frac{1}{s+3} \tag{d}$$

b. The transfer function of the system when the component is placed in a feedback loop as in Figure 9.7(a) is obtained by applying Equation (9.6):

$$\frac{C(s)}{A(s)} = \frac{\frac{1}{s+2}}{1 + \frac{0.5}{s+2}} = \frac{1}{s+2.5} \tag{e}$$

Table 9.1 Block Diagram Equivalencies

Name	Block Diagram	Equivalent Block Diagram
Series components	$A(s)$ → G_1 → G_2 → $C(s) = G_2(s)[G_1(s)A(s)]$	$A(s)$ → G_1G_2 → $C(s) = G_1(s)G_2(s)A(s)$
Parallel components	$A(s)$; G_1 (+), G_2 (+) → $C(s) = (G_1 + G_2)A(s)$	$A(s)$ → $G_1 + G_2$ → $C(s) = (G_1 + G_2)A(s)$
Feedback loop	$A(s)$ (+), (−) → G_1 → $C(s)$; G_2	$A(s)$ → $\frac{G_1}{1 + G_1G_2}$ → $C(s)$
Relocation of branch point 1	$A(s)$ → $G(s)$ → $A(s)G(s)$; $A(s)G(s)$	$A(s)$ → $G(s)$ → $A(s)G(s)$; $G(s)$ → $A(s)G(s)$
Relocation of branch point 2	$A(s)$ → $G(s)$ → $A(s)G(s)$; $A(s)$	$A(s)$ → G → $A(s)G(s)$; $\frac{1}{G}$ → $A(s)$
Summing junction	$A(s)$ → G → (+), (−) $B(s)$ → $C(s) = A(s)G(s) - B(s)$	$A(s)$ (+), (−) → G → $C(s) = A(s)G(s) - B(s)$; $\frac{1}{G}$ ← $B(s)$

Block diagram algebra is used to show that a block diagram in the right column of Table 9.1 (page 539) is equivalent to the corresponding block diagram in the middle column. These equivalencies can be used to reduce complex block diagrams and eventually obtain an equivalent transfer function.

Example 9.3

Use block diagram algebra to determine an equivalent open-loop transfer function for the system of Figure 9.8.

Solution

The block diagram of Figure 9.8 is systematically reduced using the block diagram algebra of Table 9.1. The transfer functions G_5 and G_6 are in parallel, and the combination is replaced by a single transfer function $G_5 + G_6$. The transfer functions G_1 and G_2 are in series, and the combination is replaced by a single transfer function G_1G_2. The summing junction for the feedback loop including G_4 is moved before the summing junction for the feedback loop including G_1 and G_2. These simplifications are reflected in Figure 9.9(a). The unity feedback loop involving G_1 and G_2 is in series with G_3. The + sign on the summing junction means that the open loop transfer function for the loop involving G_1, G_2, and G_3 is $\frac{G_1G_2G_3}{1 - G_1G_2}$. This is reflected in Figure 9.9(b), which also includes the elimination of this feedback loop. The resulting transfer function from elimination of the upper feedback loop is

$$\hat{G}(s) = \frac{\dfrac{G_1G_2G_3}{1 - G_1G_2}}{1 + \left(\dfrac{G_4}{G_1G_2}\right)\left(\dfrac{G_1G_2G_3}{1 - G_1G_2}\right)}$$

$$= \frac{G_1G_2G_3}{1 - G_1G_2 + G_3G_4} \quad \text{(a)}$$

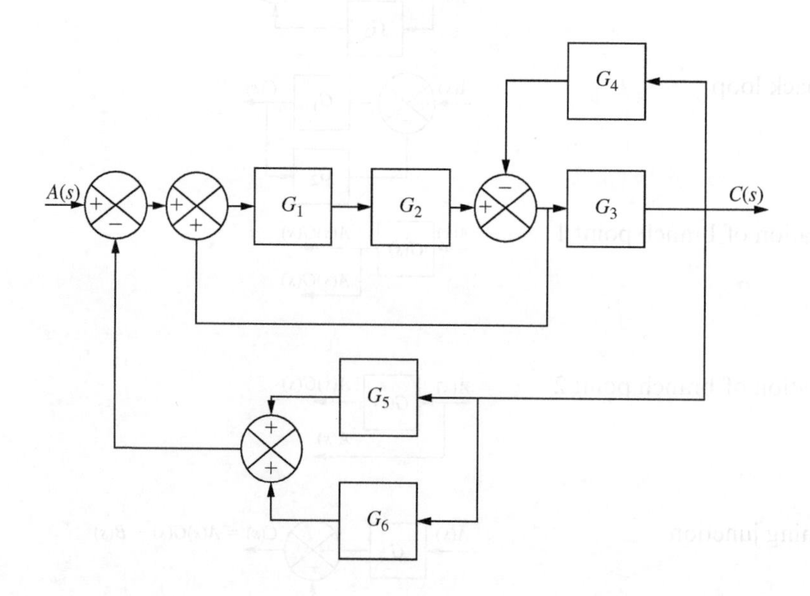

Figure 9.8

The block diagram of the system of Example 9.3.

which is reflected in Figure 9.9(c). The last feedback loop is replaced by noting that the closed-loop transfer function is

$$H(s) = \frac{\hat{G}}{1 + (G_5 + G_6)\hat{G}}$$

$$= \frac{G_1G_2G_3}{1 - G_1G_2 + G_3G_4 + (G_5 + G_6)G_1G_2G_3} \quad \text{(b)}$$

The transfer function given in Equation (b) is the transfer function for the simple block diagram of Figure 9.9(d). It is the equivalent transfer function obtained using the feedback loops. It is called the closed-loop transfer function.

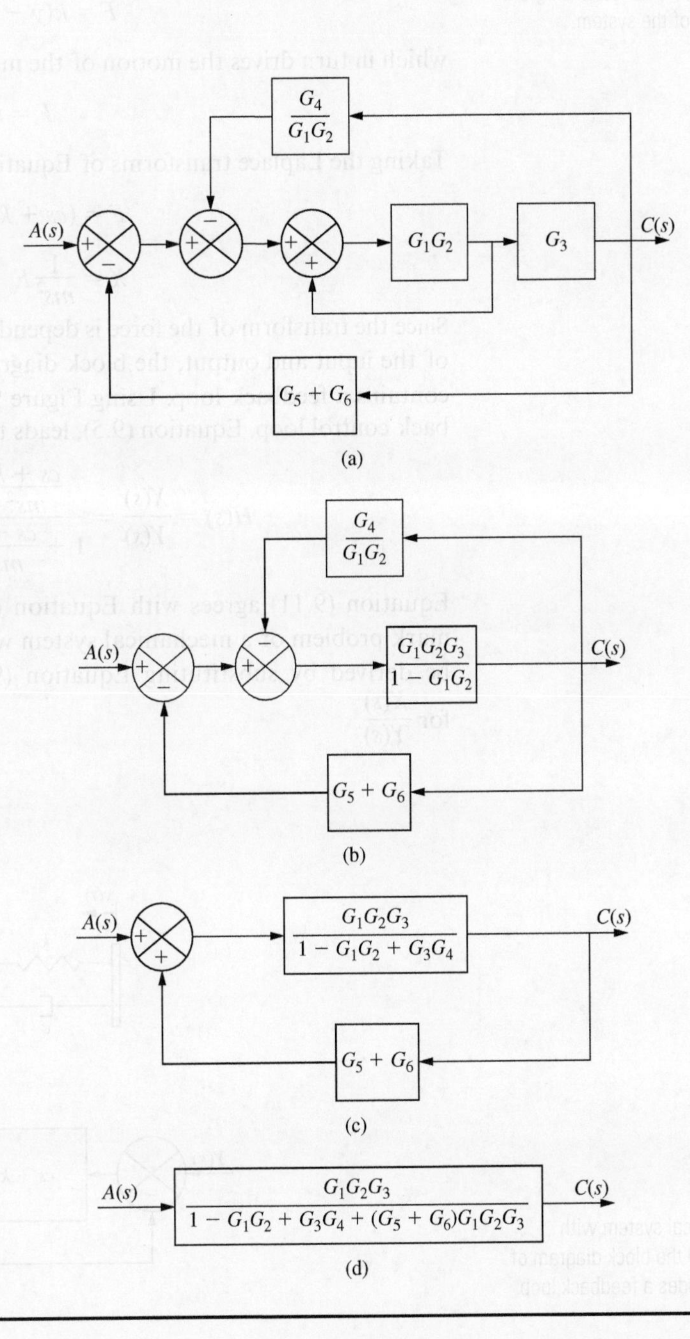

Figure 9.9

The steps in the reduction in the block diagram of Figure 9.8 to an equivalent open-loop transfer function.

9.1.2 Block Diagram Modeling of Dynamic Systems

Block diagrams can be drawn to illustrate the relationship, in the transform domain, between the input and output of a dynamic system. Consider the mass, spring, and viscous damper system of Figure 9.10(a). The transfer function relationship between the input and output is

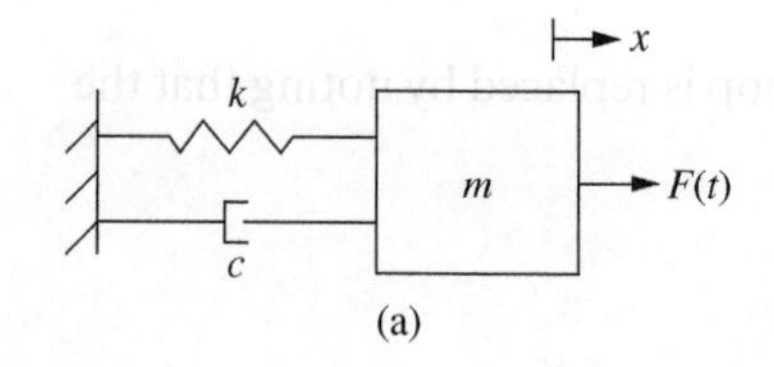

(a)

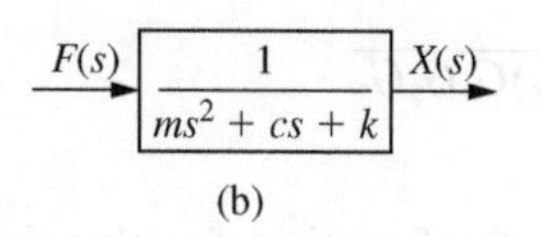

(b)

Figure 9.10

(a) A mass, spring, and viscous damper system; (b) the block diagram representation of the system.

$$X(s) = G(s)F(s) = \frac{1}{ms^2 + cs + k}F(s) \tag{9.7}$$

The block diagram representation of this system is illustrated in Figure 9.10(b).

A prescribed motion provides input to the base of the mass, spring, and viscous damper system of Figure 9.11(a). The force developed in the spring and viscous damper mechanism is

$$F = k(y - x) + c(\dot{y} - \dot{x}) \tag{9.8}$$

which in turn drives the motion of the mass

$$F = m\ddot{x} \tag{9.9}$$

Taking the Laplace transforms of Equations (9.8) and (9.9) leads to

$$F = (cs + k)(Y - X) \tag{9.10a}$$

$$X = \frac{1}{ms^2}F \tag{9.10b}$$

Since the transform of the force is dependent on the difference between the transforms of the input and output, the block diagram for the system, shown in Figure 9.11(b), contains a feedback loop. Using Figure 9.11(b) and the transfer function for a feedback control loop, Equation (9.5), leads to

$$H(s) = \frac{X(s)}{Y(s)} = \frac{\dfrac{cs + k}{ms^2}}{1 + \dfrac{cs + k}{ms^2}} = \frac{cs + k}{ms^2 + cs + K} \tag{9.11}$$

Equation (9.11) agrees with Equation (7.9), the transfer function for the benchmark problem of a mechanical system with motion input. Equation (9.11) can also be derived by substituting Equation (9.10a) into Equation (9.10b) and solving for $\frac{X(s)}{Y(s)}$.

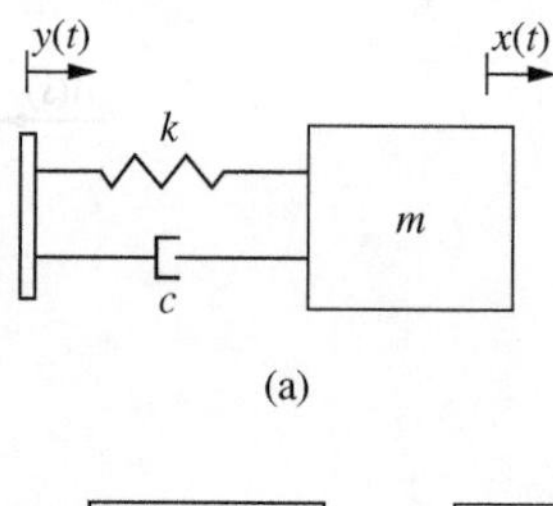

(a)

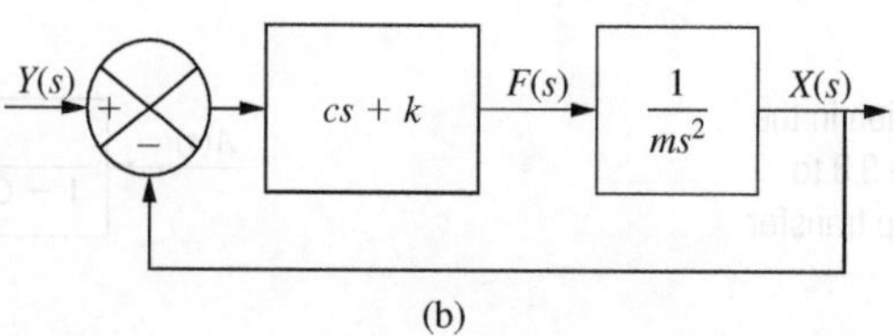

(b)

Figure 9.11

(a) The mechanical system with motion input; (b) the block diagram of the system includes a feedback loop.

Example 9.4

Draw the block diagram for an armature-controlled dc motor as illustrated in Figure 5.41. The system input is the voltage in the armature circuit and the output is the angular velocity of the motor. Use the block diagram to determine the system transfer function.

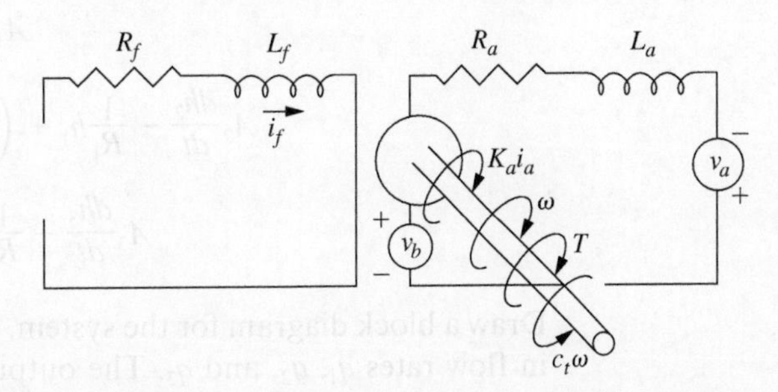

Figure 5.41

The dc servomotor of Example 5.23 and Example 9.4. (Repeated)

Solution

The differential equations for the current in the armature circuit and the angular velocity of the motor for an armature-controlled motor are Equations (a) and (b) of Example 5.23, as follows:

$$L_a \frac{di_a}{dt} + R_a i_a + K_b \omega = v_a \tag{a}$$

$$J\frac{d\omega}{dt} + c_t \omega - K_1 i_a = 0 \tag{b}$$

Taking Laplace transforms of Equations (a) and (b) leads to

$$I_a(s) = \frac{V_a(s) - K_b \Omega(s)}{L_a s + R_a} \tag{c}$$

$$\Omega(s) = \frac{K_1 I_a(s)}{Js + c_t} \tag{d}$$

Equation (d) shows that the input for the transfer function for the angular velocity is the transform of the armature current, while Equation (c) shows that the input for the transfer function of the armature current is $V_a(s) - K_b \Omega(s)$. This implies that the angular velocity output is multiplied by K_b in a feedback loop and then brought to a summing junction, where it is subtracted from the input voltage and the difference used as input for the armature current. The appropriate block diagram for the system is illustrated in Figure 9.12. The application of Equation (9.6) leads to the transfer function

$$\frac{\Omega(s)}{V_a(s)} = \frac{\dfrac{K_1}{(L_a s + R_a)(Js + c_t)}}{1 + \dfrac{K_1 K_b}{(L_a s + R_a)(Js + c_t)}} = \frac{K_1}{(L_a s + R_a)(Js + c_t) + K_1 K_b} \tag{e}$$

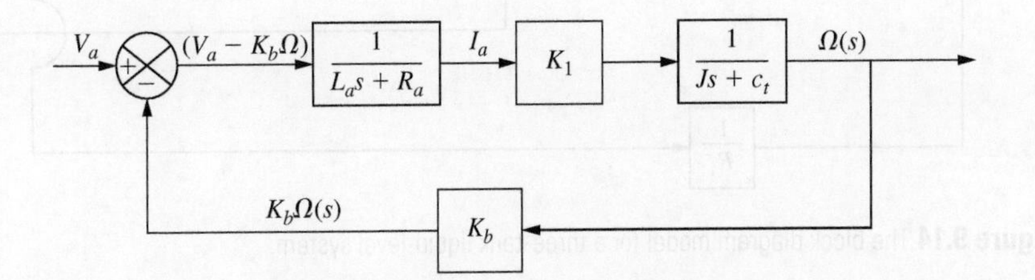

Figure 9.12

The block diagram for an armature-controlled motor.

Example 9.5

The differential equations for the flow rate perturbations in the three-tank liquid-level problem illustrated in Figure 9.13 due to perturbations of inlet flow rates are

$$A_1\frac{dh_1}{dt} + \frac{1}{R_1}h_1 - \frac{1}{R_1}h_2 = q_1(t) \tag{a}$$

$$A_2\frac{dh_2}{dt} - \frac{1}{R_1}h_1 + \left(\frac{1}{R_1} + \frac{1}{R_2}\right)h_2 - \frac{1}{R_2}h_3 = q_2(t) \tag{b}$$

$$A_3\frac{dh_3}{dt} - \frac{1}{R_2}h_2 + \left(\frac{1}{R_2} + \frac{1}{R_3}\right)h_3 = q_3(t) \tag{c}$$

Draw a block diagram for the system. The inputs to the system are the perturbations in flow rates q_1, q_2, and q_3. The outputs are the perturbations in liquid level h_1, h_2, and h_3.

Solution

Taking the Laplace transforms of Equations (a), (b), and (c) and rearranging leads to

$$H_1(s) = \frac{1}{A_1s + \frac{1}{R_1}}\left[Q_1 + \frac{1}{R_1}H_2(s)\right] \tag{d}$$

$$H_2(s) = \frac{1}{A_2s + \frac{1}{R_1} + \frac{1}{R_2}}\left[Q_2 + \frac{1}{R_1}H_1(s) + \frac{1}{R_2}H_3(s)\right] \tag{e}$$

$$H_3(s) = \frac{1}{A_3s + \frac{1}{R_2} + \frac{1}{R_3}}\left[Q_3 + \frac{1}{R_2}H_2(s)\right] \tag{f}$$

Equations (d)–(f) are used to develop the block diagram of Figure 9.14.

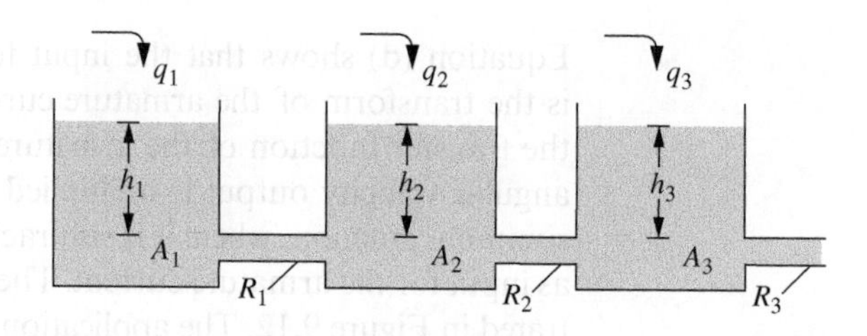

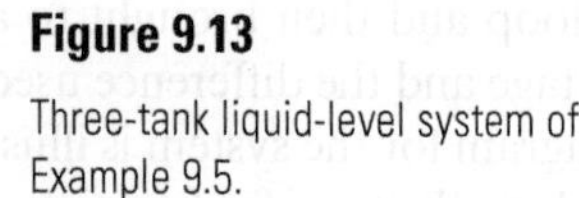

Figure 9.13
Three-tank liquid-level system of Example 9.5.

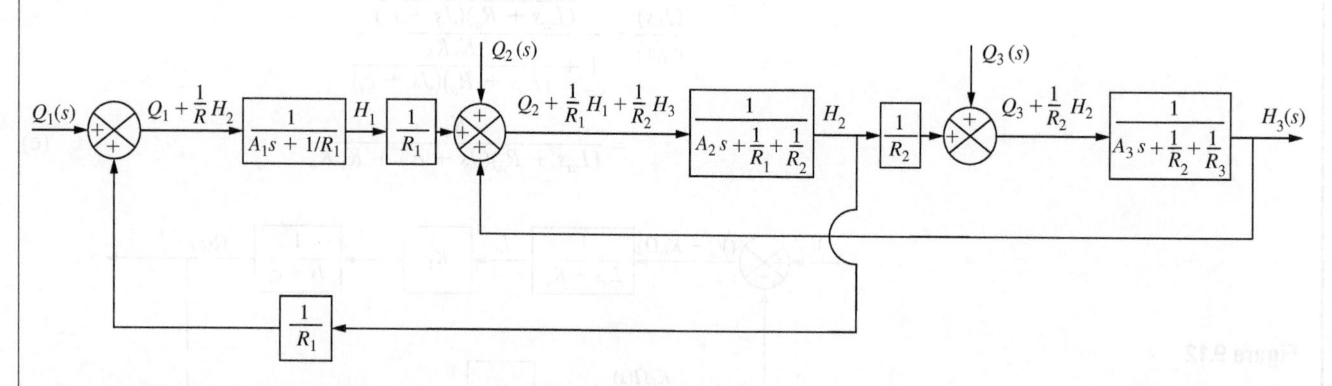

Figure 9.14 The block diagram model for a three-tank liquid-level system.

Example 9.6

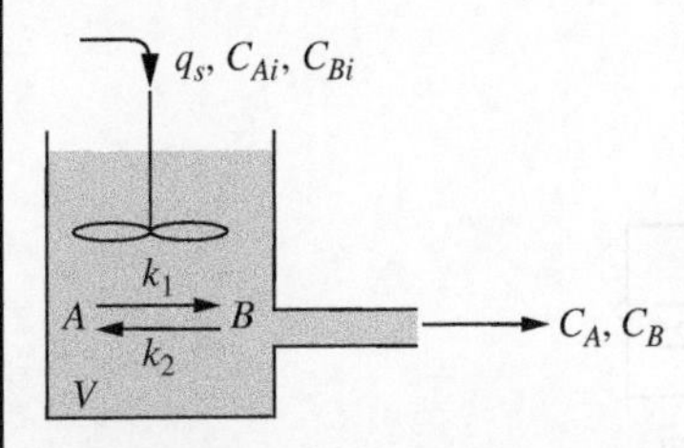

Figure 9.15

CSTR with two-way reaction.

The mathematical model for the concentrations of the reactants in a two-way reaction that occurs in a CSTR is developed in Example 6.32 and illustrated in Figure 9.15. The input to the system is the perturbation in flow rate q and the outputs are the concentrations of the reactants, C_A and C_B. The differential equations given in Equations (c) and (d) of Example 6.15 are

$$V\frac{dC_A}{dt} + (q_s + k_1V)C_A - k_2VC_B = qC_{Ai} \tag{a}$$

$$V\frac{dC_B}{dt} - k_1VC_A + (q_s + k_2V)C_B = qC_{Bi} \tag{b}$$

Draw a block diagram for this system, where q is the system input.

Solution

Taking the Laplace transforms of Equations (a) and (b) leads to

$$C_A(s) = \frac{C_{Ai}Q(s)}{Vs + q_s + k_1V} + \frac{k_2VC_B(s)}{Vs + q_s + k_1V} \tag{c}$$

$$C_B(s) = \frac{C_{Bi}Q(s)}{Vs + q_s + k_2V} + \frac{k_1VC_A(s)}{Vs + q_s + k_2V} \tag{d}$$

An appropriate block diagram is illustrated in Figure 9.16.

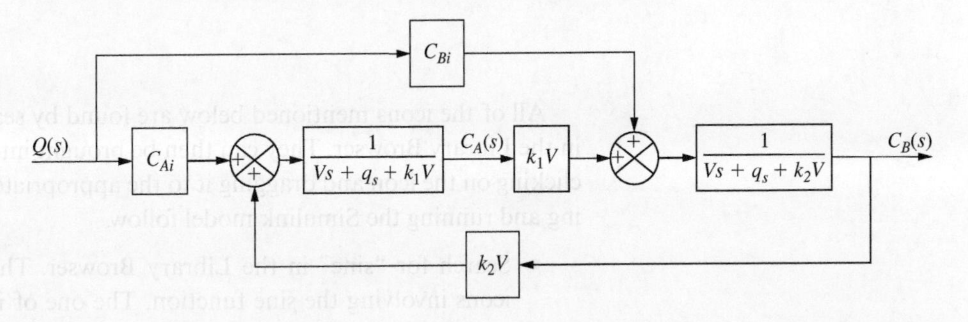

Figure 9.16

The block diagram for the CSTR system of Example 9.6.

9.2 Block Diagram Modeling Using Simulink

Simulink is a dynamic simulation software package developed in conjunction with MATLAB. Simulink has many capabilities, including modeling nonlinear and discrete time systems. The focus in this section is on building models and running dynamic simulations using Simulink.

Simulink models are developed using block diagrams that may be developed using either the transfer function method or the state-space method. Dynamic simulations are run using Simulink, which employs numerical integration software from MATLAB. The numerical methods used are discussed in Chapter 12.

The user is referred to the Simulink help function to access Simulink from the MATLAB command window, access the Simulink Library Browser, and open a model. The model is built by clicking on and dragging the appropriate icons from the Library Browser into the model workspace.

As an example, consider the mass, spring, and viscous damper system of Figure 9.10(a) with $m = 1$ kg, $k = 25$ N/m, and $c = 6\ \frac{\text{N}\cdot s}{\text{m}}$, whose block diagram is illustrated in Figure 9.10(b). The Simulink model is developed by reproducing the block diagram in the model workspace. The following steps are illustrated sequentially in Figure 9.17.

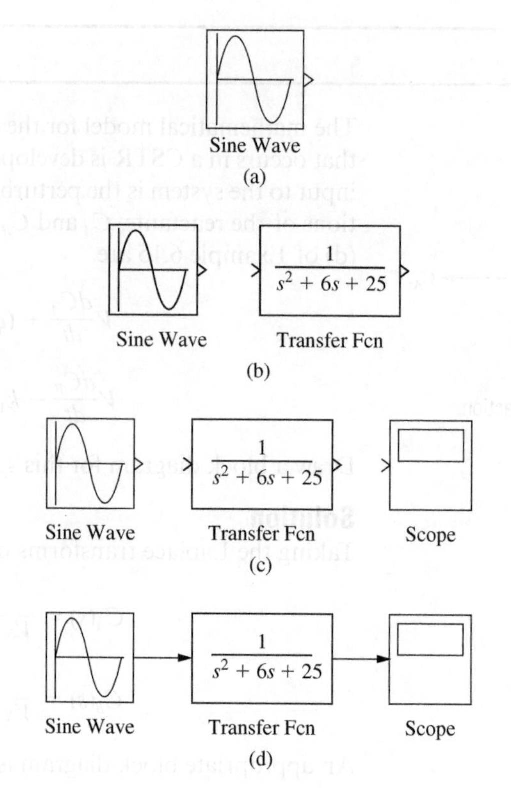

Figure 9.17

(a) Sine Wave; (b) Sine Wave and system Transfer Fcn; (c) Scope added; (d) icons connected.

All of the icons mentioned below are found by searching for the appropriate term in the Library Browser. They can then be brought into the block model workspace by clicking on the icon and dragging it to the appropriate position. The steps for generating and running the Simulink model follow.

- Search for "sine" in the Library Browser. The browser will bring up several icons involving the sine function. The one of interest is the "Sine Wave Function." Click on this and drag it into the block model. The sine wave provides a function of the form sin t to the model. If it is desired to change the amplitude, the frequency, or the phase, click on the sine wave icon to bring up a menu that allows these inputs. Also, the time source must be changed to "Use simulation time."
- Search for "transfer function" in the Library Browser. Scroll down to reach "Transfer Fcn" (with initial states). Click this icon and drag it into the block model, placing it to the right side of the "Sine Wave" icon.
- Once in the model workspace, click on the transfer function icon to bring up the block form for Block Parameters. A transfer function is defined in Simulink as it is in MATLAB; enter the coefficients of the polynomials in the numerator and denominator in a vector. For this system, leave the 1 for the numerator and enter [1 6 25] inside brackets for the denominator. The block model screen should now appear as in Figure 9.17(b).
- Find, click, and drag into the block model workspace the Scope icon. Place it to the right of the Transfer Fcn icon as in Figure 9.17(c). The scope allows the user to view the results of a simulation.
- Connect the ports on the icons. Connect the output port from the Sine Wave icon to the input port of the Transfer Fcn icon. Connect the output port from

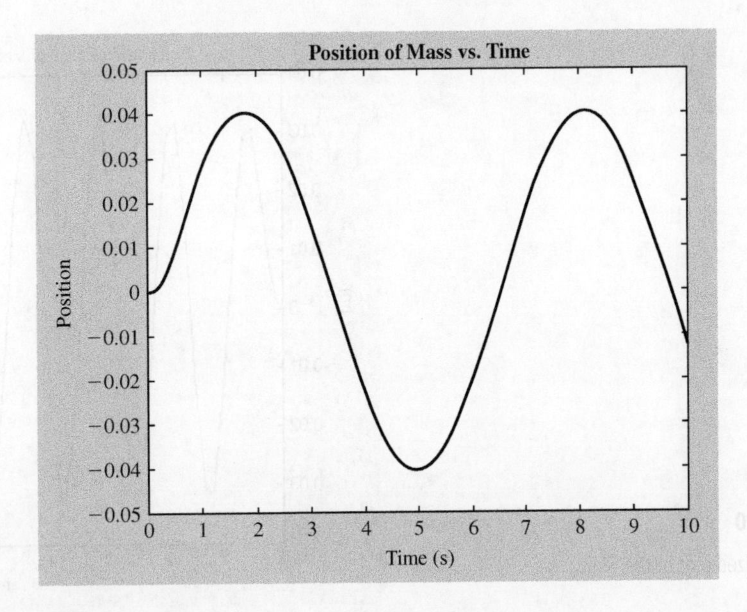

Figure 9.18

Simulink simulation of response of mass-spring-viscous damper system.

the Transfer Fcn icon to the input port of the Scope. This completes this basic model as shown in Figure 9.17(d).

- Click on Run from the Simulation menu. The output can be viewed by clicking on the Scope icon. The results are shown in Figure 9.18.

The model is complete. A simulation is run by clicking on Simulation and then Start from the model workspace toolbar. The simulation results are viewed by clicking on the Scope icon. A box titled Scope appears, showing a plot of the displacement as a function of time.

Figure 9.19 shows a Simulink model with two revisions. The Sine Wave generator has been modified by clicking on its icon and changing the frequency in the Block Parameters window to 5 r/s. A sink is obtained from the source menu in the Library Browser. This sink takes the numerical results obtained from tuning the simulation and sends them to the MATLAB command window to a vector named out. The sink is connected at a branch point after the transfer function in the block diagram. This is obtained from the source menu from the Library Browser. The "To workspace" icon is clicked and dragged to the model workspace. In the Block Parameters box that appears when the icon is clicked, the Variable Name was changed to x and the Save Format was selected as structure with time. When the simulation is run, instead of a graphical representation of the response as with a scope, the numerical values calculated are returned to the MATLAB workspace under a structure named "out." The "with time" means that the corresponding values of time are also sent to the MATLAB workspace. A customized plot is drawn beginning with the MATLAB command plot(out). The result after customization is shown in Figure 9.20.

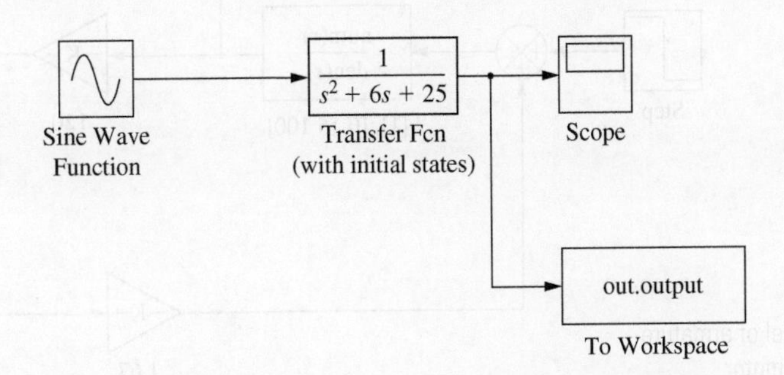

Figure 9.19

Simulink model of mass, spring, and viscous damper system with numerical values of response sent to MATLAB workspace.

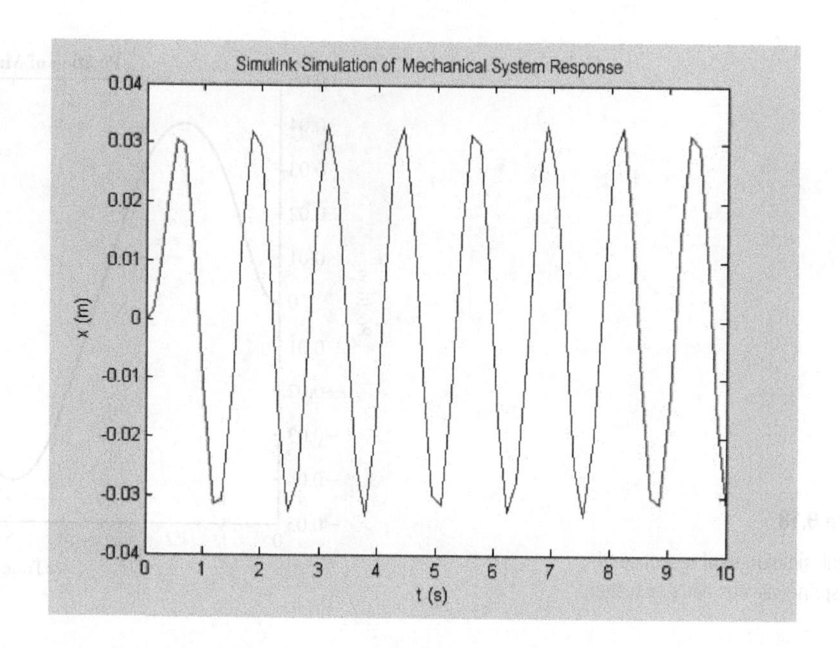

Figure 9.20

The customized plot of the Simulink simulation.

Example 9.7

Develop Simulink models and run simulations for **a.** the armature-controlled dc motor of Example 9.4 and Figure 5.41 with $J = 1.8\ \text{kg·m}^2$, $c_t = 2.15\ \text{N·m·s/r}$, $L_a = 0.15\ \text{H}$, $R_a = 100\ \Omega$, $K_1 = 120\ \text{N·m/A}$, and $K_2 = 1.67\ \text{V·s/r}$ with a unit step input for the armature voltage; **b.** the three-tank liquid-level system of Example 9.5 with $A_1 = A_2 = A_3 = 10\ \text{m}^2$, $R_1 = 5\ \text{s}^2/\text{m}$, $R_2 = 10\ \text{s}^2/\text{m}$, and $R_3 = 5\ \text{s}^2/\text{m}$ when the first tank has a unit step increase in inlet flow rate; and **c.** the CSTR with the two-way reaction of Example 9.6 with $V = 1.5 \times 10^{-3}\ \text{m}^3$, $q_s = 1.5 \times 10^{-6}\ \text{m}^3/\text{s}$, $C_{Ai} = 0.25\ \text{mol/L}$, $C_{Bi} = 0.15\ \text{mol/L}$, $k_1 = 2.0 \times 10^{-3}\ \text{s}^{-1}$, and $k_2 = 1.0 \times 10^{-3}\ \text{s}^{-1}$ when the inlet flow rate has a perturbation of $5 \times 10^{-7}\ \text{m}^3/\text{s}$.

Solution

a. The Simulink model for the armature-controlled dc servomotor is shown in Figure 9.21. The following is noted regarding this model:

- The Simulink model shown in Figure 9.21 has four sinks. There are two scopes where the current and angular velocity as functions of time can be

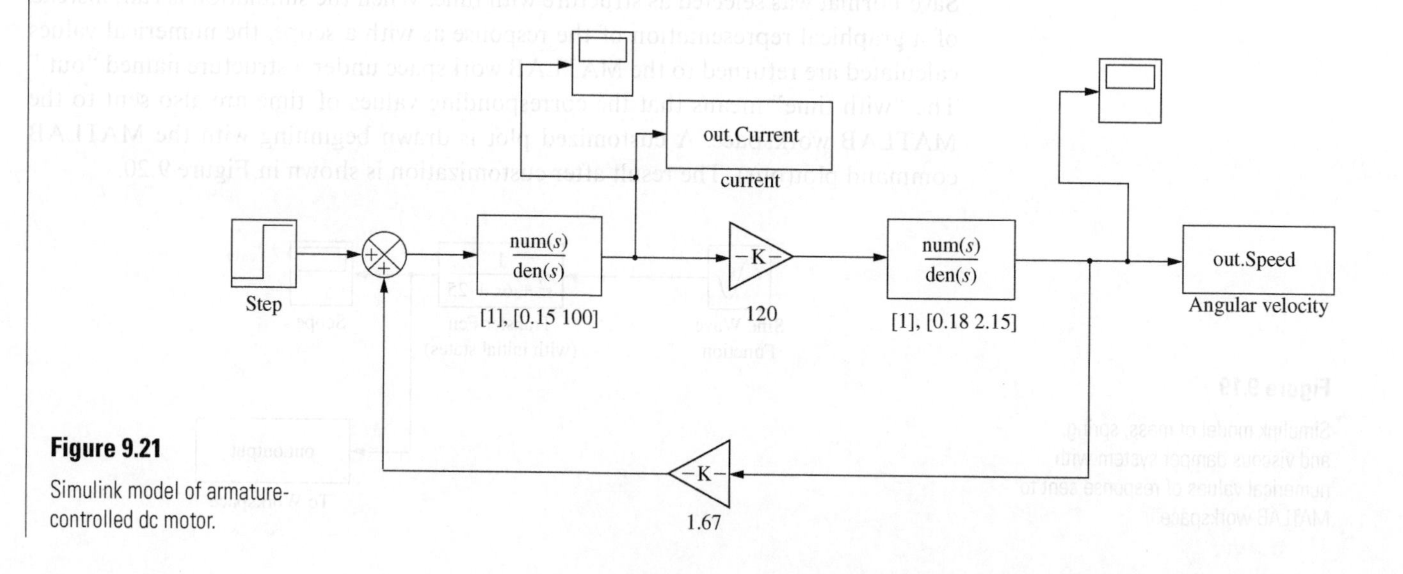

Figure 9.21

Simulink model of armature-controlled dc motor.

viewed by clicking on the appropriate scope. The current and angular velocity are also transferred to the MATLAB workspace through the out.Current sink and the out.Speed sink, respectively. This allows for the data to be plotted and annotated as in Figures 9.22(a) and 9.22(b). To access the data for the current to plot, simply type in the MATLAB workspace i=out.Current. Then type plot(i). The plot can then be customized.

- The block diagram uses the Gain block from the Simulink Library Browser. The gain block simply multiplies the signal in the line by the value of the gain.
- This model includes a gain in the feedback loop. This is done by selecting the gain icon from the Library Browser and dragging it into the Simulink model. The direction of the gain is flipped by selecting CNTL-R. The lines from the angular speed to the gain and from the gain to the summing junction are then drawn.

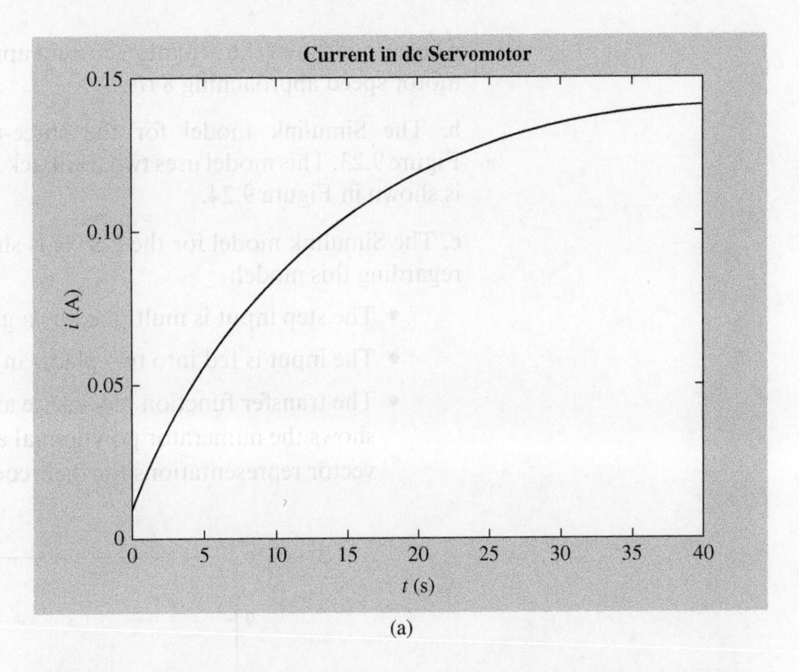

(a)

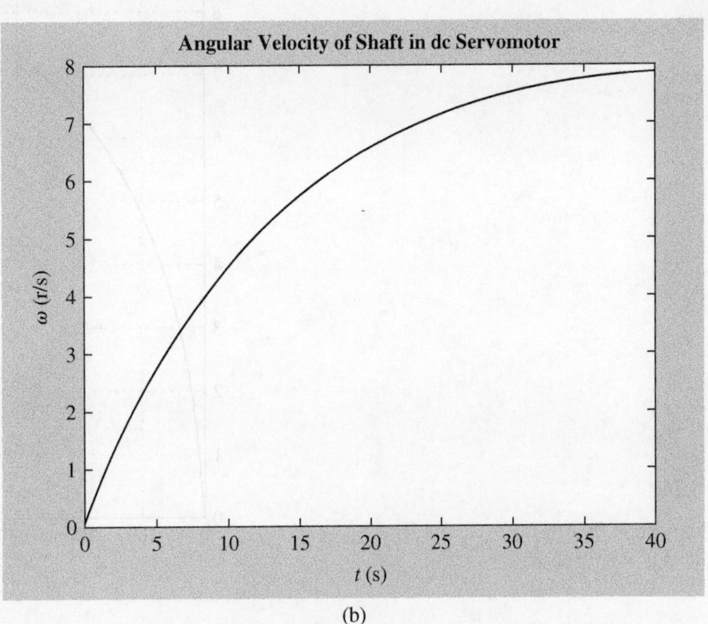

(b)

Figure 9.22

The MATLAB-generated graphs of Simulink results for the armature-controlled dc motor of Example 9.7a; (a) current; (b) angular velocity.

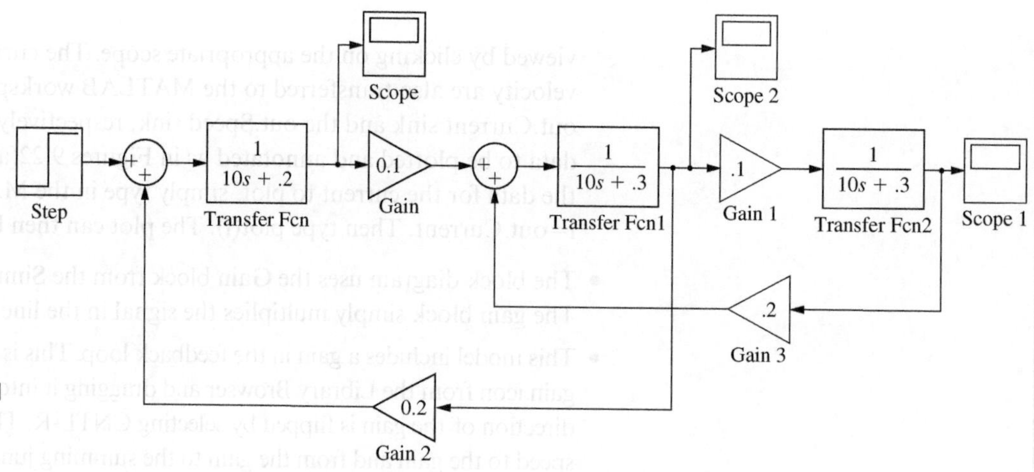

Figure 9.23
Simulink model of three-tank liquid-level system developed using block diagram of Figure 9.13.

The output shows the armature current approaching slightly more than 0.14 A and the motor speed approaching 8 r/s.

b. The Simulink model for the three-tank liquid-level system is illustrated in Figure 9.23. This model uses two feedback loops and three scopes. The system response is shown in Figure 9.24.

c. The Simulink model for the CSTR is shown in Figure 9.25. The following is noted regarding this model:

- The step input is multiplied by a gain equal to the flow rate perturbation.
- The input is fed into two places in the block diagram.
- The transfer function blocks are annotated below the block. The annotation shows the numerator polynomial and the denominator polynomial, using vector representations for their coefficients.

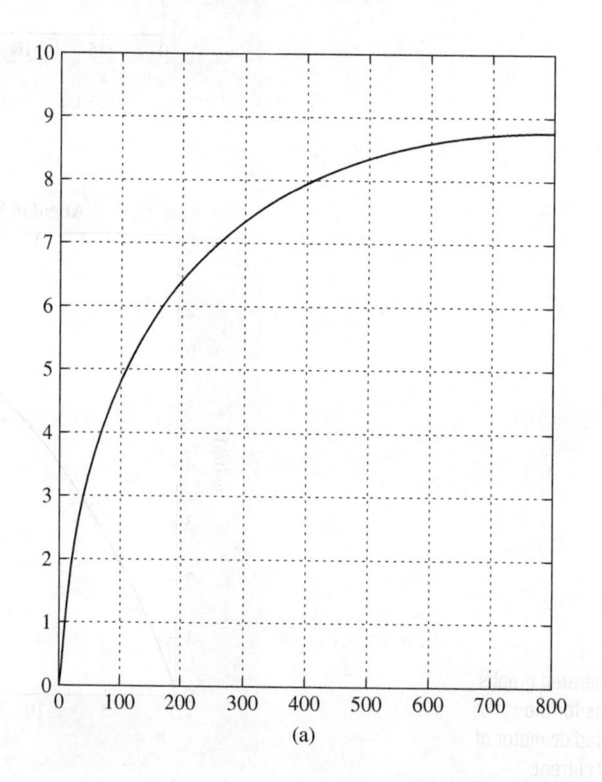

Figure 9.24
(a) $h_1(t)$ from scope; (b) $h_2(t)$ from scope 2; (c) $h_3(t)$ from scope 1.

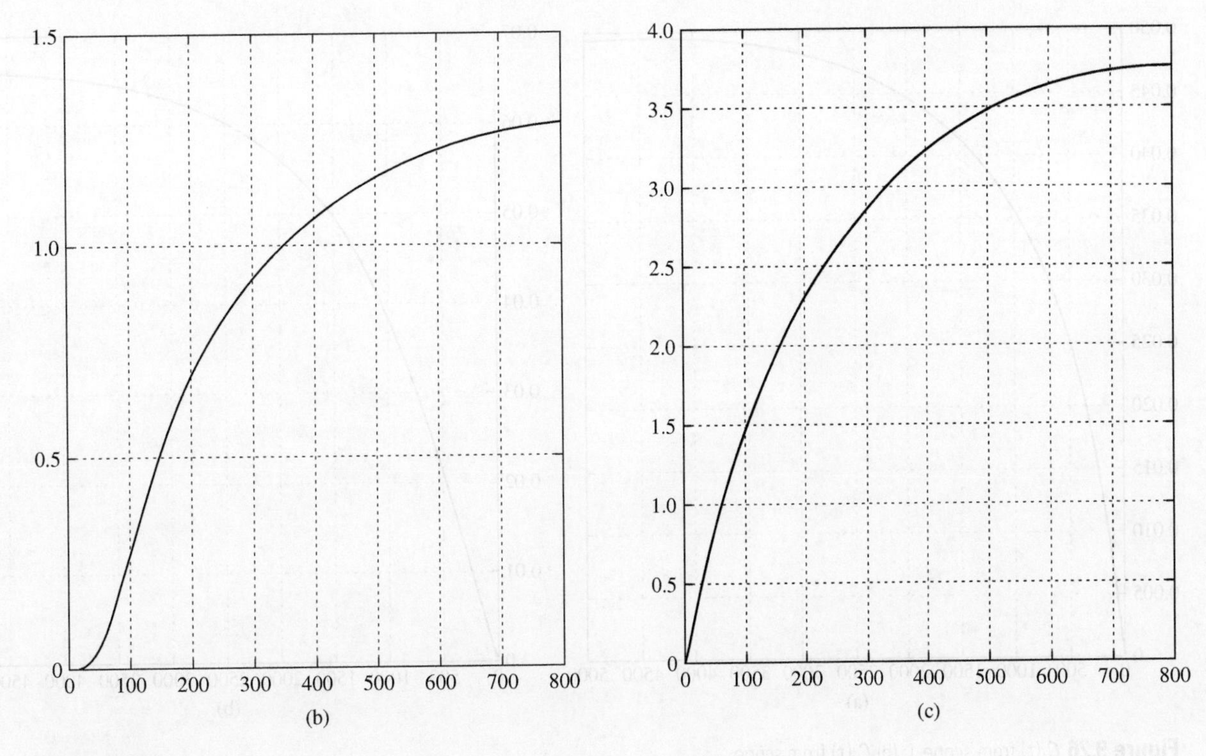

Figure 9.24
(Continued)

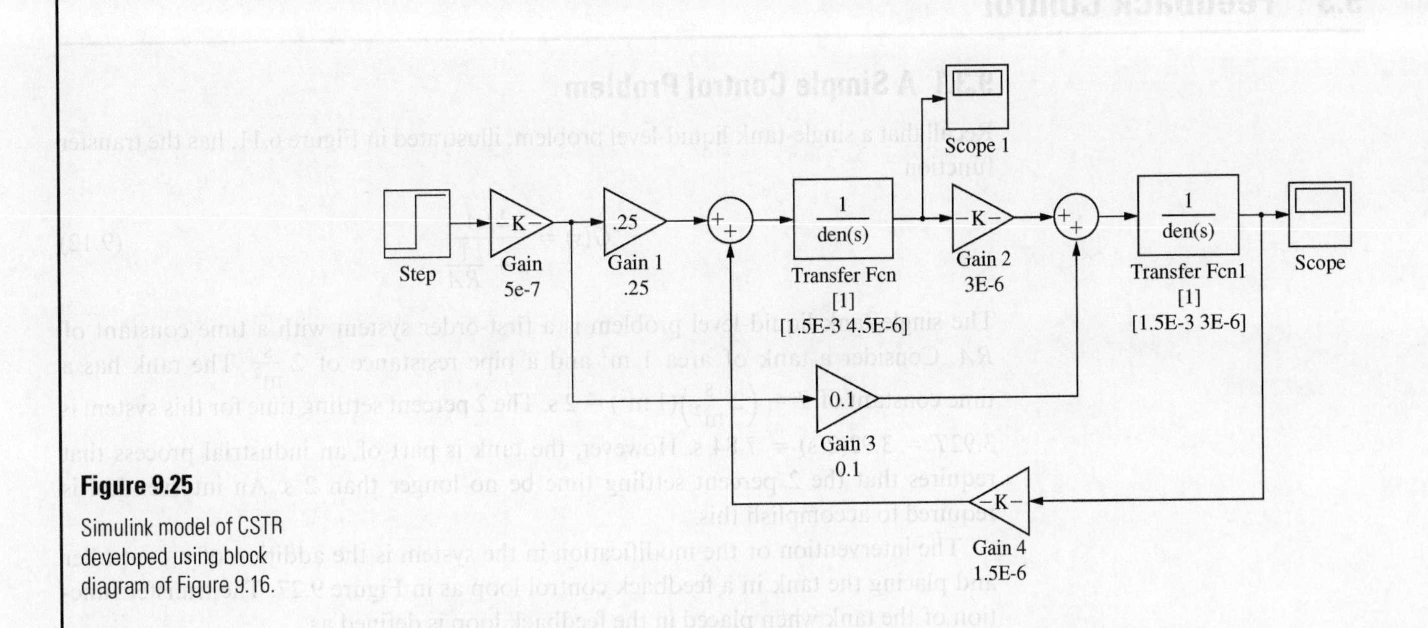

Figure 9.25
Simulink model of CSTR developed using block diagram of Figure 9.16.

The scope traces when the simulation is run are shown in Figure 9.26.

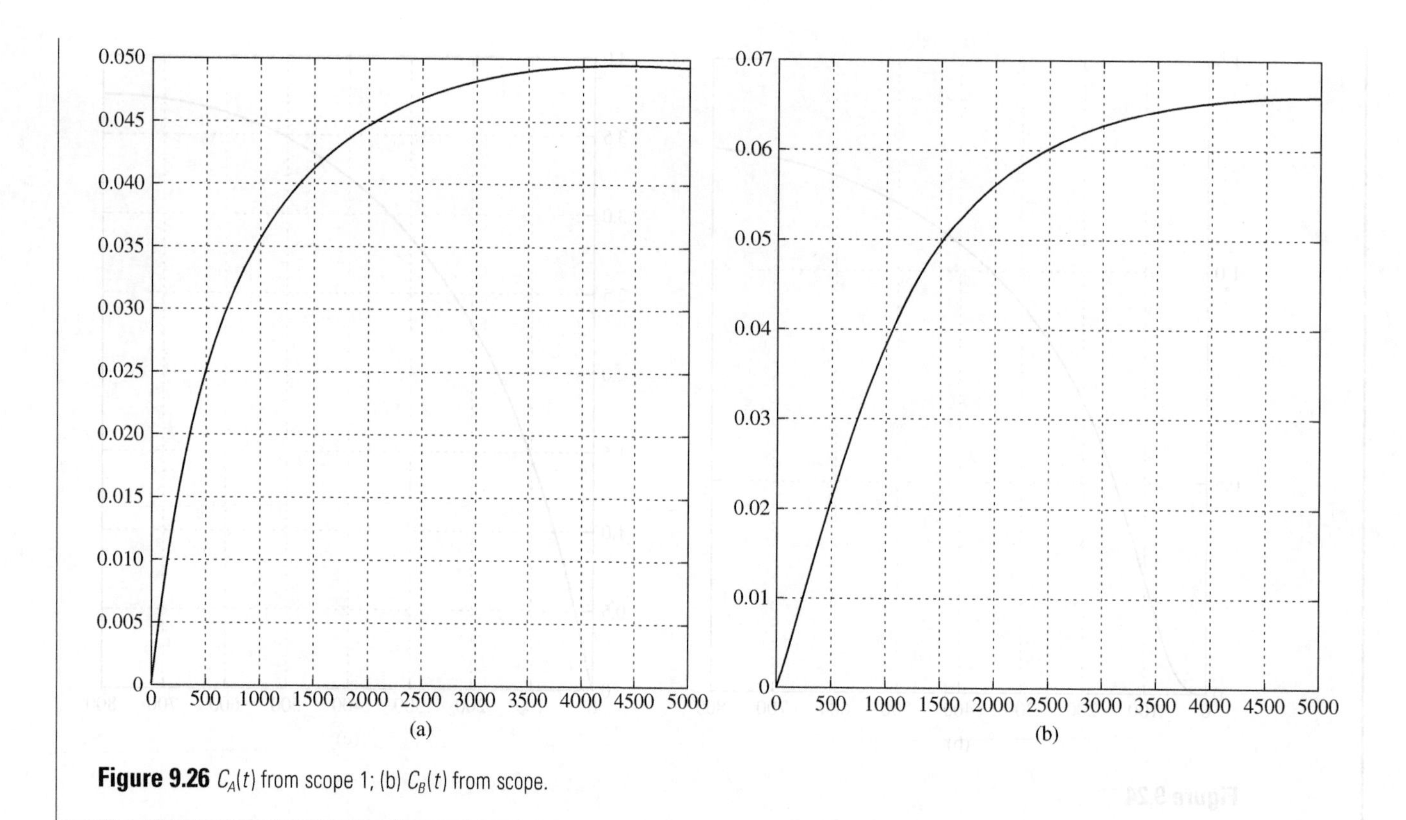

Figure 9.26 $C_A(t)$ from scope 1; (b) $C_B(t)$ from scope.

9.3 Feedback Control

9.3.1 A Simple Control Problem

Recall that a single-tank liquid-level problem, illustrated in Figure 6.11, has the transfer function

$$G(s) = \frac{1/A}{s + \frac{1}{RA}} \tag{9.12}$$

The single-tank liquid-level problem is a first-order system with a time constant of RA. Consider a tank of area 1 m^2 and a pipe resistance of $2\frac{\text{s}}{\text{m}^2}$. The tank has a time constant of $T = \left(2\frac{\text{s}}{\text{m}^2}\right)(1\ \text{m}^2) = 2$ s. The 2 percent settling time for this system is $3.92T = 3.92(2\ \text{s}) = 7.84$ s. However, the tank is part of an industrial process that requires that the 2 percent settling time be no longer than 2 s. An intervention is required to accomplish this.

The intervention or the modification in the system is the addition of a controller and placing the tank in a feedback control loop as in Figure 9.27. The transfer function of the tank when placed in the feedback loop is defined as

$$G(s) = \frac{C(s)}{A(s)} \tag{9.13}$$

where $A(s)$ is the transform of the input and $C(s)$ is the transform of the output. Let $G_c(s)$ be the transfer function for the control strategy, and using the block diagram algebra of Section 9.1,

$$[A(s) - C(s)]\frac{1}{s + 0.5}G_c(s) = C(s) \tag{9.14}$$

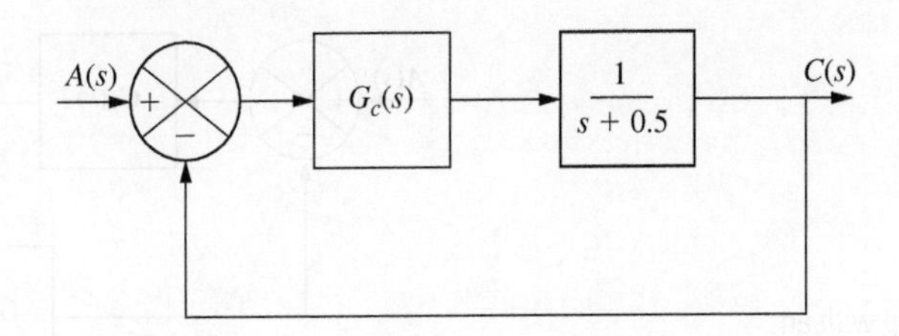

Figure 9.27

A plant with transfer function $\frac{1}{s+0.5}$ is placed in series with a controller in a feedback loop.

Solving Equation (9.14) for $G(s)$ gives

$$G(s) = \frac{\frac{1}{s+0.5}G_c(s)}{1 + \frac{1}{s+0.5}G_c(s)} \tag{9.15}$$

Suppose the transfer function of the control strategy is just a constant K. Then Equation (9.15) becomes

$$G(s) = \frac{\frac{K}{s+0.5}}{1 + \frac{K}{s+0.5}} = \frac{K}{s + 0.5 + K} \tag{9.16}$$

Using this control strategy, the transfer function for the system is that of a first-order system of time constant

$$T = \frac{1}{0.5 + K} \tag{9.17}$$

If the 2 percent settling time for the controlled system is 0.2 s, then the appropriate time constant is determined as

$$2 = 3.92\left(\frac{1}{0.5 + K}\right) \Rightarrow K = 1.46\,\text{s}^{-1} \tag{9.18}$$

Thus, a control strategy is to introduce a controller with a gain of at least 1.46 s^{-1}. The system remains a first-order system. With a gain of 1.46 s^{-1}, it has a time constant of 0.510 s^{-1}.

9.3.2 Feedback Control Systems

The diagram shown in Figure 9.28 is the simplest model for a feedback control system. The system whose transfer function is $G(s)$ is called the plant. It is in series with an actuator that serves as the control system controller. The transfer function of the actuator is $G_c(s)$. Using the rules of block diagram algebra in Section 9.1, the closed-loop transfer function $H(s)$ is defined as the ratio of the transform of the input $A(s)$ to the transform of the output $C(s)$. The output goes to the next system in line and is also fed back into the summing junction. Thus, the input to the plant in series with the actuator is $A(s) - C(s)$:

$$[A(s) - C(s)]G(s)G_c(s) = C(s) \tag{9.19}$$

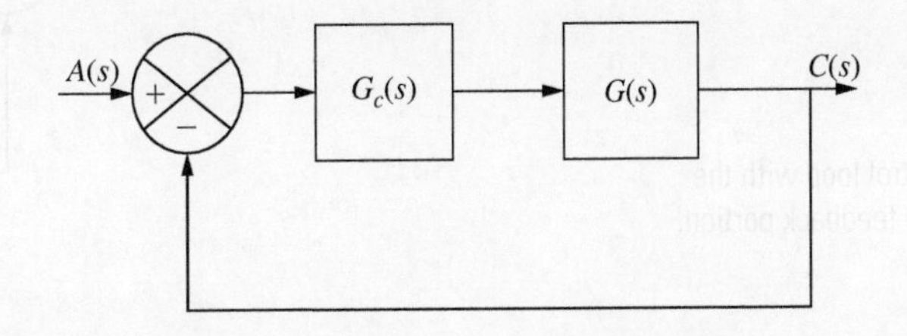

Figure 9.28

A simplified model of a plant in a feedback loop with a controller in series with the plant.

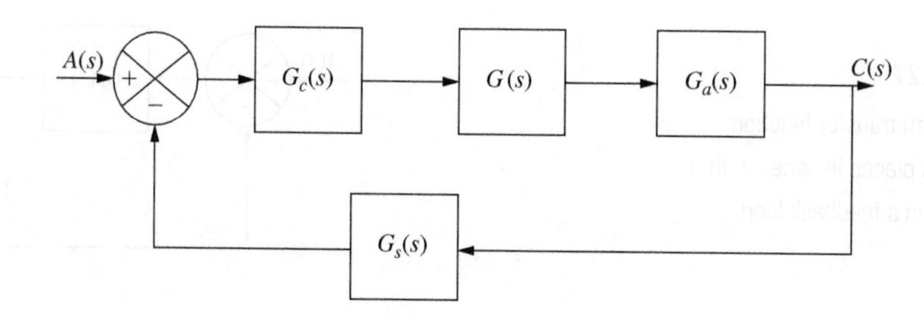

Figure 9.29

Feedback control loop with an amplifier and a sensor.

Equation (9.19) is rearranged to give

$$H(s) = \frac{C(s)}{A(s)} = \frac{G(s)\,G_c(s)}{1 + G(s)\,G_c(s)} \tag{9.20}$$

The open-loop transfer function is defined as

$$G_o(s) = G(s)\,G_c(s) \tag{9.21}$$

Equation (9.21) represents the transfer function if the plant were placed in series with the controller without a feedback loop.

The arrangement of the control system in Figure 9.28 is an idealized model. A more accurate model is illustrated in Figure 9.29. It shows the plant and the actuator as well as a sensor and an amplifier. These items generally have the effect of multiplying the signal being fed into them. The sensor picks up the signal from the series combination of the plant and the actuator and converts it into an appropriate form such that it can be subtracted from the input. The control system must satisfy the principle of dimensional homogeneity such that the input and the signal being fed into the summing junction have the same units. The difference between $A(s)$ and $G(s)\,G_a(s)$ may be small so that it must be amplified when fed back into the plant. For the purpose of this study, it is assumed that the gains from the sensor and the amplifier are part of the transfer function for the actuator. The closed-loop transfer function for a system whose block diagram is that of Figure 9.29 is

$$H(s) = \frac{K_a G(s)\,G_c(s)}{1 + K_a K_s G(s)\,G_c(s)} \tag{9.22}$$

The actuator may instead be part of the feedback loop as shown in Figure 9.30. In this case, the closed-loop transfer function becomes

$$H(s) = \frac{G(s)}{1 + G(s)\,G_c(s)} \tag{9.23}$$

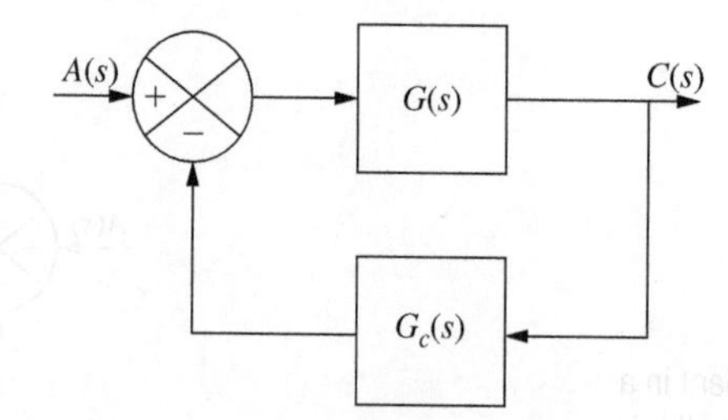

Figure 9.30

A feedback control loop with the controller in the feedback portion.

9.4 Types of Controllers

There are several types of controllers. This section catalogs these controllers and presents the forms of the transfer functions for each.

9.4.1 Proportional Control

A controller applies **proportional control** when its transfer function is a constant

$$G_c(s) = K_p \tag{9.24}$$

with K_p as the proportional gain. The substitution of Equation (9.24) into Equation (9.20) leads to

$$H(s) = \frac{K_p G(s)}{1 + K_p G(s)} \tag{9.25}$$

9.4.2 Integral Control and Proportional Plus Integral (PI) Control

The actuator transfer function for an integral controller with integral gain K_i is

$$G_c(s) = \frac{K_i}{s} \tag{9.26}$$

Substitution of Equation (9.26) into Equation (9.20) leads to

$$H(s) = \frac{\frac{K_i}{s}G(s)}{1 + \frac{K_i}{s}G(s)}$$

$$= \frac{K_i G(s)}{s + K_i G(s)} \tag{9.27}$$

The transfer function of an actuator with proportional plus integral (PI) control is

$$G_c(s) = K_p + \frac{K_i}{s} \tag{9.28}$$

Substitution into Equation (9.20) leads to

$$H(s) = \frac{\left(K_p + \frac{K_i}{s}\right)G(s)}{1 + \left(K_p + \frac{K_i}{s}\right)G(s)}$$

$$= \frac{(K_i + sK_p)G(s)}{s + (K_i + sK_p)G(s)} \tag{9.29}$$

9.4.3 Derivative Control and Proportional Plus Derivative (PD) Control

The transfer function for an actuator with derivative control is

$$G_c(s) = K_d s \tag{9.30}$$

where K_d is the derivative gain. The substitution of Equation (9.30) into Equation (9.20) leads to

$$H(s) = \frac{K_d sG(s)}{1 + K_d sG(s)} \tag{9.31}$$

A dashpot is a mechanical device that delivers proportional control. Derivative control responds to the rate of change of the signal error, not the error itself. Thus derivative control is used in conjunction with proportional control and/or integral control.

The transfer function for an actuator with proportional plus derivative (PD) control is

$$G_c(s) = K_p + K_d s \tag{9.32}$$

The substitution of Equation (9.32) into Equation (9.20) leads to

$$H(s) = \frac{(K_p + K_d s)G(s)}{1 + (K_p + sK_d)G(s)} \tag{9.33}$$

9.4.4 Proportional Plus Integral Plus Derivative (PID) Control

The transfer function for an actuator with proportional plus integral plus derivative control (PID) control is

$$G_c(s) = K_p + K_d s + \frac{K_i}{s} \tag{9.34}$$

If $E(t)$ is the input to a PID controller, then its output is

$$B(t) = K_p E(t) + K_d \frac{dE}{dt} + K_i \int_0^t E(t)dt \tag{9.35}$$

Substitution of Equation (9.34) into Equation (9.20) leads to

$$H(s) = \frac{\left(K_p + K_d s + \frac{K_i}{s}\right)G(s)}{1 + \left(K_p + K_d s + \frac{K_i}{s}\right)G(s)} = \frac{(K_d s^2 + K_p s + K_i)G(s)}{s + (K_d s^2 + K_p s + K_i)G(s)} \tag{9.36}$$

9.4.5 Compensators

When the phase of the output from the controlled system is inadequate, a compensator may be used. If a positive phase shift is required, a lead compensator is used. A lead compensator is also used to improve the system's transient response. When a negative phase shift is required, a lag compensator is used. A lag compensator is also used to improve the system's steady-state error. A lead-lag compensator is really a lead compensator in series with a lag compensator. A lead-lag compensator is used when both the system's transient response and the system's steady-state error need improvement.

All controllers are really compensators, but usually the term "compensator" is reserved for controllers that have a specific type of transfer function. A compensator has both a zero and a pole. A lead compensator has a transfer function of

$$G_c(s) = K_c \frac{s + a}{s + b} \qquad a < b \tag{9.37}$$

There are three parameters to adjust: K, a, and b. Alternately, one can introduce a time constant for this controller and write Equation (9.37) as

$$G_c(s) = K_c \frac{s + \frac{1}{T}}{s + \frac{1}{\alpha T}} \qquad \alpha < 1 \tag{9.38}$$

The value of α is called the attenuation factor. The value of K_c is called the compensator gain. The value of T is the time constant for the compensator.

A lag compensator has a transfer function of the form

$$G_c(s) = K_c \frac{s + a}{s + b} \qquad a > b \tag{9.39}$$

or, when written in terms of the time constant and an attenuation factor,

$$G_c(s) = K_c \frac{s + \frac{1}{T}}{s + \frac{1}{\alpha T}} \qquad \alpha > 1 \tag{9.40}$$

A lead-lag compensator can be thought of as a lead compensator in series with a lag compensator to give a transfer function of

$$G_c(s) = K_c \left(\frac{s + a}{s + b}\right)\left(\frac{s + c}{s + d}\right) \qquad a < b, c > d \tag{9.41}$$

Time constants for the lead compensator T_1 and lag compensator T_2 and attenuation factors α_1 for the lead compensator and α_2 for the lag compensator are introduced, and Equation (9.41) is rewritten as

$$G_c(s) = K_c \left(\frac{s + \frac{1}{T_1}}{s + \frac{1}{\alpha_1 T_1}}\right)\left(\frac{s + \frac{1}{T_2}}{s + \frac{1}{\alpha_2 T_2}}\right) \qquad \alpha_1 < 1, \alpha_2 > 1 \tag{9.42}$$

9.5 Electronic, Hydraulic, Pneumatic, and Mechanical Controllers

Many types of systems can be used as controllers if their transfer function has an appropriate form. Electric circuits, hydraulic servomotors, pneumatic systems, and mechanical systems can be used as controllers. Systems such as drug delivery systems use biological control systems. Control of populations of species and control of economic systems is achieved using non-physical controllers. This section discusses some common types of controllers.

9.5.1 Electronic Controllers

Operational amplifier circuits were introduced in Chapter 5. A circuit with two resistors and an ideal operational amplifier, illustrated in Figure 9.31, was shown to have an output of

$$v_2 = -\frac{R_2}{R_1} v_1 \tag{9.43}$$

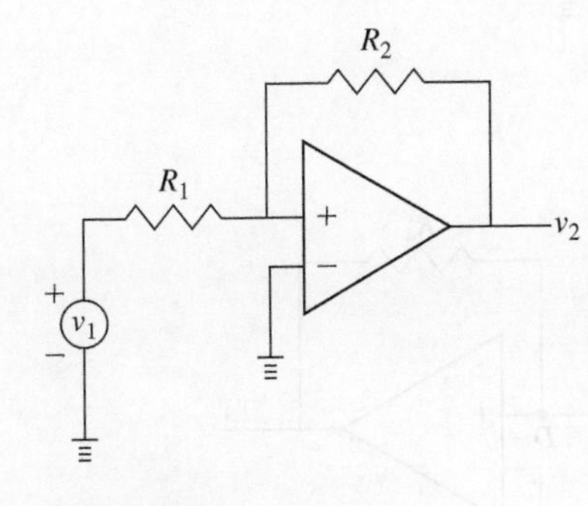

Figure 9.31

An inverting amplifier has a gain of $\frac{R_2}{R_1}$ but changes the sign of the input.

where v_1 is the potential input, R_1 is the resistance of the resistor between the input and the amplifier, and R_2 is the resistance of the resistor spanning the amplifier. If this amplifier is put in series with another amplifier, as illustrated in Figure 9.32, the sign on the output is changed and the result becomes

$$v_2 = \frac{R_4 R_2}{R_3 R_1} v_1 \tag{9.44}$$

The transfer function for this amplifier is

$$G_c(s) = \frac{V_2(s)}{V_1(s)} = \frac{R_4 R_2}{R_3 R_1} = K_p \tag{9.45}$$

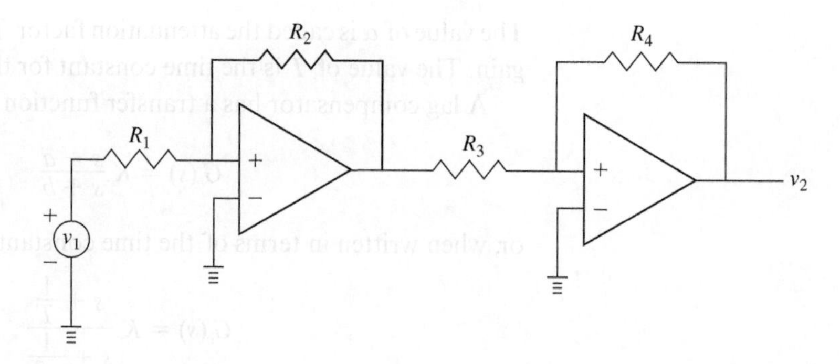

Figure 9.32

Two inverting amplifiers in series lead to a noninverting circuit.

Equation (9.45) satisfies the properties of a proportional controller. All operational amplifier circuits that serve as controllers have an inverting amplifier in series with another inverting amplifier. Then the output voltage has the same sign as the input voltage.

The amplifier circuit of Figure 9.33 with a capacitor bridging the amplifier has a transfer function of

$$G_c(s) = \frac{V_2(s)}{V_1(s)} = \frac{R_4}{R_3 R_1 Cs} = \frac{K_i}{s} \tag{9.46}$$

Equation (9.46) is the transfer function of an integral controller.

Consider the amplifier circuit of Figure 9.34. Using KCL at nodes A, B, and C and the equations for potential difference across resistors and capacitors gives

$$\frac{v_1}{R_1} + C_1\frac{dv_1}{dt} = \frac{v_2}{R_2} + C_2\frac{dv_2}{dt} \tag{9.47}$$

Taking the Laplace transform of Equation (9.47) assuming all initial conditions are zero leads to

$$\left(\frac{1}{R_1} + C_1 s\right) V_1(s) = \left(\frac{1}{R_2} + C_2 s\right) V_2(s) \tag{9.48}$$

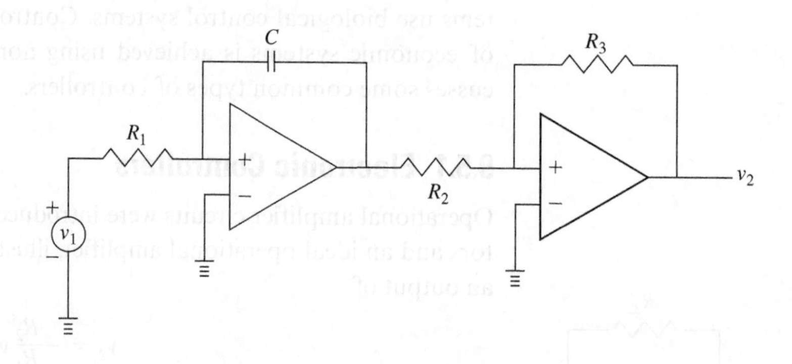

Figure 9.33

An operational amplifier circuit used as an integral controller.

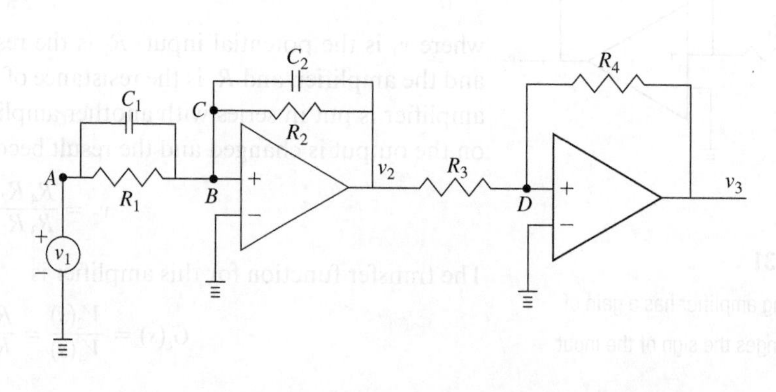

Figure 9.34

If $R_1 C_1 < R_2 C_2$, the amplifier circuit can serve as a lag compensator. If $R_1 C_1 > R_2 C_2$, the amplifier circuit can serve as a lead compensator.

The transfer function is obtained from Equation (9.48) as

$$\frac{V_2(s)}{V_1(s)} = \frac{C_1}{C_2}\frac{s + \dfrac{1}{R_1 C_1}}{s + \dfrac{1}{R_2 C_2}} \tag{9.49}$$

The use of the inverting amplifier gives a transfer function of

$$G_c(s) = \frac{R_4 C_1}{R_3 C_2}\frac{s + \dfrac{1}{R_1 C_1}}{s + \dfrac{1}{R_2 C_2}} \tag{9.50}$$

Equation (9.50) is the transfer function of a lead or lag compensator. If $R_1 C_1 > R_2 C_2$, Equation (9.50) is the transfer function of a lead compensator. In this case, Equation (9.50) can be written in the form of Equation (7.37) with a gain of

$$K_c = \frac{R_4 C_1}{R_3 C_2} \tag{9.51}$$

The compensator has a time constant of

$$T = R_1 C_1 \tag{9.52}$$

and an attenuation factor of

$$\alpha = \frac{R_2 C_2}{R_1 C_1} \tag{9.53}$$

If $R_1 C_1 < R_2 C_2$, Equation (9.50) is the transfer function of a lag compensator.

Table 9.2 (page 560) shows how operational amplifiers can serve as various types of controllers. The general principles are presented here:

- A resistor bridging an amplifier produces proportional control.
- A capacitor bridging an amplifier produces integral control.
- A parallel combination of a resistor and a capacitor upstream of an amplifier produces differential control.
- A series combination of a resistor and a capacitor bridging an amplifier produces PI type of control.
- A resistor in parallel with a capacitor bridging an amplifier produces a lead or lag compensator.
- A lead-lag compensator is constructed by having a series combination of a capacitor and a resistor in parallel with a resistor both upstream of the amplifier and bridging the amplifier.
- A second amplifier is needed in all cases to serve as an inverting amplifier.

9.5.2 Mechanical Controllers

A viscous damper provides differential control to a system. Suppose the input to the system illustrated in Figure 9.35 (page 561) is $y(t)$ and the output is $x(t)$. Assuming the junction is massless, a force balance at the joint leads to

$$c(\dot{y} - \dot{x}) - kx = 0 \tag{9.54}$$

Taking the Laplace transform of Equation (9.54), the transfer function is

$$G(s) = \frac{X(s)}{Y(s)} = \frac{cs}{cs + k} \tag{9.55}$$

Table 9.2 Electronic Controllers

Controller Action	Transfer Function $G(s) = \frac{V_2(s)}{V_1(s)}$	Parameters	Schematic Diagram
Proportional (P)	$\frac{R_4 R_2}{R_3 R_1}$	$K_p = \frac{R_4 R_2}{R_3 R_1}$	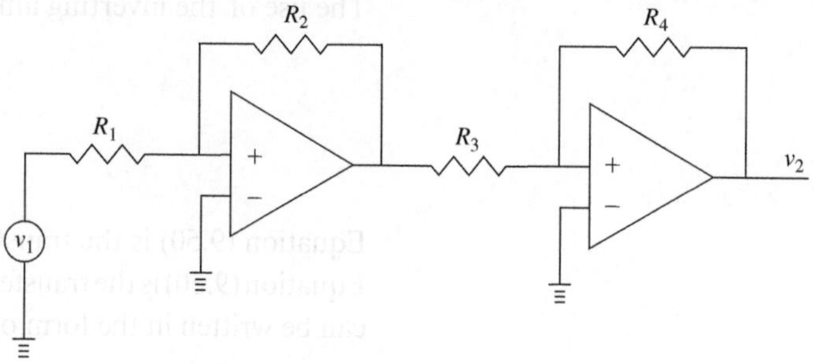
Integral (I)	$\frac{R_4}{R_3 R_1 C s}$	$K_i = \frac{R_4}{R_3 R_1 C}$	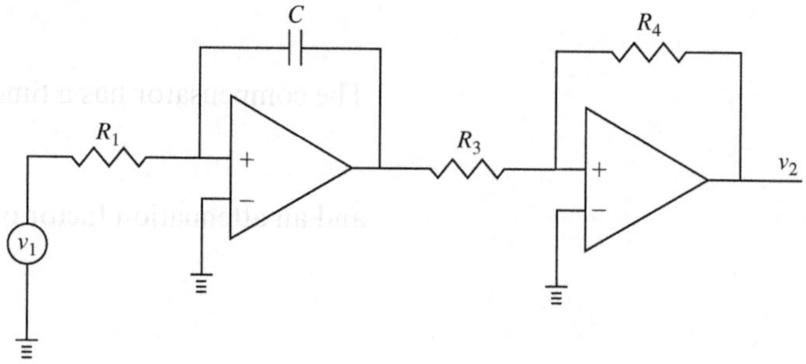
Proportional plus Integral (PI)	$\frac{R_4 R_2}{R_3 R_1}\left(1 + \frac{1}{R_2 C s}\right)$	$K_p = \frac{R_4 R_2}{R_3 R_1}$ $K_i = \frac{R_4}{R_3 R_1 C}$	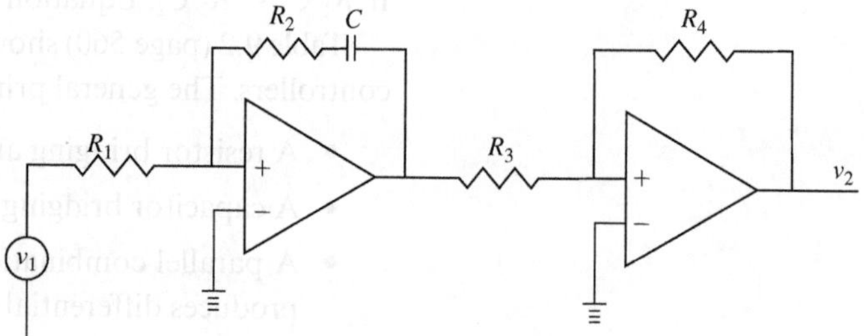
Proportional plus Differential (PD)	$\frac{R_4 R_2}{R_3 R_1}(1 + R_1 C s)$	$K_p = \frac{R_4 R_2}{R_3 R_1}$ $K_d = \frac{R_4 R_2 C}{R_3}$	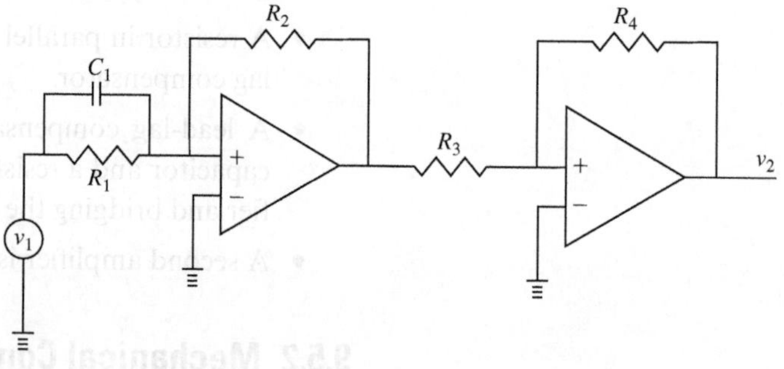
Proportional plus Integral plus Differential (PID)	$\frac{R_4 R_2}{R_3 R_1}\left[R_1 C_1 s + \left(\frac{R_1 C_1}{R_2 C_2} + 1\right) + \frac{1}{R_2 C_2 s}\right]$	$K_p = \frac{R_4 C_1}{R_3 C_2} + \frac{R_4 R_2}{R_3 R_1}$ $K_d = \frac{R_4 R_2 C_1}{R_3}$ $K_i = \frac{R_4}{R_3 R_1 C_2}$	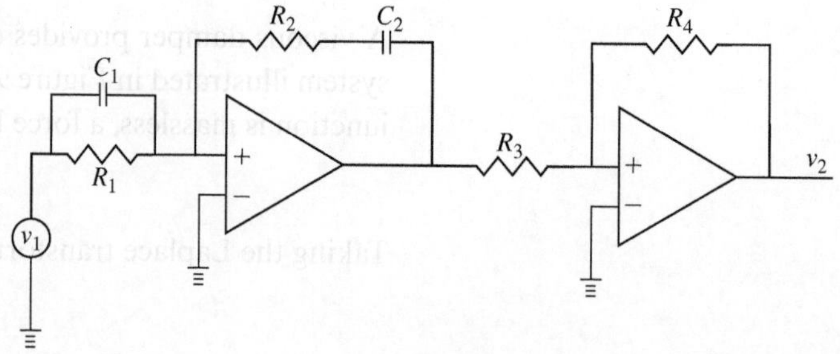

Controller Action	Transfer Function $G(s) = \frac{V_2(s)}{V_1(s)}$	Parameters	Schematic Diagram
Lag Compensator	$\frac{R_4C_1}{R_3C_2}\frac{\left(s+\frac{1}{R_1C_1}\right)}{\left(s+\frac{1}{R_2C_2}\right)}$	$K_c = \frac{R_4C_1}{R_3C_2}$ $T = R_1C_1$ $\alpha = \frac{R_2C_2}{R_1C_1} > 1$	
Lead Compensator	$\frac{R_4C_1}{R_3C_2}\frac{\left(s+\frac{1}{R_1C_1}\right)}{\left(s+\frac{1}{R_2C_2}\right)}$	$K_c = \frac{R_4C_1}{R_3C_2}$ $T = R_1C_1$ $\alpha = \frac{R_2C_2}{R_1C_1} < 1$	
Lag-Lead Compensator	$\frac{R_6R_4R_1(R_2+R_4)}{R_5R_3R_2(R_1+R_3)}\left[\frac{s+\frac{1}{(R_1+R_3)C_1}}{s+\frac{1}{R_1C_1}}\right]\left[\frac{s+\frac{1}{R_2C_2}}{s+\frac{1}{(R_2+R_4)C_2}}\right]$	$K_c = \frac{R_6R_4R_1(R_2+R_4)}{R_5R_3R_2(R_1+R_3)}$ $T_1 = (R_1+R_3)C_1$ $T_2 = R_2C_2$ $\alpha_1 = \frac{R_1}{R_1+R_3}$ $\alpha_2 = \frac{R_2+R_4}{R_2}$	

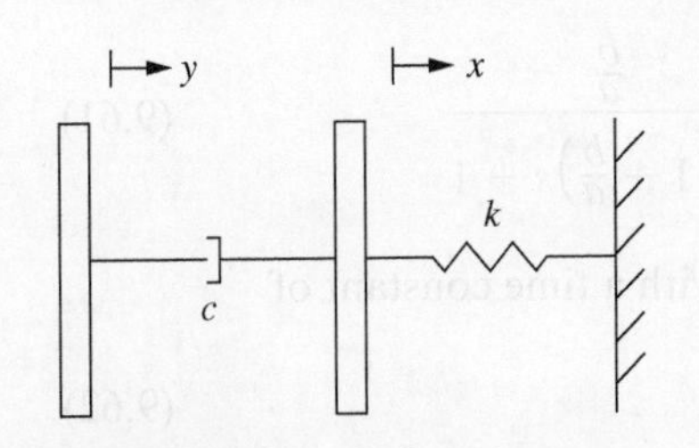

Figure 9.35

The first-order mechanical system can serve as a differential controller if $k/c \gg 1$.

If $\frac{k}{c} \gg 1$, then Equation (9.55) can be approximated by

$$G(s) = \frac{c}{k}s \tag{9.56}$$

The system of Figure 9.35 provides differential control.

Consider the system shown in Figure 9.36. The input $y(t)$ is connected to the output $x(t)$ by a system of springs and viscous dampers such that the Laplace transform of the force developed in the combination is $Z_1(s)[Y(s) - X(s)]$. The output is connected to a combination of springs and viscous dampers such that the Laplace transform of the force developed in this combination is $Z_2(s)X(s)$. Assuming the junction is massless, a force balance at the junction leads to

$$G(s) = \frac{Z_1(s)}{Z_1(s) + Z_2(s)} \tag{9.57}$$

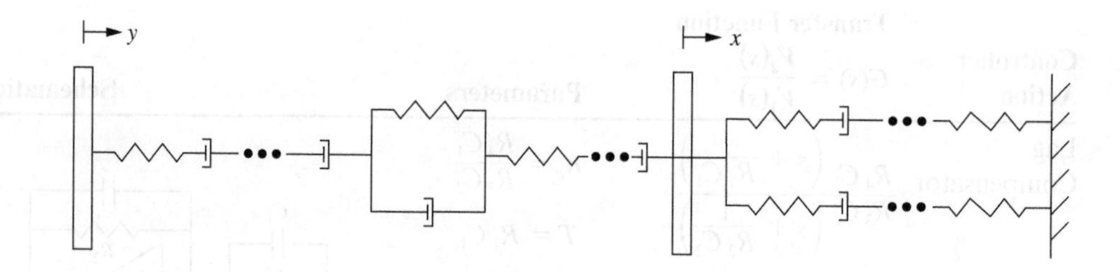

Figure 9.36

A general representation of a mechanical system. The system can serve as various types of controllers. The type is dependent upon the impedances of the system between the massless joint and the input and the massless joint and the fixed support.

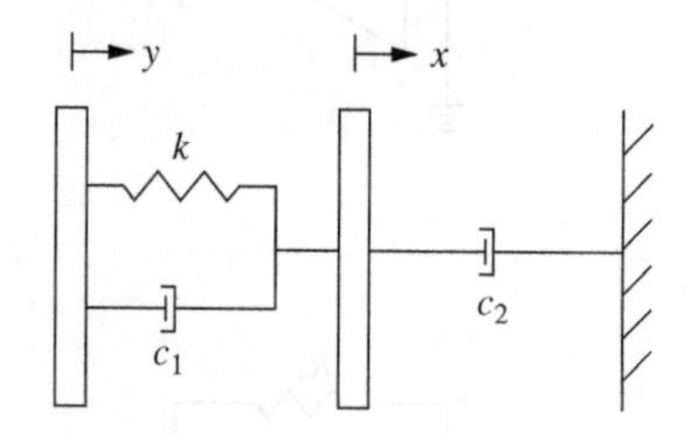

Figure 9.37

The first-order mechanical system can serve as a lag compensator.

For example, consider the system of Figure 9.37. The component whose Laplace transform leads to $Z_1(s)$ is a spring of stiffness k in parallel with a viscous damper of damping coefficient c_1. The component whose Laplace transform leads to $Z_2(s)$ is a viscous damper of damping coefficient c_2. For this combination, the transfer function is

$$G(s) = \frac{c_1 s + k}{(c_1 + c_2)s + k} = \frac{c_1}{c_1 + c_2}\left(\frac{s + \frac{k}{c_1}}{s + \frac{k}{c_1 + c_2}}\right) \tag{9.58}$$

It is noted that the constant in the denominator of Equation (9.58) is smaller than the constant in the numerator. Hence, Equation (9.58) is the transfer function for a lag compensator with $K_c = \frac{c_1}{c_1 + c_2}$, $T = \frac{c_1}{k}$, and $\alpha = \frac{c_1 + c_2}{c_1}$.

The gain in the compensator of Equation (9.58) is defined such that it is less than one. This gain would have to be amplified for the controller to be effective. For this reason, mechanical systems are rarely used as controllers.

9.5.3 Hydraulic Controllers

Consider the hydraulic servomotor in Figure 6.19. The output $y(t)$ is related to the amount the spool valve is opened $x(t)$ through the servomotor equation:

$$\hat{C}x = A_p\dot{y} \tag{9.59}$$

where $\hat{C}$ is a parameter of the motor and A_p is the area of the piston face. When a walking beam is connected to the servomotor, the equation governing the output as a function of the input is as shown in Example 6.13:

$$A_p(a + b)\dot{y} + \hat{C}ay = \hat{C}bz \tag{9.60}$$

Taking the Laplace transform of the equation with all initial conditions as zero leads to a transfer function of

$$G(s) = \frac{Y(s)}{Z(s)} = \frac{\hat{C}b}{A_p(a + b)s + \hat{C}a} = \frac{\frac{b}{a}}{\frac{A_p}{\hat{C}}\left(1 + \frac{b}{a}\right)s + 1} \tag{9.61}$$

This is the transfer function of a first-order system with a time constant of

$$T = \frac{A_p}{\hat{C}}\left(1 + \frac{b}{a}\right) \tag{9.62}$$

If $\frac{A_p}{\hat{C}} \ll 1$, then T is small and the transfer function is approximated by

$$G(s) = \frac{b}{a} \tag{9.63}$$

Then the servomotor acts as a proportional controller with a proportional gain of

$$K_p = \frac{b}{a} \tag{9.64}$$

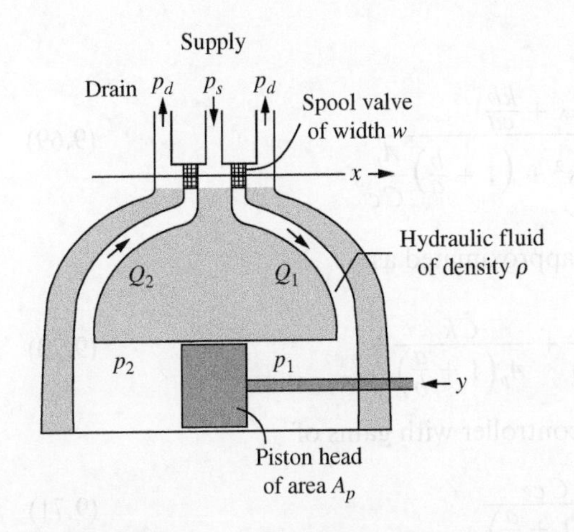

Figure 6.19

Cross section of a hydraulic servomotor. (Repeated)

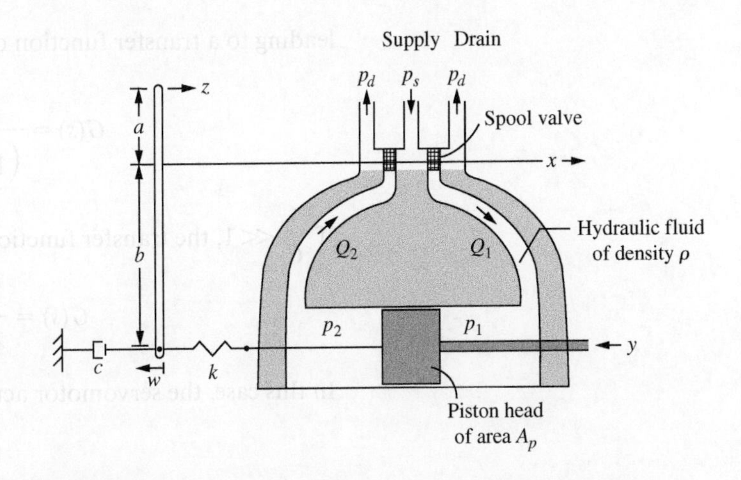

Figure 6.36

The hydraulic servomotor and walking beam system of Example 6.28. (Repeated)

Now consider the system of Example 6.28 and Figure 6.36. The differential equation relating the input and output is derived in Equation (e) of Example 6.28 as

$$\left(1+\frac{b}{a}\right)\frac{A_p}{\hat{C}}\ddot{y}+\frac{ka}{c\hat{C}}\dot{y}+\frac{ka}{c}y=b\dot{z}+\frac{kb}{c}z \tag{9.65}$$

Taking the Laplace transform to render the transfer function leads to

$$G(s)=\frac{bs+\frac{kb}{c}}{\left(1+\frac{b}{a}\right)\frac{A_p}{\hat{C}}s^2+\frac{ka}{c\hat{C}}s+\frac{ka}{c}} \tag{9.66}$$

If $\frac{A_p}{\hat{C}} \ll 1$, then the leading term in the denominator is small and the transfer function, Equation (9.66), is approximated by

$$G(s)=\frac{\frac{cb}{k}\hat{C}s+\hat{C}\frac{b}{a}}{s+\frac{1}{\hat{C}}} \tag{9.67}$$

This is the form of the transfer function of a compensator with a gain of $\frac{cb}{k}\hat{C}$. The compensator has a zero at $-\frac{k}{ca}$ and a pole at $-\frac{1}{\hat{C}}$.

Finally, consider the hydraulic servomotor of Figure 9.38. This is similar to the servomotor of Figure 6.36 with the spring and viscous damper exchanging places. The differential equation for this model is

$$\left(1+\frac{b}{a}\right)\frac{A_p}{\hat{C}}\ddot{y}+\left(1+\frac{b}{a}\right)\frac{A_p}{\hat{C}c}\dot{y}=\frac{b}{a}\dot{z}+\frac{kb}{ca}z \tag{9.68}$$

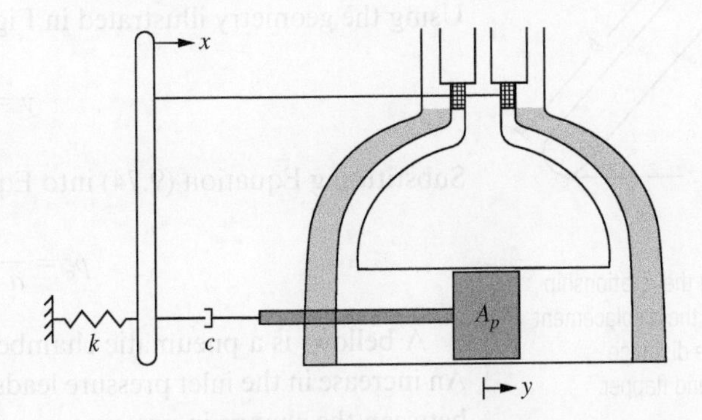

Figure 9.38

The hydraulic servomotor, in various configurations, can serve as different types of controllers. The relation between the output y and the input x is $\hat{C}x = A_p\dot{y}$.

leading to a transfer function of

$$G(s) = \frac{\frac{b}{a}s + \frac{kb}{ca}}{\left(1 + \frac{b}{a}\right)\frac{A_p}{\hat{C}}s^2 + \left(1 + \frac{b}{a}\right)\frac{A_p}{\hat{C}c}s} \tag{9.69}$$

If $\frac{A_p}{\hat{C}} \ll 1$, the transfer function may be approximated as

$$G(s) = \frac{\hat{C}c}{A_p\left(1 + \frac{a}{b}\right)} + \frac{\hat{C}k}{A_p\left(1 + \frac{a}{b}\right)}\frac{1}{s} \tag{9.70}$$

In this case, the servomotor acts as a PI controller with gains of

$$K_p = \frac{\hat{C}c}{A_p\left(1 + \frac{a}{b}\right)} \tag{9.71}$$

and

$$K_I = \frac{\hat{C}k}{A_p\left(1 + \frac{a}{b}\right)} \tag{9.72}$$

The hydraulic servomotor of Figure 6.36, in which the walking stick is connected to the piston by a spring in series with a viscous damper, serves as a PD controller. The system of Figure 9.38, in which the walking stick is connected to a fixed support by a spring and to the piston by a spring in series with a viscous damper, serves as a PID controller.

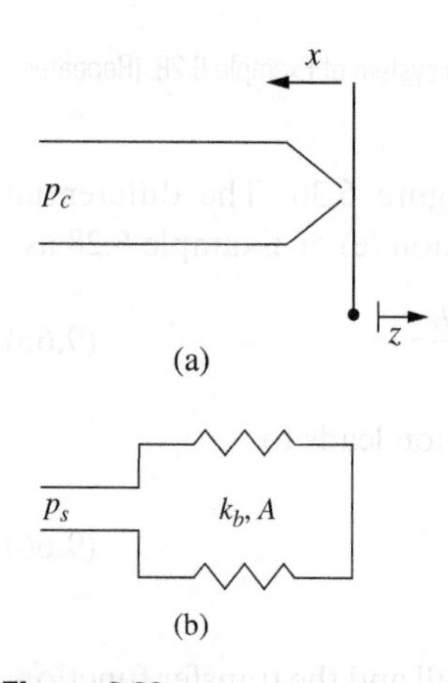

Figure 9.39

(a) The nozzle and the flapper are important components of a pneumatic controller. (b) The bellows is a chamber that expands and contracts as the upstream pressure is changed.

9.5.4 Pneumatic Controllers

Pneumatic controllers can also be used in mechanical systems. The basic elements of a pneumatic controller are a nozzle and flapper, illustrated in Figure 9.39(a), a bellows, illustrated in Figure 9.39(b), and a piping system with resistance. A flapper is a thin rigid rod, assumed to be massless. It is connected to some point in the pneumatic system but is a distance y from a nozzle. The pressure at the nozzle exit is assumed to be a linear function of the distance of the flapper from the exit. The change in the upstream pressure p_c as the exit pressure changes is also assumed to be linear. Thus

$$p_c = Ky \tag{9.73}$$

where y is the distance from the nozzle to the flapper and K is a constant of proportionality.

One end of the flapper provides input to the system x. The distance between this end and the exit of the nozzle is a. The other end is connected to a pivot that can move a distance z horizontally. The distance from the pivot point to the nozzle exit is b. Using the geometry illustrated in Figure 9.40, assuming small displacements,

$$y = \frac{bz - ax}{a + b} \tag{9.74}$$

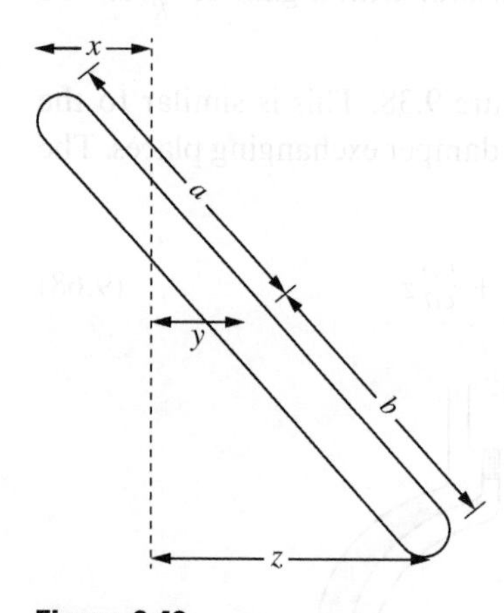

Figure 9.40

Geometry illustrating the relationship between the input x, the displacement of the pivot z, and the distance between the nozzle and flapper.

Substituting Equation (9.74) into Equation (9.73) results in

$$p_c = \frac{K}{a + b}(bz - ax) \tag{9.75}$$

A bellows is a pneumatic chamber that has the flexibility to expand and contract. An increase in the inlet pressure leads to an expansion of the bellows. The relationship between the change in pressure p and the expansion x is $pA = kx$, where A is the area

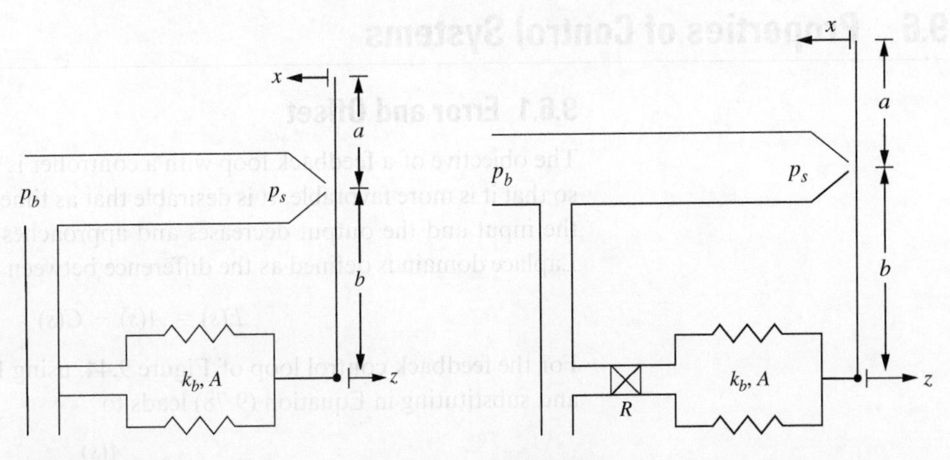

Figure 9.41

Schematic diagram of a pneumatic proportional controller.

Figure 9.42

The restriction upstream of the bellows adds derivative control to the pneumatic controller.

of the bellows and k is the bellows stiffness. The bellows behaves much like a spring. When the inlet pressure to the bellows in Figure 9.41 increases by p_c, the end of the flapper moves a distance x, causing the source pressure to increase. This leads to

$$x = \frac{p_c}{kA} \tag{9.76}$$

Using Equation (9.76) in Equation (9.75) leads to

$$p_c = \frac{KkbA}{kA(a + b) + Ka} z = \Gamma z \tag{9.77}$$

The system of Figure 9.41 may be used as a proportional controller of gain Γ.

A restriction in the flow, R, illustrated in Figure 9.42, prevents the bellows from sensing the change in pressure instantaneously. Under certain assumptions, the restriction adds derivative control to the system.

A system that has two bellows in series, as shown in Figure 9.43, with the first connected to the flapper and a restriction in the circuit for the second, acts as a proportional plus integral controller. Adding a restriction before the first bellows adds derivative control leading to a PID controller.

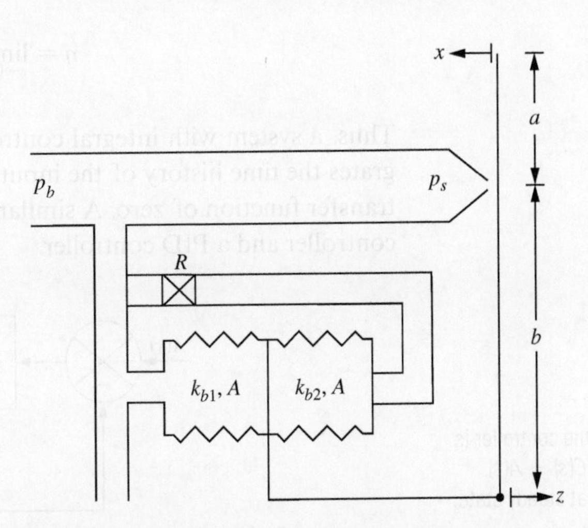

Figure 9.43

Two bellows in series and a restriction downstream of the second bellows makes this pneumatic device a PI controller. Adding a restriction upstream of the first bellows would make it a PID controller.

9.6 Properties of Control Systems

9.6.1 Error and Offset

The objective of a feedback loop with a controller is to modify the output of a system so that it is more favorable. It is desirable that as time increases, the difference between the input and the output decreases and approaches zero. Error in the system in the Laplace domain is defined as the difference between the input and the output:

$$E(s) = A(s) - C(s) \tag{9.78}$$

For the feedback control loop of Figure 9.44, using Equation (9.19), solving for $C(s)$, and substituting in Equation (9.78) leads to

$$E(s) = \frac{A(s)}{1 + G_c(s)G(s)} \tag{9.79}$$

The offset η is defined as the difference between the input and the output at steady state. Applying the final value theorem to Equation (9.79),

$$\eta = \lim_{s\to 0} sE(s)$$

$$= \lim_{s\to 0} \frac{sA(s)}{1 + G_c(s)G(s)} \tag{9.80}$$

Often the control system is designed to reduce the error from a step input. Using $A(s) = \frac{1}{s}$, the transform of a step input to Equation (9.80) leads to

$$\eta = \lim_{s\to 0} \frac{1}{1 + G_c(s)G(s)} = \frac{1}{1 + G_c(0)G(0)} \tag{9.81}$$

For a feedback loop with a proportional controller, $G_c(s) = K_p$, application of Equation (9.81) gives

$$\eta = \frac{1}{1 + K_p G(0)} \tag{9.82}$$

Equation (9.82) shows that a feedback loop employing a proportional controller has a nonzero offset. The steady-state response differs with the input. The error for a system with only a proportional controller cannot be zero because proportional control multiplies the instantaneous error signal by a constant. If the instantaneous error was zero, the output from the controller would be zero and the plant would not operate.

For a system with an integral controller, $G_c(s) = \frac{K_i}{s}$, the offset, Equation (9.81), leads to

$$\eta = \lim_{s\to 0} \frac{1}{1 + \frac{K_i}{s}G(s)} = 0 \tag{9.83}$$

Thus, a system with integral control has no offset. Since the integral controller integrates the time history of the input, an instantaneous error of zero does not lead to a transfer function of zero. A similar analysis shows that the offset is also zero for a PI controller and a PID controller.

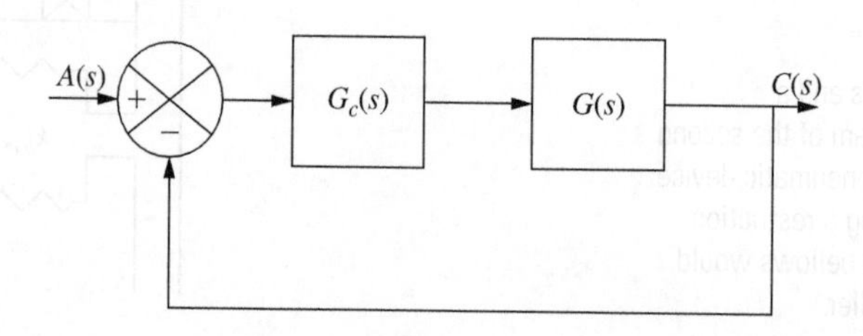

Figure 9.44

The error when using the controller is defined as the $E(s) = C(s) - A(s)$. The offset is the error at steady state.

The application of Equation (9.81) to a system with only differential control, $G_c(s) = K_d s$, leads to

$$\eta = \lim_{s\to 0} \frac{1}{1 + K_d sG(s)} = 1 \tag{9.84}$$

Thus, differential control leads to an offset that is independent of the differential gain when subject to a step input. When a PD controller, $G_a(s) = K_p + K_d s$, is used as an actuator, the offset is altered from the offset as if it were an open-loop system. When an integral, PI, or PID controller is used, the presence of the integrator eliminates the offset.

The use of a lead or lag compensator, $G_c(s) = K\dfrac{s+\frac{1}{T}}{s+\frac{1}{\alpha T}}$, gives

$$\eta = \frac{1}{1 + \alpha KG(0)} \tag{9.85}$$

Equation (9.81) applied for a lead-lag compensator gives

$$\eta = \frac{1}{1 + K\alpha_1\alpha_2 G(0)} \tag{9.86}$$

For a lead, lag, or lead-lag compensator, Equations (9.85) and (9.86) show that the steady-state response differs from the open-loop steady-state response.

Example 9.8

What is the offset when a plant whose transfer function is

$$G(s) = \frac{2s+5}{s^3 + 4s^2 + 10s + 30} \tag{a}$$

is put into a feedback loop with: **a.** a proportional controller with $K_p = 6$; **b.** a differential controller with $K_d = 3$; **c.** a proportional, integral and differential controller with $K_p = 6$, $K_i = 4$, and $K_d = 3$; and **d.** a lead-lag compensator with $K = 4$, $\alpha_1 = 0.4$, and $\alpha_2 = 2.3$?

Solution

The offset is given by Equation (9.81). Using Equation (a), it is noted that $G(0) = \frac{1}{6}$.

a. Applying Equation (9.81) with a proportional controller with $K_p = 6$ gives

$$\eta = \frac{1}{1 + G_c(0)G(0)} = \frac{1}{1 + 6\left(\frac{1}{6}\right)} = \frac{1}{2} \tag{b}$$

b. Applying Equation (9.81) with a differential controller with $K_d = 3$ gives

$$\eta = \frac{1}{1 + 3(0)6} = 1 \tag{c}$$

c. Applying Equation (9.81) with a PID controller with $K_p = 6$, $K_i = 4$, and $K_d = 3$ gives

$$\eta = \frac{1}{1 + \left[6 + 3(0) + \frac{4}{0}\right]6} = 0 \tag{d}$$

d. Applying Equation (9.81) with a lead-lag compensator with $K = 4$, $\alpha_1 = 0.4$, and $\alpha_2 = 2.3$ gives

$$\eta = \frac{1}{1 + G_c(0)G(0)} = \frac{1}{1 + 4\left[\frac{1/T_1}{1/(0.4T_1)}\right]\left[\frac{1/T_2}{1/(2.3T_2)}\right]6} = 0.0433 \tag{e}$$

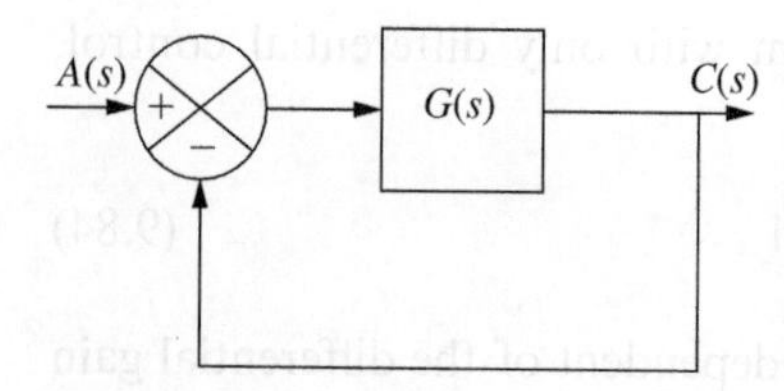

Figure 9.45
A unity feedback loop.

9.6.2 Static Error Constants

Consider the system of Figure 9.45 in which a plant is placed in a feedback loop with no controller. This is known as a unity feedback loop. The closed-loop transfer function for this feedback loop is

$$H(s) = \frac{G(s)}{1 + G(s)} \tag{9.87}$$

The error in placing a plant in a unity feedback loop is given by Equation (9.79) with $G_c(s) = 1$ and is

$$E(s) = \frac{A(s)}{1 + G(s)} \tag{9.88}$$

Application of the final value theorem to Equation (9.88) leads to the final value of the error:

$$\eta = \lim_{s \to 0} \frac{sA(s)}{1 + G(s)} \tag{9.89}$$

The final value of the error depends on the forms of $A(s)$ and $G(s)$. It is appropriate to determine the final value of the error for:

- a unit step input in which $A(s) = \frac{1}{s}$ and η is termed the static position error e_p
- a unit ramp input in which $A(s) = \frac{1}{s^2}$ and η is termed the static velocity error e_v
- a unit acceleration input in which $A(s) = \frac{1}{s^3}$ and η is termed the static acceleration error e_a

The static position error constant is defined as

$$C_p = \lim_{s \to 0} G(s) = G(0) \tag{9.90}$$

The static velocity error constant is defined as

$$C_v = \lim_{s \to 0} sG(s) \tag{9.91}$$

The static acceleration error constant is defined as

$$C_a = \lim_{s \to 0} s^2 G(s) \tag{9.92}$$

The higher the value of an error constant, the smaller the error when the plant is placed in a unity feedback loop.

Consider a plant with a transfer function of the form

$$G(s) = \frac{N(s)}{s^k D(s)} \tag{9.93}$$

The plant type is defined by the parameter k. If $k = 0$, the plant is a type 0 plant; if $k = 1$, the plant is a type 1 plant; and so on.

The static position error constant C_p is a measure of how well the system can replicate a unit step input. The larger the value of C_p, the better the replication. Applying Equation (9.89) to determine the static position error for a unit step input yields

$$e_p = \frac{1}{1 + G(0)} \tag{9.94}$$

Equation (9.94) is evaluated as for a type 0 system, leading to $e_p = \frac{1}{1 + N(0)/D(0)}$. For a type 1 or higher system, $e_p = 0$. The static position error constants are $G(0)$ for a type 0 system and ∞ for a system of higher type.

The static velocity error constant C_v is a measure of how well the system can replicate a unit ramp input. Applying Equation (9.89) to determine the static velocity error for a unit ramp input $a(t) = t$ gives

$$e_v = \frac{1}{s[1 + G(0)]} \tag{9.95}$$

Table 9.3 Steady-state Errors and Constants for a Transfer Function $G(s) = \dfrac{N(s)}{s^k D(s)}$ in a Unity Feedback Loop

Type of System (k)	e_p	C_p	e_v	C_v	e_a	C_a
Type 0	$\dfrac{1}{1 + N(0)/D(0)}$	$\dfrac{N(0)}{D(0)}$	0	∞	0	∞
Type 1	0	∞	$\dfrac{D(0)}{N(0)}$	$\dfrac{N(0)}{D(0)}$	0	∞
Type 2	0	∞	0	∞	$\dfrac{D(0)}{N(0)}$	$\dfrac{N(0)}{D(0)}$

For a type 0 system, Equation (9.95) leads to $e_v = \infty$. For a type 1 system, Equation (9.95) yields $D(0)/N(0)$. For a type 2 system, Equation (9.95) gives a value of 0. The static position error constant for a type 0 system is $C_v = 0$, for a type 1 system it is $C_v = N(0)/D(0)$, and for a type 2 or higher system it is $C_v = \infty$.

The static acceleration constant C_a is a measure of how well the system can replicate a unit parabolic input. Applying Equation (9.89) to determine the static velocity error for $a(t) = \frac{1}{2}t^2$ gives

$$e_a = \frac{1}{s^2[1 + G(0)]} \tag{9.96}$$

Applying Equation (9.96) for a type 0 or 1 system leads to $e_a = \infty$. For a type 2 system, Equation (9.96) gives $e_a = D(0)/N(0)$. The static acceleration error constant for a type 0 or 1 system is $C_a = 0$. For a type 2 system, $C_a = D(0)/N(0)$.

These results are summarized in Table 9.3

The higher an error constant, the better the performance of the closed-loop system is in replicating the steady-state behavior of the output. However, a larger value of an error constant leads to a decrease in stability.

Example 9.9

For the given transfer functions, determine the static position error constants, the static velocity error constants, and the static acceleration error constants. Discuss their stability with increasing $K > 0$.

$$\text{a. } G(s) = \frac{K}{(s+1)(s+2)} \tag{a}$$

$$\text{b. } G(s) = \frac{K}{s(s+1)(s+2)} \tag{b}$$

Solution

a. The transfer function is that of a type 0 system. Using the entry in Table 9.3, the static errors and error constants are determined as

$$e_p = \frac{1}{1 + K/2} \qquad C_p = \frac{K}{2} \tag{c}$$

$$e_v = \infty \qquad C_v = 0 \tag{d}$$

$$e_a = \infty \qquad C_a = 0 \tag{e}$$

From Equation (c) it is seen that as K increases, the static position error decreases and the ability to replicate a unit step function input increases. The closed-loop transfer function becomes

$$H(s) = \frac{K}{s^2 + 3s + 2 + K} \tag{f}$$

Equation (f) is the transfer function of a second-order system. The system is always stable if the coefficients are of the same sign. The closed-loop system is always stable as long as $K > 0$.

b. The transfer function is a type 1 transfer function. Using Table 9.3, the static errors and error constants are calculated as

$$e_p = 0 \qquad C_p = \infty \tag{g}$$

$$e_v = \frac{2}{K} \qquad C_v = \frac{K}{2} \tag{h}$$

$$e_a = \infty \qquad C_a = 0 \tag{i}$$

The closed-loop transfer function when the plant is placed in a unity feedback loop is

$$H(s) = \frac{K}{s^3 + 3s^3 + 2s + K} \tag{j}$$

Equation (j) is the transfer function of a third-order system. Equation (e) of Example 7.6, which provides the sufficient condition for the stability of a third-order system, is applied to Equation (j) of this example, leading to a system stability criterion of

$$(3)(2) > K \tag{k}$$

Thus as K increases, the static velocity constant and the static velocity error improves but the stability decreases. The system is unstable for $K > 6$.

Example 9.10

The plants whose transfer functions are

$$\textbf{a. } G(s) = \frac{2s + 3}{(s + 5)(s + 7)} \tag{a}$$

$$\textbf{b. } G(s) = \frac{2s + 3}{s(s^2 + 5s + 36)} \tag{b}$$

are placed in a feedback loop with a lag controller whose transfer function is

$$G_c(s) = K_c\frac{s + 0.025}{s + 0.0005} \tag{c}$$

Determine the value of K_c such that the appropriate static error constant is 60.

Solution

The feedback loop obtained when the controller is placed in series with a plant is illustrated in Figure 9.46. The static error constants are determined by examining the transfer function that is placed in a unity feedback loop, an equivalent transfer function. The equivalent transfer function is that of the plant placed in series with the controller:

$$G_{eq}(s) = G_c(s)G(s) \tag{d}$$

a. The plant is a type 0 plant, so the appropriate constant is the static position constant from Table 9.3:

$$C_p = G_{eq}(0) = G_c(0)G(0)$$

$$= K_c\frac{0.025}{0.0005}\frac{3}{35} = 4.286K_c \tag{e}$$

Requiring $C_p = 60$ leads to

$$4.286K_c = 60 \Rightarrow K_c = 14 \tag{f}$$

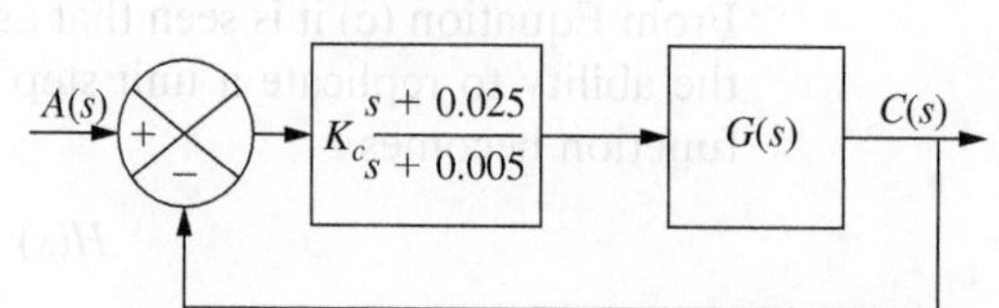

Figure 9.46
System of Example 9.10.

b. The plant is a type 1 plant, so the appropriate constant is the static velocity error constant:

$$C_v = \lim_{s \to 0} G_{eq}(s) = K_c \left(\frac{0.025}{0.0005}\right)\left(\frac{3}{36}\right) = 4.17 K_c \tag{g}$$

Requiring $C_v = 60$ leads to

$$4.17K_c = 60 \Rightarrow K_c = 14.4 \tag{h}$$

9.7 Feedback Control of First-Order Plants

The standard form of the transfer function for a first-order plant is

$$G(s) = \frac{K}{s + \frac{1}{T}} \tag{9.97}$$

where K is a constant and T is the system's time constant. The transient behavior of first-order systems is considered in Section 7.4. The response of a first-order plant when placed in a feedback loop with feedforward control is considered in this section.

9.7.1 Proportional Controller

The closed-loop transfer function for a first-order system with a proportional controller is obtained by substituting Equation (9.97) into Equation (9.25), leading to

$$H(s) = \frac{K_p \frac{K}{s + \frac{1}{T}}}{1 + K_p \frac{K}{s + \frac{1}{T}}}$$

$$= \frac{K_p K}{s + \frac{1}{T} + K_p K} \tag{9.98}$$

The transfer function of Equation (9.98) is that of a first-order system with a time constant of

$$\hat{T} = \frac{T}{1 + K_p KT} \tag{9.99}$$

and a multiplicative constant of $K_{\hat{p}}K$. For $K_p > 0$, $\hat{T} < T$.

The response of a first-order system in a feedback loop with a proportional controller added to the feedforward portion is that of a first-order system with a time constant that is less than the time constant of the original system. Thus, the settling time and the rise time are shorter than those of the original system. If the system of Section 9.3.1 was put in a feedback control loop with a proportional controller with a gain of at least 1.46, the rise time would be less than 2 s. However, Equation (9.82) shows that the offset would become $\eta = \frac{1}{1 + 1.46(2)} = 0.255$. That is, the final value with the controller is different than the final value without the controller. The final value of the error without the controller is 0.5 m, and the final value of the error with the controller is 0.255 m. The response settles faster, but to a different final value.

The output from a closed-loop system with a first-order plant actuated by a proportional controller and subject to a unit step input $A(s) = 1/s$ is

$$C(s) = \frac{K_p K}{s\left(s + \frac{1}{T} + K_p K\right)} \tag{9.100}$$

The response of a first-order plant to a step input is compared with the open-closed loop response of a first-order system with a proportional controller in Figure 9.47. The final value of the open-loop system is KT, where the final value of the closed-loop system is obtained using the final value theorem as

$$\lim_{t\to\infty} c(t) = \lim_{t\to\infty} sC(s)$$

$$= \frac{K_p K}{\frac{1}{T} + K_p K}$$

$$= K_p K\hat{T} \tag{9.101}$$

The error for the system with a unit step input is

$$E(s) = \frac{s + \frac{1}{T}}{s\left(s + \frac{1}{\hat{T}}\right)} \tag{9.102}$$

The offset is obtained using the final value theorem as

$$\eta = \lim_{s\to 0} sE(s)$$

$$= \lim_{s\to 0} \frac{s + \frac{1}{T}}{s + \frac{1}{\hat{T}}}$$

$$= \frac{\hat{T}}{T} \tag{9.103}$$

The response of the first-order system with a proportional controller due to a unit step input is obtained by inverting Equation (9.100), resulting in

$$c(t) = K_p K\hat{T}\left(1 - e^{-t/\hat{T}}\right) \tag{9.104}$$

The 2 percent settling time for the closed-loop system with the proportional controller is $t_s = 3.92\hat{T}$. The benchmark times, such as the 2 percent settling time, are lower for a first-order plant when placed in a feedback control loop with proportional controller. The benchmark times decrease with increasing values of proportional gain. However, the use of proportional control alone leads to a nonzero offset.

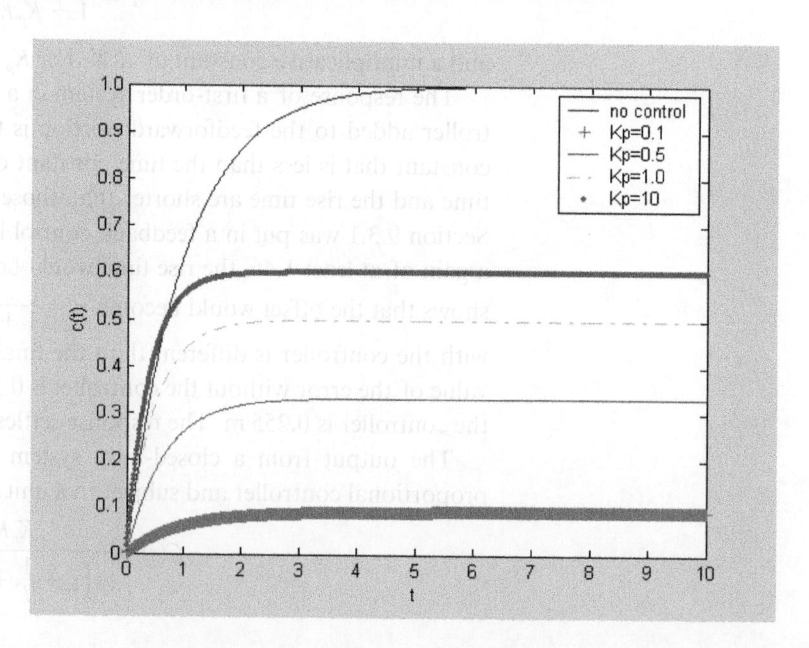

Figure 9.47

The use of a proportional controller with a first-order plant leads to offset. The offset is smaller for controllers with larger gains. The settling time is less than the settling time for a plant without a controller and is smaller for controllers with larger gains.

9.7.2 Integral Controller

The closed-loop transfer function of a first-order system with an integral controller with $G_a(s) = K_i/s$ is obtained by substituting Equation (9.97) into Equation (9.27), resulting in

$$H(s) = \frac{\dfrac{K_i K}{s\left(s + \dfrac{1}{T}\right)}}{1 + \dfrac{K_i K}{s\left(s + \dfrac{1}{T}\right)}} = \frac{K_i K}{s^2 + \dfrac{1}{T}s + K_i K} \tag{9.105}$$

Equation (9.105) is the transfer function of a second-order system with natural frequency

$$\omega_n = \sqrt{K K_i} \tag{9.106}$$

and damping ratio

$$\zeta = \frac{1}{2T\sqrt{K_i K}} \tag{9.107}$$

The larger the value of the integral gain, the smaller the value of the damping ratio and the larger the value of the natural frequency of the system. If $K_i > 1/(4KT^2)$, the system is underdamped.

The transform of the error when an integral controller is used for a first-order plant is

$$E(s) = A(s)\frac{s^2 + \dfrac{1}{T}s}{s^2 + \dfrac{1}{T}s + K_i K} \tag{9.108}$$

The transform of the response due to a unit step input is

$$C(s) = \frac{K_i K}{s\left(s^2 + \dfrac{1}{T}s + K_i K\right)} \tag{9.109}$$

As noted in Section 9.6, the offset when an integral controller is used on a first-order plant is $\eta = 0$.

The closed-loop response of a first-order system with an integral controller due to a unit step input is obtained using Table 7.5. If the closed-loop system is underdamped, the response is

$$x(t) = \frac{K_i K}{\omega_n^2}\left[1 - e^{-\zeta\omega_n t}\cos(\omega_d t) + \frac{\zeta\omega_n}{\omega_d}e^{-\zeta\omega_n t}\sin(\omega_d t)\right] \tag{9.110}$$

The use of an integral controller with a first-order plant leads to a second-order system. The system is underdamped if the integral gain is large enough. The use of the integral controller eliminates the offset that occurs when a proportional controller is used. The natural frequency of the resulting second-order system increases as the integral gain increases while the damping ratio decreases. The increase in the natural frequency decreases the benchmark times, but the decrease in the damping ratio increases the benchmark times. However, the use of an integral controller allows only one parameter, the damping ratio or the natural frequency, to be specified. The output from a closed-loop system with an integral controller and first-order plant with a unit step input is illustrated in Figure 9.48.

9.7.3 Proportional Plus Integral Controller

The use of a proportional plus integral controller in a closed-loop system with a first-order plant leads to a closed-loop transfer function of

$$H(s) = \frac{K(K_p s + K_i)}{s^2 + \left(KK_p + \dfrac{1}{T}\right)s + KK_i} \tag{9.111}$$

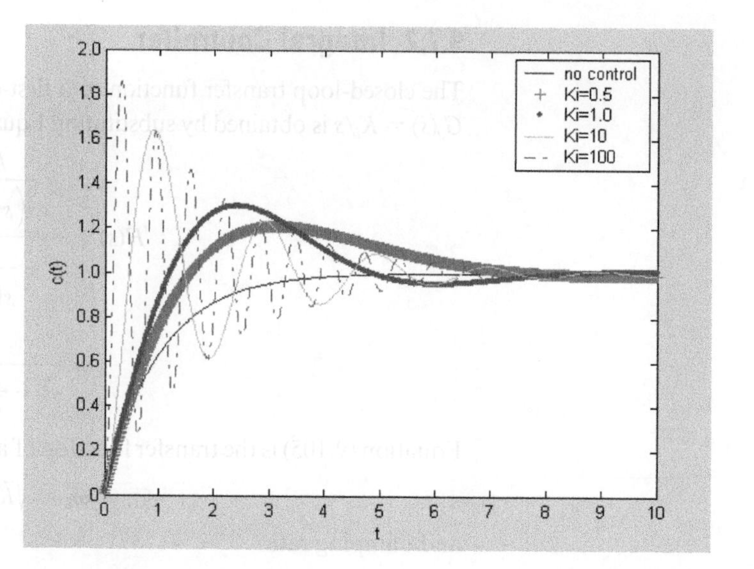

Figure 9.48
The step response of a first-order plant in a feedback loop with an integral controller in the feedforward part of the loop. There is no offset with an integral controller, but the system is underdamped for larger K_i.

The closed-loop system is a second-order system with the same natural frequency as a system with just integral control, given by Equation (9.106) and damping ratio

$$\zeta = \frac{K_p}{2}\sqrt{\frac{K}{K_i}} + \frac{1}{2T\sqrt{KK_i}} \tag{9.112}$$

The use of a PI controller with a first-order plant provides more flexibility than does the use of an integral controller. An integral controller has only one parameter, the integral gain, that can be chosen to control the system response. Thus only the natural frequency or the damping ratio can be specified when using an integral controller. When a PI controller is used, two parameters (the proportional gain and the integral gain) may be chosen. The integral gain may be chosen to specify the system's natural frequency, and the proportional gain is then chosen to specify the system's damping ratio.

The minimum value of the damping ratio for a fixed natural frequency is $\zeta_{\min} = \frac{1}{2\omega_n T}$ corresponding to $K_p = 0$. As K_p increases, so too does the damping ratio, thus decreasing the oscillation in the closed-loop response and resulting in an improved response time.

The offset for a PI controller is zero, as the integrator eliminates any error at steady state.

9.7.4 Proportional Plus Differential Controller

The use of a PD controller in a closed-loop system with a first-order plant leads to

$$H(s) = \frac{KK_d s + KK_p}{(1 + KK_d)s + \frac{1}{T} + KK_p} \tag{9.113}$$

As noted in Section 9.6, the addition of derivative control to a proportional controller does not affect the offset.

The closed-loop transfer function for a first-order system with a PD controller is that of a first-order system with a time constant of

$$\hat{T} = \frac{1 + KK_d}{\frac{1}{T} + KK_p} \tag{9.114}$$

The order of the numerator is equal to the order of the denominator. Thus, the step response is discontinuous at $t = 0$. The initial value theorem is applied to determine the initial response of the system due to a unit step input:

$$c(0) = \lim_{s\to\infty} H(s) = \frac{KK_d}{1 + KK_d} \tag{9.115}$$

The nonzero initial value is a result of the differential control. It approaches a limit of one as K_d gets larger and larger.

Application of the final value theorem to the transfer function shows that the final value of $c(t)$ is

$$c_f = H(0) = \frac{KK_p}{\frac{1}{T} + KK_p} \tag{9.116}$$

The offset of a first-order plant with a PD controller is not affected by the differential control.

The transform of the response of the system due to a unit step function is

$$C(s) = \frac{KK_p}{\frac{1}{T} + KK_p}\frac{1}{s} - \frac{\left(KK_p - \frac{KK_d}{T}\right)\hat{T}}{(1 + KK_d)^2}\frac{1}{s + \frac{1}{\hat{T}}} \tag{9.117}$$

Equation (9.117) implies that the step response is a constant if

$$K_p = \frac{K_d}{T} \tag{9.118}$$

Figure 9.49 illustrates the application of a proportional plus derivative controller to a first-order plant with a transfer function $G(s) = 1/(s + 1)$ with $K_p = 10$ and $K_d = 1$ when the system is subject to a unit step input. The addition of the derivative term does not affect the offset caused by the proportional controller, but it does lead to a discontinuity in the response at $t = 0$. This discontinuity is caused by differentiating a unit step input that is discontinuous at $t = 0$. The addition of the derivative term in the control strategy causes the system to respond more quickly, reducing the rise time.

9.7.5 PID Controller

The closed-loop transfer function of a proportional plus integral plus derivative controller used with a first-order plant is

$$H(s) = \frac{K\left(K_d s^2 + K_p s + K_i\right)}{(1 + KK_d)s^2 + \left(KK_p + \frac{1}{T}\right)s + KK_i} \tag{9.119}$$

The closed-loop system for a first-order plant with a PID controller is a second-order system. The natural frequency and damping ratio of the closed-loop system are obtained from the denominator of Equation (9.119), yielding

$$\omega_n = \sqrt{\frac{KK_i}{1 + KK_d}} \tag{9.120}$$

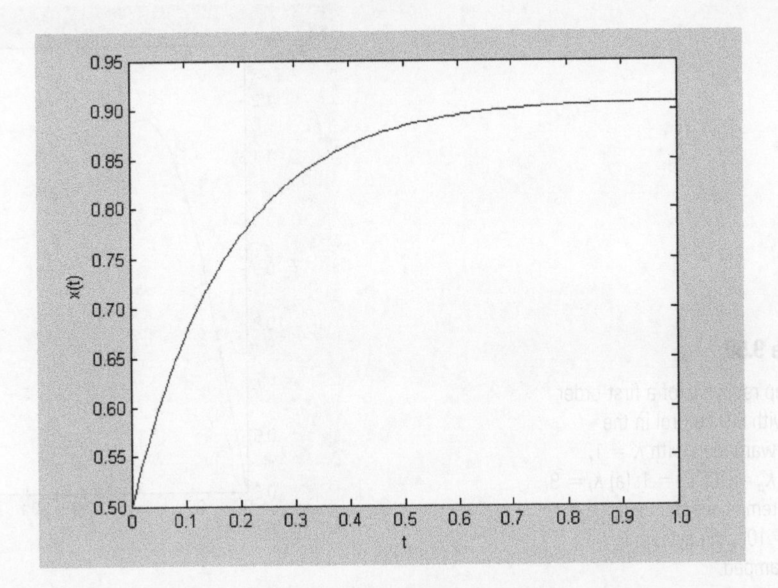

Figure 9.49

The use of a proportional plus derivative controller to a first-order plant with a step input leads to a discontinuity.

and

$$\zeta = \frac{KK_p + \frac{1}{T}}{2\sqrt{KK_i(1 + KK_d)}} \tag{9.121}$$

Comparing Equation (9.119) to the closed-loop transfer function of a first-order plant with an integral controller, Equation (9.105), shows that an effect of adding damping is to decrease the natural frequency and that an effect of adding a proportional term is to increase the damping ratio. An increase in the value of the integral gain leads to an increase in the natural frequency and a decrease in the damping ratio.

The numerator of the closed-loop transfer function is important in the rate of response due to the controller. When derivative control is present, the numerator of the closed-loop transfer function for a first-order plant is of the same order as the denominator. Thus the step response of the closed-loop system is discontinuous at $t = 0$. The rate at which the response occurs is affected by the numerator of $H(s)$. When derivative control is present, the numerator is of a higher order. This leads to a decrease in the benchmark times for the transient response of a second-order system.

Figure 9.50 illustrates the use of PID control on a first-order plant with a transfer function $G(s) = 1/(s + 1)$ with a unit step input with $K_p = 10$ and $K_d = 1$. The step

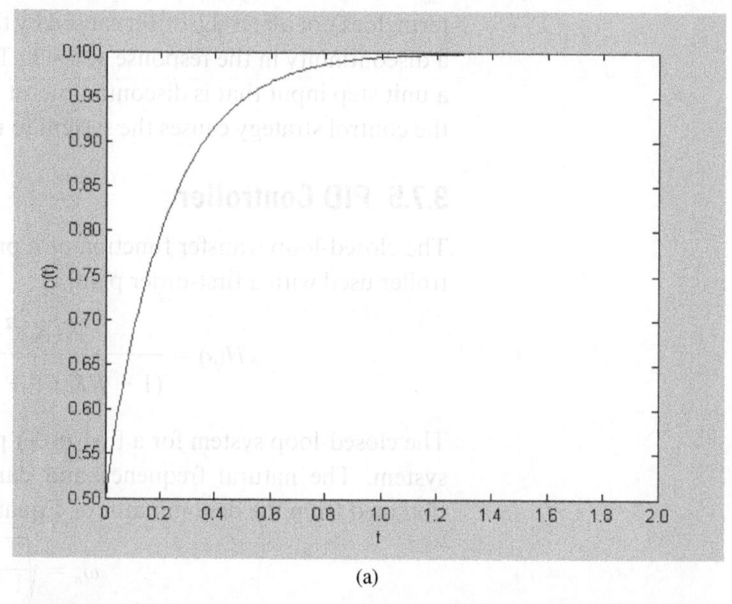

(a)

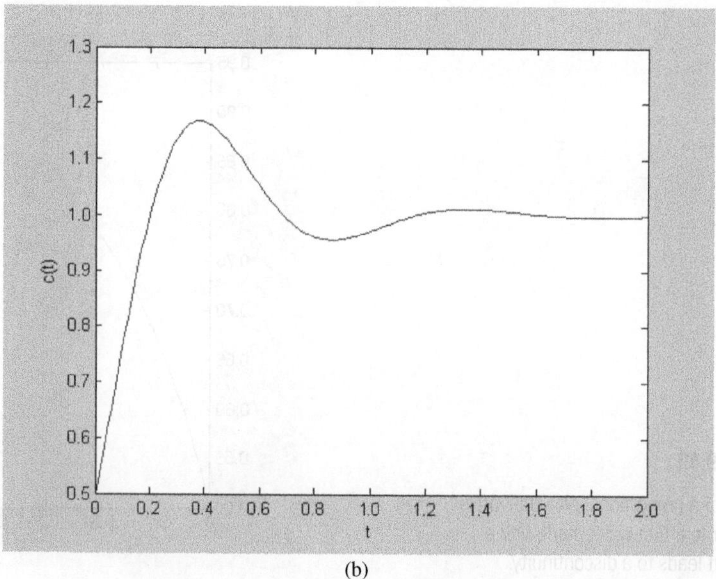

(b)

Figure 9.50

The step response of a first-order plant with PID control in the feedforward loop with $K = 1$, $T = 1$, $K_p = 10$, $K_d = 1$: (a) $K_i = 9$, the system is overdamped; (b) $K_i = 100$, the system is underdamped.

Table 9.4 Summary of Closed-Loop Response of First-Order Plants, $G(s) = K/[s + (1/T)]$ with Feedforward Actuator, $G_c(s)$

Controller	$G_c(s)$	Order of Closed-Loop System	Closed-Loop Transfer Function	Closed-Loop System Parameters	Offset
Proportional	K_p	first	Equation (9.98)	$\hat{T} = \dfrac{T}{1 + K_p KT}$	$\dfrac{\hat{T}}{T}$
Integral	$\dfrac{k_i}{s}$	second	Equation (9.105)	$\hat{\omega}_n = \sqrt{KK_i}$ $\hat{\zeta} = \dfrac{1}{2T\sqrt{KK_i}}$	0
PD	$K_p + K_d s$	first	Equation (9.113)	$\hat{T} = \dfrac{1 + KK_d}{\frac{1}{T} + KK_p}$	$\dfrac{1}{1 + K_p KT}$
PI	$K_p + \dfrac{K_i}{s}$	second	Equation (9.111)	$\hat{\omega}_n = \sqrt{KK_i}$ $\hat{\zeta} = \dfrac{K_p}{2}\sqrt{\dfrac{K}{K_i}} + \dfrac{1}{2T\sqrt{KK_i}}$	0
PID	$K_p + K_d s + \dfrac{K_i}{s}$	second	Equation (9.119)	$\hat{\omega}_n = \sqrt{\dfrac{KK_i}{1 + KK_d}}$ $\hat{\zeta} = \dfrac{KK_p + \frac{1}{T}}{2\sqrt{KK_i(1 + KK_d)}}$	0

Note: $\hat{T}$ is the time constant of a first-order closed-loop system, and $\hat{\omega}_n$ and $\hat{\zeta}$ are the natural frequency and damping ratio of a second-order closed-loop system.

response is discontinuous with $x_s(0) = KK_d/(1 + KK_d)$. The addition of the integral control eliminates the offset.

This section provides an introduction to the closed-loop response of dynamic systems with various controllers. A proportional controller is the basic controller, but its use leads to error and offset in the closed-loop response. This offset is eliminated by adding integral control. However, the use of integral control increases the order of the system and is destabilizing. While the closed-loop response of a first-order plant is inherently stable, an increase in integral gain leads to a decrease in the damping ratio, which itself is destabilizing. Derivative control is used to enhance sensitivity and increase the rate of response.

The properties of the closed-loop system of a first-order plant acted on by various controllers in the feedforward portion of the loop are summarized in Table 9.4.

Example 9.11

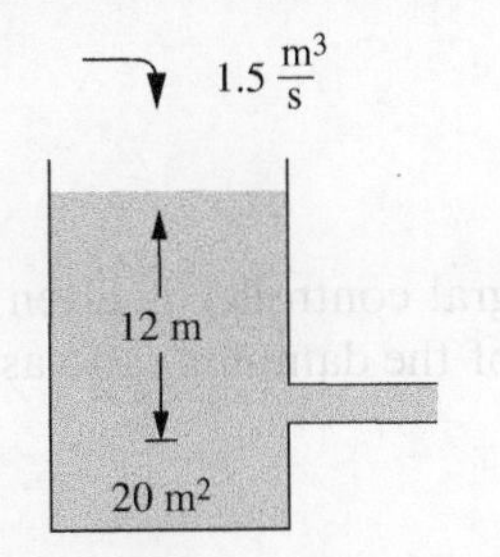

Figure 9.51

Single-tank liquid-level system of Example 9.11.

A tank of cross-sectional area 20 m^2, shown in Figure 9.51, has a steady-state level of 12 m for an incoming flow rate of 1.5 m^3/s. A closed-loop control system is being designed for the tank to improve its response when the flow rate has a step change. In all of the following cases, provide the closed-loop transfer function of the system. **a.** What is the required gain of a proportional controller such that the 2 percent settling time of the system is 180 s? **b.** What is the required gain of an integral controller such that the response is underdamped with a damping ratio of 0.8? **c.** If a proportional plus integral controller is used with the integral gain as determined in part b., what is the required proportional gain such that the system is critically damped? **d.** For the output to be compared to the input after the feedback loop, a sensor is placed in the system to convert the input from perturbation of flow rate to its resultant final value of perturbation in liquid level. This is accomplished by multiplying the perturbation flow rate by the resistance R. Plot the output for the system when this sensor is used with each of the controllers specified in parts a.–c. when the perturbation in flow rate is 0.02 m^3/s.

Solution

The resistance of the system at the steady state is calculated as

$$R = 2\frac{H}{Q}$$
$$= 2\frac{12 \text{ m}}{1.5 \text{ m}^3/\text{s}} = 16 \text{ s/m}^2 \tag{a}$$

The transfer function for the system is

$$G(s) = \frac{1}{A\left(s + \frac{1}{T}\right)} \tag{b}$$

where the time constant is

$$T = RA$$
$$= (16 \text{ s/m}^2)(20 \text{ m}^2) = 320 \text{ s} \tag{c}$$

The constant K in Equation (9.97) for this system is

$$K = \frac{1}{A} = 0.05 \text{ m}^{-2} \tag{d}$$

a. The 2 percent settling time for the closed-loop system with a proportional controller is 3.92 $\hat{T}$, where $\hat{T}$ is the time constant for the system. Requiring the settling time to be 180 s leads to

$$\hat{T} = \frac{t_s}{3.92} = \frac{180 \text{ s}}{3.92} = 45.92 \text{ s} \tag{e}$$

Equation (9.99) is rearranged to solve for the proportional gain in terms of the time constant of the plant and the time constant of the closed-loop system as

$$K_p = \frac{T - \hat{T}}{KT\hat{T}} \tag{f}$$

Substituting calculated values into Equation (f) leads to

$$K_p = \frac{(320 \text{ s}) - (45.92 \text{ s})}{(0.05 \text{ m}^{-2})(320 \text{ s})(45.92 \text{ s})} = 0.373 \text{ m}^2/\text{s} \tag{g}$$

The equivalent closed-loop transfer function for this system is obtained using Equation (9.98) as

$$H(s) = \frac{KK_p}{s + \frac{1}{\hat{T}}}$$
$$= \frac{(0.05)(0.373)}{s + \frac{1}{45.92}}$$
$$= \frac{1.87 \times 10^{-2}}{s + 2.18 \times 10^{-2}} \tag{h}$$

b. The damping ratio for a feedback system with an integral controller is given by Equation (9.107). The integral gain is determined in terms of the damping ratio as

$$K_i = \frac{1}{4KT^2\zeta^2} \tag{i}$$

Requiring the damping ratio to be 0.8 leads to

$$K_i = \frac{1}{4(0.05 \text{ m}^{-2})(320 \text{ s})^2(0.8)^2} = 7.63 \times 10^{-5}\frac{\text{m}^2}{\text{s}^2} \tag{j}$$

The transfer function when this controller is used is obtained using Equation (9.105) as

$$H(s) = \frac{(0.05)(7.63 \times 10^{-5})}{s^2 + \frac{1}{320}s + (0.05)(7.63 \times 10^{-5})}$$

$$= \frac{3.82 \times 10^{-6}}{s^2 + 3.13 \times 10^{-3}s + 3.82 \times 10^{-6}} \tag{k}$$

c. The damping ratio for a proportional plus integral controller used with a first-order plant is given by Equation (9.112). Rearranging to solve for the proportional gain leads to

$$K_p = 2\sqrt{\frac{K_i}{K}}\left(\zeta - \frac{1}{2T\sqrt{KK_i}}\right) \tag{l}$$

Noting that since the integral controller of part b. is designed for a damping ratio of 0.8, requiring a damping ratio of 1 when a PI controller is used leads to

$$K_p = 2\sqrt{\frac{7.63 \times 10^{-5}}{0.05}}(1 - 0.8)$$

$$= 1.57 \times 10^{-2} \tag{m}$$

The transfer function when a PI controller is used is given by Equation (9.111), which for this system becomes

$$H(s) = \frac{(0.05)(1.57 \times 10^{-2}s + 7.63 \times 10^{-5})}{s^2 + \left[\frac{1}{320} + (0.05)(1.57 \times 10^{-2})\right]s + (0.05)(7.63 \times 10^{-5})}$$

$$= \frac{7.85 \times 10^{-4}s + 3.82 \times 10^{-6}}{s^2 + 3.91 \times 10^{-3}s + 3.82 \times 10^{-6}} \tag{n}$$

d. The use of the sensor before the feedback control loop leads to an input of

$$A(t) = qRu(t)$$

$$= \left(0.02\frac{\text{m}^3}{\text{s}}\right)\left(16\frac{\text{s}}{\text{m}^2}\right)u(t) = 0.32u(t) \tag{o}$$

The responses of the system using the feedback controllers specified in parts a.–c. are shown in Figure 9.52. The offset when using the proportional controller is apparent.

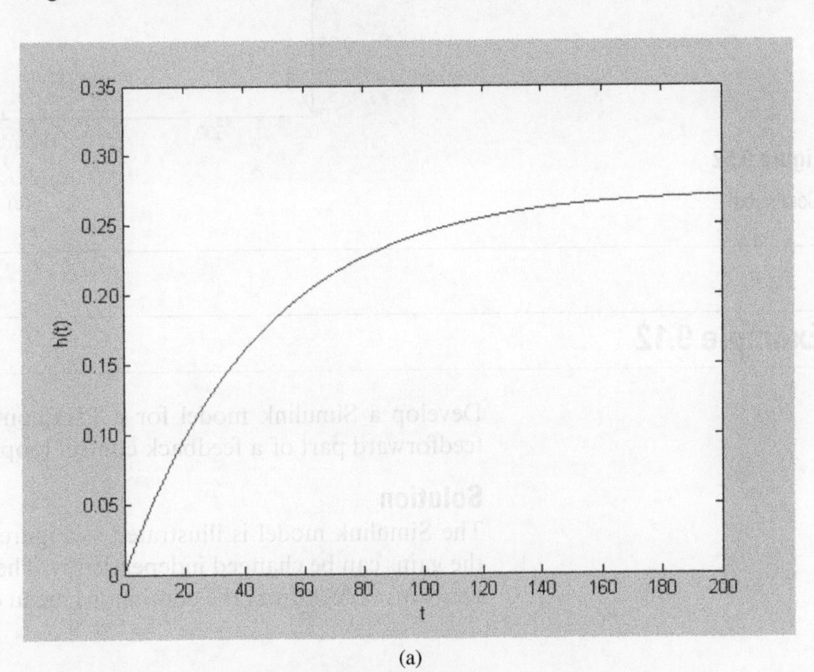

(a)

Figure 9.52

The response of the tank of Example 9.11 with feedback control using (a) the proportional controller of part a., (b) the integral controller of part b., and (c) the proportional plus integral controller of part c.

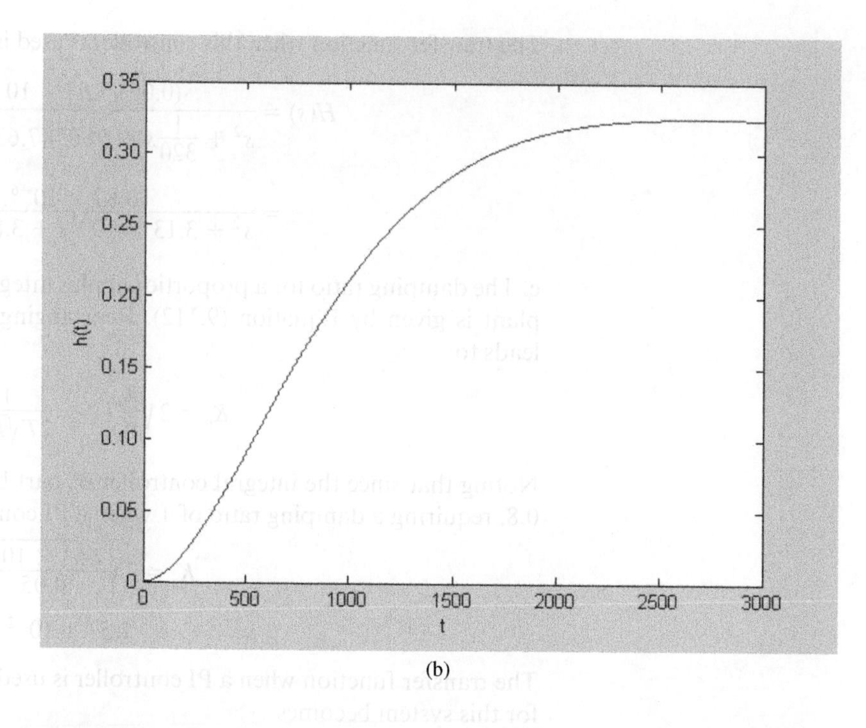

(b)

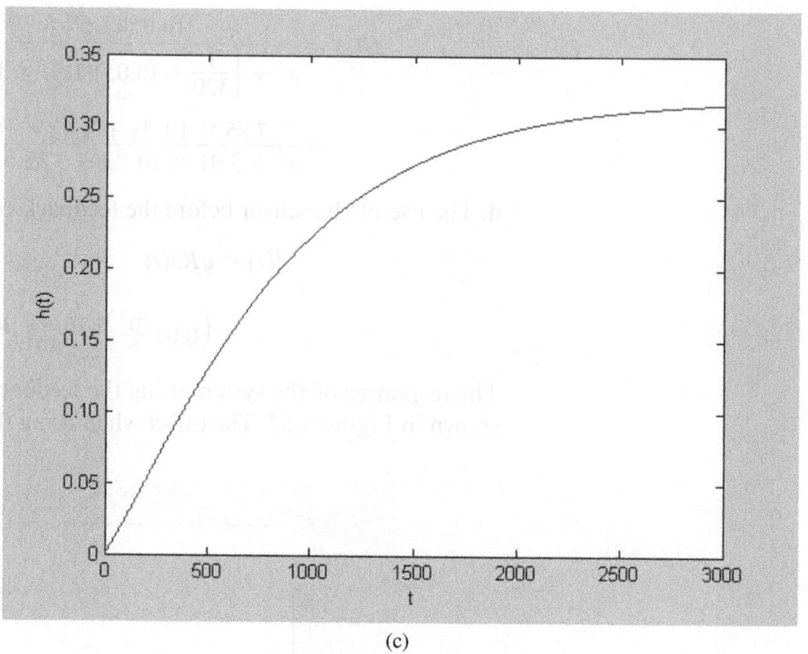

(c)

Figure 9.52
(Continued)

Example 9.12

Develop a Simulink model for a PID controller to be used as an actuator in the feedforward part of a feedback control loop for the system of Example 9.11.

Solution

The Simulink model is illustrated in Figure 9.53. The model is developed such that the gains can be changed independently. The model uses the Integrator block and the Derivative block from the continuous menu of the Simulink Library Browser.

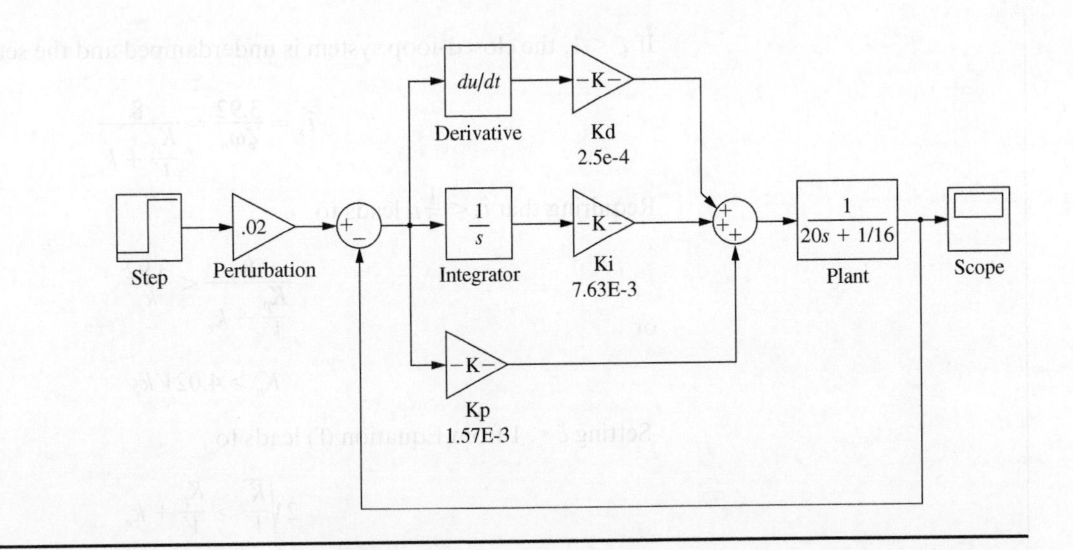

Figure 9.53
The Simulink model for Example 9.12; a first-order plant in a feedback loop with a PID controller.

Example 9.13

Recall the one-compartment model for drug delivery in Example 6.18. A drug is infused into the plasma through a mechanism such as a patch. The mathematical model for the concentration of the drug in the plasma is

$$V\frac{dC}{dt} + k_e VC = I \tag{a}$$

where V is the volume of the plasma, k_e is the rate of elimination of the drug from the plasma, and I is the rate of infusion. Discuss a control strategy for this system to speed up the settling time to half of its original value.

Solution

The transfer function for the system is obtained from Equation (a) as

$$G(s) = \frac{C(s)}{I(s)} = \frac{1/V}{s + k_e} \tag{b}$$

The time constant for this first-order model is $T = \frac{1}{k_e}$. It is desired to reduce the settling time while not having offset. Thus, a PI controller is chosen

$$G_c(s) = K_p + \frac{K_i}{s} \tag{c}$$

If the drug-delivery system is put in a feedback loop with a PI controller, the closed-loop transfer function is

$$H(s) = \frac{\frac{1}{V}(K_p s + K_i)}{s^2 + \left(\frac{K_p}{V} + k_e\right)s + \frac{K_i}{V}} \tag{d}$$

The transfer function of Equation (d) is that of a second-order system with a natural frequency of

$$\omega_n = \sqrt{\frac{K_i}{V}} \tag{e}$$

and a damping ratio of

$$\zeta = \frac{\frac{K_p}{V} + k_e}{2\sqrt{\frac{K_i}{V}}} \tag{f}$$

If $\zeta < 1$, the closed-loop system is underdamped and the settling time is

$$\hat{t}_s = \frac{3.92}{\zeta\omega_n} = \frac{8}{\frac{K_p}{V} + k_e} \tag{g}$$

Requiring that $\hat{t}_s < \frac{1}{2}t_s$ leads to

$$\frac{8}{\frac{K_p}{V} + k_e} < \frac{3.92}{k_e} \tag{h}$$

or

$$K_p > 4.02Vk_e \tag{i}$$

Setting $\zeta < 1$ from Equation (f) leads to

$$2\sqrt{\frac{K_i}{V}} > \frac{K_p}{V} + k_e \tag{j}$$

Using Equation (i) in Equation (j) leads to

$$K_i > 6.30Vk_e^2 \tag{k}$$

9.8 Control of Second-Order Plants

The transfer function of a second-order system is of the form

$$G(s) = \frac{As + B}{s^2 + 2\zeta\omega_n s + \omega_n^2} \tag{9.122}$$

where ζ is the damping ratio for the system, ω_n is its natural frequency, and A and B are system constants.

9.8.1 Proportional Controller

The use of a proportional controller on a second-order plant in the feedback control loop of Figure 9.28 leads to a closed-loop transfer function of

$$H(s) = \frac{K_p(As + B)}{s^2 + (2\zeta\omega_n + K_pA)s + K_pB + \omega_n^2} \tag{9.123}$$

Equation (9.123) is the transfer function of a second-order system of natural frequency

$$\hat{\omega}_n = \sqrt{\omega_n^2 + K_pB} \tag{9.124}$$

and damping ratio

$$\hat{\zeta} = \frac{\zeta + \frac{K_pA}{2\omega_n}}{\sqrt{1 + \frac{K_pB}{\omega_n^2}}} \tag{9.125}$$

The response of a second-order plant with proportional feedback control when subject to a unit step input is

$$C(s) = \frac{K_p(As + B)}{s\left(s^2 + 2\hat{\zeta}\hat{\omega}_n + \hat{\omega}_n^2\right)} \tag{9.126}$$

The transform of the error due to a unit step input is

$$E(s) = \frac{1}{s}[1 - H(s)]$$

$$= \frac{s^2 + 2\zeta\omega_n s + \omega_n^2}{s\left(s^2 + 2\hat{\zeta}\,\hat{\omega}_n s + \hat{\omega}_n^2\right)} \tag{9.127}$$

The offset is evaluated as

$$\lim_{t\to\infty} e(t) = \lim_{s\to 0} sE(s)$$

$$= \left(\frac{\omega_n}{\hat{\omega}_n}\right)^2$$

$$= \frac{1}{1 + \dfrac{K_p B}{\omega_n^2}} \tag{9.128}$$

Example 9.14

Plot the response of a second-order plant with $A = 0$, $B = 1$, $\zeta = 1.5$, and $\omega_n = 10$ when subject to a unit step input and placed in a feedback loop with a proportional controller for various values of the nondimensional parameter $\kappa = K_p B/\omega_n^2$.

Solution

For the given values, Equations (9.124) and (9.125) are used to determine the natural frequency and damping ratio of the closed-loop system as

$$\hat{\omega}_n = 10\sqrt{1+\kappa} \tag{a}$$

$$\hat{\zeta} = \frac{1.5}{\sqrt{1+\kappa}} \tag{b}$$

Equation (b) shows that with $A = 0$, the damping ratio for the closed-loop system is less than the damping ratio for the plant itself. The closed-loop system is over-damped $(\hat{\zeta} > 1)$ when $\kappa < 1.25$, critically damped for $\kappa = 1.25$, and underdamped for $\kappa > 1.25$. The response of the system in each case is determined using Table 7.5. Plots for different values of κ are given in Figure 9.54.

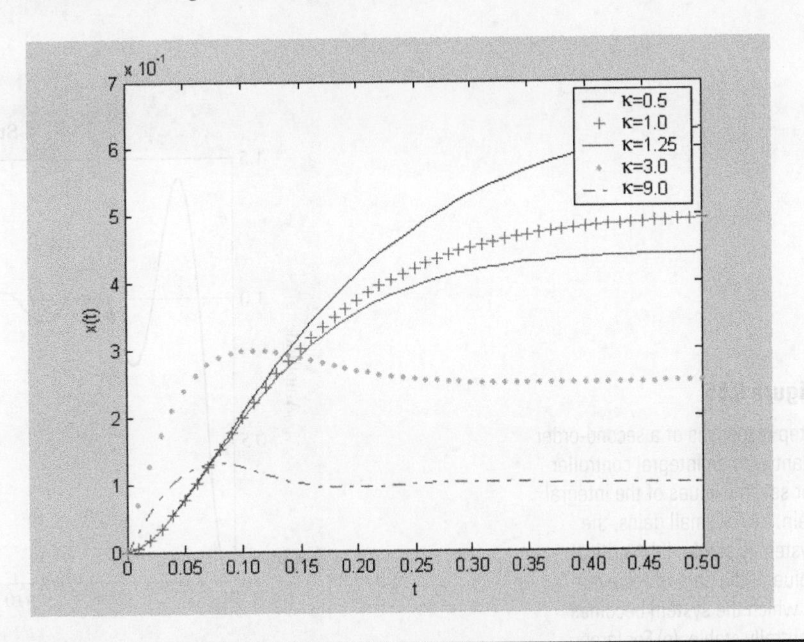

Figure 9.54

The response of a second-order plant subject to a unit step input when placed in a feedback loop with a proportional controller. The natural frequency and damping ratio of the closed-loop system vary with the proportional gain. The offset is also dependent on the gain.

9.8.2 Integral Controller

The use of an integral controller in a feedback loop with a second-order plant leads to a closed-loop transfer function of

$$H(s) = \frac{K_i(As + B)}{s^3 + 2\zeta\omega_n s^2 + (\omega_n^2 + K_i A)s + K_i B} \tag{9.129}$$

The closed-loop transfer function of Equation (9.129) is that of a third-order system. The system is stable if each of its three poles has negative real parts. The application of Routh's stability criterion to a third-order polynomial of the form

$$p(s) = s^3 + a_2 s^2 + a_1 s + a_0 \tag{9.130}$$

shows that all the roots of $p(s)$ have negative real parts if and only if

$$a_1 a_2 > a_0 \tag{9.131}$$

The application of Equation (9.131) to the transfer function of Equation (9.129) leads to a stability criterion of

$$2\zeta\omega_n(\omega_n^2 + K_i A) > K_i B$$

$$K_i < \frac{2\zeta\omega_n^3}{B - 2\zeta\omega_n A} \tag{9.132}$$

If the integral gain K_i satisfies Equation (9.132), the closed-loop system is stable. If K_i is equal to the value on the right-hand side of Equation (9.132), then one pole has a real part of zero and the system is neutrally stable. In this case the response due to a unit step input is oscillatory with constant amplitude. If K_i is greater than that value on the right-hand side of Equation (9.129), then the system is unstable and the response due to a unit step input grows without bound.

Figure 9.55 shows the step response for a system of the form of Equation (9.129) with $A = 0$, $B = 1$, $\omega_n = 2$, and $\zeta = 0.75$ in a closed loop with an integral controller for three values of K_i. The system is stable for $K_i = 6$ [Figure 9.55(a)], neutrally stable for $K_i = 12$ [Figure 9.55(b)], and unstable for $K_i = 15$ [Figure 9.55(c)], as predicted by Equation (9.132).

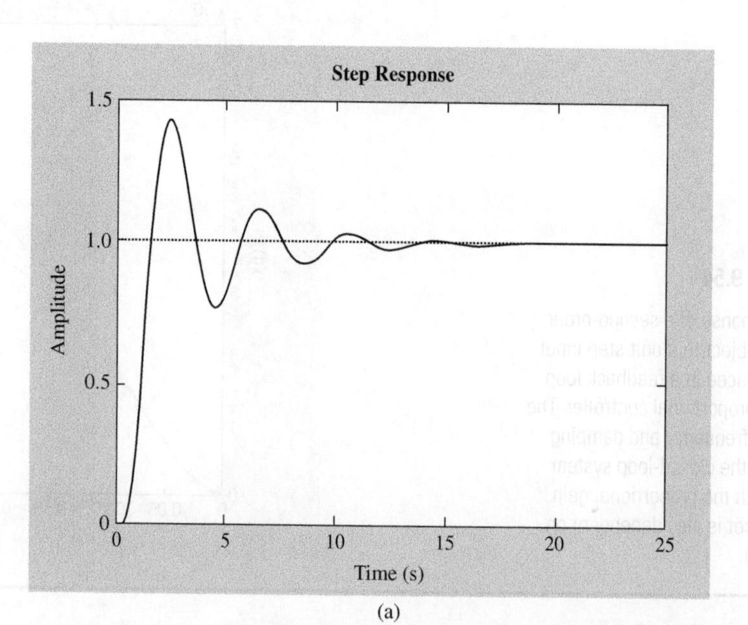

(a)

Figure 9.55

Step responses of a second-order plant with an integral controller for several values of the integral gain. (a) For small gains, the system is stable; (b) A critical value of the gain is achieved in which the system becomes neutrally stable, (c) For large gains, the system is unstable.

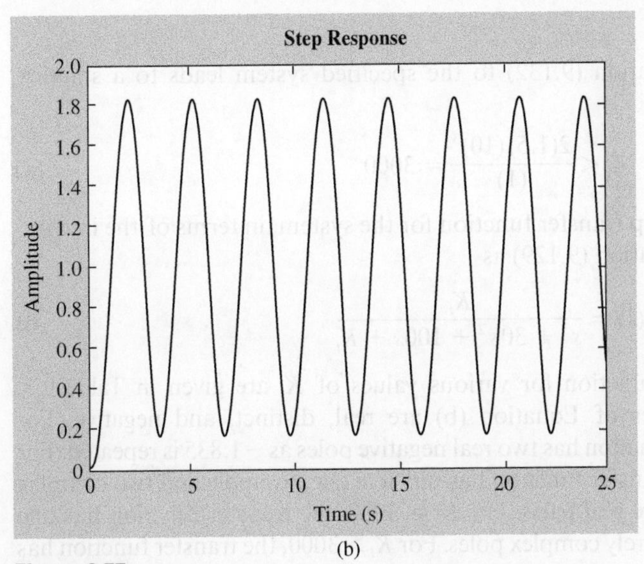

(b)

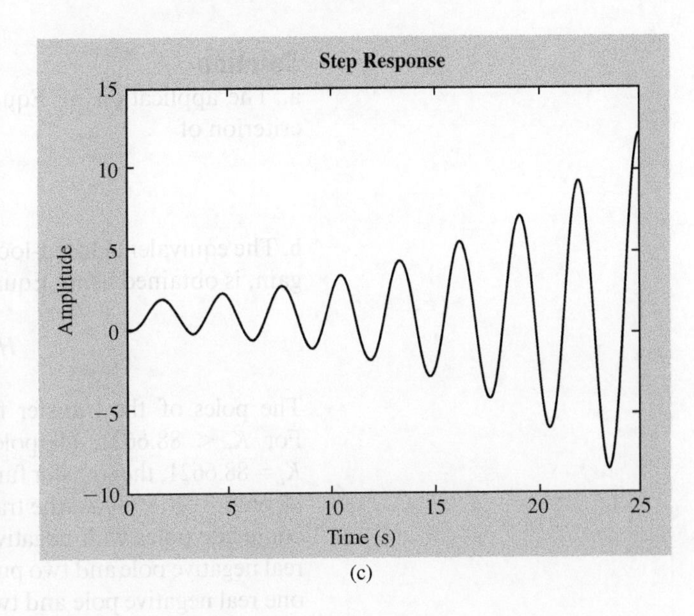

(c)

Figure 9.55
(Continued)

9.8.3 PI Controller

The use of a PI controller in a closed-loop feedback system with a second-order plant leads to a closed-loop transfer function of the form

$$H(s) = \frac{K_p A s^2 + (K_i A + K_p B)s + K_i B}{s^3 + (2\zeta\omega_n + K_p A)s^2 + (\omega_n^2 + K_p B + K_i A)s + K_i B} \tag{9.133}$$

Equation (9.133) is the transfer function for a third-order system. The application of Equation (9.131) to the transfer function of Equation (9.133) leads to a stability criterion of

$$(2\zeta\omega_n + K_p A)(\omega_n^2 + K_p B + K_i A) > K_i B \tag{9.134}$$

Equation (9.134) can be rearranged to

$$ABK_p^2 + (2\zeta\omega_n B + \omega_n^2 A + K_i A)K_p + 2\zeta\omega_n^3 + 2\zeta\omega_n K_i A - K_i AB > 0 \tag{9.135}$$

If $A = 0$, Equation (9.135) shows that if K_i satisfies Equation (9.132), then Equation (9.135) is satisfied for all positive values of K_p. If K_i does not satisfy Equation (9.132), then a value of K_p can be obtained to render the system stable. Thus the addition of proportional control enhances the stability of the system. The same result is obtained if A and B have the same sign. If $B = 0$, then A, K_i, and K_p all being of the same sign is sufficient for stability.

Example 9.15

Consider a second-order plant with $A = 0$, $B = 1$, $\zeta = 1.5$, and $\omega_n = 10$. **a.** If the plant is placed in a feedback loop with an integral controller, for what values of the integral gain is the system stable? **b.** Determine and plot the output from the feedback loop for various values of the integral gain when the system is subject to a unit step input. **c.** If the plant is placed in a feedback loop with a PI controller with an integral gain of $K_i = 4000$, for what values of the proportional gain constant is the system stable? **d.** Determine and plot the output from the feedback loop for various values of the proportional gain when the system is subject to a unit step input.

Solution

a. The application of Equation (9.132) to the specified system leads to a stability criterion of

$$K_i < \frac{2(1.5)(10)^3}{(1)} = 3000 \tag{a}$$

b. The equivalent closed-loop transfer function for the system, in terms of the integral gain, is obtained using Equation (9.129) as

$$H(s) = \frac{K_i}{s^3 + 30s^2 + 100s + K_i} \tag{b}$$

The poles of the transfer function for various values of K_i are given in Table 9.5. For $K_i < 88.6621$, all poles of Equation (b) are real, distinct, and negative. For $K_i = 88.6621$, the transfer function has two real negative poles as -1.835 is repeated. For $88.6621 < K_i < 3000$, the transfer function has one real negative pole and two complex conjugate poles with negative real parts. For $K_i = 3000$, the transfer function has one real negative pole and two purely complex poles. For $K_i > 3000$, the transfer function has one real negative pole and two complex conjugate poles with positive real parts.

The transform of the output when the input is a unit step input is

$$C(s) = \frac{1}{s}H(s)$$

$$= \frac{K_i}{s(s^3 + 30s^2 + 100s + K_i)} \tag{c}$$

For $K_i < 88.6621$, the appropriate partial fraction decomposition of Equation (c) is

$$C(s) = \frac{B_0}{s} + \frac{B_1}{s - s_1} + \frac{B_2}{s - s_2} + \frac{B_3}{s - s_3} \tag{d}$$

where the residues are evaluated as

$$B_0 = 1 \tag{e}$$

$$B_i = \lim_{s \to s_i} \frac{(s - s_i)K_i}{s(s^3 + 30s^2 + 100s + K_i)} \tag{f}$$

The inversion of Equation (d) leads to

$$c(t) = 1 + B_1 e^{s_1 t} + B_2 e^{s_2 t} + B_3 e^{s_3 t} \tag{g}$$

For $K_i = 88.6621$, the appropriate partial fraction decomposition of Equation (c) is

$$C(s) = \frac{1}{s} - \frac{5.61 \times 10^{-3}}{s + 26.33} - \frac{0.994}{s + 1.835} - \frac{1.973}{(s + 1.835)^2} \tag{h}$$

Table 9.5 Poles for Example 9.15 in Terms of the Integral Gain

K_i	s_1	s_2	s_3
1	−26.35	−3.81	−0.010
50	−26.26	−3.13	−0.609
88.6621	−26.33	−1.835	−1.835
100	−26.35	−1.853 + 0.680 *j*	−1.853 − 0.680 *j*
500	−26.98	−1.510 + 4.031 *j*	−1.510 − 4.031 *j*
1000	−27.69	−1.535 + 5.877 *j*	−1.535 − 5.877 *j*
2000	−28.94	−0.533 + 8.297 *j*	−0.533 − 8.927 *j*
2500	−29.48	−0.258 + 9.205 *j*	−0.258 − 9.205 *j*
3000	−30.00	10 *j*	−10 *j*
4000	−30.95	0.473 + 11.36 *j*	0.473 − 11.36 *j*

Inversion of Equation (h) leads to

$$c(t) = 1 - 5.61 \times 10^{-3} e^{-26.33t} - 0.994 e^{-1.835t} - 1.973 t e^{-1.835t} \tag{i}$$

For $88.6621 < K_i < 3000$, the appropriate partial fraction decomposition of Equation (c) is

$$C(s) = \frac{1}{s} + \frac{B_1}{s - s_1} + \frac{D_1 s + D_2}{s^2 - 2 s_{2r} s + \left(s_{2r}^2 + s_{2j}^2\right)} \tag{j}$$

where $s_2 = s_{2r} + j s_{2j}$, B_1 is determined using Equation (f), and D_1 and D_2 are obtained using the usual methods for quadratic factors. The inversion of Equation (j) leads to

$$c(t) = 1 + B_1 e^{s_1 t} + D_1 e^{s_{2r} t} \cos\left[\sqrt{s_{2r}^2 + s_{2j}^2}\, t\right] + \frac{D_2 + 2 s_{2r} D_1}{\sqrt{s_{2r}^2 + s_{2j}^2}} e^{s_{2r} t} \sin\left[\sqrt{s_{2r}^2 + s_{2j}^2}\, t\right] \tag{k}$$

For $K_i = 3000$, the appropriate partial fraction decomposition of Equation (c) is

$$C(s) = \frac{1}{s} - \frac{0.1}{s + 30} + \frac{0.9s - 3}{s^2 + 100} \tag{l}$$

Inversion of Equation (l) leads to

$$c(t) = 1 - 0.1 e^{-30t} + 0.9 \cos(10t) - 3 \sin(10t) \tag{m}$$

The partial fraction decomposition and the response of the system for $K_i > 3000$ are similar to those given by Equations (j) and (k) with the exception that s_{2r} is positive rather than negative.

Responses for several values of K_i are plotted in Figure 9.56.

c. An integral controller with $K_i = 4000$ leads to an unstable closed-loop system. If proportional control is added, Equation (9.135) shows that, to have a stable system,

$$K_p > \frac{K_i - 2\zeta\omega_n^3}{2\zeta\omega_n} = \frac{4000 - 2(1.5)(10)^3}{2(1.5)(10)}$$

$$K_p > \frac{100}{3} \tag{n}$$

d. The closed-loop transfer function for the system in terms of the proportional gain is obtained using Equation (9.133) as

$$H(s) = \frac{K_p s + 4000}{s^3 + 30s^2 + (100 + K_p)s + 4000} \tag{o}$$

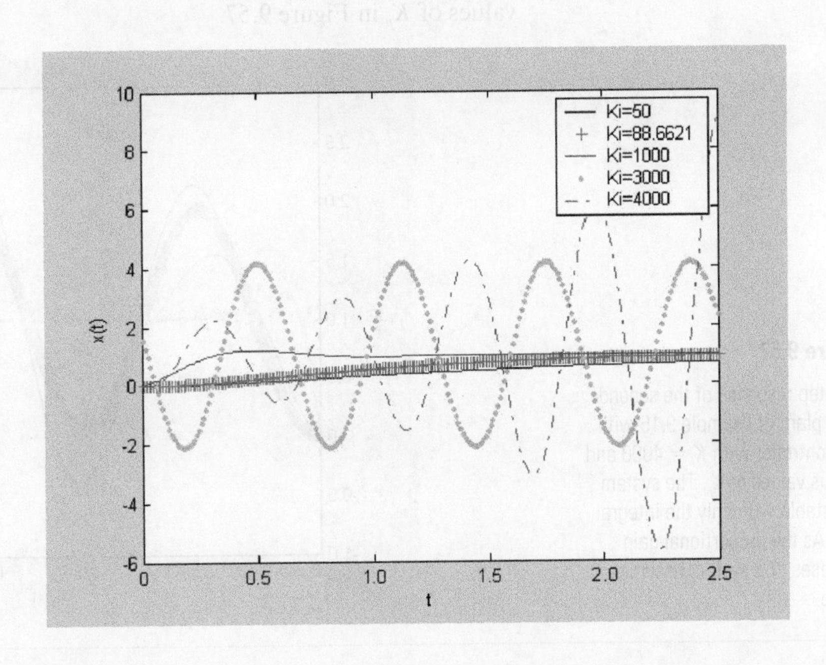

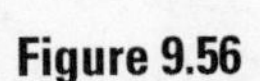

Figure 9.56

The step response of a second-order plant with an integral controller. For low values of the integral gain, the system approaches the final value asymptotically. As the integral gain increases, the response becomes oscillatory. When the integral gain reaches a critical value, the response becomes unstable and a further increase leads to unbounded growth.

Table 9.6 Poles for the System of Example 9.15 for Various Values of the Proportional Gain

K_p	s_1	s_2	s_3
0	−30.95	0.473 + 11.36j	0.473 − 11.35j
10	−30.67	0.333 + 11.42j	0.333 − 11.42j
33.33	−30	11.55j	−11.55j
50	−29.51	−0.245 + 11.64j	−0.245 − 11.64j
100	−27.96	−1.018 + 11.92j	−1.018 − 11.92j
500	−10	−10 + 17.32j	−10 − 17.32j
1000	−4.018	−12.99 + 28.76j	−12.99 − 28.76j
2000	−1.956	−14.02 + 42.99j	−14.02 − 14.99j
10,000	−0.397	−14.80 + 99.34j	−14.80 − 99.34j

Table 9.6 presents the poles of the transfer function of Equation (o) for several values of the proportional gain. For all values of K_p, the transfer function has one real and negative pole and two complex conjugate poles. For $0 \leq K_p < 100/3$, the complex poles have a positive real part and thus the closed-loop system is unstable. For $K_p = 100/3$, the complex poles have a real part equal to zero and the closed-loop system is neutrally stable. For $K_p > 100/3$, the complex poles have negative real parts and the system is stable. It is also noted that when $K_p = 500$, the real part of the complex poles equals the value of the real pole. For proportional gains greater than 500, the absolute value of the real part of the complex poles is larger than the absolute value of the real pole. For these values of the proportional gain, the settling time is dominated by the real pole.

The transform of the system output when it is subject to a unit step input is

$$C(s) = \frac{K_p s + 4000}{s\left[s^3 + 30s^2 + (100 + K_p)s + 4000\right]} \tag{p}$$

For all values of K_p, the appropriate partial fraction decomposition of Equation (p) is

$$C(s) = \frac{1}{s} + \frac{B_1}{s - s_1} + \frac{D_1 s + D_2}{s^2 + 2s_{2r}s + s_{2r}^2 + s_{2j}^2} \tag{q}$$

The system response is of the same form as Equation (j) and is plotted for several values of K_p in Figure 9.57.

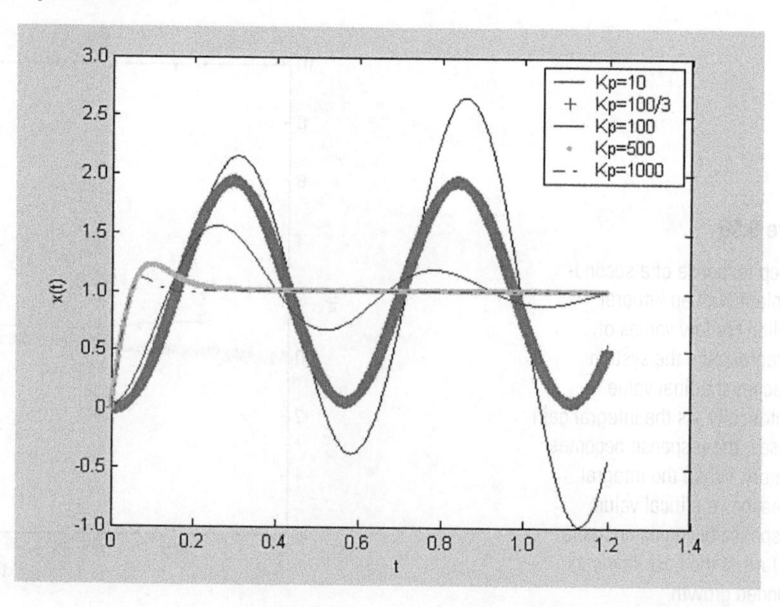

Figure 9.57

The step response of the second-order plant of Example 9.15 with a PI controller with $K_i = 4000$ and various values of K_p. The system is unstable with only the integral gain. As the proportional gain increases, the system becomes stable.

9.8.4 PD Controller

The use of a proportional plus derivative controller in a feedback loop with a second-order plant leads to a closed-loop transfer function of the form

$$H(s) = \frac{K_d A s^2 + (K_p A + K_d B)s + K_p B}{(1 + K_d A)s^2 + (K_p A + K_d B + 2\zeta\omega_n)s + K_p B + \omega_n^2} \tag{9.136}$$

Consider first the case when $A = 0$ and Equation (9.136) reduces to

$$H(s) = \frac{K_d B s + K_p B}{s^2 + (K_d B + 2\zeta\omega_n)s + K_p B + \omega_n^2} \tag{9.137}$$

Equation (9.137) is the transfer function of a second-order system. The natural frequency is unaffected by the derivative gain. It is the same as that for a system with only proportional control and is given by Equation (9.124) as

$$\hat{\omega}_n = \sqrt{\omega_n^2 + K_p B} \tag{9.124}$$

The damping ratio of the closed-loop system is

$$\hat{\zeta} = \frac{K_d B + 2\zeta\omega_n}{2\hat{\omega}_n} \tag{9.138}$$

Equation (9.138) shows that using derivative control along with proportional control increases the damping ratio beyond the damping ratio when only proportional control is used.

For a nonzero value of A, Equation (9.136) represents the transfer function for a second-order system, but with the numerator of the same order as the denominator. The response of such a system due to a unit step input is discontinuous at $t = 0$. The natural frequency and damping ratio for this system are

$$\hat{\omega}_n = \sqrt{\frac{K_p B + \omega_n^2}{1 + K_d A}} \tag{9.139}$$

$$\hat{\zeta} = \frac{K_d B + 2\zeta\omega_n}{2\hat{\omega}_n(1 + K_d A)} \tag{9.140}$$

Example 9.16

a. Consider a second-order plant with $A = 0$, $B = 1$, $\zeta = 1.5$, and $\omega_n = 10$. The plant is placed in a feedback loop with a PD controller with $K_p = 50$. Determine and plot the response of the system for various values of K_d when it is subject to a unit step input. **b.** Repeat part a. with $K_p = 500$. **c.** Repeat part a. for a second-order plant with $A = 1$, $B = 1$, $\zeta = 1.5$, and $\omega_n = 10$.

Solution

a. The response of a system with a proportional controller with $K_p = 50$ is overdamped. When derivative control is introduced in the controller, the response continues to be overdamped, as illustrated in Figure 9.58(a). The system is quicker to respond to the disturbance for higher values of the derivative gain but has a larger settling time. Overshoot occurs for larger values of K_d.

b. The response of a system with a proportional controller with $K_p = 500$ is underdamped. The system is underdamped for $K_d < 18.98$. As in part a., the system is quicker to respond for larger values of the derivative gain, but the settling time is also longer. When the system is overdamped, overshoot occurs. These responses are illustrated in Figure 9.58(b).

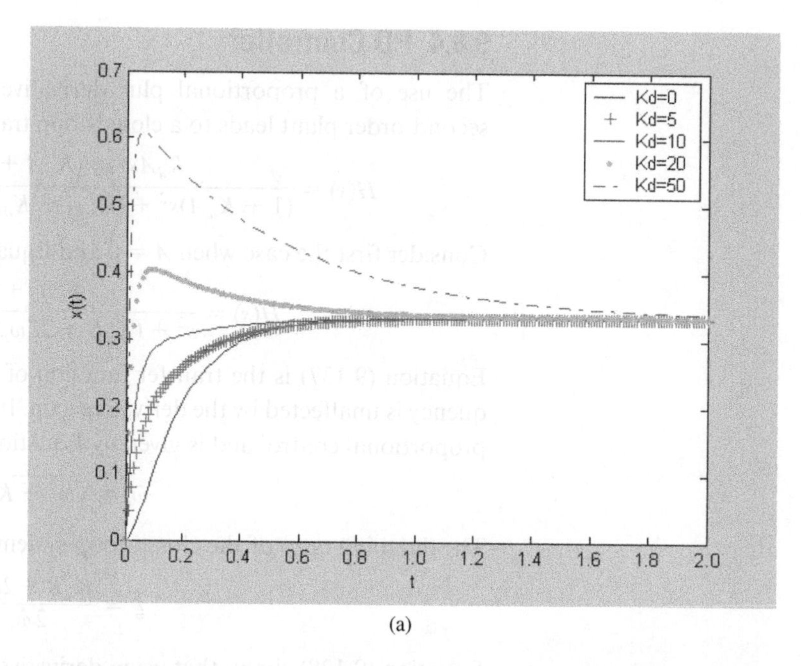

(a)

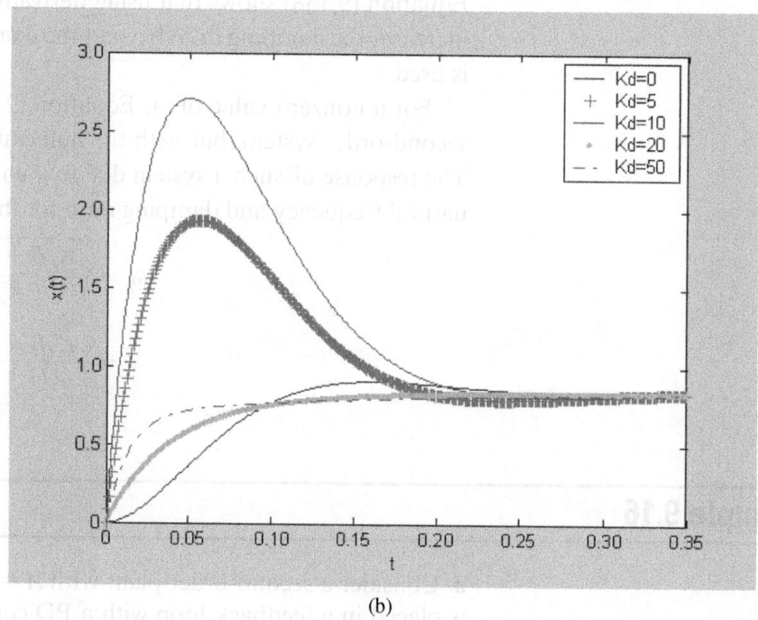

(b)

Figure 9.58

The step response of a second-order plant of Example 9.16 in a feedback loop with a PD controller for various values of the derivative gain with (a) $K_p = 50$ in which the system is overdamped without a derivative gain and (b) $K_p = 500$ in which the system is underdamped without a derivative gain.

c. For $K_p = 50$, the natural frequency and damping ratio are determined using Equations (9.124) and (9.138) as

$$\hat{\omega}_n = \sqrt{\frac{150}{1 + K_d}} \tag{a}$$

$$\zeta = \frac{K_d + 30}{2\sqrt{150(1 + K_d)}} \tag{b}$$

The natural frequency decreases as the derivative gain increases. Analysis using Equation (b) reveals that the system is underdamped when $0.556 < K_d < 539.4$, critically damped when $K_d = 0.556$ or $K_d = 539.4$, and overdamped otherwise. The damping ratio has a minimum value of 0.44 when $K_d = 28$. The responses for several values of K_d are plotted in Figure 9.59. The response of each system is discontinuous at $t = 0$ with a value of $x_s(0) = K_d/(1 + K_d)$. The addition of a derivative term does not affect the offset. For this system, the response is slower for larger values of K_d.

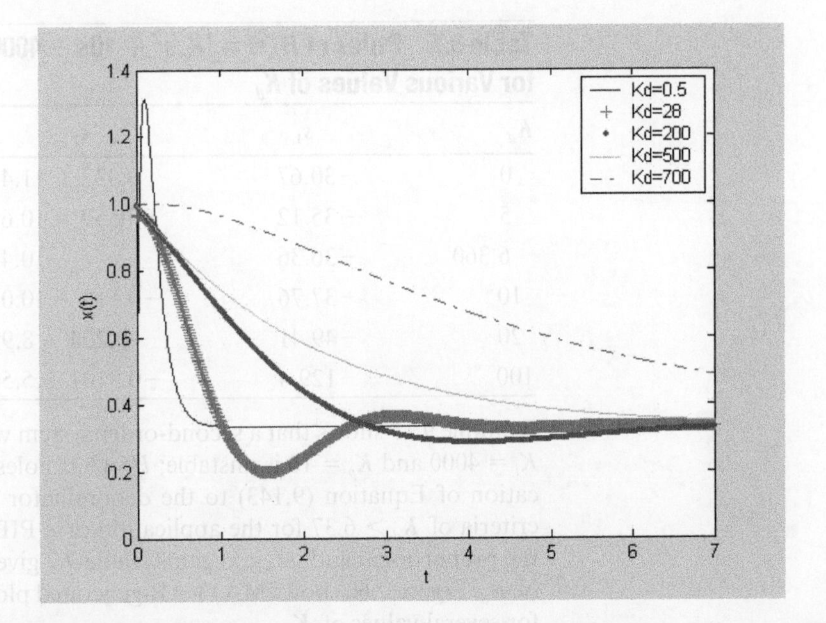

Figure 9.59

The step response of a second-order plant whose transfer function is $G(s) = (s + 1)/(s^2 + 30s + 100)$ placed in a feedback loop with a PD controller with a constant proportional gain and varying values of derivative gain.

9.8.5 PID Controller

The closed-loop transfer function when a PID controller is used with a second-order plant in a feedback loop is

$$H(s) = \frac{K_d A s^3 + (K_p A + K_d B)s^2 + (K_p B + K_i A)s + K_i B}{(1 + K_d A)s^3 + (2\zeta\omega_n + K_p A + K_d B)s^2 + (\omega_n^2 + K_p B + K_i A)s + K_i B} \quad (9.141)$$

Equation (9.141) is that of a third-order system. The application of Routh's criterion leads to a stability condition of the form

$$(2\zeta\omega_n + K_p A + K_d B)(\omega_n^2 + K_p B + K_i A) > (1 + K_d A)K_i B \quad (9.142)$$

If $A = 0$, Equation (9.142) reduces to

$$(2\zeta\omega_n + K_d B)(\omega_n^2 + K_p B) > K_i B \quad (9.143)$$

Comparison between Equations (9.134) and (9.143) shows that the addition of a derivative term to a PI controller enhances the stability of the system with a second-order plant. If A is not zero, no general conclusions can be drawn regarding the effect of the derivative gain on stability. Also, as with the PD controller, since for $A \neq 0$ the order of the polynomial in the numerator of $H(s)$ is the same as the order of the polynomial in the denominator, the response of the system due to a unit step input is discontinuous at $t = 0$.

Example 9.17

Consider a second-order plant with $A = 0$, $B = 1$, $\zeta = 1.5$, and $\omega_n = 10$. The plant is placed in a feedback loop with a PID controller with $K_i = 4000$ and $K_p = 10$. Plot the response of the system when subject to a unit step input for various values of the derivative gain.

Solution

The transfer function for this system in terms of the derivative gain is

$$H(s) = \frac{K_d s^2 + 10s + 4000}{s^3 + (30 + K_d)s^2 + 110s + 4000} \quad (a)$$

Table 9.7 Poles of $H(s) = (K_d s^2 + 10s + 4000)/(s^3 + (30 + K_d)s^2 + 110s + 4000)$ for Various Values of K_d

K_d	s_1	s_2	s_3
0	−30.67	0.337 + 11.42 *j*	0.337 − 11.42 *j*
5	−35.12	0.0559 + 10.67 *j*	0.559 − 10.67 *j*
6.360	−36.36	10.49 *j*	−10.49 *j*
10	−37.76	−0.118 + 10.03 *j*	−0.118 − 10.03 *j*
20	−49.41	−0.294 + 8.99 *j*	−0.294 − 8.99 *j*
100	−129.4	−0.3101 + 5.56 *j*	−0.3101 − 5.56 *j*

Example 9.15 shows that a second-order system with actuation by a PI controller with $K_i = 4000$ and $K_p = 10$ is unstable; $H(s)$ has poles with a positive real part. The application of Equation (9.143) to the denominator of Equation (a) leads to a stability criteria of $K_d > 6.37$ for the application of a PID controller with the same values of the proportional and integral gains. Table 9.7 gives the poles of $H(s)$ for several values of K_d. Figure 9.60 shows MATLAB-generated plots of the step response of the system for several values of K_d.

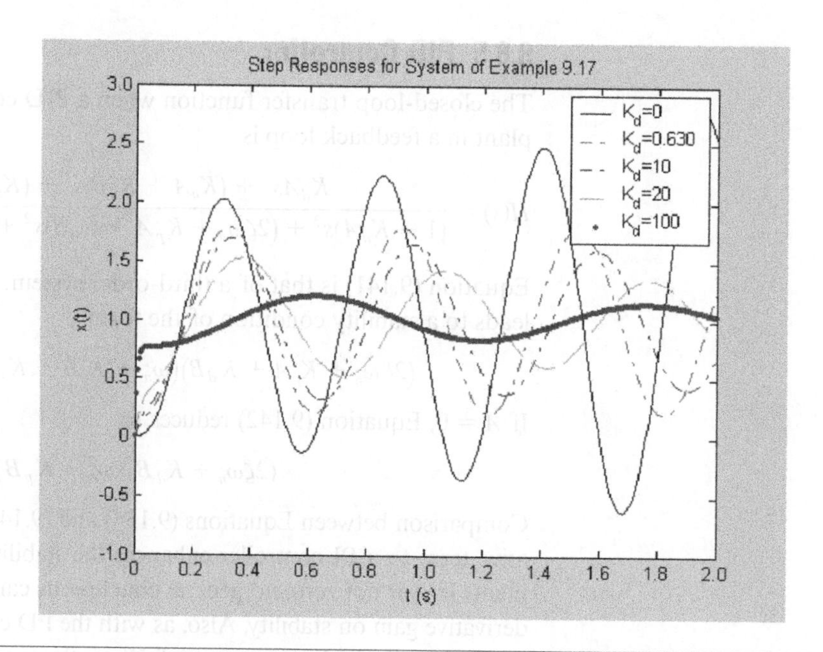

Figure 9.60

The step responses of the system of Example 9.17 for various values of K_d. The response for $K_d = 0$ is unstable and grows without bound. The response for $K_d = 6.360$ is neutrally stable. The responses for $K_d > 6.360$ are stable.

9.9 Parallel Feedback Loops

A viable control strategy is to place some of the control in a separate feedback loop and have two or more parallel feedback loops. This technique is illustrated in Figure 9.61. A special type of parallel feedback occurs when using a PID control strategy, where the derivative action is placed in a parallel feedback loop, as illustrated in Figure 9.62. This action tends to smooth an impulse that would be obtained from the differentiation that occurs when the plant is placed in a feedback loop with a derivative controller.

One type of PI control is velocity feedback control, often called a tachometer feedback. It is desired to monitor the rotational speed of a motor before it is integrated to become an angular displacement signal. The sensor is placed in a feedback loop with

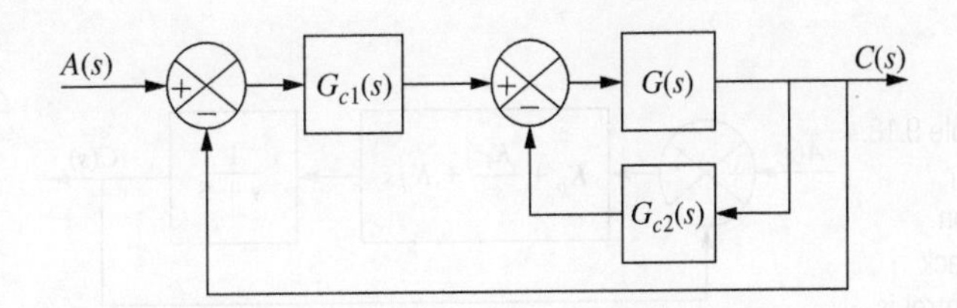

Figure 9.61
Parallel feedback loops.

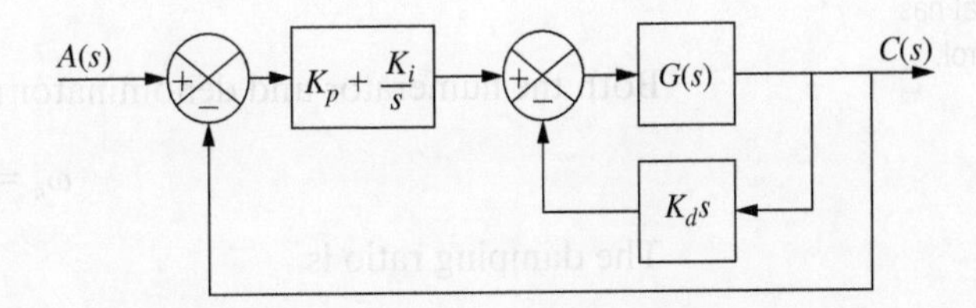

Figure 9.62
The use of PID control is often separated such that the integral plus proportional control acts on the plant with derivative control in the feedback portion of a control loop. This smooths out the effect of differentiating an impulse resulting from a differentiation.

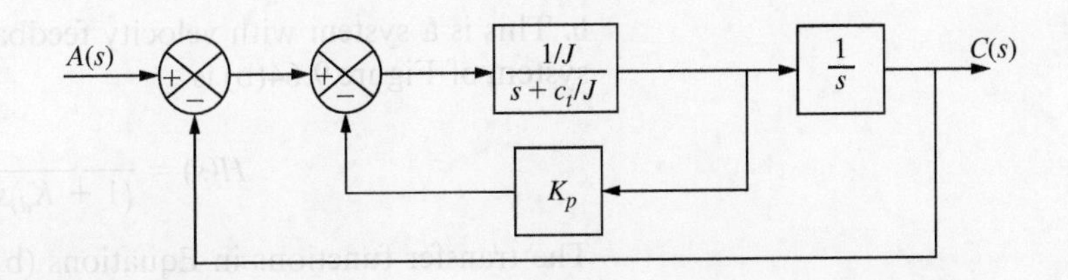

Figure 9.63
Application of velocity feedback control to a rotating shaft.

the motor and then the signal is integrated and fed back to the system. The resulting block diagram for this tachometer is shown in Figure 9.63. The closed-loop transfer function is

$$H(s) = \frac{1/J}{s^2 + \left(\frac{c_t}{J} + K_p\right)s + 1/J} \tag{9.144}$$

Equation (9.144) is that of a second-order system of natural frequency

$$\omega_n = \sqrt{\frac{1}{J}} \tag{9.145}$$

The system has a damping ratio of

$$\zeta = \frac{\sqrt{J}}{2}\left(\frac{c_t}{J} + K_p\right) \tag{9.146}$$

Example 9.18

A first-order plant with a transfer function of

$$G(s) = \frac{1}{s+5} \tag{a}$$

is to be placed in a feedback loop with PID control. Analyze and discuss the advantages of configurations (a) and (b) in Figure 9.64.

Solution

a. The configuration of Figure 9.64(a) is that of the plant in series with a PID controller and placed in a feedback loop. The closed-loop transfer function of the system is

$$H(s) = \frac{K_d s^2 + K_p s + K_i}{(1 + K_d)s^2 + (5 + K_p)s + K_i} \tag{b}$$

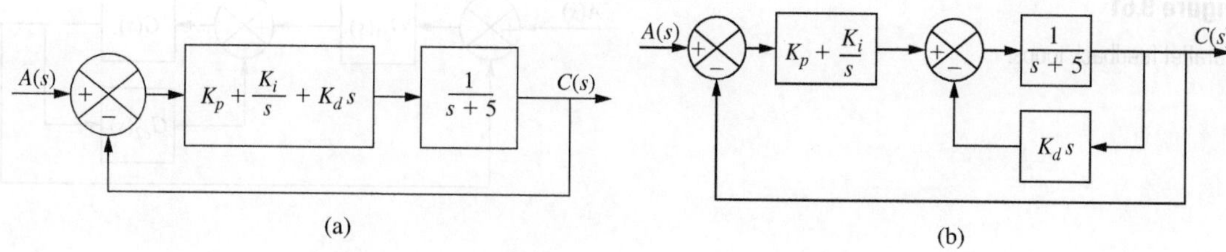

Figure 9.64

Systems of Example 9.18. (a) A PID controller is acting directly on a plant in a feedback loop. (b) The PI control is delivered separately to a feedback loop that has the derivative control.

Both the numerator and denominator are second order. The natural frequency is

$$\omega_n = \sqrt{\frac{K_i}{1 + K_d}} \tag{c}$$

The damping ratio is

$$\zeta = \frac{5 + K_p}{2\sqrt{K_i(1 + K_d)}} \tag{d}$$

b. This is a system with velocity feedback. The closed-loop transfer function of the system of Figure 9.64(b) is

$$H(s) = \frac{K_p s + K_i}{(1 + K_d)s^2 + (5 + K_p)s + K_i} \tag{e}$$

The transfer functions in Equations (b) and (e) have the same denominator but different numerators. Suppose the systems both have a natural frequency of 2 r/s and a damping ratio of $\frac{\sqrt{2}}{2}$. Choosing $K_i = 8$, $K_d = 1$, and $K_p = 0.657$ satisfies Equations (c) and (d). Equations (b) and (e) then become

$$H(s) = \frac{s^2 + 0.657s + 8}{2s^2 + 5.657s + 8} \tag{f}$$

and

$$H(s) = \frac{0.657s + 8}{2s^2 + 5.657s + 8} \tag{g}$$

Closed-loop step responses using Equations (f) and (g) are plotted in Figure 9.65. Both systems have zero offset. Application of the initial value theorem shows that the step response of the system with velocity feedback is continuous at $t = 0$. The step response of the system that has the full PID control has a discontinuity of 0.5 at $t = 0$. The rise time of the system with velocity feedback is shorter, and the settling times for both systems are approximately the same.

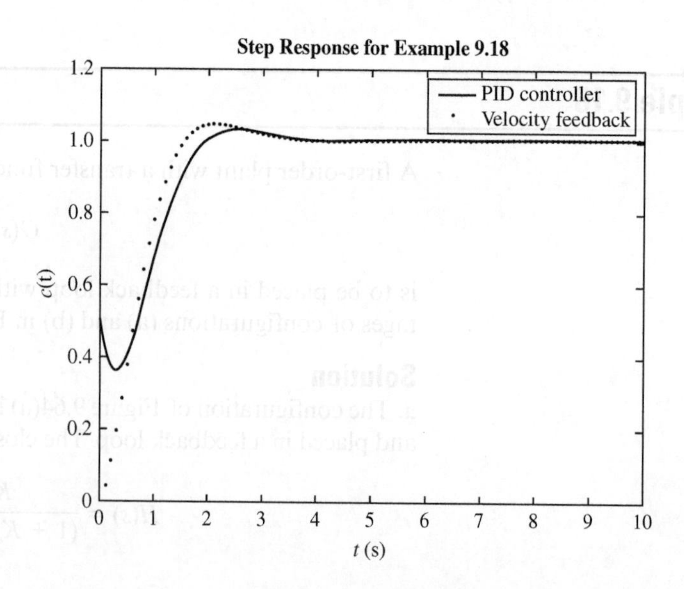

Figure 9.65

When PID control is used directly, the step response is discontinuous. When a PI controller is used in conjunction with velocity control, the response is continuous.

Example 9.19

A PID control strategy for a field-controlled dc servomotor is shown in Figure 9.66. The parameters of the dc servomotor are $R = 1\ \Omega$, $L = 0.5$ H, $J = 5 \times 10^{-3}$ kg·m², $c_t = 0.05$ N·m·s, and $K_f = 1.0\ \text{N}\cdot\frac{\text{m}}{\text{A}}$. Proportional and integral control is used in the feedforward loop, while the differential control is in a feedback loop for the torque developed in the motor. Select the control parameters such that the dominant poles have a natural frequency of 3 r/s and a damping ratio of 0.6. The third pole should be $s = -7$.

Solution

The closed-loop transfer function for the system of Figure 9.66 is

$$H(s) = \frac{(K_i + K_p s)\dfrac{K_f}{JL}}{\left(1 + \dfrac{K_d}{L}\right)s^3 + \left[\dfrac{R}{L} + \left(1 + \dfrac{K_d}{L}\right)\dfrac{c_t}{J}\right]s^2 + (K_p K_f + c_t R)\dfrac{1}{JL}s + \dfrac{K_i K_f}{JL}} \tag{a}$$

Equation (a) is the transfer function of a third-order system. Using Routh's criterion, the system is stable if

$$\frac{\left[\dfrac{R}{L} + \left(1 + \dfrac{K_d}{L}\right)\dfrac{c_t}{J}\right](K_p K_f + c_t R)\dfrac{1}{JL}}{1 + \dfrac{K_d}{L}} > \frac{K_i K_f}{JL} \tag{b}$$

The dominant poles of the system under the conditions specified are

$$s = -\zeta\omega_n \pm \omega_n\sqrt{1 - \zeta^2}\,j = -(0.6)(3) \pm (3)\sqrt{1 - 0.6^2}\,j = -1.8 \pm 2.4j \tag{c}$$

The third pole occurs at $s = -7$. The denominator with the controller should be

$$D(s) = (s + 7)(s^2 + 3.6s + 9) = s^3 + 11.6s^2 + 41.2s + 63 \tag{d}$$

Equating the denominator of Equation (a) after it has been divided by $(1 + K_d)$ with $D(s)$ from Equation (d) leads to

$$\frac{\dfrac{R}{L} + \left(1 + \dfrac{K_d}{L}\right)\dfrac{c_t}{J}}{1 + \dfrac{K_d}{L}} = 11.6 \tag{e}$$

$$\frac{(K_p K_f + c_t R)\dfrac{1}{JL}}{1 + \dfrac{K_d}{L}} = 41.2 \tag{f}$$

$$\frac{\dfrac{K_i K_f}{JL}}{1 + \dfrac{K_d}{L}} = 63 \tag{g}$$

Substituting values given in Equations (e), (f), and (g) and solving gives

$$K_d = 0.306,\ K_p = 51.03,\ K_i = 108.6 \tag{h}$$

The closed-loop transfer function is

$$H(s) = \frac{28.8s + 63}{s^3 + 11.6s^2 + 41.2s + 63} \tag{i}$$

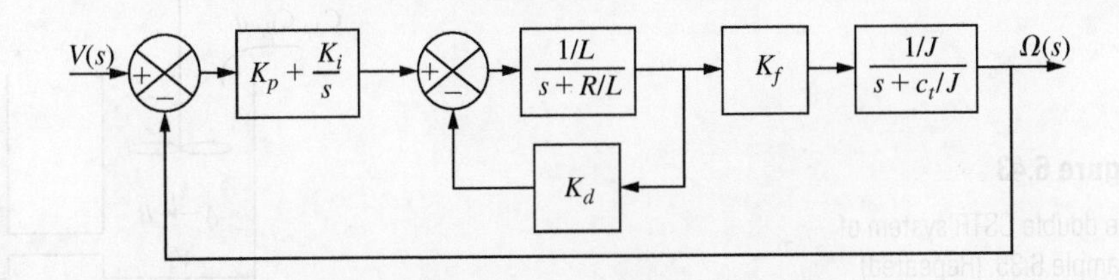

Figure 9.66

A PID control strategy is used with a field-controlled dc servomotor.

The step response for the closed-loop system is give in Figure 9.67.

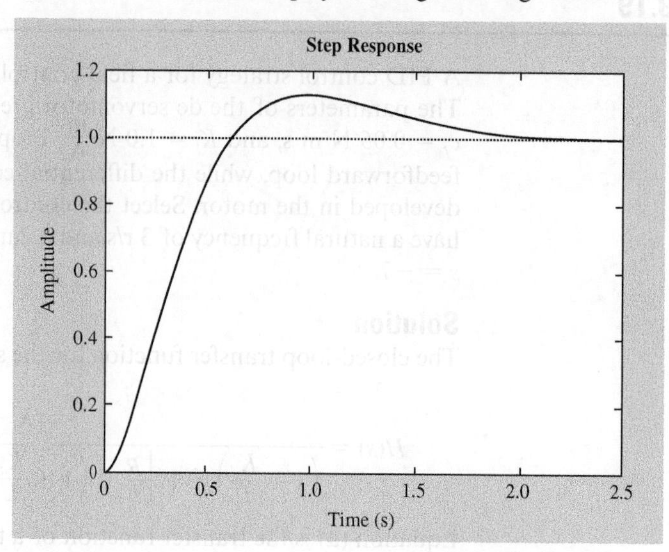

Figure 9.67

Step response for system of Example 9.19.

9.10 Further Examples

The following examples illustrate the development of block diagrams from mathematical models and transfer functions, the determination of the closed-loop response and its properties, and the stability of closed-loop systems. The actuator transfer functions are illustrative only. Applications are often more complex than illustrated here. These examples are intended to illustrate methods of analysis.

Example 9.20

a. Draw a block diagram for the two-tank CSTR problem of Example 6.35 and Figure 6.43 that includes the time delay. Include only components A and B in both tanks in the block diagram. **b.** Develop a Simulink model for this system. Run a simulation with $V_1 = V_2 = 1.5 \times 10^{-3}$ m^3, $q = 1.5 \times 10^{-6}$ m^3/s, $C_{Ai} = 0.25u(t)$ mol/L, $k_1 = 2.0 \times 10^{-3}$ s^{-1}, $k_2 = 1.0 \times 10^{-3}$ s^{-1}, and $\tau = 0.8$ s.

Solution

a. Taking the Laplace transforms of Equations (a), (b), (d), and (e) of Example 6.35 and rearranging gives

$$C_{A1p} = \frac{q\,C_{A1i}}{V_1 s + q + k_1 V_1} \tag{a}$$

$$C_{B1}(s) = \frac{k_1 C_{A1}(s)}{V_1 s + q} \tag{b}$$

$$C_{A2}(s) = \frac{q C_{A1}(s) e^{-\tau s}}{V_2 s + q + k_1 V_2} \tag{c}$$

$$C_{B2}(s) = \frac{1}{V_2 s + q + k_2 V_2}\left(q C_{B1}(s) e^{-\tau s} + k_1 V_2 C_{A2}(s)\right) \tag{d}$$

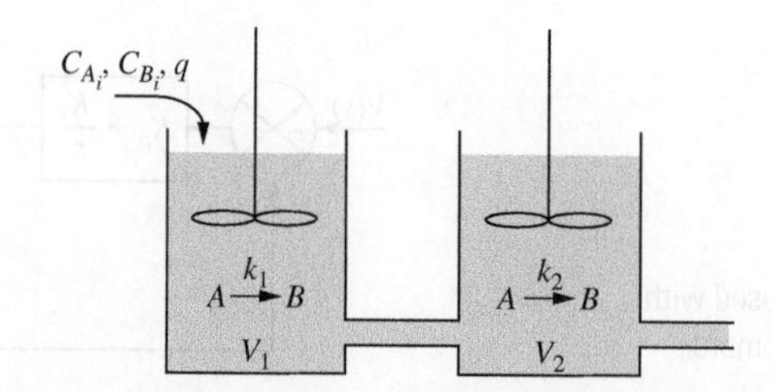

Figure 6.43

The double CSTR system of Example 6.35. (Repeated)

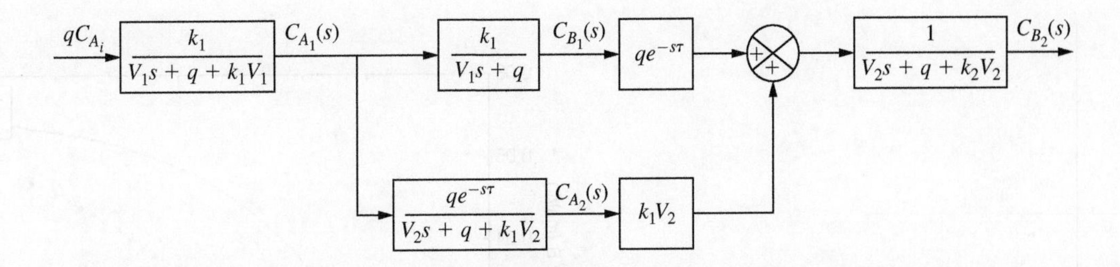

Figure 9.68

The block diagram for the two-tank CSTR with time delay.

Equations (a)–(d) are used to develop the block diagram of Figure 9.68.

b. The Simulink model of the block diagram of Figure 9.68 is shown in Figure 9.69. The concentrations of components A and B, obtained from the Simulink model, are illustrated in Figures 9.70(a) and (b), respectively. The model includes a time delay. The Time Delay block is inserted into the Simulink model from the Continuous Systems menu of the Simulink Library Browser.

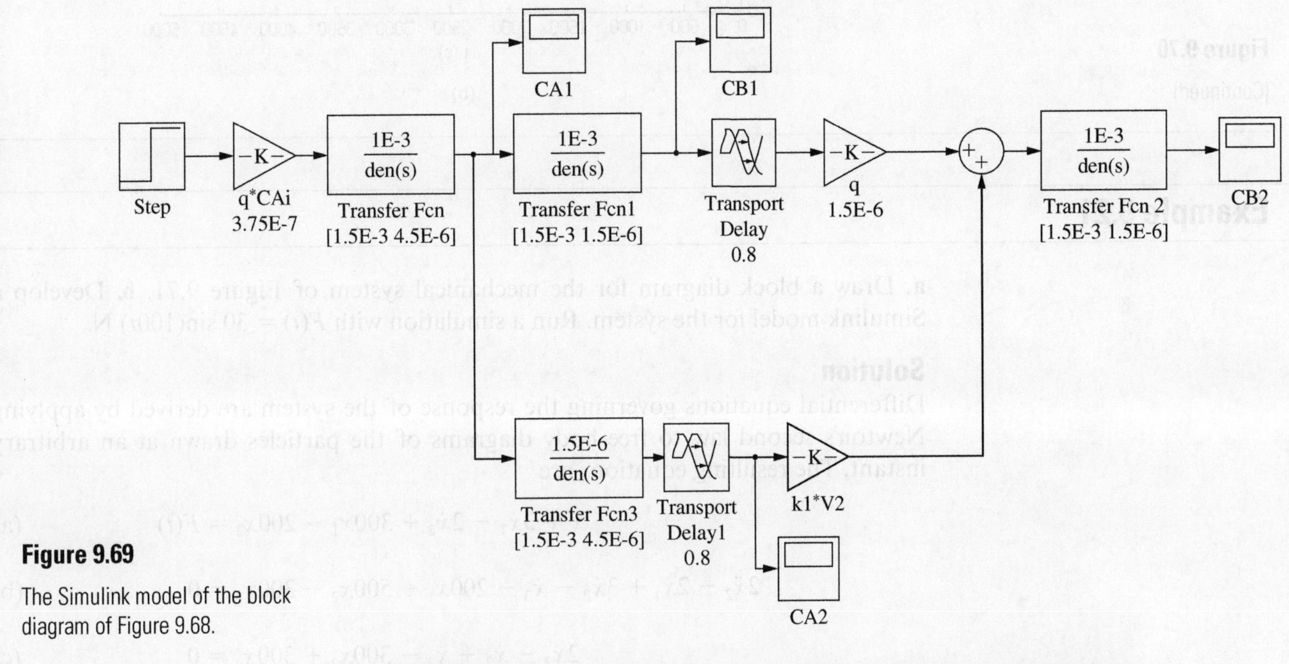

Figure 9.69

The Simulink model of the block diagram of Figure 9.68.

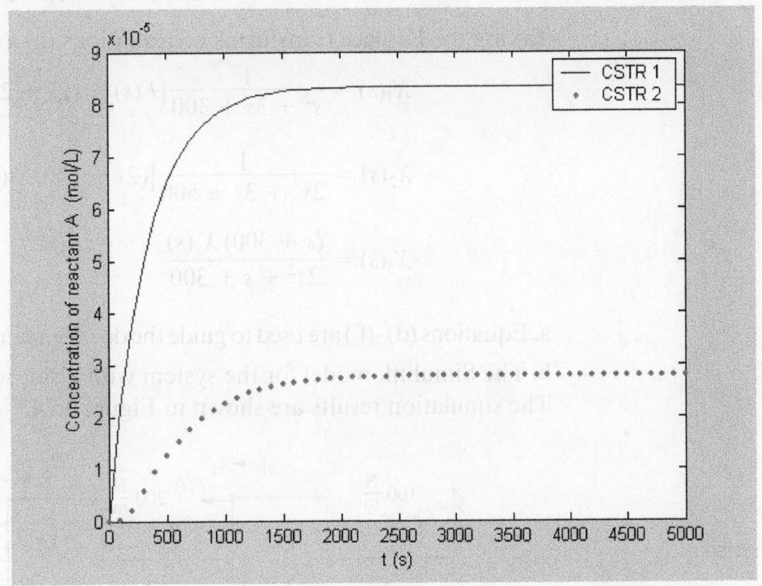

Figure 9.70

The results from running the Simulink simulation of the double CSTR of Example 9.20: (a) species A, (b) species B.

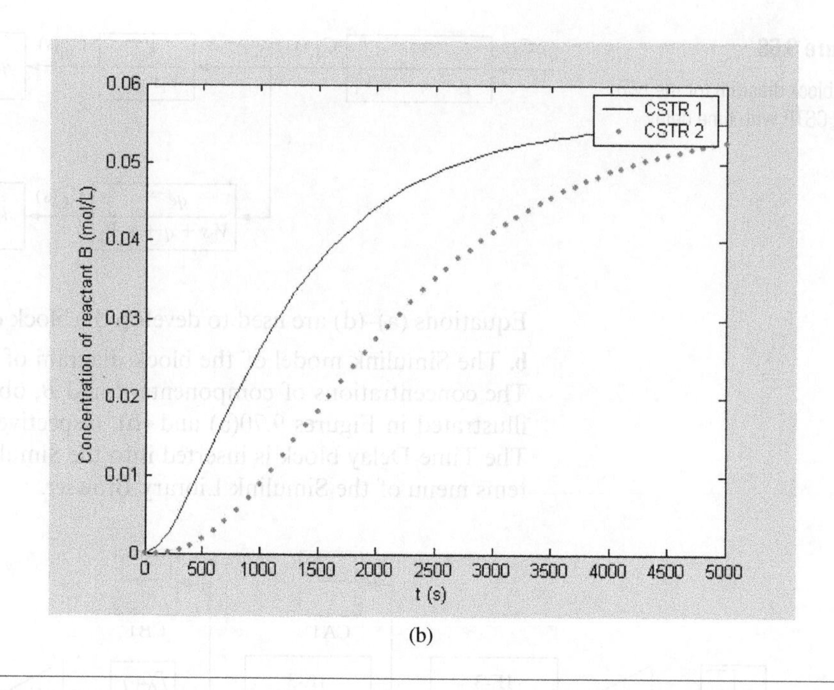

Figure 9.70

(Continued)

Example 9.21

a. Draw a block diagram for the mechanical system of Figure 9.71. **b.** Develop a Simulink model for the system. Run a simulation with $F(t) = 30 \sin(100t)$ N.

Solution

Differential equations governing the response of the system are derived by applying Newton's second law to free-body diagrams of the particles drawn at an arbitrary instant. The resulting equations are

$$\ddot{x} + 3\dot{x}_1 - 2\dot{x}_2 + 300x_1 - 200x_2 = F(t) \tag{a}$$

$$2\ddot{x}_2 - 2\dot{x}_1 + 3\dot{x}_2 - \dot{x}_3 - 200x_1 + 500x_2 - 300x_3 = 0 \tag{b}$$

$$2\ddot{x}_3 - \dot{x}_2 + \dot{x}_3 - 300x_2 + 300x_3 = 0 \tag{c}$$

Taking the Laplace transforms of Equations (a)–(c) and rearranging leads to

$$X_1(s) = \frac{1}{s^2 + 3s + 300}[F(s) + (2s + 200)X_2(s)] \tag{d}$$

$$X_2(s) = \frac{1}{2s^2 + 3s + 500}[(2s + 200)X_1(s) + (s + 300)\,X_3(s)] \tag{e}$$

$$X_3(s) = \frac{(s + 300)\,X_2(s)}{2s^2 + s + 300} \tag{f}$$

a. Equations (d)–(f) are used to guide the development of the block diagram of Figure 9.72.
b. The Simulink model for the system with a sinusoidal input is shown in Figure 9.73. The simulation results are shown in Figure 9.74.

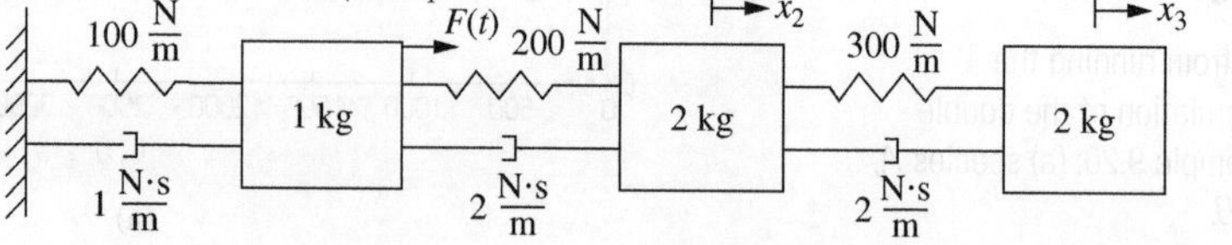

x_1 $F(t)$ x_2 x_3
$100\ \frac{\text{N}}{\text{m}}$ 1 kg $200\ \frac{\text{N}}{\text{m}}$ 2 kg $300\ \frac{\text{N}}{\text{m}}$ 2 kg
$1\ \frac{\text{N·s}}{\text{m}}$ $2\ \frac{\text{N·s}}{\text{m}}$ $2\ \frac{\text{N·s}}{\text{m}}$

Figure 9.71

The mechanical system of Example 9.21.

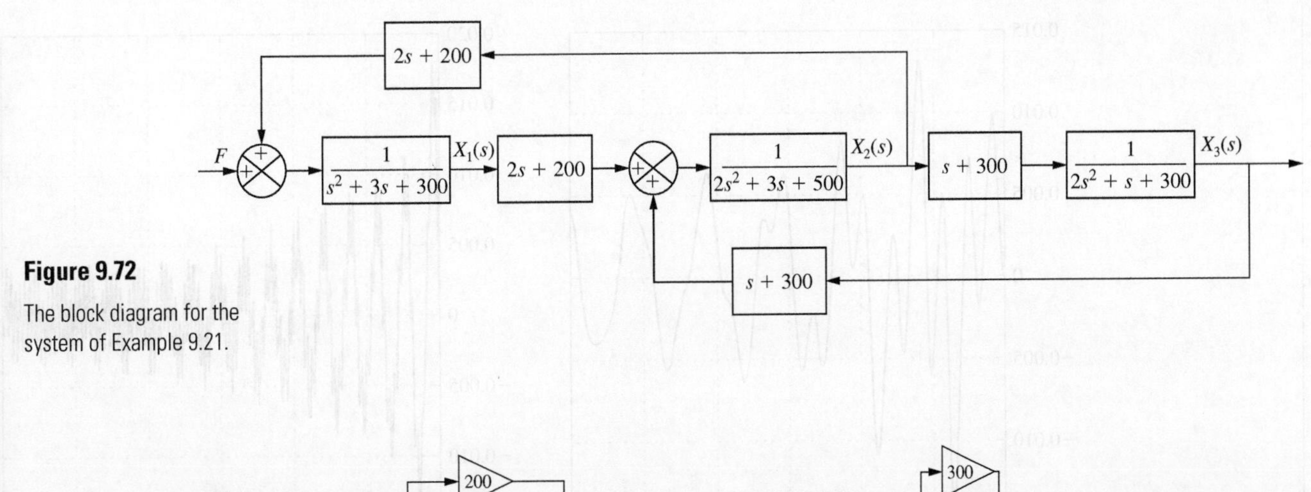

Figure 9.72

The block diagram for the system of Example 9.21.

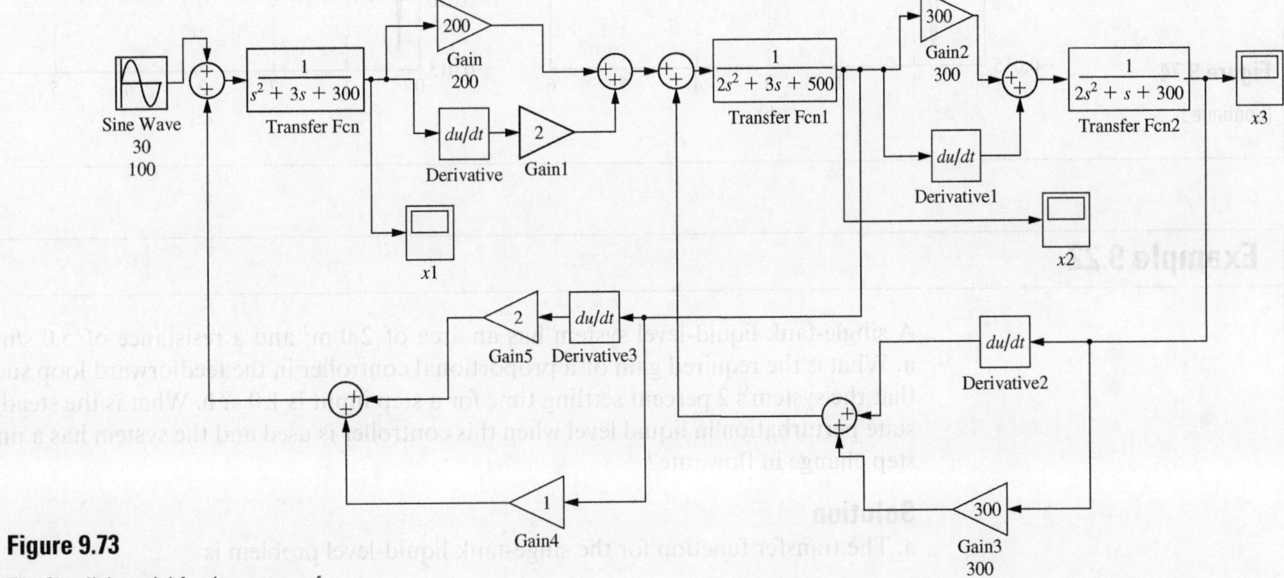

Figure 9.73

The Simulink model for the system of Example 9.21 with a sinusoidal input.

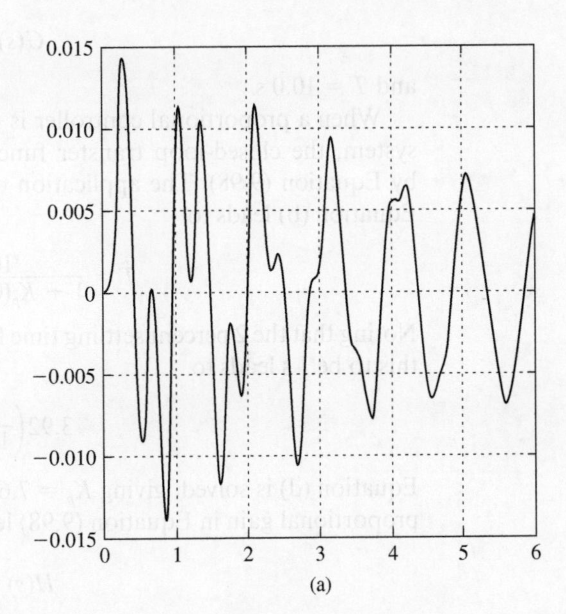

Figure 9.74

Simulation results from Simulink model: (a) $x_1(t)$; (b) $x_2(t)$; (c) $x_3(t)$.

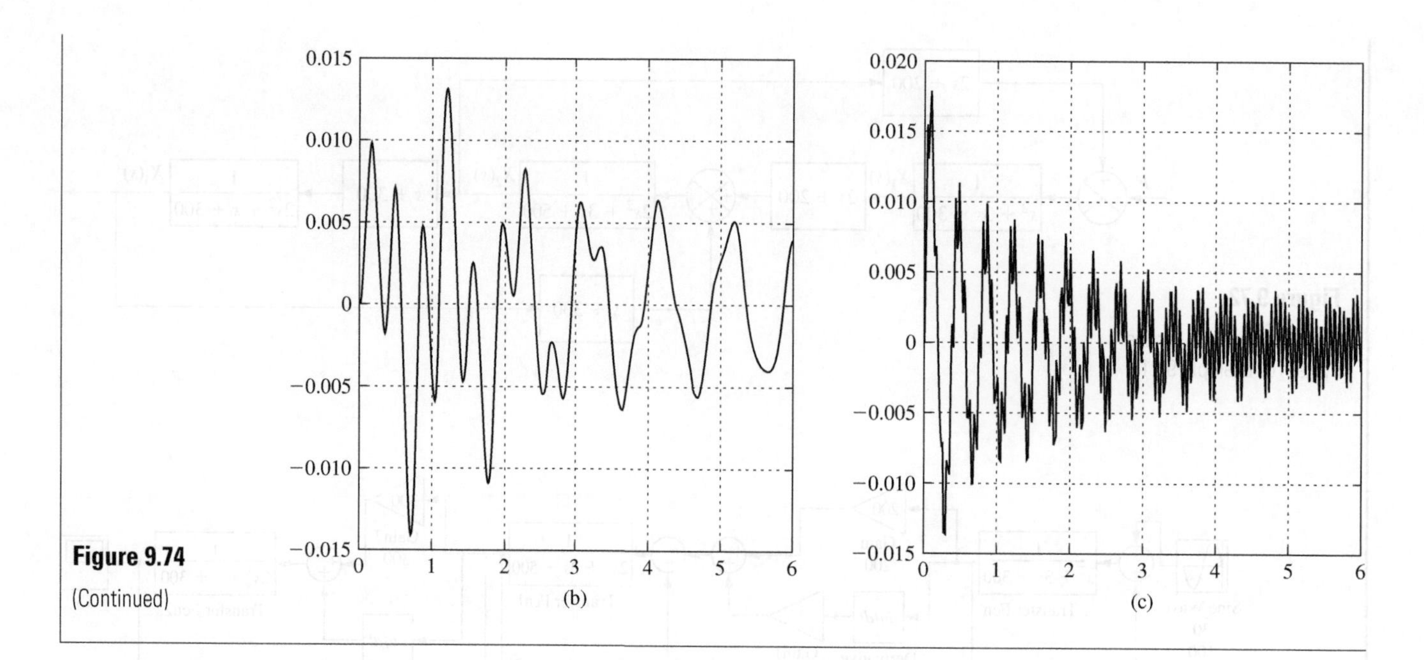

Figure 9.74
(Continued)

Example 9.22

A single-tank liquid-level system has an area of 2.0 m^2 and a resistance of 5.0 s/m^2. **a.** What is the required gain of a proportional controller in the feedforward loop such that the system's 2 percent settling time for a step input is 1.0 s? **b.** What is the steady-state perturbation in liquid level when this controller is used and the system has a unit step change in flow rate?

Solution

a. The transfer function for the singe-tank liquid-level problem is

$$G(s) = \frac{1}{As + \frac{1}{R}} = \frac{\frac{1}{A}}{s + \frac{1}{T}} \tag{a}$$

The first-order time constant is $T = RA$. The substitution of known values leads to

$$G(s) = \frac{0.5}{s + 0.1} \tag{b}$$

and $T = 10.0$ s.

When a proportional controller is used in the feedforward loop for the first-order system, the closed-loop transfer function is first order with a time constant given by Equation (9.98). The application of Equation (9.99) for the transfer function of Equation (b) leads to

$$\hat{T} = \frac{10}{1 + K_p(0.5)(10)} = \frac{10}{1 + 5K_p} \tag{c}$$

Noting that the 2 percent settling time for this first-order system is 3.92 $\hat{T}$ and requiring this to be 1 s leads to

$$3.92\left(\frac{10}{1 + 5K_p}\right) = 1 \tag{d}$$

Equation (d) is solved, giving $K_p = 7.64$ and then $\hat{T} = 0.255$ s. Using this value of the proportional gain in Equation (9.98) leads to the closed-loop transfer function of

$$H(s) = \frac{3.82}{s + 3.92} \tag{e}$$

The step responses of the system without a controller and with a proportional controller are illustrated in Figure 9.75. The offset when the controller is used is calculated using Equation (9.101) as $\hat{T}/T = 0.255$. The magnitude of the responses shown in Figure 9.75 do not compare because the units on each response are different. The uncontrolled response has meters (m) as units, whereas the controlled response must have the same units as the input (flow rate), which are cubic meters per second (m^3/s). The proportional gain must have units that allow the output to be compared with the input. In this case, $K_p = 7.64\ m^2/s$. It is clear from Figure 9.75 that the use of the controller significantly affects the dynamic response of the system, reducing the settling time and the rise time.

b. The steady-state liquid level when this controller is used is obtained using the final value theorem applied to equation (e) as

$$\begin{aligned} h_s &= \lim_{s \to 0} H(s) \\ &= \frac{3.82}{3.92} \\ &= 0.974\ \text{m} \end{aligned} \tag{f}$$

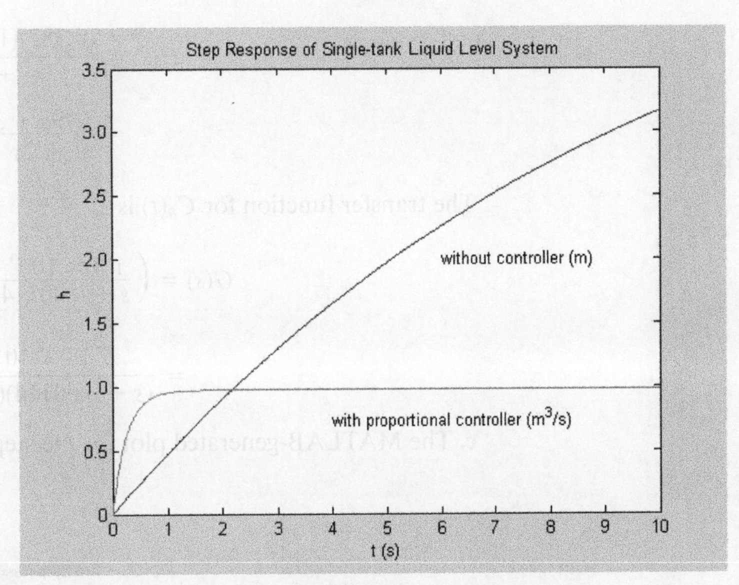

Figure 9.75

The step response of the first-order system of Example 9.22 with and without a proportional controller.

Example 9.23

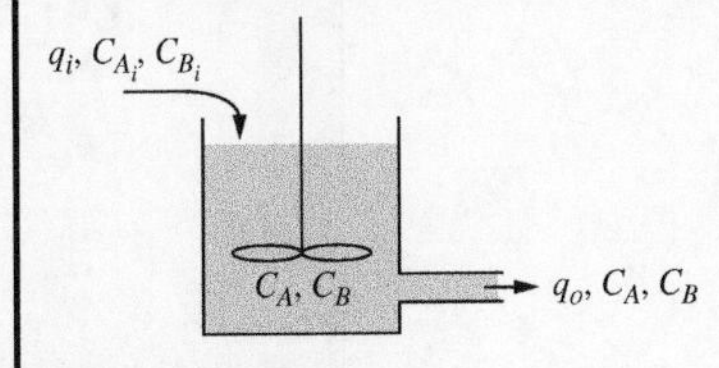

Figure 6.30

A chemical reaction occurs in a continuous stirred tank reactor (CSTR), in which component *A* is converted to component *B*. (Repeated)

Figure 9.76 shows the block diagram for the perturbation concentrations for the CSTR of Example 7.26 due to a change in the concentration of the inlet stream of a reactant A when a PI controller is used in the feedforward loop to control $C_A(t)$. **a.** What are the required gains of the PI controller such that the system is critically damped with a natural frequency of 2.0 r/s? **b.** What is the resulting transfer function $G_B(s) = C_B(s)/C_{Ai}(s)$ when this controller is used? **c.** Determine the step response for $C_B(t)$ when this controller is used.

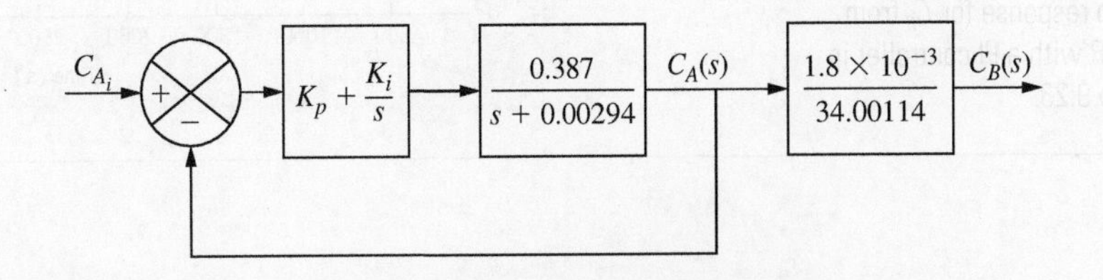

Figure 9.76

The block diagram for the system of Example 9.23, a CSTR with a PI controller to control the perturbation concentration of reactant *A*.

Solution

a. The natural frequency of the closed-loop system when a PI controller is used on a first-order system is given by Equation (9.106). Requiring the natural frequency to be 2.0 r/s leads to

$$2.0 = \sqrt{(0.387)K_i}$$

$$K_i = 10.3 \tag{a}$$

The damping ratio of the closed-loop system is given by Equation (9.112). Setting the damping ratio to 1, requiring the system to be critically damped, leads to

$$1 = \frac{K_p}{2}\sqrt{\frac{0.387}{10.3}} + \frac{1}{2\left(\frac{1}{0.00294}\right)(2.0)}$$

$$K_p = 10.3 \tag{b}$$

b. The closed-loop transfer function for $C_{Ai}(s)$ is obtained using Equation (9.111):

$$H(s) = \frac{(0.387)(10.3s + 10.3)}{s^2 + 2s + 4}$$

$$= \frac{3.99s + 3.99}{s^2 + 2s + 4} \tag{c}$$

The transfer function for $C_B(t)$ is

$$G(s) = \left(\frac{1.8 \times 10^{-3}}{s + 0.00114}\right)\left(\frac{3.99 + 3.99s}{s^2 + 2s + 4}\right)$$

$$= \frac{7.18 \times 10^{-3}(s + 1)}{(s + 0.00114)(s^2 + 2s + 4)} \tag{d}$$

c. The MATLAB-generated plot for the step response is given in Figure 9.77.

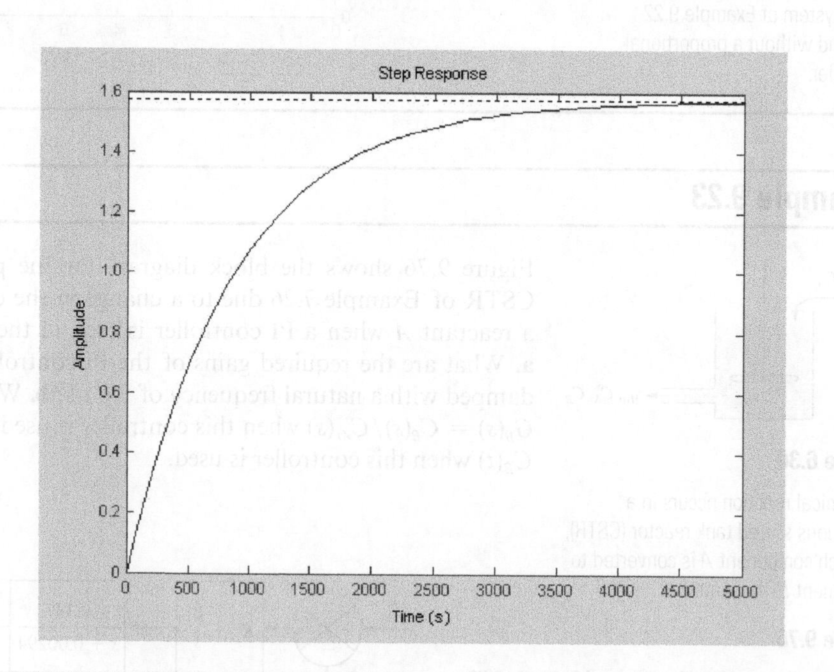

Figure 9.77

The step response for C_B from the CSTR with a PI controller in Example 9.23.

Example 9.24

A PID controller with $K_p = 50.0$ kJ/s·C, $K_i = 35.0$ MJ/s²·C, and $K_d = 6.5$ MJ/C is to be used in the feedforward loop to provide control to the thermal system of Example 6.15. When this controller is used, determine **a.** the closed-loop transfer function, **b.** the system's natural frequency and damping ratio, **c.** a MATLAB-generated plot of the step response, and **d.** the initial value of the step response.

Solution

The open-loop transfer function for the system is obtained using Equation (i) of Example 6.15 as

$$G(s) = \frac{5.95 \times 10^{-9}}{s + 1.22 \times 10^{-5}} \tag{a}$$

a. The closed-loop transfer function for a first-order system using a PID controller is given by Equation (9.119), which for this system becomes

$$H(s) = \frac{5.95 \times 10^{-9}(6.5 \times 10^6 s^2 + 5.0 \times 10^6 s + 3.5 \times 10^6)}{[1 + (5.95 \times 10^{-9})(6.5 \times 10^6)s^2] + [(5.95 \times 10^{-9})(5.0 \times 10^6) + 1.22 \times 10^{-5}]\, s + (5.95 \times 10^{-9})(3.5 \times 10^6)}$$
$$= \frac{3.87 \times 10^{-2} s^2 + 2.98 \times 10^{-2} s + 2.08 \times 10^{-2}}{1.00387 s^2 + 2.98 \times 10^{-2} s + 2.08 \times 10^{-2}} \tag{b}$$

b. Equation (b) is that of a second-order system of natural frequency $\omega_n = \sqrt{2.08 \times 10^{-2}/1.00387} = 0.144$ r/s and damping ratio $\zeta = 2.98 \times 10^{-2}/[2(1.00387)(0.144)] = 0.103$.

c. A MATLAB-generated plot of the step response is given in Figure 9.78. The amplitude of the response has units of joules per second (J/s). Since an integral term is used in the controller, the response has no offset.

d. Due to the derivative term in the controller, the step response is discontinuous at $t = 0$ with a discontinuity given by the initial value theorem of

$$\theta_s(0) = \lim_{s \to \infty} H(s) = \frac{3.87 \times 10^{-2}}{1.00387} = 3.86 \times 10^{-2} \tag{c}$$

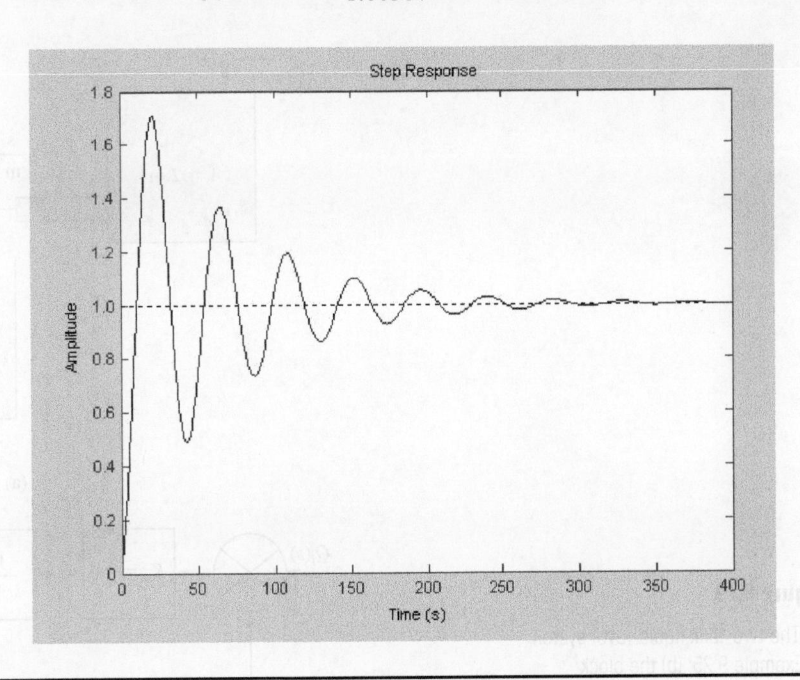

Figure 9.78

The step response of the system with a PID controller used on the tank of Example 9.24. The amplitude has units of (J/s).

Example 9.25

Figure 9.79(a) illustrates a two-tank liquid-level system. Figure 9.79(b) shows the block diagram when the system of Figure 9.79(a) is placed in a feedback loop with a proportional controller when the liquid level in the second tank is fed back into the controller.

a. Determine the transfer function of the closed-loop system. **b.** Determine the natural frequency and damping ratio of the closed-loop system. **c.** Determine the step response for both the first and second tanks. **d.** Determine the steady-state liquid-level perturbation in each tank when this controller is used and the system has a unit step change in input flow rate.

Solution

The transfer function for the system is

$$G(s) = \frac{0.05}{s^2 + 0.3s + 0.01} \tag{a}$$

Equation (a) is the transfer function of a second-order system and is of the form of Equation (9.122) with $A = 0$, $B = 0.05$, $\omega_n = 0.141$ r/s, and $\zeta = 1.06$.

a. The closed-loop transfer function for a second-order system under the action of a proportional controller is given in Equation (9.123). The application to this system leads to

$$H(s) = \frac{2.0(0.05)}{s^2 + 0.3s + 2.0(0.05) + 0.01}$$

$$= \frac{0.1}{s^2 + 0.3s + 0.11} \tag{b}$$

b. Equation (b) is the transfer function of a second-order system with $\omega_n = \sqrt{0.11} = 0.346$ r/s and $\zeta = 0.3/[2(0.346)] = 0.433$.

c. The closed-loop transform of the response for the first tank $H_1(s)$ is obtained using the block diagram and definitions of transfer functions. Note that $H_1(s)$, the output from $G_1(s)$, is

$$H_1(s) = [Q_i(s) - H_2(s)]K_pG_1(s) \tag{c}$$

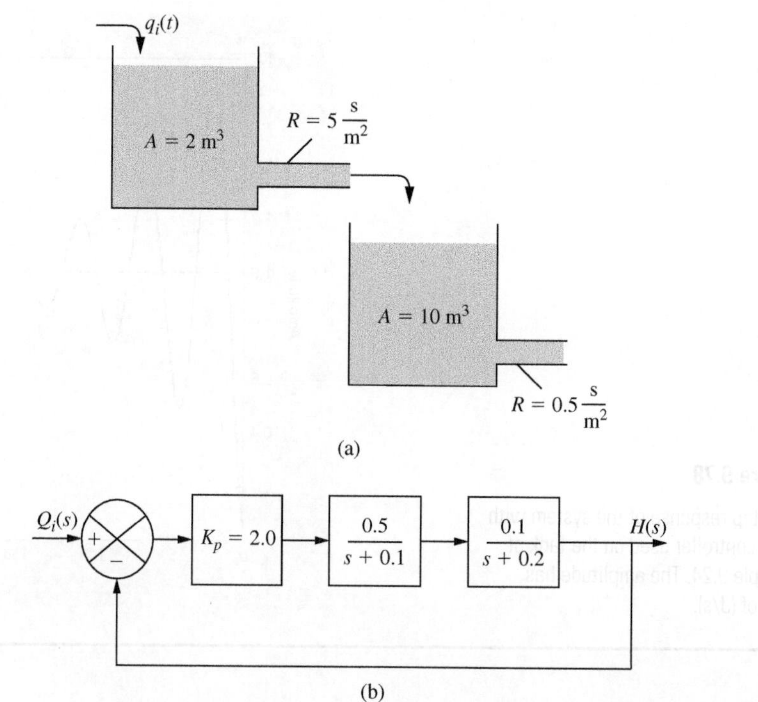

Figure 9.79

(a) The two-tank liquid-level system of Example 9.25; (b) the block diagram for the system with a proportional controller.

Since $G_1(s)$ and $G_2(s)$ are in series,

$$H_2(s) = G_2(s)H_1(s) \tag{d}$$

Eliminating $H_2(s)$ between Equations (c) and (d) leads to

$$H_1(s) = \frac{K_pG_1(s)Q_i(s)}{1 + K_pG_1(s)G_2(s)} \tag{e}$$

The substitution of known values into Equation (e) leads to

$$H_1(s) = \frac{2.0\left(\frac{0.5}{s+0.1}\right)Q_i(s)}{1 + 2.0\left(\frac{0.5}{s+0.1}\right)\left(\frac{0.1}{s+0.2}\right)}$$

$$= \frac{\left(\frac{1}{s+0.1}\right)Q_i(s)}{\frac{s^2 + 0.3s + 0.12}{(s+0.2)(s+0.1)}}$$

$$= \frac{(s+0.2)Q_i(s)}{s^2 + 0.3s + 0.12} \tag{f}$$

MATLAB-generated plots for the step responses of $H_1(s)$ and $H_2(s)$ are given in Figure 9.80. Note that the overshoot for $h_1(t)$ is much greater than the overshoot for $h_2(t)$, due to the presence of the s term in the numerator for $H_1(s)$.

d. The steady-state liquid levels are obtained using the final value theorem as

$$h_1 = \lim_{s\to\infty} H_1(s)$$

$$= \frac{0.2}{0.12}$$

$$= 1.67 \tag{g}$$

$$h_2 = \lim_{s\to\infty} H_2(s)$$

$$= \lim_{s\to\infty} G_2(s)H_1(s)$$

$$= 1.67 \lim_{s\to\infty}\left(\frac{0.1}{s+0.2}\right)$$

$$= 0.833 \tag{h}$$

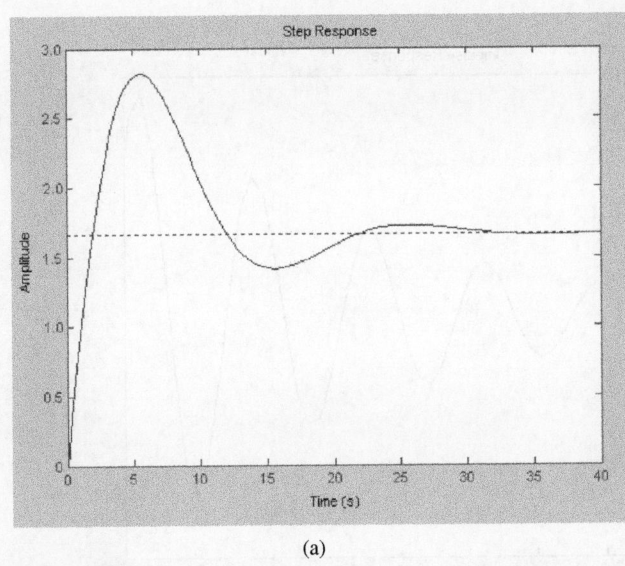

(a)

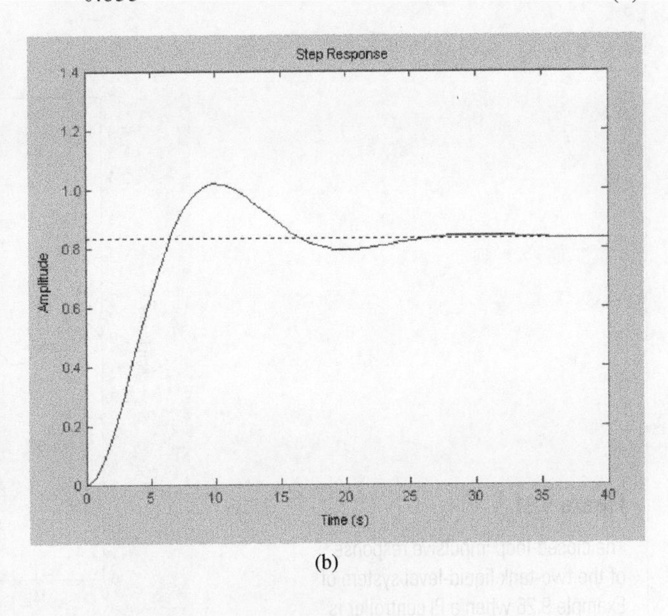

(b)

Figure 9.80

The step responses for the two-tank liquid-level problem of Example 9.25 with proportional control: (a) $h_1(t)$; (b) $h_2(t)$.

Example 9.26

The controller used for the system of Example 9.25 is being modified to a PI controller with $K_p = 2$. **a.** For what values of the integral gain K_i will the system be stable? **b.** Use MATLAB to plot the impulsive response of the system for one value of K_i for which the closed-loop system is stable and one value of K_i for which the system is unstable.

Solution

The open-loop transfer function is given by Equation (a) of Example 9.25. The closed-loop transfer function for a second-order system with a PI controller is given in Equation (9.133), which when applied to this system leads to

$$H(s) = \frac{0.2s + 0.1K_i}{s^3 + 0.3s^2 + 0.12s + 0.1K_i} \tag{a}$$

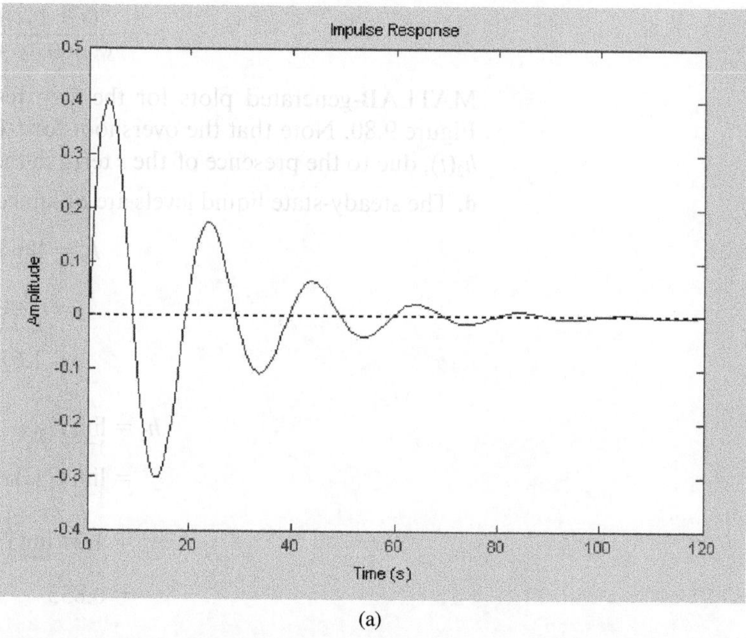

(a)

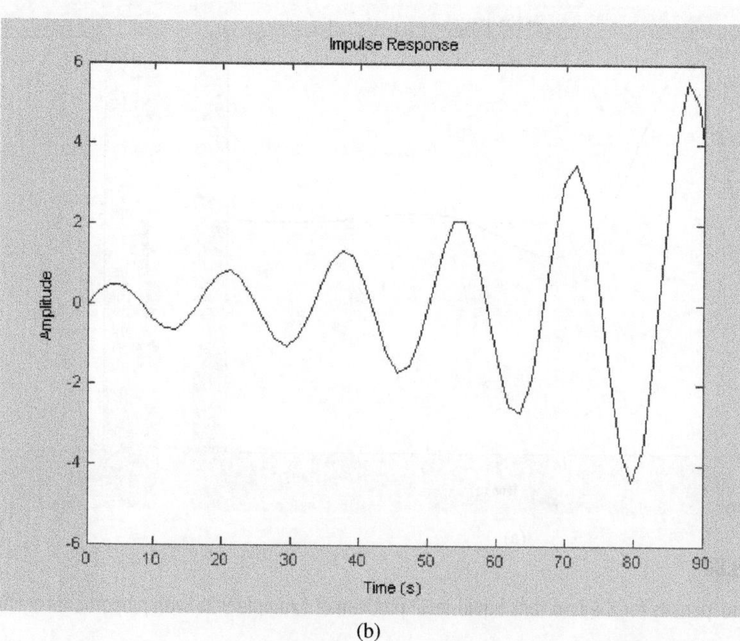

(b)

Figure 9.81

The closed-loop impulsive response of the two-tank liquid-level system of Example 9.26 when a PI controller is used with $K_p = 2.0\ m^2/s$: (a) $K_i = 0.2\ m^2/s^2$; (b) $K_i = 0.5\ m^2/s^2$.

a. The system is stable when all poles of the closed-loop transfer function have negative real parts. Application of Routh's criterion to the third-order denominator, as developed in Equation (9.131), leads to a stability criterion of

$$(0.3)(0.12) > 0.1\,K_i$$

$$K_i < 0.36 \qquad \text{(b)}$$

b. The MATLAB-generated plots are shown in Figure 9.81.

Example 9.27

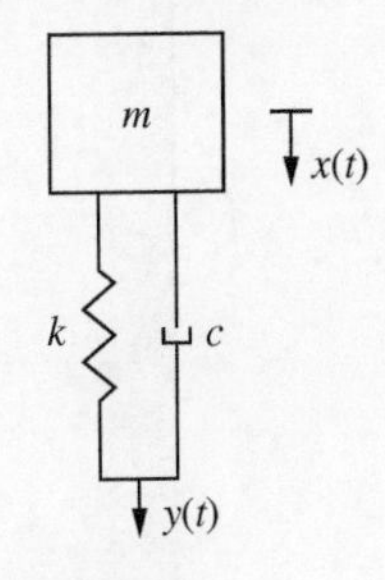

Figure 9.82

The system for Example 9.27.

An electronic PID controller is being designed to provide closed-loop feedforward control to a simplified vehicle suspension system, as illustrated in Figure 9.82. The open-loop transfer function for the system is

$$G(s) = \frac{2\zeta\omega_n + \omega_n^2}{s^2 + 2\zeta\omega_n s + \omega_n^2} \qquad \text{(a)}$$

The controller is being designed for a system with $k = 1 \times 10^5$ N/m, $c = 3000$ N·s/m, and an empty vehicle mass of 400 kg. **a.** A PID controller is to be used with $K_p = 2.0$ and $K_d = 1.5$. For what values of K_i is the system stable? **b.** For what values of K_i are all poles of the transfer function real? **c.** Compare the closed-loop and open-loop responses when subject to a step input for $K_i = 1.0$ and $K_i = 5.0$.

Solution

a. The natural frequency and damping ratio of the open-loop system are

$$\omega_n = \sqrt{\frac{1 \times 10^5 \text{ N/m}}{400 \text{ kg}}} = 15.8 \text{ r/s} \qquad \text{(b)}$$

$$\zeta = \frac{3000 \text{ N·s/m}}{2(400 \text{ kg})(15.8 \text{ r/s})} = 0.237 \qquad \text{(c)}$$

Equation (a) is of the form of the transfer function of a second-order system with $A = 2\zeta\omega_n = 7.49$ and $B = \omega_n^2 = 250.0$. Applying Equation (9.141), the closed-loop transfer function of a second-order plant in a feedback loop with a PID controller in the feedforward part of the loop leads to

$$H(s) = \frac{1.5(7.49)s^3 + [2.0(7.49) + 1.5(250)]s^2 + [2.0(250) + 7.49\,K_i]s + 250\,K_i}{[1 + 1.5(7.49)]s^3 + [7.49 + 2.0(7.49) + 1.5(250)]s^2 + [250 + 2.0(250) + 7.49\,K_i]s + 250\,K_i}$$

$$= \frac{11.24s^3 + 390.0s^2 + (500 + 7.49\,K_i)s + 250\,K_i}{12.24s^3 + 397.5s^2 + (750 + 7.49\,K_i)s + 250\,K_i} \qquad \text{(d)}$$

The application of Routh's criterion to the denominator of $H(s)$ leads to a stability criterion of

$$(397.5)(750 + 7.49\,K_i) > 12.24(250\,K_i) \qquad \text{(e)}$$

Equation (e) is rearranged, leading to $K_i < 3602.3$.

b. A trial and error solution using the MATLAB script of Figure 9.83 is used to determine that the poles of $H(s)$ are all real for $K_i < 1.504$.

Figure 9.83

The MATLAB script used to iterate on values of K such that all poles of the closed-loop transfer function are real. The script is run changing the values of K_i and inspecting the poles obtained from the residue function.

```
Ki=input ('Input value of integral gain')
num=[11.24 390 500+7.49*Ki 250*Ki]
den=[12.24 397.5 750+7.49*Ki 250*Ki]
[r,p,k]=residue(num,den)
```

c. It is clear from the MATLAB-generated plots of Figure 9.84 that the use of this PID controller does not enhance the response of the system to a step change. The settling times are longer with the controller than without the controller.

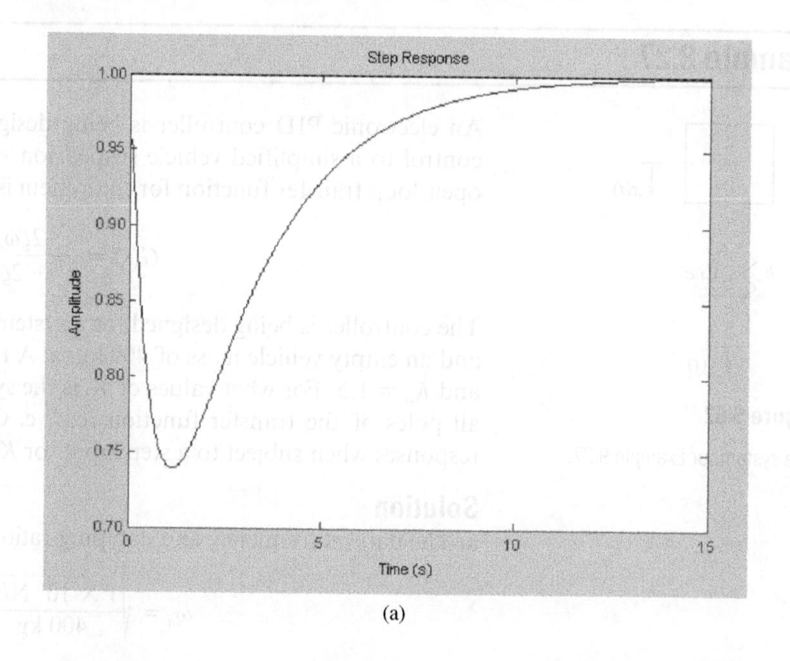

(a)

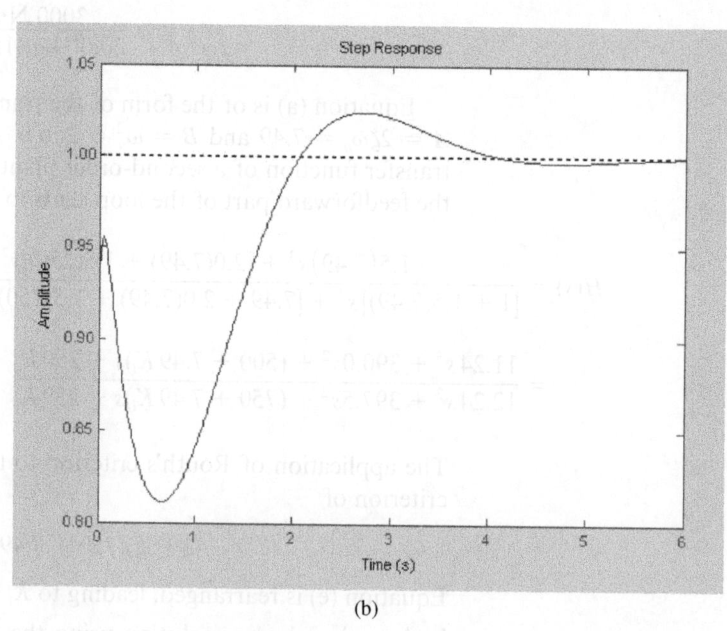

(b)

Figure 9.84

The step responses of Example 9.27: (a) PID controller with $K_i = 1$; (b) PID controller with $K_i = 5$; (c) no controller.

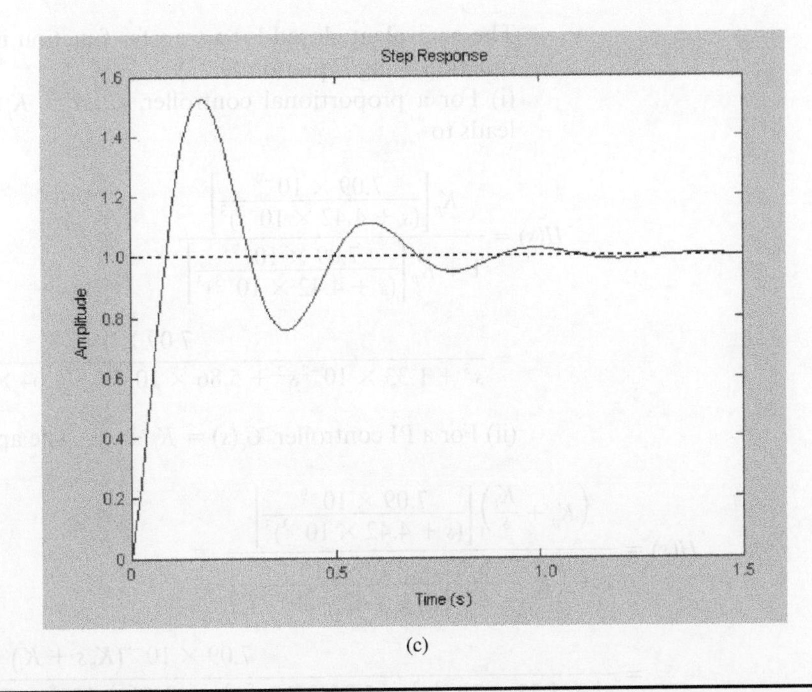

(c)

Figure 9.84

(Continued)

Example 9.28

Consider a three tank CSTR as shown in Figure 9.85. The reactors are identical with the identical reaction occurring in each. The flow rate into the first reactor is $q = 1.2 \times 10^{-6} \frac{\text{m}^3}{\text{s}}$. The volume of each reactor is $V = 6.24 \times 10^{-4}\ \text{m}^3$ and the rate of reaction in each reactor is $k = 2.5 \times 10^{-3}\ \text{s}^{-1}$. The transfer function for the concentration of reactant A in the third reactor is

$$G(s) = \frac{q^3}{(Vs + q + kV)^3} = \left(\frac{\frac{q}{V}}{s + k + \frac{q}{V}}\right)^3 \quad \text{(a)}$$

a. Determine the closed-loop transfer function when the reactors are placed in a feedback loop with (i) a proportional controller of proportional gain K_p, (ii) a PI controller with proportional gain K_p and integral gain K_i, and (iii) a PD controller with proportional gain K_p and differential gain K_d. **b.** Discuss the stability of the closed-loop response when each controller is used.

Solution

a. The substitution of the given values into Equation (a) leads to

$$G(s) = \frac{7.08 \times 10^{-9}}{(s + 4.42 \times 10^{-3})^3} \quad \text{(b)}$$

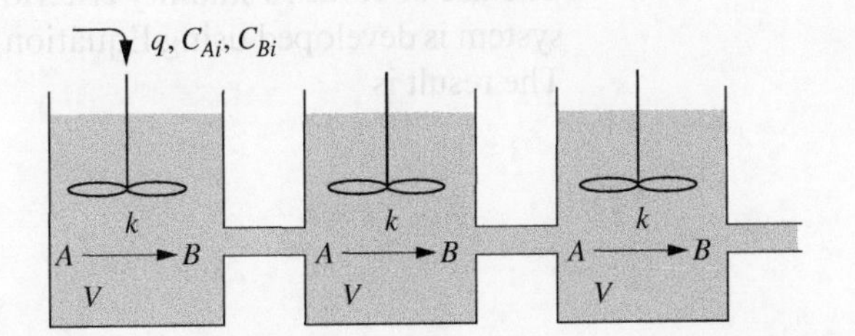

Figure 9.85

Staged CSTR system of Example 9.28.

The equivalent closed-loop transfer function is derived from the open-loop transfer function using Equation (9.23).

(i) For a proportional controller, $G_c(s) = K_p$. The application of Equation (9.23) leads to

$$H(s) = \frac{K_p\left[\dfrac{7.09 \times 10^{-9}}{(s + 4.42 \times 10^{-3})^3}\right]}{1 + K_p\left[\dfrac{7.09 \times 10^{-9}}{(s + 4.42 \times 10^{-3})^3}\right]}$$

$$= \frac{7.09 \times 10^{-9} K_p}{s^3 + 1.33 \times 10^{-2} s^2 + 5.86 \times 10^{-5} s + 8.64 \times 10^{-8} + 7.09 \times 10^{-9} K_p} \quad \text{(c)}$$

(ii) For a PI controller, $G_c(s) = K_p + \frac{K_i}{s}$. The application of Equation (9.23) leads to

$$H(s) = \frac{\left(K_p + \dfrac{K_i}{s}\right)\left[\dfrac{7.09 \times 10^{-9}}{(s + 4.42 \times 10^{-3})^3}\right]}{1 + \left(K_p + \dfrac{K_i}{s}\right)\left[\dfrac{7.09 \times 10^{-9}}{(s + 4.42 \times 10^{-3})^3}\right]}$$

$$= \frac{7.09 \times 10^{-9}(K_p s + K_i)}{s^4 + 1.33 \times 10^{-2} s^3 + 5.86 \times 10^{-5} s^2 + \left(8.64 \times 10^{-8} + 7.09 \times 10^{-9} K_p\right)s + 7.09 \times 10^{-9} K_i} \quad \text{(d)}$$

(iii) For PD controller, $G_c(s) = K_p + K_d s$. The application of Equation (9.23) leads to

$$H(s) = \frac{(K_p + K_d s)\left[\dfrac{7.09 \times 10^{-9}}{(s + 4.42 \times 10^{-3})^3}\right]}{1 + (K_p + K_d s)\left[\dfrac{7.09 \times 10^{-9}}{(s + 4.42 \times 10^{-3})^3}\right]}$$

$$= \frac{7.09 \times 10^{-9}(K_p + K_d s)}{s^3 + 1.33 \times 10^{-2} s^2 + (5.86 \times 10^{-5} + 7.09 \times 10^{-9} K_d)s + 8.64 \times 10^{-8} + 7.09 \times 10^{-9} K_p} \quad \text{(e)}$$

b. The closed-loop system is stable when all poles of the transfer function have negative real parts.

(i) The poles of the closed-loop transfer function when the proportional controller is used are roots of

$$s^3 + 1.33 \times 10^{-2} s^2 + 5.86 \times 10^{-5} s + 8.64 \times 10^{-8} + 7.09 \times 10^{-9} K_p = 0 \quad \text{(f)}$$

Routh's criterion applied to a third-order transfer function leads to the stability criterion

$$(1.33 \times 10^{-2})(5.86 \times 10^{-5}) > 8.64 \times 10^{-8} + 7.09 \times 10^{-9} K_p \quad \text{(g)}$$

Equation (g) is rearranged to $K_p < 97.7$.

(ii) The poles of the closed-loop transfer function when the PI controller is used are the roots of

$$s^4 + 1.33 \times 10^{-2} s^3 + 5.86 \times 10^{-5} s^2 + (8.64 \times 10^{-8} + 7.09 \times 10^{-9} K_p)s + 7.09 \times 10^{-9} K_i \quad \text{(h)}$$

The use of Routh's stability criterion is discussed in Section 7.3. The **B** matrix for this system is developed using Equation (7.34). The elements are given by Equation (7.35). The result is

$$\mathbf{B} = \begin{bmatrix} 1 & 5.86 \times 10^{-5} & 7.09 \times 10^{-9} K_i \\ 1.33 \times 10^{-2} & 8.64 \times 10^{-8} + 7.09 \times 10^{-9} K_p & 0 \\ 5.21 \times 10^{-5} - 5.33 \times 10^{-7} K_p & 7.09 \times 10^{-9} K_i & 0 \\ \dfrac{4.50 \times 10^{-12} - 3.23 \times 10^{-13} K_p + 3.78 \times 10^{-15} K_p^2 - 9.43 \times 10^{-11} K_i}{5.21 \times 10^{-5} - 5.33 \times 10^{-7} K_p} & 0 & 0 \\ 7.09 \times 10^{-9} K_i & 0 & 0 \end{bmatrix} \tag{i}$$

The system is stable if and only if all elements of the first column of **B** are positive. Requiring $b_{3,1} > 0$ requires $K_p < 97.7$. Requiring $b_{5,1} > 0$ requires $K_i > 0$. Noting that if $b_{3,1} > 0$ then the denominator of $b_{4,1}$ is positive; if $K_p < 97.7$ then the closed-loop response is stable if

$$4.50 \times 10^{-12} - 3.23 \times 10^{-13} K_p + 3.78 \times 10^{-15} K_p^2 - 9.43 \times 10^{-11} K_i > 0 \tag{j}$$

Equation (j) represents a parabola when plotted in the (K_i, K_p) plane, as illustrated in Figure 9.86. Stable and unstable regions are indicated in the figure with the parabola and the line $K_p = 97.7$ as boundaries of the regions between stability and instability.

(iii) The poles of the closed-loop transfer function when the PD controller is used are the roots of

$$s^3 + 1.33 \times 10^{-2} s^2 + (5.86 \times 10^{-5} + 7.09 \times 10^{-9} K_d)s + 8.64 \times 10^{-8} + 7.09 \times 10^{-9} K_p = 0 \tag{k}$$

Applying Routh's stability criterion in the same manner as for the proportional controller leads to a stability criterion of

$$(1.33 \times 10^{-2})(5.86 \times 10^{-5} + 7.09 \times 10^{-9} K_d) > 8.64 \times 10^{-8} + 7.09 \times 10^{-9} K_p \tag{l}$$

Equation (l) is rearranged to

$$97.7 + 1.33 \times 10^{-2} K_d > K_p \tag{m}$$

Equation (m) shows that if $K_p < 97.7$, then the system is stable for any positive value of K_d. The larger the value of K_d, the larger the maximum value of K_p for stability. Equation (m) is plotted in Figure 9.87, illustrating the regions of stability and instability.

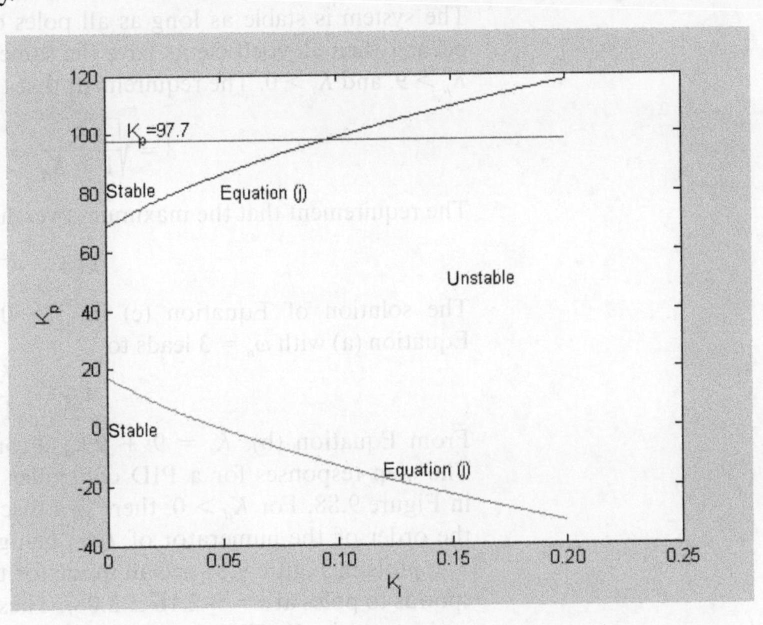

Figure 9.86

The stability regions for the PI controller of Example 9.28.

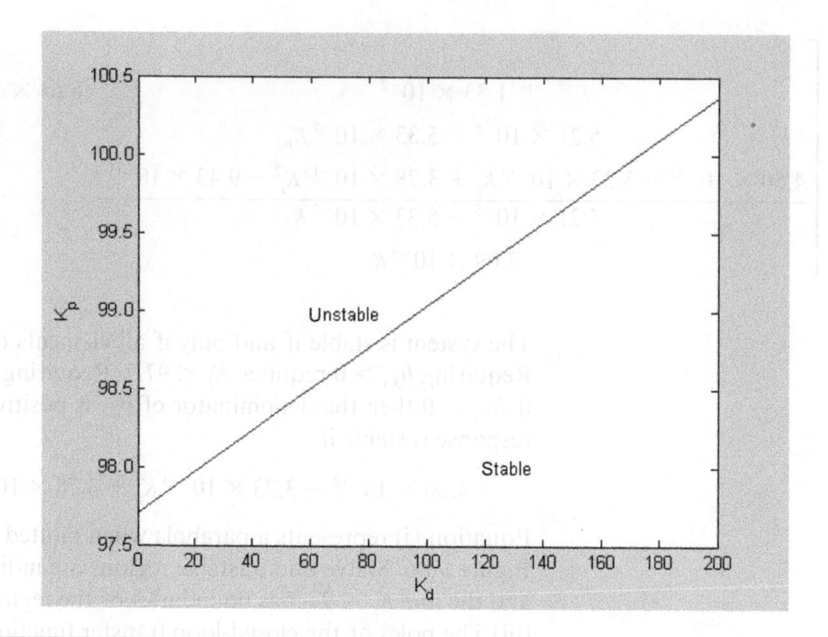

Figure 9.87
The stability regions for use of the PD controller on the three-CSTR system of Example 9.28.

Example 9.29

Design a PID controller such that the unstable plant $G(s) = \frac{4}{s-9}$ is stable when placed in a feedback loop with the PID controller. The stable system should have a natural frequency of 3 r/s and a maximum overshoot of 4 percent.

Solution

The closed-loop transfer function for the plant in a feedback loop with a PID controller is

$$H(s) = \frac{K_d s^2 + K_p s + K_i}{(1 + K_d)s^2 + (K_p - 9)s + K_i} \tag{a}$$

The system is stable as long as all poles of the transfer function are positive. This occurs when all coefficients have the same sign. The requirement leads to $K_d > -1$, $K_p > 9$, and $K_i > 0$. The requirement that the natural frequency is 3 r/s leads to

$$3 = \sqrt{\frac{K_i}{1 + K_d}} \Rightarrow K_i - 9K_d = 9 \tag{b}$$

The requirement that the maximum overshoot is 4 percent leads to

$$0.04 > e^{-\zeta\pi/\sqrt{1-\zeta^2}} \tag{c}$$

The solution of Equation (c) is $\zeta > 0.72$. Obtaining the damping ratio from Equation (a) with $\omega_n = 3$ leads to

$$4.32 < \frac{K_p - 9}{1 + K_d} \tag{d}$$

From Equation (b), $K_i = 9 + 9K_d$. From Equation (d), $K_p > 14.32 + 4.32\,K_d$. The step responses for a PID controller with several values of K_d are illustrated in Figure 9.88. For $K_d > 0$, there is a discontinuity in the response at $t = 0$ due to the order of the numerator of $H(s)$ being equal to the order of the denominator. The plots in Figure 9.88 are all made for the minimum damping ratio which corresponds to poles at $s = -2.16 \pm 2.08j$. These plots show a decrease in overshoot with an increase in K_d. The overshoot is affected by the numerator. It is noted that for

$K_d = 40$, the overshoot is 3.35 percent. For this value, the transfer function of the PID controller is

$$G_c(s) = 187.12 + \frac{369}{s} + 40s \tag{e}$$

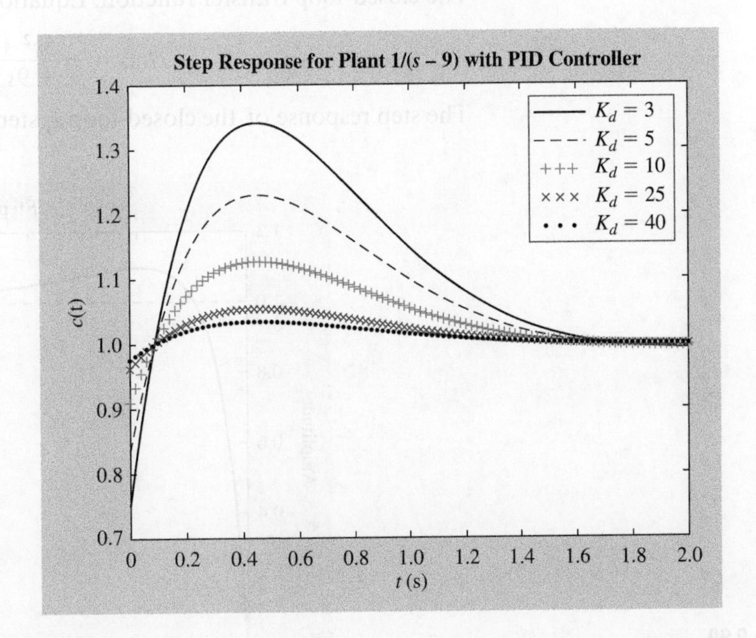

Figure 9.88

Step responses of a plant with a PID controller for several values of the derivative gain.

Example 9.30

The plant of Figure 9.89 is neutrally stable. It is placed in series with a PID controller and placed in a feedback loop. Design the PID controller such that the closed-loop system has poles at $s = -1$, $s = -3$, and $s = -5$. Plot the step response of the closed-loop system.

Solution

The closed-loop transfer function has the form of Equation (9.20):

$$H(s) = \frac{0.5\left(K_d s^2 + K_p s + K_i\right)}{s^3 + (4 + 0.5K_d)s^2 + 0.5K_p s + 0.5K_i} \tag{a}$$

In order for the poles of $H(s)$ to be at $-1, -3$, and -5, its denominator must be

$$D(s) = (s + 1)(s + 3)(s + 5) = s^3 + 9s^2 + 23s + 15 \tag{b}$$

Setting Equation (b) equal to the denominator of Equation (a) leads to

$$4 + 0.5K_d = 9 \Rightarrow K_d = 10 \tag{c}$$

$$0.5K_p = 23 \Rightarrow K_p = 46 \tag{d}$$

$$0.5K_i = 15 \Rightarrow K_i = 30 \tag{e}$$

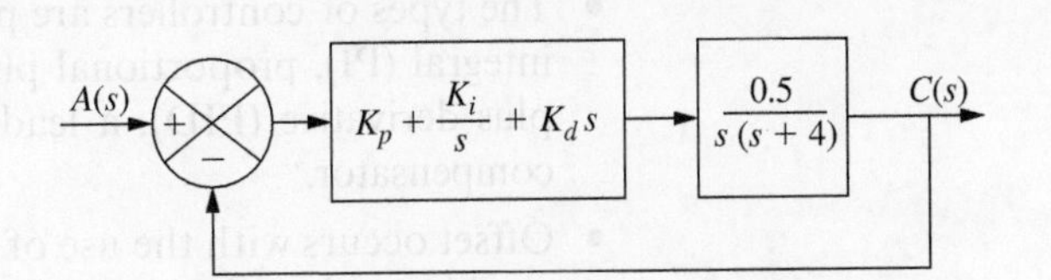

Figure 9.89

A neutrally stable plant placed in a feedback loop with a PID controller.

The PID controller is designed such that

$$G_c(s) = 46 + \frac{30}{s} + 10s \tag{f}$$

The closed-loop transfer function, Equation (a), becomes

$$H(s) = \frac{5s^2 + 23s + 15}{s^3 + 9s^2 + 23s + 15} \tag{g}$$

The step response of the closed-loop system is shown in Figure 9.90.

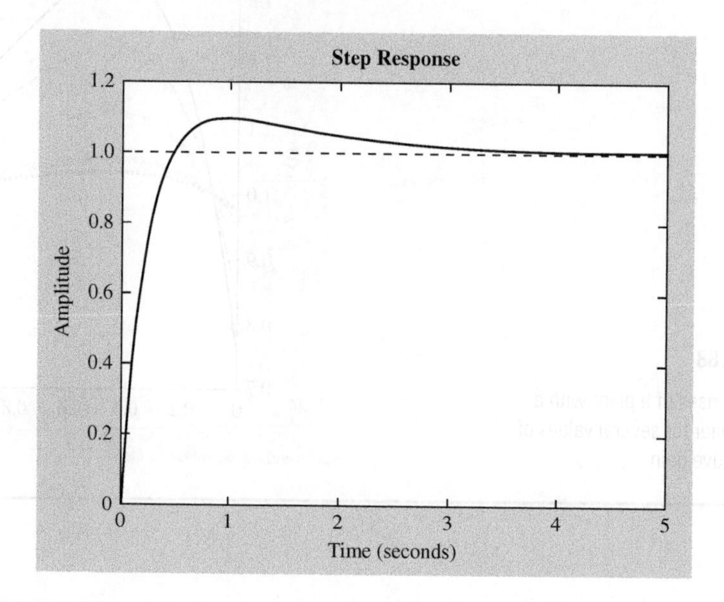

Figure 9.90

Step response for closed-loop system of Example 9.30.

9.11 Summary

9.11.1 Chapter Highlights

- Block diagrams provide a schematic representation of a dynamic system and its input and output.
- Block diagrams are used to illustrate feedback in dynamic systems.
- Simulink is a MATLAB package that simulates dynamic systems using block diagrams. Simulink uses numerical methods to run simulations for dynamic systems from block diagrams.
- Feedback control is used to modify the response of dynamic systems.
- The dynamic system response of a feedback control system is governed by the closed-loop transfer function.
- Actuators are placed in feedback control systems to modify the closed-loop transfer function and enhance system response. This study concentrates on actuators placed in the feedforward portion of the loop.
- The types of controllers are proportional, integral, derivative, proportional plus integral (PI), proportional plus derivative (PD), and proportional plus integral plus derivative (PID), a lead compensator, a lag compensator, and a lead-lag compensator.
- Offset occurs with the use of a proportional controller, but is eliminated when used in conjunction with an integral controller.

- The type of system refers to the power of s that is multiplying a function of s in the denominator of a transfer function.
- The static position error constants indicate how well a system replicates a step input.
- The static velocity error constant indicates how well a system replicates a ramp input.
- Proportional and/or derivative control of a first-order plant leads to a closed-loop transfer function for a first-order system.
- A controller with an integral gain used with a first-order plant leads to a closed-loop transfer function for a second-order system.
- Proportional and/or derivative control of a second-order plant leads to a closed-loop transfer function for a second-order system.
- A controller with an integral gain used with a second-order plant leads to a closed-loop transfer function for a third-order system.
- An integral controller used with a second-order plant could lead to an unstable system response.
- Velocity feedback is an example of a parallel control system.

9.11.2 Important Equations

- Definition of the transfer function in terms of input and output

$$C(s) = G(s)A(s) \tag{9.1}$$

- Closed-loop transfer function for a feedback loop

$$\frac{C(s)}{A(s)} = \frac{G(s)}{1 + G(s)} \tag{9.5}$$

- Closed-loop transfer function for a plant with an actuator in the feedforward portion of the loop

$$H(s) = \frac{C(s)}{A(s)} = \frac{G(s)\,G_c(s)}{1 + G(s)\,G_c(s)} \tag{9.20}$$

- Open-loop transfer function for a controller in series with a plant

$$G_o(s) = G(s)\,G_c(s) \tag{9.21}$$

- Proportional control

$$G_c(s) = K_p \tag{9.24}$$

- Integral control

$$G_c(s) = \frac{K_i}{s} \tag{9.26}$$

- Derivative control

$$G_c(s) = K_d s \tag{9.30}$$

- PID control

$$G_c(s) = K_p + K_d s + \frac{K_i}{s} \tag{9.34}$$

- Lead compensator

$$G_c(s) = K_c \frac{s + \dfrac{1}{T}}{s + \dfrac{1}{\alpha T}} \qquad \alpha < 1 \tag{9.38}$$

- Lag compensator

$$G_c(s) = K_c \frac{s + \frac{1}{T}}{s + \frac{1}{\alpha T}} \qquad \alpha > 1 \tag{9.40}$$

- Lead-lag compensator

$$G_c(s) = K_c \left(\frac{s + \frac{1}{T_1}}{s + \frac{1}{\alpha_1 T_1}} \right) \left(\frac{s + \frac{1}{T_2}}{s + \frac{1}{\alpha_2 T_2}} \right) \qquad \alpha_1 < 1, \alpha_2 > 1 \tag{9.42}$$

- Offset for a unit step input

$$\eta = \lim_{s \to 0} \frac{1}{1 + G_c(s)G(s)} = \frac{1}{1 + G_c(0)G(0)} \tag{9.81}$$

- Static position, velocity, and acceleration error constants

$$C_p = \lim_{s \to 0} G(s) = G(0) \tag{9.90}$$

$$C_v = \lim_{s \to 0} sG(s) \tag{9.91}$$

$$C_a = \lim_{s \to 0} s^2 G(s) \tag{9.92}$$

- Transfer function for a plant of type k

$$G(s) = \frac{N(s)}{s^k D(s)} \tag{9.93}$$

- Proportional control of a first-order plant

$$H(s) = \frac{K_p K}{s + \frac{1}{T} + K_p K} \tag{9.98}$$

- Integral control of a first-order plant

$$H(s) = \frac{K_i K}{s^2 + \frac{1}{T} s + K_i K} \tag{9.105}$$

- PID control of a first-order plant

$$H(s) = \frac{K\left(K_d s^2 + K_p s + K_i\right)}{(1 + KK_d)s^2 + \left(KK_p + \frac{1}{T}\right)s + KK_i} \tag{9.119}$$

- Proportional control of a second-order plant

$$H(s) = \frac{K_p(As + B)}{s^2 + (2\zeta\omega_n + K_p A)s + K_p B + \omega_n^2} \tag{9.123}$$

- Integral control of a second-order plant

$$H(s) = \frac{K_i(As + B)}{s^3 + 2\zeta\omega_n s^2 + (\omega_n^2 + K_i A)s + K_i B} \tag{9.129}$$

- PID control of a second-order plant

$$H(s) = \frac{K_d A s^3 + (K_p A + K_d B)s^2 + (K_p B + K_i A)s + K_i B}{(1 + K_d A)s^3 + (2\zeta\omega_n + K_p A + K_d B)s^2 + \left(\omega_n^2 + K_p B + K_i A\right)s + K_i B} \tag{9.141}$$

Short Answer Problems

SA9.1 What is the transfer function for the system shown in Figure SA9.1?

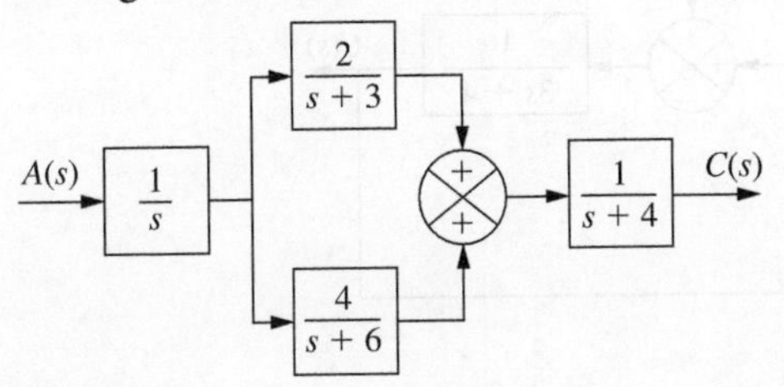

Figure SA9.1

SA9.2 What is the transfer function for the system shown in Figure SA9.2?

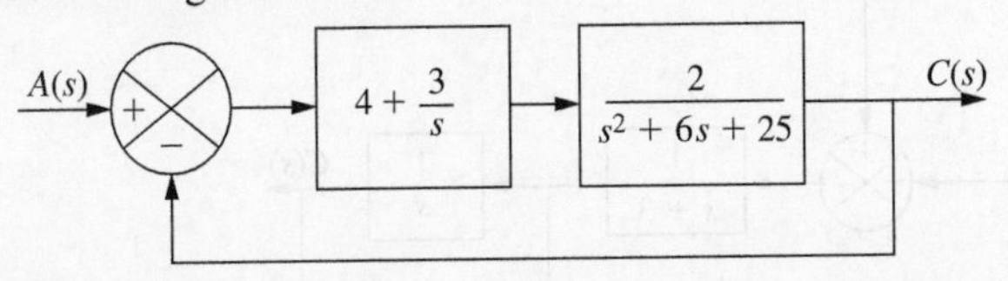

Figure SA9.2

SA9.3 What is the transfer function for the system shown in Figure SA9.3?

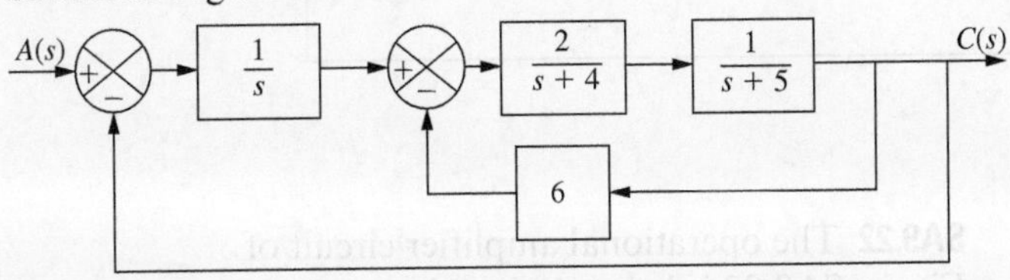

Figure SA9.3

SA9.4 What is the transfer function for the system shown in Figure SA9.4?

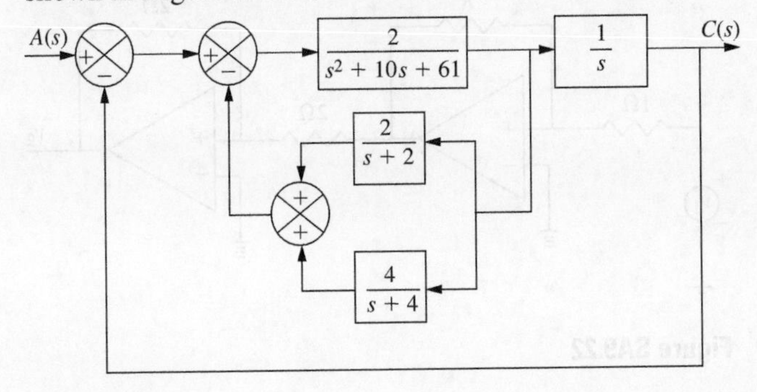

Figure SA9.4

SA9.5 If two plants with transfer functions $G_1(s)$ and $G_2(s)$ are in series, what is the output from the plants when the input to the combination is $A(s)$?

SA9.6 If two plants with transfer functions $G_1(s)$ and $G_2(s)$ are in parallel, what is the output from the plants when the input to the combination is $A(s)$?

SA9.7 What is the output from the summing junction in Figure SA9.7?

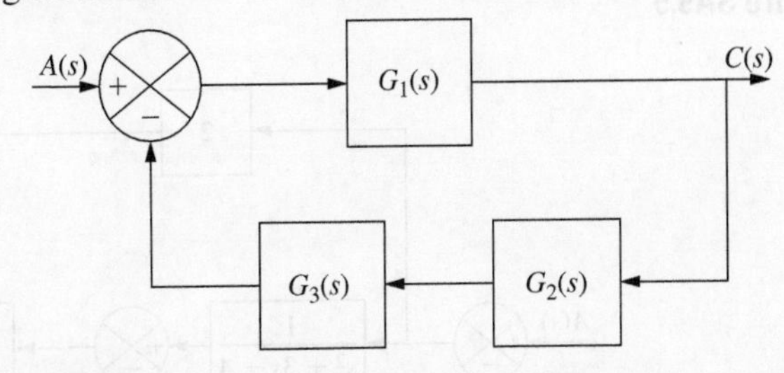

Figure SA9.7

SA9.8 What is the transfer function of the system whose block diagram model is shown in Figure SA9.8?

SA9.9 What is the transfer function of the system whose block diagram model is shown in Figure SA9.9 (page 618)?

SA9.10 What is the transfer function of the system whose block diagram model is shown in Figure SA9.10 (page 618)?

SA9.11 What is the transfer function for a proportional controller with a gain of 7.5?

SA9.12 What is the transfer function for an integral controller with a gain of 3.4?

SA9.13 What is the transfer function for a PD controller with a proportional gain of 10.4 and a differential gain of 17.5?

SA9.14 What is the transfer function for a PI controller with a proportional gain of 11.2 and an integral gain of 4.2?

SA9.15 What is the transfer function for a PID controller with a proportional gain of 12.2, an integral gain of 15.5, and a differential gain of 1.2?

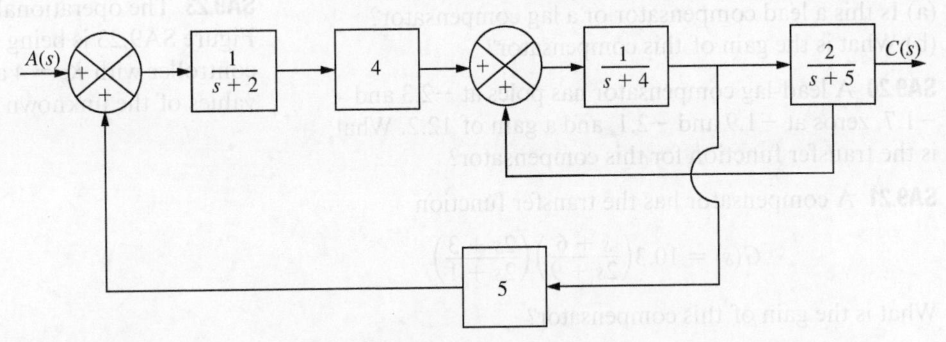

Figure SA9.8

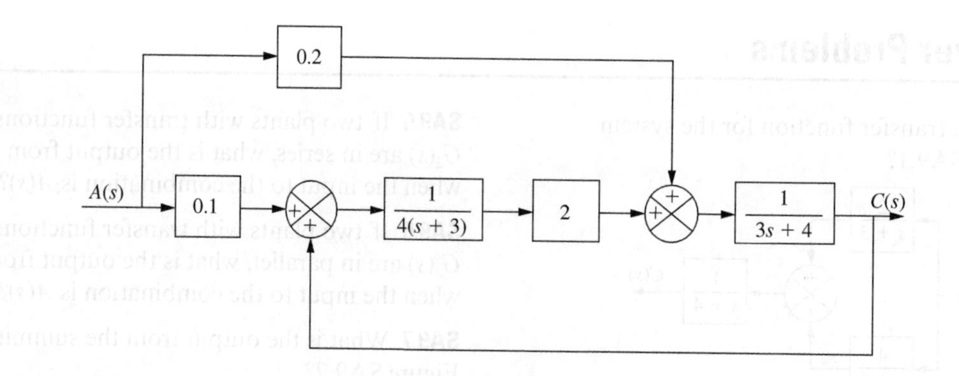

Figure SA9.9

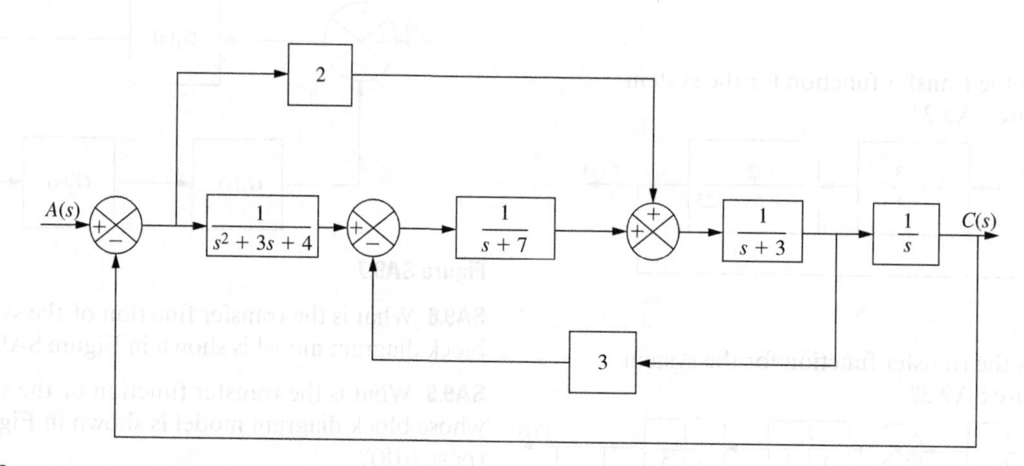

Figure SA9.10

SA9.16 A lead compensator has a pole of -2.3, a zero of -1.2, and a gain of 3.4. What is the transfer function for this compensator?

SA9.17 A lag compensator has a pole at -1.2, a zero at -2.3, and a gain of 3.4. What is the transfer function for this compensator?

SA9.18 A compensator has the transfer function

$$G(s) = 2.3\frac{3s + 6}{4s + 10}$$

(a) Is this a lead compensator or a lag compensator?
(b) What is the gain of this compensator?

SA9.19 A compensator has the transfer function

$$G(s) = 3.1\frac{3s + 15}{5s + 8}$$

(a) Is this a lead compensator or a lag compensator?
(b) What is the gain of this compensator?

SA9.20 A lead-lag compensator has poles at -2.3 and -1.7, zeros at -1.9 and -2.1, and a gain of 12.2. What is the transfer function for this compensator?

SA9.21 A compensator has the transfer function

$$G(s) = 10.3\left(\frac{s + 6}{2s + 9}\right)\left(\frac{2s + 3}{2s + 1}\right)$$

What is the gain of this compensator?

SA9.22 The operational amplifier circuit of Figure SA9.22 is being designed to serve as a proportional controller with $K_p = 3.4$. What is the necessary value of the unknown resistance?

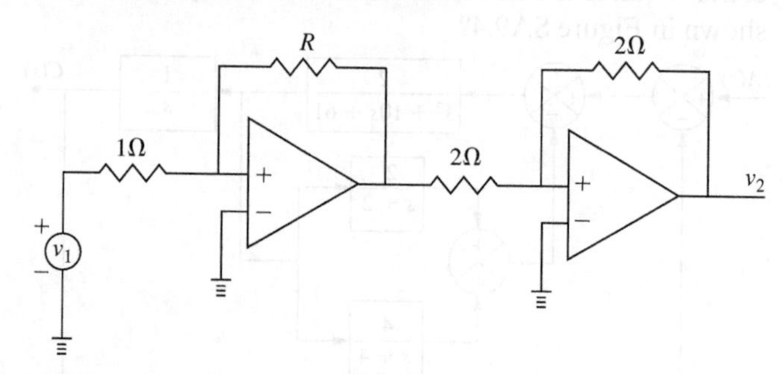

Figure SA9.22

SA9.23 The operational amplifier circuit of Figure SA9.23 is being designed to serve as a PI controller with $K_p = 4$ and $K_i = 3.2$. What are the values of the unknown resistance and the capacitance?

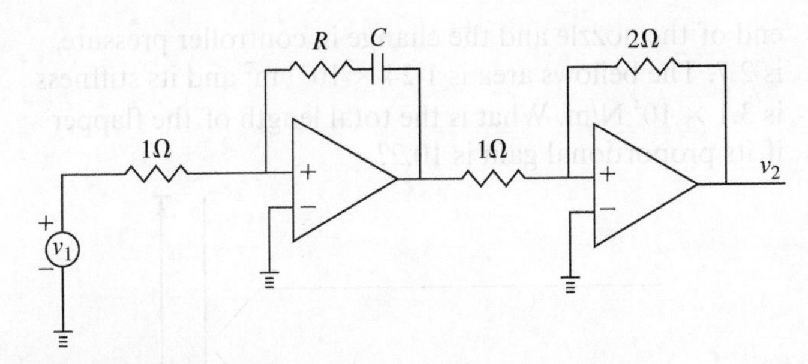

Figure SA9.23

SA9.24 The operational amplifier circuit of Figure SA9.24 is being designed to serve as a PD controller with $K_p = 4$ and $K_d = 3.2$. What are the values of the unknown resistance and the capacitance?

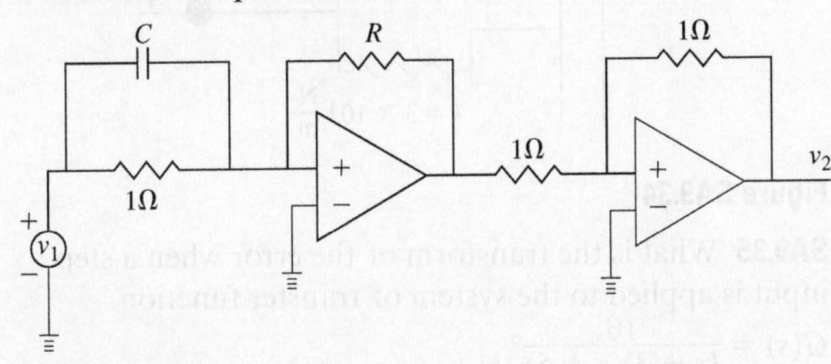

Figure SA9.24

SA9.25 The operational amplifier circuit of Figure SA9.25 is being designed to serve as a PID controller with $K_p = 4$, $K_d = 1.4$, and $K_i = 3.2$. What are the values of the unknown resistance and capacitances?

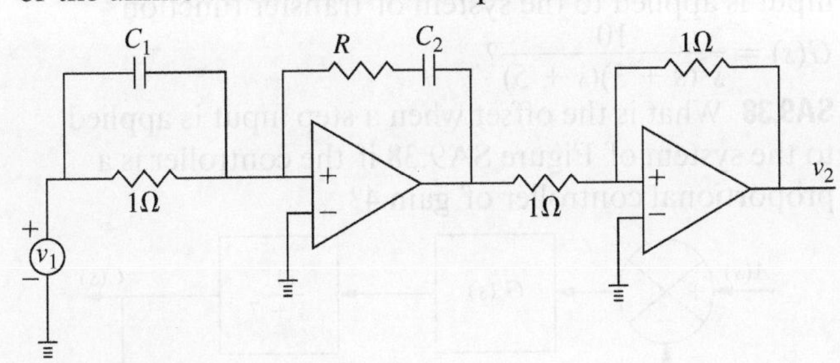

Figure SA9.25

SA9.26 What type of controller is represented by the mechanical system of Figure SA9.26?

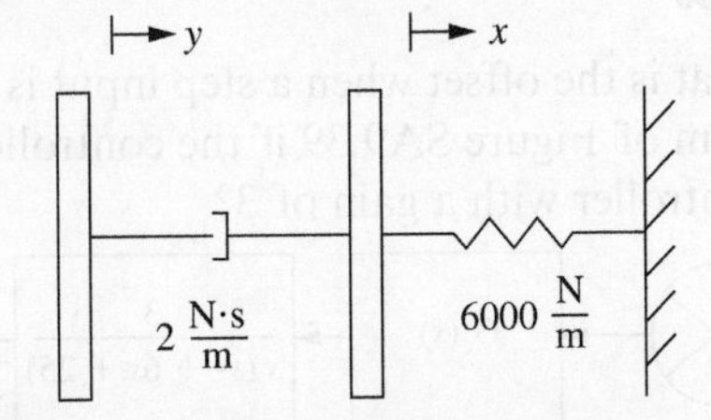

Figure SA9.26

SA9.27 What type of controller is represented by the mechanical system of Figure SA9.27?

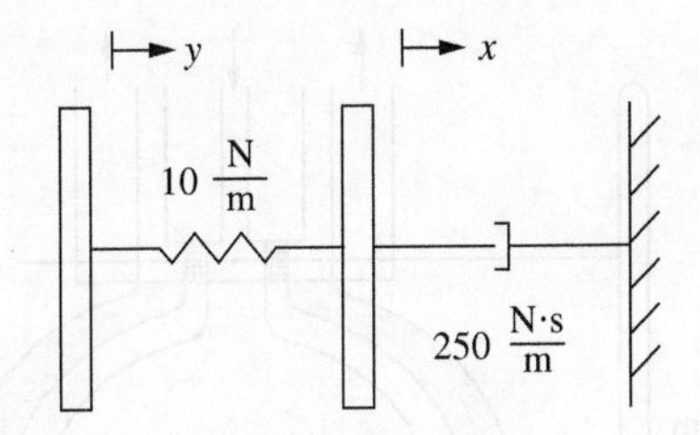

Figure SA9.27

SA9.28 What type of controller is represented by the mechanical system of Figure SA9.28?

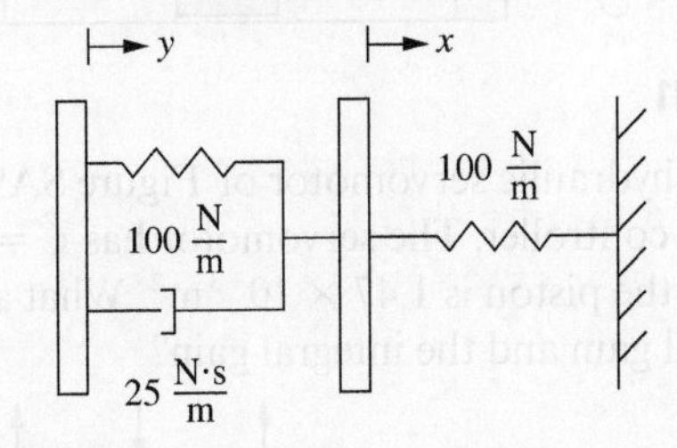

Figure SA9.28

SA9.29 What type of controller is represented by the mechanical system of Figure SA9.29?

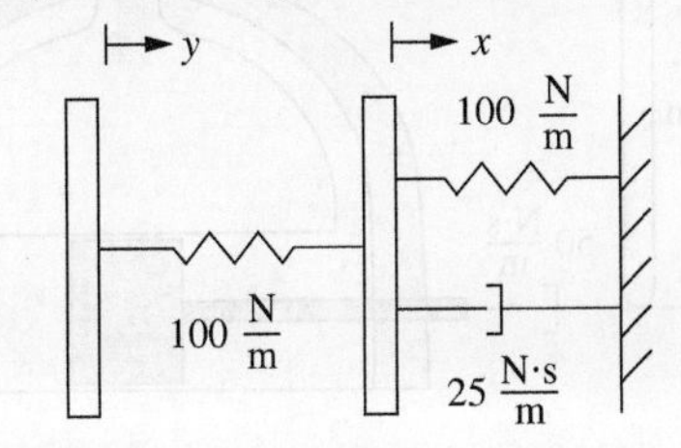

Figure SA9.29

SA9.30 What is the gain when the hydraulic servomotor of Figure SA9.30 is used as a proportional controller?

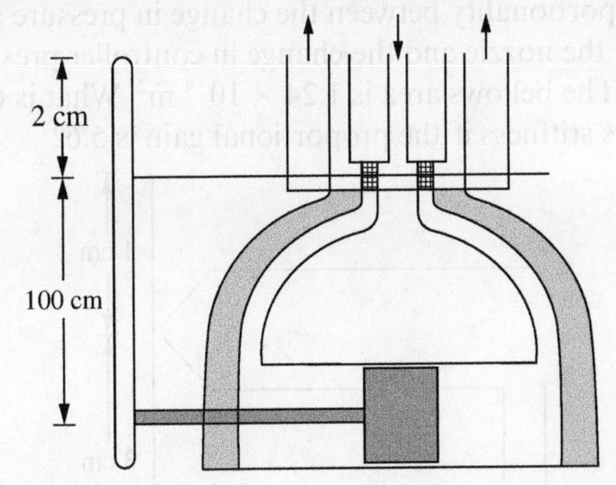

Figure SA9.30

SA9.31 The hydraulic servomotor of Figure SA9.31 is to be used as a proportional controller with a proportional gain of 3.6. If the distance between the piston and the spool valve opening is 1.3 cm, what is the total length of the walking stick?

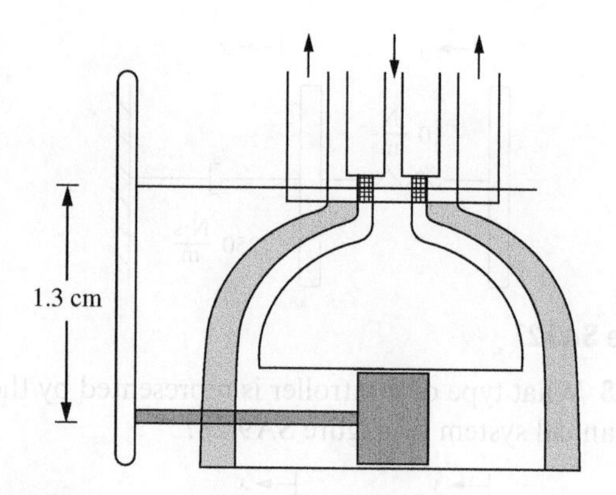

Figure SA9.31

SA9.32 The hydraulic servomotor of Figure SA9.32 is used as a PI controller. The servomotor has $\hat{C} = 2.43 \frac{\text{m}^2}{\text{s}}$. The area of the piston is $1.47 \times 10^{-3}\text{m}^2$. What are the proportional gain and the integral gain?

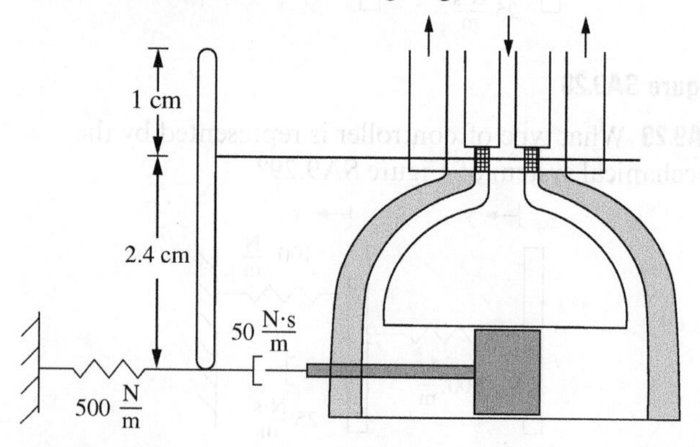

Figure SA9.32

SA9.33 The hydraulic circuit of Figure SA9.33 is used as a proportional controller. The value of K, the constant of proportionality between the change in pressure at the end of the nozzle and the change in controller pressure, is 2.7. The bellows area is 1.24×10^{-4} m^2. What is the bellows stiffness if the proportional gain is 5.6?

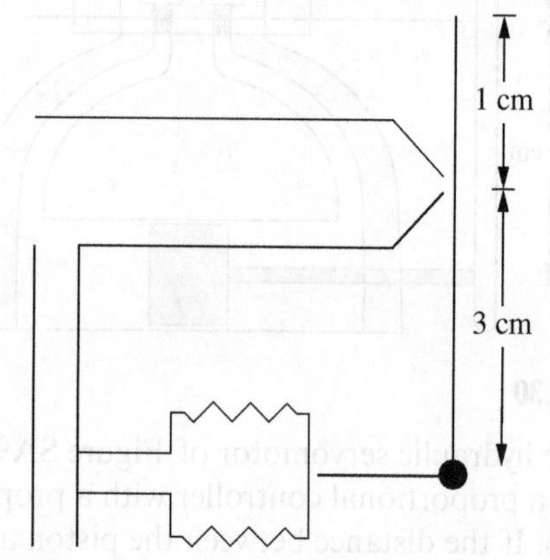

Figure SA9.33

SA9.34 The hydraulic circuit of Figure SA9.34 is used as a proportional controller. The value of K, the constant of proportionality between the change in pressure at the end of the nozzle and the change in controller pressure, is 2.7. The bellows area is 1.24×10^{-4} m^2 and its stiffness is 3.1×10^4 N/m. What is the total length of the flapper if its proportional gain is 10.2?

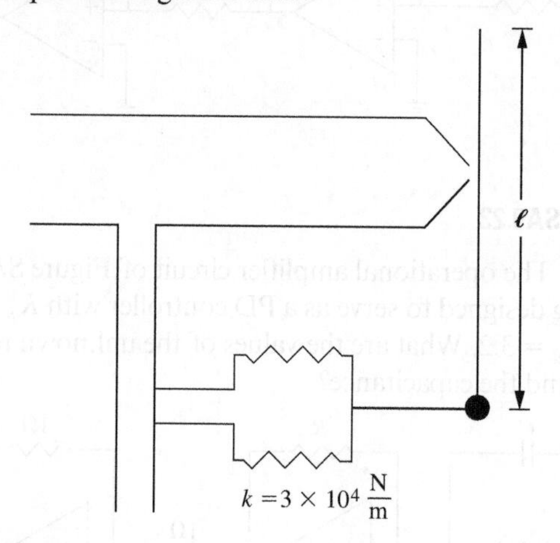

Figure SA9.34

SA9.35 What is the transform of the error when a step input is applied to the system of transfer function $G(s) = \dfrac{10}{(s+3)(s+5)}$?

SA9.36 What is the transform of the error when a step input is applied to the system of transfer function $G(s) = \dfrac{10}{s(s+3)(s+5)}$?

SA9.37 What is the transform of the error when a step input is applied to the system of transfer function $G(s) = \dfrac{10}{s^2(s+3)(s+5)}$?

SA9.38 What is the offset when a step input is applied to the system of Figure SA9.38 if the controller is a proportional controller of gain 4?

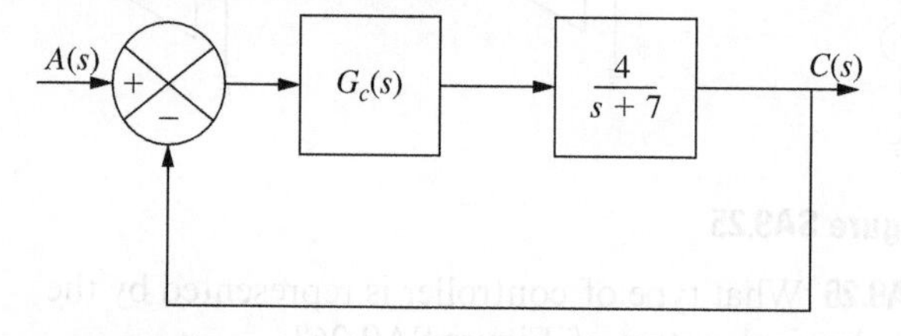

Figure SA9.38

SA9.39 What is the offset when a step input is applied to the system of Figure SA9.39 if the controller is an integral controller with a gain of 3?

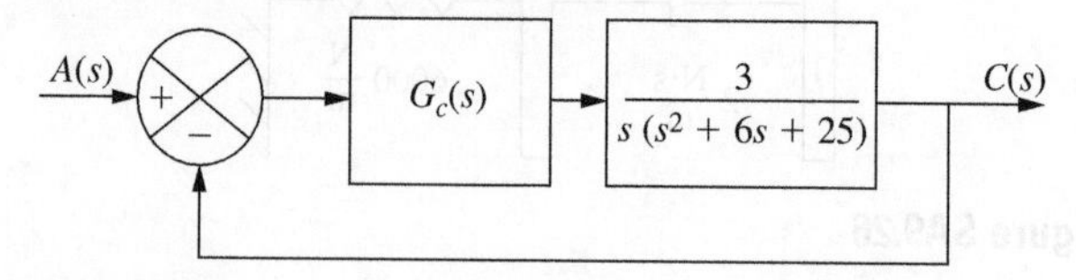

Figure SA9.39

SA9.40 What is the offset when a step input is applied to the system of Figure SA9.40 if the controller is a PI controller with a proportional gain of 4 and an integral gain of 6?

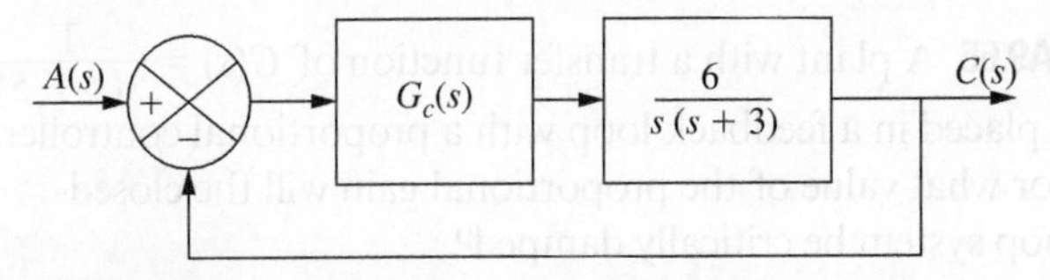

Figure SA9.40

SA9.41 What is the offset when a step input is applied to the system of Figure SA9.41 if the controller is a PD controller with a proportional gain of 4 and a differential gain of 5?

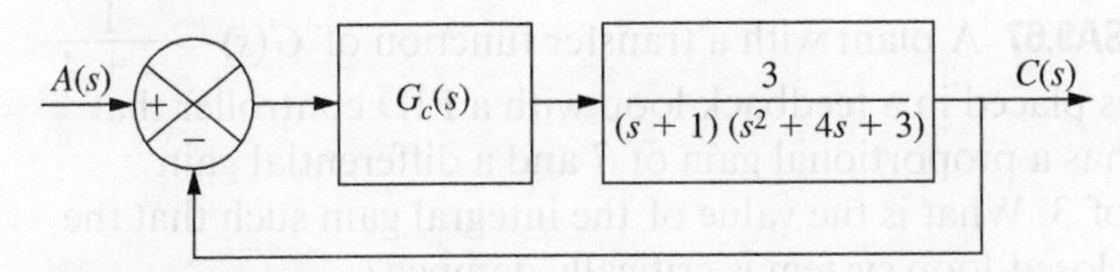

Figure SA9.41

SA9.42 What is the offset when a step input is applied to the system of Figure SA9.42 if the controller is a PID controller with a proportional gain of 4, an integral gain of 5, and a differential gain of 3?

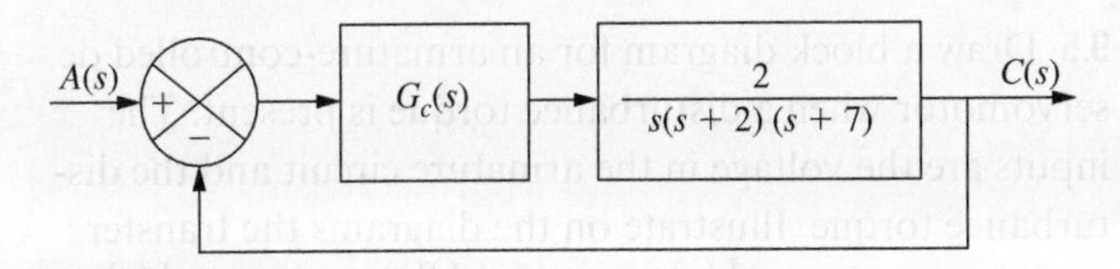

Figure SA9.42

SA9.43 What is the time constant of the closed-loop feedback system when a plant $G(s) = \frac{2}{s+6}$ is acted on by a proportional controller of gain 10?

SA9.44 What is the time constant of the closed-loop feedback system when a plant $G(s) = \frac{2}{s+6}$ is acted on by a PD controller with a proportional gain of 10 and a differential gain of 5?

SA9.45 What is the damping ratio of the closed-loop feedback system when a plant $G(s) = \frac{2}{s+6}$ is acted on by an integral controller with a gain of 5?

SA9.46 What is the natural frequency of the closed-loop feedback system when a plant $G(s) = \frac{2}{s+6}$ is acted on by a PID controller with a proportional gain of 10, an integral gain of 7, and a differential gain of 5?

SA9.47 What is the damping ratio of the closed-loop feedback system when a plant $G(s) = \frac{2}{s+6}$ is acted on by a PID controller with a proportional gain of 10, an integral gain of 7, and a differential gain of 5?

SA9.48 A fifth-order plant is placed in a feedback loop with a proportional controller. What is the order of the closed-loop transfer function?

SA9.49 A seventh-order plant is placed in a feedback loop with a PD controller. What is the order of the closed-loop transfer function?

SA9.50 A third-order plant is placed in a feedback loop with a PID controller. What is the order of the closed-loop transfer function?

SA9.51 A second-order plant is placed in a feedback control loop with a lead compensator. What is the order of the closed-loop transfer function?

SA9.52 A fifth-order plant is placed in a feedback control loop with a lead-lag compensator. What is the order of the closed-loop transfer function?

SA9.53 A plant with a transfer function of $G(s) = \frac{2}{s+6}$ is placed in a feedback loop with a proportional controller of gain 4. What is the transfer function of the closed-loop system?

SA9.54 A plant with a transfer function of $G(s) = \frac{2}{s+6}$ is placed in a feedback loop with a PI controller with a proportional gain of 4 and an integral gain of 10. What is the transfer function of the closed-loop system?

SA9.55 A plant with a transfer function of $G(s) = \frac{2}{s+6}$ is placed in a feedback loop with a PD controller with a proportional gain of 4 and a differential gain of 5. What is the transfer function of the closed-loop system?

SA9.56 A plant with a transfer function of $G(s) = \frac{2}{s+6}$ is placed in a feedback loop with a PID controller with a proportional gain of 4, a differential gain of 2, and an integral gain of 16. What is the transfer function of the closed-loop system?

SA9.57 A plant with a transfer function of $G(s) = \frac{2}{s+6}$ is placed in a feedback loop with a lag compensator whose transfer function is $G_c(s) = 3\frac{s+1}{s+2}$. What is the transfer function of the closed-loop system?

SA9.58 A plant with a transfer function of $G(s) = \frac{1}{s^2+16s+100}$ is placed in a feedback loop with a proportional controller with a gain of 200. What is the transfer function of the closed-loop system?

SA9.59 A plant with a transfer function of $G(s) = \frac{1}{s^2+16s+100}$ is placed in a feedback loop with a PI controller with a proportional gain of 100 and an integral gain of 200. What is the transfer function of the closed-loop system?

SA9.60 A plant with a transfer function of $G(s) = \frac{1}{s^2+10s+100}$ is placed in a feedback loop with an integral controller. What is the maximum gain of the controller such that the system is stable?

SA9.61 A plant with a transfer function of $G(s) = \dfrac{1}{s^2 + 10s + 100}$ is placed in a feedback loop with a PI controller. If the integral gain is 2500, for what values of the proportional gain is the system stable?

SA9.62 A plant with a transfer function of $G(s) = \dfrac{1}{s - 7}$ is placed in a feedback loop with a proportional controller. For what values of the proportional gain is the closed-loop system stable?

SA9.63 A plant with a transfer function of $G(s) = \dfrac{1}{s^2 - 16}$ is placed in a feedback loop with a proportional controller. For what values of the proportional gain is the closed-loop system stable?

SA9.64 A plant with a transfer function of $G(s) = \dfrac{1}{s(s + 5)}$ is placed in a feedback loop with a proportional controller. For what values of the proportional gain is the closed-loop system stable?

SA9.65 A plant with a transfer function of $G(s) = \dfrac{1}{s(s + 5)}$ is placed in a feedback loop with a proportional controller. For what value of the proportional gain will the closed-loop system be critically damped?

SA9.66 A plant with a transfer function of $G(s) = \dfrac{1}{s + 4}$ is placed in a feedback loop with a PI controller that has a proportional gain of 7. What is the value of the integral gain such that the closed-loop system is critically damped?

SA9.67 A plant with a transfer function of $G(s) = \dfrac{1}{s + 4}$ is placed in a feedback loop with a PID controller that has a proportional gain of 7 and a differential gain of 3. What is the value of the integral gain such that the closed-loop system is critically damped?

Problems

9.1 The differential equations for the perturbations in concentration of components A and B in a CSTR due to a perturbation in concentration of the reactant in the inlet stream are

$$V\frac{dC_A}{dt} + (k + qV)C_A = qC_{Ai} \quad \text{(a)}$$

$$V\frac{dC_B}{dt} - kC_A + qC_B = 0 \quad \text{(b)}$$

Draw a block diagram for the system.

9.2 The differential equations for the perturbations in the concentrations of components A and B in a CSTR with a two-way reaction due to perturbation in the flow rate of the inlet stream are

$$0.2\frac{dC_A}{dt} + 1.2C_A - 0.4C_B = 0.2q \quad \text{(a)}$$

$$0.2\frac{dC_B}{dt} - 0.4C_A + 0.9C_B = 0.1q \quad \text{(b)}$$

Draw a block diagram that models these differential equations.

9.3 Draw a block diagram that models the differential equations for the temperatures of the oil and water in the counterflow heat exchanger of Example 7.35.

9.4 Draw a block diagram that models the concentrations of the drug in the plasma and tissue in the two-compartment model of Example 6.19.

9.5 Draw a block diagram for an armature-controlled dc servomotor when a disturbance torque is present. The inputs are the voltage in the armature circuit and the disturbance torque. Illustrate on the diagrams the transfer functions $G_{1,1}(s) = \dfrac{I_a(s)}{V_a(s)}$, $G_{1,2}(s) = \dfrac{\Omega(s)}{V_a(s)}$, $G_{2,1}(s) = \dfrac{I_a(s)}{T(s)}$, and $G_{2,2}(s) = \dfrac{\Omega(s)}{T(s)}$.

9.6 Draw a block diagram for the mathematical model of the three-tank liquid-level problem of Figure P9.6. The system is shown at steady state before the perturbations in the flow rate occur. The inputs are the flow rate perturbations, and the outputs are the liquid-level perturbations. Identify on the block diagram the matrix of transfer functions defined as $G_{i,j}(s)$ for i and $j = 1, 2, 3$.

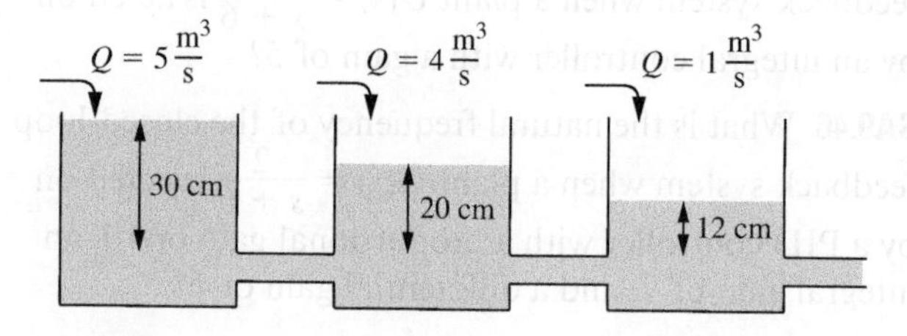

Figure P9.6

9.7 Draw a block diagram for the mechanical system of Figure P9.7. The input is the force applied to the mass whose displacement is x_3, and the outputs are the displacements of the masses.

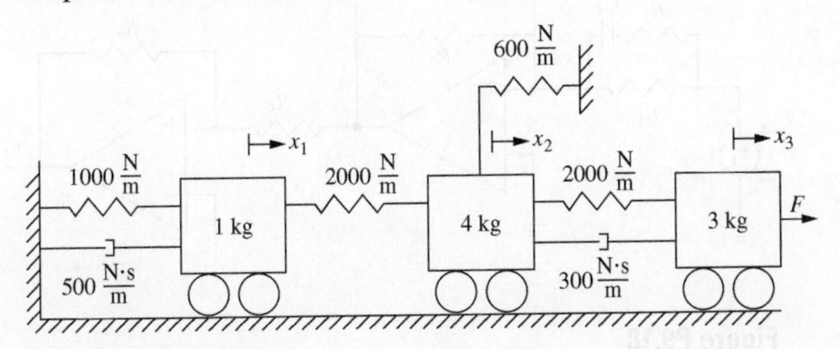

Figure P9.7

9.8 Draw a block diagram for the electric circuit of Figure P9.8. The inputs are the two voltage sources, and the outputs are the currents through the inductors.

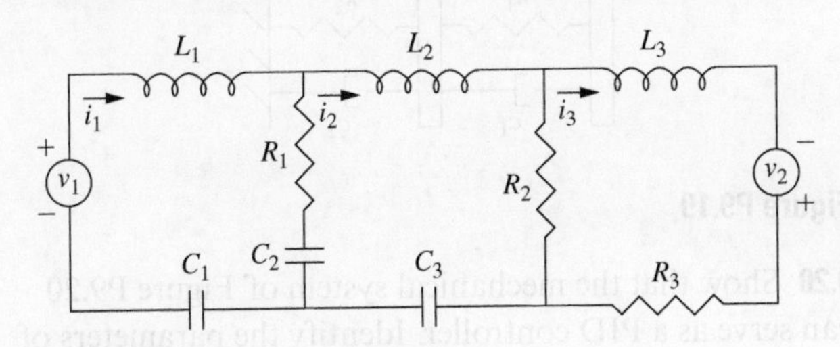

Figure P9.8

9.9 Draw a block diagram for the electric circuit of Figure P9.9. The input is the voltage source, and the output is the potential difference across the 100-Ω resistor.

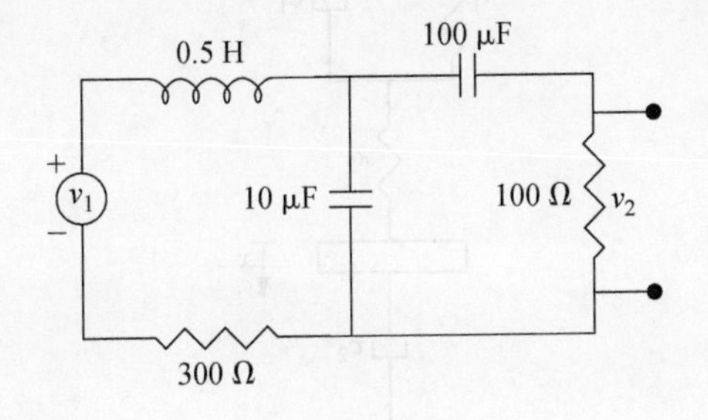

Figure P9.9

9.10 Determine the closed-loop transfer function for the block diagram of Figure P9.10.

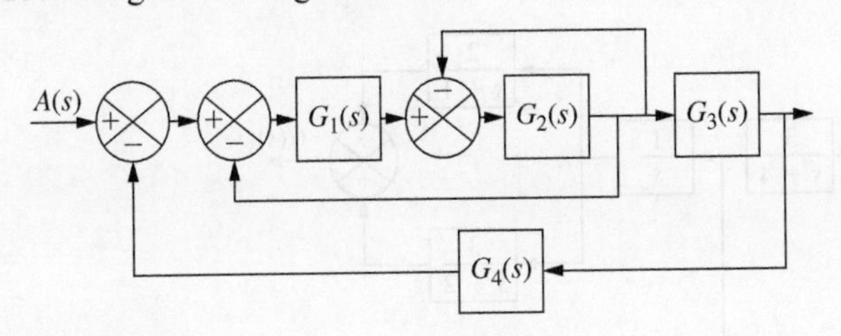

Figure P9.10

9.11 Determine the closed-loop transfer function for the block diagram of Figure P9.11.

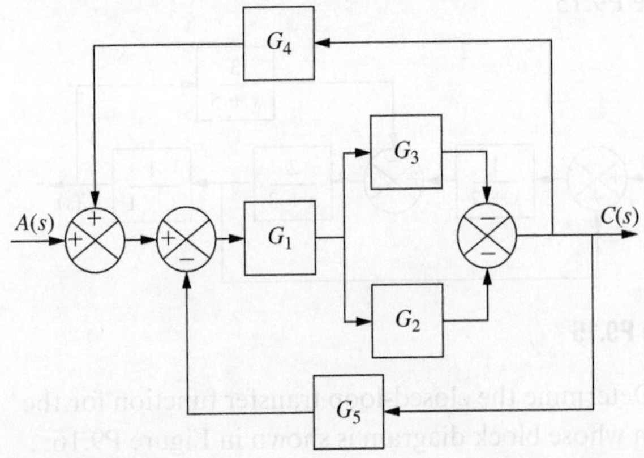

Figure P9.11

9.12 Determine the equivalent closed-loop transfer function for the system of Figure P9.12.

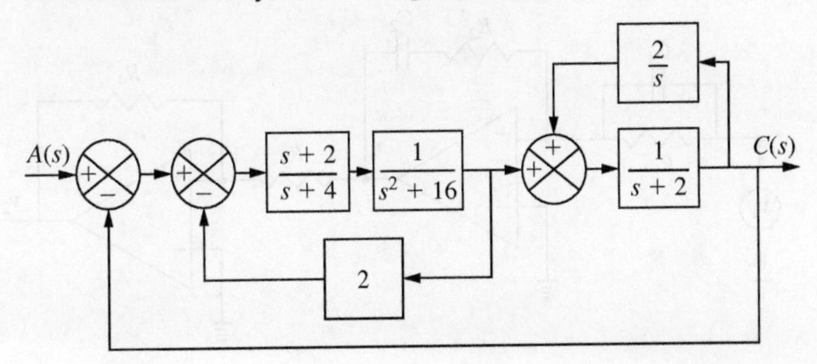

Figure P9.12

9.13 Determine the closed-loop transfer function $H(s) = C(s)/A(s)$ for the feedback system of Figure P9.13.

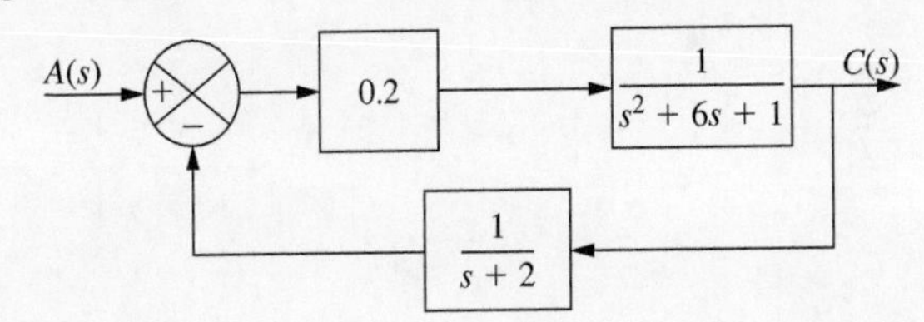

Figure P9.13

9.14 Determine the closed-loop transfer function $H(s) = C(s)/A(s)$ for the feedback system of Figure P9.14.

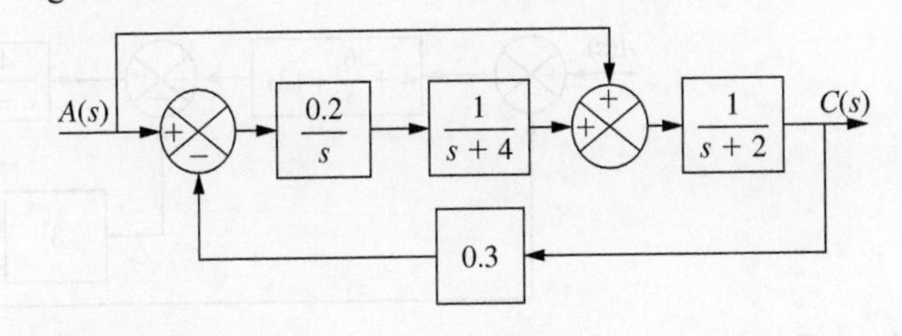

Figure P9.14

9.15 Determine the closed-loop transfer function $H(s) = C(s)/A(s)$ for the feedback system of Figure P9.15.

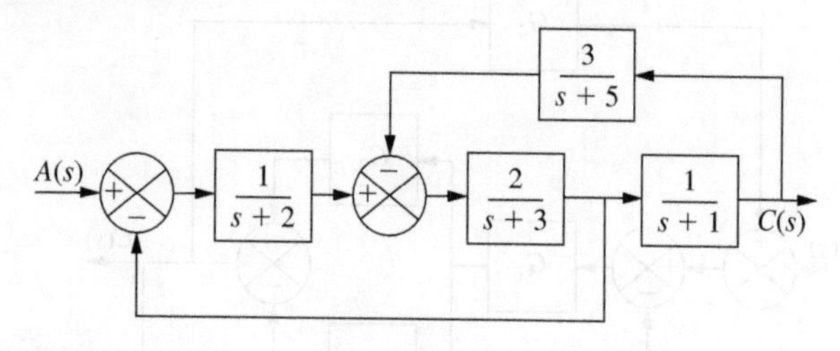

Figure P9.15

9.16 Determine the closed-loop transfer function for the system whose block diagram is shown in Figure P9.16.

9.17 Show that the operational amplifier circuit of Figure P9.17 can be used as a PID controller. Identify the gains.

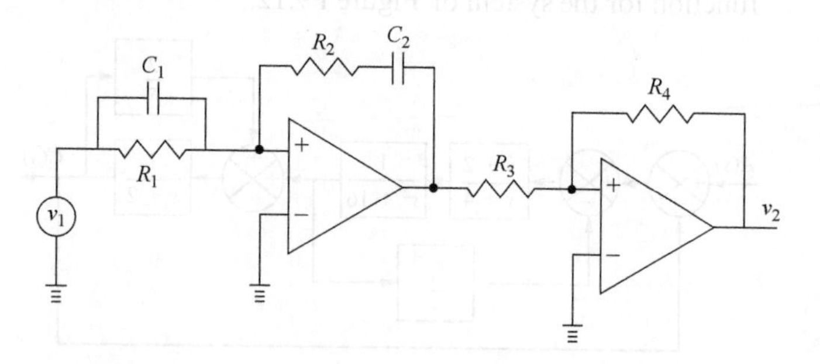

Figure P9.17

9.18 Show that the operational amplifier circuit of Figure P9.18 can be used as a lead-lag compensator. Identify the gain, T_1, T_2, α_1, and α_2.

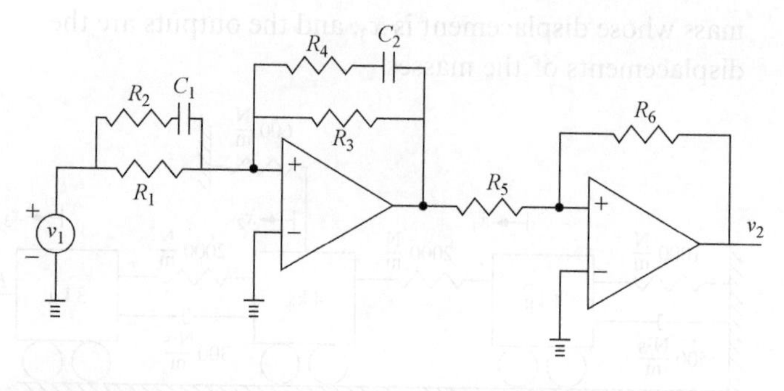

Figure P9.18

9.19 Under what conditions can the mechanical system of Figure P9.19 serve as a lag compensator?

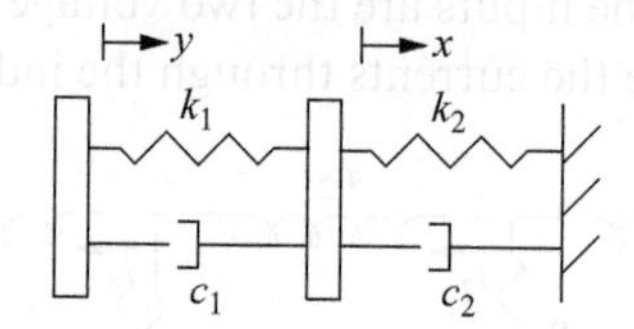

Figure P9.19

9.20 Show that the mechanical system of Figure P9.20 can serve as a PID controller. Identify the parameters of the controller in terms of the parameters of the mechanical system. The controller is connected to the particle whose displacement is x.

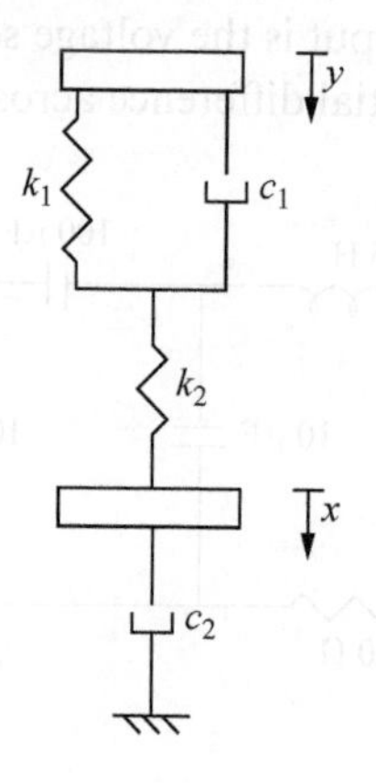

Figure P9.20

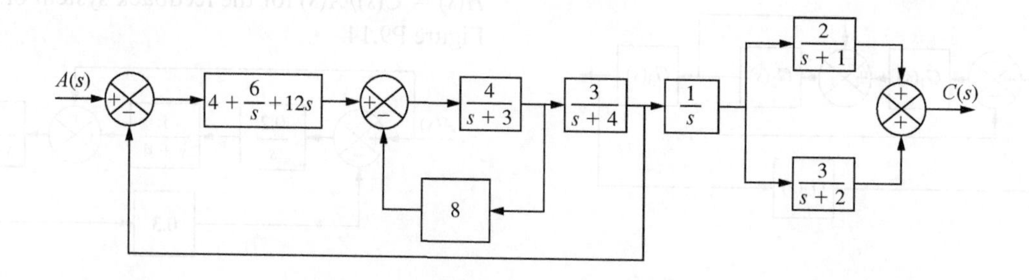

Figure P9.16

9.21 Show that the hydraulic servomotor of Figure P9.21 can serve as a PD controller if $\frac{A}{C} << 1$.

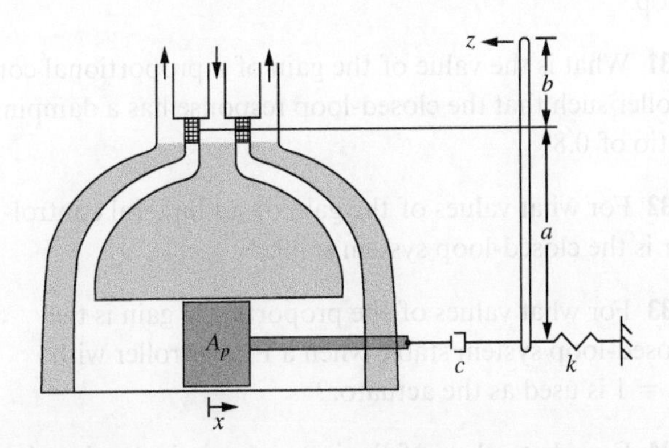

Figure P9.21

9.22 Show that the hydraulic servomotor of Figure P9.22 can serve as a PID controller if $\frac{A}{C} << 1$.

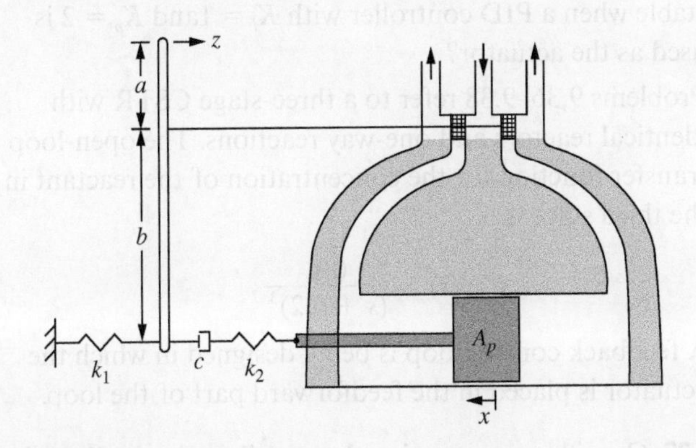

Figure P9.22

9.23 Show that the pneumatic system of Figure P9.23 can serve as a PD controller.

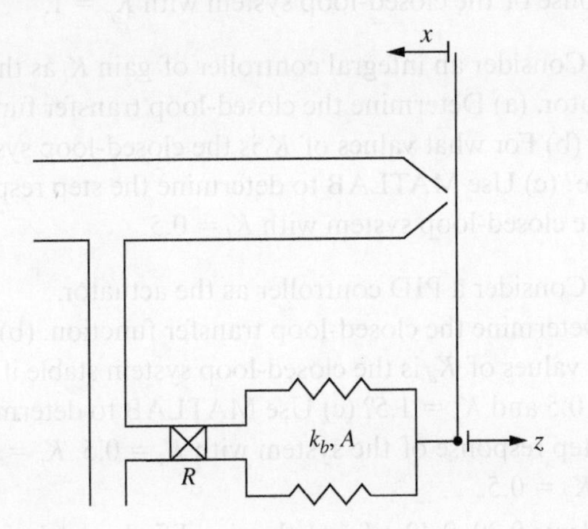

Figure P9.23

9.24 Show that the pneumatic system of Figure P9.24 can serve as a PID controller.

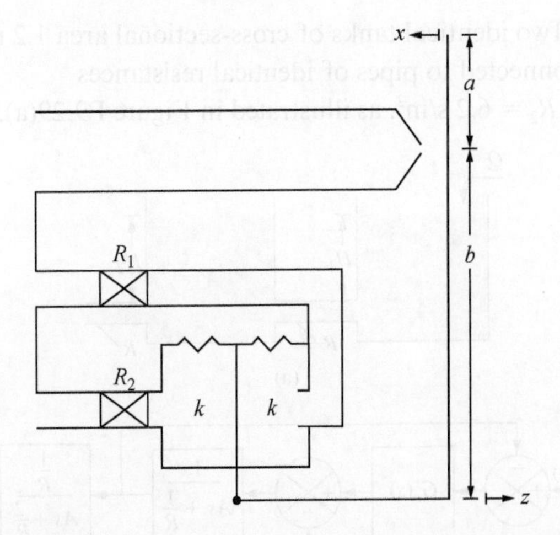

Figure P9.24

9.25 What is the minimum gain of a proportional controller required to reduce the 2 percent settling time of the thermal system of Example 6.15 by 30 percent? What is the eventual steady-state temperature when this proportional controller is used?

9.26 What is the value of the gain of an integral controller that is required such that when used on the thermal system of Example 6.15, the closed-loop response has a damping ratio of 2.5?

9.27 A PI controller with $K_p = 10$ and $K_i = 5.2$ is used with a plant whose transfer function is

$$G(s) = \frac{0.2}{s+5} \tag{a}$$

(a) Determine the closed-loop transfer function $H(s)$. (b) What are the natural frequency and damping ratio of the closed-loop system? (c) Determine the impulsive response of the closed-loop system. (d) Determine the step response of the closed-loop system.

9.28 A tank of cross-sectional area 1.2 m^2 exits into a pipe whose resistance is 6.2 s/m^2. For each of the following controllers, determine (i) the closed-loop transfer function $H(s)$, (ii) important system parameters such as time constant, or natural frequency and damping ratio, (iii) the impulsive response of the closed-loop system, and (iv) the offset when subject to a unit step input:

(a) Proportional controller of gain 0.2
(b) Integral controller of gain 1.5
(c) PI controller with $K_p = 0.2$ and $K_i = 1.5$
(d) PD controller with $K_p = 0.2$ and $K_d = 1.5$
(e) PID controller with $K_p = 0.2$, $K_i = 1.5$, and $K_d = 1.5$

9.29 Two identical tanks of cross-sectional area 1.2 m^2 are connected to pipes of identical resistances $R_1 = R_2 = 6.2$ s/m^2, as illustrated in Figure P9.29(a).

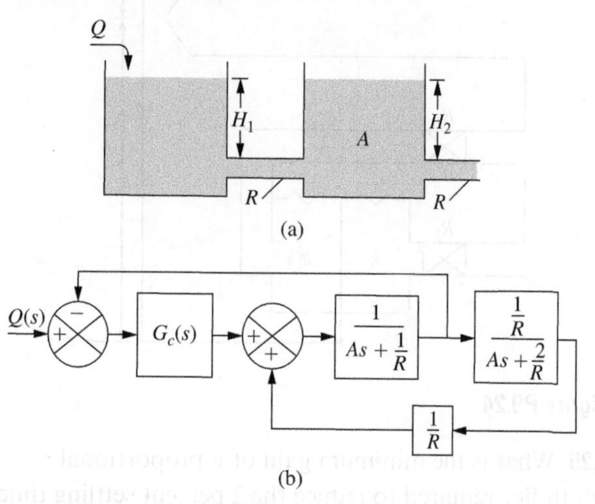

Figure P9.29

The tanks are placed in a feedback loop as shown in Figure P9.29(b). The output from the first tank is compared with the input. For each of the following actuators, determine (i) the closed-loop transfer functions pertaining to the liquid levels in both tanks, (ii) the poles of the transfer function, (iii) whether the closed-loop system is stable, (iv) the impulsive response, and (v) the step response:

(a) Proportional controller of gain 0.2
(b) Integral controller of gain 1.5
(c) PI controller with $K_p = 0.2$ and $K_i = 1.5$
(d) PD controller with $K_p = 0.2$ and $K_d = 1.5$
(e) PID controller with $K_p = 0.2$, $K_i = 1.5$, and $K_d = 1.5$

9.30 Repeat Problem 9.29, but the actuator is in a control system in which the output from the second tank is compared with the input, as in Figure P9.30.

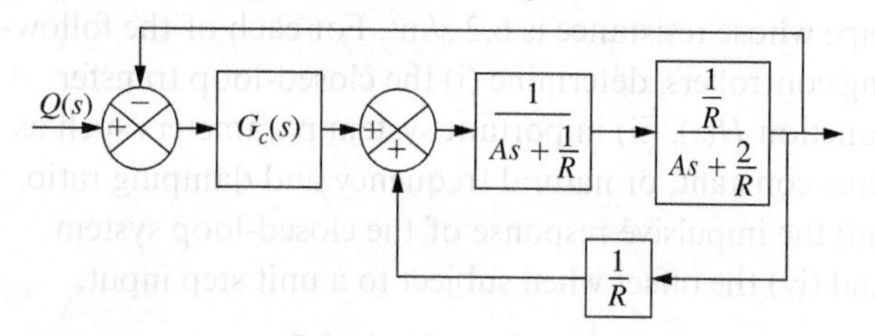

Figure P9.30

Problems 9.31–9.35 refer to a two-stage CSTR with identical reactors and one-way reactions. The transfer function for the concentration of component A in the second stage of the CSTR is

$$C_{A2}(s) = \frac{0.2}{(s + 0.4)^2}$$

A feedback control system is being designed in which the actuator is placed in the feedforward part of the loop.

9.31 What is the value of the gain of a proportional controller such that the closed-loop response has a damping ratio of 0.8?

9.32 For what values of the gain of an integral controller is the closed-loop system stable?

9.33 For what values of the proportional gain is the closed-loop system stable when a PI controller with $K_i = 1$ is used as the actuator?

9.34 For what values of the integral gain is the closed-loop system stable when a PI controller with $K_p = 2$ is used as the actuator?

9.35 For what values of K_d is the closed-loop system stable when a PID controller with $K_i = 1$and $K_p = 2$ is used as the actuator?

Problems 9.36–9.38 refer to a three-stage CSTR with identical reactors and one-way reactions. The open-loop transfer function for the concentration of the reactant in the third stage is

$$C_{A3}(s) = \frac{0.1}{(s + 0.2)^3}$$

A feedback control loop is being designed in which the actuator is placed in the feedforward part of the loop.

9.36 Consider a proportional controller of gain K_p as the actuator. (a) Determine the closed-loop transfer function. (b) For what values of K_p is the closed-loop system stable? (c) Use MATLAB to determine the impulsive response of the closed-loop system with $K_p = 1$.

9.37 Consider an integral controller of gain K_i as the actuator. (a) Determine the closed-loop transfer function. (b) For what values of K_i is the closed-loop system stable? (c) Use MATLAB to determine the step response of the closed-loop system with $K_i = 0.5$.

9.38 Consider a PID controller as the actuator. (a) Determine the closed-loop transfer function. (b) For what values of K_d is the closed-loop system stable if $K_i = 0.5$ and $K_p = 1.5$? (c) Use MATLAB to determine the step response of the system with $K_i = 0.5$, $K_p = 1.5$, and $K_d = 0.5$.

Problems 9.39–9.42 refer to the simplified model of the vehicle suspension system of Figure P9.39. The transfer function for the system is

$$G(s) = \frac{X(s)}{Y(s)} = \frac{2\zeta\omega_n s + \omega_n^2}{s^2 + 2\zeta\omega_n s + \omega_n^2}$$

where ζ is the system's damping ratio and ω_n is its natural frequency. A golf cart of empty mass 500 kg has a suspension system such that when empty its natural frequency is 20 r/s with a damping ratio of 0.7. An electronic controller is being designed to improve the performance of the suspension system.

9.39 (a) Can a proportional controller be designed such that the system is critically damped when empty? If so, what is the value of the proportional gain? (b) Consider the situation in which the golf cart is carrying golfers and their clubs. Assume the maximum added mass is 250 kg. Determine the necessary value of proportional gain as a function of the added mass such that the system is always critically damped.

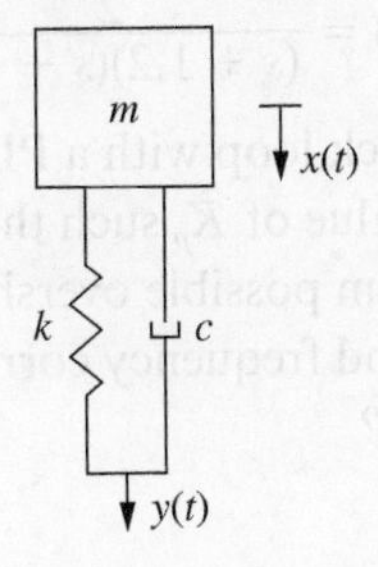

Figure P9.39

9.40 Determine the effect of the additional mass from passengers and clubs (from 0 to 250 kg) on the impulsive response of the closed-loop system when (a) a proportional controller is used, (b) an integral controller is used, and (c) a PD controller is used.

9.41 Suppose an integral controller of integral gain 1.3 is used. (a) Determine the step response of the closed-loop system. (b) Determine the 2 percent settling time. (c) Determine the 10–90 percent rise time.

9.42 Repeat Problem 9.41 but assume a differential controller of differential gain 1.2 is used.

9.43 A feedback loop for an armature-controlled dc servomotor (see Problem 9.5) is shown in Figure P9.43 when a PD controller is used. (a) Determine the closed-loop transfer functions $H_{1,1}(s) = \frac{\Omega(s)}{V(s)}$ and $H_{2,1}(s) = \frac{\Omega(s)}{T(s)}$. The parameters in this system are $L = 0.1$ H, $R_a = 20\ \Omega$, $J = 2.3 \times 10^{-4}\,\text{kg·m}^2$, and $c_t = 1.4 \times 10^{-3}$ N·m·s. (b) The closed-loop system is to be overdamped with a damping ratio of 1.5 and the rise time is less than 3 s. Determine appropriate proportional and differential gains.

9.44 Repeat Problem 9.43 if velocity feedback is used for the armature and a proportional controller is used on the entire system as in Figure P9.44.

9.45 A position control system is an armature-controlled dc servomotor (see Problem 9.5) that is then integrated to give the angular position. The position control system is placed in a feedback loop with a PD controller as shown in Figure P9.45 (page 628). (a) Determine the closed-loop transfer functions $H_{1,1}(s) = \frac{\Omega(s)}{V(s)}$ and $H_{2,1}(s) = \frac{\Theta(s)}{T(s)}$. Determine the differential and integral gains if the closed-loop system is to have a damping ratio of 0.9 and a natural frequency of 1.5 r/s.

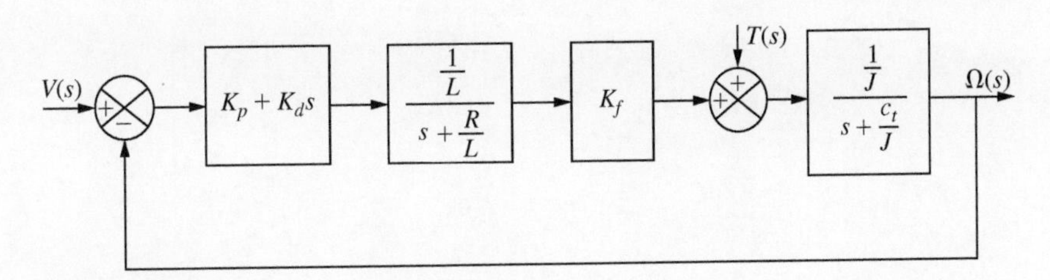

Figure P9.43

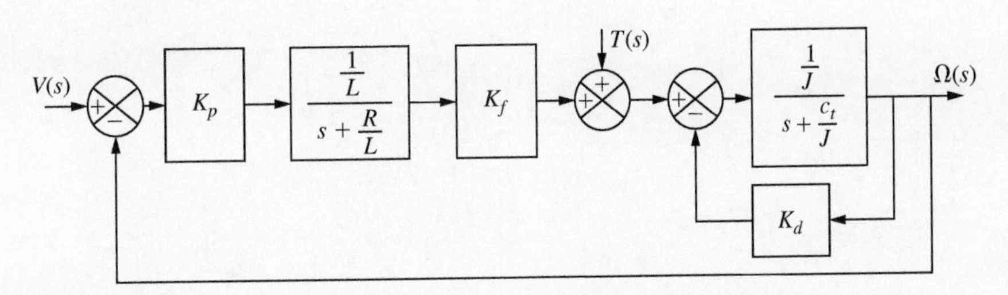

Figure P9.44

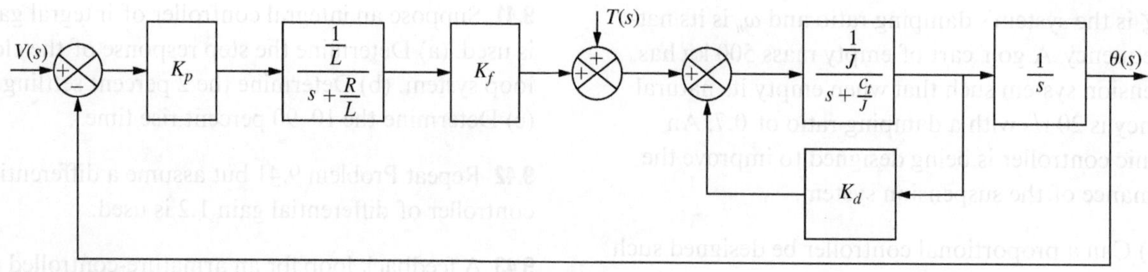

Figure P9.45

9.46 Develop a Simulink model for the system of Problem 9.2.

9.47 Develop a Simulink model for the system of Example 9.26 with a PID controller. Run the simulation for the system with the gains specified in Example 9.26. Then study the effect of changing each gain while holding the other two at the specified values. Describe the effects of changing the gains on the step response.

9.48 Develop a Simulink model for the three-CSTR system of Problem 9.38. Set up the model so that the response for the reactant concentration in each tank can be monitored.

9.49 A second-order plant whose transfer function is

$$G(s) = \frac{2}{(s + 1.2)(s + 1.8)}$$

is placed in a feedback loop with a PI controller. (a) Determine the value of K_p such that the dominant pole has the minimum possible overshoot. (b) What are the damping ratio and frequency corresponding to the minimum overshoot?

10

Root-Locus Analysis

Learning Objectives

At the conclusion of this chapter, you should be able to:

- Define a root-locus diagram and know the terminology for developing one
- Apply the magnitude criterion in developing a root-locus diagram
- Apply the angle criterion to determine which parts of the real axis lie on a branch of the root-locus diagram
- Calculate the breakaway points and the break-in points of the root-locus diagram
- Use the angle criterion to calculate the angles of departure and the angles of arrival
- Calculate the equation of the asymptotes to each branch of the root-locus diagram
- Determine the crossover frequencies and the corresponding values of K
- Sketch the root-locus diagram
- Use MATLAB to draw the root-locus diagram
- Use the root-locus diagram as a tool in the design of a compensator
- Apply the Ziegler-Nichols tuning rules for a PID controller
- Use the principles of drawing a root-locus diagram to design compensators
- Design a lead compensator to improve transient performance
- Design a lag compensator to reduce the steady-state error
- Design a lead-lag compensator to improve transient performance and reduce the steady-state error

Consider a plant placed in a feedback loop with a controller. The controller is designed with a parameter to be specified. The parameter could be the gain for a proportional controller or the gain for an integral controller. It could be one of the gains in a PI, PD, or PID controller, with the other gains specified or given in terms of the unknown gain. The parameter could be the gain of a compensator. Consider the system of Figure 10.1(a). A plant of transfer function $K\frac{R(s)}{Q(s)}$ is placed in a unity feedback loop. The closed-loop transfer function $H(s)$ illustrated in Figure 10.1(b) has a denominator that is the sum of two polynomials. One polynomial $Q(s)$ is completely specified, while the other polynomial is written as $KR(s)$, where K is the unknown parameter and $R(s)$ is a completely specified polynomial. The poles of the closed-loop system are the roots of the denominator of $H(s)$. The poles, among other things, dictate whether the system is stable. In designing a control system, the poles are a function of the parameter.

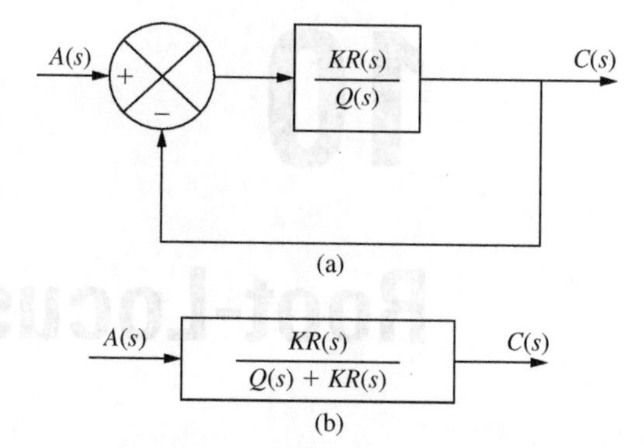

Figure 10.1

(a) Unity feedback loop; (b) Closed loop transfer function.

A root-locus diagram is a plot of the poles of $H(s)$ where K acts as a parameter. It is a parametric representation of the poles of the closed-loop transfer function when the plant is placed in a unity feedback loop.

The root-locus diagram can be drawn by following certain guidelines or by using MATLAB. Knowledge of the properties of a root-locus diagram is essential. The number of branches on a root-locus diagram is equal to the order of the polynomial $Q(s)$. Each branch represents one pole as the value of K changes. A point that is on a branch of the root-locus diagram must satisfy a magnitude criterion and an angle criterion. The number of asymptotes to the branches depends on the number of zeros and the number of poles of $H(s)$.

One method of designing a controller is to use the root-locus diagram to specify the poles of $H(s)$. The root-locus diagram can be used in the design of a lead compensator, a lag compensator, or a lead-lag compensator. One way to use the root-locus method in the design of a PID controller is to establish rules to tune the controller.

The root-locus method is focused on the denominator of $H(s)$. The numerator is included in the determination of the poles, but the numerator also affects the performance of a controller in a feedback loop. The root-locus method presented in this chapter does not consider the transient behavior of the closed-loop system, which is affected by the appropriate static error constant.

10.1 Introduction

The root-locus diagram is a plot that illustrates the effect of changing a parameter in the closed-loop transfer function on the stability of a system. The parameter may be a gain of a controller or a compensator. Although root-locus diagrams are easily drawn using MATLAB, it is instructive to understand how the diagrams are developed and how they can be sketched by hand. Such knowledge leads to a better understanding of how the poles of the closed-loop transfer function vary with a changing parameter.

Consider a plant with a transfer function of the form $G(s) = C\frac{N(s)}{D(s)}$, where $N(s)$ and $D(s)$ are polynomials in the transform variable s and C is a constant such that the leading coefficients of the polynomials are each one. It is intended to modify the behavior of the plant through placing it in a feedback loop with a controller or compensator whose transfer function is $G_c(s) = K\frac{N_c(s)}{D_c(s)}$, where $N_c(s)$ and $D_c(s)$ are polynomials in s and K is a gain. Placing the plant in a feedback loop with the controller gives a closed-loop transfer function of

$$H(s) = \frac{G_c(s)G(s)}{1 + G_c(s)G(s)}$$

$$= \frac{K\frac{N_c(s)}{D_c(s)}C\frac{N(s)}{D(s)}}{1 + K\frac{N_c(s)}{D_c(s)}C\frac{N(s)}{D(s)}}$$

$$= \frac{KCN_c(s)N(s)}{D_c(s)D(s) + KCN_c(s)N(s)} \tag{10.1}$$

The stability of the closed-loop system is determined by the poles of the denominator of $H(s)$. The pole placement of the transfer function of Equation (10.1) is essential to the behavior of the closed-loop system. It is good to know for what values of K the system will be stable and for what values of K the pole placement will be adequate for attaining the objective that the control system is designed to achieve.

The objective of drawing a root-locus diagram is to map the poles of $H(s)$ in terms of K. If the denominator of Equation (10.1) is of order n, then there are n branches of the root-locus diagram. Each branch represents the solution to

$$D_c(s)D(s) + KCN_c(s)N(s) = 0 \tag{10.2}$$

as K varies.

10.2 Definitions

A root-locus diagram shows the loci of the roots of the equation

$$Q(s) + KR(s) = 0 \tag{10.3}$$

for all K such that $0 \leq K \leq \infty$. In Equation (10.3), $Q(s)$ and $R(s)$ are polynomials with the order of $R(s)$ less than or equal to the order of $Q(s)$. If applied to a feedback control system, in accordance with Equation (10.2),

$$Q(s) = D_c(s)D(s) \tag{10.4}$$

and

$$R(s) = CN_c(s)N(s) \tag{10.5}$$

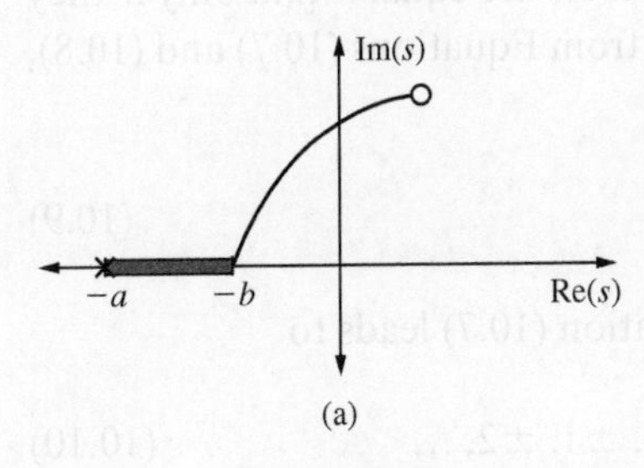

A root-locus diagram is drawn in the complex s plane, where the horizontal axis represents the real part of s and the vertical axis represents the imaginary part of s. Each locus represents the variation of one root of Equation (10.3) with the variation of K. If the order of $Q(s)$ is n, there are n roots that lead to n curves that make up the root-locus diagram. Each curve is called a branch of the diagram.

The construction of a root-locus diagram begins by determining the roots of Equation (10.3), or the roots of $Q(s)$, for $K = 0$. These values of s are called the poles of the diagram. The values of s that satisfy

$$\lim_{K\to\infty} [Q(s) + KR(s)] = 0 \tag{10.6}$$

are called the zeros of the root-locus diagram. These are the roots of $R(s)$. Each branch of the root-locus diagram must begin at a pole (corresponding to $K = 0$). The branch must either approach a zero (corresponding to $K \to \infty$) at its end or approach infinity.

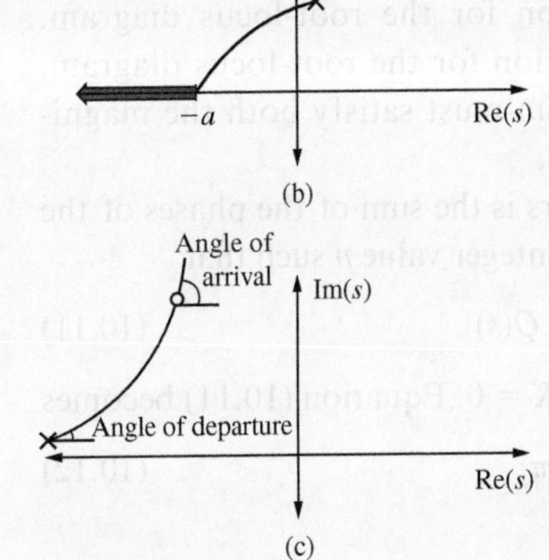

Figure 10.2

(a) The branch of the root-locus diagram begins at a real pole, $-a$, follows along the real axis until $-b$, which is a break-away point, and terminates at a complex zero; (b) this branch of a root-locus diagram begins at a complex pole and remains complex until it reaches a break-in point on the real axis at $-a$ and terminates at infinity; (c) this branch begins at a complex pole where it makes an angle, the angle of departure, with the real axis and ends at a complex zero, where it makes an angle, the angle of arrival, with the real axis.

If a branch of the root-locus diagram begins at a real pole, but for some value of K the root of Equation (10.3) becomes complex, then the point on the real axis where the branch departs from the real axis is called a breakaway point. Conversely, if a branch of the root-locus diagram begins at a complex pole, but for some value of K the solution of Equation (10.3) becomes real, the point on the real axis where this occurs is called a break-in point.

If a branch intersects the imaginary axis where the real part of the root equals zero, then the imaginary part of the root is called the crossover frequency. This term comes from the sinusoidal transfer function in which $j\omega$ is substituted for the value of s. The value of ω represents a frequency.

For a complex pole, the angle that the tangent to the root-locus diagram at the pole makes with the real direction is called the angle of departure. For a complex zero, the angle that the tangent to the root-locus diagram at the zero makes with the real direction is called the angle of arrival.

Figure 10.2(a) shows a branch on the root-locus diagram that has a real pole at $-a$, eventually breaks away from the real axis at $-b$, and crosses the imaginary axis.

Figure 10.2(b) shows a branch on the root-locus diagram that has a complex pole with a positive real part, crosses the imaginary axis, and eventually breaks into the real axis at $-a$. Figure 10.2(c) shows a complex pole and the angle of departure as well as a complex zero and the angle of arrival.

10.3 The Angle and Magnitude Criteria

Equation (10.3) may be written as

$$K\frac{R(s)}{Q(s)} = -1 \tag{10.7}$$

The number -1 has the polar representation

$$-1 = e^{j(2n+1)\pi} \tag{10.8}$$

For any integer $n = 0, \pm 1, \pm 2, \ldots$, two complex numbers are equal if and only if they have the same magnitude and the same phase. Thus, from Equations (10.7) and (10.8), the equivalence of the magnitudes gives

$$K\left|\frac{R(s)}{Q(s)}\right| = 1 \tag{10.9}$$

The equivalence of the phases of both sides of Equation (10.7) leads to

$$\phi = \angle K\frac{R(s)}{Q(s)} = (2n+1)\pi \qquad n = 0, \pm 1, \pm 2, \ldots \tag{10.10}$$

Equation (10.9) is called the magnitude criterion for the root-locus diagram. Equation (10.10) is referred to as the angle criterion for the root-locus diagram. Every point on a branch of the root-locus diagram must satisfy both the magnitude and the angle criteria.

Since the phase of a product of complex numbers is the sum of the phases of the individual parts of the product, there must exist an integer value n such that

$$(2n+1)\pi = \angle K + \angle R(s) - \angle Q(s) \tag{10.11}$$

Since K is a positive real number, $K = Ke^{j0}$ or $\angle K = 0$. Equation (10.11) becomes

$$\angle R(s) - \angle Q(s) = (2n+1)\pi \tag{10.12}$$

for some integer value of n.

Equations (10.9) and (10.12) are essential for the analytic development of the root-locus diagram. This is especially true for determining which parts of the real axis are on a branch of the root-locus diagram. The following example provides an illustration.

Example 10.1

Determine which parts of the real axis lie on some branch of the root-locus diagram for the equation

$$s(s+3)(s+5) + K(s+6) = 0 \tag{a}$$

Solution

This equation has $Q(s) = s(s+3)(s+5)$ and $R(s) = s + 6$. The poles, which are the roots of $Q(s)$, are at $s = 0, -3, -5$. The equation has one zero, the root of $R(s)$ at $s = -6$.

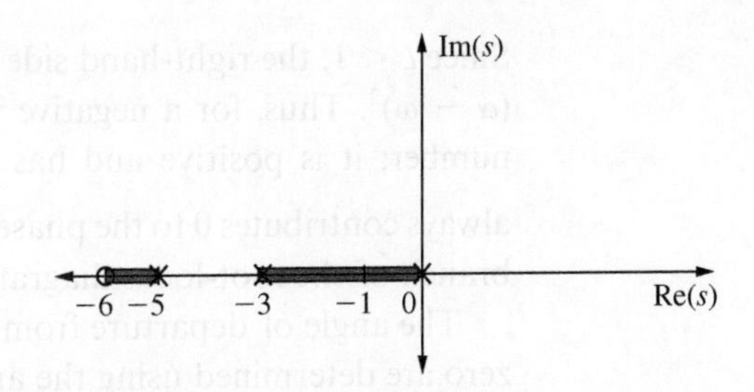

Figure 10.3

Regions which are on the real axis for the transfer function of Example 10.1.

Note that a linear factor of $s + a$, where a is a real positive value, has a polar representation of $s + a = |s + a|e^{j0}$ if $s > -a$ or $s + a = |s + a|e^{j\pi}$ if $s < -a$. Thus,

$$\angle(s + a) = \begin{cases} 0 & s > -a \\ \pi & s < -a \end{cases} \tag{b}$$

Applying the angle criterion, Equation (10.12), to Equation (a) yields

$$\angle s + \angle(s + 3) + \angle(s + 5) - \angle(s + 6) = (2n + 1)\pi \tag{c}$$

First consider a positive value of s on the real axis. Then s, $s + 3$, $s + 5$, and $s + 6$ are all positive real numbers that have a polar representation with a phase of zero. The left-hand side of Equation (c) becomes

$$0 + 0 + 0 - 0 = 0 \tag{d}$$

There is no value of n for which $(2n + 1)\pi = 0$, and thus all positive real values of s are not on any branch of the root-locus diagram.

Consider a value of s between -3 and 0. Then $\angle s = \pi$, $\angle(s + 3) = 0$, $\angle(s + 5) = 0$, and $\angle(s + 6) = 0$. The right-hand side of Equation (c) becomes $\pi + 0 + 0 - 0 = \pi = (2n + 1)\pi$ for $n = 0$. This region of the real axis between -3 and 0 is on a branch of the root-locus diagram. Of course, $s = 0$ and $s = -3$ are both poles of the root-locus diagram and are the start of branches.

Next consider a value of s between -5 and -3. Then $\angle s = \pi$, $\angle(s + 3) = \pi$, $\angle(s + 5) = 0$, and $\angle(s + 6) = 0$. The left-hand side of Equation (c) becomes $\pi + \pi + 0 - 0 = 2\pi$. There is no integer value of n such that $(2n + 1) = \pi$. Thus, the region of the real axis between -5 and -3 is not on a branch of the root-locus diagram.

Consider a value of s between -6 and -5. Then $\angle s = \pi$, $\angle(s + 3) = \pi$, $\angle(s + 5) = \pi$, and $\angle(s + 6) = 0$. The left-hand side of Equation (c) becomes $\pi + \pi + \pi - 0 = 3\pi$. If $n = 1$, then $(2n + 1) = 3\pi$. The interval on the real axis between -6 and -5 is on a branch of the root-locus diagram.

Finally, consider a value of s less than -6. All angles in Equation (c) are π. The left-hand side of Equation (c) becomes $\pi + \pi + \pi - \pi = 2\pi$. There is no value of n such that $(2n + 1) = \pi$. The range $s < -6$ is not on any branch of the root-locus diagram. The point $s = -6$ is a zero for the root-locus diagram and ends a branch. Note that as K approaches infinity in Equation (a), the equation that s must satisfy is $s + 6 = 0$, leading to $s = -6$.

Figure 10.3 illustrates the regions on the real axis that are on a branch of the root-locus diagram. The poles are designated by an x and the zero is designated by an o.

A polynomial with real coefficients can always be factored into a product of linear factors and irreducible quadratic factors of the form $s^2 + 2\zeta\omega s + \omega^2$ with $0 \le \zeta < 1$. If s is a positive real number, then an irreducible quadratic factor is positive and always has a phase of zero. If $s = -\alpha$ for a positive real value of α, then

$$s^2 + 2\zeta\omega s + \omega^2 = \alpha^2 - 2\zeta\omega\alpha + \omega^2 \tag{10.13}$$

Since $\zeta < 1$, the right-hand side of Equation (10.13) is greater than $\alpha^2 - 2\omega\alpha + \omega^2 = (\alpha - \omega)^2$. Thus, for a negative value of s, $s^2 + 2\zeta\omega s + \omega^2$ is greater than a positive number; it is positive and has a phase of 0. Then an irreducible quadratic factor always contributes 0 to the phase of $\frac{R(s)}{Q(s)}$ when determining whether a real value is on a branch of the root-locus diagram.

The angle of departure from a complex pole and the angle of arrival at a complex zero are determined using the angle criterion. The value of the pole or zero is substituted into the angle condition, Equation (10.12).

10.4 Breakaway and Break-in Points

A breakaway point occurs when a branch that started from a real pole begins to have complex roots for some value of K. This point can be thought of as the maximum value of K such that the imaginary part of the root is zero. Equation (10.3) can be interpreted as an equation to determine K as a function of a real independent variable s. In this context, $K(s)$ achieves a maximum value for a value of s that corresponds to the breakaway point. This condition requires that

$$\frac{dK}{ds} = 0 \tag{10.14}$$

Differentiating Equation (10.3) leads to

$$\frac{dQ}{ds} + K\frac{dR}{ds} + R\frac{dK}{ds} = 0 \tag{10.15}$$

Solving Equation (10.15) for $\frac{dK}{ds}$ gives

$$\frac{dK}{ds} = -\frac{1}{R}\left(\frac{dQ}{ds} + K\frac{dR}{ds}\right) \tag{10.16}$$

Using the definition of K from Equation (10.7) in Equation (10.15) gives

$$\frac{dK}{ds} = -\frac{1}{R}\left(\frac{dQ}{ds} - \frac{Q}{R}\frac{dR}{ds}\right) \tag{10.17}$$

Requiring the left-hand side of Equation (10.17) be equal to zero yields

$$R\frac{dQ}{ds} - Q\frac{dR}{ds} = 0 \tag{10.18}$$

Only values of s, obtained by solving Equation (10.18), that the angle condition has shown to be on a branch of the root-locus diagram are breakaway points.

When determining a break-in point, K is viewed as a function of a complex variable. The break-in point corresponds to the minimum value of K such that the imaginary part of the complex variable is zero. It can also be viewed as the minimum value of K such that the variable is real. Analysis of the break-in point leads to Equation (10.18) also defining a break-in point.

Example 10.2

Determine the breakaway and break-in points for the system of Example 10.1 in which

$$Q(s) = s(s + 3)(s + 5) \tag{a}$$

and

$$R(s) = s + 6 \tag{b}$$

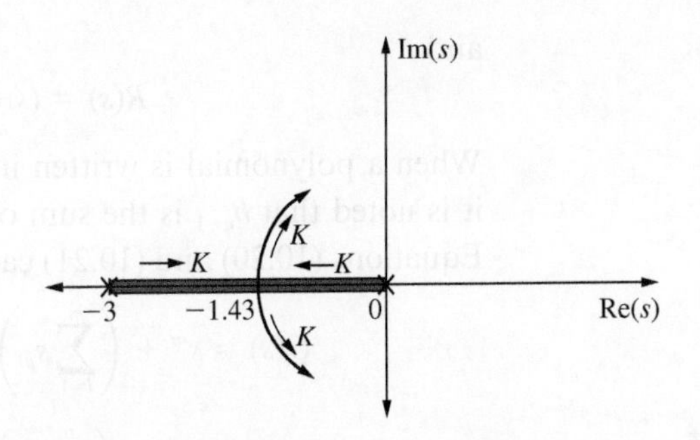

Figure 10.4
Two branches of the root-locus diagram for the transfer function of Example 10.2. One branch begins at the real pole of 0, then breaks away from the real axis at −1.43. A second branch begins at the real pole of −3 and breaks away from the real axis at −1.43. One branch follows along a path with a positive imaginary part, while the other follows with a negative imaginary part.

Solution

The breakaway and break-in points are the real values of s that are determined to be on a branch of the root-locus diagram in Example 10.1 and that are also solutions of Equation (10.18). To this end, note that

$$\frac{dQ}{ds} = 3s^2 + 16s + 15 \tag{c}$$

and

$$\frac{dR}{ds} = 1 \tag{d}$$

Substituting Equations (a)−(d) into Equation (10.18) leads to

$$(s + 6)(3s^2 + 16s + 15) - (s^3 + 8s^2 + 15s)(1) = 0 \tag{e}$$

Equation (e) simplifies to

$$2s^3 + 26s^2 + 96s + 90 = 0 \tag{f}$$

The solutions to Equation (f) are −1.43, −4.36, and −7.20. From the results of Example 10.1, the real values of −4.36 and −7.20 do not lie on any branch of the root-locus diagram. The only breakaway or break-in point is −1.43. A branch of the root-locus diagram begins at the pole of 0 and progresses along the negative real axis. Another branch begins at −3 and progresses along the real axis toward the right. Both branches intersect and break away from the real axis at $s = -1.43$, as illustrated in Figure 10.4.

10.5 Asymptotes of Branches on the Root-Locus Diagram

The branches of a root-locus diagram either end at a zero or approach infinity. To sketch the branches that end at infinity, it is helpful to know their asymptotic behavior. As s approaches infinity, the values of the poles and zeros are much smaller than s and all factors of $R(s)$ and $Q(s)$ are approximated by s. This implies that as s approaches infinity, the angles of all factors are equal. Let ϕ_a be the angle of the asymptote, n_p be the number of poles, and n_z be the number of zeros on the root-locus diagram. The angle condition Equation (10.12) requires

$$n_z\phi_a - n_p\phi_a = (2n + 1)\pi \Rightarrow \phi_a = \frac{(2n + 1)\pi}{n_z - n_p} \qquad n = 0, \pm 1, \pm 2, \ldots \tag{10.19}$$

If $n_z = n_p$, Equation (10.16) implies that there are no asymptotes on the root-locus diagram.

Using the definitions of poles and zeros,

$$Q(s) = (s - s_{p1})(s - s_{p2}) \cdots (s - s_{pm}) \tag{10.20}$$

and

$$R(s) = (s - s_{z1})(s - s_{z2}) \cdots (s - s_{zn}) \tag{10.21}$$

When a polynomial is written in the form $s^n + b_{n-1}s^{n-1} + b_{n-2}s^{n-2} + \cdots + b_1 s + b_0$, it is noted that b_{n-1} is the sum of the roots and b_0 is the product of the roots. Thus, Equations (10.20) and (10.21) can alternatively be written as

$$Q(s) = s^m + \left(\sum_{i=1}^{m} s_{pi}\right)s^{m-1} + a_{m-2}s^{m-2} + \cdots + a_1 s + \prod_{i=1}^{m} s_{pi} \tag{10.22}$$

$$R(s) = s^n + \left(\sum_{i=1}^{n} s_{zi}\right)s^{n-1} + b_{n-2}s^{n-2} + \cdots + b_1 s + \prod_{i=1}^{n} s_{zi} \tag{10.23}$$

Then

$$\frac{KR(s)}{Q(s)} = K\frac{s^n + \left(\sum_{i=1}^{n} s_{zi}\right)s^{n-1} + b_{n-2}s^{n-2} + \cdots + b_1 s + \prod_{i=1}^{n} s_{zi}}{s^m + \left(\sum_{i=1}^{m} s_{pi}\right)s^{m-1} + a_{m-2}s^{m-2} + \cdots + a_1 s + \prod_{i=1}^{m} s_{pi}} \tag{10.24}$$

Performing long division of the denominator by the numerator on the right-hand side of Equation (10.24) and rearranging leads to

$$\frac{KR(s)}{Q(s)} = \frac{K}{\left\{s + \frac{\left[\left(\sum_{i=1}^{m} s_{pi}\right) - \left(\sum_{i=1}^{n} s_{zi}\right)\right]}{m-n}\right\}^{m-n}} \tag{10.25}$$

The place on the real axis where the asymptotes intersect is obtained by setting the denominator of the right-hand side of Equation (10.25) to zero. It is

$$s = -\frac{\left[\left(\sum_{i=1}^{m} s_{pi}\right) - \left(\sum_{i=1}^{n} s_{zi}\right)\right]}{m-n} \tag{10.26}$$

Example 10.3

Determine the asymptotes for the root-locus diagram for the system of Example 10.1 in which

$$Q(s) = s(s+3)(s+5) \tag{a}$$

and

$$R(s) = s + 6 \tag{b}$$

Solution

There are three poles ($s = 0, -3, -5$) and one zero ($s = -6$) of the closed-loop transfer function. Thus, from Equation (10.19),

$$\phi_a = \frac{(2n+1)\pi}{1-3} = -(2n+1)\frac{\pi}{2} \tag{c}$$

For $n = 0$, $\phi_a = -\frac{\pi}{2}$. For $n = -1$, $\phi_a = \frac{\pi}{2}$. For $n = 1$, $\phi_a = -\frac{3\pi}{2}$. For $n = -2$, $\phi_a = \frac{3\pi}{2}$. The angle $\frac{3\pi}{2}$ is equivalent to the angle $-\frac{\pi}{2}$. The angles $\frac{\pi}{2}$ and $-\frac{3\pi}{2}$ are also equivalent. There are only two distinct angles for the asymptotes:

$$\phi_a = -\frac{\pi}{2}, \frac{\pi}{2}$$

The point where the asymptotes intersect on the real axis is given by Equation (10.26) as

$$s = \frac{1}{1-3}[(0+3+5)-(6)] = -1 \tag{d}$$

The asymptotes are vertical lines that intersect on the real axis at $s = -1$.

10.6 Summary of Steps in Drawing a Root-Locus Diagram

The procedure for drawing a root-locus diagram is outlined below. The diagram is drawn for an equation of the form of Equation (10.3) with $Q(s)$ and $R(s)$ as known functions of the transform variable s. It is assumed that $Q(s)$ is a polynomial of order m and $R(s)$ is a polynomial of order $n \leq m$. The diagram shows the roots of Equation (10.3) as a function of the parameter K. Since Equation (10.3) is a polynomial equation in terms of s and the polynomial is of order m, there will be m branches of the root-locus diagram.

1. Locate the poles, the roots of $Q(s)$, and the zeros, the roots of $R(s)$. Each branch must begin at a pole for $K = 0$. Each branch must either end at a zero or approach infinity as K approaches infinity.
2. Determine the ranges on the real axis that lie on the root-locus diagram. Example 10.1 provides a discussion on how the angle criterion is used.
3. Determine the breakaway and break-in points. Example 10.2 illustrates this procedure.
4. Determine the equations for the asymptotes of the root loci. This is illustrated in Example 10.3. The angles that the asymptotes make with the horizontal are first determined. Then their common point of intersection that lies on the real axis is located using Equation (10.26). The equations of the asymptotes are then specified.
5. The crossover frequency is obtained by substituting $s = j\omega$ into Equation (10.3). The resulting equation is broken into imaginary and real parts. These equations are solved, yielding the crossover frequencies ω and the values of K for which they occur.
6. The angles of departure from the poles and the angles of arrival at the zeros are determined using the angle criterion, Equation (10.12).
7. The values of points on a branch of the root-locus diagram can be determined using the magnitude criterion, Equation (10.9), and the angle criterion, Equation (10.12). These are necessary to determine the shape of the root-locus plots.

Example 10.4

Sketch the root-locus diagram for the system of Examples 10.1, 10.2, and 10.3. It is noted that

$$Q(s) = s(s + 3)(s + 5) \tag{a}$$

and

$$R(s) = s + 6 \tag{b}$$

Solution

The steps for drawing the diagram are listed here:

1. The poles are at $s = 0, -3, -5$. The zero is at $s = -6$.
2. In Example 10.1, it is shown that the ranges $-3 \leq s \leq 0$ and $-6 \leq s \leq -5$ are the only regions on the real axis that lie on a branch of the root-locus diagram.
3. In Example 10.2, it is shown that the only breakaway point is at $s = -1.43$. There are no break-in points. The value of K at the breakaway point is obtained by applying the magnitude criterion, Equation (10.9):

$$K\left|\frac{-1.43 + 6}{-1.43(-1.43 + 3)(-1.43 + 5)}\right| = 1 \Rightarrow K = 1.73 \tag{c}$$

This and the regions on the real axis that lie on the root-locus diagram and the fact that all branches must end at a zero or infinity imply that one branch of the root-locus diagram must start at $s = -5$ and proceed along the negative real axis that ends at $s = -6$.

4. In Example 10.3, it is shown that there are only two asymptotes that are vertical $\left(\phi_a = \pm\frac{\pi}{2}\right)$ and intersect on the real axis at $s = -1$. Along with the above, this implies that the branches that start at -3 and 0 follow the real axis until $s = -1.43$. They break away from the real axis and then grow toward infinity, one in the positive direction and one in the negative direction. They eventually approach the vertical line $\text{Re}(s) = -1$. A sketch of the root-locus diagram based on the knowledge from steps 1–4 of the outline is given in Figure 10.5(a).

5. Substituting $s = j\omega$ into Equation (10.3) with $Q(s)$ given by Equation (a) and $R(s)$ given by Equation (b) leads to

$$-j\omega^3 - 8\omega^2 + j15\omega + K(j\omega + 5) = 0 \tag{d}$$

Setting the real and imaginary parts of Equation (d) to zero leads to

$$5K - 8\omega^2 = 0 \tag{e}$$

$$-\omega^3 + (15 + K)\omega = 0 \tag{f}$$

Equation (e) implies that $K = \frac{8}{5}\omega^2$, which when substituted into Equation (f) gives

$$\left(\frac{3}{5}\omega^2 + 15\right)\omega = 0 \tag{g}$$

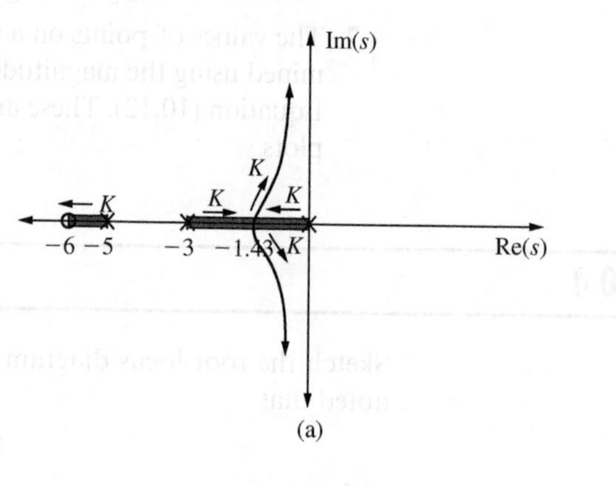

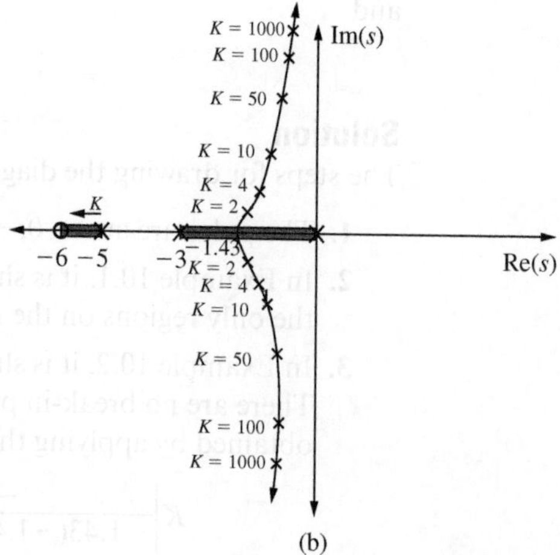

Figure 10.5

(a) A sketch of the root-locus diagram for the transfer function of Example 10.4 based upon knowledge gained from applying steps 1–4 of the construction process. (b) A sketch of the root-locus diagram with points placed from step 7.

Table 10.1 Roots of $Q(s) + KR(s)$ of Example 10.4

K	First Root	Second Root	Third Root
2	$-1.42 + 0.550j$	$-1.42 - 0.550j$	-5.15
4	$-1.37 + 1.64j$	$-1.37 - 1.64j$	-5.26
10	$-1.28 + 3.06j$	$-1.28 - 3.06j$	-5.43
50	$-1.12 + 7.13j$	$-1.12 - 7.13j$	-5.76
100	$-1.07 + 10.1j$	$-1.07 - 10.1j$	-5.87
1000	$-1.009 + 31.7j$	$-1.009 + 31.7j$	-5.98

The solutions to Equation (g) are $\omega = 0, \pm j5$. Imaginary frequencies are irrelevant. Thus, the only crossover frequency is $\omega = 0$. This is on the branch of the root-locus diagram that starts at the pole of 0.

6. There are no complex poles or complex zeros, so there is no need to calculate the angles of departure or the angles of arrival.
7. Filling in some points on the root-locus diagram by selecting a value of $K > 1.73$ and calculating the values on the branches leads to

$$K = 2 \Rightarrow s^3 + 8s^2 + 15s + 2s + 12 = s^3 + 8s^2 + 17s + 12 = 0 \Rightarrow s = -5.15, -1.42 \pm 0.550j \tag{h}$$

The remainder of the calculations are summarized in Table 10.1.

As K grows large and the real root of the equation approaches -6, the real part of the complex roots approaches -1 and the imaginary part of the complex roots approaches infinity. A sketch of the root-locus diagram is shown in Figure 10.5(b).

Example 10.5

A plant with the transfer function

$$G(s) = \frac{1}{(s+1)(s+2)(s+3)} \tag{a}$$

is in a feedback loop with a proportional controller of gain K, as shown in Figure 10.6. Draw the root-locus diagram for the closed-loop transfer function.

Solution

The closed-loop transfer function is

$$H(s) = \frac{K}{(s+1)(s+2)(s+3) + K} \tag{b}$$

The root-locus diagram is drawn for

$$Q(s) = (s+1)(s+2)(s+3) \tag{c}$$

and

$$R(s) = 1 \tag{d}$$

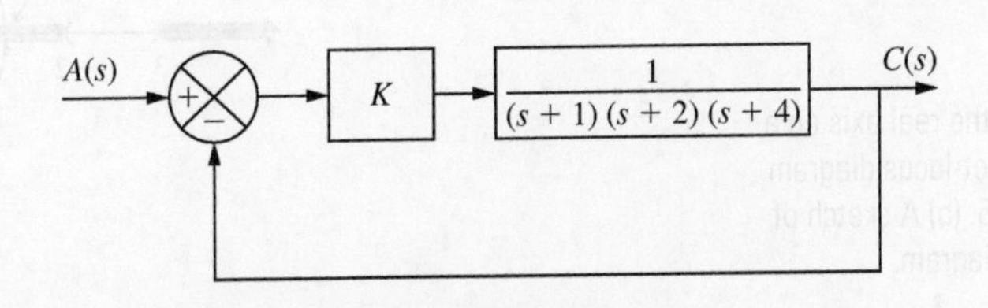

Figure 10.6

The system of Example 10.5 is a plant in a unity feedback loop with a proportional controller.

The steps in drawing the root-locus diagram follow.

1. The poles are at $s = -1, -2, -3$. There are no zeros. The closed-loop transfer function is third order. There are three branches of the root-locus diagram. They begin at the poles and, since there are no zeros, all branches approach infinity as K approaches infinity.
2. The regions in question are $s > -1$, $-2 < s < -1$, $-3 < s < -2$, and $s < -3$. A factor $s + a$ is positive if $s > -a$ and thus has an angle of 0. The same factor is negative if $s < -a$ and then its angle is π.

   For $s > -1$, all factors are positive and the angle is zero. There is no integer value of n such that $(2n + 1)\pi = 0$. Thus, the range $\mathrm{Re}(s) > -1$ does not lie on any branch of the root-locus diagram.

   For $-2 < s < -1$, only the factor $s + 1$ is negative, so the total angle is π. Choosing $n = 0$ does allow the angle to be written as $(2n + 1)\pi$. Thus, $-2 < s < -1$ for real s lies on some branch of the root-locus diagram.

   For $-3 < s < -2$, $s + 1 < 0$ and $s + 2 < 0$. The total angle is 2π. There is no integer value of n such that $(2n + 1)\pi = 2\pi$. Thus, $-3 < s < -2$ for real s does not lie on a branch of the root-locus diagram.

   For $s < -3$, all factors are negative. The total angle is 3π, which by choosing $n = 1$ can be written as $(2n + 1)\pi$. Hence, the range $s < -3$ for real s lies on a branch of the root-locus diagram.

   The ranges of the real axis that lie on a branch of the root-locus diagram are shown in Figure 10.7(a).
3. There are no break-in points as there are no complex zeros. The breakaway points are determined by substituting into Equation (10.18) and solving for s. To this end, using Equations (a) and (b),

$$0 = R\frac{dQ}{ds} - Q\frac{dR}{ds} = (1)(3s^2 + 12s + 11) - (s + 1)(s + 2)(s + 3)(0) \tag{e}$$

The solutions of Equation (e) are $s = -1.42, -2.58$. Since it is known that $s = -1.42$ lies on a branch of the root-locus diagram, it is a breakaway point. Since $s = -2.58$ does not lie on a branch of the root-locus diagram, it is not a breakaway point. The value of K at this point is determined from the magnitude criterion, Equation (10.9), as

$$K\left|\frac{1}{(-1.42 + 1)(-1.42 + 2)(-1.42 + 3)}\right| = 1 \Rightarrow K = 0.385 \tag{f}$$

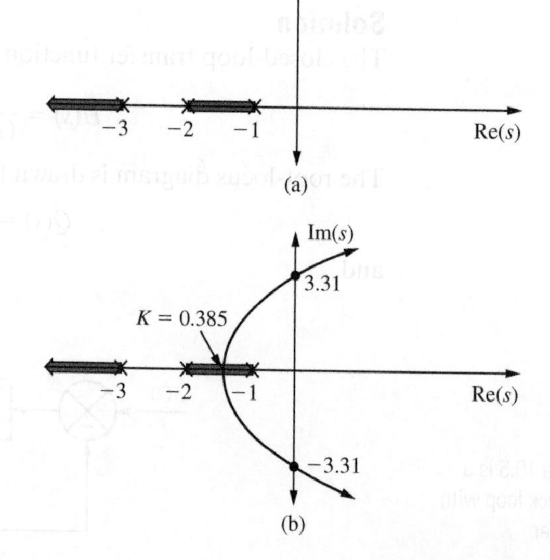

Figure 10.7

(a) The parts of the real axis on a branch of the root-locus diagram for Example 10.5. (b) A sketch of the root-locus diagram.

Table 10.2 Roots of $Q(s) + KR(s)$ of Example 10.5

K	First Root	Second Root	Third Root
0.1	−1.05	−1.12	−3.05
0.4	−0.142 + 0.43j	−0.142 − 0.43j	−3.16
10	−0.85 + 1.73j	−0.85 − 1.73j	−4.31
30	−0.39 + 2.60j	−0.39 − 2.60j	−5.21
60	0.317j	−0.317j	−6.0
100	0.36 + 3.96j	0.36 − 3.96j	−6.71

4. There are three poles, $n_p = 3$, and no zeros, $n_z = 0$. The angle of the asymptotes of the branches is obtained using Equation (10.19):

$$\phi_a = \frac{(2n+1)\pi}{0-3} = -(2n+1)\frac{\pi}{3} \tag{g}$$

For $n = 0$, $\phi_a = -\frac{\pi}{3}$. For $n = -1$, $\phi_a = \frac{\pi}{3}$. These are the only unique angles that are determined for different values of n.

The point of intersection on the real axis is determined from Equation (10.26):

$$s = \frac{1}{0-3}[(1+2+3) - 0] = -2 \tag{h}$$

The equations for the asymptotes are

$$\text{Im}(s) = \frac{\sqrt{3}}{2}\text{Re}(s) + \sqrt{3} \tag{i}$$

and

$$\text{Im}(s) = -\frac{\sqrt{3}}{2}\text{Re}(s) - \sqrt{3} \tag{j}$$

5. Substituting $s = j\omega$ into Equation (10.3) and using Equations (a) and (b) leads to

$$-j\omega^3 - 6\omega^2 + j11\omega + 6 + K = 0 \tag{k}$$

Setting the real and imaginary parts of Equation (k) to zero independently gives

$$-6\omega^2 + 6 + K = 0 \tag{l}$$

and

$$-\omega^3 + 11\omega = 0 \tag{m}$$

Equation (l) implies that

$$\omega = 0, \pm\sqrt{11} \tag{n}$$

For $\omega = 0$, Equation (l) yields $K = -6$, which is not acceptable. For $\omega = \sqrt{11}$, Equation (l) gives $K = 60$.

6. Since all the poles are real, there are no angles of departure. Since there are no zeros, there are no angles of arrival.
7. The value of K is varied and the roots are noted and reported in Table 10.2.

The root-locus diagram for $H(s)$, using the results of steps 1–7, is sketched in Figure 10.7(b).

Example 10.6

Draw the root-locus diagram for the closed-loop transfer function

$$H(s) = \frac{K(s+4)}{(s^2+6s+34)+K(s+4)} \tag{a}$$

Solution

The root-locus diagram is drawn using Equation (10.3) with

$$Q(s) = s^2 + 6s + 34 \tag{b}$$

and

$$R(s) = s + 4 \tag{c}$$

The root-locus diagram is sketched using the steps outlined at the beginning of this section:

1. The poles are determined as $s = -3 \pm j5$. The zero is $s = -4$. Since $H(s)$ is second order, there are two branches of the root-locus diagram. They begin at the poles. Since there is only one zero, one branch ends at -4; the other branch approaches infinity.
2. Recall that an irreducible second-order function does not contribute anything to the angle criterion when evaluated on the real axis. It is only necessary to consider $s > -4$ and $s < -4$ as possible regions where a branch lies on the real axis. For $s > -4$, the angle made by $s + 4$ is 0 and no integer value of n satisfies the angle criterion. For $s < -4$, the angle made by $s + 4$ is π and the angle criterion is satisfied by choosing $n = -1$. So the range $s < -4$ lies on the real axis for some branch or branches of the root-locus diagram.
3. The possible breakaway points are obtained by applying Equation (10.18):

$$(s+4)(2s+6) - (s^2+6s+34)(1) = s^2 + 8s - 10 = 0 \Rightarrow 1.10, -9.10 \tag{d}$$

Only $s = -9.10$ lies on a branch of the root-locus diagram. This must be a break-in point from the complex poles. The magnitude criterion, Equation (10.9), is used to determine the value of K at the breakaway point:

$$K\left|\frac{-9.10+4}{(-9.10)^2+6(-9.10)+34}\right| = 1 \Rightarrow K = 12.2 \tag{e}$$

4. There are two poles, $n_p = 2$, and one zero, $n_z = 1$. The angles that the asymptotes make to the root-locus branches are given by Equation (10.19):

$$\phi_a = \frac{(2n+1)\pi}{1-2} = -(2n+1)\pi \tag{f}$$

For $n = 0$, $\phi_a = -\pi$. For $n = -1$, $\phi_a = \pi$. These are the only unique values of ϕ_a.

Since the asymptotes are parallel to the real axis, they must lie on the real axis.

5. Substituting $s = j\omega$ into Equation (10.3) with $Q(s)$ as in Equation (b) and $R(s)$ as in Equation (c) leads to

$$-\omega^2 + j6\omega + 34 + K(j\omega + 4) = 0 \tag{g}$$

Table 10.3 Roots of $Q(s) + KR(s)$ of Example 10.6

K	First Root	Second Root
1	$-3.5 + 5.07j$	$-3.5 - 5.07j$
2	$-4 + 5.01j$	$-4 - 5.01j$
5	$-5.5 + 4.87j$	$-5.5 - 4.87j$
10	$-8 + 3.16j$	$-8 - 3.16j$
15	-6.47	-14.53
30	-4.96	-31.04
50	-4.55	-51.46

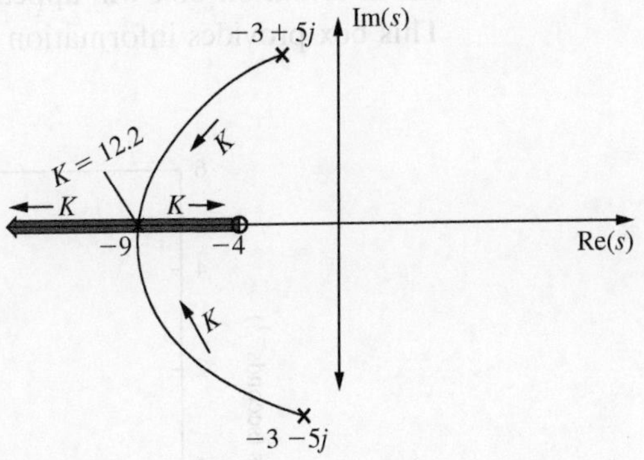

Figure 10.8

Sketch of the root-locus diagram for Example 10.6.

Setting the real and imaginary parts of Equation (f) to zero independently leads to

$$-\omega^2 + 34 + 4K = 0 \tag{h}$$

and

$$6\omega + K\omega = 0 \tag{i}$$

Equation (i) implies that $\omega = 0$. If $\omega = 0$, the real part requires $K = -\frac{17}{2}$, which is not in the range of K for which a root-locus diagram is drawn. No branches of the root-locus diagram intersect the imaginary axis.

6. The angles of departure for the complex poles are determined from the angle criterion, Equation (10.12). At the pole $-3 + j5$, the criterion gives

$$\angle[(-3 + 5j)^2 + 6(-3 + 5j) + 34] - \angle[(-3 + 5j) + 4] = (2n + 1)\pi \tag{j}$$

The tangent to the branch of the root-locus diagram beginning at $-3 + j5$ is π.

7. Calculate the points on the root-locus diagram for different values of K by substituting into Equation (10.3) using Equations (a) and (b). Partial results are given in Table 10.3.

Based on steps 1–7, the root-locus diagram for $H(s)$ is sketched in Figure 10.8. The root-locus diagram begins at $K = 0$ for two imaginary poles. The system is underdamped. It continues to be underdamped with an increasing damping ratio until K reaches 12.6 when the system is critically damped. For larger values of K, the system is overdamped, with one root approaching negative infinity and the other root approaching the zero of the transfer function $s = -4$.

10.7 Use of MATLAB

MATLAB has the capability of drawing the root-locus diagram given the vector forms of $Q(s)$ and $R(s)$. The command from the Control Systems Toolbox that is used to draw the root-locus diagram is

```
rlocus(R,Q)
```

The root-locus diagram of Figure 10.9(a) is generated for the $H(s)$ of Example 10.6 by using the commands

```
R=[1 4]
Q=[1 6 34]
rlocus(R,Q)
```

An information box will appear when a point on the root-locus diagram is clicked. This box provides information such as the real and imaginary parts of the value of

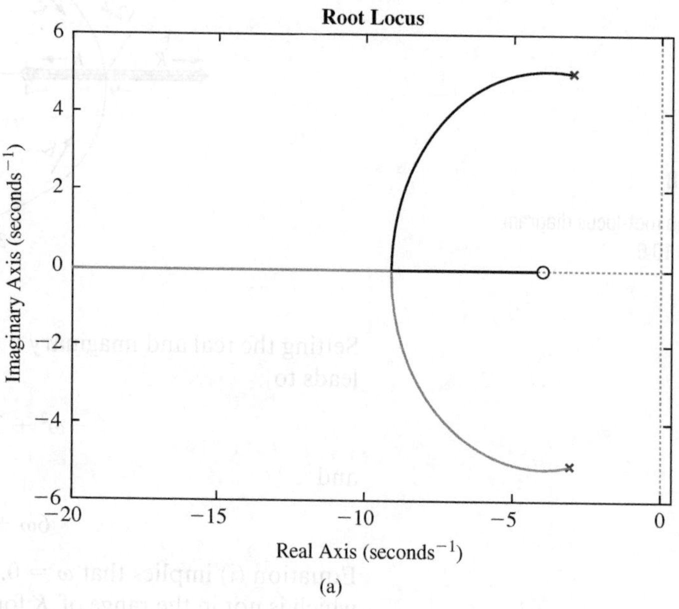

(a)

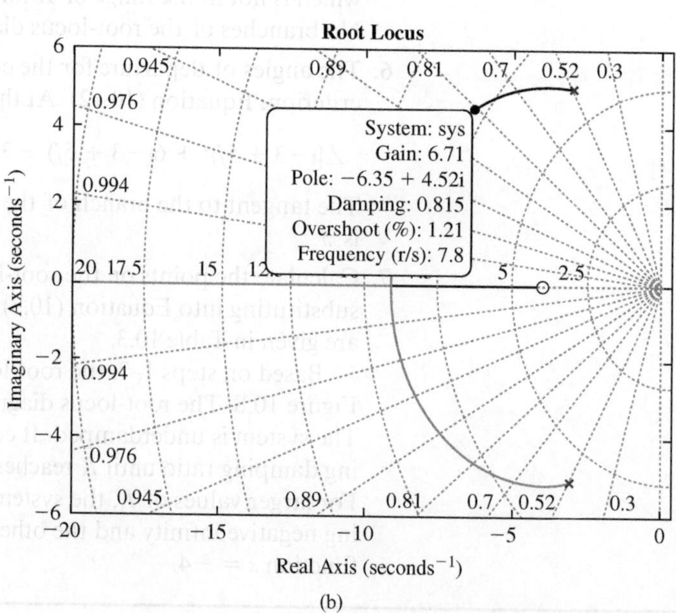

(b)

Figure 10.9

(a) MATLAB generated root-locus diagram of Example 10.6 (b) Root-locus diagram using sgrid. Clicking on a point on the root-locus diagram gives information about that point.

the point, the value of the gain that generates that point, the damping ratio, and the natural frequency for that point. This information box is made to vanish by clicking anywhere in the figure that is not on the root-locus diagram.

The values of the gain K that are used in the root-locus analysis and all of the poles p_i that correspond to the gains are obtained by using the command

```
[p,K]=rlocus(R,Q)
```

This use of the rlocus command does not draw the root-locus diagram. The command

```
sgrid
```

puts a grid of damping factors and natural frequencies (ζ, ω_n) on the plot. This command allows the user to see the effective values of the damping ratio and the natural frequency for the roots. Figure 10.9(b) illustrates the previous root-locus diagram with sgrid. The radial lines are lines of constant damping ratio, and the circles are lines of constant natural frequency. For the point highlighted, the pole has a damping ratio of 0.815 and is on a circle corresponding to a natural frequency of 7.8 r/s. The pole should be $-\zeta\omega_n + j\omega_n\sqrt{1-\zeta^2} = -(0.815)(7.8) + j7.8\sqrt{1-(0.815)^2} = -6.35 + j4.52$. This matches the value of the pole shown in Figure 10.9(b). The gain that led to this point is 6.71.

Example 10.7

A PI controller is being designed for a guidance system. The plant is

$$G(s) = \frac{1}{s^2 + 10s + 5} \tag{a}$$

Addition of the PI controller in a feedback loop to the plant yields

$$H(s) = \frac{K_p s + K_i}{s^3 + 10s^2 + (5 + K_p)s + K_i} \tag{b}$$

Use MATLAB to draw the root-locus diagram for **a.** $K_p = 20$ and **b.** $K_i = 20$.

Solution

a. For $K_p = 20$,

$$Q(s) = s^3 + 10s^2 + 25s \tag{c}$$

$$R(s) = 1 \tag{d}$$

A MATLAB .m file written to develop the root-locus diagram is given in Figure 10.10(a). The diagram drawn using the program is shown in Figure 10.10(b).

b. For $K_i = 20$,

$$Q(s) = s^3 + 10s^2 + 5s + 20 \tag{e}$$

$$R(s) = s \tag{f}$$

A MATLAB .m file written to develop the root-locus diagram is given in Figure 10.11(a). The diagram drawn using the program is shown in Figure 10.11(b).

The system of Figure 10.10(b) has poles at $s = 0$ and a double pole at $s = -5$. The branch beginning at $s = 0$ becomes more stable as K increases, and it has a breakaway

```
% Figure 10.10
% MATLAB script to draw root-locus diagram for
% Q(s)=s^3+10s^2+25s
% R(s)=1
%
% Specifying Q and R
Q=[1 10 25 0]
R=[1]
%
% Drawing root-locus diagram
%
rlocus(R,Q)
%
% Setting the grid for zeta and omegan
%
sgrid
%
% End of file
```

(a)

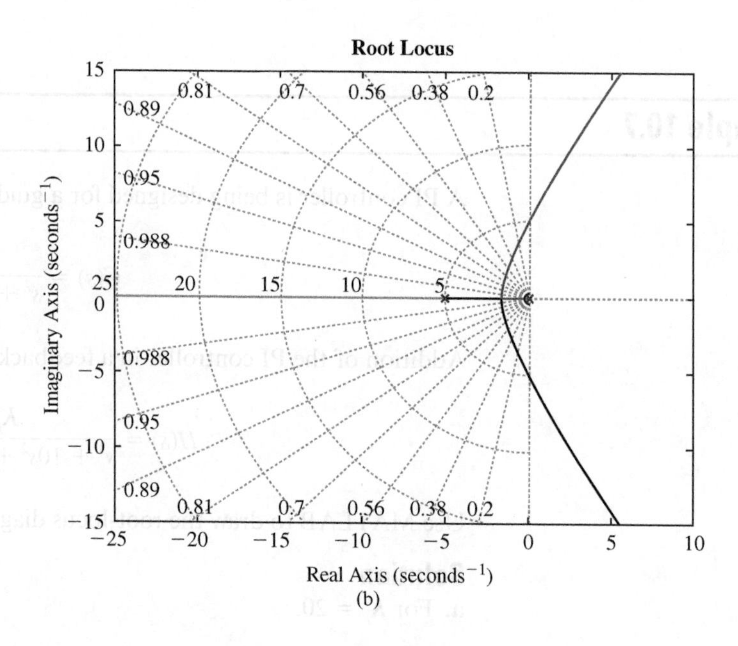

(b)

Figure 10.10
(a) MATLAB script to generate the root-locus diagram of Example 10.7a.
(b) The root-locus diagram.

point at $s = -1.67$ corresponding to $K = 18.5$. The system then loses some of its stability and becomes unstable for $K = 250$. The pole at $s = -5$ has two branches starting from it. One branch stays on the real axis and increases in stability as K increases. The other branch loses stability and breaks away from the real axis at $s = -1.67$ and becomes unstable at $K = 250$.

The system of Figure 10.11(b) has three poles: one pole at $s = -9.7$ and a pair of conjugate poles at $s = -0.151 \pm 1.43j$. The branch beginning at $s = -9.7$ loses stability until it breaks away from the real axis at $s = -4.5$ for $K = -24.2$. The branch grows large and approaches ∞ with an asymptote that makes an angle of $\pi/2$. The branches that begin at $s = -0.151 \pm 1.43j$ become more stable and become critically damped at $s = -1.71$. The system becomes overdamped, with one branch approaching $s = 0$ as K approaches ∞. The other branch meets up with the branch that began at $s = -9.7$ at the breakaway point.

```
% Figure 10.11
% MATLAB script to draw root-locus diagram for
% Q(s)=s^3+10s^2+5s+20
% R(s)=s
%
% Specifying Q and R
Q=[1 10 5 20]
R=[1 0]
%
% Drawing root-locus diagram
%
rlocus(R,Q)
%
% Setting the grid for zeta and omegan
%
sgrid
%
% End of file
```

(a)

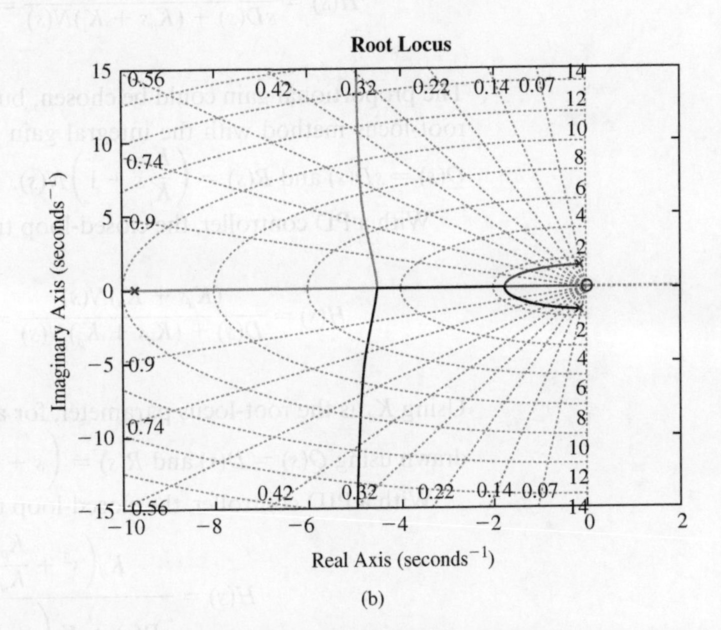

Figure 10.11
(a) MATLAB script to generate the root-locus diagram of Example 10.7b.
(b) The root-locus diagram.

10.8 Control System Design Using the Root-Locus Method

The placement of poles for the closed-loop transfer function is a key focus for a control system designer. If the dominant closed-loop pole is known, a significant amount of information can be inferred. For example, the pole leads to the value of the dominant damping ratio and natural frequency. This, in turn, leads to the maximum overshoot, the value of the settling time, and the rise time. In some cases, it could be desirable to limit the maximum overshoot and work backward to find the damping ratio. The natural frequency could be determined to specify the dominant pole.

The type of modification required for the transfer function dictates the type of controller designed. If the goal is simply to change some property of a system, such as the time constant or the damping ratio, a proportional controller can be used. However, using a proportional controller by itself results in a steady-state error. This error can be eliminated by using a PI controller. The integral action eliminates the steady-state error but can lead to oscillations and often increased overshoot, rise time, and settling time. The transient response can be better accommodated by using a PD or PID controller, where the derivative action controls the rate at which the error decreases. The derivative action leads to reduced oscillations, rise time, and settling time.

A proportional controller, a PI, PD, or PID controller, can be designed using the root-locus method. If a proportional controller is to be used, the closed-loop transfer function corresponding to a plant whose transfer function is $G(s) = \frac{N(s)}{D(s)}$ is

$$H(s) = \frac{K_p N(s)}{D(s) + K_p N(s)} \tag{10.27}$$

The root-locus method to determine the pole location for varying K_p is applied with $Q(s) = D(s)$ and $R(s) = N(s)$. The gain required to place a pole can be determined.

When integral action is to be added, two parameters must be determined: K_p, the proportional gain, and K_i, the integral gain. The closed-loop transfer function is

$$H(s) = \frac{(K_p s + K_i)N(s)}{sD(s) + (K_p s + K_i)N(s)} = \frac{K_i\left(\frac{K_p}{K_i}s + 1\right)N(s)}{sD(s) + K_i\left(\frac{K_p}{K_i}s + 1\right)N(s)} \tag{10.28}$$

The proportional gain could be chosen, but the response is tuned further by using the root-locus method with the integral gain as the root-locus parameter and choosing $Q(s) = sD(s)$ and $R(s) = \left(\frac{K_p}{K_i}s + 1\right)N(s)$.

With a PD controller, the closed-loop transfer function is

$$H(s) = \frac{(K_d s + K_p)N(s)}{D(s) + (K_d s + K_p)N(s)} = \frac{K_d\left(s + \frac{K_p}{K_d}\right)N(s)}{D(s) + K_d\left(s + \frac{K_p}{K_d}\right)N(s)} \tag{10.29}$$

Using K_d as the root-locus parameter, for a fixed value of K_p, a root-locus diagram is drawn using $Q(s) = D(s)$ and $R(s) = \left(s + \frac{K_p}{K_d}\right)N(s)$.

With a PID controller, the closed-loop transfer function is

$$H(s) = \frac{K_d\left(s^2 + \frac{K_p}{K_d}s + \frac{K_i}{K_d}\right)N(s)}{sD(s) + K_d\left(s^2 + \frac{K_p}{K_d}s + \frac{K_i}{K_d}\right)N(s)} \tag{10.30}$$

Using K_d as the root-locus parameter, for fixed values of K_p and K_i, a root-locus diagram is drawn using $Q(s) = sD(s)$ and $R(s) = \left(s^2 + \frac{K_p}{K_d}s + \frac{K_i}{K_d}\right)N(s)$. Tuning rules for PID controllers such as those presented in Section 10.9 have been developed such that this process may not be necessary.

Example 10.8

A plant has the transfer function

$$G(s) = \frac{s + 4}{s^2 + 2s + 10} \tag{a}$$

a. Draw the root-locus diagram for the closed-loop transfer function when the plant is placed in a feedback loop with a proportional controller. What is the value of the proportional gain such that the damping ratio is 0.8? **b.** Design a PI controller using

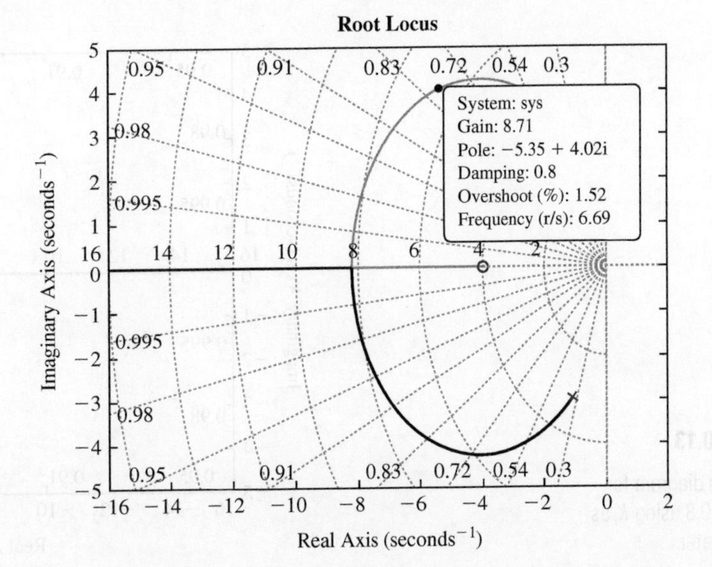

Figure 10.12

Root-locus diagram for system of Example 10.8. The root-locus diagram is used to select $K_p = 8.7$ to obtain a damping ratio of 0.8.

the K_p obtained in part a. as the value of $\frac{K_p}{K_i}$ such that the error of the closed-loop system is zero and the system has no more than 0.0004 percent overshoot. **c.** Design a PID controller using the K_p obtained in part a. as the value of $\frac{K_p}{K_d}$ and the value of K_i obtained in part b. as the value of $\frac{K_i}{K_d}$ such that there is no steady-state error and the damping ratio of the closed-loop system is 1. **d.** Compare the step responses of the closed-loop system when each controller is used.

Solution

The plant is defined by

$$N(s) = s + 4 \tag{b}$$

$$D(s) = s^2 + 2s + 10 \tag{c}$$

a. The closed-loop transfer function in a feedback loop with a proportional controller is given by Equation (10.27), with $N(s)$ and $D(s)$ given by Equations (b) and (c). These equations are substituted into Equation (10.3), with $Q(s) = D(s)$ and $R(s) = N(s)$. The MATLAB-generated result is given in Figure 10.12. The value of K_p that leads to a damping ratio of 0.8 is $K_p = 8.7$. The resulting closed-loop transfer function is

$$H(s) = \frac{8.7s + 34.8}{s^2 + 10.7s + 44.8} \tag{d}$$

b. For a PI controller with $\frac{K_p}{K_i} = 8.7$, the denominator of $H(s)$ becomes

$$s(s^2 + 10.7s + 44.8) + s\left(K_p + \frac{K_i}{s}\right)(s + 4)$$
$$= s^3 + 10.7s^2 + 44.8s + K_i\left(\frac{K_p}{K_i}s + 1\right)(s + 4) \tag{e}$$

The root-locus diagram is then drawn with K_i as the parameter and

$$Q(s) = sD(s) = s^3 + 2s^2 + 10s \tag{f}$$

$$R(s) = \left(\frac{K_p}{K_i}s + 1\right)N(s) = 8.7s^2 + 35.8s + 4 \tag{g}$$

The root-locus diagram is given in Figure 10.13. Choosing $K_i = 1.58$ leads to an overshoot of 0.0004 percent. The closed-loop transfer function is

$$H(s) = \frac{13.7s^2 + 56.6s + 6.3}{s^3 + 15.7s^2 + 66.6s + 6.3} \tag{h}$$

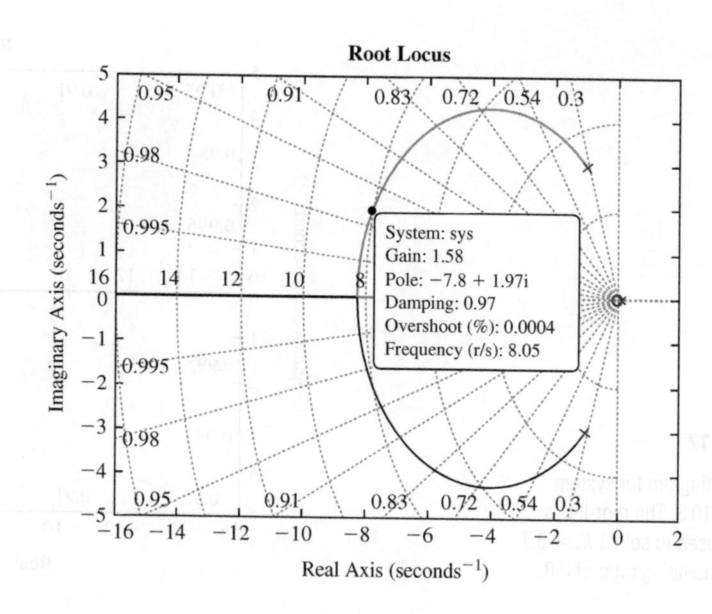

Figure 10.13

Root-locus diagram for Example 10.8 using K_i as the parameter.

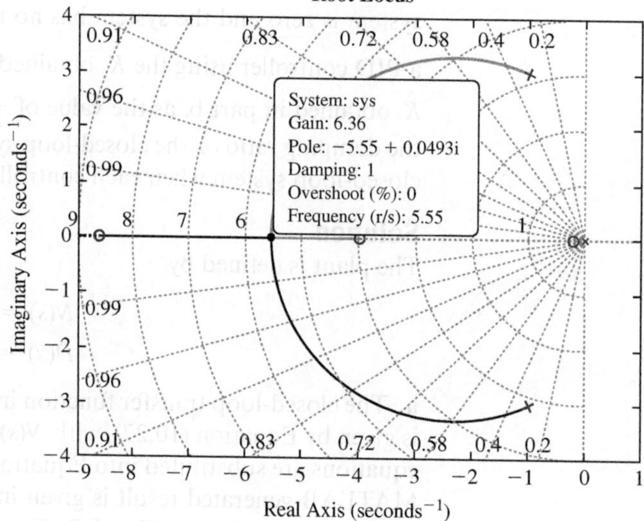

Figure 10.14

Root-locus diagram for Example 10.8 using K_d as the parameter.

c. For a PID controller with $\frac{K_p}{K_d} = 8.7$ and $\frac{K_i}{K_d} = 1.58$,

$$Q(s) = sD(s) = s^3 + 2s^2 + 10s$$

$$R(s) = \left(s^2 + \frac{K_p}{K_d}s + \frac{K_i}{K_d}\right)N(s) = (s^2 + 8.7s + 1.58)(s + 4)$$

$$= s^3 + 12.7s^2 + 36.36s + 6.32$$

The root-locus diagram for the plant with a PID controller with K_d as the parameter is shown in Figure 10.14. The value of K_d such that the system is critically damped is 6.68. The closed-loop transfer function with the PID controller is

$$H(s) = \frac{6.68s^3 + 84.84s^2 + 242.9s + 41.69}{7.68s^3 + 86.84s^2 + 252.9s + 41.69}$$

d. The step responses of the closed-loop system with the plant in series with each of the controllers is shown in Figure 10.15. The proportional controller has a steady-state error such that its final value is $\frac{(4)(8.7)}{10 + 4(8.7)} = 0.78$. The overshoot, rise time, and

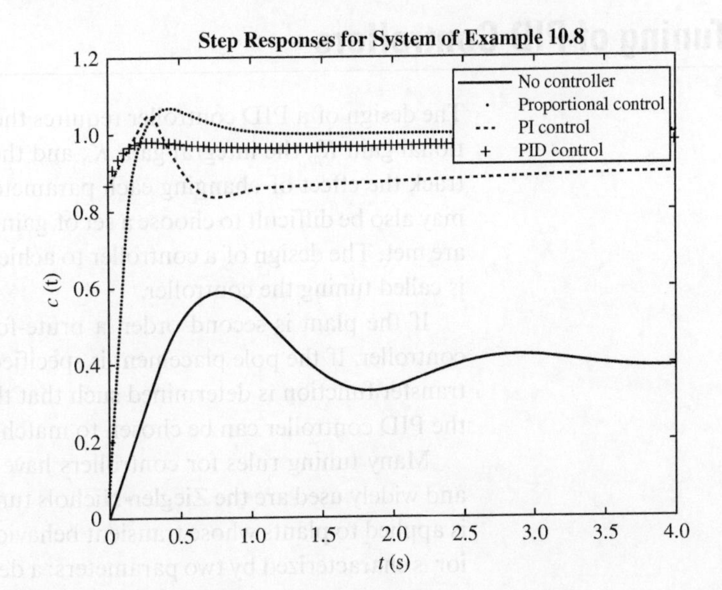

Figure 10.15

Step responses of the system of Example 10.8. The response with the PID controller has favorable rise, settling, and peak times. However, it is discontinuous at $t = 0$.

settling time are all greatly reduced by the addition of any of the designed controllers. Adding the derivative control causes the initial response to be discontinuous as the order of the numerator is equal to the order of the denominator in $H(s)$.

The step responses in part d. of Example 10.8 can be obtained either through the step function in the Control Systems Toolbox of MATLAB as discussed in Section 7.2.6 or through the use of Simulink. For example, to obtain the step response of part b. of Example 10.8, the following MATLAB commands are used:

```
N=[13.7 56.6 6.3]
D=[1 15.7 66.6 6.3]
step(N,D)
```

Multiple step responses plotted on the same graph may be obtained by saving that data to MATLAB and then plotting using MATLAB. The commands that define N and D are supplemented by

```
t=linspace(0,4,401)
[y]=step(N,D,t)
```

The last MATLAB command, which uses the Control Systems Toolbox, calculates the step response for the transfer function defined by N and D at the designated times t and stores them in the vector y.

The block diagram of Figure 10.16 is drawn using Simulink. The step input is applied to the system. Either the step response is output to the scope that plots the simulation or it can be brought into MATLAB and then plotted.

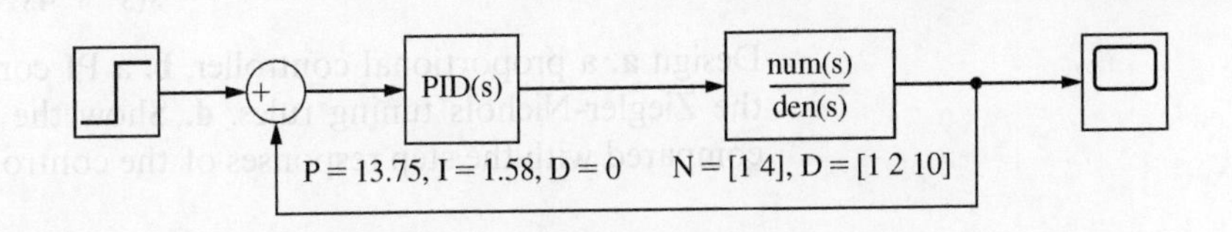

Figure 10.16

Simulink block diagram for the system of Example 10.8.

10.9 Tuning of PID Controllers

The design of a PID controller requires the selection of three parameters: the proportional gain K_p, the integral gain K_i, and the differential gain K_d. It may be difficult to track the effect of changing each parameter on the transient response of a system. It may also be difficult to choose a set of gains such that transient response specifications are met. The design of a controller to achieve specific transient response specifications is called tuning the controller.

If the plant is second order, a brute-force method may be used to design a PID controller. If the pole placement is specified, then the denominator of the closed-loop transfer function is determined such that these poles are met. Then the parameters of the PID controller can be chosen to match the polynomial.

Many tuning rules for controllers have been suggested. Perhaps the most famous and widely used are the Ziegler-Nichols tuning rules. The first Ziegler-Nichols method is applied to plants whose transient behavior is determined experimentally. The behavior is characterized by two parameters: a delay time and a time constant. These parameters are determined from the step response of the system. The first Ziegler-Nichols method is important, but it is not in the scope of this text.

The second Ziegler-Nichols method involves determining the gain of a proportional controller that corresponds to the closed-loop system's crossover frequency. The method is not applicable if the system does not have a crossover frequency. Let ω_c be the crossover frequency, and let K_c be the proportional gain at the crossover frequency. The tuning rules suggest the values of the gains for a proportional controller, a PI controller, and a PID controller in terms of ω_c and K_c. These rules are presented in Table 10.4.

Table 10.4 Ziegler-Nichols Tuning Rules

Controller	K_p	K_i	K_d
P	$0.5K_c$	0	0
PI	$0.45K_c$	$\dfrac{0.27K_c\omega_c}{\pi}$	0
PID	$0.6K_c$	$\dfrac{0.6K_c\omega_c}{\pi}$	$\dfrac{0.15K_c\pi}{\omega_c}$

Example 10.9

A PID controller is to be used for a plant whose transfer function is

$$G(s) = \frac{1}{s(s^2 + 4s + 10)} \tag{a}$$

Design **a.** a proportional controller, **b.** a PI controller, and **c.** a PID controller using the Ziegler-Nichols tuning rules. **d.** Show the step response of the original system compared with the step responses of the controlled system.

Solution

Application of the Ziegler-Nichols tuning rules requires determining the crossover frequency and the corresponding gain when a proportional controller is used. The closed-loop transfer function of the plant in a unity feedback loop under proportional control is

$$H(s) = \frac{K_p}{s^3 + 4s^2 + 10s + K_p} \tag{b}$$

Substituting $s = j\omega$ into the denominator of Equation (b) and setting it to zero leads to

$$-j\omega^3 - 4\omega^2 + 10j\omega + K_p = 0 \tag{c}$$

Setting the real and imaginary parts of Equation (c) to zero gives

$$-4\omega^2 + K_p = 0 \tag{d}$$

$$-\omega^3 + 10\omega = 0 \tag{e}$$

Equation (e) yields $\omega = 0, \pm 3.16$. Using $\omega = 0$ in Equation (d) gives $K_p = 0$. Using $\omega = 3.16$ in Equation (d) gives $K_p = 40$. Then $K_c = 40$ and $\omega_c = 3.16$ are used to determine the parameters in the design of the controllers.

a. For proportional control, according to the Ziegler-Nichols tuning rules,

$$K_p = 0.5K_c = 20 \tag{f}$$

Equation (10.27) is applied to give the closed-loop transfer function

$$H(s) = \frac{20}{s^3 + 4s^2 + 10s + 20} \tag{g}$$

b. For a PI controller, application of the Ziegler-Nichols tuning rules leads to

$$K_p = 0.45K_c = 18 \qquad K_i = \frac{0.27K_c\omega_c}{\pi} = 10.86 \tag{h}$$

The closed-loop transfer function is determined using Equation (10.29):

$$H(s) = \frac{18s + 10.86}{s^4 + 4s^3 + 10s^2 + 18s + 10.86} \tag{i}$$

c. With a PID controller, application of the Ziegler-Nichols tuning rules gives

$$K_p = 0.6K_c = 24 \qquad K_i = \frac{0.6K_c\omega_c}{\pi} = 24.14 \qquad K_d = \frac{0.15K_c\pi}{\omega_c} = 5.97 \tag{j}$$

The closed-loop transfer function for the plant of Equation (a) in a feedback loop with the PID controller whose gains are given in Equation (j) is obtained by applying Equation (10.30), yielding

$$H(s) = \frac{5.97s^2 + 24s + 24.14}{s^4 + 4s^3 + 15.97s^2 + 24s + 24.14} \tag{k}$$

d. The step responses of the plant with no feedback control and the plant in a feedback control loop with the PID controller are given in Figure 10.17(a) and (b). The original plant is unstable and has a step response that grows without bound as the time increases. The controlled response is oscillatory with a final value of 1. The overshoot for the system with a PI controller is larger than for a system with simply proportional control. The settling time is also greater. However, the use of a PID controller leads to smaller response time, settling time, and overshoot.

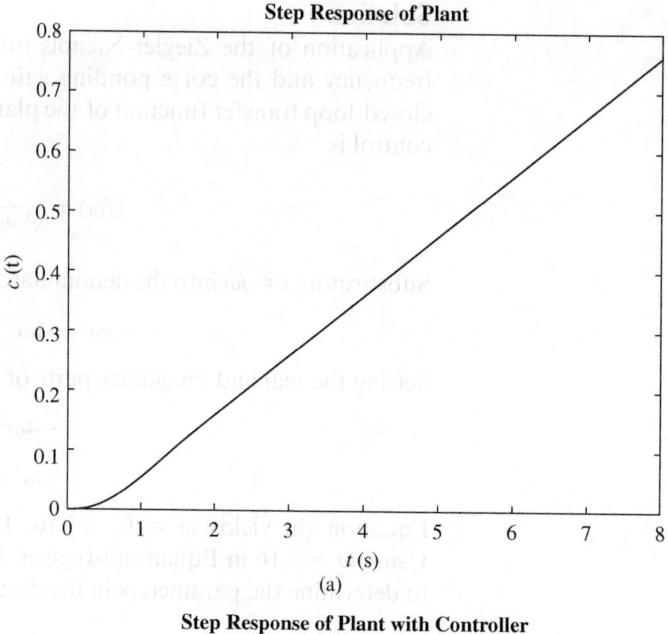

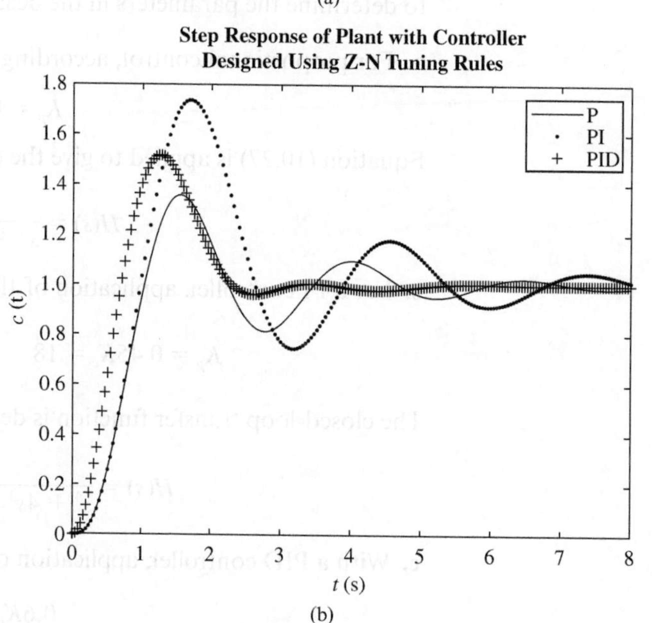

Figure 10.17

(a) Step response of the plant of Example 10.9 shows that the system is unstable if uncontrolled.
(b) Step response of system with controllers designed using the Ziegler-Nichols tuning rules.

The Ziegler-Nichols tuning rules may be applied as a first try in designing a controller such that the closed-loop response meets the desired expectations. It is noted that for a PID controller, according to the Ziegler-Nichols tuning rules,

$$K_p = 4\frac{\omega_c}{\pi}K_d \tag{10.31}$$

and

$$K_p = \frac{\pi}{\omega_c}K_i \tag{10.32}$$

If the crossover frequency is taken as the value obtained using just a proportional controller, then K_p, K_i, and K_d are related through Equations (10.31) and (10.32). One parameter may be chosen as the parameter for a root-locus analysis and the others related to it by Equations (10.31) and (10.32). Then the real and imaginary parts of the equation obtained by substituting $s = j\omega$ can be used to relate the values of K_c and ω_c. A root-locus analysis can be performed on the closed-loop transfer function with any one of the PID parameters as the parameter in the root-locus method. Then the desired performance can be obtained.

Example 10.10

Design a PID controller placed in a feedback loop with a plant whose transfer function is

$$G(s) = \frac{1}{(s+1)(s+3)(s^2+2s+5)} \quad \text{(a)}$$

such that the dominant pole has a damping ratio of 0.4 or higher.

Solution

The poles of the transfer function are $-1, -3, -1 \pm 2j$. The damping ratio of the complex poles is 0.25. When the plant is placed in a unity feedback loop, the poles are $-1.13, -2.17, -0.967 \pm 1.970j$. The dominant poles of the closed-loop transfer function have a damping ratio of 0.26. The use of just a proportional controller leads to a closed-loop transfer function of

$$H(s) = \frac{K_p}{s^4 + 6s^3 + 16s^2 + 26s + 15 + K_p} \quad \text{(b)}$$

The root-locus diagram for $H(s)$ is shown in Figure 10.18. The crossover frequency and gain are determined as $\omega_c = 2.08$ and $K_c = 35.4$. With these parameters, the PID controller designed using the Ziegler-Nichols tuning rules give

$$K_p = 0.6K_c = 21.24 \quad \text{(c)}$$

$$K_i = \frac{0.6K_c\omega_c}{\pi} = 14.06 \quad \text{(d)}$$

$$K_d = \frac{0.15K_c\pi}{\omega_c} = 8.02 \quad \text{(e)}$$

The PID controller designed using the Ziegler-Nichols tuning rules has a transfer function of

$$G_c(s) = 21.24 + \frac{14.06}{s} + 8.02s \quad \text{(f)}$$

This leads to a closed-loop transfer function of

$$H(s) = \frac{8.02s^2 + 21.24s + 14.06}{s^5 + 6s^4 + 16s^3 + 34.02s^2 + 36.24s + 14.06} \quad \text{(g)}$$

The poles are $-3.43, -0.882 \pm 0.224j, -0.402 \pm 2.19j$. This corresponds to a damping ratio of 0.183 for the dominant pole. This is lower than the desired damping ratio and even lower than the damping ratio of the plant. Using the scheme to modify the PID controller and taking K_p to be the root-locus parameter lead to

$$K_p = 4\frac{\omega_c}{\pi}K_d = 2.65K_d \Rightarrow K_d = 0.378K_p \quad \text{(h)}$$

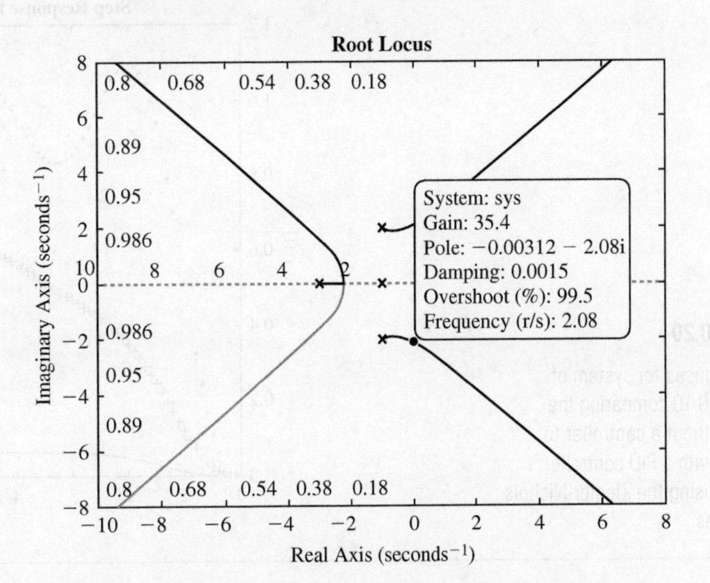

Figure 10.18

Root-locus diagram of the plant of Example 10.10 in a unity feedback loop with a proportional controller.

and

$$K_p = \frac{\pi}{\omega_c} K_i = 1.51 K_i \Rightarrow K_i = 0.662 K_p \tag{i}$$

The modified PID controller has a transfer function of

$$G_c(s) = K_p\left(1 + \frac{0.662}{s} + 0.378s\right) \tag{j}$$

The closed-loop transfer function when this controller is used is given by applying Equation (10.30):

$$H(s) = \frac{K_p(0.378s^2 + s + 0.662)}{s^5 + 6s^4 + 16s^3 + 26s^2 + 15s + K_p(0.378s^2 + s + 0.662)} \tag{k}$$

The root-locus diagram for $H(s)$ obtained using

$$Q(s) = s^5 + 6s^4 + 16s^3 + 26s^2 + 15s \tag{l}$$

and

$$R(s) = 0.378s^2 + s + 0.662 \tag{m}$$

is shown in Figure 10.19. With a gain of 3.76, the damping ratio is 0.403.

The step responses of the original system, the system designed using the Ziegler-Nichols tuning rules, and the system using the modified controller are shown in Figure 20.20.

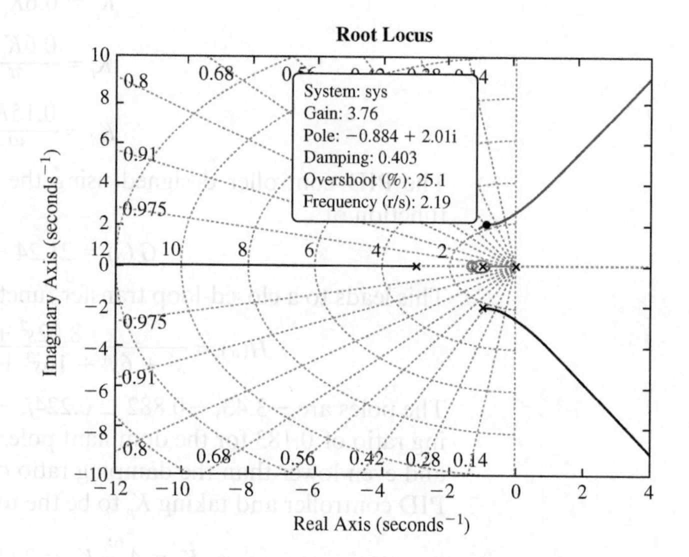

Figure 10.19

Root-locus diagram for Example 10.10 when a PID controller is designed with the proportional gain as the root-locus parameter.

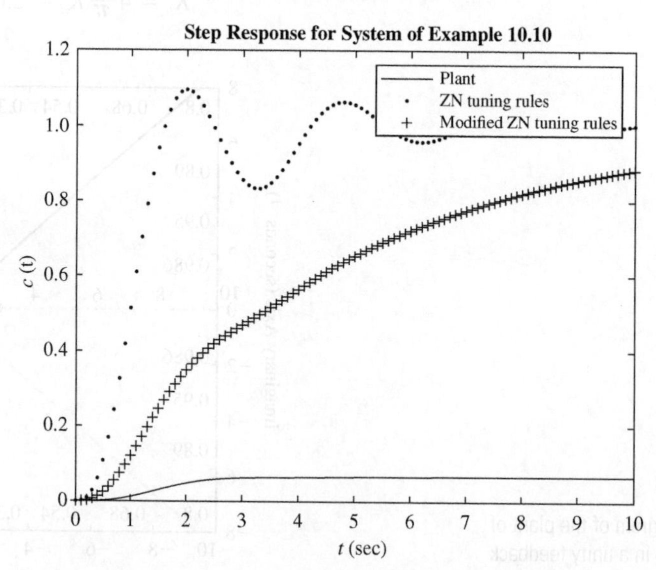

Figure 10.20

Step responses for system of Example 10.10 comparing the system without a controller to a system with a PID controller designed using the Ziegler-Nichols tuning rules.

10.10 Lead Compensators

A lead compensator is employed in a feedback loop when the gain must be increased. The transient response of a system depends heavily on the dominant poles. The farther the dominant poles are away from zero, the better the dominant response. A lead compensator is designed such that the dominant poles of the closed-loop transfer function are farther from zero than the poles when placed in a unity feedback loop. A PID controller can also be designed to improve the transient response. However, a lead compensator can do that without the integral part of the controller, which means that it can be done without an active component.

The closed-loop transfer function for a plant whose transfer function is of the form $G(s) = \frac{N(s)}{D(s)}$ placed in a feedback loop with a lead compensator is

$$H(s) = \frac{K_c\left(s + \frac{1}{T}\right)N(s)}{D(s)\left(s + \frac{1}{\alpha T}\right) + K_c\left(s + \frac{1}{T}\right)N(s)} \tag{10.33}$$

If the order of $G(s)$ is m, then the order of $H(s)$ is $m + 1$. The root-locus parameter is the compensator gain K_c. The functions defined for the development of the root-locus diagram for the closed-loop transfer function of Equation (10.33) are

$$Q(s) = D(s)\left(s + \frac{1}{\alpha T}\right) \tag{10.34}$$

$$R(s) = \left(s + \frac{1}{T}\right)N(s) \tag{10.35}$$

A lead compensator is often designed by specifying the system's closed-loop behavior. This often results in specifying the damping ratio and the natural frequency of the dominant pole. The irreducible quadratic polynomial for the dominant pole is of the form $s^2 + 2\zeta\omega_n s + \omega_n^2$. This polynomial has complex roots at $s = -\zeta\omega_n + j\omega_n\sqrt{1 - \zeta^2}$. The polynomial $Q(s) + K_cR(s)$ can have this as a factor. The parameters K_c, T, and α can be chosen accordingly. This gives two equations for three unknowns. The third equation can be obtained by choosing another root of $Q(s) + K_cR(s)$. This method works best when the plant is of second order or less. Otherwise, synthetic division of $Q(s) + K_cR(s)$ by a cubic polynomial is required.

Example 10.11

Design a lead compensator for the plant with the transfer function

$$G(s) = \frac{2}{s(s + 1)} \tag{a}$$

such that the dominant poles have a damping ratio of 0.8 and a natural frequency of 2 r/s.

Solution

Placing the plant in a unity feedback control loop gives a closed-loop transfer function of

$$H(s) = \frac{2}{s^2 + s + 2} \tag{b}$$

The unity feedback loop leads to a natural frequency of $\sqrt{2} = 1.404$ and a damping ratio of $\frac{1}{2\sqrt{2}} = 0.353$. The damping ratio and natural frequency of the dominant poles are used to determine that they are placed at

$$s = -(0.8)(2) + (2)\sqrt{1 - 0.8^2}\,j = -1.6 + 1.2j \tag{c}$$

They give rise to the quadratic factor $s^2 + 3.2s + 4$.

The closed-loop transfer function when a lead controller is placed in series with the plant in a feedback loop is

$$H(s) = \frac{2K_c\left(s + \frac{1}{T}\right)}{s^3 + \left(1 + \frac{1}{\alpha T}\right)s^2 + \left(2K_c + \frac{1}{\alpha T}\right)s + \frac{2K_c}{T}} \tag{d}$$

When the less dominant root is chosen to be located at -3, the denominator of $H(s)$ should be

$$D(s) = (s + 3)(s^2 + 3.2s + 4) = s^3 + 6.2s^2 + 13.6s + 12 \tag{e}$$

The parameters are chosen such that

$$6.2 = 1 + \frac{1}{\alpha T} \tag{f}$$

$$13.6 = \left(2K_c + \frac{1}{\alpha T}\right) \tag{g}$$

$$12 = \frac{2K_c}{T} \tag{h}$$

Equations (f), (g), and (h) are solved to give $K_c = 4.04$, $T = 0.673$, and $\alpha = 0.286$. Thus, the compensator design is

$$G_c(s) = 4.04\left(\frac{s + 1.49}{s + 5.2}\right) \tag{i}$$

The closed-loop transfer function becomes

$$H(s) = \frac{8.07(s + 1.49)}{s^3 + 6.2s^2 + 13.6s + 12} \tag{j}$$

The unit step response of the plant and the unit step response of the compensated system are shown in Figure 10.21.

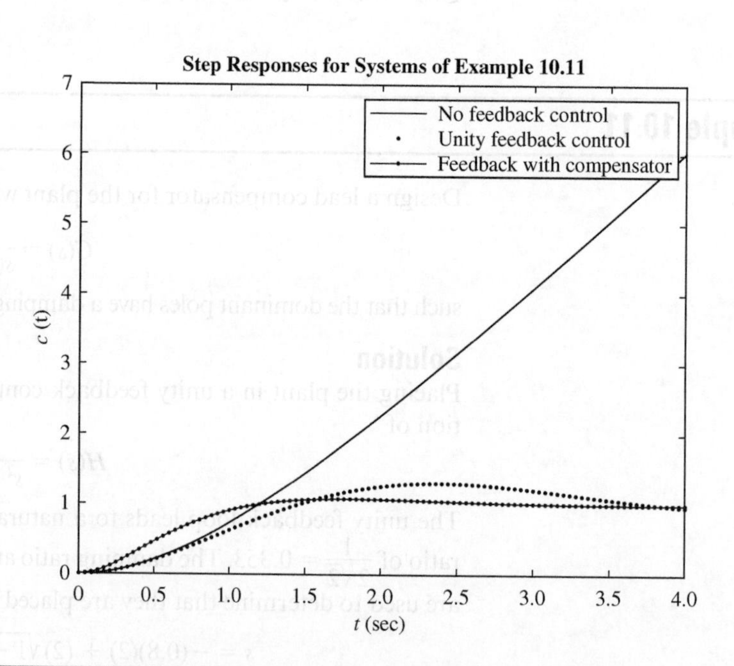

Figure 10.21

Step responses for the system of Example 10.11.

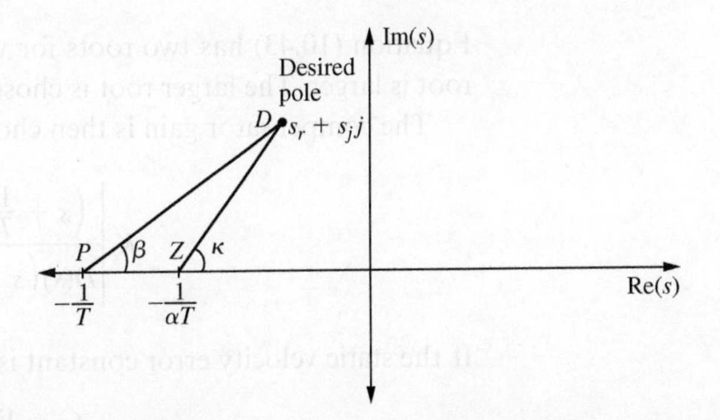

Figure 10.22

Geometry used in the design of lead compensators.

If a desired pole for $H(s)$ is given, say at $s_r + s_j j$, then the angle criterion specifies that to be on a branch of the root-locus diagram, the angle of $\frac{R(s)}{Q(s)}$ is equal to $(2k + 1)\pi$ for an integer value of k. Thus,

$$(2k + 1)\pi = \angle R(s) - \angle Q(s) = \angle N(s) + \angle s + \frac{1}{T} - \angle D(s) - \angle s + \frac{1}{\alpha T} \tag{10.36}$$

Equation (10.36) is rearranged to

$$\psi = (2k + 1)\pi + \angle D(s) - \angle N(s) = \angle s + \frac{1}{T} - \angle s + \frac{1}{\alpha T} \tag{10.37}$$

The left-hand side of Equation (10.37) is the angle that the compensator must make up and is known as the angle deficiency. The parameters α and T must be chosen to make up this deficiency. Figure 10.22 shows the placement of the compensator pole and the compensator zero. This diagram assumes that the zero is to the left of the desired pole. The angle that the zero of the compensator makes with the desired pole is κ, and the angle that the pole of the compensator makes with the desired pole is β. Using this notation in Equation (10.37) gives

$$\kappa - \beta = \psi \tag{10.38}$$

Equation (10.38) is illustrated geometrically by the triangle PZD in Figure 10.22. The angle $\angle PDZ$ in the triangle is $\pi - \beta - (\pi - \kappa) = \psi$. Using the trigonometric identity for the tangent of the sum of the angles gives

$$\tan \kappa = \tan(\psi + \beta) = \frac{\tan \psi + \tan \beta}{1 - \tan \psi \tan \beta} \tag{10.39}$$

Figure 10.22 shows that

$$\tan \kappa = \frac{s_j}{\frac{1}{T} + s_r} \tag{10.40}$$

$$\tan \beta = \frac{s_j}{\frac{1}{\alpha T} + s_r} \tag{10.41}$$

Substituting Equations (10.40) and (10.41) in Equation (10.39) leads to

$$\frac{s_j}{\frac{1}{T} + s_r} = \frac{\tan \psi + \dfrac{s_j}{\frac{1}{\alpha T} + s_r}}{1 - \dfrac{s_j}{\frac{1}{\alpha T} + s_r} \tan \psi} \tag{10.42}$$

Rearranging Equation (10.42) leads to the following quadratic equation to solve for T given the value of α:

$$\alpha\left(s_r^2 + s_j^2\right)T^2 + \left[\frac{s_j}{\tan \psi}(\alpha - 1) + s_r(\alpha + 1)\right]T + 1 = 0 \tag{10.43}$$

Equation (10.43) has two roots for values of $\alpha < 1$. One root is very small, and one root is larger. The larger root is chosen.

The compensator gain is then chosen according to the magnitude condition:

$$K_c \left| \frac{\left(s + \frac{1}{T}\right) N(s)}{D(s)\left(s + \frac{1}{\alpha T}\right)} \right|_{s=s_r+js_j} = 1 \tag{10.44}$$

If the static velocity error constant is specified, then

$$C_v = \lim_{s \to 0} s\, G_c(s) G(s) \tag{10.45}$$

must be checked. If the original plant is a type 1 system, application of Equation (10.45) leads to

$$C_v = \frac{K_c \alpha N(0)}{D(0)} \tag{10.46}$$

If the constant is not large enough for a specific application, the compensator must be redesigned by changing T and α. It is noted that the design of a compensator is not unique. There are many solutions to a compensator design. Often it is advisable to choose one that maximizes K_v.

One method of designing a lead compensator is to cancel a pole in the plant. Usually, the dominant pole is canceled. If $D(s) = (s - s_1)(s - s_2) \cdots (s - s_m)$, then choosing $T = -\frac{1}{s_1}$ causes the leading term to cancel with the numerator term of the compensator. That leaves only α to be chosen to make up the angle deficiency. This method is used when having a large K_v is not necessary.

Example 10.12

A plant has a transfer function of

$$G(s) = \frac{4}{s(s+1)} \tag{a}$$

Design a lead compensator such that the dominant poles of the closed-loop system are $-1.5 \pm 1.5j$.

Solution

The root-locus diagram when the plant is placed in a unity feedback loop shows that the poles cannot be attained by a proportional gain adjustment.

With the plant in a unity feedback loop, the closed-loop transfer function is

$$H(s) = \frac{4}{s^2 + s + 4} \tag{b}$$

The poles of the closed-loop transfer function are at $s = -0.5 \pm 1.94j$.

The angle from the pole at $s = 0$ to the desired pole of $-1 + 1.5j$ is $\angle(-1.5 + 1.5j) = 2.36$. The angle from the pole at $s = 1$ to the desired pole is $\angle 0.5 + 1.5j = 1.89$. In order for the pole to be on the root-locus diagram, the total angle must be of the form $(2k + 1)\pi$, so

$$\psi = \text{Angle deficiency} = (2k + 1)\pi + 2.36 + 1.89 \tag{c}$$

For $k = -2$, this leads to

$$\psi = 1.11 \Rightarrow \tan\psi = 2 \tag{d}$$

Substitution of Equation (d) in Equation (10.42) leads to

$$\frac{1.5}{\frac{1}{T} - 1.5} = \frac{2 + \frac{1.5}{\frac{1}{\alpha T} - 1.5}}{1 - 2\frac{1.5}{\frac{1}{\alpha T} - 1.5}} \tag{e}$$

Equation (e) is rearranged to yield

$$4.5\alpha T^2 - (2.25 + 0.75\alpha)T + 1 = 0 \tag{f}$$

Given the value of α, Equation (f) is a quadratic equation to solve for T. If $\alpha = 0.2$, then Equation (f) becomes

$$0.9T^2 - 2.44T + 1 = 0 \tag{g}$$

The smallest solution of Equation (g) is $T = 0.517$. The zero of the compensator becomes $s = \frac{1}{T} = 1.930$. The pole for the compensator is at $s = \frac{1}{\alpha T} = \frac{1}{(0.2)(0.5168)} = 9.67$. With this solution, the compensator transfer function becomes

$$G_c(s) = K_c\frac{s + 1.93}{s + 9.67} \tag{h}$$

The gain is determined using the magnitude condition:

$$K_c\left|\frac{4(-1.5 + 1.5j + 1.93)}{(-1.5 + 1.5j)(-1.5 + 1.5j + 1)(-1.5 + 1.5j + 9.67)}\right| = 1 \Rightarrow K_c = 4.46 \tag{i}$$

The static velocity error constant is evaluated using Equation (10.46):

$$C_v = \frac{K_c\alpha N(0)}{D(0)} = (4.46)(0.2)\frac{4}{1} = 3.57 \tag{j}$$

The static velocity error constant is not very high. This compensator design leads to

$$H(s) = \frac{17.84s + 34.43}{s^3 + 10.67s^2 + 27.51s + 34.43} \tag{k}$$

The poles of Equation (k) are calculated as -7.67, $-1.5 \pm 1.5j$. The step responses of the plant placed in a unity feedback loop and the plant in series with the compensator of Equation (h) are given in Figure 10.23. The step response of the system with the compensator has a shorter rise time, a shorter settling time, and less overshoot.

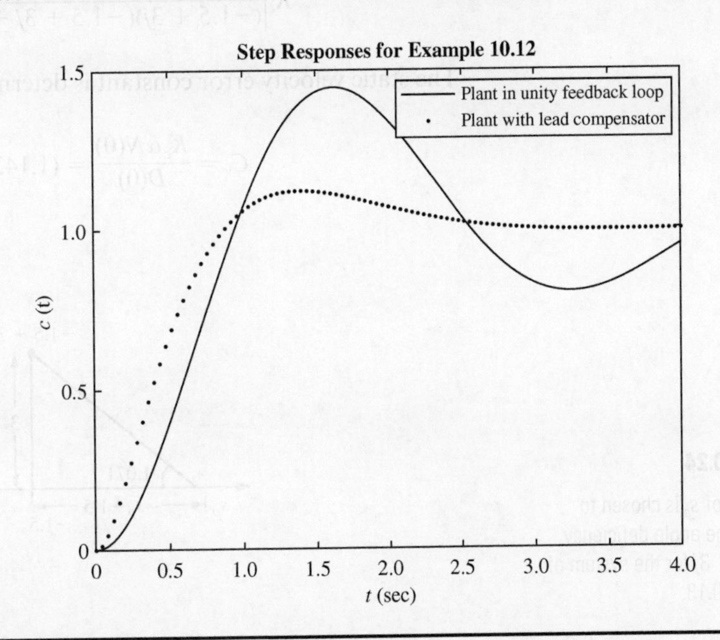

Figure 10.23

Comparison of the step response of the system of Example 10.12 in a unity feedback loop and in a feedback loop with a lead compensator.

Example 10.13

Use the method of canceling a pole to design a lead compensator for the plant with the transfer function

$$G(s) = \frac{10}{s(s+0.5)} \tag{a}$$

such that a closed-loop pole is placed at $s = -1.5 + 3j$.

Solution

The angle deficiency is obtained using Equation (10.37) as

$$\begin{aligned}\psi &= (2k+1)\pi + \angle D(s) - \angle N(s) \\ &= (2k+1)\pi + \angle(-1.5+3j) + \angle(-1+3j) \\ &= 1.071 + 1.8925\end{aligned} \tag{b}$$

Choosing $k = -1$ leads to

$$\psi = -\pi + 1.071 + 1.8925 = -0.178 \tag{c}$$

The lead compensator is designed to cancel the pole at $s = -0.5$, and it has a transfer function of

$$K_c \frac{s+0.5}{s+s_p} \tag{d}$$

The value of s_p is chosen to make up the angle deficiency of 1.071. Using Figure 10.24,

$$\tan 1.071 = \frac{3}{-s_p - 1.5} \Rightarrow s_p = -3.13 \tag{e}$$

The compensator gain is

$$K_c \left| \frac{10}{(-1.5+3j)(-1.5+3j+3.13)} \right| = 1 \Rightarrow K_c = 1.142 \tag{f}$$

The static velocity error constant is determined from Equation (10.46):

$$C_v = \frac{K_c \alpha N(0)}{D(0)} = (1.142)\left(\frac{0.5}{3.13}\right)\left(\frac{10}{1}\right) = 1.83 \tag{g}$$

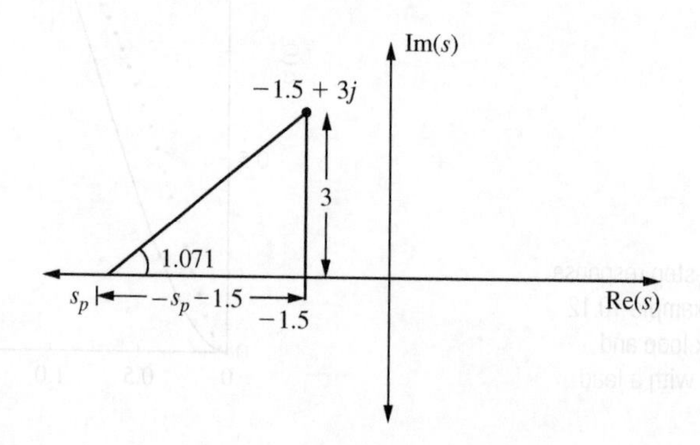

Figure 10.24
The value of s_p is chosen to make up the angle deficiency at $-1.5 + 3j$ for the system of Example 10.13.

10.11 Lag Compensators

A lag compensator is used when the damping ratio and the natural frequency of the dominant pole are acceptable but the static velocity error constant needs improvement. The static velocity error constant is $C_v = \lim_{s\to 0} sG_c(s)G(s)$, which for a compensator has the form $C_v = \frac{K_c \alpha N(0)}{D(0)}$. The larger the value of α, the larger the value of K_v. A lag compensator has values of α greater than 1. Thus, the objective of a lag compensator is to increase the static error velocity constant without appreciably affecting the locations of the poles. This means that the angle contributed by the compensator should be small.

A method to design a lag compensator for a plant such that a given static velocity error constant for the closed-loop transfer function is reached begins by determining the location of the closed-loop poles if the plant is placed in a unity feedback loop. If these poles are acceptable, then the design of a lag compensator begins by requiring the angle contributed by the lag compensator be small, say less than 5° (0.087 r). Arbitrary choices may be a large value of T (corresponding to a small zero for the compensator) and a large value of α such that the pole of the compensator is at least an order of magnitude larger than the zero. The angle provided by the compensator should be small. Then the required gain of the compensator may be calculated by choosing K_c to be large enough so that the static velocity error constant is large enough.

Example 10.14

A plant has the transfer function

$$G(s) = \frac{8}{s(s+2)(s+3)} \tag{a}$$

Design a lag compensator such that the static velocity error constant is at least 20.

Solution

When the plant is placed in a unity feedback loop, the closed-loop transfer function becomes

$$H(s) = \frac{8}{s^3 + 5s^2 + 6s + 8} = \frac{8}{(s+4)(s^2+s+2)} \tag{b}$$

The quadratic factor in the denominator of $H(s)$ is irreducible. It leads to a natural frequency of $\sqrt{2}$ and a damping ratio of 0.354. The poles of $H(s)$ are -4, $-0.5 \pm 1.32j$. The static velocity error constant is

$$K_v = \lim_{s\to 0} sG(s) = \frac{4}{3} \tag{c}$$

The transfer function of a lag compensator is

$$G_c(s) = K_c \frac{s + \frac{1}{T}}{s + \frac{1}{\alpha T}} \qquad \alpha > 1 \tag{d}$$

The system will perform as desired with these closed-loop poles. However, the static velocity constant needs to be greatly increased. The zero of the compensator is arbitrarily placed at $s = -0.01$, corresponding to $T = 100$. Choosing $\alpha = 10$, the zero is placed at $s = -\frac{1}{10(100)} = -0.001$. Then the transfer function for the compensator becomes

$$G_c(s) = K_c \frac{s + 0.01}{s + 0.001} \tag{e}$$

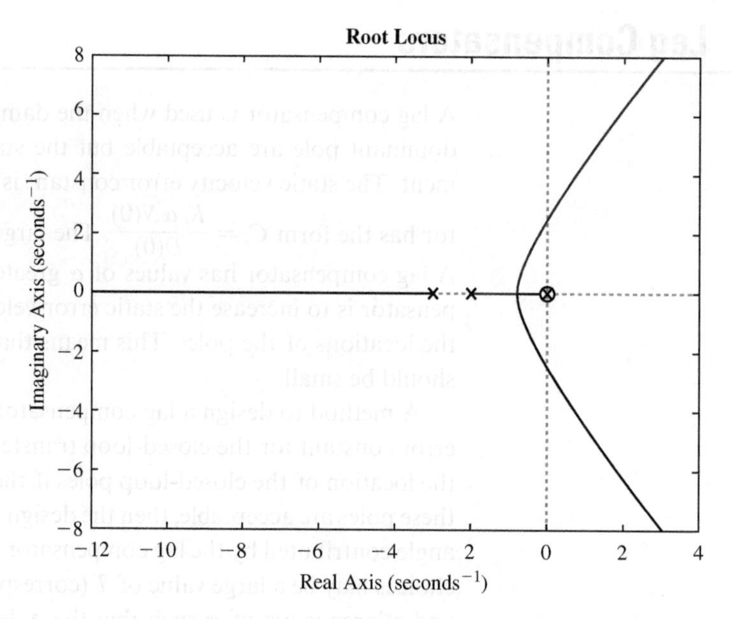

Figure 10.25

Root-locus diagram for the system of Example 10.14 using K_c as a parameter.

The closed-loop transfer function using this compensator becomes

$$H(s) = \frac{8K_c(s + 0.01)}{s(s + 2)(s + 3)(s + 0.001) + 8K_c(s + 0.01)} \tag{f}$$

The root-locus for this transfer function using K_c as the parameter is shown in Figure 10.25. Setting the static velocity error constant for the compensated system to 20 leads to

$$20 = \frac{K_c \alpha N(0)}{D(0)} = K_c(10)\frac{8}{6} \Rightarrow K_c = 1.5 \tag{g}$$

With this choice for K_c, the closed-loop transfer function becomes

$$H(s) = \frac{12s + 0.12}{s^4 + 5.001s^3 + 6.005s^2 + 12.006s + 0.12} \tag{h}$$

The poles for the compensated system are $s = -4.25, -0.01, -0.37 \pm 1.64j$. The angle contributed by the compensator at the complex pole is

$$\angle\left(\frac{-0.37 + 1.64j + 0.01}{-0.37 + 1.64j + 0.001}\right) = -0.0053 \text{ r} = 0.29° \tag{i}$$

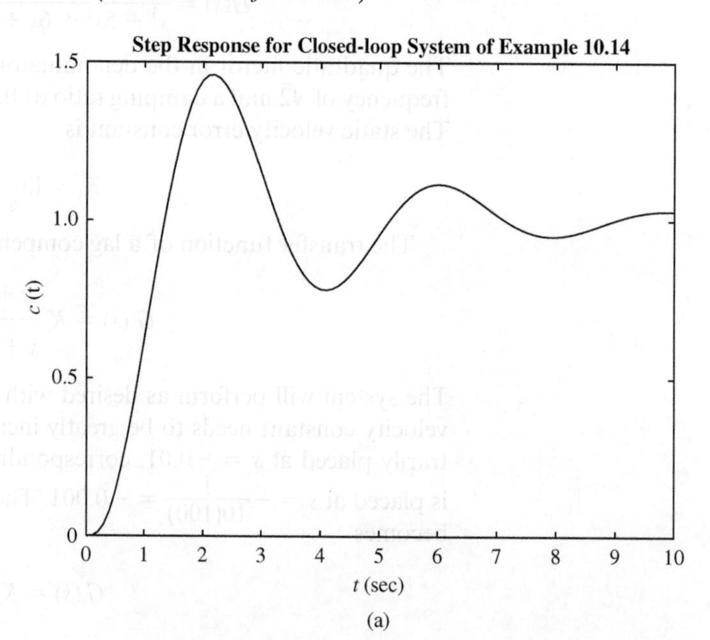

Figure 10.26

(a) Step response and (b) Ramp response for the closed-loop system of Example 10.14.

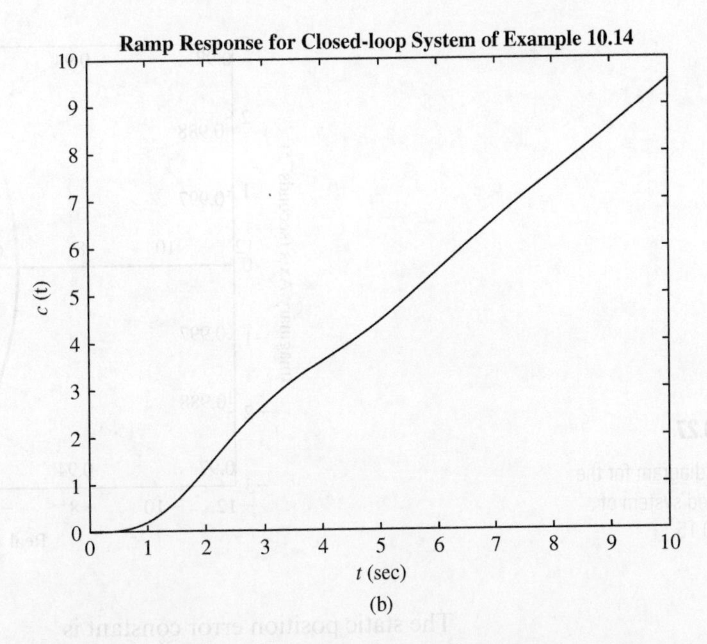

(b)

Figure 10.26
(Continued)

The step response and ramp responses of the closed-loop system are shown in Figure 10.26(a) and (b). The ramp response closely follows the ramp input.

Example 10.15

Design a compensator for a plant with the transfer function

$$G(s) = \frac{s + 4}{s^2 + 4s + 7} \tag{a}$$

such that the damping ratio of the closed-loop system is at most 0.5 and the static position error constant is 50.

Solution

When the plant is placed in a unity feedback loop, the closed-loop transfer function becomes

$$H(s) = \frac{s + 4}{s^2 + 4s + 8} \tag{b}$$

The poles of the transfer function of Equation (b) are $s = -2 \pm 2j$, which lead to a damping ratio of 0.707. The static position error constant is

$$C_p = G(0) = \frac{4}{7} \tag{c}$$

The poles are good, but the static position error constant needs to be greatly magnified. Thus, a lag compensator is used.

With $T = 50$ and $\alpha = 25$, the form of the transfer function of a lag compensator is

$$G_c(s) = K_c \frac{s + 0.02}{s + 0.0008} \tag{d}$$

With the compensator, the closed-loop transfer function becomes

$$H(s) = \frac{K_c(s + 4)(s + 0.02)}{(s^2 + 4s + 7)(s + 0.0008) + K_c(s + 0.02)} \tag{e}$$

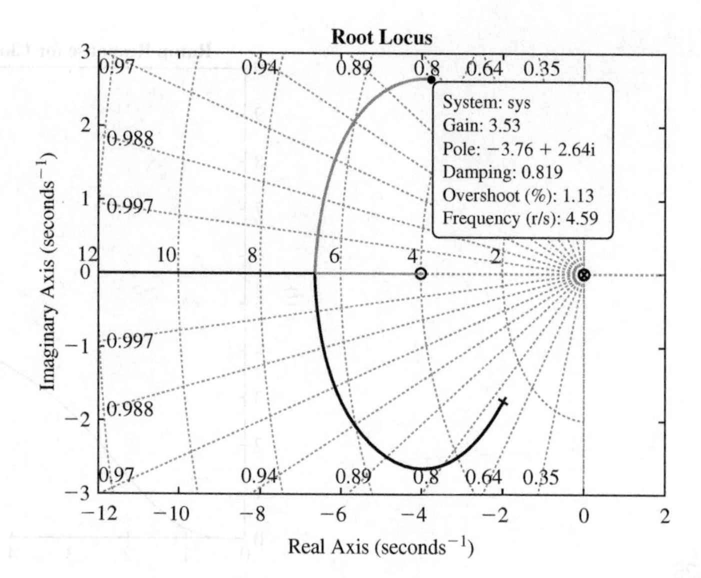

Figure 10.27

Root-locus diagram for the compensated system of Example 10.15.

The static position error constant is

$$C_p = G_c(0)G(0) = K_c\alpha\frac{4}{7} \tag{f}$$

In order for the static position error constant to be as large as 50, Equation (f) requires that K_c must be as large as 3.5. Setting $K_c = 3.5$ in Equation (e) leads to

$$H(s) = \frac{3.5s^2 + 14.07s + 0.28}{s^3 + 7.5008s^2 + 21.0732s + 0.2856} \tag{g}$$

The root-locus diagram for the compensated transfer function is shown in Figure 10.27, and the response of the compensated system due to a unit step input is given in Figure 10.28. The root-locus diagram shows that the poles of the closed-loop transfer function are at -0.0136, $-3.74 \pm 2.64j$. The angle supplied by the compensator at the dominant pole is

$$\angle\left(\frac{-1.962 + 2.56j}{-1.9608 + 2.56j}\right) = 2.594 \times 10^{-4}\,\text{r} = 0.0169° \tag{h}$$

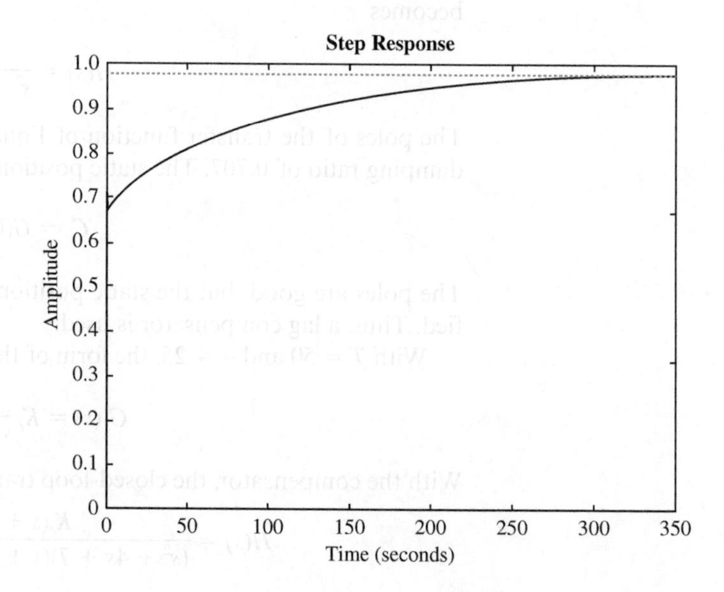

Figure 10.28

Step response for the compensated system of Example 10.15.

10.12 Lead-Lag Compensators

Lead compensators are used to improve the transient performance; lag compensators are used to reduce the steady-state error. When both are needed, a lead compensator can be placed in series with a lag compensator, or a lead-lag compensator may be employed.

The transfer function of a lead-lag compensator is

$$G_c(s) = K_c\left(\frac{s+\frac{1}{T_1}}{s+\frac{1}{\alpha_1 T_1}}\right)\left(\frac{s+\frac{1}{T_2}}{s+\frac{1}{\alpha_2 T_2}}\right) \qquad \alpha_1 < 1, \alpha_2 > 1 \tag{10.47}$$

When employed in a feedback loop with a plant whose transfer function is $G(s) = \frac{N(s)}{D(s)}$, the closed-loop transfer function becomes

$$H(s) = \frac{K_c\left(s+\frac{1}{T_1}\right)\left(s+\frac{1}{T_2}\right)N(s)}{\left(s+\frac{1}{\alpha_1 T_1}\right)\left(s+\frac{1}{\alpha_2 T_2}\right)D(s) + K_c\left(s+\frac{1}{T_1}\right)\left(s+\frac{1}{T_2}\right)N(s)} \tag{10.48}$$

The root-locus diagram can be drawn with

$$Q(s) = \left(s+\frac{1}{\alpha_1 T_1}\right)\left(s+\frac{1}{\alpha_2 T_2}\right)D(s) \tag{10.49a}$$

$$R(s) = \left(s+\frac{1}{T_1}\right)\left(s+\frac{1}{T_2}\right)N(s) \tag{10.49b}$$

If the desired dominant closed-loop poles are known, the design of a lead-lag compensator proceeds as the lead compensator does. The angle deficiency is calculated assuming that the lag compensation part of the lead-lag compensator will be designed such that it does not affect the angle very much. The magnitude condition is used to calculate the compensator gain. Then the static error constant is used to determine the value of α_2 for the lag compensator. The resulting compensator design is tested.

It is common to choose $\alpha_2 = \frac{1}{\alpha_1}$. Then the static velocity error constant becomes

$$C_v = K_c\frac{N(0)}{D(0)} \tag{10.50}$$

For this case, the value of K_c is directly determined such that the static velocity error constant is satisfied. The values of α and T_1 are determined such that the dominant poles are placed. This usually requires that the angle of deficiency is made up by the lead compensator. Since the value of K_c is already determined, the magnitude condition must be satisfied. This generates two equations to solve for α and T_1. Then a value of T_2 is chosen such that the lag compensator does not contribute much to the phase and the magnitude of the lag compensator is close to 1.

Example 10.16

Design a lead-lag compensator for a plant with the transfer function

$$G(s) = \frac{4}{s(s+1)(s+3)} \tag{a}$$

The compensated system is to have a damping ratio of 0.8 and a natural frequency of 2 r/s. The steady-state velocity error constant is to be no less than 45.

Solution

The closed-loop transfer function for the plant in a unity feedback loop is

$$H(s) = \frac{4}{s^3 + 4s^2 + 3s + 4} \tag{b}$$

The poles of the closed-loop system are $s = -3.47, -0.266 \pm 1.045j$.

The dominant poles of the system with the damping ratio of 0.8 and a natural frequency of 2 r/s are at $s = -1.6 + 1.2j$. Assuming the lag part of the compensator does not contribute sufficiently to the angle deficiency, the angle to be made up by the lead compensator is obtained using Equation (10.37) with $k = 0$:

$$\begin{aligned}\psi &= (2k+1)\pi + \angle D(s) - \angle N(s) \\ &= -\pi + \angle\left(\frac{4}{(-1.6+1.2j)(-1.6+1.2j+1)(-1.6+1.2j+3)}\right) \\ &= 0.634 \text{ r}\end{aligned} \tag{c}$$

The procedure outlined in Section 10.10 and used in Example 10.13 could be used to place the poles. However, here a lag compensator is employed to set the static velocity error constant, so the zero of the lead part of the compensator can be chosen to be an arbitrary value of s, say $s = -2$ corresponding to $T_1 = 0.5$. Using Figure 10.29, which shows the placement of the zero and the pole of the lead compensator,

$$\tan\kappa = \frac{1.2}{0.4} \Rightarrow \kappa = 1.25 \text{ r} \tag{d}$$

The angle made by the compensator pole and the desired pole must be $0.63448 - 1.25 = -0.615$ r. Then

$$\tan 0.615 = \frac{1.2}{\dfrac{1}{\alpha_1(0.5)} - 1.6} \Rightarrow \alpha_1 = 0.61 \tag{e}$$

The pole of the lead part of the compensator is $-\dfrac{1}{(0.61)(0.5)} = -3.29$.

The value of K_c is determined from the magnitude condition:

$$K_c\left|\frac{4(-1.6+1.2j+2)}{(-1.6+1.2j)(-1.6+1.2j+1)(-1.6+1.2j+3)(-1.6+1.2j+3.29)}\right| = 1 \Rightarrow K_c = 2.03 \tag{f}$$

Requiring that the static velocity error constant using the lead-lag compensator is 45 gives

$$45 = \frac{K_c\alpha_1\alpha_2 N(0)}{D(0)} = \frac{(2.03)(0.61)\alpha_2(4)}{(3)} \Rightarrow \alpha_2 = 27.3 \tag{g}$$

Choosing $T = 20$ leads to the transfer function of the compensator as

$$G_c(s) = 16.42\frac{s+2}{s+3.29}\frac{s+0.05}{s+0.00183} \tag{h}$$

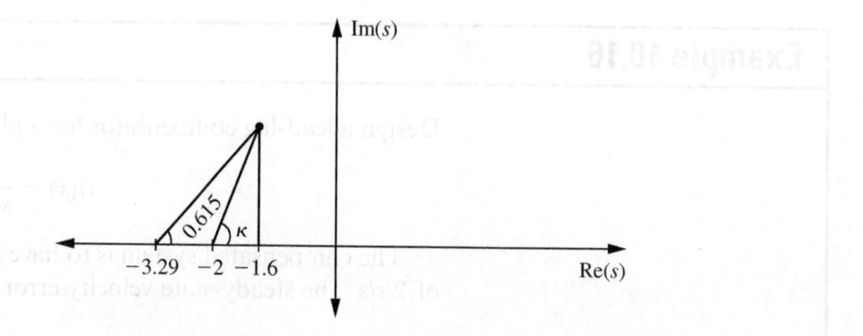

Figure 10.29
Geometry for placement of pole for Example 10.16.

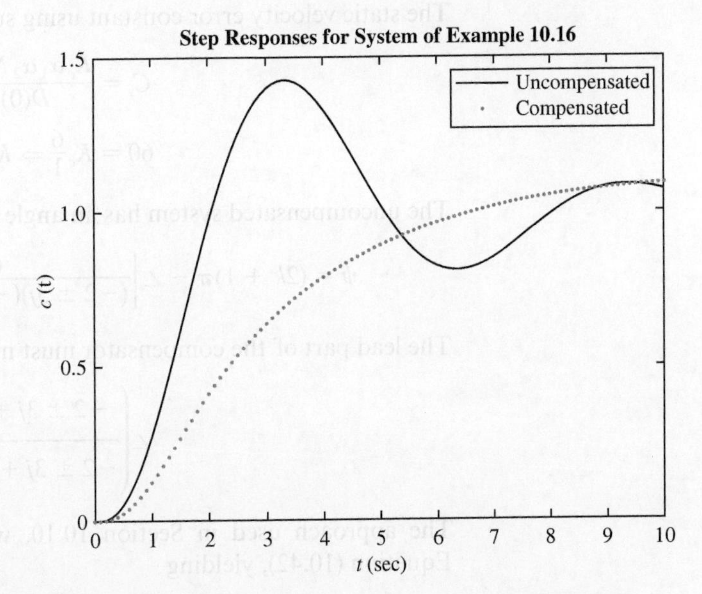

Figure 10.30

Step responses for compensated and uncompensated systems of Example 10.16.

The lag part of the compensator has a phase at the dominant poles of

$$\angle\left(\frac{-1.6 + 1.2j + 0.05}{-1.6 + 1.2j + 0.00183}\right) = -0.0148 \text{ r} = 0.848° \tag{i}$$

This is in the range for a lag compensator.

The closed-loop transfer function becomes

$$\begin{aligned} H(s) &= \frac{2.03(s+2)(s+0.05)}{(s^3 + 4s^2 + 3s)(s+3.29)(s+0.00183) + 2.03(s+2)(s+0.05)} \\ &= \frac{2.03s^2 + 4.1615s + 0.2030}{s^5 + 6.2918s^4 + 13.8815s^3 + 15.2154s^2 + 4.1856s + 0.2030} \end{aligned} \tag{j}$$

The step response of Equation (j) is shown in Figure 10.30. The lead-lag compensator greatly improves the rise time, settling time, and overshoot while tracking a ramp input with little error.

Example 10.17

Design a lead-lag compensator to use in series with a plant that has the transfer function

$$G(s) = \frac{6}{s(s+1)} \tag{a}$$

Design the compensator such that the dominant closed-loop poles are $-2 \pm 3j$. The static velocity error constant should be 60 s^{-1}.

Solution

The approach is to use a lag-lead compensator with

$$\alpha_2 = \frac{1}{\alpha_1} \tag{b}$$

The static velocity error constant using such a compensator is

$$C_v = \frac{K_c \alpha_1 \alpha_2 N(0)}{D(0)} = \frac{K_c N(0)}{D(0)} \Rightarrow$$

$$60 = K_c \frac{6}{1} \Rightarrow K_c = 10 \tag{c}$$

The uncompensated system has an angle deficiency of

$$\psi = (2k+1)\pi - \angle\left[\frac{6}{(-2 \pm 3j)(-2 \pm 3j + 1)}\right] = 0.9098 \text{ r} = 52.1° \tag{d}$$

The lead part of the compensator must make up this angle, so

$$\angle\left(\frac{-2 \pm 3j + \dfrac{1}{T_1}}{-2 \pm 3j + \dfrac{1}{\alpha T_1}}\right) = 52.1° \tag{e}$$

The approach used in Section 10.10, which was used in Example 10.12, applies Equation (10.42), yielding

$$\frac{s_j}{\dfrac{1}{T_1} + s_r} = \frac{\tan\psi + \dfrac{s_j}{\dfrac{1}{\alpha T_1} + s_r}}{1 - \left(\dfrac{s_j}{\dfrac{1}{\alpha T_1} + s_r}\right)\tan\psi} \Rightarrow \frac{4}{\dfrac{1}{T_1} - 2.5} = \frac{1.28 + \dfrac{3}{\dfrac{1}{\alpha T_1} - 2}}{1 - 1.28\left(\dfrac{3}{\dfrac{1}{\alpha T_1} - 2}\right)} \tag{f}$$

The magnitude criterion requires

$$10\left|\frac{\left(-2 \pm 3j + \dfrac{1}{T_1}\right)(-2 \pm 3j + 3)}{\left(-2 \pm 3j + \dfrac{1}{\alpha T_1}\right)(-2.5 \pm 4j)(-2.5 \pm 4j + 0.5)(-2.5 \pm 4j + 1)}\right|$$

$$= 5.26\left|\frac{\left(-2 \pm 3j + \dfrac{1}{T_1}\right)}{\left(-2 \pm 3j + \dfrac{1}{\alpha T_1}\right)}\right| = 1 \tag{g}$$

Solving Equations (f) and (g) using a trial and error procedure yields $\alpha = 0.18$ and $T_1 = 0.294$. Choosing $T_2 = 20$, the compensator transfer function becomes

$$G_c(s) = 10\left[\frac{s + \dfrac{1}{0.294}}{s + \dfrac{1}{(0.18)(0.294)}}\right]\left[\frac{s + 0.05}{s + (0.18)(0.05)}\right] = 10\left(\frac{s + 3.4}{s + 18.9}\right)\left(\frac{s + 0.05}{s + 0.009}\right) \tag{h}$$

Application of the controller in a feedback loop with the plant leads to the closed-loop transfer function

$$H(s) = \frac{60\left(\dfrac{s + 3.4}{s + 18.9}\right)\left(\dfrac{s + 0.05}{s + 0.009}\right)}{s(s+1) + 60\left(\dfrac{s + 3.4}{s + 18.9}\right)\left(\dfrac{s + 0.05}{s + 0.009}\right)}$$

$$= \frac{60s^2 + 207s + 10.2}{s^4 + 19.909s^3 + 79.0701s^2 + 207.1201s + 10.2} \tag{i}$$

The poles of the transfer function given in Equation (i) are -0.0502, -15.71, $-2.07 \pm 2.94j$. The static velocity error constant is 60. The step response of the closed-loop transfer function of Equation (i) is illustrated in Figure 10.31.

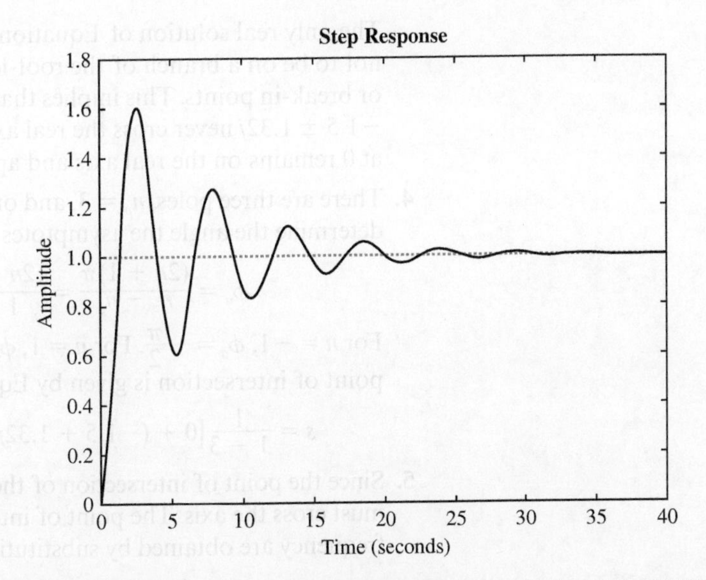

Figure 10.31
Step response for the compensated system of Example 10.17.

10.13 Further Examples

Example 10.18

Draw the root-locus diagram for a plant with the transfer function

$$H(s) = \frac{s + 6}{s(s^2 + 3s + 4) + K(s + 6)} \tag{a}$$

Solution

Comparing Equation (a) with Equation (10.3) leads to

$$Q(s) = s(s^2 + 3s + 4) \tag{b}$$

$$R(s) = s + 6 \tag{c}$$

The steps in Section 10.6 are followed for drawing the root-locus diagram.

1. The order of $Q(s)$ is 3, which implies that there are three branches on the root-locus diagram. The poles are at $s = 0, -1.5 \pm 1.32j$. The branches have starting points at these values. The zero of the root-locus diagram is at $s = -6$. One branch of the root-locus diagram ends at this point.
2. The range on the real axis corresponding to $s > 0$ is not on any branch of the root-locus diagram, as in this range both $(s + 6)$ and s are positive and have a phase of zero. The quadratic function is greater than $(s + 2)^2$ for all real s and does not contribute anything to the phase. For $-6 < s < 0$, s is negative and $(s + 6)$ is positive. The angle criterion is satisfied with $k = 0$. For $s < -6$, both s and $(s + 6)$ are negative and the phase of $\frac{R(s)}{Q(s)}$ is zero, which does not satisfy the angle criterion. The only part of the real axis that lies on a branch of the root-locus diagram is $-6 < s < 0$.
3. The breakaway and break-in points are determined by finding the values of s that satisfy Equation (10.18):

$$(s + 6)(3s^2 + 6s + 4) - (s^3 + 3s^2 + 4s)(1) = 0 \tag{d}$$

Equation (d) is simplified to

$$2s^3 + 21s^2 + 36s + 24 = 0 \tag{e}$$

The only real solution of Equation (e) is $s = -8.56$, which was shown in step 2 not to be on a branch of the root-locus diagram. Thus, there are no breakaway or break-in points. This implies that the branches that start at the complex poles $-1.5 \pm 1.32j$ never cross the real axis and the branch that starts on the real axis at 0 remains on the real axis and approaches -6 for large K.

4. There are three poles, $n_p = 3$, and one zero, $n_z = 1$. Applying Equation (10.19) to determine the angle the asymptotes for large K make with the horizontal gives

$$\phi_a = \frac{(2n+1)\pi}{n_z - n_p} = \frac{(2n+1)\pi}{1-3} = -\frac{1}{2}(2n+1)\pi \tag{f}$$

For $n = -1$, $\phi_a = -\frac{\pi}{2}$. For $n = 1$, $\phi_a = \frac{\pi}{2}$. The asymptotes are vertical lines. The point of intersection is given by Equation (10.26):

$$s = \frac{1}{1-3}[0 + (-1.5 + 1.32j) + (-1.5 - 1.32j) + 6] = \frac{3}{2} \tag{g}$$

5. Since the point of intersection of the vertical asymptotes is positive, the branches must cross the axis. The point of intersection and the corresponding crossover frequency are obtained by substituting $s = j\omega$ into Equation (10.3), leading to

$$(j\omega)\left[(j\omega)^2 + 3(j\omega) + 4\right] + K[(j\omega) + 6] = 0 \tag{h}$$

Equation (h) is rearranged as

$$6K - 3\omega^2 + j\omega(4 + K - \omega^2) = 0 \tag{i}$$

Setting the real part and the imaginary part of Equation (i) to zero leads to

$$\omega = \sqrt{8} = 2.83 \text{ r/s and } K = 4.$$

6. The angles of departure from the poles are determined by applying the angle criterion at the poles:

$$\begin{aligned}(2k+1)\pi - \angle(-1.5 + 1.32j) + \angle(-1.5 + 1.32j + 6) &= (2k+1)\pi - 2.42 + 0.2853 \\ &= 1.00 \text{ r} = 57.7^\circ\end{aligned} \tag{j}$$

7. The values of points on a branch of the root-locus diagram can be determined using the magnitude criterion, Equation (10.9), and the angle criterion, Equation (10.12). They are summarized in Table 10.5.

The hand sketch of the root-locus plot using these results is shown in Figure 10.32.

Table 10.5 Poles of Example 10.18

K	Real Pole	Complex Poles
1	−2.0	$-0.5 \pm 1.69j$
2	−2.51	$-0.24 \pm 2.17j$
5	−3.16	$0.79 \pm 3.09j$
10	−3.66	$0.33 \pm 4.04j$
15	−3.95	$0.48 \pm 4.75j$
30	−4.45	$0.73 \pm 6.33j$

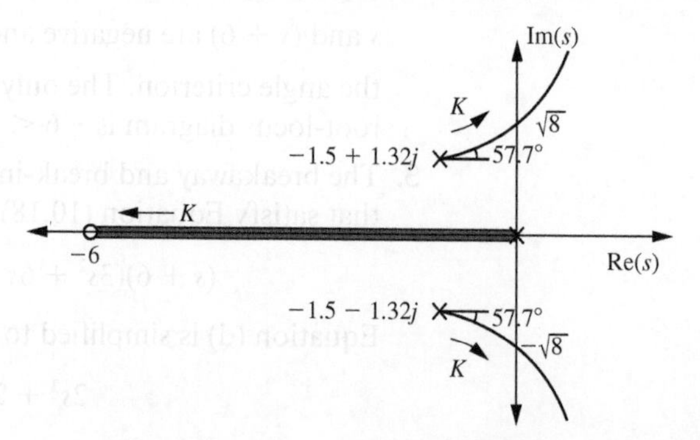

Figure 10.32

Sketch of the root-locus diagram for Example 10.18.

Example 10.19

Design a PID controller such that when it is placed in series with the plant whose transfer function is

$$G(s) = \frac{6}{(s + 1)(s + 3)} \tag{a}$$

the closed-loop transfer function has poles at $s = -1.5 \pm 3.4j, -4$.

Solution

The dominant pole has a damping ratio of 0.403 and a natural frequency of 3.70. The quadratic polynomial with these properties is $s^2 + 2\zeta\omega_n s + \omega_n^2 = s^2 + 3s + 11.6$. Thus, the denominator of $H(s)$ becomes

$$D(s) = (s + 4)(s^2 + 3s + 11.6) = s^3 + 7s^2 + 23.6s + 45.4 \tag{b}$$

The closed-loop transfer function for a PID controller in series with the plant in a feedback loop is obtained using Equation (10.30) as

$$H(s) = \frac{6\left(K_d s^2 + K_p s + K_i\right)}{s(s^2 + 4s + 3) + 6\left(K_d s^2 + K_p s + K_i\right)}$$
$$= \frac{6\left(K_d s^2 + K_p s + K_i\right)}{s^3 + (4 + 6K_d)s^2 + (3 + 6K_p)s + 6K_i} \tag{c}$$

In order for the denominator of Equation (c) to be equivalent to $D(s)$, it is required that

$$4 + 6K_d = 7 \Rightarrow K_d = 0.5 \tag{d}$$

$$3 + 6K_p = 23.6 \Rightarrow K_p = 3.43 \tag{e}$$

$$6K_i = 45.4 \Rightarrow K_i = 7.57 \tag{f}$$

Hence the transfer function for the PID controller is

$$G_c(s) = 0.5s + 3.43 + \frac{7.57}{s} \tag{g}$$

Example 10.20

The three-tank liquid-level system shown in Figure 10.33 has the transfer function

$$G_3(s) = \frac{H_3(s)}{Q(s)} = \frac{8}{12s^3 + 78s^2 + 150s + 84} \tag{a}$$

Use the Ziegler-Nichols tuning rules to design **a.** a proportional controller, **b.** a PI controller, and **c.** a PID controller for the three-tank liquid-level system.

Solution

The closed-loop transfer function for a system with a proportional controller is

$$H(s) = \frac{8K_p}{12s^3 + 78s^2 + 150s + 84 + K_p} \tag{b}$$

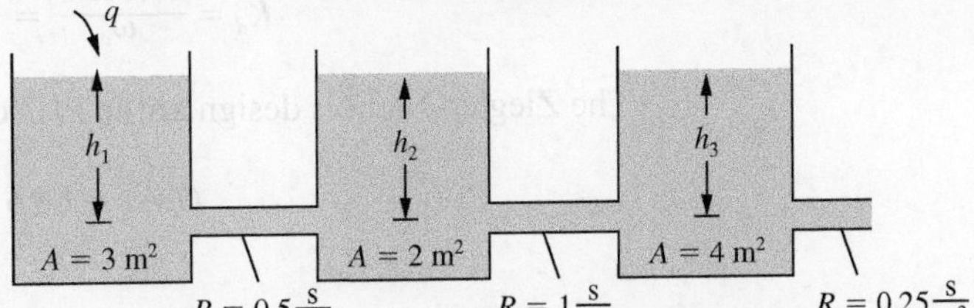

Figure 10.33

Three-tank liquid-level system of Example 10.20.

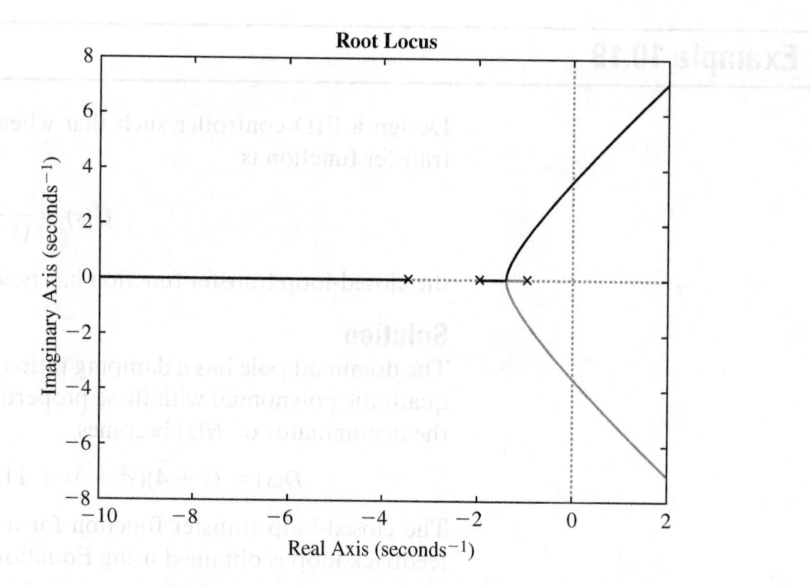

Figure 10.34

Root-locus diagram for the system of Example 10.20.

The root-locus diagram for Equation (b) is shown in Figure 10.34. The crossover frequency is $\omega_c = 3.53$ r/s. The system has a gain at this frequency of $K_c = 111$.

a. A proportional controller has a gain, according to the Ziegler-Nichols tuning rules, of

$$K_p = 0.5K_c = 55.5 \tag{c}$$

The closed-loop transfer function for the system with a proportional controller is

$$H(s) = \frac{444}{12s^3 + 78s^2 + 150s + 528} \tag{d}$$

b. The Ziegler-Nichols tuning rules state that the appropriate design of a PI controller has

$$K_p = 0.45K_c = 49.95 \tag{e}$$

$$K_i = \frac{0.27K_c\omega_c}{\pi} = 33.67 \tag{f}$$

The closed-loop transfer function when the parameters of Equations (e) and (f) are used is

$$H(s) = \frac{8(K_p s + K_i)}{s(12s^3 + 78s^2 + 150s + 84) + 8(K_p s + K_i)}$$

$$= \frac{399.6s + 269.6}{12s^4 + 78s^3 + 150s^2 + 483.6s + 269.6} \tag{g}$$

c. The parameters for the PID controller are

$$K_p = 0.6K_c = 0.6(111) = 66.6 \tag{h}$$

$$K_i = \frac{0.6K_c\omega_c}{\pi} = \frac{0.6(111)(3.53)}{\pi} = 74.8 \tag{i}$$

$$K_d = \frac{0.15K_c\pi}{\omega_c} = \frac{0.15(111)\pi}{3.53} = 14.82 \tag{j}$$

The Ziegler-Nichols design at the PID controller leads to

$$G_c(s) = 66.6 + \frac{74.8}{s} + 14.82s \tag{k}$$

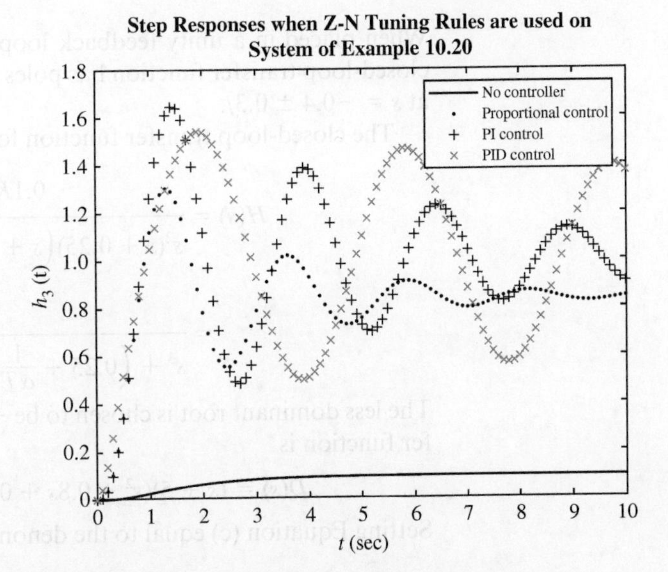

Figure 10.35

Step responses for the system of Example 10.20 with various controllers.

The closed-loop transfer function using this controller is

$$H(s) = \frac{118.6s^2 + 132.8s + 598.4}{12s^4 + 78s^3 + 268.6s^2 + 216.8s + 598.4} \tag{l}$$

The closed-loop response to a unit step input is illustrated in Figure 10.35. All controllers lead to a decrease in the rise time and increases in the settling time. The proportional controller has an error when the system is subject to a step input. The PI and PID controllers eliminate this error but result in more oscillation and a larger overshoot.

Example 10.21

Figure 10.36(a) shows a position control system with velocity feedback. Design a compensator such that the damping ratio is 0.8, the dominant roots have a natural frequency of 0.5 r/s, and the static velocity error constant is at least 0.25.

Solution

Method 1: The equivalent transfer function of velocity feedback, the system shown in Figure 10.36(b), is

$$G_{eq}(s) = \frac{0.1}{s(s + 0.25)} \tag{a}$$

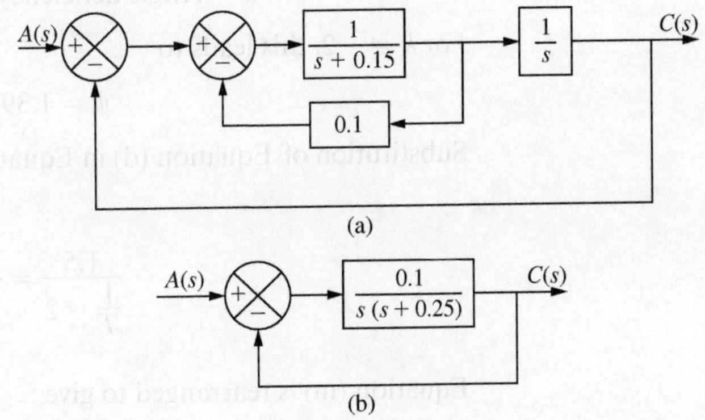

Figure 10.36

(a) Block diagram for position control system with velocity feedback; (b) Equivalent closed-loop transfer function.

When placed in a unity feedback loop with the plant of Equation (a), the resulting closed-loop transfer function has poles at $s = -0.125 \pm 0.2905j$. The desired poles are at $s = -0.4 \pm 0.3j$.

The closed-loop transfer function for this system with a lead compensator is

$$H(s) = \frac{0.1K_c\left(s + \frac{1}{T}\right)}{s^2(s + 0.25)\left(s + \frac{1}{\alpha T}\right) + 0.1K_c\left(s + \frac{1}{T}\right)}$$

$$= \frac{K_c\left(s + \frac{1}{T}\right)}{s^3 + \left(0.25 + \frac{1}{\alpha T}\right)s^2 + \left(0.1K_c + \frac{0.25}{\alpha T}\right)s + \frac{0.1K_c}{T}} \quad \text{(b)}$$

The less dominant root is chosen to be -5. The denominator of the closed-loop transfer function is

$$D(s) = (s + 5)(s^2 + 0.8s + 0.25) = s^3 + 5.8s^2 + 4.25s + 1.25 \quad \text{(c)}$$

Setting Equation (c) equal to the denominator of Equation (b) leads to

$$0.25 + \frac{1}{\alpha T} = 5.8 \quad \text{(d)}$$

$$0.1K_c + \frac{0.25}{\alpha T} = 4.25 \quad \text{(e)}$$

$$\frac{0.1K_c}{T} = 1.25 \quad \text{(f)}$$

Equations (d), (e), and (f) are solved, giving

$$K_c = 28.63 \qquad T = 2.29 \qquad \alpha = 0.078 \quad \text{(g)}$$

The lead compensator has a transfer function of

$$G_c(s) = 28.63\frac{s + 0.436}{s + 5.60} \quad \text{(h)}$$

It is noted that the static velocity error constant for the open-loop transfer function is

$$C_v = \lim_{s \to 0} 28.63\frac{s + 0.436}{s + 5.60}s\left[\frac{0.1}{s(s + 0.25)}\right] = 28.63\left(\frac{0.436}{5.60}\right)\left(\frac{0.1}{0.25}\right) = 0.89 \quad \text{(i)}$$

The closed-loop transfer function is

$$H(s) = \frac{2.863s + 1.25}{s^3 + 5.8s^2 + 4.25s + 1.25} \quad \text{(j)}$$

Method 2: The angle from the pole at $s = 0$ to the desired pole of $-0.4 + 0.3j$ is $\angle(-0.4 + 0.3j) = 2.50$. The angle from the pole at $s = -0.25$ to the desired pole is $\angle = 2.03$ r. In order for the pole to be on the root-locus diagram, the total angle must be of the form $(2k + 1)\pi$, so

$$\psi = \text{Angle deficiency} = (2k + 1)\pi + 2.50 + 2.03 \quad \text{(k)}$$

For $k = -2$, this leads to

$$\psi = 1.39 \Rightarrow \tan\psi = 5.47 \quad \text{(l)}$$

Substitution of Equation (d) in Equation (10.42) leads to

$$\frac{1.5}{\frac{1}{T} - 2} = \frac{5.47 + \frac{1.5}{\frac{1}{\alpha T} - 2}}{1 - 5.47\frac{1.5}{\frac{1}{\alpha T} - 2}} \quad \text{(m)}$$

Equation (m) is rearranged to give

$$0.25\alpha T^2 - (0.4548\alpha + 0.3452)T + 1 = 0 \quad \text{(n)}$$

Given the value of α, Equation (n) is a quadratic equation to solve for T. For $\alpha = 0.1$, Equation (f) becomes

$$0.025\,T^2 - 0.391T + 1 \tag{o}$$

The solutions of Equation (g) are $T = 3.22, 12.41$. The value of 3.22 is chosen, leading to a zero at $s = -0.311$. The pole is at $s = \frac{1}{\alpha T} = \frac{1}{(0.1)(3.22)} = 3.11$. With this solution, the compensator transfer function becomes

$$G_c(s) = K_c \frac{s + 0.311}{s + 3.11} \tag{p}$$

The gain is determined using the magnitude condition:

$$K_c \left| \frac{0.1(-0.3 + 0.4j + 0.311)}{(-0.3 + 0.4j)(-0.3 + 0.4j + 0.25)(-0.3 + 0.4j + 3.11)} \right| = 1 \Rightarrow K_c = 14.3 \tag{q}$$

The static velocity error constant is evaluated using Equation (10.46) as

$$C_v = \frac{K_c \alpha N(0)}{D(0)} = (14.3)(0.1)\left(\frac{0.1}{0.25}\right) = 0.572 \tag{r}$$

The static velocity error constant is not as high as with Method 1. The closed-loop transfer function for this compensator is

$$H(s) = \frac{1.43s + 0.445}{s^3 + 3.36s^2 + 2.21s + 0.445} \tag{s}$$

Method 3: The third method involves designing a compensator that cancels the pole at $s = -0.25$. The zero of the compensator is $\frac{1}{T} = 0.25$ or $T = 4$. The form of the transfer function for the compensator thus becomes

$$G_c(s) = K_c \frac{s + 0.25}{s + \frac{1}{\alpha T}} \tag{t}$$

The angle deficiency obtained using Equation (10.37) is evaluated in Equation (k) as 1.39 r. The pole of the compensator must make up an angle of

$$\begin{aligned} \psi &= (2k + 1)\pi + \angle D(s) - \angle N(s) \\ &= (2k + 1)\pi + \angle(-1.5 + 3j) + \angle(-1 + 3j) \\ &= (2k + 1)\pi + 1.071 + 1.8925 \end{aligned} \tag{u}$$

Choosing $k = -1$ leads to

$$\psi = -\pi + 1.071 + 1.8925 = -0.178 \text{ r} \tag{v}$$

The lead compensator is designed to cancel the pole at $s = -0.25$, and it has a transfer function of

$$K_c \frac{s + 0.25}{s + s_p} \tag{w}$$

The value of s_p is chosen to make up the angle deficiency of 2.50. Using Figure 10.37,

$$\tan 53^\circ = \frac{0.3}{-s_p - 0.4} \Rightarrow s_p = -0.626 \tag{x}$$

The compensator gain is calculated using the magnitude condition:

$$K_c \left| \frac{0.1}{(-0.4 + 0.3j)(-0.4 + 0.3j + 0.626)} \right| = 1 \Rightarrow K_c = 1.88 \tag{y}$$

The transfer function for the compensator, Equation (t), becomes

$$G_c(s) = 1.88 \frac{s + 0.25}{s + 0.626} \tag{z}$$

The static velocity error constant is determined from Equation (10.44) as

$$C_v = \frac{K_c \alpha N(0)}{D(0)} = (1.88)\left(\frac{0.25}{0.626}\right)\left(\frac{0.1}{0.25}\right) = 0.300 \tag{aa}$$

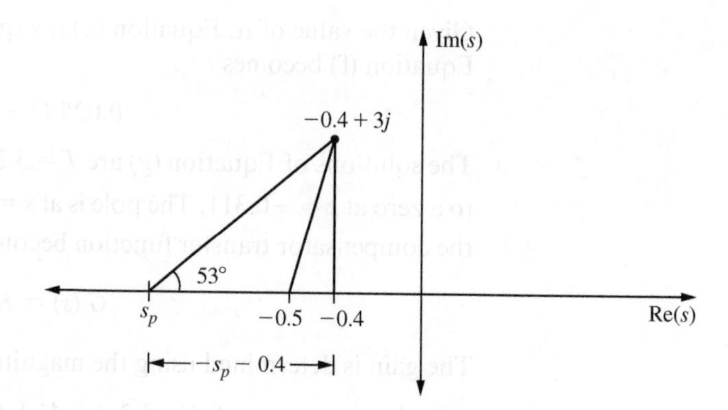

Figure 10.37

Geometry for placement of pole for Example 10.21.

The closed-loop transfer function using this compensator is

$$H(s) = \frac{0.188s + 0.047}{s^3 + 0.876s^2 + 0.353s + 0.047} \tag{ab}$$

The step responses of Equations (j), (s), and (ab) are illustrated in Figure 10.38. All compensators have a better response than simply placing the plant in a unity feedback loop. The compensator designed using Method 1, equating the denominator with the denominator for the desired poles, has the shortest rise time. However, the step response overshoots equilibrium and has a longer settling time than the one obtained using Method 2, the method of eliminating the angle deficiency. The step response of the transfer function obtained using Method 2 has the slowest settling time and does not overshoot equilibrium. Method 3 involves canceling a pole with the zero of the compensator. This method leads to the longest rise time and the longest settling time of the three methods. The step response does not overshoot equilibrium.

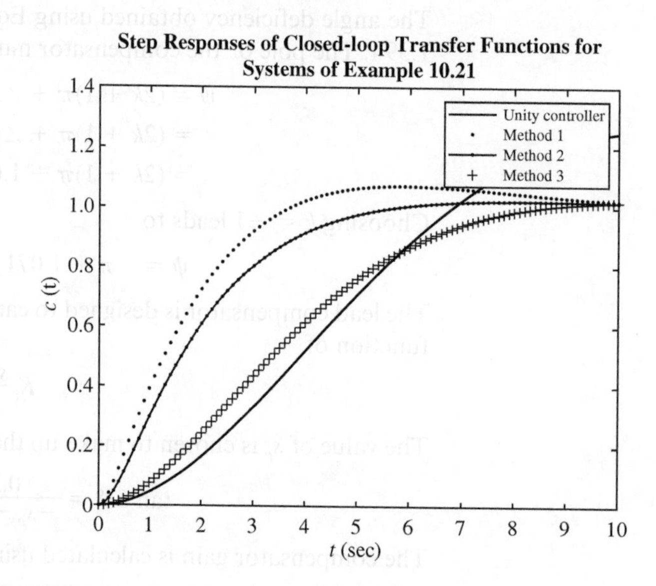

Figure 10.38

Comparison of step responses for the system of Example 10.21 with controllers designed using various methods.

Example 10.22

A plant has a transfer function of

$$G(s) = \frac{1}{s(s+4)} \tag{a}$$

When it is placed in a unity feedback loop, the closed-loop transfer function becomes

$$H(s) = \frac{1}{s^2 + 4s + 5} \tag{b}$$

Design a lag compensator such that the poles are $-2 \pm 2j$ and the static position error constant is at least 15.

Solution

The closed-loop transfer function with a proportional controller is

$$H(s) = \frac{K_p}{s^2 + 4s + 5 + K_p} \tag{c}$$

The poles are attainable if $K_p = 3$. Knowing that the poles are attainable, the lag compensator is assumed as

$$G_c(s) = K_c \frac{s + \frac{1}{T}}{s + \frac{1}{\alpha T}}, \quad \alpha > 1 \tag{d}$$

If $T = 20$ and $\alpha = 10$ are chosen,

$$G_c(s) = K_c \frac{s + 0.05}{s + 0.005} \tag{e}$$

The value of K_c is chosen by requiring that $C_v = 15$ or

$$15 = K_c(\alpha)G(0) = K_c(10)\left(\frac{1}{4}\right) = \frac{5}{2}K_c \Rightarrow K_c = 6 \tag{f}$$

The angle contributed by the compensator at the pole is

$$\angle(-2 + 2j + 0.05) - \angle(-2 + 2j + 0.005) = -0.0114 \text{ r} = -0.65° \tag{g}$$

The closed-loop transfer function with the compensator becomes

$$H(s) = \frac{15(s + 0.05)}{(s + 0.005)(s^2 + 4s + 4) + 15(s + 0.05)} = \frac{15s + 0.75}{s^3 + 4.005s^2 + 19.02s + 0.77} \tag{h}$$

The step response of the system is shown in Figure 10.39. The step response has minimal overshoot and a short rise time, but it has a long settling time.

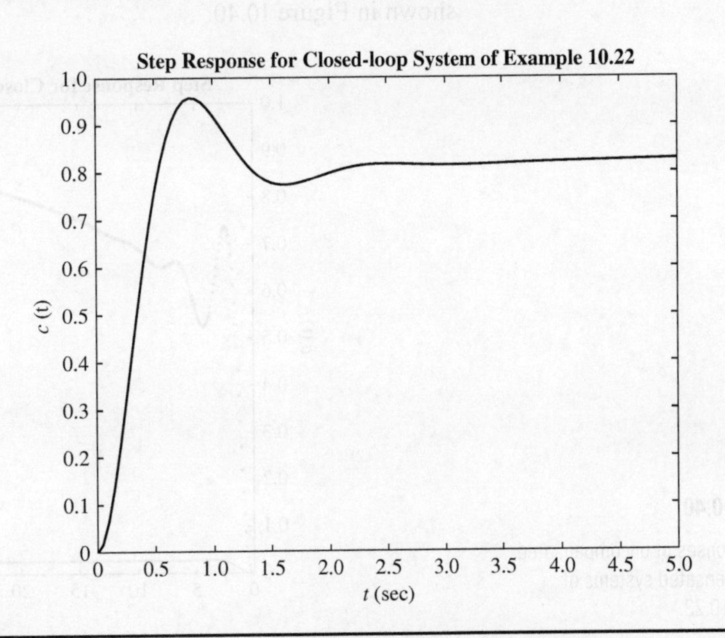

Figure 10.39

Step response of the system of Example 10.22 with a lag compensator.

Example 10.23

A plant has a transfer function of

$$G(s) = \frac{s+6}{s^5 + 17s^4 + 103s^3 + 277s^2 + 352s + 210} \tag{a}$$

Design a lag compensator such that when placed in series with the plant in a feedback loop, the static position error constant is 20.

Solution

The closed-loop transfer function when the plant is placed in a unity feedback loop is

$$H(s) = \frac{s+6}{s^5 + 17s^4 + 103s^3 + 277s^2 + 353s + 216} \tag{b}$$

The closed-loop poles are at

$$s = -6.99, -4.98, -3.07, -0.993 \pm 1.034j \tag{c}$$

The dominant poles are fine, but the static position error constant for the uncompensated system is

$$C_p = \frac{6}{216} = 0.028 \tag{d}$$

The zero for the compensator is chosen to be at $s = -0.1$ and the pole for the compensator at $s = -0.005$. That yields $\alpha = 20$. The transfer function for the compensator is

$$K_c \frac{s+0.1}{s+0.005} \tag{e}$$

The static position error of the open-loop system is

$$C_p = K_c \frac{0.1}{0.005} \frac{6}{210} = 0.571K_c \tag{f}$$

Setting $C_p = 20$ leads to $K_c = 35$.

The closed-loop transfer function is

$$\begin{aligned} H(s) &= \frac{35(s+0.1)(s+6)}{(s+0.005)(s^5 + 17s^4 + 103s^3 + 277s^2 + 352s + 210) + 35(s+0.1)} \\ &= \frac{35s^2 + 213.5s + 21}{s^6 + 17s^5 + 103.085s^4 + 277.515s^3 + 383.385s^2 + 425.2s + 22.05} \end{aligned} \tag{g}$$

The step responses of the uncompensated system and the compensated system are shown in Figure 10.40.

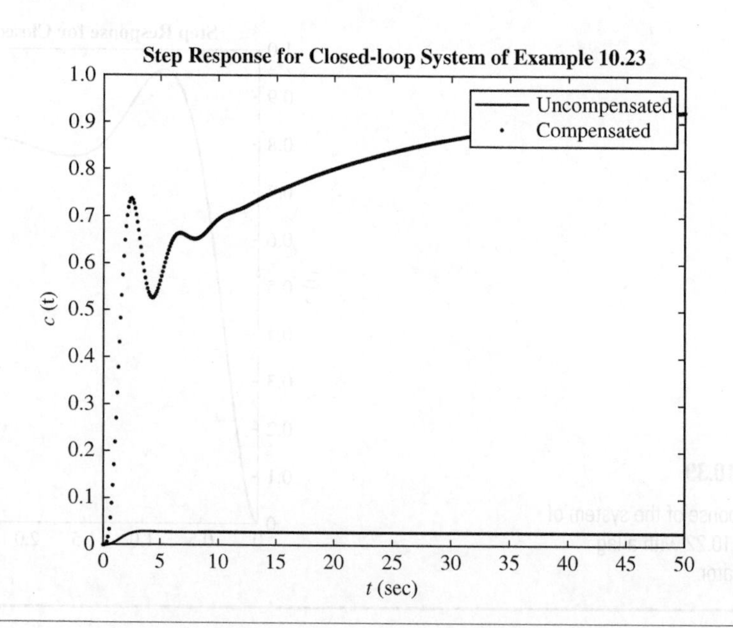

Figure 10.40

Step responses of uncompensated and compensated systems of Example 10.23.

10.14 Summary

10.14.1 Chapter Highlights

- A root-locus diagram is a plot of the poles of a transfer function in terms of a parameter. The root-locus diagram has one branch corresponding to each root. The number of branches is equal to the order of the polynomial.
- The poles are the roots of $Q(s)$ and the zeros are the roots of $R(s)$. Each branch begins at a pole and ends at either a zero or infinity.
- At every point on the root-locus diagram, the angle criterion and the magnitude criterion must be satisfied.
- The angle criterion is used to determine which parts of the real axis are on a branch of the root-locus diagram.
- The breakaway points are where a branch that starts at a real pole departs from the real axis. A break-in point is where a branch that starts at a complex pole joins the real axis.
- The crossover frequency is the frequency, if it exists, where the root-locus diagram crosses the imaginary axis.
- The asymptotes intersect at a point on the real axis.
- Pole placement is important when using the root-locus method.
- The proportional gain, the integral gain, or the differential gain may be used as the parameter in drawing the root-locus diagram for a plant in series with a PID controller.
- The Ziegler-Nichols tuning rules may be applied for a plant whose transfer function has a crossover frequency when it is placed in a feedback loop with a proportional controller.
- The angle deficiency is the angle that a lead controller needs to make up for a pole to be on the root-locus diagram.
- The gain of a lead compensator is determined using the magnitude criterion.
- One method of designing a lead compensator is through pole cancellation.
- A lag compensator is used for the pole placement when the poles achieved by a plant when placed in a unity feedback loop are adequate but the steady-state error needs improvement.
- The lag compensator does not contribute much to the angle at the pole.
- The gain of a lag compensator is determined so that an appropriate static error constant is achieved.
- A lead-lag compensator is used when the pole placement and the steady-state error both need to be improved.

10.14.2 Important Equations

- Definition of functions for drawing a root-locus diagram

$$Q(s) + KR(s) = 0 \tag{10.3}$$

- Magnitude criterion for a point to be on a branch of a root-locus diagram

$$K\left|\frac{R(s)}{Q(s)}\right| = 1 \tag{10.9}$$

- Angle criterion for a point to be on a branch of a root-locus diagram

$$\phi = \angle K\frac{R(s)}{Q(s)} = (2n + 1)\pi \qquad n = 0, \pm 1, \pm 2, \ldots \tag{10.10}$$

- Equation for determining breakaway or break-in points

$$R\frac{dQ}{ds} - Q\frac{dR}{ds} = 0 \tag{10.18}$$

- Angle of asymptotes for branches of a root-locus diagram

$$\phi_a = \frac{(2n+1)\pi}{n_z - n_p} \qquad n = 0, \pm 1, \pm 2, \ldots \tag{10.19}$$

- Point of intersection for asymptotes

$$s = \frac{\left[\left(\sum_{i=1}^{m} s_{pi}\right) - \left(\sum_{i=1}^{n} s_{zi}\right)\right]}{m - n} \tag{10.26}$$

- Closed-loop transfer function when a PID controller is in series with a plant

$$H(s) = \frac{K_d\left(s^2 + \frac{K_p}{K_d}s + \frac{K_i}{K_d}\right)N(s)}{sD(s) + K_d\left(s^2 + \frac{K_p}{K_d}s + \frac{K_i}{K_d}\right)N(s)} \tag{10.30}$$

- Ziegler-Nichols tuning rules for a PID controller

Table 10.4 Ziegler-Nichols Tuning Rules

Controller	K_p	K_i	K_d
P	$0.5K_c$	0	0
PI	$0.45K_c$	$\frac{0.27K_c\omega_c}{\pi}$	0
PID	$0.6K_c$	$\frac{0.6K_c\omega_c}{\pi}$	$\frac{0.15K_c\pi}{\omega_c}$

- Closed-loop transfer function for a system with a lead compensator

$$H(s) = \frac{K_c\left(s + \frac{1}{T}\right)N(s)}{D(s)\left(s + \frac{1}{\alpha T}\right) + K_c\left(s + \frac{1}{T}\right)N(s)} \tag{10.33}$$

- Angle deficiency for placing a desired pole when using a lead compensator

$$\psi = (2k+1)\pi + \angle D(s) - \angle N(s) = \angle s + \frac{1}{T} - \angle s + \frac{1}{\alpha T} \tag{10.37}$$

- Static velocity error constant when using a lead compensator for a type 1 system

$$C_v = \frac{K_c \alpha N(0)}{D(0)} \tag{10.46}$$

- Closed-loop transfer function for a system with a lead-lag compensator

$$H(s) = \frac{K_c\left(s + \frac{1}{T_1}\right)\left(s + \frac{1}{T_2}\right)N(s)}{\left(s + \frac{1}{\alpha_1 T_1}\right)\left(s + \frac{1}{\alpha_2 T_2}\right)D(s) + K_c\left(s + \frac{1}{T_1}\right)\left(s + \frac{1}{T_2}\right)N(s)} \tag{10.48}$$

Short Answer Problems

Problems SA10.1–SA10.15 refer to a plant with the transfer function

$$G(s) = \frac{s+2}{s(s+1)(s+5)}$$

The plant is placed in a feedback loop with a proportional controller. The root-locus diagram is drawn using the controller gain as the constant.

SA10.1 What are the functions $Q(s)$ and $R(s)$, as defined in Equation (10.3)?

SA10.2 How many branches does the root-locus diagram have?

SA10.3 Where do the branches begin?

SA10.4 Where do the branches end?

SA10.5 What are the points on the root-locus diagram that correspond to a gain of zero?

SA10.6 Determine whether $s = 3$ is on a branch of the root-locus diagram.

SA10.7 Determine whether $s = -2.5$ is on a branch of the root-locus diagram.

SA10.8 Determine whether $s = -4$ is on a branch of the root-locus diagram.

SA10.9 Determine whether $s = -7$ is on a branch of the root-locus diagram.

SA10.10 Verify that $s = -0.5$ is on a branch of the root-locus diagram and determine the value of the gain at this point.

SA10.11 Determine the breakaway points, if any.

SA10.12 Determine the break-in points, if any.

SA10.13 Determine the angle that each of the asymptotes makes with the horizontal axis.

SA10.14 Determine the point of intersection of the asymptotes on the real axis.

SA10.15 Use MATLAB to draw the root-locus diagram.

Problems SA10.16–SA10.28 refer to a plant with the transfer function

$$G(s) = \frac{s+5}{s(s^2+6s+25)}$$

The plant is placed in a feedback loop with a proportional controller. The root-locus diagram is drawn using the controller gain as the constant.

SA10.16 What are the functions $Q(s)$ and $R(s)$, as defined in Equation (10.3)?

SA10.17 How many branches does the root-locus diagram have?

SA10.18 Where do the branches begin?

SA10.19 Where do the branches end?

SA10.20 Verify that $s = -4$ is on a branch of the root-locus diagram and determine the value of the gain at this point.

SA10.21 Determine whether $s = -1$ is on a branch of the root-locus diagram.

SA10.22 Determine whether $s = -10$ is on a branch of the root-locus diagram.

SA10.23 What is the phase angle between the poles of $s^2 + 6s + 25$ and a real positive value?

SA10.24 Determine the breakaway points, if any.

SA10.25 Determine the break-in points, if any.

SA10.26 Determine the angle that each of the asymptotes makes with the horizontal axis.

SA10.27 Determine the point of intersection of the asymptotes on the real axis.

SA10.28 Use MATLAB to draw the root-locus diagram.

Problems SA10.29–SA10.42 refer to a plant with the transfer function

$$G(s) = \frac{1}{s(s+1)(s+3)}$$

The plant is placed in a feedback loop with a proportional controller. The root-locus diagram is drawn using the controller gain as the constant.

SA10.29 What are the functions $Q(s)$ and $R(s)$, as defined in Equation (10.3)?

SA10.30 How many branches does the root-locus diagram have?

SA10.31 Where do the branches begin?

SA10.32 Where do the branches end?

SA10.33 Verify that $s = -8$ is on a branch of the root-locus diagram and determine the value of the gain at this point.

SA10.34 Determine whether $s = -2$ is on a branch of the root-locus diagram.

SA10.35 The point $s = -0.202 + 1.23j$ is on a branch of the root-locus diagram. What is the gain at this point?

SA10.36 Determine the breakaway points, if any.

SA10.37 Determine the break-in points, if any.

SA10.38 Determine the angle that each of the asymptotes makes with the horizontal axis.

SA10.39 Determine the point of intersection of the asymptotes on the real axis.

SA10.40 Determine the crossover frequencies.

SA10.41 Determine the values of the gain for which branches of the root-locus diagram cross the imaginary axis.

SA10.42 Use MATLAB to draw the root-locus diagram.

Problems SA10.43–SA10.50 refer to the plant with the transfer function

$$G(s) = \frac{s+2}{s(s^2+10s+61)}$$

The plant is to be placed in a feedback control loop in series with the following controllers. Determine the forms of $Q(s)$ and $R(s)$ in each case.

SA10.43 Proportional control is used with the proportional gain as the root-locus parameter.

SA10.44 Integral control is used with the integral gain as the root-locus parameter.

SA10.45 PI control is used with (a) K_p used as the root-locus parameter and (b) K_i used as the root-locus parameter.

SA10.46 PD control is used with (a) K_p used as the root-locus parameter and (b) K_d used as the root-locus parameter.

SA10.47 PID control is used with (a) K_p used as the root-locus parameter, (b) K_d used as the root-locus parameter, and (c) K_i used as the root-locus parameter.

SA10.48 A lead compensator is used with $T = 0.5$ and $\alpha = 0.2$. Use the controller gain as the root-locus parameter.

SA10.49 A lag compensator is used with $T = 100$ and $\alpha = 2$. Use the controller gain as the root-locus parameter.

SA10.50 A lead-lag compensator is used with $T_1 = 0.5$, $T_2 = 100$, $\alpha_1 = 0.2$, and $\alpha_2 = 5$. Use the controller gain as the root-locus parameter.

SA10.51 A plant is placed in series with a proportional controller in a feedback loop. The crossover frequency for the root-locus diagram is 2.6 r/s at a gain of 3.4. What are the gains, according to the Ziegler-Nichols tuning rules, when a PI controller is used?

SA10.52 A plant is placed in series with a proportional controller in a feedback loop. The crossover frequency for the root-locus diagram is 1.3 r/s at a gain of 4.4. What are the gains, according to the Ziegler-Nichols tuning rules, when a PID controller is used?

SA10.53 A lead compensator is designed such that $T = 1.2$, $\alpha = 0.1$, and $K_c = 3.1$. What is $G_c(s)$, the transfer function for the compensator?

SA10.54 A lag compensator is designed such that $T = 100$, $\alpha = 30$, and $K_c = 30$. What is $G_c(s)$, the transfer function for the compensator?

SA10.55 A lead-lag compensator is designed such that $\alpha_1 = \frac{1}{\alpha_2} = 0.15$, $T_1 = 4$, $T_2 = 50$, and $K_c = 18$. What is $G_c(s)$, the transfer function for the compensator?

SA10.56 A lead compensator is to be designed for the plant

$$G(s) = \frac{s+7}{s(s+0.5)(s^2+10s+89)}$$

such that the pole at $s = -0.5$ is canceled by the compensator. If the compensator is designed with $\alpha = 0.1$ and $K_c = 2.4$, what is the transfer function of the compensator?

Problems SA10.57–SA10.61 refer to Figure SA10.57, which is used in the design of a lead compensator for the plant with the transfer function

$$G(s) = \frac{4}{s(s+1)(s+4)}$$

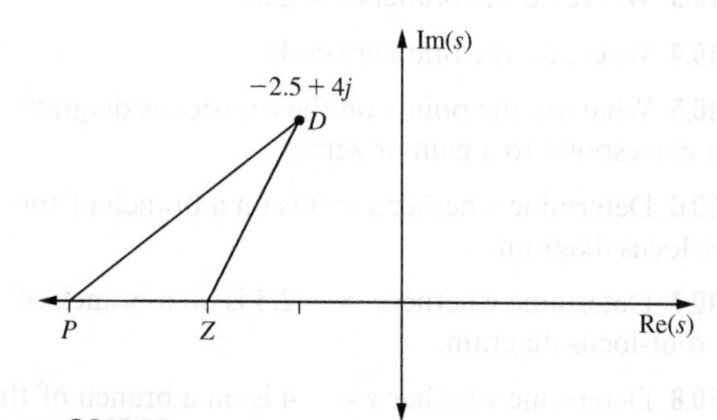

Figure SA10.57

SA10.57 What is the angle deficiency?

SA10.58 What is the angle contributed by the pole of the compensator for Figure SA10.57?

SA10.59 If the pole of the compensator is located at $s = -10$, where is the zero?

SA10.60 If the zero of the compensator is located at $s = -2$, what is the value of α?

SA10.61 If the desired pole is $s = -2.5 + 4j$, where is the zero?

Problems SA10.62–SA10.65 refer to a compensator that is being designed for a plant with the transfer function

$$G(s) = \frac{s+3}{s(s+0.4)(s+2)}$$

The desired dominant pole of the compensated system is at $s = -2 + 3.3j$.

SA10.62 What is the angle deficiency that the compensator must make up?

SA10.63 If a compensator is designed such that it has a zero at -1.7, where is the pole of the compensator?

SA10.64 What is the required gain of the compensator specified in Problem SA10.63?

SA10.65 What is the static velocity error constant of the compensator specified in Problem SA10.63?

SA10.66 An uncompensated system has a static velocity error constant of 0.6. The gain using a lead compensator with $T = 0.5$ and $\alpha = 0.2$ is determined to be 1.2. What is the static velocity error constant of the compensated system?

SA10.67 An uncompensated system has a static velocity error constant of 0.6. The gain using a lag compensator with $T = 0.05$ and $\alpha = 25$ is determined to be 17.2. What is the static velocity error constant of the compensated system?

SA10.68 An uncompensated system has a static velocity error constant of 0.6. The gain using a lead-lag compensator with $\alpha_1 = 0.2$ and $\alpha_2 = 20$ is determined to be 31.2. What is the static velocity error constant of the compensated system?

Problems SA10.69–SA10.71 refer to a lead-lag compensator that has the transfer function

$$G_c(s) = 10.2\left(\frac{s+1}{s+6}\right)\left(\frac{s+0.05}{s+0.005}\right)$$

The desired dominant pole of the system is $-2 + 3.5j$.

SA10.69 What is the angle contributed by the lag part of the compensator at the desired pole?

SA10.70 What is the magnitude of the transfer function of the plant at the desired pole, that is, $|G(-2 + 3.5j)|$?

SA10.71 If $\lim_{s \to 0} sG(s) = 2$, where $G(s)$ is the transfer function of the plant, what is the static velocity error constant of the compensated system?

Problems

For Problems 10.1–10.10, a transfer function is given. Perform the following steps to develop the root-locus diagram:

(a) Determine the functions $Q(s)$ and $R(s)$ as defined in Equation (10.3).

(b) Identify the poles and zeros of the transform.

(c) Determine which portions of the real axis are on a branch of the root-locus diagram.

(d) Determine the breakaway and break-in points.

(e) Determine the angles of departure and arrival.

(f) Determine the crossover frequency, if any. Determine the value of the imaginary part of the root-locus diagram at the crossover frequency.

(g) Determine how many asymptotes there are for the root-locus diagram.

(h) Determine the angles the asymptotes make with the real axis.

(i) Determine the values of the poles for $K = 1, 2, 5, 10, 15, 25$, and 50.

(j) Sketch the root-locus diagram.

(k) Use MATLAB to draw the root-locus diagram.

10.1 $H(s) = \dfrac{4s+2}{s^2 + 2s + 5 + K(4s+2)}$

10.2 $H(s) = \dfrac{s+5}{s^2 + 6s + 45 + K(4s+2)}$

10.3 $H(s) = \dfrac{s^2 + 6s + 5}{s(s^2 + 8s + 20) + K(s^2 + 6s + 5)}$

10.4 $H(s) = \dfrac{3}{s(s^2 + 16)(s+5) + 3K}$

10.5 $H(s) = \dfrac{s-3}{s^2 + 8s + K(s-3)}$

10.6 The closed-loop transfer function for Figure P10.6

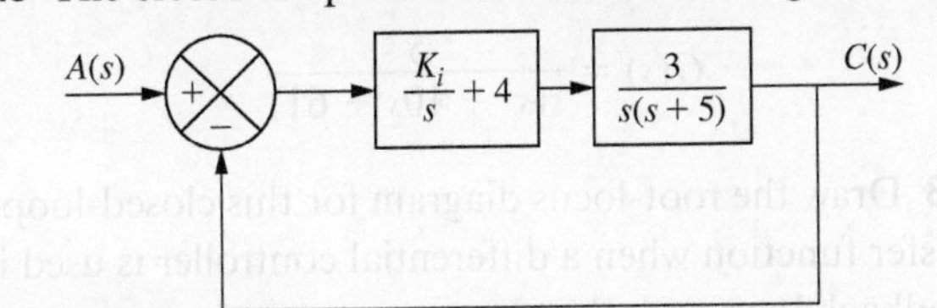

Figure P10.6

10.7 The closed-loop transfer function for Figure P10.7

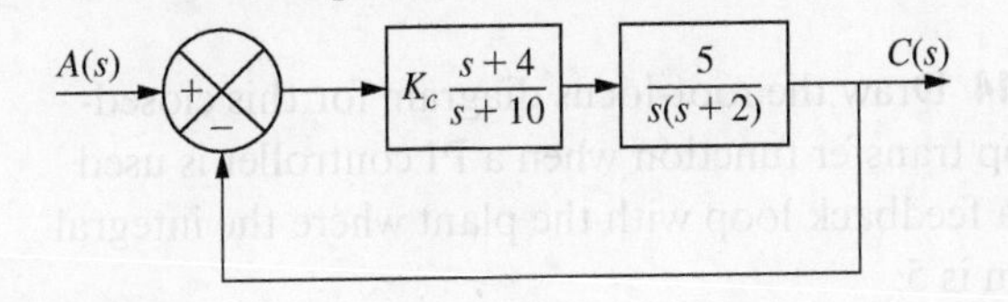

Figure P10.7

10.8 The closed-loop transfer function for Figure P10.8

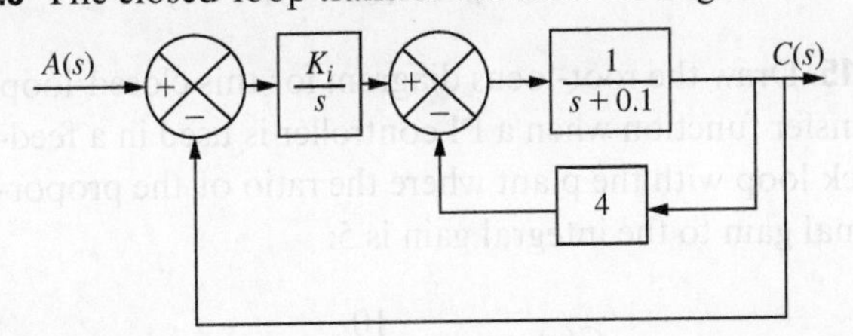

Figure P10.8

10.9 The closed-loop transfer function for Figure P10.9

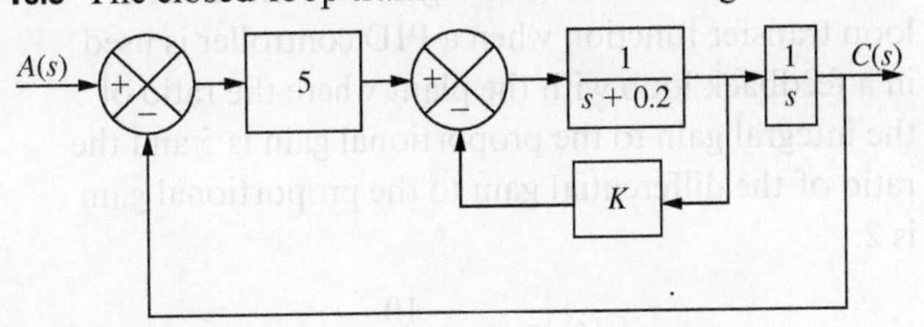

Figure P10.9

10.10 The closed-loop transfer function for Figure P10.10

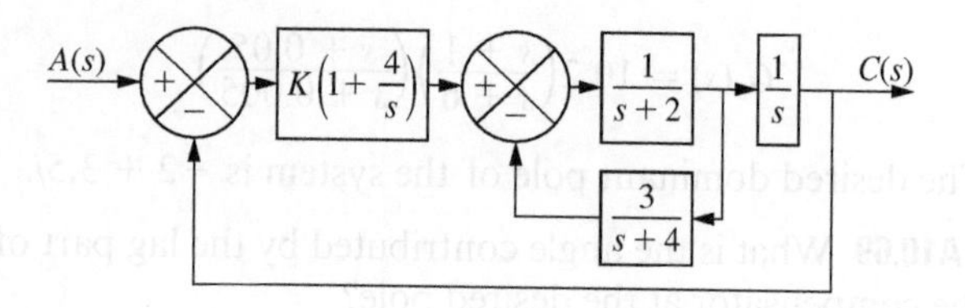

Figure P10.10

10.11 Draw the root-locus diagram for this closed-loop transfer function when an integral controller is used in a feedback loop with the plant:

$$G(s) = \frac{10}{s^2 + 10s + 61}$$

10.12 Draw the root-locus diagram for this closed-loop transfer function when a proportional controller is used in a feedback loop with the plant:

$$G(s) = \frac{5}{s(s^2 + 10s + 61)}$$

10.13 Draw the root-locus diagram for this closed-loop transfer function when a differential controller is used in a feedback loop with the plant:

$$G(s) = \frac{5}{s^2 + 10s + 61}$$

10.14 Draw the root-locus diagram for this closed-loop transfer function when a PI controller is used in a feedback loop with the plant where the integral gain is 5:

$$G(s) = \frac{10}{s^2 + 10s + 61}$$

10.15 Draw the root-locus diagram for this closed-loop transfer function when a PI controller is used in a feedback loop with the plant where the ratio of the proportional gain to the integral gain is 5:

$$G(s) = \frac{10}{s^2 + 10s + 61}$$

10.16 Draw the root-locus diagram for this closed-loop transfer function when a PID controller is used in a feedback loop with the plant where the ratio of the integral gain to the proportional gain is 5 and the ratio of the differential gain to the proportional gain is 2:

$$G(s) = \frac{10}{s^2 + 10s + 61}$$

10.17 Consider the plant in the feedback diagram of Figure P10.17. Determine the value of a such that the system has a damping ratio of 0.8.

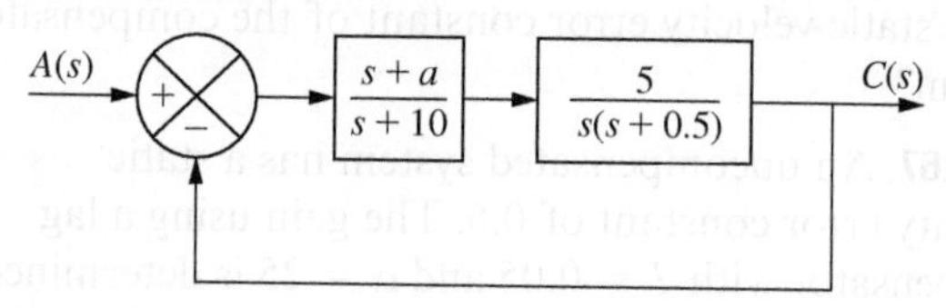

Figure P10.17

10.18 Consider the plant in the feedback diagram of Figure P10.18. Determine the value of a such that the system has a damping ratio of 0.8.

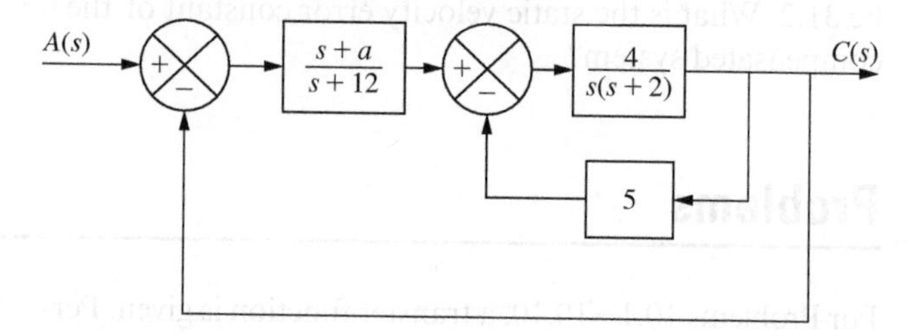

Figure P10.18

10.19 Consider the plant in the feedback diagram of Figure P10.19. Determine the value of a such that the system has a damping ratio of 0.5.

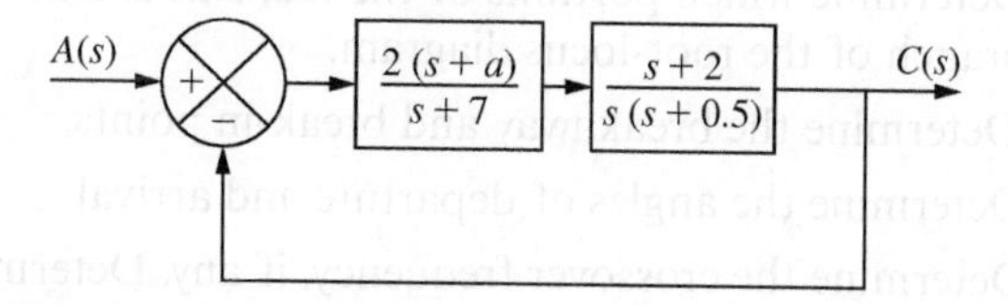

Figure P10.19

10.20 Use a root-locus diagram to determine the values of K such that the system of Figure P10.20 is stable.

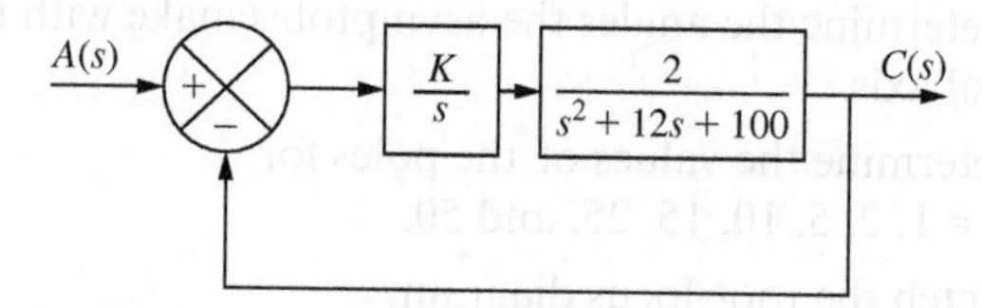

Figure P10.20

10.21 Plot the root-locus diagram for the system of Figure P10.21.

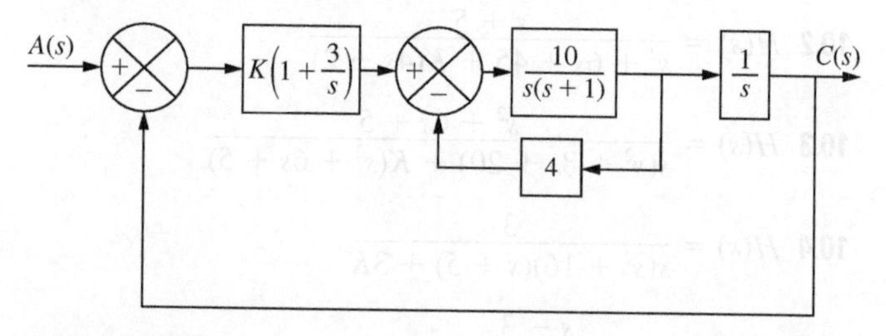

Figure P10.21

10.22 Use the root-locus diagram to design a proportional controller for the system of Problem 10.2 such that the roots have a damping ratio of 0.707.

10.23 Use the root-locus diagram to design a proportional controller for the system of Problem 10.3 such that the dominant root has a real part of -6.

10.24 Design a PID controller for the system of Figure P10.24 such that the dominant poles have a damping ratio of 0.8 and a natural frequency of 1 r/s.

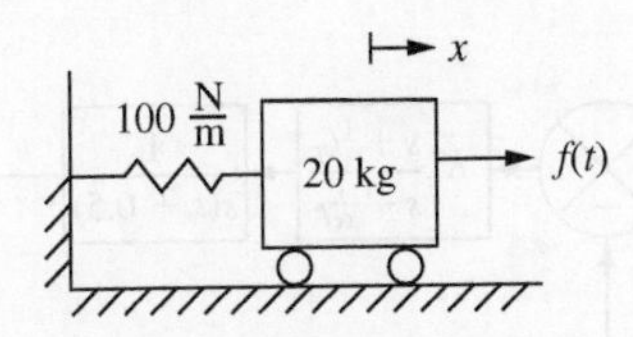

Figure P10.24

10.25 Use the Ziegler-Nichols tuning rules to design a PID controller for the system of Figure P10.25. The area of each tank is $2.7\ \text{m}^2$, $R_1 = 3\frac{\text{s}}{\text{m}^2}$, $R_2 = 1.5\frac{\text{s}}{\text{m}^2}$, and $R_3 = 1\frac{\text{s}}{\text{m}^2}$.

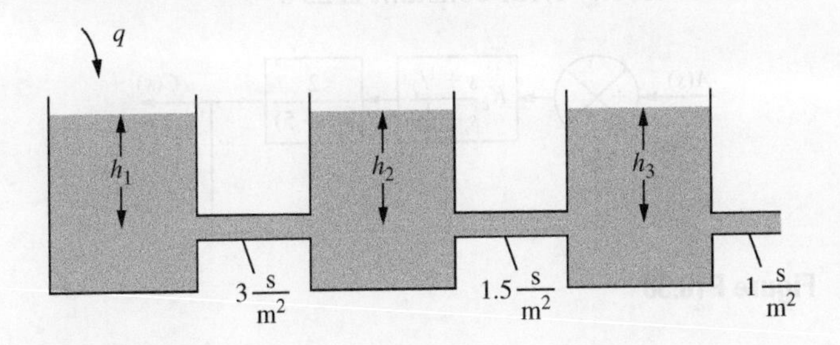

Figure P10.25

10.26 Use the Ziegler-Nichols tuning rules to design a PID controller for the model of a vehicle and simplified suspension system shown in Figure P10.26.

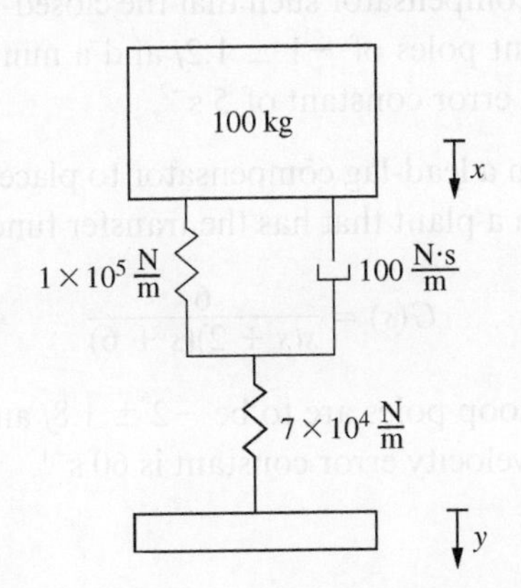

Figure P10.26

10.27 Use the Ziegler-Nichols tuning rules to design (a) a proportional controller, (b) a PI controller, and (c) a PID controller for the mechanical system of Figure P10.27, where $m_1 = 2$ kg, $m_2 = 5$ kg, $k = 10{,}000\frac{\text{N}}{\text{m}}$, and $L = 40$ cm.

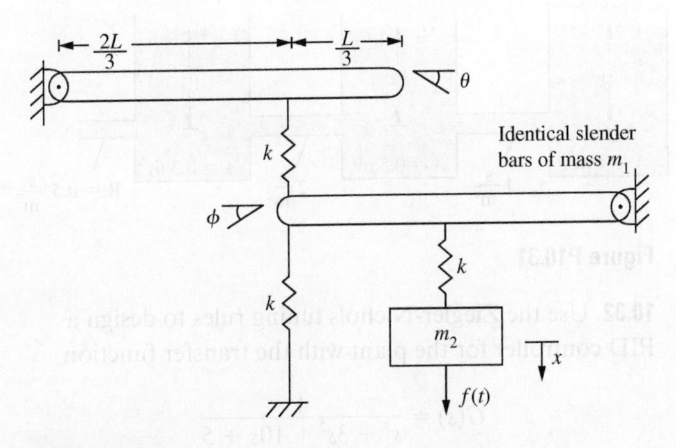

Figure P10.27

10.28 An armature-controlled dc servomotor has the following transfer function for the angular velocity of the shaft:

$$G(s) = \frac{0.5}{(s^2 + 2s + 2)(s + 4)}$$

The motor and the shaft are to be placed in a feedback control loop with a controller. Use the Ziegler-Nichols tuning rules to design (a) a proportional controller, (b) a PI controller, and (c) a PID controller.

10.29 Two consecutive reactions occur in a CSTR. The transfer function for the concentration of the product of the second reaction is

$$G(s) = \frac{0.25}{(s + 0.1)(s + 0.2)(s + 0.3)}$$

Use the Ziegler-Nichols tuning rules to design (a) a proportional controller, (b) a PI controller, and (c) a PID controller to use with this tank.

10.30 Design a PID controller for the system of Figure P10.30 such that the dominant poles have a damping ratio of 0.8 and a natural frequency of 1 r/s. Use the brute-force method.

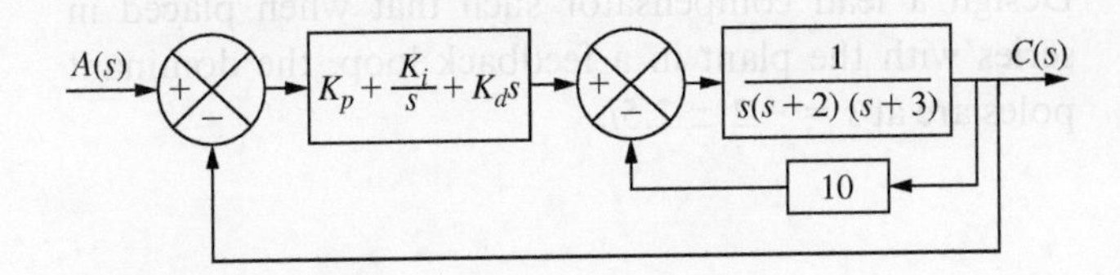

Figure P10.30

10.31 Use the root-locus method to determine the value of the gain when a proportional controller is used on the three-tank liquid-level problem of Figure P10.31 to set the dominant roots to $s = -2 \pm 2.2j$.

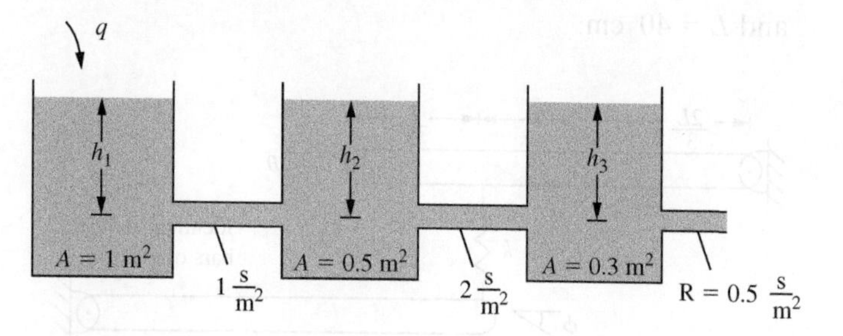

Figure P10.31

10.32 Use the Ziegler-Nichols tuning rules to design a PID controller for the plant with the transfer function

$$G(s) = \frac{4}{s^3 + 3s^2 + 10s + 5}$$

10.33 A velocity control system is illustrated in Figure P10.33. Design a proportional controller using the root-locus method such that the dominant roots are at $s = -2 \pm 3.5j$.

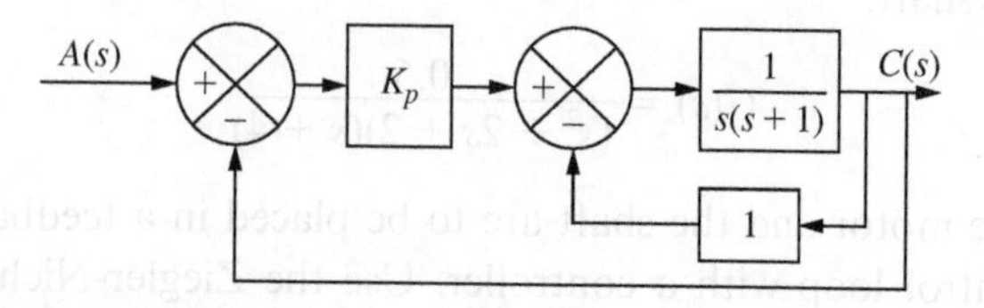

Figure P10.33

10.34 A position control system is shown in block diagram form in Figure P10.34. Design a proportional controller such that its dominant roots are at $s = -2 \pm 3j$.

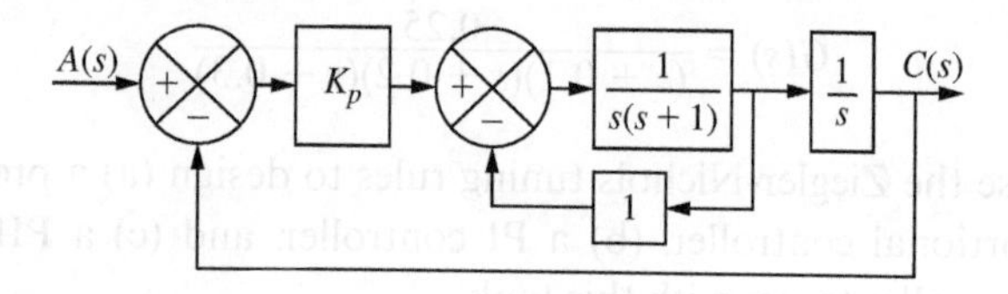

Figure P10.34

10.35 A plant has the transfer function

$$G(s) = \frac{2}{s(s+1)}$$

Design a lead compensator such that when placed in series with the plant in a feedback loop, the dominant poles are at $s = -2 \pm 2.5j$.

10.36 A plant has the transfer function

$$G(s) = \frac{2}{s(s+1)}$$

Design a lead compensator such that, when placed in series with the plant in a feedback loop, the dominant poles are at $s = -1.5 + 2j$. Design the compensator such that it cancels the pole at $s = -1$.

10.37 Consider the system of Figure P10.37. Design a lead compensator such that the natural frequency and damping ratio of the dominant poles are 1 r/s and $\frac{\sqrt{2}}{2}$, respectively.

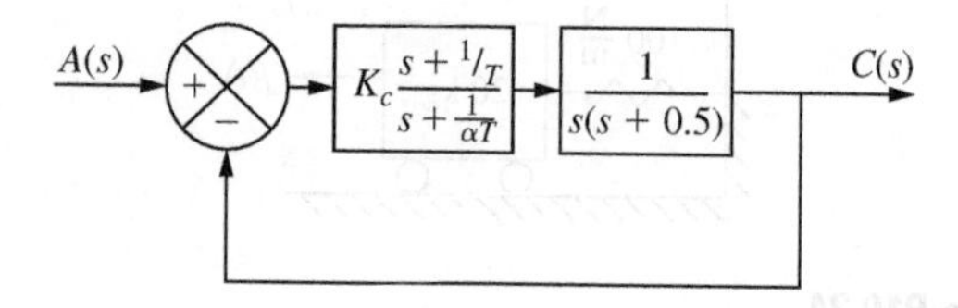

Figure P10.37

10.38 Consider the system of Figure P10.38. The closed-loop poles when the plant is placed in a unity feedback loop are acceptable. Design a lag compensator such that the static velocity error constant is 25 s^{-1}.

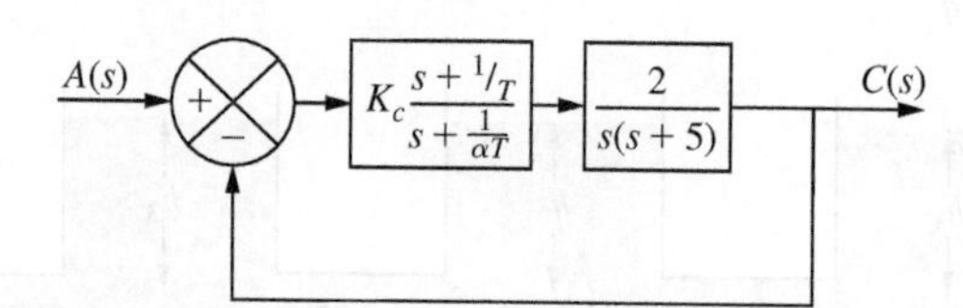

Figure P10.38

10.39 A plant with the transfer function

$$G(s) = \frac{10}{s^2(s+2)}$$

is to be placed in a feedback loop with a compensator. Design the compensator such that the closed-loop system has dominant poles of $-1 \pm 1.2j$ and a minimum static acceleration error constant of 5 s^{-2}.

10.40 Design a lead-lag compensator to place in a feedback loop in a plant that has the transfer function

$$G(s) = \frac{6}{s(s+2)(s+6)}$$

The closed-loop poles are to be $-2 \pm 1.8j$ and the minimum static velocity error constant is 60 s^{-1}.

11

Steady-State Analysis of Control Systems

Learning Objectives

At the conclusion of this chapter, you should be able to:

- Use the Nyquist diagram and the Nyquist stability criterion
- Use the definitions of gain crossover frequency, phase crossover frequency, gain margin, and phase margin in compensator design
- Apply the knowledge that the steady-state behavior of the open-loop transfer function provides information about the steady-state behavior of the closed-loop system to compensator design
- Design a PID controller using the frequency response method
- Design a lead compensator using the frequency response method
- Design a lag compensator using the frequency response method
- Design a lead-lag compensator using the frequency response method

11.1 Introduction

The steady-state behavior of a dynamic system was studied in Chapter 8. The sinusoidal transfer function is the transfer function evaluated at $s = j\omega$. The resulting analysis is said to be performed in the frequency domain. The sinusoidal transfer function $G(j\omega)$ is a complex function of a real variable. It has a real part and an imaginary part. The sinusoidal transfer function can be written in polar form as

$$G(j\omega) = |G(j\omega)|\,e^{j\phi} \tag{11.1}$$

where $|G(j\omega)|$ is the magnitude of the sinusoidal transfer function and ϕ is its phase.

Several graphical tools were developed to aid in the analysis of a system's frequency response. The Nyquist diagram is a polar plot of the real part of the sinusoidal transfer function on the horizontal axis and the imaginary part of the sinusoidal transfer function on the vertical axis. The Bode diagram has two parts. Each part is a semi-log plot with ω on the horizontal axis. The magnitude part of the Bode diagram plots $L(\omega) = 20\log|G(j\omega)|$ on the vertical axis. The phase portion plots ϕ on the vertical axis.

Figure 11.1 shows a plant in series with a controller that is placed in a unity feedback loop. The open-loop transfer function is

$$G_o(s) = G_c(s)G(s) \tag{11.2}$$

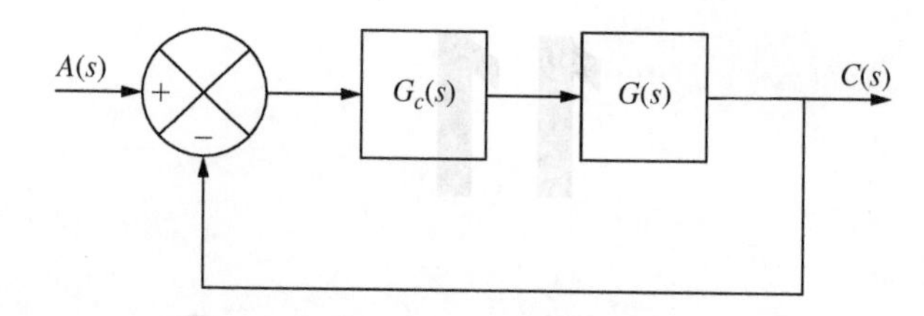

Figure 11.1
A plant in series with a controller in a unity feedback loop.

The closed-loop transfer function is

$$H(s) = \frac{G_c(s)G(s)}{1 + G_c(s)G(s)} \tag{11.3}$$

This chapter describes how the frequency domain can be used to aid in the design of controllers and compensators. The Nyquist diagram is used as a test for stability. Properties of the Bode diagram drawn for the open-loop transfer functions are used to design controllers and compensators.

11.2 The Nyquist Stability Criterion

The Nyquist stability criterion is complex. Its derivation involves complex analysis, specifically conformal mapping, and is beyond the scope of this text. Consider a plant $G(s)$ in a unity feedback loop. The open-loop transfer function is $G_o(s)$. Its closed-loop transfer function is $H(s) = \frac{G_o(s)}{1 + G(s)}$. It is assumed that $G_o(s)$ is the ratio of two polynomials and that the order of the polynomial in the denominator is at least as large as the order of the polynomial in the numerator. Note that $H(s) = \frac{G_o(s)}{Z(s)}$, where $Z(s) = 1 + G_o(s)$. Recall that the Nyquist diagram is parameterized by ω. A curve is drawn in the real and imaginary plane as ω varies from $-\infty$ to ∞. The Nyquist stability criterion, presented without proof, states:

> If the open-loop transfer function $G_o(s)$ has k poles with positive real parts, then the point $(-1, 0)$ must be encircled k times in the counterclockwise direction by the Nyquist curve in order for the closed-loop system to be stable.

The point $(-1, 0)$ refers to $-1 + j0$. If $G_o(s)$ is the transfer function of a stable system, then the Nyquist stability criterion implies that the point $(-1, 0)$ must not be encircled on the Nyquist diagram for that system in order for the closed-loop system to be stable.

Example 11.1

Determine the closed-loop stability of a plant with the transfer function

$$G(s) = \frac{1}{s^3 + 3s^2 + 2s + 1} \tag{a}$$

if it is placed in a feedback loop with a proportional controller that has the transfer function

$$G_c(s) = K_p \tag{b}$$

Determine the stability of the closed-loop system for **a.** $K_p = 2$ and **b.** $K_p = 7$.

Solution

The stability of the closed-loop system is determined by examining the transfer function of the open-loop system. The equivalent transfer function of the open-loop system is

$$G_o(s) = G_c(s)G(s) = \frac{K_p}{s^3 + 3s^2 + 2s + 1} \tag{c}$$

placed in a unity feedback loop

a. The Nyquist diagram in Figure 11.2(a) for $G_o(s)$ when $K_p = 2$ shows that the point $(-1, 0)$ is not encircled by the Nyquist curve. The closed-loop system is stable.

b. The Nyquist diagram in Figure 11.2(b) for $G_o(s)$ when $K_p = 7$ shows that the point $(-1, 0)$ is encircled by the Nyquist curve. The closed-loop system is not stable. Note that the Nyquist diagrams drawn by MATLAB identify the point $(-1, 0)$ with a + if it is on the diagram.

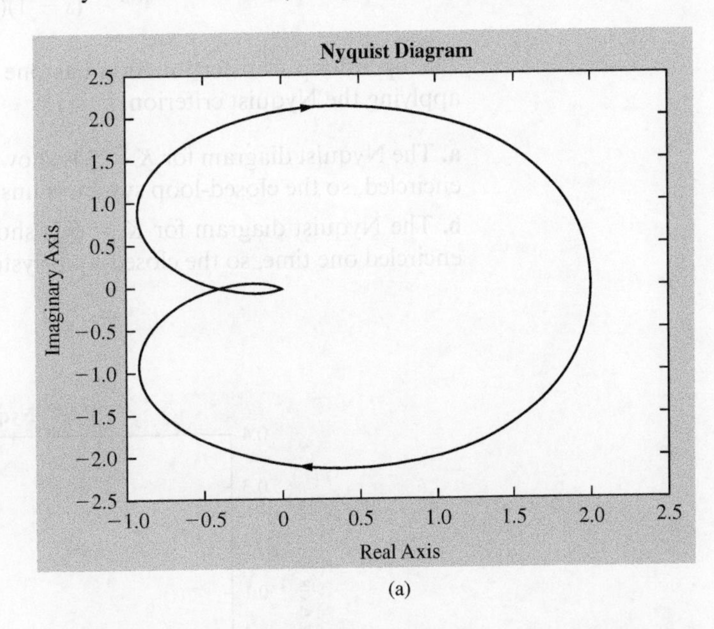

(a)

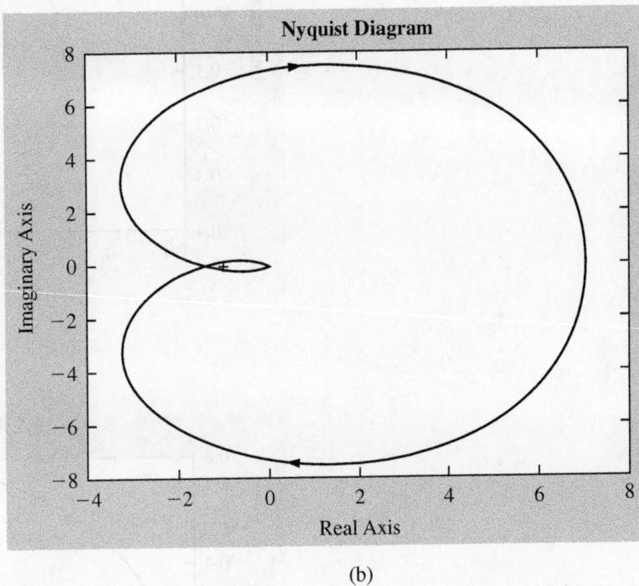

(b)

Figure 11.2

Nyquist diagrams of the plant of Example 11.1 in series with a proportional controller. (a) For $K_p = 2$, the point $(-1, 0)$ is not encircled by the Nyquist diagram and the system, when placed in a unity feedback loop, is stable. (b) For $K_p = 7$, the point $(-1, 0)$ is encircled by the Nyquist diagram and the system, when placed in a unity feedback loop, is unstable.

Example 11.2

A plant with the transfer function

$$G(s) = \frac{4}{(s + 1)(s - 1)} \tag{a}$$

is placed in a feedback loop with a lead compensator that has the transfer function

$$G_c(s) = K_c \frac{s + 2}{s + 10} \tag{b}$$

Determine the stability of the closed-loop system when **a.** $K_c = 2$ and **b.** $K_c = 5$.

Solution

The equivalent transfer function when the compensator is placed in series with the plant in a unity feedback loop is

$$G_o(s) = \frac{4K_c(s+2)}{(s-1)(s+3)(s+10)} \tag{c}$$

The open-loop transfer function has one pole with a positive real part, $s = 1$, so in applying the Nyquist criterion, $k = 1$.

a. The Nyquist diagram for $K_c = 2$ is shown in Figure 11.3(a). The point $(-1, 0)$ is not encircled, so the closed-loop system is unstable.

b. The Nyquist diagram for $K_c = 5$ is shown in Figure 11.3(b). The point $(-1, 0)$ is encircled one time, so the closed-loop system is stable.

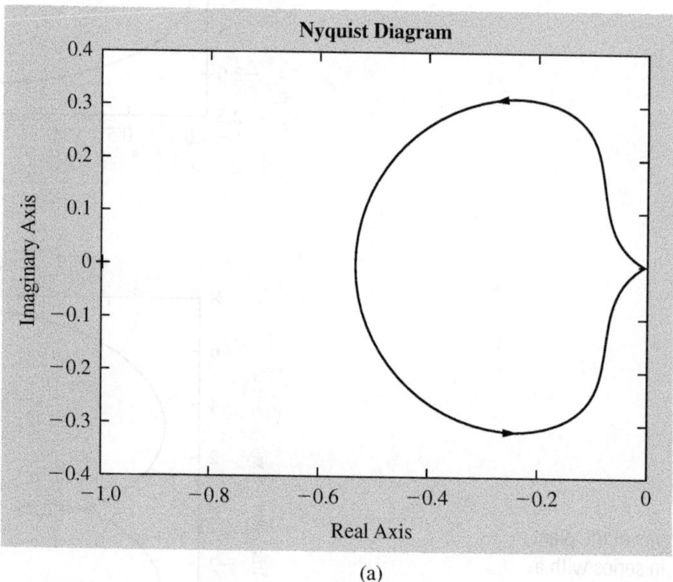

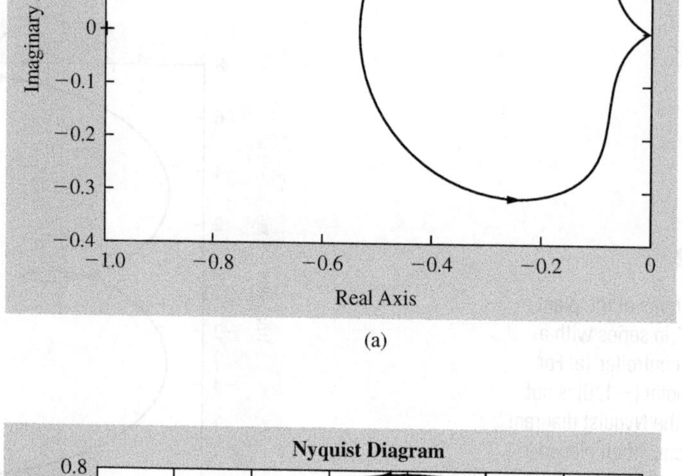

(a)

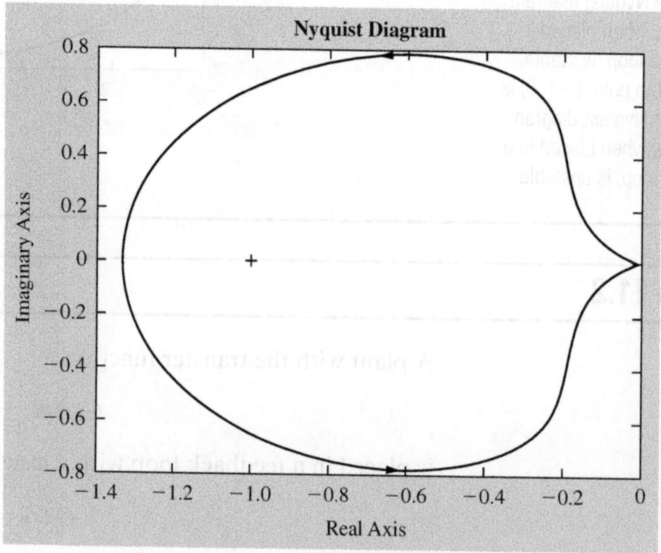

(b)

Figure 11.3

The plant of Example 11.2 has one pole with a positive real part. It is placed in series with a compensator in a feedback control loop. The Nyquist diagram for the open-loop transfer function is shown. (a) For $K_p = 2$, the point $(-1, 0)$ is not encircled by the Nyquist diagram and the system, when placed in a unity feedback loop, is unstable. (b) For $K_p = 5$, the point $(-1, 0)$ is encircled by the Nyquist diagram and the system, when placed in a unity feedback loop, is stable.

The closed-loop transfer function is

$$H(s) = \frac{4K_c(s+2)}{(s-1)(s+2)(s+10) + 4K_c(s+2)}$$

$$= \frac{4K_c(s+2)}{s^3 + 10s^2 + (17 + 4K_c)s + 8K_c - 30} \tag{d}$$

The denominator has only positive roots if all the coefficients have the same sign. This requires that $K_c > 3.75$. The system is unstable if $K_c < 3.75$. Applying Routh's criterion for a third-order system also requires that if all coefficients are positive, then the system is stable if

$$10(17 + 4K_c) > 8K_c - 30 \tag{e}$$

Equation (e) is satisfied for all positive K_c.

Example 11.3

A plant with the transfer function

$$G(s) = \frac{4}{s^6 + 12s^5 + 20s^4 + 25s^3 + 31s^2 + 24s + 20} \tag{a}$$

is placed in a feedback loop with an integral controller that has the transfer function

$$G_c(s) = \frac{5}{s} \tag{b}$$

Determine the stability of the closed-loop system.

Solution

The open-loop transfer function is

$$G_o(s) = \frac{20}{s(s^6 + 12s^5 + 20s^4 + 25s^3 + 31s^2 + 24s + 20)} \tag{c}$$

The poles of $G_o(s)$ include $0.473 \pm 0.974j$, with all other poles having non-positive real parts. In applying the Nyquist stability criterion, $k = 2$. The Nyquist curve for $G_o(s)$ is shown in Figure 11.4 and shows no encirclement of $(-1, 0)$. There must be two encirclements for stability; thus, the closed-loop system is unstable.

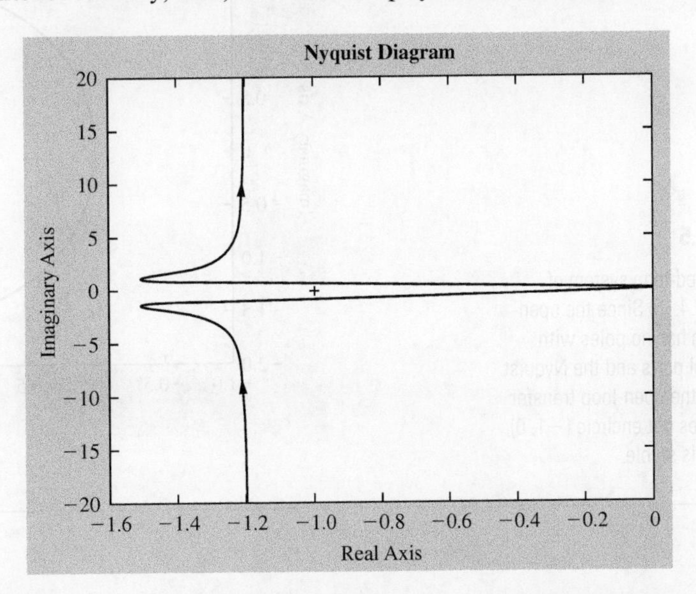

Figure 11.4

The plant of Example 11.3 has two poles with positive real parts. The plant is placed in a feedback control loop with an integral controller with $K_i = 5$. For the closed-loop system to be stable, the Nyquist diagram for the open-loop system must encircle the point $(-1, 0)$ two times. Since the Nyquist diagram does not encircle the point, the system is unstable.

Example 11.4

Determine the stability of the closed-loop system shown in Figure 11.5(a).

Solution

The equivalent open-loop transfer function of the system in Figure 11.5(a) is

$$G_o(s) = 2s\frac{\dfrac{(3s+4)(s+2)}{s(s^2+5s+25)}}{1+\dfrac{(3s+4)(s+2)}{s(s^2+5s+25)}}$$

$$= \frac{2s^3+20s^2+16s}{s^3+8s^2+14s+8} \qquad \text{(a)}$$

The poles of $G_o(s)$ are $-1.08 \pm 0.447j$, -5.84. All poles have negative real parts; thus, the parameter in the Nyquist criterion is $k = 0$. The Nyquist diagram for $G_o(s)$ is shown in Figure 11.5(b). There is no encirclement of the point $(-1, 0)$, so the closed-loop system is stable.

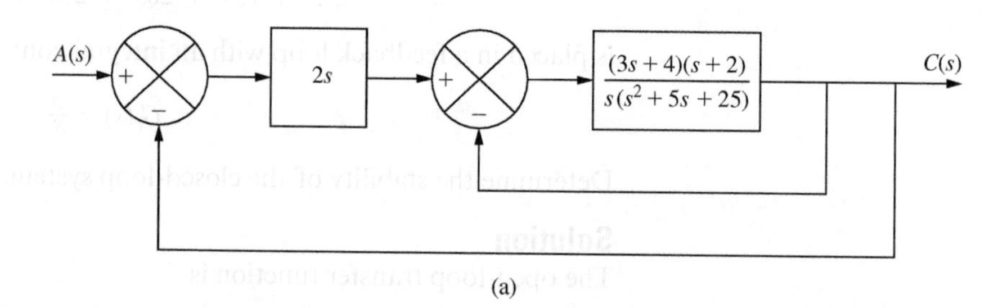

(a)

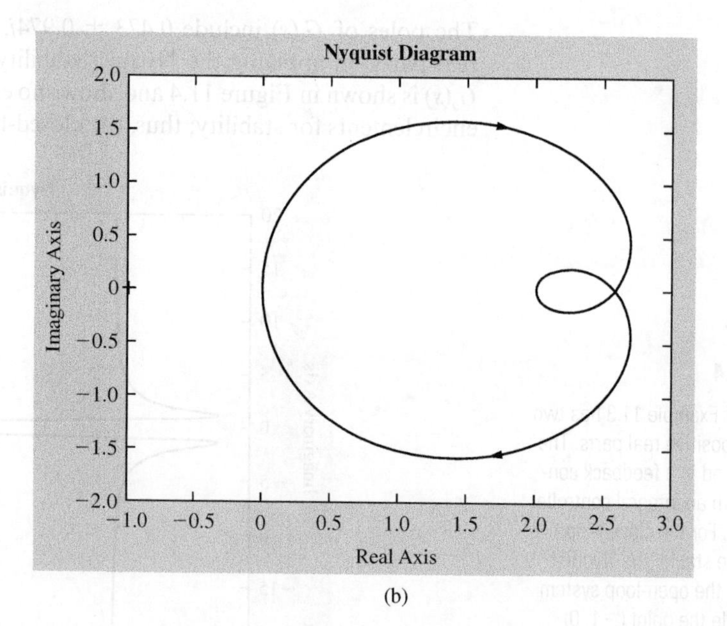

(b)

Figure 11.5

(a) The closed-loop system of Example 11.4. (b) Since the open-loop system has no poles with positive real parts and the Nyquist diagram of the open-loop transfer function does not encircle $(-1, 0)$, the system is stable.

11.3 Phase Margin and Gain Margin

The point $-1 + j0$ on the Nyquist diagram can be expressed as $(1)e^{-j\pi}$. This corresponds to points on the Bode diagram. The Bode diagram for a transfer function $G(s)$ plots $L(\omega)$ versus ω on a logarithmic scale. Recall that

$$L(\omega) = 20 \log|G(j\omega)| \tag{11.4}$$

The function $L(\omega)$ is expressed in dB.

The frequency at which the open-loop transfer function has a sinusoidal transfer function with a magnitude of 1 is called the gain crossover frequency, ω_{cg}. That is,

$$|G(j\omega_{cg})| = 1 \tag{11.5}$$

Equation (11.5) is equivalent to requiring

$$L(\omega_{cg}) = 0 \tag{11.6}$$

where $L(\omega)$ is as defined in Equation (11.4).

The phase crossover frequency ω_{cp} is the frequency at which the phase of the open-loop transfer function is equal to $-\pi$. That is,

$$\phi[G(\omega_{cp})] = -\pi \tag{11.7}$$

The gain margin (GM) is the reciprocal of the magnitude of $|G(j\omega)|$ at the phase crossover frequency. That is,

$$GM = \frac{1}{|G(\omega_{cp})|} \tag{11.8a}$$

If the gain margin is converted to dB, $\widetilde{GM}$ is the value on the Bode diagram at ω_{cp}. Then the gain margin is

$$\widetilde{GM} = -20 \log|G(j\omega_{cp})| = -L(\omega_{cp}) \tag{11.8b}$$

The phase margin (PM) is the difference between the phase angle at the gain crossover frequency and $-\pi$:

$$PM = \phi[G(\omega_{cg})] - (-\pi) = \phi[G(\omega_{cg})] + \pi \tag{11.9}$$

The closer the open-loop transfer function is to the point $-1 + j0$ on the Nyquist diagram, the more unstable the closed-loop transfer function. The gain crossover frequency is the frequency at which the real part of the Nyquist diagram is -1. The phase margin is the amount of lag when the frequency is the gain crossover frequency that is necessary to bring the closed-loop system to the verge of instability.

The phase crossover frequency is the frequency at which the imaginary part of the Nyquist diagram is 0. When the frequency is the phase crossover frequency and the closed-loop system is unstable, the gain margin is the amount of gain the system must lose in order for it to become stable. When the frequency is the phase crossover frequency and the closed-loop system is stable, the gain margin is the amount of gain the system can have before it becomes unstable. By the definition of the gain margin, this occurs only if $\widetilde{GM}$ is positive.

If either the gain margin or the phase margin is negative, the closed-loop system is unstable.

The MATLAB function margin may be used to help calculate the gain crossover frequency (omegacg), the phase crossover frequency (omegacp), the gain margin (GM), and the phase margin (PM) for a given transfer function. Given the numerator N and denominator D of the transfer function as vectors, the syntax is

```
sys=tf(N,D)
[GM,PM,omegacp,omegacg]=margin(sys)
```

The gain margin obtained using the function margin is in dB. The phase margin is measured in degrees. Both the gain crossover frequency and the phase crossover frequency are in r/s.

Example 11.5

A system has the transfer function

$$G(s) = \frac{1}{s(s+1)(s+4)} \tag{a}$$

a. Use MATLAB to generate the Bode diagram for this transfer function. Use the Bode diagram to determine the gain crossover frequency, the phase crossover frequency, the gain margin, and the phase margin for the transfer function of Equation (a). **b.** Use the margin function in MATLAB to determine the gain crossover frequency, the phase crossover frequency, the gain margin, and the phase margin for the transfer function of Equation (a).

Solution

a. The MATLAB-generated Bode diagram for the transfer function of Equation (a) is shown in Figure 11.6(a). The gain crossover frequency is the frequency at which $L(\omega) = 0$. From Figure 11.6(a), this occurs at $\omega_{cg} = 0.25$. The phase crossover frequency is the frequency at which $\phi = -\pi$. From the figure, this occurs at $\omega_{cp} = 2$. Then $\widetilde{GM} = -L(\omega_{cp}) = 20$ dB. The gain margin is calculated from

$$GM = \frac{1}{10^{\widetilde{GM}/20}} = \frac{1}{10^{20/20}} = 0.1 \tag{b}$$

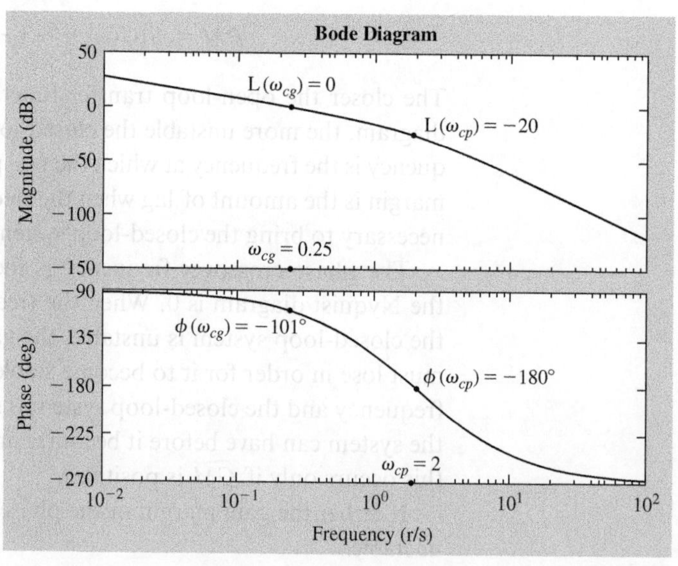

Figure 11.6

(a) Bode diagram for Example 11.5 highlighting the phase crossover frequency, the gain crossover frequency, the phase margin, and the gain margin. (b) The MATLAB code used to generate the Bode diagram. (c) Output from the MATLAB program.

```
% Figure 11.6b for Example 11.5
clear
%
% Defining numerator and denominator of transfer function
%
N=[1];
D=[1 5 4 0];
%
% Defining the transfer function from numerator and denominator
%
sys=tf(N,D)
%
% Drawing bode diagram
%
bode(N,D)
%
% Calculating using margin
%     GM=gain margin
%     PM=phase margin
%     omegacp=phase crossover frequency
%     omegacg=gain crossover frequency
%
[GM,PM,omegacp,omegacg]=margin(sys)
%
% End of file
```

(b)

```
sys=
      1
 ----------
 s^3+5s^2+4s
Continuous-time transfer function.
GM=
  20.0000
PM=
  72.8988
omegacp=
  2.0000
omegacg=
  0.2425
```

(c)

Figure 11.6 (Continued)

At the gain crossover freqeuncy, the phase is $\phi = -101°$. Thus, using Equation (11.9),

$$PM = \phi[G(\omega_{cg})] + \pi = -101° + 180° = 79° \quad \text{(c)}$$

b. The output confirms the results of part a.

Example 11.6

Determine the gain crossover frequency, the phase crossover frequency, the gain margin, and the phase margin for the transfer function

$$G(s) = \frac{100}{s(s+1)(s+4)} \tag{a}$$

Solution

From the Bode diagram in Figure 11.7 for the transfer function of Equation (a), the following are determined:

$$\omega_{cg} = 4.11 \text{ r/s}$$

$$\omega_{cp} = 2 \text{ r/s}$$

$$\widetilde{GM} = 0.2 \text{ dB}$$

$$PM = -32.14°$$

The phase margin is negative; thus, the system is unstable.

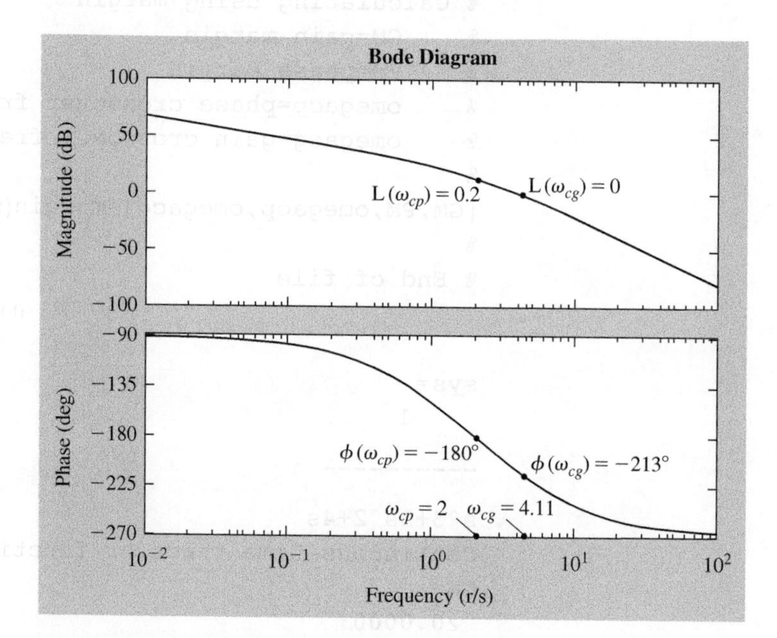

Figure 11.7

The Bode diagram for the system of Example 11.6 reveals an unstable system as the phase margin is negative. The phase margin is negative because the phase at the gain crossover frequency is less than −180°.

Example 11.7

Determine the gain crossover frequency, the phase crossover frequency, the gain margin, and the phase margin for the transfer function

$$G(s) = \frac{s+2}{s(s+1)(s+4)} \tag{a}$$

Solution

From the Bode diagram in Figure 11.8 for the transfer function of Equation (a), the following are determined:

$$\omega_{cg} = \infty$$

$$\omega_{cp} = 0.46 \text{ r/s}$$

$$\widetilde{GM} = \infty$$

$$PM = 71.6°$$

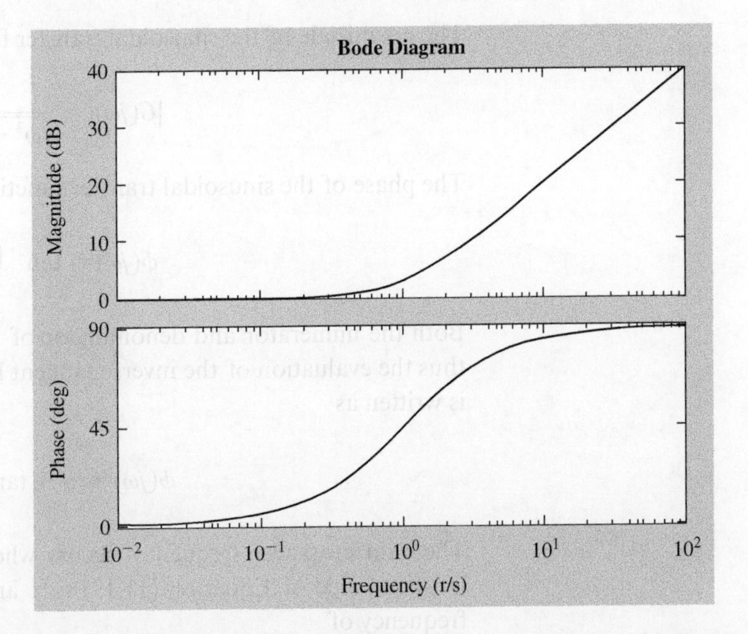

Figure 11.8

The Bode diagram for the system of Example 11.7 shows that the gain crossover frequency and the gain margin are infinite.

11.4 Steady-State Properties

Consider a plant with the transfer function

$$G(s) = \frac{a}{s(s+b)} \tag{11.10}$$

The plant is placed in a unity feedback loop as shown in Figure 11.9. The closed-loop transfer function is

$$H(s) = \frac{a}{s^2 + bs + a} \tag{11.11}$$

Equation (11.11) is the transfer function of a second-order system. It can be written in the standard form of the transfer function for a second-order system by setting

$$\omega_n = \sqrt{a} \tag{11.12}$$

and

$$\zeta = \frac{b}{2\sqrt{a}} \tag{11.13}$$

The closed-loop transfer function, Equation (11.11), is written as

$$H(s) = \frac{\omega_n^2}{s^2 + 2\zeta\omega_n s + \omega_n^2} \tag{11.14}$$

The sinusoidal transfer function for $G(s)$ is

$$G(j\omega) = \frac{\omega_n^2}{j\omega(j\omega + 2\zeta\omega_n)} = \frac{\omega_n^2}{-\omega^2 + j2\zeta\omega\omega_n} = \frac{-\omega_n^2(\omega^2 + j2\zeta\omega\omega_n)}{\omega^4 + 4\zeta^2\omega^2\omega_n^2} \tag{11.15}$$

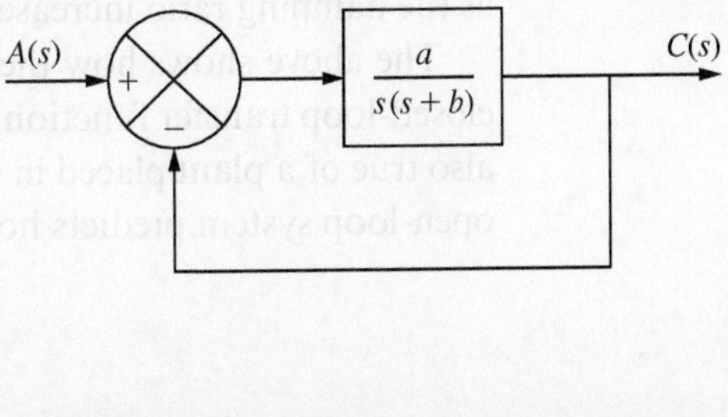

$A(s)$ + $\frac{a}{s(s+b)}$ $C(s)$

Figure 11.9

A plant $\frac{a}{s(s+b)}$ placed in a unity feedback loop.

The magnitude of the sinusoidal transfer function is

$$|G(j\omega)| = \frac{\omega_n^2}{\sqrt{\omega^4 + 4\zeta^2\omega^2\omega_n^2}} \tag{11.16}$$

The phase of the sinusoidal transfer function is

$$\phi(j\omega) = \tan^{-1}\left(\frac{-2\zeta\omega_n}{-\omega}\right) \tag{11.17}$$

Both the numerator and denominator of the ratio in Equation (11.17) are negative; thus the evaluation of the inverse tangent lies in the third quadrant. Equation (11.17) is written as

$$\phi(j\omega) = \pi + \tan^{-1}\left(\frac{2\zeta\omega_n}{\omega}\right) \tag{11.18}$$

The gain crossover frequency occurs when $|H(j\omega)| = 1$ or $L(\omega) = 0$. Setting the left-hand side of Equation (11.17) to 1 and solving for ω leads to a gain crossover frequency of

$$\omega_{cg} = \omega_n\sqrt{\sqrt{1 + 4\zeta^4} - 2\zeta^2} \tag{11.19}$$

The phase margin is obtained using Equations (11.9) and (11.18) as

$$\begin{aligned} PM = \phi[G(\omega_{cg})] + \pi &= 2\pi + \tan^{-1}\frac{2\zeta}{\sqrt{\sqrt{1 + 4\zeta^4} - 2\zeta^2}} \\ &= \tan^{-1}\frac{2\zeta}{\sqrt{\sqrt{1 + 4\zeta^4} - 2\zeta^2}} \end{aligned} \tag{11.20}$$

The gain crossover frequency is a function of ω_n and ζ, and the phase margin is a function of the damping ratio. The natural frequency and the damping ratio are written in terms of the parameters of the plant in Equations (11.12) and (11.13).

The maximum of the magnification factor for a second-order system whose transfer function is given by Equation (11.14) is

$$M_{\max} = \frac{1}{2\zeta\sqrt{1 - \zeta^2}} \tag{11.21}$$

The maximum magnification factor occurs for a frequency of

$$\omega_m = \omega_m\sqrt{1 - 2\zeta^2} \tag{11.22}$$

It is desirable to have a value of $M_{\max}$ less than 1.4. This corresponds to a damping ratio greater than 0.39.

The cutoff frequency ω_c is defined as the frequency such that

$$L(\omega_c) = L(0) - 3 \tag{11.23}$$

The bandwidth is defined as the frequency range from 0 to ω_c. The bandwidth decreases as the damping ratio increases.

The above shows how the open-loop transfer function of a plant is related to the closed-loop transfer function when the plant is placed in a unity feedback loop. This is also true of a plant placed in series with a compensator. Studying the properties of the open-loop system predicts how the closed-loop system will behave.

Example 11.8

The phase margin of a system is determined to be 31° and the gain crossover frequency is 14.7 r/s. Determine **a.** the damping ratio of the dominant roots, **b.** the natural frequency of the dominant roots, **c.** the maximum magnification factor, and **d.** the bandwidth of the system.

Solution

a. The relationship between the phase margin and the damping ratio in Equation (11.20) leads to

$$31° = \tan^{-1}\frac{2\zeta}{\sqrt{\sqrt{1+4\zeta^4}-2\zeta^2}} \tag{a}$$

or

$$\frac{2\zeta}{\sqrt{\sqrt{1+4\zeta^4}-2\zeta^2}} = \tan 31° = 0.60 \tag{b}$$

Solving Equation (b) for the damping ratio,

$$4\zeta^2 = (0.60)^2\left[\sqrt{1+4\zeta^4}-2\zeta^2\right] \tag{c}$$

$$122.7\zeta^8 + 122.7\zeta^6 + 26.68\zeta^4 - 1 = 0 \tag{d}$$

Two of the four roots of Equation (d), when written as a quartic equation for ζ^2, are complex. A third root is negative. The only positive root is 0.1456, which leads to

$$\zeta = 0.38 \tag{e}$$

b. The gain crossover frequency is related to the natural frequency and the damping ratio through Equation (11.19), which when applied to this situation leads to

$$14.7 = \omega_n\sqrt{\sqrt{1+4(0.38)^4}-2(0.38)^2} \tag{f}$$

Solving Equation (f) for the natural frequency gives

$$\omega_n = 17.0 \text{ r/s} \tag{g}$$

c. The maximum magnification factor is determined using the damping ratio and Equation (11.21):

$$M_{\max} = \frac{1}{2\zeta\sqrt{1-\zeta^2}} = \frac{1}{2(0.38)\sqrt{1-(0.38)^2}} = 1.42 \tag{h}$$

d. Assuming the dominant roots lead to a closed-loop transfer function of the form of Equation (11.14),

$$L(0) = 20\log 1 = 0 \tag{i}$$

Then the bandwidth is the frequency such that $L(\omega) = -3$ or

$$-3 = 20\log|G(j\omega)| = 20\log\frac{\omega_n^2}{\sqrt{\omega^4+4\zeta^2\omega^2\omega_n^2}}$$

$$= 20\log\frac{(17.0)^2}{\sqrt{\omega^4+4(0.38)^2(17.0)^2\omega^2}} \tag{j}$$

Rewriting Equation (j) leads to

$$52.22 = 10\log[\omega^4 + 166.9\omega^2] \tag{k}$$

Solving Equation (k) results in the cutoff frequency:

$$\omega^4 + 166.9\omega^2 - 166.9\times 10^3 = 0 \Rightarrow \omega = 18.3 \text{ r/s} \tag{l}$$

11.5 Frequency Response Design of Control Systems

The open-loop transfer function is a good predictor of how the closed-loop system will behave. The open-loop transfer function for a controller in series with a plant is given by Equation (11.2). The Nyquist stability criterion applied to the Nyquist diagram for an open-loop system can determine the closed-loop stability of a system. As evidenced by the plant of Equation (11.10), the open-loop system is used to determine the damping ratio and the natural frequency of the dominant poles of a closed-loop system. The phase margin is related to the damping ratio by Equation (11.20), and the natural frequency and damping ratio are related to the gain crossover frequency through Equation (11.19). If the natural frequency and the damping ratio corresponding to the dominant poles are specified, this leads to the determination of the phase margin and gain crossover frequency.

The static error constants specify the error when the system is subject to a benchmark input. For a type 1 system, the static velocity error constant determines how closely the system follows a unit ramp input. This determines the steady-state accuracy of the closed-loop system. Since the open-loop transfer function is a product of transfer functions and the steady-state error constants are limits of transfer functions as s approaches zero, the contribution from the controller can be taken to be one and the gain from the controller can be taken to be included with the plant. When the gain is included in the plant transfer function, the resulting transfer function is said to be the uncompensated gain-adjusted transfer function.

The frequency response approach to control systems design uses the Bode diagram for the open-loop transfer function for a plant in series with a controller or compensator. The Bode diagram for the open-loop transfer function is the sum of the Bode diagram of the uncompensated gain-adjusted transfer function of the plant and the Bode diagram of the controller. Then the phase margin that is to be contributed by the controller is determined. The required phase margin is often increased by 5° to 10° to account for the behavior contributed by other poles and the numerator. The damping ratio is specified by the phase margin, the gain margin, and the peak magnification factor. The speed of the response is specified by the crossover frequencies and the bandwidth. The static error constants are often specified to maintain steady-state accuracy.

Once a design has been achieved, it is necessary to ensure that the specifications are met. The design specified above is performed such that the dominant poles satisfy the design parameters. The remaining poles and the use of an open-loop transfer function to predict the behavior of a closed-loop system may modify the closed-loop response from what is intended. The design of a compensator is not unique. There are many designs that meet a given set of parameters.

11.6 PID Controller Design

The transfer function for a PID controller can be written as

$$G_c(s) = \frac{K_i + K_p s + K_d s^2}{s} = \frac{K_i(1 + c_1 s + c_2 s^2)}{s}$$

$$= \frac{K_i[1 + (a+b)s + abs^2]}{s} = \frac{K_i}{s}(1 + as)(1 + bs) \tag{11.24}$$

where

$$c_1 = \frac{K_p}{K_i},\ c_2 = \frac{K_d}{K_i},\ c_1 = a + b,\ c_2 = ab \tag{11.25}$$

The open-loop transfer function for a plant placed in series with a PID controller is

$$G_o(s) = \frac{K_i[1 + (a + b)s + abs^2]}{s} G(s) = \frac{K_i}{s}(1 + as)(1 + bs)G(s) \tag{11.26}$$

If a minimum value of the static velocity error constant is specified, then

$$C_v = \lim_{s \to 0} s\frac{K_i}{s}(1 + as)(1 + bs)G(s) = K_i \lim_{s \to 0} G(s) = K_i G(0) \tag{11.27}$$

Equation (11.27) can be used to specify K_i.

The open-loop transfer function can be written as

$$G_o(s) = G_1(s)\, G_2(s) \tag{11.28}$$

where

$$G_1(s) = \frac{K_i}{s} G(s) \tag{11.29}$$

and

$$G_2(s) = (1 + as)(1 + bs) \tag{11.30}$$

It is noted that $G_1(s)$ is the open-loop transfer function with an integral controller designed to satisfy the static velocity error constant requirement. From the Bode diagram for $G_1(s)$, the gain crossover frequency is determined. The gain crossover frequency for $G_o(s)$ should be the same order of magnitude as this crossover frequency. Note that $(1 + as)$ has a Bode diagram that has a corner frequency of $\frac{1}{a}$ and a slope of 20 dB per decade. The chosen crossover frequency is used as the reciprocal of the corner frequency for $(1 + as)$. The gain crossover frequency is used to specify a. The phase margin is identified at the gain crossover frequency. At the corner frequency, this portion of the compensator contributes about 45° to the phase margin.

Consider the transfer function

$$G_3(s) = (1 + as)\, G_1(s) \tag{11.31}$$

The Bode diagram of $G_3(s)$ may be used to identify the phase margin that must be supplied by $(1 + bs)$. The Bode diagram of $1 + bs$ has a corner frequency at $\frac{1}{b}$. Choosing b to be smaller than a gives a large increase in phase lead near the gain crossover frequency. Trial and error is used to choose the value of b that gives the required phase margin. An initial guess might be $b = \frac{1}{a}$. The phase margin for $b = \frac{1}{a}$ may be determined and then the value of b adjusted, if necessary.

Example 11.9

Design a PID controller for a plant with the transfer function

$$G(s) = \frac{5}{s^2 + 4} \tag{a}$$

The open-loop system must have a phase margin of at least 50°, a gain margin of at least 0.5 dB, and a static velocity error constant of at least 10 s^{-1}.

Solution

The integral gain is determined from the static velocity error constant according to Equation (11.27):

$$C_v = 10 = K_i G(0) = \frac{5}{4} K_i \Rightarrow K_i = 8 \tag{b}$$

For the transfer function

$$G_1(s) = \frac{40}{s(s^2 + 4)} \tag{c}$$

the Bode diagram is shown in Figure 11.10(a). The phase margin is $-90°$, the gain margin is zero, the phase crossover frequency is 2 r/s, and the gain crossover frequency is 3.81 r/s. The gain crossover frequency of the compensated system is chosen to be around 5 r/s, so $a = 5$.

As in Equation (11.31),

$$G_3(s) = (1 + 5s)\, G_1(s) = \frac{40(1 + 5s)}{s(s^2 + 4)} \tag{d}$$

The Bode diagram for $G_3(s)$ is shown in Figure 11.10(b). The phase margin is $-0.80°$ and the gain margin is zero. The factor $(1 + bs)$ should make up at least 51°. For a starting value of $b = \frac{1}{a} = 0.2$, the Bode diagram for the open-loop transfer function is shown in Figure 11.10(c). This system has a phase margin of 79° and a gain margin that is infinite. This is acceptable.

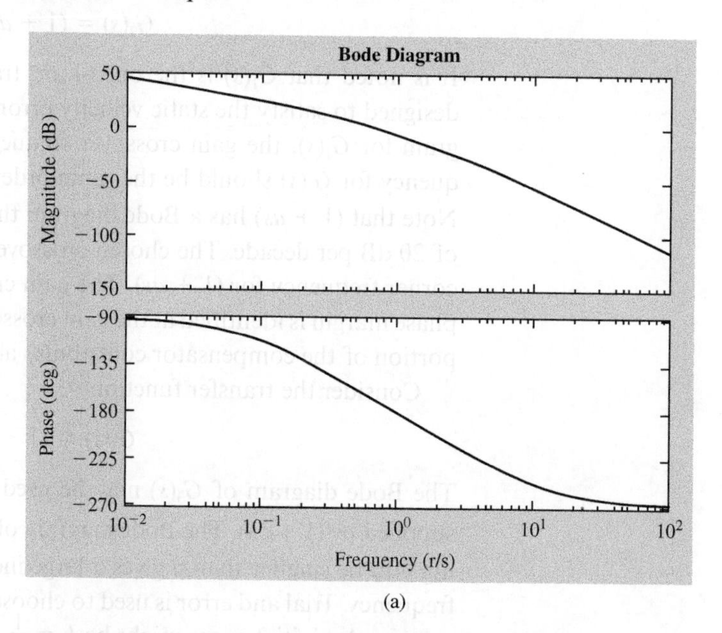

(a)

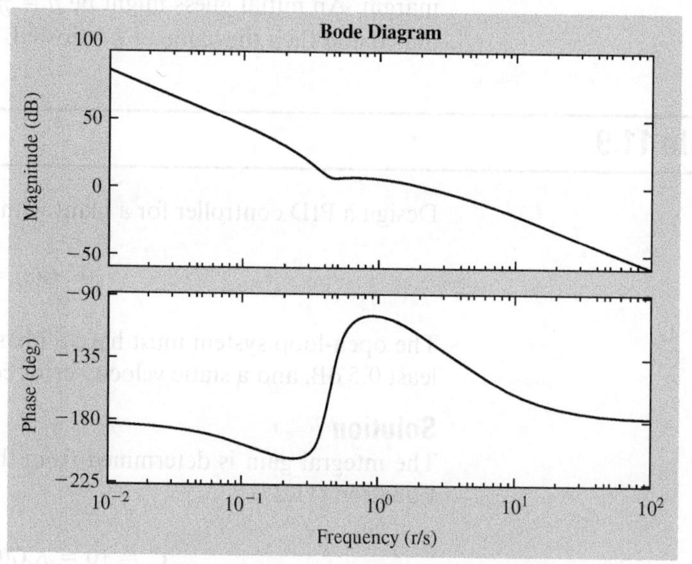

(b)

Figure 11.10

For the system of Example 11.9: (a) the Bode diagram for $G_1(s)$; (b) the Bode diagram for $G_3(s) = (1 + 5s)\, G_1(s)$ where 5 r/s is the gain crossover frequency for the compensated system; (c) the Bode diagram for the open-loop transfer function of the system compensated using a PID controller; (d) comparison of step responses for the uncompensated system and the compensated system; (e) comparison of ramp responses for the uncompensated system and the compensated system.

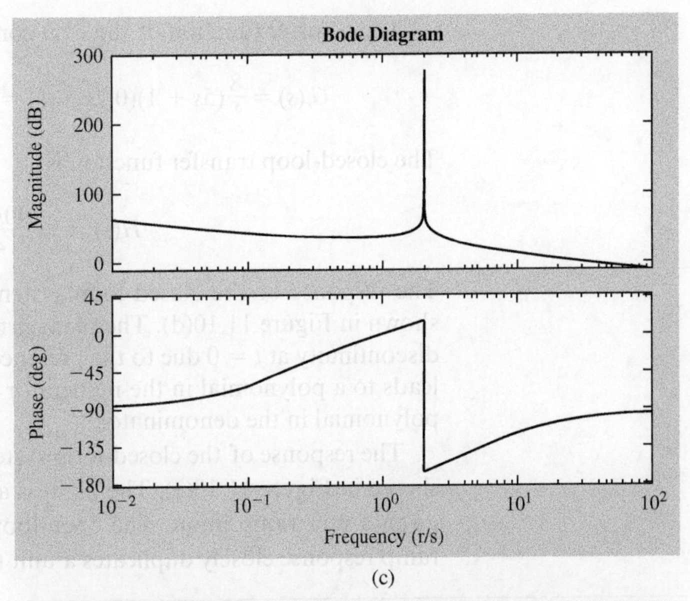

(c)

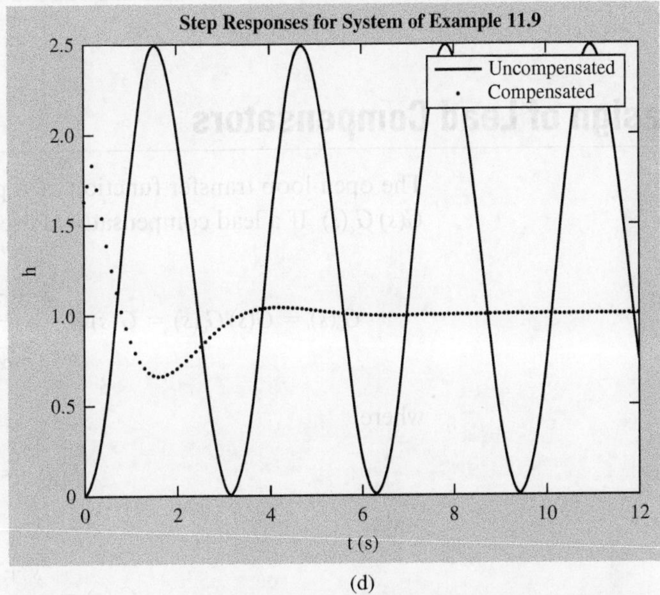

(d)

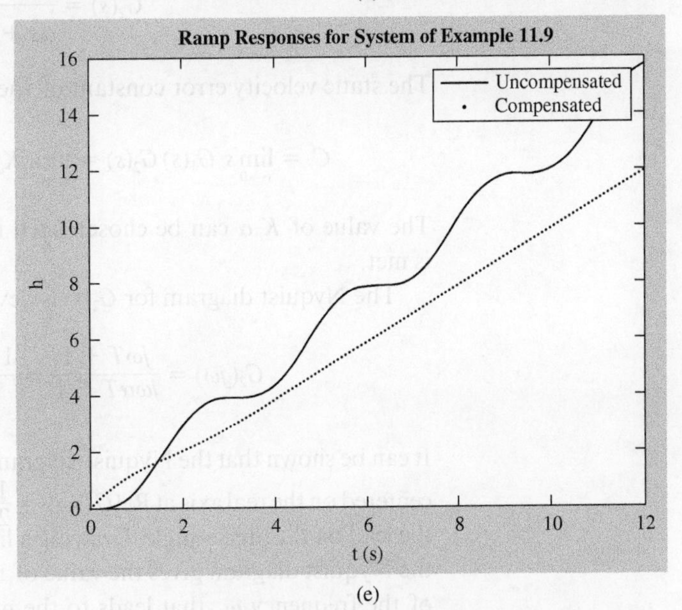

(e)

Figure 11.10
(Continued)

The transfer function of the PID controller is

$$G_c(s) = \frac{8}{s}(5s + 1)(0.2s + 1) = \frac{8}{s}(1 + 0.7s + s^2) = 5.6 + \frac{8}{s} + 8s \tag{e}$$

The closed-loop transfer function is

$$H(s) = \frac{40s^2 + 28s + 40}{s^3 + 40s^2 + 32s + 40} \tag{f}$$

The response of the closed-loop system due to a unit step input and the plant are shown in Figure 11.10(d). The plant is neutrally stable. The closed-loop system has a discontinuity at $t = 0$ due to the presence of the differential action in the controller. It leads to a polynomial in the numerator whose order is one less than the order of the polynomial in the denominator.

The response of the closed-loop system and the plant due to a unit ramp input are shown in Figure 11.10(e). The plant is a type 0 system, so there is a large error when given a unit ramp input. The open-loop transfer function is a type 1 system, so its ramp response closely duplicates a unit ramp input.

11.7 Design of Lead Compensators

The open-loop transfer function of a plant in series with a compensator is written as $G(s)\,G_c(s)$. If a lead compensator is used, then

$$G_o(s) = G(s)\,G_c(s) = G(s)\,K_c\frac{s + \frac{1}{T}}{s + \frac{1}{\alpha T}} = G(s)\,K_c\alpha\frac{s + \frac{1}{T}}{\alpha s + \frac{1}{T}} = G_1(s)\,G_2(s) \tag{11.32}$$

where

$$G_1(s) = K_c\alpha G(s) \tag{11.33}$$

and

$$G_2(s) = \frac{s + \frac{1}{T}}{\alpha s + \frac{1}{T}} = \frac{Ts + 1}{\alpha Ts + 1} \tag{11.34}$$

The static velocity error constant of the compensated system becomes

$$C_v = \lim_{s\to 0} s\ G_1(s)\,G_2(s) = \lim_{s\to 0} sK_c\alpha G(s)\frac{Ts + 1}{\alpha Ts + 1} = \lim_{s\to 0} K_c\alpha sG(s) \tag{11.35}$$

The value of $K_c\alpha$ can be chosen such that a minimum static velocity error constant is met.

The Nyquist diagram for $G_2(s)$ is developed using

$$G_2(j\omega) = \frac{j\omega T + 1}{j\omega\alpha T + 1} = \frac{1 + \alpha(T\omega)^2 + j\omega T(1 - \alpha)}{1 + (\alpha T\omega)^2} \tag{11.36}$$

It can be shown that the Nyquist diagram for $G_2(s)$ is a circle of radius $\frac{1}{2}\left(\frac{1}{\alpha} - 1\right)$ that is centered on the real axis at $\text{Re}[G(j\omega)] = \frac{1}{2}\left(\frac{1}{\alpha} + 1\right)$. The smaller the value of α, the larger the lead on the phase angle. Drawing a line from the origin that is tangent to a point on the Nyquist diagram gives the value of the maximum phase angle and hence the value of the frequency ω_m that leads to the maximum phase lead for the given value of α.

The Nyquist diagram for $T = 1$ and $\alpha = 0.25$ is shown in Figure 11.11. The maximum phase ϕ_m, using Figure 11.11 as a guide, is determined to be

$$\sin \phi_m = \frac{\frac{1}{2}\left(\frac{1}{\alpha} - 1\right)}{\frac{1}{2}\left(\frac{1}{\alpha} + 1\right)} = \frac{1 - \alpha}{1 + \alpha} \tag{11.37}$$

Equation (11.37) is rearranged, giving

$$\alpha = \frac{1 - \sin \phi_m}{1 + \sin \phi_m} \tag{11.38}$$

The design procedure for a lead compensator using the frequency response approach uses the logarithmic properties of the Bode diagram. The Bode diagram for $G_2(s)$ is the sum of two first-order Bode diagrams. One is that of a first-order lead in which the low-frequency asymptote is 0 and the high-frequency asymptote is a line of slope 20 dB per decade. It has a corner frequency of $\omega_{c1} = \frac{1}{T}$ and is shown in Figure 11.12(a). The second is the Bode diagram of a first-order lag, shown in Figure 11.12(b). A first-order lag has a low-frequency asymptote of 0 and a high-frequency asymptote of -20 dB per decade. It has a corner frequency of $\omega_{c2} = \frac{1}{\alpha T}$. The resulting Bode diagram for a compensator with $\alpha = 0.2$ and $T = 1$ is shown in Figure 11.12(c). It has a low-frequency asymptote of 0 and a high-frequency asymptote of 20 logα = 13.98 dB.

The Bode diagram for a lead compensator suggests that it acts as a high-pass filter. When added in series to a plant, the lead compensator increases the gain crossover frequency and increases the phase margin. This in turn results in improving the speed of the response.

As evidenced by Figure 11.12(c), the maximum phase angle contributed by the compensator occurs at the mean of the corner frequencies ω_c, given by

$$\log\omega_c = \frac{1}{2}(\log\omega_{c1} + \log\omega_{c2}) = \frac{1}{2}\left(\log\frac{1}{T} + \log\frac{1}{\alpha T}\right)$$
$$= -\frac{1}{2}\log(\alpha T^2) \tag{11.39}$$

Solving Equation (11.39) leads to

$$\omega_c = \frac{1}{T\sqrt{\alpha}} \tag{11.40}$$

The maximum value of ϕ occurs for a frequency of ω_c.

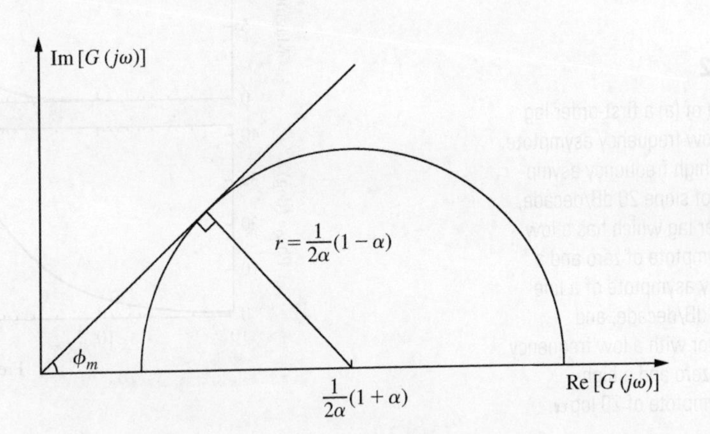

Figure 11.11

Determination of ϕ_m.

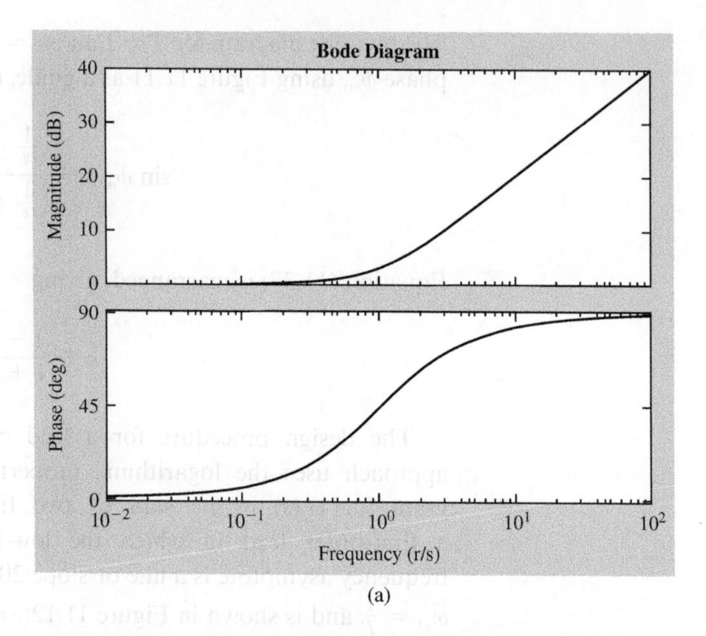

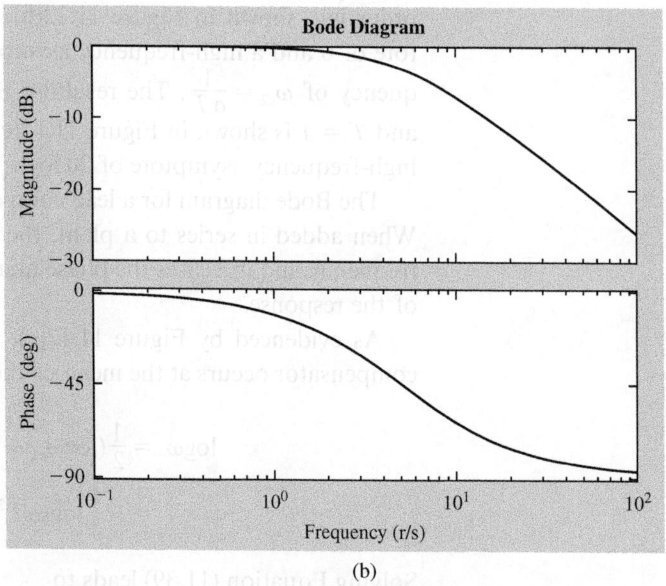

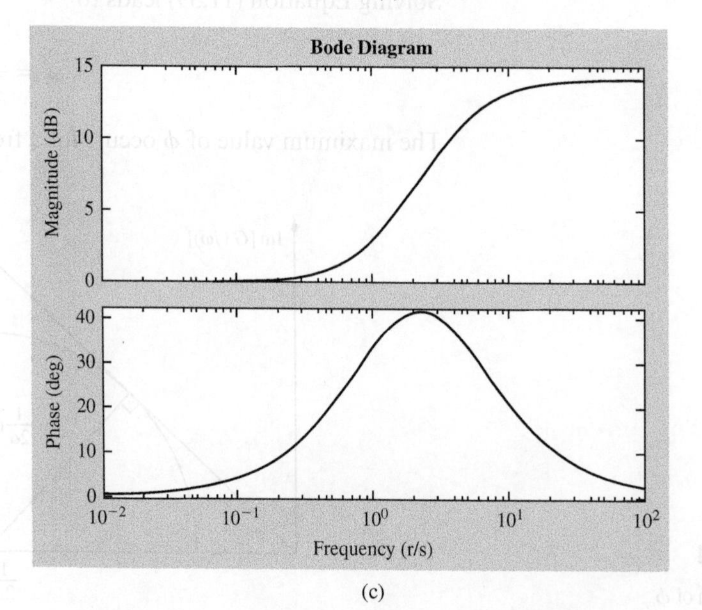

Figure 11.12

Bode diagram of (a) a first-order lag which has a low frequency asymptote of zero and a high frequency asymptote of a line of slope 20 dB/decade, (b) a first-order lag which has a low frequency asymptote of zero and a high frequency asymptote of a line of slope −20 dB/decade, and (c) compensator with a low frequency asymptote of zero and a high frequency asymptote of 20 log α.

The value of $|G_2(j\omega)|$ at the crossover frequency suggested by Equation (11.40) is

$$|G_2(j\omega_c)| = \left|\frac{j\frac{1}{T\sqrt{\alpha}}T + 1}{j\frac{1}{T\sqrt{\alpha}}\alpha T + 1}\right| = \frac{1}{\sqrt{\alpha}} \tag{11.41}$$

Since $L(\omega) = 20\log|G_1(j\omega_c)| + 20\log|G_2(j\omega_c)|$, $L(\omega_c) = 0$ requires that

$$-20\log\frac{1}{\sqrt{\alpha}} = 20\log|G_1(j\omega_c)| \tag{11.42}$$

The new gain crossover frequency is defined by Equation (11.42).

The above suggests the following procedure for designing a lead compensator using the frequency response approach:

1. Define $G_1(s)$ and $G_2(s)$ according to Equations (11.33) and (11.34). The value of $K_c\alpha$ is chosen to meet the minimum value of a static error constant as in Equation (11.35).
2. Draw the Bode diagram of $G_1(s)$. Identify the phase margin the compensator will be designed to add to the system.
3. The lead compensator decreases the gain crossover frequency and decreases the resulting phase margin. Thus, the phase margin of the compensator should be increased by 5° to 15°.
4. Determine the appropriate value of α using Equation (11.38).
5. Determine the new gain crossover frequency by requiring that Equation (11.42) is satisfied.
6. Using the new gain crossover frequency, calculate the time constant using Equation (11.40).

   The zero of the compensator s_z and the pole of the compensator s_p are

$$s_z = \frac{1}{T} \qquad s_p = \frac{1}{\alpha T} \tag{11.43}$$

   The value of K_c is determined from knowing the values of $K_c\alpha$ and α. The compensator is thus specified.

7. Test the design of the compensator by subjecting the closed-loop system to a unit step input or a unit ramp input to see if it does indeed meet specifications. If not, redesign accordingly.

Example 11.10

Design a compensator for the plant with the transfer function

$$G(s) = \frac{3}{s(s+2)} \tag{a}$$

such that the static velocity error constant is at least 24 s^{-1}, the phase margin is at least 45°, and the gain margin is at least 12 dB.

Solution

A lead compensator is used with a transfer function of the form

$$G_c(s) = K_c\frac{s + \frac{1}{T}}{s + \frac{1}{\alpha T}} \qquad \alpha < 1 \tag{b}$$

The open-loop transfer function for the plant in series with the compensator is written in the form of Equation (11.32), with $G_1(s)$ and $G_2(s)$ given by Equations (11.33) and (11.34).

The procedure follows the steps listed earlier in this section:

1. Requiring the static velocity error constant to be 24 s^{-1}, according to Equation (11.35), leads to

$$24 = \lim_{s\to 0} sK_c\alpha \frac{3}{s(s+2)} \Rightarrow K_c\alpha = 16 \tag{c}$$

Then using Equation (11.33) and Equation (c) gives

$$G_1(s) = \frac{48}{s(s+2)} \tag{d}$$

2. The Bode diagram for $G_1(s)$ is shown in Figure 11.13. From the Bode diagram, the phase margin of the uncompensated but gain-adjusted transfer function is determined to be 16.43° and the gain crossover frequency is 6.78 r/s. The gain margin is infinite, as is the phase crossover frequency. The phase margin to be added by the compensator is 45° − 16.43° = 28.57°.
3. When the compensator is added, the gain crossover frequency decreases, as does the phase margin. Taking this into account, the phase margin added by the compensator needs to be increased by at least 5°. So instead of 28.57°, 35° is used.
4. Application of Equation (11.38) gives

$$\alpha = \frac{1-\sin\phi_m}{1+\sin\phi_m} = \frac{1-\sin 35°}{1+\sin 35°} = 0.27 \tag{e}$$

5. The gain adjustment obtained by choosing $\alpha = 0.27$ is given by Equation (11.42) as

$$20\log\frac{1}{\sqrt{\alpha}} = 20\log\frac{1}{\sqrt{0.27}} = 5.7 \text{ dB} \tag{f}$$

At the new gain crossover frequency, $L(\omega) = -5.67$ dB. Using the Bode diagram in Figure 11.13, it is found that this corresponds to $\omega_{cg} = 9.6$ r/s.

6. The value of T is found from Equation (11.40):

$$T = \frac{1}{\omega_c\sqrt{\alpha}} = \frac{1}{9.6\sqrt{0.27}} = 0.20 \text{ s}$$

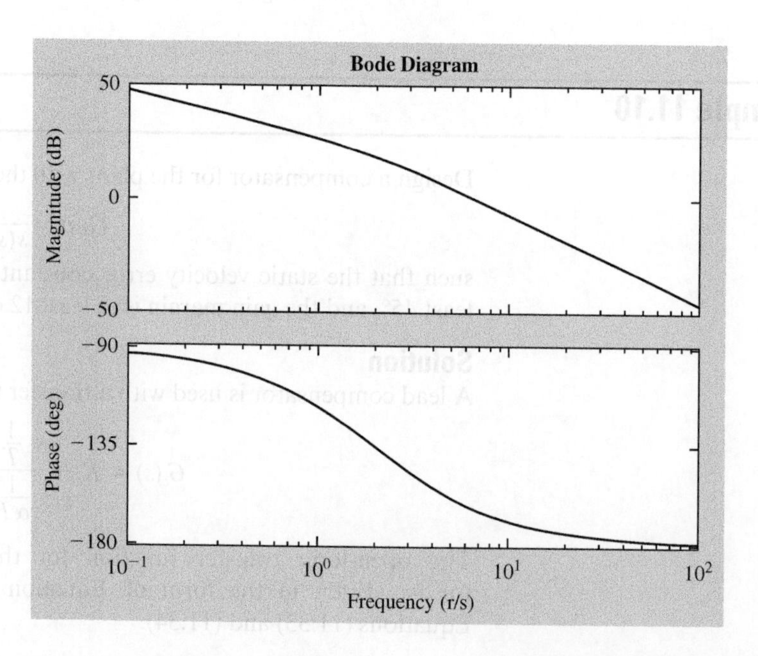

Figure 11.13
Bode diagram for $G_1(s)$ for Example 11.10.

The zero of the compensator is

$$\frac{1}{T} = 5.0 \text{ r/s}$$

The pole of the compensator is

$$\frac{1}{\alpha T} = 18.5 \text{ r/s}$$

The compensator gain is

$$K_c = \frac{K_c\alpha}{\alpha} = \frac{16}{0.27} = 59.3$$

Thus, the transfer function of the compensator becomes

$$G_c(s) = 59.3\frac{s + 5.0}{s + 18.5}$$

7. The open-loop transfer function for the compensated system is

$$G(s)\,G_c(s) = \frac{177.8(s + 5.0)}{s(s + 2)(s + 18.5)}$$

The Bode diagram for the compensated system is shown in Figure 11.14. The phase margin is 47.5°. The new gain crossover frequency is 9.5 r/s. The phase crossover frequency and the gain margin are infinite.

The closed-loop transfer function of a feedback system with the compensated system is

$$H(s) = \frac{177.9s + 889.5}{s^3 + 20.5s^2 + 206.9s + 889.5}$$

The step responses of the closed-loop transfer function of the uncompensated and compensated systems are shown in Figure 11.15(a). The ramp responses of the closed-loop transfer function of the uncompensated and compensated systems are shown in Figure 11.15(b). Both responses are adequate.

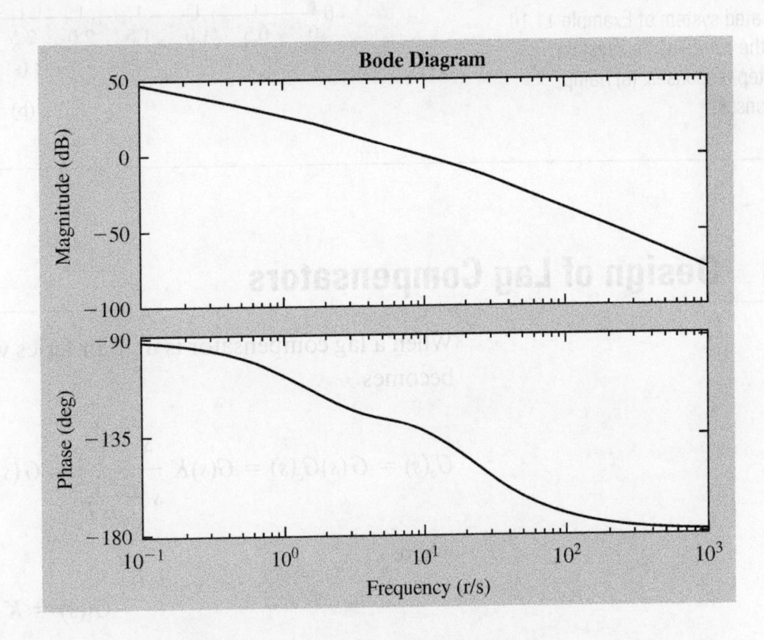

Figure 11.14

Bode diagram for compensated system of Example 11.10.

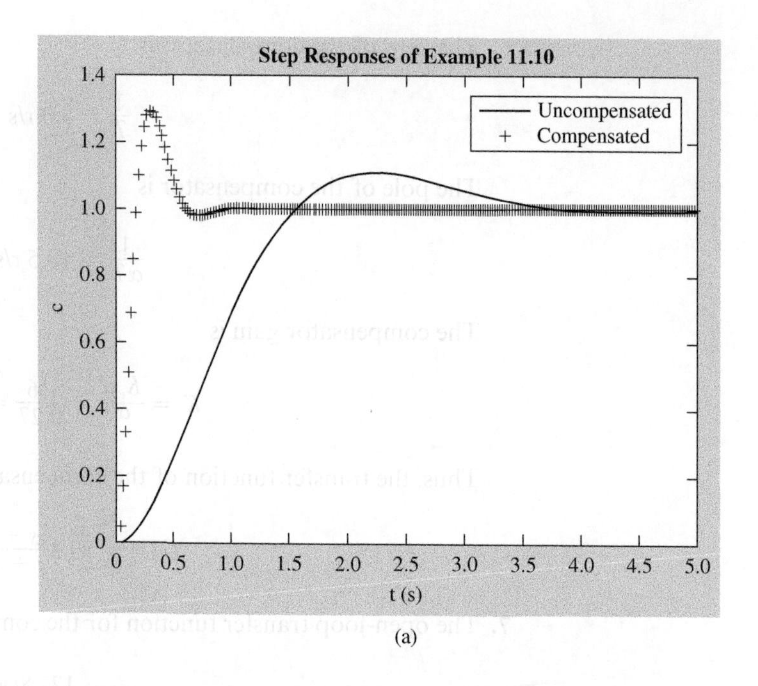

(a)

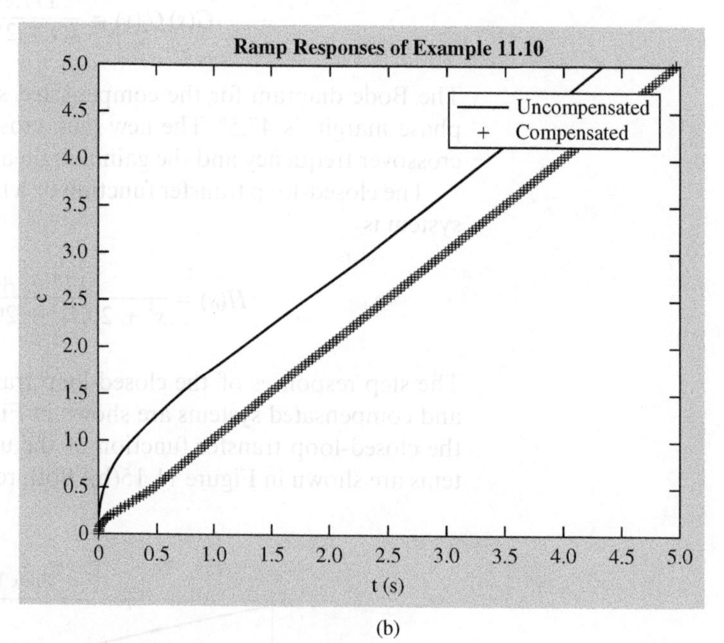

(b)

Figure 11.15

Comparison of response of uncompensated system of Example 11.10 and the compensated system. (a) Step responses. (b) Ramp responses.

11.8 Design of Lag Compensators

When a lag compensator is used in series with a plant, the open-loop transfer function becomes

$$G_o(s) = G(s)G_c(s) = G(s)K_c\frac{s+\frac{1}{T}}{s+\frac{1}{\alpha T}} = G(s)K_c\alpha\frac{s+\frac{1}{T}}{\alpha s+\frac{1}{T}} = G_1(s)\,G_2(s) \qquad \alpha > 1 \quad (11.44)$$

where

$$G_1(s) = K_c\alpha G(s) \quad (11.45)$$

and

$$G_2(s) = \frac{s + \frac{1}{T}}{\alpha s + \frac{1}{T}} = \frac{Ts + 1}{\alpha Ts + 1} \tag{11.46}$$

For a type 1 system, the static velocity error constant may be specified:

$$C_v = \lim_{s \to 0} s\ G_1(s)\, G_2(s) = \lim_{s \to 0} sK_c \alpha G(s) \frac{Ts + 1}{\alpha Ts + 1} = \lim_{s \to 0} K_c \alpha s G(s) \tag{11.47}$$

The value of $K_c\alpha$ is chosen such that the static velocity error constant is met.

The Bode diagram for $G_2(s)$ as well as a sample Bode diagram for $G_1(s)$ are shown in Figure 11.16. The Bode diagram for the compensator has a low-frequency asymptote of 0 and a high-frequency asymptote of $20 \log \alpha$. Its largest corner frequency is $\frac{1}{T}$. The lower corner frequency is $\frac{1}{\alpha T}$. Recall that for a lag compensator, the corner frequencies are very small so that the dominant poles are not affected by the transient behavior of the compensator. From the frequency response perspective, this is necessary so that the phase margin at the higher frequencies is not affected. A lag compensator is employed to minimize the steady-state error.

A lag compensator acts as a low-pass filter. It is used when the properties of the dominant poles, such as the damping ratio and natural frequency, are adequate but the steady-state error needs to be decreased.

Usually when a lag compensator is designed, the required phase margin and gain margin are specified. For the open-loop transfer function, the Bode diagram locates the point where the phase angle is the required phase margin minus 180°. The frequency where this occurs is the new gain crossover frequency. The corner frequency of the compensator $\frac{1}{T}$ is chosen to be about one decade below the new gain crossover frequency.

At the new gain crossover frequency, $L(\omega)$ for the compensated system must be zero. The value of α is chosen to make this happen. That is,

$$L(\omega_{cg}) = 0 = 20 \log|G_1(\omega_{cg})| + 20 \log|G_2(\omega_{cg})| \tag{11.48}$$

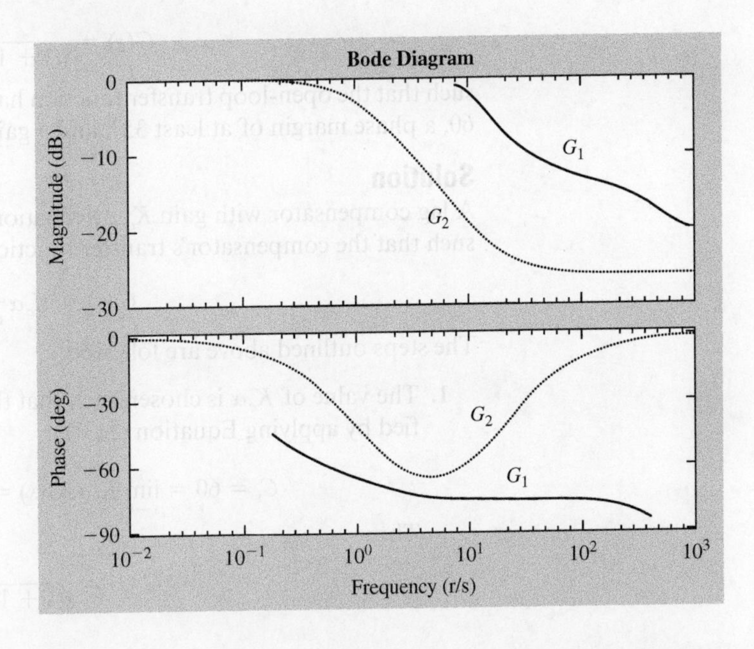

Figure 11.16

Bode diagram for $G_2(s)$ as well as a sample Bode diagram for $G_1(s)$.

However, because the corner frequency of the compensator is one decade below the new gain crossover frequency, the second term is approximately the high-frequency asymptote of the compensator. Thus,

$$20\log|G_1(\omega_{cg})| = -20\log\frac{1}{\alpha} \tag{11.49}$$

The compensator gain is then determined from the known value of $K_c\alpha$ and the known value of α from Equation (11.47).

The Bode diagram for the open-loop compensated system is drawn. The new gain margin is checked against the required value. If it is acceptable, a compensator has been designed so that it meets the requirements. If not, a new phase crossover frequency should be chosen such that the gain margin is what is required. Doing so adjusts the value of the gain crossover frequency and hence the value of α. This adjustment must take place until both the gain margin and the phase margin requirements are satisfied.

The procedure is summarized in the following steps:

1. The static error constant is used to determine an appropriate value of $K_c\alpha$.
2. The Bode diagram for $G_1(s) = K_c\alpha G(s)$ is drawn.
3. The gain crossover frequency for the compensated system is determined as the frequency where the phase for $G_1(s)$ is the required phase margin minus 180°. The required phase margin is usually adjusted by 5° to 15° to account for the phase angle introduced by the compensator.
4. The corner frequency of the compensator or the zero of the compensator is chosen to be one decade below the new gain crossover frequency. The value of α is determined using Equation (11.38).
5. The value of K_c is determined from the values of both $K_c\alpha$ and α.
6. The gain margin is checked against the required value. If necessary, the compensator design is adjusted by selecting a new phase crossover frequency.
7. The performance of the compensator is tested.

Example 11.11

Design a compensator for a plant with the transfer function

$$G(s) = \frac{10}{s(s+1)(s+1.5)} \tag{a}$$

such that the open-loop transfer function has a static velocity error constant of at least 60, a phase margin of at least 35°, and a gain margin of at least **a.** 3 dB and **b.** 5 dB.

Solution

A lag compensator with gain K_c, attenuation α, and time constant T is to be designed such that the compensator's transfer function is

$$G_c(s) = K_c\alpha\frac{Ts+1}{\alpha Ts+1} \tag{b}$$

The steps outlined above are followed:

1. The value of $K_c\alpha$ is chosen such that the static velocity error constant is satisfied by applying Equation (11.47):

$$C_v = 60 = \lim_{s\to 0} K_c\alpha sG(s) = K_c\alpha\frac{10}{1.5} \Rightarrow K_c\alpha = 9 \tag{c}$$

Then

$$G_1(s) = \frac{90}{s(s+1)(s+1.5)} \tag{d}$$

2. The Bode diagram for $G_1(s)$ is shown in Figure 11.17(a). The gain margin is 25.6 dB, the phase margin is $-58.1°$, the phase crossover frequency is 1.23 r/s, and the gain crossover frequency is 4.36 r/s. The phase margin is negative, which implies that the closed-loop system is unstable.
3. The phase margin is to be at least 35°. To account for changes in the phase part of the Bode diagram, 10° are added, so $45° - 180° = -135°$. This corresponds to a new gain crossover frequency of 0.5 r/s.
4. Choosing the corner frequency of the compensator $\frac{1}{T}$ to be one decade below the new gain crossover frequency leads to $\frac{1}{T} = 0.05$ r/s. At this value, the gain margin for the uncompensated curve is 40 dB. Applying Equation (11.49) gives

$$40 = -20 \log \frac{1}{\alpha} \Rightarrow \alpha = 100 \tag{e}$$

5. The compensator gain is obtained by applying Equation (c):

$$K_c \alpha = 9 \Rightarrow K_c = 0.09 \tag{f}$$

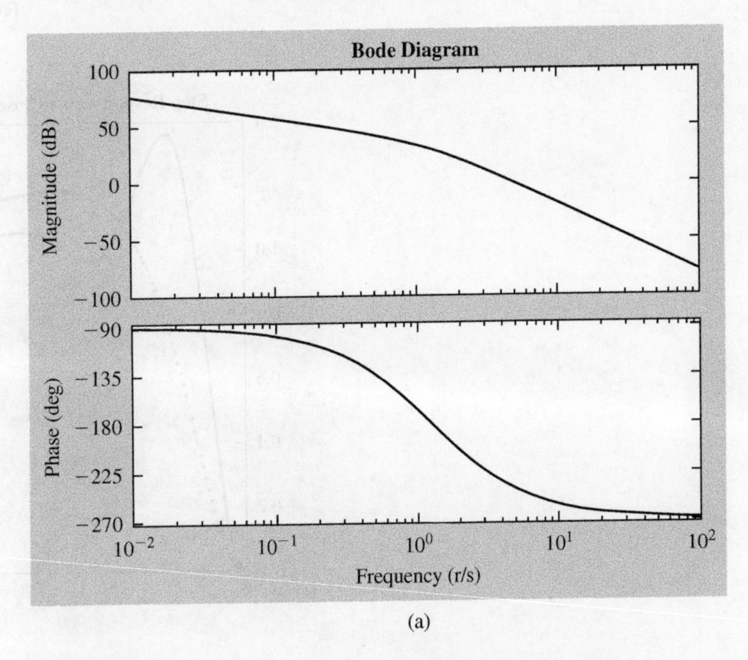

(a)

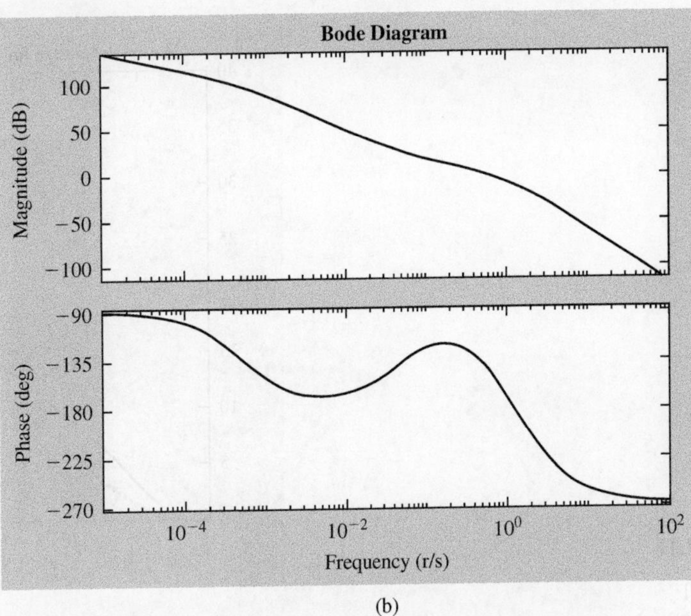

(b)

Figure 11.17

(a) Bode diagram for $G_1(s)$ for system of Example 11.11. (b) Bode diagram for compensated open-loop transfer function with a gain margin of 3.83 dB. (c) Bode diagram for compensated open-loop transfer function with a gain margin of 10.3 dB. (d) Comparison of step responses when a lag compensator of gain margin 3.83 is used and a lag compensator with a gain margin of 10.3 dB is used. (e) Comparison of ramp responses.

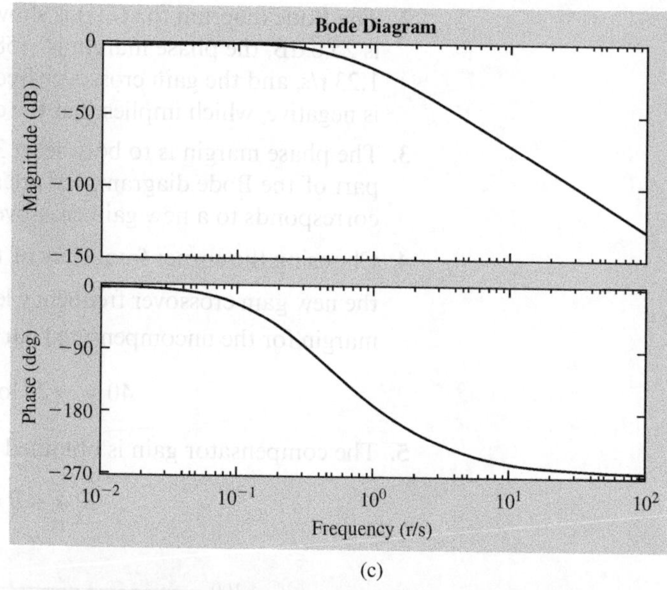

(c)

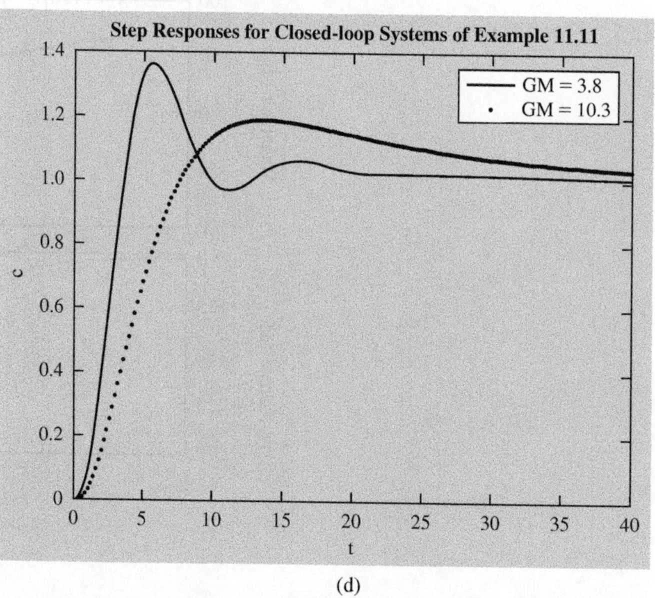

(d)

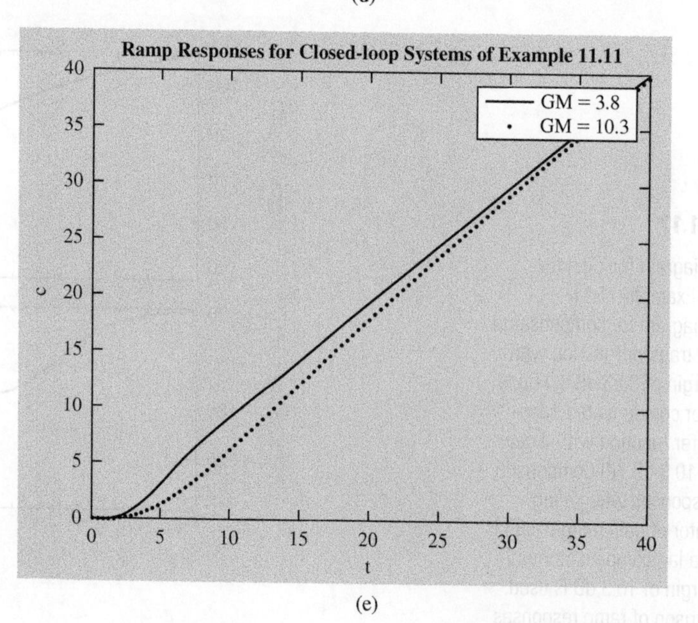

(e)

Figure 11.17

(Continued)

The transfer function for the compensator becomes

$$G_c(s) = 0.09\frac{s + 0.05}{s + 0.0005} \tag{g}$$

The open-loop transfer function becomes

$$G(s)\,G_c(s) = \left[\frac{10}{s(s+1)(s+1.5)}\right]\left(0.09\frac{s+0.05}{s+0.0005}\right)$$

$$= \frac{0.9s + 0.045}{s^4 + 2.5005s^3 + 1.50125s^2 + 0.00075s} \tag{h}$$

The Bode plot for this transfer function is shown in Figure 11.17(b).

6. The phase margin is 38.7°, which is above the required minimum. The gain margin is 3.83 dB. For part a. this is sufficient, and a compensator is designed that meets the requirements. For part b., the gain margin is below the required gain margin. The phase crossover frequency margin of the design is 1.17 r/s. The goal would be to choose a new phase crossover frequency such that the gain margin is 10 dB. However, other factors would change the gain margin. Increasing the value of α to 275 yields a compensator that leads to a gain margin for the compensated system of 10.3 dB and a phase margin of 58.1°, both of which are acceptable. The Bode diagram for the compensated system is shown in Figure 11.17(c).
7. When placed in a unity feedback loop, the plant is unstable. The step response grows without bound. The step responses of the compensated system with a gain margin of 3.8 dB and the compensated system with a gain margin of 10.3 dB are shown in Figure 11.17(d). The system with a higher gain margin is slower but has a smaller overshoot. The ramp responses of the compensated system with a gain margin of 3.8 dB and the compensated system with a gain margin of 10.3 dB are shown in Figure 11.17(e). The step responses and the ramp responses are acceptable.

11.9 Design of Lead-Lag Compensators

A lead-lag compensator has a transfer function of the form

$$G_c(s) = K_c\left(\frac{s+\frac{1}{T_1}}{s+\frac{1}{\alpha_1 T_1}}\right)\left(\frac{s+\frac{1}{T_2}}{s+\frac{1}{\alpha_2 T_2}}\right) \qquad \alpha_1 < 1, \alpha_2 > 1 \tag{11.50}$$

The open-loop transfer function when a plant with the transfer function $G(s)$ is placed in series with a lead-lag compensator is

$$G_o(s) = G(s)\,K_c\left(\frac{s+\frac{1}{T_1}}{s+\frac{1}{\alpha_1 T_1}}\right)\left(\frac{s+\frac{1}{T_2}}{s+\frac{1}{\alpha_2 T_2}}\right)$$

$$= K_c\alpha_1\alpha_2 G(s)\left(\frac{T_1 s+1}{\alpha_1 T_1 s+1}\right)\left(\frac{T_2 s+1}{\alpha_2 T_2 s+1}\right) = G_1(s)\,G_2(s) \tag{11.51}$$

where

$$G_1(s) = K_c\alpha_1\alpha_2 G(s) \tag{11.52}$$

and

$$G_2(s) = \left(\frac{T_1 s+1}{\alpha_1 T_1 s+1}\right)\left(\frac{T_2 s+1}{\alpha_2 T_2 s+1}\right) \tag{11.53}$$

The compensator is designed such that the open-loop transfer function has a minimum static error constant, a minimum phase margin, and a minimum gain margin. If the static velocity error constant is specified, then

$$C_v = \lim_{s\to 0} s\ G_1(s)\,G_2(s) = \lim_{s\to 0} sK_c\alpha_1\alpha_2 G(s)\left(\frac{T_1 s + 1}{\alpha_1 T_1 s + 1}\right)\left(\frac{T_2 s + 1}{\alpha_2 T_2 s + 1}\right)$$
$$= \lim_{s\to 0} K_c\alpha_1\alpha_2 sG(s) \tag{11.54}$$

The value of $K_c\alpha_1\alpha_2$ is specified from the static velocity error constant.

Many lead-lag compensators are designed such that $\alpha_2 = \frac{1}{\alpha_1}$. In this case, the value of K_c is specified using the static velocity error constant. The following discussion and examples all use this assumption. If the assumption is not made, then a trial-and-error procedure may be used to design the compensator. Using the root-locus method to design the compensator is an alternative technique.

The Bode diagram of a lead-lag compensator is shown in Figure 11.18 for $\alpha_1 = 0.1$, $\alpha_2 = 10, \left(\alpha_2 = \frac{1}{\alpha_1}\right), T_1 = 0.01$, and $T_2 = 100$. The Bode diagram for a lead-lag compensator has a high-frequency asymptote and a low-frequency asymptote of $L(\omega) = 0$. The corner frequencies are $\frac{1}{\alpha_2 T_2}$ and $\frac{1}{\alpha_1 T_1}$. The compensator acts as a band-reject filter. If $\alpha_2 \neq \frac{1}{\alpha_1}$, the low- and high-frequency asymptotes are $L(\omega) = C_1$ and $L(\omega) = C_2$. The minimum of the gain part of the Bode diagram for a lead-lag compensator is -20 dB.

The lead part of the compensator is used to improve the stability of the closed-loop system. The lag part of the compensator is used to give adequate steady-state performance. The design of a lead-lag compensator to meet required conditions is not unique. The following is simply a suggestion on how to proceed using the Bode diagram of the open-loop system.

After the value of $K_c\alpha_1\alpha_2$ has been specified and $G_1(s)$ is known, its Bode diagram is drawn. The phase margin to be added by the compensator is determined. From this, a new gain crossover frequency is determined. The value of α_1 is found using Equation (11.38). The corner frequency of the lag portion $\frac{1}{T_2}$ is chosen to be about one decade below the new gain crossover frequency. With $\alpha_2 = \frac{1}{\alpha_1}$, Equation (11.54) leads to

$$C_v = K_c \lim_{s\to 0} sG(s) \tag{11.55}$$

The new gain crossover frequency ω_{cg1} for $L(\omega)$ is determined as the phase crossover frequency for G_1. Suppose the gain margin for G_1 is GM_1. Notice that the gain

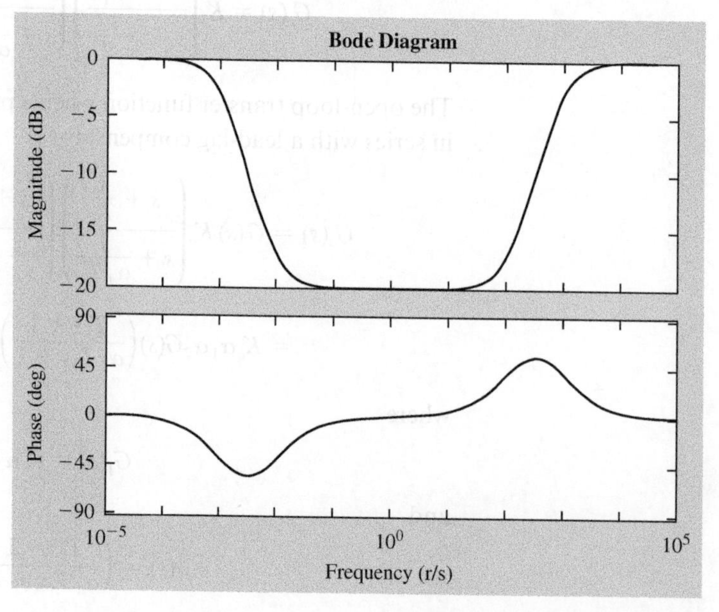

Figure 11.18

Bode diagram of lead-lag compensator.

portion of the Bode diagram for the lead part of the compensator evaluated at its corner frequency $\frac{1}{T_1}$ is

$$L_1\left(\frac{1}{T_1}\right) = 20\log\left|\frac{jT_1\frac{1}{T_1}+1}{j\alpha_1 T_1\frac{1}{T_1}+1}\right| = 20\log\left|\frac{1+j}{1+\alpha_1 j}\right| = 10\log\left|\frac{1+2\alpha_1^2}{1+\alpha_1^2}\right| \quad (11.56)$$

For small α_1, Equation (11.56) is approximated by

$$L_1\left(\frac{1}{T_1}\right) = 10\log|1| = 0 \quad (11.57)$$

The value of T_1 is determined by requiring that $L_1(\omega_{cg1}) = GM_1$ and $L_1\left(\frac{1}{T_1}\right) = 0$. This would require that a line of slope -20 dB be drawn through the point $(\omega_{cg1}, 20\log GM_1)$ and intersect the line $L = 0$ at the larger corner frequency of the compensator, $\omega = \frac{1}{\alpha T_1}$.

The corner frequency of the lag portion $\frac{1}{T_2}$ is chosen to be about one decade below the new gain crossover frequency.

The transfer function for the lead-lag compensator is then obtained. The compensator performance must be tested by simulating the closed-loop system with a step input and/or a ramp input.

Example 11.12

Design a lead-lag compensator for a plant with the transfer function

$$G(s) = \frac{4}{s(s+0.5)(s+1)} \quad \text{(a)}$$

such that the static velocity error constant is at least 40 s^{-1}, the phase margin is at least 35°, and the gain margin is at least 12 dB. Assume $\alpha_2 = \frac{1}{\alpha_1}$.

Solution

The compensator gain is specified from Equation (11.55):

$$C_v = 40\text{ s}^{-1} = \lim_{s\to 0} K_c\alpha_1\alpha_2 sG(s) = \lim_{s\to 0} K_c\alpha_1\alpha_2 s\frac{4}{s(s+0.5)(s+1)}$$

$$= \frac{4}{0.5}K_c\alpha_1\alpha_2 \Rightarrow K_c\alpha_1\alpha_2 = 5 \quad \text{(b)}$$

The Bode diagram for $G_1(s) = \frac{20}{s(s+0.5)(s+1)}$ is shown in Figure 11.19(a). The phase margin is $-58.5°$, the gain margin is 28.5 dB, the phase crossover frequency is 0.707 r/s, and the gain crossover frequency is 0.0375 = 2.64 r/s. The phase margin being negative implies that the gain-adjusted transfer function is unstable in a unity feedback loop.

The phase crossover frequency is chosen as the new gain crossover frequency for the compensated system, $\omega_{cg} = 0.707$ r/s. The phase lead angle at this frequency is to be 35°. It is chosen to add 5°, so the phase lead is chosen to be 40°. Equation (11.38) gives

$$\alpha = \frac{1-\sin\phi_m}{1+\sin\phi_m} = \frac{1-\sin 40°}{1+\sin 40°} = 0.217 \quad \text{(c)}$$

Since the gain margin of the uncompensated system is 0.188, then $L(\omega_{cg1}) = -20\log 0.0375 = 28.5$. The maximum value of T_1 is determined by drawing a line of slope 20 dB per decade in $L(\omega)$ space through the point $(0.707, -28.5)$ and determining where it intersects the line $L = 0$. This is illustrated in Figure 11.19(b) and leads to $\frac{1}{\alpha T_1} = 1.25$ r/s and $T_1 = 3.69$ s. The second corner frequency of the compensator is $\frac{1}{1.25} = 0.8$ r/s.

The value of T_2 is chosen such that $\frac{1}{T_2}$ is one decade lower than the new gain crossover frequency, which is 0.07 r/s. Thus, $T_2 = 14.3$. The pole of the lag compensator is $\frac{\alpha}{T_2} = 0.0152$ r/s.

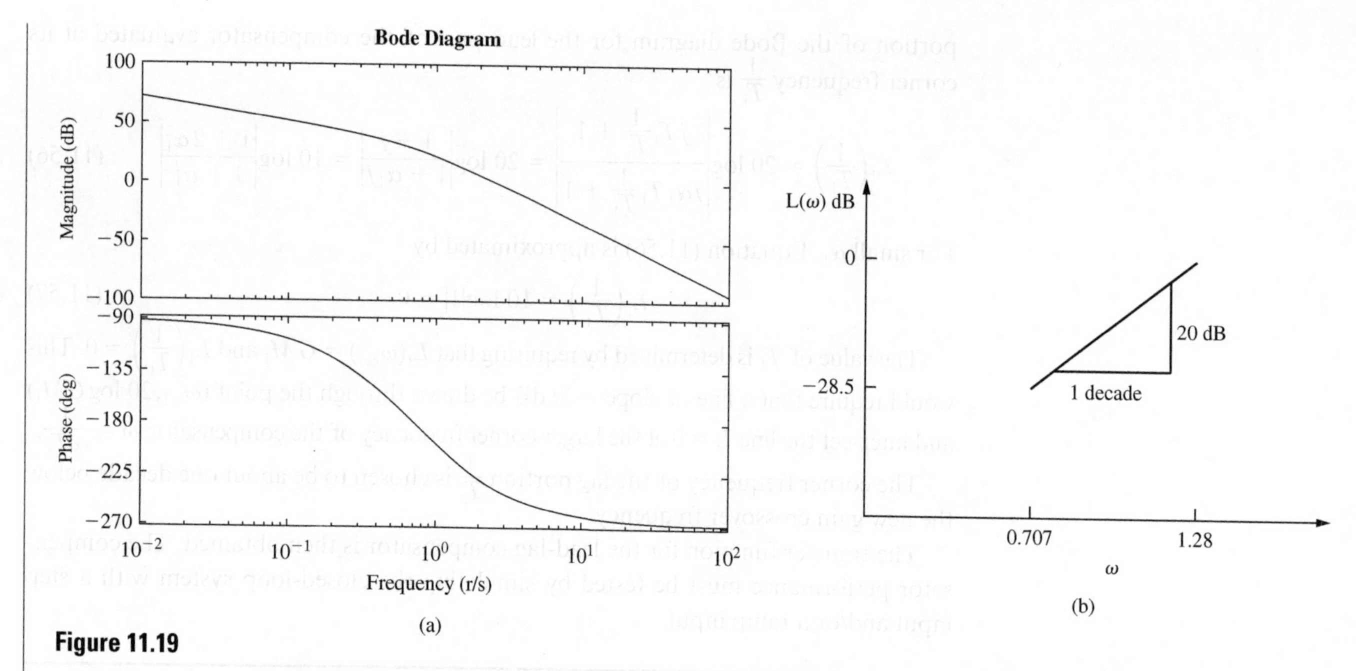

Figure 11.19

(a) Bode diagram for $G_1(s)$ of Example 11.12. The negative phase margin implies that the gain adjusted plant is unstable when placed in a unity feedback loop. (b) Geometry used for choosing T_1.

The transfer function for the compensator is

$$G_c(s) = 5\left(\frac{s+0.8}{s+1.25}\right)\left(\frac{s+0.07}{s+0.0152}\right) \tag{d}$$

The open-loop compensated transfer function is

$$G_o(s) = \frac{20(s+0.8)(s+0.07)}{s(s+0.5)(s+1)(s+1.25)(s+0.0152)}$$

$$= \frac{20s^2+17.4s+1.12}{s^5+2.7652s^4+2.4168s^3+0.6611s^2+0.0095s} \tag{e}$$

The closed-loop transfer function for the compensated system is

$$H(s) = \frac{20s^2+17.4s+1.12}{s^5+2.7652s^4+2.4168s^3+20.6611s^2+17.4094s+1.12} \tag{f}$$

11.10 Further Examples

Example 11.13

A plant has a transfer function of

$$G(s) = \frac{12(s+2)}{s^4+6s^3+12s^2+10s+45} \tag{a}$$

The plant is placed in a feedback loop with **a.** a lead compensator whose transfer function is

$$G_c(s) = 15\frac{s+3}{s+20} \tag{b}$$

and **b.** a PID controller whose transfer function is

$$G_c(s) = 1.5 + \frac{2}{s} + 6s \tag{c}$$

Use the Nyquist stability criterion to determine if the closed-loop system is stable.

Solution

a. The open-loop transfer function of the system is

$$G_o(s) = \frac{180(s+3)(s+2)}{(s+20)(s^4+6s^3+12s^2+10s+45)} \tag{d}$$

It is noted that two poles of $G_o(s)$ have a positive real part, $0.4 \pm 1.73j$. In order for the closed-loop system to be stable, the point $(-1, 0)$ must be encircled two times on the Nyquist diagram for $G_o(s)$. As evidenced by the Nyquist diagram drawn using MATLAB and shown in Figure 11.20(a), the point $(-1, 0)$ is not encircled at all. Hence, the closed-loop system is unstable.

b. The open-loop transfer function of the system is

$$G_o(s) = \frac{12(6s^2+1.5s+2)(s+2)}{s(s^4+6s^3+12s^2+10s+45)} \tag{e}$$

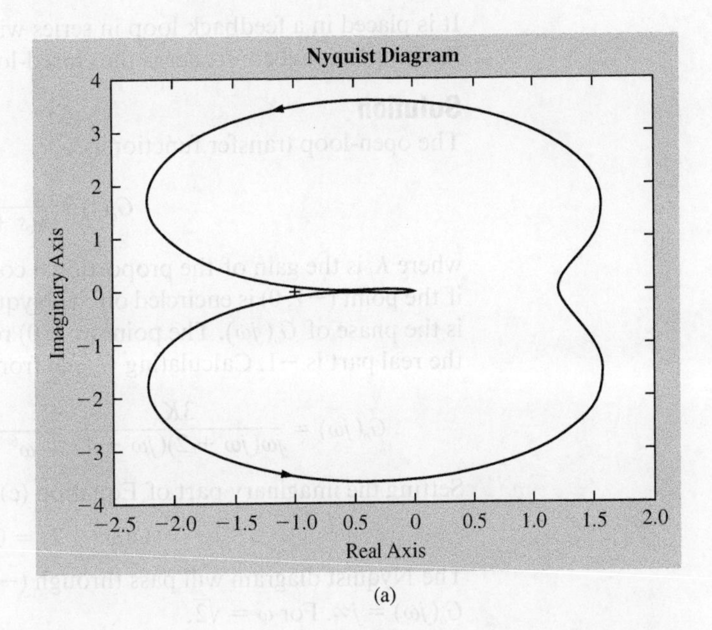

(a)

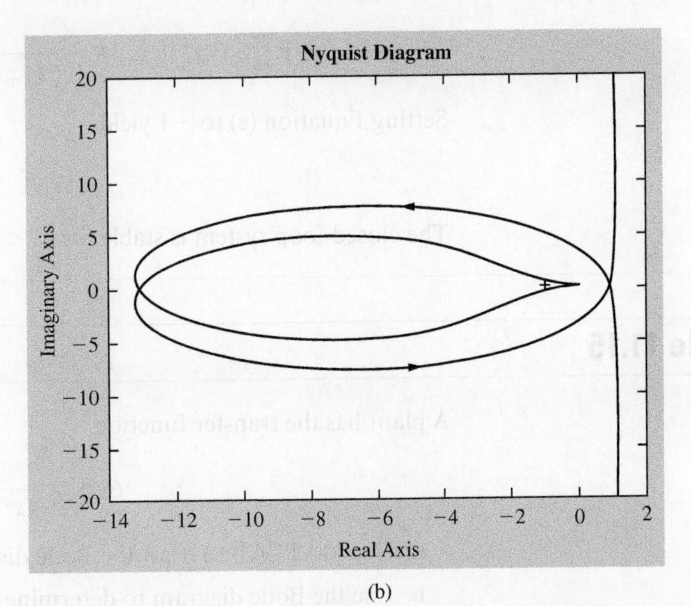

(b)

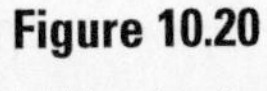

Figure 10.20

(a) Nyquist diagram for Example 11.13. Since the open-loop transfer function has two poles with positive real parts and the Nyquist diagram does not encircle the point $(-1, 0)$, the closed-loop transfer function is unstable. (b) Nyquist diagram for Example 11.13(b). The open-loop transfer function has two poles with positive real parts. The point $(-1, 0)$ is encircled two times, thus the closed-loop system is stable.

The open-loop transfer function has two roots with positive real parts and one root with a zero real part. The Nyquist diagram for the transfer function of Equation (e) is shown in Figure 11.20(b). The Nyquist stability criterion becomes more complicated when $G_o(s)$ has a root with a real part of zero. When $\omega = 0$, $|G(j0^+)| = -j\infty$ and $|G(j0^-)| = j\infty$. The Nyquist diagram is discontinuous. When a small parameter is introduced, measuring the distance from zero and varying the phase, it can be shown that these two branches connect in a very large semicircle that does not encircle the point $(-1, 0)$. Thus, because there are two poles with positive real parts, the point $(-1, 0)$ must be encircled two times for the closed-loop system to be stable. Since it does, the system is stable.

Example 11.14

A plant has the transfer function

$$G(s) = \frac{3}{(s+2)(s+1)} \tag{a}$$

It is placed in a feedback loop in series with an integral controller. For what gains of the proportional controller is the closed-loop system stable?

Solution

The open-loop transfer function is

$$G_o(s) = \frac{3K_i}{s(s+2)(s+1)} \tag{b}$$

where K_i is the gain of the proportional controller. The closed-loop system is unstable if the point $(-1, 0)$ is encircled on the Nyquist diagram for $G_o(s)$. The Nyquist diagram is the phase of $G_o(j\omega)$. The point $(-1, 0)$ represents where the imaginary part is 0 and the real part is -1. Calculating $G_o(j\omega)$ from Equation (b) gives

$$G_o(j\omega) = \frac{3K_i}{j\omega(j\omega+2)(j\omega+1)} = \frac{3K_i}{\omega^6 + 5\omega^4 + 4\omega^2}[-3\omega^2 + j\omega(\omega^2 - 2)] \tag{c}$$

Setting the imaginary part of Equation (c) to zero gives

$$\omega(\omega^2 - 2) = 0 \Rightarrow \omega, \pm\sqrt{2} \tag{d}$$

The Nyquist diagram will pass through $(-1, 0)$ for the critical value of K_i. For $\omega = 0$, $G_o(j\omega) = j\infty$. For $\omega = \sqrt{2}$,

$$\text{Re}[G_o(j\sqrt{2})] = 3K_i\frac{-3(\sqrt{2})^2}{(\sqrt{2})^6 + 5(\sqrt{2})^4 + 4(\sqrt{2})^2} = -\frac{9}{16}K_i \tag{e}$$

Setting Equation (e) to -1 yields

$$K_i = \frac{16}{9} \tag{f}$$

The closed-loop system is stable for $K_i < \frac{16}{9}$.

Example 11.15

A plant has the transfer function

$$G(s) = \frac{140}{s(s^2 + 5s + 14)} \tag{a}$$

a. Use MATLAB to draw the Bode diagram for this plant.

b. Use the Bode diagram to determine the gain crossover frequency.

c. Use the Bode diagram to determine the phase crossover frequency.

d. Use the Bode diagram to determine the phase margin.

e. Use the Bode diagram to determine the gain margin in dB.

f. Determine the gain margin, GM.

g. Use the MATLAB function margin to determine the exact values of the gain crossover frequency, phase crossover frequency, gain margin, and phase margin.

h. The plant is placed in a unity feedback loop. Is the closed-loop system stable?

i. If the closed-loop system is not stable, determine the value of the gain of a proportional controller that would make it stable.

Solution

a. The Bode diagram drawn by MATLAB is shown in Figure 11.21.

b. The gain crossover frequency is the frequency at which $L(\omega) = 0$. From the Bode diagram, this occurs at about 5 r/s.

c. The phase crossover frequency is the frequency at which the phase is $-180°$. From the phase part of the Bode diagram, this occurs at about 3.5 r/s.

d. The phase margin is the phase at the gain crossover frequency minus $-180°$. From the phase portion of the Bode diagram, the phase at the gain crossover frequency is about $-205°$. The phase margin is about $-25°$.

e. The gain margin is the value on the magnitude part of the Bode diagram at the phase crossover frequency. This appears to give 5 dB.

f. The gain margin is the value of $|G(j\omega_{cp})|$ such that the gain margin in dB is

$$\overline{GM} = -20 \log|G(j\omega_{cp})| \tag{b}$$

Solving Equation (b) for the gain margin,

$$5 \text{ dB} = -20 \log GM \tag{c}$$

Solving Equation (c) leads to

$$GM = 10^{-0.25} = 0.56 \tag{d}$$

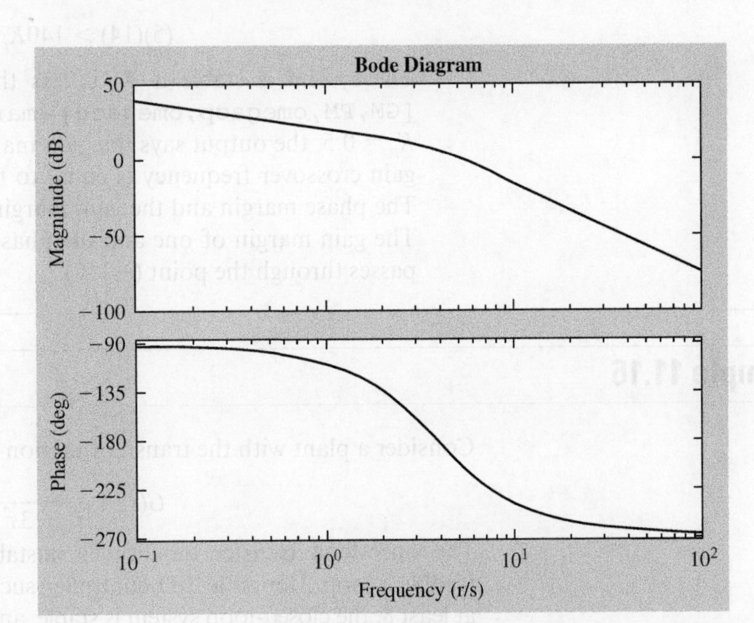

Figure 11.21

The MATLAB generated Bode diagram of the transfer function of Example 11.15.

g. The appropriate MATLAB statements are

```
N=[140]
D=[1 5 14 0]
sys(N,D)
[GM,PM,omegacp,omegacg]=margin(sys)
```

The MATLAB output using these commands is

```
Warning: The closed-loop system is unstable.
GM =
  0.5000
PM =
  -24.4667
omegacp =
  3.7417
omegacg =
  5.0483
```

h. If the plant is placed in a unity feedback loop, the closed-loop transfer function becomes

$$H(s) = \frac{140}{s^3 + 5s^2 + 14s + 140} \tag{e}$$

Applying Routh's stability criterion for a third-order system,

$$(5)(14) = 70 < 140 \tag{f}$$

The closed-loop system is unstable. This result is also determined by noting that the phase margin for the open-loop system is negative.

i. When placed with a proportional controller of gain K_p in a feedback loop, the open-loop transfer function is

$$G_o(s) = \frac{140K_p}{s(s^2 + 5s + 14)} \tag{g}$$

The closed-loop transfer function becomes

$$H(s) = \frac{140K_p}{s^3 + 5s^2 + 14s + 140K_p} \tag{h}$$

Applying Routh's stability criterion for a third-order system,

$$(5)(14) > 140K_p \Rightarrow K_p < 0.5 \tag{i}$$

The system is stable if K_p is less than 0.5. When the MATLAB statement `[GM,PM,omegacp,omegacg]=margin(sys)` for this system is used with $K_p = 0.5$, the output says the gain margin is one, the phase margin is 0, and the gain crossover frequency is equal to the phase crossover frequency of 3.74 r/s. The phase margin and the gain margin imply that the system is neutrally stable. The gain margin of one and the phase margin of 0 imply the Nyquist diagram passes through the point $(-1, 0)$.

Example 11.16

Consider a plant with the transfer function

$$G(s) = \frac{8}{s^3 + 3s^2 + 7s + 15} \tag{a}$$

The open-loop transfer function is unstable when the plant is placed in a unity feedback loop. Design a PID controller such that the static position error constant is at least 8, the closed-loop system is stable, and there is a gain margin of at least 10 dB.

Solution

The Bode diagram for the plant is shown in Figure 11.22(a). The properties are $\omega_{cg} = 2.76$ r/s, $\omega_{cp} = 2.64$ r/s, $PM = -12.1°$, and $\widetilde{GM} = -2.5$ dB.

The transfer function for the controller may be written such that the open-loop transfer function is in the form of Equation (11.26):

$$G_o(s) = \frac{K_i}{s}(1 + as)(1 + bs)G(s) \quad \text{(b)}$$

Selecting the integral gain such that the static position error constant is 8 s leads to

$$8 = \lim_{s \to 0} G_o(s) = \lim_{s \to 0} \frac{K_i}{s}(1 + as)(1 + bs)\frac{8}{s^3 + 3s^2 + 7s + 15} = \frac{8K_i}{15} \Rightarrow K_i = 15 \quad \text{(c)}$$

For

$$G_1(s) = \frac{120}{s(s^3 + 3s^2 + 7s + 15)} \quad \text{(d)}$$

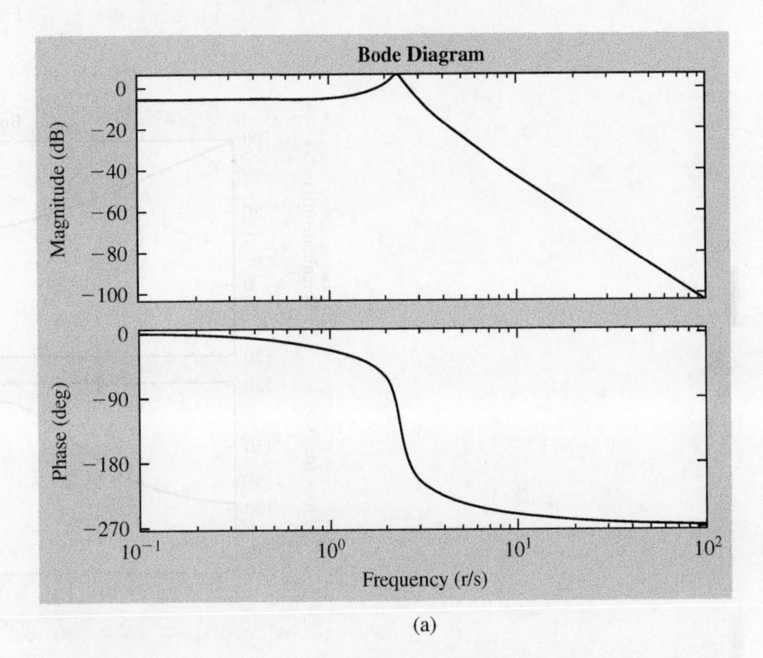

(a)

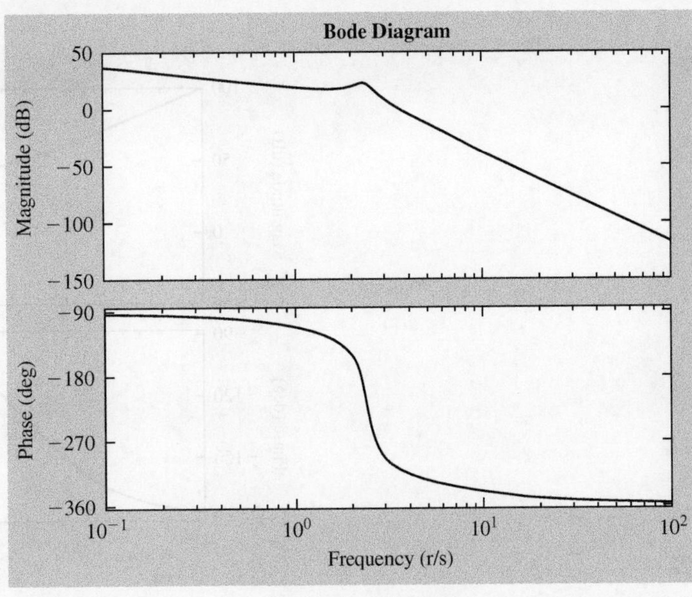

(b)

Figure 11.22

(a) The Bode diagram of the transfer function of Example 11.16. (b) The Bode diagram of $G_1(s)$. (c) The Bode diagram of $G_3(s)$. (d) The Bode diagram of $G_o(s)$ with $b = 0.2$. (e) The Bode diagram for $b = 1$ leads to an acceptable compensator design.

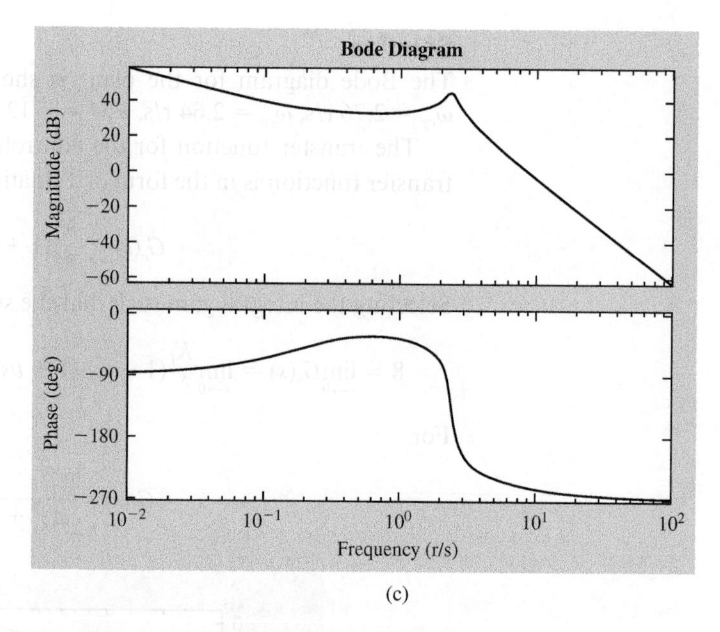

(c)

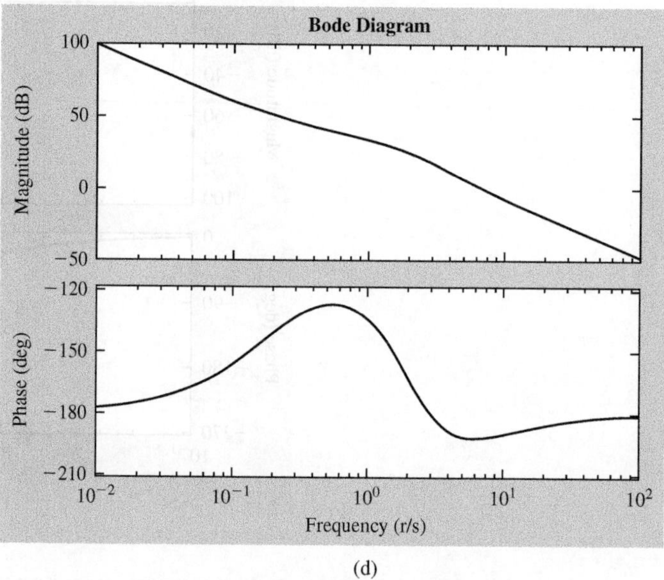

(d)

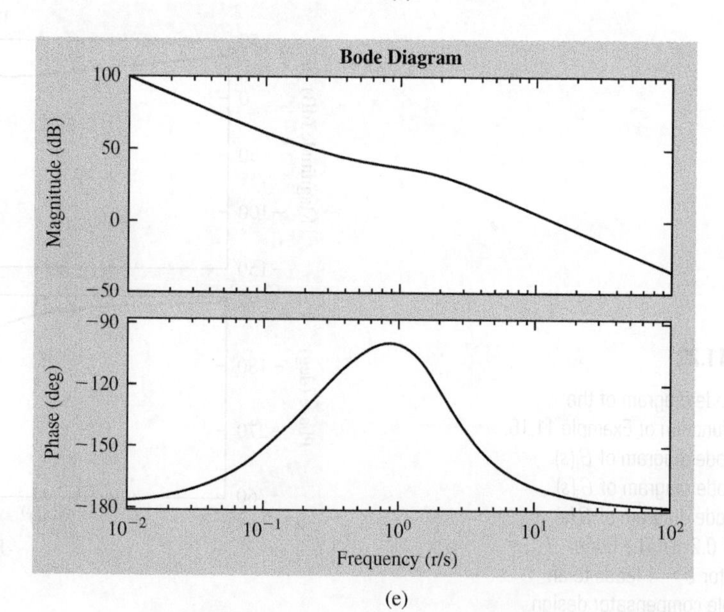

(e)

Figure 11.22

(Continued)

the Bode diagram for $G_1(s)$ is shown in Figure 11.22(b). The phase margin is $-132°$, the gain margin is 0.0833, the phase crossover frequency is 2.4 r/s, and the gain crossover frequency is 3.63 r/s. The new gain crossover frequency is chosen to be 5 r/s.

For

$$G_3(s) = \frac{120(5s + 1)}{s(s^3 + 3s^2 + 7s + 15)} \tag{e}$$

the Bode diagram for $G_3(s)$ is shown in Figure 11.22(c). The phase margin is $-71.9°$ and the gain crossover frequency is 8.55 r/s. The open-loop transfer function for an arbitrary b is

$$G_o(s) = \frac{40(5s + 1)(bs + 1)}{s^2(s^2 + 3s + 4)} \tag{f}$$

Trying $b = 0.2$ gives the Bode diagram for the open-loop transfer function shown in Figure 11.22(d). The phase margin is $-9.6°$. The closed-loop system is still unstable. Several values of b are tried, and eventually $b = 1$ leads to the Bode diagram shown in Figure 11.22(e). The phase margin is 4.2° and the gain margin is infinite. The gain crossover frequency is 2.6 r/s. An acceptable compensator design is

$$G_c(s) = \frac{15}{s}(1 + 5s)(1 + s) = 90 + \frac{15}{s} + 90s \tag{g}$$

Example 11.17

Design a lead compensator for a plant with the transfer function

$$G(s) = \frac{5}{s(s + 3)} \tag{a}$$

The static velocity error constant must be at least 50 s^{-1} and the phase margin must be at least 50°.

Solution

A lead compensator is to be employed. $G_1(s)$ and $G_2(s)$ are defined as in Equations (11.33) and (11.34). The static velocity error constant is to be at least 50 s^{-1}. Applying Equation (11.35),

$$50 = \lim_{s\to 0} s\, G_1(s)\, G_2(s) = \lim_{s\to 0} sK_c\alpha G(s)\frac{Ts + 1}{\alpha Ts + 1}$$

$$= \lim_{s\to 0} K_c\alpha sG(s) = \frac{5}{3}K_c\alpha \Rightarrow K_c\alpha = 30 \tag{b}$$

Then

$$G_1(s) = K_c\alpha G(s) = \frac{150}{s(s + 3)} \tag{c}$$

The Bode diagram for the uncompensated $G_1(s)$ is shown in Figure 11.23(a). The phase margin is 14.0°, the gain crossover frequency is 12.1 r/s, and the phase crossover frequency and the gain margin are infinite. The compensator needs to make up $50° - 14.0° = 36.0°$. Adding 5° gives a phase angle of 41.0°. The attenuation factor is calculated using Equation (11.38) as

$$\alpha = \frac{1 - \sin\phi_m}{1 + \sin\phi_m} = \frac{1 - \sin 41.0°}{1 + \sin 41.0°} = 0.207 \tag{d}$$

The new gain crossover frequency is calculated using Equation (11.39):

$$20\log\frac{1}{\sqrt{\alpha}} = 20\log\frac{1}{\sqrt{0.207}} = 6.8 \text{ dB} \tag{e}$$

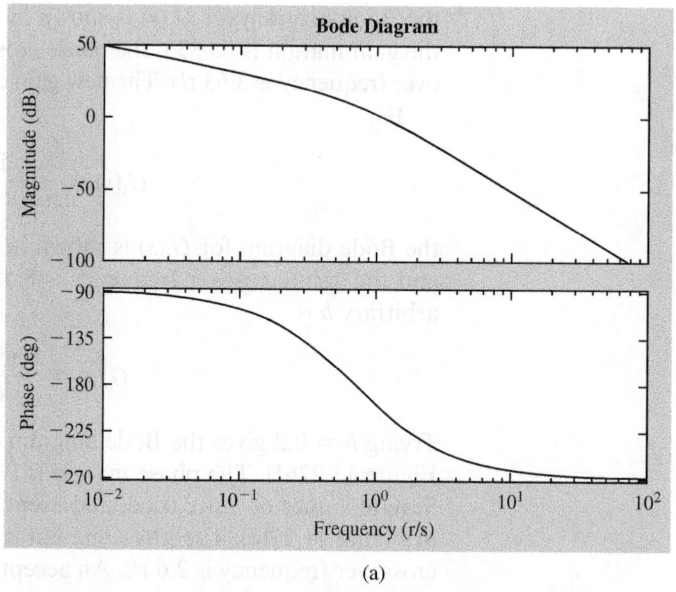

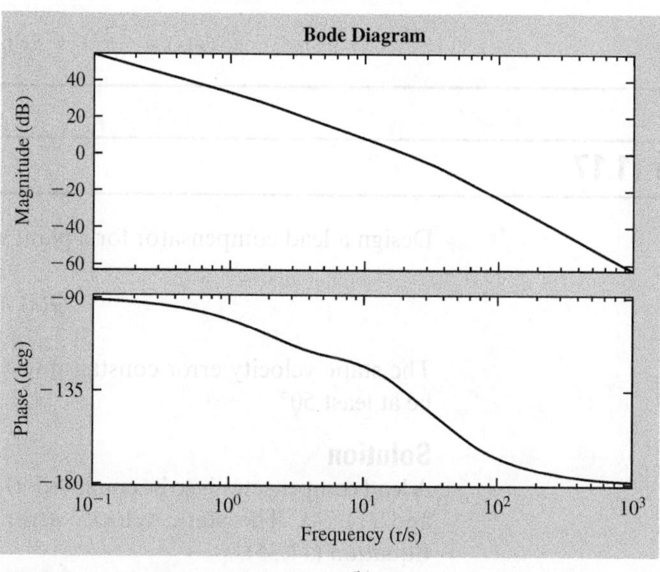

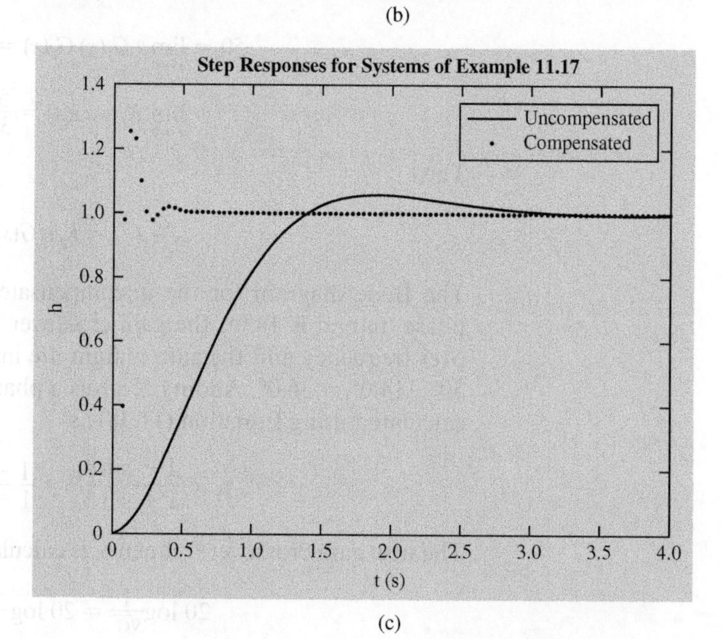

Figure 11.23

(a) Bode diagram for $G_1(s)$ of Example 11.17. (b) The Bode diagram of the open-loop transfer function. (c) Comparison of step responses of uncompensated system and compensated system. (d) Comparison of ramp responses.

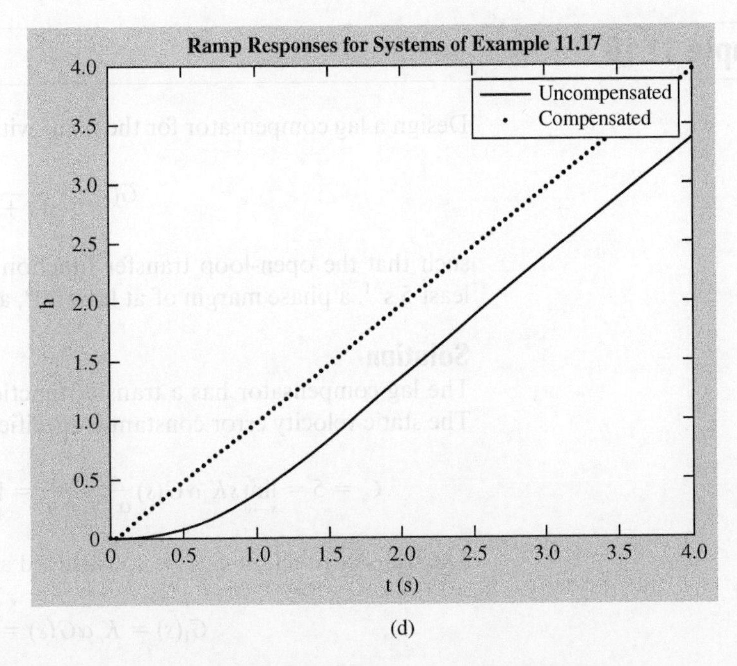

(d)

Figure 11.23
(Continued)

From Figure 11.23(a), it is found that $L(\omega) = -6.7$ dB for a frequency of 12.1 r/s. This is the new gain crossover frequency. Using Equation (11.40) leads to

$$\omega_c = \frac{1}{\sqrt{\alpha}\,T} \Rightarrow T = \frac{1}{\omega_c\sqrt{\alpha}} = \frac{1}{(12.1)\sqrt{0.207}} = 0.182 \tag{f}$$

The zero of the compensator and the pole of the compensator are calculated using Equation (11.43) as

$$s_z = \frac{1}{T} = 5.50 \quad s_p = \frac{1}{\alpha T} = 26.5 \tag{g}$$

The compensator gain is calculated using Equation (b) as

$$K_c = \frac{K_c\alpha}{\alpha} = \frac{30}{0.207} = 144.9 \tag{h}$$

The compensator design is

$$G_c(s) = 144.9\frac{s + 5.50}{s + 26.5} \tag{i}$$

The open-loop transfer function with the compensator is

$$G_o(s) = \frac{724.5(s + 5.50)}{s(s + 3)(s + 26.5)} = \frac{724.5s + 3984}{s^3 + 29.5s^2 + 79.5s} \tag{j}$$

The Bode diagram for the open-loop transfer function is shown in Figure 11.23(b). The phase margin is 44.4°, the gain crossover frequency is 21.6 r/s, and the phase crossover frequency and the gain margin are infinite.

The closed-loop transfer function is

$$H(s) = \frac{724.5(s + 5.50)}{s(s + 3)(s + 26.5)} = \frac{724.5s + 3984}{s^3 + 29.5s^2 + 804s + 3984} \tag{k}$$

The step response of the uncompensated system placed in a unity feedback loop and the compensated system are shown in Figure 11.23(c). The compensated system is much faster than the uncompensated system, but it has a larger overshoot. The ramp response of the uncompensated system placed in a unity feedback loop and the compensated system are shown in Figure 11.23(d). The ramp response of the compensated system tracks a unit ramp input much better than the uncompensated system.

Example 11.18

Design a lag compensator for the plant with the transfer function

$$G(s) = \frac{2}{s(s+1)(s+2)} \tag{a}$$

such that the open-loop transfer function has a static velocity error constant of at least 5 s^{-1}, a phase margin of at least 40°, and a gain margin of at least 8 dB.

Solution

The lag compensator has a transfer function similar to the one in Equation (11.32). The static velocity error constant is specified as in Equation (11.47):

$$C_v = 5 = \lim_{s\to 0} sK_c\alpha G(s)\frac{Ts+1}{\alpha Ts+1} = \lim_{s\to 0} K_c\alpha sG(s) = K_c\alpha \Rightarrow K_c\alpha = 5 \tag{b}$$

The transfer function $G_1(s)$ is constructed as

$$G_1(s) = K_c\alpha G(s) = \frac{10}{s(s+1)(s+2)} \tag{c}$$

The Bode diagram for $G_1(s)$ is shown in Figure 11.24(a). The phase margin is −32°, the gain margin is −10 dB, the gain crossover frequency is 2.2 r/s, and the phase crossover frequency is 1.5 r/s. The phase margin and the gain margin are both less than zero. This implies that when the plant represented by $G_1(s)$ is placed in a unity feedback loop, it is unstable.

It is noted that the required phase margin is about 40°, which corresponds to a frequency of 0.7 r/s. To account for the phase change added by the compensator, the phase margin should be about 50°. The compensated phase angle should be −180° + 50° = −130°. The frequency corresponding to this phase angle is about 0.5 r/s. This should be the new gain crossover frequency.

If a corner frequency $\frac{1}{T}$ for the compensator is chosen to be one decade below this new gain crossover frequency, then

$$\frac{1}{T} = 0.05 \Rightarrow T = 20 \text{ s} \tag{d}$$

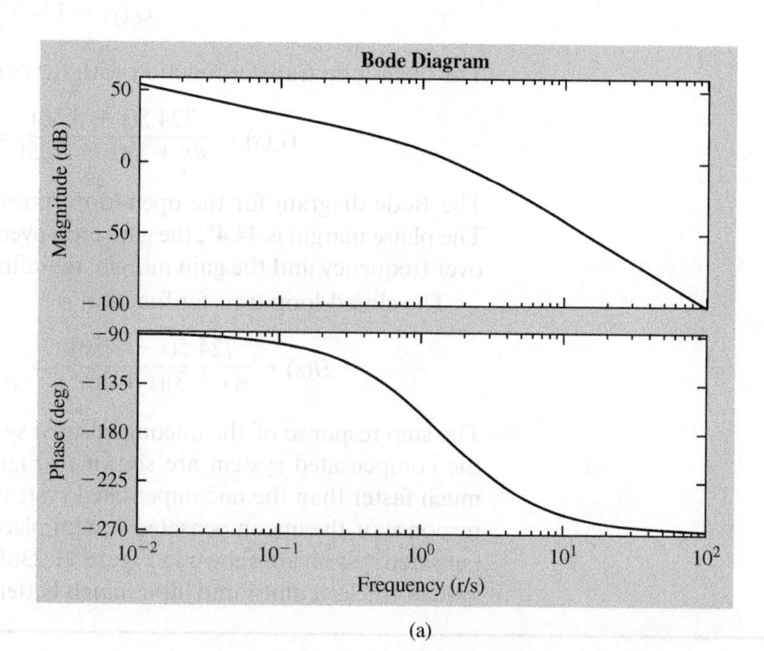

Figure 11.24
(a) Bode diagram of $G_1(s)$ for Example 11.18. (b) The Bode diagram of the open-loop transfer function. (c) Comparison of step responses of uncompensated system and compensated system. (d) Comparison of ramp responses.

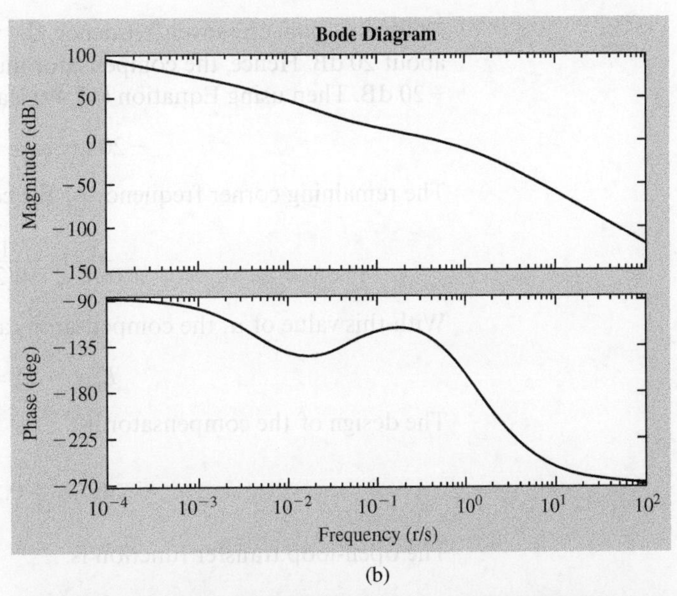

(b)

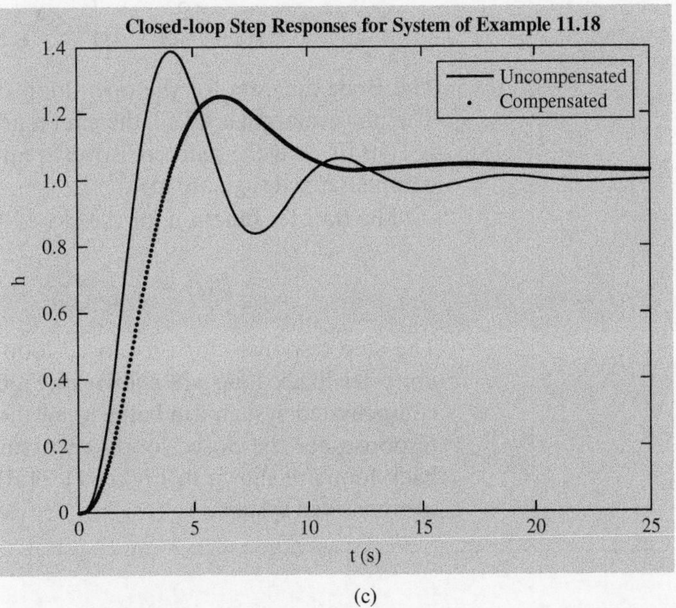

(c)

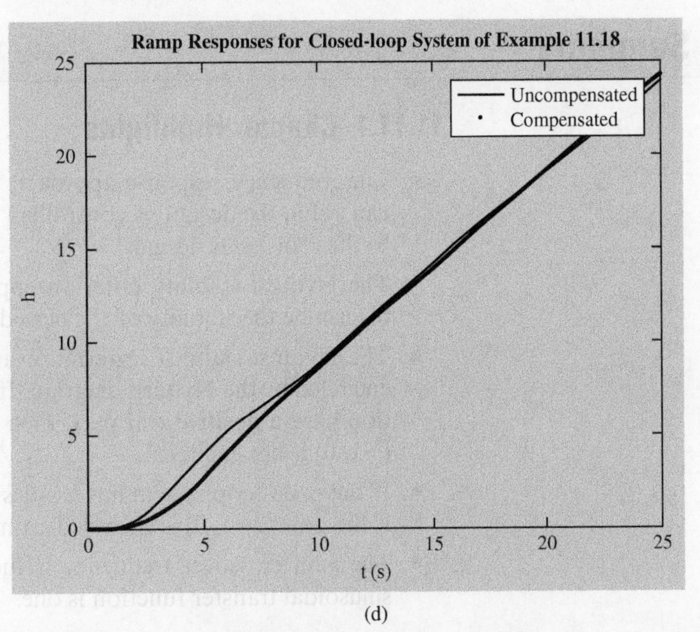

(d)

Figure 11.24

(Continued)

At the new gain crossover frequency, the value of $L(\omega)$ on the Bode diagram of $G_1(s)$ is about 20 dB. Hence, the compensator must bring the gain down to zero or contribute −20 dB. Then using Equation (11.49) leads to

$$-20 \log \alpha = -20 \Rightarrow \alpha = 10 \tag{e}$$

The remaining corner frequency of the compensator is

$$\frac{1}{\alpha T} = \frac{1}{10(20)} = 0.005 \tag{f}$$

With this value of α, the compensator gain is determined from Equation (b) as

$$K_c \alpha = 25 \Rightarrow K_c = 0.25 \tag{g}$$

The design of the compensator is

$$G_c(s) = 0.5\frac{s + 0.05}{s + 0.005} \tag{h}$$

The open-loop transfer function is

$$G_o(s) = \frac{2}{s(s+1)(s+2)}\left[0.5\frac{s + 0.05}{s + 0.005}\right] = \frac{(s + 0.05)}{s(s+1)(s+2)(s + 0.005)} \tag{i}$$

The Bode diagram for the open-loop transfer function is shown in Figure 11.24(b). The phase margin is 47.5°, the gain margin is 14.94 dB, the phase crossover frequency is 1.46 r/s, and the gain crossover frequency is 0.449 r/s. Thus, all conditions of the compensator design are met.

The transfer function for the closed-loop system is

$$H(s) = \frac{s + 0.05}{s^4 + 3.005s^3 + 2.015s^2 + 1.01s + 0.05} \tag{j}$$

The step responses for the closed-loop system and the uncompensated system in a unity feedback loop are shown in Figure 11.24(c). The transient properties for the compensated system are better than those for the uncompensated system. The ramp responses for the closed-loop system and the uncompensated system in a unity feedback loop are shown in Figure 11.24(d). The track of a ramp input is better for the compensated system.

11.11 Summary

11.11.1 Chapter Highlights

- The frequency response approach to assessing an open-loop transfer function can aid in the design of controllers and compensators. It provides an alternative to the root-locus design.
- The Nyquist stability criterion applied to an open-loop transfer function can determine the stability of the closed-loop transfer function.
- The Nyquist stability criterion examines how many times the point (−1, 0) is encircled by the Nyquist diagram. If all the poles of the open-loop transfer function have a positive real part, then the closed-loop system is stable if the point (−1, 0) is not encircled.
- If the open-loop system has k poles with a positive real part, the system is stable if the point (−1, 0) is encircled k times.
- The gain crossover frequency is the frequency at which the magnitude of the sinusoidal transfer function is one.

- The phase crossover frequency is the frequency at which the phase is −180°.
- The gain margin is the reciprocal of the sinusoidal transfer function at the phase crossover frequency.
- The phase margin is the difference between −180° and the phase at the gain crossover frequency. The phase margin is positive if the phase of the sinusoidal transfer function is greater than −180°.
- The phase margin must be positive if the closed-loop system is stable.
- The minimum of the appropriate static error constant, the minimum phase margin, and the minimum gain margin are usually specified when a controller or compensator is designed.
- The integral gain is chosen first when a PID controller is designed using the frequency response approach.
- The open-loop transfer function is written as the product of two transfer functions. One is a gain-adjusted transfer function for the plant; the other is the compensator design.
- Lead, lag, and lead-lag compensators can be designed using the frequency response approach.
- A compensator design is not unique.

11.11.2 Important Equations

- Gain crossover frequency

$$|G(j\omega_{cg})| = 1 \tag{11.5}$$

$$L(\omega_{cg}) = 0 \tag{11.6}$$

- Phase crossover frequency

$$\phi[G(\omega_{cp})] = -\pi \tag{11.7}$$

- Gain margin

$$GM = \frac{1}{|G(\omega_{cp})|} \tag{11.8a}$$

- Gain margin in dB

$$\widetilde{GM} = -20\log|G(j\omega_{cp})| = -L(\omega_{cp}) \tag{11.8b}$$

- Phase margin

$$PM = \phi[G(\omega_{cg})] - (-\pi) = \phi[G(\omega_{cg})] + \pi \tag{11.9}$$

- Gain crossover frequency in terms of natural frequency and damping ratio

$$\omega_{cg} = \omega_n\sqrt{\sqrt{1+4\zeta^4} - 2\zeta^2} \tag{11.19}$$

- Phase margin in terms of damping ratio

$$PM = \phi[G(\omega_{cg})] + \pi = \tan^{-1}\frac{2\zeta}{\sqrt{\sqrt{1+4\zeta^4} - 2\zeta^2}} \tag{11.20}$$

- PID controller design

$$G_c(s) = \frac{K_i}{s}(1+as)(1+bs) \tag{11.24}$$

- Static velocity error constant for PID controller design

$$C_v = K_iG(0) \tag{11.27}$$

- Transfer functions for PID controller design

$$G_o(s) = G_1(s)\,G_2(s) \tag{11.28}$$

$$G_1(s) = \frac{K_i}{s}G(s) \tag{11.29}$$

$$G_2(s) = (1+as)(1+bs) \tag{11.30}$$

- Lead compensator design

  Static velocity error constant

$$C_v = \lim_{s \to 0} K_c \alpha s G(s) \tag{11.35}$$

  Attenuation factor

$$\alpha = \frac{1 - \sin \phi_m}{1 + \sin \phi_m} \tag{11.38}$$

  Mean of corner frequencies of compensator

$$\omega_c = \frac{1}{T\sqrt{\alpha}} \tag{11.40}$$

  Determination of T

$$-20 \log \frac{1}{\sqrt{\alpha}} = 20 \log |G_1(j\omega_c)| \tag{11.42}$$

- Lag compensator design

  Static velocity error constant

$$C_v = \lim_{s \to 0} K_c \alpha s G(s) \tag{11.47}$$

  New gain crossover frequency

$$20 \log |G_1(\omega_{cg})| = -20 \log \frac{1}{\alpha} \tag{11.49}$$

- Lead-lag compensator design

$$C_v = \lim_{s \to 0} K_c \alpha_1 \alpha_2 s G(s) \tag{11.54}$$

Short Answer Problems

SA11.1 The Nyquist diagram for an open-loop transfer function is shown in Figure SA11.1. Is the closed-loop system stable? The open-loop transfer function has all poles with negative real parts.

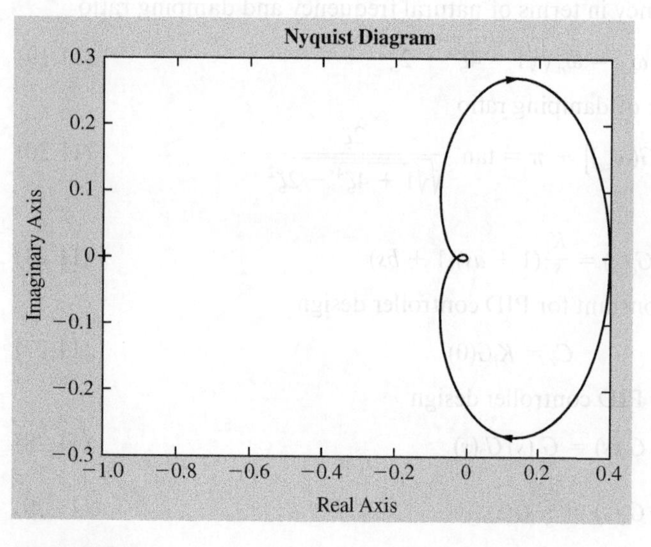

Figure SA11.1

SA11.2 The Nyquist diagram for an open-loop transfer function is shown in Figure SA11.2. Is the closed-loop system stable? The open-loop transfer function has all poles with negative real parts.

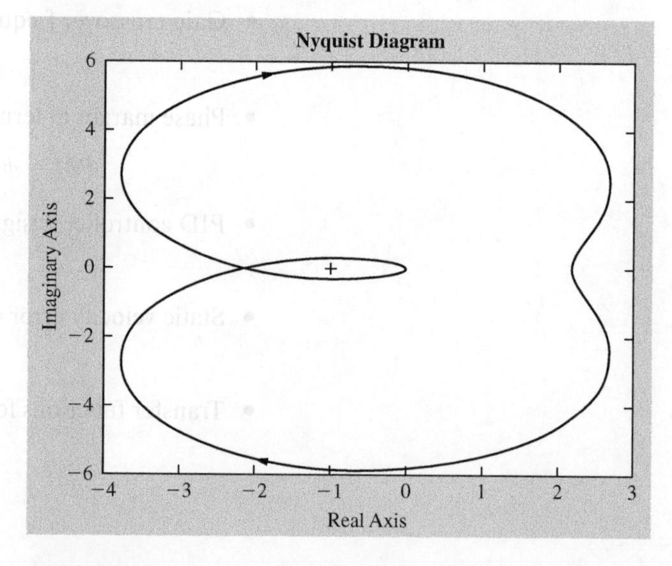

Figure SA11.2

SA11.3 The Nyquist diagram for an open-loop transfer function is shown in Figure SA11.3. Is the closed-loop system stable? The open-loop transfer function has one pole with a positive real part.

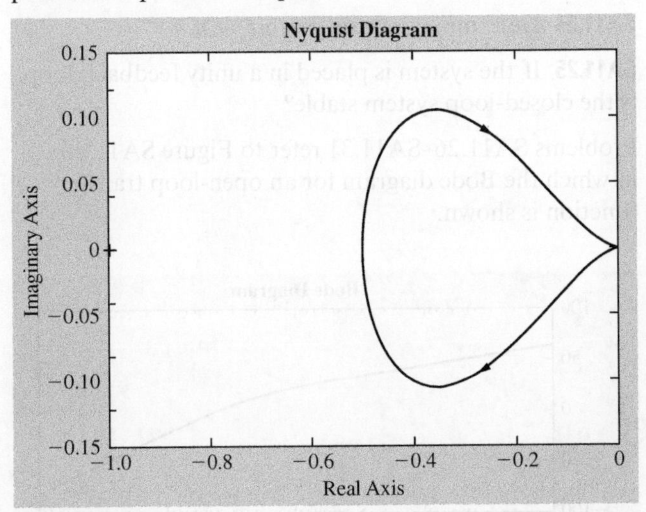

Figure SA11.3

SA11.4 The Nyquist diagram for an open-loop transfer function is shown in Figure SA11.4. Is the closed-loop system stable? The open-loop transfer function has two complex poles with a positive real part and all others with negative real parts.

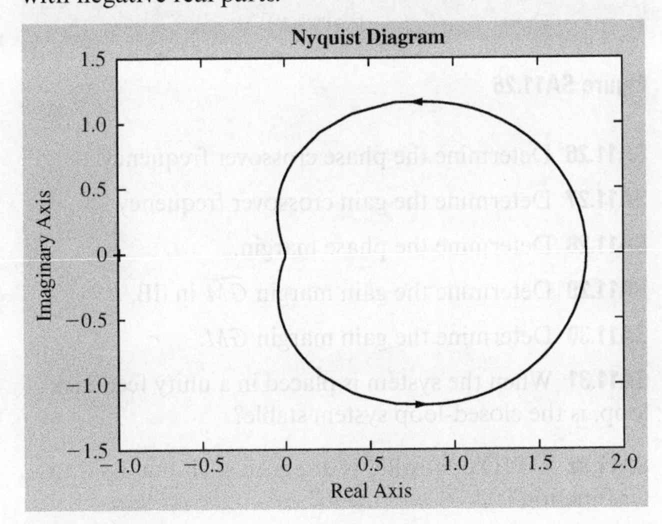

Figure SA11.4

SA11.5 A plant has the transfer function

$$G(s) = \frac{5}{s(s+3)}$$

Is the system stable when this plant is placed in a unity feedback loop?

SA11.6 A plant has the transfer function

$$G(s) = \frac{5}{s(s-3)}$$

Is the system stable when this plant is placed in a unity feedback loop?

SA11.7 A plant has the transfer function

$$G(s) = \frac{20}{s(s+3)(s+5)}$$

Is the system stable when this plant is placed in a unity feedback loop?

SA11.8 A plant with the transfer function

$$G(s) = \frac{2}{s+1}$$

is placed in series with a lead compensator that has the transfer function

$$G_c(s) = 4\frac{s+0.5}{s+7}$$

Is the closed-loop system stable when the combination is placed in a unity feedback loop?

SA11.9 A plant with the transfer function

$$G(s) = \frac{2}{s^2 + 2s + 10}$$

is placed in series with a PI controller that has the transfer function

$$G_c(s) = \frac{6}{s} + 12$$

Is the closed-loop system stable when the combination is placed in a unity feedback loop?

SA11.10 A plant with the transfer function

$$G(s) = \frac{2}{s^2 + 2s + 10}$$

is placed in series with a PID controller that has the transfer function

$$G_c(s) = \frac{15}{s} + 12 + 4s$$

Is the closed-loop system stable when the combination is placed in a unity feedback loop?

SA11.11 A plant with the transfer function

$$G(s) = \frac{2(s+5)}{(s+10)(s^2 + 6s + 34)}$$

is placed in series with a lag compensator that has the transfer function

$$G_c(s) = 15\frac{s+0.1}{s+0.005}$$

Is the closed-loop system stable when the combination is placed in a unity feedback loop?

SA11.12 A plant with the transfer function

$$G(s) = \frac{1}{s(s-2)(s+4)}$$

is in series with a proportional controller in a unity feedback loop. For what values of the proportional gain will the closed-loop system be stable?

SA11.13 A plant with the transfer function

$$G(s) = \frac{1}{(s+2)(s+4)}$$

is placed in series with an integral controller in a feedback loop. For what values of the integral gain will the closed-loop system be stable?

Problems SA11.14–SA11.19 refer to Figure SA11.14, in which the Bode diagram for an open-loop transfer function is shown.

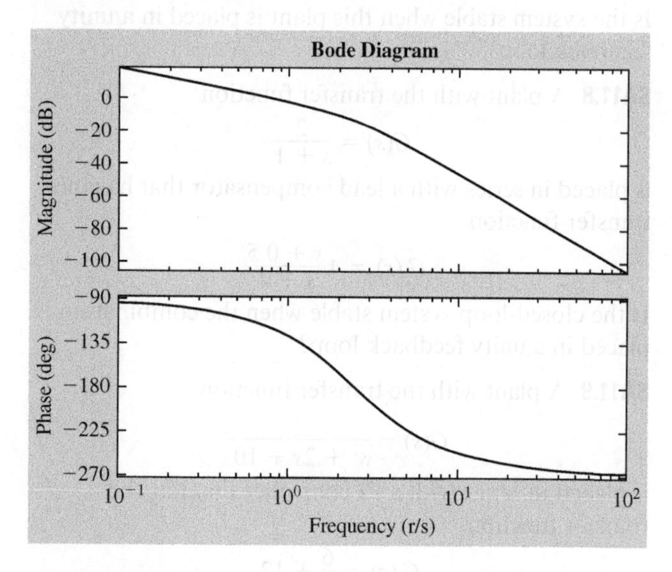

Figure SA11.14

SA11.14 Determine the phase crossover frequency.

SA11.15 Determine the gain crossover frequency.

SA11.16 Determine the phase margin.

SA11.17 Determine the gain margin $\widetilde{GM}$ in dB.

SA11.18 Determine the gain margin GM.

SA11.19 If the system is placed in a unity feedback loop, is the closed-loop system stable?

Problems SA11.20–SA11.25 refer to Figure SA11.20, in which the Bode diagram for an open-loop transfer function is shown.

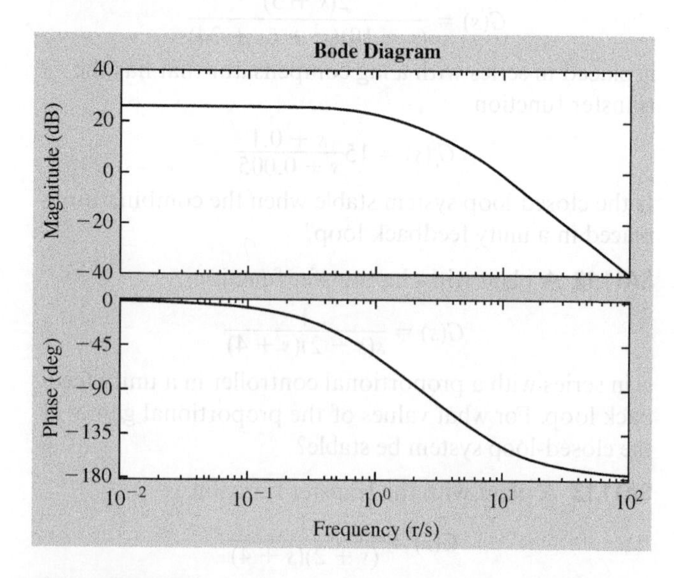

Figure SA11.20

SA11.20 Determine the phase crossover frequency.

SA11.21 Determine the gain crossover frequency.

SA11.22 Determine the phase margin.

SA11.23 Determine the gain margin $\widetilde{GM}$ in dB.

SA11.24 Determine the gain margin GM.

SA11.25 If the system is placed in a unity feedback loop, is the closed-loop system stable?

Problems SA11.26–SA11.31 refer to Figure SA11.26, in which the Bode diagram for an open-loop transfer function is shown.

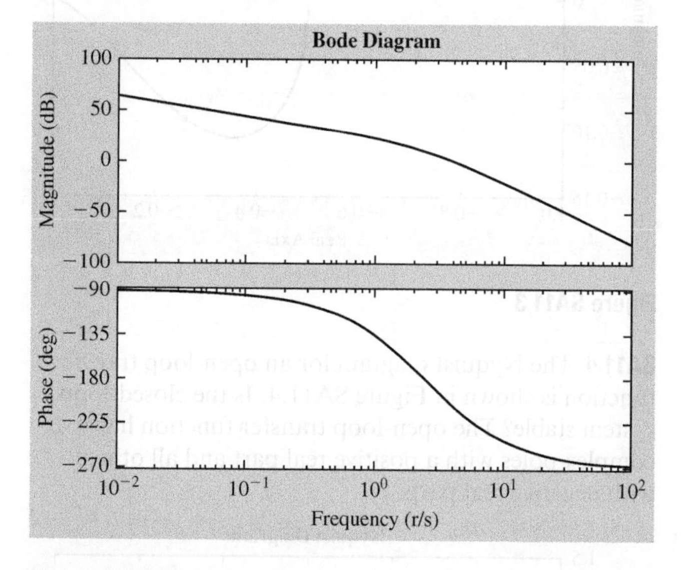

Figure SA11.26

SA11.26 Determine the phase crossover frequency.

SA11.27 Determine the gain crossover frequency.

SA11.28 Determine the phase margin.

SA11.29 Determine the gain margin $\widetilde{GM}$ in dB.

SA11.30 Determine the gain margin GM.

SA11.31 When the system is placed in a unity feedback loop, is the closed-loop system stable?

SA11.32 A PID controller is designed such that its transfer function is

$$G_c(s) = \frac{7.5(10s + 1)(0.4s + 1)}{s}$$

What are the values of the integral gain, the proportional gain, and the differential gain?

SA11.33 A PID controller is to be designed for a plant whose transfer function is

$$G(s) = \frac{s + 4}{(s + 2)(s^2 + 4s + 13)}$$

The plant is to be designed such that the static position error constant is at least 10. What is the minimum value of the integral gain?

SA11.34 A lead compensator is to be designed for a plant whose transfer function is

$$G(s) = \frac{2}{s(s+4)}$$

The static velocity error constant is to be at least 50. What is the minimum value of the compensator gain if the attenuation factor is 0.3?

SA11.35 A lead compensator is to be designed for a plant whose transfer function is

$$G(s) = \frac{s+2}{s(s+1)(s+3)}$$

The static velocity error constant is to be at least 50. What is the minimum value of the compensator gain if the attenuation factor is 0.2?

SA11.36 A lead compensator is to be designed for a plant whose transfer function is

$$G(s) = \frac{10(s+2)}{s(s+1)(s^2+2s+9)}$$

The static velocity error constant is to be at least 50. What is the minimum value of the compensator gain if the attenuation factor is 0.1?

SA11.37 A lead compensator is to be designed for a plant whose transfer function is

$$G(s) = \frac{s+2}{(s+1)(s+3)}$$

The static position error constant is to be at least 50. What is the minimum value of the compensator gain if the attenuation factor is 0.4?

SA11.38 A lag compensator is to be designed for a plant whose transfer function is

$$G(s) = \frac{3}{s(s+1)(s+3)}$$

The static velocity error constant is to be at least 50. What is the minimum value of the compensator gain if the attenuation factor is 15?

SA11.39 A lag compensator is to be designed for a plant whose transfer function is

$$G(s) = \frac{7(s+4)}{s(s+1)(s^2+4s+16)}$$

The static velocity error constant is to be at least 50. What is the minimum value of the compensator gain if the attenuation factor is 10?

SA11.40 A lag compensator is to be designed for a plant whose transfer function is

$$G(s) = \frac{s+2}{(s+1)(s+3)}$$

The static position error constant is to be at least 50. What is the minimum value of the compensator gain if the attenuation factor is 13?

SA11.41 A lead-lag compensator is to be designed for a plant whose transfer function is

$$G(s) = \frac{s+2}{s(s+1)(s+3)}$$

The static velocity error constant is to be at least 50. What is the minimum value of the compensator gain if $\alpha_2 = \frac{1}{\alpha_1}$?

SA11.42 A lead-lag compensator is to be designed for a plant whose transfer function is

$$G(s) = \frac{s+2}{(s+1)(s+3)}$$

The static position error constant is to be at least 50. What is the minimum value of the compensator gain if the lead attenuation is 0.08 and the lead attenuation factor is the reciprocal of the lag attenuation factor?

SA11.43 A lead-lag compensator is to be designed for a plant whose transfer function is

$$G(s) = \frac{s+2}{s(s+1)(s+3)}$$

The static velocity error constant is to be at least 50. What is the minimum value of the compensator gain if the lead attenuation factor is 0.08 and the lag attenuation factor is 14?

SA11.44 The maximum phase angle for a lead compensator is determined to be 37.5°. What is the attenuation factor of the compensator?

SA11.45 A lead compensator has corner frequencies of 5 r/s and 22 r/s. What is the value of the attenuation factor?

SA11.46 A lead compensator is designed for a plant whose transfer function is $G(s)$. It has corner frequencies of 5 r/s and 22 r/s. When the compensator is placed in series with the plant, what is the gain crossover frequency of the open-loop system?

SA11.47 A lead compensator has its lowest corner frequency as 2.5 r/s and an attenuation of 0.17. What is the higher corner frequency of the compensator?

SA11.48 A lead compensator has the mean between the corner frequencies of 10 r/s and an attenuation factor of 0.22. What are the compensator's corner frequencies?

SA11.49 A lag compensator is being designed for a plant whose transfer function is $G(s)$. At the new gain crossover frequency, the value of $L(\omega)$ for $G_1(s)$ is 15 dB. What is the value of the attenuation factor?

SA11.50 A lag compensator is being designed for a plant. The new gain crossover frequency for the compensated system is determined to be 2 r/s. Which of the following is appropriate for the time constant of the compensator: (a) 0.2 s, (b) 0.02 s, (c) 5 s, or (d) 50 s?

SA11.51 A compensator has a transfer function of

$$G_c(s) = 7.2\frac{s+0.05}{s+0.0004}$$

What is the high-frequency asymptote for this compensator?

SA11.52 A compensator has a transfer function of

$$G_c(s) = 7.2\frac{s + 0.05}{s + 0.0004}$$

What is the low-frequency asymptote for this compensator?

SA11.53 A compensator has a transfer function of

$$G_c(s) = 15.3\frac{s + 0.5}{s + 20}$$

What is the high-frequency asymptote for this compensator?

SA11.54 A compensator has a transfer function of

$$G_c(s) = 15.3\frac{s + 0.5}{s + 20}$$

What is the low-frequency asymptote for this compensator?

SA11.55 A compensator has a transfer function of

$$G_c(s) = 4.1\left(\frac{s + 0.5}{s + 20}\right)\left(\frac{s + 0.03}{s + 0.00075}\right)$$

What is the high-frequency asymptote for this compensator?

SA11.56 A compensator has a transfer function of

$$G_c(s) = 4.1\left(\frac{s + 0.5}{s + 20}\right)\left(\frac{s + 0.03}{s + 0.00075}\right)$$

What is the low-frequency asymptote for this compensator?

SA11.57 Does the compensator with the transfer function

$$G_c(s) = 4.1\left(\frac{s + 0.5}{s + 20}\right)\left(\frac{s + 0.03}{s + 0.00075}\right)$$

satisfy $\alpha_2 = \frac{1}{\alpha_1}$?

SA11.58 A lead compensator acts as a (choose one)

(a) low-pass filter
(b) high-pass filter
(c) band-pass filter
(d) band-reject filter

SA11.59 A lag compensator acts as a (choose one)

(a) low-pass filter
(b) high-pass filter
(c) band-pass filter
(d) band-reject filter

SA11.60 A lead-lag compensator acts as a (choose one)

(a) low-pass filter
(b) high-pass filter
(c) band-pass filter
(d) band-reject filter

Problems

11.1 A plant has the transfer function

$$G(s) = \frac{2(s - 1)}{s + 2}$$

(a) Is the closed-loop system stable if the plant is placed in a unity feedback loop?

(b) The plant is placed in series with a PI controller that has the transfer function

$$G_c(s) = 4 + \frac{3}{s}$$

Is the closed-loop system stable if the plant and the controller are placed in a unity feedback loop?

11.2 Is the closed-loop system of Figure P11.2 stable?

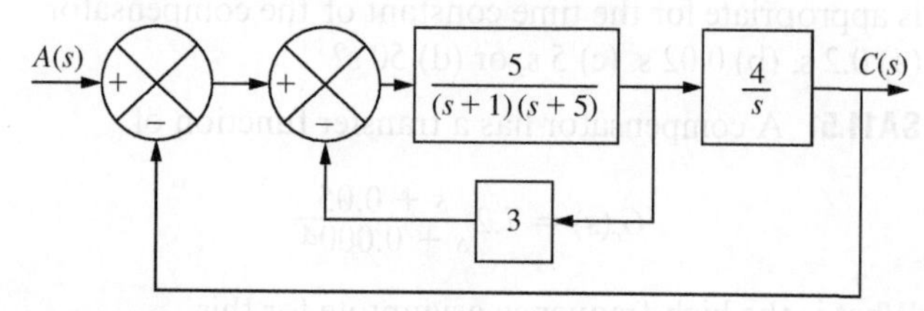

Figure P11.2

11.3 For what values of the integral gain is the system of Figure P11.3 stable?

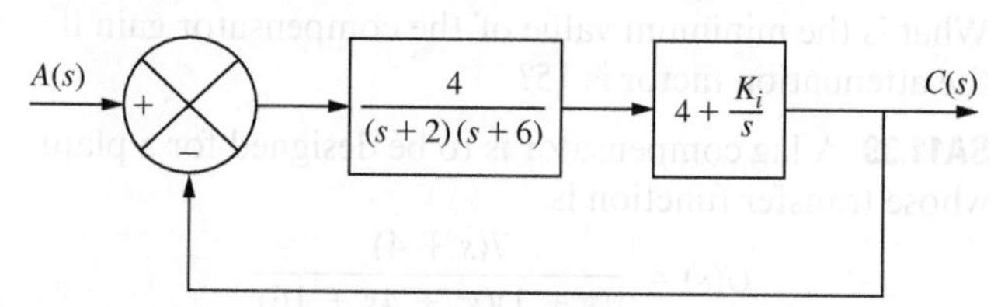

Figure P11.3

11.4 A plant has the transfer function

$$G(s) = \frac{2}{s^2(s + 2)}$$

(a) Is the closed-loop system stable if the plant is placed in a unity feedback loop?

(b) The plant is placed in series with a differentiator that has the transfer function

$$G_c(s) = 4s$$

Is the closed-loop system stable if the plant and the differentiator are placed in a unity feedback loop?

11.5 For what values of K_i is the closed-loop system of Figure P11.5 stable?

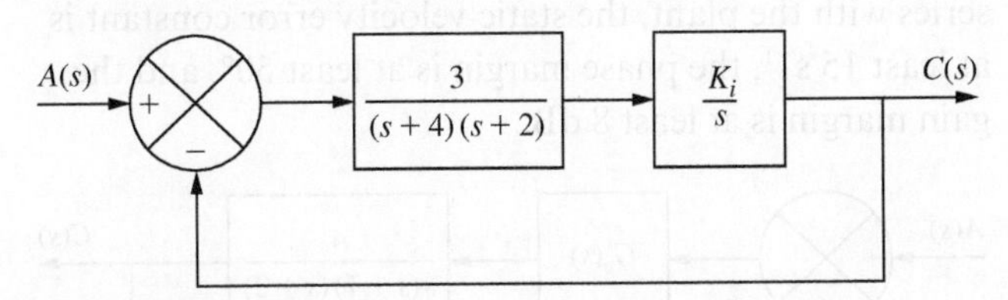

Figure P11.5

11.6 A plant with the transfer function

$$G(s) = \frac{1}{s - 0.5}$$

is placed in series with a proportional controller that has the transfer function

$$G_c(s) = \frac{K_i}{s}$$

in a unity feedback loop. For what values of the integral gain K_i is the system stable?

11.7 The open-loop transfer function of a plant is

$$G(s) = \frac{e^{-0.5s}}{s^2(s + 2)}$$

The plant is placed in series with a proportional controller that has the transfer function

$$G_c(s) = K_p$$

For what values of the proportional gain is the closed-loop system stable?

11.8 A plant has the transfer function

$$G(s) = \frac{2s^2 + 3s + 2}{s^3 + 1.5s^2 + 2s + 4}$$

It is noted that the plant is unstable. Is the closed-loop system stable when the plant is placed in a unity feedback loop?

11.9 A plant has the transfer function

$$G(s) = \frac{2s^2 + 3s + 2}{(s - 2)(s^3 + 3s^2 + 2s + 4)}$$

It is noted that the plant is unstable. Is the closed-loop system stable when the plant is placed in a unity feedback loop?

11.10 A plant has an open-loop transfer function of

$$G(s) = \frac{2s + 3}{s(s + 1)(s + 3)}$$

Determine the plant's (a) gain crossover frequency, (b) phase crossover frequency, (c) phase margin, and (d) gain margin in dB.

11.11 A plant has an open-loop transfer function of

$$G(s) = \frac{as + 3}{s(s + 1)(s + 3)}$$

Determine the value of a such that the phase margin for the plant is 50°.

11.12 A plant has an open-loop transfer function of

$$G(s) = \frac{3s + 2}{s(s + 2)(s + 3)}$$

It is placed in series with a PD controller whose transfer function is

$$K_p(1 + 3s)$$

What are the phase margin and the gain margin when $K_p = 5$?

11.13 A plant has the transfer function

$$G(s) = \frac{2}{s(s + 2)}$$

Design a lead compensator such that when placed in series with the plant, the static velocity error constant is at least 25 s^{-1}, the phase margin is at least 40°, and the gain margin is at least 8 dB.

11.14 A plant has the transfer function

$$G(s) = \frac{5}{s(s + 2)(s + 4)}$$

Design a lead compensator such that when placed in series with the plant, the static velocity error constant is at least 30 s^{-1}, the phase margin is at least 50°, and the gain margin is at least 8 dB.

11.15 A plant has the transfer function

$$G(s) = \frac{1}{s(s + 1)}$$

Design a lead compensator such that when placed in series with the plant, the static velocity error constant is at least 20 s^{-1}, the phase margin is at least 50°, and the gain margin is at least 10 dB.

11.16 A plant has the transfer function

$$G(s) = \frac{0.2}{s(s + 1)(s + 10)}$$

Design a lead compensator such that when placed in series with the plant, the static velocity error constant is at least 5 s^{-1}, the phase margin is at least 50°, and the gain margin is at least 10 dB.

11.17 Consider the plant illustrated in Figure P11.17. Design a lead compensator such that when placed in series with the plant, the static velocity error constant is at least 25 s^{-1}, the phase margin is at least 40°, and the gain margin is at least 8 dB.

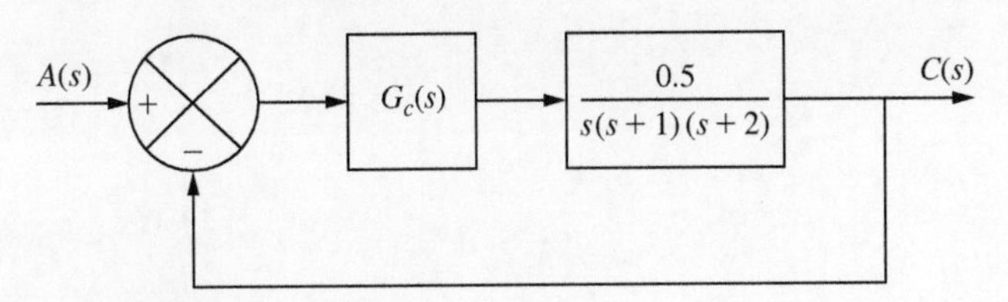

Figure P11.17

11.18 Consider the plant illustrated in Figure P11.18. Design a lead compensator such that when placed in series with the plant, the static velocity error constant is at least 25 s^{-1}, the phase margin is at least 40°, and the gain margin is at least 8 dB.

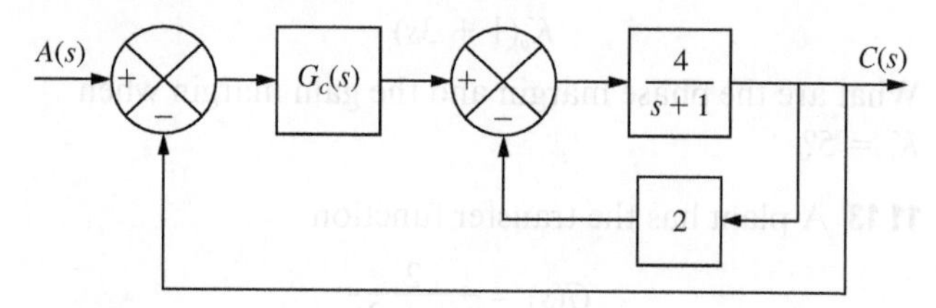

Figure P11.18

11.19 A plant has the transfer function

$$G(s) = \frac{0.5}{s(s+1)(s+3)}$$

Design a lag compensator such that the static velocity error constant is at least 6 s^{-1}, the phase margin is at least 40°, and the gain margin is at least 5 dB.

11.20 A plant has the transfer function

$$G(s) = \frac{1}{s(s+1)(s+2)}$$

Design a lead-lag compensator such that the static velocity error constant is at least 10 s^{-1}, the phase margin is at least 40°, and the gain margin is at least 10 dB.

11.21 A plant has the transfer function

$$G(s) = \frac{2(s+6)}{s(s^2+2s+5)}$$

Design a lead-lag compensator such that when placed in series with the plant, the static velocity error constant is at least 20 s^{-1}, the phase margin is at least 40°, and the gain margin is at least 8 dB.

11.22 Consider the plant illustrated in Figure P11.22. Design a lead-lag compensator such that when placed in series with the plant, the static velocity error constant is at least 15 s^{-1}, the phase margin is at least 50°, and the gain margin is at least 8 dB.

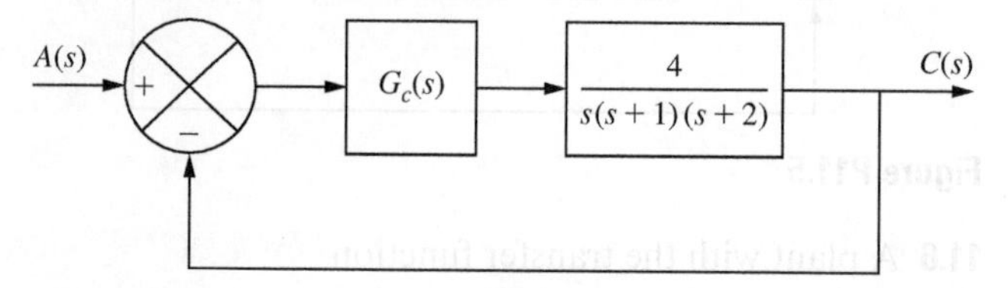

Figure P11.22

11.23 Consider the plant illustrated in Figure P11.23. Design a lead-lag compensator such that when placed in series with the plant, the static velocity error constant is at least 20 s^{-1}, the phase margin is at least 50°, and the gain margin is at least 10 dB.

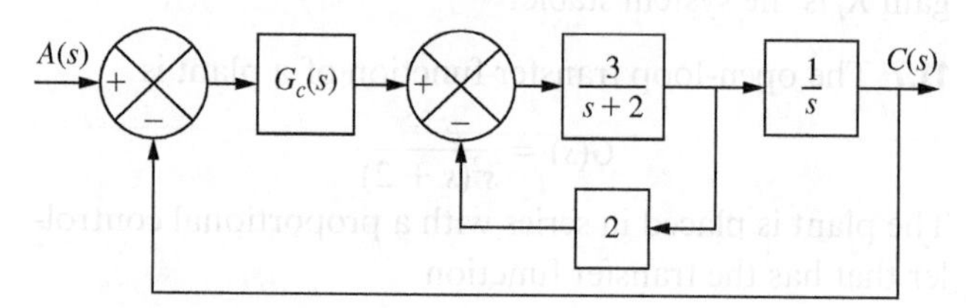

Figure P11.23

12

State-Space Methods

Learning Objectives

At the conclusion of this chapter, you should be able to:

- Develop a Simulink model of a system from the state-space formulation
- Develop a Laplace transform solution for the free response of a system in the state space
- Determine the exponential solution for the free response of a system in the state space
- Identify the matrix $e^{\mathbf{A}t}$ in terms of the eigenvalues of **A** and their corresponding eigenvectors
- Identify the state transition matrix
- Develop the Laplace transform response of a system due to any input using the state transition matrix
- Use MATLAB to develop numerical solutions from the state-space formulation
- Apply state-space methods to nonlinear systems and systems with variable coefficients
- Calculate the gains for state variable control
- Incorporate integral control into state variable control
- Calculate the Ackerman matrix

The transient and forced responses of dynamic systems are difficult to manage for complex systems. These include systems with multiple inputs and/or a large number of dependent variables. The algebra in applying these methods becomes intractable. These methods are also applicable only for linear systems and are best suited for systems whose properties remain constant. However, as with any topic in engineering and science, one needs to understand simple systems before studying complex systems. A study of second-order systems leads to the understanding of concepts of natural frequency, damping ratio, steady-state amplitude, and phase. Such concepts are generalized in the study of higher-order systems. A good knowledge of the modeling and response of linear systems is necessary before studying the behavior of nonlinear systems. Knowledge of the response of time-invariant systems is required before examining time-varying systems.

The rapid development of high-powered digital computers and sophisticated software packages such as MATLAB allow the application of a more general method for determining the response of dynamic systems. This method, called the state-space method, is applicable to all dynamic systems and is described in this chapter. The state-space formulation for a system was first introduced in Chapter 3. It involves rewriting the differential equations that govern an nth-order system as n first-order differential equations. The n independent variables in the first-order equations are referred to as state variables. The state variables usually are defined such that they are related to some form of energy.

This chapter develops the solution using the state-space formulation. This includes the transient response for both unforced systems and forced systems. The state-space formulation is convenient for numerical evaluation of these solutions. It also can be used to formulate solutions for nonlinear systems and systems with time-varying coefficients.

A Simulink model in the state space can be formulated either from the state-space formulation of the differential equations or simply by knowing the state, input, output, and transition matrices.

Control system analysis can be performed in the state space. Each state variable can be subject to a proportional control. This results in the development of a matrix, called the Ackerman matrix, that quantifies the control. A state-space formulation is made using the Ackerman matrix. The use of integral control is considered by defining a new state variable, thus increasing the order of the system.

12.1 An Example in the State Space

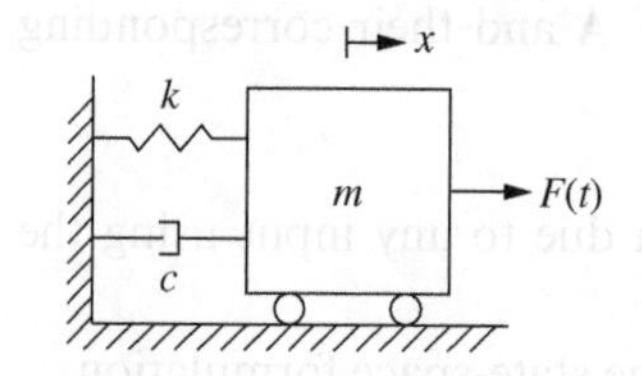

Figure 12.1

The mechanical system composed of a mass, spring, and viscous damper with a force input.

Recall that the differential equation governing the motion of the mechanical system of Figure 12.1 is

$$m\ddot{x} + c\dot{x} + kx = F(t) \tag{12.1}$$

State-space variables are defined according to

$$y_1 = x \tag{12.2a}$$

$$y_2 = \dot{x} \tag{12.2b}$$

Using the state-space variables of Equation (12.2), Equation (12.1) can be rewritten as two first-order differential equations:

$$\dot{y}_1 = y_2 \tag{12.3a}$$

$$\dot{y}_2 = \frac{1}{m}F(t) - \frac{c}{m}y_2 - \frac{k}{m}y_1 \tag{12.3b}$$

Equations (12.3) are summarized in matrix form as

$$\begin{bmatrix} \dot{y}_1 \\ \dot{y}_2 \end{bmatrix} = \begin{bmatrix} 0 & 1 \\ -\frac{k}{m} & -\frac{c}{m} \end{bmatrix} \begin{bmatrix} y_1 \\ y_2 \end{bmatrix} + \begin{bmatrix} 0 \\ \frac{1}{m} \end{bmatrix} F(t) \tag{12.4}$$

Equation (12.4) can be written as

$$\dot{\mathbf{y}} = \mathbf{A}\mathbf{y} + \mathbf{B}F(t) \tag{12.5}$$

where $\mathbf{y}$ is the state vector, $\mathbf{A}$ is the state matrix, and $\mathbf{B}$ is the input vector.

The input to the mechanical system is $F(t)$ and its output is $x(t)$. A block diagram of the state-space model for the system is illustrated in Figure 12.2. The system input is multiplied by $1/m$ and then is added to $-(k/m)y_1 - (c/m)y_2$ coming from feedback loops at a summing junction. The sum is integrated leading to $y_2(t)$, which is fed back to the summing junction. This result is integrated again leading to $y_1(t)$, which is fed back to the summing junction as well as sent as system output. The block diagram of Figure 12.2 is equivalent to block diagrams in the s domain if the integrators are replaced by $1/s$.

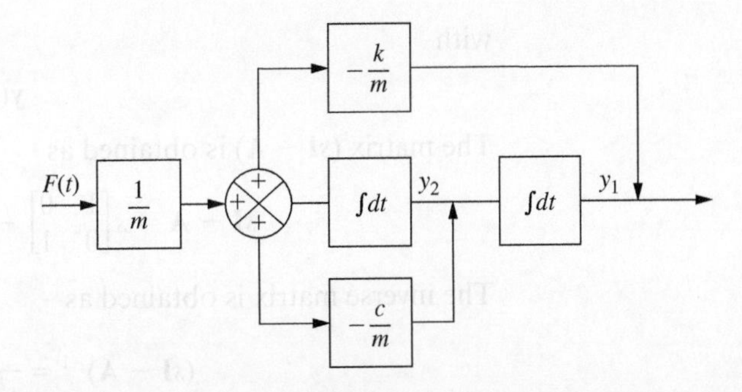

Figure 12.2

A block diagram of the state-space model for the system of Figure 12.1.

There are many methods available to determine the response of Equation (12.5) for a given $F(t)$. Given the appropriate initial conditions, Equation (12.4) is amenable to solution by single-step numerical methods such as the Runge-Kutta methods, introduced in Section 12.4. The general analytical solution is the sum of the homogeneous solution, the solution obtained when $F(t) = 0$, and the particular solution, the solution specific to $F(t)$. Many methods are available for both.

First consider the solution obtained using the Laplace transform method. To this end, define

$$\mathbf{Y}(s) = \begin{bmatrix} Y_1(s) \\ Y_2(s) \end{bmatrix} \tag{12.6a}$$

$$\mathbf{y}(0) = \begin{bmatrix} y_1(0) \\ y_2(0) \end{bmatrix} \tag{12.6b}$$

Taking the Laplace transform of Equation (12.5) using the initial conditions of Equation (12.6b) leads to

$$s\mathbf{Y}(s) = \mathbf{A}\mathbf{Y}(s) + \mathbf{y}(0) + \mathbf{B}F(s) \tag{12.7}$$

Equation (12.7) is rearranged to

$$(s\mathbf{I} - \mathbf{A})\mathbf{Y}(s) = \mathbf{y}(0) + \mathbf{B}F(s) \tag{12.8}$$

where $\mathbf{I}$ is the 2×2 identity matrix. The solution of Equation (12.8) is obtained as

$$\mathbf{Y}(s) = (s\mathbf{I} - \mathbf{A})^{-1}\,\mathbf{y}(0) + (s\mathbf{I} - \mathbf{A})^{-1}\,\mathbf{B}F(s) \tag{12.9}$$

Define

$$\mathbf{G}(t) = \mathcal{L}^{-1}\{(s\mathbf{I} - \mathbf{A})^{-1}\} \tag{12.10}$$

The convolution property, property 13 of Table 2.2, is used to invert Equation (12.9), leading to

$$\mathbf{y}(t) = \mathbf{G}(t)\mathbf{y}(0) + \int_0^t \mathbf{G}(\tau)\mathbf{B}F(t - \tau)d\tau \tag{12.11}$$

Example 12.1

Use the state-space method to determine and plot the response of a second-order mechanical system with $m = 1$ kg, $c = 5$ N·s/m, and $k = 6$ N/m subject to $F(t) = 3e^{-2t}$ N. The system's initial conditions are $x(0) = 0$ and $\dot{x}(0) = 1$ m/s.

Solution

Substitution of the given values into Equation (12.4) leads to

$$\begin{bmatrix} \dot{y}_1 \\ \dot{y}_2 \end{bmatrix} = \begin{bmatrix} 0 & 1 \\ -6 & -5 \end{bmatrix}\begin{bmatrix} y_1 \\ y_2 \end{bmatrix} + \begin{bmatrix} 0 \\ 1 \end{bmatrix} 3e^{-2t} \tag{a}$$

with

$$\mathbf{y}(0) = \begin{bmatrix} 0 \\ 1 \end{bmatrix} \tag{b}$$

The matrix $(s\mathbf{I} - \mathbf{A})$ is obtained as

$$s\mathbf{I} - \mathbf{A} = s\begin{bmatrix} 1 & 0 \\ 0 & 1 \end{bmatrix} - \begin{bmatrix} 0 & 1 \\ -6 & -5 \end{bmatrix} = \begin{bmatrix} s & -1 \\ 6 & s+5 \end{bmatrix} \tag{c}$$

The inverse matrix is obtained as

$$(s\mathbf{I} - \mathbf{A})^{-1} = \frac{1}{s^2 + 5s + 6}\begin{bmatrix} s+5 & 1 \\ -6 & s \end{bmatrix} \tag{d}$$

Noting that $s^2 + 6s + 5 = (s + 3)(s + 2)$, partial fraction decomposition leads to

$$(s\mathbf{I} - \mathbf{A})^{-1} = \begin{bmatrix} \dfrac{-2}{s+3} + \dfrac{3}{s+2} & \dfrac{1}{s+3} - \dfrac{1}{s+2} \\ -\dfrac{6}{s+3} + \dfrac{6}{s+2} & \dfrac{3}{s+3} - \dfrac{2}{s+2} \end{bmatrix} \tag{e}$$

The application of Equation (12.10) leads to

$$\mathbf{G}(t) = \begin{bmatrix} -2e^{-3t} + 3e^{-2t} & e^{-3t} - e^{-2t} \\ -6e^{-3t} + 6e^{-2t} & 3e^{-3t} - 2e^{-2t} \end{bmatrix} \tag{f}$$

Equation (12.11) becomes

$$\begin{bmatrix} y_1(t) \\ y_2(t) \end{bmatrix} = \begin{bmatrix} -2e^{-3t} + 3e^{-2t} & e^{-3t} - e^{-2t} \\ -6e^{-3t} + 6e^{-2t} & 3e^{-3t} - 2e^{-2t} \end{bmatrix}\begin{bmatrix} 0 \\ 1 \end{bmatrix}$$

$$+ \int_0^t \begin{bmatrix} -2e^{-3\tau} + 3e^{-2\tau} & e^{-3\tau} - e^{-2\tau} \\ -6e^{-3\tau} + 6e^{-2\tau} & 3e^{-3\tau} - 2e^{-2\tau} \end{bmatrix}\begin{bmatrix} 0 \\ 1 \end{bmatrix} 3e^{-3(t-\tau)} d\tau \tag{g}$$

The evaluation of the convolution integral in Equation (g) leads to

$$\begin{bmatrix} y_1(t) \\ y_2(t) \end{bmatrix} = \begin{bmatrix} e^{-3t} - e^{-2t} \\ 3e^{-3t} - 2e^{-2t} \end{bmatrix} + \begin{bmatrix} 3te^{-3t} + 3e^{-3t} - 3e^{-2t} \\ 9te^{-3t} + 6e^{-3t} - 6e^{-2t} \end{bmatrix}$$

$$= \begin{bmatrix} 3te^{-3t} + 4e^{-3t} - 4e^{-2t} \\ 9te^{-3t} + 9e^{-3t} - 8e^{-2t} \end{bmatrix} \tag{h}$$

The time-dependent response of the system is illustrated in Figure 12.3. The plot illustrates that the response satisfies the initial conditions of Equation (b). The

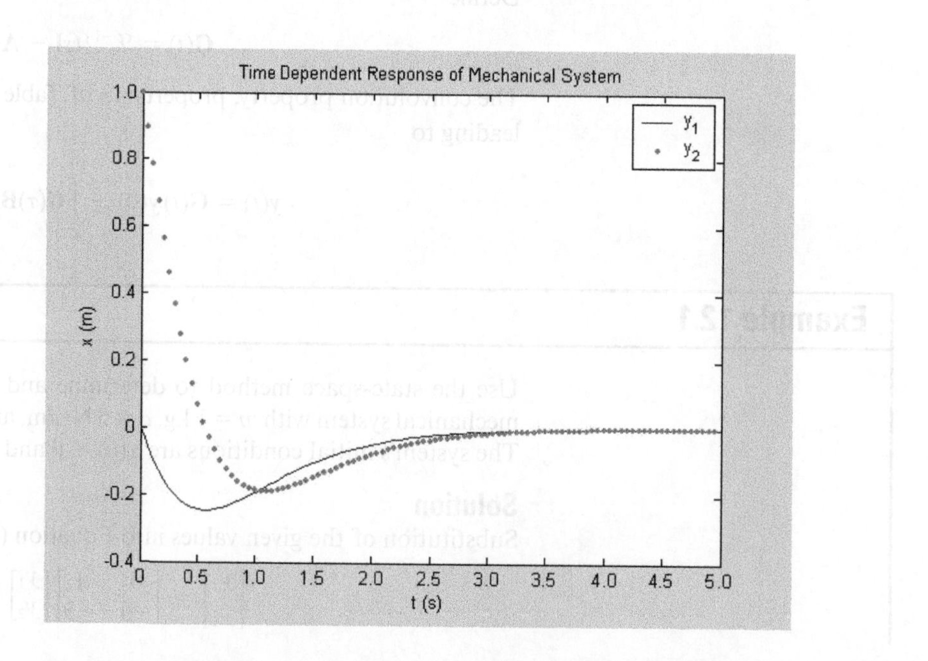

Figure 12.3

The response of the mechanical system of Example 12.1 is determined using state-space methods.

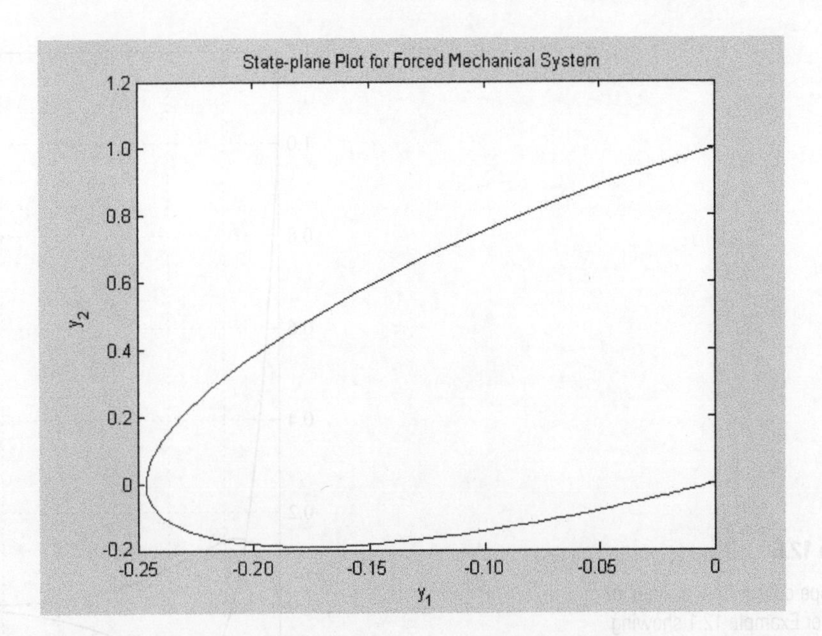

Figure 12.4

The state-plane plot of the forced response for the system of Example 12.1. The trajectory originates at (0, 1) corresponding to the system's initial conditions, is continuous, and approaches (0, 0) corresponding to the final state of the system.

state-plane plot with y_1 on the horizontal axis and y_2 on the vertical axis is illustrated in Figure 12.4. The path in the state-plane is called a **trajectory**, which is plotted using t as a parameter. The trajectory in the state-plane begins at (0,1) corresponding to the initial conditions at $t = 0$. The trajectory is continuous and approaches (0,0) for large t.

A state-plane plot shows two state variables plotted against each other. When one state variable represents displacement and the second state variable represents velocity, the state-plane plot is also called a phase-plane plot. These plots are useful for classifying an equilibrium point and determining the limiting behavior of the system.

Simulink may be used to develop state-space models of dynamic systems. The Simulink model for the system of Example 12.1 with $F(t) = 0$ is illustrated in Figure 12.5. The following is noted regarding this model:

- An initial condition of $y_2(0) = 1$ is specified using the Block Properties box for the integrator. This specifies that the variable in the block diagram after the integration has an initial value of 1.

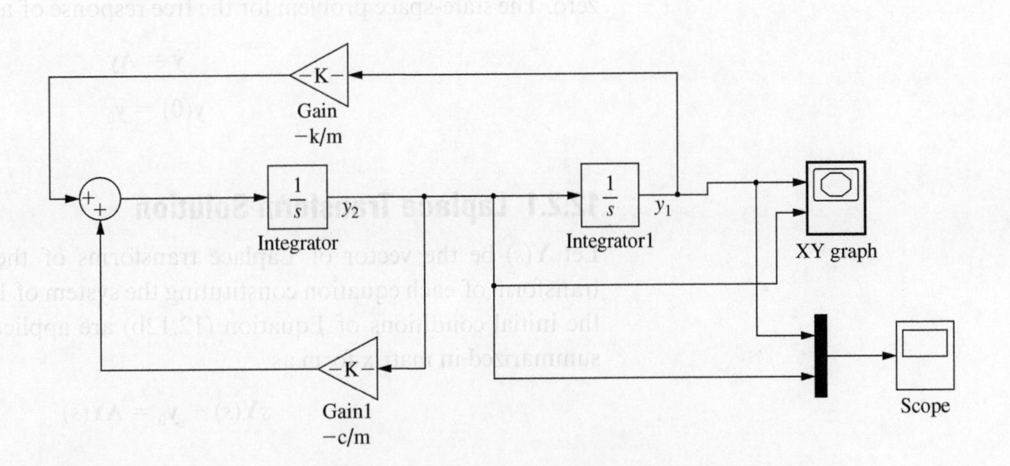

Figure 12.5

The Simulink model for the system in Example 12.1 with $F(t) = 0$.

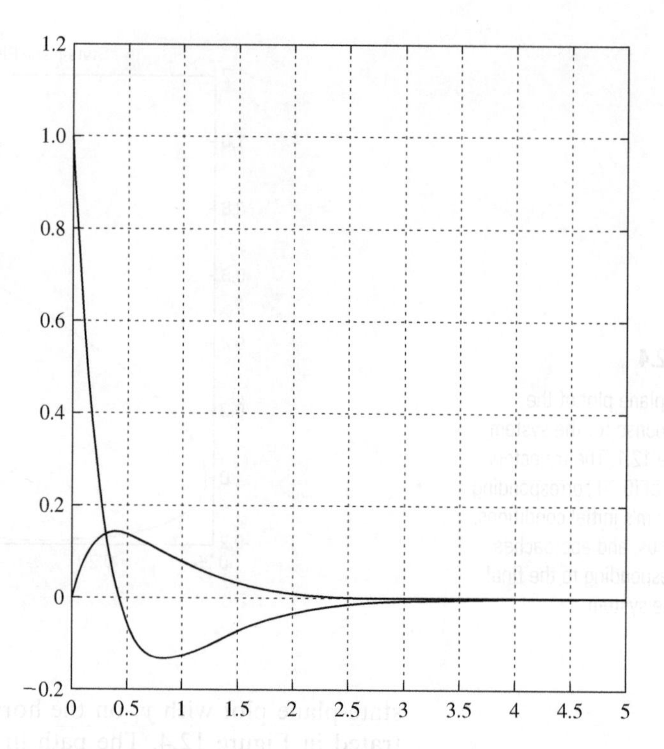

Figure 12.6

The scope output from a Simulink model for Example 12.1 showing time-dependent responses for $y_1(t)$ with an initial condition of 1 and $y_2(t)$ with an initial condition of 0.

- An XY graph, generated from the Sinks menu of the Library Browser, is used to generate a state-plane plot. The upper input to the box is y_1, which is plotted on the x axis, while the lower input is y_2, which is plotted on the y axis.
- A Mux from the Signal Routing menu of the Library Browser is used to send two signals to the same scope.
- The gains are set to $-c/m$ and $-k/m$, where m, c, and k are the current values in the MATLAB workspace.

The scope output, showing the time-dependent responses, is shown in Figure 12.6.

12.2 State-Space Solutions for Free Response

The free response of a system is its response due to initial conditions when the input is zero. The state-space problem for the free response of a system with n state variables is

$$\dot{\mathbf{y}} = \mathbf{A}\mathbf{y} \tag{12.12a}$$

$$\mathbf{y}(\mathbf{0}) = \mathbf{y}_0 \tag{12.12b}$$

12.2.1 Laplace Transform Solution

Let $\mathbf{Y}(s)$ be the vector of Laplace transforms of the state variables. The Laplace transform of each equation constituting the system of Equation (12.12a) is taken, and the initial conditions of Equation (12.12b) are applied. The resulting equations are summarized in matrix form as

$$s\mathbf{Y}(s) - \mathbf{y}_0 = \mathbf{A}\mathbf{Y}(s) \tag{12.13}$$

where $\mathbf{Y}(s) = \mathscr{L}\{\mathbf{y}(t)\}$. Equation (12.13) is rearranged as

$$(s\mathbf{I} - \mathbf{A})\mathbf{Y}(s) = \mathbf{y}_0 \tag{12.14}$$

where I represents the identity matrix. Equation (12.14) is premultiplied by $(s\mathbf{I} - \mathbf{A})^{-1}$, leading to

$$\mathbf{Y}(s) = (s\mathbf{I} - \mathbf{A})^{-1}\mathbf{y}_0 \tag{12.15}$$

The solution of Equations (12.12a) and (12.12b) is written as

$$\mathbf{y} = \mathscr{L}^{-1}\{(s\mathbf{I} - \mathbf{A})^{-1}\}\mathbf{y}_0 \tag{12.16}$$

The inverse Laplace transform of Equation (12.16) is performed on all elements of $(s\mathbf{I} - \mathbf{A})^{-1}$.

Example 12.2

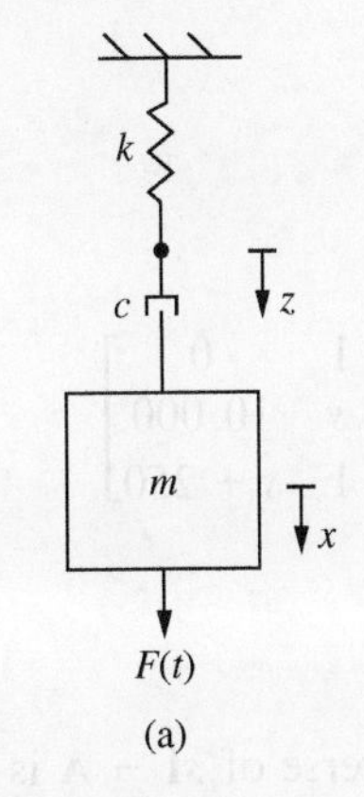

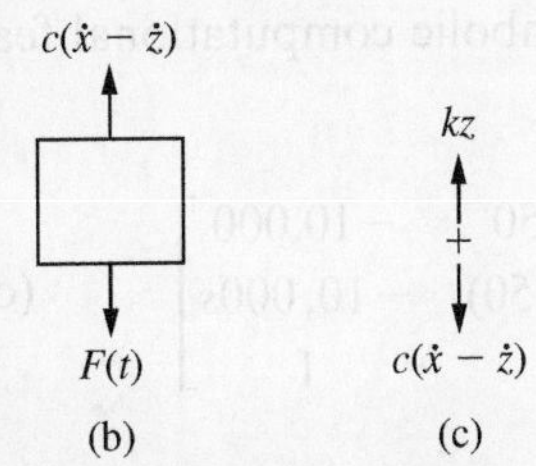

Figure 12.7

(a) The system of Example 12.2; (b) and (c) the free-body diagrams of the system of Example 12.2 at an arbitrary time.

The parameter values for the mechanical system of Figure 12.7 are $m = 10$ kg, $k = 1 \times 10^5$ N/m, and $c = 400$ N·s/m. Determine the free response of the system when $x(0) = 0$ m, $\dot{x}(0) = 0.1$ m/s, and $z(0) = 0$.

Solution

Let $z(t)$ be the displacement of the joint between the spring and the viscous damper. The application of Newton's second law to the free-body diagram of the block at an arbitrary instant, as shown in Figure 12.7(b), gives

$$F(t) - c(\dot{x} - \dot{z}) = m\ddot{x} \tag{a}$$

The application of Newton's second law to the free-body diagram of the joint between the viscous damper and the mass, as illustrated in Figure 12.7(c), noting that the mass of the joint is zero, leads to

$$-c(\dot{x} - \dot{z}) + kz = 0 \tag{b}$$

Define state variables as

$$y_1 = x \tag{c}$$

$$y_2 = \dot{x} \tag{d}$$

$$y_3 = z \tag{e}$$

The use of state variables in Equations (a) and (b) gives

$$m\dot{y}_2 - c\dot{y}_3 = -cy_2 + F(t) \tag{f}$$

$$c\dot{y}_3 = cy_2 - ky_3 \tag{g}$$

Equations (d), (f), and (g) are summarized in a matrix form as

$$\begin{bmatrix} 1 & 0 & 0 \\ 0 & m & -c \\ 0 & 0 & c \end{bmatrix}\begin{bmatrix} \dot{y}_1 \\ \dot{y}_2 \\ \dot{y}_3 \end{bmatrix} = \begin{bmatrix} 0 & 1 & 0 \\ 0 & -c & 0 \\ 0 & c & -k \end{bmatrix}\begin{bmatrix} y_1 \\ y_2 \\ y_3 \end{bmatrix} + \begin{bmatrix} 0 \\ 1 \\ 0 \end{bmatrix} F(t) \tag{h}$$

Noting that

$$\begin{bmatrix} 1 & 0 & 0 \\ 0 & m & -c \\ 0 & 0 & c \end{bmatrix}^{-1} = \begin{bmatrix} 1 & 0 & 0 \\ 0 & \frac{1}{m} & \frac{1}{m} \\ 0 & 0 & \frac{1}{c} \end{bmatrix} \tag{i}$$

premultiplication of Equation (h) by the inverse of the 3×3 matrix on its left-hand side leads to

$$\begin{bmatrix} \dot{y}_1 \\ \dot{y}_2 \\ \dot{y}_3 \end{bmatrix} = \begin{bmatrix} 1 & 0 & 0 \\ 0 & \frac{1}{m} & \frac{1}{m} \\ 0 & 0 & \frac{1}{c} \end{bmatrix} \begin{bmatrix} 0 & 1 & 0 \\ 0 & -c & 0 \\ 0 & c & -k \end{bmatrix} \begin{bmatrix} y_1 \\ y_2 \\ y_3 \end{bmatrix} + \begin{bmatrix} 1 & 0 & 0 \\ 0 & \frac{1}{m} & \frac{1}{m} \\ 0 & 0 & \frac{1}{c} \end{bmatrix} \begin{bmatrix} 0 \\ 1 \\ 0 \end{bmatrix} F(t)$$

$$= \begin{bmatrix} 0 & 1 & 0 \\ 0 & 0 & -\frac{k}{m} \\ 0 & 1 & -\frac{k}{c} \end{bmatrix} \begin{bmatrix} y_1 \\ y_2 \\ y_3 \end{bmatrix} + \begin{bmatrix} 0 \\ \frac{1}{m} \\ 0 \end{bmatrix} F(t) \tag{j}$$

The substitution of the given parameter values into Equation (j) with $F(t) = 0$ leads to

$$\begin{bmatrix} \dot{y}_1 \\ \dot{y}_2 \\ \dot{y}_3 \end{bmatrix} = \begin{bmatrix} 0 & 1 & 0 \\ 0 & 0 & -10,000 \\ 0 & 1 & -250 \end{bmatrix} \begin{bmatrix} y_1 \\ y_2 \\ y_3 \end{bmatrix} \tag{k}$$

The initial condition vector is

$$\mathbf{y}_0 = \begin{bmatrix} 0 \\ 0.1 \\ 0 \end{bmatrix} \tag{l}$$

The matrix $s\mathbf{I} - \mathbf{A}$ is evaluated as

$$s\mathbf{I} - \mathbf{A} = \begin{bmatrix} s & 0 & 0 \\ 0 & s & 0 \\ 0 & 0 & s \end{bmatrix} - \begin{bmatrix} 0 & 1 & 0 \\ 0 & 0 & -10,000 \\ 0 & 1 & -250 \end{bmatrix} = \begin{bmatrix} s & -1 & 0 \\ 0 & s & 10,000 \\ 0 & -1 & s+250 \end{bmatrix} \tag{m}$$

The determinant of the matrix is

$$|s\mathbf{I} - \mathbf{A}| = s(s^2 + 250s + 10,000) \tag{n}$$

The roots of the determinant are $s = -200, -50, 0$. The inverse of $s\mathbf{I} - \mathbf{A}$ is determined using either the methods of Appendix B or the symbolic computational features of MATLAB as

$$(s\mathbf{I} - \mathbf{A})^{-1} = \frac{1}{s(s^2 + 250s + 10,000)} \begin{bmatrix} s^2 + 250s + 10,000 & s + 250 & -10,000 \\ 0 & s(s+250) & -10,000s \\ 0 & s & 1 \end{bmatrix} \tag{o}$$

The inverse Laplace transform is determined as

$$\mathcal{L}^{-1}\{(s\mathbf{I} - \mathbf{A})^{-1}\} = \begin{bmatrix} 1 & \frac{1}{40} + \frac{1}{600}e^{-200t} + \frac{2}{75}e^{-50t} & -1 - \frac{1}{3}e^{-200t} + \frac{4}{3}e^{-50t} \\ 0 & \frac{4}{3}e^{-50t} - \frac{1}{3}e^{-200t} & -\frac{200}{3}e^{-50t} + \frac{200}{3}e^{-50t} \\ 0 & \frac{1}{150}e^{-50t} - \frac{1}{150}e^{-200t} & -\frac{1}{3}e^{-50t} + \frac{4}{3}e^{-200t} \end{bmatrix} \tag{p}$$

The state variables are determined using Equation (12.16) and the initial condition vector of Equation (l) as

$$\begin{bmatrix} y_1 \\ y_2 \\ y_3 \end{bmatrix} = 0.1 \begin{bmatrix} \frac{1}{40} + \frac{1}{600}e^{-200t} + \frac{2}{75}e^{-50t} \\ \frac{4}{3}e^{-50t} - \frac{1}{3}e^{-200t} \\ \frac{1}{150}e^{-50t} - \frac{1}{150}e^{-200t} \end{bmatrix} \tag{q}$$

12.2.2 Exponential Solution

An alternate form of the free response can be obtained by assuming an exponential solution of Equations (12.12a) and (12.12b) of the form

$$\mathbf{y} = \mathbf{Y}e^{\lambda t} \tag{12.17}$$

where $\mathbf{Y}$ is a vector of constants and λ is a parameter, both of which are to be determined through the solution process. Substitution of Equation (12.17) into Equation (12.12a) leads to

$$\lambda \mathbf{Y}e^{\lambda t} = \mathbf{A}\mathbf{Y}e^{\lambda t} \tag{12.18}$$

Equation (12.18) is rearranged to

$$(\mathbf{A} - \lambda \mathbf{I})\mathbf{Y}e^{\lambda t} = \mathbf{0}$$

$$(\mathbf{A} - \lambda \mathbf{I})\mathbf{Y} = \mathbf{0} \tag{12.19}$$

A nontrivial solution of the system of equations represented by Equation (12.19) occurs if and only if

$$|\mathbf{A} - \lambda \mathbf{I}| = 0 \tag{12.20}$$

The evaluation of the determinant of Equation (12.20) leads to a polynomial equation called the characteristic equation. The roots of the characteristic equation are the eigenvalues of $\mathbf{A}$. For each λ that is an eigenvalue of $\mathbf{A}$, a nontrivial solution of Equation (12.19) exists and is called an eigenvector of $\mathbf{A}$ corresponding to the eigenvalue λ. An eigenvector is not unique because any nonzero multiple of an eigenvector is also a nontrivial solution of Equation (12.19).

The eigenvalues of a system with n state variables are computed by finding the roots of the nth-order polynomial generated by the evaluation of the determinant of Equation (12.20). The coefficients of the polynomial are all real. Thus, if complex eigenvalues occur, they occur in complex conjugate pairs. Let $\lambda_1, \lambda_2, \ldots, \lambda_n$ be the eigenvalues of $\mathbf{A}$ ordered such that $|\lambda_1| \le |\lambda_2| \le |\lambda_3| \le \cdots \le |\lambda_n|$. Eigenvctors $\mathbf{Y}_1, \mathbf{Y}_2, \ldots, \mathbf{Y}_n$ are defined such that Y_k corresponds to the eigenvalue λ_k.

If the eigenvalues are distinct, then n linearly independent solutions of Equation (12.12a) in the form of Equation (12.17) exist. The general solution is a linear combination of such solutions. Thus it can be written as

$$\mathbf{y} = \sum_{k=1}^{n} W_k \mathbf{Y}_k e^{\lambda_k t} \tag{12.21}$$

where $W_1, W_2, \ldots, W_n$ are arbitrary constants.

The $n \times n$ matrix $\mathbf{P}(t)$ is defined such that its kth column is $e^{\lambda_k t}\mathbf{Y}_k$. Equation (12.21) is rewritten using the $\mathbf{P}$ matrix as

$$\mathbf{y} = \mathbf{P}(t)\mathbf{W} \tag{12.22}$$

where $\mathbf{W}$ is the $n \times 1$ vector whose k^{th} element is W_k. The application of Equation (12.12b) to Equation (12.22) leads to

$$\mathbf{y}_0 = \mathbf{P}(0)\mathbf{W} \tag{12.23}$$

The matrix $\mathbf{P}(0)$ is a matrix whose columns are the eigenvectors of $\mathbf{A}$, which must be linearly independent. Thus $\mathbf{P}(0)$ cannot be singular and its inverse exists. Premultiplying Equation (12.23) by $\mathbf{P}(0)^{-1}$ leads to

$$\mathbf{W} = \{\mathbf{P}(0)\}^{-1}\mathbf{y}_0 \tag{12.24}$$

Substitution for $\mathbf{W}$ from Equation (12.24) into Equation (12.22) gives

$$\mathbf{y} = \mathbf{P}(t)\{\mathbf{P}(0)\}^{-1}\mathbf{y}_0 \tag{12.25}$$

Example 12.3

Use Equation (12.25) to determine the response of the system of Example 12.2.

Solution

The matrix **A** is obtained in Equation (k) of Example 12.2 as

$$\mathbf{A} = \begin{bmatrix} 0 & 1 & 0 \\ 0 & 0 & -10{,}000 \\ 0 & 1 & -250 \end{bmatrix} \tag{a}$$

The eigenvalues of **A** are obtained using Equation (12.20):

$$|\mathbf{A} - \lambda\mathbf{I}| = 0 = \begin{vmatrix} -\lambda & 1 & 0 \\ 0 & -\lambda & -10{,}000 \\ 0 & 1 & -250-\lambda \end{vmatrix} \tag{b}$$

The determinant is evaluated using expansion by its first column:

$$-\lambda[-\lambda(-250-\lambda) - (1)(-10{,}000)] = 0$$

$$\lambda(\lambda^2 + 250\lambda + 10{,}000) = 0 \tag{c}$$

The solutions of Equation (c) are

$$\lambda_1 = 0 \qquad \lambda_2 = -50 \qquad \lambda_3 = -200 \tag{d}$$

Eigenvectors of **A** are determined by substituting the eigenvalue into Equation (12.19) and obtaining a nontrivial solution. For this problem, they are

$$\mathbf{Y}_1 = \begin{bmatrix} 1 \\ 0 \\ 0 \end{bmatrix} \qquad \mathbf{Y}_2 = \begin{bmatrix} 1 \\ -50 \\ -\frac{1}{4} \end{bmatrix} \qquad \mathbf{Y}_3 = \begin{bmatrix} 1 \\ -200 \\ -4 \end{bmatrix} \tag{e}$$

The matrix **P** is the matrix whose k^{th} column is $\mathbf{Y}_k e^{\lambda_k t}$. To this end,

$$\mathbf{P} = \begin{bmatrix} 1 & e^{-50t} & e^{-200t} \\ 0 & -50e^{-50t} & -200e^{-200t} \\ 0 & -\frac{1}{4}e^{-50t} & -4e^{-200t} \end{bmatrix} \tag{f}$$

P(0) is obtained from Equation (f) as

$$\mathbf{P}(0) = \begin{bmatrix} 1 & 1 & 1 \\ 0 & -50 & -200 \\ 0 & \frac{1}{4} & -4 \end{bmatrix} \tag{g}$$

and then

$$\mathbf{P}(0)^{-1} = \begin{bmatrix} 1 & \frac{1}{40} & -1 \\ 0 & -\frac{2}{75} & \frac{4}{3} \\ 0 & \frac{1}{600} & -\frac{1}{3} \end{bmatrix} \tag{h}$$

The response is obtained using Equation (12.25) as

$$\mathbf{y} = \begin{bmatrix} 1 & e^{-50t} & e^{-200t} \\ 0 & -50e^{-50t} & -200e^{-200t} \\ 0 & -\frac{1}{4}e^{-50t} & -4e^{-200t} \end{bmatrix} \begin{bmatrix} 1 & \frac{1}{40} & -1 \\ 0 & -\frac{2}{75} & \frac{4}{3} \\ 0 & \frac{1}{600} & -\frac{1}{3} \end{bmatrix} \begin{bmatrix} 0 \\ 0.1 \\ 0 \end{bmatrix}$$

$$\mathbf{y} = 0.1 \begin{bmatrix} \frac{1}{40} + \frac{1}{600}e^{-200t} + \frac{2}{75}e^{-50t} \\ \frac{4}{3}e^{-50t} - \frac{1}{3}e^{-200t} \\ \frac{1}{150}e^{-50t} - \frac{1}{150}e^{-200t} \end{bmatrix} \tag{i}$$

12.2.3 General Description of Free Response

As expected, the Laplace transform method applied in Example 12.2 and the exponential solution applied in Example 12.3 lead to the same free response for a system. Since the two methods must lead to identical results, the comparison of Equations (12.16) and (12.25) makes it apparent that

$$\mathcal{L}^{-1}\{(s\mathbf{I} - \mathbf{A})^{-1}\} = \mathbf{P}(t)\mathbf{P}(0)^{-1} \tag{12.26}$$

Both of these equations can be formulated as

$$\mathbf{y} = \boldsymbol{\Phi}(t)\mathbf{y}_0 \tag{12.27}$$

where $\boldsymbol{\Phi}(t)$ is called the **state transition matrix**. It is determined as either the matrix on the left-hand side of Equation (12.26) or the matrix on the right-hand side of Equation (12.26).

The inverse of a matrix is proportional to the reciprocal of the determinant of the matrix. The inverse does not exist when the determinant is zero. As such, the poles of the transfer functions are the roots of

$$|s\mathbf{I} - \mathbf{A}| = 0 \tag{12.28}$$

Thus the transfer functions have the same number of poles as state variables. The system is stable only if all poles have negative real parts. The resulting mathematical form of the free response corresponding to a pole s_i is proportional to $e^{s_i t}$. The general form of the free response is a linear combination of the terms $e^{s_i t}$ for all $i = 1, 2, \ldots, n$, where n is the order of the system. Complex poles lead to trigonometric terms in the response.

The eigenvalues of **A** are calculated as the values of λ such that $|\mathbf{A} - \lambda\mathbf{I}| = 0$. Thus the eigenvalues of **A** are identical to the poles of the system's transfer function. The system is stable only if all eigenvalues have negative real parts. The free-response is as discussed here and in Chapter 7.

Example 12.4

Determine the state transition matrix for the series *LRC* circuit of Examples 5.10 and 5.29 and Figure 5.16 when **a.** $L = 0.1$ H, $R = 1$ kΩ, $C = 0.5$ μF; **b.** $L = 0.1$ H, $R = 1$ kΩ, $C = 0.4$ μF; and **c.** $L = 0.1$ H, $R = 1$ kΩ, $C = 0.1$ μF.

Solution

From Equation (g) of Example 5.29, the state-space formulation for a series *LRC* circuit is

$$\begin{bmatrix} \dot{y}_1 \\ \dot{y}_2 \end{bmatrix} = \begin{bmatrix} 0 & 1 \\ -\frac{1}{LC} & -\frac{R}{L} \end{bmatrix} \begin{bmatrix} y_1 \\ y_2 \end{bmatrix} + \begin{bmatrix} 0 \\ \frac{1}{L} \end{bmatrix} [v] \tag{a}$$

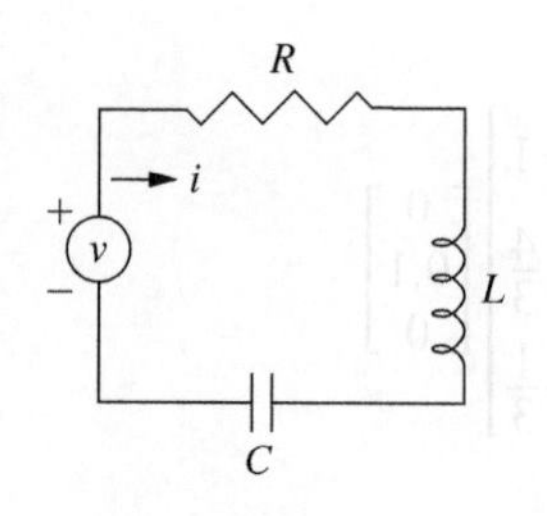

Figure 5.16

The series *LRC* circuit of Example 5.10. (Repeated)

where the state variables are

$$y_1 = \int_0^t i\,dt \tag{b}$$

and

$$y_2 = i \tag{c}$$

a. Substitution of the given values into Equation (a) leads to the state matrix

$$\mathbf{A} = \begin{bmatrix} 0 & 1 \\ -2 \times 10^7 & -1 \times 10^4 \end{bmatrix} \tag{d}$$

The eigenvalues of **A** are determined from

$$\begin{vmatrix} -\lambda & 1 \\ -2 \times 10^7 & -1 \times 10^4 - \lambda \end{vmatrix} = 0$$

$$(-\lambda)(-1 \times 10^4 - \lambda) - (1)(-2 \times 10^7) = 0$$

$$\lambda^2 + 1 \times 10^4\lambda + 210^7 = 0 \tag{e}$$

The solutions of Equation (e) are $\lambda = -8.87 \times 10^3, -1.13 \times 10^3$. The corresponding eigenvectors are

$$\mathbf{Y}_1 = \begin{bmatrix} 1 \\ -8.87 \times 10^3 \end{bmatrix} \quad \mathbf{Y}_2 = \begin{bmatrix} 1 \\ -1.13 \times 10^3 \end{bmatrix} \tag{f}$$

The matrices **P** and $\mathbf{P}(0)^{-1}$ are consequently determined as

$$\mathbf{P} = \begin{bmatrix} e^{-8.87\times10^3 t} & e^{-1.13\times10^3 t} \\ -8.87 \times 10^3 e^{-8.87\times10^3 t} & -1.13 \times 10^3 e^{-1.13\times10^3 t} \end{bmatrix} \tag{g}$$

$$\mathbf{P}(0)^{-1} = \begin{bmatrix} -0.146 & -1.29 \times 10^{-4} \\ -1.146 & 1.29 \times 10^{-4} \end{bmatrix} \tag{h}$$

The state transition matrix is

$$\boldsymbol{\Phi}(t) = \mathbf{P}\mathbf{P}(0)^{-1} = \begin{bmatrix} -0.146 & -1.29 \times 10^{-4} \\ 1.23 \times 10^3 & -1.146 \end{bmatrix} e^{-8.87\times10^3 t} + \begin{bmatrix} -0.146 & -1.29 \times 10^{-4} \\ 1.65 \times 10^2 & -0.146 \end{bmatrix} e^{-1.13\times10^3 t} \tag{i}$$

The state transition matrix of Equation (i) is an example of a state transition matrix for a second-order overdamped system.

b. The matrix **A** obtained using these parameters is

$$\mathbf{A} = \begin{bmatrix} 0 & 1 \\ -2.5 \times 10^7 & -1 \times 10^4 \end{bmatrix} \tag{j}$$

A procedure similar to that of part a. is used to obtain the eigenvalues of **A** as $\lambda = -5 \times 10^3, -5 \times 10^3$. The system has repeated eigenvectors. The exponential solution may be modified to obtain a response in the case of repeated eigenvectors, but it is beyond the scope of this text. The state transition matrix is determined as

$$\boldsymbol{\Phi}(t) = \mathcal{L}^{-1}\{(s\mathbf{I} - \mathbf{A})^{-1}\} \tag{k}$$

Use of MATLAB to perform symbolic computations leads to

$$(s\mathbf{I} - \mathbf{A})^{-1} = \begin{bmatrix} \dfrac{s + 1 \times 10^4}{(s + 5000)^2} & \dfrac{1}{(s + 5000)^2} \\ \dfrac{2.5 \times 10^7}{(s + 5000)^2} & \dfrac{s}{(s + 5000)^2} \end{bmatrix} \tag{l}$$

$$\boldsymbol{\Phi}(t) = \begin{bmatrix} 1 + 5000t & t \\ -2.5 \times 10^7 t & 1 - 5000t \end{bmatrix} e^{-5000t} \tag{m}$$

The state transition matrix of Equation (m) is an example of a state transition matrix for a second-order system with critical damping.

c. The matrix A obtained using the given parameters is

$$\mathbf{A} = \begin{bmatrix} 0 & 1 \\ -1 \times 10^8 & -1 \times 10^4 \end{bmatrix} \tag{n}$$

The eigenvalues of **A** are obtained as $\lambda_{1,2} = -5 \times 10^3 \pm j8.66 \times 10^3$. The eigenvectors of **A** are determined as

$$\mathbf{Y}_1 = \begin{bmatrix} 1 \\ -5 \times 10^3 + j8.66 \times 10^3 \end{bmatrix} \quad \mathbf{Y}_2 = \begin{bmatrix} 1 \\ -5 \times 10^3 - j8.66 \times 10^3 \end{bmatrix} \tag{o}$$

The use of MATLAB for computation leads to

$$\boldsymbol{\Phi}(t) = \begin{bmatrix} (0.5 - 0.289j)e^{\lambda_1 t} + (0.5 + 0.289j)e^{\lambda_2 t} & 5.78 \times 10^{-5}(e^{\lambda_1 t} + e^{\lambda_2 t}) \\ 5.77 \times 10^{-3}(e^{\lambda_1 t} - e^{-\lambda_2 t}) & (0.5 + 0.289j)e^{\lambda_1 t} + (0.5 - 0.289j)e^{\lambda_2 t} \end{bmatrix} \tag{p}$$

Equation (p) is further simplified by noting that $\lambda_2 = \bar{\lambda}_1$, the sum of two complex conjugates is twice the real part of one, the difference of two complex conjugates is twice the imaginary part of one, $\mathrm{Re}[e^{jbt}] = \cos(bt)$, and $\mathrm{Im}[e^{jbt}] = \sin(bt)$. The application of these simplifications leads to

$$\boldsymbol{\Phi}(t) = \begin{bmatrix} \cos(8.66 \times 10^3 t) + 0.578 \sin(8.66 \times 10^3 t) & 1.16 \times 10^{-4} \cos(8.66 \times 10^3 t) \\ 1.16 \times 10^{-2} \sin(8.66 \times 10^3 t) & \sin(8.66 \times 10^3 t) - 0.578 \cos(8.66 \times 10^3 t) \end{bmatrix} e^{-500t} \tag{q}$$

12.3 State-Space Analysis of Response Due to Inputs

The general form of the equations for the state variables when the system is subject to nonzero inputs is

$$\dot{\mathbf{y}} = \mathbf{A}\mathbf{y} + \mathbf{B}\mathbf{u} \tag{12.29}$$

Several convenient methods of solution exist for Equation (12.29). The application of the Laplace transform method leads to a solution in terms of integrals that can often be easily evaluated. State-space formulation is convenient for the application of self-starting numerical methods such as the Runge-Kutta methods.

Application of the Laplace transform to Equation (12.29), including the application of initial conditions of the form $\mathbf{y}(0) = \mathbf{y}_0$, leads to

$$s\mathbf{Y}(s) - \mathbf{y}_0 = \mathbf{A}\mathbf{Y}(s) + \mathbf{B}\mathbf{U}(s) \tag{12.30}$$

Solving Equation (12.30) for $\mathbf{Y}(s)$ gives

$$\mathbf{Y}(s) = (s\mathbf{I} - \mathbf{A})^{-1}\mathbf{y}_0 + (s\mathbf{I} - \mathbf{A})^{-1}\mathbf{B}\mathbf{U}(s) \tag{12.31}$$

Equation (12.31) shows that the matrix of transfer functions for the state variables is

$$\mathbf{G}(s) = (s\mathbf{I} - \mathbf{A})^{-1}\mathbf{B} \tag{12.32}$$

Equation (12.32) shows that the poles of the transfer functions are the roots of $|(s\mathbf{I} - \mathbf{A})^{-1}|$ and that the matrix of impulsive responses is

$$\mathbf{y}_i(t) = \mathcal{L}^{-1}\{(s\mathbf{I} - \mathbf{A})^{-1}\}\mathbf{B} \tag{12.33}$$

Recalling that the state transition matrix is defined as $\boldsymbol{\Phi}(t) = \mathcal{L}^{-1}\{(s\mathbf{I} - \mathbf{A})^{-1}\}$, Equation (12.33) is rewritten as

$$\mathbf{y}_i(t) = \boldsymbol{\Phi}(t)\mathbf{B} \tag{12.34}$$

Taking the inverse transform of Equation (12.31) leads to

$$\mathbf{y}(t) = \boldsymbol{\Phi}(t)\mathbf{y}_0 + \mathcal{L}^{-1}\{(s\mathbf{I} - \mathbf{A})^{-1}\,\mathbf{BU}(s)\} \tag{12.35}$$

The convolution property of the transforms written in its inverse form and applied to matrices of transforms that are multiplicatively compatible is

$$\mathcal{L}^{-1}\{\mathbf{F}(s)\mathbf{G}(s)\} = \mathbf{f}(t) * \mathbf{g}(t) = \int_0^t \mathbf{f}(t-\tau)\mathbf{g}(\tau)d\tau \tag{12.36}$$

The application of Equation (12.36) to Equation (12.35) and using the definition of the state transition matrix leads to

$$\mathbf{y}(t) = \boldsymbol{\Phi}(t)\mathbf{y}_0 + \int_0^t \boldsymbol{\Phi}(t-\tau)\,\mathbf{Bu}(\tau)d\tau \tag{12.37}$$

Equation (12.37) provides the general response of the state variables to any form of system input.

Example 12.5

Determine the impulsive response of the series LRC circuit of Example 12.4 for each set of circuit parameters given in parts a., b., and c.

Solution

Noting that for each set of circuit parameters of Example 12.4 $L = 0.1$ H, the input matrix determined in Example 12.4 is written as

$$\mathbf{B} = \begin{bmatrix} 0 \\ 10 \end{bmatrix} \tag{a}$$

a. The state transition matrix for these parameters is determined in Equation (i) of Example 12.4 part a. as

$$\boldsymbol{\Phi}(t) = \begin{bmatrix} -0.146 & -1.29 \times 10^{-4} \\ 1.23 \times 10^{3} & -1.146 \end{bmatrix} e^{-8.87\times 10^3 t} + \begin{bmatrix} -0.146 & -1.29 \times 10^{-4} \\ 1.65 \times 10^{2} & -0.146 \end{bmatrix} e^{-1.13\times 10^3 t}$$

Equation (12.34) and the state transition matrix are used to determine the impulsive response of this circuit as

$$\mathbf{y}_i(t) = \boldsymbol{\Phi}(t)\mathbf{B} = \begin{bmatrix} -1.29 \times 10^{-3}\left(e^{-8.87\times 10^3 t} + e^{-1.13\times 10^3 t}\right) \\ 11.46e^{-8.897\times 10^3 t} - 1.46e^{-1.13\times 10^3 t} \end{bmatrix} \tag{b}$$

Noting from Equation (c) of Example 12.4 that $i(t) = y_2(t)$ gives the current in the circuit due to an impulsive voltage:

$$i(t) = 11.46e^{-8.897\times 10^3 t} - 1.46e^{-1.13\times 10^3 t}\text{A} \tag{c}$$

b. The impulsive response is determined using Equation (12.34) along with the state transition matrix of Equation (j) of Example 12.4:

$$\mathbf{y}_i(t) = \begin{bmatrix} 10te^{-5000t} \\ 10(1 - 5000t)e^{-5000t} \end{bmatrix} \tag{d}$$

c. The impulsive response for the given parameters is determined using Equation (12.34) and the state transition matrix in Equation (n) of Example 12.4:

$$\mathbf{y}_i(t) = \begin{bmatrix} 1.16 \times 10^{-3} e^{-500t} \cos(8.66 \times 10^3 t) \\ 10e^{-500t}(\sin(8.66 \times 10^3 t) - 0.578 \cos(8.66 \times 10^3 t)) \end{bmatrix} \tag{e}$$

Example 12.6

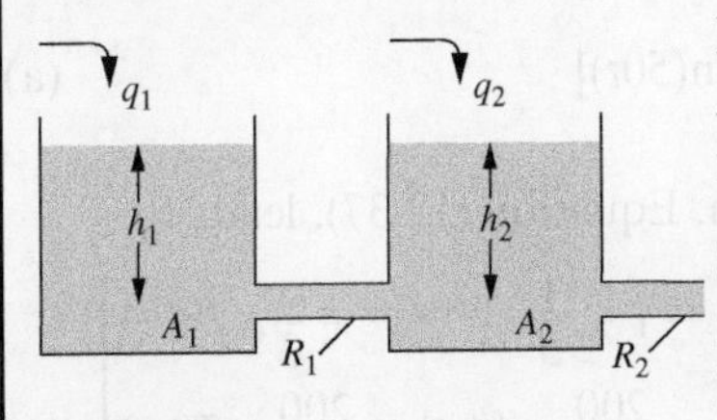

Figure 12.8

Two-tank liquid-level system of Example 12.6. The flow rates and liquid levels are perturbations from steady state.

Determine the response of the two-tank liquid-level system of Figure 12.8 when $A_1 = 10.9\ \text{m}^2$, $A_2 = 15.8\ \text{m}^2$, $R_1 = 6.5\ \text{s/m}^2$, $R_2 = 3.8\ \text{s/m}^2$, and the perturbation in flow rates are $q_1 = -0.2u(t)\ \text{m}^3/\text{s}$ and $q_2 = 0.2u(t)\ \text{m}^3/\text{s}$.

Solution

Substituting the given parameters for the perturbations in height in the two-tank liquid-level system gives a state-space model

$$\begin{bmatrix} \dot{y}_1 \\ \dot{y}_2 \end{bmatrix} = \begin{bmatrix} -0.0141 & 0.0141 \\ 0.00974 & -0.0264 \end{bmatrix} \begin{bmatrix} y_1 \\ y_2 \end{bmatrix} + \begin{bmatrix} 0.0917 & 0 \\ 0 & 0.0633 \end{bmatrix} \begin{bmatrix} -0.2u(t) \\ 0.2u(t) \end{bmatrix} \tag{a}$$

The eigenvalues of the state matrix are determined as

$$|\mathbf{A} - \lambda \mathbf{I}| = 0$$

$$\begin{vmatrix} -0.0141 - \lambda & 0.0141 \\ 0.00974 & -0.0264 - \lambda \end{vmatrix} = 0$$

$$(-0.0141 - \lambda)(-0.0264 - \lambda) - (0.00974)(0.0141) = 0$$

$$\lambda^2 + 0.0405\lambda + 1.37 \times 10^{-4} = 0$$

$$\lambda = -0.0368, -0.0186 \tag{b}$$

The eigenvectors are calculated as

$$\mathbf{Y}_1 = \begin{bmatrix} 1 \\ 1.61 \end{bmatrix} \quad \mathbf{Y}_2 = \begin{bmatrix} 1 \\ 0.319 \end{bmatrix} \tag{c}$$

The state transition matrix is determined as

$$\boldsymbol{\Phi}(t) = \begin{bmatrix} -0.2471e^{-0.0368t} + 1.2471e^{-0.0186t} & 0.775e^{-0.0368t} - 0.775e^{-0.0186t} \\ -0.398e^{-0.0368t} + 0.398e^{-0.0186t} & 1.247e^{-0.0368t} - 0.247e^{-0.0186t} \end{bmatrix} \tag{d}$$

Noting that $\mathbf{y}_0 = \mathbf{0}$, the application of Equation (12.37) leads to

$$\begin{bmatrix} y_1 \\ y_2 \end{bmatrix} = \int_0^t \begin{bmatrix} -0.2471e^{-0.0368(t-\tau)} + 1.2471e^{-0.0186(t-\tau)} & 0.775e^{-0.0368(t-\tau)} - 0.775e^{-0.0186(t-\tau)} \\ -0.398e^{-0.0368(t-\tau)} + 0.398e^{-0.0186(t-\tau)} & 1.247e^{-0.0368(t-\tau)} - 0.247e^{-0.0186(t-\tau)} \end{bmatrix}$$
$$\times \begin{bmatrix} 0.0917 & 0 \\ 0 & 0.0633 \end{bmatrix} \begin{bmatrix} -0.2u(\tau) \\ 0.2u(\tau) \end{bmatrix} d\tau \tag{e}$$

The simplification of Equation (e) gives

$$\begin{bmatrix} y_1 \\ y_2 \end{bmatrix} = \begin{bmatrix} \int_0^t (0.0143e^{-0.0368(t-\tau)} - 0.0327e^{-0.0186(t-\tau)})u(\tau)\,d\tau \\ \int_0^t (0.0231e^{-0.0368(t-\tau)} - 0.0104e^{-0.0186(t-\tau)})u(\tau)\,d\tau \end{bmatrix} \tag{f}$$

The evaluation of integrals in Equation (f) leads to

$$\begin{bmatrix} y_1 \\ y_2 \end{bmatrix} = \begin{bmatrix} 1.370 - 0.387e^{-0.0368t} + 1.76e^{-0.0186t} \\ -0.0414 - 0.601e^{-0.0368t} + 0.560e^{-0.0186t} \end{bmatrix} u(t) \tag{g}$$

Example 12.7

Determine the steady-state response of the system of Figure 12.7 and Example 12.2 when the system is subject to an input of $F(t) = 100\sin(50t)$ N.

Solution

The initial conditions are irrelevant for the steady-state response. Thus it is assumed that $\mathbf{y}_0 = 0$. The state transition matrix is given in Equation (p) of Example 12.2. **B** and **u** are determined from Example 12.2 as

$$\mathbf{B} = \begin{bmatrix} 0 \\ 0.1 \\ 0 \end{bmatrix} \quad \mathbf{u} = [100\sin(50t)] \tag{a}$$

The application of the convolution integral solution, Equation (12.37), leads to

$$\begin{bmatrix} y_1 \\ y_2 \\ y_3 \end{bmatrix} = \int_0^t \begin{bmatrix} 1 & \frac{1}{40} + \frac{1}{600}e^{-200(t-\tau)} + \frac{2}{75}e^{-50(t-\tau)} & -1 - \frac{1}{3}e^{-200(t-\tau)} + \frac{4}{3}e^{-50(t-\tau)} \\ 0 & \frac{4}{3}e^{-50(t-\tau)} - \frac{1}{3}e^{-200(t-\tau)} & -\frac{200}{3}e^{-50(t-\tau)} + \frac{200}{3}e^{-50(t-\tau)} \\ 0 & \frac{1}{150}e^{-50(t-\tau)} - \frac{1}{150}e^{-200(t-\tau)} & -\frac{1}{3}e^{-50(t-\tau)} + \frac{4}{3}e^{-200(t-\tau)} \end{bmatrix} \times \begin{bmatrix} 0 \\ 0.1 \\ 0 \end{bmatrix} 100\sin(50\tau)\,d\tau \tag{b}$$

Simplification of Equation (b) leads to

$$\begin{bmatrix} y_1 \\ y_2 \\ y_3 \end{bmatrix} = \begin{bmatrix} \int_0^t 10\left(\frac{1}{40} + \frac{1}{600}e^{-200(t-\tau)} + \frac{2}{75}e^{-50(t-\tau)}\right)\sin(50\tau)\,d\tau \\ \int_0^t 10\left(\frac{4}{3}e^{-50(t-\tau)} - \frac{1}{3}e^{-200(t-\tau)}\right)\sin(50\tau)\,d\tau \\ \int_0^t 10\left(\frac{1}{150}e^{-50(t-\tau)} - \frac{1}{150}e^{-200(t-\tau)}\right)\sin(50\tau)\,d\tau \end{bmatrix} \tag{c}$$

The system output is $x = y_1$. Thus from Equation (c),

$$x(t) = \int_0^t 10\left(\frac{1}{40} + \frac{1}{600}e^{-200(t-\tau)} + \frac{2}{75}e^{-50(t-\tau)}\right)\sin(50\tau)\,d\tau$$

$$= \frac{1}{4}\int_0^t \sin(50\tau)\,d\tau + \frac{1}{60}e^{-200t}\int_0^t e^{200\tau}\sin(50\tau)\,d\tau + \frac{4}{15}e^{-50t}\int_0^t e^{50\tau}\sin(50\tau)\,d\tau \tag{d}$$

The evaluation of the integrals in Equation (d) leads to

$$x(t) = 0.08[1 - \cos(50t)] + 2.74 \times 10^{-3}\sin(50t) + 1.96 \times 10^{-5}e^{-200t}[1 - \cos(50t)] + 2.67 \times 10^{-3}e^{-50t}[1 - \cos(50t)] \tag{e}$$

The transient response is included in Equation (e). The steady-state response is obtained by taking the limit of Equation (e) as $t \to \infty$, leading to

$$x_{ss}(t) = 0.08[1 - \cos(50t)] + 2.74 \times 10^{-3}\sin(50t) \tag{f}$$

The use of trigonometric identities allows Equation (f) to be rewritten as

$$x_{ss}(t) = 0.08 + 8.00 \times 10^{-2}\sin(50t - 1.54) \tag{g}$$

12.4 Numerical Solutions

The state-space formulation of a mathematical model is convenient for the application of numerical methods to determine a system response. Recall from Chapter 1 that the possible objectives for mathematical modeling of a dynamic system are system analysis, system design, and system synthesis. The numerical simulation of a dynamic system response can aid in achieving each of these objectives. The numerical computations involving complicated equations are often required for system analysis and system design.

Closed-form solutions for the response of dynamic systems are handy when easily obtained. Numerical solutions are still only approximate solutions and error is introduced in every computation. Thus, especially for long times, closed-form solutions are more accurate. The application of state-space methods leads to the determination of the total solution from which a steady-state response, if desired, can be extracted. Numerical solutions also determine transient behavior before a steady state is reached. System design requires analysis of the response of the system in terms of unspecified values of system parameters. Indeed, part of the design is to determine the appropriate values of these parameters. An understanding of the qualitative changes in the response of a system as a parameter changes is easier to obtain when a closed-form solution exists.

This is one reason why knowledge of both transfer function methods and state-space methods is desirable. Transfer function methods allow analysis of the system in terms of system parameters, which is useful in system design. The use of transfer function methods, as shown in Chapters 7 and 8, allows for a qualitative description of both transient and steady-state system responses. These in turn allow, as shown in Chapters 9, 10, and 11, for an understanding of how control systems are used to control and modify system response. However, the algebra leading to the determination of the response is complicated and tedious. State-space methods provide a fast and accurate determination of system response and allow for finding a quicker response from control systems. State variable control, introduced later in this chapter, is used as an alternative to compensators, and the response of these systems requires state-space analysis.

12.4.1 Runge-Kutta Methods

Numerical methods for the direct integration of state-space equations are self-starting methods; that is, only the vector of initial conditions, the matrices **A** and **B**, and the input are required to numerically determine the system response. Numerical methods approximate the system response at discrete times: $0 < t_1 < t_2 < \cdots < t_{k-1} < t_k < t_{k+1} < \cdots$. It is assumed that the difference between the discrete times is a constant h, called the **step size**, defined such that

$$t_k - t_{k-1} = h \qquad k = 1, 2, \ldots \tag{12.38}$$

The numerical approximation for $\mathbf{y}(t_k)$ is defined as $\mathbf{y}_k$. Self-starting numerical methods use a defined algorithm to approximate $\mathbf{y}_1$ using $\mathbf{y}_0$. The same algorithm is applied to approximate $\mathbf{y}_2$ using the approximation for $\mathbf{y}_1$. The algorithm is successively applied, calculating $\mathbf{y}_{k+1}$ using the approximation for $\mathbf{y}_k$ until it is terminated for a predetermined value of k.

Since numerical methods are approximate methods, there is a difference between the approximate solution $\mathbf{y}_k$ and the exact solution $\mathbf{y}(t_k)$. This difference is called the **error** of the approximation. An error, called the **local error**, occurs each time the algorithm is applied to compute $\mathbf{y}_{k+1}$ from $\mathbf{y}_k$. The local error propagates; that is, the total error in the approximation for $\mathbf{y}_{k+1}$ is the sum of the local errors from each previous step.

The error induced in a numerical approximation is determined by deriving the numerical algorithm from a Taylor series expansion for the exact solution $\mathbf{y}(t + h)$ about $\mathbf{y}(t)$. As expected, the error is strongly dependent on the step size h. The local error is usually determined to be proportional to an integer power of the step size

$$E = O(h^n) \tag{12.39}$$

Equation (12.39) is read as "the error is of the order of h^n." The equation illustrates that the error for such algorithms is reduced by decreasing the step size.

The most popular self-starting numerical methods are called **Runge-Kutta methods**. For a system with only one state variable, a Runge-Kutta algorithm is of the form

$$y_{k+1} = y_k + \phi(y_k, t_k h)h \tag{12.40}$$

where $\phi(y_k, t_k, h)$ is a function determined from an appropriate Taylor series expansion. It can generally be shown that an nth-order Runge-Kutta method has a local error of $O(h^{n+1})$ and a global error of $O(h^n)$.

The simplest of the Runge-Kutta methods is **Euler's method**, a first-order method that, when applied to a system of the form of Equation (12.29), leads to

$$\mathbf{y}_k = \mathbf{y}_{k-1} + [\mathbf{A}\mathbf{y}_{k-1} + \mathbf{B}\mathbf{u}(t_{k-1})]h \tag{12.41}$$

The local error for Euler's method is $O(h^2)$, while the global error is $O(h)$.

Example 12.8

Develop a MATLAB M-file that uses Euler's method to approximate the solution to Examples 12.2 and 12.7.

Solution

The state-space equations for the system are given in Equation (j) of Example 12.2. The values of the parameters are given in Example 12.2 and used in Example 12.7. The result is

$$\begin{bmatrix} \dot{y}_1 \\ \dot{y}_2 \\ \dot{y}_3 \end{bmatrix} = \begin{bmatrix} 0 & 1 & 0 \\ 0 & 0 & -10{,}000 \\ 0 & 1 & -250 \end{bmatrix} \begin{bmatrix} y_1 \\ y_2 \\ y_3 \end{bmatrix} + \begin{bmatrix} 0 \\ 0.1 \\ 0 \end{bmatrix} 100\sin(50t) \tag{a}$$

The application of Equation (12.41) to Equation (a) leads to

$$\begin{bmatrix} y_{1,k+1} \\ y_{2,k+1} \\ y_{3,k+1} \end{bmatrix} = \begin{bmatrix} y_{1,k} \\ y_{2,k} \\ y_{3,k} \end{bmatrix} + \left\{ \begin{bmatrix} 0 & 1 & 0 \\ 0 & 0 & -10{,}000 \\ 0 & 1 & -250 \end{bmatrix} \begin{bmatrix} y_{1,k} \\ y_{2,k} \\ y_{3,k} \end{bmatrix} + \begin{bmatrix} 0 \\ 10\sin(50\,t_k) \\ 0 \end{bmatrix} \right\} h \tag{b}$$

A MATLAB M-file used to approximate the response of this system is given in Figure 12.9. The file allows user input for two values of h and the final value for

```
% Example 12.8
% Euler's Method
%
% State matrix
A=[0 1 0; 0 0 -10000;0 1 -250]
% Input matrix
B=[0; 1; 0]
% Initial conditions
y(1,1)=0;
y(2,1)=0;
y(3,1)=0;
```

Figure 12.9

The script of Example12_8.m.

```
% Input step size and final time
h=input('Input step size  ');
tf=input('Input final time for computation  ')
t=0;
ta(1)=0;
k=1;
while t < tf
% Definition of state-vector at time t(k)
  C=[y(1,k);y(2,k);y(3,k)];
% Calculation of changes in state variables between t(k) and t(k+1)
  YN=C+h*(A*C+B*10*sin(50*t));
  k=k+1;
% Definition of state variables at t(k+1)
  y(1,k)=YN(1);
  y(2,k)=YN(2);
  y(3,k)=YN(3);
  t=t+h;
  ta(k)=t;
end
% System output is y(1)
plot(ta,y(1,:))
xlabel('t (s)')
ylabel('x (m)')
%
% End of Example12_8.m
```

Figure 12.9
(Continued)

which the response is approximated, t_∞. Figure 12.10 illustrates the response obtained from the execution of Example12_8.m for several values of h. Figure 12.10(a) shows that the numerical method is stable for $h = 0.01$, while Figure 12.10(b) shows that the numerical method is unstable for $h = 0.011$.

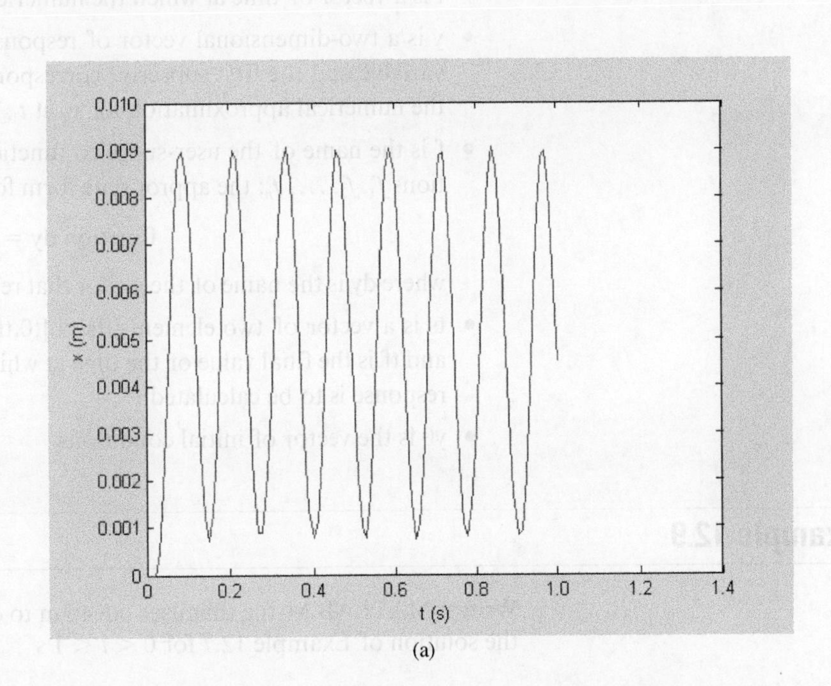

Figure 12.10

The integration of the state-space equations of Example 12.8 using Euler's method: (a) $h = 0.01$, (b) $h = 0.011$.

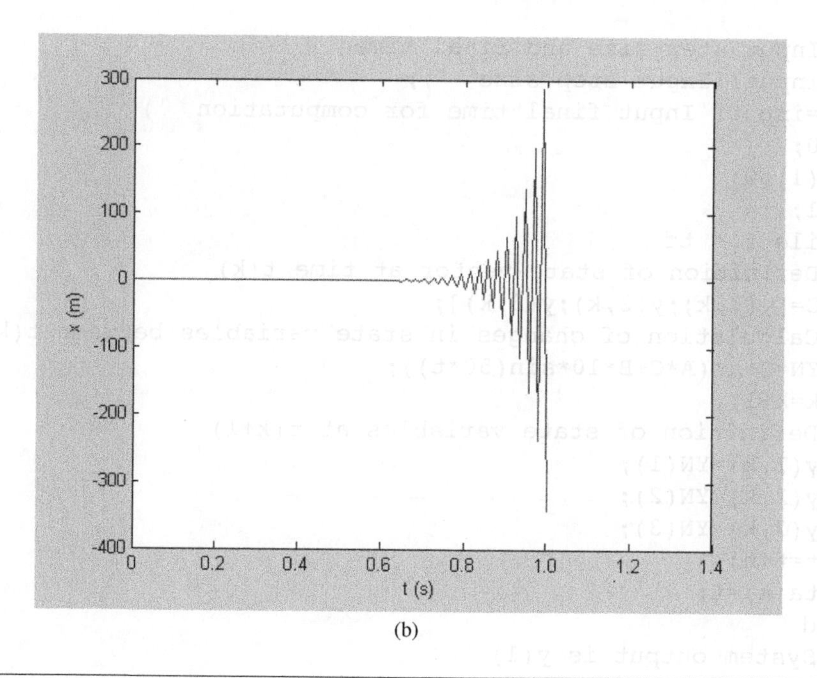

(b)

Figure 12.10

(Continued)

12.4.2 MATLAB Program ode45.m

MATLAB has several M-files in its library that use self-starting methods to numerically integrate a system of differential equations. The most widely applicable of these files is ode45.m, which is used for non-stiff equations. A stiff set of equations is one for which the numerical solution is slowly varying but can become unstable due to fast-varying solutions nearby. The basic format for using ode45.m to solve a system of n equations of the form $\dot{y}_i = f_i(y_1, y_2, \ldots, y_n, t)$ is

[t,y] = ode45(f,ts,y0)

where

- t is a vector of time at which the numerical response is returned;
- y is a two-dimensional vector of responses (the second subscript identifies the variable and the first subscript corresponds to the time; for example, $y(10,2)$ is the numerical approximation for y_2 at t_{10});
- f is the name of the user-supplied function subprogram that supplies the functions $f_1, f_2, \ldots, f_n$; the appropriate form for the function statement is

  function dy = f(t,y)

  where dy is the name of the vector that returns the values of f_i for $i = 1, 2, \ldots, n$;
- ts is a vector of two elements, ts = [t0,tf], where t0 is the initial value of time and tf is the final value of the time at which the numerical approximation to the response is to be calculated;
- y0 is the vector of initial conditions.

Example 12.9

Write a MATLAB M-file that uses ode45.m to develop a numerical approximation to the solution of Example 12.7 for $0 < t < 1$ s.

Solution

The script for Example12_9.m is presented in Figure 12.11, which also presents the script of the function subprogram required for its execution.

```
% Example 12.9
% Use of ode45.m
%
% This program uses ode45 to solve a system of 3 ODEs of the form
%      dy_i/dt=f_i(y_1,y_2,y_3,t) i=1,2,3
%
% Input initial conditions
disp('Input initial conditions')
for k=1:3
   str=['y0(',num2str(k),')= '];
   y1(k)=input(str);
end
% The initial condition vector must be a row vector
y0=[y(1) y(2) y(3)]
% Input final time
tf=input('Input final time for numerical calculation ');
ts=[0 tf];
% Calling ode45.m
%    t=vector of times at which solution is calculated
%    y=matrix containing system response
%    'f'=name of user supplied function which provides right-hand side
%        of differential equations. The format for the function is
%              function dy=f(t,y)
%        where dy is the vector of derivatives to be returned to ode45
%
[t,y]=ode45(@f,ts,y0)
plot(t,y(:,1))
xlabel('t (s)')
ylabel('x (m)')
title('ode45 solution for Example 12.9')
%
% End of Example12_9.m
```

(a)

```
% Function which provides right-hand sides of differential equations
%
%   dy is vector of derivatives to be returned to ode45
%   t is a scalar value of time for which the derivatives are to be
%      evaluated, it is supplied by ode45
%   y is the vector of values of numerical approximations at the
%      previous time step, it is supplied by ode45
%
function dy=f(t,y)
% The right-hand side is of the form Ab+c[10sin(50t)]
A=[0 1 0;0 0 -10000;0 1 -250];
b=[y(1);y(2);y(3)];
c=[0;1;0];
dy=A*b+c*10*sin(50*t);
```

(b)

Figure 12.11

(a) The script of Example12_9.m; (b) the user-supplied function required for the execution of Example12_9.m.

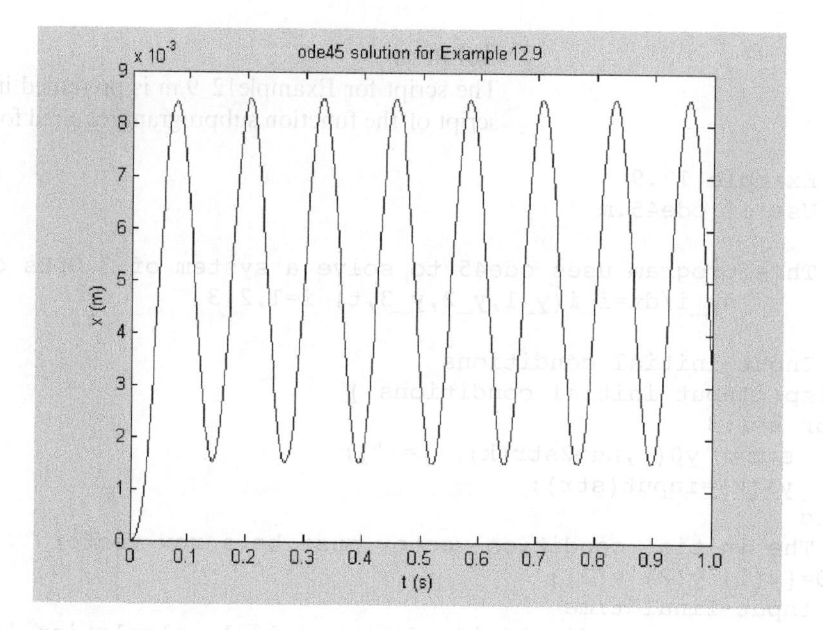

Figure 12.12

The numerical solution to Example 12.9 obtained from the execution of Example12_9.m, which uses MATLAB file ode45.m for numerical integration.

The plot of $x(t)$ generated from the execution of Example12_9.m is shown in Figure 12.12. The plot is very similar to that generated from the execution of Example12_8.m shown in Figure 12.10(a).

12.5 MATLAB and Simulink Modeling in the State Space

MATLAB is used to define transfer functions, calculate the transient response of systems, calculate the steady-state response of systems, draw Bode and Nyquist diagrams, aid in the design of control systems, and numerically solve the mathematical models of systems when developed in a state-space formulation. Simulink is used to represent block diagrams of mathematical models in both the transfer function formulation and the state-space formulation.

12.5.1 MATLAB

MATLAB and Simulink may also be used for state-space modeling of dynamic systems. A general state-space formulation for a linear system is of the form of Equations (3.20) and (3.21). These equations are used to define MATLAB and Simulink models in the state space. The matrices **A**, **B**, **C**, and **D** define the state-space model. For a given system, these matrices can be entered into the MATLAB workspace. A state-space model for the system is then defined by the statement

system = ss(A,B,C,D)

where ss is a MATLAB command from the Control Systems Toolbox that defines the state space and stores its attributes as the variable system. Figure 12.13 provides a copy of the workspace when a state-space model of Example 12.1 is entered with $m = 1$ kg, $c = 5$ N·s/m, and $k = 6$ N/m. The command

step(system)

generates the plot of the step response illustrated in Figure 12.14. Similarly, the command

impulse(system)

generates the impulsive response of the system.

```
>> A=[0 1;-6 -5];
>> B=[0;1];
>> C=[1 0];
>> D=[0];
>> system=ss(A,B,C,D)

a =
     x1  x2
x1   0   1
x2  -6  -5

b =
     u1
x1   0
x2   1

c =
     x1  x2
y1   1   0

d =
     u1
y1   0
Continuous-time model.
>>
```

Figure 12.13

The MATLAB workspace for entering the state-space model from Equation (a) of Example 12.1.

The response of a linear system due to a time-dependent input is obtained using the 'lsim' command. The times at which the simulation is to occur are defined in a vector, say t. The values of the input at these times is stored in a vector, say F. Then the command

lsim(system,F,t)

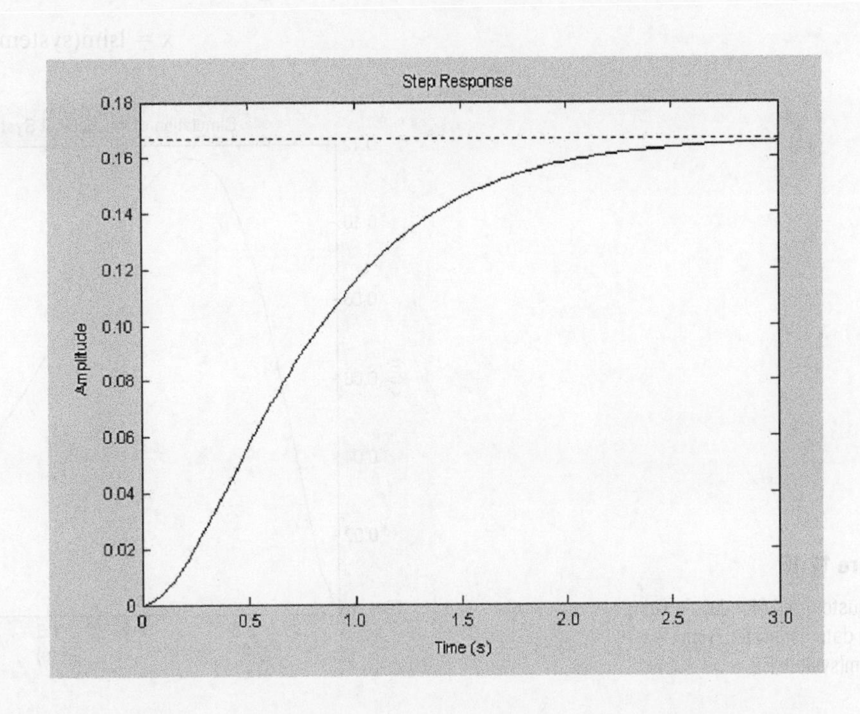

Figure 12.14

The step response for the system of Example 12.1, obtained using a state-space model.

```
>> t=0.:0.1:3;
>> F=3*exp(-3*t);
>> lsim(system,F,t)
>>
```

(a)

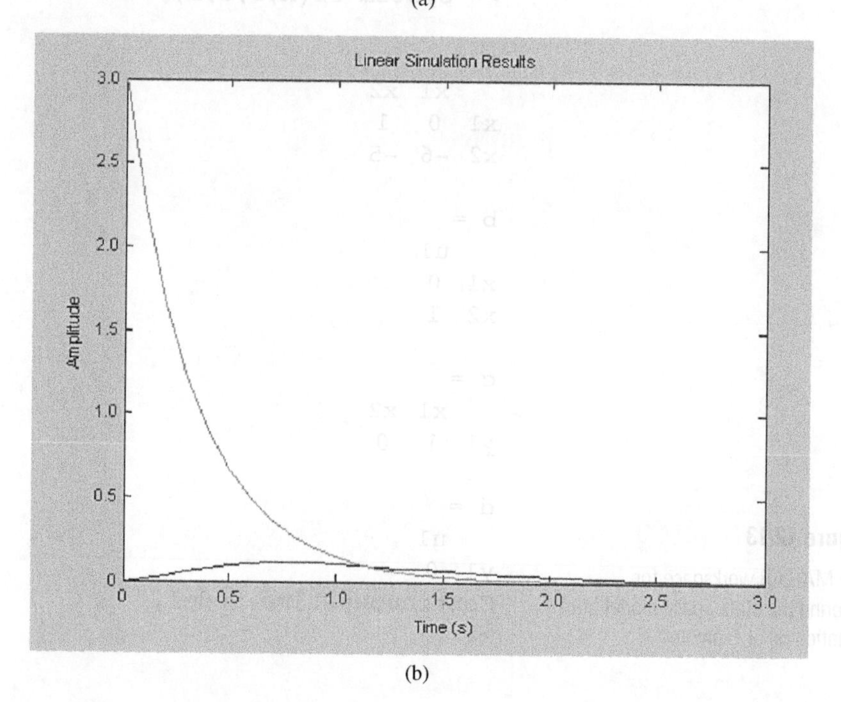

(b)

Figure 12.15

(a) The MATLAB commands to generate the simulation of the mechanical system with $F(t) = 3e^{-3t}$; (b) the resulting plot comparing the response to the input.

calculates a numerical simulation for the response of the system at the times specified in the vector t and plots the response. The MATLAB code necessary to run a simulation of the current mechanical system due to an input of the form $F(t) = 3e^{-3t}$ and the resulting simulation are illustrated in Figure 12.15. The 'lsim' command leads to a plot that compares the response to the input. It is difficult to determine much detail of the response from Figure 12.15. Thus a customized plot of the response is desirable and can be obtained by using the command

x = lsim(system,F,t)

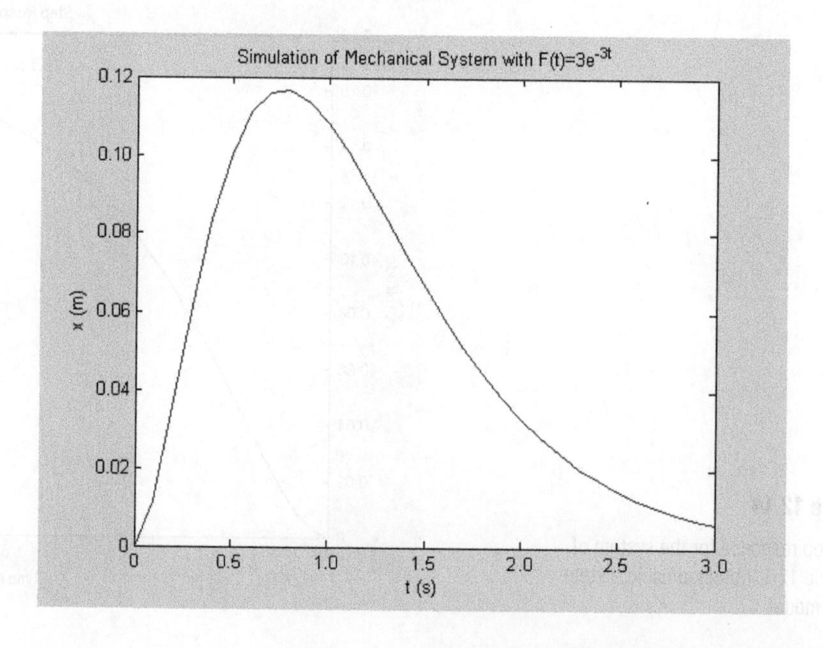

Figure 12.16

The customized plot developed using data generated from 'x=lsim(system,F,t)'.

which, instead of plotting the response, stores the numerical values of the response in the vector x. Then, a customized plot, such as that illustrated in Figure 12.16, can be developed.

It is shown in Section 3.6 that a state-space model can be used to determine the system's transfer function and that the transfer function may be used to determine a state-space model. MATLAB can perform both tasks. Given a state-space model, defined as a system, the transfer function is determined using the command

[N,D] = ss2tf(A,B,C,D)

where N and D are the vectors of coefficients defining the numerator and denominator of the system's transfer function. The matrices **A**, **B**, **C**, and **D** are the state, input, output, and transmission matrices, respectively, that are defined in the state-space formulation. The application of this command to the mechanical system of Example 12.1 is shown in Figure 12.17. Remember that the transfer function is unique, so the original N and D are expected.

Conversely, given the transfer function of a system in the form of the coefficient vectors for the numerator and denominator, a state-space model is obtained using the command

[A,B,C,D] = tf2ss(N,D)

This command is used to convert the transfer function back to a state-space model as illustrated in Figure 12.18. Remember that the transfer function for a system is unique, but the state-space model is not. The matrices A, B, C, and D could be different from the original matrices. In this case, the matrix A is different.

```
>> [N,D]=ss2tf(A,B,C,D)

N =

  0    0 1.0000

D =

1.0000 5.0000 6.0000

>> tf(N,D)

Transfer function:
  1
-------------
s^2 + 5 s + 6
```

Figure 12.17

Illustration of the development of the transfer function from the state-space model using MATLAB.

```
>> [A,B,C,D1]=tf2ss(N,D)

A =

  -5    -6
   1     0

B =

   1
   0

C =

   0     1

D1 =

   0

>>
```

Figure 12.18

Illustration of the development of a state-space model from the system's transfer function.

Example 12.10

Consider the two-degree-of-freedom mechanical system of Figure 12.19 with $m_1 = 1$ kg, $m_2 = 0.5$ kg, $c_1 = c_2 = 2$ N·s/m, $k_1 = 20$ N/m, and $k_2 = 10$ N/m. **a.** Develop a state-space model for the system in MATLAB. **b.** Use the state-space model to determine the impulsive responses for the system. **c.** Use the state-space model to determine the system response if $F_1(t) = 10 \sin(10t)$ and $F_2(t) = 5 \sin(5t)$. **d.** Use the state-space model to determine the transfer functions for the system.

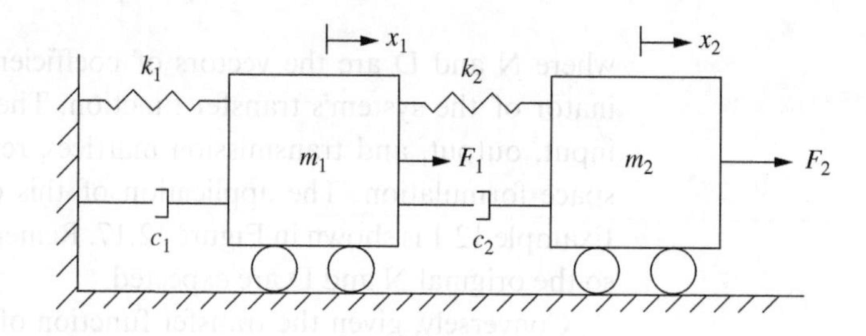

Figure 12.19

Mass-spring-viscous damper system of Example 12.10.

```
% Example 12.10
%
% Definition of parameters
m1=1;
m2=0.5;
c1=2;
c2=2;
k1=20;
k2=10;
% Definition of state-space matrices
A=[0 0 1 0;0 0 0 1;-(k1+k2)/m1 k2/m2 -(c1+c2)/m1 c2/m2;k2/m1 -k2/m2 c2/m1 -c2/m2];
B=[0 0;0 0;1/m1 0;0 1/m2];
C=[1 0 0 0;0 1 0 0];
D=[0 0; 0 0];
% State-space formulation
system=ss(A,B,C,D)
% Impulsive response
impulse(system)
% Forced responses
% Defining time range for calculation of forced response
t=0:0.02:4;
% Defining system inputs
F1=1*sin(10*t);
F2=0.5*sin(5*t);
F=[F1;F2];
figure
% Numerical simulation of forced response
lsim(system,F,t)
y=lsim(system,F,t);
figure
plot(t,y(:,1),'-',t,y(:,2),'--')
```

Figure 12.20

The script of Example12_10.m.

```
xlabel('t (s)')
ylabel('displacement (m)')
legend('x_1','x_2')
% transfer function
[N1,D1]=ss2tf(A,B,C,D,1)
[N2,D2]=ss2tf(A,B,C,D,2)
%
% End of Example12_10.m
```

Figure 12.20

(Continued)

```
> Example12_10

a =
    x1 x2 x3 x4
 x1  0  0  1  0
 x2  0  0  0  1
 x3 -30 20 -4  4
 x4 10 -20  2 -4


b =
   u1 u2
 x1  0  0
 x2  0  0
 x3  1  0
 x4  0  2


c =
    x1 x2 x3 x4
 y1  1  0  0  0
 y2  0  1  0  0


d =
   u1 u2
 y1  0  0
 y2  0  0

Continuous-time model.


N1 =

      0 -0.0000 1.0000 4.0000 20.0000
      0 -0.0000 -0.0000 2.0000 10.0000
```

Figure 12.21

(a) The workspace output from the execution of Example12_10.m; (b) the impulsive responses; (c) the forced responses plotted directly from 'lsim'; (d) the customized plots of the forced response.

```
D1 =

    1.0000 8.0000 58.0000 120.0000 400.0000

N2 =

    0 -0.0000 -0.0000 8.0000 40.0000
    0 -0.0000  2.0000 8.0000 60.0000

D2 =
    1.0000 8.0000 58.0000 120.0000 400.0000
>>
```

(a)

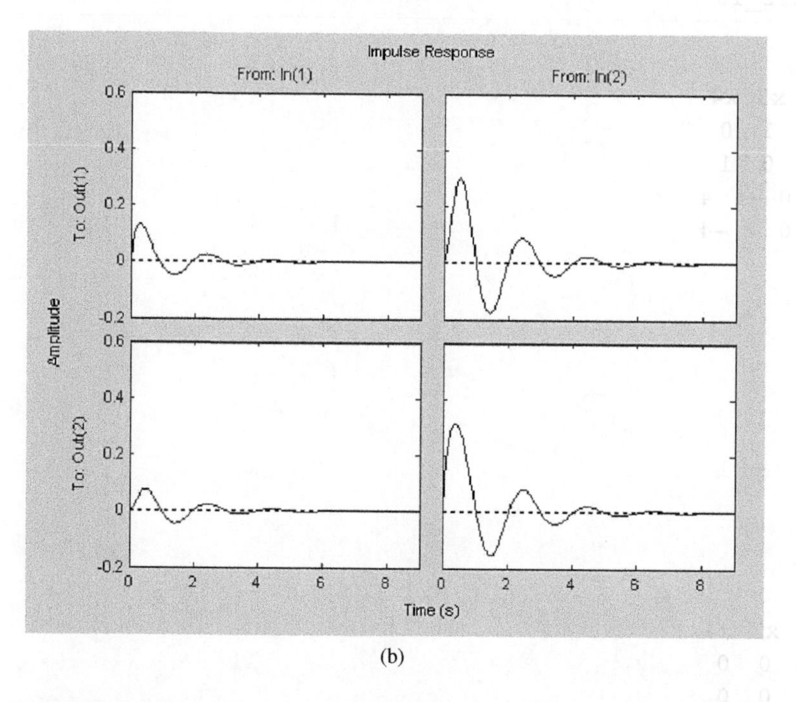

(b)

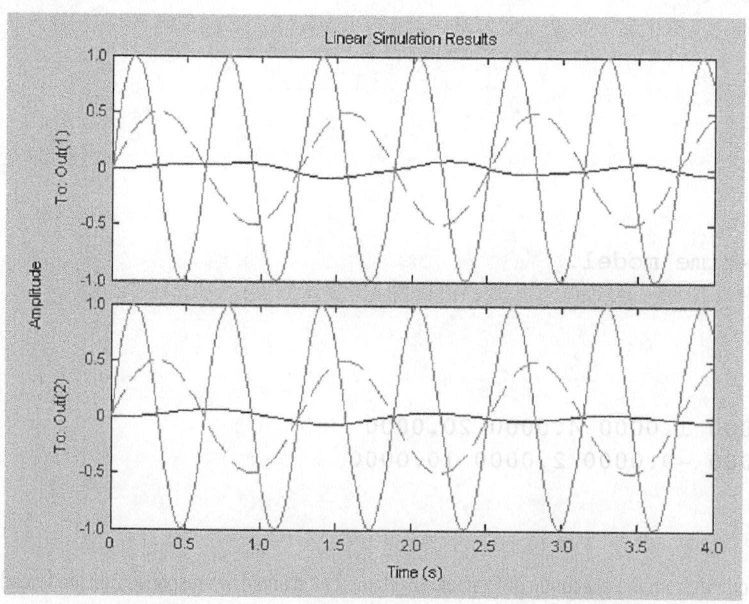

(c)

Figure 12.21
(Continued)

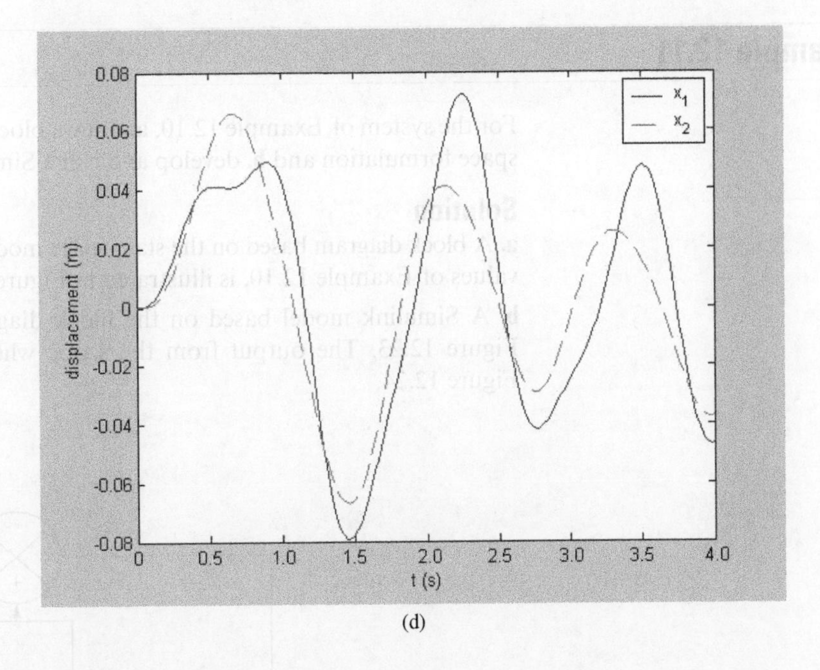

(d)

Figure 12.21

(Continued)

Solution

The script for the MATLAB program Example12_10.m is shown in Figure 12.20, while the output from its execution is shown in Figure 12.21. This system has two inputs $F_1(t)$ and $F_2(t)$ and two outputs $x_1(t)$ and $x_2(t)$. This leads to the following modifications in syntax for the state-space modeling from that of a SISO system:

- The impulsive response consists of four plots. Each plot corresponds to one input and one output. For example, the plot in the top left of Figure 12.21(b) is the response for x_1 due to $F_1 = \delta(t)$ and $F_2 = 0$.
- The variable defining the input for 'lsim' must be a matrix with two columns. Application of 'lsim' leads to two plots, one for each output variable.
- There are four transfer functions, $G_{i,j}(s) = X_i(s)/F_j(s)$. The use of 'ss2tf' for multiple output systems requires an additional argument defining which row of the transfer function matrix to determine. From the output in Figure 12.21(a), it is determined that

$$G_{1,1} = \frac{s^2 + 4s + 20}{s^4 + 8s^3 + 58s^2 + 120s + 400} \tag{a}$$

$$G_{2,2} = \frac{2s^2 + 8s + 60}{s^4 + 8s^3 + 58s^2 + 120s + 400} \tag{b}$$

12.5.2 Simulink

There are several methods of applying Simulink to simulate the response of a dynamic system from its state-space formulation. It is illustrated in Section 12.1 how a dynamic model is built in Simulink from the block diagram of the state-space formulation of the system. This is further illustrated in the following example.

Example 12.11

For the system of Example 12.10, **a.** draw a block diagram of the system from its state-space formulation and **b.** develop and run a Simulink model for the system.

Solution

a. A block diagram based on the state-space model of Figure 12.19, using the numerical values of Example 12.10, is illustrated in Figure 12.22.

b. A Simulink model based on the block diagram of Figure 12.22 is illustrated in Figure 12.23. The output from the scope when the simulation is run is shown in Figure 12.24.

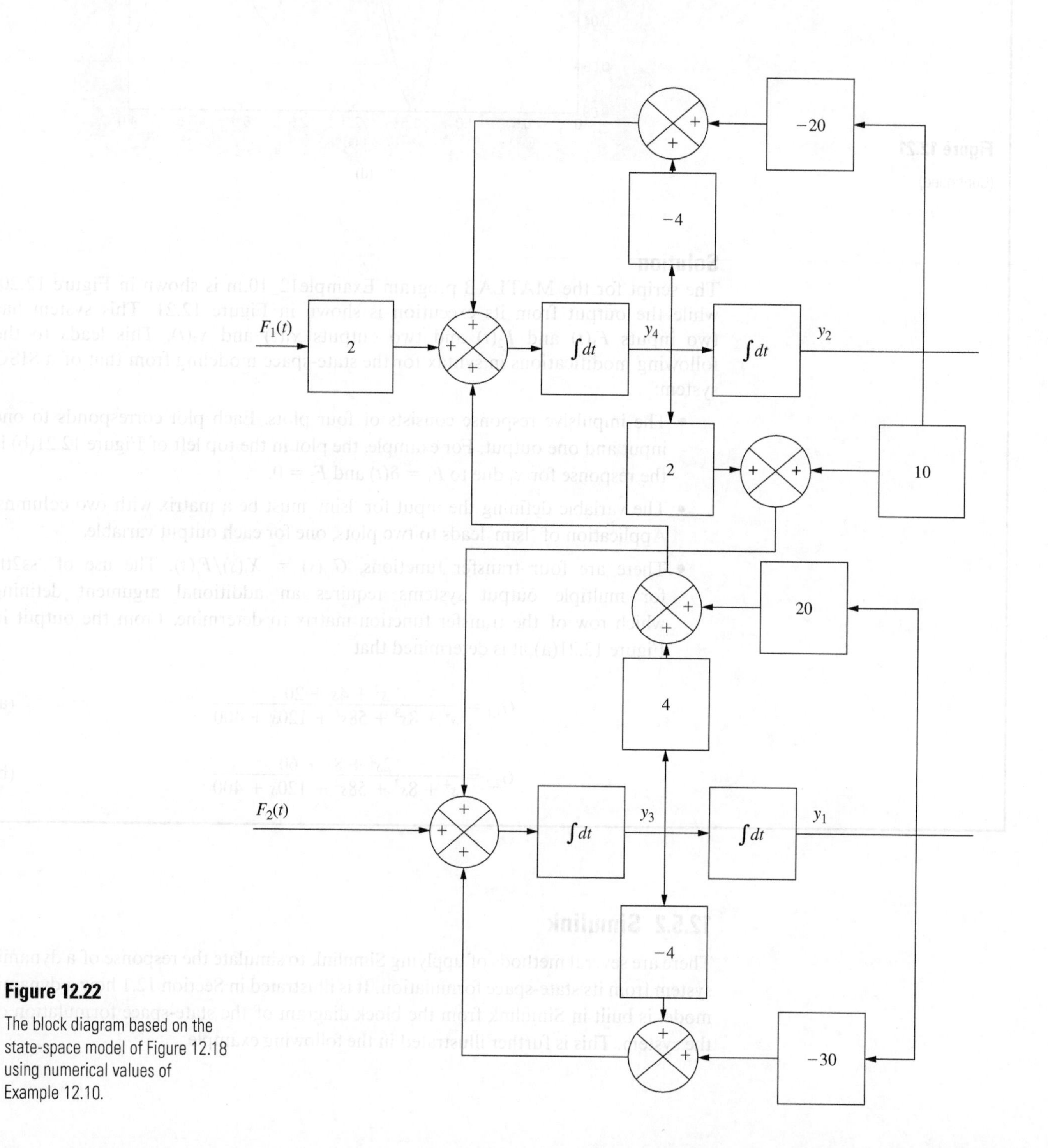

Figure 12.22
The block diagram based on the state-space model of Figure 12.18 using numerical values of Example 12.10.

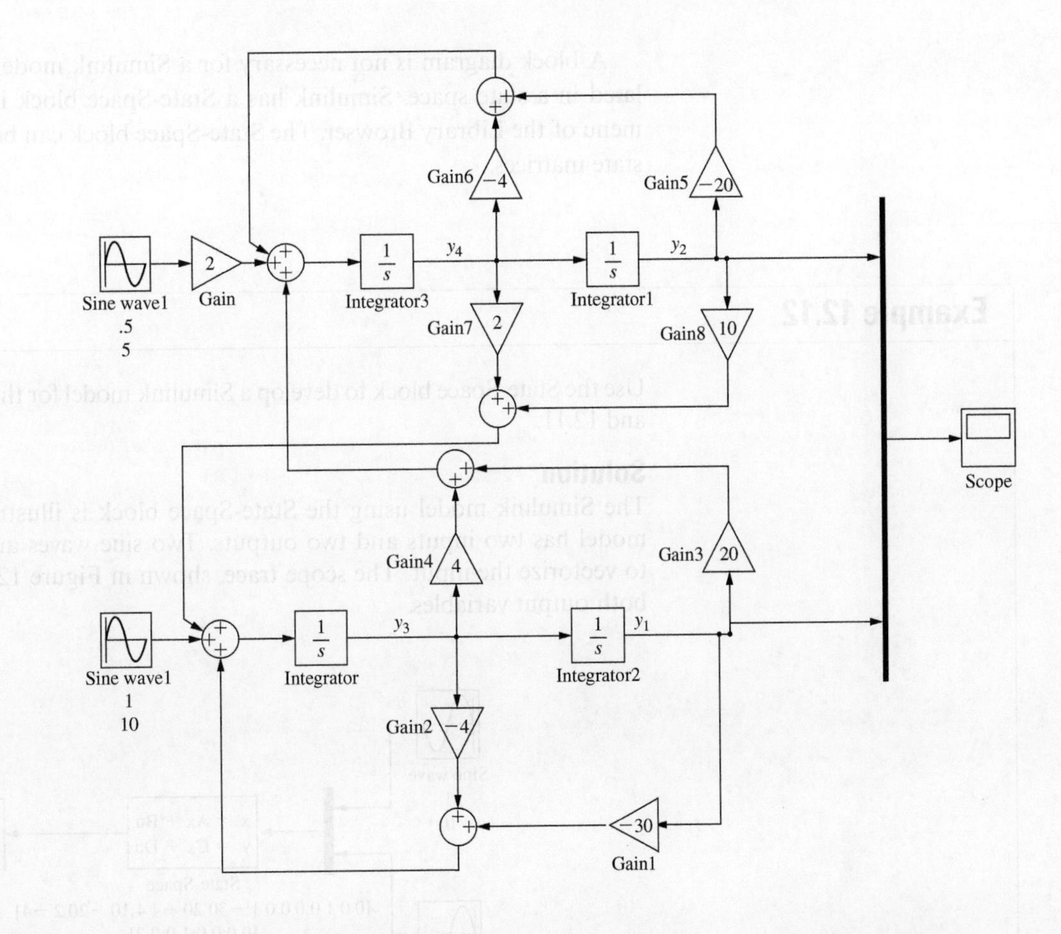

Figure 12.23

A Simulink model based on the block diagram of Figure 12.22.

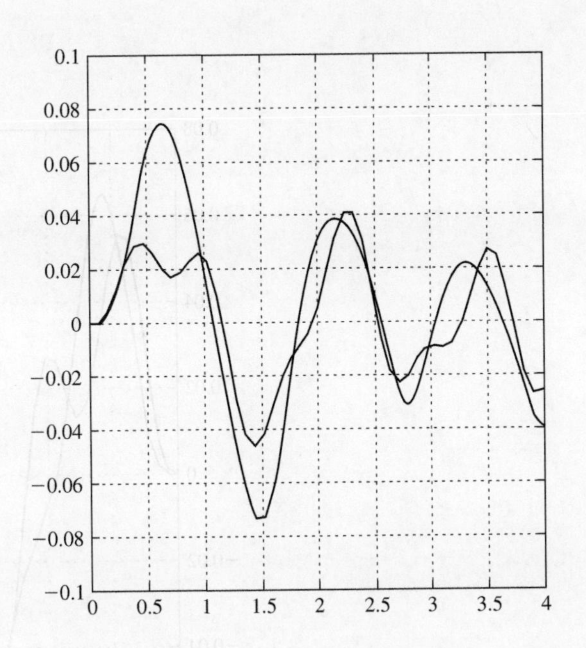

Figure 12.24

The output from the scope when the simulation is run.

A block diagram is not necessary for a Simulink model of a linear system formulated in a state space. Simulink has a State-Space block in the Continuous Systems menu of the Library Browser. The State-Space block can be directly used to define the state matrices.

Example 12.12

Use the State-Space block to develop a Simulink model for the system of Examples 12.10 and 12.11.

Solution

The Simulink model using the State-Space block is illustrated in Figure 12.25. The model has two inputs and two outputs. Two sine waves are routed through a 'Mux' to vectorize the input. The scope trace, shown in Figure 12.26, plots the response for both output variables.

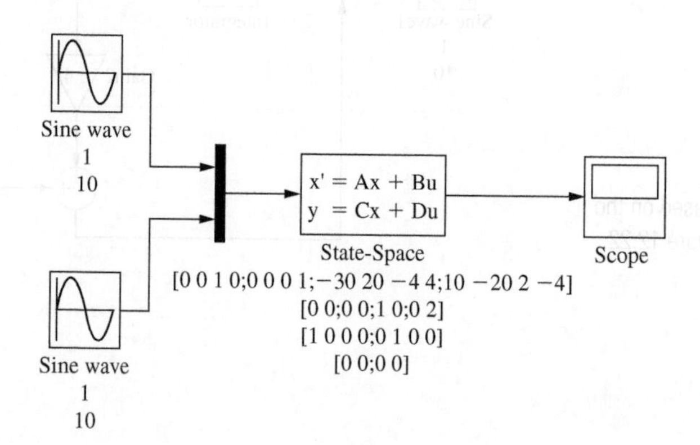

Figure 12.25

The Simulink model using the State-Space block.

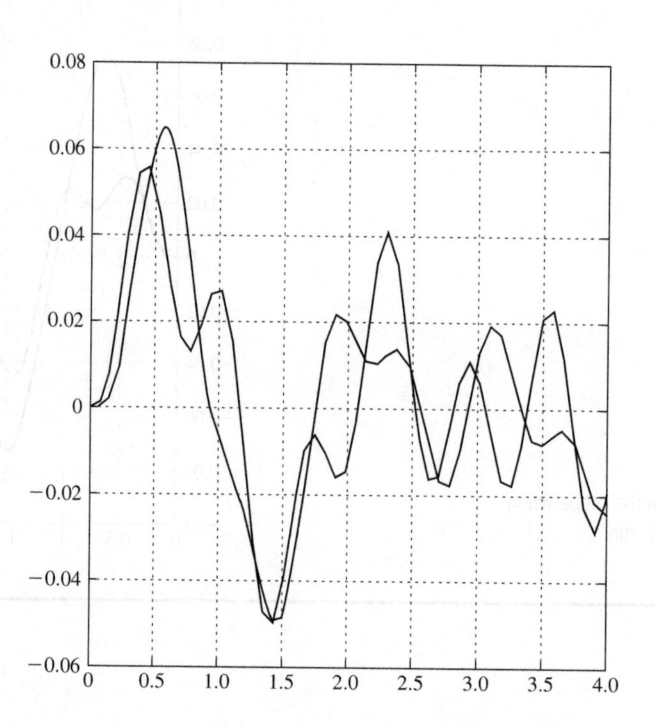

Figure 12.26

The scope trace for the responses for both output variables.

Example 12.13

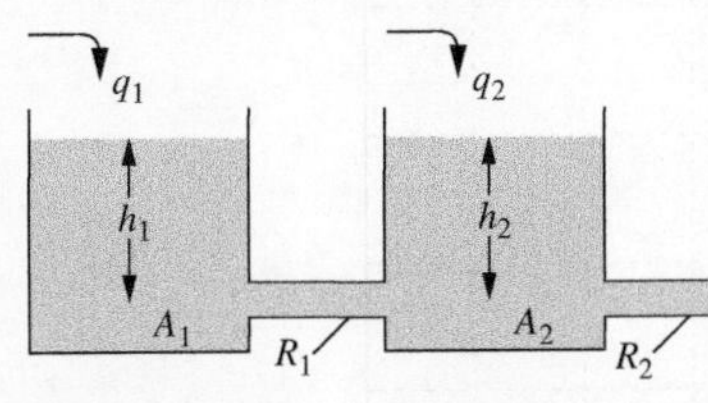

Figure 12.8

Two-tank liquid-level system of Example 12.6 and Example 12.13. The flow rates and liquid levels are perturbations from steady state. (Repeated)

Consider the two-tank liquid-level system of Figure 12.8 with $A_1 = A_2 = 2.0 \text{ m}^2$, $R_1 = R_2 = 5\frac{\text{s}}{\text{m}^2}$. **a.** Develop a state-space model for the system. **b.** Use the State-Space block to develop a Simulink model for the system. **c.** Develop a Simulink model for the closed-loop system that occurs when a PID controller is used in the feedforward portion of a feedback loop where the controller controls the level in the second tank. Run the simulation for the following proposed controllers: (i) $K_i = 0$, $K_p = 1$, $K_d = 0$; (ii) $K_i = 0.1$, $K_p = 1$, $K_d = 0$; and (iii) $K_i = 0$, $K_p = 1$, $K_d = 0.25$.

Solution

a. The differential equations governing the liquid levels in the tanks due to a perturbation in the inlet flow rate are

$$A_1\frac{dh_1}{dt} + \frac{1}{R_1}h_1 - \frac{1}{R_1}h_2 = q_i \quad \text{(a)}$$

$$A_2\frac{dh_2}{dt} - \frac{1}{R_1}h_1 + \left(\frac{1}{R_1} + \frac{1}{R_2}\right)h_2 = 0 \quad \text{(b)}$$

The state variables are $y_1 = h_1$ and $y_2 = h_2$. If the liquid levels in both tanks are considered as output, the appropriate state-space model is

$$\mathbf{A} = \begin{bmatrix} -\frac{1}{A_1R_1} & \frac{1}{A_1R_1} \\ \frac{1}{A_2R_1} & -\frac{1}{A_2}\left(\frac{1}{R_1} + \frac{1}{R_2}\right) \end{bmatrix} \quad \text{(c)}$$

$$\mathbf{B} = \begin{bmatrix} \frac{1}{A_1} \\ 0 \end{bmatrix} \quad \text{(d)}$$

$$\mathbf{C} = \begin{bmatrix} 1 & 0 \\ 0 & 1 \end{bmatrix} \quad \text{(e)}$$

$$\mathbf{D} = \begin{bmatrix} 0 \\ 0 \end{bmatrix} \quad \text{(f)}$$

b. The Simulink model is shown in Figure 12.27, while the scope trace obtained using the parameters specified is shown in Figure 12.28.

c. Since the control system monitors $h_2(t)$, the state-space model for the liquid-level system is modified such that $h_2(t)$ is the only output. Thus the output matrix and transmission matrix are modified to

$$\mathbf{C} = [0 \quad 1] \quad \text{(g)}$$

$$\mathbf{D} = [0] \quad \text{(h)}$$

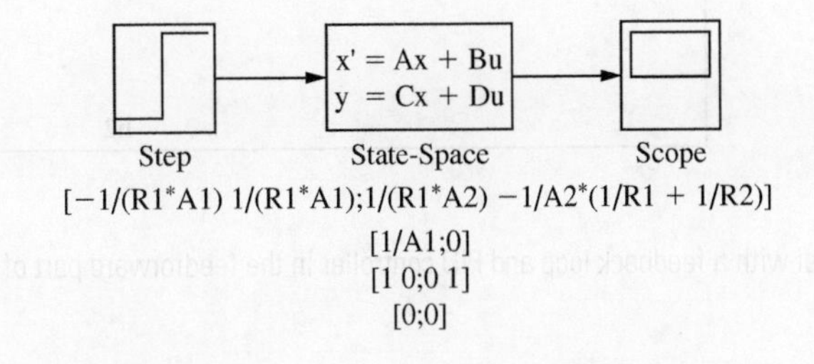

Figure 12.27

The Simulink model for the system of Example 12.13.

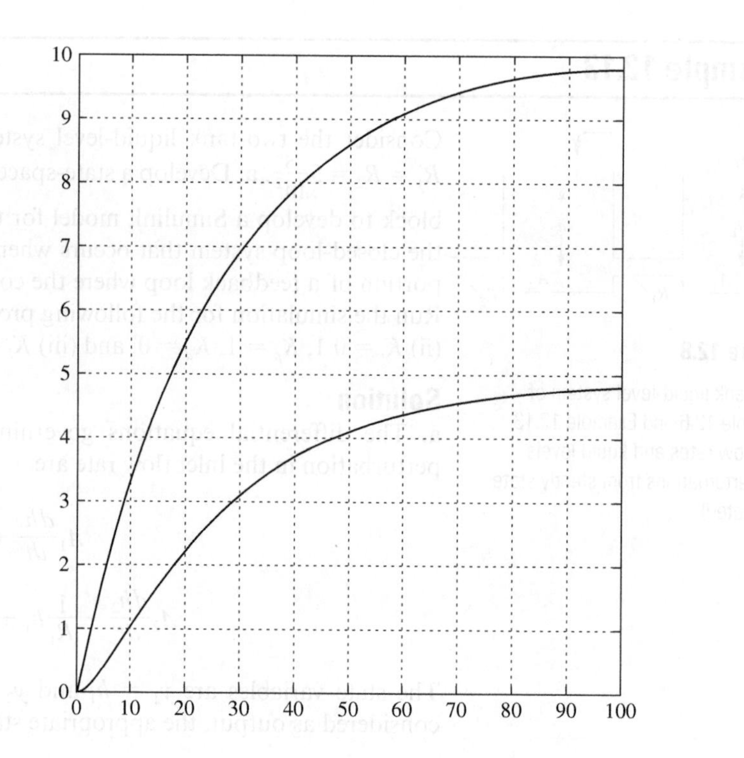

Figure 12.28
The scope trace.

The Simulink model with a feedback loop and a PID controller in the feedforward part of the loop is shown in Figure 12.29. This Simulink model uses the State-Space block for the open-loop system. The scope traces for the various controller parameters are given in Figure 12.30. The offset is clear in Figure 12.30(a), in which a proportional controller is used.

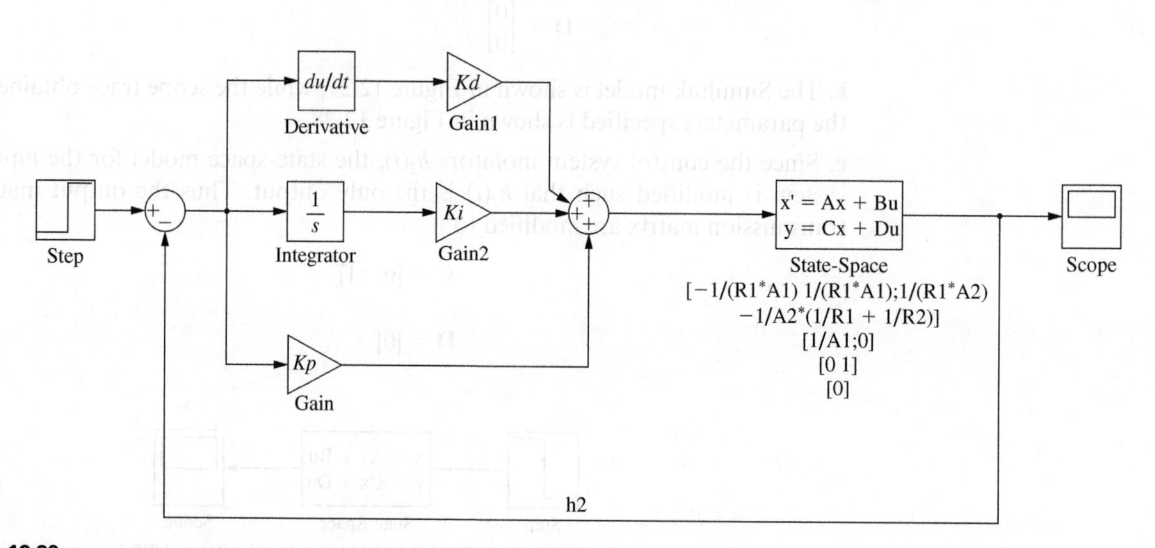

Figure 12.29
The Simulink model with a feedback loop and PID controller in the feedforward part of the loop.

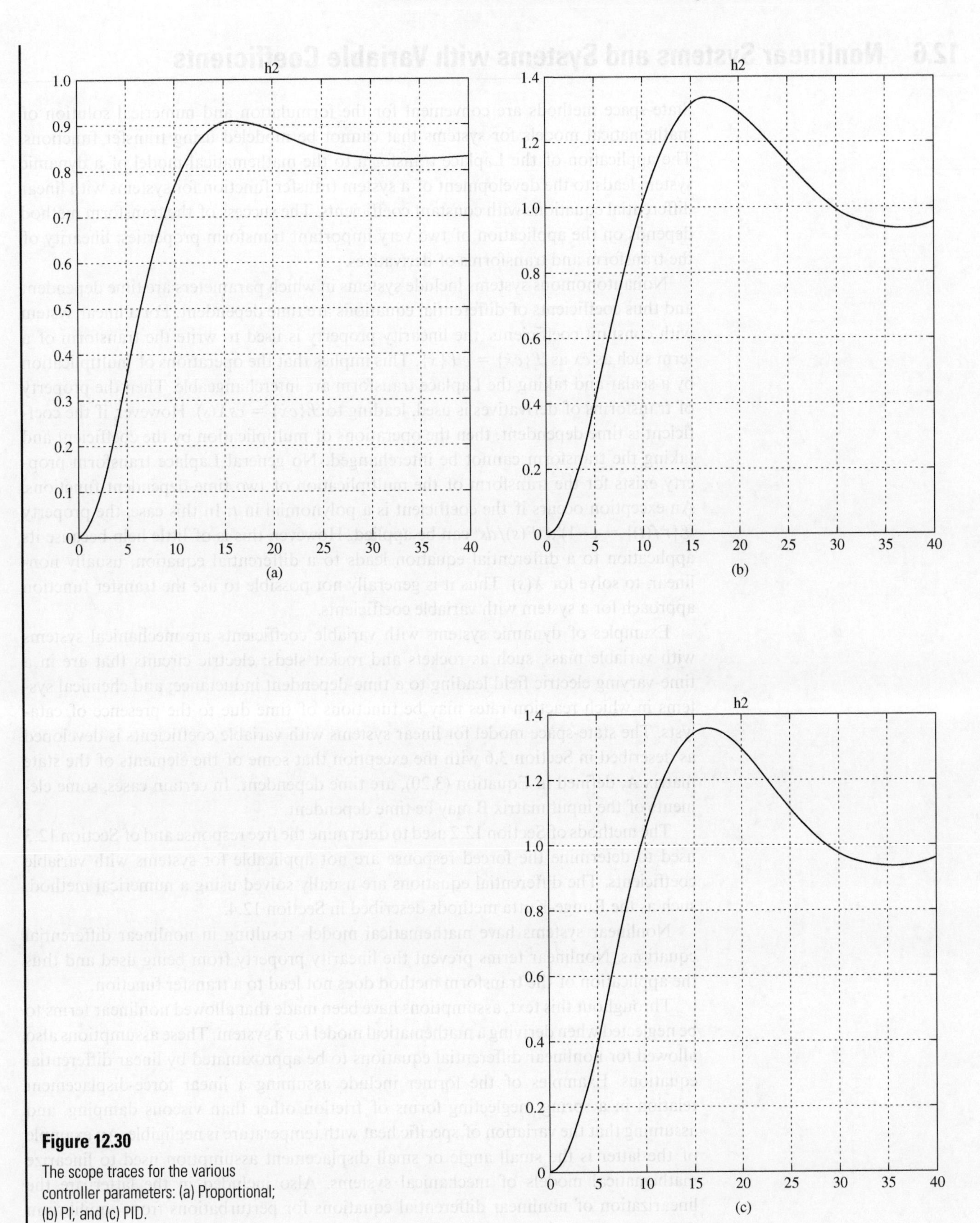

Figure 12.30

The scope traces for the various controller parameters: (a) Proportional; (b) PI; and (c) PID.

12.6 Nonlinear Systems and Systems with Variable Coefficients

State-space methods are convenient for the formulation and numerical solution of mathematical models for systems that cannot be modeled using transfer functions. The application of the Laplace transform to the mathematical model of a dynamic system leads to the development of a system transfer function for systems with linear differential equations with constant coefficients. The success of the transform method depends on the application of two very important transform properties: linearity of the transform and transforms of derivatives.

Nonautonomous systems include systems in which parameters are time dependent and thus coefficients of differential equations are time dependent. For a linear system with constant coefficients, the linearity property is used to write the transform of a term such as $c\dot{x}$ as $\mathcal{L}\{c\dot{x}\} = c\mathcal{L}\{\dot{x}\}$. This implies that the operations of multiplication by a scalar and taking the Laplace transform are interchangeable. Then the property of transforms of derivatives is used, leading to $\mathcal{L}\{c\dot{x}\} = csX(s)$. However, if the coefficient is time dependent, then the operations of multiplication by the coefficient and taking the transform cannot be interchanged. No general Laplace transform property exists for the transform of the multiplication of two time-dependent functions. An exception occurs if the coefficient is a polynomial in t. In this case, the property $\mathcal{L}\{t^n f(t)\} = (-1)^n d^n F(s)/ds^n$ can be applied. However, this is of little help because its application to a differential equation leads to a differential equation, usually nonlinear, to solve for $X(s)$. Thus it is generally not possible to use the transfer function approach for a system with variable coefficients.

Examples of dynamic systems with variable coefficients are mechanical systems with variable mass, such as rockets and rocket sleds; electric circuits that are in a time-varying electric field leading to a time-dependent inductance; and chemical systems in which reaction rates may be functions of time due to the presence of catalysts. The state-space model for linear systems with variable coefficients is developed as described in Section 3.6 with the exception that some of the elements of the state matrix **A**, defined in Equation (3.20), are time dependent. In certain cases, some elements of the input matrix **B** may be time dependent.

The methods of Section 12.2 used to determine the free response and of Section 12.3 used to determine the forced response are not applicable for systems with variable coefficients. The differential equations are usually solved using a numerical method, such as the Runge-Kutta methods described in Section 12.4.

Nonlinear systems have mathematical models resulting in nonlinear differential equations. Nonlinear terms prevent the linearity property from being used and thus the application of the transform method does not lead to a transfer function.

Throughout this text, assumptions have been made that allowed nonlinear terms to be neglected when deriving a mathematical model for a system. These assumptions also allowed for nonlinear differential equations to be approximated by linear differential equations. Examples of the former include assuming a linear force-displacement relation in a spring, neglecting forms of friction other than viscous damping, and assuming that the variation of specific heat with temperature is negligible. An example of the latter is the small angle or small displacement assumption used to linearize mathematical models of mechanical systems. Also included in the latter are the linearization of nonlinear differential equations for perturbations from equilibrium that were made in liquid-level problems by using a piping system's resistances at the original steady state or in nonisothermal CSTR systems in which the rate of reaction is dependent on temperature through the Arrhenius equation.

Linearized equations are often sufficient to meet the objectives of the modeling process. However, nonlinear systems have behaviors that cannot be predicted using linear models and that may be significant when large perturbations from equilibrium or steady state occur. Examples of nonlinear system behavior that cannot be predicted using linear models include the following:

- System parameters depend on initial conditions. The free response of a linear first-order system is characterized by the time constant, which is independent of initial conditions. The free response of a second-order system is characterized by the period and damping ratio, which are independent of initial conditions. A nonlinear first-order system may still have exponential behavior, but the time constant may depend on initial conditions.
- Resonances occur for linear systems only when an input frequency coincides with a natural frequency. In addition to these primary resonances, subharmonic and superharmonic resonances occur in nonlinear systems at frequencies different from the system's natural frequencies.
- The frequency response of a linear system is continuous, whereas the frequency response of a nonlinear system may have discrete jumps at certain frequencies.
- The amplitude of the response at a fixed frequency is proportional to the amplitude of excitation. For certain multidegree-of-freedom nonlinear systems, a mode may become saturated in that its amplitude remains constant as the amplitude of the input is increased.
- A linear system has one stable equilibrium position. A nonlinear system may have multiple equilibrium positions, some of which may be unstable.

The differential equations for a state-space model for a nonlinear system cannot be written in a traditional matrix because the state matrix cannot be formulated. The differential equations for an nth-order system with state variables $y_1, y_2, \ldots, y_n$ and system inputs $u_1, u_2, \ldots, u_k$ are formulated as

$$\begin{aligned} \dot{y}_1 &= f_1(y_1, y_2, \ldots, y_n, u_1, u_2, \ldots, u_k, t) \\ \dot{y}_2 &= f_2(y_1, y_2, \ldots, y_n, u_1, u_2, \ldots, u_k, t) \\ &\vdots \\ \dot{y}_n &= f_n(y_1, y_2, \ldots, y_n, u_1, u_2, \ldots, u_k, t) \end{aligned} \tag{12.42}$$

Applications in this study are limited to second-order systems. For such systems, a graphical representation of the solution, called the phase plane plot or the state plane plot, is used to provide useful insights into the behavior of the system. A phase plane plot for a second-order system is a plot of the variation of one state variable with the other, say with y_1 on the horizontal axis and y_2 on the vertical axis.

The formulation of the state-space model for nonlinear systems, the application of ode45 to solve the differential equations, and the development and analysis of the phase plane are illustrated in the following examples.

Example 12.14

The nonlinear differential equation governing the free response of the simple pendulum of mass m and length ℓ shown in Figure 12.31 is

$$\ddot{\theta} + \omega_n^2 \sin\theta = 0 \tag{a}$$

where $\omega_n = \sqrt{g/\ell}$ is the natural frequency of a linearized system model. **a.** Develop a state-space formulation for the system. **b.** Write a MATLAB M-file using the command

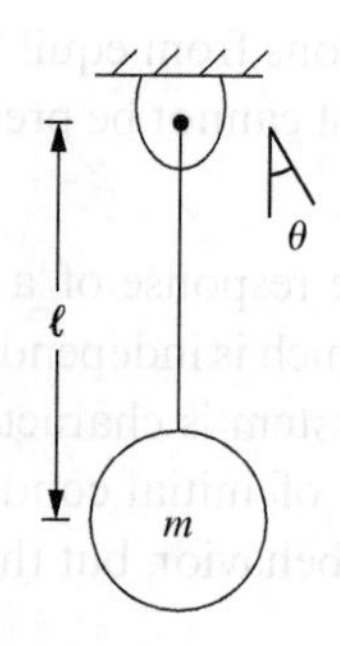

Figure 12.31

The pendulum of Example 12.14 has a nonlinear mathematical model.

'ode45' to determine a numerical solution for the system and to plot the time-dependent response and the phase plane. **c.** Use the MATLAB M-file to determine the numerical solution with $\omega_n = 1$ r/s and for initial conditions $\dot{\theta}(0) = 0$ and (i) $\theta(0) = \pi/6$, and (ii) $\theta(0) = \pi/2$. Compare and contrast the solutions.

Solution

a. Using state-space variables defined as $y_1 = \theta$ and $y_2 = \dot{\theta}$, the state-space formulation of Equation (a) is

$$\dot{y}_1 = y_2$$
$$\dot{y}_2 = \sin(y_1) \tag{b}$$

b. The script of Example12_14.m, which uses ode45 to solve Equation (b), is shown in Figure 12.32. The program plots the time-dependent response of the system as well as the state plane.

c. Plots generated from the execution of Example12_14.m are given in Figure 12.33. The comparison of Figures 12.33(b) and 12.33(c) shows that the frequency and period of a nonlinear system are dependent on initial conditions.

```
% Example12.14
% Nonlinear response of simple pendulum
% Initial conditions
theta0=input('input initial angular displacement in rad')
y0=[theta0 0];
tf=30;
% Vector of derivatives is provided from user-supplied file pend.m
[t,y]=ode45(@pend,[0 tf],y0);
% Plot time response
figure
plot(t,y(:,1))
xlabel('t (s)')
ylabel('\theta (rad)')
str=['Pendulum response for \theta_0=',num2str(theta0),'rad']
title(str)
figure
% State plane plot
plot(y(:,1),y(:,2))
xlabel('\theta (rad)')
ylabel('\omega (rad/s)')
str1=['State plane for pendulum with \theta_0=',num2str(theta0),'rad']
title(str1)
%
% End of Example12_14.m
```

(a)

```
% User supplied program to generate vector of derivatives for dynamic
%     response of simple pendulum
function dy=pend(t,y)
dy1=y(2);
dy2=sin(y(1));
dy=[dy1;dy2];
```

(b)

Figure 12.32

(a) The script of the file Example12_14.m; (b) the script of pend.m required for Example12_14.m.

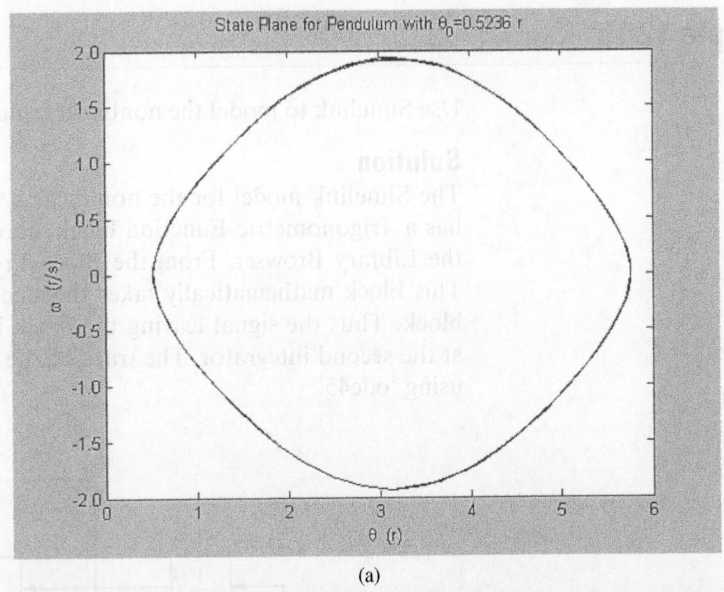

(a)

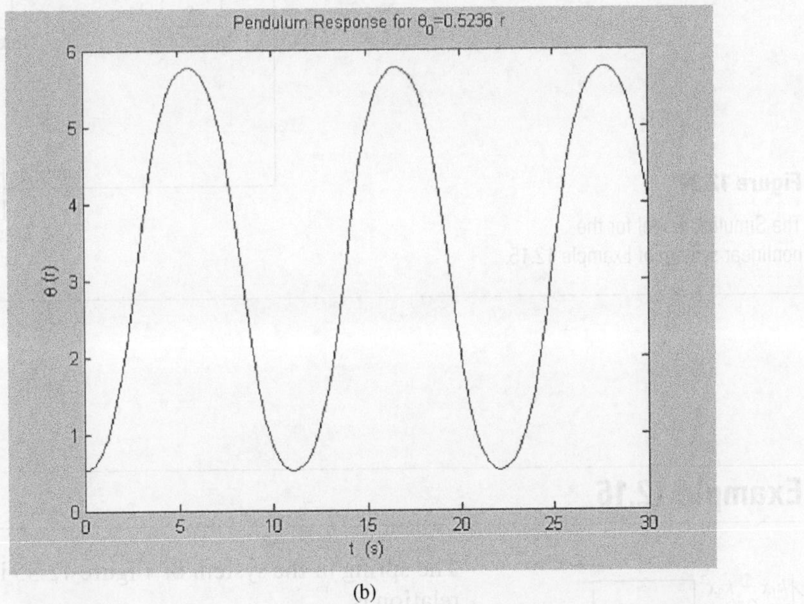

(b)

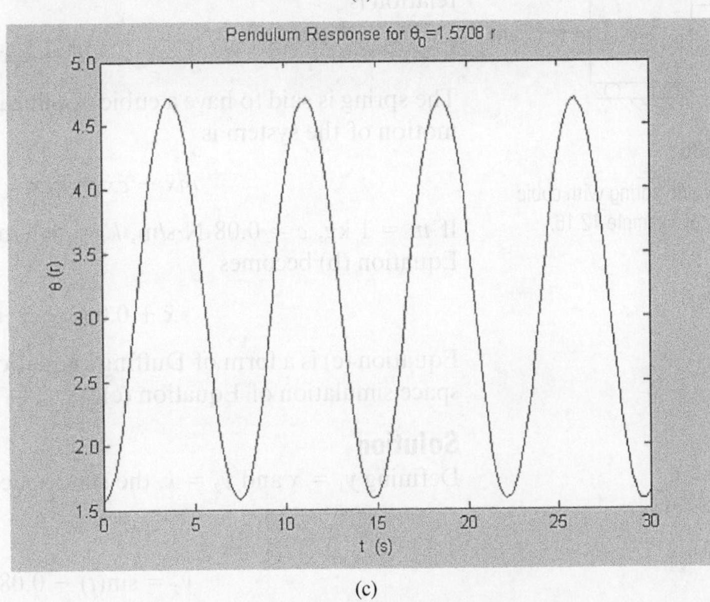

(c)

Figure 12.33

Plots obtained from the execution of Example12_14.m: (a) the state plane for the pendulum with $\theta_0 = 0.5236$ r; (b) the pendulum response for $\theta_0 = 0.5236$ r; and (c) the pendulum response for $\theta_0 = 1.5708$ r.

Example 12.15

Use Simulink to model the nonlinear pendulum of Example 12.14.

Solution

The Simulink model for the nonlinear system is shown in Figure 12.34. The model has a Trigonometric Function block, chosen from the Continuous System menu of the Library Browser. From the Block Properties menu, a choice of Sin was made. This block mathematically takes the sine of the value of the signal that enters the block. Thus the signal leaving the block is $\sin(y_1)$. The initial condition is specified at the second integrator. The trace of the scope is identical to the response obtained using 'ode45'.

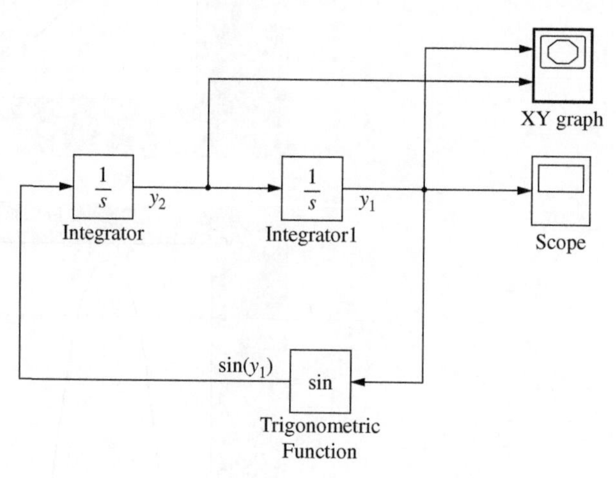

Figure 12.34

The Simulink model for the nonlinear system of Example 12.15.

Example 12.16

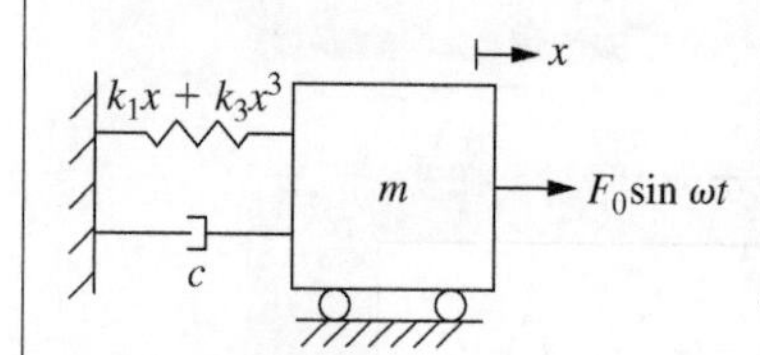

Figure 12.35

The system with spring with cubic nonlinearity of Example 12.16.

The spring in the system of Figure 12.35 is a nonlinear spring. Its force-displacement relation is

$$F = k_1 x + k_3 x^3 \tag{a}$$

The spring is said to have a cubic nonlinearity. The differential equation governing the motion of the system is

$$m\ddot{x} + c\dot{x} + k_1 x + k_3 x^3 = F_0 \sin(\omega t) \tag{b}$$

If $m = 1$ kg, $c = 0.08$ N·s/m, $k_1 = 1$ N/m, $k_3 = 0.1$ N/m^3, $F_0 = 1$ N, and $\omega = 1$ r/s, Equation (b) becomes

$$\ddot{x} + 0.08\dot{x} + x + 0.1x^3 = \sin(t) \tag{c}$$

Equation (c) is a form of Duffing's equation. Use Simulink to develop and run a state-space simulation of Equation (c).

Solution

Defining $y_1 = x$ and $y_2 = \dot{x}$, the state-space formulation of Equation (c) is

$$y_1 = y_2 \tag{d}$$

$$\dot{y}_2 = \sin(t) - 0.08 y_2 - y_1 - 0.1 y_1^3 \tag{e}$$

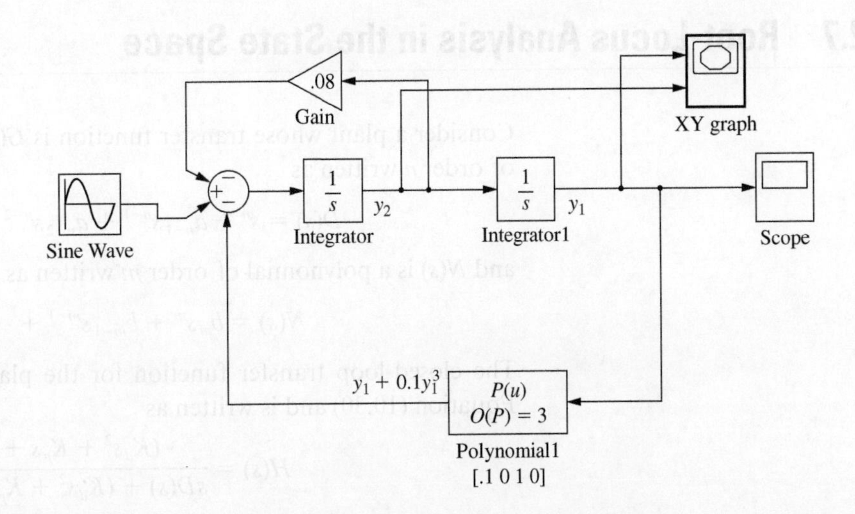

Figure 12.36

The Simulink model for the system of Example 12.16.

The Simulink model for Equations (d) and (e) is illustrated in Figure 12.36. The model uses the Polynomial block from the Math Operations menu of the Library Browser. In the Block Properties box, the coefficients of the polynomial are entered as [0.1 0 1 0] corresponding to a polynomial of the form $P(u) = u + 0.1u^3$. Thus since y_1 is the signal entering the block, the signal leaving the block has a value of $y_1 + 0.1y_1^3$. The time-dependent response and the state plane generated from running a simulation are shown in Figure 12.37.

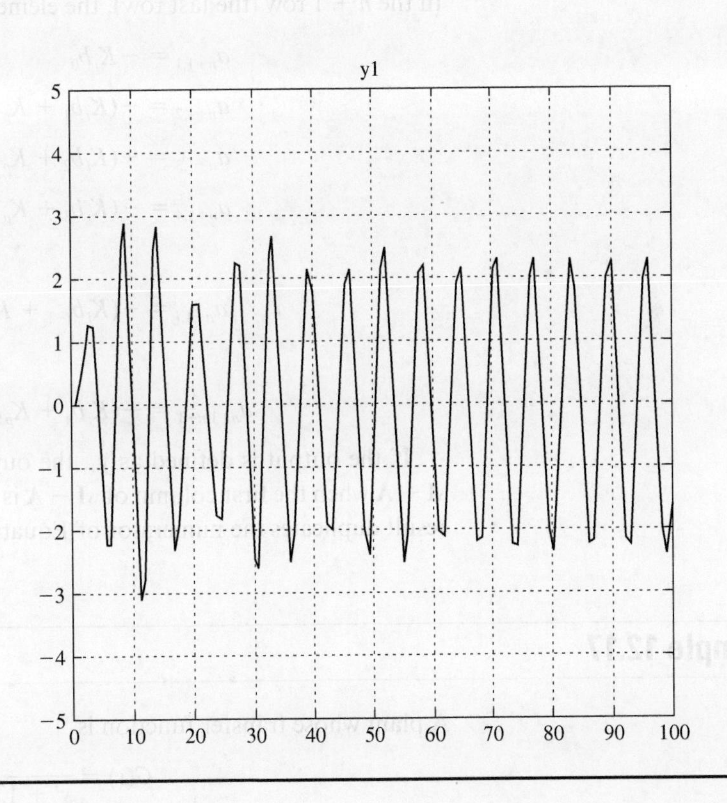

Figure 12.37

The time-dependent response and the state plane generated by the simulation of Example 12.16.

12.7 Root-Locus Analysis in the State Space

Consider a plant whose transfer function is $G(s) = \frac{N(s)}{D(s)}$, where $D(s)$ is a polynomial of order n written as

$$D(s) = s^n + a_{n-1}s^{n-1} + a_{n-2}s^{n-2} + \cdots + a_1 s + a_0 \tag{12.43}$$

and $N(s)$ is a polynomial of order m written as ...

$$N(s) = b_m s^m + b_{m-1}s^{m-1} + \cdots + b_1 s + b_0 \tag{12.44}$$

The closed-loop transfer function for the plant in series with a PID controller is Equation (10.30) and is written as

$$H(s) = \frac{(K_d s^2 + K_p s + K_i)N(s)}{sD(s) + (K_d s^2 + K_p s + K_i)N(s)} \tag{12.45}$$

The transfer function $H(s)$ has a nonunique state-space representation obtained using the methods of Section 3.7. For the transfer function of Equation (12.45), a state matrix for a state-space formulation of the system has the form of Equation (3.20). It is noted that the state matrix has $n + 1$ rows and $n + 1$ columns, meaning that $n + 1$ state variables need to be defined. If the controller is without integral action, then the state matrix has n rows and n columns. The elements in the first n rows of the state matrix for a controller with integral action have the form

$$a_{ki} = \delta_{k,i+1} = \begin{cases} 0 & k \neq i+1 \\ 1 & k = i+1 \end{cases} \tag{12.46}$$

In the $n + 1$ row (the last row), the elements are of the form

$$\begin{aligned}
a_{n+1,1} &= -K_i b_0 \\
a_{n+1,2} &= -(K_i b_1 + K_p b_0 + a_0) \\
a_{n+1,3} &= -(K_i b_2 + K_p b_1 + K_d b_0 + a_1) \\
a_{n+1,4} &= -(K_i b_3 + K_p b_2 + K_d b_1 + a_2) \\
&\vdots \\
a_{n+1,i} &= -(K_i b_{i-1} + K_p b_{i-2} + K_d b_{i-3} + a_{i-2}) \\
&\vdots \\
a_{n+1,n+1} &= -(K_i b_n + K_p b_{n-1} + K_d b_{n-2} + a_{n-2})
\end{aligned} \tag{12.47}$$

If the output is defined as y_1, the output vector is defined as the determinant of $s\mathbf{I} - \mathbf{A}$ when the first column of $s\mathbf{I} - \mathbf{A}$ is replaced by the unknown output vector. The result duplicates the numerator of Equation (12.45).

Example 12.17

A plant whose transfer function is

$$G(s) = \frac{s + 3}{s^3 + 13s^2 + 50s + 56} \tag{a}$$

is to be placed in a feedback control loop in series with a PID controller. **a.** Determine a state-space representation for the closed-loop system. **b.** Determine the values of the control parameters such that the closed-loop system has dominant poles at $-2 \pm 3j$ and another pole at -6.

Solution

a. The closed-loop transfer function, using Equation (12.45), becomes

$$H(s) = \frac{(K_d s^2 + K_p s + K_i)(s+3)}{s(s^3 + 13s^2 + 50s + 56) + (K_d s^2 + K_p s + K_i)(s+3)}$$

$$= \frac{K_d s^3 + (3K_d + K_p)s^2 + (3K_p + K_i)s + 3K_i}{s^4 + (13 + K_d)s^3 + (50 + 3K_d + K_p)s^2 + (56 + 3K_p + K_i)s + 3K_i} \quad \text{(b)}$$

The elements of the state matrix for the closed loop transfer function are defined by Equations (12.46) and (12.47). The state matrix becomes

$$\mathbf{A} = \begin{bmatrix} 0 & 1 & 0 & 0 \\ 0 & 0 & 1 & 0 \\ 0 & 0 & 0 & 1 \\ -3K_i & -(56 + 3K_p + K_i) & -(50 + 3K_d + K_p) & -(13 + K_d) \end{bmatrix} \quad \text{(c)}$$

If $\mathbf{B} = \begin{bmatrix} b_1 & b_2 & b_3 & b_4 \end{bmatrix}^T$, then the matrix $s\mathbf{I} - \mathbf{A}$ with the first column replaced by $\mathbf{B}$ is

$$\mathbf{V}_1 = \begin{bmatrix} b_1 & -1 & 0 & 0 \\ b_2 & s & -1 & 0 \\ b_3 & 0 & s & -1 \\ b_4 & (56 + 3K_p + K_i) & (50 + 3K_d + K_p) & s + (13 + K_d) \end{bmatrix} \quad \text{(d)}$$

Taking the determinant of $\mathbf{V}_1$, expanding by the first row, gives

$$|\mathbf{V}_1| = b_1\{(s)(s)[s + (13 + K_d)] - (-1)(s)(56 + 3K_p + K_i)\} - b_2\{(-1)(s)[s + (13 + K_d)]\}$$
$$+ b_3\{(-1)(-1)[s + (13 + K_d)]\} - b_4(-1)(-1)(-1)$$
$$= b_1 s^3 + [(13 + K_d)b_1 + b_2]s^2 + [(13 + K_d)b_2 + (56 + 3K_p + K_i)b_3]s + b_4 + (13 + K_d)b_3 \quad \text{(e)}$$

Setting $|\mathbf{V}_1|$ equal to the numerator of the closed-loop transfer function leads to

$$b_1 = K_d \quad \text{(f)}$$

$$(13 + K_d)b_1 + b_2 = 3K_d + K_p \Rightarrow b_2 = K_p - (10 + K_d)K_d \quad \text{(g)}$$

$$(13 + K_d)b_2 + (56 + 3K_p + K_i)b_3 = 3K_p + K_i \Rightarrow b_3 = \frac{3K_p + K_i - (13 + K_d)[K_p - (10 + K_d)K_d]}{56 + 3K_p + K_i} \quad \text{(h)}$$

$$b_4 + (13 + K_d)b_3 = 3K_i \Rightarrow b_4 = -(13 + K_d)\left\{\frac{3K_p + K_i - (13 + K_d)[K_p - (10 + K_d)K_d]}{56 + 3K_p + K_i}\right\} \quad \text{(i)}$$

b. Setting the dominant poles at $-2 \pm 3j$ yields the quadratic factor $s^2 + 4s + 9$. If another pole is to be at -6, then $s + 6$ is also a factor of $D(s)$. If the final factor of the closed-loop system is $s + \alpha$, the appropriate form of the denominator of the transfer function is

$$D(s) = (s^2 + 4s + 9)(s + 6)(s + \alpha)$$
$$= s^4 + (14 + \alpha)s^3 + (84 + 14\alpha)s^2 + (216 + 84\alpha)s + 216\alpha \quad \text{(j)}$$

Comparing Equation (j) with the denominator of Equation (b) leads to

$$13 + K_d = 14 + \alpha \quad \text{(k)}$$

$$50 + 3K_d + K_p = 84 + 14\alpha \quad \text{(l)}$$

$$56 + 3K_p + K_i = 216 + 84\alpha \quad \text{(m)}$$

$$3K_i = 216\alpha \quad \text{(n)}$$

Solving Equations (k)-(n) simultaneously leads to

$$K_p = 66.09 \quad K_i = 229.71 \quad K_d = 4.19 \quad \alpha = 3.19 \tag{o}$$

The state-space representation of the controlled system is obtained by substituting Equation (o) into Equation (c) using Equations (f)-(i). The result is

$$\begin{bmatrix} \dot{y}_1 \\ \dot{y}_2 \\ \dot{y}_3 \\ \dot{y}_4 \end{bmatrix} = \begin{bmatrix} 0 & 1 & 0 & 0 \\ 0 & 0 & 1 & 0 \\ 0 & 0 & 0 & 1 \\ -689.13 & -483.98 & -128.66 & -17.19 \end{bmatrix} \begin{bmatrix} y_1 \\ y_2 \\ y_3 \\ y_4 \end{bmatrix} + \begin{bmatrix} 4.19 \\ 6.63 \\ 0.735 \\ -12.64 \end{bmatrix} \tag{p}$$

12.8 State Variable Feedback

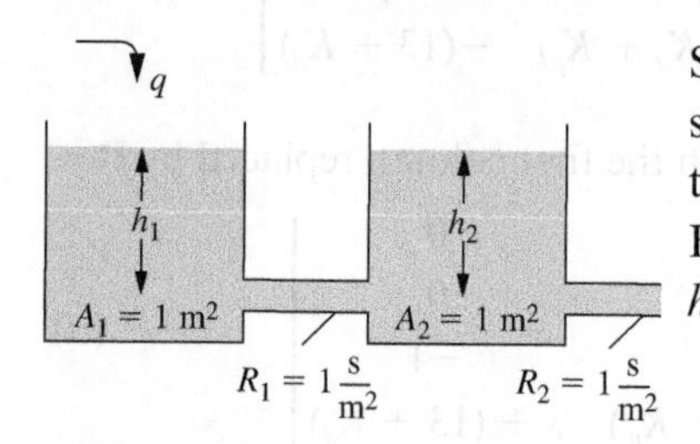

Figure 12.38

The two-tank liquid-level problem is used to illustrate state variable feedback.

State variable feedback occurs when each state variable has a feedback strategy. If all state variables can be measured, then state variable feedback provides a method for the placement of all poles. Consider the two-tank liquid-level problem represented by Figure 12.38. Taking h_1 and h_2 as state variables with $q(t)$ as the input to the system and h_1 as the output, the state-space formulation of this system is

$$\begin{bmatrix} \dot{h}_1 \\ \dot{h}_2 \end{bmatrix} = \begin{bmatrix} -1 & 1 \\ 1 & -2 \end{bmatrix} \begin{bmatrix} h_1 \\ h_2 \end{bmatrix} + \begin{bmatrix} 1 \\ 0 \end{bmatrix} [q] \tag{12.48}$$

$$[h_1] = [1 \quad 0] \begin{bmatrix} h_1 \\ h_2 \end{bmatrix} + [0][q] \tag{12.49}$$

The block diagram for the system in the s domain is shown in Figure 12.39(a). In its original configuration, the system has poles at -0.382 and -2.62. It is desired that the system has poles at -2 and -5, such that the denominator of the system's closed-loop transfer function is $D(s) = s^2 + 7s + 10$. One way to achieve this is through state-level feedback, illustrated in Figure 12.39(b), where both h_1 and h_2 are modified and fed back though a control loop. The state equations for this system are

$$\dot{h}_1 = -h_1 + h_2 + q(t) - K_1 h_1 - K_2 h_2 \tag{12.50}$$

$$\dot{h}_2 = h_1 - 2h_2 \tag{12.51}$$

The state-space formulation of the equations with the controllers has a state matrix of

$$\tilde{\mathbf{A}} = \begin{bmatrix} -1 - K_1 & 1 - K_2 \\ 1 & -2 \end{bmatrix} \tag{12.52}$$

and an output matrix of

$$\mathbf{B} = \begin{bmatrix} 1 \\ 0 \end{bmatrix} \tag{12.53}$$

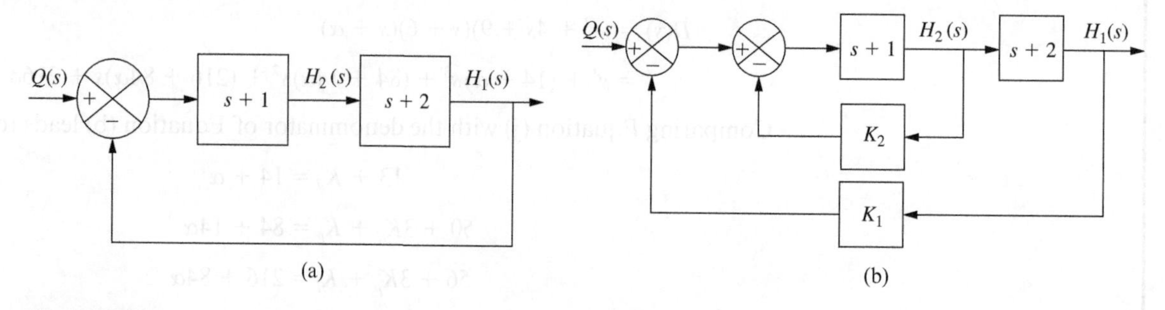

Figure 12.39

(a) Block diagram for liquid-level system of Figure 12.38. (b) Block diagram when state variable feedback is used to control the system.

Using Equation (12.25), the solution of the controlled problem is

$$\mathbf{h}(t) = e^{\widetilde{\mathbf{A}}t}\mathbf{h}(0) \tag{12.54}$$

The evaluation of $e^{\widetilde{\mathbf{A}}t}$ uses eigenvalues of $\widetilde{\mathbf{A}}$, called λ_1 and λ_2, in a linear combination. The eigenvalues are the same as the poles of the transfer function. The eigenvalues are determined from

$$\begin{vmatrix} -1 - K_1 - \lambda & 1 - K_2 \\ 1 & -2 - \lambda \end{vmatrix} = 0$$

$$(-1 - K_1 - \lambda)(-2 - \lambda) - (1 - K_2)(1) = 0$$

$$\lambda^2 + (3 + K_1)\lambda + 2K_1 + K_2 + 1 = 0 \tag{12.55}$$

Requiring the eigenvalues to be -2 and -5 leads to an equation of $\lambda^2 + 7\lambda + 10 = 0$. Equating the coefficients in this equation and Equation (12.55) gives

$$3 + K_1 = 7 \Rightarrow K_1 = 4$$

$$2K_1 + K_2 + 1 = 10 \Rightarrow K_2 = 1$$

The state-space formulation for a system is as in Equations (3.20) and (3.21), which are repeated here:

$$\dot{\mathbf{y}} = \mathbf{Ay} + \mathbf{Bu} \tag{12.56}$$

$$\mathbf{x} = \mathbf{Cy} + \mathbf{Du} \tag{12.57}$$

where $\mathbf{y}$ is the vector of state variables, $\mathbf{u}$ is the control signal, and $\mathbf{x}$ is the output signal. The matrix $\mathbf{A}$ is the state matrix; $\mathbf{B}$, which relates the state variables to the control signal, is the input matrix; $\mathbf{C}$ is the output matrix; and $\mathbf{D}$ is the transmission matrix. When state variable feedback is used, the control signal is related to the state variables by a relation of the form

$$\mathbf{u} = -\mathbf{Sy} \tag{12.58}$$

where $\mathbf{S}$ is called the state feedback gain matrix. Then Equation (12.56) becomes

$$\dot{\mathbf{y}} = (\mathbf{A} - \mathbf{BS})\mathbf{y} \tag{12.59}$$

The solution to Equation (12.58) is seen in Equation (12.54) with $\mathbf{A} - \mathbf{BS}$ replacing $\mathbf{A}$. The solution of Equation (12.58) becomes

$$\mathbf{y} = e^{(\mathbf{A}-\mathbf{BS})t}\mathbf{y}(0) \tag{12.60}$$

It is noted that $e^{(\mathbf{A}-\mathbf{BS})t}$ is obtained using Equation (12.22) as

$$e^{(\mathbf{A}-\mathbf{BS})t} = \sum_{k=1}^{n} C_k \mathbf{Y_k} e^{\lambda_k t} \tag{12.61}$$

where n is the order of the system, λ_k are the eigenvalues of $\mathbf{A} - \mathbf{BS}$, $\mathbf{Y_k}$ are the corresponding eigenvectors of $\mathbf{A} - \mathbf{BS}$, and C_k are constants. The eigenvalues of $\mathbf{A} - \mathbf{BS}$ are the poles of the closed-loop transfer function.

Specific pole placement for the system is attained by choosing the state feedback controllers to have gains such that the eigenvalues of $\mathbf{A} - \mathbf{BS}$ are the same as the required poles. For a system of order n, there are n defined state variables and n eigenvalues of $\mathbf{A} - \mathbf{BS}$. The placement of the poles can be determined with the choice of gains. This process requires state variables to be measured exactly and to be controlled individually.

For a SISO system, the matrix of gains, $\mathbf{S}$, is a $1 \times n$ column vector. For the two-tank liquid-level problem at the beginning of this section,

$$\mathbf{S} = [K_1 \;\; K_2],\ \mathbf{BS} = \begin{bmatrix} 1 \\ 0 \end{bmatrix}[K_1 \;\; K_2] = \begin{bmatrix} K_1 & K_2 \\ 0 & 0 \end{bmatrix},\ \mathbf{A} - \mathbf{BS} = \begin{bmatrix} -1 - K_1 & 1 - K_2 \\ 1 & -2 \end{bmatrix} \tag{12.62}$$

Example 12.18

The field-controlled dc servomotor of Figure 12.40 has the parameter values

$$R_f = 5\ \Omega,\ L_f = 1\ \text{H},\ J = 0.04\ \text{kg}\cdot\text{m}^2,\ c_t = 0.01\ \text{N}\cdot\text{m}\cdot\text{s},\ K_f = 0.1\ \frac{\text{N}\cdot\text{m}\cdot\text{s}}{\text{C}}$$

Design a control system with state variable feedback control such that it has poles at -2 and -10.

Solution

The differential equations modeling this system are

$$L_f \frac{di_f}{dt} + R_f i_f = v \tag{a}$$

$$J\frac{d\omega}{dt} + c_t\omega - K_f i_f = 0 \tag{b}$$

The input to the system is v and the output is ω. The state variables are $y_1 = i_f$, $y_2 = \omega$. A state-space formulation of the dc servomotor is

$$\begin{bmatrix}\dot{y}_1\\ \dot{y}_2\end{bmatrix} = \begin{bmatrix} -\dfrac{R_f}{L_f} & 0 \\ \dfrac{K_f}{J} & -\dfrac{c_t}{J}\end{bmatrix}\begin{bmatrix}y_1\\ y_2\end{bmatrix} + \begin{bmatrix}\dfrac{1}{L_f}\\ 0\end{bmatrix}[v] = \begin{bmatrix}-5 & 0\\ 2.5 & -0.25\end{bmatrix}\begin{bmatrix}y_1\\ y_2\end{bmatrix} + \begin{bmatrix}1\\ 0\end{bmatrix}[v] \tag{c}$$

The block diagram with state variable feedback is shown in Figure 12.41. The state matrix and the output vector are determined from Equation (c) as

$$\mathbf{A} = \begin{bmatrix}-5 & 0\\ 2.5 & -0.25\end{bmatrix},\quad \mathbf{B} = \begin{bmatrix}1\\ 0\end{bmatrix} \tag{d}$$

Letting $\mathbf{S} = [K_1 \quad K_2]$,

$$\mathbf{A} - \mathbf{BS} = \begin{bmatrix}-5 - K_1 & -K_2\\ 2.5 & -0.25\end{bmatrix} \tag{e}$$

The eigenvalues of $\mathbf{A} - \mathbf{BS}$ are obtained by setting the determinant of $\mathbf{A} - \mathbf{BS} - \lambda\mathbf{I}$ to zero, which leads to

$$(-5 - K_1 - \lambda)(-0.25 - \lambda) - (2.5) = \lambda^2 + (5 + K_1)\lambda + 1.25 + 0.25K_1 + 2.5K_2 = 0 \tag{f}$$

In order for the poles to be at -2 and -10, the characteristic equation should be

$$(s + 2)(s + 10) = s^2 + 12s + 20 = 0 \tag{g}$$

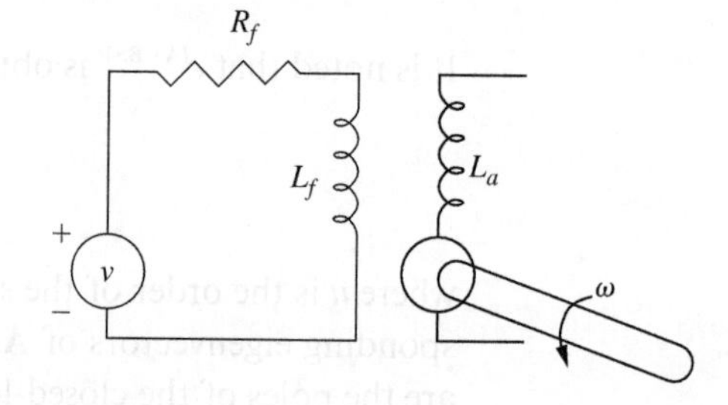

Figure 12.40
Field-controlled dc servomotor of Example 12.18.

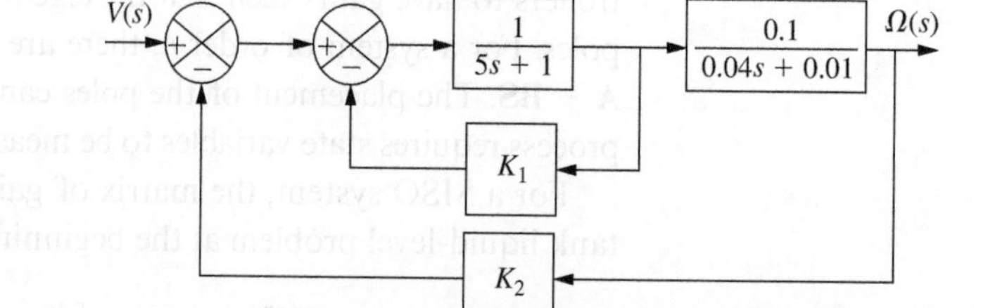

Figure 12.41
Block diagram for state variable control of dc servomotor.

Setting the coefficients of λ in Equation (f) to the coefficients of s in Equation (g) leads to

$$5 + K_1 = 12 \Rightarrow K_1 = 7 \tag{h}$$

$$1.25 + 0.25K_1 + 2.5K_2 = 20 \Rightarrow K_2 = 6.8 \tag{i}$$

The control strategy is

$$\dot{y}_1 = -12y_1 - 6.8y_2 + v \tag{j}$$

$$\dot{y}_2 = 2.5y_1 - 0.25y_2 \tag{k}$$

Example 12.19

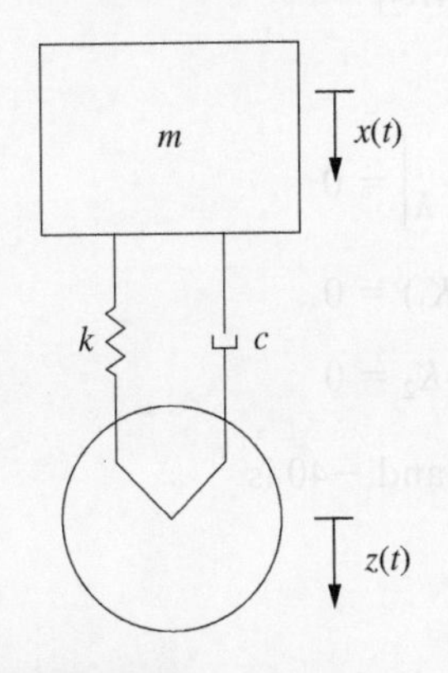

Figure 12.42

Simplified model of vehicle suspension system.

The mathematical model of the simplified vehicle suspension system illustrated in Figure 12.42 is

$$\ddot{x} + 2\zeta\omega_n\dot{x} + \omega_n^2 x = 2\zeta\omega_n\dot{z} + \omega_n^2 z \tag{a}$$

The input is $z(t)$ and the output is $x(t)$. If $\zeta = 0.3$ and $\omega_n = 10$ r/s, develop state variable feedback for the suspension system such that it has poles at -3 and -40.

Solution

Since the derivative of the input appears in the mathematical model, the state variables must be defined as in Section 3.5 as

$$y_1 = x \tag{b}$$

$$y_2 = \dot{x} - 2\zeta\omega_n z \tag{c}$$

The state-space formulation of the mathematical model is

$$\begin{bmatrix}\dot{y}_1\\ \dot{y}_2\end{bmatrix} = \begin{bmatrix}0 & 1\\ -\omega_n^2 & -2\zeta\omega_n\end{bmatrix}\begin{bmatrix}y_1\\ y_2\end{bmatrix} + \begin{bmatrix}0\\ \omega_n^2(1-4\zeta^2)\end{bmatrix}[z] \tag{d}$$

Equation (d) implies that

$$\mathbf{A} = \begin{bmatrix}0 & 1\\ -\omega_n^2 & -2\zeta\omega_n\end{bmatrix} = \begin{bmatrix}0 & 1\\ -100 & -6\end{bmatrix} \tag{e}$$

and

$$\mathbf{B} = \begin{bmatrix}0\\ \omega_n^2(1-4\zeta^2)\end{bmatrix} = \begin{bmatrix}0\\ 64\end{bmatrix} \tag{f}$$

The state-space formulation in block diagram form is shown in Figure 12.43. The state variable feedback for this system is illustrated in Figure 12.44. The state variable feedback gain vector is assumed to be

$$\mathbf{S} = [K_1 \quad K_2] \tag{g}$$

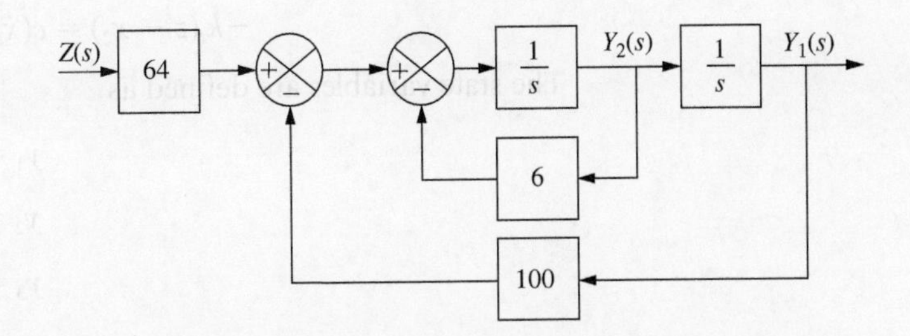

Figure 12.43

State-space formulation of vehicle suspension system in block diagram form.

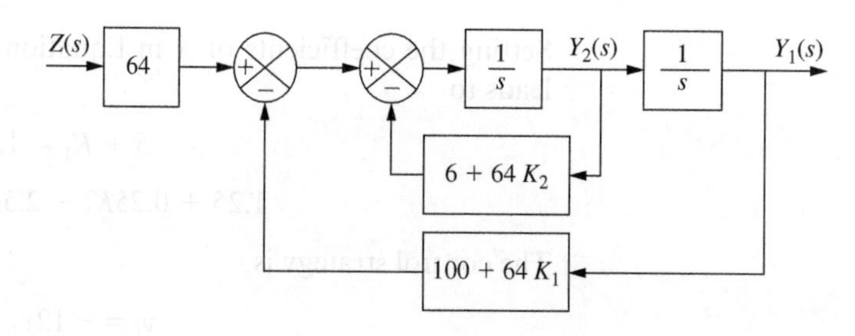

Figure 12.44
Block diagram of dc servomotor with state variable control.

Calculations show that

$$\mathbf{BS} = \begin{bmatrix} 0 \\ 64 \end{bmatrix} [K_1 \quad K_2] = \begin{bmatrix} 0 & 0 \\ 64K_1 & 64K_2 \end{bmatrix} \tag{h}$$

$$\mathbf{A} - \mathbf{BS} = \begin{bmatrix} 0 & 1 \\ -100 - 64K_1 & -6 - 64K_2 \end{bmatrix} \tag{i}$$

The eigenvalues of $\mathbf{A} - \mathbf{BS}$ are calculated from

$$\begin{vmatrix} -\lambda & 1 \\ -100 - 64K_1 & -6 - 64K_2 - \lambda \end{vmatrix} = 0$$

$$(-\lambda)(-6 - 64K_2 - \lambda) - (1)(-100 - 64K_1) = 0$$

$$\lambda^2 + (6 + 64K_1)\lambda + 100 + 64K_2 = 0 \tag{j}$$

The characteristic equation for a system with poles at -3 and -40 is

$$\lambda^2 + 43 + 120 = 0 \tag{k}$$

Comparing Equations (j) and (k) yields

$$6 + 64K_1 = 43 \Rightarrow K_1 = 0.578 \tag{l}$$

$$100 + 64K_2 = 120 \Rightarrow K_2 = 0.313 \tag{m}$$

Example 12.20

The simplified suspension system of Example 12.45(a) is attached to an axle of negligible mass that is attached to a tire of stiffness $k_t = 10{,}000$ N/m. Develop state variable feedback for this system such that the system has poles at $-2 \pm 2j$ and -5.

Solution

The differential equations modeling the system are obtained by applying Newton's law to free-body diagrams of the mass and the joint, Figure 12.45(b), as

$$m\ddot{x}_1 + c(\dot{x}_1 - \dot{x}_2) + k(x_1 - x_2) = 0 \tag{a}$$

$$-k_t(z - x_2) = c(\dot{x}_1 - \dot{x}_2) + k(x_1 - x_2) \tag{b}$$

The state variables are defined as

$$y_1 = x_1 \tag{c}$$

$$y_2 = \dot{x}_1 \tag{d}$$

$$y_3 = x_2 \tag{e}$$

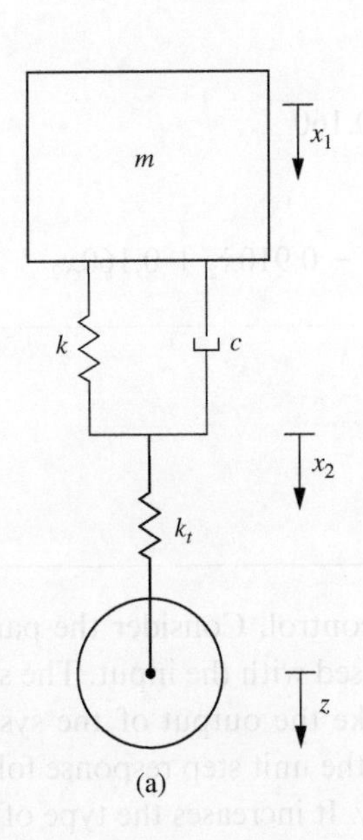

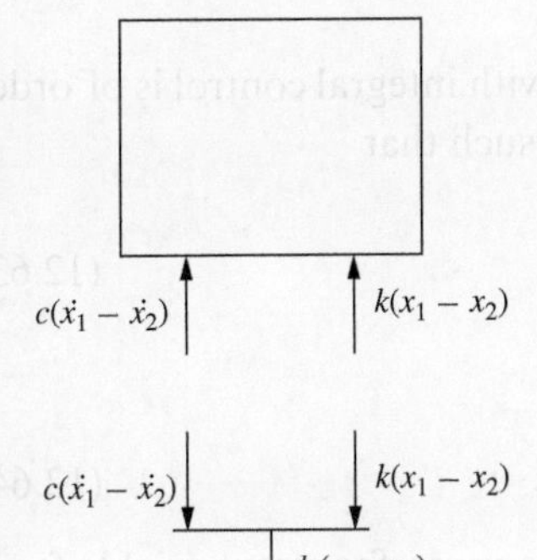

Figure 12.45

(a) Model of vehicle suspension system which includes elasticity of tire. (b) Free-body diagrams of vehicle and the joint between the axle and the tire.

Substituting the definitions of the state variables, Equations (c), (d), and (e), into Equations (a) and (b) leads to

$$\dot{y}_1 = y_2 \tag{f}$$

$$\dot{y}_2 = \frac{k_t}{m} y_3 - \frac{k_t}{m} z \tag{g}$$

$$\dot{y}_3 = \frac{k}{c} y_1 + y_2 - \frac{(k + k_t)}{c} y_3 + \frac{k_t}{c} z \tag{h}$$

The matrix form of the state-space formulation for this system is

$$\begin{bmatrix} \dot{y}_1 \\ \dot{y}_2 \\ \dot{y}_3 \end{bmatrix} = \begin{bmatrix} 0 & 0 & 0 \\ 0 & 0 & \frac{k_t}{m} \\ \frac{k}{c} & 1 & \frac{(k + k_t)}{c} \end{bmatrix} \begin{bmatrix} y_1 \\ y_2 \\ y_3 \end{bmatrix} + \begin{bmatrix} 0 \\ -\frac{k_t}{m} \\ \frac{k_t}{c} \end{bmatrix} [z] \tag{i}$$

Using state variable feedback,

$$\mathbf{A} - \mathbf{BS} = \begin{bmatrix} 0 & 1 & 0 \\ 0 & 0 & \frac{k_t}{m} \\ \frac{k}{c} & 1 & \frac{(k + k_t)}{c} \end{bmatrix} - \begin{bmatrix} 0 \\ -\frac{k_t}{m} \\ \frac{k_t}{c} \end{bmatrix} [K_1 \quad K_2 \quad K_3] \tag{j}$$

$$= \begin{bmatrix} 0 & 1 & 0 \\ \frac{k_t}{m} K_1 & \frac{k_t}{m} K_2 & \frac{k_t}{m}(K_3 + 1) \\ \frac{k}{c}(1 - K_1) & 1 - \frac{k_t}{c} K_2 & -\frac{(k + k_t)}{c} - \frac{k_t}{c} K_3 \end{bmatrix} \tag{k}$$

Substituting given values leads to

$$\mathbf{A} - \mathbf{BS} = \begin{bmatrix} 0 & 1 & 0 \\ 20K_1 & 20K_2 & 20(K_3 - 1) \\ 10(1 - K_1) & 1 - 5K_2 & -15 - 5K_3 \end{bmatrix} \tag{l}$$

The characteristic equation is

$$\begin{vmatrix} -\lambda & 1 & 0 \\ 20K_1 & 20K_2 - \lambda & 20(K_3 + 1) \\ 10(1 - K_1) & 1 - 5K_2 & -15 - 5K_3 - \lambda \end{vmatrix} = 0 \tag{m}$$

Evaluation of the determinant in Equation (m) gives

$$-\lambda[(20K_2 - \lambda)(-15 - 5K_3 - \lambda) - 20(K_3 + 1)(1 - 5K_2)]$$
$$- 1[(20K_1)(-15 - 5K_3 - \lambda) - 20(K_3 + 1)10(1 - K_1)]$$

$$= \lambda^3 - (20K_2 - 15 - 5K_3)\lambda^2 - [30K_2 + 100K_2K_3 + 20K_1 + 20(K_3 + 1)(1 - 5K_2)]\lambda$$
$$- [20K_1(15 + 5K_3) + 200(K_3 - 1)(1 - K_1)] \tag{n}$$

The required characteristic polynomial for roots of -20 and $-7 \pm 7j$ is

$$(\lambda + 20)(\lambda^2 + 14\lambda + 98) = \lambda^3 + 34\lambda^2 + 378.2\lambda + 1960 \tag{o}$$

Equating Equations (n) and (o) gives

$$-(20K_2 - 15 - 5K_3) = 34 \tag{p}$$

$$-[30K_2 + 100K_2K_3 + 20K_1 + 20(K_3 + 1)(1 - 5K_2)] = 378.2 \tag{q}$$

$$-[20K_1(15 + 5K_3) + 200(K_3 - 1)(1 - K_1)] = 1960 \tag{r}$$

Solving Equations (p), (q), and (r) gives

$$K_1 = -10.96 \quad K_2 = -0.910 \quad K_3 = 0.160 \tag{s}$$

The input to the system is

$$z(t) + 10.96y_1 - 0.910y_2 + 0.160y_3 = z(t) + 10.96x_1 - 0.910\dot{x}_2 + 0.160x_2 \tag{t}$$

12.9 State Variable Feedback with Integral Control

Integral control may be introduced with state variable control. Consider the partial system of Figure 12.46 where a reference signal is to be used with the input. The state variables with control are fed back to the input. To make the output of the system smooth, integral control is required. The idea is to make the unit step response follow the unit step response of the original system more closely. It increases the type of the system by one.

If the original system is of order n, then the system with integral control is of order $n + 1$. The output is y_k. A new state variable is defined such that

$$y_{n+1} = \int_0^t [r(t) - y_k]dt \tag{12.63}$$

This is equivalent to stating that

$$\dot{y}_{n+1} = r(t) - y_k \tag{12.64}$$

This equation is added to the state-space equations. However, for state variable feedback, the input to the system becomes

$$u(t) = f(t) - K_1y_1 - K_2y_2 - \cdots - K_ny_n + K_Iy_{n+1} \tag{12.65}$$

Using the notation of Equation (12.58), $\mathbf{u} = \mathbf{Sy}$, where

$$\mathbf{S} = [K_1 \quad K_2 \quad \cdots \quad K_n \quad -K_I] \tag{12.66}$$

Equation (12.66) can be written as

$$\mathbf{S} = [K_1\ K_2 \cdots\ K_n \quad K_{n+1}] \tag{12.67}$$

where

$$K_{n+1} = -K_I \tag{12.68}$$

State variable control is then used on the system of order $n + 1$ to place the poles.

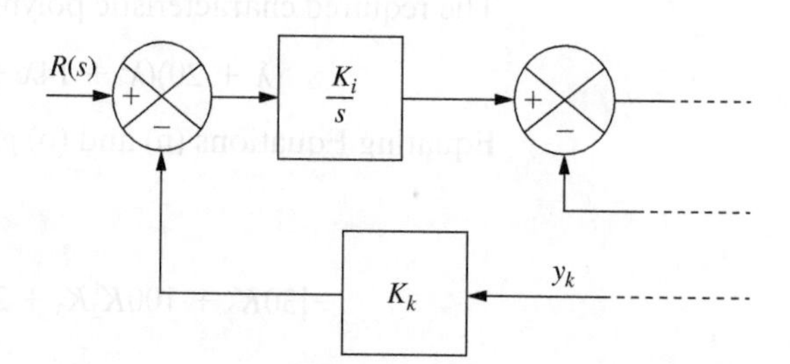

Figure 12.46

Partial control system illustrates the use of integral control along with state variable control.

Example 12.21

Consider the model for the vehicle suspension system illustrated in Example 12.19. Design a state-space system with proportional and integral control such that the poles of the system are at $s = -3 \pm 3j, -10$.

Solution

The block diagram for the system without integral control is shown in Figure 12.44. A state variable is introduced as

$$y_3 = \int_0^t z(t) - y_1\,dt \tag{a}$$

Then the input to the system is

$$u(t) = z(t) - K_1 y_1 - K_2 y_2 + K_3 y_3 \tag{b}$$

It is noted that

$$\dot{y}_3 = z - y_1 \tag{c}$$

The state-space formulation of the system with integral control is

$$\begin{bmatrix} \dot{y}_2 \\ \dot{y}_2 \\ \dot{y}_3 \end{bmatrix} = \begin{bmatrix} 0 & 1 & 0 \\ -100 & -6 & 0 \\ -1 & 0 & 0 \end{bmatrix}\begin{bmatrix} y_1 \\ y_2 \\ y_3 \end{bmatrix} + \begin{bmatrix} 0 \\ 64 \\ 1 \end{bmatrix}[z(t)] \tag{d}$$

Then

$$\mathbf{A} - \mathbf{BS} = \begin{bmatrix} 0 & 1 & 0 \\ -100 & -6 & 0 \\ -1 & 0 & 0 \end{bmatrix} - \begin{bmatrix} 0 \\ 64 \\ 1 \end{bmatrix}[K_1 \quad K_2 \quad K_3] \tag{e}$$

The eigenvalues of $\mathbf{A} - \mathbf{BS}$ are computed by

$$\begin{vmatrix} -\lambda & 1 & 0 \\ -100 - 64K_1 & -\lambda - 6 - 64K_2 & -64K_3 \\ -1 - K_1 & -K_1 & -\lambda - K_1 \end{vmatrix}$$

$$= -\lambda[(-\lambda - 6 - 64K_2)(-\lambda - K_1) - (-64K_3)(-K_1)]$$
$$1[(-100 - 64K_1)(-\lambda - K_1) - (-64K_3)(-1 - K_1)]$$
$$= -\lambda^3 - (6 + K_1 + 64K_2)\lambda^2 - (100 - 64K_1 - 64K_1K_3)\lambda + 64K_3 \tag{f}$$

In order for the poles to be at $s = -3 \pm 3j, -10$, the characteristic equation must be

$$(s + 10)(s^2 + 6s + 18) = s^3 + 16s^2 + 78s + 180 \tag{g}$$

Setting the coefficients of Equation (f) equal to the negatives of the coefficients of Equation (g) leads to

$$6 + K_1 + 64K_2 = 16 \tag{h}$$

$$100 + 64K_1 - 64K_1K_3 = 78 \tag{i}$$

$$-64K_3 = 180 \tag{j}$$

Equations (h), (i), and (j) are solved simultaneously, yielding

$$K_1 = -0.820 \quad K_2 = 0.169 \quad K_3 = -2.812 \tag{k}$$

The input to the system is

$$f(t) = 2.812\int(z - y_1)dt + 0.820y_1 - 0.169y_2$$

The feedback loop with the controllers is shown in Figure 12.47.

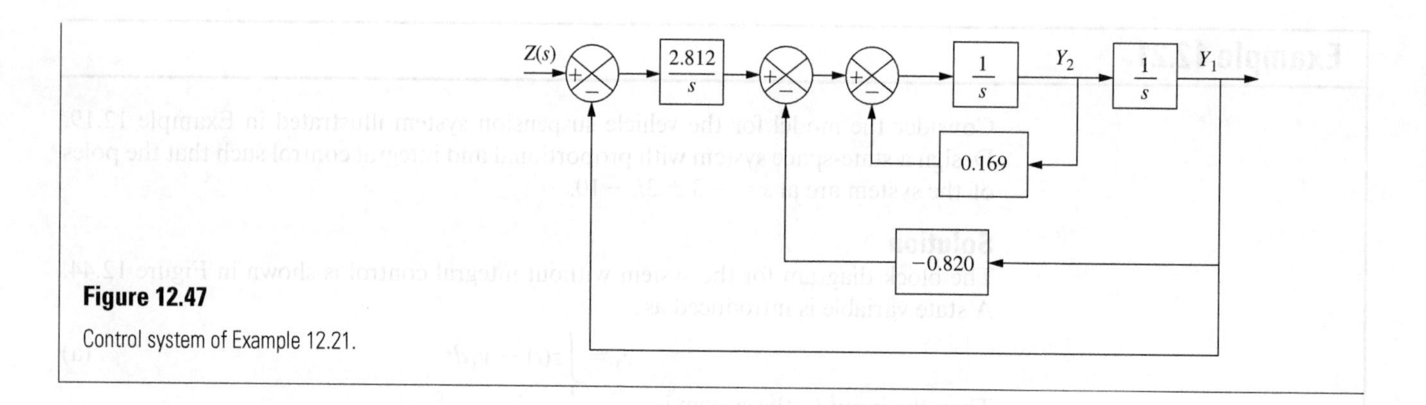

Figure 12.47

Control system of Example 12.21.

Example 12.22

Consider the field circuit-controlled dc servomotor discussed in Example 5.22. Let θ represent the position of the armature. The differential equations governing the current in the field circuit, i_f, and θ are Equations (a) and (b) of Example 5.22 written without an input torque acting on the shaft as

$$L_f\frac{di_f}{dt} + R_f i_f = v \tag{a}$$

$$J\frac{d^2\theta}{dt^2} + c_t\frac{d\theta}{dt} - K_f i_f = 0 \tag{b}$$

where L_f is the inductance of the field circuit, R_f is the resistance of the field circuit, J is the polar moment of inertia of the armature, c_t is the torsional damping coefficient of the armature bearings, K_f is the field coupling coefficient between the field circuit and the armature, and v is the voltage provided to the field circuit. The system has parameters of $R_f = 1\ \Omega$, $L_f = 0.5$ H, $J = 5 \times 10^{-3}$ kg·m², $c_t = 0.02$ N·m·s, and $K_f = 0.02\ \frac{\text{N·m}}{\text{A}}$. Design a state variable control system with integral control such that the poles are at $s = -10, -15, -20, -25$.

Solution

The state variables are defined as

$$y_1 = i_f \tag{c}$$

$$y_2 = \theta \tag{d}$$

$$y_3 = \omega = \dot{\theta} \tag{e}$$

The state equations for the field circuit dc servomotor are

$$\dot{y}_1 = -\frac{R_f}{L_f}y_1 + \frac{1}{L_f}v \tag{f}$$

$$\dot{y}_2 = y_3 \tag{g}$$

$$\dot{y}_3 = \frac{K_f}{J}y_1 - \frac{c_t}{J}y_3 \tag{h}$$

A block diagram of this system is shown in Figure 12.48 with state variable control for the angular velocity as the system output. If an integral controller is required to control the angular position of the armature, then a reference signal, $r(t)$, is sent to the entire system and the input to the system becomes

$$\int_0^t (v - y_2)\,dt - K_1y_1 - K_2y_2 - K_3y_3 \tag{i}$$

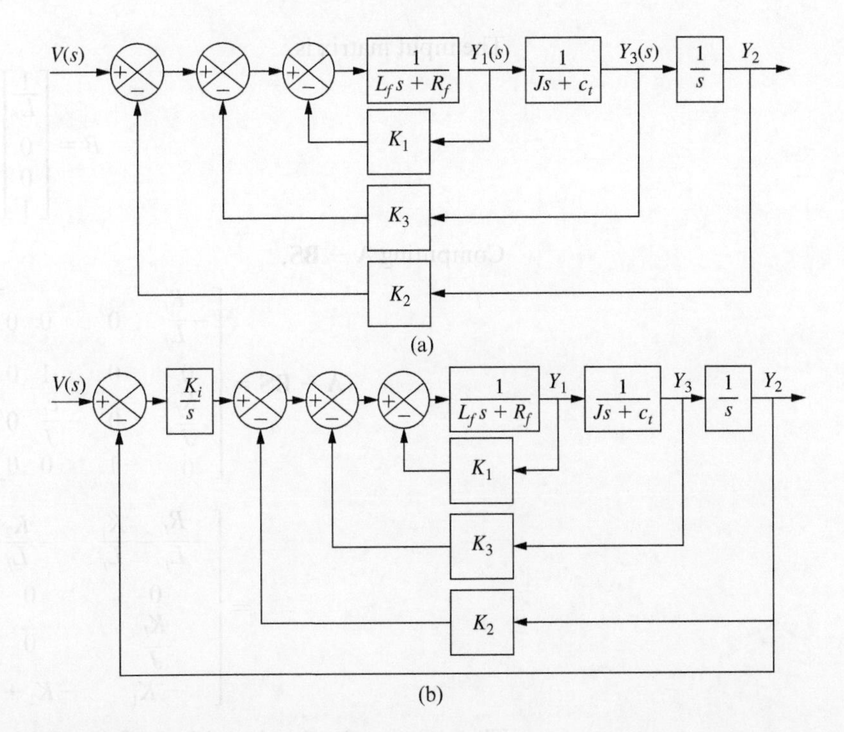

Figure 12.48

(a) Block diagram of dc servomotor of Example 12.22 with state variable control. (b) Block diagram when state variable control is used with integral control.

This is illustrated in Figure 12.48(b). The addition of the integral controller increases the denominator of the system's transfer function by one and hence the order of the system by one. This requires the definition of a new state variable:

$$y_4 = \int_0^t (v - y_2)dt \tag{j}$$

Then the input becomes

$$a(t) = K_1 y_1 - K_2 y_2 - K_3 y_3 - K_4 y_4 \tag{k}$$

where $K_4 = -K_I$. The analysis of the system proceeds as before with the introduction of a new state variable. The definitions of the state variables are

$$y_1 = i_f \tag{l}$$

$$y_2 = \theta \tag{m}$$

$$y_3 = \omega = \theta \tag{n}$$

$$y_4 = \int_0^t (v - \theta)dt \tag{o}$$

These definitions lead to

$$\dot{y}_1 = -\frac{R_f}{L_f} y_1 + \frac{1}{L_f} v \tag{p}$$

$$\dot{y}_2 = y_3 \tag{q}$$

$$\dot{y}_3 = \frac{K_f}{J} y_1 - \frac{c_t}{J} y_3 \tag{r}$$

$$\dot{y}_4 = v - y_2 \tag{s}$$

The state matrix for this formulation is

$$\begin{bmatrix} \dot{y}_1 \\ \dot{y}_2 \\ \dot{y}_3 \\ \dot{y}_4 \end{bmatrix} = \begin{bmatrix} -\frac{R_f}{L_f} & 0 & 0 & 0 \\ 0 & 0 & 1 & 0 \\ \frac{K_f}{J} & 0 & -\frac{c_t}{J} & 0 \\ 0 & -1 & 0 & 0 \end{bmatrix} \begin{bmatrix} y_1 \\ y_2 \\ y_3 \\ y_4 \end{bmatrix} + \begin{bmatrix} \frac{1}{L_f} \\ 0 \\ 0 \\ 1 \end{bmatrix} [v(t)] \tag{t}$$

The input matrix is

$$B = \begin{bmatrix} \frac{1}{L_f} \\ 0 \\ 0 \\ 1 \end{bmatrix} \tag{u}$$

Computing $\mathbf{A} - \mathbf{BS}$,

$$\mathbf{A} - \mathbf{BS} = \begin{bmatrix} -\frac{R_f}{L_f} & 0 & 0 & 0 \\ 0 & 0 & 1 & 0 \\ \frac{K_f}{J} & 0 & -\frac{c_t}{J} & 0 \\ 0 & -1 & 0 & 0 \end{bmatrix} - \begin{bmatrix} \frac{1}{L_f} \\ 0 \\ 0 \\ 1 \end{bmatrix} [K_1 \quad K_2 \quad K_3 \quad K_4]$$

$$= \begin{bmatrix} -\frac{R_f}{L_f} - \frac{K_1}{L_f} & -\frac{K_2}{L_f} & -\frac{K_3}{L_f} & -\frac{K_4}{L_f} \\ 0 & 0 & 1 & 0 \\ \frac{K_f}{J} & 0 & -\frac{c_t}{J} & 0 \\ -K_1 & -K_2 + 1 & -K_3 & -K_4 \end{bmatrix} \tag{v}$$

The equation for the eigenvalues of $\mathbf{A} - \mathbf{BS}$ is

$$\begin{vmatrix} -\frac{R_f}{L_f} - \frac{K_1}{L_f} & -\frac{K_2}{L_f} & -\frac{K_3}{L_f} & -\frac{K_4}{L_f} \\ 0 & 0 & 1 & 0 \\ \frac{K_f}{J} & 0 & -\frac{c_t}{J} & 0 \\ -K_1 & -K_2 + 1 & -K_3 & -K_4 \end{vmatrix} = 0 \tag{w}$$

Substituting given values leads to

$$\begin{vmatrix} -2 - 2K_1 - \lambda & -2K_2 & -2K_3 & -2K_4 \\ 0 & -\lambda & 1 & 0 \\ 4 & 0 & -10 - \lambda & 0 \\ -K_1 & -K_2 + 1 & -K_3 & -\lambda - K_4 \end{vmatrix} = 0 \tag{x}$$

Expanding the determinant by the first column yields

$$\begin{aligned} (-2 - 2K_1 - \lambda) &\begin{vmatrix} -\lambda & 1 & 0 \\ 0 & -4 - \lambda & 0 \\ -K_2 + 1 & -K_3 & -\lambda - K_4 \end{vmatrix} + 4 \begin{vmatrix} -2K_2 & -2K_3 & -2K_4 \\ -\lambda & 1 & 0 \\ -K_2 + 1 & -K_3 & -\lambda - K_4 \end{vmatrix} \\ &= (-2 - 2K_1 - \lambda)[(-\lambda)(-4 - \lambda)(-\lambda - K_4)] \\ &\quad + 4[(-2K_2)(1)(-\lambda - K_4) + (-2K_4)(-K_3)(-\lambda) \\ &\quad - (-2K_4)(-K_2 + 1) - (-\lambda - K_4)(-2K_3)(-\lambda)] \end{aligned} \tag{y}$$

The characteristic equation for poles at $s = -10, -15, -20, -25$ is

$$(s + 10)(s + 15)(s + 20)(s + 25) = s^4 + 70s^3 + 1775s^2 + 19{,}250s + 75{,}000 \tag{z}$$

The gains are determined by equating the coefficients in the polynomial in Equation (z) with the coefficients in the polynomial in Equation (y) and solving for the gains. The results are

$$K_1 = 89.0, K_2 = 76.75, K_3 = 98.875, K_4 = -210.0$$

12.10 Determination of the Ackerman Matrix

The matrix **S** defined by Equation (12.58) is called the Ackerman matrix. The mathematical determination of the form of the Ackerman matrix, discussed in Section 12.8, involves advanced matrix analysis and is not presented in this text. Instead, the formula for the Ackerman matrix is presented as well as its MATLAB determination.

The matrix designated as the controllability matrix is defined by

$$\mathbf{V} = [\mathbf{B} \;\; \mathbf{AB} \;\; \mathbf{A}^2\mathbf{B} \;\; \cdots \;\; \mathbf{A}^{n-1}\mathbf{B}] \tag{12.69}$$

That is, **V** is the $n \times n$ matrix whose k^{th} column is $\mathbf{A}^{k-1}\mathbf{B}$. A sufficient condition to define a vector of state variable controls given any vector of desired poles **p** is $|\mathbf{V}| \neq 0$, or **V** is nonsingular.

When the eigenvalues of $\mathbf{A} - \mathbf{BS}$ are determined, a polynomial in the eigenvalues $\tau(\lambda)$, called the characteristic polynomial, is calculated as

$$\tau(\lambda) = |\lambda\mathbf{I} - \mathbf{A} + \mathbf{BS}| \tag{12.70}$$

This is the desired characteristic equation for the controlled system. The matrix $\tau(\mathbf{A})$ is defined as the matrix obtained by substituting **A** into the characteristic polynomial of $\mathbf{A} - \mathbf{BS}$.

Ackerman's formula is

$$\mathbf{S} = [0 \;\; 0 \;\; \cdots \;\; 0 \;\; 1]\mathbf{V}^{-1}\tau(\mathbf{A}) \tag{12.71}$$

The row vector in Equation (12.71) is a vector with n elements, all zeroes except for the nth element, which is 1.

If a desired vector of poles, **p**, is specified, then the MATLAB command

```
S = acker(A,B,p)
```

computes the Ackerman matrix for the $n \times n$ state matrix **A**, the $n \times 1$ input vector **B**, and the $n \times 1$ vector of desired poles **p**.

Example 12.23

a. Use Equation (12.71) to compute the Ackerman matrix for Example 12.20 when the pole placement is specified as $s = -7 \pm 7j, -20$. **b.** Use the MATLAB command acker to calculate the Ackerman matrix for the problem in part a.

Solution

a. The state matrix for the system is

$$\mathbf{A} = \begin{bmatrix} 0 & 1 & 0 \\ 0 & 0 & \dfrac{k_t}{m} \\ \dfrac{k}{c} & 1 & -\dfrac{(k+k_t)}{c} \end{bmatrix} = \begin{bmatrix} 0 & 1 & 0 \\ 0 & 0 & 20 \\ 10 & 1 & -15 \end{bmatrix} \tag{a}$$

The input vector is

$$B = \begin{bmatrix} 0 \\ \dfrac{-k_t}{m} \\ \dfrac{k_t}{c} \end{bmatrix} = \begin{bmatrix} 0 \\ -20 \\ 5 \end{bmatrix} \tag{b}$$

The characteristic polynomial that the specified poles satisfy is

$$\tau(\lambda) = (\lambda^2 + 14\lambda + 98)(\lambda + 20) = \lambda^3 + 34\lambda^2 + 378\lambda + 1960$$

Then $\tau(\mathbf{A})$ is calculated as

$$\tau(\mathbf{A}) = \begin{bmatrix} 0 & 1 & 0 \\ 0 & 0 & 20 \\ 10 & 1 & -15 \end{bmatrix}^3 + 26\begin{bmatrix} 0 & 1 & 0 \\ 0 & 0 & 20 \\ 10 & 1 & -15 \end{bmatrix}^2 + 384\begin{bmatrix} 0 & 1 & 0 \\ 0 & 0 & 20 \\ 10 & 1 & -15 \end{bmatrix} + 1280\begin{bmatrix} 1 & 0 & 0 \\ 0 & 1 & 0 \\ 0 & 0 & 1 \end{bmatrix}$$

$$= \begin{bmatrix} 2160 & 398 & 380 \\ 3800 & 3540 & 2260 \\ 1130 & 303 & 845 \end{bmatrix} \tag{c}$$

The controllability matrix is

$$\mathbf{V} = [\mathbf{B} \quad \mathbf{AB} \quad \mathbf{A}^2\mathbf{B}] = \begin{bmatrix} 0 & -20 & 100 \\ -20 & 100 & -1900 \\ 5 & -95 & 1325 \end{bmatrix} \tag{d}$$

The inverse of $\mathbf{V}$ is

$$\mathbf{V}^{-1} = 10^{-3}\begin{bmatrix} 240 & -85 & -140 \\ -85 & 2.5 & 10 \\ -7 & 0.5 & 2 \end{bmatrix} \tag{e}$$

The Ackerman matrix is determined using Equation (12.71) as

$$\mathbf{S} = [0 \quad 0 \quad 1]\mathbf{V}^{-1}\tau(\mathbf{A}) = 10^{-3}[0 \quad 0 \quad 1]\begin{bmatrix} 240 & -85 & -140 \\ -85 & 2.5 & 10 \\ -7 & 0.5 & 2 \end{bmatrix}\begin{bmatrix} 2160 & 20 & 1060 \\ 10{,}600 & 3220 & -15{,}500 \\ -7750 & -245 & 14{,}845 \end{bmatrix}$$

$$= [-10.96 \quad -0.910 \quad 0.160] \tag{f}$$

These are the same results as in Example 12.20.

b. The MATLAB coding to obtain the Ackerman matrix is

```
% MATLAB code to obtain Ackerman matrix for Example 12.23
A=[0 1 0;0 0 20;10 1 -15]
B=[0;-20;5]
p=[-7+7j -7-7j -20]
S=acker(A,B,p)
% End of code
```

The results from the code are

```
S = -10.9600 -0.9100 0.1600
```

These are the same results found using Equation (12.71) and in Example 12.20.

12.11 Further Examples

Example 12.24

Consider the two-compartment model of the pharmokinetic system of Examples 6.19 and 7.30. **a.** Derive a state-space model for the system. **b.** Determine the state transition matrix for the system. **c.** Use the state transition matrix to derive the time-dependent tissue and plasma concentrations. **d.** Design state variable control such that the poles are at $s = -1, -2$. **e.** Determine the response of the tissue and plasma using the state variable control of part d.

Solution

a. Defining state-space variables by $y_1 = C_p$ and $y_2 = C_t$, Equations (g) and (h) of Example 6.19 are written in state-space form as

$$\begin{bmatrix} \dot{y}_1 \\ \dot{y}_2 \end{bmatrix} = \begin{bmatrix} -(k_e + k_1) & \frac{V_t}{V_p}k_2 \\ \frac{V_p}{V_t}k_1 & -k_2 \end{bmatrix} \begin{bmatrix} y_1 \\ y_2 \end{bmatrix} + \begin{bmatrix} \frac{1}{V_p} \\ 0 \end{bmatrix} I(t) \tag{a}$$

Using the numerical values of Example 7.30 in Equation (a) leads to

$$\begin{bmatrix} \dot{y}_1 \\ \dot{y}_2 \end{bmatrix} = \begin{bmatrix} -0.307 & 0.0133 \\ 2.02 & -0.133 \end{bmatrix} \begin{bmatrix} y_1 \\ y_2 \end{bmatrix} + \begin{bmatrix} 0.025 \\ 0 \end{bmatrix} I(t) \tag{b}$$

b. The state transition matrix is determined as $\Phi(t) = \mathscr{L}^{-1}\{(s\mathbf{I} - \mathbf{A})^{-1}\}$. To this end,

$$(s\mathbf{I} - \mathbf{A})^{-1} = \begin{bmatrix} s + 0.307 & -0.0133 \\ -2.02 & s + 0.133 \end{bmatrix}^{-1}$$

$$= \begin{bmatrix} \frac{s + 0.133}{s^2 + 0.440s + 0.0140} & \frac{0.0133}{s^2 + 0.880s + 0.00140} \\ \frac{2.02}{s^2 + 0.440s + 0.0140} & \frac{s + 0.307}{s^2 + 0.440s + 0.0140} \end{bmatrix} \tag{c}$$

The state transition matrix is obtained by taking the inverse Laplace transform of Equation (c). It is noted that $s^2 + 0.4405 + 0.0140 = (s + 0.4054)(s + 0.0345)$, leading to

$$\Phi(t) = \begin{bmatrix} 0.7345 & -0.03585 \\ -5.446 & 0.2655 \end{bmatrix} e^{-0.4054t} + \begin{bmatrix} 0.2655 & 0.03585 \\ 5.446 & 0.7345 \end{bmatrix} e^{-0.0345t} \tag{d}$$

c. The time-dependent infusion specified from Example 7.30 is $I(t) = 0.302 [1 - u(t - 12)]$ mg/hr. The application of Equation (12.37) to determine the system response leads to

$$\begin{bmatrix} y_1 \\ y_2 \end{bmatrix} = \int_0^t \left\{ \begin{bmatrix} 0.7345 & -0.03585 \\ -5.446 & 0.2655 \end{bmatrix} e^{-0.4054(t-\tau)} + \begin{bmatrix} 0.2655 & 0.03585 \\ 5.446 & 0.7345 \end{bmatrix} e^{-0.0345(t-\tau)} \right\}$$

$$\times \begin{bmatrix} 0.025 \\ 0 \end{bmatrix} 0.302[1 - u(\tau - 12)]d\tau \tag{e}$$

Evaluation of the integrals in Equation (e) leads to

$$y_1(t) = 0.0137[1 - e^{-0.4054t}]u(t) - 0.0137[1 - e^{-0.4054(t-12)}]u(t - 12)$$
$$+ 0.0581[1 - e^{-0.0345t}]u(t) - 0.0581[1 - e^{-0.0345(t-12)}]u(t - 12) \tag{f}$$

$$y_2(t) = -0.1014[1 - e^{-0.4054t}]u(t) - 0.1014[1 - e^{-0.4054(t-12)}]u(t - 12)$$
$$+ 1.9181[1 - e^{-0.0345t}]u(t) - 1.918[1 - e^{-0.0345(t-12)}]u(t - 12) \tag{g}$$

d. Assuming state variable control, letting $\mathbf{S} = [K_1 \quad K_2]$, the equations for the state formulation of the system response become

$$\dot{\mathbf{y}} = (\mathbf{A} - \mathbf{BS})\mathbf{y} \tag{h}$$

Substituting from Equation (b) into Equation (h) gives

$$\begin{bmatrix} \dot{y}_1 \\ \dot{y}_2 \end{bmatrix} = \left(\begin{bmatrix} -0.307 & 0.0133 \\ 2.02 & -0.133 \end{bmatrix} - \begin{bmatrix} 0.25 \\ 0 \end{bmatrix} [K_1 \quad K_2] \right) \begin{bmatrix} y_1 \\ y_2 \end{bmatrix} \tag{i}$$

Simplifying Equation (i) leads to

$$\begin{bmatrix} \dot{y}_1 \\ \dot{y}_2 \end{bmatrix} = \begin{bmatrix} -0.307 - 0.25K_1 & 0.0133 - 0.25K_2 \\ 2.02 & -0.133 \end{bmatrix} \begin{bmatrix} y_1 \\ y_2 \end{bmatrix} \tag{j}$$

The characteristic equation for $\mathbf{A} - \mathbf{BS}$ is

$$\begin{vmatrix} -\lambda - 0.307 - 0.25K_1 & 0.0133 - 0.25K_2 \\ 2.02 & -\lambda - 0.133 \end{vmatrix} = 0$$

$$= (-\lambda - 0.307 - 0.25K_1)(-\lambda - 0.133) - (0.0133 - 0.25K_2)(2.02)$$
$$= \lambda^2 + (0.440 + 0.25K_1)\lambda + 0.013105 + 0.03325K_1 + 0.505K_2 \tag{k}$$

In order for the system to have poles at $s = -1, -2$, the characteristic equation must be

$$(s + 1)(s + 2) = s^2 + 3s + 2 \tag{l}$$

The coefficients of Equation (k) must agree with the coefficients of Equation (l), which leads to

$$0.440 + 0.25K_1 = 3 \Rightarrow K_1 = 10.24 \tag{m}$$

and

$$0.013105 + 0.03325K_1 + 0.505K_2 = 2 \Rightarrow K_2 = 3.26 \tag{n}$$

e. The input is given by Equation (12.58) as

$$[u(t)] = -[10.24 \quad 3.26] \begin{bmatrix} y_1 \\ y_2 \end{bmatrix} = -10.24y_1 - 3.26y_2 \tag{o}$$

The differential equations for the state-space model with state variable control are

$$\begin{bmatrix} \dot{y}_1 \\ \dot{y}_2 \end{bmatrix} = \begin{bmatrix} -0.307 - 0.25K_1 & 0.0133 - 0.25K_2 \\ 2.02 & -0.133 \end{bmatrix} \begin{bmatrix} y_1 \\ y_2 \end{bmatrix} + \begin{bmatrix} 0.25 \\ 0 \end{bmatrix} [I(t)]$$
$$= \begin{bmatrix} -2.867 & -0.8017 \\ 2.02 & -0.133 \end{bmatrix} \begin{bmatrix} y_1 \\ y_2 \end{bmatrix} + \begin{bmatrix} 0.25 \\ 0 \end{bmatrix} [I(t)] \tag{p}$$

The state transition matrix is

$$\boldsymbol{\Phi} = \mathcal{L}^{-1}(s\mathbf{I} - \mathbf{A})^{-1} = \mathcal{L}^{-1}\left\{ \begin{bmatrix} s + 2.867 & -0.8017 \\ 2.02 & s + 0.133 \end{bmatrix}^{-1} \right\}$$
$$= \mathcal{L}^{-1}\left\{ \frac{1}{s^2 + 3s + 2} \begin{bmatrix} s + 0.133 & 0.8017 \\ -2.02 & s + 2.867 \end{bmatrix} \right\} \tag{q}$$

Taking the inverse transform of the right-hand side of Equation (q) gives

$$\boldsymbol{\Phi} = \begin{bmatrix} -0.867 & 0.8017 \\ -2.02 & 1.867 \end{bmatrix} e^{-t} + \begin{bmatrix} 1.867 & -0.8017 \\ 2.02 & -0.867 \end{bmatrix} e^{-2t} \tag{r}$$

The response of the system is obtained using Equation (12.37) as

$$\begin{bmatrix} y_1 \\ y_2 \end{bmatrix} = \int_0^t \left\{ \begin{bmatrix} -0.867 & 0.8017 \\ -2.02 & 1.867 \end{bmatrix} e^{-(t-\tau)} + \begin{bmatrix} 1.867 & -0.8017 \\ 2.02 & -0.867 \end{bmatrix} e^{-2(t-\tau)} \right\} \begin{bmatrix} 0.25 \\ 0 \end{bmatrix} 0.302[1 - u(\tau - 12)]d\tau \tag{s}$$

The convolution integrals in Equation (s) are evaluated, leading to $y_1(t)$ and $y_2(t)$.

Example 12.25

Consider the three-degree-of-freedom mechanical system of Figure 12.49. **a.** Determine a state-space model for the system. **b.** Determine the eigenvalues of the state matrix and discuss what they imply about the free response of the system. **c.** Use MATLAB to determine the impulsive response for the system.

Solution

The differential equations governing the motion of the system are

$$\begin{bmatrix} 20 & 0 & 0 \\ 0 & 30 & 0 \\ 0 & 0 & 20 \end{bmatrix}\begin{bmatrix} \ddot{x}_1 \\ \ddot{x}_2 \\ \ddot{x}_3 \end{bmatrix} + \begin{bmatrix} 3000 & -3000 & 0 \\ -3000 & 7500 & -4500 \\ 0 & -4500 & 4500 \end{bmatrix}\begin{bmatrix} \dot{x}_1 \\ \dot{x}_2 \\ \dot{x}_3 \end{bmatrix}$$

$$+ \begin{bmatrix} 4\times 10^5 & -4\times 10^5 & 0 \\ -4\times 10^5 & 1\times 10^6 & -6\times 10^5 \\ 0 & -6\times 10^5 & 6\times 10^5 \end{bmatrix}\begin{bmatrix} x_1 \\ x_2 \\ x_3 \end{bmatrix} = \begin{bmatrix} 0 \\ 0 \\ F(t) \end{bmatrix} \quad \text{(a)}$$

a. State variables for this system are $y_1 = x_1, y_2 = x_2, y_3 = x_3, y_4 = \dot{x}_1, y_5 = \dot{x}_2, y_6 = \dot{x}_3$. The resulting state-space formulation is

$$\begin{bmatrix} \dot{y}_1 \\ \dot{y}_2 \\ \dot{y}_3 \\ \dot{y}_4 \\ \dot{y}_5 \\ \dot{y}_6 \end{bmatrix} = \begin{bmatrix} 0 & 0 & 0 & 1 & 0 & 0 \\ 0 & 0 & 0 & 0 & 1 & 0 \\ 0 & 0 & 0 & 0 & 0 & 1 \\ -2\times 10^4 & 2\times 10^4 & 0 & -150 & 150 & 0 \\ 1.33\times 10^4 & -3.33\times 10^4 & 2\times 10^4 & 100 & -250 & 150 \\ 0 & 3\times 10^4 & -3\times 10^4 & 0 & 225 & -225 \end{bmatrix}\begin{bmatrix} y_1 \\ y_2 \\ y_3 \\ y_4 \\ y_5 \\ y_6 \end{bmatrix} + \begin{bmatrix} 0 \\ 0 \\ 0 \\ 0 \\ 0 \\ 0.05 \end{bmatrix} F(t) \quad \text{(b)}$$

b. The MATLAB 'eig' command is used to determine the eigenvalues of the state matrix as 0, 0, $-87.5 \pm j125.6$, $-225 \pm j96.6$. The eigenvalues of the state matrix are also the poles of the transfer function. Since 0 is a double pole of the transfer function, s^2 is a factor of the denominator of the transfer function. Thus, a partial fraction decomposition of the transfer function contains $1/s$ and $1/s^2$ terms. When inverted, the free response contains terms proportional to 1 and t. The additional terms of the free response are proportional to $e^{-87.5t}$ sin $(125.6t)$ and e^{-225t} sin $(96.6t)$.

c. The MATLAB-generated plots for the impulsive response of this unrestrained system are shown in Figure 12.50. Since the system is unrestrained, the application of an impulse imparts an initial momentum to the system. The system itself satisfies conservation of linear momentum and thus continues in motion indefinitely. While viscous damping dissipates energy associated with the sinusoidal portions of the respone, it does not dissipate energy from the systems rigid-body motion which occurs because the system is unrestrained.

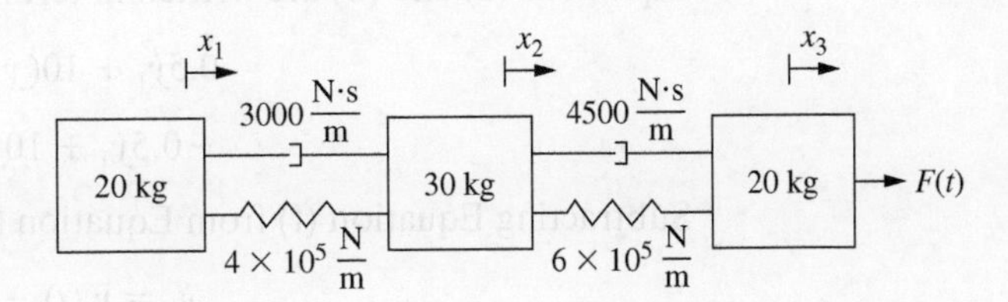

Figure 12.49

The system of Example 12.25.

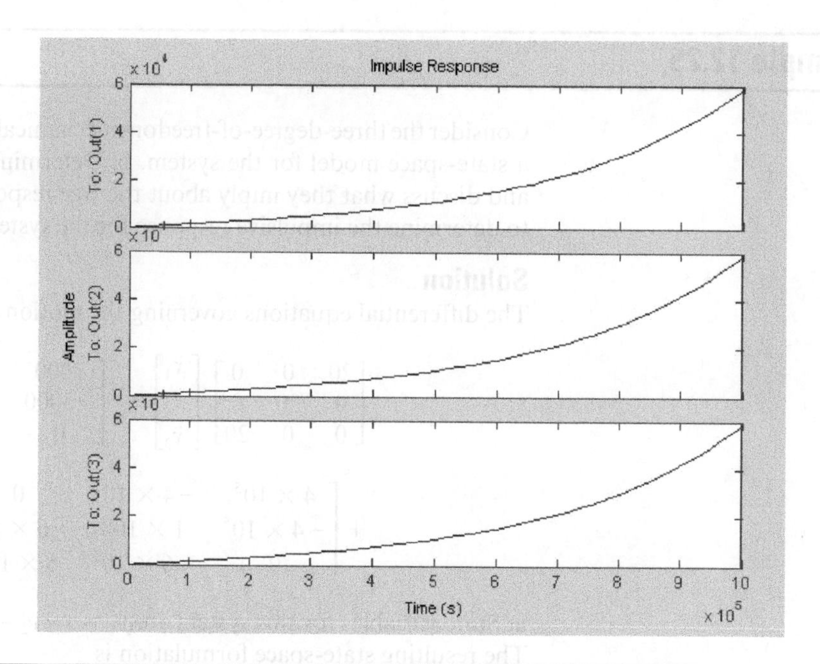

Figure 12.50

The impulsive response for the system of Example 12.25. Since the system is unrestrained, an impulse sets the system in motion indefintieily.

Example 12.26

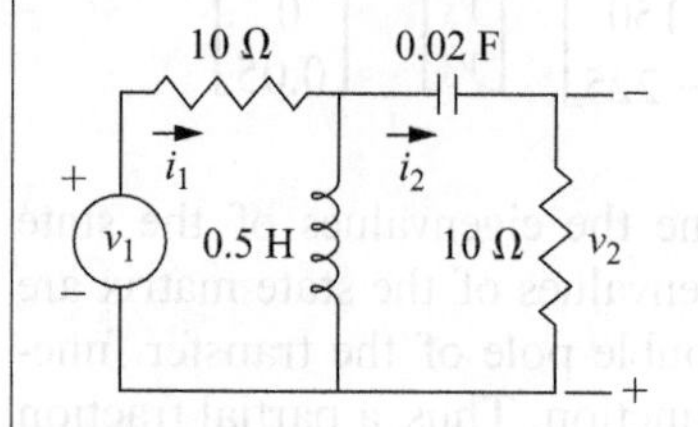

Figure 8.44

The circuit of Example 8.23, Example 8.44, and Example 12.26. (Repeated)

Consider the two-loop circuit of Example 8.23 and Figure 8.44. **a.** Determine a state-space model for the circuit where the system input is $v_1(t)$ and the system output is $v_2(t) = 10i_2(t)$. **b.** Develop a Simulink model for this system. Run a simulation for $V_1 = 120$ V and $\omega = 10$ r/s.

Solution

a. The integrodifferential equations for the circuit are Equations (a) and (b) of Example 8.23, repeated here:

$$0.5\left(\frac{di_1}{dt} - \frac{di_2}{dt}\right) + 10i_1 = v_1(t) \tag{a}$$

$$-0.5\left(\frac{di_1}{dt} - \frac{di_2}{dt}\right) + 10i_2 + 50\int_0^t i_2\,dt = 0 \tag{b}$$

Define state variables as

$$y_1 = i_1 - i_2 \tag{c}$$

$$y_2 = q_2 = \int_0^t i_2\,dt \tag{d}$$

Equations (a) and (b) are written in terms of the defined state variables as

$$0.5\dot{y}_1 + 10(y_1 + \dot{y}_2) = v_1(t) \tag{e}$$

$$-0.5\dot{y}_1 + 10\dot{y}_2 + 50y_2 = 0 \tag{f}$$

Subtracting Equation (f) from Equation (e) leads to

$$\dot{y}_1 = v_1(t) - 10y_1 + 50y_2 \tag{g}$$

Adding Equation (e) to Equation (f) leads to

$$\dot{y}_2 = \frac{1}{20}[v_1(t) - 10y_1 - 50y_2] \tag{h}$$

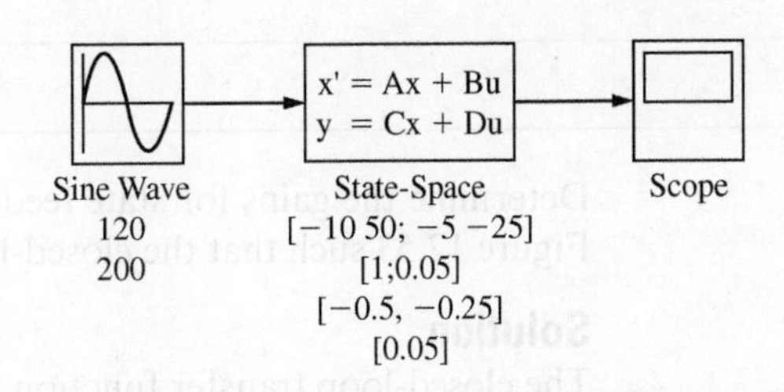

Figure 12.51

The Simulink model using the State-Space block for Example 12.26.

Equations (g) and (h) are summarized in the form of Equation (3.20) as

$$\begin{bmatrix} \dot{y}_1 \\ \dot{y}_2 \end{bmatrix} = \begin{bmatrix} -10 & 50 \\ -0.5 & -2.5 \end{bmatrix} \begin{bmatrix} y_1 \\ y_2 \end{bmatrix} + \begin{bmatrix} 1 \\ 0.05 \end{bmatrix} v(t) \tag{i}$$

The substitution of Equations (c) and (g) into Equation (a) leads to

$$0.5\,[v_1(t) - 10y_1 - 50y_2] + 10i_1 = v(t)$$

$$i_1 = 0.05v_1(t) + 0.5y_1 - 2.5y_2 \tag{j}$$

Noting from Equation (c) that $i_2 = i_1 - y_1$, use of Equation (j) leads to

$$i_2 = 0.05v_1 - 0.5y_1 - 2.5y_2 \tag{k}$$

Equation (k) is used to determine the appropriate formulation of the state-space model for this system as

$$i_2 = [-0.5 - 0.25] \begin{bmatrix} y_1 \\ y_2 \end{bmatrix} + 0.05v(t) \tag{l}$$

Equations (i) and (l) form the state-space model for the system.

b. The Simulink model developed using the State-Space block is illustrated in Figure 12.51, and the requested simulation is shown in Figure 12.52.

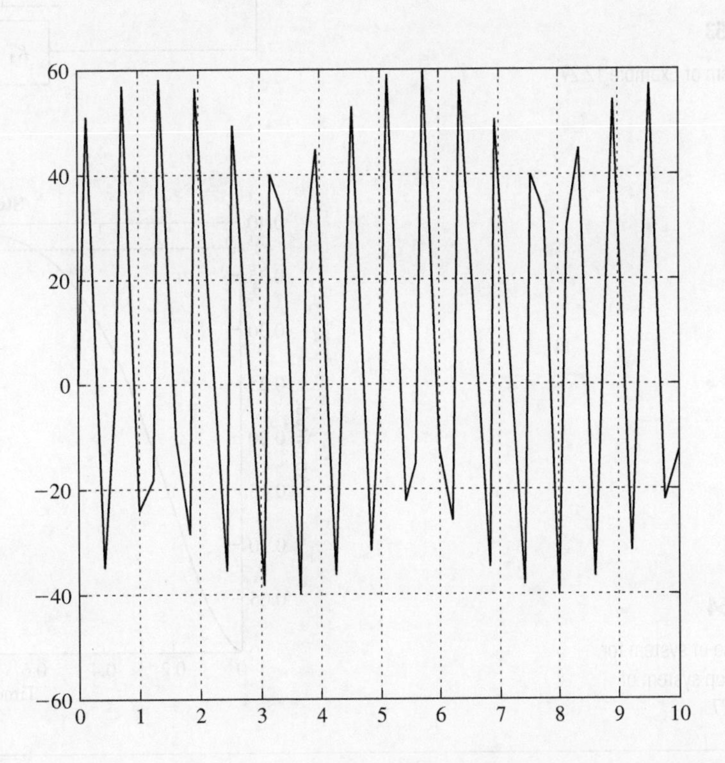

Figure 12.52

The simulation for Example 12.26.

Example 12.27

Determine the gains for state feedback control when used with the block diagram of Figure 12.53 such that the closed-loop poles are at $s = -4 \pm 4j$.

Solution

The closed-loop transfer function is

$$H(s) = \frac{C(s)}{A(s)} = \frac{12}{s^2 + (3 + 4K_2)s + 2 + 8K_2 + 12K_1} \quad \text{(a)}$$

In order to have roots at $s = -4 \pm 4j$, the denominator of the transfer function must be

$$D(s) = s^2 + 8s + 32 \quad \text{(b)}$$

Equating Equation (b) with the denominator of Equation (a) leads to

$$3 + 4K_2 = 8 \Rightarrow K_2 = 1.25 \quad \text{(c)}$$

$$2 + 8K_2 + 12K_1 = 32 \Rightarrow K_1 = 1.75 \quad \text{(d)}$$

The closed-loop transfer function is obtained from Equation (a) as

$$H(s) = \frac{12}{s^2 + 8s + 32} \quad \text{(e)}$$

The block diagram step response is shown in Figure 12.54. The final value of the step response is 0.375 instead of 1.

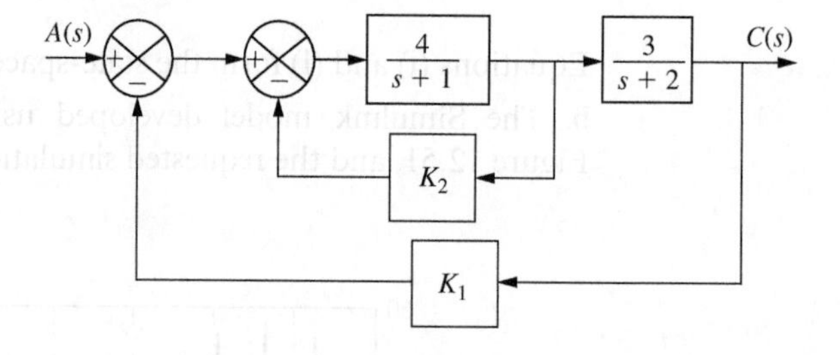

Figure 12.53

Control system of Example 12.27.

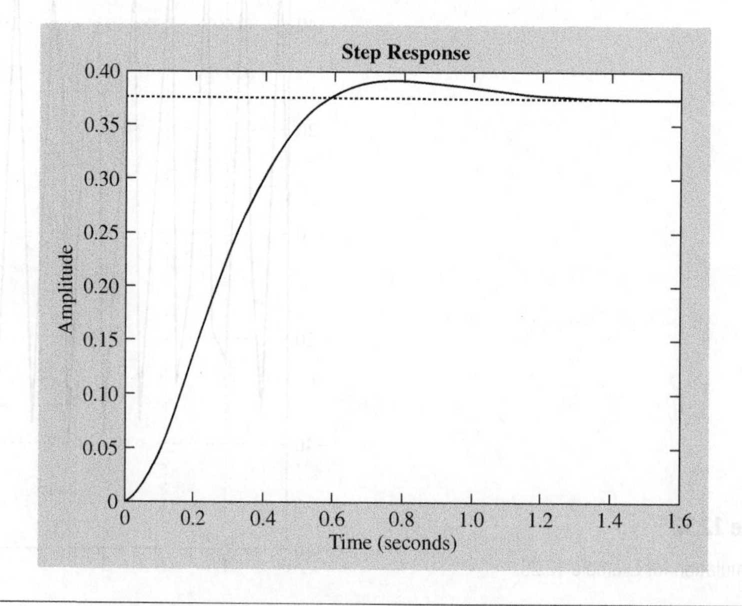

Figure 12.54

Step response of system for the closed-loop system of Example 12.27.

Example 12.28

In order to eliminate the offset of the step response that occurs when state variable control is used on the feedback system of Example 12.28, an integral control element is added. The system is illustrated in Figure 12.55. Design the system such that the dominant poles remain at $s = -4 \pm 4j$ and the third pole is at $s = -5$.

Solution

The closed-loop transfer function for the system of Figure 12.55 is

$$H(s) = \frac{12K_i}{s^3 + (3 + 4K_2)s^2 + (2 + 8K_2 + 12K_1)s + 12K_i} \tag{a}$$

The denominator of the transfer function should be

$$D(s) = (s^2 + 8s + 32)(s + 5) = s^3 + 13s^2 + 72s + 160 \tag{b}$$

Equating Equation (b) to the denominator of Equation (a) yields

$$3 + 4K_2 = 13 \Rightarrow K_2 = 2.5 \tag{c}$$

$$2 + 8K_2 + 12K_1 = 72 \Rightarrow K_1 = 8.33 \tag{d}$$

$$12K_i = 160 \Rightarrow K_i = 13.33 \tag{e}$$

The transfer function becomes

$$H(s) = \frac{160}{s^3 + 13s^2 + 72s + 160} \tag{f}$$

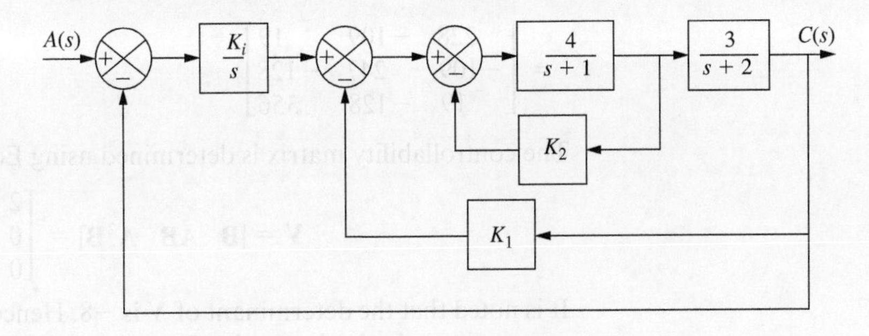

Figure 12.55

Control system of Example 12.28.

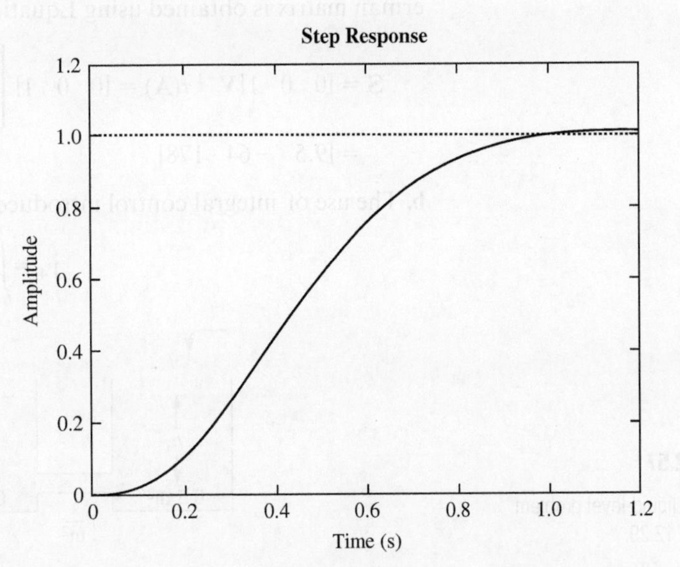

Figure 12.56

Step response of system of Example 12.28.

Comparing the step response of the system, shown in Figure 12.56, with the step response of the system of Example 12.27, it is seen that the addition of the integral controller does eliminate the offset, but it also increases the rise time, while the settling time is about the same.

Example 12.29

A three-tank liquid-level problem is illustrated in Figure 12.57. The differential equations in state-space form that govern the heights in perturbations are

$$\begin{bmatrix} \dot{y}_1 \\ \dot{y}_2 \\ \dot{y}_3 \end{bmatrix} = \begin{bmatrix} 2 & -1 & 0 \\ -1 & 2 & -1 \\ 0 & -1 & 3 \end{bmatrix} \begin{bmatrix} y_1 \\ y_2 \\ y_3 \end{bmatrix} + \begin{bmatrix} 2 \\ 0 \\ 0 \end{bmatrix} [q] \tag{a}$$

a. Specify the state variable gains if the poles of the closed-loop system are at $s = -3, -4, -5$. **b.** Specify the gains if integral control is used in conjunction with state variable control. The poles are to be at $s = -3, -4, -5, -6$.

Solution

a. The characteristic polynomial is defined by placing the poles at $-3, -4, -5$:

$$\tau(\lambda) = (\lambda + 3)(\lambda + 4)(\lambda + 5) = \lambda^3 + 12\lambda^2 + 47\lambda + 60 \tag{b}$$

Then

$$\tau(\mathbf{A}) = \begin{bmatrix} 2 & -1 & 0 \\ -1 & 2 & -1 \\ 0 & -1 & 3 \end{bmatrix}^3 + 12 \begin{bmatrix} 2 & -1 & 0 \\ -1 & 2 & -1 \\ 0 & -1 & 3 \end{bmatrix}^2 + 47 \begin{bmatrix} 2 & -1 & 0 \\ -1 & 2 & -1 \\ 0 & -1 & 3 \end{bmatrix} + 60 \begin{bmatrix} 1 & 0 & 0 \\ 0 & 1 & 0 \\ 0 & 0 & 1 \end{bmatrix}$$

$$= \begin{bmatrix} 228 & -109 & 19 \\ -109 & 247 & -128 \\ 19 & -128 & 356 \end{bmatrix} \tag{c}$$

The controllability matrix is determined using Equation (12.69):

$$\mathbf{V} = [\mathbf{B} \;\; \mathbf{AB} \;\; \mathbf{A}^2\mathbf{B}] = \begin{bmatrix} 2 & 4 & 10 \\ 0 & -2 & -8 \\ 0 & 0 & 2 \end{bmatrix} \tag{d}$$

It is noted that the determinant of $\mathbf{V}$ is -8. Hence the system is controllable. The Ackerman matrix is obtained using Equation (12.71):

$$\mathbf{S} = [0 \;\; 0 \;\; 1]\mathbf{V}^{-1}\tau(\mathbf{A}) = [0 \;\; 0 \;\; 1] \begin{bmatrix} 2 & 4 & 10 \\ 0 & -2 & -8 \\ 0 & 0 & 2 \end{bmatrix}^{-1} \begin{bmatrix} 228 & -109 & 19 \\ -109 & 247 & -128 \\ 19 & -128 & 356 \end{bmatrix}$$

$$= [9.5 \;\; -64 \;\; 178] \tag{e}$$

b. The use of integral control introduces another state variable

$$y_4 = \int_0^t (q - y_1)dt \tag{f}$$

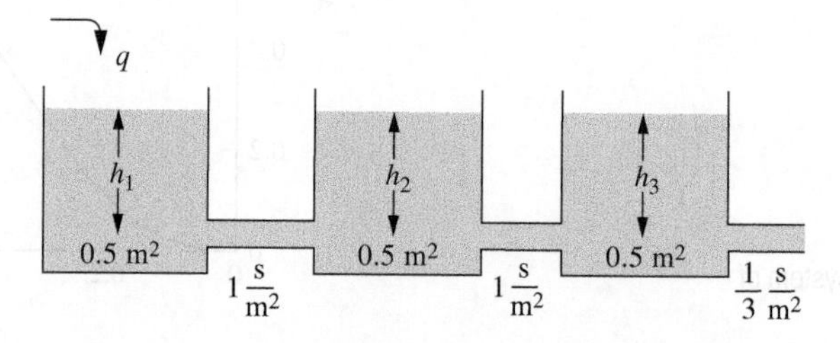

Figure 12.57

Three-tank liquid-level problem of Example 12.29.

Then

$$\dot{y}_4 = q - y_1 \tag{g}$$

The formulation of the problem in the state space assuming integral control is

$$\begin{bmatrix} \dot{y}_1 \\ \dot{y}_2 \\ \dot{y}_3 \\ \dot{y}_4 \end{bmatrix} = \begin{bmatrix} 2 & -1 & 0 & 0 \\ -1 & 2 & -1 & 0 \\ 0 & -1 & 3 & 0 \\ -1 & 0 & 0 & 0 \end{bmatrix} \begin{bmatrix} y_1 \\ y_2 \\ y_3 \\ y_4 \end{bmatrix} + \begin{bmatrix} 2 \\ 0 \\ 0 \\ 1 \end{bmatrix} [q] \tag{h}$$

The MATLAB command acker is used with

$$\mathbf{A} = \begin{bmatrix} 2 & -1 & 0 & 0 \\ -1 & 2 & -1 & 0 \\ 0 & -1 & 3 & 0 \\ -1 & 0 & 0 & 0 \end{bmatrix}, \quad \mathbf{B} = \begin{bmatrix} 2 \\ 0 \\ 0 \\ 1 \end{bmatrix}, \quad \mathbf{p} = [-3 - 4 - 5 - 6] \tag{i}$$

The code for the Ackerman matrix is

```
clear
A=[2 -1 0 0;-1 2 -1 0;0 -1 3 0;-1 0 0 0];
B=[2;0;0;1];
p=[-3 -4 -5 -6];
S=acker(A,B,p)
```

The resulting Ackerman matrix is

```
S = 23.0882   -72.6471  531.1176   -21.1765
```

Since the state variable defining integral control is Equation (12.68), the fourth parameter in the Ackerman matrix is $-K_i$. Thus the control gains are

$$K_1 = 23.1,\ K_2 = -72.6,\ K_3 = 531.1,\ K_i = 21.2 \tag{j}$$

The resulting closed-loop transfer function is

$$H(s) = \frac{60}{s^3 + 12s^2 + 47s + 60} \tag{k}$$

12.12 Summary

12.12.1 Chapter Highlights

- State-space modeling is an alternative to dynamic system modeling with transfer functions.
- State-space formulations can be developed for linear and nonlinear systems.
- The state-space model of a linear system is defined using four matrices: the state matrix, the input matrix, the output matrix, and the transmission matrix.
- The free response using the state-space formulation can be determined using either the Laplace transform method or matrix methods that involve determining the eigenvalues of the state matrix.
- The free response can be written in terms of the state transition matrix.
- The convolution integral is used to determine the forced response.
- Numerical methods such as Euler's method or Runge-Kutta methods are often employed to determine the forced response.

- The state-space formulation may be used to determine a system's transfer function, and the system's transfer function may be used to determine a state-space model.
- The state-space method can be used for MATLAB modeling of dynamic systems. Given the state-space formulation, MATLAB can be used to determine the system's impulsive response, step response, and response due to any system input.
- Simulink may be used to develop state-space models. A Simulink model can be developed from a block diagram of the state-space model, or the State-Space block can be used to define the state space.
- Simulink is well suited for nonlinear problems formulated in the state space.
- State variable control involves proportional control of each state variable.
- A block diagram can be drawn using state variable control with a feedback loop for each variable.
- The control parameters are determined by specifying the desired poles of the closed-loop system and calculating the gains to achieve this.
- A system is controllable if control variables can be specified to match any specified poles.
- A controllability matrix is defined, and the system is deemed controllable if the controllability matrix is nonsingular.
- The Ackerman matrix, for a controllable system, is the vector of state variable controls.
- If a system is to have integral control, another state variable must be introduced. The integral control parameter is essentially the gain for the new state variable.

12.12.2 Important Equations

- Free response

$$\mathbf{y} = \boldsymbol{\Phi}(t)\mathbf{y}_0 \tag{12.27}$$

- Convolution integral for the forced response

$$\mathbf{y}(t) = \boldsymbol{\Phi}(t)\mathbf{y}_0 + \int_0^t \boldsymbol{\Phi}(t-\tau)\,\mathbf{B}\mathbf{u}(\tau)d\tau \tag{12.37}$$

- Feedback gain matrix, or Ackerman matrix, relating the control signal to the state variables

$$\mathbf{u} = -\mathbf{S}\mathbf{y} \tag{12.58}$$

- Control matrix for a system with integral and state variable control

$$\mathbf{S} = [K_1 \;\; K_2 \;\cdots\; K_n \;\; -K_I] \tag{12.66}$$

- Controllability matrix

$$\mathbf{V} = [\mathbf{B} \;\; \mathbf{AB} \;\; \mathbf{A}^2\mathbf{B} \;\cdots\; \mathbf{A}^{n-1}\mathbf{B}] \tag{12.69}$$

- Ackerman's matrix

$$\mathbf{S} = [0 \;\; 0 \;\cdots\; 0 \;\; 1]\mathbf{V}^{-1}\tau(\mathbf{A}) \tag{12.71}$$

Short Answer Problems

Problems SA12.1–SA12.14 refer to the following state-space model of a system:

$$\begin{bmatrix}\dot{y}_1\\ \dot{y}_2\end{bmatrix} = \begin{bmatrix}0 & 1\\ -2 & -1\end{bmatrix}\begin{bmatrix}y_1\\ y_2\end{bmatrix} + \begin{bmatrix}0\\ 1\end{bmatrix}[u(t)]$$

$$[x] = [1 \quad 0]\begin{bmatrix}y_1\\ y_2\end{bmatrix} + [0][u(t)]$$

SA12.1 Draw a block diagram model in the time domain for this system.

SA12.2 Draw a Simulink model for this system.

SA12.3 Calculate $|s\mathbf{I} - \mathbf{A}|$.

SA12.4 Determine $\mathcal{L}\{\mathbf{y}\}$ in terms of $\mathbf{y}_0$.

SA12.5 Determine the eigenvalues of **A**.

SA12.6 Determine the eigenvector of **A** corresponding to its lowest eigenvalue.

SA12.7 Determine the eigenvector of **A** corresponding to its highest eigenvalue.

SA12.8 Determine the matrix $\mathbf{P}(t)$.

SA12.9 If $\mathbf{y}_0 = \begin{bmatrix}1\\ 0\end{bmatrix}$, determine $\mathbf{y}(t)$.

SA12.10 Is this system controllable?

SA12.11 If the system is controllable and the desired poles are at $-2, -3.5$, determine the Ackerman matrix using analytical methods.

SA12.12 If the system is controllable and the desired poles are at $-2, -3.5$, determine the Ackerman matrix using MATLAB.

SA12.13 Draw the block diagram model in the time domain if the state variable control calculated in Problem SA12.11 is used.

SA12.14 Draw and run the Simulink model for the step response if the state variable control calculated in Problem SA12.11 is used.

Problems SA12.15–SA12.28 refer to a plant with the transfer function

$$G(s) = \frac{4}{s(s+2)(s+5)}$$

A unit step input is applied to the system at the location where the second state variable is measured.

SA12.15 Draw a block diagram model in the time domain for this system.

SA12.16 Draw a Simulink model for this system.

SA12.17 Calculate $|s\mathbf{I} - \mathbf{A}|$.

SA12.18 Determine $\mathcal{L}\{\mathbf{y}\}$ in terms of $\mathbf{y}_0$.

SA12.19 Determine the eigenvalues of **A**.

SA12.20 Determine the eigenvector of **A** corresponding to its lowest eigenvalue.

SA12.21 Determine the eigenvector of **A** corresponding to its highest eigenvalue.

SA12.22 Determine the matrix $\mathbf{P}(t)$.

SA12.23 If $\mathbf{y}_0 = \begin{bmatrix}0\\ 1\\ 0\end{bmatrix}$, determine $\mathbf{y}(t)$.

SA12.24 Is this system controllable?

SA12.25 If the system is controllable and the desired poles are at $-3, -6, -9$, determine the Ackerman matrix using analytical methods.

SA12.26 If the system is controllable and the desired poles are at $-3, -6, -9$, determine the Ackerman matrix using MATLAB.

SA12.27 Draw the block diagram model in the time domain if the state variable control calculated in Problem SA12.25 is used.

SA12.28 Draw and run the Simulink model for the step response if the state variable control calculated in Problem SA12.25 is used.

Problems SA12.29–SA12.42 refer to a plant with the transfer function

$$G(s) = \frac{1}{s(s-1)(s+5)}$$

A unit step input is applied to the system at the location where the second state variable is measured.

SA12.29 Draw a block diagram model in the time domain for this system.

SA12.30 Draw a Simulink model for this system.

SA12.31 Calculate $|s\mathbf{I} - \mathbf{A}|$.

SA12.32 Determine $\mathcal{L}\{\mathbf{y}\}$ in terms of $\mathbf{y}_0$.

SA12.33 Determine the eigenvalues of **A**.

SA12.34 Determine the eigenvector of **A** corresponding to its lowest eigenvalue.

SA12.35 Determine the eigenvector of **A** corresponding to its highest eigenvalue.

SA12.36 Determine the matrix $\mathbf{P}(t)$.

SA12.37 If $\mathbf{y}_0 = \begin{bmatrix}0\\ 1\\ 0\end{bmatrix}$, determine $\mathbf{y}(t)$.

SA12.38 Is this system controllable?

SA12.39 If the system is controllable and the desired poles are at $-3, -6, -9$, determine the Ackerman matrix using analytical methods.

SA12.40 If the system is controllable and the desired poles are at $-3, -6, -9$, determine the Ackerman matrix using MATLAB.

SA12.41 Draw the block diagram model in the time domain if the state variable control calculated in Problem SA12.39 is used.

SA12.42 Draw and run the Simulink model for the step response if the state variable control calculated in Problem SA12.39 is used.

Problems SA12.43–SA12.49 refer to the three-tank liquid-level system shown in Figure SA12.43. The input to the system is the perturbation in flow rate into the first tank. The output from the system is to be the liquid level in tank 2.

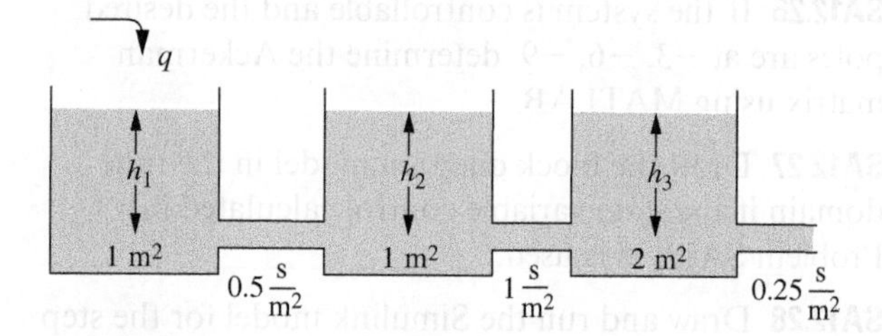

Figure SA12.43

SA12.43 Determine the state-space model for the system.

SA12.44 Draw a Simulink model for the system in the state space.

SA12.45 Use the state-space model to determine the step response of the system via the Laplace transform method.

SA12.46 Use the state-space model to determine the step response of the system using the exponential matrix.

SA12.47 Determine if the system is controllable.

SA12.48 It is desired to use state variable control to place the poles at $-3, -6, -10$. Determine the Ackerman matrix using analytical methods.

SA12.49 Develop a Simulink model for the system with the state variable control determined in Problem SA12.48.

Problems SA12.50 and SA12.51 refer to the state-space formulation for the simplified model of the vehicle suspension system shown in Figure SA12.50 where $z(t)$, which describes the road contour, is the input and $x(t)$ is the output.

SA12.50 If $z(t)$ is a unit step input, determine the response of the system using the exponential matrix.

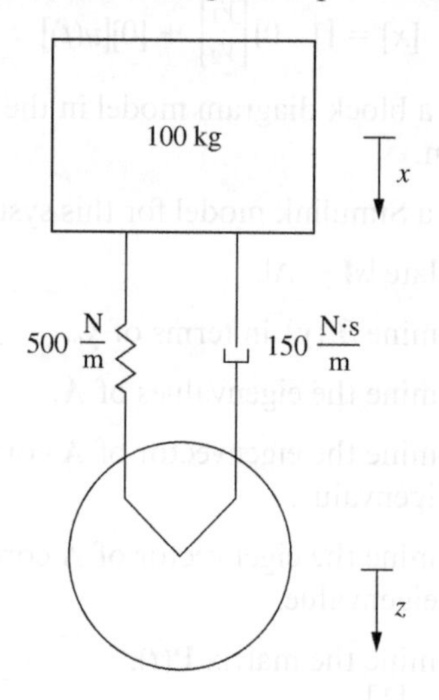

Figure SA12.50

SA12.51 Determine the Ackerman matrix such that with state variable control the poles are at $-5 \pm 5j, -10$.

Problems SA12.52–SA12.55 refer to the system whose block diagram using state variable control is shown in Figure SA12.52.

SA12.52 Determine the differential equations in the state space using state variable control.

SA12.53 The poles of this system are to be at $-2 \pm 3j$, -5. What are the required values of K_1, K_2, and K_3?

SA12.54 The system is to have integral control to better respond to a step input. Draw a block diagram in the state-space if integral control is added to the system.

SA12.55 If the poles of this system are to be at $-2 \pm 3j$, $-3, -5$, what are the required values of K_1, K_2, K_3, and K_I?

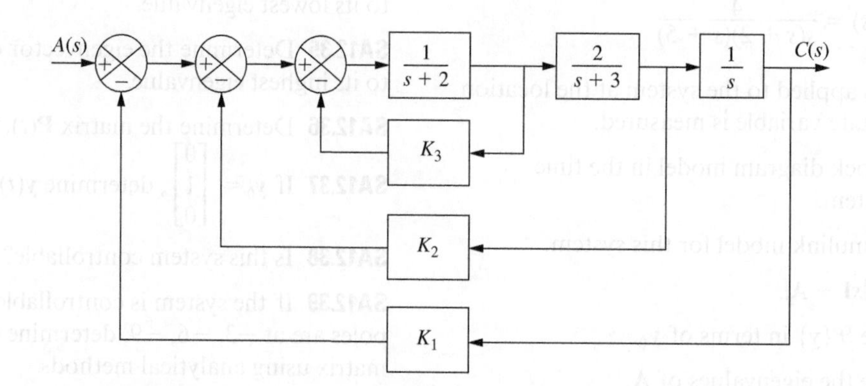

Figure SA12.52

Problems

12.1 Consider a system governed by the differential equations

$$0.2\frac{dy_1}{dt} + 2y_1 - y_2 = F_1(t)$$

$$0.25\frac{dy_2}{dt} - 2y_1 + 3y_2 = F_2(t)$$

(a) Develop a state-space formulation for the system. (b) Determine the state transition matrix. (c) If $F_1(t) = F_2(t) = 0$, determine the response of the system if $y_1(0) = 0$ and $y_2(0) = 0.5$. (d) Use state-space methods to determine the response of the system if $F_1(t) = 0.2e^{-0.2t}$ and $F_2(t) = 0.5e^{-0.2t}$.

12.2 Consider a system governed by the differential equation

$$\dddot{x} + 3\ddot{x} + 16\dot{x} + 25x = F(t)$$

(a) Develop a state-space formulation for the system. (b) Determine the state transition matrix. (c) If $F_1(t) = F_2(t) = 0$, determine the response of the system if $x(0) = 1$, $\dot{x}(0) = 0$, and $\ddot{x}(0) = 0$. (d) Use state-space methods to determine the response of the system if $F(t) = 0.5 \sin 20t$.

12.3 Use state-space methods to determine the liquid-level perturbations in the three-tank system of Figure P12.3 when the flow rate into the first tank is suddenly increased by 0.2 m³/s.

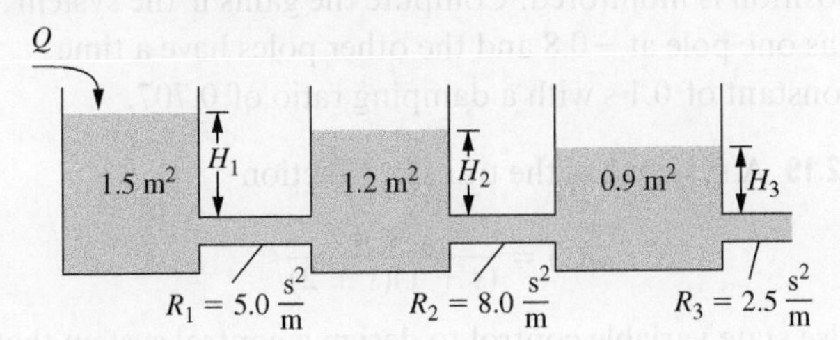

Figure P12.3

12.4 Use state-space methods to determine the response of a one-degree-of-freedom system of mass 10 kg, stiffness 4×10^5 N/m, and damping coefficient 200 N·s/m when subject to a force $F(t) = 100 \sin (200t)$ N.

12.5 Write a MATLAB M-file that determines the response of the nonlinear pendulum equation $\ddot{\theta} + \omega_n^2 \sin \theta = 0$ for $\omega_n = 10$ r/s with initial conditions of $\theta(0) = \theta_0$ and $\dot{\theta}(0) = 0$. Run the program for several values of θ_0 and discuss the dependence of the period on this initial condition.

12.6 Determine a state-space formulation for the feedback control system of Figure P12.6. Develop a Simulink model for the system. Run the model with a step input.

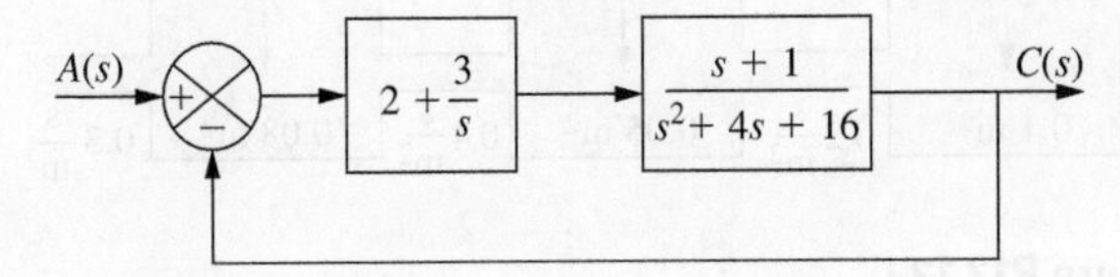

Figure P12.6

12.7 Determine a state-space formulation for the feedback control system of Figure P12.7.

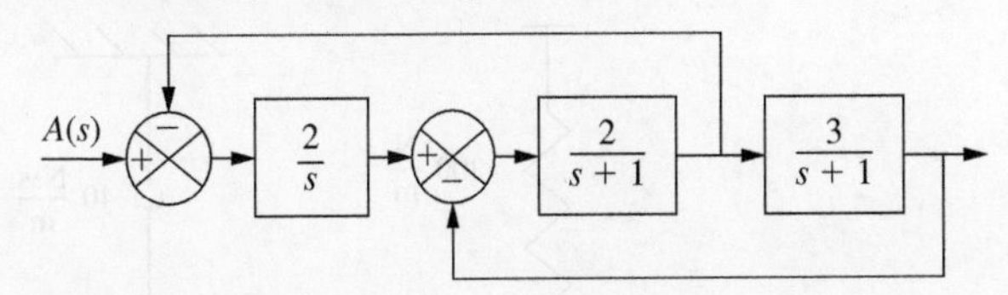

Figure P12.7

12.8 Use state-space methods to determine the response of the heat exchanger in Example 7.35.

12.9 Consider a plant with the transfer function

$$G(s) = \frac{3}{s^3 + 6s + 7}$$

Use state variable control to design a control system whose closed-loop poles are $s = -3 \pm 3j, -12$.

12.10 Consider a plant with the state-space formulation

$$\begin{bmatrix} \dot{y}_1 \\ \dot{y}_2 \\ \dot{y}_3 \end{bmatrix} = \begin{bmatrix} 0 & 1 & 0 \\ 0 & 0 & 1 \\ -1 & -4 & -5 \end{bmatrix} \begin{bmatrix} y_1 \\ y_2 \\ y_3 \end{bmatrix} + \begin{bmatrix} 0 \\ 0 \\ 1 \end{bmatrix} [u(t)]$$

$$[x] = [1 \;\; 0 \;\; 0] \begin{bmatrix} y_1 \\ y_2 \\ y_3 \end{bmatrix} + [0]\,[u(t)]$$

Use state variable control to design a control system whose closed-loop poles are $s = -2 \pm 2j, -8$.

12.11 Design a control system using state variable control for the two-tank liquid-level problem of Figure P12.11 such that it has poles at $s = -2, -10$.

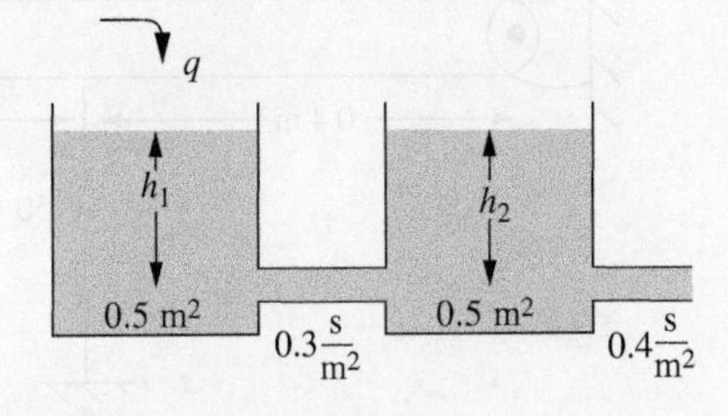

Figure P12.11

12.12 Design a control system using state variable control for the three-tank liquid-level problem of Figure P12.12 such that it has poles at $s = -2, -10, -15$.

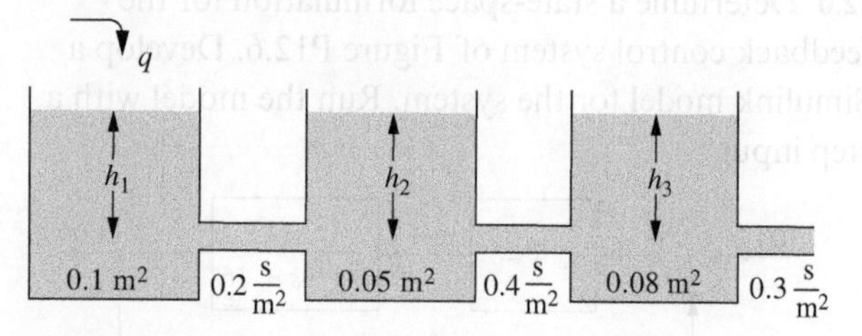

Figure P12.12

12.13 Design a control system using state variable control for the mechanical system of Figure P12.13 such that it has poles at $s = -2 \pm 2j$.

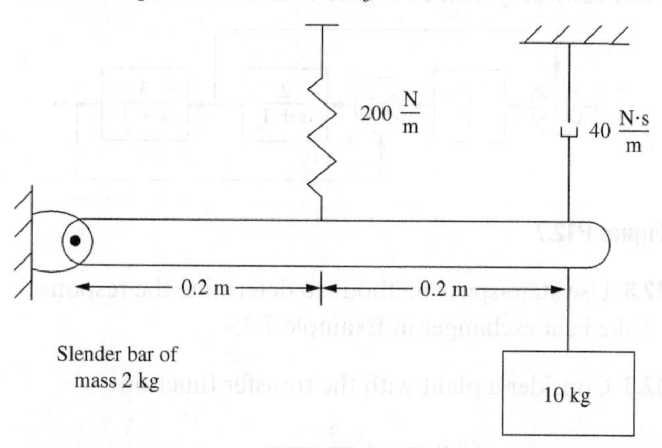

Figure P12.13

12.14 Design a control system using state variable control for the mechanical system of Figure P12.14 such that it has poles at $s = -2 \pm 2j, -6 \pm 6j$.

12.15 Design a system using state variable control for the mechanical system of Figure P12.15 such that it has poles at $s = -2 \pm 2j, -5$.

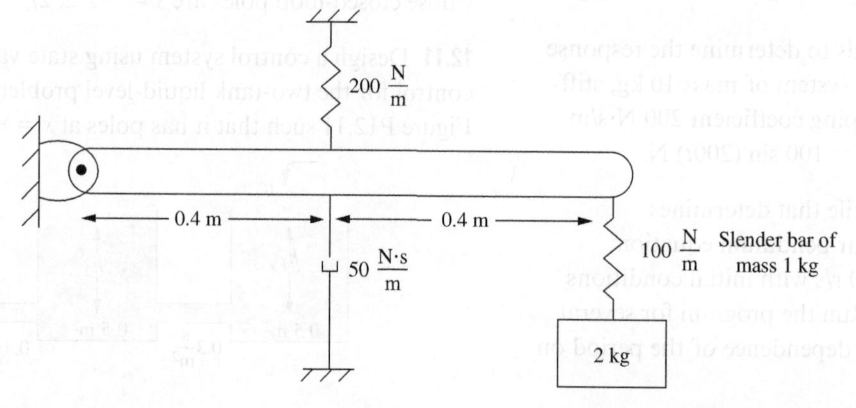

Figure P12.14

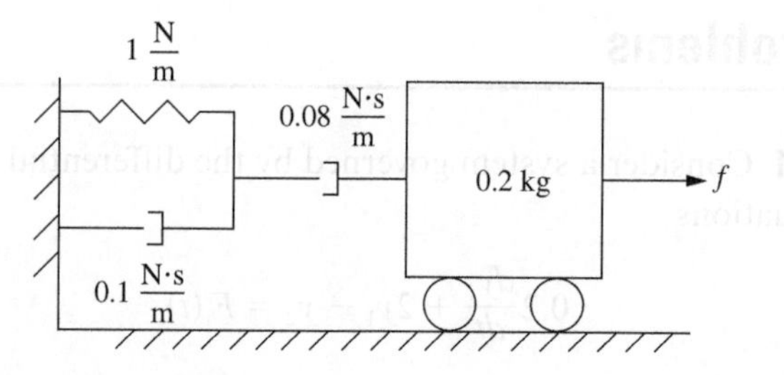

Figure P12.15

12.16 Design a system using state variable control for the CSTR system of Figure P12.16 such that it has poles at $s = -2, -3$.

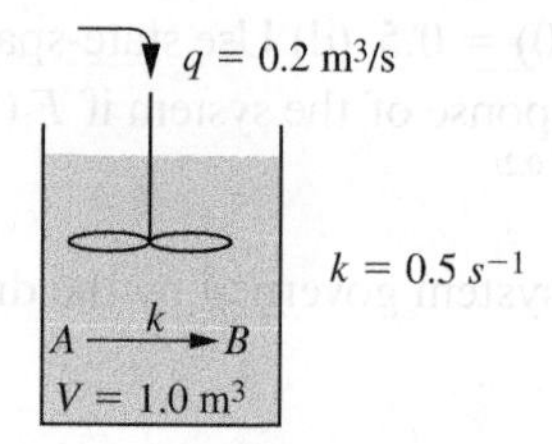

Figure P12.16

12.17 A field-controlled dc servomotor has parameters of $R_f = 1.2\ \Omega$, $L_f = 0.04$ H, $J = 4 \times 10^{-5}$ kg·m², $c_t = 0.1$ N·m·s, and $K_f = 0.02 \frac{\text{N·m}}{\text{A}}$. The servomotor is in a control loop with state variable feedback with proportional control. Compute the gains such that the system has poles with a time constant of 0.1 s with a damping ratio of 0.707.

12.18 The field-controlled dc servomotor of Problem 12.17 is placed in a feedback loop such that its position is monitored. Compute the gains if the system has one pole at -0.8 and the other poles have a time constant of 0.1 s with a damping ratio of 0.707.

12.19 A system has the transfer function

$$G(s) = \frac{s+4}{(s+1)(s+2)}$$

Use state variable control to design a control system that has poles at $-3 \pm 3.2j$.

12.20 A system has the transfer function

$$G(s) = \frac{3}{(s+1)(s+2)}$$

Use state variable control to design a control system that has poles at $-3 \pm 3.2j$.

12.21 An inverted pendulum on a cart is shown in Figure P12.21. Use as state variables $x_1 = x$, $x_2 = \dot{x}$, $x_3 = \theta$, and $x_4 = \dot{\theta}$. The motion of the pendulum is limited, so assume small θ. Design a state variable feedback system to stabilize this initially unstable system such that the closed-loop poles are at $s = -5 \pm 5j, -15, -20$. Determine the free response using the exponential matrix assuming $x_1(0) = 0$, $x_2(0) = 1\,\frac{\text{m}}{\text{s}}$, $x_3(0) = 0$, and $x_4(0) = 0$.

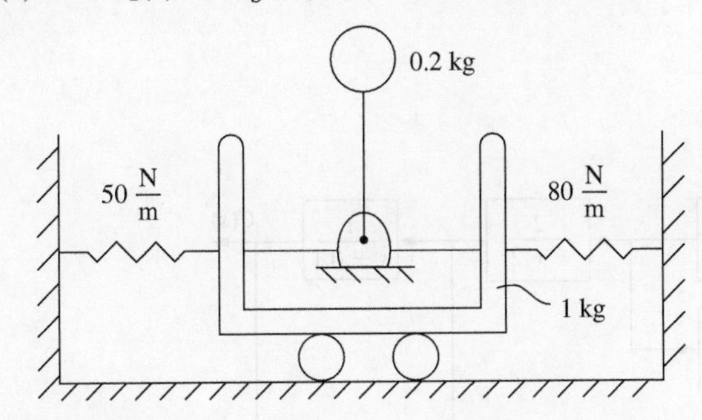

Figure P12.21

12.22 The block diagram for the concentrations in the CSTR system of Example 6.16 using state variable control is shown in Figure P12.22. Determine the state gains such that the poles are at $-2, -3$. The parameters are $V = 0.05\ \text{m}^3$, $q = 0.4\frac{\text{m}^2}{\text{s}}$, and $k = 2.3\,\frac{\text{mole}}{\text{s}}$.

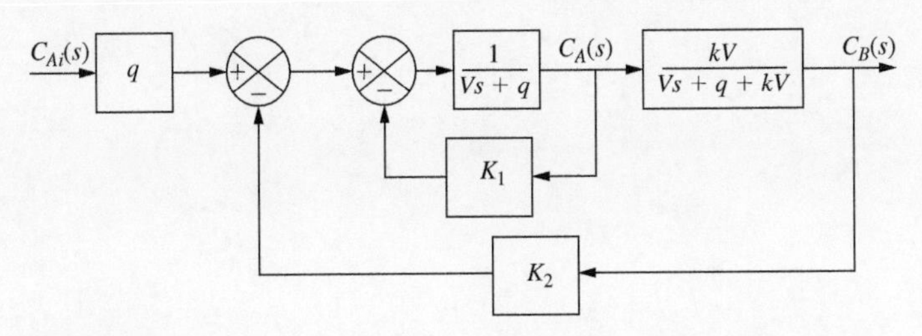

Figure P12.22

12.23 The block diagram for the concentrations in the CSTR system of Example 6.16 using state variable control with integral control is shown in Figure P12.23. Determine the gains such that the dominant roots are $-2 \pm 2j$ and the third pole is at -3. The parameters are $V = 0.05\ \text{m}^3$, $q = 0.4\frac{\text{m}^3}{\text{s}}$, and $k = 2.3\,\frac{\text{mole}}{\text{s}}$.

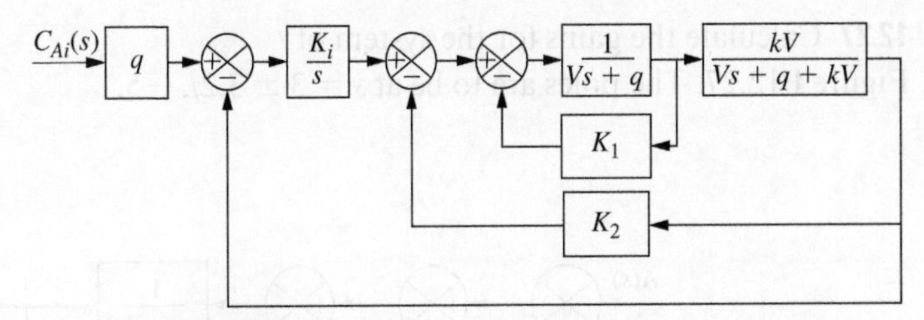

Figure P12.23

12.24 Use MATLAB to calculate the Ackerman matrix when state variable control is used on the system of Example 6.32 with parameters $V = 0.05\ \text{m}^3$, $q = 0.4\frac{\text{m}^2}{\text{s}}$, $k_1 = 2.3\,\frac{\text{mole}}{\text{s}}$, and $k_2 = 1.2\,\frac{\text{mole}}{\text{s}}$.

12.25 Calculate the gains for the system of Figure P12.25. The poles are to be at $s = 2 \pm 2j$.

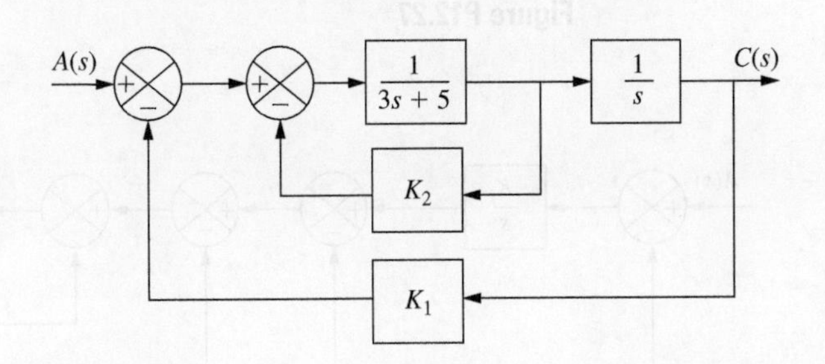

Figure P12.25

12.26 Calculate the gains for the system of Figure P12.26, the same as in Figure P12.25 but with integral control added. The poles are to be at $s = 2 \pm 2j, -4$.

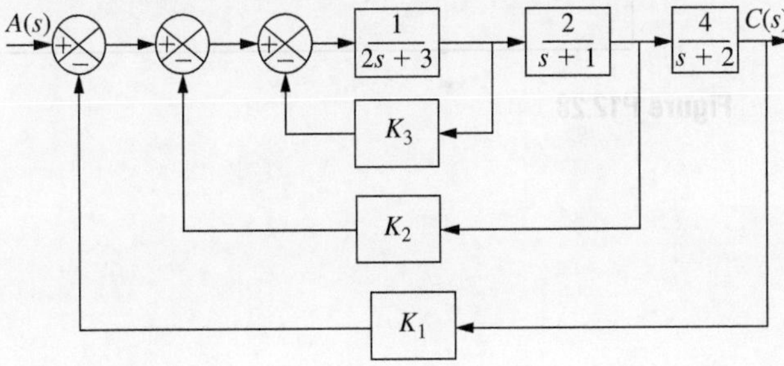

Figure P12.26

12.27 Calculate the gains for the system of Figure P12.27. The poles are to be at $s = 3 \pm 3.2j, -5$.

12.28 Calculate the gains for the system of Figure P12.28, the same as in Figure P12.27 but with integral control added. The poles are to be at $s = 3 \pm 3.2j, -5, -7$.

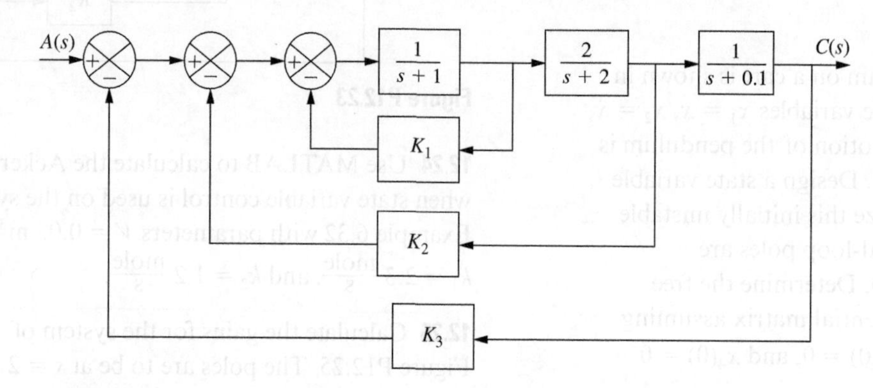

Figure P12.27

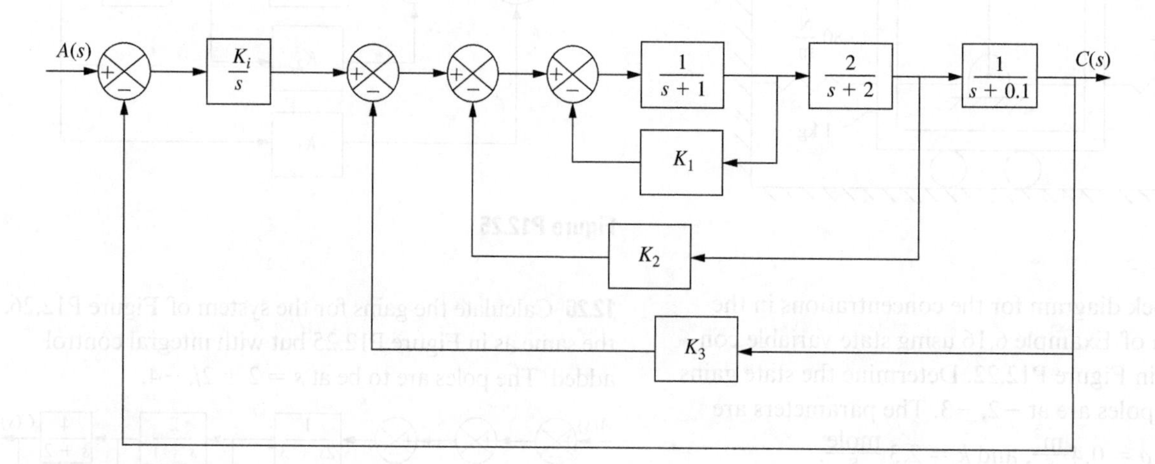

Figure P12.28

Appendix A

Complex Algebra

A **complex number** is of the form

$$c = a + jb \tag{A.1}$$

where

$$j = \sqrt{-1} \tag{A.2}$$

and a and b are real numbers. The number a is called the **real part** of c, $a = \text{Re}[c]$, and b is called the **imaginary part** of c, $b = \text{Im}[c]$. A number whose imaginary part is zero is a real number. A number whose real part is zero and whose imaginary part is not zero is an imaginary number.

Powers of j are calculated using Equation (A.2):

$$j^2 = (\sqrt{-1})(\sqrt{-1}) = -1 \tag{A.3}$$

$$j^3 = j^2(\sqrt{-1}) = -j \tag{A.4}$$

$$j^4 = jj^3 = j(-j) = -j^2 = 1 \tag{A.5}$$

The sum of two complex numbers is also a complex number. The real part of the sum is the sum of the real parts of the numbers. The imaginary part of the sum is the sum of the imaginary parts:

$$(a_1 + jb_1) + (a_2 + jb_2) = (a_1 + a_2) + j(b_1 + b_2) \tag{A.6}$$

The distributive property is used to obtain the product of two complex numbers:

$$\begin{aligned}(a_1 + jb_1)(a_2 + jb_2) &= a_1a_2 + ja_1b_2 + jb_1a_2 + j^2b_1b_2 \\ &= (a_1a_2 - b_1b_2) + j(a_1b_2 + b_1a_2)\end{aligned} \tag{A.7}$$

The **complex conjugate** of a number is a complex number such that the real parts of both numbers are the same and the imaginary parts are negatives of one another. The complex conjugate of c is denoted by $\bar{c}$ and is given by

$$\bar{c} = a - jb \tag{A.8}$$

The use of Equation (A.7) shows that the product of a number and its complex conjugate is a real number given by

$$(c)(\bar{c}) = a^2 + b^2 \tag{A.9}$$

The quotient of two complex numbers is a complex number and can be written in the form of Equation (A.1) by multiplying both the numerator and denominator of the quotient by the complex conjugate of the denominator.

Example A.1

Write the complex number

$$c = \frac{3 + j2}{3 - j4} \tag{a}$$

in the form $c = a + jb$.

Solution

Multiplying the numerator and denominator of Equation (a) by $3 + j4$ leads to

$$c = \frac{(3 + j2)(3 + j4)}{(3 - j4)(3 + 4j)} \tag{b}$$

Equation (A.7) is used to evaluate the numerator of Equation (b), while Equation (A.9) is used for the denominator, leading to

$$c = \frac{[(3)(3) - (2)(4)] + j[(3)(4) + (2)(3)]}{(3)^2 + (4)^2}$$
$$= \frac{1}{25} + j\frac{18}{25} \tag{c}$$

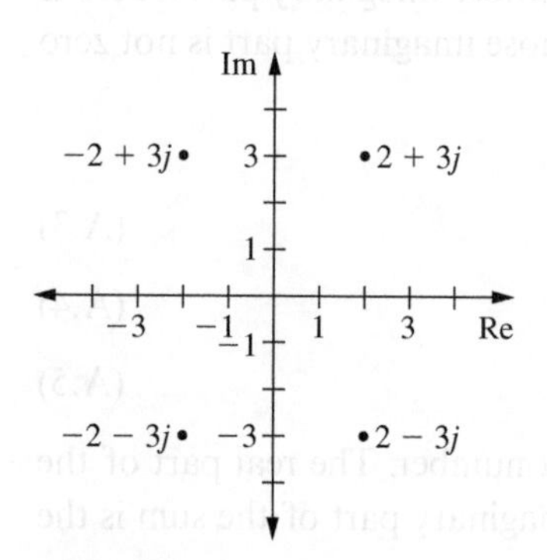

Figure A.1

Illustration of complex numbers as points in the complex plane.

A complex number can be represented geometrically as a point in the complex plane, as illustrated in Figure A.1. The horizontal axis for the complex plane is the real axis, and the vertical axis is the imaginary axis.

The polar form of a complex number, illustrated in Figure A.2, is a representation in terms of a magnitude and a phase angle. The **magnitude** is the distance of the point in the complex plane from the origin,

$$|c| = \sqrt{a^2 + b^2} \tag{A.10}$$

From Equations (A.9) and (A.10), it is seen that

$$|c| = (c\bar{c})^{1/2} \tag{A.11}$$

The phase angle θ is the angle, measured in radians, made by a line drawn from the origin to the point with the real axis. The phase angle is measured positive counterclockwise. The polar form of a complex number is

$$c = |c|e^{j\theta} \tag{A.12}$$

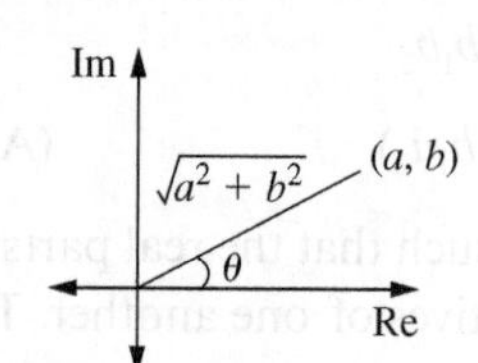

Figure A.2

The polar representation of a complex number.

Euler's identity states that

$$e^{j\theta} = \cos(\theta) + j\sin(\theta) \tag{A.13}$$

Use of Equation (A.13) in Equation (A.12) gives

$$c = |c|\cos(\theta) + j|c|\sin(\theta) \tag{A.14}$$

Noting from Equation (A.14) that Re[c]= $|c|\cos\theta$ and Im[c]=$|c|\sin\theta$ and dividing Im[c] by Re[c] leads to

$$\theta = \tan^{-1}\left\{\frac{\text{Im}[c]}{\text{Re}[c]}\right\}$$

When applying Equation (A.15), it is important to remember that $0 \le \theta < 2\pi$. Thus the value of the inverse tangent must be assigned depending on the quadrant in which c exists. The polar form of c is obtained from its real part and imaginary part by applying Equation (A.10) and Equation (A.14) to Equation (A.12).

Example A.2

Convert the complex number $c = -3 + j4$ into polar form.

Solution

The magnitude of c, obtained using Equation (A.10), is

$$|c| = \sqrt{(-3)^2 + (4)^2} = 5 \tag{a}$$

The phase angle for c is determined using Equation (A.15):

$$\theta = \tan^{-1}\left(\frac{4}{-3}\right) \tag{b}$$

The inverse tangent of Equation (b) must be evaluated corresponding to an angle in the second quadrant as the real part is negative (–3) and the imaginary part is positive (4). The appropriate value is

$$\theta = 2.21 \text{ r} \tag{c}$$

Using the numerical values from Equations (a) and (c) in Equation (A.12) leads to

$$c = 5e^{j2.21} \tag{d}$$

Appendix B
Matrix Algebra

B.1 Definitions

A **matrix** is a collection of elements arranged in rows and columns. A matrix of m rows and n columns is said to be of size $m \times n$. The number of rows equals the number of columns in a square matrix. Throughout this text, the name of a matrix is written using bold type and is identified by its elements arranged appropriately in rows and columns enclosed by square brackets. An example of a 3×3 matrix is

$$\mathbf{A} = \begin{bmatrix} 1 & 2 & -1 \\ 3 & 2 & 3 \\ 0 & 2 & -2 \end{bmatrix} \tag{B.1}$$

An element of a matrix is identified according to where it resides in the matrix. The element in the ith row and j^{th} column of matrix **A** is written as $a_{i,j}$. For example, for the matrix of Equation (B.1), $a_{3,2} = 2$.

Useful definitions regarding matrices are as follows:

- A **column vector** is a matrix with only one column, whereas a **row vector** is a matrix with only one row.
- The **diagonal elements** of a square matrix **A** are those elements $a_{i,j}$ such that $i = j$.
- The **off-diagonal elements** of a square matrix are those elements $a_{i,j}$ such that $i \neq j$.
- A **diagonal** matrix is a square matrix whose off-diagonal terms are all zero.
- The $n \times n$ **identity matrix I** is a diagonal matrix such that $a_{i,i} = 1$ for $i = 1, 2, \ldots, n$.
- The **transpose** of a matrix **A**, written $\mathbf{A}^T$, is obtained by interchanging the rows and columns of **A**. Thus if $\mathbf{B} = \mathbf{A}^T$, then $b_{i,j} = a_{j,i}$. The transpose of a column vector is a row vector and vice versa.
- A square matrix is **symmetric** if $\mathbf{A}^T = \mathbf{A}$.

B.2 Matrix Arithmetic

The **sum** of two matrices **A** and **B**, both of size $m \times n$, is the $m \times n$ matrix $\mathbf{C} = \mathbf{A} + \mathbf{B}$ whose elements are calculated as

$$c_{i,j} = a_{i,j} + b_{i,j} \qquad i = 1, 2, \ldots, m \qquad j = 1, 2, \ldots, n \tag{B.2}$$

Matrix multiplication may be defined for two matrices, written as $\mathbf{C} = \mathbf{AB}$, if the number of columns of $\mathbf{A}$ is equal to the number of rows of $\mathbf{B}$. If $\mathbf{A}$ is of size $m \times n$ and $\mathbf{B}$ is of size $n \times p$, then $\mathbf{C} = \mathbf{AB}$ is a $m \times p$ matrix such that

$$c_{i,j} = \sum_{k=1}^{n} a_{i,k} b_{k,j} \qquad i = 1, 2, \ldots, m \qquad j = 1, 2, \ldots, p \tag{B.3}$$

Matrix multiplication is not commutative: in general $\mathbf{AB} \neq \mathbf{BA}$.

B.3 Determinants

A determinant of a square matrix is a scalar quantity with great significance and is often used in computations involving matrices. The formal definition of the determinant is beyond the scope of this text, but it is a linear combination of permutations of elements of the matrix. Each permutation in the evaluation of the determinant of an $n \times n$ matrix is a product of n terms, with one element from each column and one element from each row. One permutation is a product of the diagonal elements.

The determinant of a 2×2 matrix is simply the product of the diagonal elements minus the product of the off-diagonal elements. If

$$\mathbf{A} = \begin{bmatrix} a_{1,1} & a_{1,2} \\ a_{2,1} & a_{2,2} \end{bmatrix} \qquad \text{then} \qquad \det(\mathbf{A}) = |\mathbf{A}| = a_{1,1}a_{2,2} - a_{1,2}a_{2,1} \tag{B.4}$$

The determinant of a larger matrix can be computed using a row or column expansion. For example, a row expansion of the determinant of an $n \times n$ matrix $\mathbf{A}$ using the ith row is

$$\det(\mathbf{A}) = \sum_{j=1}^{n} (-1)^{i+j} a_{i,j} \det(\mathrm{M}_{i,j}) \tag{B.5}$$

where $\mathbf{M}_{i,j}$ is the $(n-1) \times (n-1)$ matrix obtained from $\mathbf{A}$ by deleting its ith row and j^{th} column. Thus the determinant of an $n \times n$ matrix can be written as a sum of n determinants of $(n-1) \times (n-1)$ matrices. Row or column expansions can be employed to evaluate the determinant of each of the $(n-1) \times (n-1)$ matrices as a sum of $n-1$ determinants of $(n-2) \times (n-2)$ matrices. This process can be continued until 2×2 matrices are obtained whose determinants are evaluated using Equation (B.4).

Example B.1

Evaluate the determinant of the 3×3 matrix $\mathbf{A}$ of Equation (B.1) by row expansion using the second row.

Solution

Evaluation of the determinant of $\mathbf{A}$ using Equation (B.5) with $i = 2$ leads to

$$\det(\mathbf{A}) = -3\begin{vmatrix} 2 & -1 \\ 2 & -2 \end{vmatrix} + 2\begin{vmatrix} 1 & -1 \\ 0 & -2 \end{vmatrix} - 3\begin{vmatrix} 1 & 2 \\ 0 & 2 \end{vmatrix} \tag{a}$$

The 2×2 determinants are evaluated using Equation (B.4), leading to

$$\det(\mathbf{A}) = -3[(2)(-2) - (-1)(2)] + 2[(1)(-2) - (-1)(0)] - 3[(1)(2) - (2)(0)] = -4 \tag{b}$$

B.4 Matrix Inverse

The inverse of a square matrix **A**, written $\mathbf{A}^{-1}$, is the matrix such that

$$\mathbf{A}\mathbf{A}^{-1} = \mathbf{A}^{-1}\mathbf{A} = \mathbf{I} \tag{B.6}$$

$\mathbf{A}^{-1}$ exists if det (**A**) ≠ 0, Such a matrix is said to be nonsingular. If $\mathbf{B} = \mathbf{A}^{-1}$, then

$$b_{i,j} = (-1)^{i+j}\frac{\det(\mathbf{M}_{j,i})}{\det(\mathbf{A})} \tag{B.7}$$

The inverse of a diagonal matrix is a diagonal matrix with the reciprocals of the diagonal elements along the diagonal.

Example B.2

Determine $\mathbf{A}^{-1}$ for the matrix defined by Equation (B.1).

Solution

From Example B.1, det (**A**) = −4. The application of Equation (B.7) leads to

$$\mathbf{A}^{-1} = -\frac{1}{4}\begin{bmatrix} \begin{vmatrix}2 & 3\\2 & -2\end{vmatrix} & -\begin{vmatrix}2 & -1\\2 & -2\end{vmatrix} & \begin{vmatrix}2 & -1\\2 & 3\end{vmatrix} \\ \begin{vmatrix}3 & 3\\0 & -2\end{vmatrix} & \begin{vmatrix}1 & -1\\0 & -2\end{vmatrix} & \begin{vmatrix}1 & -1\\3 & 3\end{vmatrix} \\ \begin{vmatrix}3 & 2\\0 & 2\end{vmatrix} & -\begin{vmatrix}1 & 2\\0 & 2\end{vmatrix} & \begin{vmatrix}1 & 2\\3 & 2\end{vmatrix} \end{bmatrix}$$

$$= -\frac{1}{4}\begin{bmatrix} -10 & 2 & 8 \\ 6 & -2 & -6 \\ 6 & -2 & -4 \end{bmatrix} = \begin{bmatrix} \frac{5}{2} & -\frac{1}{2} & -2 \\ -\frac{3}{2} & \frac{1}{2} & \frac{3}{2} \\ -\frac{3}{2} & \frac{1}{2} & 1 \end{bmatrix} \tag{a}$$

B.5 Systems of Equations

A system of n simultaneous equations in terms of n unknowns, $x_1, x_2, \ldots, x_n$, is of the form

$$\begin{aligned} a_{1,1}x_1 + a_{1,2}x_2 + \cdots + a_{1,n}x_n &= b_1 \\ a_{2,1}x_1 + a_{2,2}x_2 + \cdots + a_{1,n}x_n &= b_2 \\ &\vdots \\ a_{n,1}x_1 + a_{n,2}x_2 + \cdots + a_{n,n}x_n &= b_n \end{aligned} \tag{B.8}$$

Equation (B.8) can be summarized in matrix form as

$$\mathbf{A}\mathbf{x} = \mathbf{b} \tag{B.9}$$

where

$$\mathbf{A} = \begin{bmatrix} a_{1,1} & a_{2,1} & \cdots & a_{n,1} \\ a_{2,1} & a_{2,2} & \cdots & a_{2,n} \\ \vdots & \vdots & \cdots & \vdots \\ a_{n,1} & a_{n,2} & \cdots & a_{n,n} \end{bmatrix} \quad \mathbf{x} = \begin{bmatrix} x_1 \\ x_2 \\ \vdots \\ x_n \end{bmatrix} \quad \mathbf{b} = \begin{bmatrix} b_1 \\ b_2 \\ \vdots \\ b_n \end{bmatrix} \tag{B.10}$$

A solution to Equation (B.8) is obtained by premultiplying both sides of Equation (B.9) by $\mathbf{A}^{-1}$, which leads to

$$\mathbf{x} = \mathbf{A}^{-1}\mathbf{b} \tag{B.11}$$

Equation (B.11) is the solution of Equation (B.8) if $\mathbf{A}^{-1}$ exists, that is, if A is nonsingular. If **A** is a singular matrix [det (**A**) = 0], then a solution of Equation (B.8) exists only for certain forms of **b**. However, when a solution does exist, it is not unique.

Example B.3

Use the matrix inverse method to determine the solution of the system of equations

$$\begin{aligned} x_1 + 2x_2 - x_3 &= 4 \\ 3x_1 + 2x_2 + 3x_3 &= -2 \\ 2x_2 - 2x_3 &= 5 \end{aligned} \tag{a}$$

Solution

The system of equations represented by Equation (a) has a matrix formulation of the form of Equation (B.9) with the matrix **A** of Equation (B.1) and $\mathbf{b} = [4 \quad -2 \quad 5]^T$. The inverse of **A** is given in Equation (a) of Example B.2. The application of Equation (B.11) leads to

$$\begin{bmatrix} x_1 \\ x_2 \\ x_3 \end{bmatrix} = \begin{bmatrix} \frac{5}{2} & -\frac{1}{2} & -2 \\ -\frac{3}{2} & \frac{1}{2} & \frac{3}{2} \\ -\frac{3}{2} & \frac{1}{2} & 1 \end{bmatrix} \begin{bmatrix} 4 \\ -2 \\ 5 \end{bmatrix} = \begin{bmatrix} 1 \\ \frac{1}{2} \\ -2 \end{bmatrix} \tag{b}$$

B.6 Cramer's Rule

Cramer's rule provides an alternate method for the solution of a set of simultaneous equations formulated in the matrix form of Equation (B.9). Cramer's rule is used to determine each element of the solution vector individually, rather than simultaneously.

Consider a set of simultaneous equations formulated as in Equation (B.9). Let V_k be the matrix formed by replacing the kth column of **A** with the vector **b**. That is,

$$(V_k)_{i,j} = \begin{cases} a_{i,j} & j \neq k \\ b_i & j = k \end{cases} \tag{B.12}$$

Cramer's rule is used to determine x_k as

$$x_k = \frac{\det(\mathbf{V}_k)}{\det(\mathbf{A})} \qquad k = 1, 2, \ldots, n \tag{B.13}$$

Example B.4

Use Cramer's rule to determine x_1, an element of the solution vector of Example B.3.

Solution

Recalling that det (**A**) = −4, Cramer's rule, Equation (B.13), is applied, giving

$$x_1 = \frac{\det(\mathbf{V}_1)}{\det(\mathbf{A})} = \frac{\begin{vmatrix} 4 & 2 & -1 \\ -2 & 2 & 3 \\ 5 & 2 & -2 \end{vmatrix}}{-4} \tag{a}$$

The determinant in the numerator of Equation (a) is evaluated using a first-row expansion and Equations (B.3) and (B.4):

$$\begin{aligned} x_1 &= -\frac{1}{4}\left[4\begin{vmatrix} 2 & 3 \\ 2 & -2 \end{vmatrix} - 2\begin{vmatrix} -2 & 3 \\ 5 & -2 \end{vmatrix} + (-1)\begin{vmatrix} -2 & 2 \\ 5 & 2 \end{vmatrix}\right] \\ &= -\frac{1}{4}[4(-10) - 2(-11) - (-14)] = 1 \end{aligned} \tag{b}$$

B.7 Eigenvalues and Eigenvectors

Consider the system of n equations represented in matrix form as

$$\mathbf{Ax} = \lambda\mathbf{x} \tag{B.14}$$

where λ is a parameter. Equation (B.14) can be rewritten as

$$(\mathbf{A} - \lambda\mathbf{I}) = \mathbf{0} \tag{B.15}$$

The application of Cramer's rule to determine any element of the solution vector of Equation (B.15) leads to

$$x_k = \frac{\det(\mathbf{V}_k)}{\det(\mathbf{A} - \lambda\mathbf{I})} \qquad k = 1, 2, \ldots, n \tag{B.16}$$

Since $\mathbf{V}_k$ is defined as in Equation (B.12), its kth column is the vector on the right-hand side of Equation (B.15), the zero vector. However, it has been noted that the determinant of a matrix with a column of zeros is itself zero. Thus Equation (B.16) becomes

$$x_k = \frac{0}{\det(\mathbf{A} - \lambda\mathbf{I})} \qquad k = 1, 2, \ldots, n \tag{B.17}$$

It is clear from Equation (B.14) that $\mathbf{x} = \mathbf{0}$, the **trivial solution**, is a solution of the set of equations. Equation (B.17) shows that the trivial solution is the only solution unless

$$\det(\mathbf{A} - \lambda\mathbf{I}) = 0 \tag{B.18}$$

Equation (B.18) is satisfied only for certain values of λ, called the eigenvalues of A.

The **eigenvalues** of a square matrix **A** are the values of the parameter λ such that $\mathbf{Ax} = \lambda\mathbf{x}$ has nontrivial solutions. The **eigenvectors** of **A** are the corresponding nontrivial solutions. An eigenvector **y** corresponding to an eigenvalue λ is not unique because any nonzero multiple of the eigenvector $c\mathbf{y}$ with $c \neq 0$ is also an eigenvector corresponding to the eigenvalue λ, $\mathrm{A}(c\mathbf{y}) = \lambda(c\mathbf{y})$.

Equation (B.18) provides a method for calculating the eigenvalues of **A**. The elements of the matrix $\mathbf{A} - \lambda\mathbf{I}$ are the same as the elements of **A** except for the diagonal elements, which are $a_{i,i} - \lambda$. Recalling that one permutation in the evaluation of the determinant is the product of the diagonal elements, it becomes clear that the evaluation of det $(\mathbf{A} - \lambda\mathbf{I})$ is an nth-order polynomial in λ. Since the eigenvalues are determined using Equation (B.18), they are thus the roots of the polynomial. Hence an $n \times n$ matrix has n eigenvalues $\lambda_1, \lambda_2, \ldots, \lambda_n$. If all elements of **A** are real, complex eigenvalues occur in complex conjugate pairs. The eigenvectors are then calculated using Equation (B.15).

Example B.5

Determine the eigenvalues and eigenvectors of the matrix **A** defined in Equation (B.1).

Solution

The eigenvalues of **A** are determined through the application of Equation (B.18), leading to

$$\begin{vmatrix} 1-\lambda & 2 & -1 \\ 3 & 2-\lambda & 3 \\ 0 & 2 & -2-\lambda \end{vmatrix} = 0 \tag{a}$$

The evaluation of the determinant in Equation (a) by column expansion using the first column leads to

$$(1-\lambda)\begin{vmatrix} 2-\lambda & 3 \\ 2 & -2-\lambda \end{vmatrix} - 3\begin{vmatrix} 2 & -1 \\ 2 & -2-\lambda \end{vmatrix} + 0\begin{vmatrix} 2 & -1 \\ 2-\lambda & 3 \end{vmatrix} = 0$$

$$(1-\lambda)[(2-\lambda)(-2-\lambda) - (3)(2)] - 3[(2)(-2-\lambda) - (-1)(2)] = 0$$

$$-\lambda^3 + \lambda^2 + 16\lambda - 4 = 0 \tag{b}$$

The solutions of Equation (b) are the eigenvalues of **A**, $\lambda_1 = -3.664$, $\lambda_2 = 0.2471$, $\lambda_3 = 4.417$. The eigenvectors are obtained by solving

$$\begin{vmatrix} 1-\lambda & 2 & -1 \\ 3 & 2-\lambda & 3 \\ 0 & 2 & -2-\lambda \end{vmatrix} \begin{bmatrix} y_1 \\ y_2 \\ y_3 \end{bmatrix} = \begin{bmatrix} 0 \\ 0 \\ 0 \end{bmatrix} \tag{c}$$

with values of λ corresponding to each of the eigenvalues. For $\lambda = -3.664$, this leads to

$$\begin{vmatrix} 4.664 & 2 & -1 \\ 3 & 5.664 & 3 \\ 0 & 2 & 1.664 \end{vmatrix} \begin{bmatrix} y_1 \\ y_2 \\ y_3 \end{bmatrix} = \begin{bmatrix} 0 \\ 0 \\ 0 \end{bmatrix} \tag{d}$$

The equations represented by the second and third rows of Equation (d) are used to obtain $y_1 = 0.571y_3$ and $y_2 = -0.832y_3$. The first equation does not yield additional information and y_3 may be taken as any nonzero constant. Thus the eigenvector corresponding to the eigenvalue $\lambda_1 = -3.664$ is $\mathbf{y}_1 = c_1\,[0.571 \quad -0.832 \quad 1]^T$. In a similar fashion, the eigenvectors corresponding to the remaining eigenvalues are $\mathbf{y}_2 = c_2\,[-1.66 \quad 1.12 \quad 1]^T$ and $\mathbf{y}_3 = c_3\,[3.21 \quad 3.59 \quad 1]^T$.

Appendix C

MATLAB

MATLAB is a special computer programming language designed for engineering and scientific computation. The name "MATLAB" is contracted from MATRIX LABORATORY. MATLAB considers all data as an array or a matrix. Scalars are considered a matrix with only one row and one column.

MATLAB can be used as a programming language or as an interactive computational worksheet. In either case, the MATLAB work environment must be open. Executable statements are entered into the work environment. The environment allocates memory for variables and follows instructions provided by executable statements. Programs developed to be executed using MATLAB are stored in M-files, which have an extension .m. These files are executed by simply typing the name of the file in the MATLAB work environment. Any data created and assigned to a variable name is stored in the work environment and may be recalled and used during a work session.

This appendix provides a summary of MATLAB commands and features used in the text. It is not exhaustive, nor does it present concepts of programming. Proficiency in programming is not required to use MATLAB. The uses of the commands presented are functional but may not be the most efficient commands to achieve the objective. MATLAB has a useful HELP feature, which the reader is encouraged to use.

MATLAB has many toolboxes that provide programs and allow special executable commands specific to the toolbox. The Symbolic Math Toolbox is included with the student edition of MATLAB and Simulink, and it is available as an add-on for the professional version. The Control System Toolbox is available as an add-on for both the student version and the professional edition.

The tables of commands and features presented in this appendix are by no means comprehensive. They are limited to the commands and features used in this text.

Table C.1 MATLAB Operators

Operator	Name	Sample Scalar Calculation	Sample Matrix Calculation
+	Addition	3 + 4 = 7	[1 2;3 4] + [5 6;7 8] = [6 8;10 12]
−	Subtraction	3 − 4 = −1	[1 2;3 4] − [5 6;7 8] = [−4 −4; −4 −4]
*	Multiplication	3*4 = 12	[1 2;3 4]*[5 6;7 8] = [19 22;43 50]
/	Division	3/4 = 0.75	
^	Exponentiation	3^4 = 81	[1 2;3 4]^2 = [7 10; 15 22]
.*	Term-by-term matrix multiplication		[1 2;3 4].*[5 6;7 8] = [5 12;21 24]
.^	Term-by-term matrix exponentiation		[1 2;3 4].^2 = [1 4;9 16]
'	Matrix transpose		[1 2;3 4]' = [1 3;2 4]
=	Assignment	A = 1	A = [1 2;3 4]

C.1 MATLAB Basics

MATLAB stores all data in matrices. A matrix is entered in MATLAB using an assignment statement with the elements of the matrix enclosed in square brackets ([]). The elements are entered row by row with each element in the row separated by a space. Rows are separated by a semicolon (;). While MATLAB considers a scalar as a matrix with one row and one column, the assignment of a scalar to a variable does not require enclosing the scalar in square brackets. Individual elements of a matrix may be defined or referenced by using subscripts to indicate the row and column in the matrix where the element resides. Since the physical use of subscripts is not possible, the subscripts are identified in parentheses after the variable name. For example, "A(1,2)" refers to the element residing in the first row and second column of the matrix **A**.

Operators are used to represent mathematical calculations. Table C.1 gives MATLAB operators and sample results of the operation. Table C.2 presents the hierarchy of operations. Some of the available MATLAB functions are listed in Table C.3. Characters with special significance are listed in Table C.4.

MATLAB allows the definition of vectors whose elements constitute an arithmetic sequence through the following command

name = i:j:k

Table C.2 Hierarchy of Operations from Highest to Lowest

1. Grouping (functions and parentheses)
2. Exponentiation
3. Multiplication and division from left to right
4. Addition and subtraction from left to right

Table C.3 MATLAB Functions

Function	Name	Mathematical Equation	Examples	Comments
abs(x)	Absolute value	$\lvert x\rvert$	abs(−4) = 4 abs (−3 + 4j) = 5	Absolute value of a real number or magnitude of a complex number
angle(x)	Angle	$\tan^{-1}\left\{\frac{\text{Im}[x]}{\text{Re}[x]}\right\}$	angle(2 − j) = −0.4636	Result in radians between $-\pi$ and π
atan(x)	Inverse tangent	$\tan^{-1}(x)$	atan(1) = 0.7854	Result in radians between $-\pi/2$ and $\pi/2$
atan2(y,x)	Four-quadrant inverse tangent	$\tan^{1}\left(\frac{y}{x}\right)$	atan2(2, −1) = 2.0344	Result in radians between $-\pi$ and π
ceil(x)	Ceiling function	Smallest integer greater than or equal to x	ceil(6.11) = 7	
conj(x)	Complex conjugate	$\bar{x}$	conj(−3 + 4j) = −3 − 4j	
cos(x)	Cosine	$\cos(x)$	cos(pi) = −1	Argument must be in radians; π is MATLAB-defined constant = 3.14159...
exp(x)	e to the x	e^x	exp(2) = 7.3891	
floor(x)	Floor function	Largest integer less than or equal to x	floor(3.15) = 3	

Table C.3 MATLAB Functions (Continued)

Function	Name	Mathematical Equation	Examples	Comments
imag(x)	Imaginary part	Im[x]	imag(−3 + 4j) = 4	Either i or j may be used to represent $\sqrt{-1}$
log(x)	Natural logarithm	ln(x)	log(10) = 2.3026	
log10(x)	Base 10 logarithm	log (x)	log10(10) = 1	
max(A)	Maximum value of elements of the vector A	$\max_{i=1...n} A_i$	max([2 1 0 5]) = 5	
min(A)	Minimum value of elements of the vector A	$\min_{i=1...n} A_i$	min([1 2 −1 −2]) = −2	
real(x)	Real part	Re[x]	real(−3 + 4j) = −3	Either i or j may be used to represent $\sqrt{-1}$
sin(x)	Sine	sin(x)	sin(pi) = 0	Argument must be in radians; π is MATLAB-defined constant = 3.14159…
tan(x)	Tangent	tan(x)	tan(pi/4) = 1	Argument must be in radians and cannot be an odd multiple of $\pi/2$

Table C.4 Characters with Special Significance

Character	Significance	Example	Comments
%	Comment statement	% Calculate natural frequency	A comment statement is a nonexecutable statement used to document a program. The character must precede a comment.
;	Suppress printing	a=3*x^2;	The presence of the semicolon at the end of an assignment statement suppresses the printing of the result.
;	Row separation	[1 2;3 4]	When used in defining a matrix, the semicolon indicates the end of one row and the beginning of the next.
' '	Literal printing	disp('The natural frequency is')	Text enclosed by single quotes will be printed as is.
:	Vector generation	G = 0:2:20	A method of generating a vector of elements that constitute an arithmetic sequence. The example generates a vector G such that G(1) = 0, G(2) = 2, G(3) = 4, …, G(11) = 20. The colons separate the first element, the constant difference between elements, and the maximum value of the last element.
…	Statement continuation	X = 3 + 6 . . . + 4 + 7	Allows for the continuation of lengthy equations on multiple lines.

where name is the vector name, i is the value of name(1), j is the constant difference between elements of name, j = name(m+1) − name(m), and k is the largest (or smallest if j is negative) possible value of an element of name.

Let t be a defined vector and let x be a vector whose elements are a prescribed function of t. A statement of the form

x = f(t)

defines the vector x for the prescribed function f. Term-by-term calculations involving matrix elements must be defined using a period (.) before an operator. For example, if t is a column vector, the statement

x = t*t

generates an error that begins with "incorrect dimensions for matrix multiplication." This statement attempts to determine the product of the $1 \times n$ vector and the $1 \times n$ vector. However, if the statement

x = t.*t

is used, a $1 \times n$ vector x is generated such that $x_i = t_i^2$.

Example C.1

A 3×3 matrix is defined as

$$\mathbf{A} = \begin{bmatrix} 4 & -1 & 2 \\ 3 & 6 & 5 \\ 2 & -1 & 0 \end{bmatrix} \quad \text{(a)}$$

a. Write the command to enter the matrix into MATLAB with the variable name A.
b. Write a MATLAB statement that calculates the product $(a_{3,2})(a_{1,3})$ and assigns it to the variable q.

Solution

a. The assignment statement to enter the matrix **A** into MATLAB's memory is

A = [4 −1 2;3 6 5;2 −1 0]

b. The assignment statement to calculate the desired product is

q = **A**(3,2)*__A__(1,3)

Example C.2

Write a MATLAB statement that evaluates the algebraic expression

$$b = \frac{3 + 4\sin(x^{2y+5})}{2 - 4\cos\left(\dfrac{x-2}{3+2x}\right)} \quad \text{(a)}$$

using the minimum number of parentheses.

Solution

Using the operators of Table C.1, the hierarchal rules of Table C.2, and the functions of Table C.3, the MATLAB assignment statement for Equation (a) using the fewest number of parentheses is

b = (3 + 4*sin(x^(2*y + 5)))/(2 − 4*cos ((x − 2)/(3 + 2*x)))

Example C.3

Write MATLAB statements that define a vector of values $t_i = 2 + 0.02 * (i - 1)$ for $i = 1, 2, \ldots, 101$ and a vector of values $x_i = e^{-2t_i} [\cos(4t_i)]^2$

Solution

It is noted that $t_{101} = 4$. The appropriate MATLAB statements are

```
t = 2:.02:4;
x = exp(-2*t).*cos(4*t).^2;
```

Note that since exp(−2*t) generates a vector and cos(4*t) generates a vector, the .* operator must be used for term-by-term multiplication. Similarly, the .^2 operator must be used to square the individual elements of a vector.

C.2 Plotting and Annotating Graphs

The development of graphs is very easy using MATLAB. A graph can be developed directly in the MATLAB work environment or from the execution of a program. A plot developed during the execution of a program may be annotated using statements in the program or from the workspace after execution of the program.

A two-dimensional plot in MATLAB is generated by plotting two vectors, using one as the abscissa and the other as the ordinate. For example, the command

```
plot(t,x)
```

constructs a graph plotting the elements of t on the horizontal axis and the elements of x on the vertical axis. The vectors t and x must be of the same length. The plot is scaled automatically.

The commands

```
t = 0:0.05:10;
x = sin(2*t);
plot(t,x)
```

generate vectors t and x and then plot x as a function of t. The resulting graph is shown in Figure C.1.

More than one graph can be displayed on one set of axes. The commands

```
t = 0:0.05:10;
x = sin(2*t);
y = exp(-.05*t).*sin(2*t);
plot(t,x,'-',t,y,'.')
```

generate vectors t, x, and y and plot x versus t and y versus t on the same set of axes. The resulting plot is shown in Figure C.2. The plot of x is continuously connected while the values of y in the generated vector are connected by a period (.), as instructed by the plot command.

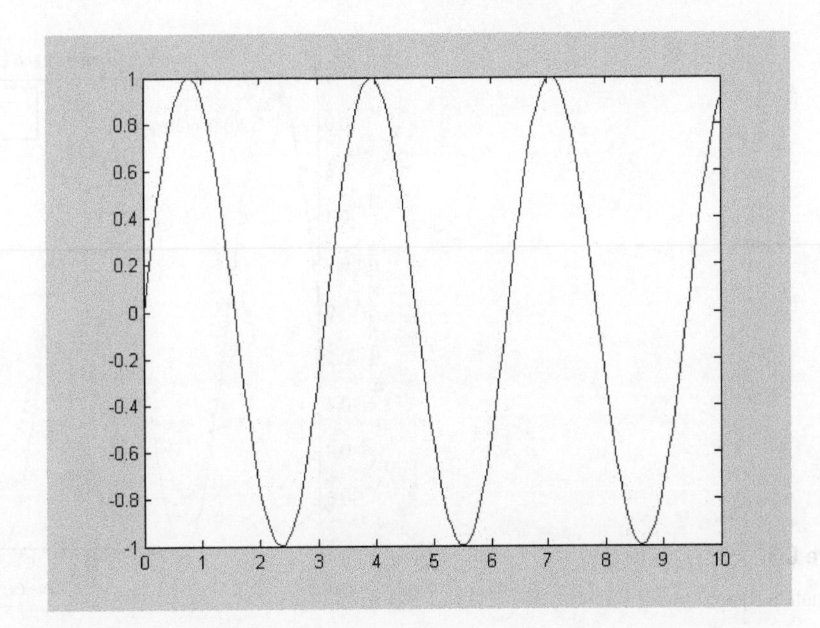

Figure C.1
The MATLAB plot generated using plot(t,x).

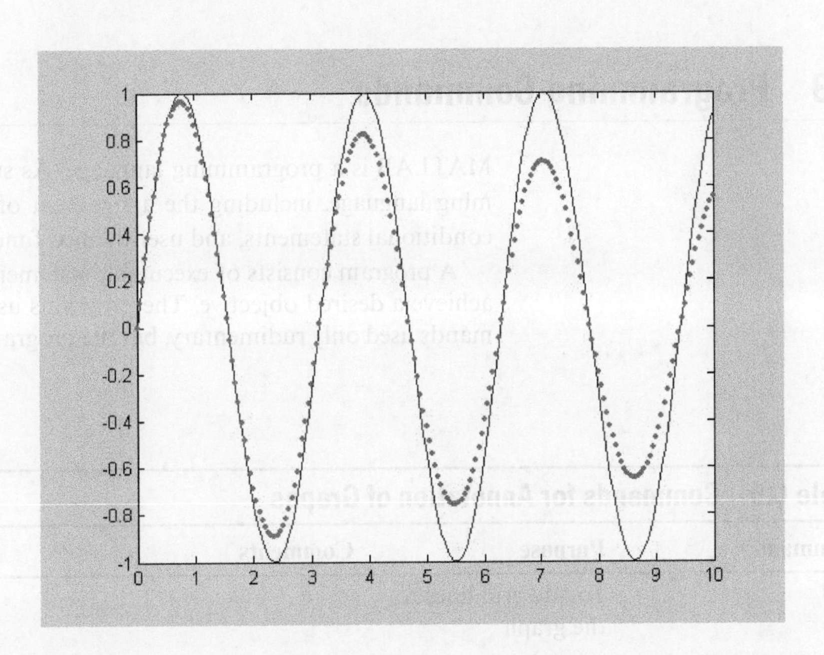

Figure C.2
Two graphs on one set of axes, generated in MATLAB.

The plot is annotated with labels on the axes, a title, a legend, and specifically placed text. The plot in Figure C.2 can be annotated using the following commands:

```
title('Comparison of Damped and Undamped Responses')
xlabel('Time (s)')
ylabel('Displacement (mm)')
legend ('undamped response','damped response\zeta =0.05')
text(6,0,'y(t) = e^-^.^0^5^tsin(2t)')
text(1.2,.8,'x(t) = sin(2t)')
```

Use of these commands leads to the graph of Figure C.3. The commands are explained in Table C.5. These commands contain features that allow exponentiation and Greek letters in the displayed text. Some of the available features are summarized in Table C.6. In each case, the trigger for the feature is used in the string inside the single quotes.

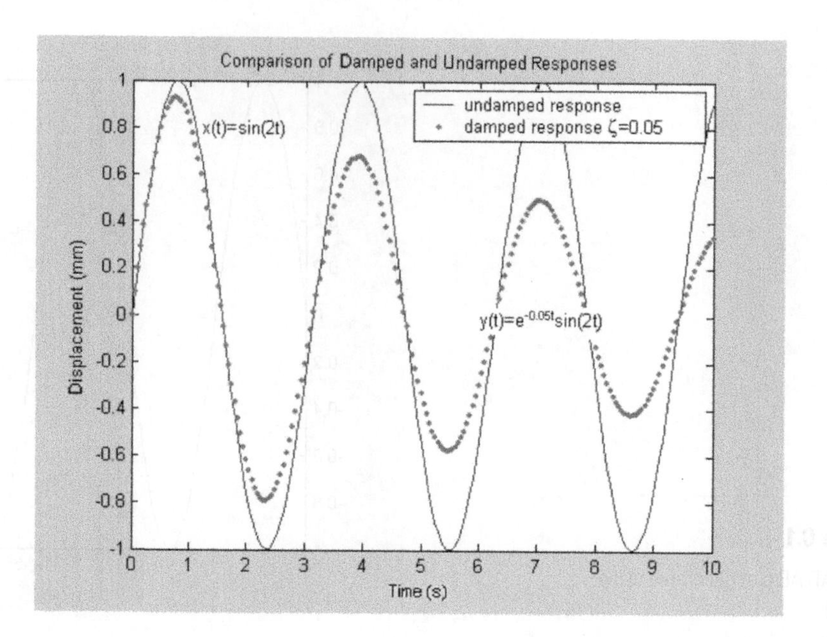

Figure C.3

An annotated graph.

C.3 Programming Commands

MATLAB is a programming language. As such, it has all the aspects of a programming language, including the assignment of variables, input and output functions, conditional statements, and user-defined functions.

A program consists of executable statements placed in a specific order designed to achieve a desired objective. The programs used in this text are modest and the commands used only rudimentary, but the programs use all these aspects of programming.

Table C.5 Commands for Annotation of Graphs

Command	Purpose	Comments
grid	To add grid lines to the graph	
legend(' ',' ',. . . .,' ')	To add a legend to the graph	Legend appears in a box in the upper right-hand corner. The number of arguments is equal to the number of graphs on the set of axes. The arguments provide the annotation for the legend for the graphs in order of listing in the plot statement.
text(x,y,' ')	To add text to the graph at a specified location	The left of the text begins at the location (x,y) on the graph, where x and y are coordinates measured along the horizontal and vertical axes of the graph. The coordinates are defined using the scale of the graph. In Figure C.3, the command that begins text(6,0,. . .) places the text beginning at the t coordinate 0 and the x coordinate 6.
title(' ')	To add a title to the graph	The title appears centered above the graph, outside the graphics box.
xlabel(' ')	To add a label to the horizontal axis	The label appears centered below the horizontal axis.
ylabel(' ')	To add a label to the vertical axis	The label appears centered to the left of the vertical axis.

Table C.6 Triggers for Special Features in Annotating Graphs

Feature	Syntax Example	Comments
Superscript	^t	The t appears as a superscript in the printed text. If more than one character is to appear as a superscript, the caret (^) must be repeated after each character. For example, ^2^t leads to a superscript of ^{2t}.
Subscript	_3	The 3 appears as a subscript in the printed text. If more than one character is to appear as a subscript, the underscore (_) must be repeated after each character. For example, _3_,_2 leads to a subscript of $_{3,2}$.
Lowercase Greek letter	\alpha	The Greek letter α is displayed. The name of the Greek letter to be displayed is spelled out in lowercase after the backslash.
Uppercase Greek letter	\Omega	The uppercase Greek letter Ω is displayed. The name of the Greek letter to be displayed is spelled out with the first letter in uppercase and the remaining letters in lowercase.

C.3.1 Input and Output

"Output" refers to how the program communicates its results. The programs in this text are all designed to output directly into the MATLAB worksheet. Not only can statements be written to physically display the value of variables in the workspace, but also once a value of a variable is calculated in the program it is stored in memory and can be recalled and used from the workspace.

By default the workspace displays the result of calculations from every assignment statement immediately upon completion of the execution of the statement. MATLAB also displays the name of the variable being displayed. This output may be suppressed by ending the statement with a semicolon (;). The value of a variable may be displayed in the workspace just by entering the name of the variable. This can be done from a statement in a program or directly from the workspace.

In lieu of the default output of results, MATLAB allows the display of more information regarding the variable through use of the 'disp' command. Suppose a program calculates a steady-state amplitude X in units of meters. Instead of simply displaying the variable name, the following 'disp' statement displays more information regarding the variable:

```
disp('The steady-state amplitude in m is')
X
```

"Input" refers to the entering of data into the program. The programs in this text allow only for data to be entered interactively from the workspace using an input statement. Suppose it is desired to enter the value of a damping ratio and assign it to the variable z. The following input statement may be used:

```
z = input ('Please enter the value of the damping ratio')
```

The user is then prompted to provide a value for z to be entered in the workspace.

C.3.2 Conditional Statements

Programs often decide how to proceed based on specified, conditions. If a condition is satisfied, the program is to do one thing; if the condition is not satisfied, the program is to do something else. Program control is exerted when the program decides which step is the next to be executed. The program control structure commonly used in this text is an 'if' statement. The basic structure of an 'if' statement is

```
if conditional statement
    statements to be executed
end
```

The conditional statement in an 'if' statement is a logical statement that is either true or false. The conditional statement is written using logical and relational operators that are summarized in Table C.7. If the conditional statement is true, then the statements following the conditional statement are executed in sequence. If the conditional statement is false, program control transfers to the 'end' statement. The 'end' statement is nonexecutable and serves as a place marker.

An expanded structure for the 'if' statement occurs when an 'else' clause is added to permit the execution of statements when the conditional is false:

```
if conditional statement
    statements to be executed
else
    statements to be executed
end
```

In this structure, if the conditional statement is true then the statements immediately following the if statement are executed until the program reaches the 'else' when control is transferred to the 'end' statement. If the conditional statement is false, control shifts to the 'else' statement.

When a false conditional statement sends the program control to an 'else' which would be followed by another conditional statement, an alternative is to use an 'elseif' statement:

```
if conditional 1
    statements to be executed
elseif conditional 2
    statements to be executed
else
    statements to be executed
end
```

Table C.7 Logical and Relational Operators

Symbol	Name	Example	Result of Example
~	Not	~(6>4)	False
==	Is equal to	3 == 2 + 1	True
~=	Is not equal to	2 ~= 2 + 1	True
<	Less than	4 < −1	False
<=	Less than or equal to	4 <= 4	True
>	Greater than	4 > 1	True
>=	Greater than or equal to	4 >= 6	False
&	And	4 < 3& 4 > −1	False
!	Or	4 < 3 ! 4 > −1	True

If conditional 1 is false, then program control shifts to the 'elseif' statement. If conditional 2 is true, then the statements following the 'elseif' statement are executed. If conditional 2 is false, then program control flows to the 'else' statement. As many 'elseif' statements as necessary may be used in a conditional.

Example C.4

A numerical value of a variable 'b' has been determined previously in the program. Write MATLAB conditional statements for the following. **a.** If b is less than 20, then its value is to be doubled. **b.** If b is less than 20, then its value is to be doubled, otherwise it is to be tripled. **c.** If b is less than 20, then its value is to be doubled; if it is between 20 and 100 inclusive it is to be tripled; otherwise it is to be quadrupled.

a. The appropriate conditional is

```
if b < 20
    b = 2*b;
end
```

b. The use of an else is required and the appropriate conditional is

```
if b < 20
    b = 2.*b;
else
    b = 3.*b;
end
```

c. The use of an elseif is convenient. One possible conditional is

```
if b < 20
    b = 2.*b;
elseif b <= 100
    b = 3.*b;
else
    b = 4.*b;
end
```

C.3.3 Looping

Looping allows statements to be executed multiple times. In some cases it is convenient to use a 'while' loop, which has the structure

```
while conditional statement
    statements to be executed
end
```

If the conditional statement is true, the immediately subsequent statements are executed. When the 'end' statement is reached, program control returns to the 'while' statement. The relational expression is again checked for validity. Once the relational expression is not true, program control is shifted to the 'end' statement and the loop ends.

A 'for' loop is executed a specified number of times. It has the structure

```
for var = i:j:k
    statements
end
```

where var is a variable name used as a counter in the loop, i is the beginning value of var, j is an incremental value, and k is the largest possible value for var. The counter is initially set to i. The following statements are executed. When 'end' is reached, program control returns to the 'for' statement. The value of var is incremented by j. Before proceeding, the new value of var is compared with k. If var is greater than k, control is shifted to the 'end' statement and the loop is exited. Otherwise the following steps are again executed and the loop continues as described.

C.3.4 User-Defined Functions

The modeling of a dynamic system focuses on being able to determine the system output given its input, which is time dependent. MATLAB is used as a tool in the determination of system response. MATLAB programs, or M-files, are written in support of this effort. A programmer developing a MATLAB code to determine the output of the dynamic system should have as an objective that the program should be able to determine the response of the system for any system input, but the program itself does not need to be modified when the system input changes. A program to determine system response for an arbitrary input should have available a function that for any value of time provides the input value(s).

The preceding objective is attained through the use of user-defined functions, which are functions defined outside a program and which the program may call whenever necessary. A user-defined function is written to return a value or values to the calling program. Information is transferred from the calling program to the user-defined function. The user-defined function uses the information to calculate the desired results, which are then transferred to the calling program. The syntax for such a user-defined function is

```
function [x,y,z] = name(a,b,c)
    assignment statements to
        calculate x, y, and z
```

The calling program sends the values of a, b, and c, the arguments of name, to the user-defined function, which uses them to calculate x, y, and z. The values of x, y, and z are then returned to the calling program.

For example, suppose that execution of an M-file requires values of two forces $F = 10 \sin(2t)$ and $G = 20e^{-2t}$ at an arbitrary time t. The name of the function is force. An appropriate form for the user-defined function is

```
function [F,G] = force(t)
        F = 10*sin(2*t);
        G = 20*exp(−2*t);
```

The user-defined function must be called from the M-file. The syntax for calling this function is

```
[A,B] = force(time)
```

MATLAB searches for force as the name of a default function or a user-defined function. The name "force" must correspond to the name of the user-defined function. The value of time is sent to the user-defined function, where it is used as t. Upon return, the value calculated in the user-defined function as F is assigned to A in the calling program, and the value calculated in the user-defined function as G is assigned to B in the calling program.

Table C.8 Symbolic Math Toolbox Commands Used In Text

Command	Syntax	Comments
syms	syms s t	Declares s and t as symbolic variables.
sym	sym('x')	Creates a symbolic representation of the value assigned to x.
laplace	laplace(F)	Takes the Laplace transform of function F defined in terms of a symbolic variable t; default is to return a function of symbolic variable s; both t and s must be declared as symbolic variables using syms.
ilaplace	ilaplace(G)	Takes the inverse Laplace transform of function G defined in terms of a symbolic variable s; default is to return a function of symbolic variable t.
simplify	simplify(F)	Algebraically simplifies the symbolic expression for F using a variety of techniques.
expand	expand(F)	Algebraically expands products of polynomials contained in F.
subs	subs(F,y)	If F is a function of a symbolic variable, say x, and y is a numerical value or an algebraic expression, subs evaluates F when y is substituted for x.
vpa	vpa(F,n)	Variable precision arithmetic computes a symbolic expression F to a decimal expression with n significant digits.

The Symbolic Math Toolbox and the Control System Toolbox are add-ons for the professional edition of MATLAB. Both toolboxes plus Simulink are bundled for students in the "MATLAB and Simulink Student Suite."

C.4 Symbolic Math Toolbox

The Symbolic Toolbox performs symbolic algebra, linear algebra, and calculus. The toolbox is used in this text for the determination of Laplace transforms, inverse Laplace transforms, and transfer functions for multiple output systems. Table C.8 summarizes the commands from the Symbolic Math Toolbox used in this regard.

C.5 Control System Toolbox

The Control System Toolbox performs calculations and develops diagrams used in the analysis and design of control systems. The functions from the Control System Toolbox are used in this text to determine poles and residues, define transfer functions, develop root-locus diagrams, draw Bode and Nyquist diagrams, and calculate the Ackerman matrix.

The use of these functions is limited to transfer functions defined as the ratio of two polynomials $G(s) = N(s)/D(s)$. However, the Symbolic Math Toolbox is not required to use these functions because the coefficients defining the polynomials in the numerator and denominator of the transfer function are entered as vectors. An nth-order polynomial is entered as a row vector of n elements. The first element in the vector is the coefficient of s^n. The command 'tf(N,D)' is used to define the transfer function.

Example C.5

Use MATLAB to define the transfer function

$$G(s) = \frac{3s^2 + 2s + 1}{s^3 + 5s^2 + 6s + 12}$$

Solution

The MATLAB worksheet for entering $N(s)$, $D(s)$, and defining $G(s)$ is shown in Figure C.4.

```
> N=[3 2 1];
>> D=[1 5 6 12];
>> G=tf(N,D)

G =

   3 s^2 + 2 s + 1
 ----------------------
 s^3 + 5 s^2 + 6 s + 12

Continuous-time transfer function.
```

Figure C.4
The MATLAB worksheet for the development of the transfer function of Example C.5.

The development of the transfer function using the function 'tf' is not often necessary. Control System Toolbox functions requiring the transfer function can also use the definitions of $N(s)$ and $D(s)$. For example, the function to plot the Bode diagram can be written as either

bode(G)

where G is the transfer function or

bode(N,D)

Some useful Control System Toolbox functions are described in Table C.9.

Table C.9 Selected Control System Toolbox Functions

Function	Use	Syntax	Comments
tf	Develop the transfer function from the numerator and denominator polynomials N(s) and D(s), respectively	G = tf(N,D)	MATLAB shows G as a function of s, but s is not defined as a symbolic variable and G is not a function of a symbolic variable.
residue	Determine the poles and residues of a transfer function	[r,p,k] = residue(N,D)	r = vector of residues p = vector of poles k = vector of coefficients obtained by dividing N by D. If the order of N is less than the order of D, then k is a null vector.
impulse	Impulsive response of system	impulse(G) impulse(N,D) impulse(G,tfinal)	Output is a plot of impulsive response. tfinal = final value at which response is calculated

Table C.9 Selected Control System Toolbox Functions (Continued)

Function	Use	Syntax	Comments
step	Step response of system	step(G) step(N,D) step(G,tfinal)	Output is a plot of step respsonse. tfinal = final value at which response is calculated
bode	Bode diagram (magnitude and phase)	bode(tf) bode(N,D)	Output is a plot of the Bode diagram.
nyquist	Nyquist diagram	nyquist(tf) nyquist(N,D)	Output is the Nyquist diagram.
rlocus	Root-locus diagram	rlocus(R,Q)	Root-locus diagram is created for the equation Q(s) + KR(s) = 0.
ss2tf	Converts a state-space model into a transfer function model	[N,D] = ss2tf(A,B,C,D)	N = vector of coefficients for numerator of transfer function D = vector of coefficients for denominator of transfer function A = state matrix B = input matrix C = output matrix D = transmission matrix
tf2ss	Converts a transfer function model into a state-space model	[A,B,C,D] = tf2ss(N,D)	A = state matrix B = input matrix C = output matrix D = transmission matrix N = vector of coefficients for numerator of transfer function D = vector of coefficients for denominator of transfer function
acker	Ackerman matrix	acker(A,B,p)	A = state matrix B = input matrix p = vector of desired poles
margin	Phase and gain margins and crossover frequencies	[GM,PM,omegacp, omegacg]=margin(G)	G = defined transfer function GM = gain margian PM = phase margin in degrees omegacp = phase crossover frequency in r/s omegacg = gain crossover frequency in r/s
stepinfo	Information about step response plots	stepinfo	Provides information about step responses including 10-90 percent rise time, 2 percent settling time, peak time, and maximum overshoot

References

Antoniou, A. (1993). *Digital Filters: Analysis, Design and Applications*, 2nd ed. New York, NY: McGraw-Hill.

Bedford, A.M. and Fowler, W. (2023). *Engineering Mechanics: Statics and Dynamics*, 6th ed. Upper Saddle River, NJ: Pearson.

Chapman, S. (2019). *MATLAB Programming for Engineers*, 6th ed. Mason, OH: Cengage.

Chapra, S.C. and Canale, R.P. (2020). *Numerical Methods for Engineers*, 8th ed. New York, NY: McGraw-Hill.

Churchill, R.V. (1972). *Operational Mathematics*, 3rd ed. New York, NY: McGraw-Hill.

Fournier, R.L. (2017). *Basic Transport Phenomena in Biomedical Engineering*, 4th ed. Boca Raton, FL: CRC Press.

Irwin, J.D. and Nelms, R.M. (2021). *Basic Engineering Circuit Analysis*, 12th ed. Hoboken, NJ: Wiley.

James, J. (2020). *MATLAB Programming Fundamentals*, Natick, MA: MathWorks.

Kelly, S.G. (2023). *Mechanical Vibrations: Theory and Applications*, 2nd ed. San Diego, CA: Cognella.

__________. (1996). *Schaum's Outline for Mechanical Vibrations*. New York, NY: McGraw-Hill.

Luyben, W.L. (1989). *Process Modeling, Simulation and Control for Chemical Engineers*, 2nd ed. New York, NY: McGraw-Hill.

Marshall, J.E., Gorecki, H., Korytowski, A., and Walton, K. (1992). *Time-Delay Systems: Stability and Performance Criteria with Applications*. Upper Saddle River, NJ: Pearson.

Navhi, M. and Edminister, J. (2017). *Schaum's Outline of Electric Circuits*, 7th ed. New York, NY: McGraw-Hill.

Newton, C.M. (1984). "Modern Model-Based Methods and Software for Pharmokinetics Research" in C. Nicolin, ed. *Modeling and Analysis in Biomedicine*. Singapore: World Scientific.

Ogata, K.O. (2009). *Modern Control Engineering*, 5th ed. Upper Saddle River, NJ: Pearson.

Palm, W.J. (2020). *System Dynamics*, 4th ed. New York, NY: McGraw-Hill.

Pelesko, J.A. and Bernstein, D.H. (2002). *Modeling MEMS and NEMS*. Boca Raton, FL: CRC Press.

Raven, F. (1995). *Automatic Control Engineering*, 5th ed. New York, NY: McGraw-Hill.

Simon, L. (2012). *Control of Biological and Drug-Delivery Systems for Chemical, Biomedical, and Pharmaceutical Engineering*. Hoboken, NJ: Wiley.

Smith, J.M. (2014). *Chemical Engineering Kinetics*, 3rd ed. New York, NY: McGraw-Hill.

Trietley, H. (1986). *Transducers in Mechanical and Electronic Design*. Boca Raton, FL: CRC Press.

Watton, J. (2009). *Fundamentals of Fluid Power Control*. New York, NY: Cambridge University Press.

White, F.M. and Xue, H. (2020). *Fluid Mechanics*, 9th ed. New York, NY: McGraw-Hill.

Index

D

E

F

G

H

I

K

L

M

T